CHAP. 301. *Of Ducks meat.*

Lens palustris.
Ducks meat.

¶ *The Description.*

DVcks meat is as it were a certain green mosse, with very little round leaues of the bignes of Lentils : out of the midst whereof on the nether side grow downe very fine threds like haires, which are to them in stead of roots : it hath neither stalke, floure, nor fruit.

¶ *The Place.*

It is found in ponds, lakes, city ditches, & other standing waters euery where.

¶ *The Time.*

The time of Ducks meat is known to all.

¶ *The Names.*

Duckes meat is called in Latine, *Lens lacustris, Lens aquatilis,* and *Lens palustris* : of the Apothecaries it is named *Aqua Lenticula* : in high-Dutch, **Meerlinsen** : in low-Dutch, **Waterlinsen**, & more vsually **Enden gruen**, that is to say, *Anatum herba,* Ducks herb, becaufe Ducks do feed thereon; whereupon alfo in English it is called Ducks meat: some term it after the Greek, water Lentils; and of others it is named Graines. The Italians call it *Lent di palude* : in French, *Lentille d'eaue* : in Spanish, *Lenteias de agua.*

¶ *The Nature.*

Galen sheweth that it is cold and moist after a fort in the second degree.

¶ *The Vertues.*

Dioscorides faith that it is a remedie against all A manner of inflammations, S. *Anthonies* fire, and hot Agues, if they be either applied alone, or else vfed with parched barly meale. It alfo knitteth ruptures in yong children.

Ducks meat mingled with fine wheaten floure, and applied, preuaileth much against hot Swel-B lings, as Phlegmons, Erifipela's, and paines of the joints.

The fame helpeth the fundament fallen downe in yong children.   C

CHAP. 302. *Of Water Crow-foot.*

1 *Ranunculus aquatilis.*
Water Crow-foot.

¶ *The Defcription.*

WAter Crow-foot hath flender branches trailing far abroad, whereupon grow leaues vnder the water, moft finely cut and jagged like thofe of Cammomill : thofe aboue the water are fomwhat round, indented about the edges, in forme not vnlike the fmal tender leaues of the Mallow, but leffer : among which doe grow the floures, fmall, and white of colour made of fine little leaues, with fome yellowneffe in the middle like the floures of the Strawberry, and of a fweet fmell : after which there come round rough and prickly knaps like thofe of the field **Crow**foot. The roots be very fmall hairy ftrings.

‡ There is fomtimes to be found a varietie of this, with the leaues leffe, and diuided into three parts after the maner of an Iuy leafe, and the floures alfo are much leffer, but white of colour, with a yellow bottom. I queftion whether this be not the *Ranunculus hederaceus Dalefchampij, pag.* 1031. of the *Hiſt. Lugd.* ‡

2 There is another plant growing in the water, of fmal moment, yet not amiffe to be remembred, called *Hederula aquatica*, or water Iuy ; the which is very rare to finde, neuertheleffe I found it once in a ditch by Bermondfey houfe neere London, and neuer elfewhere : it hath fmall threddie ftrings in ftead of roots and ftalks, rifing from the bottome of the water to the top, whereunto are faftned fmall leaues fwimming or floting vpon the water, triangled or three cornered like thofe of barren Iuy, or rather noble Liuerwort, barren of floures and feeds.

2 *Hederula aquatica.*	‡ 3 *Stellaria aquatica.*
Water Iuy.	Water Star-wort.

3 There is likewife another herb of fmall reckoning that floteth vpon the water, called *Stellaria aquatica* or water Starwort, which hath many fmall graffy ftems like threds comming from the bottom of the water, vnto the vpper face of the fame; whereupon grow fmal double flours of a greenifh or herby colour. ‡ I take this *Stellaria* to be nothing elfe but a water Chickeweed, growing almoft in euery ditch, with long narrow leaues at each joint, and halfe a dozen or more lying clofe together at the top of the water, in fafhion of a ftar : in which fhape it may be feen in the end of Aprill and beginning of **May**. I haue not yet obferued either the floure or feed thereof. ‡

¶ The

¶ *The Place.*

Water Crowfoot growes by ditches and shallow springs, and in other moist and plashy places.

¶ *The Time.*

It floureth in Aprill and May, and somtimes in Iune.

¶ *The Names.*

Water Crow-foot is called in Latine *Ranunculus aquatilis*, and *Polyanthemum aquatile*: in English, water Crow-foot, and white water Crow-foot. Most Apothecaries and Herbarists doe erroniously name it *Hepatica aquatica*, and *Hepatica alba*; and with greater error they mix it in medicines in stead of *Hepatica alba* or grasse of Parnassus. ‡ I know none that commit this great error here mentioned, neither haue I knowne either the one or the other euer vsed or appointed in medicine with vs in England; though *Dodonæus* (from whom our Author had this and most else) blame his country-men for this mistake and error.

¶ *The Temperature and Vertues.*

Water Crow-foot is hot, and like to common Crow-foot.

CHAP. 303.   Of Dragon.

1 *Dracontium majus.*
Great Dragons.

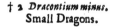

† 2 *Dracontium minus.*
Small Dragons.

¶ *The Description.*

1   THe great Dragon riseth vp with a straight stalke a cubit and a halfe high or higher, thicke, round, smooth, sprinkled with spots of diuers colours, like those of the adder or snake: the leaues are great and wide, consisting of seuen or more ioyned together in order, euery one of which is long and narrow, much like to the leaues of Dock, smooth and slippe-rie: out of the top of the stalke groweth a long hose or husk greater than that of the Cuckow pint, of a greenish colour without, and crimson within, with his pestell which is blackish, long, thicke, and pointed like a horne; the skinne or filme whereof, when the seed waxeth big, beeing stretched or broken

3 *Dracunculus aquaticus.*
Water Dragons.

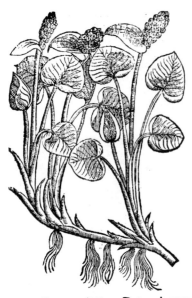

broken aſunder,there appeareth the fruit, like to a bunch or cluſter of grapes ; the berries whereof at the firſt are green, afterwards red and ful of juice; in which is contained ſeed that is ſomewhat hard : the root continueth freſh, thicke like to a knob, white,couered with a thinne pilling,oftentimes of the bigneſſe of a meane apple, full of white little threds appendant thereunto.

2   The leſſer Dragon is like Aron or Wake-Robin in leaues, hoſe or huske, peſtle and berries; yet are not the leaues ſprinkled with blacke, but with whitiſh ſpots, which periſh not ſo ſoone as thoſe of Wake-Robin, but endure together with the berries euen vntill winter: theſe berries alſo be not of a deepe red,but of a colour inclining to ſaffron. The root is not vnlike to the Cuckow-pintle, hauing the forme of a bulb,full of ſtrings, with diuers rude ſhapes of new plants, whereby it greatly increaſeth.

‡   The figure which our Author here gaue by the title of *Dracuntium minus*,was no other than of Aron which is deſcribed in the firſt place of the next Chapter : neither is the deſcription of any other plant than of that ſort thereof which hath leaues ſpotted either with white or blacke ſpots, though our Author ſay only with white. I haue giuen you *Cluſius* his figure of *Arum Byzantinum*, in ſtead of that which our Author gaue. ‡

3   The root of Water Dragon is not round like a bulbe,but very long,creeping,jointed,and of mean bigneſſe ; out of the joints whereof ariſe the ſtalks of the leaues, which are round, ſmooth, and ſpongie within,certaine white and ſlender ſtrings growing downewards. The fruit ſprings forth at the top vpon a ſhort ſtalke,together with one of the leaues,being at the beginning couered with little white threds,which are in ſtead of the floures : after that it groweth into a bunch or cluſter, at the firſt greene,and when it is ripe red,leſſer than that of Cuckow-pint,but not leſſe biting : The leaues are broad, greeniſh, glib,and ſmooth, in faſhion like thoſe of Ivy, yet leſſer than thoſe of Cuckow-pint ; and that thing whereunto the cluſtered fruit growes,is alſo leſſer, and in that part which is towards the fruit (that is to ſay the vpper part) is white.

4   The great Dragon of *Matthiolus* his deſcription is a ſtranger not onely in England,but elſewhere for any thing that we can learne : my ſelfe haue diligently enquired of moſt ſtrangers skilful in plants,that haue reſorted vnto me for conference ſake,but no man can giue mee any certainetie thereof ; and therefore I hold it vnfit to giue you his figure or any deſcription,for that I take it for a feigned picture.

¶ *The Place.*

The greater and leſſer Dragons are planted in gardens. The water Dragons grow in watery and mariſh places, for the moſt part in fenny and ſtanding waters.

¶ *The Time.*

The berries of theſe plants are ripe in Autumne.

¶ *The Names.*

The Dragon is called in Greeke Δϱακόντιον : in Latine,*Dracunculus*. The greater is named *Serpentaria major* : of ſome,*Biſaria*,and *Colubrina* : *Cordus* calleth it *Dracunculus Polyphyllos*,and *Luph Criſpum:* in high Dutch,**Schlangenkraut :** in low-Dutch, **Speerwoꝛtele :** in French,*Serpentaire* : in Italian,*Dracontea* : in Spaniſh,*Taragontia* : in Engliſh,Dragons,and Dragonwort. *Apuleius* calls Dragon *Dracontea* ; and ſetteth downe many ſtrange names thereof,which whether they agree with the greater or the leſſer,or both of them,he doth not expound ; as *Python*, *Anchomanes*, *Sauchromaton*, *Therion*,*Schœnos*,*Dorcadion*,*Typhonion*,*Theriophonon*,and *Eminion*. *Athenæus* ſheweth, That Dragon is called *Aronia*,becauſe it is like to Aron.

¶ *The Temperature.*

Dragon,as *Galen* ſaith, hath a certaine likeneſſe with Aron or Wake-Robin,both in leaues and alſo in root,yet more biting and more bitter than it,and therefore hotter,and of thinner parts : it is
alſo

alfo fomething binding, which by reafon that it is adjoyned with the two former qualities, that is to fay,biting and bitter, it is made in like manner a fingular medicine of very great efficacy.

¶ *The Vertues.*

The root of Dragons doth clenfe and fcoure all the entrailes, making thinne, efpecially thicke A and tough humours ; and it is a fingular remedy for vlcers that are hard to be eured, named in Greeke, ʊϰειϗϞ.

It fcoureth and cleanfeth mightily, afwell fuch things as haue need of fcouring, as alfo white B and blacke morphew, being tempered with vinegre.

The leaues alfo by reafon that they are of like quality are good for vlcers and greene wounds : C and the leffe dry they are, the fitter they be to heale ; for the dryer ones are of a more fharpe or bi-ting quality than is conuenient for wounds.

The fruit is of greater operation than either the leaues or the root : and therefore it is thought D to be of force to confume and take away cankers and proud flefh growing in the nofthrils,called in Greeke, *Polypus* : alfo the juyce doth cleanfe away webs and fpots in the eies.

Furthermore, *Diofcorides* writeth, that it is reported that they who haue rubbed the leaues or E root vpon their hands, are not bitten of the viper.

Pliny faith,that ferpents will not come neere vnto him that beareth dragons about him,and thefe F things are read concerning both the Dragons, in the two chapters of *Diofcorides.*

Galen alfo hath made mention of Dragon in his booke of the faculties of nourifhments ; where G he faith,that the root of Dragon being twice or thrice fod,to the end it may lofe all his acrimony or fharpeneffe,is fomtimes giuen as Aron, or wake-Robin is,when it is needfull to expell the more forceable thicke and clammie humors that are troublefome to the cheft and lungs.

And *Diofcorides* writeth,that the root of the leffer Dragon being both fodde and roft with hony, H or taken of it felfe in meate,caufeth the humors which fticke faft in the cheft to be eafily voided.

The juyce of the Garden Dragons, as faith *Diofcorides*,being dropped into the eies, doth clenfe I them,and greatly amend the dimneffe of the fight.

The diftilled water hath vertue againft the peftilence or any peftilentiall feuer or poyfon,being K drunke bloud warme with the beft treacle or mithridate.

The fmell of the floures is hurtfull to women newly conceiued with child.   L

CHAP. 304.   *Of Cuckow pint, or wake-Robin.*

¶ *The Defcription.*

1   A Rum or Cockow pint hath great,large, fmooth, fhining,fharpe pointed leaues,befpot-ted here and there with blackifh fpots,mixed with fome blewneffe : among which rifeth vp a ftalke nine inches long, befpeckled in many places with certaine purple fpots. It beareth alfo a certaine long hofe or hood, in proportion like the eare of an hare : in the middle of which hood commeth forth a peftle or clapper of a darke murry or pale purple colour : which being paft, there fucceedeth in place thereof a bunch or clufter of berries in manner of a bunch of grapes,greene at the firft, but after they be ripe of a yellowifh red like corall, and full of juyce, wherein lie hid one or two little hard feeds. The root is tuberous, of the bigneffe of a large Oliue,white and fucculent,with fome threddy additaments annexed thereto.

2   There is in Ægypt a kinde of *Arum* which alfo is to be feene in Africa, and in certaine pla-ces,of Lufitania,about riuers and flouds, which differeth from that which groweth in England and other parts of Europe. This plant is large and great, and the leaues thereof are greater than thofe of the water Lillie : the root is thicke and tuberous, and toward the lower end thicker and broader, and may be eaten. It is reported to be without floure and feed, but the increafe that it hath is by the fibres which runne and fpread from the roots.   ‡   This plant hath alfo peftels and clufters of berries as the common Aron,but fomewhat different,the leaues are not cut into the ftalke, but joy-ned before the fetting thereto : the root alfo is very large. Thofe that defire to fee more of this plant,and the queftion which fome haue mooued, whether this be the *Colocafia*, or *Faba Ægyptia* of the Antients ? let them haue recourfe to the firft chapter of *Fabius Columna*,his *Minus cognitarum ftirpium pars altera,*and there they fhall finde fatisfaction. ‡

1 *Arum vulgare.*
Cockow pint.

‡ 2 *Arum Ægyptiacum.*
Ægyptian Cockow pint.

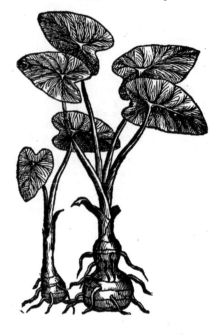

¶ *The Place.*

Cockow pint groweth in woods neere vnto ditches vnder hedges, euery where in ſhadowie places.

¶ *The Time.*

The leaues appeare preſently after Winter:the peſtell ſheweth it ſelfe out of his huske or ſheath in Iune,whileſt the leaues are in withering : and when they are gone,the bunch or cluſter of berries becommeth ripe,which is in Iuly and Auguſt.

¶ *The Names.*

There groweth in Ægypt a kinde of Aron or Cuckow pint which is found alſo in Africa, and likewiſe in certaine places of Portingale neere vnto riuers and ſtreames, that differeth from thoſe of our countries growing, which the people of Caſtile call *Manta de naeſtra ſenora* : moſt would haue it to be called *Colocaſia*, but *Dioſcorides* ſaith that *Colocaſia* is the root of *Faba Ægyptia*, or the Beane of Ægypt. ‡ *Fabius Columna* (in the place formerly alledged) prooues this not to be the true *Colocaſia*,and yet *Proſper Alpinus* ſince in his ſecond booke *de plantis exoticis,cap.*17.and 18. labours to prooue the contrary : let the curious haue recourſe to theſe, for it is to tedious for me in this place to inſiſt vpon it, being ſo large a point of controuerſie, which hath ſo much troubled all the late Writers. ‡

The common Cuckow pint is called in Latine, *Arum* : in Greeke, *ἄρον*: in ſhops *Iarus*,and *Barba-Aron* : of others, *Pes vituli* : of the Syrians, *Lupha* : of the men of Cyprus, *Colocaſia*, as wee finde among the baſtard names. *Pliny* in his 24 booke, 16.chapter,doth witneſſe, that there is great difference betweene *Aron* and *Dracontium*,although there hath been ſome controuerſie about the ſame among the old Writers, affirming them to be all one : in high Dutch it is called, **Paſſen pint :** in Italian, *Gigora:* in Spaniſh, *Yaro:* in low Dutch, **Calfsuoet :** in French, *Pied d'veau :* in Engliſh, Cuckow pint, and Cuckow pintle, wake-Robin,Prieſts pintle, Aron,Calfes foot,and Rampe ; and of ſome Starchwort.

¶ *The Temperature.*

The faculties of Cuckow pint doe differ according to the varietie of countries : for the root hereof, as *Galen* in his booke of the faculties of nouriſhments doth affirme, is ſharper and more biting in ſome countries than in others, almoſt as much as Dragons, contrariwiſe in Cyren a city in Africke,it is generally in all places hot and dry,at the leaſt in the third degree.

¶ *The*

¶ *The Vertues.*

If any may would haue thicke and tough humours which are gathered in the cheſt and lungs to A
be cleanſed and voided out by coughing, then that Cuckow-pint is beſt that biteth moſt.

It is eaten being ſodden in two or three waters, and freſh put to, whereby it may looſe his acri- B
monie ; and being ſo eaten, they cut thicke humours meanely, but Dragons is better for the ſame
purpoſe.

Dioſcorides ſheweth, that the leaues alſo are preſerued to be eaten, and that they muſt be eaten af- C
ter they be dried and boiled ; and writeth alſo, that the root hath a peculiar vertue againſt the gout,
being laied on ſtamped with Cowes dung.

Beares after they haue lien in their dens forty daies without any manner of ſuſtenance, but what D
they get with licking and ſucking their owne feet, doe as ſoone as they come forth eat the herbe
Cuckow-pint, through the windie nature thereof the hungry gut is opened and made fit againe to
receiue ſuſtenance : for by abſtaining from food ſo long a time, the gut is ſhrunke or drawne ſo
cloſe together, that in a manner it is quite ſhut vp, as *Ariſtole, Ælianus, Plutarch, Pliny,* and others do
write.

The moſt pure and white ſtarch is made of the roots of Cuckow-pint ; but moſt hurtfull to the E
hands of the Laundreſſe that hath the handling of it, for it choppeth, bliſtereth, and maketh the
hands rough and rugged, and withall ſmarting.

CHAP. 305. *Of Friers Cowle, or hooded Cuckowpint.*

1 *Ariſarum latifolium.*
Broad leaued Friers Cowle.

2 *Ariſarum anguſtifolium.*
Narrow leated Friers Cowle.

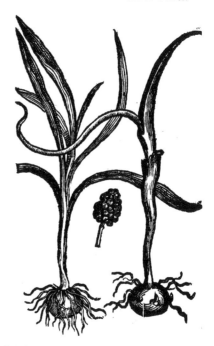

¶ *The Deſcription.*

1   BRoad leaued Friers hood hath a leafe like Iuie, broad and ſharpe pointed but far leſſe approching neere to the forme of thoſe of Cuckowpint : the ſtalke thereof is ſmall and
ſlender : the huske or hoſe is little ; the peſtell ſmall, and of a blacke purpliſh colour ;
the cluſter when it is ripe is redde ; the kernels ſmall ; the root white, hauing the forme of Aron or
Cuckowpint, but leſſer, whereof doubtleſſe it is a kinde.

The ſecond Friers hood hath many leaues, long and narrow, ſmooth and glittering: The huske or hoſe is narrow and long; the peſtell that commeth forth of it is ſlender, in forme like a great earth worme, of a blackiſh purple colour, as hath alſo the inſide of the hoſe, vpon which, hard to the ground, and ſometimes a little within the ground, groweth a certaine bunch or cluſter of berries, greene at the firſt, and afterwards red: the root is round and white like the others.

¶ The Place.

Theſe plants are ſtrangers in England, but common in Italy, and eſpecially in Tuſcane about Rome, and in Dalmatia, as *Aloiſius Anguillara* witneſſeth: notwithſtanding I haue them in my Garden.

¶ The Time.

The floures and fruit of theſe come to perfection with thoſe of Cuckowpint and Dragons.

¶ The Names.

Friers hood is called *Dioſcorides*, Ἀρίσαρον: in Latine, *Ariſarum*: but *Pliny* calleth it Ἄρις, or *Aris*; for in his twenty fourth booke, *cap.* 16. hee ſaith, That *Aris* which groweth in Ægypt is like Aron or Cuckowpint: it may be called in Engliſh after the Latine name *Ariſarum*; but in my opinion it may be more fitly called Friers hood, or Friers cowle, to which the floures ſeeme to be like; whereupon the Spaniards name it *Fraililloſ*, as *Daleſcampius* noteth.

¶ The Temperature.

Friers Cowle is like in power and faculty to the Cuckow-pint, yet is it more biting, as *Galen* ſaith.

¶ The Vertues.

A　There is no great vſe of theſe plants in Phyſicke; but it is reported that they ſtay running or eating ſores or vlcers: and likewiſe that there is made of the roots certaine compoſitions called in Greeke *Collyria*, good againſt fiſtula's: and being put into the ſecret part of any liuing thing, it rotteth the ſame, as *Dioſcorides* writeth.

† That which was formerly figured and deſcribed in the third place, vnder the title of *Ariſarum latifolium Matthioli*, was the ſame with that deſcribed by the name of *Dracontium minus*, in the precedent chapter, and therefore here omitted.

CHAP. 306.　Of Aſarabacca.

1 *Aſarum.*
Aſarabacca.

2 *Aſarina Matthioli.*
Italian Aſarabacca.

¶ *The Deſcription.*

1   THe leaues of Aſarabacca are ſmooth, of a deepe greene colour, rounder, broader, and ten
derer than thoſe of Iuie, and not cornered at all, not vnlike to thoſe of Sow bread. he
floures lie cloſe to the roots, hid vnder the leaues, ſtanding vpon ſlender foot-ſtalkes, of
an ill-fauoured purple colour, like to the floures and huskes of Henbane, but leſſe, wherein are contained ſmall ſeeds, cornered, and ſomewhat rough : the roots are many, ſmall and ſlender, growing
aſlope vnder the vpper cruſt of the earth, one folded within another, of an vnpleaſant taſte, but of a
moſt ſweet and pleaſing ſmell, hauing withall a kinde of biting quality.

2   This ſtrange kinde of Aſarabacca, which *Matthiolus* hath ſet forth creeping on the ground,
in manner of our common Aſtrabacca, hath leaues ſomewhat rounder and rougher, ſleightly indented about the edges, and ſet vpon long ſlender foot-ſtalkes : the floures grow hard vnto the ground
like vnto thoſe of Cammomill, but much leſſer, of a mealy or duſty colour, and not without ſmell.
The roots are long and ſlender, creeping vnder the vpper cruſt of the earth, of a ſharpe taſte, and bitter withall. ‡ This *Aſarina* of *Matthiolus, Cluſius* (whoſe opinion I here follow) hath iudged to be
the *Tuſſilago Alpina* 2. of his deſcription, wherefore I giue you his figure in ſtead of that of our
Author, which had the floures expreſt, which this wants. ‡

¶ *The Place.*

It delighteth to grow in ſhadowie places, and is very common in moſt Gardens.

¶ *The Time.*

The herbe is alwaies greene ; yet doth it in the Spring bring forth new leaues and floures.

¶ *The Names.*

It is called in Greeke Ἄσαρον, *Aſarum* : in Latine, *Nardus ruſtica* : and of diuers, *Perpenſa* : *Perpenſa*
is alſo *Baccharis* in *Pliny*, *lib.* 21. *cap.* 21. *Macer* ſaith, That *Aſarum* is called *Vulgago*, in theſe words :

Eſt Aſaron Gracè Vulgago dicta Latinè.

This herbe, *Aſaron* do the Græcians name ;
Whereas the Latines *Vulgago* clepe the ſame.

It is found alſo amongſt the baſtard names, that it was called of the great learned Philoſophers
Αἷμα ἄρεος : that is, *Martis ſanguis*, or the bloud of *Mars* : and of the French men, *Baccar* ; and thereupon
it ſeemeth that the word *Aſarabacca* came, which the Apothecaries vſe, and likewiſe the common
people : but there is another *Baccharis* differing from *Aſarum*, yet notwithſtanding *Crateuas* doth alſo call *Baccharis, Aſarum*.

This confuſion of both the names hath been the cauſe, that moſt could not ſufficiently expound
themſelues concerning *Aſarum* and *Baccharis* ; and that many things haue been written amiſſe in
many copies of *Dioſcorides*, in the chapter of *Aſarum* : for when it is ſet downe in the Greeke copies
a ſweet ſmelling garden herbe, it belongeth not to the deſcription of this *Aſarum*, but to that of
Baccharis : for *Aſarum* (as *Pliny* ſaith) is ſo called, becauſe it is not put into garlands : and ſo by
that meanes it came to paſſe, that oftentimes the deſcriptions of the old Writers were found corrupted and confuſed : which thing, as it is in this place manifeſt, ſo oftentimes it cannot ſo eaſily
be marked in other places. Furthermore, *Aſarum* in called in French *Cabaret* : in high Dutch, Ha-
zelwurtz : in low Dutch, Mans ooren : in Engliſh, Aſarabacca, Fole-foot, and Hazel wort.

¶ *The Temperature.*

The leaues of Aſarabacca are hot and dry, with a purging qualitie adjoyned thereunto, yet not
without a certaine kinde of aſtriction or binding. The roots are alſo hot and dry, yet more than
the leaues ; they are of thin and ſubtill parts : they procure vrine, bring downe the deſired ſickneſſe,
and are like in faculty, as *Galen* ſaith, to the roots of *Acorus*, but yet more forceable ; and the roots
of *Acorus* are alſo of a thinne eſſence, heating, attenuating, drying, and prouoking vrine, as he affirmeth : which things are happily performed by taking the roots of Aſarabacca, either by themſelues are mixed with other things.

¶ *The Vertues.*

The leaues draw forth by vomit, thicke phlegmaticke and cholericke humours, and withall   A
mooue the belly ; and in this they are more forceable and of greater effect than the roots themſelues.

They are thought to keepe in hard ſwelling cankers that they encreaſe not, or come to exulcera-   B
tion, or creeping any farther, if they be outwardly applied vpon the ſame.

The roots are good againſt the ſtoppings of the liuer, gall, and ſpleene, againſt wens and hard   C
ſwellings, and agues of long continuance : but being taken in the greater quantitie, they purge
flegme and choler not much leſſe than the leaues (though *Galen* ſay no) by vomit eſpecially, and
alſo by ſiege.

D   One dram of the pouder of the roots giuen to drinke in ale or wine, groſſely beaten, prouoketh
vomit for the purpoſes aforeſaid ; but being beaten into fine pouder, and ſo giuen, it purgeth very
little by vomit, but worketh moſt by procuring much vrine; therefore the groſſer the pouder is, ſo
much the better.

E   But if the roots be infuſed or boiled, then muſt two, three, or foure drams be put to the infuſion;
and of the leaues eight or nine be ſufficient : the juyce of which ſtamped with ſome liquid thing,
is to be giuen. The roots may be ſteeped in wine, but more effectually in whay or honied water, as
Meſues teacheth.

F   The ſame is good for them that are tormented with the Sciatica or gout in the huckle bones, for
thoſe that haue the dropſie, and for ſuch as are vexed with a quartane ague, who are cured and made
whole by vomiting.

C H A P. 307.   *Of Sea Binde-weed.*

1 *Soldanella marina.*
Sea Binde-weed.

‡ 2 *Soldanella Alpina major.*
Mountaine Binde-weed.

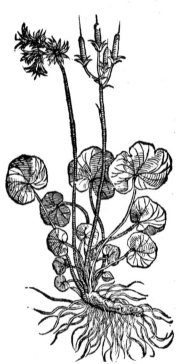

¶ *The Deſcription.*

1   Soldanella or Sea Binde-weed, hath many ſmall branches, ſomewhat red, trailing vpon the
ground, beſet with ſmall and round leaues, not much vnlike Aſarabacca, or the leaues
of Ariſtolochia, but ſmaller; betwixt which leaues and the ſtalkes come forth floures
formed like a bell, of a bright red incarnate colour, in euery reſpect anſwering the ſmal Bindeweed,
whereof it is a kinde, albeit I haue here placed the ſame, for the reaſons rendred in my Proëme.
The ſeed is blacke, and groweth in round huskes : the root is long and ſmall, thruſting it ſelfe far
abroad. and into the earth like the other Binde-weeds.

2   *Soldanella* or mountaine Binde-weed hath many round leaues ſpred vpon the ground, not
much vnlike the former, but rounder, and more full of veines, greener, of a bitter taſte like ſea Binde-
weed : among which commeth forth a ſmall and tender ſtalke a handfull high, bearing at the top
little floures like the ſmall Bell-floure, of a sky colour. The root is ſmall and threddy.

‡ 3 There

‡ 3 *Soldanella Alpina minor.*
Small Mountaine Bindweed.

‡ 3 There is of this kinde another hauing all the parts ſmaller, and the leaues redder and rounder : the floures alſo blew, and compoſed of one leafe diuided into fiue parts, and ſucceeded by a longiſh cod, round and ſharp pointed. ‡

¶ *The Place.*

The firſt grows plentifully by the Sea ſhore in moſt places of England, eſpecially neere ro Lee in Eſſex, at Merſey in the ſame county, in moſt places of the Iſle of Thanet, and Shepey, and in many places along the Northerne coaſt.

The ſecond groweth vpon the mountaines of Germany, and the Alpes ; it groweth vpon the mountaines of Wales, not farre from Commers Meare in North-Wales.

¶ *The Time.*

Theſe herbes doe floure in Iune, and are gathered in Auguſt to be kept for medicine.

¶ *The Names.*

The firſt called *Soldanella* is of the Apothecaries and the Antients called *Marina Braſſica,* that is to ſay, Sea Colewoort : but what reaſon hath moued them ſo to do I canot conceiue, vnleſſe it be penury and ſcarſitie of names, and becauſe they know not otherwiſe how to terme it : of this I am ſure, that this plant and *Braſſica* are no more like than things which are moſt vnlike ; for *Braſſica Marina* is the Sea Colewort, which doth much reſemble the garden Cabbage or Cole, both in ſhape and in nature, as I haue in his due place expreſſed. A great fault and ouerſight therefore it hath been of the old writers and their ſucceſſors, which haue continued the cuſtome of this error, not taking the paines to diſtinguiſh a Binde-weed from a Cole-wort. But to auoid controuerſies, the truth is, as I haue before ſhewed, that this *Soldanella* is a Binde-weed, and cannot be eſteemed for a *Braſſica,* that is, a Colewort. The later Herbariſts call it *Soldana,* and *Soldanella :* in Dutch, Zeewind, that is to ſay, *Convolvulus Marinus :* of *Dioſcorides* κράμβη θαλασσία, (i) *Braſſica Marina* in Engliſh, Sea Withwinde, Sea Binde-weed, Sea-Bels, Sea-Coale, of ſome, Sea-Fole-foot, and Scottiſh Scuruie-graſſe.

The ſecond is called *Soldanella Montana :* in Engliſh, Mountaine Bind-weed.

¶ *The Nature.*

Sea Bind-weed is hot and dry in the ſecond degree : the ſecond is bitter and very aſtringent.

¶ *The Vertues.*

Soldanella purgeth downe mightily all kinde of watriſh humours, and openeth the ſtoppings of **A** the liuer, and is giuen with great profit againſt the dropſie : but it muſt be boiled with the broth of ſome fat meat or fleſh, and the broth drunke, or elſe the herbe taken in pouder worketh the like effect.

Soldanella hurteth the ſtomack, and troubleth the weake and delicate bodies which do receiue it **B** in pouder, wherefore aduice muſt be taken to mix the ſaid pouder with Anniſe ſeeds, Cinnamon, ginger, and ſugar, which ſpices do correct his malignitie.

Practitioners about Auſpurge and Rauiſpurge (cities of Germany) doe greatly boaſt that they **C** haue done wonders with this herbe *Soldanella Montana ;* ſaying, that the leaues taken and emplaiſtred vpon the nauell and ſomewhat lower, draw forth water from their bellies that are hydropicke, that is, troubled with water or the dropſie : this effect it worketh in other parts without heating.

It doth alſo wonderfully bring fleſh in wounds, and healeth them.   **D**

Dioſcorides witneſſeth, that the whole herbe is an enemy to the ſtomacke, biting and extremely **E** purging (both ſodden, and taken with meat) and bringeth troubleſome gripings thereunto, and doth oftentimes more hurt than good.

‡ My friend Mr. *Goodyer* hath told me, that in Hampſhire at Chicheſter and thereabout they **F** make vſe of this for Scuruie-graſſe, and that not without great errour, as any that know the qualities may eaſily perceiue. ‡

CHAP.

CHAP. 308.   *Of the Grasse of Parnassus.*

† 1 *Gramen Parnaßi.*
Grasse of Parnassus.

‡ 2 *Gramen Parnaßi flore duplici.*
Grasse of Parnassus with double floures.

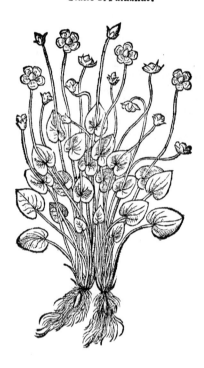 

¶ *The Description.*

THe Grasse of Parnassus hath small round leaues, very much differing from any kinde of Grasse, much resembling the leaues of Iuie, or Asarabacca, but smaller, and not of so darke a colour: among these leaues spring vp small stalkes a foot high, bearing little white floures consisting of fiue round pointed leaues; which being falne and past, there come vp round knops or heads, wherein is contained a reddish seed. The root is somewhat thicke, with many strings annexed thereto.

2   The second kinde of *Gramen Parnaßi* doth answer the former in each respect, sauing that the leaues are somewhat larger, and the floures double, otherwise very like.

¶ *The Place.*

The first groweth very plentifully in Lansdall and Crauen, in the North parts of England; at Doncaster, and in Thornton fields in the same country: moreouer, in the Moore neer to Linton, by Cambridge, at Hesset also in Suffolke, at a place named Drinkstone, in the medow called Butchers mead. ‡ M^r. *Goodyer* found it in the boggy ground below the red well of Wellingborough in Northampton shire: and M^r. *William Broad* obserued it to grow plentifully in the Castle fields of Berwicke vpon Tweed. ‡

The second is a stranger in England.

¶ *The Time.*

These herbes do floure in the end of Iuly, and their seed is ripe in the end of August.

¶ *The Names.*

Valerius Cordus hath among many that haue written of these herbes said something of them to good purpose, calling them by the name of *Hepatica alba* (whereof without controuersie they are kindes) in English, white Liuerwoort: although there is another plant called *Hepatica alba*, which
for

for diſtinction ſake I haue thought good to Engliſh, Noble white Liuerwort.

The ſecond may be called Noble white Liuerwort the double floure.

¶ *The Temperature.*

The ſeed of Parnaſſus Graſſe, or white Liuerwort, is dry, and of ſubtill parts.

¶ *The Vertues.*

The decoction of the leaues of Parnaſſus Graſſe drunken, doth dry and ſtrengthen the feeble A and moiſt ſtomacke, ſtoppeth the belly, and taketh away the deſire to vomit.

The ſame boiled in wine or water, and drunken, eſpecially the ſeed thereof, prouoketh vrine, brea- B keth the ſtone, and driueth it forth.

† The figure that was formerly in the firſt place of this Chapter was of *Vnifolium*, deſcribed before, *cap. 60. pag. 406.* that which was in the ſecond place belonged to the firſt deſcription.

CHAP. 309.　*Of white Saxifrage, or Golden Saxifrage.*

¶ *The Deſcription.*

1 THe white Saxifrage hath round leaues ſpred vpon the ground, and ſomewhat jagged about the edges, not much vnlike the leaues of ground Ivie, but ſofter and ſmaller, and of a more faint yellowiſh greene : among which riſeth vp a round hairie ſtalke a cubit high, bearing at the top ſmall white floures, almoſt like Stocke-gilloſloures: the root is compact of a number of blacke ſtrings, whereunto are faſtened very many ſmall reddiſh graines or round roots as bigge as pepper cornes, which are vſed in medicine, and are called *Semen Saxifraga albæ*; that is, the ſeed of white Saxifrage, or Stone-breake, although (beſides theſe foreſaid round knobbes) it hath alſo ſmall ſeed contained in little huskes, following his floure as other herbes haue.

1 *Saxifraga alba.*
White Saxifrage.

2 *Saxifraga aurea.*
Golden Saxifrage.

‡ 3 *Saxifraga alba petræa.*
White Rocke Saxifrage.

2   Golden Saxifrage hath round compaſſed leaues,bluntly indented about the borders
like the former, among which riſe vp ſtalkes a
handfull high, at the top whereof grow two or
three little leaues together: out of the middle
of them ſpring ſmall floures of a golden color;
after which come little husks, wherein is contained the red ſeed, not vnlike the former: the
root is tender, creeping in the ground with
long threds or haires.

‡ 3   *Pona* hath ſet forth this plant by the
name of *Saxifraga alba petræa*, and therefore I
haue placed it here; though I thinke I might
more fitly haue ranked him with *Paronychia ru-
taceo folio* formerly deſcribed. It hath a ſmall
ſingle root from which ariſe diuers fat longiſh
leaues,ſomewhat hairy,and diuided into three
parts: amongſt thoſe riſeth vp a round knotty
ſtalke,roughiſh,and of a purpliſh colour,ſome
halfe foot high, diuided into ſundry branches
which carry white floures, conſiſting of fiue
leaues apiece, with ſome yellowiſh threds in
their middles: theſe falling, there remaines a
cup containing a very ſmall ſeed. It floures at
the end of Iune in the ſhadowie places of the
Alpes,whereas *Pona* firſt obſerued it. ‡

¶ *The Place.*
The white Saxifrage groweth plentifully
in ſundry places of England, and eſpecially in
a field on the left hand of the high way, as you
goe from the place of execution called Saint
Thomas Waterings vnto Dedford by London. It groweth alſo in the great field by Iſlington called the Mantles: alſo in the greene places by the ſea ſide at Lee in Eſſex,among the ruſhes, and in
ſundry other places thereabout, and elſe where. ‡ It alſo growes in Saint Georges fields behinde
Southwarke. ‡

The golden Saxifrage groweth in the moiſt and mariſh grounds about Bathe and Wels, alſo in
the Moores by Boſton and Wisbich in Lincolneſhire : ‡ and Mr.*George Bowles* hath found it growing in diuers woods at Chiſſelhurſt in Kent: Mr.*Goodyer* alſo hath obſerued it abundantly on the
ſhadowie moiſt rockes by Mapledurham in Hampſhire: and I haue found it in the like places in
Yorkſhire. ‡

¶ *The Time.*
The white Saxifrage floureth in May and Iune : the herbe with his floure are no more ſeene vn-
till the next yeare.
The golden Saxifrage floureth in March and Aprill.

¶ *The Names.*
The firſt is called in Latine *Saxifraga alba* : in Engliſh,white Saxifrage,or white Stone-breake.
The ſecond is called Golden Saxifrage,or Golden Stone-breake.

¶ *The Nature.*
The firſt of theſe, eſpecially the root and ſeed thereof, is of a warme or hot complexion.
Golden Saxifrage is of a cold nature,as the taſte doth manifeſtly declare.

¶ *The Vertues.*
A   The root of white Saxifrage boiled in wine and drunken, prouoketh vrine,clenſeth the kidnies
and bladder, breaketh the ſtone and driueth it forth, and is ſingular good againſt the ſtrangury,and
all other griefes and imperfections in the reines.

B   The vertues of golden Saxifrage are yet vnto vs vnknowne, notwithſtanding I am of this minde,
that it is a ſingular wound herbe,equall with Sanicle.

CHAP.

CHAP. 310.　Of Sow-bread.

¶ *The Description.*

1　THe first being the common kinde of Sow-bread, called in shops *Panis porcinus*, and *Arthanita*, hath many greene and round leaues like vnto Asarabacca, sauing that the vpper part of the leaues are mixed here and there confusedly with white spots, and vnder the leaues next the ground of a purple colour : among which rise vp little stemmes like vnto the stalkes of violets, bearing at the top small purple floures, which turne themselues backward (being full blowne) like a Turks cap, or Tulepan, of a small sent or sauour, or none at all : which being past, there succeed little round knops or heads which containe slender browne seeds : these knops are

1　*Cyclamen orbiculato folio.*
Round Sow-bread.

2　*Cyclamen folio Hederæ.*
Ivie Sow-bread.

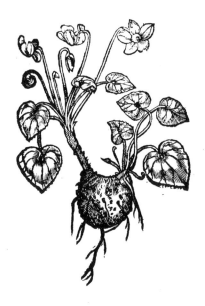
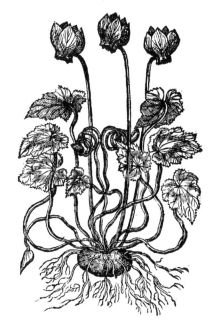

‡ 3　*Cyclamen Vernum.*　　Spring Sow-bread.

wrapped

wrapped after a few daies in the ſmall ſtalkes, as thred about a bottome, where it remaineth ſo defended from the iniurie of Winter cloſe vpon the ground, couered alſo with the greene leaues aforeſaid, by which meanes it is kept from the froſt, euen from the time of his ſeeding, which is in September, vntill Iune : at which time the leaues doe ſade away, the ſtalkes & ſeed remaining bare and naked, whereby it inioyeth the Sun (whereof it was long depriued) the ſooner to bring them vnto maturitie: the root is round like a Turnep, blacke without and white within, with many ſmall ſtrings annexed thereto.

2   The ſecond kinde of Sowbread hath broad leaues ſpred vpon the ground, ſharp pointed, ſomwhat indented about the edges, of a darke greene colour, with ſome little lines or ſtrakes of white on the vpper ſide, and of a darke reddiſh colour on that ſide next the ground : among which riſe vp ſlender foot-ſtalkes of two or three inches long : at the tops whereof ſtand ſuch floures as the precedent, but of a ſweeter ſmel, and more pleaſant colour. The ſeed is alſo wrapped vp in the ſtalk, for his further defence againſt the iniurie of winter. The root is ſomewhat greater, and of more ver-tue, as ſhall be declared.

‡ 4 *Cyclamen Vernum album.*
White floured Sowbread.

‡ 5 *An Cyclaminos altera, hederaceis folijs planta ?*

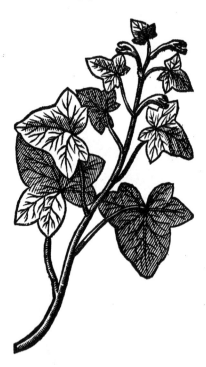

3   There is a third kinde of Sow-bread that hath round leaues without peaked corners, as the laſt before mentioned, yet ſomewhat ſnipt about the edges, and ſpeckled with white about the brims of the leaues, and of a blackiſh colour in the middle ; the floures are like to the reſt, but of a deeper purple : the root alſo like, but ſmaller, and this commonly floures in the Spring.

‡   4   This in leaues and roots is much like the laſt deſcribed; but the flours are ſmaller, ſnow white, and ſweet ſmelling. There are diuers other varieties of theſe plants, which I thinke it not neceſſary for me to inſiſt vpon : wherefore I referre the curious to the Garden of floures ſet forth by M[r] *Iohn Parkinſon*, where they ſhall finde ſatisfaction. ‡

5   There is a plant which I haue ſet forth in this place that may very well be called into que-ſtion, and his place alſo, conſidering that there hath been great contention about the ſame, and not fully determined on either part, which hath moued me to place him with thoſe plants that moſt do reſemble one another, both in ſhape and name : this plant hath greene cornered leaues like to Iuie,
long

long and small gaping floures like the small Snapdragon : more hath not been said of this plant, either of stalke or root, but is left vnto the consideration of the learned.

‡ The plant which our Author would here acquaint you with,is that which *Lobel* figures with this title which I here giue, and saith it was gathered among other plants on the hils of Italy, but in what part or place, or how growing he knew not ; and he only questions whether it may not be the *Cyclaminos altera* of *Dioscorides,lib.2.cap.195.* ‡

¶ The Place.

Sow-bread groweth plentifully about Artoies and Vermandois in France, and in the Forest of Arden, and in Brabant : but the second groweth plentifully in many places of Italie.

It is reported vnto mee by men of good credit, that *Cyclamen* or Sow-bread groweth vpon the mountaines of Wales ; on the hils of Lincolnshire, and in Somerset shire by the house of a gentleman called M^r. *Hales* ; vpon a Fox-borough also not far from M^r.*Bamfields*,neer to a town called Hardington. The first two kindes grow in my garden, where they prosper well. ‡ I cannot learne that this growes wilde in England. ‡

¶ The Time.

Sow-bread floureth in September when the plant is without leafe,which doth afterwards spring vp,continuing green all the Winter,couering and keeping warme the seed vntil Midsommer next, at what time the seed is ripe as aforesaid. The third floureth in the spring, for which cause it was called *Cyclamen vernum* : and so doth also the fourth.

¶ The Names.

Sow-bread is called in Greek, ϰυϰλάμινος : in Latine,*Tuber terra*, and *Terra rapum* : of *Marcellus* ,*Orbicularis* : of *Apuleius*,*Palalia,Rapum porcinum,*& *Terra malum* : in shops, *Cyclamen,Panis Porcinus,*and *Arthanita* : in Italian,*Pan Porcino* : in Spanish,*Mazan de Puerco* : in High Dutch, **Schweinbrot:**in Low Dutch, **Uerckins broot:**in French,*Pain de Porceau* : in English, Sow-bread. *Pliny* calleth the colour of this floure in Latine,*Colossinus color* : in English, Murrey colour.

¶ The Nature.

Sow-bread is hot and drie in the third degree.

¶ The Vertues.

The root of Sow-bread dried into pouder and taken inwardly in the quantitie of a dram and a A halfe, with mead or honied water, purgeth downeward tough and grosse flegme, and other sharpe humors.

The same taken in wine as aforesaid, is very profitable against all poison, and the bitings of ve- B nomous beasts,and to be outwardly applied to the hurt place.

The pouder taken as aforesaid, cureth the jaundise and the stoppings of the liuer, taketh away C the yellow colour of the bodie, if the patient after the taking hereof be caused to sweat.

The leaues stamped with honie, and the juice put into the eies,cleareth the sight taketh away D all spots and webs, pearle or haw, and all impediments of the sight, and is put into that excellent ointment called *Vnguentum Arthanita.*

The root hanged about women in their extreme trauell with childe, causeth them to be deliue- E red incontinent, and taketh away much of their paine.

The leaues put into the place hath the like effect, as my wife hath prooued sundry times vpon F diuers women,by my aduise and commandement, with good successe.

The juice of Sow-bread doth open the Hemorrhoids, and causeth them to flow being applied G with wooll or flocks.

It is mixed with medicines that consume or waste away knots, the Kings euill, and other hard H swellings : moreouer it clenseth the head by the nosthrils,it purgeth the belly being anointed therwith,and killeth the childe. It is a strong medicine to destroy the birth,being put vp as a pessary.

It scoureth the skin, and taketh away the Sun-burning, and all blemishes of the face,pilling off I the haire,and marks also that remaine after the small pocks & mesels : and giuen in wine to drink, it maketh a man drunke.

The decoction thereof serueth as a good and effectuall bath for members out of joint, the gout, K and kibed heeles.

The root being made hollow and filled with oile, closed with a little wax, and rosted in the hot L embers,maketh an excellent ointment for the griefes last rehearsed.

Being beaten and made vp into trochisches, or little flat cakes, it is reported to be a good amo- M rous medicine to make one in loue,if it be inwardly taken.

¶ The Daunger.

It is not good for women with childe to touch or take this herbe, or to come neere vnto it, or stride ouer the same where it groweth ; for the naturall attractiue vertue therein contained is such, that without controuersie they that attempt it in manner abouesaid,shall be deliuered before their

time :

time : which danger and inconuenience to auoid, I haue (about the place where it groweth in my garden) faſtned ſticks in the ground, and ſome other ſtickes I haue faſtned alſo croſſe-waies ouer them, leaſt any woman ſhould by lamentable experiment finde my words to be true, by their ſtepping ouer the ſame.

‡   I iudge our Author ſomthing to womaniſh in this, that is, led more by vaine opinion than by any reaſon or experience, to confirme this his aſſertion, which frequent experience ſhews to be vaine and friuolous, eſpecially for the touching, ſtriding ouer, or comming neere to this herbe. ‡

Chap. 311. *Of Birthwoorts.*

¶ *The Kindes.*

Birthwort, as *Dioſcorides* writeth, is of three ſorts, long, round, and winding : *Pliny* hath added a fourth kinde called *Piſtolochia*, or little Birthwoort. The later writers haue joined vnto them a fifth named Saracens Birthwoort.

1   *Ariſtolochia longa.*   2   *Ariſtolochia rotunda.*
    Long Birthwoort.   Round Birthwoort.

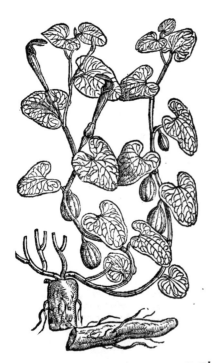 

¶ *The Deſcription.*

1   LOng Birthwoort hath many ſmall long ſlender ſtalkes creeping vpon the ground, tangling one with another very intricately, beſet with round leaues not much vnlike Sowbread or Iuie, but larger, of a light or ouerworne green colour, and of a grieuous or lothſome ſmell and ſauour : among which come forth long hollow floures, not much vnlike the flours of Aron, but without any peſtell or clapper in the ſame ; of a darke purple colour : after which follow ſmall fruit like vnto little peares, containing triangled ſeeds of a blackiſh colour. The root is long, thicke, of the colour of box, of a ſtrong ſauour and bitter taſte.

2   The round Birthwoort in ſtalkes and leaues is like the firſt, but his leaues are rounder : the floures differ onely in this, that they be ſomwhat longer and narrower, and of a faint yellowiſh colour, but the ſmall flap or point of the floure that turneth backe againe, is of a darke or blacke purple

ple colour. The fruit is formed like a peare, ſharp toward the top, more ribbed and fuller than the former: the root is round like vnto Sow-bread, in taſte and ſauor like the former.

3 *Ariſtolochia clematitis.* Climing Birthwort.

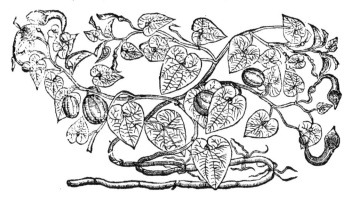

‡ 4 *Ariſtolochia Saracenica.*
Saracens Birthwort.

‡ 5 *Piſtolochia.*
Small Birthwort.

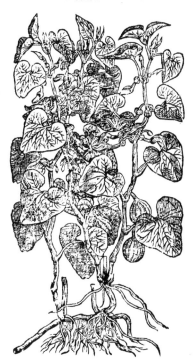

　3　Climing Birthwort taketh hold of any thing that is next vnto it, with his long and claſping ſtalks, which be oftentimes branched, and windeth it ſelfe like Bindeweed: the ſtalks of the leaues are longer, whoſe leaues be ſmooth, broad, ſharp pointed, as be thoſe of the others; the flours likewiſe hollow, long, yellow, or of a blackiſh purple color: the fruit differs not from that of the other, but the roots be ſlender and very long, ſometimes creeping on the top of the earth, and ſometimes growing deeper, being of like colour with the former ones.
　4　There is a fourth kinde of Birthwort reſembling the reſt in leaues and branched ſtalkes, yet

higher

higher,and longer than either the long or the round : the leaues thereof be greater than thoſe of *A.ſarabacca*,the floures hollow,long, and in one ſide hanging ouer, of a yellowiſh colour : the fruit is long and round like a peare,in which the ſeeds lie ſeuered, of forme three ſquare, of an ilfauoured blackiſh colour : the root is ſomwhat long,oftentimes of a mean thickneſſe, yellow like Box, not inferior in bitterneſſe either to the long or round Birthwort : and ſometimes theſe are found to bee ſmall and ſlender,and that is when they were but lately digged vp and gathered : for by the little parcels of the roots which are left, the yong plants bring forth at the beginning tender and branched roots.

5　　Small Birthwort is like to the long and round Birthwort both in ſtalkes and leaues, yet is it leſſer and tenderer : the leaues thereof are broad, and like thoſe of Iuy ; the floure is long, hollow in the vpper part,and on the outſide blackiſh ; the fruit ſomething round like the fruit of round Birthwort : in ſtead of roots there grow forth a multitude of ſlender ſtrings.

‡ 6 *Piſtolochia Cretica, ſiue Virginiana.*　Virginian Snake-root.

‡ 6　*Cluſius* figures and deſcribes another ſmal *Piſtolochia*,by the name of *Piſtolochia Cretica*,to which I thought good to adde the Epithit *Virginiana* alſo, for that the much admired Snakeweed of Virginia ſeems no otherwiſe to differ from it than an inhabitant of Candy from one of the Virginians,which none I think wil ſay to differ in *ſpecie*.I wil firſt giue *Cluſius* his deſcription,and then expreſſe the little varietie that I haue obſerued in the plants that were brought from Virginia,and grew here with vs : It ſends forth many ſlender ſtalks a foot long,more or leſſe,and theſe are cornered or indented,creſted,branched,tough,and bending towards the ground,or ſpred thereon,and of a darke greene colour ; vpon which without order grow leaues neruous, and like thoſe of the laſt deſcribed,but much ſharper pointed,and after a ſort reſembling the ſhape of thoſe of *Smilax aſpera* ; but leſſe;and of a darke and laſting greene colour,faſtned to longiſh ſtalks ; out of whoſe boſomes grow long and crooked floures in ſhape like thoſe of the long Birthwort,but of a darker red on the outſide,but ſomewhat yellowiſh within ; and theſe are alſo faſtned to pretty long ſtalkes, and are ſucceeded by fruit not vnlike, yet leſſe than that of the long Birthwort. This hath aboundance of roots like as the former,but much ſmaller,and more fibrous,and of a ſtronger ſmel:it flours in Iuly and Auguſt. Thus *Cluſius* deſcribes his : To which that Snake-weed that was brought from *Virginia*,and grew with Mᵗ *Iohn Tradeſcant* at South Lambeth, *An.*1632, was agreeable in all points, but here and there one of the lower leaues were ſomewhat broader, and rounder pointed than the reſt : the floure was long, red, crooked, and a little hairy, and it did not open the top or ſhew the inner ſide,which I iudge was by reaſon of the coldneſſe and vnſeaſonableneſſe of the later part of the ſomer when it floured : the ſtalks in the figure ſhould haue bin expreſt more crooking or indenting, for they commonly grow ſo. How hard it is to iudge of plants by one particle or faculty may very well appeare by this herb I now treat of : for ſome by the ſimilitude the root had with *Aſarum*, and a vomiting quality which they attributed to it(which certainly is no other than accidental)would forthwith pronounce and maintaine it an *Aſarum* : ſome alſo refer it to other things,as to Primroſes,*Vincetoxicum*,&c. Others more warily name it *Serpentaria Virginiana*,& *Radix Virginiana*, names as it were offering themſelues,and eaſily fitted and impoſed vpon ſundry things,but yet too general,and therfore not fit any more to be vſed,ſeeing the true and ſpecifique denomination is found.‡

¶ *The Place.*

Pliny ſheweth, That the Birthworts grow in fat and Champian places : The fields of Spaine are full

full of these three long and round Birthworts : they are also found in Italy and Narbone or Languedoc a country of France. *Petrus Bellonius* writeth, That he found branched Birthwort vpon Ida a mountaine in Candy. *Clusius* saith he found this about Hispalis, & in many other places of Granado in Spain among bushes and brambles. They grow all in my garden.

¶ *The Time.*

They floure in May, Iune, and Iuly.

¶ *The Names.*

Birthwort is called in Greeke Ἀριστολοχία : in Latine likewise *Aristolochia*, because it is ἀρίστη ταῖς λόχοις that is to say good for women newly brought to bed or deliuered with childe : in English, Birthwort, Hartwort, and of some *Aristolochia*.

The first is called *Aristolochia longa*, or long Birthwort, of the forme of his root, and also *Aristolochia mas* or male Birthwort : the second is thought to be *Fæmina*, or female Birthwort, and is called *Rotunda Aristolochia*, or round Birthwort ; of diuers also, *Terra Malum*, the Apple of the earth : yet *Cyclaminus* is also called *Terra Malum*, or Earth Apple.

¶ *The Temperature.*

All these Birthworts are of temperature hot and dry, euen in the third degree, hauing power to clense.

¶ *The Vertues.*

Dioscorides writeth, That a dram weight of long Birthwort drunke with wine and so applied, is **A** good against serpents and deadly things : and that being drunke with Myrrhe and pepper, it expels whatsoeuer is left in the matrice after the childe is deliuered, the flowrs also and dead childe ; and that being put in a pessarie it performeth the same.

Round Birthwort serues for all these things, and for the rest of the other poisons: it preuails also **B** against the stuffing of the lungs, the hicket, the shakings or shiuerings of agues, hardnesse of the milt or spleen, burstings, cramps, convulsions, and pain of the sides, if it be drunk with water.

It plucketh out thornes, splinters, or shiuers, and being mixed in plaisters or pulteßes, it drawes **C** forth scales or bones, remoueth rottennesse or corruption, mundifieth and scoureth foul and filthy vlcers, and filleth them vp with new flesh, if it be mixed with Ireos and hony.

Galen saith, That branched Birthwort is of a more sweet and pleasant smell, and therefore is vsed **D** in ointments ; but it is weaker in operation than the former.

Birth-wort, as *Pliny* writeth, being drunke in water, is a most excellent remedie for cramps, con- **E** vulsions, bruises, and for such as haue fallen from high places.

It is good against short windednesse and the Falling sicknesse.   **F**

The round *Aristolochia* doth beautifie, clense, and fasten the teeth, if they be often fretted or rub- **G** bed with the pouder thereof.

‡ The root of the Virginian *Pistolochia*, which is of a strong and aromatick sent, is a singular and **H** much vsed antidote against the bite of the Rattle-snake, or rather Adder or Viper, whose bite is very deadly; and therefore by the prouidence of the Creator hee hath vpon his taile a skinny dry substance parted into cels, which contain some loose hard dry bodies that rattle in them (as if one should put little stones or pease into a stiffe and very dry bladder) that so he may by this noise giue warning of his approch, the better to be auoided: but if any be bitten, they know nor stand in need of no better antidote than this root, which they chew and apply to the wound, & also swallow some of it downe, by which means they quickly ouercome the malignitie of this poisonous bite, which otherwise in a very short time would proue deadly. Many also commend the vse of this against the plague, small pox, measels, and such like maligne and contagious diseases. ‡

CHAP. 312.   *Of Violets.*

¶ *The Kindes.*

THere might be described many kindes of floures vnder this name of Violets, if their differences should be more curiously looked into than is necessarie : for we might ioine hereunto the stock Gillofloures, Wall-floures, Dames Gillofloures, Marian violets, & likewise some of the bulbed floures, because some of them by *Theophrastus* are termed Violets. But this was not our charge, holding it sufficient to distinguish and diuide them as neere as may be in kindred and neighbourhood ; addressing my selfe vnto the Violets called the blacke or purple violets, or March Violets of the garden, which haue a great prerogatiue aboue others, not only because the mind conceiueth a certain pleasure and recreation by smelling and handling those most odoriferous floures, but also for that very many by these violets receiue ornament and comely grace; for there be made of them garlands for the head, nosegaies and poesies, which are delightfull to looke on and pleasant to smel to, speaking nothing of their appropriat vertues; yea gardens themselues receiue by these the greatest

test ornament of all, chiefest beauty, and most excellent grace, and the recreation of the minde which is taken hereby cannot be but very good and honest ; for they admonish and stirre vp a man to that which is comely and honest ; for floures through their beauty, variety of colour, and exquisit forme, do bring to a liberall and gentle manly minde, the remembrance of honestie, comlinesse, and all kindes of vertues : for it would be an vnseemly and filthy thing (as a certaine wise man saith) for him that doth looke vpon and handle faire and beautiful things, to haue his mind not faire, but filthy and deformed.

¶ *The Description.*

1   THe blacke or purple Violet doth forthwith bring from the root many leaues, broad, sleightly indented in the edges, rounder than the leaues of Iuy; among the midst whereof spring vp fine slender stems, and vpon euery one a beautifull flour sweetly smelling, of a blew darkish purple, consisting of fiue little leaues, the lowest whereof is the greatest: after them do appeare little hanging cups or knaps, which when they be ripe do open and diuide themselues into three parts. The seed is smal, long, and somwhat round withall: the root consisteth of many threddy strings.

1 *Viola nigra siue purpurea.*
   The purple garden Violet.

2 *Viola flore albo.*
   The white garden Violet.

2   The white garden Violet hath many milke white floures, in forme and figure like the precedent ; the colour of whose floures especially setteth forth the difference.

3   The double garden Violet hath leaues, creeping branches, and roots like the garden single Violet ; differing in that, that this Violet bringeth forth most beautifull sweet double floures, and the other single.

4   The white double Violet likewise agrees with the other of his kinde, differing onely in the colour ; for as the last described bringeth double blew or purple flours, contrariwise this plant beareth double white floures, which maketh the difference.

5   The yellow Violet is by nature one of the wilde Violets, for it groweth seldom any where but vpon most high and craggy mountaines, from whence it hath been diuers times brought into the garden, but it can hardly be brought to culture or grow in the garden without great industrie. And by the relation of a gentleman often remembred, called Master *Thomas Hesketh*, who found it

growing

3 *Viola martia purpurea multiplex.*
The double garden purple Violet.

5 *Viola martia lutea.*
Yellow Violets.

† 6 *Viola canina ſylueſtris.*
Dogs Violets, or wilde Violets.

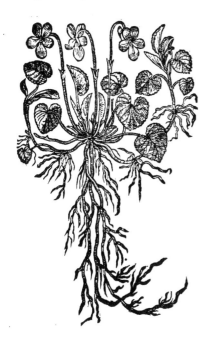

growing vpon the hils in Lancaſhire, neere
vnto a village called Latham ; and though
he brought them into his garden, they withered and pined. The whole plant is deſcribed to be like vnto the field Violet, and
differeth from it, in that this plant brings
forth yellow floures, yet like in forme and
figure, but without ſmell.

6   The wilde field Violet with round
leaues riſeth forth of the ground from a
fibrous root , with long ſlender branches,
whereupon do grow round ſmooth leaues.
The floures grow at the top of the ſtalkes,
of a light blew colour : ‡ and this growes
commonly in woods and ſuch like places,
and floures in Iuly and Auguſt. There is
another varietie of this wilde Violet , hauing the leaues longer, narrower, and ſharper pointed: and this was formerly figured
& deſcribed in this place by our Author. ‡

7   There is found in Germany about
Noremberg and Strasburg, a kind of Violet altogether a ſtranger in theſe parts. It
hath (ſaith my Author) a thick tough root
of a wooddy ſubſtance, from which riſeth
vp a ſtalk diuiding it ſelf into diuers branches of a woody ſubſtance : whereon grow
long jagged leaues like thoſe of the panſy:
the floures grow at the top, compact of fine
leaues apiece, of a watchet colour.

¶ *The*

¶ *The Place.*

The Violet groweth in gardens almoſt euery where : the others which are ſtrangers haue beene touched in their deſcriptions.

¶ *The Time.*

The floures for the moſt part appeare in March, at the fartheſt in Aprill.

¶ *The Names.*

The Violet is called in Greeke, ʹɪ: of *Theophraſtus*, both ʹɪοπμίλιω, and μλαιον : in Latine, *Nigra viola* or blacke Violet, of the blackiſh purple colour of the floures. The Apothecaries keepe the Latine name *Viola*, but they call it *Herba Violaria*, and *Mater Violarum* : in high-Dutch, **Blaw Viel** : in low Dutch **Violetten** : in French, *Violette de Mars* : in Italian, *Viola mammola* : in Spaniſh, *Violeta* : in Engliſh, Violet. *Nicander* in his Geoponicks beleeueth (as *Hermolaus* ſheweth) that the Grecians did call it ʹɪν becauſe certaine Nymphs of Iönia gaue that floure firſt to *Iupiter*. Others ſay it was called ʹɪν, becauſe when *Iupiter* had turned the yong damoſell *Iö*, whom he tenderly loued, into a Cow, the earth brought forth this floure for her food ; which being made for her ſake, receiued the name from her : and thereupon it is thought that the Latines alſo called it *Viola*, as though they ſhould ſay *Vitula*, by blotting out the letter *t*. *Seruius* reporteth, That for the ſame cauſe the Latines alſo name it *Vaccinium*, alledging the place of *Virgil* in his Bucolicks ;

Alba liguſtra cadunt, vaccinia nigra leguntur.

Notwithſtanding *Virgil* in his tenth Eclog ſheweth, that *Vaccinium* and *Viola* do differ:

Et nigræ violæ ſunt, & vaccinia nigra.

† *Vitruuius* alſo, *lib.* 7. of Architecture or Building, doth diſtinguiſh *Viola* from *Vaccinium* : for he ſheweth, that the colour called *Sile Atticum* or the Azure of Athens, is made *ex Viola* ; and the gallant purple, *ex Vaccinio*. The Dyers, ſaith he, when they would counterfeit *Sile* or Azure of Athens, put the dried violets into a fat, kettle, or caldron, and boile them with water : afterwards when it is tempered, they poure it into a linnen ſtrainer, and wringing it with their hands, receiue into a mortar the liquor coloured with the Violets ; and ſteeping earth of *Erethria* in it, and grinding it, they make the Azure colour of Athens. After the ſame maner they temper *Vaccinium*, and putting milk vnto it do make a gallant purple colour. But what *Vaccinia* are we will elſewhere declare.

¶ *The Temperature.*

The floures and leaues of Violets are cold and moiſt.

¶ *The Vertues.*

A   The floures are good for all inflammations, eſpecially of the ſides and lungs ; they take away the hoarſeneſſe of the cheſt, the ruggedneſſe of the winde-pipe and iawes, allay the extreme heate of the liuer, kidnies, and bladder, mitigate the fiery heate of burning agues, temper the ſharpneſſe of choler, and take away thirſt.

B   There is an oile made of Violets which is likewiſe cold and moiſt. The ſame being anointed vpon the teſticles doth gently prouoke ſleep which is hindred by an hot and dry diſtemper: mixed or labored together in a woodden diſh with the yelke of an egge, it aſſwageth the pain of the fundament and hemorrhoids : it is likewiſe good to be put into cooling cliſters, and into pulteſſes that coole and eaſe paine.

C   But let the oile in which the Violets be ſteeped be either of vnripe Oliues, called *Omphacinum*, or of ſweet almonds, as *Meſues* ſaith ; and the Violets themſelues muſt be freſh and moiſt, for being dry and hauing loſt their moiſture they do not coole, but ſeem to haue gotten a kinde of heate.

D   The later phyſitions thinke it good to mix dry Violets with medicines that are to comfort and ſtrengthen the heart.

E   The leaues of Violets inwardly taken do coole, moiſten, and make the belly ſoluble. Being outwardly applied, they mitigate all kinde of hot inflammations, both taken by themſelues, and alſo applied with Barley floure dried at the fire, after it hath lien ſoking in the water. They are likewiſe laid vpon a hot ſtomacke, and on burning eies, as *Galen* witneſſeth. *Dioſcorides* writeth, that they be alſo applied to the fundament that is fallen out.

F   They may help the fundament that is fallen out, not as a binder keeping backe the fundament, but as a ſuppler and a mollifier. Beſides, *Pliny* ſaith that Violets are as well vſed in garlands, as for ſmell, and are good againſt ſurfeting, heauineſſe of the head ; and being dried in water & drunk, they remoue the Squinancie or inward ſwellings of the throat. They cure the falling ſickeneſſe, eſpecially in yong children, and the ſeed is good againſt the ſtinging of ſcorpions.

G   There is a ſyrrup made of Violets and ſugar, whereof three or foure ounces taken at one time, ſoften the belly and purge choler. The manner to make it is as followes:

H   Firſt make of clarified ſugar by boiling, a ſimple ſyrrup of good conſiſtence or meane thickeneſſe, whereunto put the floures cleane picked from all manner of filth, as alſo the white ends nipped away, a quantitie according to the quantity of the ſyrrup, to your owne diſcretion, wherein let them infuſe or ſteep foure and twenty houres, and ſet vpon a few warme embers ; then ſtrain it, and put more violets into the ſame ſyrrup : thus do three or foure times, the oftner the better, then ſet them vpon a gentle fire to ſimper, but not to boile in any wiſe : ſo haue you it ſimply made of a moſt perfect purple colour, and of the ſmell of the floures themſelues. Some do adde therto a little of the juice of the floures in the boiling, which maketh it of better force and vertue. Likewiſe ſome do put a little quantitie of the juice of Lymmons in the boiling, which doth greatly increaſe the beauty thercof, but nothing at all the vertue.

There is likewiſe made of Violets and ſugar certaine plates called Sugar violet, Violet tables, I or Plate, which is moſt pleaſant and wholeſome, eſpecially it comforteth the heart and the other inward parts.

The decoction of Violets is good againſt hot feuers, and the inflammation of the liuer, and all K other inward parts : the like propertie hath the juyce, ſyrrup or conſerue of the ſame.

Syrrup of Violetts is good againſt the inflammation of the lungs and breſt, againſt the pluriſie L and cough, againſt feuers and agues in yong children, eſpecially if you put vnto an ounce of ſyrrup eight or nine drops of oyle of Vitrioll, and mix it together, and giue it to the childe a ſpoonefull at once.

The ſame giuen in manner aforeſaid is of great efficacie in burning feuers and peſtilent diſea- M tes, greatly cooling the inward parts : and it may ſeeme ſtrange to ſome that ſo ſharpe a corroſiue as oile of Vitrioll ſhould be giuen into the bodie ; yet being delayed and giuen as aforeſaid, ſuck-ing children may take it without any perill.

The ſame taken as aforeſaid cures all inflammations of the throat, mouth, uvula, ſquinancy, and N the falling euill in children.

Sugar-Violet hath power to ceaſe inflammations, roughneſſe of the throat, comforts the heart, O aſſwageth paines of the head, and cauſeth ſleep.

The leaues of Violets are vſed in cooling plaiſters, oiles, and comfortable cataplaſmes or pul- P teſſes ; and are of greater efficacie among other herbs, as Mercurie, Mallowes, and ſuch like, in cli-ſters for the purpoſes aforeſaid.

CHAP. 313.　　*Of Hearts-eaſe, or Panſies.*

¶ *The Deſcription.*

1　THe Hearts-eaſe or Panſie hath many round leaues at the firſt comming vp ; afterward they grow ſomewhat longer, ſleightly cut about the edges, trailing or creeping vpon the ground : the ſtalks are weake and tender, whereupon grow floures in form & figure like the Violet, and for the moſt part of the ſame bigneſſe, of three ſundry colours, whereof it tooke the ſyrname *Tricolor*, that is to ſay, purple, yellow, and white or blew ; by reaſon of the beauty and brauerie of which colours they are very pleaſing to the eye, for ſmel they haue little or none at all. The ſeed is contained in little knaps of the bigneſſe of a Tare, which come forth after the floures be fallen, and do open of themſelues when the ſeed is ripe. The root is nothing elſe but as it were a bundle of threddy ſtrings.

2　The vpright Panſie bringeth forth long leaues deeply cut in the edges, ſharpe pointed, of a bleake or pale green colour, ſet vpon ſlender vpright ſtalkes, cornered, jointed, or kneed a foot high or higher ; whereupon grow very faire floures of three colours, viz. of purple, blew, and yellow, in ſhape like the common Hearts-eaſe, but greater and fairer ; which colours are ſo excellently and orderly placed, that they bring great delight to the beholders, though they haue little or no ſmell at all : for oftentimes it hapneth that the vppermoſt floures are differing from thoſe that grow vpon the middle of the plant, and thoſe vary from the lowermoſt, as Nature liſt to dally with things of ſuch beauty. The ſeed is like that of the precedent.

3　The wilde Panſie differeth from that of the garden, in leaues, roots, and tender branches : the floures of this wilde one are of a bleak and pale colour, far inferior in beauty to that of the garden, wherein conſiſteth the difference.

4　Stony Hearts-eaſe is a baſe and low plant : the leaues are rounder, not ſo much cut about the edges as the others : the branches are weak and feeble, trailing vpon the ground : the flours are likewiſe of three colours, that is to ſay, white, blew, and yellow, void of ſmell : the root periſheth ha-uing perfected his ſeed.

5　There is found in ſundry places of England a wilde kinde hereof, hauing floures of a feint yellow colour, without mixture of any other colour, yet hauing a deeper yellow ſpot in the lowest

leaſe,

1 *Viola tricolor.*
Hearts eaſe.

2 *Viola aſſurgens tricolor.*
Vpright Hearts eaſe.

3 *Viola tricolor ſylveſtris.*
Wilde Paunſies.

4 *Viola tricolor petræa.*
Stony Hearts eaſe.

leaſe with foure or fiue blackiſh purple lines, wherein it differeth from the other wilde kinde : and this hath been taken of ſome young Herbariſts to be the yellow Violet.

¶ *The Place.*

The Hearts caſe groweth in fields in many places, and in gardens alſo, and that oftentimes of it ſelfe : it is more gallant and beautifull than any of the wilde ones.

Matthiolus reporteth, that the vpright Paunſie is found on mount **Baldus in Italy**. *Lobel* ſaith that it groweth in Languedocke in France, and on the tops of ſome hills in England ; but as yet I haue not ſeene the ſame.

Thoſe with yellow floures haue beene found by a village in Lancaſhire called Latham , foure miles from Kyrcham, by Mr. *Thomas Hesketh* before remembred.

¶ *The Time.*

They floure not onely in the Spring, but for the moſt part all Summer thorow, euen vntill Autumne.

¶ *The Names.*

Hearts eaſe is named in Latine *Viola tricolor*, or the three coloured Violet ; and of diuers, *Iacea* ; (yet there is another *Iacea* ſyrnamed *Nigra* : in Engliſh, Knap-weed, Bull-weed, and Matfellon) of others, *Herba Trinitatis*, or herbe Trinitie, by reaſon of the triple colour of the floures : of ſome others, *Herba Clauellata* : in French, *Penſees* : by which name they became knowne to the Brabanders and others of the Low-countries that are next adioyning. It ſeemeth to be *Viola flammea*, which *Theophraſtus* calleth φλόγα, which is alſo called φλόγιον : in Engliſh, Hearts-eaſe, Paunſies, Liue in idleneſſe, Cull me to you, and Three faces in a hood.

The vpright Panſie is called not vnproperly *Viola aſſurgens*, or *Surrecta*, and withall *Tricolor*, that is to ſay, ſtraight or vpright Violet three coloured : of ſome, *Viola arboreſcens*, or Tree Violet, for that in the multitude of branches and manner of growing, it reſembles a little Tree.

¶ *The Nature.*

It is of temperature obſcurely cold, but more euidently moiſt, of a tough and ſlimie juice, like that of the Mallow ; for which cauſe it moiſtneth and ſuppleth , but not ſo much as the Mallow doth.

¶ *The Vertues.*

It is good, as the later Phyſitians write, for ſuch as are ſicke of an ague, eſpecially children and **A** infants, whoſe convulſions and fits of the falling ſickneſſe it is thought to cure.

It is commended againſt inflammations of the lungs and cheſt, and againſt ſcabs and itchings **B** of the whole bodie, and healeth vlcers.

The diſtilled water of the herbe or floures giuen to drinke for ten or more daies together, three **C** ounces in the morning, and the like quantitie at night , doth wonderfully eaſe the paines of the French diſeaſe, and cureth the ſame, if the patient be cauſed to ſweat ſundry times, as *Coſtæus* reporteth, in his booke *de natura Vniuerſ. ſtirp.*

Chap. 314.　*Of Ground-Iuy, or Ale-hoofe.*

¶ *The Deſcription.*

‡ GRound Ivy is a low or baſe herbe ; it creepeth and ſpreads vpon the ground hither and thither all about, with many ſtalkes of an vncertaine length, ſlender, and like thoſe of the Vine, ſomething cornered, and ſometimes reddiſh : whereupon grow leaues ſomething broad and round, wrinckled, hairy, nicked in the edges, for the moſt part two out of euerie joint : amongſt which come forth the floures gaping like little hoods, not vnlike to thoſe of Germander, of a purpliſh blew colour : the roots are very threddy : the whole plant is of a ſtrong ſmell and bitter taſte.

‡ 2 Vpon the rockie and mountainous places of Prouince and Daulphine grows this other kinde of Ale-hoofe, which hath leaues, ſtalkes, floures, and roots like in ſhape to thoſe of the former, but the floures and leaues are of a light purple colour, and alſo larger and longer. This by *Lobel* is called *Aſarina, ſiue ſaxatilis hedera*. ‡

¶ *The Place.*

It is found as well in tilled as in vntilled places, but moſt commonly in obſcure and darke places, vpon banks vnder hedges, and by the ſides of houſes.

¶ *The Time.*

It remaineth greene not onely in Summer, but alſo in Winter at any time of the yeare : it floureth from Aprill till Summer be far ſpent.

¶ *The*

1 *Hedera terreftris.*
Ale-hoofe.

‡ 2 *Hedera faxatilis.*
Rocke Ale-hoofe.

¶ *The Names.*

It is commonly called *Hedera terreftris :* in Greeke, χαμαιϰισσος : alfo *Corona terra :* in High-Dutch, **Gundelreb :** in Low-Dutch, **Onderhaue :** in French, *Lierre terreftre : Hedera humilis* of fome, and *Chamæcißum :* in Englifh, Ground-Ivy, Ale-hoofe, Gill go by ground, Tune-hoof, and Cats-foot. ‡ Many queftion whether this be the *Chamæcißus* of the Antients : which controuerfie *Dodonæus* hath largely handled, *Pempt.3. lib.3.cap.4.* ‡

¶ *The Temperature.*

Ground Ivie is hot and dry, and becaufe it is bitter it fcoureth, and remoueth ftoppings out of the intrals.

¶ *The Vertues.*

A   Ground-Ivy is commended againft the humming noyfe and ringing found of the eares, being put into them, and for them that are hard of hearing.

B   *Matthiolus* writeth, That the juice being tempered with Verdugreafe, is good againft fiftulaes and hollow vlcers.

C   *Diofcorides* teacheth, That halfe a dram of the leaues being drunk in foure ounces and a halfe of faire water, for fortie or fiftie daies together, is a remedie againft the Sciatica, or ache in the huckle bone.

D   The fame taken in like fort fix or feuen daies doth alfo cure the yellow jaundife. *Galen* hath attributed (as we haue faid) all the vertue vnto the floures : Seeing the floures of Ground-Ivy (faith he) are very bitter, they remoue ftoppings out of the liuer, and are giuen to them that are vexed with the Sciatica.

E   Ground-Ivy, Celandine, and Daifies, of each a like quantitie, ftamped and ftrained, and a little fugar and rofe water put thereto, and dropped with a feather into the eies, taketh away all manner of inflammation, fpots, webs, itch, fmarting, or any griefe whatfoeuer in the eyes, yea although the fight were nigh hand gone : it is proued to be the beft medicine in the world.

F   The herbes ftamped as aforefaid, and mixed with a little ale and honey, and ftrained, take away the pinne and web, or any griefe out of the eyes of horfe or cow, or any other beaft, being fquirted intothe fame with a fyringe, or I might haue faid the liquor injected into the eies with a fyringe. But I lift not to be ouer eloquent among Gentlewomen, to whom efpecially my Workes are moft neceffarie.

G   The women of our Northerne parts, efpecially about Wales and Chefhire, do turne the herbe Ale-hoof into their Ale ; but the reafon thereof I know not : notwithftanding without all controuerfie

uerſie it is moſt ſingular againſt the griefes aforeſaid ; being tunned vp in ale and drunke, it alſo purgeth the head from rheumaticke humors flowing from the braine.

— *Hedera terreſtris* boyled in water ſtaieth the termes ; and boyled in mutton broth it helps weake   I and aking backes.

They haue vſed to put it into ointments againſt burning with fire, gunpouder, and ſuch like.   K

Hedera terreſtris, being bound in a bundle, or chopt as herbes for the pot, and eaten and drunke in   L thin broth ſtaieth the flux in women.

Chap. 315. *Of Iuy.*

¶ *The Kindes.*

THere be two kindes of Iuie, as *Theophraſtus* witneſſeth, reckoned among the number of thoſe plants which haue need to be propped vp ; for they ſtand not of themſelues, but are faſtned to ſtone walls, trees, and ſuch like, and yet notwithſtanding both of a wooddy ſubſtance, and yet not to be placed among the trees, ſhrubs, or buſhes, becauſe of the affinitie they haue with climbing herbes ; as alſo agreeing in forme and figure with many other plants that climbe, and are indeed ſimply to be reckoned among the herbes that clamber vp. But if any will cauill, or charge me with my promiſe made in the beginning of this hiſtory, where we made our diuiſion, namely, to place each plant as neere as may be in kindred and neighbourhood ; this promiſe I haue fulfilled, if the curious eie can be content to reade without raſhneſſe thoſe plants following in order, and not only this climbing Iuie that lifteth her ſelfe to the tops of trees, but alſo the other Iuie that creepeth vp-on the ground.

Of the greater or the climing Iuie there are alſo many ſorts ; but eſpecially three, the white, the blacke, and that which is called *Hedera Helix*, or *Hedera ſterilis*.

1 *Hedera corymboſa.*
Clymbing or berried Iuie.

2 *Hedera Helix.*
Barren or creeping Iuie.

¶ *The Deſcription.*

1   THe greater Iuie climbeth on trees, old buildings, and walls : the ſtalkes thereof are wooddy, and now and then ſo great as it ſeemes to become a tree ; from which it ſen-deth a multitude of little boughes or branches euery way, whereby as it were with

armes it creepeth and wandereth far about: it alſo bringeth forth contiually fine little roots, by which it faſtneth it ſelfe and cleaueth wonderfull hard vpon trees, and vpon the ſmootheſt ſtone walls: the leaues are ſmooth, ſhining eſpecially on the vpper ſide, cornered with ſharpe pointed corners. The floures are very ſmall and moſſie ; after which ſucceed bundles of black berries, euery one hauing a ſmall ſharpe pointall.

There is another ſort of great Iuie that bringeth forth white fruit, which ſome call *Acharnicam irriguam*: and alſo another leſſer, the which hath blacke berries. This *Pliny* calleth *Selinitium*.

We alſo finde mentioned another ſort hereof ſpred abroad, with a fruit of a yellow Saffron colour, called of diuers, *Dionyſias*, as *Dioſcorides* writeth : others *Bacchica*, of which the Poëts vſed to make garlands, as *Pliny* teſtifieth, *lib. 16.cap.34.*

2   Barren Iuie is not much vnlike vnto the common Iuie aforeſaid, ſauing that his branches are both ſmaller and tenderer, not lifting or bearing it ſelfe vpward, but creeping along by the ground vnder moiſt and ſhadowie ditch bankes. The leaues are moſt commonly three ſquare, cornered, of a blackiſh greene colour, which at the end of Summer become browniſh red vpon the lower ſide. The whole plant beareth neither floures nor fruit, but is altogether barren and fruitleſſe.

‡ 3   There is kept for noueltes ſake in diuers gardens a Virginian, by ſome (though vnfitly) termed a Vine, being indeed an Iuie. The ſtalkes of this grow to a great heighth, if they be planted nigh any thing that may ſuſtaine or beare them vp : and they take firſt hold by certaine ſmall tendrels, vpon what body ſoeuer they grow, whether ſtone, boords, bricke, yea glaſſe, and that ſo firmely, that oftentimes they will bring pieces with them if you plucke them off. The leaues are large, conſiſting of foure, fiue, or more particular leaues, each of them being long, and deepely notched about the edges, ſo that they ſomewhat reſemble the leaues of the Cheſnut tree : the floures grow cluſtering together after the manner of Iuie, but neuer with vs ſhew themſelues open, ſo that we cannot iuſtly ſay any thing of their color, or the fruit that ſucceeds them. It puts forth his leaues in Aprill, and the ſtalkes with the rudiments of the floures are to be ſeene in Auguſt. It may as I ſaid be fitly called *Hedera Virginiana.* ‡

¶ *The Place.*

Iuie groweth commonly about walls and trees ; the white Iuie groweth in Greece, and the barren Iuie groweth vpon the ground in ditch bankes and ſhadowie woods.

¶ *The Time.*

Iuie flouriſheth in Autumne : the berries are ripe after the Winter Solſtice.

¶ *The Names.*

Iuie is called in Latine, *Hedera* : in Greeke, χίϑος, and χιϲϲος : in high Dutch, **Epheu** : in low Dutch, **Weple** : in Spaniſh, *Yedra* : in French, *Liarre.*

The great Iuie is called of *Theophraſtus*, ὑψος χιϲϲος : in Latine, *Hedera attollens*, or *Hedera aſſurgens* : *Gaza* interpreteth it *Hedera excelſa*. The later Herbariſts would haue it to be *Hedera arborea*, or tree Iuie, becauſe it groweth vpon trees, and *Hedera muralis*, which hangeth vpon walls.

Creeping or barren Iuie is called in Greeke, χϑὶμαιχιϲϲος : in Engliſh, Ground-Iuie : yet doth it much differ from *Hedera terreſtris* or Ground-Iuie before deſcribed : of ſome it is called *Clauicula*, *Hedera Helix*, and *Hedera ſterilis* ; and is that herbe wherein the Bore delighteth, according to *Iohannes Khuenius*.

¶ *The Temperature.*

Iuie, as *Galen* ſaith, is compounded of contrarie faculties ; for it hath a certaine binding earthy and cold ſubſtance, and alſo a ſubſtance ſomewhat biting, which euen the very taſte doth ſhew to be hot. Neither is it without a third faculty, as being of a certaine warme watery ſubſtance, and that is if it be greene : for whileſt it is in drying, this watery ſubſtance being earthy, cold, and binding conſumeth away, and that which is hot and biting remaineth.

¶ *The Vertues.*

A   The leaues of Iuie freſh and greene boyled in wine, do heale old vlcers, and perfectly cure thoſe that haue a venomous and malitious quality joyned with them ; and are a remedy likewiſe againſt burnings and ſcaldings.

B   Moreouer, the leaues boiled in vinegre are good for ſuch as haue bad ſpleenes ; but the floures or fruit are of more force, being very finely beaten and tempered with vinegre, eſpecially ſo vſed they are commended againſt burnings.

C   The juyce drawne or ſnift vp into the noſe doth effectually purge the head, ſtaieth the running of the eares that hath beene of long continuance, and healeth old vlcers both in the eares and alſo in the noſthrils : but if it be too ſharpe, it is to be mixed with oile of Roſes, or ſallad oyle.

D   The gum that is found vpon the trunke or body of the old ſtocke of Iuie, killeth nits and lice, and taketh away haire : it is of ſo hot a quality, as that it doth obſcurely burne : it is as it were a

cettaine

certaine waterifh liquor congealed of thofe gummie drops. Thus farre *Galen*.

The very fame almoſt hath *Dioſcorides*, but yet alſo fomewhat more: for ouer and befides he faith, **E** that fiue of the berries beaten fmall, and made hot in a Pomegranat rinde, with oyle of rofes, and dropped into the contrary eare, doth eafe the tooth-ach; and that the berries make the haire blacke.

Iuie in our time is very feldome vfed, faue that the leaues are laid vpon little vlcers made in the **F** thighes, legs, or other parts of the body, which are called iffues ; for they draw humors and wate-rifh fubftance to thofe parts, and keepe them from hot fwellings or inflammations, that is to fay, the leaues newly gathered, and not as yet withered or dried.

Some likewife affirme that the berries are effectuall to procure vrine ; and are giuen vnto thofe **G** that be troubled with the ftone and difeafes of the kidnies.

The leaues laid in fteepe in water for a day and a nights fpace, helpe fore and fmarting waterifh **H** eies, if they be bathed and wafhed with the water wherein they haue beene infufed.

CHAP. 316.　*Of rough Binde-weed.*

1　*Smilax Peruuiana, Salſa parilla*.
Rough Binde-weed of Peru.

2　*Smilax aſpera*.
Common rough Bindeweed.

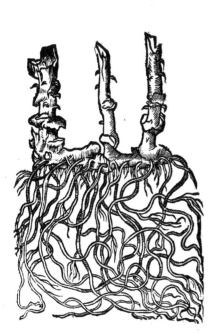

¶ *The Deſcription.*

1　ALthough we haue great plenty of the roots of this Binde-weed of Peru, which we vfu-ally cally *Zarʒa*, or *Sarſa parilla*, wherewith diuers griefes and maladies are cured, and that thefe roots are very well knowne to all ; yet fuch hath beene the carelefneffe and fmall prouidence of fuch as haue trauelled into the Indies, that hitherto not any haue giuen vs in-ftruction fufficient, either concerning the leaues, floures, or fruit : onely *Monardus* faith, that it hath long roots deepe thruſt into the ground : which is as much as if a great learned man fhould tell the fimple, that our common carrion Crow were of a blacke colour. For who is fo blinde that feeth the root it felfe, but can eafily affirme the root to be very long ? Notwithſtanding, there is in the reports of fuch as fay they haue feene the plant it felfe growing, fome contradiction or con-trarietie : fome report that it is a kind of Bindweed, and efpecially one of thefe rough Bind-weeds :

　　　　others

3 *Smilax aſpera Luſitanica.*
Rough Bind-weed of Portugall.

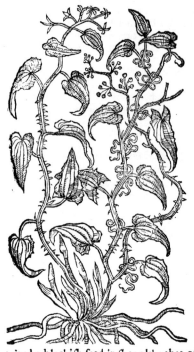

others,as one Mr.*White* an excellent painter,
who carried very many people into Virginia
(or after ſome Norembega)there to inhabit,
at which time he did ſee thereof great plenty,as he himſelfe reported vnto me,with this
bare deſcription; It is (ſaith he) the root of
a ſmall ſhrubby tree, or hedge tree, ſuch as
are thoſe of our country called Haw-thorns,
hauing leaues reſembling thoſe of Iuy, but
the floures or fruit he remembreth not. ‡ It
is moſt certaine, that *Sarſa parilla* is the root
of the Americane *Smilax aſpera*,both by conſent of moſt Writers, and by the relation of
ſuch as haue ſeene it growing there. ‡
   2   The common rough Bind-weed hath
many branches ſet full of little ſharpe prickles, with certaine claſping tendrels, wherewith it taketh hold vpon hedges, ſhrubs,and
whatſoeuer ſtandeth next vnto it, winding
and claſping it ſelfe about from the bottome
to the top;whereon are placed at euery joint
one leafe like that of Iuie,without corners,
ſharpe pointed, leſſer and harder than thoſe
of ſmooth Binde-weed, oftentimes marked
with little white ſpots, and garded or borde-
red about the edges with crooked prickles.
The floures grow at the top of crooked ſtalks
of a white colour,and ſweet of ſmell. After
commeth the fruit like thoſe of the wilde
Vine, greene at the firſt, and red when they
be ripe, and of a biting taſte ; wherein is con-

tained a blackiſh ſeed in ſhape like that of hempe. The root is long, ſomewhat hard,and parted
into very many branches.
   3   This rough Binde-weed,found for the moſt part in the barren mountaines of Portugall,diffe-
reth not from the precedent in ſtalkes and floures, but in the leaues and fruit ; for the leaues are
ſofter, and leſſe prickly, and ſometimes haue no prickles at all, and they are alſo oftentimes much
narrower : the fruit or berry is not red but blacke when as it commeth to be ripe. The root hereof
is one ſingle root of a wooddy ſubſtance, with ſome fibres annexed thereto, wherein conſiſteth the
difference.

¶ *The Place.*

Zarza Parilla, or the prickly Binde-weed of America, groweth in Peru a prouince of America,
in Virginia, and in diuers other places both in the Eaſt and Weſt Indies.
   The others grow in rough and vntilled places, about the edges and borders of fields, on mountaines and vallies,in Italy,Languedoc in France, Spaine, and Germany.

¶ *The Time.*

They floure and flouriſh in the Spring : their fruit is ripe in Autumne,or a little before.

¶ *The Names.*

It is named in Greeke, ςμιλαξ τραχεια. *Gaza* (*Theophraſtus* his tranſlator) names it *Hedera Cilicia* ; as
likewiſe *Pliny*,who *lib.*24.*cap.*10.writeth,that it is alſo ſyrnamed *Nicophoron.* Of the Hetrurians,
Hedera ſpinoſa, and *Rubus Cervinus* : of the Caſtilians in Spaine, as *Lacuna* ſaith, *Zarza parilla,* as
though they ſhould ſay *Rubus viticula,* or Bramble little Vine. *Sarſa* as *Matthiolus* interpreteth it,
doth ſignifie a Vine ; and *Parilla,* a ſmall or little Vine.
   Diuers affirme that the root (brought out of Peru a Prouince in America)which the later Herbariſts do call *Zarza,* is the root of this Bindeweed. *Garcias Lopius Luſitanus* granteth it to be like
thereunto, but yet he doth not affirme that it is the ſame. Plants are oftentimes found to be like
one another, which notwithſtanding are proued not to be the ſame by ſome little difference ; the
diuers conſtitutions of the weather and of the ſoile maketh the difference.
   Zarza parilla of Peru is a ſtrange plant, and is brought vnto vs from the Countries of the new
world called America ; and ſuch things as are brought from thence, although they alſo ſeeme and
are like to thoſe that grow in Europe, notwithſtanding they do often differ in vertue and operation : for the diuerſitie of the ſoile and of the weather doth not only breed an alteration in the forme
                                                                               but

but doth moſt of all preuaile in making the vertues and qualities greater or leſſer. Such things as grow in hot places be of more force, and greater ſmell; and in cold, of leſſer. Some things that are deadly and pernitious, being remoued wax milde, and are made wholeſome: ſo in like manner, although *Zarza parilla* of Peru be like to rough Bind-weed, or to Spaniſh *Zarza parilla*, notwithſtanding by reaſon of the temperature of the weather, and alſo through the nature of the ſoile, it is of a great deale more force than that which groweth either in Spaine or in Africke.

The roots of *Zarza parilla* of Peru, which are brought alone without the plant, be long and ſlender, like to the leſſer roots of common liquorice, very many oftentimes hanging from one head, in which roots the middle ſtring is hardeſt. They haue little taſte, and ſo ſmall a ſmell that it is not to be perceiued. Theſe are reported to grow in Honduras a prouince of Peru. They had their name of the likeneſſe of rough Binde-weed, which among the inhabitants it keepeth; ſignifying in Spaniſh, a rough or prickly vine, as *Garcias Lopius* witneſſeth.

¶ *The Temperature.*

The roots are of temperature hot and dry, and of thinne and ſubtill parts, inſomuch as their decoction doth very eaſily procure ſweat.

¶ *The Vertues.*

The roots are a remedie againſt long continuall paine of the joynts and head and againſt cold A diſeaſes. They are good for all manner of infirmities wherein there is hope of cure by ſweating, ſo that there be no ague joyned.

The cure is perfected in few daies, if the diſeaſe be not old or great; but if it be, it requireth a B longer time of cure. The roots here meant are as I take it thoſe of *Zarza parilla*, whereof this *Smilax aſpera* or rough Binde weed is holden for a kinde: notwithſtanding this of Spaine and the other parts of Europe, though it be counted leſſe worth, yet is it commended of *Dioſcorides* and *Pliny* againſt poyſons. The leaues hereof, ſaith *Dioſcorides*, are a counterpoyſon againſt deadly medicines, whether they be drunke before or after.

† The ſecond and fourth were both formerly of one plant, I meane the hiſtorie; for the figure in the fourth place ſhould haue beene in the third, and the figure in the third was the ſame with the ſecond, and ſhould haue beene in the fourth place.

Chap. 317.   *Of ſmooth or gentle Binde-weed.*

1 *Smilax lenis ſiue læuis major.*
Great ſmooth Binde-weed.

2 *Smilax lenis minor.*
Small Binde-weed.

¶ *The Deſcription.*

1   IT is a ſtrange thing vnto me, that the name of *Smilax* ſhould be ſo largely extended, as that it ſhould be aſſigned to thoſe plants that come nothing neere the nature, and ſcarſly vnto any part of the forme of *Smilax* indeed. But we will leaue controuerſies to the further conſideration of ſuch as loue to dance in quag-mires, and come to this our common ſmooth *Smilax*, called and knowne by that name among vs, or rather more truly by the name of *Convolvulus major*, or *Volubilis major* : it beareth the long branches of a Vine, but tenderer, and for the length and great ſpreading thereof it is very fit to make ſhadows in arbors : the leaues are ſmooth like Iuie, but ſomwhat bigger, and being broken are full of milke : amongſt which come forth great white and hollow floures like bells. The ſeed is three cornered, growing in ſmall huskes couered with a thinne skin. The root is ſmall, white, and long, like the great Dogs graſſe.

2   *Smilax læuis minor* is much like vnto the former in ſtalkes, leaues, floures, ſeed, and roots, ſauing that in all reſpects it is much ſmaller, and creepeth vpon the ground. The branches are ſmall and ſmooth : the little leaues tender and ſoft : the floures like vnto little bells, of a purple colour : the ſeed three cornered like vnto the others.

3 *Convolvulus minimus ſpica-folius.*
Lauander leafed Binde-weed.

† 4 *Convolvulus argenteus Altheæ folio.*
Siluer leafed Binde-weed.

3   This Bindweed *Pena* ſaith he neuer ſaw but in the brinks of quicke-ſets and Oliuets in Prouince, Sauoy, and Narbone ; notwithſtanding I found it growing in the corne fields about great Dunmow in Eſſex, in ſuch abundance, that it doth much hurt vnto their corne. This kind of Bindweed or *Volubilis* is like vnto the ſmall Bindweed before mentioned, but it hath a finer floure, plaited or folded in the compaſſe of a bell very orderly, eſpecially before the Sun riſe (for after it opens it ſelfe the welts are not ſo much perceiued) and it is of a darke purple colour : the ſeed is not vnlike the reſt, cornered and flat, growing out of ſlender branches which ſtand vpright and thicke together, proceeding out of a wooddy white root. The leaues are long and narrow, reſembling *Linaria* both in colour and hairineſſe, in taſte drying, and ſomewhat heating.

‡ 4 The

5 *Volubilis nigra.*
Blacke Bindweed.

‡ 4 The ſtalkes and branches of this are ſome cubite long, ſlender, weake and hairy, ſo that they lie vpon the ground, if they haue nothing to ſuſtaine them:vpon theſe without any order grow leaues, ſhaped like thoſe of Ivie, or the Marſh Mallow, but leſſe,and couered ouer with a ſiluer-like downe or hairineſſe,and diuided ſomewhat deepe on the edges, ſometimes alſo curled,and otherwhiles onely ſnipt about. The floure growes vpon a long ſtalke like as in other plants of this kinde, and conſiſts of one folding leafe, like as that of the laſt mentioned, and it is either of a whitiſh purple, or els abſolute purple colour: The root is ſmall and creeping. It growes in many places of Spaine, and there floures in March and Aprill. *Cluſius* calls this *Convolvulus Althea folio*,and ſaith that the Portugals name it *Verdezilla*, and commend it as a thing moſt effectuall to heale wounds. Our Author gaue the figure hereof (how fitly let the Reader iudge) by the name of *Papauer cornutum luteum minus*,making it a horned Poppy, as you may ſee in the former Edition, *pag.* 294. ‡

† 5 This kind of Bindweed hath a tough root full of thready ſtrings, from which riſe vp immediatly diuers trailing branches, whereon grow leaues like the common field Bindweed, or like thoſe of Orach,of a black green colour, whereof it tooke his name:the floures are ſmal, and like thoſe of Orach:the ſeed is black,three ſquare, like, but leſſe than that of Buck-wheat.

The whole plant is not onely a hurtfull weed, but of an euill ſmell alſo, and too frequently found amongſt corne. *Dodonæus* calls this *Convolvulum nigrum* : and *Helxine Ciſſampelos* : *Tabernamontanus, Volubilis nigra* : and *Lobel, Helxine Ciſſampelos altera Atriplicis effigie.*

¶ *The Place.*

All theſe kindes of Bindweeds do grow very plentifully in moſt parts of England. ‡ The third and fourth excepted. ‡

¶ *The Time.*

They do floure from May to the end of Auguſt.

¶ *The Names.*

The great Bindweed is called in Greeke, ϲμιλαξ λεία: in Latine, *Smilax Lævis* : of *Galen* and *Paulus Ægineta,* μίλαξ λεία: it is ſurnamed *Lævis* or ſmooth,becauſe the ſtalkes and branches thereof haue no prickes at all. *Dolichus* called alſo *Smilax hortenſis,* or Kidney beane, doth differ from this : and likewiſe *Smilax* the tree,which the Latines call *Taxus* : in Engliſh,the Yew tree. The later Herbariſts do call this Bindweed *Volubilis maior,Campanella,Funis arborum,Convolvulus albus,*and *Smilax lævis major* : in like manner *Pliny* in his 21.booke,5.chap.doth alſo name it *Convolvulus.*It is thought to be *Liguſtrum,*not the ſhrub priuet,but that which *Martial* in his firſt booke of Epigrams ſpeaketh of, writing againſt *Procillus.*

The ſmall Bindweed is called *Convolvulus minor,*and *Smilax lævis minor, Volubilis minor* : in high Dutch, **Weindkraut :** in low Dutch, **Wranghe :** in French, *Liſeron* : in Italian, *Vilucchio* : in Spaniſh, *Campanella Yerua* : in Engliſh,Withwinde,Bindeweed,and Hedge-bels.

¶ *The Temperature.*

Theſe herbes are of a hot and dry temperature.

¶ *The Vertues.*

The leaues of blacke Bindeweed called *Helxine Ciſſampelos,* ſtamped and ſtrained, and the juyce   A drunken,doth looſe and open the belly exceedingly.

The leaues pound and laid to the grieued place, diſſolue, waſte, and conſume hard lumps and   B ſwellings,as *Galen* ſaith.

The

D   The rest of the Bindweeds are not fit for medicine, but vnprofitable weeds, and hurtfull vnto each thing that groweth next vnto them.

   †   The description which our Author intended in the first place for *Volubilis nigra*, and tooke out of the 274. page of the *Aduersaria*, but so confusedly and imperfectly, neither agreeing with that he intended, I haue omitted as impertinent, and made his later, though also imperfect description, somewhat more compleat and agreeable to the plant figured and intended.

Chap. 318. *Of Blew Bindweed.*

¶ *The Description.*

1   **B**Lew Bindweed bringeth forth long, tender, and winding branches, by which it climeth vpon things that stand neere vnto it, and foldeth it selfe about them with many turnings and windings, wrapping it selfe against the Sun, contrary to all other things whatsoeuer, that with their clasping tendrels doe embrace things that stand neere vnto them; whereupon doe grow broad cornered leaues very like vnto those of Iuie, something rough and hairy, of an ouer-worne russet greene colour: among which come forth most pleasant floures bell fashion, somthing cornered as are those of the common Bindweed, of a most shining azure colour tending to purple: which being past, there succeed round knobbed seed vessels, wherein is contained long blackish seed of the bignesse of a Tare, and like vnto those of the great hedge Bindweed. The root it thready, and perisheth at the first approach of Winter.

1 *Convolvulus Cæruleus.*
Blew Binde-weed.

‡ 2 *Convolvulus cæruleus folio rotundo.*
Round leaued blew Bindweed.

‡ 2   There are also kept in our Gardens two other blew floured Bindweeds. The one a large and great plant, the other a lesser. The great sends vp many large and long winding branches, like those of the last described, and a little hairie: the leaues are large and roundish, ending in a sharpe point: the floures are as large as those of the great Bindweed, and in shape like them, but blew of colour, with fiue broad purplish veines equally distant each from other: and these floures commonly grow three neere together vpon three seuerall stalkes some inch long, fastened to another
stalke

‡ 3 *Convolvulus cæruleus minor folio oblongo.*
Small blew Bind-weed.

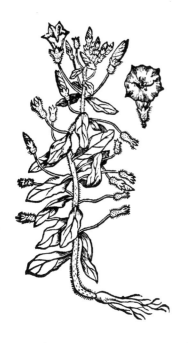

stalke some handfull long: the cup wh ch holds the floures, and afterwards becomes the seed vessell, is rough and hairie: the seed is blacke, and of the bignesse of a Tare: the root is stringie, and lasts no longer than to the perfecting of the seed. I haue onely giuen the figure of the leafe and floure largely exprest, because for the root and manner of growing it resembles the last described.

3 This small blew Bind-weed sendeth forth diuers long slender creeping hairie branches, lying flat vpon the ground vnlesse there be something for it to rest vpon: the leaues be longish and hairy, and out of their bosomes (almost from the bottome to the tops of the stalkes) come small foot-stalkes carrying beautifull floures of the bignesse and shape of the common smal Bindweed, but commonly of three colours; that is, white in the very bottome, yellow in the middle, and a perfect azure at the top; and these twine themselues vp, open and shut in fiue plaits like as most other floures of this kinde doe. The seed is contained in round knaps or heads, and is blacke and cornered: the root is small, and perishes euery yeare. *Bauhine* was the first that set this forth, and that by the name of *Convolvulus peregrinus cæruleus folio oblongo.* ‡

¶ *The Place.*

The seed of this rare plant was first brought from Syria and other remote places of the world, and is a stranger in these Northerne parts, yet haue I brought vp and nourished it in my Garden vnto flouring, but the whole plant perished before it could perfect his seed.

¶ *The Time.*

The seed must be sowne as Melons and Cucumbers are, and at the same time: it floured with me at the end of August.

¶ *The Names.*

It is called *Campana Lazura*, and *Lazura*: of the later Herbarists *Campana Cærulea*, and also *Convolvulum Cæruleum*: it is thought to be the *Ligustrum nigrum*; of which *Columella* in his tenth booke hath made mention:

Fer calathis violam, & nigro permista ligustro
Balsama cum Cassia nectens, &c.

In Baskets bring thou Violets, and blew Bindweed withall,
But mixed with pleasant Baulme, and Cassia medicinall.

For if the greater smooth Withwinde, or Bindweed be *Ligustrum*, then may this be not vnproperly called *Ligustrum nigrum* · for a blew purple colour is oftentimes called blacke, as hath beene said in the blacke Violet. But there be some that would haue this Bindweed to be *Granum nil Auicenne*, of which he writes in the 306. chapter; the which differeth from that *Nil* that is described in the 512. chapter. For this is *Isatis Græcorum*, or the Grecian Woad: but that is a strange plant, and is brought from India, as both *Avicen* and *Serapio* do testifie: *Avicen* in this manner: what is *Granum Nil*? It is *Cartamum Indum*: and *Serapio* thus, *Habal Nil*, is *Granum Indicum*, in *cap.*283. where the same is described in these words: [The plant hereof is like to the plant of *Leblab*, that is to say of *Convolvulus*, or Bind-weed, taking hold of trees with his tender stalke: it hath both green branches and leaues, and there commeth out by euery leafe a purple floure, in fashion of the Bell-floures: and when the floure doth fall away, it yeeldeth a seed in small cods (I read little heads) in which

which are three graines,leſſer than the ſeeds of Staueſaker] to which deſcription this blew Binde-weed is anſwerable.

There be alſo other ſorts of Bindweeds,which be reſerred to *Nil Auicenna*; which no doubt may be kindes of *Nil*; for nothing gaine-ſaith it why they ſhould not be ſo. Therefore to conclude,this beautifull Bind-weed,which we call *Convolvulus Caruleus*,is called of the Arabians *Nil*: of *Serapio, Habab Nil*: about Alepo and Tripolis in Syria the inhabitants call it *Haſmiſen*:the Italians,*Campana laʒurea*: of the beautifull azured floures, and alſo *Fior de notte*,becauſe his beauty appeares moſt in the night.

¶ The Temperature.

Convolvulus Caruleus,or *Nil*,as *Auicen* ſaith,is hot and dry in the firſt degree:but *Serapio* maketh it to be hot and dry in the third degree.

¶ The Vertues.

A   It purgeth and voideth forth raw, thicke, flegmaticke,and melancholy humours : it driues out all kinde of wormes,but it troubleth the belly,and cauſeth a readineſſe to vomit,as *Auicen* ſaith : it worketh ſlowly,as *Serapio* writeth; in whom more hereof may be found,but to little purpoſe,wherefore we thinke good to paſſe it ouer.

CHAP. 319.   Of Scammonie, or purging Bindweed.

1 *Scammonium Syriacum.*
Syrian Scammonie.

† 2 *Scammonea Valentina.*
Scammonie of Valentia.

¶ The Deſcription.

1   SCammonie of Syria bath many ſtalkes riſing from one root, which are long,ſlender, and like the claſping tendrels of the vine,by which it climeth and taketh hold of ſuch things as are next vnto it. The leaues be broad, ſharpe pointed like thoſe of the ſmooth or hedge Bind-weed:among which come forth very faire white floures tending to a bluſh colour,bell-faſhion. The root is long,thicke, and white within: out of which is gathered a juyce that being
hardned

hardned,is greatly vfed in Phyficke : for which confideration, there is not any plant growing vpon the earth,the knowledge whereof more concerneth a Phyfition, both for his fhape and properties, than this Scammonie, which *Pena* calleth *Lactaria fcanforiaque volvula*, that is, milky and climbing Windweed,whereof it is a kinde; although for diftinction fake I haue placed them as two feverall kinds. And although this herbe be fufpected, and halfe condemned of fome learned men,yet there is not any other herbe to be found,whereof fo fmall a quantity will do fo much good : neither could thofe which haue carped at it,and reproued this herbe, finde any fimple in refpect of his vertues to be put in his roome : and hereof enfueth great blame to all practitioners,who haue not endeuoured to be better acquainted with this herbe, chiefely to auoid the deceit of the crafty Drug-feller and Medicine-maker of this confected Scammony, brought vs from farre places, rather to be called I feare infected Scammony,or poyfoned Scammony,·than confected. But to auoid the inconuenien-ces hereof, by reafon of the counterfeiting and ill mixing thereof : I haue therefore thought good to fet downe what I haue taken out of the diligent, and no leffe learned obferuations of *Pena*, con-cerning this plant,*Anno* 1561. or 1562.*Vid.adnerf pag*.272.

‡ 3 *Scammonum Monfpelienfe.*
French Scammonie.

Sequinus Martinellus an Apothecary of Ve-nice,being a moft diligent fearcher of Simples, that he might haue the right Scammony of An-tioch,trauelled into Syria,where from the citie of Aleppo he fent an 100. weight of the juvce of Scammony of Antioch, prepared and hard-ned into a lumpe,at the making whereof he was prefent himfelfe. This man fent alfo of the feeds thereof, which in all points anfwered the cornered feed of *Volubilis*; which being fowne in the beginning of the Spring at Padua and Venice,grew vp to the form of a braue & good-ly *Convolvulus*,in leaues,floures,and fhew fo like vnto our *Ciffampelos*,that a man would haue ta-ken it for the fame without controuerfie,fauing that the root was great, and in bigneffe equall to the great Brionie,as alfo in tenderneffe. The outward bark of the root was of a dusky colour, and white within : the inner pith beeing taken forth feemeth in all mens iudgements to be the fame and the beft allowed *Turbith officinarum* : and yet it differeth from Turbith,in that,that it is more brittle, and will more eafily bee bro-ken, though the pith in Scammony be no leffe gummie and full of milkie juyce,than Turbith. Further *Pena* reporteth, that afterward he fent of this feed vnto Antwerpe,where it grew very brauely,the climing ftrings and branches grow-ing vp to the height of fiue or fix cubits,not differing from that which was fowne in Italy. Alfo *William Dries* of Antwerpe, a moft excel-lent Apothecary,did cut of the branches of his Antwerpian Scammonie from the root,and dri-ed them, planted the feeds in his Garden, and

conferred the fuperfluous branched roots with the Turbith of Alexandria, and could not find them to differ or difagree the one from the other in any point. But he that will know more concerning the making, difference, choife and vfe of Scammony, let him read *Pena* in his chapter of Scammo-nie,in the place formerly cited, where he fhall finde many excellent fecrets worthy the noting of thofe which would know how to vfe fuch rare and excellent medicines.

2   Scammony of Valentia (whereof I haue plenty in my Garden) is alfo a kinde of Bindweed, growing naturally by the fea fide vpon the grauelly fhore, by the mouth of the riuer Rhodanus, at the waters called *Aquas Marianas*, where the Apothecaries of Montpellier gather of it great plen-tie,who haue attempted to harden the milkie juyce thereof, to vfe it in ftead of Scammony of An-tioch. This plant bringeth forth many flender branches,which will climbe and very well run vpon a pole; as being fupported therewith, and mounteth to the height of fiue or fix cubits, climbing and ramping like the firft kinde of Scammony. The leaues are greene, fmooth, plaine, and fharpe

pointed,

pointed, which being broken do yeeld abundance of milke : the floures are white, ſmall, and ſtarre-faſhion : the roots white and many, ſhooting forth ſundry other roots, whereby it mightily increaſeth.

† 3   This ſtrange kinde of Scammony, which *Cluſius* maketh rightly to be *Periploca ſpecies*, hath very many long branches ramping and taking hold of ſuch things as do grow neer vnto them, of a darkiſh aſh colour : whereupon do grow leaues ſharp pointed, crooked at the ſetting on of the ſtalke like thoſe of the blacke Bryony, and likewiſe of an aſhe colour, ſet together by couples : from the boſome whereof thruſt forth ſmall tender foot-ſtalkes, whereon are placed ſmall white floures ſtar-faſhion : the ſeeds are contained in long cods, and are wrapped in downe, like as thoſe of Swallow-wort. The root is very long, ſlender, and creeping, like that of the ſmall Bindweed, ſo that if it once take in any ground, it can hardly be deſtroied.

¶ The Place.

It doth grow in hot regions, in a fat ſoile, as in Miſia, Syria, and other like countries of Aſia; it is likewiſe found in the Iſland of Candia, as *Bellon.* witneſſeth; from whence I had ſome ſeeds, of which ſeed I recciued two plants that proſpered exceeding well ; the one whereof I beſtowed vpon a learned Apothecary of Colcheſter, which continueth to this day, bearing both floures and ripe ſeed. But an ignorant weeder of my garden plucked mine vp, and caſt it away in my abſence, in ſtead of a weed : by which miſchance I am not able to write hereof ſo abſolutely as I determined : it likewiſe groweth neere vnto the ſea ſide about Tripolis in Syria, where the inhabitants doe call it *Meudheudi.*

¶ The Time.

It floured in my Garden about S.*Iames* tide, as I remember, for when I went to Briſtow Faire I left it in floure ; but at my returne it was deſtroied as aforeſaid.

¶ The Names.

The Greekes call it, ἐσκαμμωνία : the Latines, *Scammonium,* ſo naming not onely the plant it ſelfe, but alſo the hard and condenſed juyce : of the Apothecaries, *Scammonea;* and when it is prepared, *Diagridium* : as though they ſhould ſay, δακρύδιον : which ſignifieth a little teare : both the herbe and juyce are named *Scamony* : of *Rhaſis, Coriziola.*

¶ The Temperature.

The juyce doth mightily purge by the ſtoole, and is the ſtrongeſt purge whatſoeuer ; for as *Oribaſius* ſaith, it is in no part ouercome by thoſe things which ſtir and moue the body. It worketh the ſame not vehemently by any hot quality, but by ſome other hid and ſecret property of the whole ſubſtance : for there is no extremity of heat perceiued in it by taſte : for with what liquor or thing ſoeuer it is mixed, it giueth vnto it no bitterneſſe, biting, or other vnpleaſant taſte at all, and therefore it is not to be accounted among the extreme hot medicines, but amongſt thoſe that are moderately hot and dry.

¶ The Vertues.

A   It cleanſeth and draweth forth eſpecially choler : alſo thinne and wateriſh humours, and oftentimes flegme, yet is it as *Paulus* teacheth, more hurtfull to the ſtomacke than any other medicine.

B   *Meſues* thinketh that it is not onely troubleſome and hurtfull to the ſtomacke, but alſo that it ſhaueth the guts, gnawing and fretting the intrails; openeth the ends of the veines, and through the eſſence of his whole ſubſtance, it is an enemy to the heart, and to the reſt of the inward parts: if it be vſed immoderately and in time not conuenient, it cauſeth ſwounings, vomitings, and ouerturnings of the ſtomacke, ſcouring, the bloudy flux and vlcers in the lower gut, which bring a continuall deſire to the ſtoole.

C   Theſe miſchiefes are preuented if the Scammonie be boiled in a Quince and mixed with the ſlime or mucilage of *Pſillium,* called Fleawoort, the pap or pulp of Prunes, or other things that haue a ſlimie jnyce, with a little maſticke added, or ſome other caſie binding thing.

D   *Pliny* affirmeth that the hurt thereof is taken away if Aloes be tempered with it : [Scammonie (ſaith he) ouerthroweth the ſtomack, purgeth choler, looſeth the belly vnleſſe two drams of Aloes be put vnto one ſcruple of it] which alſo *Oribaſius* alloweth of in the firſt booke of his *Synopſes,* and the ſeuenth booke of his medicinall Colleſtions.

E   The old Phyſitions were alſo woont to boile Scammonie in a quince, and to giue the quince to be eaten, hauing caſt away the Scammony : and this quince ſo taken doth moue the belly without any hurt vnto the ſtomacke, as *Galen* in his firſt booke of the Faculties of Nouriſhments doth ſet downe, and likewiſe in his third booke of the Faculties of ſimple Medicines.

   The Apothecaries do vſe Scammonie prepared in a Quince, which as we haue ſaid they name *Diagridium,* and do mix it in diuers compoſitions.

They

They keepe vſually in their ſhops two compoſitions, or electuaries, the one of *Pſyllium* or Flea- G
woort, ſet downe by *Meſue*: the other of Prunes, fathered vpon *Nicolaus*, which were deuiſed for the
tempering and correction of Scammony, and be commended for hot burning agues, and tertians,
and for what diſeaſes ſoeuer that proceed of choler.

Galen hath taken Maſticke and *Bdellium* out of the pilles called *Cochia*, which alſo containe in H
them a great quantitie of Scammony, as we may read in his firſt Booke of medicines according
to the places affected, which alſo we meane to touch in the chapter of Coloquintida, where we intend to treate at large concerning maſtick, and other binding things, that are accuſtomed to bee
mixed for the correction of ſtrong and violent purgers.

The quantitie of Scammonie or of *Diagridium* it ſelfe, to be taken at one time, as *Meſue* writeth, I
is from fiue grains to ten or twelue : it may be kept as the ſame Author ſheweth, foure yeres: *Pliny*
iudgeth it to be little worth after 2 yeares: it is to be vſed, ſaith he, when it is two yeres old, and it
is not good before, nor after. The mixing or otherwiſe the vſe therof, more than is ſet down, I think
it not expedient to ſet forth in the Phyſicall vertues of Scammony, vpon the receipt whereof many times death inſueth: my reaſons are diuers, for that the ſame is very daungerous, either if too
great a quantitie thereof be taken, or if it be giuen without correction ; or taken at the hands of
ſome runnagat phyſick-monger, quackſaluer, old women-leaches, and ſuch like abuſers of phiſick,
and deceiuers of people. The vſe of Scammony I commit to the learned, vnto whom it eſpecially
and onely belongeth, who can very carefully and curiouſly vſe the ſame.

‡ The titles of the ſecond and third were formerly tranſpoſed, and both the figures belonged to the ſecond deſcription, which was of the *ſcammonium Mon-*
ſpeluſe of the *Aduerſ,* being the ſame with the *Scammonea Valentina* of *Cluſius*.

Chap. 320.　*Of Briony, or the white Vine.*

¶ *The Kindes.*

THere be two kinds of Bryony, the one white, the other blacke: of the white Briony as followeth.

Bryonis alba.
White Bryonie.

¶ *The Deſcription.*

WHite Briony brings forth diuers long
and ſlender ſtalkes with many claſping tendrels like the vine, wherwith
it catcheth hold of thoſe things that are next to
it. The leaues are broad, fiue cornered, and indented like thoſe of the vine; but rougher, more
hairy and whiter of colour: the floures be ſmall
and white, growing many together. The fruit
conſiſteth in little cluſters, the berries whereof
are at the firſt green, and red when they be ripe.
The root is very great, long, and thicke, growing deepe in the earth, of a white yellowiſh colour, extreame bitter, and altogether of an vnpleaſant taſte. The Queens chiefe Surgion Mr.
William Godorous, a very curious & learned gentleman, ſhewed me a root herof, that waied half
an hundred weight, and of the bignes of a child
of a yeare old.

¶ *The Place.*
Briony groweth almoſt euery where among
pot-herbes, hedge-buſhes, and ſuch like places.

¶ *The Time.*
It floureth in May, and bringeth forth his
grapes in Autumne.

¶ *The Names.*
Briony is called in Greek, ἀμπελολευϰη in Latine,

Dddd　　　　　　　　　　　　　　　　*Vitis*

Vitis alba, or white Vine, and it is named, ἀμπελ. becauſe it is not onely like the vine in leaues, but alſo for that it bringeth forth his fruit made vp after the likeneſſe of a little cluſter, although the berries ſtand not cloſe together : it is called of *Pliny*, *Bryonia*, and *Madon* : of the Arabians *Alpheſera* : of *Mattheus*, *Syluaticus*, *Viticella* : in the poore mans Treaſure, *Roraſtrum* : of *Apuleius*, *Apiaſtellum*, *Vitis Taminia*, *Vitis alba*, and *Vitalba* : in high Dutch, **Suchwurtz** : in low Dutch, **Brionie** : in Engliſh, Briony, white Briony, and tetter Berrie : in French *Couleuree* : in Italian, *Zucca ſyluatica* : in Spaniſh, *Nueza blanca*.

¶ *The Temperature.*

White Briony is in all parts hot and dry, exceeding the third degree, eſpecially of heate, with an exceeding great force of clenſing and ſcouring, by reaſon wherof it purgeth and draweth forth, not onely cholericke and flegmaticke humors, but alſo watrie.

¶ *The Vertues.*

A　*Dioſcorides* writeth, that the firſt ſprings or ſproutings being boiled and eaten, do purge by ſiege, and vrine. *Galen* ſaith, that all men vſe accuſtomably to eat of it in the ſpring time, and that it is a nouriſhment wholeſome, by reaſon of the binding quality that it hath ; which is to be vnderſtood of thoſe of the wild vine, called in Latine, *Tamus* ; and not of the ſproutings of this plant ; for the ſproutings of the firſt ſprings of white Bryony are nothing binding at all, but do mightily purge the belly, and torment the ſtomacke.

B　*Dioſcorides* alſo affirmeth, That the juice of the root being preſſed out in the ſpring, and drunke with meade or honied water, purgeth flegme : and not only the juice, but alſo the decoction of the root draweth forth flegme, choler, and wateriſh humors, and that very ſtrongly ; and it is withall oftentimes ſo troubleſome to the ſtomacke, as it procureth vomite.

C　This kinde of ſtrong purgation is good for thoſe that haue the dropſie, the falling ſickneſſe, and the dizzineſſe and ſwimming of the braine and head, which hath continued long, and is hardly to be remoued : yet notwithſtanding it is not daily to be giuen (as *Dioſcorides* admoniſheth) to them that haue the falling ſickneſſe, for it will be troubleſome enough to take it now and then : and it is (as we haue ſaid) an exceeding ſtrong medicine, purging with violence, and very forceable for mans nature.

D　The root put vp in maner of a peſſary bringeth forth the dead childe and afterbirth : being boiled for a bath to ſit in, it worketh the ſame effect.

E　It ſcoureth the skin, and taketh away wrinkles, freckles, ſun burning, blacke markes, ſpots, and ſcars of the face, being tempered with the meale of vetches or Tares, or of Fenugreeke : or boiled in oile till it be conſumed ; it taketh away blacke and blew ſpots which come of ſtripes : it is good againſt Whitlowes : being ſtamped with wine and applied it breaketh biles, and ſmall apoſtumes, it draweth forth ſplinters and broken bones, if it be ſtamped and laid thereto.

F　The ſame is alſo fitly mixed with eating medicines, as *Dioſcorides* writeth.

G　The fruit is good againſt ſcabs and the leprie, if it be applied and anointed on, as the ſame Author affirmeth.

H　*Galen* writeth, that it is profitable for Tanners to thicken their leather hides with.

I　Furthermore, an electuary made of the roots and hony or ſugar, is ſingular good for them that are ſhort winded, troubled with an old cough, paine in the ſides, and for ſuch as are hurt and burſten inwardly : for it diſſolueth and ſcattereth abroad congealed and clottered bloud.

K　The root ſtamped with ſalt is good to be laid vpon filthy vlcers and ſcabbed legges. The fruit is likewiſe good to the ſame intent, if it be applied in manner aforeſaid.

L　The root of Briony and of wake-Robin ſtamped with ſome ſulphur or brimſtone, and made vp into a maſſe or lump and wrapped in a linnen clout, taketh away the morphew, freckles, and ſpots of the face, if it be rubbed with the ſame being dipped firſt in vineger.

CHAP. 321.　*Of blacke Brionie, or the wilde Vine.*

¶ *The Deſcription.*

1　THe black Briony hath long flexible branches of a woody ſubſtance, couered with a gaping or clouen bark growing very far abroad, winding it ſelfe with his ſmall tendrels about trees, hedges, and what elſe is next vnto it, like vnto the branches of the vine : the leaues are like vnto thoſe of Iuie or garden Nightſhade, ſharpe pointed, and of a ſhining greene colour : the floures are white, ſmall, and moſſie ; which being paſt, there ſucceed little cluſters of red berries

1 *Bryonia nigra.*
Blacke Bryonie.

berries, fomewhat bigger than thofe of the fmall Raifins or Ribes, which wee call Currans or fmall Raifins. The root is very great and thicke, often-times as big as a mans leg, blackifh without, and very clammy or flimy within; which beeing but fcraped with a knife or any other thing fit for that purpofe, it feemes to be a matter fit to fpred vpon cloath or leather in manner of a plaifter of Sear-cloath: which being fo fpred and vfed, it fer-ueth to lay vpon many infirmities, and vnto verie excellent purpofes, as fhal be declared in the pro-per place.

2   The wild black Bryonie refembles the for-mer as well in flender Vine-ftalkes, as leaues; but clafping tendrels haue they none, neuerthelefſe by reafon of the infinitenefſe of the branches, and the tendernefſe of the fame, it takes hold of thofe things that ftand next vnto it, although eafie to be loofed, contrarie vnto the other of his kinde. The berries hereof are blacke of colour when they be ripe; the root alfo is blacke without, and within of a pale yellow colour like Box.   ‡ This which is here defcribed is the *Bryonia nigra* of *Dodonæus*; but *Bauhine* calleth it *Bryonia alba*, and faith it dif-fereth from the common white Bryonie, onely in that the root is of a yellowifh Box colour on the in-fide, and the fruit or berries are blacke when as they come to ripenefſe.

Bryonia nigra florens non fructum ferens.
3   This is altogether like the firft defcribed in roots, branches, and leaues, only the foot-ftalks whereon the floures grow are about eight or nine inches long: the floures are fomething greater, hauing neither before or after their flouring any betries or fhew thereof, but the flours & footftalks do foon wither and fall away. This I haue heretofore and now this Summer 1621, diligently ob-ferued, becaufe it hath not bin mentioned or obferued by any that I know.   *Iohn Goodyer.*   ‡

¶ *The Place.*
The firft of thefe plants groweth in hedges and bufhes almoft euery where.
The fecond growes in Heffen, Saxonie, Weftphalia, Pomerland, and Mifnia, where white Bryo-nie groweth not, as *Valerius Cordus* hath written; who faith that it growes vnder hafell trees neere a city of Germany called Argentine or Strausborough.

¶ *The Time.*
They fpring in March, bring forth the floures in May, and their ripe fruit in September.

¶ *The Names.*
Blacke Bryonie is called in Greeke, Ἄμπελος ἄγρια: in Latine, *Bryonia nigra*, and *Vitis fyluestris*, or wild Vine; notwithftanding it much differs from *Labrufca*, or *Vitis vinifera fyluestris*, that is, the wild Vine which bringeth fo th wine called *Ampelos agria*. Why both thefe were called by one name I.ny was the caufe, who could not fufficiently expound them; *Lib.23. cap.1.* but confounded and made them all one, in which error are alfo the Arabians.
This wilde Vine is alfo called in Latine, *Tamus*; and the fruit thereof *Vua Taminia. Pliny* nameth it alfo *Salicaftrum. Ruellius* faith that in certain fhops it is called *Sigillum B. Mariæ*; it is alfo called *Cyclaminus altera*, but not properly: in Englifh, Blacke Bryonie, wilde Vine, and our Ladies feale.

¶ *The Nature.*
The roots of the wilde Vine are hot and dry in the third degree: the fruit is of like temper, but not fo forceable: both of them fcoure and wafte away.

¶ The

¶ *The Vertues.*

A   *Dioſcorides* ſaith that the roots do purge wateriſh humors,& are good for ſuch as haue the drop-ſie,if they be boiled in wine,adding vnto the wine a little ſea water,and be drunke in three ounces of faire freſh water: he ſaith furthermore,that the fruit or berries doth take away the Sunne-burne and other blemiſhes of the skin.

B   The berries do not only clenſe and remoue ſuch kinde of ſpots, but do alſo very quickely waſte and conſume away blacke and blew marks that come of bruiſes and dry-beatings:which thing alſo the roots performe being laid vpon them.

C   The yong and tender ſproutings are kept in pickle,and reſerued to be eaten with meat,as *Dioſco-rides* teacheth. *Matthiolus* writeth,that they are ſerued at mens tables in our age alſo in Tuſcanie, others alſo report the like to be done in Andoloſia one of the kingdomes of Granado.

D   It is ſaid that ſwine ſeeke after the roots hereof, which they dig vp and eat with no leſſe delight than they do the roots of *Cyclaminus* or *Panis Porcinus* ; whereupon it was called *Cyclaminus altera*,or Sow-bread.If this reaſon ſtand for good ; then may we in like manner ioine hereunto many other roots,and likewiſe call them *Cyclaminus altera* or Sow-bread ; for ſwine do not ſeeke after the roots of this only,but alſo of diuers other plants,of which none are of the kinds of Sow-bread. It would therefore be a point of raſhneſſe to affirme *Tamus* or our Ladies ſeale to be a kinde of Sow-bread, becauſe the roots hereof are a pleaſant meat to ſwine.

E   The root ſpred vpon ſheeps leather in manner of a plaiſter,whileſt it is yet freſh and greene, ta-keth away blacke and blew marks,all ſcars and deformitie of the skin,breaks hard apoſtems,drawes forth ſplinters and broken bones,diſſolueth congealed bloud,and being laid on and vſed vpon the hip or huckle bones,ſhoulders,armes,or any other part where there is great paine and ache,it takes it away in ſhort ſpace,and worketh very effeを̌ually.

†   The figure that was formerly in the ſecond place of this Chapter did no way agree with the deſcription,for it was of the *Vitrus* or *Trauellors* Ioy, hereafter to be mentioned ; which *Tabernamontanus* (whoſe figures our Author made vſe of) calls *Vitis nigra ſecunda.*

C H A P. 322.

Of Bryonie of Mexico;

¶ *The Deſcription.*

1   THat plant which is now called *Mechoacan* or Bryonic of Mexico commeth very neere the kinds of Bindweeds in leaues and trailing branches, but in roots like the Bryonies: for there ſhooteth from the root thereof many long ſlender tendrels,which do infinite-ly graſpe or claſpe about ſuch things as grow or ſtand next vnto them ; whereupon do grow great broad leaues ſharp poinred,of a darke green colour, in ſhape like thoſe of our Ladies ſeale, ſome-what rough and hairy,and a little biting the tongue :among the leaues come forth the floures (as *Nicolaus Monardus* writeth) not vnlike thoſe of the Orenge tree, but rather of the golden Apple of Loue,conſiſting of fiue ſmall leaues,out of the midſt whereof comes forth a little clapper or peſtil in manner of a round lump as big as an haſell nut ; which being diuided with a thin skin or mem-brane that commeth through it, openeth into two parts, in each whereof are contained two ſeeds as big as Peaſe,in colour blacke and ſhining.The root is thicke and long,very like vnto the root of white Bryonie,whereof we make this a kinde,although in the taſte of the roots there is ſome dif-ference ; for the root of white Bryonie hath a bitter taſte,and this hath little or no taſte at all.

2   The Bryonie or *Mechoacan* of Peru groweth vp with many long trailing flexible branches, interlaced with diuers Viny tendrels which take hold of ſuch things are next or neere vnto them, euen in ſuch manner of claſping or climing as doth the blacke Bryonie or wine Vine,whereunto it is very like almoſt in each reſpeを̌,ſauing that his moſſie floures do ſmell very ſweetly. The fruit as yet I haue not obſerued, by reaſon that the plant which growes in my garden did not perfeを̌ the ſame,by occaſion of the great raine and intemperat weather that hapned *An. 1596.* but I am in good hope to ſee it in his perfeを̌ion,and then we ſhall eaſily iudge whether it be that right *Mecho-acan* that hath bin brought from Mexico and other places of the Weſt Indies, or no. The root by the figure ſhould ſeeme to anſwer that of the wilde Vine,but as yet thereof I cannot write cer-tainly.

‡   3   There is brought to vs and into vſe of late time,the root of another plant which ſeemes

1 *Mechoacan.*
Bryonie of Mexico.

2 *Mechoacan Peruviana.*
Bryonie of Peru.

haue much affinity with *Mechoacan*,and therefore *Bauhine* hath called it *Bryonia Mechoacan nigricans*
and thus deliuers the historie thereof: [It is a root like *Mechoacan*, but couered with a blackish
barke,and reddish,or rather grayish on the inside, and cut into slices ; it was brought some yeares
agon out of India,by the name of *Chelapa* or *Gelapa*. It is called by those of Alexandria and Marsei-
les,*Ialapium*,or *Gelapum* ; and of those of Marseilles it is thought the blacke or male *Mechoacan*:the
taste is not vngratefull,but gummy,and by reason of the much gumminesse,put to the fire it quickly flames : it in facultie exceeds the common *Mechoacan*,for by reason of the great gumminesse it
more powerfully purgeth serous humours with a little griping, also it principally strengthens the
liuer and stomacke ; wherefore it is safely giuen ʒj. and performs the operation without nauseous-
nesse. It is vsually giuen in Succorie water or some thin broth, three houres before meat.] Thus
much *Bauhine*,who saith it was first brought to these parts eleuen yeres before he set forth his *Prodromus*,which was about 1611. It hath bin little vsed here till within these ten yeares. ‡

¶ The Place.

Some write,that *Mechoacan* was first found in the prouince of New Spain,neer the city of Mexico or Mexican,whereof it tooke his name. It groweth likewise in a prouince of the West Indies
called Nicaragua and Quito,where the best is thought to grow.

¶ The Names.

It beareth his name,as is said,of the prouince wherein it is found. Some take it for a kinde of
Bryonie : but seeing the root is nothing bitter,but rather without taste, it agrees little with Bryonie,for the root of Bryonie is very bitter. Diuers name it *Rha album* or white Rubarb,but vnproperly,being nothing like. It comes neere vnto Scammonie,and if I might yeeld my censure,it seems
to be *Scammonium quoddam Americanum*,or a certaine Scammonie of America. Scammony creepeth,as we haue said,like Binde-weed : the root is both white and thicke : the juice hath but little
taste,as also hath this of *Mechoacan*. It is called in English, Mechoca, and Mechocan, and may be
called Indian Bryonie.

¶ The

¶ *The Temperature.*

The root is of a mean temper between hot and cold, but yet dry.

¶ *The Vertues.*

It purgeth by ſiege eſpecially flegme, and then wateriſh humors. It is giuen from one full dram weight to two, and that with wine, or with ſome diſtilled water (according as the diſeaſe requires) or elſe in fleſh broth.

It is giuen with good effect to all whoſe diſeaſes proceed of flegme and cold humors: it is good againſt head-ache that hath continued long, old coughs, hardneſſe of breathing, the colique, paine of the kidnies and joints, the diſeaſes of the reins and belly.

CHAP. 323.   *Of the manured Vine.*

¶ *The Kindes.*

THe Vine may be accounted among thoſe plants that haue need of ſtayes and props, and canot ſtand by themſelues ; it is held vp with poles and frames of wood, and by that meaus it ſpreds all about and climbeth aloft : it ioyneth it ſelfe vnto trees or whatſoeuer ſtandeth neere vnto it.

Of Vines that bring forth wine ſome be tame and husbanded, others wild : of tame Vines there are many that be greater, and likewiſe another ſort that are leſſer.

¶ *The Deſcription.*

THe trunke or body of the Vine is great and thicke, very hard, couered with many barks, which are full of cliffes or chinks ; from which grow forth branches as it were armes, many wayes ſpreading ; out of which come forth jointed ſhoots or ſprings ; and from the boſom of thoſe joints, leaues and claſping tendrels, and likewiſe bunches or cluſters full of grapes : the leaues be broad, ſomething round, fiue cornered, and ſomewhat indented about the edges : amongſt which come forth many claſping tendrels, that take hold of ſuch props or ſtaies as ſtand next vnto it. The grapes differ both in colour and greatneſſe, and alſo in many other things, which to diſtinguiſh ſe-uerally were impoſſible, conſidering the infinite ſorts or kinds, and alſo thoſe which are tranſplan-ted from one region or clymat to another, do likewiſe alter both from the forme and taſte they had before : wherefore it ſhall be ſufficient to ſet forth the figure of the manured grape, and ſpeak ſom-what of the reſt.

There is found in Grecia and the parts of Morea, as *Pantalarea, Zante, Cephalonia,* and *Petras* (wher-of ſome are Iſlands, and others of the continent) a certain Vine that hath a trunk or body of a woo-dy ſubſtance, with a ſcaly or rugged barke of a grayiſh colour, whereupon grow faire broad leaues ſleightly indented about the edges, not vnlike thoſe of the marſh Mallow : from the boſom wherof come forth many ſmall claſping tendrels, and alſo tough and pliant foot-ſtalks, whereon grow very faire bunches of grapes of a Watchet blewiſh colour : from the which fruit commeth forth long tender laces or ſtrings ſuch as is found among Sauorie, whereupon we call that plant which hath it laced Sauorie, not vnlike that growing among and vpon flax, which we cal Dodder, or *Podagra lini,* wherof is made a blacke wine called Greeke wine, yet of the taſte of Sack. The laced fruit of this Vine may be fitly termed *Vva barbata,* laced or bearded Grapes.

The plant that beareth thoſe ſmall Raiſins which are commonly called Corans or Currans, or rather Raiſins of Corinth, is not that plant which among the vulgar people is taken for Currans, it being a ſhrub or buſh that brings forth ſmall cluſters or berries, differing as much as may be from Corans, hauing no affinitie with the Vine or any kinde thereof. The Vine that beareth ſmall Rai-ſins hath a body or ſtock as other Vines haue, branches and tendrels likewiſe. The leaues are larger than any of the others, ſnipt about the edges like the teeth of a ſaw : among which come forth clu-ſters of grapes in forme like the other, but ſmaller, of a blewiſh colour ; which beeing ripe are ga-thered and laid vpon hurdles, carpets, mats, and ſuch like, in the Sun to dry : then are they caried to ſome houſe and laid vpon heaps, as wee lay apples and corne in a garner, vntill the merchants buy them : then do they put them into large buts or other woodden veſſels, and tread them downe with their bare feet, which they call Stiuing, and ſo are they brought into theſe parts for our vſe. ‡ And they are commonly termed in Latine, *Vvæ Corinthiacæ,* and *Paſſulæ minores.*

There is alſo another which beareth exceeding faire Grapes whereof they make Raiſins, whiter coloured, and much exceeding the bigneſſe of the common Raiſin of the Sunne ; yet that grape whereof

Vitis Vinifera.
The manured Vine.

wherof the Raiſin of the Sun is made is a large one, and thought to bee the *Vva Zibibi* of the Arabians; and it is that which *Tabernamontanus* figured vnder that name, who therin was followed by our Author: but the figures being little to the purpoſe, I haue thought good to omit them. ‡

There is another kinde of Vine which hath great leaues very broad, of an ouerworne colour; where-upon grow great bunches of grapes of a blewiſh colour, the pulpe or meat whereof ſticketh or clea-ueth ſo hard to the graines or little ſtones, that the one is not eaſily diuided from the other, reſembling ſome ſtarued or withered berry that hath been bla-ſted, whereof it was named *Duracina.*

There be ſome Vines that bring forth grapes of a whitiſh or reddiſh yellow colour; others of a deep red, both in the outward skin, juice, and pulpe within.

There be others whoſe grapes are of a blew co-lour, or ſomething red, yet is the juice like thoſe of the former. Theſe grapes doe yeeld forth a white wine before they are put into the preſſe, and a red-diſh or paller wine when they are trodden with the husks, & ſo left to macerate or ferment, with which if they remain too long, they yeeld forth a wine of a higher colour.

There be others which make a black and obſcure red wine, whereof ſome bring bigger cluſters, and conſiſt of greater grapes, others of leſſer; ſome grow more cluſtered or cloſer together, others looſer; ſome haue but one ſtone, others more; ſome make a more auſtere or harſh wine, others a more ſweet : of ſome the old wine is beſt, of diuers the firſt yeres wine is moſt excellent : ſome bring forth fruit foure ſquare, of which kindes we haue great plenty.

¶ *The Place.*

A fit ſoile for Vines, ſaith *Florentinus*, is euery blacke earth, which is not very cloſe nor clammy, hauing ſome moiſture. But *Columella* ſaith great regard is to be had, what kinde or ſort of Vine you would nouriſh, according to the nature of the country and ſoile.

A wiſe husbandman will commit to a fat and fruitfull ſoile a leane Vine, of his own nature not too fruitfull : to a lean ground a fruitfull Vine : to a cloſe and, compact earth a ſpreading vine, and that is full of matter to make branches of: to a looſe and fruitful ſoile a Vine of few branches. The ſame *Columella* ſaith, that the Vine delighteth not in dung, of what kinde ſoeuer it be; but freſh mold mixed with ſome ſhauings of horne is the beſt to be diſpoſed about the roots to cauſe ferti-litie.

¶ *The Time.*

Columella ſaith Vines muſt be pruned before the yong branches ſpring forth. *Palladius* writeth, in Februarie : if they be pruned later they loſe their nouriſhment with weeping.

¶ *The Names.*

The Vine is called in Greeke, ᾽Αμπλος ἠμιρεψε : as much to ſay in Latine, as *Vitis Vinifera*, or the vine which beareth wine; and ᾽Αμπλος ἡμερος : that is, *Vitis manſuefacta, ſiue cultiva*, Tame or manured Vine. And it is called ἡμερος, that it may differ from both the Bryonies, the white and blacke, and from *Tamus* or our Ladies ſeale, which be likewiſe named ᾽Αμπλος. It is called *Vitis*, becauſe *inuitatur ad uvas pariendas*. It is cheriſhed to the intent to bring forth full cluſters, as *Varro* ſaith.

Pliny, lib. 14. *cap.* 3. maketh *Vva Zibeba, Alexandrina vitis*, or Vine of Alexandria; deſcribing the ſame by thoſe very words that *Theophraſtus* doth. *Dioſcorides* ſets it down to be *altera ſpecies Vitis ſyl-veſtris*, or a ſecond kinde of wilde Vine; but we had rather retaine it among the tame Vines. We may name it in Engliſh, Raiſin Vine. The fruits hereof are called in ſhops, *Paſſularum de Corintho*: in Engliſh, Corans, or ſmall Raiſins.

Sylueſtris

Sylvestris Vitis or wilde Vine is called in Greek, ʾλμπλοϲ ἀγεια : and in Latine *Labrusca* ; as in *Virgils* Eclogs.

—— *Adspice ad antrum*
Sylvestris raris sparsit labrusca racemis.

—— See how the wilde Vine
Bedecks the caue with sparsed clusters fine.

To this wilde Vine doth belong those which *Pliny, lib. 16. cap. 27.* reporteth to be called *Trisera* or that bring three sundry fruits in a yeare, as *Insana* and mad bearing Vines ; because in those some clusters are ripe and full growne, some in swelling, and others but flouring.

The fruit of the Vine is called in Greeke ϲϊϱυϲ, and ϲτϼυλὴ : in Latine, *Racemus*, and *Vva* : in English a bunch or cluster of Grapes.

The cluster of grapes that hath bin withered or dried in the Sun is named in Greeke ϼϣϊϲ : in Latine, *Vva Passa* : in shops, *Passula* : in English, Raisins of the Sun.

The berry or grape it selfe is called in Latine *Acinus*, and also *Granum*, as *Democritus* saith, speaking of the berry.

The seeds or stones contained within the berries are called in Latine *Vinacea*, and somtimes *Nuclei* : in shops, *Arilli*, as though they should say *Ariduli*, because they are dry and yeeld no juice: notwithstanding *Vinacea* are taken in *Columella* for the drosse or remnant of the grapes after they bee pressed.

The stalke which is in the middle of the clusters, and vpon which the grapes doe hang, is called of *Galen*, ϲϊϼυϠϲ : of *Varro*, *Scapus vvarum*.

¶ *The Nature and Vertues.*

A   The tender clasping branches of the Vine and the leaues do coole and mightily binde: they stay bleeding in any part of the body : they are good against the laske, the bloudy flix, the heart burne, heate of the stomacke, or readinesse to vomit. It stayeth the lusting or longing of women with childe, though but outwardly applied, and also taken inwardly after any maner: they be moreouer a remedie for the inflammation of the mouth and almonds of the throat, if they be gargled, or the mouth washed therewith.

B   Of the same facultie be also the clusters gathered before they be ripe; and likewise the bunches of the wilde Grape, which is accounted to be more effectuall against all those infirmities.

C   *Dioscorides* saith, That the liquor which falleth from the body and branches beeing cut, and that somtimes is turned as it were into a gum (which driueth stones out of the kidnies and bladder, if it be drunke in wine) healeth ring-wormes, scabs, and lepry, but the place is first to be anointed with Nitre. Being often anointed or layd on it taketh away superfluous haires : but yet he saith that the same is best which issueth forth of the green and smaller sticks, especially that liquor which falls away whilest the branches are burning, and it takes away Nits being laid on them.

D   The stones and other things remaining after the pressing are good against the bloudy flix, the laske of long continuance, and for those that are much subiect to vomiting.

E   The ashes made of the sticks and drosse that remain after the pressing, being laid vpon the piles and hard swellings about the fundament, cure the same : beeing mixed with oile of Rue or Herbe-Grace and Vineger, as the same Authour affirmeth, it helpeth to strengthen members out of joynt, and such as are bitten with any venomous beast, and easeth the pain of the spleen or milt, being ap-in manner of a plaister.

F   The later age do vse to make a Lye with the ashes of Vine stickes, in composition of causticke and burning medicines, which serue in stead of an hot iron : the one wee call a potentiall cauterie, and the other actuall.

¶ *Of Grapes.*

G   OF Grapes, those that are eaten raw do trouble the belly, and fil the stomacke full of wind, especially such as are of a soure and austere taste ; such kindes of Grapes do very much hinder the concoction of the stomacke ; and while they are dispersed through the liuer and veins they ingender cold and raw juice, which cannot easily be changed into good bloud.

H   Sweet Grapes and such as are thorow ripe are lesse hurtfull : the juice is hotter, and easilier dispersed. They also sooner passe through the belly, especially being moist, and most of all if the liquor with the pulp be taken without the stones and skin, as *Galen* saith.

I   The substance of the stones, although it be drier, and of a binding quality, doth descend thorow:

all.

all the bowels, and is nothing changed; as also the skins, which are little or nothing at all altered in the body.

Those grapes which haue a strong taste of wine are in a mean between soure and sweet.   A

Such grapes as haue little juice do nourish more, and those lesse that haue more juice: but these B do sooner descend; for the body receiueth more nourishment by the pulp than by the juice; by the juice the belly is made more soluble.

Grapes haue the preheminence among the Autumne fruits, and nourish more than they all, but C yet not so much as figges; and they haue in them little ill juice, especially when they bee thorow ripe.

Grapes may be kept the whole yeare, being ordered after the same manner that *Ioachimus Came-* D *rarius* reporteth. You shall take, saith hee, the meale of mustard seed, and strew in the bottome of any earthen pot well leaded; whereupon you shall lay the fairest bunches of the ripest grapes, the which you shall couer with more of the foresaid meale, and lay vpon it another sort of Grapes, so doing vntill the pot be full: then shall you fill vp the pot to the brimme with a kind of sweet wine called Must. The pot being very close couered shall be set into some cellar or other cold place: the grapes you may take forth at your pleasure, washing them with faire water from the pouder.

¶ Of Raisins.

OF Raisins most are sweet, some haue an austere or harsh taste: sweet Raisins are hotter, austere E colder; both of them do moderatly bind, but the austere somwhat more, which do strengthen the stomacke more. The sweet ones do neither slacken the stomack, nor make the belly soluble, beeing taken with their stones, which are of a binding qualitie: otherwise the stones taken forth, they do make the belly loose and soluble.

Raisins yeeld good nourishment to the body, they haue in them no ill juice at all, but ingender F somwhat a thick juice, which notwithstanding nourisheth the more.

There comes of sweet and fat Raisins most plenty of nourishment; of which they are the best G that haue a thin skin.

There is in the sweet ones a temperat and smoothing qualitie, with a power to clense moderate- H ly: they are good for the chest, lungs, winde-pipe, kidnies, bladder, and stomacke; for they make smooth the roughnesse of the winde-pipe, and are good against hoarsenesse, short windednesse, or difficultie of breathing: they serue to concoct the spittle, and cause it to rise more easily in any disease whatsoeuer of the chest, sides, or lungs, and do mitigate the pain of the kidnies and bladder which hath joined with it heate and sharpenesse of vrine: they dull and allay the malice of sharpe and biting humors that hurt the mouth of the stomacke.

Moreouer, Raisins are good for the liuer, as *Galen* writeth in his seuenth booke of Medicines ac- I cording to the places affected; for they be of force to concoct raw humors and restraine their ma-lignitie, and they themselues do hardly putrifie: besides, they are properly and of their owne sub-stance familiar to the intrals, cure any distemper, and nourish much, wherein is their chief vertue; for they strengthen, resist putrifaction, and helpe without hurt any distemperature of moisture or coldnesse, as the said *Galen* affirmeth.

The old Physitions haue taught vs to take out the stones, as we may see in diuers compositions K of the antient writers; as namely in that called by *Galen, Arteriaca Mithridatis,* which hath the seeds taken out: for seeing that Raisins containe in them a thicke substance, they canot easil passe tho-row the veins, but are apt to breed obstructions and stoppings of the intrals; which happeneth the rather by reason of the seeds, for they so much the harder passe through the body, and do quickely and more easily cause obstructions, in that they are more astringent or binding. Wherefore the seeds are to be taken out, for so shall the juice of the Raisins more easily passe, and the sooner bee distributed through the intrals.

Dioscorides reporteth, That Raisins being chewed with pepper draw flegme and water out of the L head.

Of Raisins is made a pultesse good for the gout, rottings about the joynts, gangrens, and morti- M fied vlcers: beeing stamped with the herb All-heale it quickely takes away the loose nailes in the fingers and toes, being laid thereon.

¶ Of Must.

MVst, called in Latine *Mustum,* that is to say, the liquor newly issuing out of the grapes when they be trodden or pressed, doth fill the stomacke and intrals with winde; it is hardly dige-sted, it is of a thicke juice, and if it doe not speedily passe through the body, it becommeth more hurtfull.

hurtfull. It hath only this one good thing in it(as *Galen* ſaith)that it maketh the body ſoluble.

A   That which is ſweeteſt and preſſed out of ripe grapes doth ſooneſt paſſe through; but that which is made of ſoure and auſtere grapes is worſt of all; it is more windy,it is hardly concoɛ̃ed, it ingendreth raw humors,and although it doth deſcend with a looſeneſſe of the belly, notwith-ſtanding it oftentimes bringeth withall the Colique and pain of the ſtone; but if the belly be not moued,all things are the worſe and more troubleſome,and it oft times brings an extreme laske and the bloudy flix.

B   The firſt part of the wine that comes forth of it ſelfe before the grapes be preſſed,is anſwerable to the Grape it ſelf,and doth quickly deſcend; but that which iſſues forth afterward,hauing ſome part of the nature of the ſtones,ſtalks,and skins,is far worſe.

¶ *Of Cute.*

C   OF Cute that is made of Muſt,which the Latines call *Sapa,* and *Defrutum,* is that liquor called in Engliſh Cute,which is made of the ſweeteſt Muſt,by boiling it to a certaine thickneſſe, or to a third part,as *Columella* writeth.

D   *Pliny* affirmeth,That *Sapa* and *Defrutum* differ in the maner of the boiling,and that *Sapa* is made when the new Wine is boiled away till onely a third part remaines; and *Defrutum* till halfe bee boiled.

E   *Siræum* (ſaith he,*lib.*14.*cap.*17.) which others call ἕψημα. and we *Sapa,* a worke of wit,and not of Nature, is made of new Wine boyled to a third part; which beeing boyled to halfe, wee call *Defrutum.*

F   *Palladius* ioyneth to theſe *Carænum,*which as hee ſaith is made when a third part is boiled away, and two remaine.

G   *Leontius* in his Geoponicks ſheweth,That *Hepſema* muſt be made of eight parts of new Wine, and an hundred of wine it ſelfe boiled to a third.

H   *Galen* teſtifieth,That *Hepſema* is new wine very much boiled.The later phyſitians call *Hepſema* or *Sapa* boiled wine.

I   Cute or boiled wine is hot,yet not ſo hot as wine,but thicker,not ſo eaſily diſtributed or caried through the body,and it ſlowly deſcendeth by vrine,but by the belly oftentimes ſooner, for it moderatly looſens it.

K   It nouriſheth more,and filleth the body quickly; yet doth it by reaſon of his thickneſſe ſticke in the ſtomack for a time,and is not ſo fit for the liuer or ſpleen. Cute alſo digeſteth raw humours that ſticke in the cheſt and lungs,and raiſeth them vp ſpeedily;wherfore it is good for the cough and ſhortneſſe of breath.

L   The Vintners of the Low-Countries (and perhaps of London)do make of Cute and wine mixt in a certain proportion, a compound and counterfeit wine which they ſell for Candy wine, vſually called Malmſey.

M   *Pliny,lib.*14.*cap.*9. ſaith,that Cute was firſt deuiſed for a baſtard hony.

¶ *Of Wine.*

N   TO ſpeake of wine the juice of Grapes,which being newly preſſed forth is called (as wee haue ſaid) *Muſtum* or new wine;after the dregs and droſſe are ſetled, and it appeareth pure and cleer, it is called in Greek ὄινος: in Latine,*Vinum* : in Engliſh,Wine,and that not vnproperly.For certain other juices,as of Apples,Pomegranats,Peares,Medlars,Seruices,or ſuch otherwiſe made(for examples ſake)of Barley and Graine,be not at all ſimply called wine,but with the name of the thing added whereof they do conſiſt. Hereupon is the wine which is preſſed forth of the Pomegranate berries named *Rhoites,*or wine of Pomegranats; out of Quinces,*Cydonites,*or wine of Quinces: out of Peares, *Apyites* or Perry; and that which is compounded of Barley is called *Zythum,* or Barley wine : in Engliſh,Ale or Beere.

O   And other certain wines haue borrowed ſyrnames of the plants that haue bin infuſed or ſteeped in them; and yet all wines of the Vine,as Wormwood wine,Myrtle wine,and Hyſſop wine,which are all called artificiall wines.

P   That is properly and ſimply called wine which is preſſed out of the grapes of the wine, and is without any manner of mixture.

Q   The kindes of Wines are not of one nature,nor of one facultie or power, but of many, differing one from another; for there is one difference thereof in taſte, another in colour, the third is referred to the conſiſtence or ſubſtance of the Wine; the fourth conſiſteth in the vertue and ſtrength thereof. *Galen* addeth that which is found in the ſmell, which belongs to the vertue and ſtrength of the Wine.

That

That may also bee joyned vnto them, which respecteth the age : for by age wines become hot- A
ter and sharper, and doe withall change oftentimes the colour, the substance, and the smell : for
some wines are sweet of taste ; others austere or somthing harsh ; diuers of a rough taste, or altogether harsh ; and most of them sufficient sharp : there be likewise wines of a middle sort, inclining to one or other qualitie.

Wine is of colour either white or reddish, or of a blackish deep red, which is called blacke, or B
of some middle colour betweene these.

Some wine is of substance altogether thin ; other some thick and fat ; and many also of a mid- C
dle consistence.

One wine is of great strength, and another is weake, which is called a waterish wine : a full wine D
is called in Latine *Vinosum*. There be also among these very many that be of a middle strength.

There is in all wines, be they neuer so weake, a certaine winie substance thin and hot. There be E
likewise waterie parts, and also diuers earthy : for wine is not simple, but (as *Galen*, testifieth *lib.* 4.
of the faculties of medicines) consisteth of parts that haue diuers faculties.

Of the sundry mixture and proportion of these substances one with another, there rise diuers and F
sundry faculties of the wine.

That is the best and fullest wine in which the hot and winie parts doe most of all abound : and G
the weakest is that wherein the waterie haue the preheminence.

The earthy substance abounding in the mixture causeth the wine to be austere or somthing H
harsh, as a crude or raw substance doth make it altogether harsh. The earthy substance being seuered falleth downe, and in continuance of time sinketh to the bottome, and becomes the dreggs
or lees of the wine : yet it is not alwaies wholly seuered, but hath both the tast and other qualities
of this substance remaining in the wine.

All wines haue their heate, partly from the proper nature and inward or originall heate of the I
vine, and partly from the Sun : for there is a double heate which ripeneth not only the grapes, but
also all other fruits, as *Galen* testifieth ; the one is proper and naturall to euery thing ; the other is
borrowed of the Sun ; which if it be perceiued in any thing, it is vndoubtedly best and especially
in the ripening of grapes.

For the heate which proceeds from the Sun concocteth the grapes and the juice of the grapes, K
and doth especially ripen them, stirring vp and increasing the inward and naturall heat of the wine,
which otherwise is so ouerwhelmed with abundance of raw and waterish parts, as it seemes to bee
dulled and almost without life.

For vnlesse wine had in it a proper and originall heat, the grapes could not be so concocted by L
the force of the Sunne, as that the wine should become hot ; no lesse than many other things naturally cold, which although they be ripened and made perfect by the heat of the Sun, do not for
all that lose their originall nature ; as the fruits, iuices, or seeds of Mandrake, Nightshade, Hemlocke, Poppy, and of other such like, which though they be made ripe, and brought to full perfection, yet still retaine their owne cold qualitie.

Wherefore seeing that wine through the heate of the Sunne is for the most part brought to his M
proper heate, and that the heate and force is not all alike in all regions and places of the earth ;
therefore by reason of the diuersitie of regions and places, the wines are made not a little to differ
in facultie.

The stronger and fuller wine groweth in hot countries and places that lie to the Sun ; the rawer N
and weaker in cold regions and prouinces that lie open to the North.

The hotter the Summer is, the stronger is the wine ; the lesse hot or the moister it is, the lesse O
ripe is the wine. Notwithstanding not only the manner of the weather and of the Sun, maketh the
qualities of the wine to differ, but the natiue propertie of the soile also ; for both the tast and other
qualities of the wine are according to the manner of the Soile. And it is very well knowne, that
not only the colour of the wine, but the tast also dependeth vpon the diuersity of the grapes.

Wine (as *Galen* writeth) is hot in the second degree, and that which is very old, in the third ; but P
new wine is hot in the first degree : which things are especially to bee vnderstood concerning the
meane betweene the strongest and the weakest ; for the fullest and mightiest (being but *Horna*, that
is as I take it, of one yeare old) are for the most part hot in the second degree The weakest and
the most waterish wines, although they be old, do seldome exceed the second degree.

The drinesse is answerable to the heate in proportion, as *Galen* saith in his book of Simples : but Q
in his books of the gouernment of health he sheweth, that wine doth not only heate, but also moisten our bodies, and that the same doth moisten and nourish such bodies as are extreame dry ; and
both these opinions be true.

For the faculties of wine are of one sort as it is a medicine, and of another as it is a nourish- R
ment ; which *Galen* in his book of the faculties of nourishments doth plainly shew, affirming that
those qualities of the wine which *Hippocrates* writeth of in his booke of the manner of diet, be not

as

as a nouriſhment, but rather as of a medicine. For wine as it is a medicine doth dry, eſpecially being outwardly applied ; in which caſe, for that it doth not nouriſh the body at all, the drines doth more plainely appeare, and is more manifeſtly perceiued.

A   Wine is a ſpeciall good medicine for an vlcer, by reaſon of his heate and moderate drying, as *Galen* teacheth in his fourth booke of the method of healing.

B   *Hippocrates* writeth, That vlcers, what manner of ones ſoeuer they are, muſt not be moiſtned vnleſſe it be with wine : for that which is dry (as *Galen* addeth) commeth neerer to that which is whole, and the thing that is moiſt, to that which is not whole.

C   It is manifeſt, that Wine is in power or facultie dry, and not in act ; for Wine actually is moiſt and liquid, and alſo cold : for the ſame cauſe it likewiſe quencheth thirſt, which is an appetite or deſire of cold and moiſt, and by this actuall moiſture (that we may ſo terme it) it is if it be inwardly taken, not a medicine, but a nouriſhment ; for it nouriſheth, and through his moiſture maketh plenty of bloud ; and by increaſing the nouriſhment it moiſtneth the body, vnleſſe peraduenture it be old and very ſtrong : for it is made ſharpe and biting by long lying, and ſuch kinde of Wine doth not only heate, but alſo conſume and dry the body, for as much as it is not now a nouriſhment, but a medicine.

D   That wine which is neither ſharp by long lying, nor made medicinable, doth nouriſh and moiſten, ſeruing as it were to make plenty of nouriſhment and bloud, by reaſon that through his actuall moiſture it more moiſtneth by feeding, nouriſhing, and comforting, than it is able to dry by his power.

E   Wine doth refreſh the inward and naturall heate, comforteth the ſtomacke, cauſeth it to haue an appetite to meat, moueth concoction, and conueyeth the nouriſhment through all parts of the body, increaſeth ſtrength, inlargeth the body, maketh flegme thin, bringeth forth by vrine cholericke and waterie humors, procureth ſweating, ingendreth pure bloud, maketh the body well coloured, and turneth an ill colour into a better.

F   It is good for ſuch as are in a conſumption, by reaſon of ſome diſeaſe, and that haue need to haue their bodies nouriſhed atd refreſhed (alwaies prouided they haue no feuer,) as *Galen* ſaith in his ſeuenth booke of the Method of curing. It reſtoreth ſtrength moſt of all other things, and that ſpeedily : It maketh a man merry and ioyfull : It putteth away feare , care , troubles of minde, and ſorrow : It moueth pleaſure and luſt of the body , and bringeth ſleepe gently.

G   And theſe things proceed of the moderate vſe of wine : for immoderate drinking of wine doth altogether bring the contrarie. They that are drunke are diſtraughted in minde, become fooliſh, and oppreſſed with a drowſie ſleepineſſe, and be afterward taken with the Apoplexy, the gout, or altogether with other moſt grieuous diſeaſes ; the braine, liuer, lungs, or ſome other of the intrals being corrupted with too often and ouermuch drinking of wine.

H   Moreouer, wine is a remedy againſt taking of Hemlock or green Coriander, the juice of black Poppy, Wolfes bane, and Leopards-bane, Tode-ſtooles, and other cold poiſons, and alſo againſt the biting of Serpents, and ſtings of venomous beaſts, that hurt and kill by cooling.

I   Wine alſo is a remedy againſt the ouer-fulneſſe and ſtretching out of the ſides, windy ſwellings, the green ſickneſſe, the dropſie, and generally all cold infirmities of the ſtomack, liuer, milt, and alſo of the matrix.

K   But Wine which is of colour and ſubſtance like water, through ſhining bright, pure, of a thinne ſubſtance, which is called white, is of all wines the weakeſt ; and if the ſame ſhould bee tempered with water it would beare very little : and hereupon *Hippocrates* calleth it ἰνγόνεον, that is to ſay, bearing little water to delay it withall.

L   This troubleth the head and hurteth the ſinewes leſſe than others do, and is not vnpleaſant to the ſtomacke : it is eaſily and quickely diſperſed thorow all parts of the body : it is giuen with far leſſe danger than any other wine, to thoſe that haue the Ague (except ſome inflammation or hot ſwelling be ſuſpected) and oftentimes with good ſucceſſe to ſuch as haue intermitting feuers ; for as *Galen, lib.*8. of his Method ſaith, it helpeth concoction, digeſteth humors that be halfe raw, procureth vrine and ſweat, and is good for thoſe that cannot ſleep, and that be full of care and ſorrow, and for ſuch as are ouerwearied.

M   Black wine, that is to ſay, wine of a deep red colour, is thick, and hardly diſperſed, and doth not eaſily paſſe through the bladder : it quickely taketh hold of the braine and makes a man drunk : it is harder of digeſtion : it remaineth longer in the body ; it eaſily ſtoppeth the liuer and ſpleen ; for the moſt part it bindes, notwithſtanding it nouriſheth more, and is more fit to engender bloud : it filleth the body with fleſh ſooner than others do.

N   That which is of a light crimſon red colour is for the moſt part more delightfull to the taſte, fitter for the ſtomacke ; it is ſooner and eaſier diſperſed : it troubleth the head leſſe, it remains not ſo long vnder the ſhort ribs, and eaſilier deſcendeth to the bladder than blacke wine doth : it doth

also

alſo make the belly coſtiue, if ſo be that it be not ripe. For ſuch crude and rough wines do often-times moleſt weake ſtomackes, and are troubleſome to the belly.

Reddiſh yellow wine ſeemeth to be in a meane betweene a thinne and thicke ſubſtance : other-wiſe it is of all vines the hotteſt ; and ſuffereth moſt water to be mixed with it, as *Hippocrates* wri-teth.   A

The old vine of this kinde, being of a thin ſubſtance and good ſmell, is a ſingular medicine for all thoſe that are much ſubject to ſwouning, although the cauſe thereof proceed of choler that hurteth the mouth of the ſtomacke, as *Galen* teſtifieth in the 12. booke of his method.   B

Sweet wine the leſſe hot it is, the leſſe doth it trouble the head, and offend the minde ; and it bet-ter paſſeth through the belly, making it oftentimes ſoluble : but it doth not ſo eaſily paſſe or de-ſcend by vrine.   C

Againe, the thicker it is of ſubſtance, the harder and ſlowlier it paſſeth through: it is good for the lungs and for thoſe that haue the cough. It ripeneth raw humors that ſticke in the cheſt, and cauſeth them to be eaſilier ſpit vp ; but it is not ſo good for the liuer, whereunto it bringeth no ſmall hurt when either it is inflamed, or ſchirrous, or when it is ſtopped. It is alſo an enemy to the ſpleene, it ſticketh vnder the ſhort ribs, and is hurtfull to thoſe that are full of choler. For this kind of wine, eſpecially the thicker it is, is in them very ſpeedily turned into choler : and in others when it is well concocted, it increaſeth plenty of nouriſhment.   D

Auſtere wine, or that which is ſomewhat harſh in taſte, nouriſheth not much ; and if ſo be that it be thin and white, it is apt to prouoke vrine, it leſſe troubleth the head, it is not quickly digeſted, for which cauſe it is the more to be ſhunned, as *Galen* ſaith in 12. booke of his method.   E

That wine which is altogether harſh or rough in taſte, the leſſe ripe it is, the neerer it commeth to the qualities of Verjuyce made of ſower grapes, being euidently binding. It ſtrengtheneth a weake ſtomacke ; it is good againſt the vnkindely luſting or longing of women with childe; it ſtai-eth the laske, but it ſticketh in the bowels : breedeth ſtoppings in the liuer and milt ; it ſlowly deſ-cendeth by vrine, and ſomething troubleth the head.   F

Old wine which is alſo made ſharpe by reaſon of age, is not only troubleſome to the braine, but alſo hurteth the ſinewes : it is an enemy to the entrailes, and maketh the body leane.   G

New wine, and wine of the firſt yeere, doth eaſily make the body to ſwell, and ingendreth winde, it cauſeth troubleſome dreames, eſpecially that which is not throughly refined, or thicke, or very ſweet : for ſuch do ſooner ſticke in the entrailes than others do. Other wines that are in a meane in colour, ſubſtance, taſte or age, as they doe decline in vertues and goodneſſe from the extreames ; ſo alſo they be free from their faults and diſcommodities. They come neere in faculties to thoſe wines whereunto they be next, either in colour, taſte, or ſubſtance, or elſe in ſmell or in age.   H

Wine is fitteſt for thoſe that be of nature cold and dry ; and alſo for old men, as *Galen* ſheweth in his fifth booke of the gouernment of health : for it heateth all the members of their bodies, and purgeth away the watery part of the bloud, if their be any.   I

The beſt wines are thoſe that be of a fat ſubſtance : for thoſe both increaſe bloud, and nouriſh the body; both which commodities they bring to old men, eſpecially at ſuch time as they haue no ſerous humour in their veines, and haue need of much nouriſhment. It happeneth that oftentimes there doth abound in their bodies a wateriſh excrement, and then ſtand they in moſt need of all of ſuch wines as do prouoke vrine.   K

As wine is beſt for old men, ſo it is worſt for children : by reaſon that being drunke, it both moi-ſteneth and drieth ouermuch, and alſo filleth the head with vapours, in thoſe who are of a moiſt and hot complexion, or whoſe bodies are in a meane betweene the extreames, whom *Galen* in his booke of the gouernment of health doth perſuade, that they ſhould not ſo much as taſte of wine for a very long time : for neither is it good for them to haue their heads filled, nor to be made moiſt and hot, more than is ſufficient, becauſe they are already of ſuch a heate and moiſture, as if you ſhould but little increaſe either quality, they would forthwith fall into the extreme.   L

And ſeeing that euery exceſſe is to be ſhunned, it is expedient moſt of all to ſhun this, by which not only the body, but alſo the minde receiueth hurt.   M

Wherefore we thinke, that wine is not fit for men that be already of full age, vnleſſe it be mode-rately taken, becauſe it carrieth them headlong into fury and luſt, and troubleth and dulleth the rea-ſonable part of the minde.   N

¶ Of the delaying, or tempering of Wine.

IT was an ancient cuſtome, and of long continuance in oldtime, for wines to be mixed with wa-ter, as it is plaine and euident not onely by *Hippocrates*, but alſo by other old mens Writings. Wine firſt began to be mixed with water for health and wholeſomeneſſe ſake : for as *Hippocra-tes* writeth in his booke of antient Phyſicke, beeing ſimply and of it ſelfe much drunke, it ma-   O

keth

keth a man in ſome ſort weake and feeble: which thing *Ouid* ſeemeth alſo to allow of, writing thus:

Vt Venus eneruat vires,ſic copia vini
Et tentat greſſas, debilitatque pedes.

As Venery the vigour ſpends, ſo ſtore of Wine
Makes man to ſtagger,makes his ſtrength decline.

A   Moreouer, wine is the ſweeter, hauing water poured into it, as *Athenæus* ſaith. *Homer* likewiſe commendeth that wine which is well and fitly allaied. *Philocorus* writeth(as *Athenæus* reporteth) that *Amphictyon* king of Athens was the firſt that allaied wine, as hauing learned the ſame of *Dionyſius* : wherefore he ſaith,that thoſe who in that manner drunke it remained in health, that before had their bodies feeble and ouerweakened with pure and vnmixed wine.

B   The manner of mingling or tempering of wine was diuers: for ſometimes to one part of wine, there were added two,and ſometimes three or foure of water;or to two parts of wine three of water: of a leſſe delay was that which conſiſted of equall parts of wine and water.

C   The old Comedians did thinke that this leſſer mixture was ſufficient to make men mad,among whom was *Mneſitheus*,whoſe words be extant in *Athenæus.*

D   *Hippocrates* in the ſeuenth booke of his Aphoriſmes ſaith,that this manner of tempering of wine and water by equall parts bringeth as it were a light pleaſant drunkenneſſe, and that it is kinde of remedy againſt diſquietneſſe,yawnings,and ſhiuerings ; and this mingling belongeth to the ſtrongeſt wines.

E   Such kinde of wines they might be which in times paſt the Scythians were reported of the old writers to drinke, who for this cauſe doe call vnmixed wine the Scythians drinke. And they that drinke ſimple wine ſay, that they will *Scithyzare*, or do as the Scythians do,as we may reade in the tenth booke of *Athenæus.*

F   The Scythians, as *Hippocrates* and diuers other of the old writers affirme, be people of Germany beyond the floud Danubius, which is alſo called Iſter : Rhene is a riuer of Scythia : and *Cyrus* hauing paſſed ouer Iſter is reported to haue come into the borders of the Scythians.

G   And in this our age all the people of Germany do drinke vnmixed wine,which groweth in their owne country,and likewiſe other people of the North parts, who make no ſcruple at all to drinke of the ſtrongeſt wines without any mixture.

¶ *Of the liquor which is diſtilled out of Wine, commonly called,* Aqua vitæ.

H   THere is drawne out of Wine a liquor,which in Latine is commonly called *Aqua vitæ,* or water of life, and alſo *Aqua ardens*,or burning water, which as diſtilled waters are drawne out of herbes and other things, is after the ſame manner diſtilled out of ſtrong wine, that is to ſay, by certaine inſtruments made for this purpoſe,which are commonly called Limbeckes.

I   This kinde of liquor is in colour and ſubſtance like vnto waters diſtilled out of herbes,and alſo reſembleth cleere ſimple water in colour,but in faculty it farre differeth.

K   It beareth the ſyrname of life, becauſe that it ſerueth to preſerue and prolong the life of man.

L   It is called *Ardens*, burning, for that it is eaſily turned into a burning flame: for ſeeing it is not any other thing than the thinneſt and ſtrongeſt part of the wine, it being put to the flame of fire, is quickly burned.

M   This liquor is very hot, and of moſt ſubtill and thin parts,hot and dry in the later end of the third degree, eſpecially the pureſt ſpirits thereof: for the purer it is, the hotter it is, the dryer,and of thinner parts: which is made more pure by often diſtilling.

N   This water diſtilled out of wine is good for all thoſe that are made cold either by a long diſeaſe, or through age,as for old and impotent men : for it cheriſheth and increaſeth naturall heate ; vpholdeth ſtrength,repaireth and augmenteth the ſame : it prolongeth life, quickeneth all the ſenſes,and doth not only preſerue the memory, but alſo recouereth it when it is loſt : it ſharpeneth the ſight.

O   It is fit for thoſe that are taken with the Catalepſie (which is a diſeaſe in the braine proceeding of drineſſe and cold) and are ſubject to dead ſleepes, if there be no feuer joyned ; it ſerueth for the weakeneſſe,trembling,and beating of the heart ; it ſtrengtheneth and heateth a feeble ſtomacke ; it conſumeth winde both in the ſtomacke, ſides, and bowels ; it maketh good concoction of meate, and is a ſingular remedy againſt cold poyſons.

P   It hath ſuch force and power,in ſtrengthening of the heart,and ſtirreth vp the inſtruments of the
ſenſes

ſenſes, that it is moſt effectuall, not onely inwardly taken to the quantity of a little ſpoonefull, but alſo outwardly applied : that is to ſay, ſet to the noſthrils, or laid vpon the temples of the head, and to the wreſts of the armes; and alſo to foment and bath ſundry hurts and griefes.

Being held in the mouth it helpeth the toothache : it is alſo good againſt cold cramps and con- A
vulſions, being chafed and rubbed therewith.

Some are bold to giue it in quartaines before the fit, eſpecially after the height or prim of the B
diſeaſe.

This water is to be giuen in wine with great iudgement and diſcretion; for ſeeing it is extreme C
hot, and of moſt ſubtill parts, and nothing elſe but the very ſpirit of the wine, it moſt ſpeedily pierceth through, and doth eaſily aſſault and hurt the braine.

Therefore it may be giuen to ſuch as haue the apoplexie and falling ſickneſſe, the megrim, the D
headache of long continuance, the Vertigo, or giddineſſe proceeding through a cold cauſe : yet can
it not be alwaies ſafely giuen; for vnleſſe the matter the efficient cauſe of the diſeaſe be ſmall, and
the ſicke man of temperature very cold, it cannot be miniſtred without danger : for that it ſpredeth and diſperſeth the humours, it filleth or ſtuffeth the head, and maketh the ſicke man worſe :
and if the humors be hot, as bloud is, it doth not a little increaſe inflammations alſo.

This water is hurtfull to all that be of nature and complexion hot, and moſt of all to cholericke E
men : it is alſo offenſiue to the liuer, and likewiſe vnprofitable for the kidnies, being often and plentifully taken.

If I ſhould take in hand to write of euery mixture, of each infuſion, of the ſundry colours, and
euery other circumſtance that the vulgar people doe giue vnto this water, and their diuers vſe, I
ſhould ſpend much time but to ſmall purpoſe.

¶ Of Argall, Tartar, or wine Lees.

THe Lees of wine which is become hard like a cruſt, and ſticketh to the ſides of the veſſell, and F
wine casks, being dried, hard, ſound, and well compact, and which may be beaten into pouder, is
called in ſhops *Tartarum* : in Engliſh, Argall, and Tartar.

Theſe Lees are vſed for many things : the ſiluer-Smiths poliſh their ſiluer herewith : the Diers G
vſe it : and it is profitable in medicine.

It doth greatly dry and waſte away, as *Paulus Ægineta* ſaith : it hath withall a binding faculty, H
proceeding from the kinde of wine, of which it commeth.

The ſame ſerueth for moiſt diſeaſes of the body : it is good for thoſe that haue the greene ſicke- I
neſſe and the dropſie, eſpecially that kinde that lieth in the fleſh, called in Latine, *Leucophlagmatica:*
being taken euery day faſting halfe a penny weight or a full penny weight (which is a dram and
nine graines after the Romanes computation) doth not onely dry vp the wateriſh excrements, and
voideth them by vrine, but it preuaileth much to clenſe the belly by ſiege.

It would worke more effectually, if it were mixed either with hot ſpices, or with other things K
that breake winde, or elſe with diuretickes, which are medicines that prouoke vrine; likewiſe to
be mixed with gentle purgers, as the ſicke mans caſe ſhall require.

The ſame of it ſelfe, or tempered with oile of Myrtles, is a remedy againſt ſoft ſwellings, as *Di-* L
ſcorides teacheth : it ſtaieth the laske, and vomiting, being applied outwardly vpon the region of
the ſtomacke in a pultis; and if it be laid to the bottome of the belly and ſecret parts, it ſtoppeth
the whites, waſteth away hot ſwellings of the kernels in the flankes, and other places, which be
not yet exulcerated : it aſſwageth great breſts, and drieth vp the milke, if it be annointed on with
vinegre.

Theſe Lees are oftentimes burnt : if it become all white, it is a ſigne of right and perfect bur- M
ning, for till then it muſt be burned : being ſo burnt the Grecians terme it, σφίαν, as *Ægineta* ſaith,
the Apothecaries call it, *Tartarum vſtum*, and *Tartarum calcinatum* : that is to ſay, burnt or calcined
Tartar.

It hath a very great cauſticke or burning quality : it clenſeth and throughly heateth, bindeth, N
eateth, and very much drieth, as *Dioſcorides* doth write : being mixed with Roſin, it maketh rough
and ill nailes to fall away : *Paulus* ſaith, That it is mixed with cauſtickes or burning medicines to
increaſe their burning quality : it muſt be vſed whileſt it is new made, becauſe it quickly vaniſheth,
for the Lees of wine burned, doe ſoone relent or wax moiſt, and are ſpeedily reſolued into liquor :
therefore he that would vſe it dry, muſt haue it put in a glaſſe, or glaſſed veſſell well ſtopped, and
ſet in a hot and dry place. It melteth and is turned into liquor if it be hanged in a linnen bag in
ſome place in a celler vnder the ground.

The Apothecaries call this liquor that droppeth away from it, oile of Tartar. It retaineth a cau- O
ſticke and burning quality, and alſo a very dry faculty : it very ſoon taketh away lepric, ſcabs, ret-
ters, and other filth and deformity of the skin and face : with an equall quantity of Roſe water

added,

added,and as much Ceruſe as is ſufficient for a linament,wherewith the blemiſhed or ſpotted parts muſt be annointed ouer night.

¶ *The briefe ſumme of that hath beene ſaid of the Vine.*

A   THe juyce of the greene leaues, branches, and tendrels of the Vine drunken, is good for thoſe that vomit and ſpit bloud,for the bloudy flix, and for women with child that vomit ouermuch. The kernell within the grapes boiled in water and drunke hath the ſame effect.

B   Wine moderately drunke profiteth much, and maketh good digeſtion, but it hurteth and diſtempereth them that drinke it ſeldome.

C   White wine is good to be drunke before meat ; it preſerueth the body,and pierceth quickely into the bladder : but vpon a full ſtomacke it rather maketh opilations or ſtoppings,becauſe it doth ſwiftly driue downe meate before Nature hath of her ſelfe digeſted it.

D   Claret wine doth greatly nouriſh and warme the body, and is wholeſome with meate, eſpecially vnto phlegmaticke people ; but very vnwholeſome for young children, as *Galen* ſaith, becauſe it heateth aboue nature,and hurteth the head.

E   Red wine ſtops the belly,corrupteth the bloud,breedeth the ſtone,is hurtfull to old people, and good or profitable to few,ſaue to ſuch as are troubled with the laske,bloudy flix,or any other looſeneſſe of the body.

F   Sacke or Spaniſh wine hath been vſed of a long time to be drunke after meat, to cauſe the meat the better to digeſt ; but common experience hath found it to be more beneficiall to the ſtomacke to be drunke before meat.

G   Likewiſe Malmſey,Muskadell,Baſtard,and ſuch like ſweet wines haue been vſed before meat, to comfort the cold and weake ſtomacke, eſpecially being taken faſting : but experience teacheth, that Sacke drunke in ſtead thereof is much better,and warmeth more effectually.

H   Almighty God for the comfort of mankinde ordained Wine ; but decreed withall, That it ſhould be moderately taken,for ſo it is wholſome and comfortable:but when meaſure is turned into exceſſe, it becommeth vnwholeſome, and a poyſon moſt venomous, relaxing the ſinewes, bringing with it the palſey and falling ſickneſſe:to thoſe of a middle age it bringeth hot feuers,frenſie, and lecherie ; it conſumeth the liuer and other of the inward parts : beſides,how little credence is to be giuen to drunkards it is euident ; for though they be mighty men, yet it maketh them monſters,and worſe than brute beaſts. Finally in a word to conclude ; this exceſſiue drinking of Wine diſhonoreth Noblemen, beggereth the poore, and more haue beene deſtroied by ſurfeiting therewith,than by the ſword.

Cʜᴀᴘ. 324. *Of Hops.*

¶ *The Kindes.*

THere bee two ſorts of Hops : one the manured or the Garden Hop ; the other wilde or of the hedge.

¶ *The Deſcription.*

1   THe Hop doth liue and flouriſh by embracing and taking hold of poles, pearches, and other things vpon which it climeth. It bringeth forth very long ſtalkes, rough, and hairie ; alſo rugged leaues broad like thoſe of the Vine, or rather of Bryony, but yet blacker, and with fewer dented diuiſions : the floures hang downe by cluſters from the tops of the branches,puffed vp, ſet as it were with ſcales like little canes,or ſcaled Pine apples,of a whitiſh colour tending to yellowneſſe, ſtrong of ſmell : the roots are ſlender, and diuerſly folded one within another.

2   The wilde Hop differeth not from the manured Hop in forme or faſhion, but is altogether leſſer,as well in the cluſters of floures, as alſo in the franke ſhoots, and doth not bring forth ſuch ſtore of floures,wherein eſpecially conſiſteth the difference.

¶ *The Place.*

The Hop joyeth in a fat and fruitfull ground : it proſpereth the better by manuring : alſo it groweth among briers and thornes about the borders of fields,I meane the wilde kinde.

                      ¶ *The*

1 *Lupus ſalictarius.*
Hops.

¶ *The Time.*

The floures of hops are gathered in Auguſt and September, and reſerued to be vſed in beere: in the Spring time come forth new ſhoots or buds: in the Winter onely the roots remaine aliue.

¶ *The Names.*

It is called in ſhops and in all other places, *Lupulus:* of ſome, *Lupus ſalictarius,* or *Lupulus ſalictarius:* in high Dutch, **Hopſſen:** in low Dutch, **Hoppe:** in Spaniſh, *Hombreζillos:* in French, *Houblon:* in Engliſh, Hops.

Pliny, lib. 2 1. cap. 15. maketh mention of Hops among the prickly plants.

¶ *The Temperature.*

The floures of the hop are hot and dry in the ſecond degree: they fill and ſtuffe the head, and hurt the ſame with their ſtrong ſmell. Of the ſame temperature alſo are the leaues themſelues, which doe likewiſe open and clenſe.

¶ *The Vertues.*

The buds or firſt ſprouts which come forth in A the Spring are vſed to be eaten in ſallads; yet are they, as *Pliny* ſaith, more toothſome than nouriſhing, for they yeeld but very ſmall nouriſhment: notwithſtanding they be good for the intrals, both in opening and procuring of vrine, and likewiſe in keeping the body ſoluble.

The leaues and little tender ſtalkes, and alſo the B floures themſelues remoue ſtoppings out of the liuer and ſpleene, purge by vrine, helpe the ſpleene, clenſe the bloud, and be profitable againſt long lingring Agues, ſcabs, and ſuch like filth of the skin, if they be boyled in whay.

The juyce is of more force, and doth not onely remoue obſtructions out of the intrals, but it is C alſo thought to auoid choler and flegme by the ſtoole. It is written, that the ſame dropped into the eares, taketh away the ſtench and corruption thereof.

The floures are vſed to ſeaſon Beere or Ale with, and too many do cauſe bitterneſſe thereof, and D are ill for the head.

The floures make bread light, and the lumpe to be ſooner and eaſilier leauened, if the meale be E tempered with liquor wherein they haue been boiled.

The decoction of Hops drunke openeth the ſtoppings of the liuer, the ſpleene, and kidnies, and F purgeth the bloud from all corrupt humors, cauſing the ſame to come forth with the vrine.

The juyce of Hops openeth the belly, and driueth forth yellow and cholericke humors, and pur- G geth the bloud from all filthineſſe.

The manifold vertues of Hops do manifeſt argue the wholeſomeneſſe of beere aboue ale; for the H hops rather make it a phyſicall drinke to keepe the body in health, than an ordinary drinke for the quenching of our thirſt.

Chap. 325.   *Of Trauellers-Ioy.*

¶ *The Deſcription.*

1   THe plant which *Lobel* ſetteth forth vnder the title of *Viorna, Dodonæus* makes *Vitis alba;* but not properly; whoſe long wooddy and viny branches extend themſelues very far, and into infinite numbers, decking with his claſping tendrels and white ſtarre-like floures (being very ſweet) all the buſhes, hedges, and ſhrubs that are neere vnto it. It ſends forth many branched ſtalkes, thicke, tough, full of ſhoots and claſping tendrels, wherewith it foldeth it ſelfe vpon the hedges, and taketh hold and climeth vpon euery thing that ſtandeth neere vnto it.

The

The leaues are fastned for the most part by fiues vpon one rib or stem, two on either side, and one in the midst or point standing alone ; which leaues are broad like those of Iuy, but not cornered at all: among which come forth clusters of white floures, and after them great tufts of flat seeds, each seed hauing a fine white plume like a feather fastned to it, which maketh in the Winter a goodly shew, couering the hedges white all ouer with his feather-like tops. The root is long, tough, and thicke, with many strings fastned thereto.

2   *Clusius* hath set forth a kinde of *Clematis*, calling it *Clematis Bætica*, hauing a maruellous long small branch full of joynts, with many leaues indented about the edges like those of the peare tree, but stiffer and smaller, comming from euery joynt ; from whence also at each joynt proceed two small clasping tendrels, as also the small foot-stalkes whereon the seeds do stand, growing in great tufted plumes or feathers, like vnto the precedent, whereof it is a kinde. The floures are not expressed in the figure, nor seene by the Author, and therefore what hath been said shall suffice.

<table>
<tr><td>1 Viorna.
The Trauellers-Ioy.</td><td>2 Clematis Bætica.
The Spanish Trauellers-Ioy.</td></tr>
</table>

 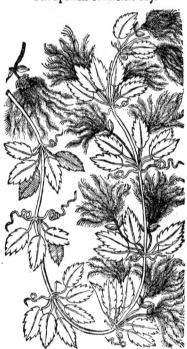

¶ *The Place.*

The Trauellers-Ioy is found in the borders of fields among thornes and briers, almost in euery hedge as you go from Grauesend to Canturbury in Kent; in many places of Essex, and in most of these Southerly parts about London, but not in the North of England that I can heare of.

The second is a stranger in these parts : yet haue I found it in the Isle of Wight, and in a wood by Waltham abbey.

¶ *The Time.*

The floures come forth in Iuly : the beautie thereof appeares in Nouember and December.

¶ *The Names.*

The first is commonly called *Viorna, quasi vias ornans*, of decking and adorning waies and hedges, where people trauel; and thereupon I haue named it the Tauellers-Ioy : of *Fuchsius* it is called *Vitis nigra* : of *Dodonæus, Vitalba* : of *Matthiolus, Clematis altera* : of *Cordus, Vitis alba* : of *Dioscorides, Vitis syluestris* : of *Theophrastus, Atragene* : in Dutch, **Linen :** in French, as *Ruellius* writeth, *Viorne.*

¶ *The Temperature and Vertues.*

These plants haue no vse in physicke as yet found out, but are esteemed onely for pleasure, by reason of the goodly shadow which they make with their thicke bushing and clyming, as also for the beauty of the floures, and the pleasant sent or sauour of the same.

CHAP. 326.   Of Ladies Bower, or Virgins Bower.

¶ The Description.

1   THat which *Lobel* describeth by the name *Clematis peregrina*, hath very long and slender
stalks like the Vine, which are joynted, of a darke colour; it climeth aloft, and taketh
hold with his crooked claspers vpon euery thing that standeth nere vnto it: it hath many leaues diuided into diuers parts; among which come the floures that hang vpon slender toot-
stalkes, something like to those of Peruinkle, consisting onely of foure leaues, of a blew colour, and
sometimes purple, with certaine threds in the middle: the seeds be flat, plaine amd sharpe pointed.
The roots are slender, and spreading all about.

1 2 *Clematis peregrina Carulea, siue rubra.*
   Blew or red floured Ladies-Bower.

‡ 3 *Clematis Carulea flore pleno.*
   Double floured Virgins-Bower.

2   The second differeth not from the other, in leaues, stalkes, branches nor seed. The only difference consisteth in that, that this plant bringeth forth red floures, and the other blew.
‡ 3   There is preserued also in some Gardens another sort of this *Clematis*, which in roots,
leaues, branches, and manner of growing differs not from the former: but the floure is much different, being composed of abundance of longish narrow leaues, growing thicke together, with foure
broader or larger leaues lying vnder, or bearing them vp, and these leaues are of a darke blewish purple colour. *Clusius* calls this *Clematis altera flore pleno.* ‡
¶ The Place.
These plants delight to grow in Sunnie places: they prosper better in a fruitfull soile than in
barren. They grow in my garden, where they flourish exceedingly.
¶ The Time.
They floure in Iuly and August, and perfect their seed in September.
¶ The Names.
Ladies Bower in called in Greeke, ꙁꙁꙁꙁ: in Latine, *Ambuxum*: in English you may call it, La-
dies

dies bower, which I take from his aptneſſe in making of Arbors, Bowers, and ſhadie couertures in gardens.

¶ *The Temperature and Vertues.*

The faculties and vſe of theſe in Phyſicke is not yet knowne.

CHAP. 327. *Of purging Peruinkle.*

¶ *The Deſcription.*

1   A Mong theſe plants which are called *Clematides,* theſe be alſo to be numbred, as hauing certaine affinitie becauſe of the ſpreading branching, and ſemblance of the Vine; and this is called *Flammula vrens,* by reaſon of his fiery and burning heate, becauſe that being laid vpon the skin, it burneth the place, and maketh an eſchar, euen as our common cauſticke or corroſiue medicines do. The leaues hereof anſwer both in colour and ſmoothneſſe, *Vinca, Peruinca,* or Peruinkle, growing vpon long clambring tender branches, like the other kindes of climbing plants. The flours are very white, ſtar-faſhion, and of an exceeding ſweet ſmell, much like vnto the ſmell of Hawthorne floures, but more pleaſant, and leſſe offenſiue to the head: hauing in the middle of the floures certaine ſmall chiues or threds. The root is tender, and diſperſeth it ſelfe far vnder the ground.

1 *Clematis vrens.*
Virgins Bower.

2 *Flammula Iovis ſurrecta.*
Vpright Virgins Bower.

2   Vpright Clamberer or Virgins Bower is alſo a kinde of *Clematis,* hauing long tough roots not vnlike to thoſe of Licorice; from which riſeth vp a ſtraight vpright ſtalke, of the height of three or foure cubits, ſet about with winged leaues, compoſed of diuers ſmall leaues, ſet vpon a middle rib, as are thoſe of the aſhe tre, or Valerian, but fewer in number: at the top of the ſtalkes come forth ſmall white floures, very like the precedent, but not of ſo pleaſant a ſweet ſmell; afterwhich come the ſeeds, flat and ſharpe pointed.

                 3 There

3   There is another *Clematis* of the kinde of the white *Clematis* or burning *Clematis*,which I haue recouered from ſeed,that hath been ſent me from a curious and learned citiſen of Strawsborough, which is like vnto the other in each reſpeɛt, ſauing that, that the floures hereof are very double, wherein conſiſteth the eſpeciall difference.

4   Amongſt the kindes of climbing or clambering plants, *Carolus Cluſius*, and likewiſe *Lobel* haue numbred theſe two, which approch neere vnto them in leaues and floures, but are far different in claſping tendrels, or climbing otherwiſe, being low and baſe plants in reſpeɛt of the others of their kinde. The firſt hath for his roots a bundle of tough tangling threddes, in number infinite, and thicke thruſt together ; from which riſe vp many ſmall ſtalkes, of a browniſh colour, foure ſquare, and of a wooddy ſubſtance : whereupon do grow long leaues, of a biting taſte, ſet together by couples, in ſhape like thoſe of *Aſclepias*, or ſilken Swallow-wort. The floures grow at the top of the ſtalkes, of a faire blew or sky colour, conſiſting of foure parts in manner of a croſſe, hauing in the middle a bunched pointell, like vnto the head of field Poppie when it is young, of a whitiſh yellow colour,hauing little or no ſmell at all. The floures being paſt, then commeth the ſeed,ſuch as is to be ſeene in the other kindes of *Clematis*. The whole plant dieth at the approch of Winter, and recouereth it ſelfe againe from the root,which endureth, whereby it greatly increaſeth.

4 *Clematis Pannonica.*
Buſh Bower.

5 *Clematis major Pannonica.*
Great Buſh Bower.

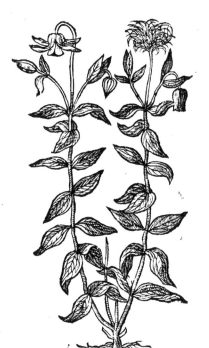

†   5   The great Buſh Bower differeth from the former laſt deſcribed in greatneſſe, as alſo in the ſhape and colour of the floures, which are leſſer, ſtare faſhioned, of a deepe purple colour, and edged with greene. †

‡   6   Of theſe there is another,whoſe bending creſted ſtalkes are ſome three cubits high, which ſend forth ſundry ſmall branches, ſet with leaues growing together by threes vpon ſhort foot-ſtalkes, and they are like myrtle leaues, but bigger, more wrinkled, darke coloured,and ſnipt about the edges : the floure reſembles a croſſe,with foure ſharpe pointed rough leaues of a whitiſh blew colour,which containe diuers ſmall looſe little leaues in their middles. The root is long and laſting.It groweth vpon the rocky places of mount Baldus in Italy, where *Pona* found it,and he calls it *Clematis cruciata Alpina.* ‡

¶ *The*

‡ *6 Clematis cruciata Alpina.*
Virgins Bower of the Alps.

A

¶ *The Place.*

Theſe plants do not grow wilde in England, that I can as yet learne; notwithſtanding I haue them all in my garden, where they flouriſh exceedingly.

¶ *The Time.*

Theſe plants doe floure from Auguſt to the end of September.

¶ *The Names.*

There is not much more found of their names than is expreſſed in their ſeuerall titles, notwithſtanding there hath beene ſomewhat ſaid, as I thinke, by hearſay, but nothing of certaintie : wherefore let that which is ſet downe ſuffice. We may in Engliſh call the firſt, Biting Clematis, or white Clematis, Biting Peruinkle or purging Peruinkle, Ladies Bower, and Virgins Bower.

¶ *The Temperature.*

The leafe hereof is biting, and doth mightily bliſter, being, as *Galen* ſaith, of a cauſticke or burning quality : it is hot in the beginning of the fourth degree.

¶ *The Vertues.*

Dioſcorides writeth, that the leaues being applied do heale the ſcurfe and lepry, and that the ſeed beaten, and the pouder drunke with faire water or with mead, purgeth flegme and choler by the ſtoole.

Cʜᴀᴘ. 328. *Of Wood-binde, or Hony-ſuckle.*

¶ *The Kindes.*

THere be diuers ſorts of Wood-bindes, ſome of them ſhrubs with winding ſtalkes, that wrap themſelues vnto ſuch things as are neere about them. Likewiſe there be other ſorts or kindes found out by the later Herbariſts, that clime not at all, but ſtand vpright, the which ſhall be ſet forth among the ſhrubby plants. And firſt of the common Wood-binde.

¶ *The Deſcription.*

1　WOod-binde or Hony-ſuckle climeth vp aloft, hauing long ſlender wooddy ſtalkes, parted into diuers branches : about which ſtand by certaine diſtances ſmooth leaues, ſet together by couples one right againſt another ; of a light greene colour aboue, vnderneath of a whitiſh greene. The floures ſhew themſeluas in the tops of the branches, many in number, long, white, ſweet of ſmell, hollow within; in one part ſtanding more out, with certaine threddes growing out of the middle. The fruit is like little bunches of grapes, red when they be ripe, wherein is contained ſmall hard ſeed. The root is wooddy, and not without ſtrings.

2　This ſtrange kind of Woodbind hath leaues, ſtalks, and roots like vnto the common Woodbinde or Honiſuckle, ſauing that neere vnto the place where the floures come forth, the ſtalkes do grow through the leaues, like vnto the herbe Thorow-wax, called *Perfoliata*, which leaues do reſemble little ſaucers : out of which broad round leaues proceed faire, beautifull, and well ſmelling floures, ſhining with a whitiſh purple colour, and ſomewhat daſht with yellow, by little and little ſtretched out like the noſe of an Elephant, garniſhed within with ſmall yellow chiues or threddes : and when the floures are in their flouriſhing, the leaues and floures do reſemble ſaucers filled with
the

the floures of Woodbinde : many times it falleth out, that there is to be found three or foure ſaucers one aboue another, filled with floures, as the firſt, which hath cauſed it to be called double Hony-ſuckle, or Woodbinde.

1 *Periclymenum.*
Woodbinde or Honiſuckles.

2 *Periclymenum perfoliatum.*
Italian Woodbinde.

¶ The Place.

The Woodbinde groweth in woods and hedges, and vpon ſhrubs and buſhes, oftentimes winding it ſelfe ſo ſtraight and hard about, that it leaueth his print vpon thoſe things ſo wrapped.

The double Honiſuckle groweth now in my Garden, and many others likewiſe in great plenty, although not long ſince, very rare and hard to be found, except in the garden of ſome diligent Herbariſts.

¶ The Time.

The leaues come forth betimes in the ſpring : the floures bud forth in May and Iune : the fruit is ripe in Autumne.

¶ The Names.

It is called in Greeke, περικλύμενον in Latine, *Volucrum majus* : of *Scribonius Largus*, *Sylua mater* : in ſhops, *Caprifolium*, and *Matriſylva* : of ſome, *Lilium inter ſpinas* : in Italian, *Vinciboſco* : in High Dutch, **Geyſbladt** : in Low Dutch, **Gheytenbladt**, and **Mammekens Cruit** : in French, *Cheurefueille* : in Spaniſh, *Madreſelva* : in Engliſh, Woodbinde, Honiſuckle, and Capriſoly.

¶ The Temperature.

There hath an errour in times paſt growne amongſt a few, and now almoſt paſt recouery to bee called againe, being growne an errour vniuerſall, which errour is, the decoction of the leaues of Honiſuckles, or the diſtilled water of the floures, are raſhly giuen for the inflammations of the mouth and throte, as though they were binding and cooling. But contrariwiſe Honiſuckle is neither cold nor binding ; but hot, and attenuating or making thinne. For as *Galen* ſaith, both the fruit of Woodbinde, and alſo the leaues, do ſo much attenuate and heate, as if ſomewhat too much of them be drunke, they will cauſe the vrine to be as red as bloud, yet do they at the firſt onely prouoke vrine.

¶ The

¶ *The Vertues.*

A   *Dioſcorides* writeth, that the ripe ſeed gathered and dried in the ſhadow, and drunke vnto the quantitie of one dram weight, fortie daies together, doth waſte and conſume away the hardneſſe of the ſpleene, remoueth weariſomeneſſe, helpeth the ſhortneſſe and difficulty of breathing, cureth the hicket, procureth bloudy vrine after the ſixth day, and cauſeth women to haue ſpeedy trauell in childe bearing.

B   The leaues be of the ſame force : which being drunke thirty daies together, are reported to make men barren, and deſtroy their naturall ſeed.

C   The floures ſteeped in oile, and ſet in the Sun, are good to annoint the body that is benummed, and growne very cold.

D   The diſtilled water of the floures are giuen to be drunke with good ſucceſſe againſt the piſſing of bloud.

E   A ſyrrup made of the floures is good to be drunke againſt the diſeaſes of the lungs and ſpleene that is ſtopped, being drunke with a little wine.

F   Notwithſtanding the words of *Galen* (or rather of *Dodonæus*) it is certainly found by experience, that the water of Honiſuckles is good againſt the ſoreneſſe of the throat and uvula : and with the ſame leaues boyled, and the leaues or flours diſtilled, are made diuers good medicines againſt cankers, and ſore mouthes, as well in children as in elder people, and likewiſe for vlcerations and ſcaldings in the priuie parts of man or woman ; if there be added to the decoction hereof ſome allome or Verdigreace if the ſore require greater clenſing outwardly, prouided alwaies that there be no Verdigreace put into the water that muſt be injected into the ſecret parts.

Chap. 329. *Of Iaſmine, or Gelſemine.*

1 *Iaſminum album.*
White Geſſemine.

2 *Iaſminum Candiflorum majus.*
Great white Geſſemine.

¶ *The Deſcription.*

1   IAſmine, or Gelſemine, is of the number of thoſe plants which haue need to be ſupported or propped vp, and yet notwithſtanding of it ſelfe claſpeth not or windeth his ſtalkes about

3 *Iaſminum luteum.*
Yellow Iaſmine.

bout ſuch things as ſtand neere vnto it, but onely leaneth and lieth vpon thoſe things that are prepared to ſuſtain it about arbors and banqueting houſes in gardens, by which it is held vp: the ſtalkes thereof are long, round, branched, jointed or kneed, and of a green colour, hauing within a white ſpongeous pith. The leaues ſtand vpon a middle rib, ſet together by couples like thoſe of the aſh tree, but much ſmaller, of a deepe greene colour: the floures grow at the vppermoſt part of the branches, ſtanding in a ſmall tuft far ſet one from another, ſweet in ſmell, of colour white: the ſeed is flat and broad like thoſe of Lupines, which ſeldom come to ripeneſſe: the root is tough and thready.

2　　*Lobel* reporteth, That hee ſaw in a garden at Bruſſels, belonging to a reuerend perſon called Mr. *Boiſot*, a kinde of Gelſemine very much differing from our Iaſmine, which he nouriſhed in an earthen pot: it grew not aboue (ſaith he) the height of a cubit, diuided into diuers branches, wherupon did grow leaues like thoſe of the common white Iaſmine, but blacker and rounder. The floures in ſhew were moſt beautifull, ſhaped like thoſe of the common Iaſmine, but foure times bigger, gaping wide open, white on the vpper ſide, and of a bright red on the vnder ſide.

3　　There is a kinde hereof with yellow floures; but ſome doe deſcribe for the yellow Iaſmine, the ſhrubby Trefoil, called of ſome, *Trifolium fruticans;* and of others, *Polemonium:* but this yellow Iaſmine is one, and that is another plant, differing from the kindes of Iaſmine, as ſhall be declared in his proper place. The yellow Iaſmine differs not from the common white Geſmine in leaues, ſtalks, nor faſhion of the flours: the only difference is, that this plant brings forth yellow flours, and the other white.

4　　There is likewiſe another ſort that differs not from the former in any reſpect but in the color of the floure; for this plant hath floures of a blew colour, and the others not ſo, wherein conſiſteth the difference.

Gelſemine is foſtred in gardens, and is vſed for arbors, & to couer banqueting houſes in gardens: it growes not wild in England, that I can vnderſtand of, though Mr *Lyte* be of another opinion: the white Iaſmine is common in moſt places of England: the reſt are ſtrangers, and not ſeene in theſe parts as yet.

They bring forth their pleaſant floures in Iuly and Auguſt.

Among the Arabians *Serapio* was the firſt that named Geſſemine, *Zambach:* it is called *Iaſminum, Ieſeminum,* and alſo *Geſſeminum:* in Engliſh, Iaſmine, Geſſemine, and Geſſe.
There is in *Dioſcorides* a compoſition of oile of Iaſmine, which he ſaith is made in Perſia of the white floures of Violets, which Violets ſeem to be none other than the floures of this Geſſemine; for *Dioſcorides* hath often reckoned faire and elegant floures among the Violets; ſo that it muſt not ſeem ſtrange that he calleth the floures of Ieſſemine Violets, eſpecially ſeeing the plant it ſelf was vnknown to him, as is euident.

Geſſemine, and eſpecially the floures thereof, be hot in the beginning of the ſecond degree, as *Serapio* reporteth out of *Meſue.*

The oile which is made of the flours hereof waſteth away raw humors, and is good againſt cold　A rheums; but in thoſe that are of a hot conſtitution it cauſeth head-ache, and the ouermuch ſmell thereof maketh the noſe to bleed, as the ſame Author affirmeth. It is vſed (as *Dioſcorides* writeth, and after him *Aetius*) of the Perſians in their banquets for pleaſures ſake : it is good to be anointed after baths, in thoſe bodies that haue need to be ſuppled and warmed, but for ſmell it is not much vſed.

B   The leaues boiled in wine vntill they be ſoft, and made vp to the forme of a pultis, and applied, diſſolue cold ſwellings, wens, hard lumps, and ſuch like out-goings.

CHAP. 330.   *Of Peruinkle.*

¶ *The Deſcription.*

1   PErunkle hath ſlender and long branches trailing vpon the ground, taking hold here and there as it runneth, ſmall like ruſhes, with naked or bare ſpaces between joint and joint. The leaues are ſmooth, not vnlike to the Bay leafe, but leſſer : the floures grow hard by the leaues, ſpreading wide open, compoſed of fiue ſmall blew leaues.

We haue in our London gardens a kinde hereof bearing white floures, wherein it differs from the former.

1 *Vinca Peruinca minor.*
Peruinkle.

‡ 2 *Clematis Daphnoides, ſiue Peruinca major.*
Great Peruinkle.

There is another with purple floures, doubling it ſelfe ſomewhat in the middle, with ſmaller leaues, wherein is the difference.

2   There is another ſort greater than any of the reſt, called of ſome, *Clematis Daphnoides*, of the likeneſſe the leaues haue with thoſe of Bay : the leaues and floures are like thoſe of the precedent, but altogether greater, wherein conſiſts the difference.

¶ *The Place.*
They grow in moſt of our London gardens, they loue moiſt and ſhadowie places ; the branches remain alwaies green.

¶ *The Time.*
They floure and flouriſh in March, Aprill, and May, and ſometimes later.

¶ *The*

¶ *The Names.*

Peruinkle is called in Greeke, χαμαιτ δαφνιδι, becauſe it bringeth forth ſtalkes which creepe like thoſe of the Vine ; and *Daphnoeides*, by reaſon that the leaues are like thoſe of the Bay, as aforeſaid. *Pliny* calls it *Vinca peruinca*, and *Chamædaphne* : notwithſtanding there is another *Chamædaphne*, of which in his place. The ſame Author likewiſe calleth it *Centunculus* : in high-Dutch, **Jngrun:** in low-Dutch, **Wincoozte, maegden cruyt :** in French, *Pucellage*, *Vauche*, and *Peruanche* : in Italian, *Prouenca* : in Spaniſh, *Peruinqua* : in ſhops, *Clematis peruinca* : in Engliſh, Peruinkle, Pervinkle, and Periwinkle.

¶ *The Nature.*

Peruinkle is hot within the ſecond degree, ſomthing dry and aſtringent.

¶ *The Vertues.*

The leaues boiled in wine and drunke ſtop the laske and bloudy flix.   A

An handfull of the leaues ſtamped, and the juice giuen to drinke in red wine, ſtoppeth the laske B and bloudy flix, ſpitting of bloud, which neuer faileth: it likewiſe ſtops the inordinat courſe of the monethly ſickneſſe.

Cʜᴀᴘ. 331. *Of Capers.*

¶ *The Kindes.*

THere be two ſorts of Capers eſpecially ; one with broad leaues ſharp pointed : the other with rounder leaues. The Brabanders haue alſo another ſort called *Capparis fabago*, or Bean Capers.

1 *Capparis folio acuto.*
Sharp leafed Capers.

2 *Capparis rotundiore folio.*
Round leafed Capers.

¶ The Deſcription.

1   THe Caper is a prickly ſhrub,the ſhoots or branches whereof be full of ſharpe prickly
thornes,trailing vpon the ground,if they be not ſupported or propped vp : whereupon
grow leaues like thoſe of the Quince tree, but rounder : among the which come forth
long ſlender foot ſtalks,whereon grow round knops,which open or ſpred abroad into faire floures :
after which come in place long fruit like to an Oliue,and of the ſame colour, wherein is contained
flat rough ſeeds,of a dusky colour.   The root is wooddy,and couered with a thicke barke or rinde,
which is much vſed in Phyſicke.

2   The ſecond kinde of Caper is likewiſe a prickly plant much like the bramble buſh,hauing
many ſlender branches ſet full of ſharp prickles. The whole plant traileth vpon the place where it
groweth,beſet with round blackiſh leaues diſorderly placed,in ſhape like thoſe of Aſtrabacca,but
greater,approching to the forme of Fole-foot : among which commeth forth a ſmall and tender
naked twig,charged at the end with a ſmall knap or bud,which openeth it ſelfe to a ſmall ſtar-like
floure,of a pleaſant ſweet ſmell ; in place whereof comes a ſmal fruit long and round like the Cornell berry,of a brown colour.   The root is long and wooddy,couered with a thicke barke or rinde,
which is likewiſe vſed in medicine.

¶ The Place.

The Caper groweth in Italy,Spaine,and other hot regions,without manuring, in a lean ſoile,in
rough places amongſt rubbiſh,and vpon old walls,as Dioſcorides reporteth.

Theophraſtus writeth,That it is by nature wilde, and refuſeth to be husbanded, yet in theſe our
dayes diuers vſe to cheriſh the ſame,and to ſet it in dry and ſtony places. My ſelfe at the impreſſion hereof planted ſome ſeeds in the bricke walls of my garden, which as yet doe ſpring and grow
green : the ſucceſſe I expect.

¶ The Time.

The Caper floureth in Summer euen vntil Autumne. The knops of the flours before they open
are thoſe Capers we eat as ſauce,which are gathered and preſerued in pickle or ſalt.

¶ The Names.

It is called in Greeke, κάππαρις : in Latine alſo Capparis : of ſome, Cynoſbatos,yet Cynoſbatos or Canirubus is properly taken for the wild Roſe : it is generally called Cappers in moſt languages ; in Engliſh,Cappers,Caper,and Capers.

¶ The Temperature.

Capers,or the floures not yet fully growne,be of temperature hot, and of thin parts : if they be
eaten green they yeeld very little nouriſhment, and much leſſe if they be ſalted : and therefore are
rather a ſauce and medicine,than a meat.

¶ The Vertues.

A   They ſtir vp an appetite to meat, are good for a moiſt ſtomacke, and ſtay the watering thereof,
clenſing away the flegme that cleaueth vnto it.They open the ſtoppings of the liuer and milt:with
meat they are good to be taken of thoſe that haue a quartan Ague and ill ſpleens. They are eaten
boiled(the ſalt firſt waſhed off) with oile and vineger,as other ſallads be,and ſomtimes are boiled
with meat.

B   The rinde or bark of the root conſiſteth of diuers faculties, it heateth,clenſeth,purgeth,cutteth
and digeſteth.

C   This barke is a ſingular remedie for hard ſpleenes, being outwardly applied,or inwardly taken :
and the ſame boiled with vineger or Oxymel, or beaten and mixed with other ſimples, expelleth
thicke groſſe humors,and conueyeth away the ſame mixed with bloud,by vrine and ſiege, wherby
the milt or ſpleen is helpled,and the pain of the huckle bones taken away : moreouer, it bringeth
downe the deſired ſickneſſe,purgeth and draweth flegme out of the head,as Galen writeth.

D   The ſame bark (as Dioſcorides teacheth) clenſeth old filthy ſores,and ſcoureth away the thicke
lips and cruſts about the edges,and being chewed it takes away the tooth-ache.

E   Being ſtamped with vineger it ſcoureth away tettars and ring-wormes,hard ſwellings,and cures
the Kings euill.

F   The barke of the roots of Capers is good againſt the hardneſſe and ſtopping of the ſpleen, and
profiteth much if it be giuen in drink,to ſuch as haue the ſciatica,palſie,or that are burſten or bruſed by falling from ſome high place : it mightily prouokes vrine,inſomuch that if it be vſed ouermuch,or giuen in too great a quantitie,it draweth out bloud with the vrin.

CHAP.

Chap. 332. Of Beane Capers.

Capparis Fabago.
Beane Capers.

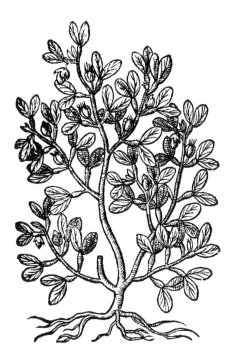

¶ The Deſcription.

THis plant which the Germanes call **fabago**, and *Dodonæus* ſauouring of Dutch, calleth it in his laſt Edition *Capparis Fabago*, and properly: *Lobel* calleth it *Capparis leguminoſa*: between which there is no great difference, who labour to refer this plant vnto the kinds of Capers, which is but a low and baſe herb, and not a ſhrubbie buſh, as are the true Capers. It bringeth forth ſmooth ſtalkes tender and branched, wherupondo grow long thick leaues, leſſer than thoſe of the true Capers, & not vnlike to the leaues of Purſlane, comming out of the branches by couples, of a light greene colour. The floures before they bee opened are like to thoſe of the precedent, but when they be come to maturity & ful ripeneſſe they waxe white, with ſome yellow chiues in the midſt: which being paſt, there appeare long cods, wherein is contained ſmall flat ſeed. The root is tender, branching hither and thither.

¶ The Place.

It groweth of it ſelfe in corne fields of the low Countries, from whence I haue receiued ſeeds for my garden, where they flouriſh.

¶ The Time.

It floureth when the Caper doth.

¶ The Names.

It is called in Latine of the later Herbariſts *Capparis fabago*: of moſt *Capparis Leguminoſa*: it is thought to be that herbe which *Avicen*, deſcribeth in his 28 chapter, by the name of *Ardifrigi*: we may content our ſelues that *Capparis fabago* retaine that name ſtill, and ſeek for none other, vnleſſe it be for an Engliſh name, by which it may be called after the Latine, Beane Caper.

¶ The Temperature and Vertues.

Touching the faculties thereof we haue nothing left in writing worth the remembrance.

Chap. 333. Of Swallow-wort.

¶ The Deſcription.

1 SWallow-wort with white floures hath diuers vpright branches of a browniſh colour, of the height of two cubits, beſet with leaues not vnlike to thoſe of *Dulcamara* or wooddy Night-ſhade, ſomewhat long, broad, ſharpe pointed, of a blackiſh greene colour, and ſtrong ſauor: among which come forth very many ſmall white floures ſtar-faſhion, hanging vpon little ſlender foot-ſtalkes: after which come in place thereof long ſharp pointed cods, ſtuffed full of a moſt perfect white cotton reſembling ſilk, as well in ſhew as handling; (our London Gentlewomen haue named it Silken Ciſlie) among which is wrapped ſoft browniſh ſeed. The roots are very many, white, threddie, and of a ſtrong ſauour.

2 The ſecond kinde is oftentimes found with ſtalks much longer, climing vpon props or ſuch things as ſtand neere vnto it, attaining to the height of fiue or ſix cubites, wrapping it ſelfe vpon them with many and ſundry foldings: the floures hereof are black: the leaues, cods, and roots be like thoſe of the former.

    ¶ The

1 *Aſclepias flore albo.*
White Swallow-woort.

2 *Aſclepias flore nigro.*
Blacke Swallow-woort.

¶ *The Place.*

Both theſe kindes do grow in my garden, but not wilde in England; yet I haue heard it reported that it groweth in the fields about Northampton, but as yet I am not certaine of it.

¶ *The Time.*

They floure about Iune, in Autumne the downe hangeth out of the cods, and the ſeed falleth to the ground.

¶ *The Names.*

It is called of the later Herbariſts *Vincetoxicum* : of *Ruellius, Hederalis* : in High Dutch, **Swalwe woztele,** that is to ſay in Latine, *Hirundinaria* : in Engliſh, Swallow-woort : of our Gentlewomen it is called Silken Ciſlie; *Æſculapius* (who is ſaid to be the firſt inuentor of phyſick, whom therefore the Greekes and Gentiles honored as a god) called it after his owne name *Aſclepias,* or *Æſculapius* herbe, for that he was the firſt that wrote thereof, and now it is called in ſhops *Hirundinaria.*

¶ *The Temperature.*

The roots of Swallow-woort are hot and dry; they are thought to be good againſt poyſon.

¶ *The Vertues.*

A   *Dioſcorides* writeth, That the roots of *Aſclepias* or Swallow-woort boiled in wine, and the decoction drunke, are a remedy againſt the gripings of the belly, the ſtingings of Serpents, and againſt deadly poyſon, being one of the eſpecialleſt herbes againſt the ſame.

B   The leaues boiled and applied in forme of a pultis, cure the euill ſores of the paps or dugs, and matrix, that are hard to be cured.

CHAP. 334. *Of Iudian Swallow-woort.*

¶ *The Deſcription.*

THere groweth in that part of Virginia, or Norembega, where our Engliſh men dwelled (intending there to erect a certaine Colonie) a kinde of *Aſclepias,* or Swallow-woort, which
the

the Sauages call *Wiſanck* : there riſeth vp from a ſingle crooked root, one vpright ſtalk a foot high,
ſlender, and of a greeniſh colour:whereupon do grow faire broad leaues ſharp pointed, with many
ribs or nerues running through the ſame like thoſe of Ribwort or Plaintaine, ſet together by couples at certaine diſtances. The floures come forth at the top of the ſtalks, which as yet are not obſerued,by reaſon the man that brought the ſeeds & plants hereof did not regard them:after which,
there come in place two cods(ſeldome more) ſharp pointed like thoſe of our Swallow-woort, but
greater, ſtuffed full of a moſt pure ſilke of a ſhining white colour : among which ſilke appeareth a
ſmall long tongue (which is the ſeed) reſembling the tongue of a bird, or that of the herbe called
Adders tongue. The cods are not only full of ſilke,but euery nerue or ſinew wherewith the leaues
be ribbed are likewiſe moſt pure ſilke ; and alſo the pilling of the ſtems, euen as flax is torne from
his ſtalks.This conſidered,behold the juſtice of God, that as he hath ſhut vp thoſe people and nations in infidelity and nakednes, ſo hath he not as yet giuen them vnderſtanding to couer their na-
kedneſſe,nor mater wherewith to do the ſame;notwithſtanding the earth is couered ouer with this
ſilke,which daily they tread vnder their feet, which were ſufficient to apparell many kingdomes,
if they were carefully manured and cheriſhed.

Wiſanck, ſiue Vincetoxicum Indianum.　　　　　　　　‡ *Apocynum Syriacum Cluſij.*
Indian Swallow-wort.

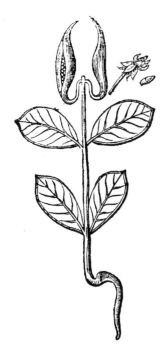

‡　This plant,which is kept in ſome gardens by the name of Virginia Silke graſſe,I take to be
the ſame,or very like the *Beidelſar* of *Alpinus*;and the *Apocynum Syriacum* of *Cluſius* : at Padua they
call it *Eſula Indica*, by reaſon of the hot milky juice. *Bauhinus* hath very vnfitly named it *Lapathum
Ægyptiacum lacteſcens ſiliqua Aſclepiadis*. But he is to be pardoned ; for *Iohannes Carolus Roſenbergus,
cap.* 16.*p.*46.of his *Animad.& Exerc. Medica*, or *Roſa nobilis iatrica*, hath taken vpon him the credit
and inuention of this abſurd denomination:I may call it abſurd, for that neither any way in ſhape
or qualitie it reſembleth or participateth any thing with a Docke. I haue giuen you the figure
of our Author with his title , and that of *Cluſius* with his : in the former the cods are onely
well expreſt ; in the later the leaues and floures reaſonably well, but that they are too few in number,and ſet too far aſunder. Vpon the ſight of the growing and flouring plant I took this deſcription : The root is long and creeping ; the ſtalks two or three cubits high, ſquare, hollow, a finger
thicke, and of a light green colour, ſending out towards the top ſome few branches : vpon this at
certaine

certaine ſpaces grow by couples, leaues ſome halfe foot long, and three inches broad, darke greene on their vpper ſides , more whitiſh below, and full of large and eminent veines : at the top of the ſtalke and branches it carries moſt commonly an hundred or more floures, growing vpon footſtalks ſome inch long, all cloſe thruſt together after the manner of the Hyacinth of Peru at the firſt flouring : each floure is thus compoſed ; firſt it hath fiue ſmall greene leaues bending backe, that ſerue for the cup : then hath it other fiue leaues foure times larger than the former, which bend back and couer them ; and theſe are green on the vnder ſide, and of a pale colour with ſome redneſſe aboue : then are there fiue little graines (as I may ſo terme them) of a pleaſant red colour, and on their outſide like corns of Millet, but hollow on their inſides, with a little thred or chiue comming forth of each of them : theſe fiue ingirt a ſmall head like a button, greeniſh vnderneath, and whitiſh aboue. I haue giuen you the figure of one floure by the ſide of our Authors figure. The leaues and ſtalkes of this plant are very full of a milky juice. ‡

¶ *The Place.*

It groweth, as before is rehearſed, in the countries of Norembega, now called Virginia, by the honourable Knight Sir *Walter Raleigh*, who hath beſtowed great ſums of money in the diſcouerie thereof ; where are dwelling at this preſent Engliſh men.

¶ *The Time.*

It ſpringeth vp, floureth, and flouriſheth both Winter and Summer, as do many or moſt of the plants of that countrey. ‡ It dies down with vs euery winter and comes vp in the Spring, and flours in Auguſt, but neuer bringeth forth the cods with vs, by reaſon of the coldnes of our Climate. ‡

¶ *The Names.*

The ſilke is vſed of the people of Pomeioc and other of the prouinces adioyning, being parts of Virginia, to couer the ſecret parts of maidens that neuer taſted man ; as in other places they vſe a white kinde of moſſe Wiſanck : we haue thought *Aſclepias Virginiana*, or *Vincetoxicum Indianum* fit and proper names for it : in Engliſh, Virginia Swallow-wort, or the Silke-wort of Norembega.

¶ *The Temperature and Vertues.*

A   We finde nothing by report, or otherwiſe of our owne knowledge, of his phyſicall vertues , but only report of the abundance of moſt pure ſilke, wherewith the whole plant is poſſeſſed.

B   ‡ The leaues beaten either crude, or boiled in water, and applied as a pulteſſe, are good againſt ſwellings and paines proceeding of a cold cauſe.

C   The milkie juice, which is very hot, purges violently ; and outwardly applied is good againſt tettars, to fetch haire off skins, if they be ſteeped in it, and the like. *Alpinus.* ‡

Cʜᴀᴘ. 335.
Of the Bombaſte or Cotton-Plant.

¶ *The Deſcription.*

THe Cotton buſh is a low and baſe Plant, hauing ſmall ſtalkes of a cubit high, and ſometimes higher ; diuided from the loweſt part to the top into ſundry ſmall branches, whereupon are ſet confuſedly or without order a few broad leaues, cut for the moſt part into three ſeaions , and ſometimes more, as Nature liſt to beſtow, ſomwhat indented about the edges, not vnlike to the leafe of the Vine, or rather the Veruaine Mallow, but leſſer, ſofter, and of a grayiſh colour : among which come forth the floures, ſtanding vpon ſlender foot-ſtalks, the brims or edges whereof are of a yellow colour, the middle part purple : after which appeareth the fruit, round, and of the bigneſſe of a Tenniſe ball, wherein is thruſt together a great quantity of fine white Cotton wooll ; among which is wrapped vp blacke ſeed of the bigneſſe of peaſen, in ſhape like the trettles or dung of a cony. The fruit being come to maturity or ripeneſſe, the husk or cod opens it ſelfe into foure parts or diuiſions, and caſteth forth his wooll and ſeed vpon the ground, if it bee not gathered in his time and ſeaſon. The root is ſmall and ſingle, with lew threds anexed therto, and of a wooddy ſubſtance, as is all the reſt of the plant.

¶ *The Place.*

It groweth in India, in Arabia, Egypt, and in certaine Iſlands of the Mediterranean ſea, as Cyprus, Candy, Malta, Sicilia, and in other prouinces of the continent adiacent. It groweth about Tripolis and Aleppo in Syria, from whence the factor of a worſhipfull merchant in London, Maſter *Nicholas Lete* before remembred, did ſend vnto his ſaid maſter diuers pounds weight of the ſeed ; whereof ſome were committed to the earth at the impreſſion hereof, the ſucceſſe we leaue to
the

Goſsipium, ſiue Xylon.
The Cottonbuſh.

the Lord. Notwithſtanding my ſelf three yeares paſt did ſow of the ſeed, which did grow verie frankely,but periſhed before it came to perfecti-on,by reaſon of the cold froſts that ouertooke it in the time of flouring.

¶ *The Time.*

Cotton ſeed is ſowne in plowed fields in the ſpring of the yeare, and reaped and cut downe in harueſt, euen as corne with vs ; and the ground muſt be tilled and ſown new again the next yere, and vſed in ſuch ſort as we do the tillage for corn and grain ; for it is a plant of one yeare, and peri-ſheth when it hath perfected his fruit, as many other plants do.

¶ *The Names.*

Cotton is called in Greeke ἔυλον, and Γοσσίπιον : in Latine, *Xylum*, and *Goſsipium*, after the Greeke : in ſhops, *Lanugo, Bombax*, and *Cotum* : in Italian, *Bom-bagia*: in Spaniſh, *Algodon*: in high Dutch, Baum-wool : in Engliſh and French, Cotton, Bombaſt, and Bombace.

Theophraſtus, lib.4. cap. 9. hath made mention hereof,but without a name ; and hee ſaith it is a tree in Tylus which beares wooll. Neither is it any maruell if he took an vnknown ſhrub or plant growing in countries far off,for a tree ; ſeeing al-ſo in this age (in which very many things come to be better knowne than in times paſt)the Cot-ton or wooll hereof is called of the Germanes(as we haue ſaid) Baum wool, that is, Wooll of a tree,whereas indeed it is rather an herb or ſmall ſhrub,and not to be numbred among trees.

Of this *Theophraſtus* writeth thus ; It is reported that the ſame Iſland *(viz.* Tylus) brings forth many trees that beare wooll,which haue leaues like thoſe of the Vine,&c.

Pliny, lib. 19. cap. 1. writing of the ſame, ſaith thus : The vpper part of Ægypt towards Arabia brings forth a ſhrub called *Goſsipion* or *Xylon*, and therefore the linnen that is made thereof is called *Xylina*. It is(ſaith he)the plant that beareth that wooll wherewith the garments are made which the prieſts of Egypt do weare.

¶ *The Temperature.*

The ſeed of Cotton(according to the opinion of *Serapio*) is hot and moiſt,the wooll it ſelfe is hot and dry.

¶ *The Vertues.*

The ſeed of Cotton is good againſt the cough and ſhort windedneſſe : it alſo ſtirreth vp luſt of A the body by increaſing naturall ſeed ; wherein it ſurpaſſeth.

The oile preſſed out of the ſeed takes away freckles,ſpots,and other blemiſhes of the skin. B

The aſhes of the wooll burned ſtanch the bleeding of wounds, vſed in reſtrictiue medicines, as C Bole Armonicke,and is more reſtrictiue than Bole it ſelfe.

To ſpeake of the commodities of the wooll of this plant were ſuperfluous,common experience D and the daily vſe and benefit we receiue by it ſhew them. So that it were impertinent to our hiſto-rie,to ſpeak of the making of Fuſtian,Bombaſies,and many other things that are made of the wool thereof.

CHAP. 336. *Of Dogs-bane.*

¶ *The Kindes.*

THere be two kindes of Dogs-bane : the one a clymbing or clambering plant ; the other an vp-right ſhrub.

¶ *The*

¶ *The Description.*

1   DOgs-bane riseth vp like vnto a small hedge bush, vpright and straight, vntil it haue attained to a certain height; then doth it clasp and clymbe with his tender branches as do the Bindeweeds, taking hold vpon props or poles, or whatsoeuer stands next vnto it: whereupon grow faire broad leaues sharp pointed like those of the Bay tree, of a deep green colour. The floures come forth at the top of the stalks, consisting of fiue small white leaues: which beeing past, there succeed long cods set vpon a slender foot-stalk by couples, ioyning themselues together at the extreme point, and likewise at the stalke, making of two pieces knit together one intire cod; which cod is full of such downy matter and seed as that of *Asclepias*, but more in quantitie, by reason the cods are greater: which beeing dry and ripe, the silken cotton hangeth forth, and by little and little sheddeth, vntill the whole be fallen vpon the ground. The whole plant yeelds that yellow stinking milky iuice that the other doth, and sometimes it is of a white colour, according to the clymat where it groweth; for the more cold the country is, the whiter is the iuice; and the hotter, the more yellow. The root is long and single, with some threds anexed thereto.

<table>
<tr><td>1 Periploca repens angustifolia.
Climing Dogs-bane.</td><td>‡ 2 Periploca latifolia.
Broad leafed Dogs-bane.</td></tr>
</table>

2   There is another Dogs-bane that hath long slender stalkes like those of the Vine, but of a browne reddish colour, wherewith it windeth it selfe about such things as stand neere vnto it, in manner of a Bindeweed: whereupon are set leaues not vnlike those of the Iuy, but not so much cornered, of a darke green colour, and of a rank smell being bruised betwixt the fingers, yeelding forth a stinking yellow milky iuice when it is so broken: amongst which come forth little white floures standing scatteringly vpon little huskes: after the floures come long cods very like *Asclepias* or Swallow-wort, but greater, stuffed with the like soft downy silke, among which downe is wrapped vp flat blacke seed. The roots are many and threddy, creeping all about within the ground, budding forth new shoots in sundry places, whereby it greatly increaseth.

¶ *The Place.*

They grow naturally in Syria, and also in Italy, as *Matthiolus* reporteth. My louing friend *Iohn Robin*, herbarist in Paris, did send me plants of both the kinds for my garden, where they floure and flourish: but whether they grow in France, or that hee procured them from some other region, as yet I know not.

The

¶ *The Time.*

They begin to bud forth their leaues in the beginning of May, and ſhew their floures in September.

¶ *The Names.*

Dogs-bane is called by the learned of our age, *Periploca*: it is euident that they are to be referred to the *Apocynum* of *Dioſcorides*. The former of the two hath beene likewiſe called κυνόμορον; and *Braſſica canina*, or Dogs-Cole:notwithſtanding there is another Dogs Cole which is a kind of wild Mercurie. We may call the firſt Creeping Dogs bane; and the other, Vpright or Syrian Dogsbane.

¶ *The Nature.*

Theſe plants are of the nature of that peſtilent or poyſonous herbe *Thora*, which being eaten of dogs or any other liuing creature doth certainly kill them,vnleſſe there be in readines an antidote or preſeruatiue againſt poiſon,and giuen,which by probability is the herb deſcribed in the former chapter,called *Vincetoxicum*; euen as *Anthora* is the antidote and remedy againſt the poiſon of *Thora*; and *Herba Paris* againſt *Pardalianches.*

¶ *The Vertues.*

Dogs-bane is a deadly and dangerous plant,eſpecially to foure footed beaſts; for as *Dioſcorides* A writes,the leaues hereof mixed with bread and giuen, kill dogs,wolues,foxes,and leopards, the vſe of their legs and huckle bones being preſently taken from them,and death it ſelf following incontinent : wherefore it is not to be vſed in medicine.

CHAP. 337.　*Of Solomons Seale.*

1 *Polygonatum.*
Solomons Seale.

2 *Polygonatum minus.*
Small Solomons Seale.

¶ *The Deſcription.*

1　THe firſt kinde of Solomons Seale hath long round ſtalks,ſet for the moſt part with long leaues ſomewhat furrowed and ribbed, not much vnlike Plantain,but narrower,which for the moſt part ſtand all vpon one ſide of the ſtalke,and hath ſmall white floures reſembling the

floures

floures of Lilly Conual : on the other ſide when the floures be vaded, there come forth round berries, which at the firſt are green and of a blacke colour tending to blewneſſe, and being ripe, are of the bigneſſe of Iuy berries, of a very ſweet and pleaſant taſte. The root is white and thicke, full of knobs or joints, in ſome places reſembling the marke of a ſeale, whereof I thinke it tooke the name *Sigillum Solomonis* ; it is ſweet at the firſt, but afterward of a bitter taſte with ſome ſharpneſſe.

2   The ſecond kinde of *Polygonatum* doth not much vary from the former, ſauing in the leaues, which be narrower, and grow round about the ſtalke like a ſpur, in faſhion like vnto Woodroofe or red Madder : amongſt the leaues come forth floures like the former, but of a greener white colour ; which being paſt, there ſucceed berries like the former, but of a reddiſh colour : the roots are thick and knobby like the former, with ſome fibres anexed thereto.

3 *Polygonatum latifolium 2. Cluſij.*   4 *Polygonatum ramoſum.*
Sweet ſmelling Solomons ſeale.   Branched Solomons ſeale.

3   The third kinde of Solomons Seale, which *Cluſius* found in the wooddy mountains of Leitenberg, aboue Manderſtorf, and in many other mountains beyond the riuer Danubius, eſpecially among the ſtones, he ſent to London to Mr *Garth*, a worſhipful gentleman, and one that greatly delighteth in ſtrange plants, who very louingly imparted the ſame vnto me. This plant hath ſtalkes very like to the common Solomons ſeale, a foot high, beſet with leaues vpon one ſide of the ſtalke like the firſt and common kinde, but larger, and more approching to the bigneſſe of the broad leafed Plantain ; the taſt whereof is not very pleaſant : from the boſome of which leaues come forth ſmall well ſmelling greeniſh white flours not much vnlike the firſt : which being paſt, there follow ſeeds or berries that are at the firſt green, but afterwards blacke, containing within the ſame berries a ſmall ſeed as big as a Vetch, and as hard as a ſtone. The roots are like the other of his kinde, yet not ſo thicke as the firſt.

4   The fourth kinde according to my account, but the third of *Cluſius* (which he found alſo in the mountains aforeſaid) groweth a foot high, but ſeldom a cubit, differing from all the others of his kinde ; for his ſtalkes diuide themſelues into ſundry other branches, which are garniſhed with goodly leaues, larger and ſharper pointed than any of the reſt, which do embrace the ſtalkes about after the manner of *Perfoliata* or Thorow-wax, yet very like vnto the kindes of Solomons Seale in
ſhew,

shew, saue that they are somwhat hoarie vnderneath the leaues ; which at the first are sweet in taste, but somewhat acride or biting towards the later end. From the backe part of the leaues shoot forth small long tender and crooked stems, bearing at the end little gaping white flours not much vnlike *Lillium conuallium*, sauouring like Hawthorne floures, spotted on the inner side with blacke spots : which being past, there come forth three cornered berries like the narrow leafed Solomons seale, greene at the first, and red when they be ripe, containing many white hard graines. The roots differ from all the other kindes, and are like vnto the crambling roots of *Thalictrum*, which the grauer hath omitted in the figure.

5 *Polygonatum angustifolium ramosum.*
Narrow leaued Solomons seale.

5 This rare sort of Solomons Seale rises vp from his tuberous or knobby root, with a straight vpright stalke joynted at certaine distances, leauing betweene each joynt a bare and naked stalke, smooth, and of a greenish colour tending to yellownes; from the which joynts thrust forth diuers smal branches, with foure narrow leaues set about like a star or the herbe Woodroofe : vpon which tender branches are set about the stalkes by certaine spaces long narrow leaues inclosing the same round about: among which leaues come forth small whitish floures of little regard. The fruit is small, and of a red colour, full of pulpe or meate ; among which is contained a hard stony seed like that of the first Solomons seale.

‡ 6   There is kept in our gardens, and said to be brought from some part of America another *Polygonatum*, which sends vp a stalk some foot and more high, and it hath leaues long, neruous, and very greene and shining, growing one by another without any order vpon the stalke, which is somewhat crested, crooked, and very greene ; bearing at the very top thereof, aboue the highest leafe, vpon little foot-stalks, some eight or nine little white floures, consisting of six leaues apiece, which are succeeded by berries, as in the former. This floures in May, and is vulgarly named *Polygonatum Virginianum*, or Virginians Solomons seale. ‡

¶ *The Place.*

The first sort of Solomons seale growes naturally wilde in Somerset-shire, vpon the North side of a place called Mendip, in the parish of Shepton Mallet : also in Kent by a village called Crayford, vpon Rough or Row hill : also in Odiam parke in Hampshire ; in Bradfords wood, neere to a towne in Wiltshire foure miles from bath ; in a wood neere to a village called Horsley, fiue miles from Gilford in Surrey, and in diuers other places.

That sort of Solomons seale with broad leaues groweth in certaine woods in Yorkeshire called Clapdale woods, three miles from a village named Settle.

¶ *The Time.*

They spring vp in March, and shew their floures in May : the fruit is ripe in September.

¶ *The Names.*

Solomons seale is called in Greeke, πολύγονον : in Latine likewise, *Polygonatum*, of many, Knees, for so the Greeke word doth import : in shops, *Sigillum Salomonis*, and *Scala cæli* : in English likewise, Scala cœly, Solomons seale, and White woort, or White-root : in high Dutch, weisswurtz : in French, *Seau de Solomon* : of the Hetrurians, *Frasinella*, and *Fraxinella*.

¶ *The Temperature.*

The roots of Solomons seale, as *Galen* saith, haue both a mixt faculty and quality also : For they haue (saith he) a certaine kinde of astriction or binding, and biting withall, and likewise a certaine loathsome bitternesse, as the same Author affirmeth: which is not to be found in those that do grow in our climate.

¶ *The Vertues.*

A   *Dioscorides* writeth, That the roots are excellent good for to seale or close vp greene wounds, being stamped and laid thereon; whereupon it was called *Sigillum Salomonis*, of the singular vertue that it hath in sealing or healing vp wounds, broken bones, and such like. Some haue thought it tooke the name *Sigillum* of the markes vpon the roots: but the first reason seemes to be more probable.

B   The root of Solomons seale stamped while it is fresh and greene, and applied, taketh away in one night, or two at the most, any bruise, blacke or blew spots gotten by fals or womens wilfulnesse, in stumbling vpon their hasty husbands fists, or such like.

C   *Galen* saith, that neither herbe nor root hereof is to be giuen inwardly: but note what experience hath found out, and of late daies, especially among the vulgar sort of people in Hampshire, which *Galen, Dioscorides*, or any other that haue written of plants haue not so much as dreamed of; which is, That if any of what sex or age soeuer chance to haue any bones broken, in what part of their bodies soeuer; their refuge is to stampe the roots hereof, and giue it vnto the patient in ale to drinke: which sodoreth and glues together the bones in very short space, and very strangely, yea although the bones be but slenderly and vnhandsomely placed and wrapped vp. Moreouer, the said people do giue it in like manner vnto their cattell, if they chance to haue any bones broken, with good successe; which they do also stampe and apply outwardly in manner of a pultesse, as well vnto themselues as their cattell.

D   The root stamped and applied in manner of a pultesse, and laid vpon members that haue beene out of joynt, and newly restored to their places, driueth away the paine, and knitteth the joynt very firmely, and taketh away the inflammation, if there chance to be any.

E   The same stamped, and the juyce giuen to drinke with ale or white wine, as aforesaid, or the decoction thereof made in wine, helps any inward bruise, disperseth the congealed and clotted bloud in very short space.

F   That which might be wrttten of this herbe as touching the knitting of bones, and that truely, would seeme vnto some incredible; but common experience teacheth, that in the world there is not to be found another herbe comparable to it for the purposes aforesaid: and therefore in briefe, if it be for bruises inward, the roots must be stamped, some ale or wine put thereto, strained, and giuen to drinke.

G   It must be giuen in the same manner to knit broken bones, against bruises, blacke or blew marks gotten by stripes, falls, or such like; against inflammation, tumors or swellings that happen vnto members whose bones are broken, or members out of joynt, after restauration: the roots are to be stamped small, and applied pultesse or plaisterwise, wherewith many great workes haue beene performed beyond credit.

H   *Matthiolus* teacheth, That a water is drawne out of the roots, wherewith the women of Italy vse to scoure their faces from Sunne-burning, freckles, morphew, or any such deformities of the skinne.

† That which our Authour formerly figured and described in the fifth place of this chapter, by the name of *Polygonatum acutum Clusii*, was that described by him in the fourth place; but the figure was not so well exprest.

CHAP. 338.   *Of Knee-holme, or Butchers broome.*

¶ *The Description.*

K Nee-holme is a low wooddy plant, hauing diuers small branches, or rather stems. rising immediately from the ground, of the height of a foot; whereupon are set many leaues like vnto those of the Box-tree, or rather of the Myrtle, but sharpe and pricking at the point. The fruit groweth vpon the middle rib of the leafe, greene at the first, and red as Corall when it is ripe, like those of *Asparagus*, but bigger. The roots are white, branched, of a meane thickenesse, and full of tough sprouting shoots thrusting forth in other places, whereby it greatly increaseth.

¶ *The Place.*

It groweth plentifully in most places in England in rough and barren grounds, especially vpon Hampsted heath foure miles from London; in diuers places of Kent, Essex, and Barkshire, almost in euery copse and low wood.

¶ *The Time.*

The young and tender sprouts come forth at the first of the Spring, which are eaten in some places, as the young tender stalkes of Asparagus and such like herbes. The berries are ripe in August.

Ruscus, sive Bruscus.
Knee-holme, or Butchers broome.

¶ *The Names.*

It is called in Greeke, ἰξυμυρσίνη, as though they should say *Acuta Myrtus*, or pricking Myrtle; and *Myrtus syluestris*, or wild Myrtle, in Latine, *Ruscum*, or *Ruscus*: in shops, *Bruscus*, of diuers, *Scopa regia*, as testifieth *Marcellus Empericus* an old Writer: in high Dutch, **Muessdorn**: in low Dutch, **Stekende palm**: in Italian, *Rusco*, and *Pontogopi*: in Spanish, *Gilberbaira* in English, Kace-holme, Knee-huluer, Butchers broome, and Petigree.

There be some (saith *Pliny*, *lib.25. cap.13*) that call it *Oxymyrsine*.

Serapio, *cap.288.* supposeth that *Myrtus Agria*, or wilde Myrtle, is the same that *Cubebæ* are: he alleageth a reason, because *Galen* hath not described *Myrtus Agria*, or Knee-holme; neither *Dioscorides Cubebæ*. Which as it is a reason of no account, so is it also without truth: for *Galen* doth no where make mention of *Cubebæ*; and be it that he had, it should not therefore follow that Knee-holme is *Cubebæ*. *Galen* speaketh of *Carpesium*, which *Auicen* in his 137 chapter makes to be *Cubebæ*: and that *Carpesium* doth much differ from Kneeholme, those things do euidently declare which *Galen* hath left written hereof in his first booke of Counterpoysons. *Carpesium* (saith he) is an herbe like in kind to that which is called *Phu*, or Setwall, but of greater force, and more aromaticall or spicie. This groweth very plentifully in Sida a city of Pamphilia. Also he saith further, that some of the stickes of Carpesium are like to those of Cinnamon: there be two kindes thereof, one which is named *Laërtium* ; and another that is called *Ponticum*. They both take their names of the mountaines on which they grow: but *Ponticum* is the better, which is put into medicines in which the herbe *Phu* ought to bee put. For *Carpesium*, as I haue said, is like vnto *Phu*, or Setwall, yet is it stronger, and yeeldeth a certain aromaticall qualitie both in taste and smell. Thus far *Galen*. By which it plainely appeareth, that Knee-holme is not *Carpesium*, that is to say, *Auicenna* his *Cubebæ*, as shall be further declared in the chapter of *Cubeba*.

Herein *Serapio* was likewise deceiued, who suspected it to be such a like thing ; saying, There be certaine fruits or graines called *Cubeba*, not sticks : yet do they neither agree with Knee-holme, neither yet were they knowne vnto *Galen*.

Isaac in the second booke of his practise doth number it among the graines : and likewise *Haliabbas* in the second booke of his practise also, *num.162.* The later Græcians, among whom is *Nicolaus Myrepsus*, call them *Cubeba*.

¶ *The Temperature.*

The roots of Knee-holme, which be chiefely vsed, are of temperature hot, and meanely dry, with a thinnesse of essence.

¶ *The Vertues.*

The decoction of the roots of Knee-holme made in wine and drunken, prouoketh vrine, breaketh **A** the stone, driueth forth grauell and sand, and easeth those that make their water with great paine.

Dioscorides writeth the same things of the leaues and berries, which moreouer (saith hee) bring **B** downe the desired sickenesse, helpe the head-ache and the yellow jaundise. Ouer and besides, the roots do serue to raise vp gently tough and grosse flegme which sticketh in the lungs and chest, and do concoct the same.

CHAP. 339.   Of Horse-tongue, or Double-tongue.

¶ *The Description.*

1   HOrse-tongue sendeth forth round stalkes of a span long ; whereupon are set long broad and sharpe pointed leaues, but not pricking as are those of Knee-holme, not vnlike to the leaues of the Bay-tree, but lesser ; greater than those of Knee-holme : out of the middle rib whereof commeth forth another leafe, sharpe pointed also, but small, and of the bignesse of the leafe of Knee-holme, resembling a little tongue. From the bosome of which two leaues commeth forth a berry of the bignesse of a pease, of colour red when it is ripe, which is sometimes in a manner all hid vnder the leafe. The root is white, long, and tough, and of a sweet and pleasant smell.

1 Hippoglossum mas.
The male Horse-tongue.

2 Hippoglossum fæmina.
The female Horse-tongue.

2   The female Horse-tongue differeth not from the precedent but in stature and colour of the fruit : it riseth vp (saith my Author) foure or fiue handfulls high : the berries come forth of the middle part of the greater leafe, and the setting on of the lesser, of a feint yellowish red colour, wherein consisteth the difference. ‡ This is all one with the former. ‡

3   There is likewise another sort of Double-tongue set forth by *Matthiolus*, which seemes vnto some not to differ from the first described or best known Horse-tongue, being in truth the self same plant without any difference : notwithstanding I haue set forth the figure, that it may appeare to be the same, or very little different, and that not to be distinguished : but *Matthiolus* may not escape without reprehension, who knowing the vntrue translation of *Ruellius*, would set forth so false a figure in his Commentaries.

‡ Our Author here, as in many other places, mistakes himselfe ; for *Matthiolus* did not set forth that figure that our Author giues in this place, for *Hippoglossum*, but by the title of *Laurus Alexandrina altera* : and it thus differs from the common Horse-tongue ; it hath shorter and rounder leaues, yet sharpe pointed, and the berries are not couered with little leaues as in the other, neither haue they any apparant stalkes at all, but grow close to the leaues, as you may see them exprest in the figure. ‡

¶ *The*

3 Hippogloſſum Matthioli.
Italian Horſe-tongue.

¶ *The Place.*

They are found on the Alps of Liguria, and on the mountaines of Auſtria. *Bellonius* writeth, that they do grow very plentifully about the hil Athos.

The firſt of the Horſe-tongues growes in my garden very plentifully.

¶ *The Time.*

That which groweth in my Garden floured in the beginning of May : the fruit is ripe in the fall of the leafe.

¶ *The Names.*

Horſe-tongue is called in Greeke, ἱπτόγλωσσα : of the later Herbariſts, *Bonifacia, Vvularia, Biſlingua, Lingua Pagana,* and *Victoriola.* The ſame is alſo named δάφνη ἰδαία, of Ida a mountaine of Troy, which is called *Alexanders* Troy : of ſome, *Laurus Alexandrina,* or the Bay of Alexandria, and *Laurus Idaa.*

This *Hippogloſſum,* or *Bonifacia* is called in high Dutch, **Zapflinkraut** : in low Dutch, **Tongenblabt** : in Spaniſh, *Lengua de Cauallo*: in Engliſh , Horſe-tongue, Tongue-blade, Double-tongue, and Laurel of Alexandria.

¶ *The Temperature.*

Horſe-tongue is euidently hot in the ſecond degree, and dry in the firſt.

¶ *The Vertues.*

A The roots of Double-tongue boiled in wine, and the decoction drunke, helpeth the ſtrangury, prouoketh vrine, eaſeth women that haue hard trauell in childe-bearing. It expelleth the ſecondine or afterbirth. The root beaten to pouder, whereof ſix drams giuen in ſweet wine, do helpe the diſeaſes aforeſaid: it bringeth down the termes, as *Dioſcorides* teacheth. The like writeth *Pliny* alſo : adding further, That it cauſeth women to haue ſpeedy deliuerance, eſpecially if halfe an ounce of the pouder of the root be giuen to drinke in a draught of ſweet wine.

B *Baptiſta Sardus* doth notably commend this herbe for the diſeaſes of the mother; by giuing, ſaith he, a little ſpoonfull of the pouder either of the herbe, the fruit, or of the root, to her that is troubled with the mother, ſhe is thereby forthwith recouered. He alſo writeth, that the ſame is a ſingular good medicine for thoſe that be burſten, if a ſpoonefull of the pouder of the root be drunke in the broth of fleſh certaine daies together.

Chap. 340. *Of Cucumbers.*

¶ *The Kindes.*

THere be diuers ſorts of Cucumbers; ſome greater, others leſſer; ſome of the garden, ſome wilde; ſome of one faſhion, and ſome of another, as ſhall be declared in the following chapter.

¶ *The Deſcription.*

1 THe Cucumber creepes alongſt vpon the ground all about, with long rough branches ; whereupon doe grow broad rough leaues vneuen about the edges : from the boſome whereof come forth crooked claſping tendrels like thoſe of the Vine. The floures ſhoot forth betweene the ſtalkes and the leaues, ſet vpon tender foot-ſtalkes compoſed of fiue ſmall yellow leaues : which being paſt, the fruit ſucceedeth, long, cornered, rough, and ſet with certaine bumpes or riſings, greene at the firſt, and yellow when they be ripe, wherein is contained a firme and ſollid pulpe or ſubſtance tranſparent or thorow-ſhining, which together with the ſeed is eaten a little before they be fully ripe. The ſeeds be white, long, and flat.

2 There

1 *Cucumis vulgaris.*
Common Cucumber.

2 *Cucumis Anguinus.*
Adders Cucumber.

4 *Cucumis ex Hiſpanico ſemine natus.*
Spaniſh Cucumber.

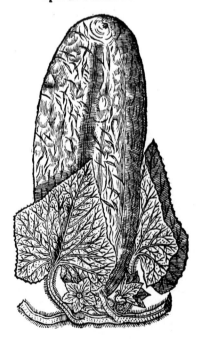

2 There be alſo certaine long cucumbers, which were firſt made (as is ſaid) by art and manuring, which Nature afterwards did preſerue: for at the firſt, when as the fruit is very little, it is put into ſome hollow cane, or other thing made of purpoſe, in which the cucumber groweth very long, by reaſon of that narrow hollowneſſe, which being filled vp, the cucumber encreaſeth in length. The ſeeds of this kinde of cucumber being ſowne bringeth forth not ſuch as were before, but ſuch as art hath framed; which of their own growth are found long, and oftentimes very crookedly turned: and thereupon they haue beene called *Anguini,* or long Cucumbers.

3 The peare faſhioned Cucumber hath many trailing branches lying flat vpon the ground, rough and prickly; whereon do ſtand at each joynt one rough leafe, ſharpe pointed, and of an ouerworn green colour; among which come forth claſping tendrels, and alſo ſlender foot-ſtalks, whereon do grow yellow ſtarre-like floures. The fruit ſucceeds, ſhaped like a peare, as big as a great Warden. The root is threddy.

4 There hath bin not long ſince ſent out of Spain ſome ſeeds of a rare & beautiful cucumber, into Strausburg a city in Germany, which there brought forth long trailing branches, rough & hairy, ſet with very large rough leaues ſharp pointed, faſhioned like vnto the leaues of
the

the great Bur-docke, but more cut in or diuided : amongſt which come forth faire yellow floures growing nakedly vpon their tender foot-ſtalkes : the which being paſt, the fruit commeth in place, of a foot in length, greene on the ſide toward the ground, yellow to the Sunward, ſtraked with many ſpots and lines of diuers colours. The pulpe or meat is hard and faſt like that of our Pompion.

¶ *The Place.*

Theſe kindes of Cucumbers are planted in gardens in moſt countries of the world.

¶ *The Time.*

According to my promiſe heretofore made, I haue thought it good and conuenient in this place to ſet downe not onely the time of ſowing and ſetting of Cucumbers, Muske-melons, Citruls, Pompions, Gourds, and ſuch like, but alſo how to ſet or ſow all manner and kindes of other colde ſeeds, as alſo whatſoeuer ſtrange ſeeds are brought vnto vs from the Indies, or other hot Regions : *videl.*

Firſt of all in the middeſt of Aprill or ſomewhat ſooner (if the weather be any thing temperate) you ſhall cauſe to be made a bed or banke of hot and new horſedung taken from the ſtable (and not from the dunghill) of an ell in breadth, and the like in depth or thickneſſe, of what length you pleaſe, according to the quantity of your ſeed:the which bank you ſhall couer with hoops or poles, that you may the more conueniently couer the whole bed or banke with Mats, old painted cloth, ſtraw or ſuch like, to keepe it from the injurie of the cold froſty nights, and not hurt the things planted in the bed : then ſhall you couer the bed all ouer with the moſt fertileſt earth finely ſifted, halfe a foot thicke, wherein you ſhall ſet or ſow your ſeeds:that being done, caſt your ſtraw or other couerture ouer the ſame; and ſo let it reſt without looking vpon it, or taking away of your couering for the ſpace of ſeuen or eight daies at the moſt. for commonly in that ſpace they will thruſt themſelues vp nakedly forth of the ground : then muſt you caſt vpon them in the hotteſt time of the day ſome water that hath ſtood in the houſe or in the Sun a day before, becauſe the water ſo caſt vpon them newly taken forth of a well or pumpe, will ſo chill and coole them being brought and nouriſhed vp in ſuch a hot place, that preſently in one day you haue loſt all your labour; I mean not only your ſeed, but your banke alſo ; for in this ſpace the great heate of the dung is loſt and ſpent, keeping in memory that euery night they muſt be couered and opened when the day is warmed with the Sun beames : this muſt be done from time to time vntill that the plants haue foure or ſix leaues a piece, and that the danger of the cold nights is paſt : then muſt they be replanted very curiouſly with the earth ſticking to the plant, as neere as may be vnto the moſt fruitfull place, and where the Sun hath moſt force in the garden; prouided that vpon the remouing of them you muſt couer them with ſome Docke leaues or wiſpes of ſtraw, propped vp with forked ſtickes, as well to keepe them from the cold of the night, as alſo the heat of the Sun : for they cannot whileſt they be young and newly planted, endure neither ouermuch cold nor ouermuch heate, vntill they are well rooted in their new place or dwelling.

Oftentimes it falleth out that ſome ſeeds are more franker and forwarder than the reſt, which commonly do riſe vp very nakedly with long necks not vnlike to the ſtalke of a ſmall muſhrome, of a night old. This naked ſtalke muſt you couer with the like fine earth euen to the greene leaues, hauing regard to place your banke ſo that it may be defended from the North-windes.

Obſerue theſe inſtructions diligently, and then you ſhall not haue cauſe to complaine that your ſeeds were not good, nor of the intemperancie of the climat (by reaſon wherof you can get no fruit) although it were in the furtheſt parts of the North of Scotland.

¶ *The Names.*

The Cucumber is named generally *Cucumis* : in ſhops, *Cucumer* : and is taken for that which the Grecians call ~~...~~ : in Latine, *Cucumis ſatiuus*, or garden Cucumber : in high Cutch, **Cucumer** : in Italian, *Concomero* : in Spaniſh, *Cogombro* : in French, *Concombre* : in low Dutch, **Concommeren** : in Engliſh, Cowcumbers and Cucumbers.

¶ *The Temperature and Vertues.*

All the Cucumbers are of temperature cold and moiſt in the ſecond degree. They putrifie ſoon   A
in the ſtomacke, and yeeld vnto the body a cold and moiſt nouriſhment, and that very little, and the ſame not good.

Thoſe Cucumbers muſt be choſen which are greene and not yet ripe : for when they are ripe and   B
yellow they be vnfit to be eaten.

The ſeed is cold, but nothing ſo much as the fruit. It openeth and clenſeth, prouoketh vrine, o-   C
peneth the ſtoppings of the liuer, helpeth the cheſt and lungs that are inflamed ; and being ſtamped and outwardly applied in ſtead of a clenſer, it maketh the skin ſmooth and faire.

Cucumber (ſaith my Author) taken in meats, is good for the ſtomacke and other parts troubled   D
with heat. It yeeldeth not any nouriſhment that is good, inſomuch as the vnmeaſurable vſe thereof filleth the veines with naughty cold humors.

The ſeed ſtamped and made into milke like as they do with Almonds, or ſtrained with milke or   E
ſweet

ſweet wine and drunke, looſeth the belly gently, and is excellent againſt the exulceration of the bladder.

F   The fruit cut in pieces or chopped as herbes to the pot, and boiled in a ſmall pipkin with a piece of mutton, being made into potage with Ote-meale, euen as herb potage are made, whereof a meſſe eaten to break-faſt, as much to dinner, and the like to ſupper; taken in this manner for the ſpace of three weekes together without intermiſſion, doth perfectly cure all manner of ſauce flegme and copper faces, red and ſhining fierie noſes (as red as red Roſes) with pimples, pumples, rubies, and ſuch like precious faces.

G   Prouided alwaies that during the time of curing you doe vſe to waſh or bathe the face with this liquor following.

H   Take a pinte of ſtrong white wine vinegre, pouder of the roots of Ireos or Orrice three dragmes, ſearced or bolted into moſt fine duſt, Brimmeſtone in fine pouder halfe a ounce, Camphire two dragmes, ſtamped with two blanched Almonds, foure Oke apples cut thorow the middle, and the juyce of foure Limons: put them all together in a ſtrong double glaſſe, ſhake them together very ſtrongly, ſetting the ſame in the Sunne for the ſpace of ten daies: with which let the face be waſhed and bathed daily, ſuffering it to drie of it ſelfe without wiping it away. This doth not onely helpe fierie faces, but alſo taketh away lentils, ſpots, morphew, Sun-burne, and all other deformities of the face.

     † That which formerly was in the ſecond place by the name of *Cucumis Turcicus*, was the ſame with the fifth of the former Edition (now the fourth) and is there-fore omitted.

Chap. 341.   *Of Wilde Cucumber.*

Cucumis Aſininus.
Wilde Cucumber.

¶ *The Deſcription.*

THe wilde Cucumber hath many fat hairie branches, very rough and full of juyce, cree-ping or trailing vpon the ground, whereupon are ſet very rough leaues, hairy, ſharpe pointed, & of an ouerworne grayiſh greene colour: from the boſome of which come forth long tender foot-ſtalkes: on the ends whereof doe grow ſmall floures compoſed of fiue ſmall leaues of a pale yellow colour: after which commeth forth the fruit, of the bignes of the ſmalleſt pullets egge, but ſomewhat longer, very rough and hairy on the outſide, and of the colour and ſubſtance of the ſtalkes, wherein is contained very much wa-ter and ſmal hard blackiſh ſeeds alſo, of the big-neſſe of tares; which being come to maturitie and ripeneſſe, it caſteth or ſquirteth forth his water with the ſeeds, either of it owne accord, or being touched with the moſt tender or delicate hand neuer ſo gently, and oftentimes ſtriketh ſo hard againſt thoſe that touch it (eſpecially if it chance to hit againſt the face) that the place ſmarteth long after: whereupon of ſome it hath beene called *Noli me tangere*, Touch me not. The root is thicke, white and long laſting.

¶ *The Place.*

It is found in moſt of the hot countries among rubbiſh, grauell, and other vntilled places: it is planted in gardens in the Low-countries, and being once planted, ſaith *Dodonæus*, it eaſily commeth vp againe many yeares after (which is true:) and yet ſaith he further, that it doth not ſpring againe of the root, but of the ſeeds ſpirted or caſt about: which may likewiſe be true where he hath obſerued it, but in my garden it is otherwiſe, for as I ſaid before, the root is long laſting, and continueth from yeare to yeare.

     ¶ *The*

¶ *The Time.*

It fpringeth vp in May, it floureth and is ripe in Autumne, and is to be gathered at the fame time, to make that excellent compofition called *Elaterium.*

¶ *The Names.*

It is called in Greeke ϭιϰυς εγϱιος: in Latine, *Agreſtis,* and *Errraticus Cucumis* : in fhops, *Cucumer afini-nus* : in Italian, *Cocomero falvatico* : in Spanifh, *Cogumbrillo amargo* : in Englifh, wilde Cucumber, fpirting Cucumbers, and touch me not : in French, *Concombres ſauvages.*

¶ *The Temperature.*

The leaues of wilde Cucumbers, roots and their rindes, as they are bitter in tafte, fo they be like-wife hot and clenfing. The juyce is hot in the fecond degree, as *Galen* witneffeth, and of thin parts. It clenfeth and wafteth away.

¶ *The Vertues.*

The juyce called *Elaterium* doth purge forth choler, flegme, and watery humours, and that with   A force and not onely by fiege, but fometimes alfo by vomit.

The quantity that is to be taken at one time is from fiue graines to ten, according to the ftrength   B of the patient.

The juyce dried or hardened, and the quantity of halfe a fcruple taken, driueth forth by fiege   C groffe flegme, cholericke humours, and preuaileth mightily againft the dropfie, and fhortneffe of breath.

The fame drawne vp into the nofthrils mixed with a little milke, taketh away the redneffe of the   D eies.

The juyce of the root doth alfo purge flegme cholericke and waterifh humours, and is good for   E the dropfie: but not of fuch force as *Elaterium,* which is made of the juyce of the fruit : the making whereof I commend to the learned and curious Apothecaries : among which number M^r. *William Wright* in Bucklers Burie my louing friend hath taken more paines in curious compoſing of it, and hath more exactly performed the fame than any other whatfoeuer that I haue had knowledge of.

CHAP. 342.   *Of Citrull Cucumbers.*

1 *Citrullus officinarum.*
Citrull Cucumber.

‡ 2 *Citrullus minor.*
Small Citrull.

¶ *The*

¶ *The Deſcription.*

1   THe Citrull Cucumber hath many long, flexible, and tender ſtalkes trailing vpon the
ground, branched like vnto the Vine, ſet with certaine great leaues deeply cut, and very
much jagged: among which come forth long claſping tendrels, and alſo tender foot-
ſtalkes, on the ends whereof do grow floures of a gold yellow colour: the fruit is ſomewhat round,
ſtraked or ribbed with certaine deepe furrowes alongſt the ſame, of a greene colour aboue, and vnderneath on that ſide that lyeth vpon the ground ſomething white: the outward skin whereof is
very ſmooth; the meat within is indifferent hard, more like to that of the Pompion than of the Cucumber or Muske-Melon: the pulpe wherein the ſeed lieth, is ſpungie, and of a ſlimie ſubſtance:
the ſeed is long, flat, and greater than thoſe of the Cucumbers: the ſhell or outward barke is blackiſh, ſometimes of an ouerworne reddiſh colour. The fruit of the Citrull doth not ſo eaſily rot or
putrifie as doth the Melon, which being gathered in a faire dry day may be kept a long time, eſpecially being couered in a heape of Wheat, as *Matthiolus* ſaith; but according to my practiſe you
may keepe them much longer and better in a heape of dry ſand.
2   The ſecond kinde of Citrull differeth not from the former, ſauing that it is altogether leſſer,
and the leaues are not ſo deeply cut or jagged, wherein conſiſteth the difference.

¶ *The Place and Time.*

The Citrull proſpereth beſt in hot Regions, as in Sicilia, Apulia, Calabria, and Syria, about Alepo and Tripolis. We haue many times ſowne the ſeeds, and diligently obſerued the order preſcri-
bed in planting of Cucumbers.

¶ *The Names.*

The later Herbariſts do call it *Anguria*: in ſhops, *Citrullus*, and *Cucumis Citrullus*: in Engliſh,
Citruls, and Cucumber Citruls, and the ſeed is knowne by the name of *Semen Citrulli*, or Citrull
ſeed. But if *Cucumis Citrullus* be ſo called of the yellow colour of the Citron, then is the common
Cucumber properly *Cucumis Citrullus*: which is knowne vnto all to be contrary.

¶ *The Temperature and Vertues.*

A   The meat or pulpe of Cucumer Citrull which is next vnto the bark is eaten raw, but more commonly boyled: it yeeldeth to the body little nouriſhment, and the ſame cold: it ingendreth a wateriſh bloud, mitigateth the extremity of heat of the inner parts, and tempereth the ſharpneſſe and
feruent heat of choler: being raw and held in the mouth, it takes away the roughneſſe of the tongue
in Agues, and quencheth thirſt.

B   The ſeeds are of the like faculty with thoſe of Cucumbers.

Cʜᴀᴘ. 343.   *Of the wilde Citrull called* Colocynthis.

¶ *The Deſcription.*

1   COloquintida hath beene taken of many to be a kind of the wilde Gourd, it lieth along
creeping on the ground as do the Cucumbers and Melons, comming neereſt of all to
that which in thoſe daies of ſome Herbariſts is called Citrull Cucumber: it bringeth forth vpon his long branches ſmall crooked tendrels like the Vine, and alſo very great broad
leaues deeply cut or jagged: among which come forth ſmall floures of a pale yellow colour; then
commeth the fruit round as a bowle, couered with a thin rinde, of a yellow colour when it is ripe,
which when it is pilled or pared off, the white pulpe or ſpungie ſubſtance appeareth full of ſeeds,
of a white or elſe an ouerworne browne colour; the fruit ſo pared or pilled, is dryed for medicine;
the which is moſt extreame bitter, and likewiſe the ſeed, and the whole plant it ſelfe in all his
parts.
2   The ſecond kinde of Coloquintida hath likewiſe many long branches and claſping tendrels,
wherewith it taketh hold of ſuch things as are neere vnto it. It bringeth forth the like leaues, but
not ſo much jagged. The floures are ſmall and yellow: the fruit is faſhioned like a peare, and the
other ſort round, wherein the eſpeciall difference conſiſteth.

¶ *The Place.*

Coloquintida is ſowne and commeth to perfection in hot regions, but ſeldome or neuer in theſe
Northerly and cold countries.

¶ *The*

1 *Colocynthis.*
The wilde Citrull or Coloquintida.

2 *Colocynthis pyriformis.*
Peare fashioned Coloquintida.

¶ *The Time.*

It is sowne in the Spring, and bringeth his fruit to perfection in August.

It hath beene diuers times deliuered vnto me for a truth, that they doe grow in the sands of the Mediterranean sea shore, or very neere vnto it, wilde, for euery man to gather that list, especially on the coast of Barbarie, as also without the mouth of the Streights neer to *Sancta Crux* and other places adjacent, from whence diuers Surgions of London that hath trauelled thither for the curing of sicke and hurt men in the ship haue brought great quantities thereof at their returne.

¶ *The Names.*

It is vulgarly called *Coloquintida*: in Greeke, κολοκυνθίς: the Latine translators for *Colocinthis* doe oftentimes set downe *Cucurbita syluestris*: notwithstanding there is a *Cucurbita syluestris* that differeth from *Colocinthis*, or Coloquintida: for *Cucurbita syluestris* is called in Greeke κολοκύνθα ἀγρία: or wilde Coloquintida, whereof shall be set forth a peculiar chapter next after the *Cucurbita* or Gourd: in English it is called Coloquintida, or Apple of Coloquintida.

¶ *The Temperature.*

Coloquintida as it is in his whole nature and in all his parts bitter, so it is likewise hot and drie in the later end of the second degree; and therefore it purgeth, clenseth, openeth and performeth all those things that most bitter things do: but that the strong qualitie which it hath to purge by the stoole, is, as *Galen* saith, of more force than the rest of his operations.

¶ *The Vertues.*

Which operation of purging it worketh so violently, that it doth not onely draw forth flegme **A** and choler maruellous speedily, and in very great quantitie: but oftentimes fetcheth forth bloud and bloudy excrements, by snauing the guts, and opening the ends of the meseraicall veines.

So that therefore the same is not to bee vsed either rashly, or without some dangerous and ex- **B** treme disease constraine thereunto: neither yet at all, vnlesse some tough and clammie thing bee mixed therewith, whereby the vehemency thereof may be repressed, the hurtfull force dulled, and the same speedily passing through the belly, the guts be not fret or shaued. *Mesues* teacheth to mixe with it either Mastich, or gum Tragacanth.

There be made of it Trochises, or little flat cakes, with Mastich, gum Arabick, Tragacanth and **C** Bdellium,

Bdellium, of thefe, Maftich hath a manifeft binding quality : but tough and clammy things are much better, which haue no aftriction at all in them, or very little.

D   For by fuch binding or aftringent things, violent medicines being reftrained and brideled, do afterward work their operation with more violence and trouble : but fuch as haue not binding things mixed with them do eafilier worke, and with leffer paine, as be thofe pils which *Rhafis* in his ninth booke of *Almanzor* calleth *Illiaca* : which are compounded of Coloquintida and Scamony, two of the ftrongeft medicines that are ; and of a third called gum *Sagapene*, which through his clamminefle doth as it were daube the intrails and guts, and defend them from the harme that might haue come of either of them.

E   The which compofition, although it be wonderfull ftrong, and not to be vfed without very great neceffitie vrge thereunto, doth notwithftanding eafily purge, aad without any great trouble, and with leffer torment than moft of the mildeft and gentleft medicines which haue Maftich and other things mixed with them that are aftringent.

F   And for this caufe it is very like that *Galen* in his firft booke of Medicines, according to the places affected, would not fuffer Maftich aud Bdellium to be in the pilles, which are furnamed *Cochia* : the which notwithftanding his Schoolemafter *Quintus* was alfo woont before to adde vnto the fame.

G   But Coloquintida is not onely good for purgations, in which it is a remedy for the diffineffe or the turning fickneffe, the megrim, continuall head-ache, the Apoplexie, the falling fickeneffe, the ftuffing of the lungs, the gnawings and gripings of the guts and intrailes, and other moft dangerous difeafes, but alfo it doth outwardly worke his operations, which are not altogether to be rejected.

H   Common oile wherein the fame is boyled, is good againft the finging in the eares, and deafeneffe: the fame killeth and driueth forth all manner of wormes of the belly, and doth oftentimes prouoke to the ftoole, if the nauell and bottome of the belly be therewith annointed.

I   Being boiled in vinegre, and the teeth wafhed therewith, it is a remedy for the tooth-ache, as *Mefues* teacheth.

K   The feed is very profitable to keepe and preferue dead bodies with ; efpecially if Aloes and Myrrhe be mixed with it.

L   The white pulpe or fpungious pith taken in the weight of a fcruple openeth the belly mightily, and purgeth groffe flegme, and cholericke humors.

M   It hath the like force if it be boyled and laid to infufe in wine or ale, and giuen to drink.

N   Being taken after the fame manner it profiteth the difeafes before remembred, that is, the Apoplexie, falling fickneffe, giddineffe of the head, the collicke, loofeneffe of finewes, and places out of joynt, and all difeafes proceeding of cold.

O   For the fame purpofes it may be vfed in clifters.

P   The fame boiled in oile, and applied with cotton or wooll, taketh away the paine of the Hemorrhoides.

Q   The decoction made in wine, and vfed as a fomentation or bathe, bringeth downe the defired fickeneffe.

CHAP. 344. *Of Muske-Melon, or Million.*

¶ *The Kindes.*

THere be diuers forts of Melons found at this day, differing very notably in fhape and proportion, as alfo in tafte, according to the climate and country where they grow : but of the Antients there was onely one and no more, which is that *Melopepo* called of *Galen*, *Cucumis*, or *Galens* Cucumber : notwithftanding fome haue comprehended the Muske-Melons vnder the kindes of Citruls, wherein they haue greatly erred : for doubtleffe the Muske-Melon is a kinde of Cucumber, according to the beft approued Authors.

¶ *The Defcription.*

1.   THat which the later Herbarifts do call Muske-Melons is like to the common Cucumber in ftalks, lying flat vpon the ground, long, branched, and rough. The leaues be much alike, yet are they leffer, rounder, and not fo cornered : the floures in like manner bee yellow : the fruit is bigger, at the firft fomwhat hairy, fomthing long, now and then fomwhat round ; oftentimes greater, and many times leffer : the barke or rinde is of an ouerworne ruffet greene
colour,

1 *Melo*,
The Muske Melon.

2 *Melo Saccharinus.*
Sugar Melon.

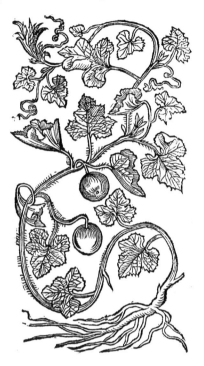

4 *Melo Hispanicus.*
Spanish Melons.

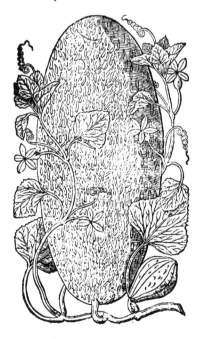

colour, ribbed and furrowed very deepely, hauing often chaps or chinks, and a confused roughnesse: the pulp or inner substance which is to be eaten is of a feint yellow colour; the middle part whereof is full of a slimie moisture: amongst which is contained the seed, like to those of the Cucumber, but lesser, and of a browner colour.

2   The sugar Melon hath long trailing stalks lying vpon the ground, whereon are set small clasping tendrels like those of the Vine, and also leaues like vnto the common Cucumber, but of a greener colour: the fruit commeth forth among those leaues, standing vpon slender foot-stalkes, round as the fruit of Coloquintida, and of the same bignesse, of a most pleasant taste like sugar, whereof it tooke the syrname *Saccharatus*.

3   The peare-fashioned Melon hath many long viny branches, whereupon grow cornered leaues like those of the Vine, and likewise great store of long tendrels, clasping and taking hold of each thing that it toucheth: the fruit groweth vpon sleeder foot-stalks fashioned like a peare, of the bignesse of a great Quince.

4   The Spanish Melon brings forth long trailing branches, wheron are set broad leaues slightly indented about the edges, not diuided at all, as are all the rest of the Melons. The fruit groweth neere vnto the stalke, like vnto the common Pompion, very long, not crested or furrowed at all, but

Hhhh   spotted

ſpotted with very many ſuch marks as are on the backe of the Hearts-tongue leafe. The pulpe or meat is not ſo pleaſing in taſte as the other.

¶ The Place.

They delight in hot regions,notwithſtanding I haue ſeen at the Queens houſe at S. *Iames* many of the firſt ſort ripe,through the diligent and curious nouriſhing of them by a ſkilfull gentleman the keeper of the ſaid houſe called Mr *Fowle* : and in other places neere the right honourable Lord of Suſſex his houſe of Bermondſey by London,where yearely there is very great plenty,eſpecially if the weather be any thing temperat.

¶ The Time.

They are ſet or ſowne in Aprill,as I haue already ſhewne in the chapter of Cucumbers. Their fruit is ripe in the end of Auguſt,and ſomtimes ſooner.

¶ The Names.

The Muske Melon is called in Latine, *Melo* : in Italian, *Mellone* : in Spaniſh, *Melon* : in French, *Melons* : in high-Dutch,**Melaun** : in low-Dutch,**Meloenen** : in Greeke, Μῆλον, which ſignifieth an apple ; and therefore this kinde of Cucumber is more truly called μηλοπέπων, or *Melopepon*, by reaſon that *Pepo* ſmelleth like an apple,whereto the ſmell of this fruit is like, hauing withall the ſmell as it were of Muske ; and for that cauſe are alſo named *Melones Muſchatellini*,or Muske Melons.

¶ The Nature.

The meat of the Muske Melon is very cold and moiſt.

¶ The Vertues.

A　It is harder of digeſtion than is any of the Cucumbers, and if it remaine long in the ſtomack it putrifieth,and is occaſion of peſtilent feuers. Which thing alſo *Aëtius* witneſſeth , in his firſt booke of *Tetrabibles* ; writing,That the vſe of *Cucumeres* or Cucumbers breedeth peſtilent feuers : for hee alſo taketh *Cucumis* to be that which is commonly called a Melon ; which is vſually eaten of the Italians and Spaniards rather to repreſſe the rage of luſt,than for any other phyſicall vertue.

B　The ſeed is of like operation with that of the former Cucumber.

Chap. 345. *Of Melons or Pompions.*

¶ The Kindes.

THere be found diuers kindes of Pompions which differ either in bigneſſe or forme; it ſhal be therefore ſufficient to deſcribe ſome one or two of them,and refer the reſt to the view of the figures, which moſt liuely doe expreſſe their differences; eſpecially becauſe this Volume waxeth great,the deſcription of no moment,and I haſten to an end.

¶ The Deſcription.

1　THe great Pompion bringeth forth thicke and rough prickely ſtalkes,which with their claſping tendrels take hold vpon ſuch things as are neere vnto them, as poles, arbours, pales,and ledges, which vnleſſe they were neere vnto them would creep along vpon the ground : the leaues be wilde and great,very rough,and cut with certain deep gaſhes,nicked alſo on the edges like a ſaw : the floures be very great like vnto a bell cup,of a yellow colour like gold, hauing fiue corners ſtanding out like teeth : the fruit is great,thicke,round,ſet with thicke ribbs like edges ſticking forth ; the pulpe or meat whereof next vnder the rinde is white,and of a mean hardneſſe : the pith or ſubſtance in the middle is ſpongie and ſlimy ; the ſeed is great,broad, flat,ſomthing white,much greater than that of the Cucumber,otherwiſe not differing at all in forme. The colour of the barke or rinde is oftentimes of an obſcure green, ſometimes gray. The rinde of the green Pompion is harder,and as it were of a wooddy ſubſtance : the rinde of the gray is ſofter and tenderer.

2　The ſecond kinde of Melons or Pompions is like vnto the former in ſtalkes and leaues, and alſo in claſping tendrels ; but the gaſhes of the leaues are not ſo deepe, and the ſtalkes bee tenderer : the floures are in like manner yellow, gaping, cornered at the top, as be thoſe of the former : but the fruit is ſomewhat rounder, ſometimes greater, and many times leſſer ; and oftentimes

of

of a green colour,with an harder barke,now and then ſofter and whiter. The meat within is like the former : the ſeeds haue alſo the ſame forme,but are ſomewhat leſſe.

1 *Pepo maximus oblongus.*
The great long Pompion.

2 *Pepo maximus rotundus.*
The great round Pompion.

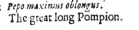

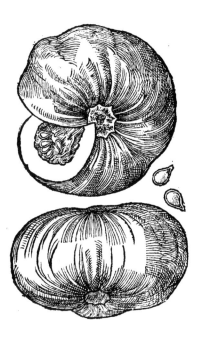

3　Of this kinde there is alſo another Pompion like vnto the former in rough ſtalks,and in ga-ſhed and nicked leaues : the floure is alſo great and yellow like thoſe of the others : the fruit is of a great bigneſſe, whoſe barke is full of little bunnies or billy welts, as is the rinde of the Citron, which is in like manner yellow when it is ripe.

4　The fourth Pompion doth very much differ from the others in form : the ſtalks, leaues, and flours are like thoſe of the reſt,but the fruit is not long or round,but altogether broad,and in a ma-ner flat like vnto a ſhield or buckler, thicker in the middle, thinner in the compaſſe, and curled or bumped in certain places about the edges,like the rugged or vneuen barke of the Pomecitron, the which rinde is very ſoft, thin, and white ; the meat within is meetly hard and durable : the ſeed is greater than that of the common Cucumber,in forme and colour all one.

‡ *Macocks Virginiani, ſiue Pepo Virginianus.*
The Virginian Macock or Pompion.

‡　This hath rough cornered ſtraked trailing branches proceeding from the root,eight or nine foot long or longer,and thoſe again diuided into other branches of a blackiſh greene colour, trai-ling,ſpreading,or running alongſt the earth,couering a great deale of ground,ſending forth broad cornered rough leaues,on great groſſe long rough hairy footſtalks,like & fully as big as the leaues of the common Pompion,with claſping tendrels and great broad ſhriueled yellow flours alſo like thoſe of the common Pompion : the fruit ſucceedeth growing alongſt the ſtalks, commonly not neere the root,but towards the vpper part or tops of the branches, ſomewhat round, not extending in length, but flat like a bowle, but not ſo big as an ordinarie bowle, beeing ſeldome foure inches broad,and three inches long, of a blackiſh green colour when it is ripe.　The ſubſtance or eatable part is of a yellowiſh white colour, containing in the middeſt a great deale of pulp or ſoft matter, wherein the ſeed lieth in certain rowes alſo, like the common Pompion, but ſmaller.　The root is made of many whitiſh branches creeping far abroad in the earth,and periſh at the firſt approch of Winter.

3 *Pepo maximus compressus*. The great flat bottomed Pompion.

4 *Pepo maximus clypeatus*. The great Buckler Pompion.

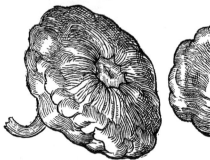

5 *Pepo Indicus minor rotundus*.
The small round Indian Pompion.

6 *Pepo Indicus angulosus*.
The cornered Indian Pompion.

Melones aquatici edules Virginiani.
The Virginian water-Melon.

This Melon or Pompion is like and fully as big as the common Pompion, in spreading running creeping branches, leaues, floures, and clasping tendrels : the fruit is of a very blackish greene colour, and extendeth it selfe in length neere foure inches, and three inches broad, no bigger nor longer than a great apple, and grow alongst the branches forth of the bosomes of the leaues, not farre from the root, euen to the tops of the branches, containing a substance, pulp, and flat seed like the ordinarie Pompion : the root is whitish, and disperseth it selfe very far abroad in the earth, and perisheth about the beginning of Winter. Octob. 10. 1621. *Iohn Goodyer.* ‡

¶ *The Place.*
All these Melons or Pompions be garden plants : they ioy best in a fruitfull soile, and are common in England, except the last described, which is as yet a stranger.

¶ *The Time.*
They are planted in the beginning of Aprill : they floure in August : the fruit is ripe in September.

¶ *The Names.*
The great Melon or Pompion is named in Greek πέπων: in Latine likewise *Pepo*: the fruits of them all when they be ripe are called by a common name in Greeke, πέπονες: in English, Melons, or Pompions. Whereupon certain Physitions (saith *Galen*) haue contended, that this fruit ought to be called σικυοπέπων, that is to say in Latine, *Pepo Cucumeralis*, or Cucumber. *Pliny, lib. 9. cap. 5.* writes, That *Cucumeres* when they exceed in greatnesse are called *Pepones* : it is called in high-Dutch, **Pluker**: in low-Dutch, **Pepoenen**: in French, *Pompons.*

¶ *The Nature and Vertues.*
All the Melons are of a cold nature, with plenty of moisture : they haue a certain clensing qualitie, by means whereof they prouoke vrine, and do more speedily passe through the body than doe either the Gourd, Citron, or Cucumber, as *Galen* hath written.　　A

The pulp of the Pompion is neuer eaten raw, but boiled ; for so it doth more easily descend, making the belly soluble. The nourishment which comes hereof is little, thin, moist, and cold, bad, as *Galen* saith, and that especially when it is not well digested ; by reason whereof it maketh a man apt and ready to fall into the disease called the Cholericke passion, or Felonie.　　B

The seed clenseth more than the meat, it prouoketh vrin, and is good for those that are troubled with the stone of the kidnies.　　C

The fruit boiled in milke and buttered, is not only a good and wholsome meat for mans bodie, but so prepared, it is also a physical medicine for such as haue an hot stomack, and the inward parts inflamed.　　D

The flesh or pulp of the same sliced and fried in a pan with butter, is also a good and wholsome meat : but baked with apples in an ouen, it fills the body with flatuous or windy belchings, and is food vtterly vnwholsome for such as liue idlely ; but vnto robustious and rusticke people nothing hurteth that filleth the belly.　　E

CHAP. 346. *Of wilde Pompions.*

¶ *The Description.*

1　AS there is a wilde sort of Cucumbers, of Melons, Citrons, and Gourds, so likewise are there certain wilde Pompions that be so of their own nature. These bring forth rough stalks set with sharp thorny prickles : the leaues be likewise rough, the floures yellow, as be those of the garden Melon, but euery part is lesser : the root is thicke, round, and sharpe pointed, hauing a hard green rinde ; the pulpe or meat whereof, and the middle pith, with the seed, are like those of the garden Pompion, but very bitter in taste.

2　The second is like vnto the former, but it is altogether lesse, wherein consists the difference.

¶ *The Place.*
These Melons grow wilde in Barbarie, Africa, and most parts of the East and West Indies : they grow not in these parts, except they be sowne.

¶ *The Time.*
Their time of flouring and flourishing answers that of the garden Pompion.

　　　　　　　　¶ *The*

1 *Pepo major ſylueſtris.*
The great wilde Pompion.

2 *Pepo minor ſylueſtris.*
The ſmall wilde Pompion.

¶ *The Names.*

Although the antient phyſitions haue made no mention of this plant, yet the thing it ſelf doth ſhew that there be ſuch, and ought to be called in Greeke πίπωνε ἄγριοι : in Latine, *Pepones ſylveſtres:* in Engliſh, wilde Melons or Pompions.

¶ *The Temperature.*

Like as theſe wild Melons be altogether of their owne nature very bitter, ſo be they alſo of temperature hot and dry, and that in the later end of the ſecond degree. They haue likewiſe a clenſing facultie, not inferior to the wilde Cucumbers.

¶ *The Vertues.*

A   The wine which when the pith and ſeed is taken forth is poured into the rinde, and hath remained ſo long therein till ſuch time as it becommeth bitter, doth purge the belly, and bringeth forth flegmaticke and cholericke humors. To be briefe, the juice herof is of the ſame operation that the wilde Cucumber is, and being dried it may be vſed in ſtead of *Elaterium*, which is the dried juice of the wilde Cucumber.

CHAP. 347:   *Of Gourds.*

¶ *The Kindes.*

THere diuers ſorts of Gourds, ſome wilde, others tame of the garden : ſome bearing fruit like vnto a bottle ; others long, bigger at the end, keeping no certain form or faſhion; ſome greater, others leſſe.

‡   I will only figure and deſcribe two or three of the chiefeſt, and ſo paſſe ouer the reſt, becauſe each one vpon the firſt ſight of them knowes to what kinde to refer them. ‡

¶ *The*

¶ *The Deſcription.*

1   THe Gourd bringeth forth very long ſtalks as be thoſe of the Vine, cornered and parted into diuers branches, which with his claſping tendrels taketh hold and clymeth vpon ſuch things as ſtand neere vnto it : the leaues bee very great, broad, and ſharpe pointed, almoſt as great as thoſe of the Clot-burre, but ſofter, and ſomwhat couered as it were with a white freeſe, as be alſo the ſtalkes and branches, like thoſe of the mariſh Mallow : the floures be white, and grow forth from the boſome of the leaues : in their places come vp the fruit, which are not all of one faſhion, for oftentimes they haue the forme of flagons and bottles, with a great large belly and a ſmall necke. The Gourd (ſaith *Pliny, lib. 19. cap. 5.*) groweth into any forme or faſhion that you would haue it ; either like vnto a wreathed Dragon, the leg of a man, or any other ſhape, accor-ding to the mold wherein it is put while it is yong : being ſuffered to clyme vpon any arbor where the fruit may hang, it hath bin ſeen to be nine foot long, by reaſon of his great weight which hath ſtretched it out in the length : the rinde when it is ripe, is very hard, woody, and of a yellow colour : the meat or inward pulpe is white ; the ſeed long, flat, pointed at the top, broad, below, with two peaks ſtanding out like hornes, white within, and ſweet of taſte.

2   The ſecond differs not from the precedent in ſtalks, leaues, or floures : the fruit hereof is for the moſt part faſhioned like a flagon or bottle, wherein eſpecially conſiſteth the difference.

1 *Cucurbita anguina.* Snakes Gourd.	2 *Cucurbita lagenaria.* Bottle Gourds.

¶ *The Place.*

Gourds are cheriſhed in the gardens of theſe cold regions rather for pleaſure than profit : in the hot countries where they come to ripenes they are ſomtimes eaten, but with ſmall delight, eſpeci-ally they are kept for the rinds, wherein they put turpentine, oile, hony, and alſo ſerue them as pailes to fetch water in, and many other the like vſes.

¶ *The Time.*

They are planted in a bed of horſe dung in April, like as we haue taught the planting of cucum-bers : they flouriſh in Iune and Iuly, the fruit is ripe in the end of Auguſt.

¶ *The Names.*

The Gourd is called in Greeke, κολοκύντη ἡμερος : in Latine, *Cucurbita edulis, Cucurbita ſatiua* : of *Pliny,*

Cucurbita

*Cucurbita Cameraria,*becauſe it climeth vp,and is a couering for arbors,walking places,and banqueting houſes in gardens : he calls the other which climeth not vp,but lieth crawling on the ground,
Cucurbita plebeia: in Italian,*Zucca* : in Spaniſh,*Calabazza:* in French,*Courge:* in high-Dutch,**Kurbs:**
in low-Dutch,**Cauwoozden :** in Engliſh,Gourds.

¶ *The Temperature.*

The meat or inner pulp of the Gourd is of temperature cold and moiſt, euen in the ſecond degree.

¶ *The Vertues.*

A   The juice being dropped into the eares with oile of Roſes is good for the paine thereof proceeding of an hot cauſe.

B   The pulp or meat thereof mitigateth all hot ſwellings,if it be laid thereon in manner of a pultis,
and vſed in like manner it takes away the head-ache and inflammation of the eies.

C   The ſame Author affirmeth,that a long Gourd or Cucumber being laid in the cradle or bed by
the yong infant whileſt it is aſleep and ſicke of an ague,it ſhall be very quickly made whole.

D   The pulp alſo is eaten ſod,but becauſe it hath in it a wateriſh and thinne juice,it yeeldeth ſmall
nouriſhment to the body,and the ſame cold and moiſt;but it eaſily paſſeth thorow,eſpecially being ſodden,which by reaſon of the ſlipperineſſe and moiſtneſſe alſo of his ſubſtance, mollifies the
belly.

E   But being boiled in an ouen or fried in a pan it loſeth moſt part of his naturall moiſture,where-
fore it more ſlowly deſcendeth,and doth not mollifie the belly ſo ſoon.

F   The ſeed allayes the ſharpneſſe of vrin,and brings down the ſame.

CHAP. 348.   Of the wild Gourd.

1 *Cucurbita lagenaria ſylueſtris.*
  Wilde Bottle Gourd.

2 *Cucurbita ſylueſtris fungiformis.*
  Muſhrom wilde Gourd.

¶ *The Description.*

1 THere is besides the former ones a certaine wilde Gourd : this is like the garden Gourd in clyming stalks, clasping tendrels, and soft leaues, and as it were downy; all and euery one of which things being far lesse : this also clymbeth vpon arbours and banqueting houses : the fruit representeth the great bellied Gourd, and those that be like vnto bottles in form, but in bignesse it is very far inferior, for it is small, and scarce so great as an ordinarie Quince, and may be held within the compasse of a mans hand: the outward rinde at the first is green, afterward it is as hard as wood, and of the colour thereof : the inner pulp is moist and very full of juice, wherin lies the seed. The whole is as bitter as Coloquintida, which hath made so many errors, one especially, in taking the fruit Coloquintida for the wilde Gourd.

2 The second wilde Gourd hath likewise many trailing branches and clasping tendrels, wherwith it taketh hold of such things as be neere vnto it : the leaues be broad, deepely cut into diuers sections like those of the Vine, soft and very downy, whereby it is especially knowne to be one of the Gourds : the flours are very white, as are also those of the Gourds : the fruit succeeds, growing to a round forme, flat on the top like the head o a Mushrome, whereof it tooke his syr-name.

¶ *The Place.*

They grow of themselues wilde in hot regions : they neuer come to perfection of ripenesse in these cold countries.

¶ *The Time.*

The time answereth those of the garden.

¶ *The Names.*

The wilde Gourd is called in Greeke, Κολοκύνθα ἀγρία : in Latine, *Cucurbita syluestris*, or wilde Gourd. *Pliny, lib. 20. cap. 3.* affirmeth, That the wilde Gourd is named of the Grecians, σίκυος, which is hollow, an inch thicke, not growing but among stones, the juice whereof being taken is very good for the stomacke. But the wilde Gourd is not that which is so described, for it is aboue an inch thick, neither is it hollow, but full of juice, and by reason of the extreme bitternesse offensiue to the stomacke.

Some also there be that take this for Coloquintida, but they are far deceiued ; for Colocynthis is the wilde Citrull Cucumber, whereof we haue treated in the chapter of Citruls.

¶ *The Temperature.*

The wilde Gourd is as hot and dry as Coloquintida, namely in the second degree.

¶ *The Vertues.*

The wild Gourd is extreme bitter, for which cause it opens and scoures the stopped passages of   A the body: it also purgeth downwards as do wilde Melons.

Moreouer, the Wine which hath continued all night in this Gourd likewise purgeth the belly   B mightily, bringing forth cholericke and flegmaticke humors.

Chap. 349.   *Of Potato's.*

Sisarum Peruuianum, siue Batata Hispanorum.
Potatus, or Potato's.

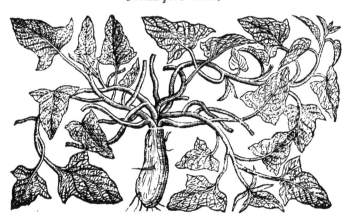

¶ *The Deſcription.*

THis Plant (which is called of ſome, *Siſarum Peruvianum*, or Skyrrets of Peru) is generally of vs called Potatus or Potato's. It hath long rough flexible branches trailing vpon the ground like vnto thoſe of Pompions, whereupon are ſet greene three cornered leaues very like thoſe of the wilde Cucumber. There is not any that haue written of this plant, haue ſaid any thing of the floures; therefore I refer their deſcription vnto thoſe that ſhall hereafter haue further knowledge of the ſame. Yet haue I had in my garden diuers roots that haue flouriſhed vnto the firſt approch of Winter, and haue growne vnto a great length of branches, but they brought forth no floures at all; whether becauſe the Winter cauſed them to periſh before their time of flouring, or that they be of nature barren of floures, I am not certaine. The roots are many, thicke, and knobby, like vnto the roots of Peonies, or rather of the white Aſphodill, ioined together at the top into one head, in maner of the Skyrret, which being diuided into diuers parts and planted, do make a great increaſe, eſpecially if the greateſt roots be cut into diuers goblets, and planted in good and fertile ground.

¶ *The Place.*

The Potato's grow in India, Barbarie, Spaine, and other hot regions; of which I planted diuers roots (which I bought at the Exchange in London) in my garden, where they flouriſhed vntil winter, at which time they periſhed and rotted.

¶ *The Time.*

It flouriſheth to the end of September: at the firſt approch of great froſts the leaues together with the roots and ſtalks do periſh.

¶ *The Names.*

Cluſius calleth it *Batata, Camotes, Amotes,* and *Ignames* : in Engliſh, Potatoes, Potatus, and Potades.

¶ *The Nature.*

The leaues of Potato's are hot and dry, as may euidently appeare by the taſte: the roots are of a temperat qualitie.

¶ *The Vertues.*

A The Potato roots are among the Spaniards, Italians, Indians, and many other nations, ordinarie and common meat; which no doubt are of mighty and nouriſhing parts, and doe ſtrengthen and comfort nature; whoſe nutriment is as it were a mean between fleſh and fruit, but ſomewhat windie; yet being roſted in the embers they loſe much of their windineſſe, eſpecially being eaten ſopped in wine.

B Of theſe roots may be made conſerues no leſſe toothſome, wholeſome, and dainty, than of the fleſh of Quinces; and likewiſe thoſe comfortable and delicate meats called in ſhops, *Morſelli, Placentulæ,* and diuers other ſuch like.

C Theſe roots may ſerue as a ground or foundation whereon the cunning Confectioner or Sugar-Baker may worke and frame many comfortable delicat Conſerues and reſtoratiue ſweet-meats.

D They are vſed to be eaten roſted in the aſhes. Some when they be ſo roſted infuſe and ſop them in wine: and others to giue them the greater grace in eating, do boile them with prunes and ſo eat them: likewiſe others dreſſe them (being firſt roſted) with oile, vineger, and ſalt, euery man according to his owne taſte and liking. Notwithſtanding howſoeuer they be dreſſed, they comfort, nouriſh, and ſtrengthen the body, vehemently procuring bodily luſt.

C H A P. 350: *Of Potato's of Virginia.*

¶ *The Deſcription.*

VIrginian Potato hath many hollow flexible branches trailing vpon the ground, three ſquare, vneuen, knotted or kneed in ſundry places at certaine diſtances: from the which knots commeth forth one great leafe made of diuers leaues, ſome ſmaller, and others greater, ſet together vpon a fat middle rib by couples, of a ſwart greene colour tending to redneſſe; the whole leafe reſembling thoſe of the Winter-Creſſes, but much larger; in taſte at the firſt like graſſe, but afterward ſharp and nipping the tongue. From the boſome of which leaues come forth long
round

round flender footftalkes, whereon grow very faire and pleafant floures, made of one entire whole leafe, which is folded or plaited in fuch ftrange fort, that it feemes to be a floure made of fiue fundry fmal leaues, which canot eafily be perceiued, except the fame be pulled open. The whole floure is of a light purple colour, ftriped downe the middle of euery fold or welt with a light fhew of yellowneffe, as if purple and yellow were mixed together. In the middle of the floure thrufteth forth a thicke flat pointall yellow as gold, with a fmall fharp green pricke or point in the midft thereof. The fruit fucceeds the floures, round as a ball, of the bigneffe of a little Bulleffe or wilde plumme, green at the firft, and blacke when it is ripe, wherein is contained fmall white feed leffer than thofe of Muftard: the root is thicke, fat, and tuberous, not much differing either in fhape, colour, or tafte, from the common Potato's, fauing that the roots hereof are not fo great nor long, fome of them are as round as a ball, fome oual or egge-fafhion, fome longer, and others fhorter; the which knobby roots are faftned vnto the ftalks with an infinite number of threddy ftrings.

Battata Virginiana fiue Virginianorum, & Pappus,
Virginian Potatoes.

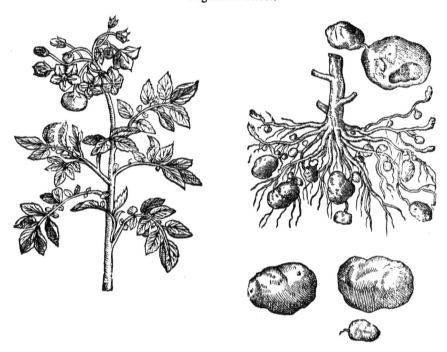

¶ The Place.

It groweth naturally in America, where it was firft difcouered, as reporteth *Clufius*, fince which time I haue receiued roots hereof from Virginia, otherwife called Norembega, which grow & profper in my garden as in their owne natiue country.

¶ The Time.

The leaues thruft forth of the ground in the beginning of May; the flours bud forth in Auguft, the fruit is ripe in September.

¶ The Names.

The Indians call this plant *Pappus*, meaning the roots; by which name alfo the common Potatoes are called in thofe Indian countries. Wee haue it's proper name mentioned in the title. Becaufe it hath not only the fhape and proportion of Potato's, but alfo the pleafant tafte and vertues of the fame, we may call it in Englifh, Potatoes of America or Virginia.

‡ *Clufius* queftions whether it be not the *Arachidna* of *Theophraftus*. *Bauhine* hath referred it to the Nightfhades, and calls it *Solanum tuberofum efculentum*; and largely figures and defcribes it in his *Prodromus, pag. 89.* ‡

¶ *The Temperature and Vertues.*

The temperature and vertues be referred to the common Potato's, being likewise a food, as also a meat for pleasure, equall in goodnesse and wholesomnesse to the same, being either rosted in the embers, or boiled and eaten with oile, vineger and pepper, or dressed some other way by the hand of a skilfull Cooke.

‡ *Bauhine* faith, That he heard that the vse of these roots was forbidden in Bourgondy (where they call them Indian Artichokes) for that they were persuaded the too frequent vse of them cau-fed the leprosie. ‡

CHAP. 351.
Of the garden Mallow called Hollyhocke.

¶ *The Kindes.*

THere be diuers sorts or kindes of Mallowes; some of the garden; some of the marish or sea shore, others of the field, both which are wilde. And first of the garden Mallow or Hollihock.

1 *Malua hortensis.*
Single garden Hollihock.

2 *Malua rosea simplex peregrina.*
Iagged strange Hollihock.

¶ *The Description.*

1 THe tame or garden Mallow bringeth forth broad round leaues of a whitish greene co-lour, rough, and greater than those of the wilde Mallow: the stalke is streight, of the height of foure or six cubits; whereon do grow vpon slender foot-stalks single floures, not much vnlike to the wilde Mallow, but greater, consisting only of fiue leaues, sometimes white or red, now and then of a deep purple colour, varying diuersly as Nature list to play with it: in their places groweth vp a round knop like a little cake, compact or made vp of a multitude of flat seeds like little cheefes. The root is long, white, tough, easily bowed, and groweth deep in the ground.

2 The

3 *Malua purpurea multiplex*.
Double purple Hollihocke.

2 The second being a strange kinde of Hollihocke hath likewise broad leaues, rough and hoarie, or of an ouerworne ruffet colour, cut into diuers sections euen to the middle ribbe, like those of Palma Christi. The floures are very single, but of a perfect red colour, wherein consisteth the greatest difference. ‡ And this may be called *Malua rosea simplex peregrina folio Ficus*, Iagged strange Hollihocke. ‡

3 The double Hollihocke with purple floures hath great broad leaues, confusedly indented about the edges, and likewise toothed like a saw. The stalke groweth to the height of foure or fiue cubits. The floures are double, and of a bright purple colour.

4 The Garden Hollihocke with double floures of the colour of skarlet, groweth to the height of fiue or six cubits, hauing many broad leaues cut about the edges. The stalke and root is like the precedent. ‡ This may be called *Multea hortensis rubra multiplex*, Double red Hollihockes, or Rose mallow. ‡

5 The tree mallow is likewise one of the Hollihockes; it bringeth forth a great stalke of the height of ten or twelue foot, growing to the forme of a small tree, whereon are placed diuers great broad leaues of a ruffet greene colour, not vnlike to those of the great Clot Burre Docke, deepely indented about the edges. The floures are very great and double as the greatest Rose, or double Peiony, of a deepe red colour tending to blackenesse. The root is great, thicke, and of a wooddy substance, as is the rest of the plant. ‡ This may be called *Malua hortensis atrorubente multiplici flore*. ‡

¶ *The Place.*

These Hollihockes are sowne in gardens, almost euery where, and are in vaine sought elsewhere.

¶ *The Time.*

The second yeere after they are sowne they bring forth their floures in Iuly and August, when the seed is ripe the stalke withereth, the root remaineth and sendeth forth new stalkes, leaues and floures, many yeares after.

¶ *The Names.*

The Hollihocke is called in Greeke, ᵐᵃᶜᵉᵒⁿ: of diuers, *Rosa vltra-marina*, or outlandish Rose, and *Rosa hyemalis*, or Winter Rose. And this is that Rose which *Pliny* in his 21. book, 4. chapter, writes to haue the stalke of a mallow, and the leaues of a pot-herbe, which they cal *Mosceuton*: in high Dutch, **Garten pappelen**: in low Dutch, **winter Roosen**: in French, *Rose d'outre mer*: in English, Hollihocke, and Hockes.

¶ *The Temperature.*

The Hollihocke is meetely hot, and also moist, but not so much as the wilde Mallow: it hath likewise a clammy substance, which is more manifest in the seed and root, than in any other part.

¶ *The Vertues.*

The decoction of the floures, especially those of the red, doth stop the ouermuch flowing of the   A monethly courses, if they be boiled in red wine.

The roots, leaues, and seeds serue for all those things for which the wilde Mallowes doe, which   B are more commonly and familiarly vsed.

Chap. 352.   Of the wilde Mallowes.

¶ *The Description.*

1 THe Wilde Mallow hath broad leaues somewhat round and cornered, nickt about the edges, smooth, and greene of colour: among which rise vp many slender tough stalkes,

with the like leaues, but smaller. The floures grow vpon little foot-stalkes of a reddish colour
ed with purple strakes, consisting of fiue leaues, fashioned like a bell : after which commeth vp
vp or round button, like vnto a flat cake, compact of many small seeds. The root is white, tough,
full of a slimie juyce, as is all the rest of the plant.

The dwarfe wilde Mallow creepeth vpon the ground : the stalkes are slender and weake, yet
tough and flexible. The leaues be rounder, and more hoary than the other. The floures are small,
and of a white colour.

3   The crispe or curled Mallow, called of the vulgar sort French Mallowes, hath many small
vpright stalkes, growing to the height of a cubit, and sometimes higher ; whereon doe grow broad
leaues somewhat round and smooth, of a light greene colour, plaited or curled about the brims like
a ruffe. The floures be small and white. The root perisheth when it hath perfected his seed.

1 *Malua syluestris.*   2 *Malua syluestris pumila.*
The field Mallow.   The wilde dwarfe Mallow.

4   The Veruaine Mallow hath many straight stalkes, whereon doe grow diuers leaues deepely
cut and jagged euen to the middle ribbe, not vnlike to the leaues of Veruaine, whereof it tooke his
name : among which come forth faire and pleasant floures like vnto those of the common Mallow
in forme, but of a more bright red colour, mixed with stripes of purple, which setteth forth the
beautie. The root is thicke, and continueth many yeares. ‡ This is sometimes though more rarely
found with white floures. ‡

‡ 5   This annuall Mallow, called by *Clusius, Malua trimestris,* is very like our common Mallow,
sending vp slender branched stalkes some three foot high ; the bottome leaues are round, those on
the stalkes more sharpe pointed, greene aboue, and whiter vnderneath : the floures consist of fiue
leaues of a light carnation colour, the seed is like that of the ordinary Mallow, but smaller ; and
such also is the root which perishes euery yeare as soone as the seed is ripe : it is sowne in some gardens, and growes wilde in Spaine. ‡

¶ *The Place.*

The two first Mallowes grow in vntoiled places among pot-herbes, by high waies, and the borders of fields.

The French mallow is an excellent pot-herbe, for the which cause it is sowne in gardens, and is
not to be found wilde that I know of.

The

3 *Malua crispa.*
The French curled Mallow.

4 *Malua verbenaca.*
Veruaine Mallow.

‡ 5 *Malua æstiua Hispanica.*
The Spanish Mallow.

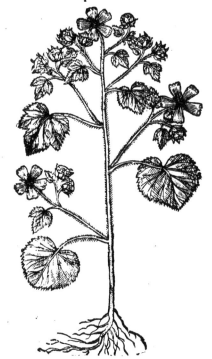

The Veruaine Mallow groweth not euerie where: it growes on the ditch sides on the left hand of the place of execution by London, called Tyborn: also in a field neere vnto a village fourteene miles from London called Bushey, on the backe-side of a Gentlemans house named Mr. *Robert Wilbraham*: likewise amongst the bushes and hedges as you go from London to a bathing place called the Old Foord: and in the bushes as you go to Hackny a village by London, in the closes next the town, and in diuers other places, as at Bassingburne in Hartfordshire, three miles from Roiston.

‡ Mr. *Goodyer* found the Veruain Mallow with white floures growing plentifully in a close neere Maple-durham in Hampshire, called Aldercrofts. ‡

¶ *The Time.*

These wilde Mallowes do floure from Iune till Summer be well spent: in the meane time their seed also waxeth ripe.

¶ *The Names.*

The wilde Mallow is called in Latine *Malua syluestris*: in Greeke, Μαλάχη ἄγρια, or χορτάια: and ἄλμος, as though they should say a mitigator of paine: of some, *Osiriaca*: in high Dutch, **Pappelin**: in low Dutch, **Maluwe**, and **Keelkens cruit**: in English, Mallow.

Iiii 2   The

The Veruaine Mallow is called of *Dioſcorides, Alcea :* in Greeke, ἀλκαία: of ſome, *Herba Hungarica,* and *Herba Simeonis,* or Simons Mallow : in Engliſh, Varuaine Mallow, and jagged Mallow.

The name of this herbe *Malua* ſeemeth to come from the Hebrewes, who call it in their tongue חלמות *Malluach,* of the ſaltneſſe, becauſe the Mallow groweth in ſaltiſh and old ruinous places, as in dung-hills and ſuch like, which in moſt abundant manner yeeldeth forth Salt-peter and ſuch like matter : for חלמ *Melach* ſignifieth ſalt, as the learned know. I am perſuaded that the Latine word *Malua* commeth from the Chaldee name *Mallucha,* the gutturall letter מ *Ch,* being left out for good ſounds ſake : ſo that it were better in this word *Malüa* to reade *u* as a vowell, that as a con-ſonant : which words are vttered by the learned Doctor *Rabbi Dauid Kimhi,* and ſeeme to carrie a great ſhew of truth: in Engliſh it is called, Mallow ; which name commeth as neere as may be to the Hebrew word.

¶ *The Temperature.*

The wilde Mallowes haue a certaine moderate and middle heate, and moiſtneſſe withall : the juyce thereof is ſlimie, clammie, or gluing, the which are to be preferred before the garden Mallow or Hollihocke, as *Diphilus Siphinus* in *Athenæus* doth rightly thinke ; who plainely ſheweth, that the wilde mallow is better than that of the garden : although ſome doe prefer the Hollihocke, whereunto we may not conſent, neither yet yeeld vnto *Galen,* who is partly of that minde, yet ſtandeth he doubtfull : for the wilde Mallow without controuerſie is fitter to be eaten, and more pleaſant than thoſe of the garden, except the French Mallow, which is generally holden the wholſomeſt, and amongſt the pot-herbes not the leaſt commended by *Heſiodo:* whoſe opinion was *Horace,* writing in his ſecond Ode of his *Epodon,*

  —— *& graui*
  Malua ſalubres corpori.

The Mallow (ſaith *Galen*) doth nouriſh moderately, ingendreth groſſe bloud, keepeth the body ſoluble, and looſeth the belly that is bound. It eaſily deſcendeth, not onely becauſe it is moiſt, but alſo by reaſon it is ſlimie.

¶ *The Vertues.*

A The leaues of Mallowes are good againſt the ſtinging of Scorpions, Bees, Waſps, and ſuch like : and if a man be firſt annointed with the leaues ſtamped with a little oile, he ſhal not be ſtung at all, as *Dioſcorides* ſaith.

B The decoction of mallowes with their roots drunken are good againſt all venome and poyſon, if it be incontinently taken after the poyſon, ſo that it be vomited vp againe.

C The leaues of mallowes boiled till they be ſoft and applied, doe mollifie tumors and hard ſwellings of the mother, if they do withall ſit ouer the fume thereof, and bathe themſelues therewith.

D The decoction vſed in cliſters is good againſt the roughneſſe and fretting of the guts, bladder, and fundament.

E The roots of the Veruaine-mallow do heale the bloudy flix and inward burſtings, being drunke with wine and water, as *Dioſcorides* and *Paulus Ægineta* teſtifie.

CHAP. 353.   Of Marſh Mallow.

¶ *The Deſcription.*

1 Marſh-Mallow is alſo a certaine kinde of wilde Mallow : it hath broad leaues, ſmall toward the point, ſoft, white and freeſed or cottoned, and ſlightly nicked about the edges : the ſtalkes be round and ſtraight, three or foure foot high, of a whitiſh gray colour ; whereon do grow floures like vnto thoſe of the wilde Mallowes, yet not red as they are, but commonly white, or of a very light purple colour out of the white: the knop or round button wherein the ſeeds lie is like that of the firſt wilde mallow. The root is thicke, tough, white within, and containeth in it a clammy and ſlimy juyce.

† 2 This ſtrange kinde of Mallow is holden amongſt the beſt writers to be a kinde of Marſh mallow : ſome excellent Herbariſts haue ſet it downe for *Sida Theophraſti,* whereto it doth not fully anſwer : it hath ſtalks two cubits high, whereon are ſet without order many broad leaues hoarie and whitiſh, not vnlike thoſe of the other Marſh mallow : the floures conſiſt of fiue leaues, and are larger than thoſe of the marſh mallow, and of a purple colour tending to redneſſe : after which there come round bladders of a pale colour, in ſhape like the fruit or ſeeds of round *Ariſtochia,* or Birthwort, wherein is contained round blacke ſeed. The root is thicke and tough, much like that of the common mallow.

1 *Althæa, Ibiſcus.*
Marſh Mallow.

2 *Althæa paluſtris.*
Water Mallow.

3 *Althæa Arboreſcens.*
Tree-Mallow.

4 *Althæa frutex Cluſij.*
Shrubby Mallow.

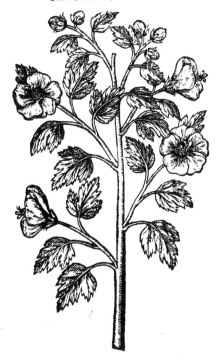

‡ 5 *Alcea fruticofa cannabina.*
Hempe-leaued Mallow.

3　This wilde Mallow is likewiſe referred vnto the kindes of marſh mallow, called generally by the name of *Althæa*, which groweth to the form of a ſmal hedge tree, approching neerer to the ſubſtance and nature of wood than any of the other; wherewith the people of Olbia and Narbone in France doe make hedges, to ſeuer or diuide their gardens and vine-yards (euen as wee doe with quick-ſets of priuet or thorne) which continueth long: the ſtalke whereof groweth vpright, very high, comming neere to the Willow in wooddineſſe and ſubſtance. The floures grow alongſt the ſame, in faſhion and colour of the common wilde mallow.

4　The ſhrubby mallow riſeth vp like vnto a hedge buſh, and of a wooddy ſubſtance, diuiding it ſelfe into diuers tough and limber branches, couered with a barke of the colour of aſhes; whereupon doe grow round pointed leaues, ſomewhat nickt about the edges, very ſoft, not vnlike to thoſe of the common marſh mallow, and of an ouerworne hoary colour. The floures grow at the top of the ſtalkes, of a purple colour, conſiſting of fiue liues, very like to the common wilde mallow, and the ſeed of the marſh mallow.

5　We haue another ſort of mallow, called of *Pena*, *Alcea fruticoſior petaphylla*: it bringeth forth in my Garden many twiggy branches, ſet vpon ſtiffe ſtalkes of the bigneſſe of a mans thumbe, growing to the height of ten or twelue foot: whereupon are ſet very many leaues deepely cut euen to the middle rib, like vnto the leaues of hempe: the floures and ſeeds are like vnto the common mallow: the root is exceeding great, thicke, and of a wooddy ſubſtance. ‡ *Cluſius* calls this *Alcea fruticoſa cannabino folio*: and it is with good reaſon thought to be the *Cannabis ſylueſtris* deſcribed by *Dioſcorides, lib. 3. cap. 166.* ‡

¶ *The Place.*
The common marſh mallow groweth very plentifully in the marſhes both on the Kentiſh and Eſſex ſhore alongſt the riuer of Thames, about Woolwych, Erith, Greenhyth, Graueſend, Tilbury, Lee, Colcheſter, Harwich, and in moſt ſalt marſhes about London: being planted in Gardens it proſpereth well, and continueth long.
The ſecond groweth in the moiſt and fenny places of Ferraria, betweene Padua in Italy, and the riuer Eridanus.
The others are ſtrangers likewiſe in England: notwithſtanding at the impreſſion hereof I haue ſowen ſome ſeeds thereof in my garden, expecting the ſucceſſe.

¶ *The Time.*
They floure and flouriſh in Iuly and Auguſt: the root ſpringeth forth afreſh euery yeare in the beginning of March, which are then to be gathered, or in September.

¶ *The Names.*
The common marſh mallow is called in Greeke, Αλθαία, and ἰβίσκος: the Latines retaine the names *Althæa* and *Ibiſcus*: in ſhops, *Biſmalua*, and *Maluauiſcus*; as though they ſhould ſay *Malua Ibiſcus*: in high Dutch, **Ibiſch**: in low Dutch, **witte Maluwe**, and **witten Hemſt**: In Italian and Spaniſh, *Maluauiſco*: in French, *Guimaulue*: in Engliſh, marſh mallow, mooriſh mallow, and white mallow.
The reſt of the mallowes retaine the names expreſſed in their ſeuerall titles.

¶ *The Temperature.*
Marſh mallow is moderately hot, but drier than the other mallowes: the roots and ſeeds hereof are more dry, and of thinner parts, as *Galen* writeth; and likewiſe of a digeſting ſoftning or mollifying nature.

¶ *The*

¶ *The Vertues.*

The leaues of Marſh mallow are of the power to digeſt, mitigate paine, and to concoct.　A

They be with good effect mixed with fomentations and pulteſſes againſt paines of the ſides, of　B the ſtone, and of the bladder; in a bath alſo they ſerue to take away any manner of paine.

The decoction of the leaues drunke doth the ſame, which doth not onely aſſwage paine, which　C proceedeth of the ſtone, but alſo is very good to cauſe the ſame to deſcend more eaſily, and to paſſe forth.

The roots and ſeeds are profitable for the ſame purpoſe : moreouer, the decoction of the roots　D helpeth the bloudy flix, yet nor by any binding quality, but by mitigating the gripings and frettings thereof : for they doe not binde at all, although *Galen* otherwiſe thought, but they cure the bloudy flix, by hauing things added vnto them, as the roots of Biſtort, Tormentill, the floures and rindes of Pomegranates and ſuch like.

The mucilage or ſlimie iuyce of the roots, is mixed very effectually with all oiles, ointments, and　E plaiſters that ſlacken and mitigate paine.

The roots boyled in wine, and the decoction giuen to drinke, expell the ſtone and grauell, helpe　F the bloudy flix, ſciatica, crampes and convulſions.

The roots of Marſh mallows, the leaues of common Mallowes, and the leaues of Violets, boiled　G in water vntill they be very ſoft, and that little water that is left drained away, ſtamped in a ſtone morter, adding thereto a certaine quantitie of Fenugreeke, and Lineſeed in pouder : the root of the black Bryony, and ſome good quantitie of Barrowes greaſe, ſtamped all together to the forme of a pultis, and applied very warme, mollifie and ſoften Apoſtumes and hard ſwellings, ſwellings in the joynts, and ſores of the mother : it conſumeth all cold tumors, blaſtings, and windie outgrowings ; it cureth the rifts of the fundament ; it comforteth, defendeth, and preſerueth dangerous greene wounds from any manner of accidents that may happen thereto, it helpeth digeſtion in them, and bringeth old vlcers to maturitie.

The ſeeds dried and beaten into pouder and giuen to drinke, ſtop the bloudy flix and laske, and　H all other iſſues of bloud.

Chap. 354.　*Of the yellow Mallow.*

Althæa Lutea.
Yellow Mallow.

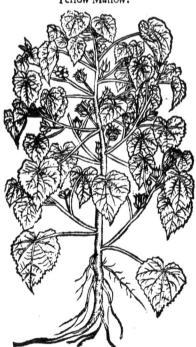

¶ *The Deſcription.*

THe yellow Mallow riſeth vp with a round ſtalke, ſomthing hard and wooddy, three or foure cubits high, couered with broad leaues ſomething round, but ſharpe pointed, white, ſoft, ſet with very fine haires like to the leaues of gourds, hanging vpon long tender foot-ſtalkes : from the boſome of which leaues come forth yellow floures, not vnlike to thoſe of the common Mallow in forme : the knops or ſeed veſſels are blacke, crooked, or wrinkled, made vp of many ſmall cods, in which is black ſeed : the root is ſmall, and dieth when it hath perfected his ſeed.

¶ *The Place.*

The ſeed hereof is brought vnto vs from Spaine and Italy : we doe yearely ſow it in our gardens, the which ſeldome or neuer doth bring his ſeed to ripeneſſe : by reaſon whereof, we are to ſeeke for ſeeds againſt the next yeare.

¶ *The Time.*

It is ſowne in the midſt of Aprill, it brings forth his floures in September.

¶ *The Names.*

Some thinke this to be *Abutilon* : whereupon that agreeth which *Auicen* writeth, that it is like to the Gourd, that is to ſay, in leaſe, and to be named *Abutilon*, and *Arblutilon* : diuers

take

take it to be that *Althæa* or Marſh mallow, vnto which *Theophraſtus* in his ninth booke of the Hiſtorie of Plants doth attribute *Florem* μιλον, or a yellow floure : for the floure of the common Marſh mallow is not yellow, but white ; yet may *Theophraſtus* his copy, which in diuers places is faultie, and hath many emptie and vnwritten places, be alſo faultie in this place ; therefore it is hard to ſay, that this is *Theophraſtus* Marſh mallow, eſpecially ſeeing that *Theophraſtus* ſeemeth alſo to attribute vnto the roote of Marſh mallow ſo much ſlime, as that water may be thickened therewith, which the roots of common Marſh mallow can very well doe : but the root of *Abutilon* or yellow mallow not at all : it may be called in Engliſh, yellow Mallow, and *Auicen* his Mallow.

¶ The Temperature.

The temperature of this Mallow is referred vnto the Tree-mallow.

¶ The Vertues.

A   *Auicen* ſaith, that *Abutilon* or yellow mallow, is held to be good for greene wounds, and doth preſently glew together and perfectly cure the ſame.

B   The ſeed drunke in wine preuaileth mightily againſt the ſtone.

C   *Bernardus Paludanus* of Anchuſen reporteth, that the Turks do drinke the ſeed to prouoke ſleep and reſt.

CHAP. 3ᵢ5.   *Of Venice Mallow, or Good-night at Noone.*

1 *Alcea Peregrina.*
Venice Mallow.

2 *Sabdariſa.*
Thorny Mallow.

¶ The Deſcription.

† 1   THe Venice mallow riſeth vp with long, round, feeble ſtalkes, whereon are ſet vpon long ſlender foot-ſtalkes, broad jagged leaues, deepely cut euen to the middle rib : amongſt which come forth very pleaſant and beautifull floures, in ſhape like thoſe of the common mallow,
fomething

‡ 3 *Alcea Ægyptia.*
The Ægyptian Codded Mallow.

fomething white about the edges, but in the middle of a fine purple: in the middeſt of this floure ſtandeth forth a knap or peſtel, as yellow as gold: it openeth it ſelfe about eight of the clocke, and ſhutteth vp againe at noone, about twelue a clock when it hath receiued the beams of the Sun, for two or three houres, whereon it ſhould ſeeme to rejoyce to looke. and for whoſe departure, being then vpon the point of declenſion, it ſeemes to grieue. and ſo ſhuts vp the floures that were open, and neuer opens them againe ; whereupon it might more properly be called *Malua horaria,* or the Mallow of an houre: and this *Colamella* ſeemeth to call *Moloche,* in this verſe;

——— *Et Moloche, Prono ſequitur quæ vertice ſolem.*

The ſeed is contained in thicke rough bladders, whereupon *Dodonæus* calleth it *Alcea Veſicaria :* within theſe bladders or ſeed veſſels are contained blacke ſeed, not vnlike to thoſe of *Nigella Romana.* The root is ſmall and tender, and periſheth when the ſeed is ripe, and muſt be increaſed by new and yearely ſowing of the ſeed, carefully reſerued.

2 Thorn Mallow riſeth vp with one vpright ſtalk of two cubits high, diuiding it ſelfe into diuers branches, whereupon are placed leaues deeply cut to the middle rib, and likewiſe ſnipt about the edges like a ſaw: in taſte like Sorrel: the floures for the moſt part thruſt forth of the trunke or body of the ſmall ſtalke, compact of fiue ſmall leaues, of a yellowiſh colour; the middle part whereof is of a purple tending to redneſſe: the huske or cod wherein the floure doth ſtand is ſet or armed with ſharpe thornes: the root is ſmall, ſingle, and moſt impatient of our cold climate, inſomuch that when I had with great induſtry nouriſhed vp ſome plants from the ſeed, and kept them vnto the midſt of May; notwithſtanding one cold night chancing among many, hath deſtroied them all.

‡ 3   This alſo is a ſtranger cut leaued mallow, which *Cluſius* hath ſet forth by the name of *Alcea Ægyptia:* and *Proſper Alpinus* by the title of *Bammia :* the ſtalke is round, ſtraight, green, ſome cubit and halfe high: vpon which without order grow leaues at the bottome of the ſtalke, like thoſe of Mallow, cornered and ſnipt about the edges; but from the middle of the ſtalke to the top they are cut in with fiue deepe gaſhes like as the leaues of the laſt deſcribed: the floures grow forth by the ſides of the ſtalke, in forme and colour like thoſe of the laſt mentioned, to wit, with fiue yellowiſh leaues: after theſe follow long thicke fiue cornered hairy and ſharpe pointed ſeed veſſels, containing a ſeed like *Orobus,* couered with a little downineſſe: this growes in Ægypt, where they eat the fruit thereof as we do Peaſe and Beanes: *Alpinus* attributes diuers vertues to this plant, agreeable to thoſe of the common Marſh-mallow. ‡

¶ *The Place.*

The ſeeds hereof haue been brought out of Spaine and other hot countries. The firſt proſpereth well in my garden from yeare to yeare.

¶ *The Time.*

They are to be ſowne in the moſt fertill ground and ſunnie places of the garden, in the beginning of May, or in the end of Aprill.

¶ *The Names.*

Their names haue beene ſufficiently touched in their ſeuerall deſcriptions. The firſt may be called in Engliſh, Venice-mallow, Good-night at noone, or the Mallow flouring but an houre: of *Matthiolus* it is called *Hypecoon,* or Rue Poppy, but vnproperly.

¶ *The Temperature and Vertues.*

There is a certaine clammie juyce in the leaues of the Venice-mallow, whereupon it is thought **A** to come neere vnto the temperature of the common Mallow, and to be of a mollifying faculti e:but his vſe in Phyſicke is not yet knowne, and therefore can there be no certainty affirmed.

CHAP. 356. *Of Cranes-bill.*

¶ *The Kindes.*

THere be many kindes of Cranes-bil, whereof two were known to *Dioſcorides*, one with the knobby root, the other with the Mallow leafe.

Geranium Columbinum.
Doues foot, or Cranes-bill.

¶ *The Deſcription.*

DOues-foot hath many hairy ſtalks, trailing or leaning toward the ground, of a brownifh colour, ſomewhat kneed or joynted; wherupon do grow rough leaues of an ouerworn green color, round, cut about the edges, and like vnto thoſe of the common Mallow: amongſt which come forth the floures of a bright purple colour: after which is the feed, ſet together like the head and bill of a bird; wherupon it was called Cranes-bill, or Storks-bill, as are alfo all the other of his kind. The root is ſlender, with ſome fibres annexed thereto.

‡ 2   There is another kinde of this with larger ſtalkes and leaues, alſo the leaues are more deepely cut in or diuided, and the floures are either of the fame colour as thoſe of the common kinde, or elſe ſomewhat more whitiſh. This may be called *Geranium columbinum majus diſſeĉtis folius*, Great Doues foot.

3   To this kinde may alſo fitly be referred the *Geranium Saxatile* of *Thalius*: the root is ſmal and thready, the leaues are ſmoother, redder, more bluntly cut about the edges, and tranſparent than thoſe of the firſt deſcribed, yet round, and otherwiſe like them: the floures are ſmall and red, and the bills like thoſe of the former. Mr *Goodyer* found it growing plentifully on the bankes by the high way leading frow Gilford towards London, neere vnto the townes end. ‡

¶ *The Place.*
It is found neere to common high waies, deſart places, vntilled grounds, and ſpecially vpon mud walls almoſt euery where.

¶ *The Time.*
It ſpringeth vp in March and Aprill: floureth in May, and bringeth his feed to ripeneſſe in Iune.

¶ *The Names.*
It is commonly called in Latine, *Pes Columbinus*: in High Dutch, **Scatter kraut**: in Low Dutch, **Duyuen boet**: in French, *Pied de Pigeon*: hereupon it may be called *Geranium Columbinum*: in Engliſh, Doues-foot, and Pigeons foot: of *Dioſcorides, Geranium alterum*: of ſome *Pulmonia*, and *Gruina*.

¶ *The Temperature.*
Doues foot is cold and ſomewhat drie, with ſome aſtriĉtion or binding, hauing power to ſoder or joyne together.

¶ *The Vertues.*
A   It ſeemeth, ſaith my Author, to be good for greene and bleeding wounds, and aſſwageth inflammations or hot ſwellings.

The

The herbe and roots dried,beaten into moſt fine pouder,and giuen halfe a ſpoonfull faſting, and **B** the like quantitie to bedwards in red wine,or old claret,for the ſpace of one and twenty daies toge-ther,cure miraculouſly ruptures or burſtings,as my ſelfe haue often proued,whereby I haue gotten crownes and credit : if the ruptures be in aged perſons,it ſhall be needfull to adde thereto the pow-der of red ſnailes(thoſe without ſhels)dried in an ouen in number nine,which fortifieth the herbes in ſuch ſort, that it neuer faileth, although the rupture be great and of long continuance : it like-wiſe profiteth much thoſe that are wounded into the body, and the decoction of the herbe made in wine,preuaileth mightily in healing inward wounds,as my ſelfe haue likewiſe proued.

Chap. 357. *Of Herbe Robert.*

Geranium Robertianum.
Herbe Robert.

¶ *The Deſcription.*

HErbe Robert bringeth forth ſlender weake and brittle ſtalks,ſomewhat hairie,and of a reddiſh colour,as are oftentimes the leaues al-ſo,which are jagged and deepely cut, like vnto thoſe of Cheruile, of a moſt loathſome ſtin-king ſmell. The flours are of a moſt bright pur-ple colour;which being paſt, there follow cer-taine ſmal heads, with ſharpe beaks or bils like thoſe of birds : the root is ſmall and thteddy.

¶ *The Place.*

Herb Robert groweth vpon old wals,as wel thoſe made of brick and ſtone,as thoſe of mud or earth:it groweth likewiſe among rubbiſh,in the bodies of trees that are cut downe, and in moiſt and ſhadowie ditch banks.

¶ *The Time.*

It floureth from Aprill till Summer be al-moſt ſpent : the herbe is green in Winter alſo, and is hardly hurt with cold.

¶ *The Names.*

It is called in high Dutch,**Rupꝛechts kraut:** in low Dutch, **Robꝛechts kruit:**and thereup-on it is named in Latine, *Ruberta,* and *Roberti herba: Ruellius* calleth it *Robertiana* ; and we *Ro-bertianum: of Tabernamontanus,Rupertianum :* in Engliſh, Herbe Robert.He that conferreth this Cranes-bill with *Dioſcorides* his third *Sideritis* ſhall plainely perceiue,that they are both one, and that this is moſt apparantly *Sideritis* 3. *Di-oſcoridis* ; for *Dioſcorides* ſetteth downe three *Sideritides,* one with the leafe of Horehound ; the next with the leafe of Fearne ; and the third groweth in walls and Vineyards : the natiue ſoile of Herbe Robert agree thereunto, and likewiſe the leaues, being like vnto Cheruile, and not vnlike to thoſe of Corianders,according to *Dioſcorides* deſcription.

¶ *The Temperature.*

Herbe Robert is of temperature ſomewhat cold : and yet both ſcouring and ſomewhat binding, participating of mixt qualities.

¶ *The Vertues.*

It is good for wounds and vlcers of the dugs and ſecret parts ; it is thought to ſtanch bloud, **A** which thing *Dioſcorides* doth attribute to his third *Sideritis :* the vertue of this,ſaith he,is applied to heale vp bloudy wounds.

Chap.

Chap. 358. *Of knobbed Cranes-bill.*

Geranium tuberofum.
Knobbie Cranes-bill.

¶ *The Defcription.*

THis kinde of Cranes-bill hath many flexi-ble branches, weake and tender, fat, and full of moifture, wheron are placed very great leaues cut into diuers fmall fections or diuifions, re-fembling the leaues of the tuberous *Anemone*, or Wind-floure, but fomewhat greater, of an ouer-worn greenifh colour: among which come forth long foot-ftalkes, whereon do grow faire flours, of a bright purple color, and like vnto the fmal-left brier Rofe in forme: which being paft, there fucceed fuch heads and beaks as the reft of the Cranes-bill haue : the root is thick, bumped or knobbed, which we call tuberous.

¶ *The Place.*
This kinde of Cranes-bill is a ftranger in England, notwithftanding I haue it growing in my Garden.

¶ *The Time.*
The time anfwereth the reft of the Cranes-bills.

¶ *The Names.*
Cranes-bill is called in Greeke, γεράνιον: in La-tine, *Gruinalis*, commonly *Roftrum Gruis*, or *Ro-ftrum Ciconia*, of the likeneffe of a Cranes-bill, or ftorkes bill : of fome, *Acus mofcata* : but that name doth rather belong to another of this kind: it is alfo called *Acus Paftoris*: in Italian, *Ro-ftro di grua* : in French, *Bec de Grue* : in Spanifh, *Pico di Ciquena, pico del grou* : in high Dutch, **Storckenfchnabel** : in low Dutch, **Otteuoers beck** : in Englifh, Storks-bill, Cranes-bill, Herons-bill, and Pincke-needle : this is alfo called for diftin-ctions fake, *Geranium tuberofum*, and *Geranium bulbofum* : it is likewife *Geranium primum Diofcoridis*, or *Diofcorides* his firft Cranes-bill.

¶ *The Temperature.*
The roots of this Cranes-bill haue a little kinde of heat in them.

¶ *The Vertues.*

A   *Diofcorides* faith that the roots may be eaten, and that a dram weight of them drunk in wine doth wafte and confume away the windineffe of the Matrix.

B   Alfo *Pliny* affirmeth, that the root hereof is fingular good for fuch as after weakeneffe craue to be reftored to their former ftrength.

C   The fame Author affirmeth that the weight of a dram of it drunke in wine three times in a day, is excellent good againft the Ptificke, or confumption of the lungs.

Chap. 359. *Of Musked Cranes-bill.*

¶ *The Defcription.*

MVsked Cranes-bill hath many weake and feeble branches trailing vpon the ground, whereon doe grow long leaues, made of many fmaller leaues, fet vpon a middle rib, fnipt or cut about the edges, of a pleafant fweet fmell, not vnlike to that of Muske: among which come forth the floures fet vpon tender foot-ftalkes, of a red colour, compact of fiue fmall leaues apiece : after which appeare fmall heads and pointed beakes or bills like the other kindes of Cranes-bills : the root is fmall and threddy.

¶ *The*

Geranium moſchatum.
Musked Cranes-bill.

¶ *The Place.*

It is planted in gardens for the ſweet ſmel that the whole plant is poſſeſſed with, ‡ but if you rub the leaues and then ſmel to them, you ſhall find them to haue a ſent quite contrary to the former. ‡

¶ *The Time.*

It floureth and flouriſheth al the ſummer long.

¶ *The Names.*

It is called *Myrrhida Plinij, Roſtrum Ciconiæ, Acus moſchata,* in ſhops, and *Acus paſtoris,* & likewiſe *Geranium moſchatum:* in Engliſh, Musked Storks bill, and Cranes bill, *Muſchatum,* and of the vulgar ſort Muſchata, and alſo Pick-needle.

¶ *The Temperature.*

This Cranes bill hath not any of his faculties found out or knowne : yet it ſeemeth to be cold and a little dry, with ſome aſtriction or binding.

¶ *The Vertues.*

The vertues are referred vnto thoſe of A Doues foot, and are thought of *Dioſcorides* to be good for greene and bloudy wounds, and hot ſwellings that are newly begun.

CHAP. 360. *Of Crow-foot Cranes-bill,* or Gratia Dei.

¶ *The Deſcription.*

1　CRow-foot Cranes bill hath many long and tender branches tending to rednes, ſet with great leaues deeply cut or jagged, in forme like thoſe of the field Crow-foot, whereof it tooke his name ; the floures are pretty large, and grow at the top of the ſtalkes vpon tender foot-ſtalks, of a perfect blew colour : which being paſt, there ſucceed ſuch heads, beaks, and bils as the other Cranes bils haue.

I haue in my garden another ſort of this Cranes bill, bringing forth very faire white floures, which maketh it to differ from the precedent ; in other reſpects there is no difference at all.

‡　2　This which is the *Geranium 2. Batrachioides minus* of *Cluſius* hath large ſtalks and leaues, and thoſe very much diuided or cut in; the ſtalks alſo are diuided into ſundry branches, which vpon long foot-ſtalkes carry floures like in ſhape, but leſſe than thoſe of the formerly deſcribed, and not blew, but of a reddiſh purple color, hauing ten threds and a pointall comming forth of the middle of the floure; the beaks or bils which are the ſeed ſtand vpright, and hang not downe their points as moſt others do. The root is large and liues many yeares.

3　The ſtalkes of this are ſtiffe, greene, and hairy, diuided at their tops into ſundry branches which end in long foot-ſtalks, vpon which grow floures commonly by couples, and they conſiſt of fiue leaues apeece, and theſe of a darke red colour. The leaues are large, ſoft, and hairy, diuided into ſix or ſeuen parts, and ſnipt about the edges ; the roots are large and laſting. It is kept with vs in gardens and floures in May. *Cluſius* calls it *Geranium 1. pullo flore.*

4　This alſo hath ſtalkes and leaues much like thoſe of the laſt deſcribed, but ſomewhat leſſe : the flours are as large as thoſe of the laſt deſcribed, but of a more light red, and they are contained in thicker and ſhorter cups, and ſucceeded by ſhorter ſeeds or bills, and are commonly of a ſweet muske-like ſmell : The root is very long, red, and laſting. It floures in the middeſt of May, and is

† 1 *Geranium Batrachioides.*
Crow-foot Cranes-bill.

2 *Geranium Batrachioides alternum.*
Small Crow-foot Cranes-bil.

‡ 3 *Geranium Batrachioides pullo flore.*
Duskie Cranes-bill.

‡ 4 *Geranium Batrachioides longius radicatum.*
Long rooted Cranes-bill.

called by *Gefner, Geranium montanum*:by *Dodonæus, Geranium batrachioides alterum*:and by *Lobel, Geranium batrachioides longius radicatum*. ‡

¶ *The Place.*

Thefe Cranes bils are wilde of their own nature, and grow in barren places,and in vallies rather than in mountaines;both of them do grow in my garden.

¶ *The Time.*

They floure,flourifh,and grow greene moft part of the **Summer.**

¶ *The Names.*

It is called in Greek, βατραχιώδης and *Geranium batrachioides*,which name it taketh from the likeneffe of Crow-foot:of fome it is called *Ranunculus cæruleus*, or blew Crow-foot:*Fuchfius* calleth it **Gottes Gnad**,that is in Latine,*Gratia Dei*:in Englifh alfo Gratia dei, blew Cranes bil,or Cranes bill with the blew floures,or blew Crowfoot Cranes bill.

¶ *The Temperature.*

The temperature is referred to the other Cranes bils.

¶ *The Vertues.*

None of thefe plants are now in vfe in Phyfick;yet *Fuchfius* faith that Cranes bill with the blew floure is an excellent thing to heale wounds.

CHAP. 361. *Of Candy Cranes bill.*

1 *Geranium Creticum.*
Candy Cranes bill.

2 *Geranium Malacoides.*
Baftard Candy Cranes bill.

¶ *The Defcription.*

1   THe Cranes bill of Candie hath many long tender ftalks,foft,and full of juice : diuiding it felfe into diuers branches,whereon are fet great broad leaues, cut,or jagged in diuers

sections or cuts: among which come forth floures composed of fiue leaues apiece, of a blewish or watchet colour, in the middle part whereof come forth a few chiues, and a small pointell of a purplish colour: the head and beake is like to the rest of the Cranes bills, but greater: the root dieth when it hath perfected his seed.

2　This Cranes-bill, being a bastard kinde of the former, hath long slender branches growing to the height of two or three cubits, set about with very great leaues, not vnlike to those of Holihocks, but somewhat lesser, of an ouerworne greene colour: among which rise vp little foot-stalks, on the ends whereof do grow small floures, lesser than those of the precedent, and of a murrey colour: the head and seeds are like also, but much lesser: the roots likewise die at the first approach of Winter.

¶ The Place.

These are strangers in England, except in the gardens of some Herbarists: they grow in my garden very plentifully.

¶ The Time.

The time answereth the rest of the Cranes-bils, yet doth that of Candy floure for the most part with me in May.

¶ The Names.

There is not more to be said of the names than hath beene remembred in their seuerall titles: they may be called in English, Cranes-bils, or Storkes-bils.

¶ The Nature.

Their temperature answereth that of Doues-foot.

¶ The Vertues.

Their faculties in working are equall to those of Doues-foot, and vsed for the same purposes, (and rightly) especially beeing vsed in wound drinkes, for the which it doth far excell any of the Cranes-bils, and is equall with any other herbe whatsoeuer for the same purpose.

CHAP. 362. Of diuers wilde Cranes-bills.

¶ The Kindes.

THere be diuers sorts or kindes of Cranes-bils which haue not been remembred of the antient, nor much spoken of by the later writers, all which I meane to comprehend vnder this chapter. making as it were of them a Chapter of wilde Cranes-bills, although some of them haue place in our London gardens, and that worthily, especially for the beautie of the floures: their names shall be expressed in their seuerall titles, their natures and faculties are referred to the other cranes-bils, or if you please to a further consideration.

¶ The Description.

1　SPotted Cranes-bill, or Storks-bill, the which *Lobel* describeth in the title thus, *Geranium Fuscum flore liuido purpurante, & medio Candicante*, whose leaues are like vnto Crowfoot (being a kind doubtlesse of Cranes-bil, called *Gratia Dei*) of an ouerworne dusty colour, and of a strong sauor, yet not altogether vnpleasant: the stalks are dry and brittle, at the tops whereof doe grow pleasant floures of a darke purple colour, the middle part of them tending to whitenesse: from the stile or pointell thereof, commeth forth a tuft of small purple hairy threds. The root is thicke and very brittle, lifting it selfe forth of the ground, insomuch that many of the said roots lieaboue the ground naked without earth, euen as the roots of Floure-de-luces do.

2　Of these wilde ones I haue another sort in my garden, which *Clusius* in his Pannonicke obseruations hath called *Geranium Hæmatoides*, or sanguine Cranes-bill: and *Lobel, Geranium Gruinum*, or *Gruinale*: it hath many flexible branches creeping vpon the ground: the leaues are much like vnto Doues foot in forme, but cut euen to the middle rib: the floures are like those of the small wilde mallow, and of the same bignesse, of a perfect bright red colour, which if they be suffered to

grow

1. *Geranium maculatum, siue fuscum.*
Spotted Cranes bill.

2. *Geranium sanguinarium.*
Bloudy Cranes bill.

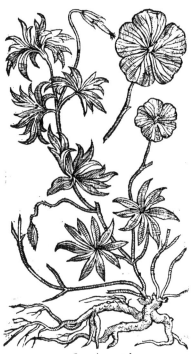

3. *Geranium Cicutæ folio inodorum.*
Vnsauorie field Cranes-bill.

5. *Geranium Violaceum.*
Violet Storks bill.

grow and stand vntill the next day wil be a murry colour;and if they stand vnto the third day,they will turn into a deep purple tending to blewnesse : their change is such, that you shall finde at one time vpon one branch flours like in forme,but of diuers colours. The root is thick,and of a woody substance.

3   This wilde kinde of musked Cranes bill,being altogether without sauor or smell, is called *Myrrhida inodorum*,or *Geranium aruense inodorum*,which hath many leaues spred flat on the ground, euery leafe made of diuers small leaues,and those cut or jagged about the edges,of no smel at all: amongst which rise vp slender branches,whereon grow small floures of a light purple colour. The root is long and fibrous.

4   This is also one of the wilde kindes of Cranes-bills, agreeing with the last described in ech respect,except the floures, for as the other hath purple floures, so this plant bringeth forth white floures,other difference there is none at all.

5   The Cranes-bill with violet coloured flours hath a thick woody root,with some few strings anexed thereto : from which rise immediatly forth of the ground diuers stiffe stalks,which diuide themselues into other small branches , whereupon are set confusedly broad leaues made of three leaues apiece,and those jagged or cut about the edges:the floures grow at the top of the branches, of a perfect violet colour,whereof it tooke his name : after come such beaks or bils as the other of his kinde.

‡   The figure that was put vnto this description was the same with *Geranium Robertianum*, and therefore I thought it not much amisse to put it here againe. ‡

6   I haue likewise another sort that was sent me from *Robinus* of Paris, whose figure was neuer set forth nor described of any : it bringeth forth a thick tough root with many branches of a brownish colour ; whereupon grow leaues not vnlike those of Gratia Dei, but not so deepely cut, somewhat cornered,and of a shining green colour : the floures grow at the top of the tender branches, composed of six small leaues of a bright skarlet colour.

¶ The Place.

The third and fourth of these Cranes bills grow of themselues about old walls, and about the borders of fields,woods,and copses ; and most of the rest we haue growing in our gardens.

¶ The Time.

Their time of flouring and seeding answers the rest of the Cranes bills.

¶ The Names.

Their seuerall titles shall serue for their names,referring what might haue bin said more to a farther consideration.

¶ The Nature and Vertues.

There hath nothing been found as yet either of their temperature or faculties,but may be referred to the other of their kinde.

CHAP. 363.

‡ Of certaine other Cranes-bills.

‡   1   THis,which *Clusius* receiued from Doctor *Thomas Penny* of London , and setteth forth by the same title you finde it heere expressed, hath a root consisting of sundry long and small bulbes, and which is fibrous towards the top : the stalk is a Cubit high, jointed, and red neere vnto the root and about the joints : out of each of these joynts come two leaues, which are fastened vnto somewhat long foot-stalks, and diuided into fiue parts, which also are snipt about the edges : out of each of which joints, by the setting on of the
foot-stalks

‡ 1 *Geranium bulboſum Pennæi.*
Pennies bulbous Cranes bill.

‡ 2 *Geranium nodoſum, Plateau.*
Knotty Cranes bill.

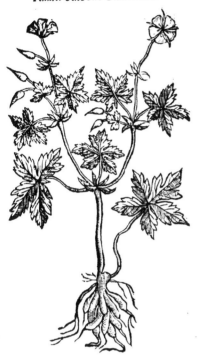

‡ 3 *Geranium argenteum Alpinum.*
Siluer leaued mountaine Cranes-bill.

foot-ſtalkes come forth fiue little ſharpe pointed leaues : the floures grow by couples vpon
the tops of the ſtalkes, and are of a reddiſh purple colour. It grows wilde in Denmark, whence
Dꝛ *Turner* brought it, and beſtowed it vpon Doctor *Penny* before mentioned.

2   This hath ſtalkes ſome foot high , jointed, and of a purpliſh colour ; vpon which grow
leaues diuided into three parts , but thoſe below are cut into fiue, and both the one and the
other are ſnipt about the edges : the floures are
compoſed of fiue reddiſh purple leaues of a
pretty largeneſſe, with a reddiſh pointall in the
middle, and falling, the ſeed followeth , as in
other plants of this kinde : the root is knotty,
and jointed with ſome fibres. It flours in May,
and continueth a great part of the Summer af-
ter. *Cluſius* calls this, *Geranium* 5 *nodoſum, Plateau.*
This is ſometimes found to carry tuberous excreſcences vpon the ſtalks, toward the later end
of Summer ; whence *Plateau* diſtinguiſhed it
from the other, but afterward found it to be the
ſame : and *Cluſius* figures and deſcribeth this later varietie alſo by the name of *Geranium* 6 *tu-
beriferum Plateau.*

3   The root of this is ſome 2 handfuls long,
black without, and white within, and toward the
top diuided into ſundry parts, whence put forth
leaues

leaues couered ouer with a fine filuer downe ; and they are diuided into fiue parts, each of which again is diuided into three others, which are faftned to long flender & round footftalks:the flours grow vpon footftalks fhorter than thofe of the leaues, and in colour and fhape are like thofe of the Veruaine Mallow, but much leffe; and after it is vaded there followes a fhort bill as in the other plants of this kinde. It floures in Iuly, and growes vpon the Alpes, where *Pona* found it, and fet it forth by the name of *Geranium Alpinum longius radicatum.*

4 The ftalks of this pretty Cranes bill are fome foot or better high, whereon grow leaues parted into fiue or fix parts like thofe of the *Geranium fufcum,* but of a lighter green colour:the floures are large, compofed of fiue thin and foon fading leaues of a whitifh colour, all ouer intermixt with fine veins of a reddifh colour, which adde a great deale of beauty to the floure; for thefe veines are very fmall, and curiously difperfed ouer the leaues of the floure : it floures in Iune, and is preferued in diuers of our gardens. Some call it *Geran. Romanum ftriatum :* in the *Hortus Eftettenfis* it is fet fo. th by the name of *Geranium Anglicum variegatum. Bauhine* calls it *Geranium batrachioides flore variegato.* We may call it variegated or ftriped Cranes bill.

5 There is of late brought into this kingdome, and to our knowledge, by the induftry of Mr. *Iohn Tradefcant,* another more rare and no leffe beautifull than any of the former; and hee had it by the name of *Geranium Indicum noctu odoratum :* this hath not as yet beene written of by any that I know ; therefore I will giue you the defcription thereof, but cannot as yet giue you the figure, becaufe I omitted the taking thereof the laft yeare, and it is not as yet come to his perfection. The leaues are larger, being almoft a foot long, compofed of fundry little leaues of an vnequall bignes, fet vpon a thick and ftif middle rib; and thefe leaues are much diuided or cut in, fo that the whole leafe fomewhat refembles that of *Tanacetum inodorum;* and they are thick, green, and fomwhat hairie : the ftalke is thicke , and fome cubit high; at the top of each branch vpon foot-ftalkes fome inch long grew fome eleuen or twelue floures , and each of thefe floures confifteth of fiue round pointed leaues, of a yellowifh colour, with a large blacke purple fpot in the middle of each leafe, as if it were painted, which giues the floure a great deale of beauty; and it alfo hath a good fmell. I did fee it in floure about the end of Iuly, 1632. being the firft time that it floured with the owner thereof. We may fitly call it Sweet Indian Storks bill; or painted Storks bill : and in Latine, *Geranium Indicum odoratum flore maculato.* ‡

CHAP. 364. *Of Sanicle.*

Sanicula, fiue Diapenfia. Sanicle.

¶ *The Defcription.*

S Anicle hath leaues of a blackifh greene colour, fmooth and fhining, fomewhat round, diuided into fiue parts like thofe of the Vine, or rather thofe of the Maple : among which rife vp flender ftalks of a brown colour, on the tops wherof ftand white moffie flours : in their places come vp round feed , rough, cleauing to mens garments as they paffe by, in manner of little burs : the root is black and full of threddy ftrings.

¶ *The Place.*

It groweth in fhadowie Woods and Copfes almoft euery where : it joyeth in a fat and fruitfull moift foile.

¶ *The Time.*

It floureth in May and Iune : the feed is ripe in Auguft : the leaues of the herbe are greene all the yeare, and are not hurt with the cold of winter.

¶ *The Names.*

It is commonly called *Sanicula :* of diuers, *Diapenfia.* In high and low Dutch it is named Sanikel : in French, *Sanicle :* in Englifh, Sanicle, or Sanikel : and it is fo called, *à fanandis vulneribus,* or of healing of wounds, as *Ruellius* faith : there be alfo
other

other Sanicles, so named of most Herbarists, as that which is described by the name of *Dentaria*, or Coral-wort, and likewise *Auricula vrsi*, or Beares eare, which is a kind of Cowslip; and likewise another set forth by the name of *Sanicula guttata*, whereof we haue intreated among the kindes of Beares eares.

¶ *The Temperature*

Sanicle as it is in taste bitter, with a certaine binding qualitie; so besides that it clenseth, and by the binding facultie strengtheneth, it is hot and dry, and that in the second degree, and after some Authors, hot in the third degree, and astringent.

¶ *The Vertues.*

The juice being inwardly taken is good to heale wounds.

The decoction of it also made in wine or water is giuen against spitting of bloud, and the bloudie flix: also foule and filthy vlcers be cured by being bathed therewith. The herbe boiled in water, and applied in manner of a pultesse, doth dissolue and wast away cold swellings: it is vsed in potions which are called Vulnerarie potions, or wound drinkes, which make whole and sound all inward wounds and outward hurts: it also helps the vlcerations of the kidnies, ruptures, or burstings.

CHAP. 365.   *Of Ladies Mantle, or great Sanicle.*

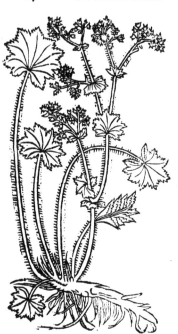

Alchimilla.
Lyons-foot, or Ladies Mantle.

¶ *The Description.*

LAdies mantle hath many round leaues, with fiue or six corners finely indented about the edges, which before they be opened are plaited and folded together, not vnlike to the leaues of Mallowes, but whiter, and more curled: among which rise vp tender stalks set with the like leaues but much lesser: on the tops whereof grow small mossie floures clustering thicke together, of a yellowish greene colour. The seed is small and yellow, inclosed in green husks. The root is thicke, and full of threddie strings.

¶ *The Place.*

It groweth of it selfe wilde in diuers places, as in the towne pastures of Andouer, and in many other places in Barkshire and Hampshire, in their pastures and copses, or low woods, and also vpon the banke of a mote that incloseth a house in Bushey called Bourn hal, fourteene miles from London, and in the high-way from thence to Watford, a small mile distant from it.

¶ *The Time.*

It floureth in May and Iune: it flourisheth in Winter, as well as in Summer.

¶ *The Names.*

It is called of the later Herbarists *Alchimilla*: and of most, *Stellaria, Pes Leonis, Pata Leonis*, and *Sanicula major*: in high-Dutch, **Synnauw**, and **Onser frauwen mantel**: in French, *Pied de Lion*: in English, Ladies Mantle, great Sanicle, Lyons foot, Lyons paw; and of some, Padelyon.

¶ *The Temperature.*

Ladies mantle is like in temperature to little Sanicle, yet is it more drying and more binding.

¶ *The Vertues.*

It is applied to wounds after the same manner that the smaller Sanicle is, being of like efficacy: it stoppeth bleeding, and also the ouermuch flowing of the naturall sicknesse: it keeps downe maidens paps or dugs, and when they be too great or flaggy it maketh them lesser or harder.

CHAP. 366. *Of Neeſe-woort Sanicle.*

Elleborine Alpina.
Neeſe wort Saniċle.

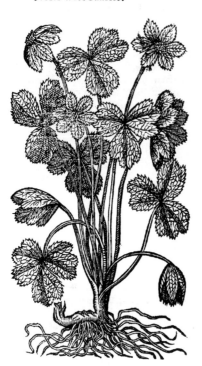

¶ *The Deſcription.*

WHen I made mention of *Helleborus albus*, I did alſo ſet downe my cenſure concerning *Elleborine* or *Epipactis* : but this *Elleborine* of the Alpes I put in this place, becauſe it approcheth neerer vnto Sanicle and *Ranunculus*, as participating of both: it groweth in the mountaines and higheſt parts of the Alpiſh hills, and is a ſtranger as yet in our Engliſh gardens. The root is compact of many ſmall twiſted ſtrings like blacke Hellebor : from thence ariſe ſmall tender ſtalkes, ſmooth, and eaſie to bend; in whoſe tops grow leaues with fiue diuiſions, ſomewhat nickt about the edges like vnto Sanicle : the floures conſiſt of ſix leaues ſomwhat ſhining, in taſte ſharpe, yet not vnpleaſant. This is the plant which *Pena* found in the forreſt of Eſens, not far from Iupiters mount, and ſets forth by the name of *Alpina Elleborine Sanitula & Ellebori nigri facie.*

¶ *The Temperature and Vertues.*

I haue not as yet found any thing of his nature or vertues.

CHAP. 367. *Of Crow-feet.*

¶ *The Kindes.*

THere be diuers ſorts or kinds of theſe pernitious herbes comprehended vnder the name of *Ranunculus*, or Crowfoot, whereof moſt are very dangerous to be taken into the body, and therefore they require a very exquiſite moderation, with a moſt exact and due maner of tempering, not any of them are to be taken alone by themſelues, becauſe they are of moſt violent force, and therefore haue the greater need of correction.

The knowledge of this herbe is as neceſſarie to the Phyſitian as of other herbes, to the end they may ſhun the ſame, as *Scribonius Largus* ſaith, and not take them ignorantly : or alſo, if neceſſitie at any time require, that they vſe them, and that with ſome deliberation and ſpeciall choice, and with their proper correctiues. For theſe dangerous ſimples are likewiſe many times of themſelues beneficial, & oftentimes profitable: for ſome of them are not ſo daungerous, but that they may in ſome ſort, and oftentimes in fit and due ſeaſon profit and doe good, if temperature and moderation bee vſed: of this there be foure kindes, as *Dioſcorides* writeth ; one with broad leaues, another that is downy, the third very ſmall, and the fourth with a white floure: the later Herbariſts alſo haue obſerued many moe: all theſe may be brought into two principall kinds, ſo that one be a garden or tame one, and the other wild; and of theſe ſome are common, and others rare, or forreign. Moreouer, there is a difference both in the roots and in the leaues ; for one hath a bumped or knobby root, another a long leafe as Speare-woort: and firſt of the wilde or field Crow-feet, referring the reader vnto the end of the ſtock and kindred of the ſame, for the temperature and vertues.

1 *Ranunculus pratensis, etiamque hortensis.*
Common Crow-foot.

2 *Ranunculus surrectis cauliculis.*
Right Crow-foot.

3 *Ranunculus aruorum.*
Crow-foot of the fallowed field.

4 *Ranunculus Alpinus albus.*
White mountaine Crow-foot.

¶ The

¶ *The Deſcription.*

1　THe common Crow-foot hath leaues diuided into many parts, commonly three, ſometimes fiue, cut here and there in the edges, of a deep green colour, in which ſtand diuers
white ſpots: the ſtalks be round, ſomthing hairie, ſome of them bow downe toward the
ground, and put forth many little roots, whereby it taketh hold of the ground as it traileth along:
ſome of them ſtand vpright, a foot high or higher, on the tops whereof grow ſmall ſloures with fiue
leaues apiece, of a yellow glittering colour like gold: in the middle part of theſe floures ſtand certaine ſmall threads of like colour: which being paſt, the ſeeds follow, made vp in a rough ball: the
roots are white and thready.

2　The ſecond kind of Crow-foot is like vnto the precedent, ſauing that his leaues are fatter,
thicker, and greener, and his ſmall twiggy ſtalks ſtand vpright, otherwiſe it is like: of which kinde
it chanced, that walking in the field next to the Theatre by London, in the company of a worſhipfull Merchant named Mr. *Nicholas Lete*, I found one of this kind there with double floures, which
before that time I had not ſeene.

¶ *The Place.*

They grow of themſelues in paſtures and medowes almoſt euerie where.

¶ *The Time.*

They floure in May and many moneths after.

¶ *The Names.*

Crow-foot is called of *Lobel, Ranunculus prateꝼſis* : of *Dodonæus, Ranunculus hortenſis*, but vnproperly: of *Pliny, Polyanthemum*, which he ſaith diuers name *Batrachion* : in high-Dutch, **Schmalk-**
bluom : in low-Dutch, **Boter bloemen:** in Engliſh, King Kob, Gold cups, Gold knobs, Crow-foot,
and Butter-floures.

¶ *The Deſcription.*

3　The third kinde of Crow-foot, called in Latine *Ranunculus aruorum*, becauſe it growes commonly in fallow fields where corne hath beene lately ſowne, and may be called Corne Crow-foot,
hath for the moſt part an vpright ſtalke of a foot high, which diuides it ſelfe into other branches:
whereon do grow fat thick leaues very much cut or jagged, reſembling the leaues of Sampire, but
nothing ſo green, but rather of an ouerworne colour. The floures grow at the top of the branches,
compact of fiue ſmall leaues of a faint yellow colour: after which come in place cluſters of rough
and ſharp pointed ſeeds. The root is ſmall and thready.

4　The fourth Crow-foot, which is called *Ranunculus Alpinus*, becauſe thoſe that haue firſt written thereof haue not found it elſewhere but vpon the Alpiſh mountains (notwithſtanding it groweth in England plentifully wilde, eſpecially in a wood called Hampſted Wood, and is planted in
gardens) hath diuers great fat branches two cubits high, ſet with large leaues like the common
Crow-foot, but greater, of a deepe greene colour, much like to thoſe of the yellow Aconite, called
Aconitum luteum Ponticum. The floures conſiſt of fiue white leaues, with ſmal yellow chiues in the
middle, ſmelling like the floures of May or Haw-thorne, but more pleaſant. The roots are greater
than any of the ſtocke of Crow-feet.

¶ *The Place and Time.*

Their place of growing is touched in their deſcription: their time of flouring and ſeeding anſwereth the other of their kindes.

¶ *The Names.*

The white Crow-foot of the Alpes and French mountaines is the fourth of *Dioſcorides* his deſcription; for he deſcribeth his fourth to haue a white floure: more hath not bin ſaid touching the
names, yet *Tabern.* calls it *Batrachium album* : in Engliſh, white Crow-foot.

¶ *The Deſcription.*

5　Among the wilde Crow-feet there is one that is ſyrnamed *Illyricus*, which brings forth ſlender ſtalks, round, and of a meane length: wherupon do grow long narrow leaues cut into many long
gaſhes, ſomthing white, and couered with a certain downineſſe: the floures be of a pale yellow colour: the root conſiſteth of many ſmall bumpes as it were graines of corne, or little long bulbes
growing cloſe together like thoſe of Pilewort. It is reported, that it was firſt brought out of Illyria into Italy, and from thence into the Low-countries: notwithſtanding we haue it growing very
common in England. ‡ But onely in gardens that I haue ſeene. ‡

6　The ſixth kinde of Crow-foot, called *Ranunculus bulboſus*, or Onion rooted Crow-foot, and
round rooted Crow-foot, hath a round knobby or onion-faſhioned root, like vnto a ſmall Turnep,
and of the bigneſſe of a great Oliue: from the which riſes vp many leaues ſpred vpon the ground,
like thoſe of the field Crow-foot, but ſmaller, and of an ouerworne greene colour: amongſt which
riſe vp ſlender ſtalks of the height of a foot: whereupon do grow floures of a faint yellow colour.
‡ This growes wilde in moſt places, and floures at the beginning of May. ‡

¶ *The*

¶ *The Place.*

It is alſo reported to be found not only in Illyria and Sclauonia, but alſo in the Iſland Sardinia, ſtanding in the midland, or Mediterranean ſea.

¶ *The Names.*

This Illyrian Crow-foot is named in Greeke ʍʍθϵω, that is, *Apium ſylueſtre*, or wilde Smallage, alſo *Herba Sardoa*: it may be, ſaith my Author, that kinde of Crow-foot called *Apium riſus*, anʊ γɩκατʊ· ʍ; and this is thought to be that *Gelotophyllis*, of which *Pliny* maketh mention in his 24 booke, 17 chap. which being drunke, ſaith he, whith wine and myrrhe, cauſeth a man to ſee diuers ſtrange ſights, and not to ceaſe laughing till he hath drunke Pine apple kernells with Pepper in wine of the Date-tree, (I thinke he would haue ſaid vntill he be dead) becauſe the nature of laughing Crow-foot is thought to kill laughing, but without doubt the thing is cleane contrary; for it cauſeth ſuch convulſions, cramps, and wringings of the mouth and jawes, that it hath ſeemed to ſome that the parties haue died laughing, whereas in truth they haue died in great torment.

5 *Ranunculus Illyricus.*   6 *Ranunculus bulboſus.*
Crow-foot of Illyria.   Rround rooted Crow-foot.

¶ *The Deſcription.*

7   The ſeuenth kinde of Crow-foot, called *Auricomus* of the golden lockes wherewith the floure is thrummed, hath for his root a great buſh of blackiſh hairy ſtrings; from which ſhoot forth ſmall jagged leaues, not much vnlike to Sanicle, but diuided onely into three parts, yet ſometimes into fiue; among which riſe vp branched ſtalkes of a foot high, whereon are placed the like leaues but ſmaller, ſet about the top of the ſtalkes, whereon do grow yellow floures, ſweet ſmelling, of which it hath beene called *Ranunculus dulcis, Tragi,* or *Tragus* his ſweet Crowfoot. ‡ It growes in medowes and about the ſides of woods, and floures in Aprill. ‡

† 8   Frogge Crow foot, called of *Pena, Aconitum Batrachioides*: of *Dodonæus, Batrachion Apulei,* is that formerly deſcribed in the fourth place, whereto this is much alike, but that the ſtalkes and leaues are larger, as alſo the floures, which are white: the root is tough and threddy.

9   The ninth Crow-foot hath many graſſie leaues, of a deepe greene tending to blewneſſe, ſomewhat long, narrow, and ſmooth, very like vnto thoſe of the ſmall Biſtort, or Snake-weed:

7 *Ranunculus auricomus.*
Golden-haired Crow-foot.

† 8 *Ranunculus Aconiti folio.*
Frog Crow-foot.

9 *Ranunculus gramineus Lovelÿ.*
Graſſie Crow-foot.

10 *Ranunculus Aatumnalis Cluſÿ.*
Winter Crow-foot.

among which rise vp slender stalkes, bearing at the top small yellow floures like the other Crow-feet : the root is small and threddy. ‡ There is a variety of this hauing double floures;and I haue giuen you the figure thereof in stead of the single that was formerly in this place. ‡

10   The Autumne or Winter Crow-foot hath diuers broad leaues spred vpon the ground,snipt about the edges, of a bright shining greene colour on the vpper side, and hoary vnderneath, full of ribs or sinewes as are those of Plantaine, of an vnpleasant taste at the first, afterward nipping the tongue : among which leaues rise vp sundry tender foot-stalkes, on the tops whereof stand yellow floures consisting of six small floures apiece : after which succeed little knaps of seed like to a dry or withered straw-berry. The root is compact of a number of limber roots, rudely thrust together in manner of the Asphodill.

11   The Portugall Crow-foot hath many thicke cloggy roots fastned vnto one head,very like those of the yellow Asphodill : from which rise vp three leaues,seldome more, broad, thicke, and puffed vp in diuers places, as if it were a thing that were blistered,by meanes whereof it is very vn-euen. From the middle of which leaues riseth vp a naked stalke,thicke, fat,very tender,but yet fra-gile, or easie to breake : on the end whereof standeth a faire single yellow floure, hauing in the mid-dle a naked rundle of a gold yellow tending to a Saffron colour.

1 1   *Ranunculus Lusitanicus Clusij.*   1 2   *Ranunculus globosus.*
      Portugall Crow-foot.   Locker Gowlons,or Globe Crow-foot.

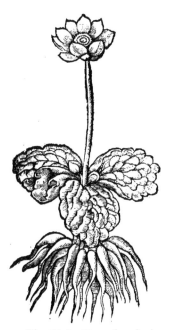 

12   The Globe Crow-foot hath very many leaues deepely cut or jagged, of a bright greene colour like those of the field Crow-foot : among which riseth vp a stalke, diuided towards the top into other branches, furnished with the like leaues of those next the ground, but smaller : on the tops of which branches grow very faire yellow floures, consisting of a few leaues folded or rolled vp together like a round ball or globe : whereupon it was called *Ranunculus globosus*, or the Globe Crow-foot, or Globe-floure : which being past, there succeed round knaps,wherein is a blackish seed. The root is small and threddy.

‡ 13   This hath large leaues like those of the last described,but rough and hairy : the stalke is some foot high : the floures are pretty large, composed of fiue white sharpish pointed leaues. It floures in Iuly,and growes in the Alps : it is the *Ranunculi montani alteri 2. species* of *Clusius.*

14   This other hath leaues not vnlike those of the precedent, and such stalkes also ; but the floures consist of 5 round leaues,purplish beneath; the edges of the vpper side are of a whitish pur-ple,and the residue wholly white,with many yellow threds in the middle : it grows in the mountain

‡ 13 *Ranunculus hirſutus Alpinus flo.albo.*
Rough white floured mountaine Crow-foot.

‡ 14 *Ranunculus montanus hirſutus purpureus.*
Rough purple floured mountain Crow-foot.

Iura, againſt the city of Geneua, whereas it floures in Iune, and ripens the ſeed in Auguſt. *Cluſius* had the figure and deſcription hereof from Dr.*Penny* and he calls it *Ranunculus montanus* 3. ‡

¶ *The Place.*

The twelfth kind of Crow-foot groweth in moſt places of York-ſhire and Lancaſhire, and other bordering ſhires of the North countrey, almoſt in euery medow, but not found wilde in theſe Southerly or Weſterly parts of England that euer I could vnderſtand of.

¶ *The Time.*

It floureth in May and Iune : the ſeed is ripe in Auguſt.

¶ *The Names.*

The Globe-floure is called generally *Ranunculus globoſus* : of ſome, *Flos Trollius,* and *Ranunculus Alpinus* : in Engliſh, Globe Crow-foot, Troll floures, and Lockron gowlons.

CHAP. 368.   *Of Double yellow and white Batchelors Buttons.*

¶ *The Deſcription.*

1   THe great double Crow-foot, or Batchelors button hath many jagged leaues of a deepe greene colour : among which riſe vp ſtalkes, whereon do grow faire yellow floures exceeding double, of a ſhining yellow colour, oftentimes thruſting forth of the middeſt of the ſaid floures one other ſmaller floure : the root is round, or faſhioned like a Turnep; the form whereof hath cauſed it to be called of ſome S.Anthonies Turnep, or Rape Crow-foot. The ſeed is wrapped in a cluſter of rough knobs, as are moſt of the Crow-feet.

2   The double yellow Crow-foot hath leaues of a bright greene colour, with many weake branches trailing vpon the ground ; whereon do grow very double yellow floures like vnto the precedent but altogether leſſer. The whole plant is likewiſe without any manifeſt difference, ſauing that theſe floures doe neuer bring forth any ſmaller floure out of the middle of the greater, as the other doth, and alſo hath no Turnep or knobby root at all, wherein conſiſts the greateſt difference.

3 The

† 1 *Ranunculus maximus Anglicus.*
Double Crow-foot,or Batchelors buttons.

2 *Ranunculus dulcis multiplex.*
Double wilde Crow-foot.

3 *Ranunculus albus multiflorus.*
Double white Crow-foot.

3   The white double Crow-foot hath many
great leaues deeply cut with great gaſhes,and thoſe
ſnipt about the edges.The ſtalks diuide themſelues
into diuers brittle branches,on the tops whereof do
grow very double floures as white as ſnow, and of
the bigneſſe of our yellow Batchelors button. The
root is tough, limber, and diſperceth it ſelfe farre a-
broad,whereby it greatly increaſeth.

¶ *The Place.*

The firſt and third are planted in gardens for the
beauty of the floures,and likewiſe the ſecond,which
hath of late beene brought out of Lancaſhire vnto
our London gardens, by a curious gentleman in the
ſearching forth of Simples, Mr.*Thomas Hesketh*,who
found it growing wilde in the towne fields of a ſmal
village called Hesketh, not farre from Latham in
Lancaſhire.

¶ *The Time.*

They floure from the beginning of May to the
end of Iune.

¶ *The Names.*

Dioſcorides hath made no mention thereof;but *A-
puleius* hath ſeparated the firſt of theſe from the
others,intreating of it apart, and naming it by a peculiar name *Batrachion*; whereupon it is alſo called
Apuleij Batrachion,or *Apuleius* Crow-foot.

It is commonly called *Rapum D.Anthonij*,or Saint
Anthonies Rape:it may be called in Engliſh, Rape
Crow-foot: it is called generally about London,
Batchelors buttons , and double Crowfoot : in

                       Dutch,

Dutch, **S.Anthony Rapkin.** ‡ These names and faculties properly belong to the *Ranunculus bulbosus*, described in the sixt place of the last chapter; and also to the first double one here described; for they vary little but in colour, and the singlenesse and doublenesse of their floures. ‡

The third is called of *Lobel Ranunculus niueus Polyanthos* : of *Tabern.Ranunculus albus multiflorus* : in English, Double white Crow-foot, or Batchelors buttons.

¶ *The Temperature.*

These plants do bite as the other Crow-feet do.

¶ *The Vertues.*

A   The chiefest vertue is in the root, which being stamped with salt is good for those that haue a plague sore, if it be presently in the beginning tied to the thigh, in the middle betweene the groine or flanke and the knee : by meanes whereof the poyson and malignitie of the disease is drawn from the inward parts, by the emunctory or clensing place of the flanke, into those outward parts of lesse account : for it exulcerateth and presently raiseth a blister, to what part of the body soeuer it is applied. And if it chance that the sore hapneth vnder the arme, then it is requisite to apply it to the arme a little aboue the elbow. My opinion is, that any of the Crow-feet will do the same : my reason is, because they all and euery of them do blister and cause paine; wheresoeuer they be applied, and paine doth draw vnto it selfe more paine; for the nature of paine is to resort vnto the weakest place, and where it may finde paine; and likewise the poyson and venomous quality of that disease is to resort vnto that painefull place.

B   *Apuleius* saith further, That if it be hanged in a linnen cloath about the necke of him that is lunaticke, in the waine of the Moone, when the signe shall be in the first degree of *Taurus* or *Scorpio*, that then he shall forthwith be cured. Moreouer, the herbe *Batrachion* stamped with vinegre, root and all, is vsed for them that haue blacke skars or such like marks on their skins, it eats them out, and leaues a colour like that of the body.

† The figure that formerly was in the first place of this chapter was the double one mentioned in the second description of the foregoing chapter, where also you may finde a double floure exprest by the side of the figure.

CHAP. 369.   *Of Turkie or Asian Crow-foot.*

1 *Ranunculus sanguineus multiplex.*
The double red Crow-foot.

‡ 2 *Ranunculus Asiaticus flo.pleno miniato.*
The double Asian skarlet Crow-foot.

‡ 3 *Ranunculus Asiaticus flore pleno prolifero.*
The double buttoned scarlet Asian Crow-foot.

4 *Ranunculus Tripolitanus.*
Crow-foot of Tripolie.

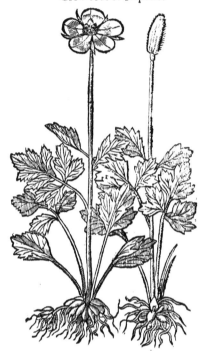

‡ 5 *Ranunculus grumosa radice ramosus.*
Branched red Asian Crowfoot.

‡ 6 *Ranunculus Asiaticus grumosa radice flo. albo.*
White floured Asian Crowfoot.

‡ 7 *Ranunculus Aſiaticus grumoſa radice flore flauo vario.*
Aſian Crow-foot with yellow ſtriped floures.

¶ *The Deſcription.*

1 THe double red **Crow**-foot hath a few leaues riſing immediatly forth of the ground, cut in the edges with deepe gaſhes, ſomewhat hollow and of a bright ſhining green colour. The ſtalke riſeth vp to the height of a foot, ſmooth and very brittle, diuiding it ſelfe into other branches, ſometimes two, ſeldome three: whereon doe grow leaues confuſedly ſet without order: the floures grow at the top of the ſtalks, very double, and of great beautie, of a perfect ſcarlet colour, tending to redneſſe. The root is compact of many long tough roots, like thoſe of the yellow Aſphodill.

‡ 2 Of this kinde there is alſo another, or other the ſame better expreſt; for *Cluſius* the Author of theſe neuer ſee the former, but makes it onely to differ, in that the floures are of a ſanguine colour, and thoſe of this of a kind of ſcarlet, or red lead colour.

3 This differs nothing from the former, but that it ſends vp another floure ſomwhat leſſer, out of the middle of the firſt floure, which happens by the ſtrength of the root, and goodneſſe of the ſoile where it is planted. ‡

4 The Crow-foot of Tripolis or the ſingle red *Ranunculus* hath leaues at the firſt comming vp like vnto thoſe of Groundſwell: among which riſeth vp a ſtalke of the height of halfe a cubit, ſomwhat hairy, wheron grow broad leaues deeply cut, euen to the middle rib, like thoſe of the common Crowfoot, but greener: the floure groweth at the top of the ſtalke, conſiſting of fiue leaues, on the outſide of a darke ouerworne red colour, on the inſide of a red lead colour, bright and ſhining, in ſhape like the wilde corne Poppy: the knop or ſtile in the middle which containeth the ſeed is garniſhed or bedeckt with very many ſmall purple thrummes tending to blackeneſſe: the root is as it were a roundell of little bulbes or graines like thoſe of the ſmall Celandine or Pilewort.

‡ 5 There be diuers other Aſian Crow-feet which *Cluſius* hath ſet forth, and which grow in the moſt part in the Gardens of our prime Floriſts, and they differ little in their roots, ſtalkes, or leaues, but chiefely in the floures; wherefore I will onely briefely note their differences, not thinking it pertinent to ſtand vpon whole deſcriptions, vnleſſe they were more neceſſarie: this fifth differs from the fourth in that the ſtalkes are diuided into ſundry branches, which beare like, but leſſe floures than thoſe which ſtand vpon the main ſtalke: the colour of theſe differs not from that of the laſt deſcribed.

6 This is like the laſt deſcribed, but the floures are of a pure white colour, and ſometimes haue a few ſtreaks of red about their edges.

7 This in ſtalkes and manner of growing is like the ′precedent: the ſtalke ſeldome parting it ſelfe into branches; but on the top thereof it carries a faire floure conſiſting commonly of round topped leaues of a greeniſh yellow colour, with diuers red veines here and there diſperſed and running alongſt the leaues, with ſome purple thrums, and a head ſtanding vp in the middle as in the former. ‡

¶ *The Place.*

The firſt groweth naturally in and about Conſtantinople, and in Aſia on the further ſide of Boſphorus, from whence there hath beene brought plants at diuers times, and by diuers perſons, but they haue periſhed by reaſon of their long journey, and want of skill of thoſe bringers, that haue ſuffered them to lie in a box or ſuch like ſo long, that when we haue receiued them they haue beene as dry as ginger; notwithſtanding *Cluſius* ſaith he receiued a plant freſh and greene, the which a domeſtical theefe ſtole forth of his garden. My Lord and Maſter the right Honorable the Lord Treaſurer

surer had diuers plants sent him from thence which were drie before they came, as aforesaid. The other groweth in Aleppo and Tripolis in Syria naturally, from whence we haue receiued plants for our gardens, where they flourish as in their owne country.

¶ *The Time.*

They bring forth their pleasant floures in May and Iune, the seed is ripe in August.

¶ *The Names.*

The first is called *Ranunculus Constantinopolitanus* : of *Lobel*, *Ranunculus sanguineus multiplex*, *Ranunculus Bizantinus, siue Asiaticus :* in the Turkish tongue, *Torobolos, Catamer laile :* in English, the double red Ranunculus or Crow foot.

The fourth is called *Ranunculus Tripolitanus*, of the place from whence it was first brought into these parts : of the Turks, *Taroboles Catamer*, without that addition *laile :* which is a proper word to all floures that are double.

¶ *The Temperature and Vertues.*

Their temperature and vertues are referred to the other Crow-feet, whereof they are thought to be kindes.

CHAP. 370. *Of Speare-woort, or Bane-wort.*

¶ *The Description.*

1   SPeare-wort hath an hollow stalke full of knees or joynts, whereon do grow long leaues, a little hairy, not vnlike those of the willow, of a shining green colour: the floures are very large, and grow at the tops of the stalks, consisting of fiue leaues of a faire yellow colour, very like to the field gold cup, or wilde Crow-foot : after which come round knops or seed vessels, wherein is the seed : the root is compact of diuers bulbes or long clogs, mixed with an infinite number of hairy threds.

1 *Ranunculus flammeus maior.*
Great Speare-wort.

2 *Ranunculus flammeus minor.*
The lesser Speare-wort.

2 The

2   The common Spearewort being that which we haue called the leſſer, hath leaues, floures, and ſtalkes like the precedent, but altogether leſſer: the root conſiſteth of an infinite number of threddy ſtrings.

3   Iagged Speare-wort hath a thicke, fat, hollow ſtalke, diuiding it ſelfe into diuers branches, whereon are ſet ſomtimes by couples two long leaues, ſharpe pointed, and cut about the edges like the teeth of a ſaw. The floures grow at the top of the branches, of a yellow colour, in form like thoſe the field Crow-foot: the root conſiſteth of a number of hairy ſtrings.

3 Ranunculus flammeus ſerratus.
Iagged Speare-wort.

4 Ranunculus paluſtris rotundi folius.
Mariſh Crow-foot, or Speare-wort.

4   Marſh Crow-foot, or Speare-wort (whereof it is a kinde, taken of the beſt approued authors to be the true *Apium riſus*, though diuers thinke that *Pulſatilla* is the ſame: of ſome it is called *Apium hæmorrhoidarum*) riſeth forth of the mud or wateriſh mire from a threddy root, to the height of a cubit, ſometimes higher. The ſtalke diuideth it ſelfe into ſundry branches, whereupon doe grow leaues deeply cut round about like thoſe of Doues-foot, and not vnlike to the cut Mallow, but ſomwhat greater, and of a moſt bright ſhining greene colour: the floures grow at the top of the branches, of a yellow colour, like vnto the other water Crow-feet.

¶ *The Place.*

They grow in moiſt and dankiſh places, in brinkes or water courſes, and ſuch like places almoſt euery where.

¶ *The Time.*

They floure in May when other Crowfeet do.

¶ *The Names.*

Speare-wort is called of the later Herbariſts *Flammula*, and *Ranunculus Flammeus*; of *Cordus*, *Ranunculus* παχύυπιε, or broad leaued Crow-foot: of others, *Ranunculus longifolius*, or long leafed Crowfoot: in low Dutch, 𝕰𝕘𝕖𝕝𝕔𝕠𝕠𝕝𝕖𝕟: in Engliſh, Speare Crow-foot, Speare-wort, and Banewort, becauſe it is dangerous and deadly for ſheepe; and that if they feed of the ſame it inflameth their liuers, fretteth and bliſtereth their guts and intrails.

¶ *The Temperature of all the Crow-feet.*

Speare-wort is like to the other Crow-feet in faculty, it is hot in the mouth or biting, it exulcerateth

cerateth and raifeth blifters, and being taken inwardly it killeth remedileffe. Generally all the Crow-feet, as *Galen* faith, are of a very fharpe and biting quality, infomuch as they raife blifters with paine : and they are hot and dry in the fourth degree.

¶ *The Vertues of all the Crow-feet.*

The leaues or roots of **Crow-feet** ftamped and applied to any part of the body, coufeth the skin A to fwell and blifter, and raifeth vp wheales, bladders, caufeth skars, crufts, and ouglie vlcers : it is laid vpon cragged warts, corrupt nailes, and fuch like excrefcenfes, to caufe them to fall away.

The leaues ftamped and applied vnto any peftilentiall or plague fore, or carbuncle, ftaieth the B fpreading nature of the fame, and caufeth the venomous or peftilentiall matter to breath forth, by opening the parts and paffages in the skin.

It preuaileth much to draw a plague fore from the inward parts, being of danger, vnto other re- C mote places further from the heart, and other of the fpirituall parts, as harh beene declared in the defcription.

Many do vfe to tie a little of the herbe ftamped with falt vnto any of the fingers, againft the pain D of the teeth; which medicine feldome faileth ; for it caufeth greater paine in the finger than was in the tooth, by the meanes whereof the greater paine taketh away the leffer.

Cunning beggers do vfe to ftampe the leaues, and lay it vnto their legs and arms, which caufeth E fuch filthy vlcers as we dayly fee (among fuch wicked vagabonds) to moue the people the more to pittie.

The kinde of Crow-foot of Illyria, being taken to be *Apium rifus* of fome, yet others thinke *Aco-* F *nitum Batrachioides* to be it. This plant fpoileth the fences and vnderftanding, and draweth together the finewes and mufcles of the face in fuch ftrange manner, that thofe who beholding fuch as died by the taking hereof, haue fuppofed that they died laughing ; fo forceably hath it drawne and con- tracted the nerues and finewes, that their faces haue beene drawne awry, as though they laughed, whereas contrariwife they haue died with great torment.

‡ CHAP. 371. *Of diuers others Crow-feet.*

‡ 1 *Ranunculus Creticus latifolius.*
Broad leaued Candy Crow-foot.

‡ 2 *Ranunculus folio Plantaginis.*
Plantaine leaued Crow-foot.

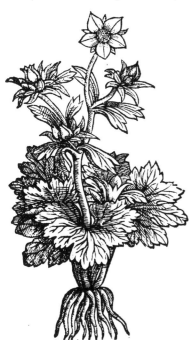 

¶ *The Description.*

‡ 1   THe roots of this are somewhat like those of the Asian *Ranunculus*: the leaues are very
large and roundish, of a light greene colour, cut about the edges, and here and there
deeply diuided : the stalke is thicke, round, and stiffe, diuided into two or three branches ; at the setting on of which grow longish leaues a little nickt about the end : the floures are of
an indifferent bignesse, and consist of fiue longish round pointed leaues, standing a little each from
other, so that the green points of the cups shew themselues between them : there are yellow threds
in the middle of these floures, which commonly shew themselues in February, or March. It is found
only in some gardens, and *Clusius* only hath set it forth by the name we here giue you.

2   This also that came from the Pyrenæan hills is made a Denizen in our gardens : it hath a
stalke some foot high, set with neruous leaues, like those of Plantaine, but thinner, and of the colour
of Woad, and they are something broad at their setting on, and end in a sharpe point : at the top of
the stalke grow the floures ; each consisting of fiue round slender pure white leaues, of a reasonable
bignesse, with yellowish threds and a little head in the middle : the root is white and fibrous. It
floures about the beginning of May. *Clusius* set forth this by the title of *Ranunculus Pyrenæus albo
flore.*

3   The same Author hath also giuen vs the knowledge of diuers other plants of this kinde, and
this hee calls *Ranunculus montanus* 1. It hath many round leaues, here and there deepely cut in, and
snipt about the edges, of a darke greene colour, and shining, pretty thicke, and of a very hot taste :
amongst which riseth vp a slender, single, and short stalke, bearing a white floure made of fiue little
leaues with a yellowish thrum in the middle : which falling, the seeds grow clustering together as
in other plants of this kinde : the root is white and fibrous.

‡ 3 *Ranunculus montanus flo. minore.*   ‡ 4 *Ranunculus montanus flore majore.*
Mountain Crow-foot with the lesser floure,   Mountain Crow-foot with the bigger floure.

4   This also is nothing else but a variety of the last described, and differs from it in that the
floures are larger, and it is sometimes found with them double. Both these grow on the tops of the
Alpes, and there they floure as soone as the snow is melted away, which is vsually in Iune : but
brought into gardens they floure very early, to wit, in Aprill.

5   The leaues of this are cut or diuided into many parts, like those of Rue, but softer, & greener
(whence *Clusius* names it *Ranunculus Ruta folio*) or not much vnlike those of Coriander (whereupon

‡ 5 *Ranunculus præcox rutacco folio.*
Rue leaued Crow-foot.

‡ 6 *Ranunculus præcox Thalictri folio.*
Columbine Crow-foot.

‡ 7 *Ranunculus parvus echinatus.*
Small rough leaued Crow-foot.

Pona calls it *Ranunculus Coriandri folio:*) amongſt, or
rather before theſe, comes vp a ſtalke ſome handfull high, bearing at the top thereof one floure of
a reaſonable bigneſſe, on the outſide before it be
throughly open of a pleaſing red colour, but white
within, compoſed of twelue or more leaues.

6   This hath a ſtalke ſome foot high, ſmal and
reddiſh, whereon grow ſundry leaues like thoſe of
the greater *Thalictrum*, or thoſe of Columbines,
but much leſſe, and of a bitter taſte: out of the boſomes of theſe leaues come the floures, at each
ſpace one, white, and conſiſting of fiue leaues a-
piece : which falling, there ſucceed two or three
little horns containing a round reddiſh ſeed. The
root is fibrous, white, very bitter, and creeps heere
and there, putting vp new ſhoots. It growes in diuers woods of Auſtria, and floures in Aprill, and
the ſeed is ripe in May or Iune. *Cluſius* calls it *Ranunculus præcox* 2. *Thalietri folio*. It is the *Aquilegia
minor Daleſchampÿ*, in the *Hiſt. Lugd*.

7   This, which (as *Cluſius* ſaith) ſome call the
Ranunculus of *Apuleius*, hath alſo a fibrous root,
with ſmall leaues diuided into three parts, & cut
about the edges, and they grow vpon ſhort foot-
ſtalks: the ſtalks are ſome two handfuls high, commonly leaning on the ground, and on them grow
ſuch leaues as the former; and out of their boſoms
come little foot-ſtalkes bearing floures of a pale
yellow colour, made of fiue leaues apiece, which

M m m m   fallen,

fallen, there ſucceed fiue or ſix ſharpe pointed rough cods, containing ſeed almoſt like that of the former. ‡

Chap. 372: Of Wolfes-bane.

¶ *The Kindes.*

THere be diuers ſorts of Wolfs-banes, whereof ſome bring forth flours of a yellow color, others of a blew or tending to purple: among the yellow ones there are ſome greater, others leſſer; ſome with broader leaues, and others with narrower.

1 *Thora Valdenſis.*
Broad leafed Wolfs-bane.

2 *Thora montis Baldi, ſiue Subaudica.*
Mountain Wolfs-bane.

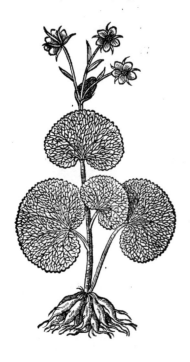

¶ *The Deſcription.*

1   THe firſt kinde of Aconite, of ſome called *Thora*, others adde thereto the place where it groweth in great aboundance, which is the Alps, and call it it *Thora Valdenſium*. This plant tooke his name of the Greeke word ✿✿, ſignifying corruption, poiſon, or death, which are the certaine effeCts of this pernitious plant: for this they vſe very much in poiſon, and when they mean to infeCt their arrow heads, the more ſpeedily and deadly to diſpatch the wilde beaſts which greatly annoy thoſe Mountaines of the Alpes. To which purpoſe alſo it is brought into the Mart townes neere thoſe places, to be ſold vnto the hunters, the juyce thereof being prepared by preſſing forth, and ſo kept in hornes and hoofes of beaſts for the moſt ſpeedy poyſon of the Aconites: for an arrow touched therewith leaues the wound vncurable (if it but fetch bloud where it entred in) vnleſſe that round about the wound the fleſh bee ſpeedily cut away in great quantitie: this plant therefore may rightly be accounted as firſt and chiefe of thoſe called Sagittaries or Aconites, by reaſon of the malignant qualities aforeſaid. This that hath beene ſayd,
argueth

argueth alſo that *Matthiolus* hath vnproperly called it *Pſeudoaconitum,* that is,falſe or baſtard Aco-nite;for without queſtion there is no worſe or more ſpeedie venome in the world,nor no Aconite or toxible plant comparable hereunto. And yet let vs conſider the fatherly care and prouidence of God,who hath prouided a conquerour and triumpher ouer this plant ſo venomous, namely his *Antigoniſt,Antithora,*or to ſpeake in ſhorter and fewer ſyllables, *Anthora,*which is the very antidote or remedie againſt this kinde of Aconite. The ſtalke of this plant is ſmal and ruſhie,very ſmooth, two or three handfuls high:wherupon do grow two,three,or foure leaues,ſeldome more, which be ſomthing hard,round,ſmooth,of a light greene colour tending to blewneſſe,like the colour of the leaues of Woad,nicked in the edges. The floures grow at the top of the ſtalks,of a yellow colour, leſſer than thoſe of the field Crowfoot,otherwiſe alike,in the place therof grows a knop or round head,wherin is the ſeed:the root conſiſteth of nine or ten ſlender clogs,with ſome ſmall fibres alſo, and they are faſtned together with little ſtrings vnto one head,like thoſe of the white Aſphodill.

2　Wolfes-bane of the mount Baldus hath one ſtalke,ſmooth and plain,in the middle where-of come forth two leaues and no more,wherein it differeth from the other of the Valdens, hauing likewiſe three or foure ſharp pointed leaues, narrow and ſomewhat jagged at the place where the ſtalke diuideth it ſelfe into ſmaller branches ; whereon do grow ſmall yellow floures like the pre-cedent,but much leſſer.

¶ The Place.

Theſe venomous plants do grow on the Alpes, and the mountaines of Sauoy and Switzerland : the firſt growes plentifully in the countrey of the Valdens, who inhabite part of thoſe mountains towards Italy. The other is found on Baldus,a mountaine of Italy. They are ſtrangers in England.

¶ The Time.

They floure in March and Aprill,their ſeed is ripe in Iune.

¶ The Names.

This kinde of Aconite or Wolfs-bane is called *Thora,Taura,*and *Tura,* it is ſyrnamed *Valdenſis,* that it may differ from *Napellus,* or Monkes-hood,which is likewiſe named *Thora.*

Auicen maketh mention of a certaine deadly herbe in his fourth book, ſixt Fen, called *Farſiun ;* it is hard to affirme the ſame to be *Thora Valdenſis.*

‡　*Geſner* iudges this to be the *Aconitum Pardalianches* of *Dioſcorides,*and herein is followed by *Bauhine.* ‡

¶ The Temperature and Vertues.

The force of theſe Wolfes-banes,is moſt pernitious and poiſonſome, and (as it is reported) ex-　A ceedeth the malice of *Napellus,*or any of the other Wolfes-banes,as we haue ſaid.

They ſay that it is of ſuch force, that if a man eſpecially, and then next any foure footed beaſt　B be wounded with an arrow or other inſtrument dipped in the juice hereof, they die within halfe an houre after remedileſſe.

† There were formerly four figures in this chapter,with as many deſcriptions,though the plants figured and deſcribed were but two,to which number they are now reduced.The two former,which were by the names of *Thora Valdenſirmas* and *famina,* thus differed, the male had onely two large round leaues and the female foure. The other two being alſo of one plant are more deeply cut in vpon the top of the leaues,which are fewer and leſſer than thoſe of the former.

Chap. 373. Of Winter Wolfes-bane.

¶ The Deſcription.

THis kind of Aconite is called *Aconitum hyemale Belgarum,*of *Dodonæus,Aconitum luteum minus:* in Engliſh,Wolfs-bane,or ſmal yellow Wolfs-bane,whoſe leaues come forth of the ground in the dead time of winter,many times bearing the ſnow on the heads of his leaues & flours; yea the colder the weather is,and the deeper that the ſnow is,the fairer and larger is the floure;and the warmer that the weather is,the leſſer is the floure, and worſe coloured:theſe leaues I ſay come forth of the ground immediatly from the root, with a naked, ſoft, and ſlender ſtem, deeply cut or jagged on the leaues, of an exceeding faire greene colour, in the midſt of which commeth forth a yellow flour,in ſhew or faſhion like vnto the common field Crow-foot:after which follow ſundry cods full of browne ſeeds like the other kinds of Aconites:the root is thick,tuberous,and knottie, like to the kindes of Anemone.

Aconitum hyemale.
Winter Woolfes-bane.

A

¶ *The Place.*

It groweth vpon the mountains of Germany:we haue great quantitie of it in our London gardens.

¶ *The Time.*

It floureth in Ianuary ; the ſeed is ripe in the end of March.

¶ *The Names.*

It is called *Aconitum hyemale* or *Hibernum*, or winter Aconite:that it is a kind of Aconite or Wolfs-bane,both the form of the leaues and cods, and alſo the daungerous faculties of the herb it ſelfe declare.

It is much like to *Aconitum Theophraſti:* which he deſcribes in his ninth book,ſaying,it is a ſhort herb hauing no ἀκτὶν,or ſuperfluous thing growing on it,and is without branches as this plant is:the root,ſaith he,is like to ῥαφα̃,or to a nut,or els to ἰσχνῆ a dry fig, only the leafe ſeemeth to make againſt it , which is nothing at all like to that of Succory,which he compares it to.

¶ *The Temperature and Vertues.*

This herb is counted to be very dangerous and deadly, hot and dry in the fourth degree,as *Theophraſtus* in plain words doth teſtifie concerning his own Aconite ; for which he ſaith that there was neuer found his Antidote or remedie:wherof *Athenæus* and *Theopompus* write, that this plant is the moſt poiſonous herbe of all others,which moued *Ouid* to ſay,*Quæ quia naſcuntur dura viuacia caute:*notwithſtanding it is not without his peculiar vertues. *Ioachimus Camerarius* now liuing in Noremberg ſaith, the water dropped into the eies ceaſeth the pain & burning:it is reported to preuaile mightily againſt the bitings of ſcorpions, and is of ſuch force,that if the Scorpion paſſe by where it groweth and touch the ſame,preſently he becommeth dull,heauy,and ſenceleſſe,and if the ſame ſcorpion by chance touch the white Hellebor, he is preſently deliuered from his drowſineſſe.

Chap. 374. *Of Mithridate Wolfes-bane.*

¶ *The Deſcription.*

THis plant called *Anthora*, being the antidote againſt the poiſon of *Thora, Aconite* or Wolfes bane, hath ſlender hollow ſtalkes,very brittle,a cubit high, garniſhed with fine cut or jagged leaues,very like to *Nigella Romana*,or the common Larks ſpurre,called *Conſolida regalis* : at the top of the ſtalks grow faire floures,faſhioned like a little helmet, of an ouerworne yellow colour ; afterwhich come ſmall blackiſh cods,wherein is contained blacke ſhining ſeed like thoſe of Onions:the root conſiſteth of diuers knobs or tuberous lumps,of the bigneſſe of a mans thumbe.

¶ *The Place.*

This plant which in Greeke we may terme Ἀντιθώρα groweth abundantly in the Alpes, called *Rhetici*,in Sauoy,and in Liguria.The Ligurians of Turnin,and thoſe that dwell neer the lake Lemane, haue found this herbe to be a preſent remedie againſt the deadly poiſon of the herb *Thora*, and the reſt of the Aconits,prouided that when it is brought into the garden there to be kept for phiſicks vſe,it muſt not be planted neere to any of the Aconites:for through his attractiue qualitie, it will
draw

Anthora ſiue Aconitum ſalutiferum.
Wholeſome Wolfes-bane

draw vnto it ſelfe the maligne and venomous poiſon of the Aconite,wherby it wil become of the like qualitie, that is, to become poiſonous likewiſe : but being kept far off,it retaineth his owne naturall qualitie ſtill.

¶ *The Time.*

It floureth in Auguſt,the ſeed is ripe in the end of September.

¶ *The Names.*

The inhabitants of the lake of Geneva,and the Piemontoiſe do call it *Anthora,* and the common people *Anthoro. Auicen* calleth a certaine herbe which is like to Monks hood,as a remedy againſt the poiſon thereof , by the name of *Napellus Moyſis,* in the 500 chapter of his ſecond booke;and in the 745 chapter hee ſaith,that *Zedoaria* doth grow with *Napellus* or Monks hood, and that by reaſon of the neerneſſe of the ſame, the force and ſtrength therof is dulled,and made weaker, and that it is a treacle, that is , a counterpoiſon againſt the Viper,Monks hood,and all other poiſons:and hereupon it followeth, that it is not only *Napellus Moyſis,*but alſo *Zedoaria Auicennæ*:notwithſtanding the Apothecaries doe ſell another *Zedoaria* differing from *Anthora,*which is a root of a longer forme , which not without cauſe is thought to be *Auicens* and *Seraptos Zerumbeth,*or *Zurumbeth.*

It is called *Anthora,*as though they ſhould ſay *Antithora,*becauſe it is an enemy to *Thora,* and a counterpoyſon to the ſame. *Thora* and *Anthora,*or *Tura* and *Antura,*ſeem to be new words,but yet they are vſed in *Marcellus Empericus,* an old writer,who teaches vs a medicine to be made of *Tura* and *Antura,* againſt the pin and web in the eies:in Engliſh,yellow Monks-hood,yellow Helmet floure,and Aconites Mithridate.

¶ *The Vertues.*

The root of *Anthora* is wonderfull bitter,it is an enemy to all poiſons :it is good for purgations ; A for it voideth by the ſtoole both watery and ſlimie humors,killeth and driueth forth all manner of wormes of the belly.

Hugo Solerius ſaith,that the roots of *Anthora* do largely purge,not only by the ſtool, but alſo by B vomit.and that the meaſure thereof is taken to the quantity of *Faſelus* (which is commonly called a beane)in broth or wine,and is giuen to ſtrong bodies.

Antonius Guanerius doth ſhew in his treatie of the plague, the ſecond difference,the third chap- C ter,that *Anthora* is of great force,yea & that againſt the plague:and the root is of like vertues,giuen with Dittanie, which I haue ſeene, ſaith he,by experience: and he further ſaith, it is an herbe that groweth hard by that herbe *Thora,*of which there is made a poyſon, wherewith they of Sauoy and thoſe parts adjacent,do enuenome their arrowes,the more ſpeedily to kill the wild Goats,& other wilde beaſts of the Alpiſh mountaines. And this root *Anthora,* is the *Bezoar* or counterpoiſon to that *Thora,*which is of ſo great a venom as that it killeth all liuing creatures with his poiſonſome qualitie:and thus much *Guanerius.*

Simon Ianuenſis hath alſo made mention of *Anthora,* and *Arnoldus Villanouanus* in his treatie of D poiſons : but their writings do declare that they did not well know *Anthora.*

Chap. 375. *Of yellow Wolfes-bane.*

¶ *The Deſcription.*

THe yellow kinde of Wolfes-bane called *Aconitum luteum Ponticum,* or according to *Dodonæus Aconitum Lycoctonon luteum majus* · in Engliſh,yellow Wolfs-bane,whereof this our age hath found out ſundry ſorts not knowne to *Dioſcorides,* although ſome of the ſorts ſeeme to ſtand indifferent

Aconitum luteum Ponticum.
Yellow Woolfes bane.

indifferent betweene the kindes of *Ranunculus, Helleborus,* & *Napellus:*) this yellow kind I ſay hath large ſhining green leaues faſhioned like a vine, and of the ſame bigneſſe, deeply indented or cut, not much vnlike the leaues of *Geranium Fuſcum,* or black Cranebill: the ſtalks are bare or naked, not bearing his leaues vpon the ſame ſtalks, one oppoſit againſt another, as in the other of his kind: his ſtalks grow vp to the height of three cubits, bearing very fine yellow floures, fantaſtically faſhioned, and in ſuch manner ſhaped, that I can very hardly deſcribe them to you. They are ſomewhat like vnto the helmet Monkes hood, open and hollow at one end, firme and ſhut vp at the other: his rootes are many, compact of a number of threddie or black ſtrings, of an ouerworne yellow color, ſpreading far abroad euery way, folding themſelues one within another very confuſedly. This plant groweth naturally in the darke hilly forreſts, & ſhadowie woods, that are not trauelled nor haunted, but by wilde and ſauage beaſts, and is thought to bee the ſtrongeſt and next vnto *Thora* in his poiſoning qualitie, of all the reſt of the Aconites, or Woolfes banes; inſomuch that if a few of the floures be chewed in the mouth, and ſpit forth againe preſently, yet forthwith it burneth the jaws and tongue, cauſing them to ſwell, and making a certain ſwimming or giddineſſe in the head. This calleth to my remembrance an hiſtory of a certain Gentleman dwelling in Lincolneſhire, called *Mahewe,* the true report whereof my very good friend Mr. *Nicholas Belſon,* ſomtimes Fellow of Kings Colledge in Cambridge, hath deliuered vnto me: Mr. *Mahewe* dwelling in Boſton, a ſtudent in phyſick, hauing occaſion to ride through the fens of Lincolnſhire, found a root that the hogs had turned vp, which ſeemed vnto him very ſtrange and vnknowne, for that it was in the ſpring before the leaues were out: this he taſted, and it ſo inflamed his mouth, tongue, and lips, that it cauſed them to ſwell very extremely, ſo that before he could get to the towne of Boſton, he could not ſpeake, and no doubt had loſt his life if that the Lord God had not bleſſed thoſe good remedies which preſently he procured and vſed. I haue here thought good to expreſſe this hiſtory, for two ſpeciall cauſes; the firſt is, that ſome induſtrious and diligent obſeruer of nature may be prouoked to ſeeke forth that venomous plant, or ſome of his kindes: for I am certainly perſuaded that it is either the *Thora Valdenſium,* or *Aconitum luteum,* whereof this gentleman taſted, which two plants haue not at any time bin thought to grow naturally in England: the other cauſe is, for that I would warne others to beware by that gentlemans harme. ‡ I am of opinion that this root which Mr. *Mahewe* taſted was of the *Ranunculus flammeus major,* deſcribed in the firſt place of the 370. chapter aforegoing; for that grows plentifully in ſuch places, and is of a very hot taſte and hurtfull qualitie. ‡

¶ *The Place.*

The yellow Wolfes-bane groweth in my garden, but not wilde in England, or in any other of theſe Northerly regions.

¶ *The Time.*

It floureth in the end of Iune, ſomwhat after the other Aconites.

¶ *The Names.*

This yellow Wolfs-bane is called of *Lobel, Aconitum luteum Ponticum,* or Pontick Wolfs-bane. There is mention made in *Dioſcorides* his copies of three Wolfes-banes, of which the hunters vſe one, and Phyſitians the other two. *Marcellus Virgilius* holdeth opinion that the vſe of this plant is vtterly to be refuſed in medicine.

¶ *The Nature and Vertues.*

A   The facultie of this Aconite, as alſo of the other Wolfes-banes, is deadly to man, and likewiſe to all other liuing creatures.

It is vſed among the hunters which ſeeke after wolues,the juice whereof they put into raw fleſh B which the wolues deuoure,and are killed.

CHAP. 376.   Of other Wolfes-banes and Monkes hoods.

¶ The Deſcription.

1   THis kinde of Wolfes-bane(called *Aconitum Lycoctonum :*and of *Dodonæus, Aconitum Ly-cottonon flore Delphinij,*by reaſon of the ſhape and likenes that the floure hath with *Del-phinium,* or Larks-ſpur:and in Engliſh it is called black Wolfs-bane)hath many large leaues of a very deep green or ouerworne colour,very deepely cut or jagged:among which riſeth vp a ſtalk two cubits high;wherupon grow floures faſhioned like a hood,of a very ill fauoted blewiſh colour,and the thrums or threds within the hood are black : the ſeed is alſo black,and three corne-red,growing in ſmall husks:the root is thick and knobby.

† 1 *Aconitum Lycoctonon flore Delphinij.*   † 2 *Aconitum Lycoctonon cæruleum parvum.*
Larks-heele Woolfs-bane.   Small blew Wolfes-bane.

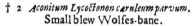

2   This kind of Wolfes-bane, called *Lycoctonon cæruleum parvum,facie Napelli:*in Engliſh,ſmal Wolfes-bane, or round Wolfs-bane, hath many ſlender brittle ſtalks two cubits high, beſet with leaues very much jagged,and like vnto *Napellus,*called in Engliſh,Helmet-floure. The floures doe grow at the top of the ſtalkes,of a blewiſh colour, faſhioned alſo like a hood, but wider open than any of the reſt:the cods and ſeed are like vnto the other:the root is round and ſmal, faſhioned like a Peare or ſmall Rape or Turnep : which moued the Germanes to call the ſame **Rapen-bloemen**, which is in Latine, *Flos rapaceus:*in Engliſh,Rape-floure.

3   This kinde of Wolfes-bane, called *Napellus verus,* in Engliſh, Helmet-floure, or the great Monkes-hood, beareth very faire and goodly blew floures in ſhape like an Helmet ; which are ſo beautifull,that a man would thinke they were of ſome excellent vertue,but *non eſt ſemper fides ha-benda fronti.*This plant is vniuerſally knowne in our London gardens and elſewhere ; but naturally
it

it groweth in the mountaines of Rhetia, and in ſundry places of the Alps, where you ſhall find the graſſe that groweth round it eaten vp with cattell, but no part of the herbe it ſelfe touched, except by certaine flies, who in ſuch abundant meaſure ſwarme about the ſame that they couer the whole plant: and (which is very ſtraunge) although theſe flies do with great delight feed hereupon, yet of them there is confected an Antidot or moſt auaileable medicine againſt the deadly bite of the ſpider called *Tarantala*, or any other venomous beaſt whatſoeuer; yea, an excellent remedy not only againſt the Aconites, but all other poiſons whatſoeuer. The medicine of the foreſaid flies is thus made: Take of the flies which haue fed themſelues as is aboue mentioned, in number twentie, of *Ariſtolechia rotunda*, and bole Armoniack, of each a dram.

3 *Napellus verus cæruleus*.
Blew Helmet-floure, or Monks-hood.

‡ 4 *Aconitum Lycoctonum ex Cod. Cæſareo*.

4 There is a kind of wolfes-bane which *Dodonæus* reports he found in an old written greek book in the Emperors Librarie at Vienna, vnder the title of *Aconitum Lycoctonum*, that anſwereth in all points vnto *Dioſcorides* his deſcription, except in the leaues. It hath leaues (ſaith he) like vnto the Plane tree, but leſſer, and more full of jagges or diuiſions; a ſlender ſtalke as Ferne, of a cubit high, bearing his ſeed in long cods: it hath blacke roots in ſhape like Creauiſes. Hereunto agreeth the Emperors picture in all things ſauing in the leaues, which are not ſo large, nor ſo much diuided, but notched or toothed like the teeth of a ſaw.

‡ 5 Beſides theſe mentioned by our Author there are ſundry other plants belonging to this pernitious Tribe, whoſe hiſtorie I wil briefly run ouer: The firſt of theſe is that which *Cluſius* hath ſet forth by the name of *Aconitum Lycoctonum flo. Delphinij Sileſiacum*: it hath ſtalkes ſome two or three cubits high, ſmooth and hollow, of a greeniſh purple colour, and couered with a certain mea-lines: the leaues grow vpon long ſtalks, being rough, and faſhioned like thoſe of the yellow Wolfs bane, but of a blacker colour: the top of the ſtalk ends in a long ſpike of ſpur-floures, which before they be open reſemble locuſts or little Lyzards, with their long and crooking tailes; but opening they ſhew fiue leaues, two on the ſides, two below, and one aboue, which ends in a crooked taile or horne: all theſe leaues are wrinckled, and purple on their outſides, but ſmooth, and of an elegant blew within. After the floures are paſt ſucceed three ſquare cods, as in other Aconites, wherein is contained an vnequall browniſh wrinckled ſeed: the root is thick, black, and tuberous. This grows naturally inſome mountaines of Sileſia, and floures in Iuly and Auguſt.

6 The

‡ 5 *Aconitum Lycoct. hirsutum flo. Delphinii.*
Rough Larks-heele Wolfes-bane.

‡ 6 *Aconitum violaceum.*
Violet coloured Monks-hood.

‡ 7 *Aconitum purpureum Newbergense.*
Purple Monks-hood of Newburg.

‡ 8 *Aconitum maximum Iudenbergense.*
Large floured Monks-hood.

6   The leaues of this are ſomwhat like, yet leſſe than thoſe of our common Monks-hood, blackiſh on the vpper ſide and ſhining. The ſtalk is ſome cubit and half high, firm, full of pith, ſmooth, and ſhining; diuided towards the top into ſome branches carrying few flours like in form to thoſe of the vulgar Monks-hood, of a moſt elegant and deep violet colour: the ſeeds are like the former, and roots round, thicke, and ſhort, with many fibres. It growes vpon the hills nigh Saltsburg, where it floures in Iuly: but brought into gardens it floures ſooner than the reſt of this kinde, to wit in May. *Cluſius* cals this *Aconitum Lycoctonum 4. Tauricum.*

7   This hath leaues broader than thoſe of our ordinary Monks-hood, yet like them: the ſtalk is round, ſtraight, and firme, and of ſome three cubits height, and oft times toward the top diuided into many branches, which cary their floures ſpike-faſhion, of a purple colour, abſolutely like thoſe of the common ſort, but that the thrummie matter in the midſt of the flours is of a duskier colour. The root and reſt of the parts are like thoſe of the common kind: it grows naturally vpon the Stirian Alpes, whereas it floures ſomewhat after the common kinde, to wit, in Iuly. *Cluſius* hath it by the name of *Aconitum Lycoctonum 5. Neubergenſe.*

‡ 9 *Aconitum maximum nutante coma.*
Monkes-hood, with the bending or
nodding head.

8   The leaues of this are alſo diuided into fiue parts, and ſnipt about the edges, and doe very much reſemble thoſe of the ſmal Wolfs-bane deſcribed in the ſecond place, but that the leaues of that ſhine, when as theſe do not: the ſtalke is two cubits high, not very thicke, yet firm and ſtraight, of a greeniſh purple colour; and at the top carries fiue or ſix floures, the largeſt of all the Monks-hoods, conſiſting of foure leaues, as in the reſt of this kind, with a very large helmet ouer them, beeing ſometimes an inch long, of an elegant blewiſh purple colour: the ſeed-veſſels, ſeeds, and roots are like the reſt of this kinde. This grows on Iudenberg, the higheſt hill of all Styria, and floures in Auguſt; in gardens about the end of Iuly. *Cluſius* names it *Aconitum Lycoct. 9. Iudenbergenſe.*

9   This riſes vp to the height of three cubits, with a ſlender round ſtalke which is diuided into ſundry branches, and commonly hangeth downe the head, whence *Cluſius* cals it *Aconitum Lycoctonum 8. coma nutante.* The flours are like thoſe of the common Monks-hood, but of ſomwhat a lighter purple colour. The leaues are larger and long, and much more cut in or diuided than any of the reſt. The roots, ſeeds, and other particles are not vnlike thoſe of the reſt of this kinde. ‡

¶ *The Place.*
Diuers of theſe Wolfs-banes grow in ſome gardens, except *Aconitum lycoctonon* taken forth of the Emperors booke.

¶ *The Time.*
Theſe plants do floure from May vnto the end of Auguſt.

¶ *The Names.*
The firſt is *Lycoctoni ſpecies,* or a kinde of Wolfes-bane, and is as hurtfull as any of the reſt, and called of *Lobel, Aconitum flore Delphinij,* or Larke-ſpur Wolfes-bane. *Auicen* ſpeaketh hereof in his ſecond booke, and afterwards in his fourth booke, Fen. 6. the firſt Treatiſe; hauing his reaſons why and wherefore he hath ſeparated this from *Canach adip,* that is to ſay, the Wolfes ſtrangler, or the Wolfes-bane.

The later and barbarous Herbariſts call the third Wolfes-bane in Latine *Napellus,* of the figure and ſhape of the roots of *Napns,* or *Nauet,* or Nauew gentle: it is likewiſe *Aconiti lycoctoni ſpecies,* or a kinde of Wolfes-bane: alſo it may be called *Toxicum*; for *Toxicum* is a deadly medicine, wherewith the Hunters poyſon their ſpeares, darts, and arrowes, that bring preſent death: ſo named of arrowes which the Barbarians call *Toxeumata* and *Toxa. Dioſcorides* ſetting downe the ſymptomes or accidents cauſed by *Toxicum*, togetherwith the remedies, reckoneth vp almoſt the very ſame that *Aviceu* doth concerning *Napellus* : notwithſtanding *Auicen* writes of *Napellus* and *Toxicum* ſeuerally,but not knowing what *Toxicum* is,as he himſelfe confeſſeth : ſo that it is not to be maruelled at,that hauing written of *Napellus*,he ſhould afterward entreat againe of *Toxicum*.

¶ *The Nature and Vertues.*

All theſe plants are hot and dry in the fourth degree,and of a moſt venomous qualitie.

The force and facultie of Wolfs-bane is deadly to man and all kindes of beaſts : the ſame was A tried of late in Antwerpe, and is as yet freſh in memorie,by an euident experiment, but moſt lamentable ; for when the leaues hereof were by certaine ignorant perſons ſerued vp in ſallads, all that did eat thereof were preſently taken with moſt cruell ſymptomes,and ſo died.

The ſymptomes that follow thoſe that doe eat of theſe deadly Herbs are theſe ; their lipps and B tongue ſwell forthwith, their eyes hang out, their thighes are ſtiffe, and their wits are taken from them,as *Auicen* writes,*lib.*4. The force of this poiſon is ſuch,that if the points of darts or arrowes be touched therewith,it brings deadly hurt to thoſe that are wounded with the ſame.

Againſt ſo deadly a poiſon *Auicen* reckoneth vp certain remedies, which help after the poyſon C is vomited vp : and among theſe he maketh mention of the Mouſe (as the copies euery where haue it) nouriſhed and fed vp with *Napellus*,which is altogether an enemie to the poiſonſome nature of it,and deliuereth him that hath taken it from all perill and danger.

Antonius Guanerius of Pauia,a famous phyſition in his age,in his treaty of poiſons is of opinion, D that it is not a mouſe which *Avicen* ſpeaketh of,but a fly : for he telleth of a certaine Philoſopher who did very carefully and diligently make ſearch after this mouſe, and neither could find at any time any mouſe,nor the roots of Wolfs-bane gnawn or bitten, as he had read : but in ſearching he found many flies feeding on the leaues, which the ſaid Philoſopher tooke , and made of them an antidote or counterpoiſon, which hee found to be good and effectuall againſt other poiſons,but eſpecially againſt the poiſon of Wolfs-bane.

The compoſition conſiſteth of two ounces of *Terra lemnia*, as many of the berries of the Bay E tree, and the like weight of Mithridate, 24 of the flies that haue taken their repaſt vpon Wolfes-bane,of hony and oile Oliue a ſufficient quantitie.

The ſame opinion that *Guanerius* is of, *Pena* and *Lobel* do alſo hold ; who affirme, that there was F neuer ſeene at any time any mouſe feeding thereon, but that there bee flies which reſort vnto it by ſwarmes,and feed not only vpon the floures,but on the herb alſo.

¶ *The Danger.*

There hath bin little heretofore ſet down concerning the Vertues of Aconites,but much might be ſaid of the hurts that haue come hereby,as the wofull experience of the lamentable example at Antwerp yet freſh in memorie,doth declare,as we haue ſaid.

† The figure that was in the firſt place formerly, was of the *Aconitum luteum Ponticum* ; and that in the ſecond place was of a *Napellus*.

CHAP. 377.   *Of blacke Hellebore.*

¶ *The Deſcription.*

1   THe firſt kind of blacke Hellebor *Dodonæus* ſets forth vnder this title,*Veratrum nigrum*, and it may properly be called in Engliſh,blacke Hellebor,which is a name moſt fitly agreeing vnto the true and vndoubted blacke Hellebor, for the kindes and other ſorts hereof which hereafter follow are falſe and baſtard kindes thereof. This plant hath thicke and fat leaues of a deep green colour,the vpper part whereof is ſomewhat bluntly nicked or toothed, hauing ſundry diuiſions or cuts,in ſome leaues many,in others fewer,like to the femal Peony or *Smyrnium Creticum*. It beareth Roſe-faſhioned floures vpon ſlender ſtems, growing immediatly out of the ground an handfull high,ſomtimes very white,and oftentimes mixed with a little ſhew of purple : which being vaded, there ſucceed ſmall husks full of blacke ſeeds : the roots are many , with long blacke ſtrings comming from one head.

2   The ſecond kinde of blacke Hellebor,called of *Pena,Helleboraſtrum*; and of *Dodonæus,Veratrum ſecundum*, (in Engliſh,Baſtard Hellebor) hath leaues much like the former, but narrower and blacker,each leafe beeing much jagged or toothed about the edges like a ſaw : the ſtalkes grow to the height of a foot or more,diuiding themſelues into other branches toward the top; whereon grow floures not much vnlike to the former in ſhew, ſaue that they are of a greeniſh herby colour. The roots are ſmall and threddy,but not ſo blacke as the former.

3 The

1 *Helleborus niger verus.*
The true blacke Hellebor.

2 *Helleboraſtrum.*
Wilde blacke Hellebor.

3 *Helleboraſter maximus.*
The great Oxe-heele.

4 *Conſiligo Ruel. & Seſamoides maz. Cordi.*
Setter-wort, or Beare-foot.

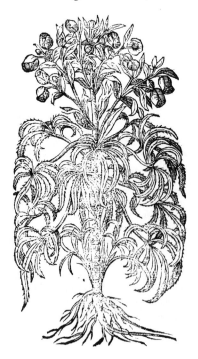

3 . The third kinde of blacke Hellebor, called of *Pena*, *Helleboraſter maximus*, with this adition, *flore & ſemine prægnans*, that is, full both of floures and ſeed, hath leaues ſomewhat like the former wilde Hellebor,(ſaue that they be greater,more jagged, and deeply cut. The ſtalkes grow vp to the height of two cubits,diuiding themſelues at the top into ſundry ſmall branches, whereupon grow little round and bottle-like hollow greene floures ; after which come forth ſeeds which come to perfect maturitie and ripeneſſe. The root conſiſteth of many ſmall black ſtrings,inuoloued or wrapped one within another very intricately.

4   The fourth kinde of blacke Hellebor (called of *Pena* and *Lobel*, according to the deſcription of *Cordus* and *Ruellius*, *Saſamoides magnum*, and *Conſiligo*: in Engliſh,Ox-heele,or Setter-wort,which names are taken from his vertues in curing Oxen and ſuch like cattell,as ſhall be ſhewed afterward in the names thereof) is ſo well knowne vnto the moſt ſort of people by the name of Beare-foot, that I ſhall not haue cauſe to ſpend much time about the deſcription. ‡ Indeed it was not much need full for our Author to deſcribe it, for it was the laſt thing he did; for both theſe two laſt are of one plant, both figures and deſcriptions ; the former of theſe figures expreſſing it in floure, and the later in ſeed : but the former of our Author was with ſomewhat broader leaues, and the later with narrower. ‡

¶ *The Place.*

Theſe Hellebors grow vpon rough and craggy mountains: the laſt growes wilde in many woods and ſhadowie places in England : we haue them all in our London gardens.

¶ *The Time.*

The firſt floureth about Chriſtmaſſe,if the Winter be milde and warme : the others later.

¶ *The Names.*

It is agreed among the later writers, that theſe plants are *Veratra nigra* : in Engliſh,blacke Hel-leboros : in Greeke, ισσετωνες : in Italian,*Elleboro nero* : in Spaniſh,*Verde gambe negro* :of diuers,*Me-Lampodium*, becauſe it was firſt found by *Melampos*, who was firſt thought to purge therewith *Prœ-tus* his mad daughters,and to reſtore them to health. *Dioſcorides* writeth, that this man was a ſhep-heard : others, a Sooth-ſayer. In high Dutch it is called,**Chriſtwurtz**, that is, Chriſts herbe, or Chriſtmaſſe herbe : in low Dutch,**Heylich kerſt cruyt**,and that becauſe it floureth about the birth of our Lord Ieſus Chriſt.

The third kind was called of *Fuchſius*,*Pſeudobelleborus*, and *Veratrum nigrum adulterinum*,which is in Engliſh, falſe or baſtard blacke Hellebor. Moſt name it *Conſiligo*, becauſe the husbandmen of our time doe herewith cure their cattell, no otherwiſe than the old Farriers or horſe-leeches were wont to doe,that is,they cut a ſlit or hole in the dewlap,as they terme it(which is an emptie ſkinne vnder the throat of the beaſt)wherein they put a piece of the root of Setter-wort or Beare-foot, ſuf-fering it there to remaine for certaine daies together : which manner of curing they do call Sette-ring of their cattell, and is a manner of rowelling, as the ſaid Horſe-leeches doe their horſes with horſe haire twiſted,or ſuch like,and as in Surgerie we doe vſe with ſilke,which in ſtead of the word *Seton*, a certaine Phyſitian called it by the name Rowell ; a word very vnproperly ſpoken of a lear-ned man,becauſe there would be ſome difference betweene men and beaſts. This manner of Sette-ring of cattell helpeth the diſeaſe of the lungs, the cough, and wheeſing. Moreouer, in the time of peſtilence or murraine,or any other diſeaſes affecting cattell,they put the root into the place afore-ſaid,which draweth vnto it all the venomous matter,and voideth it forth at the wound.The which *Abſyrtus* and *Hierocles* the Greeke Horſe-leeches haue at large ſet downe. And it is called in Eng-liſh,Beare-foot,Setterwort,and Setter-graſſe.

The ſecond is named in the German tongue,**Lowſzkraut**, that is, *Pedicularis*, or Lowſie graſſe : for it is thought to deſtroy and kill lice, and not onely lice but ſheepe and other cattell : and may be reckoned among the Beare-feet,as kindes thereof.

¶ *The Temperature.*

Blacke Hellebor, as *Galen* holdeth opinion, is hotter in taſte than the white Hellebor : in like manner hot and dry in the third degree.

¶ *The Vertues.*

Blacke Hellebor purgeth downeward flegme,choler,and alſo melancholy eſpecially,and all me-   A
lancholy humors,yet not without trouble and difficulty:therefore it is not to be giuen but to robu-ſtious and ſtrong bodies,as *Meſues* teacheth. A purgation of Hellebor is good for mad and furious men,for melancholy,dull and heauie perſons, for thoſe that are troubled with the falling ſickenes, for lepers, for them that are ſicke of a quartaine Ague, and briefly for all thoſe that are troubled with blacke choler,and moleſted with melancholy.

The manner of giuing it(meaning the firſt blacke Hellebor)ſaith *Arctuarius* in his firſt booke,is   B
three ſcruples little more or leſſe.

It is giuen with wine of raiſins or oxymel,but for pleaſantneſſe ſake ſome ſweet and odoriferous   C ſeeds muſt be put vnto it : but if you would haue it ſtronger, adde thereunto a graine or two of Sca-monie. Thus much *Actuarius*.

D   The firſt of theſe kindes is beſt, then the ſecond ; the reſt are of leſſe force.

E   The roots take away the morphew and blacke ſpots in the skin, tetters, ring-wormes, leproſies, and ſcabs.

F   . The root ſodden in pottage with fleſh, openeth the bellies of ſuch as haue the dropſie.

G   The root of baſtard Hellebor, called among our Engliſh women Beare-foot, ſteeped in wine and drunken, looſeth the belly euen as the true blacke Hellebor, and is good againſt all the diſeaſes whereunto blacke Hellebor ſerueth, and killeth wormes in children.

H   It doth his operation with more force and might, if it be made into pouder, and a dram thereof be receited in wine.

I   The ſame boyled in water with Rue and Agrimonie, cureth the jaundiſe, and purgeth yellow ſu-perfluities by ſiege.

K   The leaues of baſtard Hellebor dried in an ouen, after the bread is drawne out, and the pouder thereof taken in a figge or raiſin, or ſtrawed vpon a piece of bread ſpred with hony, and eaten, killeth wormes in children exceedingly.

Chap. 378.   Of Dioſcorides *his blacke Hellebor.*

Aſtrantia nigra, ſiue veratrum nigrum Dioſcoridis, Dod.
Blacke Maſter-worts, or *Dioſcorides* his blacke Hellebor.   ¶ *The Deſcription.*

THis kinde of blacke Hellebor, ſet forth by *Lobel* vnder the name of *Aſtrantia nigra*, a-greeth very well in ſhape with the true *A-ſtrantia*, which is called *Imperatoria* : neuertheles by the conſent of *Dioſcorides* and other Authors, who haue expreſſed this plant for a kinde of *Ve-ratrum nigrum*, or blacke Hellebor, it hath many blackiſh green leaues parted or cut into foure or fiue deepe cuts, after the maner of the vine leafe very like vnto thoſe of Sanicle, both in greennes of colour and alſo in proportion. The ſtalke is euen, ſmooth, and plaine: at the top whereof grow floures in little tufts or vmbels, ſet together like thoſe of Scabious, of a whitiſh light greene co-lour, daſhed ouer as it were with a little darke purple : after which come the ſeed like vnto *Car-thamus* or baſtard Saffron. The roots are many blackiſh thredsknit to one head or maſter root.

¶ *The Place.*

Blacke Hellebor is found in the mountains of Germany, and in other vntilled and rough pla-ces : it proſpereth in gardens.

Dioſcorides writeth, That blacke Hellebor groweth likewiſe in rough and dry places : and that is the beſt which is taken from ſuch like places ; as that (ſaith hee) which is brought out of Anticyra a city in Greece. It groweth in my garden.

¶ *The Time.*

This blacke Hellebor floureth not in Winter, but in the Summer moneths. The herbe is greene all the yeare thorow.

¶ *The Names.*

It is called of the later Herbariſts, *Aſtrantia nigra* : of others, *Sanicula fœmina* : notwithſtanding it differeth much from *Aſtrantia*, an herbe which is alſo named *Imperatoria*, or Maſter-wort. The vulgar people call it Pellitorie of Spaine, but vntruly : it may be called blacke Maſter-wort, yet doubtleſſe a kinde of Hellebor, as the purging facultie doth ſhew : for it is certaine, that diuers experienced phyſitians can witneſſe, that the roots hereof do purge melancholy and other humors,

and

and that they themſelues haue perfectly cured mad melancholy people being purged herewith. And that it hath a purging quality, *Conradus Geſnerus* doth likewiſe teſtifie in a certaine Epiſtle written to *Adolphus Occo*, in which he ſheweth, that *Aſtrantia nigra* is almoſt as ſtrong as white Hellebor, and that he himſelfe was the firſt that had experience of the purging faculty therof by ſiege: which things confirme that it is *Dioſcorides* his blacke Hellebor.

Dioſcorides hath alſo attributed to this plant all thoſe names that are deſcribed to the other black Hellebors. He ſaith further, that the ſeed therof in Anticyra is called *Seſamoides*, the which is vſed to purge with, if ſo be that the Text be true, and not corrupted. But it ſeemeth not to be altogether perfect; for if *Seſamoides*, as *Pliny* ſaith, and the word it ſelfe doth ſhew, hath his name of the likeneſſe of *Seſamum*, the ſeed of this blacke Hellebor ſhall vnproperly be called *Seſamoides*; being not like that of *Seſamum*, but of *Cnicus* or baſtard Saffron. By theſe proofes we may ſuſpect, that theſe words are brought into *Dioſcorides* from ſome other Author.

¶ *The Temperature and Vertues.*

The faculties of this plant wee haue already written to be by triall found like to thoſe of the A other blacke Hellebor: notwithſtanding thoſe that are deſcribed in the former chapter are to be accounted of greater force.

† This whole Chapter (as moſt beſides) was out of *Dodonæus*, who, *Pempt. 3. lib. 2. cap. 30.* labours to proue this plant to be the true blacke Hellebor of *Dioſcorides*. There was alſo another deſcription thruſt by our Author into this chapter, being of the *Perſicaria ſiliquoſa*, or *Noli me tangere* formerly deſcribed in the fourth place of the *114. chap. pag. 446.*

C H A P. 379.　*Of Herbe Chriſtopher.*

Chriſtophoriana.
Herbe Chriſtopher.

¶ *The Deſcription.*

ALthough Herbe Chriſtopher be none of the Binde-weeds, or of thoſe plants that haue need of ſupporting or vnderpropping, wherewith it may clime or rampe, yet becauſe it beareth grapes, or cluſters of berries, it might haue been numbred among the *Cyman*, or thoſe that grow like Vines. It brings forth little tender ſtalkes a foot long, or not much longer; whereupon doe grow ſundry leaues ſet vpon a tender foot-ſtalke, which doe make one leafe ſomewhat jagged or cut about the edges, of a light greene colour: the floures grow at the top of the ſtalkes, in ſpokie tufts conſiſting of foure little white leaues apiece: which being paſt, the fruit ſucceeds, round, ſomewhat long, and blacke when it is ripe, hauing vpon one ſide a ſtreaked furrow or hollowneſſe growing neere together as doe the cluſters of grapes. The root is thicke, blacke without, and yellow within like box, with many trailing ſtrings anexed thereto, creeping far abroad in the earth, whereby it doth greatly increaſe, and laſteth long.

¶ *The Place.*

Herbe Chriſtopher groweth in the North parts of England, neere vnto the houſe of the right worſhipfull Sir *William Bowes*. I haue receiued plants thereof from *Robinus* of Paris, for my garden, where they flouriſh.

¶ *The Time.*

It floureth and flouriſheth in May and Iune, and the fruit is ripe in the end of Summer.

　　　　　¶ *The*

¶ *The Names.*

It is called in our age *Christophoriana,* and *S. Christophori herba* : in English, Herbe Christopher : some there be that name it *Costus niger* : others had rather haue it *Aconitum bacciferum* : it hath no likenes at all nor affinitie with *Costus,* as the simplest may perceiue that do know both. But doubt-lesse it is of the number of the Aconites, or Wolfs-banes, by reason of the deadly and pernicious quality that it hath, like vnto Wolfes-bane, or Leopards-bane.

¶ *The Temperature.*

The temperature of Herbe Christopher answereth that of the Aconites, as we haue said.

¶ *The Vertues.*

I finde little or nothing extant in the antient or later writers, of any one good propertie where-with any part of this plant is possessed : therefore I wish those that loue new medicines to take heed that this be none of them, because it is thought to be of a venomous and deadly quality.

CHAP. 380. *Of Peionie.*

¶ *The Kindes.*

THere be three Peionies, one male, and two females, described by the Antients : the later writers haue found out foure more ; one of the female kinde, called *Paonia pumila,* or dwarfe Peionie ; and another called *Paonia promiscua siue neutra,* Bastard, Mis-begotten, or neither of both, but as it were a plant participating of the male and female ; one double Peionie with white floures, and a fourth kinde bearing single white floures.

1 *Paonia mas.*
Male Peionie.

Paonia mas cum semine.
Male Peionie in seed.

¶ *The Description.*

1   THe first kinde of Peionie (being the male, called *Paonia mas :* in English, Male Peiony) hath thicke red stalkes a cubit long : the leaues be great and large, consisting of diuers leaues growing or joyned together vpon one slender stemme or rib, not much vnlike the leaues of
the

the VVall-nut tree both in faſhion and greatneſſe : at the top of the ſtalkes grow faire large redde floures very like roſes, hauing alſo in the midſt, yellow threds or thrums like them in the roſe called *Anthera* which being vaded and fallen away; there come in place three or foure great cods or husks, which do open when they are ripe ; the inner part of which cods is of a faire red colour, wherein is contained blacke ſhining and poliſhed ſeeds, as big as a Peaſe, and betweene euery blacke ſeed is couched a red or crimſon ſeed, which is barren and empty. The root is thicke, great, and tuberous, like vnto the common Peionie.

2   There is another kinde of Peionie called of *Dodonæus*, *Pæonia fæmina prior* : of *Lobel*, *Pæonia fæmina* : in Engliſh, female Peionie, which is ſo well knowne vnto all that it needeth not any deſcription.

3   The third kinde of Peionie (which *Pena* ſetteth forth vnder the name *Pæonia fæmina Polyanthos* : *Dodonæus*, *Pæonia fæmina multiplex* : in Engliſh, Double Peionie) hath leaues, roots, and floures like the common female Peionie, ſaue that his leaues are not ſo much jagged, and are of a lighter greene colour : the roots are thicker and more tuberous, and the floures much greater, exceeding double, of a very deepe red colour, in faſhion very like the great double roſe of Prouince, but greater and more double.

2 *Pæonia fæmina.*
Female Peionie.

3 *Pæonia fæmina multiplex.*
Double red Peionie.

4   There is found another ſort of the double Peionie, not differing from the precedent in ſtalks, leaues, or roots : this plant bringeth forth white floures, wherein conſiſteth the difference.

5   There is another kinde of Peionie (called of *Dodonæus*, *Pæonia fæmina altera* : but of *Pena*, *Pæonia promiſcua, ſiue neutra* : in Engliſh, Maiden or Virgin Peiony) that is like to the common Peiony, ſauing that his leaues and floures are much leſſe, and the ſtalkes ſhorter : it beareth red floures and ſeed alſo like the former.

6   VVe haue likewiſe in our London gardens another ſort bearing floures of a pale whitiſh colour very ſingle, reſembling the female wilde Peionie, in other reſpects like the double white Peiony, but leſſer in all the parts thereof.

† 7   *Cluſius* by ſeed ſent him from Conſtantinople had two other varieties of ſingle Peionies ; the one had the leaues red when they came out of the ground ; and the floure of this was of a deep red colour : the other had them of a whitiſh greene, and the floures of this were ſomwhat larger, and of a lighter colour. In the leaues and other parts they reſembled the common double Peiony. ‡

4 *Pæonia fæmina polyanthos flore albo,*
The double white Peionie.

‡ 5 *Pæonia promiscua.*
Maiden Peionie.

‡ 6 *Pæonia fæmina pumila.*
Dwarfe female Peionie.

‡ 7 *Pæonia Byzantina.*
Turkish Peionie.

¶ *The Place.*

All tHd ſorts of Peionies do grow in our London gardens, except that double Peiony with white floures, which we do expect from the Low countries or Flanders.

The male Peionie groweth wild vpon a conny berry in Betſome, being in the pariſh of South-fleet in Kent, two miles from Grauefend, and in the ground ſometimes belonging to a farmer there called *Iohn Bradicy*.

‡ I haue beene told that our Author himſelfe planted that Peionie there, and afterwards ſee-med to finde it there by accident: and I doe beleeue it was ſo, becauſe none before or ſince haue euer ſeene or hard of it growing wilde in any part of this kingdome. ‡

¶ *The Time.*

They floure in May : the ſeed is ripe in Iuly.

¶ *The Names.*

The Peionie is called in Greeke, παιωνία : in Latine alſo, *Pæonia*, and *Dulciſida* : in ſhops, *Pionia* : in high Dutch, Peonien bloemen : in low Dutch, Maſt bloemen; in French, *Pinoine*: in Spaniſh, *Roſa del monte* : in Engliſh, Peionie: it hath alſo many baſtard names, as *Roſa fatuina, Herba Caſta* : of ſome, *Luᵣaris*, or *Lunaria Pæonia*: becauſe it cureth thoſe that haue the falling ſickneſſe, whom ſome men call *Lunaticos*, or Lunaticke. It is called *Idæus Dactylus* : which agreeth with the female Peio-nie; the knobby roots of which be like to *Dactyli Idæi*, and *Dactyli Idai* are certaine precious ſtones of the forme of a mans finger, growing in the Iſland of Candie : it is called of diuers *Aglaophotis*, or brightly ſhining, taking his name of the ſhining and glittering graines, which are of the colour of ſcarlet.

There be found two *Aglaophotides*, deſcribed by *Ælianus* in his 14 booke ; one of the ſea, in the 24. Chapter : the other of the earth, in the 27. chapter. That of the ſea is a kinde of *Fucus*, or ſea moſſe, which groweth vpon high rocks, of the bigneſſe of Tamarisk, with the head of Poppy; which opening in the Summer Solſtace doth yeeld in the night time a certaine fierie, and as it were ſparkling brightneſſe or light.

That of the earth, ſaith he, which by another name is called *Cynoſpaſtus*, lieth hid in the day time among other herbes, and is not knowne at all, and in the night time it is eaſily ſeene : for it ſhineth like a ſtar, and glittereth with a fiery brightneſſe.

And this *Aglaophotis* of the earth, or *Cynoſpaſtus*, is *Pæonia* ; for *Apuleius* ſaith, that the ſeeds or graines of Peionie ſhine in the night time like a candle, and that plenty of it is in the night ſeaſon found out and gathered by the ſhepheards. *Theophraſtus* and *Pliny* do ſhew that Peionie is gathered in the night ſeaſon ; which *Ælianus* alſo affirmeth concerneth *Aglaophotis*.

This *Aglaophotis* of the earth, or *Cynoſpaſtus*, is called of *Ioſephus* the writer of the Iewes warre, in his ſeuenth booke, 25. chapter, *Baaras*, of the place wherein it is found; which thing is plaine to him that conferreth thoſe things which *Ælianus* hath written of *Aglaophotis* of the earth, or *Cynoſpaſtus*, with thoſe which *Ioſephus* hath ſet downe of *Baaras* : for *Ælianus* ſaith, that *Cynoſpaſtus* is not pluck-ked vp without danger ; and that it is reported how he that firſt touched it, not knowing the nature thereof, periſhed. Therefore a ſtring muſt be faſtned to it in the night, and a hungry dog tied ther-to, who being allured by the ſmell of roſted fleſh ſet towards him, may plucke it vp by the roots. *Ioſephus* alſo writeth, that *Baaras* doth ſhine in the euening like the day ſtar, and that they who come neere, and would plucke it vp, can hardly do it, except that either a womans vrine, or her menſes be poured vpon it, and that ſo it may be pluckt vp at the length.

Moreouer, it is ſet downe by the ſaid Author, as alſo by *Pliny* and *Theophraſtus*, that of neceſſitie it muſt be gathered in the night; for if any man ſhall pluck off the fruit in the day time, being ſeene of the Wood-pecker, he is in danger to loſe his eies ; and if hee cut the root, it is a chance if his fundament fall not out. The like fabulous tale hath been ſet forth of Mandrake, the which I haue partly touched in the ſame chapter. But all theſe things be moſt vaine and friuolous : for the root of Peionie, as alſo the Mandrake, may be remoued at any time of the yeare, day or houre whatſoe-uer.

But it is no maruell, that ſuch kindes of trifles, and moſt ſuperſtitious and wicked ceremonies are found in the books of the moſt Antient Writers; for there were many things in their time very vainly feined and cogged in for oſtentation ſake, as by the Ægyptians and other counterfeit mates, as *Pliny* doth truly teſtifie : an imator of whom in times paſt, was one *Andreas* a Phyſition, who as *Galen* ſaith, conueied into the art of Phyſick, lies and ſubtill deluſions. For which cauſe *Galen* com-manded is Schollers to refraine from the reading of him, and of all ſuch like lying and deceitfull ſycophants. It is reported that theſe herbes tooke the name of Peionie, or *Pæan*, of that excellent Phyſition of the ſame name, who firſt found out and taught the knowledge of this herbe vnto po-ſteritie.

¶ *Thɛ*

¶ *The Temperature.*

The root of Peionie, as *Galen* ſaith, doth gently binde with a kinde of ſweetneſſe: and hath alſo joyned with it a certaine bitteriſh ſharpneſſe: it Is in temperature not very hot, little more than meanely hot; but it is dry, and of ſubtill parts.

¶ *The Vertues.*

A   *Dioſcorides* writeth, that the root of the Male Peionie being dried, is giuen to women that be not well clenſed after their deliuerie, being drunke in Mead or honied water to the quantitie of a beane; for it ſcoureth thoſe parts, appeaſeth the griping throwes and torments of the belly, and bringeth downe the deſired ſickeneſſe.

B   *Galen* addeth, that it is good for thoſe that haue the yellow jaundiſe, and paine in the kidnies and bladder, it clenſeth the liuer and kidnies that are ſtopped.

C   It is found by ſure and euident experience made by *Galen*, that the freſh root tied about the necks of children, is an effectuall remedy againſt the falling ſickeneſſe; but vnto thoſe that are growne vp in more yeares, the root thereof muſt alſo be miniſtred inwardly.

D   It is alſo giuen, ſaith *Pliny*, againſt the diſeaſe of the minde. The root of the male Peionie is preferred in this cure.

E   Ten or twelue of the red berries or ſeeds drunke in wine that is ſomething harſh or ſower, and red, do ſtay the inordinate flux, and are good for the ſtone in the beginning.

F   The blacke graines (that is the ſeed) to the number of fifteene taken in wine or mead, helpes the ſtrangling and paines of the matrix or mother, and is a ſpeciall remedy for thoſe that are troubled in the night with the diſeaſe called *Ephialtes* or night Mare, which is as though a heauie burthen were laid vpon them, and they oppreſſed therewith, as if they were ouercome by their enemies, or ouerpreſt with ſome great weight or burthen; and they are alſo good againſt melancholicke dreams.

G   Syrrup made of the floures of Peionie helpeth greatly the falling ſickneſſe: likewiſe the extraction of the roots doth the ſame.

C H A P. 381.   *Of toothed Violets, or Corall worts.*

1 *Dentaria Bulbifera.*
Toothed Violet.

2 *Dentaria Coralloide radice, ſiue Dent. Enneaphyllos.*
The Corall toothed Violet.

3 *Dentaria Heptaphyllos Clusij.*
The seuen leafed toothed Violet.

4 *Dentaria Pentaphyllos Clusij.*
Fiue leafed toothed Violet.

‡ 5 *Dentaria Pentaphyllos alter.*
The other fiue leaued Corall-wort.

¶ *The Description.*

1 THe firſt kinde of *Dentaria* (called in Latine *Dentaria baccifera* : of *Dodonæus*, *Dentaria prior* : in Engliſh, Dogs tooth violet) hath a tuberous and knobby root, toothed, or as it were kneed like vnto the crags of Corall, of an vnpleaſant ſauor, and ſomewhat ſharp in taſte: from which ſpring forth certaine ſmall and ſlender ſtalkes a foot high, which haue leaues very much cut or jagged, like vnto thoſe of Hempe, of the forme and faſhion of Aſhen leaues: at the top of the ſtalkes doe grow ſmall white floures, in ſhape like *Viola matronales*, that is, Queenes Gilloſloures, or rather like ſtocke Gilloſloures, of a white yellow colour, laid ouer with a light ſprinkling of purple : among which come forth ſmall knobs growing vpon the ſtalks among the leaues, ſuch as are to be ſeen vpon the *Chimiſts Martagon*, which being ripe do fall vpon the ground, where of many other plants are ingendred.

2 The ſecond kinde of Dogs-tooth Violet bringeth forth ſmall round ſtalks, firm, and ſtiffe, a foot high, beſet with leaues much broader, rounder, and greener than the former, bearing at the top many little floures conſiſting of foure ſmall leaues, of a pale herbie colour ; which being paſt, there ſucceed long and ſlender coddes
ſomewhat

ſomewhat like the cods of Queenes Gillofloures, wherein is contained ſmall blackiſh ſeed : the root is like the former, but not in euery reſpeƈt, much reſembling Corall, yet white and tuberous notwithſtanding.

3　The third kinde of Dogs-tooth Violet is called of *Cluſius, Dentaria heptaphyllos,* that is, con-ſiſting of ſeuen leaues faſtened vpon one rib,ſinew,or ſmall ſtem ; of *Lobel* with this title, *Alabaſtri-tes altera,*or *Dentaria altera* : but *Cordus* calleth it *Coralloides altera* : in Engliſh, Corall violet : it hath ſtalkes,floures,and roots like vnto the firſt of his kinde,ſauing that the floures are much fairer, and white of colour,and the roots haue a greater reſemblance of Corall than the other.

4　The fourth kinde of Dogs-tooth violet,called in Engliſh Codded violet (which *Cluſius* ſet-teth forth vnder the title *Dentaria Matthioli Pentaphyllos* ; which *Pena* doth alſo expreſſe vnder the ti-tle of *Nemoralis alpina Herbariorum Alabaſtrites , Cordus* calleth it *Coralloides,* and may very well be called in Engliſh Cinkfoile violet)hath leaues ſo like the greater Cinkfoile,that it is hard to know one from another; therefore it might very well haue beene reckoned among the herbes called *Pen-taphylla,*that is,fiue leaued herbes. This plant groweth in the ſhadowie forreſt about Turin, and the mountaine Sauena called Calcaris, and by the Rhene not far from Baſill. The ſtalkes grow to the height of a cubit,beſet with a tuft of flours at the top like to that of the firſt, but of a deeper purple colour:which being vaded,there ſucceed long and flat cods like vnto Rocket, or the great Celan-dine, wherein is contained a ſmall feed. All the whole plant is of a hot and bitter taſte. The roots are like vnto Corall, of a pale whitiſh colour : the leaues are rough and harſh in handling, and of a deepe greene colour.

‡　5　*Cluſius* giues vs another variety of *Dentaria pentaphyllos,* whoſe roots are more vneuen and knobby than the laſt deſcribed : the ſtalke is ſome foot high: the leaues fiue vpon a ſtalke, but not ſo rough, nor ſo deepe a greene as thoſe of the former ; yet the floures are of a deepe purple colour, like thoſe of the laſt deſcribed. ‡

¶ *The Place.*

They grow on diuers ſhadowie and darke hills. *Valerius Cordus* writeth, that they are found a-bout the forreſt Hercinia,not far from Northuſium,moſt plentifully,in a fat ſoile that hath quaries of ſtone in it. The firſt I haue in my garden.

¶ *The Time.*

They floure eſpecially in Aprill and May:the ſeed commeth to perfeƈtion in the end of Auguſt.

¶ *The Names.*

The toothed Violet, or after ſome, Dogs-tooth violet,is commonly called *Dentaria:*of *Cordus,Co-ralloides,*of the root that is in forme like to Corall. *Matthiolus* placeth it *inter Solidagines & Symphy-ta,*among the Conſounds and Comfries. We had rather call them *Viola Dentaria,* of the likeneſſe the floures haue with Stocke gillo-floures. They may be called in Engliſh, Toothed Violets, or Corall-worts.

¶ *The Nature and Vertues.*

A　I haue read of few or no vertues contained in theſe herbes,ſauing thoſe which ſome women haue experienced to be in the firſt kinde thereof, and which *Matthiolus* aſcribeth vnto *Pentaphylla denta-ria,*the fourth kinde, in the fourth booke of his Commentaries vpon *Dioſcorides,* and in the chapter concerning *Symphytum,*where he ſaith that the root is vſed in drinkes which are made againſt *En-terocele* and inward wounds, but eſpecially thoſe wounds and hurts which haue entred into the hol-lowneſſe of the breſt.

Cʜᴀᴘ. 382.　*Of Cinkefoile, or fiue finger Graſſe.*

¶ *The Deſcription.*

1　THe firſt kinde of Cinkefoile is ſo common and ſo vniuerſally knowne, that I thinke it a needleſſe trauell to ſtand about the deſcription. ‡ It hath many long ſlender ſtalks,ly-ing ſpred vpon the ground, out whereof grow leaues made of fiue longiſh ſnipt leaues faſtened to one long foot-ſtalke : the floures alſo grow vpon the like foot-ſtalkes,and are compoſed of fiue yellow leaues.The root is pretty large,of a reddiſh colour,and round ; but dried,it becomes ſquare. ‡

2　The ſecond kind of Cinkefoile or Quinquefoile hath round and ſmall ſtalks of a cubit high ; the leaues are large, and very much jagged about the edges,very like the common Cinkefoile : the floures grow at the top of the ſtalkes,in faſhion like the common kinde, but much greater, and of a pale or bleake yellow or elſe whitiſh colour : the root is blacke without,and full of ſtrings annexed thereto,and of a wooddy ſubſtance.

1 *Quinquefolium vulgare.*
Common Cinkfoile.

† 2 *Quinquefolium majus rectum.*
Great vpright Cinkfoile.

3 *Pentaphyllum purpureum.*
Purple Cinkfoile.

4 *Pentaphyllum rubrum paluſtre,*
Marſh Cinkfoile.

 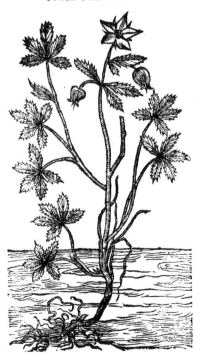

5 *Pentaphyllum petroſum, Heptaphyllum Cluſij.*
Stone Cinkfoile.

† 6 *Pentaphyllon ſupinum Potentillæ facie.*
Siluerweed Cinkfoile.

7 *Quinquefolium Tormentillæ facie.*
Wall Cinkfole.

8 *Pentaphyllum Incanum.*
Hoarie Cinkfoile.

† 3 The third kinde of Cinkfoile hath leaues like thofe of the laft defcribed, and his floures are of a purple colour; which boing paft, there fucceedeth a round knop of feed like a Strawberry before it be ripe: the ftalks are creeping vpon the ground; the root is of a wooddy fubftance, full of blacke ftrings appendant thereto. ‡ This differs not from the laft defcribed, but in the colour of the floures. ‡

4 The fourth kinde of Cinkefoile is very like vnto the other, efpecially the great kinde: The ftalkes are a cubit high, and of a reddifh colour: the leaues confift of fiue parts, fomewhat fnipped about the edges: the floures grow at the top of the ftalks like vnto the other Cinkfoiles, faue that they be of a dark red colour; the root is of a woody fubftance, with fome fibres or threddy ftrings hanging thereat.

9 *Pentaphyllum incanum minus repens.*
Small hoary creeping Cinkfoile.

Wood Cinkfoile with white floures.

⸪ The fifth kinde of Cinkfoile groweth vpon the cold mountaines of Sauoy, and in the vally of Auftenfie, and in Narbone in France, and (if my memorie faile me not) I haue feene it growing vpon Beefton caftle in Chefhire: the leaues hereof are few, and thin fet, confifting of fiue parts like the other Cinkfoiles, oftentimes fix or feuen fet vpon one foot-ftalke, not fnipt about the edges as the other, but plaine and fmooth: the leaues are of a bright white filuer colour, very foft and fhining: the floures grow like ftars vpon flender ftalkes by tufts and bunches, of a white colour, and fometimes purple, in fafhion like the floures of *Alchimilla* or Ladies mantle: the root is thicke and full of ftrings, of a browne purple colour.

‡ 6 This plant, whofe figure out Authour formerly gaue for *Fragaria fterilis*, and in his Defcription confounded with it, to auoid confufion I think fit to giue you here among the Cinkfoils, and in that place the *Fragaria fterilis*, as moft agreeable thereto. This feems to challenge kinred of three feuerall plants, that is, Cinkfoile, Tormentill, and Siluer-weed; for it hath the vpper leaues, the yellow flours, creeping branches, and root of Cinkfoile, but the lower leaues are of a dark green and grow many vpon one middle rib like thofe of Siluer-weed: the fruit is like an vnripe ftrawberty. *Lobel* calls this, *Pentaphyllum fupinum Tormentillæ facie:* and *Tabernamontanus, Quinquefolium fragiferum repens.* ‡

7 The feuenth kinde of Cinkfoile *Pena* that diligent fearcher of Simples found in the Alps of Rhetia neere Clauena, and at the firft fight fuppofed it to be a kinde of *Tormentilla* or *Pentaphyllum,*

ſaue that it had a more threddy root rather like *Geranium* ; it is of a dark colour outwardly,hauing ſome ſweet ſmell repreſenting *Gariophyllata* in the ſauor of his roots : in leaues and floures it reſembles Cinkfoile and Tormentil,and in ſhape of his ſtalks and roots,*Auens*,or *Gariophyllata*,participating of them all ; notwithſtanding it approcheth neereſt vnto the Cinkfoiles,hauing ſtalks a foot high, whereupon grow leaues diuided into fiue parts , and jagged round about the edges like the teeth of a Saw, hauing the pale yellow floures of *Pentaphylla* or *Tormentilla* ; within which are little moſſie or downy threds of the colour of Saffron,but leſſer than the common Auens.

8　　The eighth kinde of Cinkfoile (according to the opinion of diuers learned men who haue had the view thereof,and haue iudged it to be the true *Leucas* of *Dioſcorides*,agreeable to *Dioſcorides* his deſcription) is all hoary,whereupon it tooke the addition *Incanum*.The ſtalks are thick,wooddy,and ſomewhat red,wrinkled alſo,and of a brown colour,which riſe vnequall from the root, ſpreding themſelues into many branches,ſhadowing the place where it groweth,beſet with thicke and notched leaues like *Scordium* or water Germander, which according to the iudgment of the Learned is thought to be of no leſſe force againſt poiſon,than *Pentaphyllon* or *Tormentilla*,being of an aſtringent and drying qualitie : hereupon it may be that ſome tɪying the force hereof,haue yeelded it vp for *Leucas Dioſcoridis*. This rare plant I neuer found growing naturally,but in the hollowneſſe of the Peakiſh mountaines and dry grauelly vallies.

‡ 11　*Quinquefolium ſyluaticum minus flo.albo.*　　‡ 12　*Quinquefolium minus flo. aureo.*
Small white floured wood Cinkfoile.　　　　　Small golden floured Cinkfoile.

‡ 9　This hath the like creeping purple branches as the laſt deſcribed : the leaues are narrower, more hairy,and deeper cut in ; the floures are alſo of a more golden colour : in other reſpeƈts they are alike. ‡

†　10　The wood Cinkfoile hath many leaues ſpred vpon the ground,conſiſting of fiue parts; among which riſe vp other leaues ſet vpon very tall foot-ſtalks and long in reſpeƈt of thoſe that did grow by the ground,and ſomwhat ſnipt about the ends,and not all alongſt the edges. The floures grow vpon ſlender ſtalks,conſiſting of fiue white leaues.The root is thick,with diuers fibres comming from it.

‡　11　This alſo from ſuch a root as the laſt deſcribed ſends forth many ſlender branches not creeping,but ſtanding vpright,and ſet with little hoary leaues, ſnipped onely at the ends like vnto
thoſe

‡ 13 *Pentaphyllum fragiferum.*
Straw-berry Cink-foile.

thoſe of the laſt deſcribed : the tops of the branches carry pretty white floures like thoſe of the laſt deſcribed, whereof it ſeems to be a kinde, yet leſſe in each reſpect.

12   This from a blacke and fibrous root ſends forth creeping branches, ſet with leaues like the common Cinkfoile, but leſſe, ſomewhat hoary and ſhining; the ſtalkes are ſome handfull high, and on their tops carry large flours in reſpect of the ſmalneſſe of the plant, and theſe of a faire golden colour, with ſaffron coloured threds in their middle; the ſeeds grow after the manner of other Cinke-foiles : this floures in Iune, and it is *Cluſius* his *Quinquefolium 3. aureo flore.* ‡

13   There is one of the mountain Cink-foiles that hath diuers ſlender brittle ſtalkes, riſing immediatly out of the ground, where-upon are ſet by equall diſtances certaine jag-ged leaues, not vnlike to the ſmalleſt leaues of Auens : the floures are white and grow at the top, hauing in them threds yellow of co-lour, and like to the other Cinkefoiles, but altogether leſſer. The root is thicke, tough, and of a wooddy ſubſtance. ‡ The ſeeds grow cluſtering together like little Straw-berries, whence *Cluſius* calls it *Quinquefolium fragiferum.* ‡

¶ *The Place.*

They grow in low and moiſt medowes, vp-on bankes and by high-way ſides : the ſecond is onely to be found in gardens.

The third groweth in the woods of Saue-nia and Narbon, but not in England. The fourth groweth in a marſh ground adjoyning to the land called Bourne ponds, halfe a mile from Colcheſter; from whence I brought ſome plants for my Garden, where they flouriſh and proſper well.

The fifth groweth vpon Beeſton caſtle in Cheſhire : the ſixth vpon bricke and ſtone walls about London, eſpecially vpon the bricke wall in Liuer-lane.

The place of the ſeuenth and eight is ſet forth in their deſcriptions.

¶ *The Time.*

Theſe plants do floure from the beginning of May to the end of Iune.

¶ *The Names.*

Cinke-foile is called in Greeke, πεντάφυλλον : in Latine, *Quinquefolium* : the Apothecaries vſe the Greeke name *Pentaphyllon* : and ſometime the Latine name. There be very many baſtard names, wherewith I will not trouble your eares : in high Dutch, **Iunff fingerkraut** : in low Dutch, **Uiiff Unger krutit** : in Italian, *Cinquefoglie* : in French, *Quintefueille* : in Spaniſh, *Cinco en rama*. in Eng-liſh Cink-foile, Fiue finger Graſſe, Fiue leaued graſſe, and Sinkfield.

¶ *The Temperature.*

The roots of Cink-foile, eſpecially of the firſt do vehemently dry, and that in the third degree, but without biting : for they haue very little apparant heat or ſharpneſſe.

¶ *The Vertues.*

The decoction of the roots of Cinke-foile drunke, cureth the bloudy flix, and all other fluxes of A the belly, and ſtancheth all exceſſiue bleeding.

The juyce of the roots while they be young and tender, is giuen to be drunke againſt the diſeaſes B of the liuer and lungs and all poyſon.

The ſame drunke in mede or honied water, or wine wherein ſome pepper hath been mingled, cu- C reth the tertian or quartaine feuers : and being drunken after the ſame manner for thirty daies to-gether, it helpeth the falling ſickeneſſe.

The leaues vſed among herbes appropriate for the ſame purpoſe, cure ruptures and burſtings of D the rim, and guts falling in the cods.

E   The juyce of the leaues drunken doth cure the jaundiſe, and comforteth the ſtomacke and liuer.

F   The decoction of the roots held in the mouth doth mitigate the paine of the teeth, ſtaieth putrifaction, and all putrified vlcers of the mouth, helpeth the inflammations of the almonds, throat and the parts adjoyning, it ſtaieth the laske, and helpeth the bloudy flix.

G   The root boyled in vinegre is good againſt the ſhingles, appeaſeth the rage of fretting ſores, and cankerous vlcers.

H   It is reported that foure branches hereof cure quartaine agues, three tertians, and one branch quotidians : which things are moſt vaine and friuolous, as likewiſe many other ſuch like, which are not onely found in *Dioſcorides*, but alſo in other Authors, which we willingly withſtand.

I   *Ortolpho Morolto* a learned Phyſition, commended the leaues being boyled with water, and ſome *Lignum vitæ* added thereto, againſt the falling ſickeneſſe, if the patient be cauſed to ſweat vpon the taking thereof. He likewiſe commendeth the extraction of the roots againſt the bloudy flix.

† Our Author formerly in his deſcription, title, and place of growing mentioned tnat plant which he figured, and is yet kept in the ſecond place ; and in the firſt place he figured the common Cinke-foile, and made mention of it, yet without deſcription in the ſecond. That which formerly was in the ſixth place, by the name of *Pentaphyllum ſupinum*, was the ſame with that in the fifth place.

CHAP. 383. *Of Setfoile, or Tormentill.*

¶ *The Deſcription.*

Tormentilla.
Setfoile.

¶ *The Deſcription.*

THis herbe Tormentill or Setfoile is one of the Cinke-foiles, it brings forth many ſtalkes, ſlender, weake, ſcarſe able to lift it ſelfe vp, but rather lies down vpon the ground: the leaues be leſſer than Cinkfoile, but moe in number, ſometimes fiue, but commonly ſeuen, whereupon it tooke his name Setfoile, which is ſeuen leaues, and thoſe ſomwhat ſnipt about the edges: the floures grow on the tops of ſlender ſtalkes, of a yellow colour, like thoſe of the Cinkfoiles. The root is blacke without, reddiſh within, thicke, tuberous or knobby.

¶ *The Place.*

This plant loueth woods and ſhadowie places, and is likewiſe found in paſtures lying open to the Sun, almoſt euery where.

¶ *The Time.*

It floureth from May, vnto the end of Auguſt.

¶ *The Names.*

It is called of the later Herbariſts *Tormentilla*: ſome name it after the number of the leaues ἑπτάφυλλον, and *Septifolium*: in Engliſh, Setfoile and Tormentill: in high Dutch, **Birckwurtz** : moſt take it to be *Chryſogonon*; whereof *Dioſcorides* hath made a briefe deſcription.

¶ *The Temperature.*

The root of Tormentill doth mightily dry and that in the third degree, and is of thin parts: it hath in it very little heate, and is of a binding quality.

¶ *The Vertues.*

A   Tormentill is not onely of like vertue with Cinkefoile, but alſo of greater efficacie: it is much vſed againſt peſtilent diſeaſes : for it ſtrongly reſiſteth putrifaction, and procureth ſweat.

The

The leaues and roots boiled in wine, or the iuice thereof drunk, prouoke ſweat, & by that means B
driue out all venom from the heart, expell poiſon, and preſerue the body from infection in time of
peſtilence, and from all other infectious diſeaſes.

The roots dried, made into pouder, and drunke in wine, do the ſame.   C

The ſame pouder taken as aforeſaid, or in the water of a ſmiths forge, or rather in water wherein D
hot ſteele hath bin often quenched of purpoſe, cureth the laske and bloudy flix, yea although the
Patient haue adioyning to his ſcouring a grieuous feuer.

It ſtoppeth the ſpitting of bloud, piſsing of bloud, and all other iſſues of bloud in man or wo- E
man.

The decoction of the leaues and roots, or the iuice thereof drunke, is good for all wounds both F
outward and inward : it alſo openeth and healeth the ſtoppings of the liuer and lungs, and cureth
the iaundice.

The root beaten into pouder, tempered or kneaded with the white of an egge and eaten, ſtayeth G
the deſire to vomit, and is good againſt choler and melancholy.

Chap. 384.
Of wilde Tanſie or Siluer-weed.

Argentina.
Siluer-weed, or wilde Tanſie.

¶ *The Deſcription.*

Wilde Tanſie creepeth along on the ground
with fine ſlender ſtalkes and claſping tendrels : the leaues are long, made vp of many ſmall leaues like vnto thoſe of the garden Tanſy,
but leſſer ; on the vpper ſide greene, and vnder very
white. The flours be yellow, and ſtand vpon ſlender
ſtems as do thoſe of Cinkfoile.

¶ *The Place.*
It groweth in moiſt places neere highwayes and
running brooks almoſt euerie where.

¶ *The Time.*
It floureth in Iune and Iuly.

¶ *The Names.*
The later Herbariſts call it *Argentina*, of the ſiluer drops that are to bee ſeene in the diſtilled water
thereof when it is put into a glaſſe, which you ſhall
eaſily ſee rolling and tumbling vp and downe in the
bottome. ‡ I iudge it rather ſo called of the fine
ſhining ſiluer colored leaues. ‡ It is likewiſe called *Potentilla* : of diuers, *Agrimonia ſylueſtris*, *Anſerina*,
and *Tanacetum ſylueſtre* : in high-Dutch, **Genſerich**:
in low-Dutch, **Ganſerick** : in French, *Argentine* :
in Engliſh, wilde Tanſie, and Siluer-weed.

¶ *The Temperature.*
It is moderatly cold, and dry almoſt in the third
degree, hauing withall a binding facultie.

¶ *The Vertues.*
Wilde Tanſie boiled in wine and drunke, ſtops the lask and bloudy flix, and all other fluxes of A
bloud in man or woman.

The ſame boiled in water and ſalt and drunk diſſolues clotted and congealed bloud in ſuch as B
are hurt or bruiſed by falling from ſome high place.

The decoction hereof made in water cures the vlcers and cankers of the mouth, if ſome honey C
and allom be added thereto in the boiling.

Wilde Tanſie hath many other good vertues, eſpecially againſt the ſtone, inward wounds, and D
wounds of the ſecret parts, and cloſeth vp all green and freſh wounds.

The

E   The diſtilled water takes away freckles,ſpots,pimples in the face,and ſun-burning;but the herb laid to infuſe or ſteep in white wine is far better: but the beſt of all is to ſteepe it in ſtrong white wine vineger,the face being often bathed or waſhed therewith.

Chap. 385. Of Auens or herb Bennet.

1 *Caryophyllata.*
Auens, or herbe Bennet.

2 *Caryophyllata montana.*
Mountaine Auens.

¶ *The Deſcription.*

1   THe common Auens hath leaues not vnlike to Agrimonie, rough, blackiſh , and much clouen or deepely cut into diuers gaſhes : the ſtalke is round and hairy,a foot high, diuiding it ſelfe at the top into diuers branches,whereupon do grow yellow flours like thoſe of Cinkfoile or wilde Tanſie : which beeing paſt,there follow round rough reddiſh hairy heads or knops full of ſeed,which beeing ripe will hang vpon garments as Burres do.The root is thick, reddiſh within , with certaine yellow ſtrings faſtned therunto, ſmelling like Cloues or the roots of Cyperus.

2   The mountaine Auens hath greater and thicker leaues than the precedent,rougher & more hairy,not parted into three,but rather round,nicked on the edges : amongſt which riſeth vp ſlender ſtalkes,whereupon doe grow little longiſh ſharpe pointed leaues: on the top of each ſtalke doth
grow

3 *Caryophyllata Alpina pentaphyllæa.*
Fiue leaued Auens.

‡ 4 *Caryophyllata montana purpurea.*
Red floured mountain Auens.

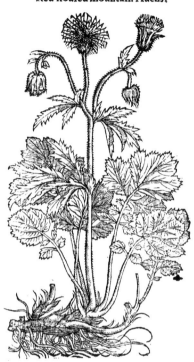

‡ 5 *Caryophyllata Alpina minima.*
Dwarfe Mountain Auens.

grow one floure greater than that of the former, which confifteth of fiue little leaues as yellow as gold : after which growes vp the feeds among long hairy threds. The root is long, growing aflope, fomewhat thicke, with ftrings annexed thereto.

3   Fiue finger Auens hath many fmall leaues fpred vpon the ground, diuided into fiue parts, fomewhat fnipt about the edges like Cinke-foile, whereof it tooke his name. Among which rife vp flender ftalkes diuided at the top into diuers branches, whereon doe grow fmall yellow floures like thofe of Cinke-foile : the root is compofed of many tough ftrings of the fmell of Cloues, which makes it a kinde of Auens, otherwife doubtles it muft of neceffitie be one of the Cinkfoiles.

‡ 4   This hath joynted ftringy roots fome finger thicke, from whence rife vp many large and hairy leaues, compofed of diuers little leaues, with larger at the top, and thefe fnipt about the edges like as the common Auens : amongft thefe leaues grow vp fundry ftalkes fome foot or better high, whereon grow floures hanging downe their heads, and the tops of the ftalkes and cups of the flours are commonly of a purplifh colour : the floures themfelues are of a pretty red colour, and are of diuers fhapes, and grow diuers waies ; which hath beene the reafon that *Clufius* and others haue iudged them feuerall plants, as may be feene in *Clufius* his Workes, where he giues you the floures, which you here finde expreft, for a different kinde. Now fome of thefe floures euen the greater part of them grow with fiue red round pointed leaues, which neuer lie faire open, but only ftand ftraight out, the middle part being filled with an hairy matter and yellowifh threds : other-fome confift of feuen, eight, nine, or more leaues ; and fome againe lie wholly open, with greene leaues growing clofe vnder the cup of the floure, as you may fee them reprefented in the figure ; and fome few now and then may be found compofed of a great many little leaues thicke thruft together, making a very double floure. After the floures are falne come fuch hairy heads as in other plants of this kind, among which lies the feed. *Gefner* calls this *Geum rivale : Thalius, Caryophyllata major purpurea : Camerarius, Caryophyllata aquatica : Clufius, Caryophyllata montana prima, & tertia.*

5   The root of this is alfo thicke, fibrous and whitifh ; from which arife many leaues three fingers high, refembling thofe of Agrimonie, the little leaues ftanding directly oppofite each againft other, fnipt about the edges, hairy, a little curld, and of a deepe greene colour : out of the midft of thofe, vpon a fhort ftalke growes commonly one fingle floure of a gold yellow colour, much like the mountaine Auens defcribed in the fecond place. It floures at the beginning of Iuly, and groweth vpon the Alpes. *Pona* was the firft that defcribed it, and that by the name of *Caryophyllata Alpina omnium minima.* ‡

¶ *The Place.*

Thefe kindes of Auens are found in high mountaines and thicke woods of the North parts of England : we haue them in our London gardens, where they flourifh and increafe infinitely.

‡   The red floured mountaine Auens was found growing in Wales by my much honoured friend Mr. *Thomas Glynn,* who fent fome plants thereof to our Herbarifts, in whofe Gardens it thriueth exceedingly. ‡

¶ *The Time.*

They floure from the beginning of May to the end of Iuly.

¶ *The Names.*

Auens is called *Caryophyllata,* fo named of the fmell of Cloues which is in the roots, and diuers call it *Sanamunda, Herba benedicta,* and *Nardus ruftica* in high Dutch, **Benedicten woꝛtꜩ:** in French, *Galiot :* of the Wallons, *Gloria filia :* in Englifh, Auens, and herbe Benet : it is thought to be *Geum Plinij,* which moft doe fufpect by reafon he is fo briefe. *Geum,* faith *Pliny, lib. 26. cap. 7.* hath little flender roots, blacke, and of a good fmell.

The other kinde of Auens is called of the later Herbarifts, *Caryophyllata montana,* Mountaine Auens : it might agree with the defcription of *Baccharis,* if the floures were purple tending to whiteneffe ; which as we haue faid are yellow, and likewife differ in that, that the roots of Auens fmell of Cloues, and thofe of *Baccharis* haue the fmell of Cinnamon.

¶ *The Temperature.*

The roots and leaues of Auens are manifeftly dry, and fomething hot, with a kinde of fcouring quality.

¶ *The Vertues.*

A   The decoction of Auens made in wine is commended againft cruditie or rawneffe of the ftomacke, paine of the Collicke, and the biting of venomuus beafts.

B   The fame is likewife a remedy for ftiches and griefes in the fides, for ftopping of the liuer ; it concocteth raw humours, fcoureth away fuch things as cleaue to the intrals, wafteth and diffolueth winde, efpecially being boyled with wine : but if it be boyled with pottage or broth it is of great efficacie, and of all other pot-herbes is chiefe, not only in Phyficall broths, but commonly to be vfed in all.

C   The leaues and roots taken in this manner diffolue and confume clottered bloud in any inward

part

part of the body; and therefore they are mixed with potions which are drunk of thoſe that are bruiſed, that are inwardly broken, or that haue falne from ſome high place.

The roots taken vp in Autumne and dried, do keepe garments from being eaten with moths, and   F
make them to haue an excellent good odour, and ſerue for all the phyſicall purpoſes that Cinke-
foiles do.

Chap. 386.   Of Straw-berries.

¶ The Kindes.

THere be diuers ſorts of Straw-berries; one red, another white, a third ſort greene, and likewiſe
a wilde Straw-berry, which is altogether barren of fruit.

1 Fragaria & Fraga.
Red Straw-berries.

2 Fragaria & Fraga ſubalba.
White Straw-berries.

¶ The Deſcription.

1   THe Straw-berry hath leaues ſpred vpon the ground, ſomewhat ſnipt about the edges,
three ſet together vpon one ſlender foot-ſtalke like the Trefoile, greene on the vpper
ſide, and on the nether ſide more white: among which riſe vp ſlender ſtems, whereon
do grow ſmall floures, conſiſting of fiue little white leaues, the middle part ſomewhat yellow, after
which commeth the fruit, not vnlike to the Mulberrie, or rather the Raſpis, red of colour, hauing the
taſte of wine, the inner pulpe or ſubſtance whereof is moiſt and white, in which is contained little
ſeeds: the root is threddy, of long continuance, ſending forth many ſtrings, which diſperſe themſelues far abroad, whereby it greatly increaſeth.

2   Of theſe there is alſo a ſecond kinde, which is like the former in ſtems, ſtrings, leaues, and
floures. The fruit is ſomething greater, and of a whitiſh colour, wherein is the difference.

There is another ſort, which brings forth leaues, floures, and ſtrings like the other of his kinde.

The

† 3 *Fragaria minime veſca, ſive ſterilis.*
Wilde or barren Straw-berry.

The fruit is greene when it is ripe, tending to
redneſſe vpon that ſide that lieth to the Sunne,
cleauing faſter to the ſtems, and is of a ſweeter
taſte, wherein only conſiſteth the difference.

‡ There is alſo kept in our gardens (onely
for variety) another Strawberry which in leaues
and growing is like the common kinde; but the
floure is greeniſh, and the fruit is harſh, rough
and prickely, beeing of a greeniſh colour, with
ſome ſhew of redneſſe. Mr. *Iohn Tradeſcant* hath
told me that he was the firſt that took notice of
this Straw-berry, and that in a womans garden
at Plimoth, whoſe daughter had gathered and
ſet the roots in her garden in ſtead of the common Straw-berry: but ſhe finding the fruit not
to anſwer her expectation, intended to throw it
away: which labor he ſpared her, in taking it and
beſtowing it among the louers of ſuch varieties,
in whoſe gardens it is yet preſerued. This may
be called in Latine, *Fragaria fructu hiſpido,* The
prickly Straw-berry. ‡

† 3 This wild Strawberry hath leaues like
the other Straw-berry, but ſomewhat leſſe, and
ſofter, ſlightly indented about the edges, and of
a light green colour: among which riſe vp ſlender ſtems bearing ſuch floures as the common
Straw-berries doe, but leſſer, which doe wither
away, leauing behinde a barren or chaffie head,
in ſhape like a Straw-berry, but of no worth or
value: the root is like the others.

¶ *The Place.*

Straw-berries do grow vpon hills and vallies,
likewiſe in woods and other ſuch places that bee ſomewhat ſhadowie: they proſper well in Gardens, the firſt euery where, the other two more rare, and are not to be found ſaue onely in gardens.
‡ The barren one growes in diuers places, as vpon Blacke-heath, in Greenewich parke, &c. ‡

¶ *The Time.*

The leaues continue greene all the yeare: in the Spring they ſpred further with their ſtrings, and
floure afterward: the berries are ripe in Iune and Iuly. ‡ The barren one floures in Aprill and May,
but neuer carries any berries. ‡

¶ *The Names.*

The fruit or berries are called in Latine by *Virgil* and *Ovid, Fraga*: neither haue they any other
name commonly knowne: the are called in high Dutch, **Erdbeeren**: in low Dutch, **Eertbeſſen**:
in French, *Fraiſes*: in Engliſh, Straw-berries.

¶ *The Temperature.*

The leaues and roots do coole and dry, with an aſtriction or binding quality: but the berries be
cold and moiſt.

¶ *The Vertues.*

A   The leaues boyled and applied in manner of a pultis taketh away the burning heate in wounds:
the decoction thereof ſtrengthneth the gummes, faſtneth the teeth, and is good to be held in the
mouth, both againſt the inflammation or burning heate thereof, and alſo of the Almonds of the
throat: they ſtay the ouermuch flowing of the bloudy flix, and other iſſues of bloud.

B   The berries quench thirſt, and do allaw the inflammation or heate of the ſtomacke: the nouriſhment which they yeeld is little, thin, and wateriſh, and if they happen to putrifie in the ſtomacke,
their nouriſhment is naught.

C   The diſtilled water drunke with white Wine is good againſt the paſſion of the heart, reuiuing
the ſpirits, and making the heart merry.

D   The diſtilled water is reported to ſcoure the face, to take away ſpots, and to make the face faire
and ſmooth; and is likewiſe drunde with good ſucceſſe againſt the ſtone in the kidnies.

E   The leaues are good to be put into Lotions or waſhing waters, for the mouth and the priuie
parts.

The

The ripe Strawberries quench thirſt, coole heate of the ſtomack and inflammation of the liuer, F take away, if they be often vſed, the redneſſe and heate of the face.

† The figure that formerly was in this place, and ſome part of the deſcription were (as I haue formerly noted) of the *Pentaphyllum ſupinum Potentilla facie*, which you may finde deſcribed among the Cinkfoiles in the ſixt place.

CHAP. 387. Of Angelica.

¶ The Kindes.

THere be diuers kindes of Angelica's ; the garden Angelica, that of the water, and a third ſort wilde growing vpon the land.

<table>
<tr><td>1 Angelica ſativa.
Garden Angelica.</td><td>2 Angelica ſylueſtris.
Wilde Angelica.</td></tr>
</table>

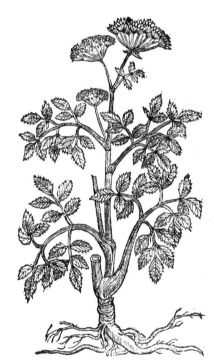

¶ The Deſcription.

1 COncerning this plant Angelica there hath bin hertofore ſome contention and contro-uerſie ; *Cordus* calling it *Smyrnium* ; ſome later writers, *Coſtus niger :* but to auoid cauil, the controuerſie is ſoone decided, ſith it and no other doth aſſuredly retaine the name Angelica. It hath great broad leaues, diuided again into other leaues, which are indented or ſnipt about much like to the vppermoſt leaues of *Sphondylium*, but lower, tenderer, greener, and of a ſtron-ger ſauor : among which leaues ſpring vp the ſtalks, very great, thicke, and hollow, ſix or ſeuen foot high, jointed or kneed : from which joints proceed other arms or branches, at the top wherof grow tufts of whitiſh floures like Fennell or Dill : the root is thicke, great, and oilous, out of which iſſu-eth, if it be cut or broken, an oily liquor : the whole plant, as well leaues, ſtalks, as roots, are of a rea-ſonable pleaſant ſauor, not much vnlike *Petroleum*.

There is another kinde of true Angelica found in our Engliſh gardens (which I haue obſerued) being like vnto the former, ſauing that the roots of this kinde are more fragrant, and of a more aro-maticke ſauor, and the leaues next the ground of a purpliſh red colour, and the whole plant leſſer.

2 The

‡ 3 *Archangelica.*
Great wilde Angelica.

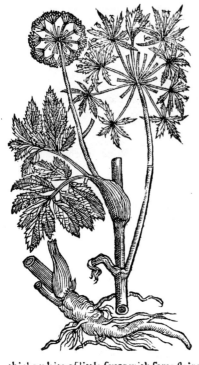

2   The wild Angelica,which ſeldom growes in gardens , but is found to grow plentifully in water-ſoken grounds and cold moiſt medowes,is like to that of the garden,ſaue that his leaues are not ſo deeply cut or jagged : they be alſo blacker and narrower : the ſtalkes are much ſlenderer and ſhorter, and the floures whiter : the root is much ſmaller, and hath more thredy ſtrings appendant thereunto,and is not ſo ſtrong of ſauor by a great deale.

3   *Matthiolus* and *Geſner* haue made mention of another kind of Angelica,but we are very ſlenderly inſtructed by their inſufficient deſcriptions: notwithſtanding for our better knowledge and more certaine aſſurance I muſt needs record that which my friend Mr *Bredwel* related to mee concerning his ſight therof,who found this plant growing by the mote which compaſſes the houſe of Mr *Munke* of the pariſh of Iuer two miles from Colbrooke ; and ſince that I haue ſeene the ſame in low fenny and marſhy places of Eſſex about Harwich. This plant hath leaues like vnto the garden Angelica, but ſmaller, and fewer in number, ſet vpon one rib a great ſtalk groſſe & thick, whoſe joints and that ſmall rib wheron the leaſe grows are of a reddiſh colour, hauing many long branches comming forth of an husk or caſe, ſuch as is in the common garden Parſnep : the floures grow at the top of the branches,and are of a white colour and tuft faſhion : which being paſt, there ſucceed broad long and thicke ſeeds, longer and thicker than garden Angelica : the root is great, thicke,white,of little ſauor,with ſome ſtrings appendant thereto.

‡ This of our Authors deſcription ſeems to agree with the *Archangelica* of *Lobel,Dodonæus,*and *Cluſius* ; wherefore I haue put their figure to it. ‡

¶ *The Place.*

The firſt is very common in our Engliſh gardens ; in other places it growes wild without planting,as in Norway,and in an Iſland of the North called Iſland,where it groweth very high:it is eaten of the inhabitants, the bark being pilled off,as we vnderſtand by ſome that haue trauelled into Iſland,who were ſometimes compelled to eat hereof for want of other food ; and they report that it hath a good and pleaſant taſte to them that are hungry.It groweth likewiſe in diuers mountains of Germanie,and eſpecially of Bohemia.

¶ *The Time.*

They floure in Iuly and Auguſt,and the roots for the moſt part periſh after the ſeed is ripe : yet haue I with often cutting the plant kept it from ſeeding,by which means the root and plant haue continued ſundry yeares together.

¶ *The Names.*

It is called of the later Age *Angelica* : in high-Dutch,𝕬𝖓𝖌𝖊𝖑𝖎𝖈𝖐,𝕭𝖟𝖚𝖘𝖙𝖜𝖚𝖗𝖙𝖟,or 𝖉𝖊𝖘 𝖍𝖊𝖎𝖑𝖎𝖌𝖍𝖊𝖓 𝕲𝖊𝖞𝖘𝖙 𝖜𝖚𝖗𝖙𝖟𝖊𝖑,that is, *Spiritus ſancti radix,*the root of the holy Ghoſt, as *Fuchſius* witneſſeth : in low-Dutch,𝕬𝖓𝖌𝖊𝖑𝖎𝖐𝖆 : in French, *Angelic* : in Engliſh alſo Angelica.

It ſeems to be a kind of *Laſerpitium* : for if it be compared with thoſe things which *Theophraſtus* at large hath written concerning *Silphium* or *Laſerpitium, lib.6.* of the hiſtorie of Plants,it ſhall appeare to be anſwerable thereunto. But whether wilde Angelica be that which *Theophraſtus* calleth *Magydaris,*that is to ſay,another kinde of *Laſerpitium,*we leaue to be examined and conſidered by the learned Phyſitians of our London Colledge.

¶ *The Temperature.*

Angelica,eſpecially that of the garden,is hot and dry in the third degree,wherefore it openeth, attenuateth,or maketh thin,digeſteth,and procureth ſweat.

¶ *The*

¶ *The Vertues.*

The root of garden Angelica is a fingular remedy againft poyfon, and againft the plague, and   A all infections taken by euill and corrupt aire ; if you doe but take a piece of the root and hold it in your mouth, or chew the fame betweene your teeth, it doth moft certainely driue away the peftilentiall aire, yea although the corrupt aire haue poffeffed the hart, yet it driueth it out againe by vrine and fweat, as Rue and Treacle, and fuch like *Antipharmaca* do.

Angelica is an enemy to poyfons : it cureth peftilent difeafes if it bee vfed in feafon : a dramme   B weight of the pouder hereof is giuen with thin wine, or if the feuer be vehement, with the diftilled water of *Carduus benedictus*, or of *Tormentill*, and with a little vinegre, and by it felf alfo, or with treacle of Vipers added.

It openeth the liuer and fpleene : draweth downe the termes, driueth out or expelleth the fecon-   C dine.

The decoction of the root made in wine, is good againft the cold fhiuering of agues.   D

It is reported that the root is auaileable againft witchcraft and inchantments, if a man carry the   E fame about them, as *Fuchfius* faith.

It attenuateth and maketh thin, groffe and tough flegme : the root being vfed greene, and while   F it is full of iuyce, helpeth them that bee afthmaticke, diffoluing and expectorating the ftuffings therein, by cutting off and clenfing the parts affected, reducing the body to health againe ; but when it is dry it worketh not fo effectually.

It is a moft fingular medicine againft furfeting and loathfomeneffe to meate : it helpeth conco-   G ction in the ftomacke, and is right beneficiall to the heart : it cureth the biting of mad dogges, and all other venomous beafts.

The wilde kindes are not of fuch force in working, albeit they haue the fame vertues attributed   H vnto them.

CHAP. 388.   *Of* Masterworts *and herbe* Gerard.

1 *Imperatoria.*   2 *Herba Gerardi.*
Mafterworts.   Herbe Gerard, or Aifh-weed.

¶ *The Deſcription.*

1   *Imperatoria* or Maſterwort hath great broad leaues not much vnlike wilde Angelica, but ſmaller, and of a deeper greene colour, in fauor like Angelica, and euery leafe diuided into ſundry other little leaues : the tender knotted ſtalks are of a reddiſh colour, bearing at the top round ſpokie tuſts with white floures : the ſeed is like the ſeed of Dill : the root is thicke, knotty and tuberous of a good fauour, and hot or biting vpon the tongue, which hath mooued the vnskilfull to call it Pellitory of Spaine, but very vnfitly and vntruly.

2   *Herba Gererda,* which *Pena* doth alſo call *Imperatoria* and *Oſtrutium : the Germaines Podagraria,* that is, Gout-wort : in Engliſh, herbe Gerard, or wilde Maſterwort, and in ſome places after *Lyte,* Aſhweed is very like the other in leaues, floures, and roots, ſauing that they be ſmaller, growing vpon long ſtems : the roots tenderer, whiter, and not ſo thicke or tuberous. The whole plant is of a reſonable good fauour, but not ſo ſtrong as Maſterwort.

¶ *The Place.*

Imperatoria groweth in darke woods and deſerts ; in my Garden and ſundry others very plentifully.

Herbe Gerard groweth of it ſelfe in gardens without ſetting or ſowing, and is ſo fruitfull in his increaſe, that where it hath once taken root, it will hardly be gotten out againe, ſpoiling and getting euery yeere more ground, to the annoying of better herbes.

¶ *The Time.*

The floure from the beginning of Iune to the beginning of Auguſt.

¶ *The Names.*

Imperatoria, or *Aſtrantia,* is called in Engliſh, Maſterwort, or baſtard Pellitory of Spaine.

Herba Gerardi is called in Engliſh, Herbe Gerard, Aiſhweed, and Gout-wort : in Latine alſo *Podagraria Germanica.*

¶ *The Nature.*

Imperatoria, eſpecially the root, is hot and dry in the third degree. The wilde *Imperatoria,* or herbe Gerard, is almoſt of the ſame nature and quality, but not ſo ſtrong.

¶ *The Vertues.*

A   *Imperatoria* is not onely good againſt all poiſon, but alſo ſingular againſt all corrupt and naughty aire and infection of the peſtilence, if it be drunken with wine.

B   The roots and leaues ſtamped, diſſolue and cure peſtilentiall carbuncles and botches, and ſuch other apoſtumations and ſwellings, being applied thereto.

C   The root drunke in wine cureth the extreme and rigorous cold fits of agues, and is good againſt the dropſie, and prouoketh ſweat.

D   The ſame taken in manner aforeſaid, comforteth and ſtrengthneth the ſtomacke, helpeth digeſtion, reſtoreth appetite, and diſſolueth all ventoſities or windineſſe of the ſtomacke and other parts.

E   It greatly helpeth ſuch as haue taken great ſquats, bruſes, or falls from ſome high place, diſſoluing and ſcattering abroad congealed and clotted bloud within the body : the root with his leaues ſtamped and laid vpon the members infected, cureth the bitings of mad dogs, and of all other venomous beaſts.

F   Herbe Gerard with his roots ſtamped, and laid vpon members that are troubled or vexed with the gout, ſwageth the paine, and taketh away the ſwelling and inflammations thereof, which occaſioned the Germanes to giue it the name *Podagraria* becauſe of his vertues in curing the gout.

G   It cureth alſo the Hemorrhoids, if the fundament be bathed with the decoction of the leaues and roots, and the ſoft and tender ſodden herbes laid thereon very hot.

H   Falſe Pellitory of Spaine attenuateth or maketh thinne, digeſteth, prouoketh ſweat and vrine, concocteth groſſe and cold humors, caſteth away windineſſe of the entrailes, ſtomacke and matrix : it is good againſt the collike and ſtone.

I   One dram of the root in pouder giuen certaine daies together, is a remedy for them that haue the dropſie, and alſo for thoſe that are troubled with convulſions, cramps, and the falling ſickeneſſe.

K   Being giuen with wine before the fit come, it cureth the quartaine ague, and is a remedy againſt peſtilent diſeaſes.

L   The ſame boyled in ſharpe or ſower wine, eaſeth the tooth-ache, if the mouth be waſhed therewith very hot.

M   Being chewed it draweth forth water and flegme out of the mouth (which kinde of remedies in Latine are called *Apophlegmatiſmi*) and disburdeneth the braine of phlegmaticke homours, and are likewiſe vſed with good ſucceſſe in apoplexies, drowſie ſleepes, and other like infirmities.

CHAP.

CHAP. 389.
Of Hercules Wound-wort, or All-heale.

¶ The Kindes.

Pᴀɴᴀx is of ſundry kindes, as witneſſeth Theophraſtus in his ninth booke; one groweth in Syria, and likewiſe other three, that is to ſay, Chironium, Heracleum, and Æſculapium; or Chirons All-heale, Hercules All-heale, and Æſculapius All-heale. Beſides theſe there is one Platyphyllon, or broad leafed; ſo that in Theophraſtus there are ſix kindes of Panax: but Dioſcorides deſcribeth onely three, Heracleum, Aſclepium, and Chironium: whereunto we haue added another ſort, whoſe vertues wee found out by meanes of an husbandman, and for that cauſe haue named it Panax Coloni, or Clownes-wort.

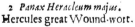

1 Panax Heracleum.   2 Panax Heracleum majus.
Hercules All-heale.   Hercules great Wound-wort.

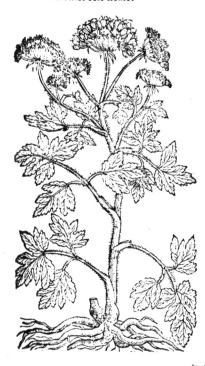

¶ The Deſcription.

1   Hᴇʀᴄᴜʟᴇs All-heale or Wound-wort hath many broad leaues ſpred vpon the ground, very rough and hairy, of an ouerworne greene colour, and deepely cut into diuers ſe-ctions like thoſe of the Cow Parſenep, and not vnlike to the fig leaues: among the which riſeth vp a very ſtrong ſtalke couered ouer with a rough hairineſſe, of the height of foure or fiue cubits. Being wounded it yeeldeth forth a yellow gummie juyce, as doth euery part of the plant, which is that precious gum called Opopanax: at the top of which ſtalkes ſtand great tufts or vmbels of yellowiſh floures, ſet together in ſpoky rundles like thoſe of Dill, which turne into ſeed of a ſtraw colour, ſharpe and hot in taſte, and of a pleaſing ſauour: the root is very thicke, fat, and full of juyce, and of a white colour.

2   The great Wound-wort, which the Venetians nouriſh in their Gardens, hath great large leaues ſomewhat rough or hairie, conſiſting of diuers ſmall leaues ſet together vpon a middle rib, which make one entire leafe joyned together in one, whereof each collaterall or ſide leafe is long

and ſharpe pointed : among which riſeth vp a knotty ſtalke three or foure cubits high, diuiding it ſelfe into diuers branches ; on the tops whereof do grow ſpokie tufts or rundles like the precedent, but the floures are commonly white : the ſeed is flat and plaine : the root long, thicke, and white, which being broken or wounded, yeeldeth forth liquor like that of the former, of a hot and biting taſte.

¶ The Place.

Theſe plants grow in Syria ; the firſt of them alſo in my Garden : but what *Panax* of Syria is, *Theophraſtus* doth not expreſſe. *Pliny* in his 12 booke, chap 26 ſaith, that the leaues are round, and of a great compaſſe : but it is ſuſpected that theſe are drawne from the deſcription of Hercules Panax.

Proad leafed Panax is thought to be the great Centorie : for *Pliny* witneſſeth, that Panax which *Chiron* found out is ſyrnamed *Centaurium*, Centorie.

Matthiolus ſaith it growes of it ſelfe in the tops of the hills Apennini, in the Cape Argentaria, in the ſea coaſts of Siena, and it is cheriſhed in the Gardens of Italy : but hee cannot affirme, That the liquor hereof is gathered in Italy ; for the liquor *Opopanex* which is ſold in Venice is brought, ſaith he, out of Alexandria a city in Ægypt : it groweth alſo in Syria, and about Phocea Citie of Bœotia.

¶ The Time.

They floure and flouriſh from the firſt of May vnto the end of September.

¶ The Names.

That which is called πάναξ in Greeke, is likewiſe named *Panax* in Latine : and that *Panax Heraclium* which *Dioſcorides* ſetteth downe is called in Latine, *Panax Heraculanum,* or *Herculeum,* or Hercules Panax : it may be called in Engliſh, Hercules his Wound-wort or All-heale, or Opopanax-wort, of the Greeke name.

¶ The Temperature.

The barke of the root of Hercules Wound-wort is hot and dry, yet leſſe than the juyce, as *Galen* teacheth.

¶ The Vertues.

A   The ſeed beat to pouder and drunke in Wormwood wine is good againſt poyſon, the biting of mad dogs, and the ſtinging of all manner of venomous beaſts.

B   The leafe or root ſtamped with honey, and brought to the forme of an Vnguent or Salue, cureth wounds and vlcers of great difficultie, and couereth bones that are bare or naked without fleſh.

Chap. 390.   Of Clownes Wound-wort, or All-heale.

¶ The Deſcription.

CLownes All-heale, or the Husbandmans Wound-wort, hath long ſlender ſquare ſtalkes of the height of two cubits, furrowed or chamfered along the ſame as it were with ſmall gutters, and ſomewhat rough or hairy : whereupon are ſet by couples one oppoſite to another, long rough leaues ſomewhat narrow, bluntly indented about the edges like the teeth of a ſaw, of the forme of the leaues of Speare-mint, and of an ouerworne greene colour : at the top of the ſtalkes grow the floures ſpike faſhion, of a purple colour mixed with ſome few ſpots of white, in forme like to little hoods. The root conſiſteth of many ſmall threddy ſtrings, whereunto are annexed or tied diuers knobby or tuberous lumpes, of a white colour tending to yellowneſſe : all the whole plant is of an vnpleaſant fauour like *Stachys* or ſtinking Horehound. ‡ The root in the Winter time and the beginning of the ſpring is ſomewhat knobby, tuberous, and joynted, which after the ſtalkes grow vp become flaccide and hollow, and ſo the old ones decay, and then it putteth forth new ones. ‡

¶ The Place.

It groweth in moiſt medowes by the ſides of ditches, and likewiſe in fertile fields that are ſomewhat moiſt, almoſt euery where ; eſpecially in Kent about South-fleet, neer to Graueſend, and likewiſe in the medowes by Lambeth neere London.

¶ The Time.

It floureth in Auguſt, and bringeth his ſeed to perfection in the end of September.

¶ The Names.

That which hath been ſaid in the deſcription ſhall ſuffice touching the names, as well in Latine as Engliſh.

‡ This

Panax Coloni.
Clownes All-heale.

‡ This plant by *Gesner* was called *Stachys paluſtris*, and *Betonica fœtida*, and thought to be of the kinde of *Herba Indaica*, or *Sideritis*; to which indeed I ſhould, and *Thalius* hath referred it, calling it *Sideritis* t. *grauis odoris: Caſalpinus* calls it *Tertiola*; and giues this reaſon, *quod Tertianas ſanet*, becauſe it cures Tertians. *Tabernamont.* called it *Stachys aquatica*, whoſe figure with a deſcription our Authour in the former edition gaue, *pag.* 565. by the name of *Marrubium aquaticum acutum*; yet (as it ſeemeth) either not knowing, or forgetting what he had formerly done, he here againe ſetteth it forth as a new thing, vnder another title: but the former figure of *Tabern.* being in my iudgment the better, I haue here giuen you, with addition of the joynted tuberous roots as they are in Winter: yet by the Caruers fault they are not altogether ſo exquiſitely expreſt as I intended. ‡

¶ *The Temperature.*

This plant is hot in the ſecond degree, and dry in the firſt.

¶ *The Vertues.*

The leaues hereof ſtamped with *Axungia* or A
hogs greaſe, and applied vnto greene wounds in manner of a pulteſſe, heale them in ſhort time, and in ſuch abſolute manner, that it is hard for any that haue not had the experience thereof to beleeue: for being in Kent about a Patient, it chanced that a poore man in mowing of Peaſon did cut his leg with a ſithe, wherein hee made a wound to the bones, and withall very large and wide, and alſo with great effuſion of bloud; the poore man crept vnto this herbe, which he bruiſed with his hands, and tied a great quantitie of it vnto the wound with a piece of his ſhirt, which preſently ſtanched the bleeding, and eeaſed the paine, inſomuch that the poore man preſently went to his daies worke againe, and ſo did from day to day, without reſting one day vntill he was perfectly whole; which was accompliſhed in a few daies, by this herbe ſtamped with a little hogs greaſe, and ſo laid vpon it in manner of a pulteſſe, which did as it were glew or ſodder the lips of the wound together, and heale it according to the firſt intention, as wee terme it, that is, without drawing or bringing the wound to ſuppuration or matter; which was fully performed in ſeuen daies, that would haue required forty daies with balſam it ſelfe. I ſaw the wound and offered to heale the ſame for charity; which he refuſed, ſaying that I could not heale it ſo well as himſelfe: a clowniſh anſwer I confeſſe, without any thankes for my good will: whereupon I haue named it Clownes Wound-wort, as aforeſaid. Since which time my ſelfe haue cured many grieuous wounds, and ſome mortall, with the ſame herbe; one for example done vpon a Gentleman of Grayes Inne in Holborne, M^r. *Edmund Cartwright*, who was thruſt into the lungs, the wound entring in at the lower part of the *Thorax*, or the breſt-blade, euen through that cartilaginous ſubſtance called *Mucronata Cartilago*, inſomuch that from day to day the frothing and puffing of the lungs did ſpew forth of the wound ſuch excrements as it was poſſeſſed of, beſides the Gentleman was moſt dangerouſly vexed with a double quotidian feuer; whom by Gods permiſſion I perfectly cured in very ſhort time, and with this Clownes experiment, and ſome of my foreknowne helpes, which were as followeth.

Firſt I framed a ſlight vnguent hereof thus: I tooke foure handfulls of the herbe ſtamped, and B
put them into a pan, whereunto I added foure ounces of Barrowes greaſe, halfe a pinte of oyle Oliue, wax three ounces, which I boyled vnto the conſumption of the juyce (which is knowne when the ſtuffe doth not bubble at all) then did I ſtraine it, putting it to the fire againe, adding thereto two ounces of Turpentine, the which I ſuffered to boile a little, reſeruing the ſame for my vſe.

The which I warmed in a ſawcer, dipping therein ſmall ſoft tents, which I put into the wound, C
defending the parts adjoyning with a plaiſter of *Calcitheos*, relented with oyle of roſes: which manner of dreſſing and preſeruing I vſed euen vntill the wound was perfectly whole: notwithſtanding once in a day I gaue him two ſpoonfulls of this decoction following.

I tooke a quart of good Claret wine, wherein I boyled an handfull of the leaues of *Solidago* D
Saracenica,

Saracenica, or Saracens confound, and foure ounces of honey, whereof I gaue him in the morning two Spoonefulls to drinke in a ſmall draught of wine tempered with a little ſugar.

E   In like manner I cured a Shoo-makers ſeruant in Holborne, who intended to deſtroy himſelfe for cauſes knowne vnto many now liuing: but I deemed it better to couer the fault, than to put the ſame in print, which might moue ſuch a graceleſſe fellow to attempt the like: his attempt was thus; Firſt, he gaue himſelfe a moſt mortall wound in the throat, in ſuch ſort, that when I gaue him drinke it came forth at the wound, which likewiſe did blow out the candle: another deepe and grieuous wound in the breſt with the ſaid dagger, and alſo two others in *Abdomine* or the nether belly, ſo that the *Zirbus* or fat, commonly called the caule, iſſued forth, with the guts likewiſe: the which mortall wounds, by Gods permiſſion, and the vertues of this herbe, I perfectly cured within twenty daies: for the which the name of God be praiſed.

Chap. 391.   *Of Magydare, or Laſer-wort.*

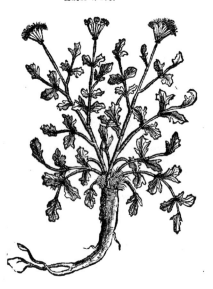

¶ *The Deſcription.*

IT ſeemeth that neither *Dioſcorides* nor yet *Theophraſtus* haue euer ſeene *Laſerpitium, Sagapenum,* or any other of the gummiferous roots, but haue barely and nakedly ſet downe their iudgments vpon the ſame, either by heare-ſay, or by reading of other mens Workes. Now then ſeeing the old Writers be vnperfect herein, it behooueth vs in this caſe to ſearch with more diligence the truth hereof; and the rather, for that very few haue ſet forth the true deſcription of that Plant which is called *Laſerpitium,* that is indeed the true *Laſerpitium,* from the roots whereof flow that ſap or liquor called *Laſer.* This plant, as *Pena* and *Lobel* themſelues ſay, was found out not far from the Iſles which *Dioſcorides* calls Stœchades, ouer againſt Maſſilia, among ſundry other rare plants. His ſtalke is great and thicke like Ferula, or Fennell gyant: The leaues are like vnto the common Smallage, and of an vnpleaſant ſauour. The floures grow at the top of the ſtalkes, tuft-faſhion like Ferula or Fennell: which being paſt, there ſucceed broad and flat ſeeds like Angelica, of a good ſauour, and of the colour of Box. The roots are many, comming from one head or chiefe root, and are couered ouer with a thicke and fat barke. Theſe roots and ſtalks being ſcarified or cut, there floweth out of them a ſtrong liquor, which being dried is very medicinable, and is called *Laſer.*

¶ *The Place.*

There be ſundry ſorts of Laſer, flowing from the roots and ſtalkes of *Laſerpitium,* the goodneſſe or quality whereof varieth according to the country or clymate wherein the plant groweth. For the beſt groweth vpon the high mountaines of Cyrene and Africa, and is of a pleaſant ſmell: in Syria alſo, Media, Armenia, and Lybia; the liquour of which plant growing in theſe places is of a moſt ſtrong and deteſtable ſauour. *Lobel* reporteth, that *Iacobus Rainaudus* an Apothecarie of Maſſilia was the firſt that made it knowne, or brought the plants thereof to Montpellier in France, vnto the learned *Rondeletius,* who right well beholding the ſame, concluded, that of all the kindes of Ferula that he had euer ſeene, there was not any ſo anſwerable vnto the true *Laſerpitium* as this onely plant.

¶ *The Time.*

This Plant floureth in Montpellier about Midſummer.

¶ *The*

¶ *The Names.*

It is called in Latine, *Laſerpitium*: in Engliſh, Laſerwort, and Magydare : the gum or liquor that iſſueth out of the ſame is called *Laſer*, but that which is gathered from thoſe plants that grow in Media and Syria, is called *Aſa fœtida*.

¶ *The Nature.*

Laſerpitium, eſpecially the root, is hot and dry in the third degree : *Laſer* is alſo hot and dry in the third degree, but it exceedeth much the heat of the leaues and ſtalkes of *Laſerpitium*.

¶ *The Vertues.*

The root of *Laſerpitium* well pounded, or ſtamped with oyle, ſcattereth clotted bloud, taketh a- A way blacke and blew markes that come of bruiſes or ſtripes, cureth and diſſolueth the Kings-euill, and all hard ſwellings and botches, the places being annointed or plaiſtered therewith.

The ſame root made into a plaiſter with the oyle of Ireos and wax, doth both aſſwage and cure B the Sciatica, or gout of the hip or huckle bone.

The ſame holden in the mouth and chewed, doth aſſwage the tooth-ache; for they are ſuch roots C as draw from the braine a great quantitie of humors.

The liquour or gum of *Laſerpitium*, eſpecially the *Laſer* of Cyrene broken and diſſolued in wa- D ter and drunken, taketh away the hoarſeneſſe that commeth ſuddenly : and being ſupt vp with a reare egge, cureth the cough : and taken with ſome good broth or ſupping, is good againſt an old pleuriſie.

Laſer cureth the jaundiſe and dropſie, taken with dried figs : alſo being taken in the quantitie of E a ſcruple, with a little pepper and Myrrhe, is very good againſt ſhrinking of ſinewes, and members out of joynt.

The ſame taken with hony and vinegre, or the ſyrrup of vinegre, is very good againſt the falling F ſickeneſſe.

It is good againſt the flux of the belly comming of the debilitie and weakeneſſe of the ſtomacke G (called in Latine *Cœliacus morbus*) if it be taken with raiſons of the Sunne.

It driueth away the ſhakings and ſhiuerings of agues, being drunke with wine, pepper, and white H Frankincenſe. Alſo there is made an electuarie thereof called *Antidotus ex ſucco Cyrenaico*, which is a ſingular medicine againſt feuer quartaines.

It is excellent againſt the bitings of all venomous beaſts, and venomous ſhot of darts or arrows, I not onely taken inwardly, but alſo applied outwardly vpon wounds.

It bringeth to maturation, and breaketh all peſtilentiall impoſtumes, botches and carbuncles, K being applied thereto with Rue, Salt-peter, and hony : after the ſame manner it taketh away cornes after they haue beene ſcarified with a knife.

Being laid to with Copperas and Verdigreaſe, it taketh away all ſuperfluous outgrowings of L the fleſh, the Polypus that happeneth in the noſe, and all ſcuruie mangineſſe.

If it be applied with vinegre, pepper, and wine, it cureth the naughty ſcurfe of the head, and fal- M ling off of the haire.

The gum or liquor of *Laſerpitium* which groweth in Armenia, Lybia, and ſundry other places, is N that ſtinking and lothſome gum called of the Arabian Phyſitions *Aſa* and *Aſſa*, as alſo with vs in ſhops *Aſa fœtida* : but the *Laſerpitium* growing in Cyrene is the beſt, and of a reaſonable pleaſant ſmell, and is called *Laſer* to diſtinguiſh and make difference betweene the two juyces; though *Aſa fœtida* be good for all purpoſes aforeſaid, yet is it not ſo good as *Laſer* of Cyrene : it is good alſo to ſmell vnto, and to be applied vnto the nauels of women vexed with the choking, or riſing of the mo-ther.

✝ That figure which formerly was in this place, was of the common Lovage deſcribed in the following chapter.

C ʜ ᴀ ᴘ. 392.　*Of common Louage.*

¶ *The Deſcription.*

A Ntient writers haue added vnto this common kinde of Lovage, a ſecond ſort, yet knowing that the plant ſo ſuppoſed is the true *Siler montanum*, and not *Leuiſticum*, though others haue alſo deemed it *Laſerpitium* ; theſe two ſuppoſitions are eaſily anſwered, ſith they bee ſundry kindes of plants, though they be very neere in ſhape and faculties one vnto another. This plant being

† *Leuiſticum vulgare.*
Common Louage.

A

B

being our common garden Louage, hath large and broad leaues, almoſt like to ſmallage. The ſtalks are round, hollow and knotty, 3. cubits high, hauing ſpokie tufts, or buſhy rundles; and at the top of the ſtalkes of a yellow colour, a round, flat, and browne ſeed, like the ſeed of Angelica : the root is long and thicke, and bringeth forth euery yeare new ſtems.

¶ *The Place.*

The right *Leuiſticum* or 'Louage groweth in ſundry gardens, and not wild (as far as I know) in England.

¶ *The Time.*

Louage floureth moſt commonly in Iuly and Auguſt.

¶ *The Names.*

It is called in Latine, *Leuiſticum* : and by ſome, *Liguſticum* : of other ſome, *Siler montanum,* but not truely : in high Dutch, **Libſtockel** : in French, *Liuiſche* : in low Dutch, **Lauetſe** : in Engliſh, Louage.

¶ *The Temperature.*

This plant is hot and dry in the third de-gree.

¶ *The Vertues.*

The roots of Louage are very good for all inward diſeaſes, driuing away ventoſities or windineſſe, eſpecially of the ſtomacke.

The ſeed thereof warmeth the ſtomack, hel-peth digeſtion; wherefore the people of Gennes in times paſt did vſe it in their meates, as wee doe pepper, according to the teſtimony of *Ant. Muſa.*

C　　The diſtilled water of Louage cleareth the ſight, and putteth away all ſpots, lentils, freckles, and redneſſe of the face, if they be often waſhed therewith.

　† The figure that was here was of the *Siler montanum,* or *Seſeli officinarum.*

CHAP. 393. *Of Cow Parſnep.*

¶ *The Deſcription.*

THis plant *Sphondilium* groweth in all Countries, and is knowne by the name of wilde Parſnep or Sphondylium, whereunto it effectually anſwereth, both in his grieuous and ranke ſauour, as alſo in the likeneſſe of the root, whereupon it was called *Sphondilium* ; and of the Germans, *Acanthus,* but vntruly : the leaues of this plant are long and large, not much vnlike the leaues of wilde Parſnep, or *Panax Heracleum* ; deeply notched or cut about the edges like the teeth of a ſaw, and of an ouerworne greene colour. The floures grow in tufts or rundles, like vnto wilde Parſneps : the root is like to Henbane : this herbe in each part thereof hath an euill ſauour, and differeth from the right *Acanthium,* not onely in faculties, but euen in all other things.

¶ *The Place.*

This plant groweth in fertile moiſt medowes, and feeding paſtures, very commonly in all parts of England, or elſewhere in ſuch places as I haue trauelled.

¶ *The Time.*

Sphondylium floureth in Iune and Iuly.

¶ *The Names.*

It is called in Greeke σπονδύλιον : in Latine likewiſe *Sphondyliam* : in the ſhops of high and low Ger-many

† *Sphondylium.*
Cow Parſenep.

many *Branca vrſina*, who vnaduiſedly in times paſt haue vſed it in clyſters, in ſtead of Brancke Vrſine, and thereupon haue named it **Bernclaw**: in Engliſh, Cow Parſnep, medow Parſenep, and Madnep.

¶ *The Temperature.*

Cow Parſnep is of a manifeſt warme complexion.

¶ *The Vertues..*

The leaues of this plant doe conſume and **A** diſſolue cold ſwellings if they bee bruiſed and applied thereto.

The people of Polonia and Lituania vſe **B** to make drinke with the decoction of this herbe, and leuen or ſome other thing made of meale, which is vſed in ſtead of beere and other ordinary drinke.

The ſeed of Cow Parſenep drunken, ſcou-**C** reth out flegmaticke matter through the guts, it healeth the jaundiſe, the falling ſickneſſe, the ſtrangling of the mother, and them that are ſhort winded.

Alſo if a man bee falne into a dead ſleepe, **D** or a ſwoune, the fume of the ſeed will waken him againe.

If a phrenticke or melancholicke mans **E** head bee annointed with oyle wherein the leaues and roots haue beene ſodden, it helpeth him very much, and ſuch as bee troubled with the head-ache and the lethargie, or ſickneſſe called the forgetfull euill.

‖ The figure formerly was of the *Paſtinacà Syluestris, or E.aphobosceum* of *Tabernamontanus*, and the figure that ſhould haue beene here was afterwards vnder the title *ri Hippoſelinum.*

C H A P. 394. *Of Herbe Frankincenſe.*

¶ *The Deſcription.*

1　THere hath beene from the beginning diuers plants of ſundry kindes, which men haue termed by this glorious name *Libanotis*, onely in reſpect of the excellent and fragrant ſmell which they haue yeelded vnto the ſences of man, ſomewhat reſembling Frankincenſe. The ſent and ſmell *Dioſcorides* doth aſcribe to the root of this firſt kind, which bringeth forth a long ſtalke with joynts like Fennell, whereon grow leaues almoſt like Cheruill or Hemlocks, ſauing that they be greater, broader, and thicker: at the top of the ſtalkes grow ſpokie taſſels bearing whitiſh floures, which do turne into ſweet ſmelling ſeed, ſomewhat flat, and almoſt like the ſeed of Angelica. The root is blacke without, and white within, hairie aboue, at the parting of the root and ſtalke like vnto *Meum* or *Peucedanum*, and ſauoureth like vnto Roſine, or Frankincenſe.

2　The ſecond kinde of *Libanotis* hath alſo a ſtraight ſtalke, full of knots and joynts: the leaues are like vnto Smallage: the floures grow in taſſels like vnto the former, and bring forth great, long and vneuen ſeed, of a ſharpe taſte: the root is like the former, and ſo is the whole plant very like, but leſſer.

3　The third kinde of *Libanotis* differeth ſomewhat from the others in forme and ſhape, yet it agreeth with them in ſmell, which in ſome ſort is like Frankinſence: the leaues are whiter, longer, and rougher than the leaues of Smallage: the ſtalkes do grow to the height of two cubits, bearing at the top the ſpokie tufts of Dill, ſomwhat yellow: the root is like the former, but thicker, neither wanteth it hairy taſſels at the top of the root; which the others alſo haue before rehearſed.

4　T car-

1 *Libanotis Theophrafti major.*
Great herbe Frankinfence.

2 *Libanotis Theophrafti minor.*
Small herbe Frankinfence.

3 *Libanotis Theophrafti nigra.*
Blacke herbe Frankinfence.

4 *Libanotis Galeni, Cachrys verior.*
Rofemary Frankinfence.

4 I cannot finde among all the plants called *Libanotides*, any one more agreeable to the true and right *Libanotis*, of *Dioſcorides* than this herbe, which ariſeth vp to the height of fiue or ſix cubits with the cleere ſhining ſtalkes of *Ferula* ; diuiding it ſelfe from his knotty joynts into ſundry arms or branches, ſet full of leaues like Fennell, but thicker and bigger, and fatter than the leaues of *Cotula fœtida*, of a grayiſh greene colour, bearing at the top of the ſtalks the tufts of *Ferula*, or rather of Carrots, full of yellow floures: which being paſt there ſucceedeth long flat ſeed like the ſeed of the Aſh tree, ſmelling like Roſin, or Frankincenſe, which being chewed filleth the mouth with the taſt of Frankincenſe, but ſharper: all the reſt of the plant is tender, and ſomewhat hot, but not vnpleaſant: the plant is like vnto *Ferula*, and aboundeth with milke as *Ferula* doth, of a reaſonable good ſauour.

¶ *The Place.*

I haue the two laſt kindes growing in my Garden; the firſt and ſecond grow vpon the high Deſerts and mountaines of Germanie.

¶ *The Time.*

Theſe herbes do floure in Iuly and Auguſt.

¶ *The Names.*

This herbe is called in Greeke, Λιβανωτίς, becauſe their roots do ſmell like incenſe, which is called in Greeke, λιβανός in Latine, *Roſmarinus*; the firſt may be Engliſhed great Frankincenſe Roſemarie ; the ſecond ſmall Frankincenſe Roſemarie ; M^r. *Lite* calleth the third in Engliſh, blacke Hart-root, the fourth white Hart-root: the ſeed is called *Cachrys* or *Canchrys*.

¶ *The Temperature.*

Theſe herbes with their ſeeds and roots are hot and dry in the ſecond degree, and are of a digeſting, diſſoluing and mundifying quality.

¶ *The Vertues.*

The leaues of *Libanotis* pounded, ſtop the flux of the Hemorrhoides or piles, and ſupple the ſwellings and inflammations of the fundament called *Condylomata*, concoct the ſwellings of the throat called *Struma*, and ripen botches that will hardly be brought to ſuppuration or to ripeneſſe. A

The juice of the leaues and roots mixed with hony, and put into the eies, doth quicken the ſight, and cleereth the dimneſſe of the ſame. B

The ſeed mingled with honey, doth ſcoure and clenſe rotten vlcers, and being applied vnto cold and hard ſwellings conſumeth and waſteth them. C

The leaues and roots boyled vntill they be ſoft, and mingled with the meale of Darnell and vineger, aſſwage the paine of the gout, if they be applied thereto. D

Moreouer being receiued in wine and pepper, it helpeth the jaundiſe, and prouoketh ſweat, and being put into oile aud vſed as an ointment, it cureth ruptures alſo. E

It purgeth the diſeaſe called in Greeke, Λέπρα: in Latine, *Vitiligo*, or *Impetigo*, that is, the white ſpottines of the skin, chaps, or rifts in the palms of the hands and ſoles of the feet, and by your patience couſin german to the ſcab of Naples, tranſported or transferred into France, and prettily well ſprinkled ouer our Northern coaſts. F

When the ſeed of *Libanotis* is put into receits, you muſt vnderſtand, that it is not meant of the ſeed of Cachris, becauſe it doth with his ſharpeneſſe exaſperate or make rough the gullet; for it hath a very heating quality, and doth dry very vehemently, yea this ſeed being taken inwardly, or the herbe it ſelfe, cauſeth to purge vpward and downeward very vehemently. G

C H A P. 395. *Of Corianders.*

¶ *The Deſcription.*

1 THe firſt or common kinde of Coriander is a very ſtinking herbe, ſmelling like the ſtinking worme called in Latine *Cimex* : it hath a round ſtalke full of branches, two foot long. The leaues are of a faint greene colour, very much cut or jagged : the leaues that grow loweſt, and ſpring firſt, are almoſt like the leaues of Cheruill or Parſley, but thoſe which come forth afterward, and grow vpon the ſtalkes, are more jagged, almoſt like the leaues of Fumitorie, though a great deale ſmaller, tenderer, and more jagged. The floures are white, and doe grow in round taſſels like vnto Dill. The ſeed is round, hollow within, and of a pleaſant ſent and ſauour when it is dry. The root is hard, and of a wooddy ſubſtance, which dieth when the fruit is ripe, and ſoweth it ſelfe from yeare to yeare, whereby it mightily increaſeth.

There

2　There is a ſecond kinde of Coriander very like vnto the former, ſauing that the bottome leaues and ſtalks are ſmaller : the fruit thereof is greater, and growing together by couples, it is not ſo pleaſant of ſauour nor taſte, being a wilde kinde thereof, vnfit either for meat or medicine.

1　*Coriandrum.*
Coriander.

‡ 2　*Coriandrum alterum minus odorum.*
Baſtard Coriander.

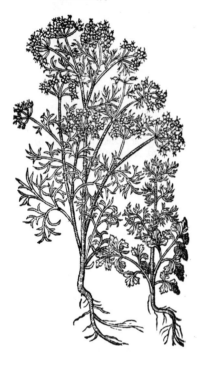 

¶ *The Place.*

Coriander is ſowne in fertile fields and gardens, and the firſt doth come of it ſelfe from time to time in my garden, though I neuer ſowed the ſame but once.

¶ *The Time.*

They floure in Iune and Iuly, and deliuer their ſeed at the end of Auguſt.

¶ *The Names.*

The firſt is called in Latine *Coriandrum :* in Engliſh, Corianders. The ſecond, *Coriandrum alterum,* wilde Corianders.

¶ *The Temperature.*

The greene and ſtinking leaues of Corianders are of complexion cold and dry, and very naught, vnwholeſome, and hurtfull to the body.

The dry and pleaſant well ſauouring ſeed is warme, and very conuenient to ſundry purpoſes.

¶ *The Vertues.*

A　Coriander ſeed prepared and couered with ſugar, as comfits, taken after meat cloſeth vp the mouth of the ſtomacke, ſtaieth vomiting, and helpeth digeſtion.

B　The ſame parched or roſted, or dried in an ouen, and drunke with wine killeth and bringeth forth wormes, ſtoppeth the laske, and bloudy flix, and all other extraordinary iſſues of bloud.

The manner how to prepare Coriander, both for meat and medicine.

C　Take the ſeed well and ſufficiently dried, whereupon poure ſome wine and vinegre, and ſo leaue them to infuſe or ſteepe foure and twenty houres, then take them forth and dry them, and keepe them for your vſe.

D　The greene leaues of Coriander boiled with the crums of bread or barley meale, conſume all
hot

hot fwellings and inflammations; and with Beane meale diffolue the Kings euill, wens, and hard lumpes.

The juice of the leaues mixed and laboured in a leaden mortar, with Cerufe, Litharge of filuer, F vineger, and oile of Rofes, cureth S. Anthonies fire, and takes away all inflammations whatfoeuer.

The juice of the greene Coriander leaues, taken in the quantitie of foure dragmes, killeth and G poifoneth the body.

The feeds of Coriander prepared with fugar, preuaile much againft the gout taken in fome fmal H quantitie before dinner vpon a fafting ftomack, and after dinner the like without drinking immediately after the fame, or in three or foure houres. Alfo if the fame be taken after fupper, it preuaileth the more, and hath more fuperiority ouer the difeafe.

Alfo if it be taken with meat fafting, it caufeth good digeftion, and fhuts vp the ftomacke, kee- I peth away fumes from rifing vp out of the fame: it taketh away the founding in the eares, drieth vp the rheume, and eafeth the fquinancie.

Chap. 396. Of Parfley.

Apium hortenfe.
Garden Parfley.

¶ The Defcription.

1 THe leaues of garden Parfley are of a beautifull greene, confifting of many little ones faftned together, diuided moft commonly into three parts, and alfo fnipt round about the edges: the ftalke is aboue one cubit high, flender, fomthing chamfered, on the top wherof ftand fpoked rundles, bringing forth very fine little floures, and afterwards fmall feeds fomwhat of a fiery taft: the root is long & white, & good to be eaten.

2 There is another garden Parfley in taft and vertue like vnto the precedent : the onely difference is , that this plant bringeth forth leaues very admirably crifped or curled like fannes of curled feathers , whence it is called *Apium crifpum, fiue multifidum,* Curl'd Parfley.

‡ 3 There is alfo kept in fome gardens another Parfley called *Apium fiue Petrofelinum Virginianū,* or Virginian parfley; it hath leaues like the ordinary, but rounder, and of a yellowifh greene colour, the ftalkes are fome three foot high, diuided into fundry branches wheron grow vmbells of whitifh floures : the feeds are like, but larger than thofe of the common Parfley, & when they are ripe they commonly fow themfelues, and the old roots die, and the young ones beare feed the fecond yeare after their fowing. ‡

¶ The Place.

It is fowne in beds in gardens ; it groweth both in hot and cold places, fo that the ground be either by nature moift, or bee oftentimes watered : for it profpereth in moift places, and is delighted with water, and therefore it naturally commeth vp neere to fountaines or fprings : *Fuchfius* writeth that it is found growing of it felfe in diuers fenny grounds in Germany.

¶ The Time.

It may be fown in time but it flowly commeth vp: it may oftentimes be cut and cropped: it bringeth forth his ftalks the fecond yeare: the feeds be ripe in Iuly or Auguft.

¶ The Names.

Euery one of the Parfleyes is called in Greeke ꜱꜱꜱ but this is named, ꜱꜱꜱꜱ, that is to fay, *Apium hortenfe:* the Apothecaries and common Herbarifts name it *Petrofelinum :* in high Dutch, ꜱꜱꜱꜱ terſilgen : in low Dutch, **Crimen Peterſelie :** in French, *du Perſil :* in Spaniſh, *Perexil Iuliuert,* and *Salſa :* in Italian, *Petroſello :* in Engliſh, Perſele, Parſely, common Parſley, and garden Parſley. Yet is it not the true and right *Petroſelinum* which groweth vpon rockes and ſtones, whereupon it tooke his name, and whereof the beſt is in Macedonia: therefore they are deceiued who think that garden Parſley doth not differ from ſtone Parſley, and that the only difference is, for that Garden Parſley is of leſſe force than the wild; for wild herbs are more ſtrong in operation than thoſe of the garden.

¶ *The Temperature.*

Garden Parſley is hot and dry, but the ſeed is more hot and dry, which is hot in the ſecond degree, and dry almoſt in the third : the root is alſo of a moderate heate.

¶ *The Vertues.*

A   The leaues are pleaſant in ſauces and broth, in which beſides that they giue a pleaſant taſt, they be alſo ſingular good to take away ſtoppings, and to prouoke vrine: which thing the roots likewiſe do notably performe if they be boiled in broth : they be alſo delightfull to the taſte, and agreeable to the ſtomacke.

B   The ſeeds are more profitable for medicine, they make thinne, open, prouoke vrine, diſſolue the ſtone, break and waſt away winde, are good for ſuch as haue the dropſie, draw downe menſes, bring away the birth, and after-birth : they be commended alſo againſt the cough, if they bee mixed or boiled with medicines made for that purpoſe : laſtly they reſiſt poiſons , and therefore are mixed with treacles.

C   The roots or the ſeeds of any of them boiled in ale and drunken, caſt forth ſtrong venom or poiſon, but the ſeed is the ſtrongeſt part of the herbe.

D   They are alſo good to be put into clyſters againſt the ſtone or torments of the guts.

CHAP. 397.   *Of water Parſley, or Smallage.*

Eleoſelinum, ſiue Paludapium.
Smallage.

¶ *The Deſcription.*

SMallage hath green ſmooth and glittering leaues , cut into very many parcels , yet greater & broader than thoſe of common Parſley: the ſtalkes be chamfered and diuided into branches, on the tops whereof ſtand little white floures; after which do grow ſeeds ſomething leſſer than thoſe of common Parſley: the root is faſtned with many ſtrings.

¶ *The Place.*

This kind of Parſley delighteth to grow in moiſt places, and is brought from thence into gardens. ‡ It growes wilde abundantly vpon the bankes in the ſalt marſhes of Kent and Eſſex. ‡

¶ *The Time.*

It flouriſhes when the garden Parſley doth, and the ſtalke likewiſe commeth vp the next yeare after it is ſown, and then alſo it bringeth forth ſeeds which are ripe in Iuly & Auguſt.

¶ *The Names.*

It is called in Greeke ἑλειοσέλινον : of *Gaza, Paludapium :* in ſhoppes, *Apium,* abſolutely without any addition: in Latine, *Paluſtre Apium,* and *Apium ruſticum :* in high Dutch **Epffich:** in low Dutch, **Eppe,** & of diuers **Jouffroumerck :** in Spaniſh and Italian, *Apio :* in French *de L'ache :* in Engliſh, Smallage, Marſh Parſley, or water Parſley.

¶ *The Temperature.*

This Parſley is like in temperature and vertues to that of the garden, but it is both hotter and

drier

drier,and of more force in moſt things: this is ſeldome eaten,neither is it counted good for ſauce, but is very vnprofitable in medicine.

¶ *The Vertues.*

The juice thereof is good for many things,it clenſeth,openeth,attenuateth,or maketh thin;it re- **A** moueth obſtructions and prouoketh vrine,and therefore thoſe ſyrrups which haue this mixt with them,as that which is called *Syrupus Byzantinus*,open the ſtoppings of the liuer and ſpleen,and are a remedie for long laſting agues,whether they be tertians or quartans,and all other which proceed both of a cold cauſe and alſo of obſtructions or ſtoppings , and are very good againſt the yellow jaundiſe.

The ſame juice doth perfectly cure the malitious and venomous vlcers of the mouth and of the **B** almonds of the throat,with the decoction of Barly and *Mel Roſarum* or hony of Roſes added, if the parts be waſhed therewith : it likewiſe helpeth all outward vlcers and foule wounds : with hony it is profitable alſo for cankers exulcerated,for although it cannot cure them yet it keeps them from putrifaction,and preſerueth them from ſtinking : the ſeed is good for thoſe things for which that of the garden Parſly is;yet is not the vſe thereof ſo ſafe, for it hurteth thoſe that are troubled with the falling ſickneſſe,as by euident proofes it is very well knowne.

Smallage,as *Pliny* writeth,hath a peculiar vertue againſt the biting of venomous ſpiders. **G**

The juice of Smallage mixed with hony and bean floure,doth make an excellent mundificatiue **D** for old vlcers and malignant ſores,and ſtayeth alſo the weeping of the cut or hurt ſinues in ſimple members,which are not very fatty or fleſhy,and bringeth the ſame to perfect digeſtion.

The leaues boiled in hogs greaſe,and made into the forme of a pultis, take away the pain of fe- **E** lons and whitlowes in the fingers,and ripen and heale them.

Chap.398.　*Of Mountaine Parſley.*

† *Oreoſelinum.*
Mountaine Parſley.

¶ *The Deſcription.*

THe ſtalke of mountaine Parſley,as *Dioſcorides* writeth, is a ſpan high, growing from a ſlender root ; vpon which are branches and little heads like thoſe of Hemlocke , yet much ſlenderer : on which ſtalks grow the ſeed, which is long,of a ſharp or biting taſte,ſlender,and of a ſtrong ſmell like vnto Cumin:but we canot find that this kinde of mountain Parſly is knowne in our age: the leaues of this we here giue are like thoſe of common Parſly, but greater & broader, conſiſting of many ſlender footſtalks faſtned vnto them : the ſtalke is ſhort, the floures on the ſpoked tufts be white, the ſeed ſmall;the root is white, and of a meane length or bigneſſe,in taſte ſomewhat biting and bitteriſh , and of a ſweet ſmell.

¶ *The Place.*

† *Dioſcorides* writeth,That mountain parſley groweth vpon rocks and mountains. And *Dodonæus* affirmeth that this herb deſcribed grows vpon the hills which diuide Sileſia from Morauia, called in times paſt the countrey of the Marcomans : alſo it is ſaid to be found on other mountains and hils in the North parts of England.

¶ *The Names.*

The Grecians do name it of the mountaines, Ὀρεοσέλινον, which the Latines alſo for that cauſe do call *Apium montanum*,and *Montapium*: in Engliſh, mountaine Parſley : in Latine, *Apium* : but *Dioſcorides* maketh *Petroſelinum* or ſtone Parſley to differ from mountaine Parſley ; for, ſaith hee, we muſt not be deceiued, taking mountaine Parſley to be that which groweth on rocks, for Rock Parſley is another plant: of ſome it is called **Veelgutta :** in Latine, *Multi-bona :* in Engliſh, Much Good, ſo named becauſe it is good and profitable for many things : and this is not altogether vn-properly named *Oreoſelinum*, or mountaine Parſley, for it growes, as we haue ſaid, on mountains, and is not vnlike to ſtone Parſley : the ſeed is not like that of Cumin, for if it were ſo, who would deny it to be *Oreoſelinum*, or *Dioſcorides* his mountaine Parſley.

¶ *The Temperature and Vertues.*

A   *Oreoſelinum* or mountaine Parſley is, as *Galen* ſaith, like in facultie to Smallage, but more effe-ctuall : *Dioſcorides* writeth, that the ſeed and root being drunke in wine prouoke vrine, bring downe the Menſes : and that they are mixed with counterpoiſons, diureticke medicines , and medicines that are hot.

B   The root of *Veelgutta* or Much-good is alſo hot and dry, euen in the later end of the ſecond de-gree, it maketh thin, it cutteth, openeth, prouoketh, breaketh the ſtone and expelleth it, openeth the ſtopping of the liuer and ſpleen, and cureth the yellow Iaundice : being chewed it helps the tooth-ache, and brings much water out of the mouth.

† This whole Chapter was wholly taken from *Dodonæus, Tempt. 5. lib. 4 cap.3.* wherefore I haue giuen his figure, which was agreeable to the hiſtorie : for the figure our Author here gaue was of the *Selinum montanum pumilum*, far different from this, as I ſhall hereafter ſhew you in the Chap. of *Peucedanum*.

CHAP. 399.   *Of ſtone Parſley or Macedonie.*

† 1 *Petroſelinum Macedonicum Fuchſij.*
Baſtard ſtone-Parſley.

† 2 *Petroſelinum Macedonicum verum.*
The true Parſley of Macedonia.

¶ *The Deſcription.*

O F ſtone-Parſley there is very little ſpoken by the antient Writers : onely *Dioſcorides* ſaith, That it hath ſeed like to that of Ameos, but of a more pleaſant ſmell, ſharpe, aromaticall, or ſpiced : touching the forme of the leaues, the colour of the floures , and the faſhion of the
roots

root he writeth nothing at all:and *Pliny* is more briefe ; as for *Theophraſtus* he doth not ſo much as name it,making mention only of Parſley,Alexander,Smallage,and mountaine Parſley.

1   For ſtone Parſley *Leonhartus Fuchſius* hath ſet down a plant,hauing leaues not ſpred and cut after the maner of garden Parſley,but long,and ſnipped round about, made vp and faſtened to a rib or ſtemme in the midſt,ſomthing like,but yet not altogether, to the firſt leaues of the leſſer Saxifrage;the ſtalke is ſlender, and a cubit and a halfe high ; the floures on the ſpokie tufts are white ; the ſeed ſo nthing blacke,like to that of Ameos,and garden Parſley,very ſweet of ſmell,ſomthing ſharp or biting:the root is ſlender and full of ſtrings.

2   *Lobel* alſo in ſtead of the right ſtone Parſley deſcribeth another, which the Venetians call ſtone Parſley of Macedonia:this hath leaues like thoſe of garden parſley,or rather of the Venetian Saxifrage which is the blacke herbe Frankincenſe formerly deſcribed : the ſtalke is a cubit high ; the ſpokie tufts ſomthing white : the ſeed ſmall, quickly vading (as he ſaith) inferiour to that of garden Parſley in temperature and vertues:but whether this be the true and right ſtone Parſley, he addeth,he is ignorant.

¶ *The Place.*

It groweth on craggie rockes, and among ſtones : but the beſt in Macedonia,whereupon it beareth the ſyrname *Macedonicum,*of Macedonia.

¶ *The Time.*

It floureth in the ſummer moneths.

¶ *The Names.*

It is called in Greeke,πετρι λινον,of the ſtony places where it groweth:in Latine,*Petrapium.*and *Petroſelinum Macedonicum.* in Engliſh,ſtone Parſley:the Apothecaries know it not:they are far deceiued that would haue the herbe which *Fuchſius* pictureth to be *Amomum* : for *Amomum* differeth from this,as it is very plaine by the deſcription thereof in *Dioſcorides :*but we hold this for the true ſtone Parſley,till ſuch time as we may learne ſome other more like in leaues to the Parſleyes, and in ſeed,ſuch as that of ſtone Parſley ought to be:and the very ſeed it ſelfe may cauſe vs to hold this opinion,being ſo agreeing to the deſcription as no herbe more ; for it is ſharp and biting,and of a ſweeter ſmell than is that of Ameos,and of a more ſpicy ſent ; yet doe not the leaues gaine ſay it, which though they haue not the perfect form of other Parſleyes,yet notwithſtanding are not altogether vnlike.   ‡ The firſt of theſe is thought by *Anguillara,Turner,Geſner,Cordus,* and others to be the *Siſon* of *Dioſcorides,*& *Tragus* cals it *Amomum Germanicum,*& the ſeeds in ſhops retain the name of *Sem. Amomi.*The ſecond is thought by *Columna* to be the ſecond *Daucus* of *Dioſcorides.* ‡

¶ *The Nature.*

The ſeed of ſtone Parſley,which is moſt commonly vſed, is hot and dry, hauing withall a cutting quality.

¶ *The Vertues.*

It prouoketh vrine,and bringeth down the floures:it is profitable againſt wind in the ſtomack, A and collicke gut,and gripings in the belly:for it is,as *Galen* ſaith, εφιμμ,that is to ſay,a waſter or conſumer of wind:it is a remedy againſt pain in the ſides,kidnies,and bladder,it is alſo mixed in counterpoiſons : *Dioſcorides.*

The firſt figure that was formerly in this chapter ſhould haue beene in the ſecond place, and that in the ſecond place was of Alexanders, and ſhould haue ceen put in the following chapter.

---

‡ C H A P. 400.   *Of Corne Parſley, or Hone-wort.*

¶ *The Deſcription.*

THis Herbe commeth vp at the firſt from ſeed like Parſley,with two ſmall long narrow leaues, the next that ſpring are two ſmall round ſmooth leaues nickt about the edges, and ſo for two or three couples of leaues of the next growth there are ſuch round leaues growing on a middle rib by couples,and one round one,alſo at the top;after as more leaues ſpring vp,ſo the faſhion of them alſo change,that is to ſay, euery leafe hath about eight or nine ſmal ſmooth green leaues, growing on each ſide of a middle rib one oppoſit againſt another,and one growing by it ſelf at the top,and are finely ſnipt or indented about the edges, and in forme reſembling thoſe of *Sium oderatum Tragi,*but not ſo big,long, or at all browniſh;amongſt which riſe vp many ſmall round ſtraked ſtalks or branches,about two foot long,now and then aboue twenty from one root,ſomtimes growing vpright, ſometimes creeping not farre from the ground, joynted or kneed,and diuiding themſelues

‡ *Selinum Sÿ folÿs.*
Hone-wort.

ſelues into very many branches;at euerie joint groweth one leafe ſmaller than the former , which together with the lowermoſt periſh,ſo that ſeldom one green leaf is to be ſeen on this herb when the ſeed is ripe;the floures are white, and grow moſt commonly at the tops of the branches, ſomtimes at moſt of the joints euen from the earth, in vneuen or vnorderly vmbells, euery floure hauing fiue exceeding ſmall leaues,flat,and broad at the top,and in the middle very ſmall cheiues with purple tops,the whole flour not much exceeding the bigneſſe of a ſmall pins head, which being paſt there comes vp in the place, of euery floure two ſmall gray crooked ſtraked ſeeds,like parſley ſeeds,but bigger,in taſt hot and aromaticall. The rot is ſmall and whitiſh, with many threds not ſo big as Parſley roots. It begins to floure about the beginning of Iuly, and ſo continueth flouring a long time ; part of the ſeede is ripe in Auguſt, and ſome ſcarſe in the beginning of October,mean while ſome fals whereby it renues it ſelfe,and grows with flouriſhing greene leaues all the winter.

    I took the deſcription of this herb the yere,1620.but obſerued it long afore, not knowing any name for it:firſt I referred it to *Sium*,calling it,*Sium terreſtre*,& *Sium ſegetum & agrorum*;afterwards vpon ſight of *Selinum peregrinum primum Cluſij*,which in ſome reſpects reſembleth this herbe,I named it *Selinum Sÿ folÿs*;yet wanting an Engliſh name, at length about the yeare 1625. I ſaw Miſtris *Vrſula Leigh* (then ſeruant to Miſtris *Bilſon* of Mapledurham in Hampſhire, and now (5. *Martÿ* 1632.wife to Mr.*William Mooring* Schoolemaſter of Petersfield,a town neer the ſaid Mapledurham)gather it in the wheat erſhes about Mapledurham aforeſaid(where in ſuch like grounds it ſtill groweth, eſpecially in clay grounds) who told mee it was called Hone-wort, and that her Mother miſtris *Charitie Leigh* late of Brading in the Iſland of Wight deceaſed,taught her to vſe it after the manner here expreſſed, for a ſwelling which ſhe had in her left cheeke,which for many yeares would once a yeare at the leaſt ariſe there,and ſwell with great heate,redneſſe,and itching, vntill by the vſe of this herbe it was perfectly cured,and roſe no more nor ſwelled,being now(5. *Martÿ* 1632)about twenty yeres ſince, only the ſcar remaineth to this day.This ſwelling her mother called by the name of a Hone,but asking whether ſuch tumors were in the ſaid Iſle vſually called Hones ſhe could not tell, by reaſon ſhe was brought from Brading aforeſaid young,and not being aboue twelue yeares old when ſhe vſed this medicine.

¶ *The Vertues.*

'A   Take one handfull of the greene leaues of this Hone-wort,and ſtamp them,put to it about halfe a pint or more of beer,ſtraine it,and drink it,and ſo continue to drink the like quantity euery morning faſting , till the ſwelling doth abate,which with or in her was performed in the ſpace of two weeks at the moſt.Auguſt,18.1620. *Iohn Goodyer.* ‡

Chap. 401. *Of Alexander.*

¶ *The Deſcription.*

THe leaues of Alexander are cut into many parcels like thoſe of Smallage, but they be much greater and broader,ſmooth alſo, and of a deep greene colour : the ſtalke is thick,oftentimes a cubit high:the floures be white,and grow vpon ſpokie tufts:the ſeed is thicke, long, blacke, ſomthing

ſomthing bitter,and of an aromaticall or ſpicy ſmell:the root is thicke, blacke without,and white within,and like to a little Radiſh, and is good to be eaten, out of which being broken or cut, there iſſues forth a juice that quickly waxeth thick, hauing in it a ſharp bitterneſſe,like in taſt to Myrrh ; which thing alſo *Theophraſtus* hath noted,there iſſueth out of it,ſaith he,a juice like Myrrhe.

† *Hippoſelinum.*
Alexanders.

¶ *The Place.*

Alexanders or great Parſley groweth in moſt places of England.

¶ *The Time.*

The ſeed waxeth ripe the ſecond yeare,in the Moneth of Auguſt.

¶ *The Names.*

It is called in Greeke, of the greatneſſe wherein it excelleth the other Parſleyes-ꭐꝙꝹ-ꝗꝪꝧor Horſe Parſley;of *Gaza,Equapium* : it is alſo named *Olus atrum,*or the blacke pot-herbe;and of diuers *Sylueſtre Apium,* or wild Parſley;of *Galen* and certain others, ꝙꝗꝧꝹ, by reaſon of the juice that iſſueth forth therof, that is, as wee haue ſaid, like vnto Myrrhe, which is called in Greeke ꝙꝗꝧꝹꝧthere is alſo another *Smyrnium* of mount Aman, of which we do write in the 404.chapter:the Apothe-caries cal it *Petroſelinum Macedonicum:*others *Petroſelinum Alexandrinum :* the Germaines, **Groſz Epffich:**the Low-country-men,**Pe-terſelie van Macedonion :** in Spaniſh, *Pe-rexil Macedonico :* the French, and Engliſh-men, Alexandre, Alexanders.

¶ *The Temperature.*

The ſeede and root of Alexanders, are no leſſe hot & dry than are thoſe of the garden Parſley, they clenſe and make thinne,being hot and dry in the third degree.

¶ *The Vertues.*

Dioſcorides ſaith,that the leaues and ſtalks are boiled and eaten, and dreſſed alone by themſelues **A** or with fiſhes : that they are preſerued raw in pickle:that the root eaten both raw and ſod,is good for the ſtomack:the root hereof is alſo in our age ſerued to the table raw for a ſallad herbe.

The ſeeds bring down the floures,expel the ſecondine,break and conſume wind,prouoke vrine, **B** and are good againſt the ſtrangury:the decoction alſo of the root doth the ſame,eſpecially if it be made with wine.

‡ The figure formerly here was of *Sphondilium,*and that belonging to this place was put in the foregoing chapter.

CHAP.402.
Of wilde Parſley.

¶ *The Deſcription.*

THis is like to the kindes of Parſleyes in the ſundry cuts of the leaues, and alſo in the bignes ; for they be broad,and cut into diuers parcels:the ſtalks are round, chamfred ſet with certain joints,hollow within,a cubit high or higher,two or three comming forth together out of one root,and in the nether part many times of a darke reddiſh colour. The floures be white and grow vpon ſpokie tufts:the ſeed is round, flat,like that of Dill:the root is white within,and diuided into many branches and ſtrings.This plant in what part ſoeuer it be cut or broken,yeelds forth a milky juice.

¶ *The*

† *Apium ſylueſtre ſiue Thiſſelium.*
Wilde Parſley.

¶ *The Place.*

It is found by ponds ſides in moiſt and dankiſh places, in ditches alſo, hauing in them ſtanding waters, and oftentimes by old ſtocks of Alder trees.

‡ I haue not as yet obſerued this plant growing wilde with vs. ‡

¶ *The Time.*

It floureth and bringeth forth ſeed in Iune and Iuly.

¶ *The Names.*

The ſhops of the Low countries haue miſcalled it in times paſt by the name of *Meum,* and vſed it for the right Mew, or Spiknel wort. The Germains name it **Olſenich:** *Valerius Cordus, Olſenichium :* diuers in the Low countries call it **wilde Eppe:** that is to ſay, in Latine, *Apium ſylueſtre,* or wild Parſley: and ſome, **water Eppe :** that is, *Hydroſelinon,* or *Apium aquatile,* water Parſley : and oftentimes it is named, as we haue already written, *Eleoſelinum,* & *Sium.* It may be more rightly termed in Latine, *Apium ſylueſtre,* and in Engliſh, wilde Parſley.

Dioſcorides hath made mention of wild Parſley in the chapter of *Daucus* or wilde Carrot : and *Theophraſtus, lib.* 7. where hee maketh the Parſleyes to differ both in leaues and ſtalkes, and ſhews that ſome haue white ſtalks, others, purple, or elſe of ſundry colours, and that there is alſo a certaine wilde Parſley; for he ſaith that thoſe which haue the purple ſtalks, and the ſtalks of diuers colours, come neereſt of all to the wilde Parſley. And therefore ſeeing that *Olſenichium* or wilde Parſley, hath the lower part of the ſtalke of a purpliſh colour, and like in leaues to Parſley, which in times paſt we thought good rather to call *Apium ſylueſtre,* or wild Parſly, than to erre with the Apothecaries, and to take it for Mew. And after when we now know that it was held to bee *Thyſſelium Plinij,* and that wee could alleadge nothing to the contrary, wee alſo ſettled our ſelues to be of their opinion ; and the rather, becauſe the faculties are agreeable. *Thyſſelium,* ſaith *Plin.lib.* 25.*cap.*11.is not vnlike to Parſley: the root hereof purgeth flegme out of the head ; which thing alſo the root of *Olſenichium* doth effectually performe, as we will forthwith declare: the name alſo is agreeable, for it ſeemeth to be called ϑυσϑλιον, becauſe it extendeth it ſelfe, in Greek, ϑλω, thorow ϑλιϑυς, or mariſh places.

¶ *The Temperature.*

The root hereof is hot and dry in the third degree.

¶ *The Vertues.*

A   The root being chewed, bringeth by the mouth flegme out of the head, and is a remedy for the tooth-ach, and there is no doubt but that it alſo makes thin, cutteth and openeth, prouoketh vrine, and bringeth downe the floures, and doth likewiſe no leſſe but more effectually performe thoſe things that the reſt of the Parſleyes do.

† The figure formerly put in this place was of the *Cereſolium ſylueſtre* of *Tabernamontanus,* whoſe hiſtory I intend hereafter to giue you.

CHAP. 403.   *Of baſtard Parſley.*

¶ *The Deſcription.*

1   THe firſt kind of baſtard Parſley is a rough hairy herb, not much vnlike to Carrots: the leaues are like to thoſe of Corianders, but parted into many ſmall jags : at the top of the branches grow ſhadowie vmbels, or ſpokie rundles conſiſting of many ſmall white floures :

1 *Caucalis albis floribus.*
Baftard Parfley with white floures.

‡ 2 *Caucalis Apij folijs flore rubro.*
Baftard Parfley with red floures.

‡ 3 *Caucalis Peucedani folio.*
Hogs Parfley.

‡ 4 *Caucalis major Cluf.*
Great rough Parfley.

floures : the ſeed is long and rough, like the ſeed of Carrots, but greater : the root is ſtraight and ſingle, growing deepe into the ground, of a white colour, and in taſte like the Parſnep.

2　　There is another ſort like vnto the former, ſauing that the leaues thereof are broader, and the floures are of a reddiſh colour : there hath great controuerſie riſen about the true determination of *Caucalis*, becauſe the Latine interpretation of *Dioſcorides* is greatly ſuſpected, containing in it ſelfe much ſuperfluous matter, not pertinent to the hiſtory : but wee deeme that this plant is the true *Caucalis*, the notes ſet downe declare it ſo to be : the floures, ſaith he, are reddiſh : the ſeeds couered with a rough huſke ſet about with prickles, which cleaue vnto garments that it toucheth, as doe Burs, which roughneſſe being pilled off, the ſeed appeares like vnto hulled Otes, not vnpleaſant in taſte, all which do ſhew it to be the ſame.

3　　There is likewiſe another ſort that hath a long ſingle root, thrummed about the vpper end with many thrummie threds of a browne colour : from which riſeth vp diuers ſtalkes full of joynts or knees, couered with a ſheath or skinnie filme, like vnto that of *Meum* : the leaues are finely cut or jagged, reſembling the leaues of our Engliſh Saxifrage : the floures grow at the top of the ſtalks in ſpokie rundles like Fennell : the ſeed is ſmall like that of Parſley.

‡ 5 *Caucalis minor floſculis rubentibus.*　　　　‡ 6 *Caucalis nodoſa echinoto ſemine.*
Hedge Parſley.　　　　　　　　　　　　　Knotted Parſley.

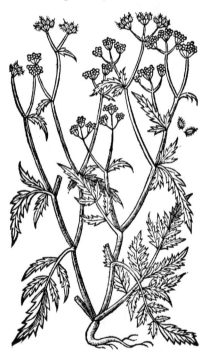

‡ 4　*Cluſius* vnder the name of *Caucalis major* hath deſcribed and figured this, which hath many creſted ſtraight ſtalkes ſome two cubits high or more, which are diuided into ſundry branches, and at each joynt ſend forth large and winged leaues ſomwhat like thoſe of Angelica, but rougher, and of a darker green ; at the tops of the branches grow vmbels of whitiſh floures, being of ſomewhat a purpliſh or fleſh colour vnderneath ; and theſe are ſucceeded by broad ſeed almoſt like thoſe of the Cow-Parſnep, but that they are rougher, and forked at the top, and prickly : the root is white, hard and wooddy. It floures in Iune, ripens the ſeed in Iuly and Auguſt, and then the root dies, and the ſeed muſt be ſowne in September, and ſo it will come vp and continue greene all the Winter.

5　　Beſides theſe formerly deſcribed there are two others growing wilde with vs ; the firſt of theſe, which I haue thought good to call Hedge, or field Parſley, (becauſe it growes about hedges, and in plowed fields very plentifully euery where) hath creſted hollow ſtalkes growing vp to ſome cubit and halfe high, whereon ſtand winged leaues made of ſundry little longiſh ones, ſet one

againſt

againſt another, ſnipt about the edges, and ending in a long and ſharp pointed leaſe: theſe leaues as alſo the ſtalks are ſomewhat rough and harſh, and of a darke greene colour: the floures are ſmall and reddiſh, and grow in little vmbels, and are ſucceeded by longiſh little rough ſeed of ſomewhat a ſtrong and aromaticke taſt and ſmell: It is an annuall plant, and floures commonly in July, and the feeds are ripe in Auguſt. *Cordus* and *Thalius* cal it *Daucoides minus*; and *Bauhine, Caucal: ſemine aſpero floſculis ſubrubentibus.* There is a bigger and leſſer varietie or ſort of this plant, for you ſhall finde it growing to the height of two cubits, with leaues and all the vpper parts anſwerable, and you may againe obſerue it not to exceed the height of halfe a foot.

6 This other, which *Bauhine* hath firſt ſet forth in writing by the name of *Caucalis nodoſa echinato ſemine*, hath a white & long root, from which it ſends vp ſundry ſmall creſted and rough branches which commonly lie along vpon the ground, and are of an vnequal length, ſome a cubit long, other-ſome ſcarſe two handfulls: the leaues are ſmall, rough, winged, and deeply jagged, and at the ſetting on of each leafe, cloſe to the ſtalkes vſually vpon very ſhort foot-ſtalkes grow ſmall little flours of colour white, or reddiſh, and made of fiue little leaues apeece: after theſe follow the ſeed, round, ſmall, and rough, and they grow cloſe to the ſtalkes. It floures in Iune and Iuly, and growes wilde in ſundry places, as in the fields, and vpon the banks about S. Iames, and Pickadilla. *Fabius Columna* iudges it to be the true *Scandix* of the Antients. ‡

There is likewiſe one of theſe found in Spaine, called *Caucalis Hiſpanica*, like the firſt: but it is an annuall plant, which periſhes at the firſt approch of winter, the which I haue ſowne in my garden, but it periſhed before the ſeed was perfected.

¶ The Place.

Theſe plants do grow naturally vpon rocks and ſtony grounds: we haue the firſt and the third in our paſtures in moſt places of England: that with red floures is a ſtranger in England.

‡ I haue not heard that the third growes wilde with vs, but the ſecond was found growing in the corne fields on the hils about Bathe, by Mr. *Bowles*. ‡

¶ The Time.

They floure and flouriſh from May to the end of Auguſt.

¶ The Names.

Baſtard Parſley is called in Greeke ϰαυκαλίς: in Latine alſo *Caucalis* : of ſome, *Daucus ſylueſtris* : among the baſtard names of *Democritus*, βρίον : in Latine, *Pes Gallinaceus, Pes Pulli* : the Egyptians name it *Seſelis* : the countrey-men of Hetruria, *Petroſello ſaluatico* : in Engliſh, baſtard Parſley, and Hennes foot.

¶ The Temperature and Vertues.

Dioſcorides ſaith, that baſtard Parſley is a pot-herbe which is eaten either raw or boiled, and prouoketh vrine. **A**

Pliny doth reckon it vp alſo among the pot-herbes : *Galen* addeth, that it is preſerued in pickle for ſallades in winter. **B**

The ſeed of baſtard Parſley is euidently hot and dry, and that in the ſecond degree: it prouoketh vrine, and bringeth downe the deſired ſickneſſe : it diſſolueth the ſtone and driueth it forth. **C**

It taketh away the ſtoppings of the liuer, ſpleen, and kidnies: it cutteth and concocteth raw and flegmaticke humors : it comforteth a cold ſtomacke, diſſolueth winde, quickneth the ſight, and refreſheth the heart, if it be taken faſting. **D**

Matthiolus in his Commentaries vpon *Dioſcorides, Lib. 2.* attributeth vnto it many excellent vertues, to prouoke venery and bodily luſt, and erection of the parts. **E**

† The figure which belonged to the third deſcription in this chapter was formerly put for Engliſh Saxifrage.

CHAP. 404. *Of Candy Alexanders.*

¶ The Deſcription.

Dioſcorides and *Pliny* haue reckoned *Smyrnium* among the kindes of Parſley, whoſe judgement, while this plant is young, and not growne vp to a ſtalke, may ſtand with very good reaſon, for that the young leaues next the ground are like to Parſley, but ſomwhat thicker and larger : among which riſeth vp a ſtalke a cubit high, and ſomewhat more, garniſhed with round leaues, farre different from thoſe next the ground , incloſing the ſtalke about like Thorow wax, or *Perfoliata*, which leaues are of a yellow colour, and doe rather reſemble the leaues of Fole-foot than Parſley :

at

Smyrnium Creticum.
Candy Alexander.

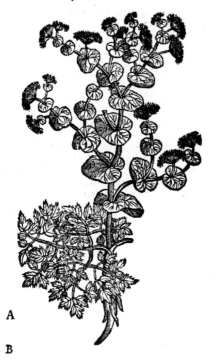

A

B

at the top of the ſtalkes doe grow round ſpoky tufts of a yellow color, after which comes round and blacke ſeed like Cole-worts, of a ſharp & bitter taſt like Myrrh: the root is white and thicke, contrary to the opinion of *Dodonæus*, who ſaith it is blacke without, but I ſpeak that which I haue ſeen and proued.

¶ *The Place.*

Smyrnium groweth naturally vpon the hils and mountaines of Candy, and in my garden alſo in great plenty: alſo vpon the mountain Amanus in Cilicia.

¶ *The Time.*

Smyrnium floureth in Iune, and the ſeed is ripe in Auguſt.

¶ *The Names.*

This plant is called in Latine, *Smyrnium* : in Greek, ϲμύρνιον: in Cilicia, *Petroſelinū*, and as *Galen* teſtifies, ſome haue called it, *Hippoſelinum agreſte*: in Engliſh, Candy Alexanders, or Thorow bored Parſley.

¶ *The Temperature.*

Smyrnium is hot and dry in the third degree.

¶ *The Vertues.*

The leaues of *Smyrnium* diſſolue wens and hard ſwellings, dry vp vlcers and ex-coriations, and glew wounds together.

The ſeeds are good againſt the ſtop-pings of the ſpleene, kidnies, and bladder.

C   Candy Alexander hath force to digeſt and waſt away hard ſwellings, in other things it is like to garden Parſley, and ſtone Parſley, and therefore we vſe the ſeed hereof to prouoke the deſired ſick-neſſe, and vrine, and to help thoſe that are ſtuffed in the lungs, as *Galen* writeth,

D   The root is hot, ſo is the herbe and ſeed, which is good to be drunke againſt the biting of ſer-pents : it is a remedy for the cough, and profitable for thoſe that cannot take their breath vnleſſe they do ſit or ſtand vpright : it helpeth thoſe that can hardly make their water : the ſeed is good a-gainſt the infirmities of the ſpleen or milt, the kidnies and bladder: it is likewiſe a good medicine for thoſe that haue the dropſie, as *Dioſcorides* writeth.

C H A P. 405. *Of Parſneps.*

¶ *The Deſcription.*

1   THe leaues of the tame or Garden Parſneps are broad, conſiſting of many ſmall leaues faſtned to one middle rib like thoſe of the aſh tree: the ſtalk is vpright, of the height of a man: the floures ſtand vpon ſpokie tufts, of colour yellow; after which commeth the ſeed flat and round, greater than thoſe of Dil : the root is white, long, ſweet, and good to be eaten.

2   The wild Parſnep is like to that of the Garden, in leaues, ſtalke, tuft, yellow floures, flat and round ſeed, but altogether leſſer : the root is ſmall, hard, wooddy, and not fit to be eaten.

¶ *The Place.*

The garden Parſnep requireth a fat and looſe earth, and that that is digged vp deep.

The wilde Parſnep groweth in vntoiled places, eſpecially in the ſalt marſhes, vpon the bankes and borders of the ſame: the ſeed whereof being gathered and brought into the garden, and ſowed

in

1 *Paſtinaca latifolia ſatiua.*
Garden Parſneps.

2 *Paſtinaca latifolia ſylueſtris.*
Wilde Parſneps.

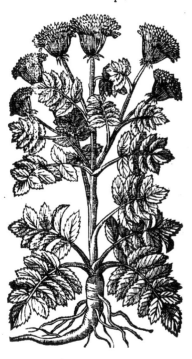

in fertill ground, do proue better roots, ſweeter and greater than they that are ſowne of ſeeds gathered from thoſe of the garden.

They floure in Iuly and Auguſt, and feed the ſecond yeare after they be ſowne.

¶ The Names.

The Herbariſts of our time do call the garden Parſneps ꜱꜩꜵꜹꜽꜩꜷ, and *Paſtinaca*, and therefore wee
haue ſurnamed it *Latifolia*, or broad leafed, that it may differ from the other garden Parſenep with
narrow leaues, which is truly and properly called *Stathylinus*, that is, the garden Carrot. Some Phyſitians doubting, and not knowing to what herbe of the Antients it ſhould be referred, haue fained
the wilde kinde hereof to bee *Panacis ſpecies*, or a kinde of All-heale : diuers haue named it *Baucia* ;
others, *Branca Leonina*, but if you diligently marke, and confer it with *Elaphoboſcum* of *Dioſcorides*, you
ſhall hardly finde any difference at all : but the plant called at Montpellier *Pabulum Ceruinum* : in
Engliſh, Harts fodder, ſuppoſed there to bee the true *Elaphoboſcum*, differeth much from the true
notes thereof. Now *Baucia*, as *Iacobus Manlius* reporteth in *Luminari maiore*, is *Dioſcorides*, and the old
Writers *Paſtinaca*, that is to ſay, *Tenuifolia*, or Carrot : but the old Writers, and eſpecially, *Dioſcorides* haue called this wilde Parſnep by the name of *Elaphoboſcum* : and we do call them Parſneps and
Mypes.

¶ The Temperature.

The Parſnep root is moderately hot, and more dry than moiſt.

¶ The Vertues.

The Parſneps nouriſh more than do the Turneps or the Carrots, and the nouriſhment is ſome- A
what thicker, but not faulty nor bad; notwithſtanding they be ſomewhat windie: they paſſe through
the body neither ſlowly nor ſpeedily : they neither binde nor looſe the belly : they prouoke vrine,
and luſt of the body : they be good for the ſtomacke, kidnies, bladder, and lungs.

There is a good and pleaſant food or bread made of the roots of Parſneps, as my friend Mr. *Plat* B
hath ſet forth in his booke of experiments, which I haue made no tryall of, nor meane to do.

The ſeed is hotter and dryer euen vnto the ſecond degree, it mooueth vrine, and conſumeth C
winde.

D   It is reported, ſaith *Dioſcorides*, that Deare are preſerued from bitings of Serpents, by eateing of the herbe *Elaphoboſcum*, or wilde Parſnep, whereupon the ſeed is giuen with wine againſt the bitings and ſtingings of Serpents.

† Both the figures that formerly were in this chapter were of the Garden Parſneps ; the firſt being that of *Lobel*, and the ſecond that of *Tabernæmont.* that which ſhould haue beene in the ſecond place, was formerly put for *Sphondylium.*

C h a p . 406. *Of Skirrets.*

Siſarum.   Skirrets.

¶ *The Deſcription.*

THe leaues of the Skirret do likewiſe conſiſt of many ſmall leaues faſtened to one rib, euery particular one whereof is ſomething nicked in the edges, but they are leſſer, greener, & ſmoother than thoſe aſ the Parſnep. The ſtalkes be ſhort, and ſeldome a cubit high ; the floures in the ſpokie tufts are white, the roots bee many in number, growing out of one head an hand breadth long, moſt commonly not a finger thick, they are ſweet, white, good to be eaten, and moſt pleaſant in taſte.

¶ *The Place and Time.*

This ſkirret is planted in Gardens, and eſpecially by the root, for the greater and thicker ones being taken away, the leſſer are put into the earth againe: which thing is beſt to be done in March or Aprill, before the ſtalkes come vp, and at this time the roots which bee gathered are eaten raw, or boyled.

¶ *The Names.*

This herb is called in Latine, *Siſarum,* and alſo in Greeke, σίσαρον ; the Latines do likewiſe call it *Siſer*; and diuers of the later Herbariſts *Seruillum* or *Cheruillam*, or *Seruilla*, the Germans name it, **Sterlin:** *Tragus*, **Zam gatten Rapunkelen:** in the Low-countries, **Suycker wortelen,** that is to ſay, Sugar roots, and oftentimes, **Serillen:** in Spaniſh, *Cherinia*: in Italian, *Siſaro*: in French, *Cheruy* : in Engliſh, Skirret and Skirwort. And this is that *Siſer* or Skirret which *Tiberius* the Emperour commaunded to bee conueied vnto him from Gelduba a caſtle about the riuer of Rhene, as *Pliny* reporteth in *lib.* 16. *cap.* 5. The Skirret is a medicinable herbe, and is the ſame that the foreſaid Emperour did ſo much commend, inſomuch that he deſired the ſame to be brought vnto him euery yeare out of Germany. It is not, as diuers ſuppoſe, *Serapio* his *Secacul,* of which he hath written in his 89. chapter : for *Secacul* is deſcribed by the leaſe of *Iulben*, that is to ſay, of the peaſe, as *Matthiolus Syluaticus* expoundeth it : and it bringeth forth a blacke fruit of the bigneſſe of a Cich-peaſe, full of moiſture, and of a ſweet taſte, which is called *Granum Culcul* : But the Skirret hath not the leaſe of the peaſe, neither doth it bring forth fruit like to the Ciche-peaſe; whereupon it is maniſeſt, that the Skirret doth very much differ from *Serapio* his *Secacul* : ſo farre is it from beeing the ſame.

¶ *The Temperature and Vertues.*

A   The roots of the Skirret be moderately hot and moiſt ; they be eaſily concoſted ; they nouriſh meanly, and yeeld a reaſonable good juyce : but they are ſomething windie, by reaſon whereof they alſo prouoke luſt.

B   They be eaten boiled, with vineger, ſalt, and a little oyle, after the manner of a ſallad, and oftentimes they be fried in oile and butter, and alſo dreſſed after other faſhions, according to the ſkill of the cooke, and the taſte of the eater.

The

· The women in Sueuia,ſaith *Hieronymus Heroldus*,prepare the roots hereof for their husbands,and C know full well wherefore and why, &c.

The juyce of the roots drunke with goats milke ſtoppeth the laske. The ſame drunke with wine D putteth away windineſſe out of the ſtomacke, and gripings of the belly, and helpeth the hicket or yeoxing. They ſtirre vp appetite, and prouoke vrine.

CHAP. 407.　*Of Carrots.*

¶ *The Deſcription.*

1　THe leaues of the Garden Carrots are of a deepe greene colour, compoſed of many fine Fennell-like leaues,very notably cut or jagged;among which riſeth vp a ſtalke ſtraight and round,foure cubits high,ſomewhat hairie and hollow,hauing at the top round ſpo-ked tuſts, in which do grow little white floures : in their places commeth the ſeed,rough and hai-rie,of a ſweet ſmell when it is rubbed.The root is long,thicke and ſingle, of a faire yellow colour, pleaſant to be eaten, and very ſweet in taſte.

1 *Paſtinaca ſativa tenuifolia.*
Yellow Carrot.

‡ 2 *Paſtinaca ſativa atro-rubens.*
Red Carrot.

2　There is another kinde hereof like to the former in all parts,and differeth from it only in the colour of the root,which in this is not yellow, but of a blackiſh red colour.

¶ *The Place.*

Theſe Carrots are ſowne in the fields, and in gardens where other pot-herbes are : they require a loofe and well manured ſoile.

¶ *The Time.*

They are to be ſowne in Aprill ; they bring forth their floures and ſeed the yeare after they be ſowne.

¶ *The Names.*

The Carrot is properly called in Greeke, σταφυλῖνος, for that which we haue termed in Latine by the name of *Paſtinaca latioris folij*, or the Garden Parſenep, is deſcribed of the old Writers by another name : this Carrot is called in Latine likewiſe, *Paſtinaca ſativa*, but with this addition *tenuifolia*, that it may differ from the garden Parſnep with broad leaues, and white roots. *Theophraſtus* in the ninth booke of his Hiſtory of Plants nameth this *Staphylinus*, or Carrot, ﹍﹍, and writeth that it groweth in Arcadia, and ſaith that the beſt is found in *Spartenſi Achaia*, but doubtleſſe hee meant that *Daucus* which we call *Cretenſis*, that may be numbred among the Carrots : *Galen* in his booke of the faculties of Simple medicines doth alſo make it to be *Daucus*, but yet not ſimply *Daucus* ; for hee addeth alſo *Staphilinus* or *Paſtinaca* : in high Dutch it is called 𝔊𝔢𝔢𝔩 𝔯𝔲𝔟𝔢𝔫 : in low Dutch, 𝔊𝔢𝔢𝔩 𝔓𝔢𝔢𝔫, 𝔊𝔢𝔢𝔩 𝔓𝔬𝔬𝔱𝔢𝔫, and 𝔊𝔢𝔢𝔩 𝔴𝔬𝔷𝔱𝔢𝔩𝔢𝔫 : in French, *Carrotte*, and *Racine jaulne* : in Italian, *Paſtinaca*: in Spaniſh, *Canahoria* : in Engliſh, Yellow Carrots : the other is called red Carrot, and blacke Carrot.

¶ *The Temperature and Vertues.*

A   The root of the yellow Carrot is moſt commonly boyled with fat fleſh and eaten : it is temperately hot and ſomething moiſt. The nouriſhment which commeth hereof is not much, and not very good : it is ſomething windie, but not ſo much as be the Turneps, and doth not ſo ſoon as they paſſe through the body.

B   The red Carrot is of like facultie with the yellow. The ſeed of them both is hot and dry, it breaketh and conſumeth windineſſe, prouoketh vrine, as doth that of the wilde Carrot.

C<small>HAP</small>. 408.   *Of Wilde Carrot.*

Paſtinaca ſylueſtris tenuifolia.
Wilde Carrot, or Bees-neſt.

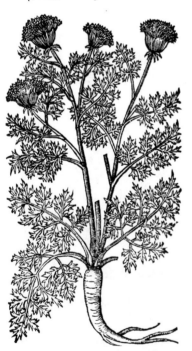

¶ *The Deſcription.*

THe leaues of the wilde Carrot are cut into diuers ſlender narrow parcels, very like vnto thoſe of the garden Carrots, but they bee ſomewhat whiter and more hairy : the ſtalkes be likewiſe hairie and ſomewhat rough : the floures are little, and ſtand vpon broad ſpoked tufts, of a white color, of which tuft of floures the middlemoſt part is of a deepe purple : the whole tuft is drawne together when the ſeed is ripe , reſembling a birds neſt ; whereupon it hath beene named of ſome, Birds-neſt : the root is ſlender, and of a meane length.

¶ *The Place.*

It groweth of it ſelfe in vntoyled places, in fields, and in the borders thereof, almoſt euery where.

¶ *The Time.*

It floures and flouriſhes in Iune and Iuly, the ſeed is ripe in Auguſt.

¶ *The Names.*

The wilde Carrot is called in Greeke, ςαφυλι- νυ αγριου : in Latine, *Paſtinaca ſylueſtris tenuifolia* : in ſhops, *Daucus* : and it is vſed in ſtead of the true *Daucus*, and not amiſſe nor vnprofitably : for *Galen* alſo in his time doth teſtifie that it was taken for *Dancus*, or baſtard Parſly, and is without doubt *Dauci ſylueſtris genus*, or a wilde kinde of baſtard parſly, ſo called of *Theophraſtus* : in high Dutch it is named, 𝔚𝔦𝔩𝔡 𝔓𝔞ſ𝔱𝔢𝔫𝔢𝔫, 𝔘𝔬𝔤𝔬𝔩 𝔫𝔢ſ𝔱 : in low Dutch, 𝔘𝔬𝔤𝔢𝔩𝔰 𝔫𝔢ſ𝔱, and 𝔴𝔦𝔩𝔡𝔢 𝔠𝔞𝔯𝔬𝔱𝔢𝔫 𝔠𝔯𝔬𝔬𝔨𝔢𝔫𝔰 𝔠𝔯𝔲𝔶𝔱 : in French, *Paſtena de Sauvage* : in Engliſh, wilde Carrot, and after the Dutch, Birds-neſt, and in ſome places Bees-neſt.

Athenæus citing *Diphilus* for his Author ſaith, That the Carrot is called φιλτρον, becauſe it ſerueth for loue matters ; and *Orpheus*, as *Pliny* writeth, ſaid, that the vſe hereof winneth loue : which things be be written of wilde Carrot, the root whereof is more effectuall than that of the garden, and containeth in it, as *Galen* ſaith, a certaine force to procure luſt.

¶ *The Temperature and Vertues.*

The ſeed of this wilde Carrot, and likewiſe the root is hot and dry in the ſecond degree, and doth A
withall open obſtructions.

The root boiled and eaten, or boiled in wine, and the decoction drunke prouoketh vrine, expel- B
leth the ſtone, bringeth forth the birth ; it alſo procureth bodily luſt.

The ſeed drunke bringeth downe the deſired ſickneſſe, it is good for them that can hardly make C
water, it breaketh and diſſolueth winde, it remedieth the dropſie, it cureth the collick and ſtone, being drunke in wine.

It is alſo good for the paſſions of the mother, and helpeth conception : it is good againſt the bi- D
tings of all manner of venomous beaſts : it is reported, ſaith *Dioſcorides*, that ſuch as haue firſt taken of it are not hurt by them.

CHAP. 409.　*Of Candie Carrots.*

Daucus Cretenſis verus.
Candie Carrots.

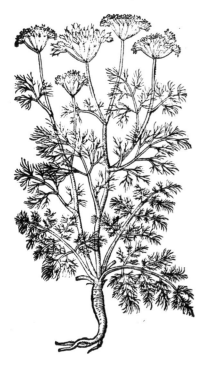

¶ *The Deſcription.*

THis *Daucus Cretenſis*, being the true *Daucus* of *Dioſcorides*, doth not grow in Candy only, but is found vpon the mountains of Germany, and vpon the hills and rockes of Iura about Geneua, from whence it hath beene ſent and conueied by one friendly Herbariſt vnto another , into ſundry regions : it beareth leaues which are ſmall, and very finely jagged, reſembling either Fennel or wild Carrot : among which riſeth vp a ſtalke of a cubit high, hauing at the top white ſpokie tufts, and the floures of Dill, which being paſt, there come great plentie of long ſeed, well ſwelling, not vnlike the ſeed of Cumin, ſaue that it is whitiſh, with a certaine moſſineſſe, and a ſharpe taſte, and is in greater vſe than any part of the plant. The root alſo is right good in medicine, being leſſer than the root of a Parſenep, but hotter in taſte, and of a fragrant ſmell.

¶ *The Time.*
This floures in Iune and Iuly, his ſeed is ripe in Auguſt.

¶ *The Names.*
There is ſufficient ſpoken in the deſcription as touching the name.

¶ *The Nature.*
Theſe plants are hot and dry , eſpecially the ſeed of *Daucus Creticus*, which is hot and dry in the third degree : but the ſeed of the wilde Carrot is hot and dry in the ſecond degree.

¶ *The Vertues.*

The ſeed of *Daucus* drunken is good againſt the ſtangurie, and painefull making of water, it pre- A
uaileth againſt the grauell and ſtone, and prouoketh vrine.

It aſſuageth the torments and gripings of the belly, diſſolueth windineſſe, cureth the collick, and B
ripeneth an old cough.

The ſame beeing taken in Wine , is very good againſt the bitings of beaſts, and expelleth C
poiſon.

The ſeed of *Daucus Creticus* is of great efficacy and vertue being put into Treacle, Mithridate, D
or any antidotes againſt poyſon or peſtilence.

E   The root thereof drunke in wine ſtoppeth the laske, and is alſo a ſoueraigne remedy againſt venome and poyſon.

C H A P. 410. *Of ſtinking and deadly Carrots.*

¶ *The Deſcription.*

1   THe great ſtinking Carrot hath very great leaues, ſpred abroad like wings, reſembling
thoſe of Fennell gyant (whereof ſome haue taken it to be a kinde, but vnproperly) of a
bright greene colour, ſomewhat hairy : among which riſeth vp a ſtalke of the height of
two cubits, and of the bigneſſe of a mans finger, hollow and full of a ſpungious pith; whereupon are
ſet at certaine joynts, leaues like thoſe next the ground, but ſmaller. The floures are yellow, ſtanding
at the top of the ſtalke in ſpokie rundles, like thoſe of Dill : after which commeth the ſeed, flat and
broad like thoſe of the Parſnep, but much greater and broader. The root is thicke, garniſhed at the
top with certaine capillaments or hairy threds, blacke without, white within, full of milky juyce, of
a moſt bitter, ſharpe, and lothſome taſte and ſmell, inſomuch that if a man do ſtand where the winde
doth blow frow the plant, the aire doth exulcerate and bliſter the face, and euery other bare or naked place that may be ſubject to his venomous blaſt, and poiſonous quality.

1 *Thapſia latifolia Cluſij.*
Stinking Carrots.

2 *Thapſia tenuifolia.*
Small leafed ſtinking Carrot.

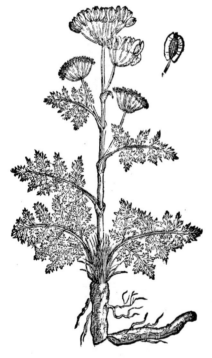

2   This ſmall kind of ſtinking or deadly Carrot is like to the laſt deſcribed in each reſpect, ſauing that the leaues are thinner and more finely minced or jagged, wherein conſiſts the difference.

3   The common deadly Carrot is like vnto the precedent, ſauing that he doth more neerely reſemble the ſtalkes and leaues of the garden carrot, and is not garniſhed with the like buſh of haire
about the top of the ſtalkes : otherwiſe in ſeed, root, and euill ſmell, taſte and quality like.

¶ *The Place.*

Theſe pernicious plants delight in ſtony hills and mountaines : they are ſtrangers in England.

¶ *The*

3 *Thapſia vulgaris.*
Deadly Carrots.

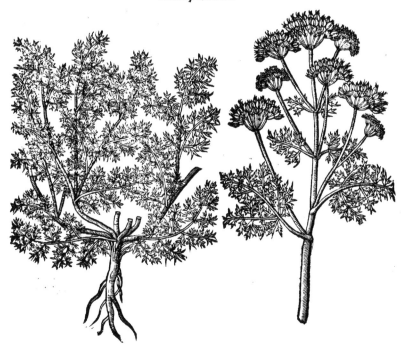

¶ *The Time.*

They floure in Auguſt or ſomewhat after.

¶ *The Names.*

The French Phyſitians haue accepted the root of *Thapſia* for a kinde of Turbith, calling it *Turpetum Cineritium* ; notwithſtanding vpon better conſideration they haue left the vſe thereof, eſpecially in purging, for it mightily hurteth the principall parts, and doth often cauſe cruell gripings in the guts and belly, with convulſions and cramps : neuertheleſſe the venomous quality may be taken away with thoſe correctiues which are vſed in mitigating the extreme heate and virulent quality of *Sarcocolla, Hammoniacum,* and *Turpetum :* but where there be ſo many wholeſome Simples, and likewiſe compounds they are not to be vſed.

Of ſome it is called *Turpetum Griſeum :* it is called *Thapſia,* as ſome thinke, of the Iſland Thapſus, where it was firſt found.

Of the people of Sicilia and Apulia it is called *Ferulacoli,* where it doth grow in great aboundance.

¶ *The Temperature and Vertues.*

The temperature and faculties in working haue beene touched in the deſcription, and likewiſe in the names.

CHAP. 411. *Of Fennell.*

¶ *The Deſcription.*

1   THe firſt kinde of Fennell. called in Latine, *Fœniculum :* in Greeke, Μάραθον, is ſo well knowne amongſt vs, that it were but loſt labour to deſcribe the ſame.

    2   The ſecond kinde of Fennell is likewiſe well knowne by the name of Sweet Fennell, ſo called becauſe the ſeeds therof are in taſte ſweet like vnto Anniſe ſeeds, reſembling the common Fennell, ſauing that the leaues are larger and fatter, or more oleous : the ſeed greater and whiter, and the whole plant in each reſpect greater.

¶ *The*

Fœniculum vulgare.
Common Fennell.

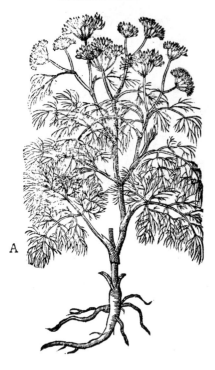

A

¶ *The Place.*

Theſe herbs are ſet and ſowne in gardens; but the ſecond doth not proſper well in this Countrey: for being ſowne of good and perfect ſeed, yet in the ſecond yeare after his ſowing it will degenerate from the right kinde, and become common Fennell.

¶ *The Time.*

They floure in Iune and Iuly, and the ſeed is ripe in the end of Auguſt.

¶ *The Names.*

Fennell is called in Greeke, Μάραθρον: in Latine, *Marathrum*, and *Fœniculum*: in high Dutch, 𝕱𝖊𝖓𝖈𝖐𝖊𝖑𝖑: in low Dutch, 𝖂𝖊𝖓𝖈𝖐𝖊𝖑𝖑: in Italian, *Finocchio*: in Spaniſh, *Hinoio*: in French, *Fenoil*: in Engliſh, Fennell, and Fenckell.

¶ *The Temperature.*

The ſeed of Fennel is hot and dry in the third degree.

¶ *The Vertues.*

The pouder of the ſeed of Fennell drunke for certaine daies together faſting preſerueth the eye-ſight: whereof was written this Diſtichon following:

Fœniculum, Roſa, Verbena, Chelidonia, Ruta,
　Ex his fit aqua quæ lumina reddit acuta.

Of Fennell, Roſes, Veruain, Rue, and Celandine, Is made a water good to cleere the ſight of eine.

B　The greene leaues of Fennell eaten, or the ſeed drunke made into a Ptiſan, do fill womens breſts with milke.

C　The decoction of Fennell drunke eaſeth the paines of the kidnies, cauſeth one to auoid the ſtone, and prouoketh vrine.

D　The roots are as effectuall, and not onely good for the intents aforeſaid, but againſt the dropſie alſo, being boiled in wine and drunken.

E　Fennell ſeed drunke aſſwageth the paine of the ſtomacke, and wambling of the ſame, or deſire to vomit, and breaketh winde.

F　The herbe, ſeed, and root of Fennell are very good for the lungs, the liuer, and the kidnies, for it openeth the obſtructions or ſtoppings of the ſame, and comforteth the inward parts.

G　The ſeed and herbe of ſweet Fennell is equall in vertues with Anniſe ſeed.

Chap. 412. *Of Dill.*

¶ *The Deſcription.*

DIll hath a little ſtalke of a cubit high, round and joynted: whereupon doe grow leaues very finely cut, like to thoſe of Fennell, but much ſmaller: the floures be little and yellow, ſtanding in a ſpokie tuft or rundle: the ſeed is round, flat and thin: the whole plant is of a ſtrong ſmell: the root is threddy.

¶ *The Place.*

It is ſowne in gardens, and is alſo ſometimes found wilde.

¶ *Thi*

Anethum.
Dill.

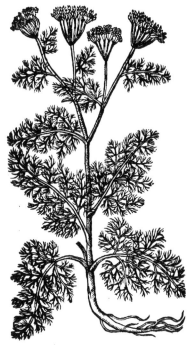

¶ *The Time.*

It bringeth forth floures and ſeed in Auguſt.

¶ *The Names.*

Dil is called in Greek, *Κπδωι* : in Latine likewiſe *Anethum*, and *Anetum* : in high Dutch, **Dyllen** : in low Dutch, **Dille** : in Italian *Anetho* : in Spaniſh, *Eneldo* : in French, *Anet* : in Engliſh, Dill, and Anet.

¶ *The Temperature.*

Dill, as *Galen* ſaith, is hot in the end of the ſecond degree, and dry in the beginning of the ſame, or in the end of the firſt degree.

¶ *The Vertues.*

The decoction of the tops of dried Dil, and A likewiſe of the ſeed, being drunke, engendreth milke in the breſts of nurſes, allaieth gripings and windineſſe, prouoketh vrine, increaſeth ſeed, ſtayeth the yeox, hicket, or hicquet, as *Dioſcorides* teacheth.

The ſeed likewiſe if it be ſmelled vnto ſtay- B eth the hicket, eſpecially if it bee boyled in wine, but chiefely if it be boyled in Wormewood Wine, or Wine and a few branches of Worme-wood, and Roſe-leaues, and the ſtomacke bathed therewith.

Galen ſaith, that being burnt and layd vpon C moiſt vlcers, it cureth them, eſpecially thoſe in the ſecret parts, and likewiſe thoſe *ſub Praputio*, though they be old and of long continuance.

Common oyle, in which Dill is boyled or ſunned, as we do oyle of Roſes, doth digeſt, mitigate D paine, procureth ſleepe, bringeth raw and vnconcocted humors to perfect digeſtion, and prouoketh bodily luſt.

Dill is of great force or efficacie againſt the ſuffocation or ſtrangling of the mother, if the wo- E man doe receiue the fume thereof being boiled in wine, and put vnder a cloſe ſtoole or hollow ſeat fit for the purpoſe.

CHAP. 413. *Of Caruwaies.*

¶ *The Deſcription.*

C Aruwaies haue an hollow ſtalke foure ſquare, of two cubits high, full of knots or joynts, from which proceed ſundry other ſmall branches, ſet full of leaues very finely cut or jagged, like vnto thoſe of Carrots or Dill : at the top of the ſtalkes grow ſpokie white tufts like thoſe of Dill : after which commeth the ſeed, ſharpe in eating, yet of a pleaſant taſte : the root is like that of Parſley, often white, ſeldome yellow, and in taſte like vnto the Carrot.

¶ *The Place.*

It groweth almoſt euery where in Germany and in Bohemia, in fat and fruitfull fields, and in medowes that are now and then ouer-run with water : it groweth alſo in Caria, as *Dioſcorides* ſheweth, from whence it tooke his name.

¶ *The Time.*

It floureth and ſeedeth from May to the end of Auguſt.

¶ *The*

Carum, ſive Careum.
Caruwaie.

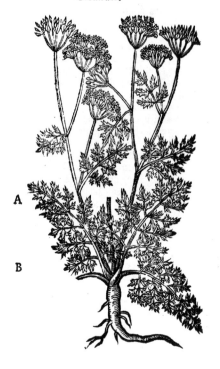

A

B

¶ *The Names.*

It is called in Greeke, ϰάρον: in Latine, *Carum* and *Careum* : in ſhops, *Carui.* Simeon Zethy calleth it *Carnabadion* : in high Dutch, **kym**, and **kymmel** : in low Dutch, **Carup ſaet** : in French, *du Caruy* : in Italian, *Caro* : in Spaniſh, *Carauea*, and an article being ioyned vnto it, *Alkarauea* : in Engliſh, Caruwaie, and the ſeed is called Caruwaie ſeed.

¶ *The Temperature.*

The ſeed of Caruwaies, as *Galen* ſaith, is hot and dry in the third degree, and hath a moderate biting qualitie.

¶ *The Vertues.*

It conſumeth winde, it is delightfull to the ſtomacke and taſte, it helpeth concoction, prouoketh vrine, and is mixed with counterpoyſons : the root may be ſodden, and eaten as the Parſenep or Carrot is.

The ſeeds confected, or made with ſugar into Comfits, are very good for the ſtomacke they helpe digeſtion, prouoke vrine, aſſwage and diſſolue all windineſſe : to conclude in a word, they are anſwerable to Aniſe ſeed in operation and vertues.

Cʜᴀᴘ. 414. *Of Aniſe.*

¶ *The Deſcription.*

1   THe ſtalke of Aniſe is round and hollow, diuided into diuers ſmall branches, ſet with leaues next the ground ſomewhat broad and round : thoſe that grow higher are more jagged, like thoſe of young Parſley, but whiter : on the top o ̔ the ſtalkes do ſtand ſpokie rundles or tuſts of white floures, and afterward ſeed, which hath a pleaſant taſte as euery one doth know.

‡ 2 This other Aniſe (whoſe vmbels *Cluſius* had out of England from Maſter *Morgan* the Queenes Apothecarie, and *Iames Garret*, and which were brought from the Philippines by Mʳ. *Tho. Candiſh* in his voyage when he incompaſſed the world) is thus deſcribed by *Cluſius* : The vmbels were large, no leſſe than thoſe of the Archangelica, made of diuers thicke ſtiffe foot-ſtalkes, each whereof carried not double ſeed as the common Aniſe, but more, in a round head ſome inch ouer, made of cods ſet ſtar faſhion, ſix, eight, or more, of a dusky colour, wrinkled, diuided into two equall parts, and open aboue : moſt of theſe huskes were empty, yet ſome of them contained one ſmooth ſhining aſh coloured ſeed, of the bigneſſe of that of *Orobus* ; the taſte and ſmell was the fame with our common Aniſe ſeed, where ̔ore they which ſent it to *Cluſius* called it Aniſe : yet in the place where it grew it was called *Damor* ; for Mʳ. *Candiſh* had the name ſo written in the China characters, after their manner of writing. ‡

¶ *The Place.*

It groweth plentifully in Candy, Syria, Ægypt, and other countries of the Eaſt. I haue often ſowne it in my garden, where it hath brought forth his ripe ſeed when the yeare hath fallen out to be temperate.

¶ *The*

1 *Aniſum.*
Aniſe.

‡ 2 *Aniſum Indicum ſtellatum,*
Starry headed Aniſe.

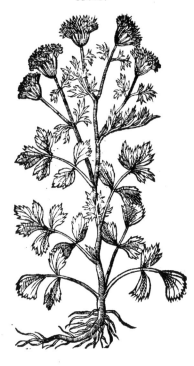 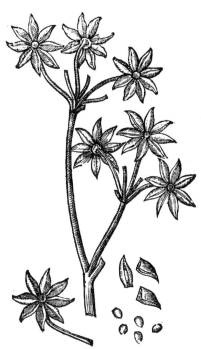

¶ *The Time.*

It is to be ſowne in theſe cold regions in the moneth of May: the ſeed is ripe in Auguſt.

¶ *The Names.*

It is called in Latine, *Aniſum* : in Greeke, ᾿Άνισον : in high Dutch, **Aniß** : in low Dutch, **Aniſſaet** : in Italian, *Aniſo* : in Spaniſh, *Matahalua* : in French, *Anis* : in Engliſh, Aniſe, and Anniſe ſeed.

¶ *The Temperature.*

Galen writeth, That the ſeed of Aniſe is hot and dry in the third degree: after others, it is hot in the ſecond degree, and much leſſe than dry in the ſecond degree ; for it ingendreth milke, which it could not doe if it were very dry, as *Galen* in his chapter of Fennell doth whether hee will or no declare and teſtifie ; in that it doth ingender milke, his opinion is that it is not hot aboue the firſt degree : which thing alſo may be in Aniſe ſeed, both by this reaſon, and alſo becauſe it is ſweet. Therefore to conclude, Aniſe ſeed is dry in the firſt degree, and hot in the ſecond.

¶ *The Vertues.*

The ſeed waſteth and conſumeth winde, and is good againſt belchings and vpbraidings of the **A** ſtomacke, allaieth gripings of the belly, prouoketh vrine gently, maketh abundance of milke, and ſtirreth vp bodily luſt : it ſtayeth the laske, and alſo the white flux in women.

Being chewed it makes the breath ſweet, and is good for them that are ſhort winded, and quen- **B** cheth thirſt, and therefore it is fit for ſuch as haue the dropſie : it helpeth the yeoxing or hicket, both when it is drunken or eaten dry : the ſmell thereof doth alſo preuaile very much.

The ſame being dried by the fire and taken with honey clenſeth the breſt very much from fleg- **C** maticke ſuperfluities : and if it be eaten with bitter Almonds it doth helpe the old cough.

It is to be giuen to young children and infants to eat which are like to haue the falling ſickneſ, **D** or to ſuch as haue it by patrimony or ſucceſſion.

It taketh away the ſquinancie or Quincie, (that is, a ſwelling in the throat) being gargled with **E** hony, vinegre, and a little Hyſſop gently boyled together.

Chap.

CHAP. 415.
Of Biſhops Weed, Herbe-William, or Ameos.

¶ *The Deſcription.*

1   THe common Ameos,eſpecially with vs here in England,hath round greene ſtalks,with diuers boughes and branches, and large long leaues, diuided into diuers other narrow long and ſmall leaues, dented or ſnipt about the edges, hauing at the top of the ſtalke white ſloures in great ſpoky tufts, which bring forth a little ſharpe and bitter ſeed : the root there-of is white and threddy.

2   This excellent and aromaticall Ameos of Candy hath tufts and leaues like *Daucus Creticus,* and a root like vnto the garden Carrot, of a yellow colour, and hot ſeed like *Origanum,* of an excel-lent ſpicie ſauour or ſmell, growing in ſpokie tufts or roundles like *Carum :* it hath beene brought from Candy and Syria into Venice, and from Venice into France, Flanders, and England, where we haue often ſowne it ; but without doubt we haue beene beguiled therein by the deceitfull drug-maſters,who haue firſt boyled it,or vſed ſome other falſe and deceitfull deuiſe,to bring greater ad-miration vnto the Venice treacle, for the confeƈtion whereof this ſeed is a chiefe and moſt princi-pall ingredient.

      1 *Ammi vulgare.*   † 2 *Ammi Creticum.*
     Common Biſhops-weed.   Candy Biſhops-weed.

3   There is another kinde of Ameos, which is an herbe very ſmall and tender, hauing ſtalkes a foot and a halfe high,very ſmall and tender,beſet with leaues like vnto Dill,finely jagged,and ſom-what ſlender ; and at the top of the ſtalkes grow little tufts or ſpokie white rundles, which after-wards do turne into ſmall gray ſeed,hot,and ſharpe in taſte. The root is ſmall and ſlender.

¶ *The Place.*

    Theſe plants do all grow in my garden, except *Ammi Creticum,* whereof hath beene ſufficiently ſpoken in the deſcription.

¶ *The*

‡ 3 *Ammi perpuſillum.*
Small Biſhops-weed.

¶ *The Time.*

They floure in Iune and Iuly, and yeeld their ſeed in the end of Auguſt.

¶ *The Names.*

The Græcians call it κμμι: the Latines alſo *Ammi*: diuers call it *Cuminum Æthiopicum*: others, *Cuminum Reginum*, or Comin Ro all: in ſhops, *Ammios*, or *Ameos* in the Genitiue caſe: the Germanes, **Amep**: in Engliſh, Ameos, or Ammi: of ſome, Herbe-William, Bullwort, and Biſhops-weed.

¶ *The Temperature.*

The ſeed of Ameos is hot and dry in the later end of the third degree.

¶ *The Vertues.*

It auaileth againſt gripings of the belly in **A** making of vrine, againſt the biting of ſerpents taken in wine, and alſo it bringeth downe the floures: beeing applied with honey it taketh away blacke and blew ſpots which come of ſtripes: the ſeed of *Siſon* doth alſo the like, for it is hot and dry, and that in the third degree; likewiſe of thin parts, prouoking vrine, and bringing downe the deſired ſickneſſe.

The ſeed of Ameos is good to bee drunken **B** in wine againſt the biting of all manner of venomous beaſts, and hath power againſt all maner of poyſon & peſtilent feuers, or the plague, and is vſed in the correcting of Cantharides, whereby thoſe flies are made medicinable to be applied to the body without danger.

Ameos brayed and mingled with hony ſcattereth congealed bloud, and putteth away blacke and **C** blew markes which come by ſtripes or falls, if it be applied thereto in manner of a plaiſter.

† The figure which was formerly in the ſecond place was of the *Hippomarathrum album* of *Tabernamontanus*.

Chap. 416. *Of Cheruill.*

¶ *The Deſcription.*

1　THe leaues of Cheruill are ſlender, and diuerſly cut, ſomething hairy, of a whitiſh greene: the ſtalkes be ſhort, ſlender, round, and hollow within, which at　　firſt together with the leaues are of a whitiſh green, but tending to a red when the ſeeds are ripe: the floures be white, and grow vpon ſcattered tufts. The ſeed is long, narrow, ſlender, ſharpe pointed: the root is full of ſtrings.

‡ 2　There is found in Iune and Iuly, almoſt in euery hedge, a certaine plant which *Tabernamontanus* and *Bauhine* fitly call *Chærophyllum*, or *Cerefolium ſylueſtre*, and the figure was vnfitly giuen by our Author for *Thyſſelinum*: It hath a whitiſh wooddy root, from which ariſe round red and hairy ſtalkes ſome two cubits high, ſometimes more, and oft times ſomewhat big and ſwolne about the joynts, and they are not hollow but full of pith: toward the top it is diuided into ſundry branches which on their tops carry vmbels of ſmall pure white little floures, which are ſucceeded by longiſh ſeeds. The leaues are vſually parted into three chiefe parts, and theſe againe ſubdiuided into fiue, and they are ſnipt about the edges, ſoft and hairy, of a darke greene or elſe reddiſh colour. It floureth in Iune and Iuly, and then ripens the ſeed. ‡

3　Great Cheruill hath large leaues deepely cut or jagged, in ſhew very like vnto Hemlockes, of a very good and pleaſant ſmell and taſte like vnto Cheruill, and ſomething hairy, which hath cauſed vs to call it ſweet Cheruill. Among theſe leaues riſeth vp a ſtalke ſomewhat creſted or furrowed, of the height of two cubits, at the top whereof grow ſpokie tufts or rundles with white

floures,

floures, which doe turne into long browne crested and shining seed, one seed being as big as foure Fennell seeds, which being greene do taste like annise seed. The root is great, thicke, and long, as big as *Enula Campana*, exceeding sweet in smell, and tasting like vnto Anise seeds.

1 *Cerefolium vulgare satiuum.*
Common Cheruill.

† 2 *Cerefolium syluestre.*
Hedge Cheruill.

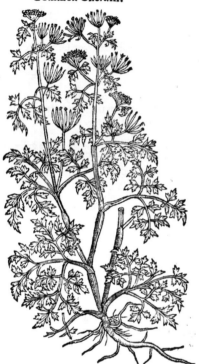 

‡ 4 There is found in some part of the Alps, **as about Genua and in other places,** another *Myrrhis*, which in the leaues and vmbels is like that of the last described, but the whole plant is lesse; the seed is long, smooth, and shaped like an Oat, and in taste somewhat like that of the *Daucus Creticus, Lobel* ha this by the same name as we here giue it you.

5 About mud walls, high-waies, and such places, here about London, and in diuers other places, is found growing a small plant, which in all things but the smell and height agrees with that referred to this kinde by *Fabius Columna*, and called *Myrrhis Æquicolorum nova*. The root hereof is small and white, perishing euery yeare when it hath perfected his seed: the stalkes are slender, hollow, smooth, and not hairy, seldome exceeding the height of a cubit, or cubit and halfe it is diuided into sundry branches, vpon the seeds whereof against the setting on of the leaues, or out of their bosomes, grow forth the stalkes, which carry vmbels of small white floures: after which follow the seeds, growing two together, and these longish, rough, round, and hairy, about the bignesse of Anise seeds. The leaues are small, and finely cut or diuided like those of Hemlocke, but of a whitish colour, and hairy: it comes vp in March, floures in May, and ripens his seed in Iune. In Italy they eate the young leaues in sallads, and call it wilde Cheruile: we may in English for distinction sake call it small Hemlocke Cheruill.

6 To these we may fitly adde that plant which in the *Hist. Lugd.* is called *Cicutaria alba*, and by *Camerarius, Cicutaria palustris*; for it floures at the same time with the last mentioned, and is found in floure and seed in May and Iune very frequently al.most in all places; but afterwards his stalkes die downe, yet his roots liue, and the leaues are greene all the yeare. The root of this is very large, and diuided into sundry parts, white also and spungie, of a pleasant strong smell, with a hot and biting taste: the stalkes grow vp in good ground to be some three cubits high, and they are hollow, joynted, pretty thicke, greene, and much crested, sending forth of the bosomes of the leaues many branches, which vpon their tops carry vmbels composed of many white floures, each floure consisting of fiue little leaues, whereof the lowest is twice as big as the rest, the two side-ones lesse, and the vppermost the least of all. The leaues are large like those of *Myrrhis*, but of a dark greene colour,

and

and thoſe that grow vpon the tops of the ſtalks are commonly diuided into three parts, and theſe ſubdiuided into ſundry long ſharp pointed and ſnipt leaues like as in *Myrrhis*. The ſeeds grow two together, being longiſh, round, ſharp pointed, blacke, and ſhining. We may fitly terme this plant wilde Cicely, for that it ſo much reſembles the *Myrrhis* or garden Cicely, not only in ſhape, but if I be not deceiued, in vertues alſo.

3 *Cerefolium magnum, ſiue Myrrhis.*
Great Cheruill or Myrrhe.

‡ 4 *Myrrhis altera parua.*
Small ſweet Cheruill.

¶ *The Place.*
The common Cheruill groweth in gardens with other pot-herbes: it proſpers in a ground that is dunged and ſomewhat moiſt. The great ſweet Cheruill groweth in my garden, and in the gardens of other men who haue bin diligent in theſe matters.

¶ *The Time.*
Theſe herbs floure in May, and their ſeed is ripe in Iuly.

¶ *The Names.*
Cheruill is commonly called in Latine *Cerefolium*, and as diuers affirme, *Chærofolium*, with *o* in the ſecond ſyllable. *Columella* nameth it *Chærephyllum*, and it is thought to be ſo called, becauſe it delights to grow with many leaues, or rather in that it cauſeth ioy and gladneſſe: in high-Dutch, **Roꝛffelkraut**: in low-Dutch, **Keruell**: in Italian, *Cerefoglio*: in French, *Du Cerfueil*: in Engliſh, Cheruell and Cheruill.

Myrrhis is alſo called *Myrrha*, taken from his pleaſant ſauor of Myrrh: of ſome, *Conila*, as it is found noted among the baſtard names. It is alſo by reaſon of the ſimilitude it hath with Hemlocke, called by moſt later writers *Cicutaria*. Of this *Pliny* maketh mention, *lib.24.cap.16.* where he reporteth that it is called *Smyrrhiza*: in Engliſh it is called Cheruill, ſweete Cheruill, or ſweete Cicely.

¶ *The Nature and Vertues.*
Cheruill is held to be one of the pot-herbs, it is pleaſant to the ſtomack and taſte, it is of a temperat heate and moderat drineſſe, but nothing ſo much as the Parſlies.　**A**

It prouoketh vrine, eſpecially being boiled in Wine, and applied hot to the ſhare or nethermoſt part of the belly, and the wine drunke in which it was boiled.　**B**

It hath in it a certain windineſſe, by means whereof it procureth luſt.　**C**

It is vſed very much among the Dutch people in a kinde of Loblolly or hotchpot which they do eat, called Warmus.　**D**

E   The leaues of ſweet Cheruill are exceeding good,wholeſome and pleaſant among other ſallad herbs,giuing the taſte of Aniſe ſeed vnto the reſt.

F   The root,ſaith *Galen*,is hot in the ſecond degree,hauing a thinneſſe of ſubſtance joyned with it.

G   *Dioſcorides* teacheth,That the root drunke in wine is a remedie againſt the bitings of the venomous ſpiders called in Latine *Phalangia*; and that it bringeth downe the menſes and ſecondines; and being boiled and drunk it is good for ſuch as haue the Ptyſick and conſumption of the lungs.

H   The ſeeds eaten as a ſallad whiles they are yet green, with oile, vineger, and pepper, exceed all other ſallads by many degrees,both in pleaſantneſſe of taſte,ſweetneſſe of ſmell,and wholſomneſſe for the cold and feeble ſtomacke.

I   The roots are likewiſe moſt excellent in a ſallad,if they be boiled and afterwards dreſſed as the cunning Cooke knoweth how better than my ſelfe : notwithſtanding I vſe to eat them with oile and vineger,being firſt boiled ; which is very good for old people that are dull and without courage : it rejoiceth and comforteth the heart,and increaſeth their luſt and ſtrength.

CHAP. 417.   *Of Shepheards Needle or wilde Cheruill.*

¶ *The Deſcription.*

1   SCandix or *Pecten Veneris* doth not much differ in the quantitie of the ſtalkes, leaues, and floures,from Cheruill ; but *Scandix* hath no ſuch pleaſant ſmell as Cheruill hath. The leaues be leſſer,more finely cut,and of a brown green colour : the flours grow at the top of the ſtalks in ſmall white tufts ; after which come vp long ſeeds very like vnto pack-needles,orderly ſet one by another like the great teeth of a combe,whereof it tooke the name *Pecten Veneris*, Venus combe,or Venus needle : the root is white,a finger long.

1 *Pecten Veneris, ſiue Scandix.*
Shepheards Needle or Venus Combe.

‡ 2 *Scandix minor, ſiue Anthriſcus.*
Small Shepheards Needle.

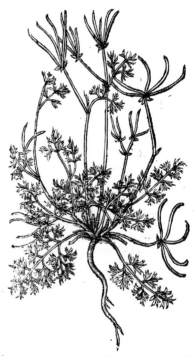

‡ 2 This from a ſlender long and whitiſh root ſends vp many ſmal leaues like thoſe of the laſt deſcribed,but of a pleaſing ſmell,and taſte ſomething like that of the common Cheruill:amongſt theſe leaues grow vp ſlender ſtalks a little hairy,diuided into ſhort green and ſlender branches carrying little vmbels,conſiſting of fiue,ſix,ſeuen,or eight ſmal white floures,compoſed of fiue leaues apiece,

apiece,with a darke purplish chiue in the middle;the floures are succeeded by,or rather grow vpon long slender ends,which become some inch long,and resemble those of the last described:it flours in Iune, as *Clusius* affirmeth, who giues vs the history of it ; and he receiued it from *Honorius Bellus* outof Candy ; who writes,that in the Spring time it is much vsed in sallads, and desired,for that it much excites to Venery. He also thinks this plant to be the *Anthriscus* of *Pliny*,and by the same name *Clusius* sets it forth.*Columna* hath called it *Anifo-marathrum*,because the smell and taste is be-tweene that of Anise and Fennell. ‡

    ¶ *The Place.*

It groweth in most corne fields in England,especially among wheat and barley.

    ¶ *The Time.*

It floureth in May : the seed is ripe in August with corne.

    ¶ *The Names.*

The Latines call it *Scandix*,hauing borrowed that name of the Grecians,who call it Σκάνδιξ we finde among the bastard words,that the Romans did call it *Scanaria*, and *Acula*,of the seed that is like vnto a needle.*Ruellius* describeth it vnder the name *Pecten Veneris* : of others, *Acus Veneris*,and *Acus pastoris*, or Shepheards Needle, wilde Cheruill,and Ladies combe : in high Dutch,**Nadel Kärnel:** This is that herbe (saith *Pliny,lib.22.cap.22.*) which *Aristophanes* obiected in sport to the Poet *Euripides*,that his mother was wont to sell no right pot-herbe but *Scandix*,or Shepheards nee-dle;meaning,as I take it,*Visnaga*,wherwith the Spaniards do picke their teeth when they haue ea-ten no meat at all except a few oranges or such a like trifle,called also *Scandix*.

    ¶ *The Temperature.*

Shepheards needle,saith *Galen*,is an herbe somewhat binding, and bitter in taste,insomuch that it is hot and dry either in the later end of the second degree,or in the beginning of the third.

    ¶ *The Vertues.*

Dioscorides saith it is eaten both raw and boiled, and that it is an vnwholsome pot-herbe among A the Greekes ; but in these dayes it is of small estimation or value, and taken but for a wilde Wort, as appeareth by *Aristophanes* taunting of *Euripides*,as aforesaid.

The decoction thereof is good for the bladder, kidneyes, and liuer ; but as I deeme hee meant B Cheruill,when he set the same downe to be vsed in physicke.

Chap. 418. *Of Tooth-picke Cheruill.*

    ¶ *The Description.*

1   THe first of these Tooth-pick Cheruils beareth leaues like wild Turneps,a round stalke furrowed,joynted,blackish,and hairy,diuided into many branches,on the tops wherof grow spokie tufts,beset round about with many small leaues. The floures therof are whitish : after commeth the seed, which being once ripe doe cluster, and are drawne together, in a round thick tuft like a small birds nest,as be those of the wild Carrot;whose seeds whole toucheth, they will cleaue and sticke to his fingers,by reason of the glutinous or slimie matter they are pos-sessed with. The root is small and whitish,bitter in taste,as is all the rest of the plant.

2   The Spanish Tooth-picke hath leaues, floures, and knobby stalkes like vnto wilde carrots, sauing that the leaues are somwhat finer,cut or jagged thicker,and tenderer,but not rough or hairy at all as is the former, of a bitter taste, and a reasonable good smell : among which rise vp bushie rundles or spokie tufts like those of the wilde Carrot or Birds nest, closely drawne together when the seed is ripe ; at what time also the sharpe needles are hardned, fit to make Tooth-pickes, and such like,for which purpose they do very fitly serue.

    ¶ *The Place.*

Both of them grow in Syria, and most commonly in Cilicia : the later likewise is to be found in Spaine almost euery where ; and I haue it likewise in my garden in great plentie.

    ¶ *The Time.*

They floure in my garden about August,and deliuer their seed in October.

    ¶ *The Names.*

That which the Grecians call τοχίδιον,the Latines do likewise name *Gingidium* : and it is called in Syria,*Lepidium* : yet is there another *Lepidium*. It is reported among the bastard names to be cal-led by the Romans, *Bifacutum* : of which name some shew remaines among the Syrians, who com-monly call the later,*Gingidium*,*Visnaga* : this is named in English, Tooth-picke Cheruill,

1 *Gingidium latifolium.*
Broad Tooth-picke Cheruill.

2 *Gingidium Hiſpanicum.*
Spaniſh Tooth-picke Cheruill.

¶ *The Temperature and Vertues.*

A　There is, ſaith *Galen*, great increaſe of *Gingidium* in Syria, and it is eaten no otherwiſe than *Scandix* is with vs at Pergamum : it is, ſaith he, very wholſome for the ſtomack, whether it be eaten raw or boiled ; notwithſtanding it is euident that it is a medicine rather than a nouriſhment. As it is bitter and binding, ſo is it likewiſe of a temperate heat and drines. The heat is not very apparant, but it is found to be dry in the later end of the ſecond degree, as alſo the ſaid Author alledgeth in his diſcourſe of the faculties of ſimple medicines.

B　*Dioſcorides* doth alſo write the ſame : This pot-herbe (ſaith he) is eaten raw, ſodden, and preſerued with great good to the ſtomack; it prouoketh vrine, and the decoction therof made with wine and drunke, is profitable to ſcoure the bladder, prouoketh vrine, and is good againſt the grauell and ſtone.

C　The hard quils whereon the ſeeds do grow are good to clenſe the teeth and gums, and do eaſily take away all filth and baggage ſticking in them, without any hurt vnto the gums, as followeth after many other Tooth-picks, and they leaue a good ſent or ſauor in the mouth.

CHAP. 419.　*Of Mede-ſweet, or Queene of the Medowes.*

¶ *The Deſcription.*

1　THis herbe hath leaues like Agrimony, conſiſting of diuers leaues ſet vpon a middle rib like thoſe of the aſh tree, euery ſmall leaf ſleightly ſnipt about the edges, white on the inner ſide , and on the vpper ſide crumpled or wrinkled like vnto thoſe of the Elme tree ; whereof it tooke the name *Vlmaria*, of the ſimilitude or likeneſſe that the leaues haue with the Elme leaues. The ſtalke is three or foure foot high, rough, and very fragile or caſie to bee broken, of a reddiſh purple colour : on the top whereof are very many little floures cluſtering and growing together , of a white colour tending to yellowneſſe , and of a pleaſant ſweet
ſmell,

1 Regina prati.
Queene of the Medow

ſmell, as are the leaues likewiſe : after which come the ſeeds, ſmall, crookedly turning or winding one with another, made into a fine little head:the root hath a ſweet ſmell,ſpreading far abroad, black without,& of a darkiſh red color within.

‡ 2 There is alſo another which by *Fuchſius, Tragus Lonicerus, Geſner,* and o-thers, is called *Barba Capri* : it hath large wooddy roots,leaues of the bigneſſe, and growing ſomwhat after the maner of the wild Angelica:the ſtalks are creſted, and diuided into ſundry branches,which cary long bending ſpikes or eares of white floures and ſeeds ſomewhat like thoſe of the common kind:this flours at the ſame time as the former,& I haue not yet heard of it wild with vs, but only ſeene it grow-ing with Mr.*Tradeſcant.* ‡

¶ *The Place.*

It groweth in the brinkes of waterie ditches and riuers ſides, and alſo in me-dowes:it liketh watery and moiſt places, and groweth almoſt euery where.

¶ *The Time.*

It floureth and floureſheth in Iune,Iu-ly,and Auguſt.

¶ *The Names.*

It is called of the later age *Regina prati,* and *Barba Capri* : of ſome, *Vlmaria,à folio-rum Vlmi ſimilitudine,*for the likeneſſe it hath with the Elme tree leafe:in high Dutch, 𝕾𝖈𝖎𝖋𝖇𝖆𝖗𝖙. It is called *Barba Hirci,*which name belongeth to the plant which the Grecians do cal *Tragopogon:* of *Anguillara,Potentilla major.*It hath ſome likeneſſe with *Rhodora Plinij,* but yet we cannot affirme it to be the ſame.It is called in low Dutch,𝕽𝖊𝖋𝖎𝖓𝖊𝖙𝖙𝖊:in French, *Barbe de Cheure, Reine des Praiz* : in Engliſh,Meadſ-ſweet,Medow-ſweet,and Queen of the medowes. *Camerarius* of Noremberg ſaith it is called of the Germanes his countrimen,𝖂𝖚𝖗𝖒𝖊 𝖐𝖗𝖆𝖚𝖙 : becauſe the roots, ſaith he, ſeeme to be eaten with wormes.I rather ſuppoſe they call it ſo, becauſe the antient hackny men and horſe-leaches do giue the decoction thereof to their horſes and aſſes,againſt the bots and worms,for the which it is greatly commended.

¶ *The Nature.*

Mede-ſweet is cold and dry,with an euident binding qualitie adioined.

¶ *The Vertues.*

The root boiled,or made into pouder and drunke,helpeth the bloudy flix,ſtaieth the lask, and all other fluxes of bloud in man or woman.   A

It is reported,that the floures boiled in wine and drunke,do take away the fits of a quartaine a-gue,and make the heart merrie.   B

The leaues and floures farre excell all other ſtrowing herbes, for to decke vp houſes, to ſtraw in chambers,halls,and banqueting houſes in the Summer time;for the ſmell thereof makes the heart   C merrie, delighteth the ſenſes : neither doth it cauſe head-ache,or lothſomeneſſe to meat, as ſome other ſweet ſmelling herbes do.

The diſtilled water of the floures dropped into the eies, taketh away the burning and itching   D thereof, and cleareth the ſight.

CHAP. 420. *Of Burnet Saxifrage.*

¶ *The Deſcription.*

1 THis great kinde of Pimpinell,or rather Saxifrage, hath great and long roots, faſhioned like a Parſnep, of an hot and biting taſte like Ginger : from which riſeth vp an hollow
ſtalke

ſtalke with joints and knees two cubits high, beſet with large leaues, which do more neerely repreſent Smallage than Pimpernell, or rather the garden Parſnep. This plant conſiſteth of many ſmall leaues growing vpon one ſtem, ſnipt or dented about the edges like a ſaw : the floures doe grow at the top of the ſtalkes in white round tufts : the ſeed is like the common Parſley, ſauing that it is hotter and biting vpon the tongue.

‡ There is a bigger and leſſer of this kinde, which differ little, but that the ſtalkes and veins of the leaues of the leſſer are of a purpliſh colour, and the root is hotter. Our Author formerly gaue the figure of the leſſer in the ſecond place, in ſtead of that of *Bipinella*. ‡

1 *Pimpinella Saxifraga.*
Burnet Saxifrage.

‡ 2 *Bipinella, ſiue Saxifraga minor.*
Small Burnet Saxifrage.

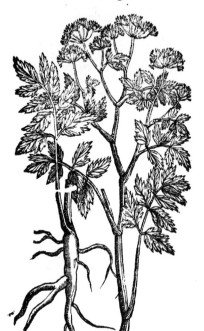 

2   *Bipinella* is likewiſe a kind of Burnet or Pimpinell, vpon which *Pena* hath beſtowed this addition *Saxifraga minor :* vnder which name *Saxifraga* are comprehended diuers herbes of diuers kinds, and the one very vnlike to the other : but that kinde of Saxifrage which is called *Hircina*, which is rough or hairy Saxifrage, of others *Bipinella*, is beſt known, and the beſt of all the reſt, like vnto the ſmall Burnet, or common Parſley, ſauing that it is void of haires, as may appeare by the old Latine, verſe,

Pimpinella pilos, Saxifraga non habet vllos.
Pimpinell hath haires ſome, but Saxifrage hath none.

Notwithſtanding, I haue found a kinde hereof growing in our paſtures adioining to London, the leaues whereof if you take and tenderly breake with your hands, you may draw forth ſmall threds. like the web of a ſpider, ſuch as you may draw from the leaues of Scabious. The ſtalke is hollow, diuiding it ſelfe from the joints or knees, into ſundry other ſmall branches, at the top whereof do grow ſmall tufts or ſpokie rundles, of a white colour : after which commeth the ſeed like to *Carui*, or Caruwaies, of a ſharp taſte : the root is alſo ſharp and hot in taſte.

¶ *The Place.*

Theſe plants do grow in dry paſtures and medowes in this countrey very plentifully.

¶ *The Time.*

They floure from Iune to the end of Auguſt.

¶ *The Names.*

That which *Fuchſius* calleth *Pimpinella major*, *Dodonæus* termeth *Saxifragia major*, which kinde of Saxifrage more abſolutely anſwereth the true *Phellandrium* of *Pliny*, than any other plant whatſoeuer :

ſoeuer : wherein the Phyſitions of Paris haue bin deceiued, calling or ſuppoſing the medow Rue to be the right *Phellandrium*, whereunto it is not like either in ſhape or faculty : for it is nothing ſo effectuall in breaking the ſtone, or prouoking of vrine, as either of theſe plants, eſpecially *Pimpinella Hircina*, which is not ſo called becauſe it hath any rammiſh ſmell of a Goat, but becauſe practitioners haue vſed to feed goats with it, whoſe fleſh and bloud is ſingular good againſt the ſtone: but we rather take it to be named *Hircina*, of *Hircinea ſylua*, where it growes in great aboundance, the fauor of the herb not being vnpleaſant, ſomewhat reſembling the ſmell and taſt of *Daucus, Liguſtrum*, and *Paſtinaca* : ſo to conclude, both theſe are called *Saxifragia*. The ſmaller is called of ſome, *Petraſindula, Bipinella*, and *B. peſula* : of *Baptiſta Sardus*, and alſo of *Leonardus Fuchſius, Pimpinella maior* : Diuers call it *P. mpinella ſaxifraga* ; for there is alſo another *Pimpinella*, called *Pimpinella ſanguiſorba* notwithſtanding the verſe before rehearſed ſheweth a difference between *Pimpinella* and *Saxifraga* : in high Dutch it is called **Bibernel** : in low-Dutch, **Bauenaert** : in Engliſh the greater may be called great Saxifrage, and the other ſmall Saxifrage.

B. pinella is called *Saxifragia minor* : in Engliſh, ſmall Saxifrage, as *Pimpinella* is called great Saxifrage. ‡ *Columna* iudges it to be the *Tragium* of *Dioſcorides*. ‡

¶ *The Nature.*

Saxifrage of both kinds, with their ſeed, leaues, and roots are hot and dry in the third dagree, and of thin and ſubtill parts.

¶ *The Vertues.*

The ſeed and root of Saxifrage drunk with wine, or the decoction thereof made with wine, cauſeth to piſſe well, breaketh the ſtone in the kidneis and bladder, and is ſingular againſt the ſtrangurie and ſtoppings of the kidneis and bladder ; whereof it took the name *Saxifragia*, or Break-ſtone.   A

The juice of the leaues of Saxifrage doth clenſe and take away all ſpots and freckles of the face and leaueth a good colour.   B

The diſtilled water thereof mingled with ſome vineger in the diſtillation, cleareth the ſight, and taketh away all obſcuritie and darkneſſe of the ſame.   C

Chap. 421. *Of Burnet.*

1 *Pimpinella hortenſis.*
Garden Burnet.

2 *Pimpinella ſylueſtris.*
Wilde Burnet.

¶ *The Kindes.*

BVrnet of which we will intreat, doth differ from *Pimpinella*, which is alſo called *Saxifraga*. One of the Burnets is leſſer, for the moſt part growing in gardens, notwithſtanding it groweth in barren fields, where it is much ſmaller: the other greater, is altogether wilde.

¶ *The Deſcription.*

1   GArden Burnet hath long leaues made vp together of a great many vpon one ſtem, euery one whereof is ſomething round, nicked on the edges, ſomewhat hairy: among theſe riſeth a ſtalke that is not altogether without leaues, ſomething chamfered: vpon the tops whereof grow little round heads or knaps, which bring forth ſmall floures of a browne purple colour, and after them cornered ſeeds, which are thruſt vp together. The root is long: the whole plant doth ſmell ſomething like a Melon, or Cucumber.

2   Wilde Burnet is greater in all parts, it hath wider and bigger leaues than thoſe of the former: the ſtalke is longer, ſometimes two cubits high: the knaps are greater, of a darke purple colour, and the ſeed is likewiſe cornered and greater: the root longer, but this Burnet hath no pleaſant ſmell at all.

‡ 3   There is kept in ſome gardens another of this kinde, with very large leaues, ſtalkes, and heads, for the heads are ſome inch and halfe long, yet but ſlender conſidering the length, and the floures (as I remember) are of a whitiſh colour: in other reſpects it differs not from the precedent: it may fitly be called *Pimpinella ſanguiſorba hortenſis maxima*, Great Garden Burnet. ‡

¶ *The Place.*

The ſmall Pimpinell is commonly planted in Gardens, notwithſtanding it doth grow wild vpon many barren heaths and paſtures.

The great wilde Burnet groweth (as Mr. *Lyte* ſaith) in dry medowes about Viluord, and my ſelfe haue found it growing vpon the ſide of a cauſey which croſſeth the one halfe of a field, whereof the one part is earable ground, and the other part medow, lying betweene Paddington and Lyſſon green neere vnto London, vpon the high way.

¶ *The Time.*

They floure from Iune, vnto the end of Auguſt.

¶ *The Names.*

The later herbariſts doe call Burnet *Pimpinella ſanguiſorba*, that it may differ from the other, and yet it is called by ſeuerall names, *Sanguiſorba*, and *Sanguinaria*: *Geſner* had rather it ſhould be called *Peponella* of the ſmell of Melons or Pompions, to which it is like, as we haue ſaid: of others it is named *Pimpinella*, or *Bipennula*: of moſt men, *Solbaſtrella*: in high Dutch, **Solbleſkraut, her Gots Bartlin, Blutkraut, megelkraut**: in French, *Pimpennelle, Sanguiſorbe*: in Engliſh, Burnet. It agreeth *cum altera Dioſcoridis Sideritide*, that is to ſay, with *Dioſcorides* his ſecond Iron-wort: the leaſe (and eſpecially that of the leſſer ſort) which we haue written to conſiſt of many nicks in the edges of the leaues; and this may be the very ſame which *Pliny* in his 24 booke, chapter 17. reporteth to be named in Perſia, *Siſitieptris*, becauſe it made them merry; he alſo calleth the ſame *Protomedia*, and *Caſigneta*, and likewiſe *Dionyſionymphæs*, for that it doth maruellouſly agree with wine, to which alſo this *Pimpinella* (as we haue ſaid) doth giue a pleaſant ſent: neither is that repugnant, that *Pliny* in another place hath written, *De Sideritibus*, of the Iron-worts; for it often falleth out that he intreateth of one and the ſelfe ſame plant in diuers places, vnder diuers names: which thing then hapneth ſooner when the writers themſelues do not well know the plant, as that *Pliny* did not well know *Sideritis*, or Iron-wort, it is euen thereby manifeſt, becauſe hee ſetteth not downe his owne opinion hereof, but other mens.

¶ *The Temperature.*

Burnet, beſides the drying and binding faculty that it hath, doth likewiſe meanly coole: and the leſſer Burnet hath likewiſe withall a certaine ſuperficiall, ſleight, and temperate ſent, which when it is put into the wine it doth leaue behind it: this is not in the dry herbe, in the juyce, nor in the decoction.

¶ *The Vertues.*

A   Burnet is a ſingular good herb for wounds (which thing *Dioſcorides* doth attribute to his ſecond Iron-wort) and commended of a number: it ſtancheth bleeding, and therefore it was named *Sanguiſorba*, as well inwardly taken, as outwardly applied.

B   Either the juyce is giuen, or the decoction of the pouder of the dry leaues of the herbe, beeing bruiſed, it is outwardly applied, or elſe put among other externall medicines.

C   It ſtaieth the laske and bloudy flix: it is alſo moſt effectuall to ſtop the monethly courſe.

D   The leſſer Burnet is pleaſant to be eaten in ſallads, in which it is thought to make the heart mery and glad, as alſo being put into wine, to which it yeeldeth a certaine grace in the drinking.

The

The decoction of Pimpinell drunken, cureth the bloudy flix, the ſpitting of bloud, and all other F fluxes of bloud in man or woman.

The herbe and ſeed made into pouder, and drunke with wine, or water wherein yron hath beene G quenched doth the like.

The leaues of Pimpinell are very good to heale wounds, and are receiued in drinks that are made H for inward wounds.

The leaues of Burnet ſteeped in wine and drunken, comfort the heart, and make it merry, and are I good againſt the trembling and ſhaking thereof.

CHAP. 422.　Of Engliſh Saxifrage.

¶ The Deſcription.

1 THis kinde of Saxifrage our Engliſh women Phyſitions haue in great vſe, and is familiarly knowne vnto them, vouchſaſing that name vnto it of his vertues againſt the ſton: it hath the leaues of Fennell, but thicker and broader, very like vnto Seſeli pratenſe Mo. ſpelienſium (which addition Pena hath beſtowed vpon this our Engliſh Saxifrage) among which riſeth vp a ſtalke, of a cubit high or more, bearing at the top ſpokie rundles beſet with whitiſh yellow floures: the root is thicke, blacke without, and white within, and of a good ſauour.

† 1 Saxifraga Anglicana facie Seſeli pratenſis.　　　‡ 2 Saxifraga Pannonica Cluſij.
　English Saxifrage.　　　　　　　　　　　　Auſtrian Saxifrage.

‡ 2 Cluſius hath ſet forth another plant not much different from this our common Saxifrage and called it Saxifraga Pannonica, which I haue thought fit here to inſert: the leaues, ſaith hee, are much ſhorter than thoſe of Hogs-Fennell, & ſomwhat like to thoſe of Fumitory: the ſtalks are ſome foot high, ſlender, hauing ſome few ſmall leaues, and at the top carrying an vmbell of white floures: the root is not much vnlike that of Hogs-Fennell, but ſhorter and more acride; it is hairy at the top
thereof

thereof, whence the ftalkes and leaues come forth: it growes vpon fome hills in Hungary and Au-ftria, and floures in Iuly. ‡

¶ *The Place.*

Saxifrage groweth in moft fields and medowes euery where throughout this our Kingdome of England.

¶ *The Time.*

It floureth from the beginning of May to the end of Auguft.

¶ *The Names.*

Saxifraga Anglicana is called in our mother tongue Stone-breake or Englifh Saxifrage: *Pena* and *Lobel* call it by this name *Saxifraga Anglicana:* for that it groweth more plentifully in England than in any other countrey.

¶ *The Temperature.*

Stone-breake is hot and dry in the third degree.

¶ *The Vertues.*

A   A decoction made with the feeds and roots of Saxifrage, breaketh the ftone in the bladder and kidnies, helpeth the ftrangury, and caufeth one to piffe freely.

B   The root of ftone-breake boiled in wine, and the decoction drunken, bringeth downe womens fickeneffe, expelleth the fecondine and dead childe.

C   The root dried and made into pouder, and taken with fugar, comforteth and warmeth the fto-macke, cureth the gnawing and griping paines of the belly.

D   It helpeth the collicke, and driueth away ventofities or windineffe.

E   Our Englifh women vfe to put it in their running or rennet for cheefe, efpecially in Chefhire (where I was borne) where the beft cheefe of this Land is made.

† I haue formerly Chap. 188 deliuredthe hiftory of the *Saxifraga maior* of *Matthiolus,* and *Saxifraga Antiquorum* of *Lobel;* not thinking that our Author had but their defcriptions here amongft the *Vmbellifera,* for if I had, I fhould haue fpared my labour there beftowed, and haue giuen the figures here to the defcriptions of our Author, which are now omitted. The figure formerly here was of the *Caucalis,* defcribed in the third place of the 403. chapter.

C H A P. 423. *Of Siler Mountaine, or baftard Louage.*

† 1 *Siler montanum Officinarum.*   † 2 *Sefeli pratenfe Monfpelienfium.*
Baftard Louage.   Horfe Fennell.

¶ *The Description.*

1   THe naturall plants of *Seseli*, being now better knowne than in times past, especially a-
mong our Apothecaries, is called by them *Siler montanum*, and *Seseleos* : this plant they
haue retained to very good purpose and consideration ; but the errour of the name hath
caused diuers of our late writers to erre, and to suppose that *Siler montanum*, called in shops *Seseleos*,
was no ther then *S. seli Massiliensium* of *Dioscorides*. But this plant containeth in his substance much
more acrimony, sharpenesse, and efficacy in working, than any of the plants called *Seselios*. It hath
stalkes like Ferula, two cubits high. The root smelleth like *Ligusticum* : the leaues are very much
cut or diuided, like the leaues of Fennell or *Seseli Massilicnse*, and broader than the leaues of *Peucedanum*. At the top of the stalkes grow spokie tufts like Angelica, which bring forth a long and leafie
seed like Cumine, of a pale colour ; in taste seeming as though it were condited with sugar, but
withall somewhat sharpe, and sharper than *Seseli pratense*.

2   There is a second kinde of *Siler* which *Pena* and *Lobel* set forth vnder the title of *Seseli pratense
Montpeliensium*, which *Dodonæus* in his last edition calleth *Siler pratense alterum*, that is in shew very
like the former. The stalkes thereof grow to the height of two cubits, but his leaues are somewhat
broader and blacker ; there are not so many leaues growing vpon the stalke, and they are lesse diuided than the former, and are of a little sauour. The seed is smaller than the former, and sauouring
very little or nothing. The root is blacke without, and white within, diuiding it selfe into sundry
diuisions.

¶ *The Place.*

It groweth of it selfe in Liguria, not far from Genua in the craggy mountaines, and in the gardens of diligent Herbalists.

¶ *The Time.*

These plants do floure from Iune to the end of August.

¶ *The Names.*

It is called commonly *Siler Montanum* : in French and Dutch by a corrupt name *Ser-Montain* :
in diuers shops, *Seseleos*, but vntruly : for it is not *Seseli*, nor a kinde thereof : in English, Siler mountaine, after the Latine name, and bastard Louage. ‡ The first is thought to be the *Ligusticum* of the
Antients, and it is so called by *Matthiolus* and others. ‡

¶ *The Temperature.*

This plant with his seed is hot and dry in the third degree.

¶ *The Vertues.*

The seeds of *Siler* drunke with Wormewood Wine, or Wine wherein Wormewood hath beene   **A**
sodden, moueth womens diseases in great abundance : cureth the suffocation and strangling of the
matrix, and causeth it to returne vnto the naturall place againe.

The root stamped with hony, and applied or put into old sores, doth cure them, and couer bare   **B**
and naked bones with flesh.

Being drunke it prouoketh vrine, easeth the paines of the guts or entrails proceeding of crudity   **C**
or rawnesse, it helpeth concoction, consumeth winde, and swelling of the stomacke.

The root hath the same vertue or operation, but not so effectuall, as not being so hot and dry.   **D**

† The figure which formerly was here was of the *Seseli Massiliense* described in the next chapter in the fourth place, and that which belonged to this place was
put for our common Louage. Also that figure which belonged to the second description was formerly vnder the title of *Fæniculum dulce*.

CHAP. 424.   *Of Seselios, or Hart-worts of Candy.*

¶ *The Description.*

1   THis plant beeing the *Seseli* of Candy, and in times past not elsewhere found, tooke his
surname of that place where it was first found, but now adaies it is to be seene in the
corne fields about Narbone in France, from whence I had seeds, which prosper well
in my garden. This is but an annuall plant, and increaseth from yeare to yeare by his owne sowing.
The leaues grow at the first euen with the ground, somewhat hairie, of an ouerworne greene colour,
in shape much like vnto Cheruill, but thicker : among which riseth vp an hairy rough stalke, of
the height of a cubit, bearing at the top spokie tufts with white floures : which being vaded, there
followeth round and flat seed, compassed and cunningly wreathed about the edges like a ring.

The ſeed is flat like the other, joyned two together in one, as you may ſee in the ſeed of Ferula or Angelica, in ſhape like a round target, in taſte like *Myrrhis*. *Matthiolus* did greatly miſtake this plant.

     2   There is a kinde of *Seſeli Creticum*, called alſo *Tordylion* : and is very like vnto the former, ſauing that his leaues are more like vnto common Parſneps than Cheruil, and the whole plant is bigger than the former.

   1 *Seſeli Creticum minus.*
Small Seſeleos of Candie.

   ‡ 2 *Seſeli Creticum majus.*
Great Seſelios of Candie.

     3   There is likewiſe a kinde of *Seſeli* that hath a root as bigge as a mans arme, eſpecially if the plant be old, but the new and young plants beare roots an inch thicke, with ſome knobs and tuberous ſprouts, about the lower part, the root is thicke, rough, and couered ouer with a thicke barke, the ſubſtance whereof is firſt gummie, afterward ſharpe, and as it were full of ſpattle; from the vpper part of the root proceed many knobs or thicke ſwelling roots, out of which there iſſueth great and large wings or branches of leaues, ſome whereof are notched or dented round about, growing vnto one ſide or rib of the leaſe, ſtanding alſo one oppoſite vnto another, of a darke and delaid green colour, and ſomwhat ſhining aboue, but vnderneath of a grayiſh or aſh-colour : from amongſt theſe leaues there ariſeth a ſtraked or guttered ſtalke, a cubit and a halfe high, ſomtimes an inch thick, hauing many joynts or knees, and many branches growing about them, and vpon each joynt leſſer branches of leaues. At the top of the ſtalkes, and vpper ends of the branches grow little cups or vmbels of white floures ; which being vaded, there commeth in place a ſeed, which is very like *Siler montanum*. ‡ I take this here deſcribed to be the *Seſeli montanum* 1, of *Cluſius*, or *Liguſticum alterum Belgarum* of *Lobel* : and therefore I haue giuen you *Cluſius* his figure in this place. ‡

     There is alſo a kind of *Seſeli*, which *Pena* ſetteth forth for the firſt kinde of *Daucus*, whereof I take it to be a kinde, growing euery where in the paſtures about London, that hath large leaues, growing for a time euen with the earth, and ſpred thereupon, and diuided into many parts, in manner almoſt like to the former for the moſt part in all things, in the round ſpokie tufts or vmbels, bearing ſtiffe and faire white floures in ſhape like them of Cinkefoile ; in ſmell like *Sambucus* or Elder. When the floure is vaded, there commeth in place a yellow guttered ſeed, of a ſpicie and very hot taſte. The root is thicke, and blacke without, which rotteth and periſheth in the ground (as wee
                    may

may ſee in many gummie or Ferulous plants) after it hath ſeeded, neither will it floure before the ſecond or third yeare after it is ſowne. ‡ I am ignorant what our Author meanes by this deſcription. ‡

‡ 3 *Seſeli montanum majus.*
Mountaine Seſelios.

† 4 *Seſeli Maſsilienſe.*
Seſelios of *Marſeilles.*

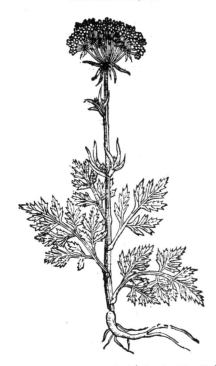 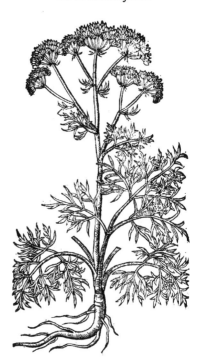

4   There is likewiſe a kinde of *Seſeli* called *Seſeli Maſsilienſe,* which hath leaues very much clouen or cut, and finely jagged, very much like vnto the leaues of ſweet Fennell, greater and thicker than the common Fennell. The ſtalke groweth to the height of three cubits, hauing knotty joynts, as it were knees; bearing at the top thereof tufts like vnto Dill, and ſeed ſomewhat long and cornered, of a ſharpe and biting taſte. The root is long and thicke like vnto great Saxifrage, of a pleaſant ſmell and ſharpe in taſte.

There is another *Seſeli* of Maſſilia, which hath large and great leaues like vnto Ferula, and not much vnlike *Siler Montanum :* among which riſe vp ſtalkes foure cubits high, bearing at the tops ſpokie tufts like vnto the laſt before rehearſed, of a good ſauour. The root is like vnto the former in ſhape, ſubſtance, and ſauour, but that it is greater.

¶ *The Place.*

Theſe plants are ſtrangers in England, notwithſtanding I haue them in my garden.

¶ *The Time.*

They floure and flouriſh in September.

¶ *The Names.*

Their names haue beene touched in their ſeuerall deſcriptions.

¶ *The Temperature and Vertues.*

It prouoketh vrine, and helpeth the ſtrangury, bringeth downe the ſickeneſſe and dead birth : it   A helpeth the cough and ſhortneſſe of breath, the ſuffocation of the mother, and helpeth the falling ſickeneſſe.

The ſeed drunke with wine concoĉteth raw humours, taketh away the griping and torments of   B the belly, and helpeth the ague, as *Dioſcorides* ſaith.

The juyce of the leaues is giuen to Goats and other cattell to drinke, that they may the ſooner   C be deliuered of their young ones, as the ſame Author reporteth.

          Chap.

Chap. 425. *Of Spignell, Spicknell, or Mewe.*

¶ *The Deſcription.*

1   SPignell hath ſtalks riſing vp to the height of a cubit and a halfe, beſet with leaues reſembling Fennell or Dill, but thicker, more buſhie, and more finely jagged ; and at the top of
the ſtalkes do grow ſpokie tufts like vnto Dil. The roots are thicke, and full of an oleous
ſubſtance, ſmelling well, and chafing or heating the tongue, of a reaſonable good ſauour.

1 *Meum.*
Spignell.

‡ 2 *Meum alterum Italicum.*
Italian Spignell.

2   There is a baſtard kinde of Spignell like vnto the former, ſauing that the leaues are not ſo
finely cut or jagged : the floures are tufted more thicker than the former: the roots are many, thicke,
and full of ſap.

¶ *The Place.*

Mew, or Meon groweth in Weſtmerland, at a place called Round-twhat betwixt Aplebie and
Kendall, in the pariſh of Orton.

Baſtard Mewe, or *Meum*, groweth in the waſte mountaines of Italy, and the Alpes, and (as it hath
been told me) vpon Saint Vincents rocke by Briſtow, where I ſpent two daies to ſeeke it, but it was
not my hap to finde it, therefore I make ſome doubt of the truth thereof.

¶ *The Time.*

Theſe herbes do floure in Iune and Iuly, and yeeld their ſeed in Auguſt.

¶ *The Names.*

It is called of the Græcians, μήϊν or μάον: likewiſe of the Latines, *Meum:* of the Italians, *Meo* : in
Apulia, as *Matthiolus* declareth, it is called *Imperatrix :* in diuers places of Spaine, *Siſtra :* in others,
Pinello : in high Dutch, **Beerewurtz:** in French, *Siſtre : Ruellius* ſaith that it is named in France *Ane-
thum tortuoſum*, and *ſylueſtre*, or writhed Dill, and wilde Dill : alſo it is called in Engliſh, Spignell, or
Spicknell, of ſome Mew, and Bearewort.

The ſecond may be called baſtard Spicknell.

¶ *The*

¶ *The Temperature.*

Theſe herbes, eſpecially the roots of right Meon, is hot in the third degree, and dry in the ſecond.

¶ *The Vertues.*

The roots of Meon, boyled in water and drunke, mightily open the ſtoppings of the kidnies and **A** bladder, prouoke vrine and bodily luſt, eaſe and helpe the ſtrangury, and conſume all windineſſe and belchings of the ſtomacke.

The ſame taken with hony doth appeaſe the griefe of the belly, and is exceeding good againſt **B** all Catarrhes, rheumes, and aches of the joynts, as alſo any phlegme which falls vpon the lungs.

If the ſame be laied plaiſterwiſe vpon the bellies of children, it maketh them to piſſe well.   **C**

They clenſe the entrals, and deliuer them of obſtructions or ſtoppings: they prouoke vrine, **D** driue forth the ſtone, and bring downe the floures: but if they be taken more than is requiſite, they cauſe the head-ache; for ſeeing they haue in them more heat than drineſſe, they carry to the head raw moiſture and windie heat, as *Galen* ſaith.

C H A P. 426.   *Of Horeſtrange, or Sulpherwort.*

¶ *The Deſcription.*

1   SVlphurwort or Hogs-fennell hath a ſtiffe and hard ſtalke full of knees or knots, beſet with leaues like vnto Fennell, but greater, comming neerer vnto Ferula, or rather like the leaues of wilde Pine-tree, and at the top of the ſtalkes round ſpokie tufts full of little yollow floures, which doe turne into broad browne ſeed. The root is thicke and long: I haue digged vp roots thereof as big as a mans thigh, blacke without, and white within, of a ſtrong and grieuious ſmell, and full of yellow ſap or liquor, which quickely waxeth hard or dry, ſmelling not much vnlike brimſtone, called *Sulphur*; which hath induced ſome to call it Sulphurwort; hauing alſo at the top toward the vpper face of the earth, a certaine buſh of haire, of a browne colour, among which the leaues and ſtalkes do ſpring forth.

2   The ſecond kinde of *Peucedanum* or Hogs-fennell is very like vnto the former, ſauing that the leaues be like Ferula: the roots are nothing ſo great as the former, but all the reſt of the plant doth far exceed the other in greatneſſe.

3   There is another kinde of *Peucedanum* or Hogs-Fennell, which *Pena* found vpon Saint Vincents rock by Briſtow, whoſe picture he hath ſet forth in his *Aduerſaria*, which that famous Engliſh Phyſition of late memory, Dr. *Turner* found there alſo, ſuppoſing it to be the right and true *Peucedanum*, whereof no doubt it is a kinde: it groweth not aboue a foot high, and is in ſhape and leaues like the right *Peucedanum*, but they be ſhorter and leſſer; growing ſomwhat like the writhed Fennell of Maſſilia, but the branches are more largely writhed, and the leaues are of the colour of the branches, which are of a pale greene colour. At the top of the branches grow ſmall white tufts, hauing ſeed like Dill, but ſhorter and ſlenderer, of a good taſte, ſomewhat ſharpe. The root is thicker than the ſmalneſſe of the herbe will well beare. Among the people about Briſtow, and the rocke aforeſaid, this hath beene thought good to eat.

‡   The figure of this our Author formerly gaue (yet vnfitly, it not agreeing with that deſcription) for *Oreoſelinum*: it may be he thought it the ſame with that of *Dodon*. his deſcription, becauſe he found it vnder the ſame title in *Tabernamontanus*. This is the *Selinum montanum pumilum* of *Cluſius*; and the *Peucedani facie puſilla planta* of *Pena* and *Lobel*; wherefore *Bauhine* was miſtaken in his *Pinax*, whereas he refers that of *Lobel* to his third *Peucedanum*: the root of this is blacke without, and white within, but ſhort, yet at the top about the thickeneſſe of ones finger: the leaues are ſmall and greene, commonly diuided into fiue parts; and theſe againe ſubdiuided by threes: the ſtalke is ſome ſix inches or halfe a foot high, diuided into ſundry branches, creſted, broad, and at the toppes of the branches, euen when they firſt ſhoot vp, appeare little vmbels of white floures very ſmall, and conſiſting of fiue leaues apiece. The ſeed is blacke, ſhining and round, two being joyned together, as in moſt vmbelliferous plants. It floures in May, and ripens the ſeed in Iuly: I receiued in Iuly 1632, ſome plants of this from Briſtow, by the meanes of my oft mentioned friend Maſter *George Bowles*, who gathered it vpon Saint Vincents Rocke, whereas the Authors of the *Adeerſaria* report it to grow. ‡

1 *Peucedanum*.
Sulphurwort.

2 *Peucedanum majus*.
Great Sulphurwort.

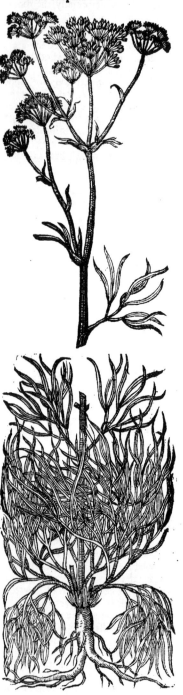

‡ 3 *Peucedanum pumilum*.
Dwarfe Hogs-Fennell.

¶ *The Place.*

The firſt kinde of *Peucedanum* or Hogs Fennell groweth very plentifully on the South ſide of a wood belonging to Waltham, at the Naſe in Eſſex by the high-way ſide; alſo at Whitſtable in Kent, in a medow neere to the ſea ſide, ſometime belonging to Sir *Henry Criſpe*, and adjoyning to his houſe there. It groweth alſo in great plenty at Feuerſhaam in Kent, neere vnto the hauen vpon the bankes thereof, and in the medowes adjoyning.

The ſecond kinde groweth vpon the ſea coaſts of Montpellier in France, and in the coaſts of Italy.

¶ *The Time.*

Theſe plants do floure in Iune, Iuly, and Auguſt.

¶ *The Names.*

The Græcans call it, ꝑꝏ: the Latines in like manner, *Peucedanos*, or *Peucedanum*, and alſo *Pinaſtellum* : moſt of the ſhops, and likewiſe the common people name it *Fœniculum Porcinum* : of diuers, *Stataria* : of the Prophets, ꝑꝏ: that is to ſay, a good Angell or Ghoſt: in high Dutch, **Harſtrang, Schweffel wurkel, Sewfenckel** : in Italian and French, *Peucedano* : in Spaniſh, *Herbatum* : in Engliſh, Hore-ſtrange, and Hore-ſtrong, Sow-Fennell, or Hogs Fennell, Sulpher-wort, or Brimſtone-wort. It is called *Peucedanum* and *Pinaſtellum*, of the Greeke and Latine words, ꝑꝏ and *Pinus*.

¶ *The Temperature.*

Theſe herbes, eſpecially the yellow ſap of the root, is hot in the ſecond degree, and dry in the beginning of the third.

¶ *The Vertues.*

The yellow ſap of the root of Hogs Fennell, or as they call it in ſome places of England, Hore-ſtrange, taken by it ſelfe, or with bitter Almonds and Rue, is good againſt the ſhortneſſe of breath, it aſſwageth the griping paines of the belly, diſſolueth and driueth away ventoſitie or windineſſe of the ſtomacke; it waſteth the ſwelling of the milt or ſpleene, looſeth the belly gently, and purgeth by ſiege both flegme and choler.   A

The ſame taken in manner aforeſaid prouoketh vrine, eaſeth the paine of the kidnies and bladder, cauſeth eaſie deliuerance of childe, and expelleth the ſecondine, or after-birth, and the dead childe.   B

The ſap or juyce of the root mixed with oyle of Roſes, or Vineger, and applied, eaſeth the palſie, crampes, contraction or drawing together of ſinewes, and all old cold diſeaſes, eſpecially the Sciatica.   C

It is vſed with good ſucceſſe againſt the rupture or burſtings in young children, and is very good to be applied vnto the nauels of children that ſtand out ouer much.   D

The decoction of the root drunke is of like vertue vnto the juyce, but not altogether ſo effectuall againſt the foreſaid diſeaſes.   E

The root dried and made into pouder doth mundifie and clenſe old ſtinking and corrupt ſores and vlcers, and healeth them: it alſo draweth forth the corrupt and rotten bones that hinder the ſame from healing, and likewiſe ſplinters and other things fixed in the fleſh.   F

The ſaid pouder or juyce of the root mixed with oyle of Roſes, cauſeth one to ſweat, if the body be annointed therewith, and therefore good to be put into the vnction or ointment for the French diſeaſe.   G

The congealed liquor tempered with oile of Roſes, and applied to the head after the manner of an ointment, is good for them that haue the Lethergie, that are franticke, that haue dizzineſſe in the head, that are troubled with the falling ſickeneſſe, that haue the palſie, that are vexed with convulſions and cramps, and generally it is a remedy for all infirmities of the ſinewes, with vineger and oyle as *Dioſcorides* teacheth.   H

The ſame being ſmelt vnto reuiueth and calleth them againe that be ſtrangled with the mother, and that lie in a dead ſleepe.   I

Being taken in a reare egge it helpeth the cough and difficulty of breathing, gripings and windineſſe, which, as *Galen* addeth, proceedeth from the groſſeneſſe and clammineſſe of humours.   K

It purgeth gently, it diminiſheth the ſpleene, by cutting, digeſting and making thin humours that are thicke: it cauſeth eaſie trauell, and openeth the matrix.   L

A ſmall piece of the root holden in the mouth is a preſent remedy againſt the ſuffocation of the mother.   M

Chap. 472.
Of Herbe Ferula, or Fennell Gyant.

¶ The Kindes.

Dioſcorides maketh mention of a Ferula, out of which is gathered the Gum Sagapene; and alſo he declareth, that the Gums Galbanum and Ammoniacum are liquors of this herb Ferula: but what difference there is in the liquors, according to the clymate or countrey where it groweth, he doth not ſet downe; for it may be that out of one kinde of Ferula ſundry juyces may be gathered, that is to ſay, according to the diuerſitie of the countries where they grow, as we haue ſaid: for as in Laſer, the juyce of Laſerwort that groweth in Cyrene doth differ from that liquor which groweth in Media and Syria; ſo it is likewiſe that the herbe Ferula doth bring forth in Media Sagapenum, in Cyrene Ammoniacum, and in Syria Galbanum. Theophraſtus ſaith that the herbe Ferula is diuided into mo kindes, and he calleth one great, by the name of Ferula; and another little, by the name Ferulago.

1 Ferula.
Fennell-Gyant.

‡ 2 Ferulago.
Small Fennell-Gyant.

¶ The Deſcription.

1   Ferula, or Fennell Gyant, hath very great and large leaues of a deepe greene colour, cut and jagged like thoſe of Fennell, ſpreading themſelues abroad like wings: amongſt which riſeth vp a great hollow ſtalke, ſomewhat reddiſh on that ſide which is next vnto the Sun, diuided into certaine ſpaces, with joynts or knees like thoſe of Hemlocks or Kexes, of the bigneſſe of a mans arme in the wreſt, of the height of foure or fiue cubits where it groweth naturally, as in Italy, Greece, and other hot countries; notwithſtanding it hath attained to the height of foureteene or fifteene foot in my Garden, and likewiſe groweth fairer and greater than from whence it came, as it fareth with other plants that come hither from hot regions: as for example our great Artichoke, which firſt was brought out of Italy into England, is become (by reaſon of the great moiſture which our countrey is ſubject vnto) greater and better than thoſe of Italy;
                                                       infomuch

inſomuch that diuers Italians haue ſent for ſome plants of our Artichokes, deeming them to be of another kinde ; neuertheleſſe in Italy they are ſmall and dry as they were before. Euen ſo it happeth to this *Ferula*, as we haue ſaid. This foreſaid ſtalke diuideth it ſelfe toward the top into diuers other ſmall branches, whereon are ſet the like leaues that grow next the ground, but much leſſer. At the top of the branches at the firſt budding of the floures appeare certaine bundles incloſed in thin skins, like the yolke of an egge, which diuers call *Corculum Ferula,* or the little heart of *Ferula,* which being brought to maturitie, open themſelues into a tuft or vmbell like that of Dill, of a yellowiſh colour : afterwhich come the ſeed, in colour and faſhion like thoſe of the Parſnep, but longer and greater, alwaies growing two together, ſo cloſely joyned, that it cannot be diſcerned to be more than one ſeed vntill they be diuided : the root is very thicke and great, full of a certaine gummie juyce, that floweth forth, the root being bruiſed, broken, or cut ; which being dried or hardned, is that gum which is called *Sagapenum,* and in ſome ſhops *Serapinum.*

‡ 3 *Panax Aſclepium Ferula facie.*
Æſculapius his All-heale.

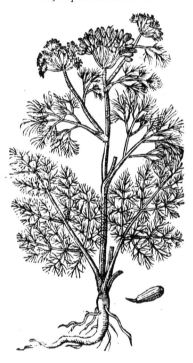

2　There is likewiſe another ſmaller *Ferula* like vnto the former in each reſpect, ſauing that it is altogether leſſe: the root likewiſe being wounded yeeldeth forth a ſap or juyce, which when it is hardened is called *Galbanum :* of the Aſſyrians, *Metopium.*

I haue likewiſe another ſort ſent mee from Paris, with this title *Ferula nigra* ; which proſpereth exceeding well in my Garden, but difference I cannot finde any from the former, ſauing that the leaues are of a more blacke or ſwart colour.

‡ 3 I know not where more fitly than in this place to giue you the Hiſtory of that *Ferula* or Ferulaceous plant that *Dodonæus, Lobel,* and others haue ſet downe vnder the name of *Panax Aſclepium.* The ſtalke hereof is ſlender, a cubit high, creſted and joynted, and from theſe joynts proceed leaues bigger than thoſe of Fennell, and alſo rougher, and of a ſtrong ſmell: at the tops of the branches grow vmbels of yellow floures : the ſeed is flattiſh, like that of the other *Ferula :* the root long, white, and of a ſtrong ſmell. This growes naturally in Iſtria. ‡

¶ *The Place.*
Theſe plants are not growing wilde in England ; I haue them all in my garden.

¶ *The Time.*
They floure in Iune and Iuly ; they perfect their ſeed in September ; not long after, the ſtalke with his leaues periſh : the root remaineth freſh and greene all Winter.

¶ *The Names.*
The firſt is called in Greeke, νάρθηξ : in Latine, *Ferula :* in Italian, *Ferola :* in Spaniſh, *Cananheia :* in Engliſh, Herbe Ferula, and Fennell Gyant.

¶ *The Temperature.*
Theſe plants with their Gums are hot in the third degree, and dry in the ſecond.

¶ *The Vertues.*
The pith or marrow, called *Corculum Ferula,* as *Galen* teacheth, is of an aſtringent or binding qua- A
litie, and therefore good for them that ſpit bloud, and that are troubled with the flix.

Dioſcorides ſaith, that beeing put into the noſthrils it ſtaieth bleeding, and is giuen in Wine to B
thoſe that are bitten with Vipers.

It is reported to be eaten in Apulia roſted in the embers, firſt wrapped in leaues or in old clouts, C
with pepper and ſalt ; which, as they ſay, is a pleaſant ſweet food, that ſtirreth vp luſt, as they report.

The ſeed doth heate, and attenuate or make thinne : it is a remedy againſt cold fits of an Ague, D
by procuring ſweat, being mixed with oyle, and the body annointed therewith.

A dram of the juyce of *Ferula* which beareth *Sagapenum,* purgeth by ſiege tough and ſlimie hu- E
mors,

mors, and all groſſe flegme and choler, and is alſo good againſt all old and cold diſeaſes which are hard to be cured; it purgeth the braine, and is very good againſt all diſeaſes of the head, againſt
F   the Apoplexie and Epilepſie.

Being taken in the ſame manner, it is good againſt crampes, palſies, ſhrinkings and paines of the ſinewes.

G   It is good againſt the ſhortneſſe of breath, the cold and long cough, the paine in the ſide and breſt, for it mundifieth and cleanſeth the breſt from all cold flegme and rheumaticke humors.

H   *Sagapenum* infuſed or ſteeped in vineger all night, and ſpread vpon leather or cloath, ſcattereth, diſſolueth, and driueth away all hard and cold ſwellings, tumors, botches, and hard lumps growing about the joynts or elſewhere, and is excellent good to be put into or mingled with all oyntments or emplaiſters which are made to mollifie or ſoften.

I   The juyce of *Ferula Galbanifera*, called *Galbanum*, drunke in wine with a little myrrhe, is good againſt all venome or poyſon that hath beene taken inwardly, or ſhot into the body with venomous darts, quarrels, or arrowes.

K   It helps womens painfull trauell, if they do take thereof in a cup of wine the quantitie of a bean.

L   The perfume of *Galbanum* helpeth women that are grieued with the riſing of the mother, and is good for thoſe that haue the falling ſickeneſſe.

M   *Galbanum* ſoftneth, mollifieth, and draweth forth thornes, ſplinters, or broken bones, and conſumeth cold and phlegmaticke humors, ſeruing in ſundry oyntments and emplaiſters for the vſe of Surgerie, and hath the ſame phyſicall vertues that are attributed vnto *Sagapenum*.

Chap. 428. *Of Drop-wort, or Filipendula.*

1 *Filipendula.*
Drop-wort.

2 *Filipendula montana.*
Mountaine Drop-wort.

¶ *The Kindes.*

THere be diuers ſorts of Drop-worts, ſome of the champion or fertill paſtures, ſome of more moiſt and dankiſh grounds, and ſome of the mountaine.

¶ *The*

¶ *The Deſcription.*

1　THe firſt kinde of Filipendula hath leaues growing and ſpred abroad like feathers, each leafe conſiſting of ſundry ſmall leaues dented or ſnipt round about the edges, growing to the ſtalke by a ſmall and ſlender ſtem : theſe leaues reſemble wilde Tanſie or Burnet, but that they be longer and thicker, ſet like feathers, as is aforeſaid : among theſe riſe vp ſtalkes a cubit and a halfe high, at the top whereof grow many faire white floures, each ſmall floure conſiſting of ſix ſlender leaues, like a little ſtar, buſhing together in a tuft like the floures of Medeſweet, of a ſoft ſweet ſmell : the ſeed is ſmall, and groweth together like a button : the roots are ſmall and blacke, whereupon depend many little knops or blacke pellets, much like the roots of the female Peonie, ſauing that they be a great deale ſmaller.

2　The ſecond kinde of Filipendula, called of *Pena* in his Obſeruations, *Oenanthe, ſiue Philipendula altera montana*, is neither at this day very well knowne, neither did the old Writers heretofore once write or ſpeake of it : but *Pena* that painefull Herbariſt found it growing naturally in Narbone in France, neere vnto Veganium, on the top of the high hills called *Paradiſus Dei*, and neere vnto the mountaine Calcaris : this rare plant hath many knobby long roots, in ſhape like to *Aſphodelus luteus*, or rather like the roots of *Corrada*, or wilde Aſparagus ; from which riſeth vp a ſtalke a foot high, and more, which is thicke, round, and chanelled, beſet full of leaues like thoſe of common Filipendula, but they be not ſo thicke ſet or winged, but more like vnto the leaues of a Thiſtle, conſiſting of ſundry ſmall leaues, in faſhion like to *Coronopus Ruellij*, that is, *Ruellius* his Buckes-horne : round about the tops of the ſtalke there groweth a very faire tuft of white floures, reſembling fine ſmall hoods, growing cloſe and thicke together like the floures of *Pedicularis*, that is, Red Rattle, called of *Carolus Cluſius*, *Alectorolophos* ; whereof he maketh this plant a kinde, but in my judgement and opinion it is rather like *Cynoſorchis*, a kinde of Satyrion.

3 *Filipendula anguſtifolia.*
Narrow leaued Drop-wort.

† 4 *Filipendula Cicute facie.*
Hemlocke Drop-wort.

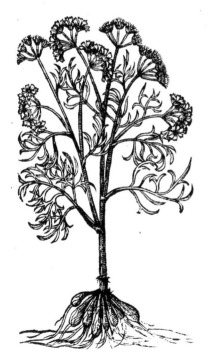

3　There is another kinde of Filipendula ſet forth vnder the name of *Oenanthe*, that hath many tuberous and thicke roots like thoſe of Drop-wort, but white of colour, and euery one of thoſe knobs hath a certaine ſtring or fibre annexed thereto, from whence ariſeth a creſted ſtalke two foot high,

5 *Filipendula aquatica.*
Water Drop-wort.

high, diuiding it selfe toward the top into sundry arms or branches: from the hollow place or bosome of euery joynt (out of which doe grow those branches) the leaues doe also proceed, very much cut or jagged like Fennell: at the top of those branches come forth spokie rundles of white floures fashioned like stars.

† 4 The fourth kinde of Filipendula is as strange a plant as the former, especially with vs here in England, except in the watery places and rills in the North, where *Paludapium* or water Smallage groweth; whereunto in leaues it is not vnlike, but more like *Ruta pratensis:* it hath many large branches, a naughty sauour, and in color & shape like *Cicuta,* that is, Hemlocke. The stalkes are more than two cubits high, comming from a root which exceedingly multiplies it selfe into bulbes, like *Asphodelus albus.* The smell of this plant is strong and grieuous; the taste hot and biting, it being full of a juyce, at first milky, but afterward turning yellow. The spoky tufts or rundles growing at the top are like *Cicuta,* yea, it much resembleth Hemlocke in property and qualities, and so do they affirme that haue proued and seene the experience of it: for being eaten in sallads it did well nigh poyson those which ate of it, making them giddie in their heads, waxing very pale, staggering and reeling like drunken men. Beware and take good heed of this and such like Simples; for there is no Physitian that wil giue it, because there be many other excellent good Simples which God hath bestowed vpon vs from the preuenting and curing of diseases. ‡ Pernitious and not excusable is the ignorance of some of our time, that haue bought and (as one may probably conjecture) vsed the roots of this plant in stead of those of Peionie; and I know they are daily by the ignorant women in Cheape-side sold to people more ignorant than themselues, by the name of water Louage; *Caueat Emptor.* The danger that may ensue by vsing them may be gathered by that which our Author hath here set downe, being taken out of the *Aduersaria, pag.* 326. ‡

5 The fifth and last kinde of Filipendula, which is the fourth according to *Matthiolus* his account, hath leaues like water Smallage, which *Pliny* calleth *Sylaus,* the leaues very much resembling those of *Lauer Cratena:* among which riseth vp a small stalke deepely furrowed or crested, bearing at the top whereof spokie or bushie rundles of white floures thicke thrust together. The roots are compact of very many filaments or threds; among which come forth a few tuberous or knobbie roots like vntothe second.

¶ *The Place and Time.*

The first groweth plentifully vpon stonie rockes or mountaines, and rough places, and in fertile pastures. I found great plenty thereof growing in a field adjoyning to Sion house, somtime a Nunnerie, neere London, on the side of a medow called Sion Medow.

The second hath beene sufficiently spoken of in the description. The third groweth neere vnto brookes and riuers sides. The fourth groweth betweene the plowed lands in the moist and wet furrowes of a field belonging to Battersey by London. ‡ It also groweth in great abundance in many places by the Thames side; as amongst the oysiers against Yorke house, a little aboue the Horse-ferrey, against Lambeth, &c. ‡ The fifth groweth neere the sides of riuers and water-streames, especially neere the riuer of Thames or Tems, as in S. Georges fields, and about the Bishop of Londons house at Fulham, and such like places.

They floure from May to the end of Iune.

¶ *The Names.*

They are commonly called *Filipendula.* The first is called of *Nicolaus Myrepsus, Philipendula:* of some, *Saxifraga rubra,* and *Millefolium syluestre:* of Pliny, *Molon:* in Italian and Spanish, *Filipendula:* in English, Filipendula and Drop-wort. Water Filipendula is called *Filipendula aquatica,* *Oenanthe aquatica,* and *Silaus Plinij.*

The

The fourth,whoſe leaues are like to Homlocks,is as ſome thinke called of *Cordus, Olſenichium :* in Engliſh,Homlocke Filipendula.

¶ *The Nature.*

Theſe kinds of Filipendula are hot and dry in the third degree,opening and clenſing,yet with a little aſtriction or binding. All the kindes of Onanthes haue the ſame facultie,except the fourth, whoſe pernitiouſneſſe we haue formerly touched.

¶ *The Vertues.*

The root of common Filipendula boiled in wine and drunken,is good againſt all pains of the A bladder, cauſeth one to make water, and breaketh the ſtone. The like *Dioſcorides* hath written of Oenanthe : the root,ſaith he,is good for them that piſſe by drops.

The pouder of the roots of Filipendula often vſed in meat, will preſerue a man from the falling B ſickneſſe.

† The figures that were formerly in the fourth and fifth places were both of the plant deſcribed in the fifth place. I haue giuen you in the fourth place the figure which *Lobel* and others haue giuen for the plant there deſcribed ; but it is not well expreſt, for the leaues are large like thoſe of Smallage, the ſtalke, branches, and vmbels very large,and like thoſe of Hemlocke,but rather bigger.

CHAP. 429. *Of Homlocks or herb Bennet.*

¶ *The Deſcription.*

1 THe firſt kind of Hemlocke hath a long ſtalke fiue or ſix foot high,great and hollow,ful of joints like the ſtalks of Fennell,of an herby colour,poudered with ſmall red ſpots almoſt like the ſtems of Dragons : the leaues are great,thick,and ſmal cut or jagged like the leaues of Cheruill,but much greater,and of a very ſtrong vnpleaſant ſauor:the flours are white, growing by tufts or ſpoky tops,which do change and turne into a white flat ſeed : the root is ſhort and ſomwhat hollow within.

1 *Cicuta.*
Hemlocks.

2 The Apothecaries in times paſt not knowing the right *Seſeli Peloponnenſe* , haue erroniouſly taken this *Cicuta latifolia* for the ſame : the leaues whereof are broad, thicke, and like vnto *Cicutaria*, but not the ſame. They called it *Seſeli Peloponnenſe cum folio Cicuta,* the faculties whereof deny and refute that aſſertion and opinion ; yea, and the plant it ſelfe, which beeing touched, yeeldeth or breatheth out a moſt virulent or loathſome ſmell. Theſe things ſufficiently argue , that it is not a kinde of *Seſeli* ; beſides the reaſons following : *Seſeli* hath a reaſonable good ſauour in the whole plant : the root is bare and ſingle, without fibres, like a Carrot. But *Cicuta* hath not onely a loathſome ſmell, but his roots are great, thicke, and knobbie, like the roots of *Myrrhis :* The whole plant doth in a manner reſemble the leaues, ſtalks, and floures of *Myrrhis odorata* , whoſe ſmall white floures do turn into long and crooked ſeeds, growing at the top of the branches , three cubits high.

‡ 3 This in leaues, ſtalkes, and roots is larger than the laſt deſcribed,the ſtalkes equalling or exceeding the height of a man : the ſmel is ſtrange and grieuous, and in all the parts thereof it is like vnto the other plants of this kinde. *Lobel* figures it by the name of *Cicutaria maxima Brancionis,* and queſtioneth whether it be not the *Thapſia tertia Salamanticenſium* of *Cluſius :* but *Cluſius* denies it ſo to bee ‡

2 *Cicuta latifolia fætidiſsima .*
Broad leaſed ſtinking Hemlocke.

‡ 3 *Cicuta latifolia altera.*
Gyant Hemlocke.

¶ *The Place.*

Common Hemlock growes plentifully about towne walls and villages, in ſhadowie places and fat ſoiles neere ditches.

The ſecond groweth vpon mountains and deſart places, and is a ſtranger in England : yet I haue plants thereof in my garden.

¶ *The Time.*
They flouriſh and ſeed in September.

¶ *The Names.*
Hemlocke is called in Greeke *κώνειον* : in Latine, *Cicuta :* in high-Dutch, **Schirling :** in Low-Dutch, **Scheerlinck :** in Spaniſh , *Cegutay,* Ca-*naheia :* in French, *Cigue :* in Engliſh, Hemlocke, Homlock, Kexe, and herb Bennet.

The ſecond is called *Cicuta latifolia,* and *Ci-cutaria latifolia,* and *Seſeli Peloponnenſe quorundam :* In Engliſh, great Hemlockes, and garden Hem-locks.

¶ *The Temperature.*
Galen ſaith, that Hemlocke is extreme cold in operation, euen in the fourth degree of coldneſſe.

¶ *The*

¶ *The Vertues.*

It is therefore a very rafh part to lay the leaues of Hemlocke to the ftones of yong boyes, or vir- A
gins brefts, by that means to keep thofe parts from growing great; for it doth not only eafily caufe
thofe members to pine away, but alfo hurteth the heart and liuer, being outwardly applied, & muft
of neceffitie hurt more being inwardly taken; for it is one of the deadly poifons which killeth by
his cold qualitie, as *Diofcorides* writeth, faying, Hemlocke is a very euill, dangerous, hurtfull, and
poyfonous herbe, infomuch that whofoeuer taketh of it into his body, dieth remedileffe, except the party drinke fome wine, that is naturally hot, before the venom hath taken the heart, as *Pliny* faith: but being drunke with wine, the poifon is with greater fpeed carried to the heart, by reafon whereof it killeth prefently: therefore not to be applied outwardly, much leffe taken inwardly into the body.

The great Hemlocke doubtleffe is not poffeffed with any good faculty, as appeares by his loth- B
fome fmell and other apparant fignes, and therefore not to be vfed in phyficke.

CHAP. 430.   *Of wilde and water Hemlocks.*

¶ *The Defcription.*

† 1   THis wilde kinde of Hemlocke hath a fmall tough white root, from which arife vp
diuers ftiffe ftalks, hollow, fomwhat reddifh toward the Sunne, jointed or kneed at
certain diftances: from which joints fpring forth long leaues very green, and finely minced or jagged like the common Cheruil or Parfly: the floures ftand at the tops of the ftalks
in fmall fpoky vmbels, with little longifh greene leaues about them: the feed followes like thofe
of Hemlocke, or as they grow together on the tops of the ftalkes they refemble Coriander feeds,
but leffer: the whole plant is of a naughty fmell.

† 1 *Cicutaria tenuifolia.*   2 *Cicutaria paluftris.*
Then leafed wilde Hemlocke.   Wilde water Hemlocke.

2   Water Hemlock, which *Lobel* calleth *Cicutaria paluftris*; *Clufius* and *Dodonæus*, *Phellandrium*,
rifeth vp with a thicke fat and empty hollow ftalke, full of knees or joints, crefted, chamfered, or

furrowed,of a yellowiſh green colour : the leaues ſhoot forth of the joints and branches,like wilde Hemlocke, but much thicker, fatter, and oilous, very finely cut or jagged almoſt like thoſe of the ſmalleſt *Viſnaga* or Spaniſh Tooth-picks : the floures ſtand at the top of the ſtalkes in ſmall white tufts : the ſeed followes,blackiſh,of the bigneſſe of Aniſe ſeed,& of a ſweet ſauor:the root is thick and long,within the water,very ſoft and tender,with very many ſtrings faſtned thereto.

¶ *The Place.*

1   This groweth amongſt ſtones and rubbiſh by the walls of cities and townes almoſt euerie where.

The other groweth in the middeſt of water ditches and ſtanding pooles and ponds, in moſt places of England : it groweth very plentifully in the ditches by a cauſey as you go from Redriffe to Detford neere London,and in many other places.

¶ *The Time.*

They floure and flouriſh in Iuly and Auguſt.

¶ *The Names.*

‡   1   This is *Petroſelini vitium* of *Tragus* ; and *Dauci inutilis genus* of *Geſner* : *Thalius* calls it *Apium cicutarium* : *Lobel*,*Cicutaria fatua* : *Tabern. Petroſelinum Caninum* ; which we may fitly Engliſh,Dogs Parſly.

2   This is *Liguſticum ſyl.& Fœminicum ſyl.*of *Tragus* : *Cicutaria paluſtris* of *Lobel* and others.*Dodonæus* thinks it *Plinies Phellandrion* ; and *Caſalpinus* iudges it his *Silaus.* ‡

¶ *The Nature and Vertues.*

Their temperature and faculties are anſwerable to the common Hemlocke, and haue no vſe in Phyſicke.

†   The figure formerly in the firſt place was of *Myrrhis*; the deſcription I thinke was intended, but not fully agreeing with that I here giue you, wherefore I haue a little altered it.

Chap. 431.
Of Earth-Nut, Earth Cheſtnut,or Kippernut.

† 1 *Bulbocaſtanon minus.*   2 *Bulbocaſtanon majus.*
   Small Earth-nut.   Great Earth-nut.

¶ *The Deſcription.*

1   EArth-nut or Kipper-nut, called after *Lobelius Nucula terreſtris* , hath ſmall euen creſted
ſtalks a foot or ſomewhat more high : whereon grow next the ground, leaues like thoſe
of Parſley, and thoſe that do grow higher like vnto thoſe of Dill ; the white floures do
ſtand on the top of the ſtalks in ſpokie rundles, like the tops of Dill, which turne into ſmall ſeed,
growing together by couples, of a very good ſmell, not vnlike to thoſe of Fennell, but much ſmaller : the root is round, knobbed, with certaine eminences or bunchings out; browne without, white
within, of a firme and ſollid ſubſtance, and of a taſte like the Cheſſe-nut, or Cheſt-nut, whereof it
tooke his name.

2   There is alſo another Earth-nut that hath ſtalkes a foot high, whereon grow jagged leaues
like thoſe of Engliſh Saxifrage, of a bright green colour : the floures grow at the top of the branches, in ſmall ſpokie tufts conſiſting of little white floures : the root is like the other, bulbous faſhion, with ſome few ſtrings hanging at the bottome, of a good and pleaſant taſte. ‡ This differs
from the former , in that the leaues are larger and greener: the root alſo is not ſo far within the
ground, and it alſo ſends forth ſome leaues from the bulb it ſelfe ; whereas our common kind hath
onely the end of a ſmall root that carries the ſtem and leaues vpon it, faſtned vnto it as you ſee it
expreſt in the former figure. ‡

¶ *The Place.*

Theſe herbes do grow in paſtures and corne fields almoſt euery where : there is a field adjoining
to High-gate, on the right ſide of the middle of the village, couered ouer with the ſame ; and likewiſe in the next field to the conduit heads by Maribone, neer the way that leads to Padington by
London, and in diuers other places. ‡ I haue not yet obſerued the ſecond to grow wild with vs. ‡

¶ *The Time.*

They floure in Iune and Iuly : the ſeed commeth to perfection afterward.

¶ *The Names.*

Alexander Traſlianus hath made mention of βικβοκανανο, *Lib.* 7. reckoning it vp among thoſe kindes
of meat or ſuſtenances which be good for ſuch as haue rotten lungs : of ſome it is called *Agrioca-
ſtanon.*

Gwinterius thought the word was corrupted, and that *Balanocaſtanon* ſhould be read : but this is as
ſtrange a word as *Bolbocaſtanon*, which was deriued of the form of a bulb, and the taſt of a Cheſt-nut:
of ſome, *Nucula terriſtris*, or the little Earth-nut : it is thought to be *Bunium Dioſcoridis* of ſome; but
we think not ſo : of Dr. *Turner, Apios* : yet there is another *Apios*, being a kind of *Tithymale* : of *Matthiolus, Oenanthe*, making it a kind of Filipendula : in high-Dutch, **Erdnuſʒ** : in low Dutch, **Erd-
noten** : the people of Sauoy cal it *Fauerottes* : in Engliſh, Earth-nuts, Kipper-nuts, & Earth cheſtnuts.

¶ *The Nature.*

The roots of Earth-nuts are moderately hot and dry, and alſo binding : but the ſeed is both hotter and drier.

¶ *The Vertues.*

The ſeed openeth and prouoketh vrine, and ſo doth the root likewiſe.   A

The root is good for thoſe that ſpit and piſſe bloud, if the root be eaten raw, or roſted in the em-   B
bers.

The Dutch people doe vſe to eat them boiled and buttered, as we do Parſeneps and Carrots,   C
which ſo eaten comfort the ſtomacke, and yeeld nouriſhment that is good for the bladder and kidneyes.

There is a plaiſter made of the ſeeds hereof, whereof to write in this place were impertinent to   D
our hiſtorie.

CHAP. 432.   *Of Cumin.*

¶ *The Deſcription.*

THis garden Cumin is a low or baſe herbe of a foot high : the ſtalk diuideth it ſelf into diuers
ſmall branches, whereon doe grow little jagged leaues very finely cut into ſmall parcels, like
thoſe of Fennell, but more finely cut, ſhorter and leſſer, the ſpoky tufts grow at the top of the
branches and ſtalkes, of a red or purpliſh colour : after which come the ſeed, of a ſtrong or rancke
ſmell, and biting taſte : the root is ſlender, which periſheth when it hath ripened his ſeed.

¶ *The*

Cuminum ſativum Dioſcoridis.
Garden Cumin.

¶ *The Place*

Cumin is husbanded and ſowne in Italy and Spaine, and is very common in other hot countries, as in Æthiopia, Egypt, Cilicia, and all the leſſer Aſia.

It delights to grow eſpecially in putrified and hot ſoiles : I haue proued the ſeeds in my garden,where they haue brought forth ripe ſeed much fairer and greater than any that comes from beyond the ſeas.

¶ *The Time.*

It is to be ſown in the middle of the ſpring; a ſhowre of rain preſently following much hindereth the growth thereof, as *Ruellius* ſaith.

My ſelf did ſow it in the midſt of May,which ſprung vp in ſix daies after : and the ſeed was ripe in the end of Iuly.

¶ *The Names.*

It is called in Greeke ᾤμινον ἥμερον that is, tame or garden Cumin, that it may differ from the wild ones: it is named in Latine *Cuminum* : in ſhops, *Cyminum* : in high-Dutch, **Roomiſche kymmel** : in Italian,*Comino* in Spaniſh,*Cominchos* : in French, *Comin* : in Engliſh, Cumin.

¶ *The Temperature.*

The ſeed of garden Cumin, as *Galen* ſaith, is hot and dry in the third degree:*Dioſcorides* ſaith that it hath in it alſo a binding quality.

¶ *The Vertues.*

A   The ſeed of Cumin ſcattereth and breaketh all the windineſſe of the ſtomacke, belly,guts, and matrix:it is good againſt the griping torments,gnawing or fretting of the belly,not only receiued inwardly by the mouth, but alſo in clyſters,and outwardly applied to the belly with wine and barley meale boiled together to the forme of a pultis.

B   Being handled according to art,either in a cataplaſm,pultis, or plaiſter,or boiled in wine and ſo applied,it taketh away blaſtings,ſwellings of the cods or genitors:it conſumeth windie ſwellings in the joints,and ſuch like.

C   Being taken in a ſupping broth it is good for the cheſt and cold lungs,and ſuch as are oppreſſed with abundance of raw humors.

D   It ſtancheth bleeding at the noſe, being tempered with vineger and ſmelt vnto.

E   Being quilted in a little bag with ſome ſmall quantitie of Bay ſalt , and made hot vpon a bed-pan with fire or ſuch like,and ſprinkled with good wine vineger,and applied to the ſide very hot, it taketh away the ſtitch and paines thereof,and eaſeth the pleuriſie very much.

Chap. 420. *Of wilde Cumin.*

¶ *The Kindes.*

THere be diuers plants differing very notably one from another in ſhape , and yet all comprehended vnder the wilde Cumin.

¶ *The Deſcription.*

1   THe wild Cumin hath ſmall white roots with ſome fibres thereto appendant; from the which ariſe ſundry little jagged leaues, conſiſting of many leſſer leaues, finely dented about the edges,in faſhion like the ſmalleſt leaues of wild parſnep:among which ſpringeth vp a ſlender bending ſtalk a foot high,like vnto *Pecten Veneris*,bearing at the top therof white round

1 *Cuminum ſylueſtre.*
Wilde Cumin.

2 *Cuminum ſiliquoſum.*
Codded wilde Cumin.

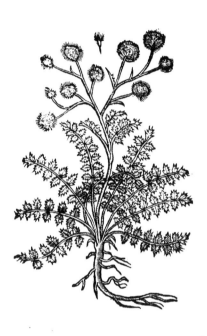

3 *Cuminum corniculatum, ſiue Hypecoum Cluſij.*
Horned wild Cumin.

round and hairie buttons or knops, like *Ar-
ction*, as *Dioſcorides* hath right wel obſerued:
within which knops is contained a tender
downie ſubſtance, among which is the ſeed,
like the ſeed of *Dens Leonis*, but much leſſer.

2   The ſecond kinde of Cumin is verie
like to the foreſaid wilde Cumin, ſaue that
it beareth a number of horned or crooked
cods, after the maner of *Scorpioides*, but thicker, and leſſe crooked, and the ſeeds within
the cods are ſeuerally diſtinct & ſeparated
one from another by equall partitions, in
ſmall croſſes, yellow of color, and ſomwhat
long : the ſtalkes are little and tender beſet
with leaues much like vnto the ſmall leaues
of *Carui*, or *Pecten Veneris* : and at the top of
the ſtalks there grow pretty yellow floures,
like thoſe of great Celandine or Rocket, ſauing that they be ſomwhat leſſer.

† 3   The third kind of Cumin is very
like vnto the laſt before mentioned, but the
leaues are much greater, more ſlender, and
more finely cut or jagged, like the leaues of
Seſeli of *Maſſylia* : among which riſeth vp a
ſtalke a cubit high or ſomewhat more, very
ſmooth & whitiſh: at the top wherof ſpring
forth fine yellow flours, not like the former,
but conſiſting of ſix leaues apiece, wherof
two are large, and edged with greene on the
outſide :

outfide : the other foure are fmall ones,and grow two on a fide between the two larger leaues:thefe floures being vaded, there fucceed crooked cods, greater, and more full of knots or diuifions than the former,wherein is contained a fmall and flat yellow feed like *Galega*:the root is long,thick, and fingle.

¶ *The Place.*

Thefe wilde Cumins do grow in Lycia,and Galatia,a prouince of Afia,and in Carthage a citie of Spaine ; feldome feen in thefe Northerne parts : notwithftanding at the impreffion hereof, the laft did floure and flourifh in my garden. ‡ Thefe grow in Prouince in France,and in diuers parts of Spaine. ‡

¶ *The Time.*

They floure in Auguft,and perfect their feed in September.

¶ *The Names.*

Their names hath been touched in their titles in as ample manner as hath been fet down by any Author.

¶ *The Nature and Vertues.*

Their temperature and vertues are referred to the garden Cumin,notwithftanding I cannot read in any Author of their vfe in Phyficke.

CHAP. 434. *Of Flixweed.*

i *Sophia Chirurgorum.*
Flixweed.

¶ *The Defcription.*

1   FLixweed hath round & hard ftalks, a cubit and a halfe high,whereon do grow leaues moft finely cut and diuided into innumerable fine jags, like thofe of the fea Wormwood called *Seriphium*,or *Abfinthium tenuifolium* , but much finer and fmaller, drawing neer vnto the fmalleft leaues of Corianders,of an ouerworne green col   :the flours grow alongft the tops of the fpriggy branches, of ad ark yellow colour:after which come long cods full of fmall red feeds : the root is long, ftraight,and of a wooddie fubftance.

2   The fecond fort differeth not from the precedent, fauing that the leaues of this plant are broader, wherein efpecially confifteth the difference ; notwithftanding in mine opinion *Tabernamontanus* found this fecond fort growing in fome fertil place,whereby the leaues did grow broader and greater,which moued him to make of this a fecond fort , whereas in truth they are both but one and the felfe fame plant.

¶ *The Place.*

This Flixweede groweth in moft places of England , almoft euery where in the ruines of old buildings,byhigh waies,and in filthie obfcure bafe places.

¶ *The Time.*

It floureth and feedeth from Iune tothe end of September.

¶ *The Names.*

Flixweed is called *Thalietrum* ; and of fome,*Thalictrum*,but vnproperly ; for *Thalictrum* belongs to Englifh Rubarbe : the Paracelfians doe vaunt and brag very much of an herbe called *Sophia*, adding thereto the fyrname *Paracelfi* , wherewith they imagine to doe wonders, whether this be the fame plant it is difputable, the controuerfie not as yet decided ; neuerthelefe we muft be content

to

to accept of this for the true *Sophia*, vntill ſome diſciple or other of his doe ſhew or ſet forth the plant wherewith their Maſter *Paracelſus* did ſuch great matters : in Engliſh we call it Fliſweed, of his faculties againſt the ſtix.

❡ The Temperature.

Sophia drieth without any manifeſt ſharpneſſe or heate.

❡ The Vertues.

The ſeed of *Sophia* or Flix-weed drunke with with wine or ſmiths water, ſtoppeth the bloudy A flix, the laske, and all other iſſues of bloud.

The herb bruiſed or put into vnguents cloſeth and healeth vlcers, or old ſores and wounds, as P *Paracelſus* ſaith ; and that becauſe it drieth without acrimonie or ſharpneſſe.

Chap. 435.　Of the great Celandine or Swallow-wort.

❡ The Deſcription.

1　The great Celandine hath a tender brittle ſtalke, round, hairy, and full of branches, each whereof hath diuers knees or knotty ioints ſet with leaues not vnlike to thoſe of Columbine, but tenderer, and deeper cut or iagged, of a grayiſh green vnder, and greene on the other ſide tending to blewneſſe : the floures grow at the top of the ſtalks, of a gold yellow colour, in ſhape like thoſe of the Wal-floure: after which come long cods full of bleak or pale ſeeds: the whole plant is of a ſtrong vnpleaſant ſmell, and yeeldeth a thicke iuice of a milky ſubſtance, of the colour of Saffron : the root is thicke and knobby, with ſome threds anexed thereto, which beeing broken or bruiſed, yeeldeth a ſap or iuice of the colour of gold.

1　*Chelidonium majus.*
　Great Celandine.

‡ 2　*Chelidonium majus folio magis diſſecto.*
　Great Celandine with more cut leaues.

‡　2　This other doth not in forme and magnitude differ from the former, but in the leaues, which are finelier cut and iagged, and ſomewhat in their ſhape reſemble an Oken leafe: the floures alſo

alſo are a little jagged or cut about the edges; and in theſe two particulars conſiſts the whole difference. *Cluſius* calls it *Cheledonium majus laciniato flore* : and *Bauhine*, *Cheledonium majus folys quernis.* ‡

¶ *The Place.*

It groweth in vntilled places by common way ſides, among briers and brambles, about old wals, and in the ſhade rather than in the Sun.

¶ *The Time.*

It is greene all the yeare: it floureth from Aprill to a good part of Summer: the cods are perfected in the mean time.

¶ *The Names.*

It is called in Greeke χελιδόνιον : in Latine, *Chelidonium majus*, and *Hirundinarium major* : amongſt the Apothecaries, *Chelidonia* : diuers miſcall it by the name *Chelidonium* : it is named in Italian, *Celidonia*: in Spaniſh, *Celiduhenha*, *Yerva de las golundrinhas* : in high-Dutch, **Groſz Scholtwurtz :** in low-Dutch, **Stinkende Goutve :** in French, *Eſclere*, or *Eſclayre*, and *Celidoine* : in Engliſh, Celandine, or great Celandine, Swallow-wort, and Tetter-wort.

It is called Celandine not becauſe it firſt ſpringeth at the comming in of Swallowes, or dieth when they go away, (for as we haue ſaid, it may be found all the yere) but becauſe ſome hold opinion, that with this herb the dams reſtore ſight to their yong ones when they cannot ſee. Which things are vain and falſe ; for *Cornelius Celſus, lib. 6.* witneſſeth, That when the ſight of the eies of diuers yong birds is put forth by ſome outward means, it will after a time be reſtored of it ſelfe, and ſooneſt of all the ſight of the Swallow: whereupon (as the ſame Author ſaith) the tale grew, how thorow an herb the dams reſtore that thing which healeth of it ſelfe. The very ſame doth *Ariſtotle* alledge, *lib. 6. de Animal.* The eies of Swallowes (ſaith he) that are not fledge, if a man do pricke them out, do afterwards grow againe and perfectly recouer their ſight.

¶ *The Nature.*

The great Celandine is manifeſtly hot and dry, and that in the third degree, and withall ſcoures and clenſeth effectually.

¶ *The Vertues.*

A   The juice of the herbe is good to ſharpen the ſight, for it clenſeth and conſumeth away ſlimie things that cleaue about the ball of the eye, and hinder the ſight, and eſpecially being boiled with hony in a braſen veſſell, as *Dioſcorides* teacheth.

B   The root cureth the yellow jaundice which comes of the ſtopping of the gall, eſpecially when there is no ague adjoined with it, for it opens and deliuers the gall and liuer from ſtoppings.

C   The root being chewed is reported to be good againſt the tooth-ache.

D   The juice muſt be drawn forth in the beginning of Summer, and dried in the ſunne, ſaith *Dioſcorides.*

E   The root of Celandine boiled with Aniſe ſeed in white wine, opens the ſtoppings of the liuer, and cureth the jaundice very ſafely, as hath been often proued.

F   The root cut into ſmall pieces is good to be giuen vnto Hauks againſt ſundry diſeaſes, wherunto they are ſubiect, as wormes, Cray, and ſuch like.

G   ‡ I haue by experience found (ſaith *Cluſius*) that the juice of the great Celandine dropped into ſmall greene wounds of what ſort ſoeuer, wonderfully cures them. ‡

C H A P. 436.

Of Cocks-combe or yellow Rattle.

¶ *The Deſcription.*

CHriſta Galli, or *Griſta Gallinacea*, hath a ſtrait vpright ſtalke ſet about with narrow leaues ſnipt round about the edges : the floures grow at the top of the ſtemmes, of a yellow colour; after which come vp little flat pouches or purſes, couered ouer or contained within a little bladder or flat skin, open before like the mouth of a fiſh, wherin is contained flat yellowiſh ſeed, which being ripe and dry will make a noiſe or rattling when it is ſhaken or moued, of which property it took the name yellow Rattle.

¶ *The.*

Chriſti Galli. Yellow rattle or Coxcomb.

¶ *The Place.*

It growes in dry medowes and paſtures and is to them a great anoyance.

¶ *The Time.*

It floureth moſt part of the Summer.

¶ *The Names.*

It is called in low-Dutch, **Ratelen**, and **Geele Ratelen**: commonly in Latine, *Criſta Galli*, and *Gallinacea Chriſti*: In Engliſh, Coxcombe, Penny-graſſe, yellow or white Rattle: in high-Dutch it is called **geel Rodel**: in French, *Creſte de Coc:* diuers take it to be the old writers *Alectorolophos*.

‡ Some thinke it to be the *Mimmulus*: or as others (and that more fitly) reade it, *Nummulus*, mentioned by *Pliny, lib.* 18. *cap.* 28. ‡

¶ *The Nature and Vertues.*

What temperature or vertua this herb is of, men haue not as yet been carefull to know, ſeeing it is accounted vnprofitable.

C н а р. 437.　*Of Red Rattle or Louſe-wort.*

Pedicularis.
Louſe-wort, or red-Rattle.

¶ *The Deſcription.*

R Ed Rattle (of *Dodonæus* called *Fiſtularia*, and according to the opinion and cenſure of *Cluſius, Pena*, and others, the true *Alectorolophos*) hath very ſmall rent or jagged leaues of a brown red colour, and weake ſmall and tender ſtalks, wherof ſome lie along trailing vpon the ground: in very mooriſh medows they grow a cubit high and more: but in moiſt and wet heaths and ſuch like barren grounds not aboue an handfull high: the floures grow round about the ſtalke, from the midſt thereof euen to the top, of a brown colour, in ſhape like the floures of dead Nettle: which being paſt, there ſucceed little flat pouches, wherein is contained flat and blackiſh ſeed, in ſhew very like to the former: the root is ſmall, white, and tender.

¶ *The Place.*

It growes in moiſt and mooriſh medows: the herb is not only vnprofitable, but alſo hurtful to medowes

¶ *The Time.*

It is found with the floures and ſtalks in May and Iune.

¶ *The*

¶ *The Names.*

It is called in Greeke, ꜵꜷꜵ : in high-Dutch,**Bꝛaun Rodel**: in Latine, *Pedicularis*, of the effect, becauſe it filleth ſheep and other cattell that feed in medowes where this groweth, full of lice:diuers of the later herbariſts call it *Piſtularia* : of ſome, *Criſti Galla* : and ſome take it to be *Mimmulus herba* : in Engliſh, Rattle-graſſe, red Rattle-graſſe, and Louſe-wort.

¶ *The Nature.*

It is cold and dry, and aſtringent.

¶ *The Vertues.*

A   It is held to be good for fiſtulaes and hollow vlcers , and to ſtay the ouermuch flowing of the menſes or any other flux of bloud, if it be boiled in red wine and drunke.

Chap. 438. *Of Yarrow or Noſe-bleed.*

¶ *The Deſcription.*

1   COmmon Yarrow hath very many ſtalks comming vp a cubit high, round and ſomwhat hard, about which ſtand long leaues cut in the ſides ſundry wiſe, and as it were made vp of many ſmal jagged leaues, euery one of which ſeem to reſemble the ſlender leaues of Coriander : there ſtand at the top tufts or ſpoked rundles, the floures whereof are either white or purple, which being rubbed do yeeld a ſtrong vnpleaſant ſmel : the root ſends down many ſtrings.

1 *Millefolium terreſtre vulgare.*   2 *Millefolium flore rubro.*
  Common Yarrow.   Red floured Yarrow.

2   The ſecond kinde of Milfoile or Yarrow hath ſtalkes, leaues, and roots like the former, ſaue that his ſpoky tufts are of an excellent faire red or crimſon colour, and being a little rubbed in the hand, of a reaſonable good ſauor.

¶ *The Place.*

The firſt groweth euery where in dry paſtures and medowes : red Milfoile growes in a field by Sutton in Kent, called Holy-Deane, from whence I brought thoſe plants that grow in my garden : but it is not common euery where as the other is.

¶ *The Time.*

They floure from May to the end of October.

¶ *The*

¶ *The Names.*

Yarrow is called of the Latine Herbarists *Millefolium*: it is *Dioscorides* his ἐχίλλεον : in Latine, *Achillea*, and *Achillea sideritis* ; which thing he may very plainely see that will compare with that description which *Dioscorides* hath set downe : this was found out, saith *Pliny* in his 25.booke,chap.5.by *Achilles, Chirons* disciple,which for that cause is named *Achilleios*: of others, *Sideritis*: among vs, *Millefolium* · yet be there other *Sideritides*, and also another *Panaces Heracleion*, whereof we haue intreated in another place : *Apuleius* setteth downe diuers names hereof, some of which are also found among the bastard names in *Dioscorides* : in Latine it is called *Militaris*, *Supercilium Veneris*, *Acrum*, or *Acorum syluaticum* : of the French-men, *Millefueille* : in high Dutch, **Garben, scharffgras** : in low Dutch, **Geruwe**: in Italian, *Millefoglio* : in Spanish, *Milhoyas yerua* : in English, Yarrow, Nosebleed, common Yarrow, red Yarrow, and Milfoile.

¶ *The Temperature.*

Yarrow as *Galen* saith, is not vnlike in temperature to the *Sideritides*, or Iron-worts, that is to say, clensing, and meanely cold, but it most of all bindeth.

¶ *The Vertues.*

The leaues of Yarrow do close vp wounds, and keepe them from inflammation, or fiery swelling :   A it stancheth bloud in any part of the body, and it is likewise put into bathes for women to sit in : it stoppeth the laske, and being drunke it helpeth the bloudy flix.

Most men say that the leaues chewed, and especially greene are a remedy for the tooth-ache.   B

The leaues being put into the nose, do cause it to bleed, and ease the paine of the megrim.   C

It cureth the inward excorations of the yard of a man, comming by reason of pollutions or ex-   D treme flowing of the seed, although the issue doe cause indammation and swelling of those secret parts, and though the spermaticke matter do come downe in great quantity, if the juyce be injected with a syringe, or the decoction. This hath been proued by a certaine friend of mine, sometimes a Fellow of Kings Colledge in Cambridge, who lightly brused the leaues of common Yarrow, with Hogs-grease, and applied it warme vnto the priuie parts, and thereby did diuers times helpe himselfe, and others of his fellowes, when he was a student and a single man liuing in Cambridge.

One dram in pouder of the herbe giuen in wine, presently taketh away the paines of the colicke.   E

CHAP. 439. *Of yellow Yarrow, or Milfoile.*

1 *Millefolium luteum.*   † 2 *Achillea, siue Millefolium nobile.*
   Yellow Yarrow.   Achilles Yarrow.

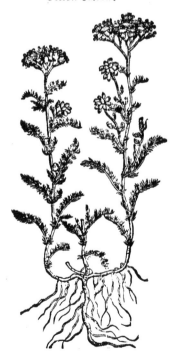

¶ *The Deſcription.*

1   YEllow Yarrow is a ſmall plant ſeldome aboue a ſpan high : the ſtalkes whereof are couered with long leaues, very finely cut in the edges like feathers in the wings of little
birds : the tufts or ſpokie rundles bring forth yellow floures, of the ſame ſhape and
forme of the common Yarrow : the root conſiſteth of thready ſtrings.

2   Achilles Yarrow, or noble Milfoile, hath a thicke and tough root, with ſtrings faſtened thereto : from which immediatly riſe vp diuers ſtalkes, very greene and creſted, whereupon do grow long
leaues compoſed of many ſmall jagges, cut euen to the middle rib : the floures ſtand on the top of
the ſtalkes with ſpokie vmbels or tufts, of a whitiſh colour, and pleaſant ſmell.

¶ *The Place.*

Theſe kinds of Yarrow are ſeldome found : they grow in a fat and fruitfull ſoile, and ſometimes
in medowes, and are ſtrangers in England.

¶ *The Time.*

They floure from May vntill Auguſt.

¶ *The Names.*

Dioſcorides deſcription doth ſufficiently declare, that this herbe is *Stratiotes Millefolium* : in
Greeke, ςρατιωτε χιλιοφυλλοs : the height of the herbe ſheweth it, the forme of the leaues agree ; there is
ſome ambiguitie or doubt in the colour of the floures, which *Dioſcorides* deſcribeth to be white, as
the vulgar copies haue ; but *Andreas Lacuna* addeth out of the old booke of a yellow colour : it is
named of the later age, *Millefolium minus*, or little Yarrow, and *Millefolium luteum*, yellow Yarrow,
or Noſe bleed : the Apothecaries and common people know it not.

¶ *The Temperature.*

Yarrow is meanely cold and ſomewhat binding.

¶ *The Vertues.*

A   It is a principall herbe for all kinde of bleedings, and to heale vp new and old vlcers and greene
wounds : there be ſome, ſaith *Galen,* that vſe it for fiſtulaes.

B   This plant *Achillea* is thought to be the very ſame wherewith *Achilles* cured the wounds of his
ſouldiers, as before in the former chapter.

† The plant here figured and deſcribed in the ſecond place, was alſo figured and deſcribed formerly in the fifth place of the 209 chapter of this booke, by the
title of *Tanacetum minus album*, but the figure of *Lobell* which is put there being ſomewhat imperfect, I thought it not amiſſe here to giue you that of *Dodonæus* which
is ſomewhat more exquiſite, otherwiſe both the figure and hiſtory might in this place haue beene omitted.

CHAP. 440.   *Of Valerian, or Setwall.*

¶ *The Deſcription.*

1   THe tame or garden Valerian hath his firſt leaues long, broad, ſmooth, greene, and vndiuided ; and the leaues vpon the ſtalkes greater, longer, and deepely gaſhed on either
ſide, like the leaues of the greater Parſenep, but yet leſſer : the ſtalke is aboue a cubit
high, ſmooth, and hollow, with certaine joynts farre diſtant one from another : out of which joynts
grow forth a couple of leaues, and in the tops of the ſtalkes vpon ſpokie rundles ſtand floures hea-
ped together, which are ſmall, opening themſelues out of a long little narrow necke, of colour whitiſh, and ſometimes withall of a light red : the root is an inch thicke, growing aſlope, faſtned on the
vpper part of the earth by a multitude of ſtrings, the moſt part of it ſtanding out of the ground, of
a pleaſant ſweet ſmell when it is broken.

2   The greater wilde Valerian hath leaues diuided and jagged as thoſe of the former ; thoſe
about the ſtalke hereof are alſo ſmooth, hollow, and joynted, and aboue a cubit high : the floures
ſtand on ſpokie rundles like to thoſe of the former, but of a light purple colour : the roots are ſlender, and full of ſtrings and ſmall threds, not altogether without ſmell.

3   The other wilde one is much like in forme to the garden Valerian, but farre leſſer : the firſt
leaues thereof be vndiuided, the other are parted and cut in ſunder : the ſtalkes a ſpan long : the
floures which ſtand on ſpokie rundles are like to thoſe of the others, of a light whitiſh purple colour : the roots be ſlender, growing aſlope, creeping, and full of fine ſmall threds, of little ſmell.

4   There is a ſmall Valerian growing vpon rockes and ſtony places, that is like vnto the laſt deſcribed, ſauing it is altogether leſſe. ‡ The ſtalk is ſome halfe foor high, and ſtait, diuiding it ſelfe
into branches toward the top, and that alwaies by couples : the bottome leaues are whole, the top

leaues

1 *Valeriana hortenſis.*
Garden Valerian, or Setwall.

2 *Valeriana major ſylueſtris,*
Great wilde Valerian.

3 *Valeriana minor.*
Small Valerian.

4 *Valeriana Petræa.*
Stone Valerian.

leaues much diuided, the floures are ſmall, of a whitiſh purple colour, parted into fiue, and ſtanding vpon round rough heads, which when the floures are falne, become ſtar-faſhioned, diuided into ſix parts : it floures in Iune, and is an annuall plant. ‡

5   The fifth ſort of Valerian hath diuers ſmall hollow ſtalkes, a foot high and ſomewhat more, garniſhed with leaues like vnto thoſe that do grow on the vpper part of the ſtalks of common Valerian, but ſmaller, cut or jagged almoſt to the middle rib : at the top of the ſtalks do grow the floures cluſtering together, of a blew colour, conſiſting of fiue leaues apiece, hauing in the middle thereof ſmall white threds tipped with yellow : the ſeed is ſmall, growing in little huskes or ſeed veſſels : the root is nothing elſe but as it were all of threds.

<table>
<tr><td>5   Valeriana Græca.
Greekiſh Valerian.</td><td>‡ 6   Valeriana Mexicana.
Indian Valerian.</td></tr>
</table>

6   I haue another ſort of Valerian (the ſeed whereof was ſent me from that reuerend Phyſition *Bernard Paludane*, vnder the title of *Valeriana Maxicana*:) hauing ſmall tender ſtalkes trailing vpon the ground, very weake and brittle: whereupon doe grow ſmooth greeniſh leaues like thoſe of Corne Sallade (which wee haue ſet forth amongſt the Lettuce, vnder the title *Lactuca Agnina*, or Lambs Lettuce:) among the leaues come forth the floures cluſtering together, like vnto the great Valerian in forme, but of a deepe purple colour: the root is very ſmall and threddy, which periſheth with the reſt of the plant, when it hath brought his ſeed to maturity or ripeneſſe, and muſt bee ſowne anew the next yeare in May, and not before.

7   There is alſo another ſort or kinde of Valerian called by the name *Phyteuma*, of the learned Phyſitions of Montpellier and others (ſet forth vnder the ſtocke or kindred of the Valerians, reſembling the afore-ſaid Corn-ſallad, which is called of ſome *Prolifera*, from the Greeke title *Phyteuma* ; as if you ſhould ſay, good to make conception, and to procure loue:) the loweſt leaues are like thoſe of the ſmall Valerian, of a yellowiſh colour : the vpper leaues become more jagged : the ſtalks are an handfull high : on the tops whereof doe ſtand ſmall round ſpokie tufts of white floures ; which being paſt, the ſeeds appeare like ſmall round pearles, which being ripe, grow to be ſomewhat flat, hauing in the middle of each ſeed the print of an hole, as it were grauen or bored therein. The root is ſmall and ſingle, with ſome fibres annexed thereto.

‡ 8   This ſends forth from a white and wooddy root many leaues ſpred vpon the ground, green,

and

‡ 8 *Valeriana annua, Cluf.*
Annuall Valerian.

‡ 9 *Valeriana Alpina latifolia.*
Broad leaued Setwall of the Alps.

‡ 10 *Valeriana Alpina angustifolia.*
Small Alpine Set-wall.

and not vnlike thofe of the Star-Thiftle:among thefe rife vp fome round hollow branched ftalks two cubits high : at each joynt grow forth two leaues leffer, yet like the lower : at the tops of the branches grow the floures as it were in little vmbels confifting of fiue leaues apiece ; and thefe of a light red, or flefh colour : and then thefe as it were vmbels grow into longifh branches bearing feed almoft like, yet leffe than the red Valerian : it floures in Iuly, and perifheth when ir hath ripened the feed. *Clufius* hath fet this forth by the name of *Valeriana annua altera.*‡

9   The fame Author hath alfo giuen vs the hiftorie of fome other Plants of this kinde; and this he cals *Valeriana fyl. Alpina* 1 *latifol.* the ftalk hereof is fome foot high, round, greene, and crefted : vpon which ftand leaues longifh, fharpe pointed, and cut in with two or three deepe gafhes : but the bottome leaues are more round and larger, comming neere to thofe of *Trachelium,* yet leffer, flenderer, and bitter of tafte : the floures which are white of colour, and the feed, are like thofe of the other Valerians : the root is fmall, creeping, fibrous, white and aromatick : it growes vpon the Alpes, and floures in Iune and Iuly.

10   This fends forth leaues like thofe of the mountaine Daifie: out of the midft of which

riſeth vp a ſtalk ſome foot high, joynted, and at the top diuided into little branches, carrying white floures like the other Valerians : the root is as aromaticke as that of the laſt mentioned, and grows in the chinkes of the Alpine rockes, where it floures in Iune and Iuly. *Cluſius* hath it by the name of *Valeriana ſylueſtris Alpin.2.Saxatilis,* ‡

¶ *The Place.*

• The firſt and likewiſe the Greeke Valerian are planted in gardens ; the wilde ones are found in moiſt places hard to riuers ſides, ditches, and watery pits ; yet the greater of theſe is brought into gardens where it flouriſheth, but the leſſer hardly proſpereth.

¶ *The Time.*

Theſe floure in May, Iune, and Iuly, and moſt of the Summer moneths.

¶ *The Names.*

Generally the Valerians are called by one name, in Latine, *Valeriana* : in Greeke, φῦ· in ſhoppes alſo *Phu*, which for the moſt part is meant by the garden Valerian, that is called of *Dioſc.* νάρδος ἀγρία: in Latine, *Sylueſtris*, or *Ruſtica Nardus* : of *Pliny*, *Nardus Cretica* : which names are rather referred to thoſe of the next chapter, although theſe be reckoned as wilde kindes thereof : of certaine in our age, *Marinella, Amantilla, Valentiana, Genicularis, Herba Benedicta*, and *Theriacaria* : in moſt ſhops, *Valeriana Domeſtica* : of *Theophraſtus Paracelſus, Terdina* : in high Dutch, **Groß baldrian :** in low Dutch, **Speercruyt, S. Joris cruyt,** and **Valeriane :** in Engliſh, Valerian, Capons taile, and Setwall; but vnproperly, for that name belongeth to *Zedoaria*, which is not Valerian : what hath beene ſet downe in the titles ſhall ſerue for the diſtinctions of the other kindes.

¶ *The Temperature.*

The garden Valerian is hot, as *Dioſcorides* ſaith, but not much, neither the green root, but the dried ones; for the greene is eaſily perceiued to haue very little heate, and the dried to be hotter, which is found by the taſte and ſmell.

¶ *The Vertues.*

A　The dry root as *Dioſcorides* teacheth, prouoketh vrine, bringeth downe the deſired ſickeneſſe, helpeth the paine in the ſides ; and is put into counterpoyſons and medicines preſeruatiue againſt the peſtilence, as are treacles, mithridates, and ſuch like : whereupon it hath been had (and is to this day among the poore people of our Northerne parts) in ſuch veneration amongſt them, that no broths, pottage or phyſicall meats are worth any thing, if Setwall were not at an end: whereupon ſome woman Poët or other hath made theſe verſes.

They that will haue their heale,
Muſt put Setwall in their keale.

B　It is vſed generally in ſleight cuts, wounds, and ſmall hurts.
C　The extraction of the roots giuen, is a moſt ſingular medicine againſt the difficulty of making water, and the yellow jaundiſe.
D　Wilde Valerian is thought of the later Herbariſts to be good for them that are burſten, for ſuch as be troubled with the crampe and other convulſions, and alſo for all thoſe that are bruiſed with falls.
E　The leaues of theſe and alſo thoſe of the garden, are good againſt vlcers and foreneſſe of the mouth and gums, if the decoction thereof be gargarized or held in the mouth.
F　Some hold opinion that the roots of wilde Valerian dried and poudered, and a dramme weight thereof taken with wine, do purge vpward and downeward.

Chap. 441. *Of Mountaine Setwall, or Nardus.*

¶ *The Deſcription.*

1　THe *Nardus* named *Celtica*, but now by ſome, *Liguſtica Nardus*, flouriſheth in high mountaines. The Valleſians in their mother tongue call it *Selliga*; whence *Geſner* thought it to be *Saliunca* ; neither do I doubt, but that it is the ſame which *Virgil* ſpeaketh of in theſe verſes :

Puniceis humiliæ quantum Saliunca roſetis,
Iudicio noſtro tantum tibi cedit Amintas.

For it is a very little herbe creeping on the ground, and afterward lifting vp it ſelfe with a ſtalke of a handfull high; whereupon from the lower part grow ſmall thin leaues, firſt greene, but afterwards ſomewhat yellowiſh : vpon the roughneſſe of the root there are many ſcales, platted one vpon another ; but vnder the root there are many browne ſtrings and hairy threds, in ſmell like the roots of *Aſtrabacca*, or rather the wilde mountaine Valerian, whereof it ſeemes to be a kinde, in taſte ſharpe and bitter. The floures grow along the vpper branches, white or yellowiſh, and very ſmall.

2 The

1 *Nardus Celtica.*
Celticke Spikenard.

3 *Hirculus.*
Vrine-wort.

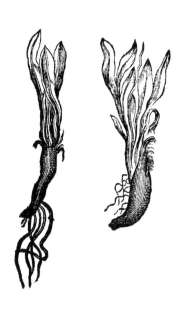

‡ 4 *Nardus montana germinans.*
Mountaine Nard at the firſt ſpringing vp.

4 *Nardus montana.*
Mountaine Spikenard.

2   The ſecond ſort of Spike-nard
hath many threddy roots, from the
which riſe vp many ſcaly rough and
thicke ſtalkes, hauing at the top certain
flat hoary leaues growing vpon ſmall
and tender foot-ſtalks. The whole plant
is of a pleaſant ſweet ſmell.

3   *Hirculus* is a plant very rare, which
as yet I neuer ſaw, notwithſtanding wee
are greatly beholding to *Carolus Cluſius*
the father of forreine Simples, who finding this plant among many bunches
or handfulls of mountaine Spikenard,
hath made it knowne vnto poſteritie, as
he hath done many other rare plans, in
                 tranſlating

tranſlating of *Garcias* the Luſitanian Phyſitian, he ſetteth it forth with a light deſcription, ſaying, It is a baſe and low herbe two handfuls high, bringing forth leaues without any ſtalks at all, ‡ very hairy about the root, and blackiſh, hauing no pleaſant ſent at all. The leaues chewed yeeld no aromaticke taſte, but are clammie, or viſcide ; whereas the leaues of Celticke Narde are hot, with a little aſtriction, and of a pleaſant ſmell and taſte. ‡

4 Mountaine Spikenard hath a great thicke knobbed root, ſet here and there with ſome tender fibres, of a pleaſant ſweet ſmell; from the which come forth three or foure ſmooth broad leaues, and likewiſe jagged leaues deeply cut euen to the middle rib : among which riſe vp naked ſtalkes, garniſhed in the middle with a tuſt of jagged leaues. The floures grow at the top of the ſtalkes, in an vmbel or tuſt like thoſe of the wilde Valerian in ſhape and colour, and ſuch alſo is the ſeed. ‡ I haue giuen you the figure of the root and whole leaues as they ſhew themſelues when they firſt appeare, as it was taken by *Cluſius*. ‡

5 The Spikenard of India is a low plant, growing cloſe vnto the ground, compoſed of many rough browne hairy cloues, of a ſtrong, yet not vnpleaſant ſmell. The root is ſmall and threddie. ‡ It hath certainely ſtalkes, floures, and ſeeds ; but none of our Indian Writers or Trauellers haue as yet deſcribed them. I haue ſeene little pieces of ſlender hollow ſtalkes ſome two inches long faſtned to the roots that are brought to vs. ‡

| 5 *Nardus Indica.* | 6 *Nardus Narbonenſis.* |
| Indian Spikenard. | French Spikenard. |

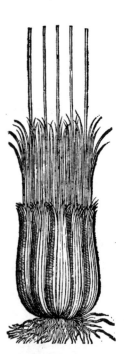

† 6 This French Spikenard, being a baſtard kinde, groweth cloſe vpon the ground like the precedent, compact of ſcaly rough leaues : in the middle whereof commeth forth a great buſh of round greene ſtiffe and ruſhy leaues : among the which ſhoot vp diuers round ſtalkes a cubit high, ſet from the middle to the top with greeniſh little cods, ſtanding in chaffie huskes like thoſe of Schœnanth. The root is ſmall and threddy : the whole plant is altogether without ſmell, which ſheweth it to be a baſtard kinde of Spikenard.

¶ *The Place.*

Theſe plants [the firſt foure] are ſtrangers in England, growing in great plenty vpon the mountaines of Iudenberg and Heluetia, on the rockes among the moſſe, and in the mountaines of Tiroll and Saltsburg.

The firſt and ſecond, if my memorie faile me not, doe grow in a field in the North part of England

land,called Crag cloſe, and in the foot of the mountaine called Ingleborow Fels. ‡ The fourth may be found in ſome gardens with vs. The fifth growes in the Eaſt Indies, in the Prouinces of Mando and Chiten in the kingdome of Lengala and Decan. The laſt growes in Prouince in France,neere a little city called Gange. ‡

¶. *The Time.*

The leaues grow to withering in September, at which time they ſmell more pleaſantly than when they flouriſhed and were greene.

¶ *The Names.*

Nardus is called in Pannonia or Hungary,of the countrey people,*Speick :* of ſome,*Becht fix ;* that is,the herbe of Vienna, becauſe it doth grow there in great abundance, from whence it is brought into other countries : of *Geſner,Saliunca :* in Engliſh,Celticke Spikenard : of the Valletians, *Selliga,* and *Nardus Celtica.*

¶ *The Temperature and Vertues.*

Celticke Narde mightily prouokes vrine, as recordeth *Rondeletius ;* who trauelling through the **A** deſart countrey , chanced to lodge in a monaſtery where was a Chanon that could not make his water, but was preſently helped by the decoction of this herbe, through the aduice of the ſaid *Rondeletius.*

‡ The true Spikenard or Indian Nard hath a heating and drying facultie, being (according to **B** *Galen)* hot in the firſt degree [yet the Greeke copy hath the third] and dry in the ſecond. It is compoſed of a ſufficiently aſtringent ſubſtance, and not much acride heate, and a certaine light bitterneſſe. Conſiſting of theſe faculties, according to reaſon, both inwardly and outwardly vſed it is conuenient for the liuer and ſtomacke.

It prouoketh vrine,helps the gnawing paines of the ſtomacke, dries vp the defluxions that trou- **C** ble the belly and intrails ,as alſo that moleſt the head and breſt.

It ſtaies the fluxes of the belly, and thoſe of the wombe, being vſed in a peſſarie, and in a bath it **D** helpes the inflammation thereof.

Drunke in cold water, it helpes the nauſeouſneſſe, gnawings, and windineſſe of the ſtomacke, **E** the liuer,and the diſeaſes of the kidnies,and it is much vſed to be put into Antidotes.

It is good to cauſe haire to grow on the eye-lids of ſuch as want it,and is good to be ſtrewed vp- **F** on any part of the body that abounds with ſuperfluous moiſture,to dry it vp.

The Celticke-Nard is good for all the forementioned vſes, but of leſſe efficacie, vnleſſe in the **G** prouoking of vrine. It is alſo much vſed in Antidotes.

The mountaine Nard hath alſo the ſame faculties,but is much weaker than the former, and not **H** in vſe at this day that I know of. ‡

Cʜᴀᴘ. **442.**　*Of Larks heele or Larks claw.*

¶ *The Deſcription.*

1　　THe garden Larks ſpur hath a round ſtem ful of branches,ſet with tender jagged leaues very like vnto the ſmall Sothernwood : the floures grow alongſt the ſtalks toward the tops of the branches,of a blew colour,conſiſting of fiue little leaues which grow together and make one hollow floure,hauing a taile or ſpur at the end turning in like the ſpur of Todeflax. After come the ſeed, very blacke, like thoſe of Leekes : the root periſheth at the firſt approch of Winter.

2　The ſecond Larks ſpur is like the precedent, but ſomewhat ſmaller in ſtalkes and leaues : the floures are alſo like in forme,but of a white colour,wherein eſpecially is the difference. Theſe floures are ſometimes of a purple colour, ſometimes white, murrey, carnation, and of ſundry other colours,varying infinitely,according to the ſoile or country wherein they liue.

‡ 3　Larks ſpur with double floures hath leaues, ſtalkes,roots, and ſeeds like the other ſingle kinde, but the floures of this are double ; and hereof there are as many ſeuerall varieties as there be of the ſingle kinde,to wit,white,red,blew,purple,bluſh,&c.

4　There is alſo another variety of this plant, which hath taller ſtalkes and larger leaues than the common kinde : the floures alſo are more double and larger, with a leſſer heele : this kinde alſo yeeldeth vſually leſſe ſeed than the former. The colour of the floure is as various as that of the former,being either blew,purple,white,red,or bluſh,and ſometimes mixed of ſome of theſe. ‡

5　The wilde Larks ſpur hath moſt fine jagged leaues, cut and backt into diuers parts, confuſedly ſet vpon a ſmall middle tendrell : among which grow the floures,in ſhape like the others, but

a great

1 *Conſolida regalis ſatiua.*
Garden Larks heele.

2 *Conſolida ſatiua flore albo vel rubro.*
White or red Larks ſpur.

‡ 3 *Conſolida regalis flore duplici.*
Double Larks ſpur.

‡ 4 *Conſolida reg. elatior flo. pleno.*
Great double Larks ſpur.

a great deale leſſer, ſometimes purple, otherwhiles white, and often of a mixt colour. The root is ſmall and threddy.

5 *Conſolida regalis ſylueſtris.*
Wilde Larkes heele.

¶ *The Place.*

Theſe plants are ſet and ſowne in gardens: the laſt groweth wilde in corne fields, and where corn hath grown, ‡ but not with vs, that I haue yet obſerued ; though it be frequently found in ſuch places in many parts of Germanie. ‡

¶ *The Time.*

They floure for the moſt part all Summer long, from Iune to the end of Auguſt, and oft-times after.

¶ *The Names.*

Larks heele is called *Flos Regius* : of diuers, *Conſolida regalis* : who make it one of the Conſounds or Comfreyes. It is alſo thought to be the *Delphinium* which *Dioſcorides* deſcribes in his third booke ; wherewith it may agree. It is reported by *Gerardus* of Veltwijcke, who remained Lieger with the great Turke from the Emperor *Charles* the fifth, That the ſaid *Gerard* ſaw at Conſtantinople a copy which had in the chap. of *Delphinium*, not leaues but floures like Dolphins : for the floures, and eſpecially before they be perfeſted, haue a certaine ſhew and likeneſſe of thoſe Dolphins, which old piſtures and armes of certain antient families haue expreſſed with a crooked and bending fiｇure or ſhape, by which ſigne alſo the heauenly Dolphine is ſet forth. And it skilleth not, though the chapter of *Delphinium* be thought to be falſified and counterfeited ; for although it bee ſo ne other mans, and not of *Dioſcorides*, it is notwithſtanding ſome one of the old Writers, out of who n it is taken, and foiſted into *Dioſcorides* his bookes : of ſome it is called *Bucinus, or Bucinum* : in Engliſh Larks ſpur, Larks heele, Larks toes, and Larks claw : in high Dutch **Ridder ſpooren**, that is, *Equitis Calcar*, Knights ſpur : in Italian, *Sperone* : in French, *Pied d'alouette*.

¶ *The Temperature.*

Theſe herbes are temperate and warme of nature.

¶ *The Vertues.*

We finde little extant of the vertues of Larks heele, either in the antient or later writers, worth A the noting, or to be credited; yet it is ſet downe, that the ſeed of Larks ſpur drunken is good againſt the ſtingings of Scorpions ; whoſe vertues are ſo forceable, that the herbe onely throwne before the Scorpion or any other venomous beaſt, couſeth them to be without force or ſtrength to hurt, inſomuch that they cannot moue or ſtirre vntill the herbe be taken away : with many other ſuch trifling toyes not worth the reading.

Chap. 443. *Of Gith, or Nigella.*

¶ *The Kinds.*

THere be diuers ſorts of Gith or Nigella, differing ſome in the colour of the floures, others in the doubleneſſe thereof, and in ſmell of the ſeed.

¶ *The Deſcription.*

1 THe firſt kind of Nigella hath weake and brittle ſtalks of the height of a foot, full of branches, beſet with leaues very much cut or jagged, reſembling the leaues of Fumiterie, but much greener : the floures grow at the top of the branches, of a whittiſh blew colour, each floure
being

1 *Melanthium.*
Garden Nigella.

2 *Melanthium ſylueſtre.*
Wilde Nigella.

3 *Melanthium Damaſcenum.*
Damaske Nigella.

‡ 4 *Melanthium Damaſcenum flo.pleno.*
Double floured Damaske Nigella.

being parted into fiue ſmall leaues, ſtarre-faſhion: the floures being vaded, there come vp ſmall knobs or heads, hauing at the end thereof fiue or ſix little ſharpe hornes or pointals, and euery knob or head is diuided into ſundry ſmall cels or partitions, wherein the ſeed is contained, which is of a blackiſh colour, very like vnto Onion ſeed, in taſte ſharpe, and of an excellent ſweet ſauour.

2   The wilde Nigella hath a ſtreaked ſtalke a foot or more high, beſet full of grayiſh leaues, very finely jagged, almoſt like the leaues of Dill: the floures are like the former, ſaue that they are blewer: the cods or knops are like the heads or huskes of Columbines, wherein is conteined the ſweet and pleaſant ſeed, like the former.

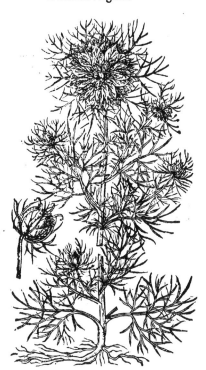

5 *Nigella flore albo multiplici.*
Damaske Nigella.

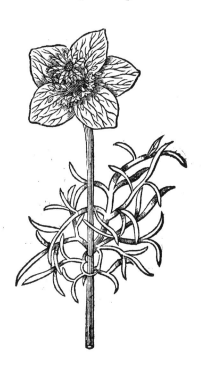

‡ 6 *Nigella Hiſpanica flore amplo.*
Great Spaniſh Nigella.

3   The third kinde of Nigella, which is both faire and pleaſant, called Damaske Nigella, is ve-ry like vnto the wilde Nigella in his ſmall cut and jagged leaues, but his ſtalke is longer: the floures are like the former, but greater, and euery floure hath fiue ſmall greene leaues vnder him, as it were to ſupport and beare him vp: which floures being gone, there ſucceed and follow knops and ſeed like the former, but without ſmell or ſauour.

‡ 4   This in the ſmalneſſe, and ſhape of the leaues and the manner of growing is like to the laſt deſcribed hauing ſmall leaues growing vnder the floure, which is not ſingle, as in the laſt deſcri-bed, but double, conſiſting of fiue or more rankes of little blewiſh leaues, which are ſucceeded by ſuch cornered heads as thoſe of the former, hauing in them a blacke ſeed without any manifeſt ſmell. ‡

5   The fifth kinde of Nigella hath many ſmall and ſlender ſtalkes, ſet full of ſlender and thinne leaues deepely cut or jagged, of a faint yellowiſh greene colour: the floures grow at the top of the ſtalkes, of a whitiſh colour, and exceeding double: which being vaded, there ſucceed bowles or knobs, full of ſweet blacke ſeed like the former: the root is ſmall and tender.

‡ 6   The root of this is ſlender, and yellowiſh; the ſtalke is ſome cubit high, round, greene, cre-ſted, and toward the top diuided into ſundry branches, the leaues toward the bottome are ſome-what ſmall cut, but ſomewhat larger vpon the ſtalkes. The floure is much larger than any of the former, compoſed of fiue leaues, of a light blew aboue, and ſomewhat whitiſh vnderneath, with

large veines running about them : in the middle ſtands vp the head, encompaſſed with blackiſh threds, and ſome 7.or 8. little gaping blewiſh floures at the bottomes of them ; the leaues of the floures decaying the head becomes bigger, hauing at the tops thereof 6.7. or 8. longiſh twined hornes growing, in a ſtar faſhion ; the inſide is parted into cels conteining a yellowiſh green, or elſe blackiſh ſeed. It is ſet forth in the *Hortus Eyſtettenſis* by the name of *Melanthium Hiſpanicum majus* ; by Mr. *Parkinſon* it is called *Nigella Hiſpanica flore ſimplici* ; and *Bauhine* in his *Prodromus* hath it by the name of *Nigella latifolia flore maiore ſimplici cæruleo.* It is an annuall plant, and floures in Iuly ; it is ſometimes to be found in the gardens of our Floriſts.

¶ *The Place.*

The tame are ſowne in gardens : the wilde ones doe grow of themſelues among corne and other graine, in diuers countries beyond the ſeas.

¶ *The Time.*

The ſeed muſt be ſowne in Aprill : it floureth in Iuly and Auguſt.

¶ *The Names.*

Gith is called in Greeke, μελάνθιον : in Latine alſo *Melanthium* : in ſhops, *Nigella*, and *Nigella Romana* : of diuers, *Gith*, and *Saluſandria*, and ſome among the former baſtard names, *Papauer nigrum* : in high Dutch, ***Swartzkymmich*** : in low Dutch, ***Narduſ ſaet*** : in Italian, *Nigella* : in Spaniſh, *Axenuz, Alipiure* : in French, *Nielle odorante* : in Engliſh, Gith, and Nigella Romana, in Cambridgeſhire, Biſhops wort : and alſo *Dinæ Catherinæ flos*, Saint Katharines floure.

¶ *The Temperature.*

The ſeed of the garden Nigella is hot and dry in the third degree, and of thin parts.

¶ *The Vertues.*

A　The ſeed of *Nigella Romana* drunke with wine, is a remedy againſt the ſhortneſſe of breath, diſſolueth and putteth forth windineſſe, prouoketh vrine, the menſes, increaſeth milke in the breſts of nurſes if it be drunke moderately ; otherwiſe it is not onely hurtfull to them, but to any that take thereof too often, or in too great a quantity.

B　The ſeed killeth and driueth forth wormes, whether it be taken with wine or water, or laid to the nauell in manner of a plaiſter.

C　The oyle that is drawne forth thereof hath the ſame property.

D　The ſeed parched or dried at the fire, brought into pouder, and wrapped in a piece of fine lawne or ſarcenet, cureth all murs, catarrhes, rheumes, and the poſe, drieth the braine, and reſtoreth the ſence of ſmelling vnto thoſe which haue loſt it, being often ſmelled vnto from day to day, and made warme at the fire when it is vſed.

E　It takes away freckles, ſcurfs, and hard ſwellings, being laid on mixed with vineger. To be briefe, as *Galen* ſaith, it is a moſt excellent remedy where there is need of clenſing, drying, and heating.

F　It ſerueth well among other ſweets to put into ſweet waters, bagges, and odoriferous pouders.

† The figures of the third and fourth of the former edition were tranſpoſed.

CHAP. 444. *Of Cockle.*

¶ *The Deſcription.*

COckle is a common and hurtfull weed in our Corne, and very well knowne by the name of Cockle, which *Pena* calleth *Pſeudomelanthium*, and *Nigellaſtrum*, by which name *Dodonæus* and *Fuchſius* do alſo terme it ; *Mutonus* calleth it *Lolium* ; and *Tragus* calleth it *Lychnoides ſegetum.* This plant hath ſtraight, ſlender and hairy ſtems, garniſhed with long hairy and grayiſh leaues, which grow together by couples, incloſing the ſtalke round about : the floures are of a purple colour, declining to redneſſe, conſiſting of fiue ſmall leaues, in proportion very like to wilde Campions ; when the floures be vaded there follow round knobs or heads full of blackiſh ſeed, like vnto the ſeed of *Nigella*, but without any ſmell or ſauour at all.

¶ *The Place and Time.*

The place of his growing and time of his flouring, are better knowne than deſired.

¶ *The Names.*

Cockle is called *Pſeudomelanthium*, and *Nigellaſtrum*, wilde or baſtard Nigella ; of *Fuchſius*, *Lolium* : of *Mouton*, *Lychnoides ſegetum* : of *Tragus*, *Githago* : in high Dutch, ***Kornegele*** : in low Dutch, ***Corne rooſen*** : in French, *Nielle des Bledʒ* : in Engliſh, Cockle, field Nigella, or wilde Nigella : in Italian, *Githone* ; whereupon moſt Herbariſts being moued with the likeneſſe of the word, haue thought it to be the true Gith or *Melanthium* ; but how farre they are deceiued it is better knowne, than needfull to be confuted : for it doth not onely differ in leaues from the true Gith, but alſo in other properties, and yet it is called Gith or *Melanthium*, and that is of the blackeneſſe of the ſeed, yet not properly, but with a certaine addition, that it may differ from the true *Melanthium* : for

Hippocrates

Pſeudomelanthium.
Baſtard Nigella, or Cockle.

Hippocrates calleth it *Melanthium ex tritico*, of
Wheat: *Octavius Horatianus* calleth that Gith
which groweth among corne: and for the ſame
cauſe it is named of the Learned of this our
time, *Nigellaſtrum, Gigatho*, and *Pſeudomelanthium*.
Ruellius ſaith it is called in French *Niele*, & *Flos
Micancalus*.

¶ *The Temperature.*

The ſeed of Cockle is hot and dry in the later
end of the ſecond degree.

¶ *The Vertues.*

The ſeed made in a peſſarie or mother ſuppo-   A
ſitorie, with hony put vp, bringeth down the deſired ſickneſſe, as *Hippocrates* witneſſeth in his
booke of womens diſeaſes.

Octavius Horatianus giues the ſeed parched and   B
beaten to pouder, to be drunk againſt the yellow
Iaundice.

Some ignorant people haue vſed the ſeed of   C
this for the ſeed of Darnel, to the great danger
of thoſe who haue receiued the ſame. What hurt
it doth among corne, the ſpoile of bread, as well
in colour, taſte, and vnwholeſomneſſe, is better
knowne than deſired.

C H A P. 445.   *Of Fumitorie.*

¶ *The Kindes.*

THere be diuers herbs comprehended vnder the title of Fumitorie; ſome wilde, and others of
the garden; ſome with bulbous or tuberous roots, and others with fibrous or threddy roots:
and firſt of thoſe whoſe roots are nothing but ſtrings.

¶ *The Deſcription.*

1   FVmitorie is a very tender little herbe: the ſtalkes thereof are ſlender, hauing as it were
little knots or joints ful of branches, that ſcarce grow vp from the ground without prop-
pings, but for the moſt part grow ſidelong: the leaues round about are ſmall, cut on the
edges as thoſe of Coriander, which as well as the ſtalks are of a whitiſh green: the flours be made
vp in cluſters at the tops of the ſmall branches, of a red purple colour: then riſe vp huskes round
and little, in which lieth the ſmall ſeed, the root is ſlender, and groweth ſtraight downe. ‡ This
is alſo found with floures of a purple violet colour, and alſo ſometimes with them white. ‡

2   The ſecond kinde of Fumitorie hath many ſmall long and tender branches, wherupon grow
little leaues vſually ſet together by threes or fiues, in colour and taſte like the former, hauing at the
tops of the branches many ſmall claſping tendrels, with which it taketh hold vpon hedges, buſhes,
and whatſoeuer groweth next to it: the flours are ſmall, and cluſtering together, of a white colour,
with a little ſpot in their middles; after which ſucceed cods containing the ſeed: the root is ſingle, and of a fingers length.

3   The third kinde of Fumitorie hath a very ſmal root, conſiſting of diuers little ſtrings, from
which ariſe ſmall and tender branches trailing here and there vpon the ground, beſet with many
ſmall and tender leaues moſt finely cut and jagged like the little leaues of Dill, of a deepe greene
colour tending to blewneſſe, the floures ſtand at the tops of the branches in bunches or cluſters,
thicke thruſt together, like thoſe of the medow Claver or three leaued graſſe, of a moſt bright red
colour, and very beautifull to behold: the root is very ſmall and threddy.

1 *Fumaria purpurea.*
Common or purple Fumitory.

† 2 *Fumaria alba latifolia claviculata.*
White broad leafed Fumitorie.

3 *Fumaria tenuifolia.*
Fine leafed Fumitory.

4 *Fumaria lutea.*
Yellow Fumitorie.

4   The yellow Fumitorie hath many crambling threddy roots, fomwhat thicke, groffe, and fat, like thofe of *Afparagus* : from which rife diuers vpright ftalkes a cubit high, diuiding themfelues toward the top into other fmaller branches , whereon are confufedly placed leaues like thofe of *Thalictrum*, or Englifh Rubarb, but leffer and tninner: alongft the tops of the branches grow yellow floures, refembling thofe of Sage : which being paft, there followeth fmall feed like vnto duft.

¶ *The Place.*

The Fumitories grow in corne fields among Barley and other graine ; in vineyards, gardens, and fuch like manured places. I found the fecond and third growing in a corne field betweene a fmall village called Charleton and Greenwhich.

¶ *The Time.*

Fumitory is found with his floure in the beginning of May, and fo continues to the end of fummer. When it is in floure is the beft time to gather it to keep dry, or to diftill.

¶ *The Names.*

Fumitorie is called in Greeke Καπνίτ, and Κάπνιος, and often Καπνίτης : in Latine, *Fumaria* : of *Pliny*, *Cap-nos* : in fhops, *Fumus terræ* : in high Dutch, **Erdrauch** : in low Dutch, **Grijffecrom**, **Dupuen**, **Ker-nel** : in Spanifh, *Palomilha* : in French and Englifh, Fumiterre.

¶ *The Temperature.*

Fumitorie is not hot, as fome haue thought it to be, but cold and fomething dry ; it openeth and clenfeth by vrine.

¶ *The Vertues.*

It is good for all them that haue either fcabs or any other filthe growing on the skinne, and for **A** them alfo that haue the French difeafe.

It remoueth ftoppings from the liuer and fpleene : it putrifieth the blond, and is oft times good **B** for them that haue a quartane ague.

The decoction of the herbe is vfed to be giuen, or elfe the fyrrup that is made of the juyce : the **C** diftilled water thereof is profitable againft the purpofes aforefaid.

It is oftentimes boyled in whay, and in this manner it helpeth in the end of the Spring and in **D** Summer time thofe that are troubled with fcabs.

Paulus Ægineta faith that it plentifully prouoketh vrine, and taketh away the ftoppings of the **E** liuer, and feeblenelfe thereof ; that it ftrengthneth the ftomacke, and maketh the belly foluble.

Diofcorides affirmeth, that the juice of Fumitorie, of that which groweth among Barley, as *Æ-* **F** *gineta* addeth, with gum Arabicke, doth take away vnprofitable haires that pricke the eyes, grow-ing vpon the eye lids, the haires that pricke being firft plucked away, for it will not fuffer others to grow in their places.

The decoction of Fumitorie drunken driueth forth by vrine and fiege all hot chollericke burnt **G** and hurtfull humors, and is a moft fingular digefter of falt and pituitous humors.

† There were formerly fix figures and defcriptions in this chapter ; whereof the two firft figures were of the common Fumitorie, the one with purple, the other with white floures; and the two later were of the *Fumaria latifolia clauiculata*, differing only in the largeneffe and fmallneffe of the leafe. The defcription in the fe-cond place belonged to the *Fumaria clauiculata*, which alfo was againe defcribed in the fifth and fixth places, yet not to much purpofe; wherefore I haue put the fi-gure to the fecond, and omitted the other as fuperfluous.

CHAP. 446.   *Of bulbous Fumitorie, or Hollow-root.*

¶ *The Defcription.*

1   THe leaues of great Hollow-root are jagged and cut in funder, as be thofe of Corian-der, of a light greenifh colour, that is to fay, like the gray colour of the leaues of Co-lumbine, whereunto they be alfo in forme like, but leffer : the ftalks be fmooth, round, and flender , an handfull long ; about which on the vpper part ftand little floures orderly placed, long, with a little horne at the end like the floures of Tode-flax, of a light red tending to a purple colour: the feed lieth in flat cods, very foft and greenifh when it is ready to yeeld vp his black fhi-ning ripe feed : the root is bumped or bulbous, hollow within, and on the vpper part preffed down fomewhat flat, couered ouer with a dark yellow skin or barke, with certaine ftrings faftned thereto, and of a bitter and auftere taft.

2   The fecond is like vnto the firft in each refpect , fauing that it bringeth floures of a white colour, and the other not fo.

3   The fmall purple Hollow-root hath roots, leaues, ftalkes, floures, and feeds like the prece-dent, the efpeciall difference is, that this plant is fomwhat leffe.

4 The

4   The ſmall white Hollow-root likewiſe agreeth with the former in each reſpect, ſauing that this plant bringeth white floures, and the other not ſo.

1 *Radix caua major purpurea.*   2 *Radix caua major alba.*
Great purple Hollow-root.   Great white Hollow-root.

 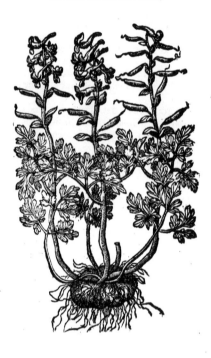

5   This kinde of Hollow-root is alſo like the laſt deſcribed, ſauing that the floures hereof are mixed with purple and white, which maketh it to differ from the others.

6   There is no difference in this, that can poſſibly be diſtinguiſhed from the laſt deſcribed, ſauing that the floures hereof are of a mixt colour, white and purple, with ſome yellow in the hollowneſſe of the ſame, wherein conſiſteth the difference from the precedent.

7   This thin leafed Hollow-root hath likewiſe an hollow-root, couered ouer with a yellow pilling, of the bigneſſe of a tenniſe ball: from which ſhoot vp leaues ſpred vpon the ground, very like vnto the leaues of Columbines, as well in forme as colour, but much thinner, more jagged, and altogether leſſer: among which riſe vp ſmall tender ſtalkes, weake and feeble, of an handfull high, bearing from the middle thereof to the top very fine floures, faſhioned vnto one peece of the Columbine floure, which reſembleth a little bird of a purple colour.

8   This other thin leafed Hollow-root is like the precedent, ſauing that this plant brings forth white floures tending to yellowneſſe, or as it were of the colour of the field Primroſe.

9   **Bunnykens holwortele,** as the Dutch men do call it, hath many ſmall jagged leaues growing immediately from the ground, among which riſe vp very ſlender ſtalks, whereon do grow ſuch leaues as thoſe next the ground: on the top of the branches ſtand faire purple floures like vnto the others of his kinde, ſauing that the floures hereof are as it were ſmall birds, the bellies or lower parts whereof are of a white colour: wherein it differeth from all the reſt of the Hollow-roots.

10   The laſt and ſmall Hollow-root is like the laſt deſcribed, ſauing that it is altogether leſſe, and the floures hereof are of a green colour, not vnlike in ſhape to the floures of Cinkfoile. ‡ This plant, whoſe figure our Author here gaue with this ſmall deſcription, is that which from the ſmell of musk is called *Moſchatella,* by *Cordus* and others: it is the *Denticulata* of *Daleſchampius*: the *Fumaria bulboſa tuberoſa minima* of *Tabernamontanus*: and the *Ranunculus minimus ſeptentrionalium herbido muſcoſo flore* of *Lobel*. The root hereof is ſmal and toothed, or made of little bulbs reſembling teeth and ending in white hairy fibres: it ſends vp diuers little branches ſome two or three inches high: the leaues are ſomwhat like thoſe of the yellow Fumitorie, or *Radix caua,* but much leſſe: the flours
grow

grow cluſtering on the top of the ſtalke, commonly fiue or ſeuen together, each of them made oſ foure yellowiſh greene leaues with ſome threds in them ; it floures in Aprill, and is to be found in diuers places among buſhes at that time, as in Kent about Chiſlehurſt, eſpecially in *Pits* his wood, and at the further end of Cray heath, on the left hand vnder a hedge among bryers and brambles, which is his proper ſeat. ‡

9 *Radix caua minor.*
Bunnikens Holwoort.

10 *Radix caua minima viridi flore*
Small Bunnikens Holwoort,

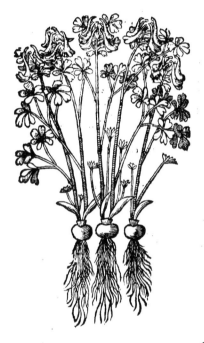

¶ The Place.

Theſe plants do grow about hedges, brambles, and in the borders of fields and vineyards, in low and fertile grounds, in Germanie and the Low-countries , neuertheleſſe the two firſt, and alſo the two laſt deſcribed do grow in my garden.

¶ The Time.

Theſe do floure in March, and their ſeed is ripe in Aprill : the leaues and ſtalks are gone in May, and nothing remaining ſaue only the roots, ſo little a while do they continue.

¶ The Names.

Hollow root is called in high Dutch **Holwurtz** : in low Dutch, **Hoolewoortele**, that is, *Radix caua* : in Engliſh, Hollow root, and Holewoort : it is vſed in ſhops in ſtead of *Ariſtolochia*, or round Birthwnort ; which errour is better knowne than needfull to be confuted : and likewiſe their error is apparant, who raſhly judge it to be *Piſtolochia*, or little Birthwort. It ſhould ſeem the old Writers knew it not ; wherefore ſome of our later Authors haue made it *Leontopetali ſpecies*, or a kind of Lions Turnep : others, *Eriphium*: and otherſome, *Theſium* : moſt men, *Capnos Chelidonia* : it ſeemeth to agree with *Leontopetalon* in bulbed roots, and ſomewhat in leaues, but in no other reſpects, as may be perceiued by *Dioſcorides* and *Plinies* deſcription of *Leontopetalon*. And if *Eriphium* haue his name ἀπὸ τῆ ἦρος, that is to ſay of the Spring, then this root may be not vnproperly *Eriphium*, and *Veris Planta* : or the plant of the Spring : for it is euident that it appeareth and is green in the Spring onely : ſome think it hath been called *Eriphium, ab Hædo*, or of the Goat: but this *Eriphion* is quite another plant, as both *Apuleius* writeth, and that book alſo mentioneth which is attributed to *Galen*, and dedicated to *Paternianus*. In the booke which is dedicated to *Paternianus*, there be read theſe words ; [*Eriphion* is an herbe which is found vpon high mountaines , it hath leaues like Smallage, a fine floure like the Violet, and a root as great as an onion: it hath likewiſe other roots which ſend forth

roots

roots after roots. Whereby it is euident that this root whereof we intreat is not this kinde of *Eriphium.* Concerning *Theſium* the old Writers haue written but little: *Theophraſtus* ſaith, that the root thereof is bitter, and being ſtamped purgeth the belly. *Pliny, lib. 2 1. cap. 17.* ſheweth, that the root which is called *Theſium,* is like the bulbed plants, and is rough in taſte: *Athenæus* citing *Timachida* for an Author ſaith, that *Theſium* is called a floure, of which *Ariadnes* garland was made. Theſe things ſeem well to agree with Hollow root, for it is bumped or bulbous, of taſt bitter, and auſtere, or ſomthing rough, which is alſo thought to purge: but what certainty can be affirmed, ſeeing the old Writers are ſo briefe? what manner of herbe *Capnos Chelidonia* is, which groweth by hedges, and hereupon is ſyrnamed ερλημ, *Aetius* doth not expound, only the name thereof is found in his ſecond Tetrab. the third booke, chap. 1 10. in *Martianus* his *Collyrium,* and in his Tetrab. *lib. 3. cap. 2.* among ſuch things as ſtrengthen the liuer. But if *Capnos Chelidonia* be that which *Pliny* in his 2 5. booke, chap. 1 3, doth call *Prima Capnos,* or the firſt *Capnos,* and commendeth it for the dimneſſe of the ſight, it is plaine enough that *Radix caua,* or the Hollow root, is not *Capnos Chelidonia:* for *Plinies* firſt *Capnos* is branched, and foldeth it ſelfe vpon hedges: but Hollow root hath no ſuch branches growing on it, and is a low herbe, and is not held vp with props, nor needeth them. But if *Aetius* his *Capnos Chelidonia* be another herb differing from that of *Pliny* (which thing perchance was the cauſe why it ſhould be ſyrnamed *Chelidonia*) there is ſome reaſon why it ſhould be called *Capnos Chelidonia;* for it is ſomwhat like Fumitorie in leaues, though greater, and commeth vp at the firſt Spring, which is about the time when the Swallows do come in; neuertheleſſe it doth not follow, that it is true, and right *Capnos Chelidonia,* for there be alſo other herbes comming vp at the ſame ſeaſon, and periſh in ſhort time after, which notwithſtanding are not called *Chelidonia.*

¶ *The Temperature.*

Hollow root is hot and drie, yet more drie than hot, that is to ſay, dry in the third degree, and hot in the ſecond; it bindeth, clenſeth, and ſomwhat waſteth.

¶ *The Vertues.*

A   Hollow root is good againſt old and long laſting ſwellings of the Almonds in the throat, and of the jawes: it likewiſe preuaileth againſt the pains of the hemorrhoides, which are ſwoln and painfull being mixed with the ointment of Poplar buds, called *Vng. Populeon.*

B   It is reported that a dram weight hereof being taken inwardly, doth purge by ſiege, and drawes forth ſlegme.

† Ihaue reduced the eight figures which were formerly here put to the firſt 8. deſcriptions, beeing all of one and the ſame plant, to two, yet haue I left the deſcriptions, which in my opinion might haue been as well ſpared as the figures, for excepting the various colour of the floures, there are but two diſtinct differences of the *Fumaria bulboſa maier,* the one hauing a hollowneſſe in the bottome of the root, and the other wanting it; and this which hath the ſollid root hath alſo the greene leaues betweene the floures cut in or diuided, the floures alſo are leſſe, more in number, and of an elegant red purple colour; and ſeldome found of any other colour, whereas the other varies much in the colour of the floures.

CHAP. 447. *Of Columbine.*

¶ *The Deſcription.*

1   THe blew Columbine hath leaues like the great Celandine, but ſomewhat rounder, indented on the edges, parted into diuers ſections, of a blewiſh green colour, which beeing broken, yeeld forth little juice or none at all: the ſtalke is a cubit and a halfe high, ſlender, reddiſh, and ſleightly haired: the ſlender ſprigs whereof bring forth euerie one one floure with fiue little hollow hornes, as it were hanging forth, with ſmall leaues ſtanding vpright, of the ſhape of little birds: theſe floures are of colour ſomtimes blew, at other times of a red or purple, often white, or of mixt colours, which to diſtinguiſh ſeuerally were to ſmall purpoſe, being things ſo familiarly known to all: after the floures grow vp cods, in which is contained little black and glittering ſeed: the roots are thicke, with ſome ſtrings thereto belonging, which continue many yeres.

2   The ſecond doth not differ ſauing in the colour of the floures; for like as the others are deſcribed to be blew, ſo thoſe are of a purple red, or horſe-fleſh colour, which maketh the difference.

3   The double Columbine hath ſtalks, leaues, and roots, like the former: the floures hereof are very double, that is to ſay, many of thoſe little floures (hauing the forme of birds) are thruſt one into the belly of another, ſometimes blew, often white, and otherwhiles of mixt colours, as nature liſt to play with her little ones, differing ſo infinitely, that to diſtinguiſh them apart would require

more

1 *Aquilegia cærulea.*
Blew Columbines.

2 *Aquileia rubra.*
Red Columbines.

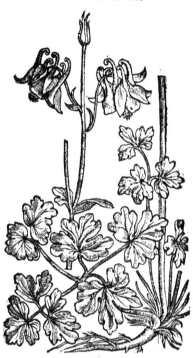

3 *Aquilina multiplex.*
Double Columbines.

‡ 4 *Aquilegia variegata.*
Variegated Columbine.

‡ 5 *Aquilegia flo. inverſo rubro.*
Columbine with the inverted red floure.

‡ 6 *Aquilegia flo. inverſo albo.*
Inverted Columbine with the white floure.

‡ 7 *Aquilegia flore roſeo.*
Roſe Columbine.

‡ 8 *Aquilegia degener.*
Degenerate Columbine.

more time than were requiſite to leeſe : and therefore it ſhall ſuffice what hath beene ſaid for their deſcriptions.

‡ 4　There are alſo other varieties of this double kinde, which haue the floures of diuers or partie colours, as blew and white, and white and red variouſly marked or ſpotted.

5　This kinde hath the floures with their heeles or ſpurtes turned outward or in the middle of the floure, whence it is called *Aquilina inuerſa:* the floures of this are commonly reddiſh, or of a light or darke purple colour, and double.

6　This differs from the laſt in the colour of the floures which are white, yet double, and inuerted as the former.

7　The roots, leaues, and ſtalks of this are not vnlike that of the precedent, but the floure is much different in ſhape; for it hath no heels or ſpurs, but it is made of ſundry long leaues lying flat open, being ſometimes more ſingle, and otherwhiles more double. The colour of the floure is either red, white, blew, or variouſly mixt of theſe as the former.

8　This though it be termed degenerate, is a kinde of it ſelfe, and it differs from the laſt deſcribed in that the vtmoſt leaues are the largeſt, and the colour thereof is commonly greene, or greene ſomewhat inclining to purple. ‡

¶ *The Place.*

They are ſet and ſowne in gardens for the beautie and variable colour of the floures.

¶ *The Time.*

They floure in May, Iune, and Iuly.

¶ *The Names.*

Columbine is called of the later Herbariſts, *Aquileia, Aquilina,* and *Aquilegia:* of *Coſteus, Pothos :* of *Geſner, Leontoſtomum :* of *Daleſchampius, Iouis flos :* of ſome, *Herba Leonis,* or the herbe, wherein the Lion doth delight : in high Dutch, **Agley :** in low Dutch, **Akelepen :** in French, *Ancoiles :* in Engliſh, Columbine. ‡ *Fabius Columna* iudges it to be the *Iſopyrum* deſcribed by *Dioſcorides.* ‡

¶ *The Temperature.*

Columbines are thought to be temperate betweene heate and moiſture.

¶ *The Vertues.*

Notwithſtanding what temperature or vertues Columbines haue is not yet ſufficiently known; A for they are vſed eſpecially to decke the gardens of the curious, garlands and houſes : neuertheleſſe *Tragus* writeth, that a dram weight of the ſeed, with halfe a ſcruple or ten graines of Saffron giuen in wine, is a good and effectuall medicine for the ſtopping of the liuer, and the yellow jaundiſe; but he ſaith, that whoſo hath taken it muſt be well couered with cloathes, and then ſweat.

Moſt in theſe daies following others by tradition, do vſe to boyle the leaues in milke againſt the B ſoreneſſe of the throat, falling and excoriation of the vuula : but the antient writers haue ſaid nothing hereof : *Ruellius* reporteth, that the floures of Columbines are not vſed in medicine : yet ſome there be that do affirme they are good againſt the ſtopping of the liuer, which effect the leaues doe alſo performe.

‡ *Cluſius* ſaith, that Dr. *Francis Rapard* a Phyſition of Bruges in Flanders, told him that the ſeed C of this common Columbine very finely beaten to pouder, and giuen in wine, was a ſingular medicine to be giuen to women to haſten and facilitate their labour, and if the firſt taking it were not ſufficiently effectuall, that then they ſhould repeate it againe. ‡

CHAP. 448.　*Of Wormewood.*

¶ *The Deſcription.*

1　THe firſt kind being our common and beſt knowne Wormwood, hath leaues of a grayiſh colour, very much cut or jagged, and very bitter : the ſtalks are of a wooddy ſubſtance, two cubits high, and full of branches, alongſt which doe grow little yellowiſh buttons, wherein is found ſmall ſeed like the ſeed of Tanſey, but ſmaller : the root is likewiſe of a wooddy ſubſtance, and full of fibres.

2　The ſecond kind of Wormewood bringeth forth ſlender ſtalkes about a foot high or ſomewhat more, garniſhed with leaues like the former, but whiter, much leſſer, and cut or jagged into moſt fine and ſmall cuts or diuiſions : the floures are like the former, hanging vpon ſmall ſtemmes with their heads downeward : the roots are whitiſh, ſmall and many, crawling and crambling one ouer another, and thereby infinitely do increaſe, of ſauour leſſe pleaſant than the common Wormwood.

wood. Some haue termed this plant *Abſinthium ſantonicum*, but they had ſlender reaſon ſo to do: for if it was ſo called becauſe it was imagined to grow in the Prouince of Saintoinge, it may very well appeare to the contrary ; for in the Alpes of Galatia, a countrey in Aſia *minor*, it groweth in great plenty, and therefore may rather be called *Galatium Sardonicum*, and not *Santonicum* : but leauing controuerſies impertinent to the Hiſtory, it is the Ponticke Wormwood of *Galens* deſcriptio n, and ſo holden of the learned *Paludane* (who for his ſingular knowledge in plants is worthy triple honor) an d likewiſe many others.

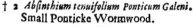

1 *Abſinthium latifolium ſiue Ponticum.*	† 2 *Abſinthium tenuifolium Ponticum Galeni.*
Broad leafed Wormwood.	Small Ponticke Wormwood.

¶ *The Place.*

This broad leafed Wormewood delighteth to grow on rockes and mountaines, and in vntilled places ; it groweth moſt vpon dry bankes, it is common euery where in all countries : the beſt, ſaith *Dioſcorides*, is found in Pontus, Cappadocia, and on mount Taurus : *Pliny* writeth, that Ponticke Wormwood is better than that of Italy : *Ouid* in theſe words doth declare the Ponticke Worme-wood to be extreme bitter.

Turpia deformes gignunt Abſinthia campi,
Terraque defructu, quam ſit amara docet.

Vntilled barren ground the lothſome Wormwood yeelds,
And knowne it's by the fruit how bitter are the fields.

And *Bellonius* in his firſt booke of Singularities, chap.76. doth ſhew, that there is alſo a broad leafed Wormwood like vnto ours, growing in the Prouinces of Pontus, and is vſed in Conſtantino-ple by the Phyſitions there, it is likewiſe found in certaine cold places of Switzerland, which by reaſon of the chilneſſe of the aire riſeth not vp, but creepeth vpon the ground, whereupon diuers call it creeping Wormewood.

¶ *The Time.*

The little floures and ſeeds are perfected in Iuly and Auguſt, then may Wormewood be gathe-red and laied vp for profitable vſes.

¶ *The Names.*

It is called in Greeke, ἀψίσιον : it is named of *Apuleius*, *Abſinthium ruſticum*, country Wormwood,

or

or pesants Wormewood : we haue named it *Absinthium latifolium*,broad leafed Wormewood, that it may differ from the rest : the interpreters of the Arabians call the better sort, which *Dioscorides* nameth Ponticke Wormewood, *Romanum Absinthium*,Roman Wormewood:and after these, the barbarous Ph sitions of the later age : the Italians name Wormewood,*Assenso* the Spaniards, *Axenxios*,*Assinsios*,most of them *Donzell* : the Portingales,*Alosua*:in high Dutch,ꝏertonmut,ꝏertmut: in French, *Aluyne* : in English, common Wormewood.

Victor Trincauilla,a singular Physition,in his practise tooke it for *Absinthium Ponticum*.

2 This is commonly called *Absinthium Romanum*:and in low Dutch,Roomsche Alsene:by which name it is knowne to very many Physitons and Apothecaries, who vse this in stead of Ponticke wormwood:furthermore,it hath a leafe and floure far lesse that the other wormwoods : likewise the smell of this is not onely plesant,but it yeeldeth also a spicie sent,whereas all the rest haue a strong and lothsome smell : and this Ponticke Wormewood doth differ from that which *Dioscorides* commendeth : for *Dioscorides* his Pontick Wormwood is accounted among them of the first kind,or of broad leafed wormwood;which thing also *Galen* affirmeth in his sixt booke of the Faculties of medicines in the chapter of Sothernwood. There be three kinds of Wormwood(saith he) whereof they vse to call one by the generall name,and that is especially Pontick : whereby it is manifest that *Galen* in this place hath referred Ponticke to no other than to the first Wormwood;and therefore many not without cause maruell, that *Galen* hath written in his booke of the Method of curing, how Ponticke Wormwood is lesse in floure and leafe : many excuse him, and lay the fault vpon the coruption of the booke,and in his 9. booke of Method, the lesser they would haue the longer : therefore this wormwood with the lesser leafe is not the right Ponticke Wormwood,neither againe the Arabians Romane wormewood,who haue no other Romane than Ponticke of the Græcians. Also many beleeue that this is called *Santonicum*,but this is not to be sought for in Mysia, Thracia,or other countries Eastward, but in France beyoud the Alps, if we may beleeue *Dioscorides* his copies, there be that would haue it grow not beyond the Alps of Italy,but in Galatia a country in Asia,& in the region of the Sardines,which is in the lesser Asia;whereupon it was called in Greeke ϲαρδόνιον, which was changed into the name *Santonicum* through the errour of the translators : *Dioscorides* his copies keep the word *Sardonium*,& *Galens* copies *Santonicum*,which came to posterity as it seemeth, is called in English,Romane Wormewood, garden or Cypres Wormewood, and French Wormewood.

¶ *The Temperature.*

Wormewood is of temperature hot and dry, hot in the second degree and dry in the third : it is bitter and clensing,and likewise hath power to binde or strengthen.

¶ *The Vertues.*

It is very profitable to a weake stomacke that is troubled with choler, for it clenseth it through **A** his bitternesse, purgeth by siege and vrine : by reason of the binding quality, it strengthneth and comforteth the stomacke, but helpeth nothing at all to remoue flegme contained in the stomacke, as *Galen* addeth.

If it be taken before a surfeit it keepeth it off,and remoueth lothsomnesse, saith *Dioscorides*, and **B** it helpeth not only before a surfeit, but also it quickly refresheth the stomack and belly after large eating and drinking.

It is oftentimes a good remedy against long and lingring agues, especially tertians : for it doth **C** not onely strengthen the stomacke and make an appetite to meat, but it yeeldeth strength to the liuer also,and riddeth it of obstructions or stoppings,clensing by vrine naughty humours.

Furthermore, Wormewood is excellent good for them that vomit bloud from the spleene, the **D** which hapneth when the spleene being ouercharged and filled vp with grosse bloud doth vnburden it selfe,and then great plenty of bloud is oftentimes cast vp by vomite. It happeneth likewise that store of blacke and corrupt bloud mixed with excrements passeth downewards by the stoole, and it oftentimes hapneth that with violent and large vomiting the sicke man fainteth or swouneth, or when he is reuiued doth fall into a difficult and almost incurable tympanie, especially when the disease doth often happen ; but from these dangers Wormewood can deliuer him,if when he is refreshed after vomite,and his strength any way recouered, he shall a good while vse it,in what manner soeuer he himselfe shall thinke good.

Againe, Wormewood voideth away the wormes of the guts, not onely taken inwardly, but ap- **E** plied outwardly : it withstandeth all putrifactions ; it is good against a stinking breath ; it keepeth garments also from the mothes ; it driueth away gnats, the body being annointed with the oyle thereof.

Likewise it is singular good in pultesses and fomentations to binde and to dry. **F**

Besides all this,*Dioscorides* declareth, that it is good also against windinesse and griping pains of **G** the stomacke and belly,with Seseli and French Spikenard : the decoction cureth the yellow jaundies or the infusion,if it be drunke thrise a day some ten or twelue spoonfuls at a time.

H   It helpeth them that are strangled with eating of Mushromes, or toad stools,if it be drunke with vinegre.

I   And being taken in wine,it is good against the poyson of *Ixia* (being a viscous matter procee-ding from the thistle *Chamælion*) and of Hemlocke, and against the biting of the shrew mouse, and of the Sea Dragon : it is applied to the squincie or inflammations of the throat with hony and ni-ter, and with water to night wheales, and with hony to swartish markes that come vpon brises.

K   It is applied after the same manner to dim eies,and to mattering eares.

L   *Ioachimus Camerarius* of Noremberg commendeth it greatly against the iaundise, giuing of the floures of Wormewood,Rosemarie, Sloes, of each a small quantitie, and a little Saffron, boiled in wine, the body first being purged and prepared by the learned Physition.

† The figure which formerly was in the second place,was of a small Wormewood not different from the common kinde,but onely in the smalnesse,and more aro-maticke taste,it growes on mountainous places,and *Gesner* calls it *Absinthium commune minus, vel Alpinum :* now our Authors description was intended for this,whose figure we haue giuen you,for it is the *Absinthium santonicum,* of some, as *Ruellius* and *Cesalpinus ;* and the *Galatium Sardonium* of *Pena* and *Lobel.*

C H A P. 449. *Of Small leafed Worme-wood.*

Absinthium tenuifolium Austriacum.
Austrian Wormewood.

¶ *The Description.*

SMall leafed Wormwood bringeth forth ve-ry many little branches, slender, a span or a foot high, full of leaues , lesse by a great deale,and tenderer than the former,most finely and nicely minced:the floures like those of the former,hang vpon the little branches & sprigs: the roots are small,creeping ouerthwart, from whence do rise a great number of yong sprouts: this Wormwood also is somewhat white, and no lesse bitter than the broad leafed one, and hath not so ranke, or so vnpleasant a smell,but rather delightfull.

¶ *The Place.*

It grows plentifully in Mysia,Thracia,Hun-garie, and Austria, and in other regions neere, adjoyning : it is also found in Bohemia, and in many vntilled places of Germany, it is a gar-den plant in the low Countries,and in Eng-land.

¶ *The Time.*

It bringeth forth floures and seed in Au-tumne : a little while after when Winter com-meth, the herbe withereth away, but the root remaineth aliue,from which leaues and stalkes do come againe in the Spring.

¶ *The Names.*

‡   This *Lobel* calls *Absinthium Ponticum Tridentinum Herbariorum : Clusius, Absinthium tenuifolium Austriacum : Tabernamontanus, Absin-thium Nabathæum Auicennæ :* wee may call it in English,small leafed Wormewood. ‡

¶ *The Nature.*

Small leafed Wormewood is of facultie hot and dry, it is as bitter also as the broad leafed one, and of like facultie.

¶ *The Vertues.*

The faculties are referred vnto the common Wormewood.

CHAP. 450. *Of Sea Worme-wood.*

¶ *The Deſcription.*

1 THe white or common Sea Wormwood hath many leaues cut and diuided into infinite fine jags, like thoſe of Sothernwood, of a white hoarie colour and ſtrong ſmell, but not vnpleaſant : among which riſe vp tough hoarie ſtalks ſet with the like leaues, on the top whereof do grow ſmall yellowiſh floures; the root is tough, and creepeth far abroad, by means where of it greatly increaſeth.

1 *Abſinthium marinum album.*
White Sea Wormwood.

2 *Abſinthium marinum repens.*
Creeping Sea Wormwood.

2 The broad leafed Sea Wormwood hath very many ſoft leaues, growing cloſe by the ground, of a darke ſwart colour, nothing ſo finely cut or jagged as the other of his kinds : the floures grow vpon the tops of the ſtalks, of a yellowiſh colour : the root is tough and creeping. ‡ This hath many weake ſlender branches commonly two foot long at their ful growth, red of colour and creeping vpon the ground : the leaues are ſmall, narrow, long and jagged, or parted towards their ends into ſundry parcels : they are greene aboue, and graviſh vnderneath : the toppes of the branches are ſet with many little ſtalkes, ſome inch long : which vpon ſhort foot-ſtalkes comming out of the boſomes of little longiſh narrow leaues carry ſmall round knops, like as in other plants of this kinde: the taſte is a little bitteriſh, and the ſmell not vnpleaſant: this growes with Mʳ *Parkinſon* and others, and (as I remember) it was firſt ſent ouer from the Iſle of Rees by Mʳ *Iohn Tradeſcant*. *Lobel* in his Obſeruations mentions it by the name of *Abſinth. Ponticum ſupinum Herbariorum* ; and *Tabern.* ſets it forth by the title of *Abſinthium repens*. ‡

¶ *The Place.*

Theſe Wormwoods doe grow vpon the raiſed grounds in the ſalt marſhes neere vnto the ſea, in moſt places of England ; which being brought into gardens doth there flouriſh as in his naturall place, and retaineth his ſmell, taſte and naturall qualitie, as hath beene often proued. ‡ I haue not heard that the later growes wilde with vs in England. ‡

¶ *The Time.*

These bring forth floures and seeds when the other Wormwoods do. ‡ The later scarce seeds with vs, it floures so late in the yeare. ‡

¶ *The Names.*

Sea Wormwood is called in Grkke, ἀψίνθιον θαλάσσιον : in Latine, *Absinthium marinum*, and likewise *Seriphium* : in Dutch, **See Alsene** : of diuers, *Santonicum*, as witnesseth *Dioscorides* : neuerthelesse there is another *Santonicum* differing from Sea Wormewood : in English of some Women in the country, Garden Cypresse.

¶ *The Temperature.*

Sea Wormwood is of nature hot and dry, but not so much as the common.

¶ *The Vertues.*

A   *Dioscorides* affirmeth, that being taken of it selfe, or boyled with Rice, and eaten with hony, it killeth the small wormes of the guts, and gently looseth the belly, the which *Pliny* doth also affirme.

B   The juyce of sea Wormwood drunke with wine resisteth poison, especially the poison of Hemlockes.

C   The leaues stamped with figs, salt-peter, and the meale of Darnel, and applied to the belly, sides, or flankes, helpe the dropsie, and such as are spleneticke.

D   The same is singular against all inflammations, and heat of the stomacke and liuer, exceeding all the kindes of Wormwood for the same purposes that common Wormwood serueth.

E   It is reported by such as dwell neere the sea side, that the cattell which doe feed where it groweth become fat and lusty very quickly.

F   The herbe with his stalks laid in chests, presses, and ward-robes, keepeth clothes from moths and other vermine.

CHAP. 451.   *Of Holy Wormewood.*

Sementina.
Holy Wormewood.

¶ *The Description.*

THis Wormewood called *Sementina*, and *Semen sanctum*, which we haue Englished, Holy, is that kinde of Wormwood which beareth that seed which we haue in vse, called Wormeseed : in shops, *Semen Santolinum* : about which there hath been great controuersies amongst writers : some holding that the seed of *Santonicum Galatium* to be the true Wormeseed : others deeming it to be that of *Romanum Absinthium* : it doth much resemble the first of the sea Wormwoods in shape and proportion : it riseth vp with a wooddy stalke, of the height of a cubite, diuided into diuers branches and wings ; whereupon are set very small leaues : among which are placed clusters of seeds in such abundance, that to the first view it seemeth to be a plant consisting all of seed.

¶ *The Place.*

It is a forreine plant: the seeds being sowne in the gardens of hot regions doe prosper well ; in these cold countries it will not grow at all. Neuertheles there is one or two companions about London, who haue reported vnto me that they had great store of it growing in their Gardens yearely, which they sold at a great price vnto our London Apothecaries, and gained much mony thereby ; one of the men dwelling by the Bagge and Bottle neere London, whose name is *Cornewall* ; into whose garden I was brought to see the thing that I would not beleeue ; for being often

told .

told that there it did grow, I ſtill perſiſted it was not true : but when I did behold this great quantitie of Wormewood, it was nothing elſe but common *Amcos*. How many Apothecaries haue been
deceiued, how many they haue robbed of their mony, and how many children haue beene nothing
the better for taking it, I refer it to the iudgement of the ſimpleſt, conſidering their owne report, to
haue ſold many hundreth pounds weight of it ; the more to their ſhame be it ſpoken, and the leſſe
wit or skill in the Apothecaries : therefore haue I ſet downe this as a caueat vnto thoſe that buy
of theſe ſeeds, firſt to taſte and trie the ſame before they giue it to their children, or commit it to
any other vſe. ‡ Certainly our Author was either miſinformed, or the people of theſe times were
very ſimple, for I dare boldly ſay there is not any Apothecary, or ſcarce any other ſo ſimple as to be
thus deceiued now. ‡

¶ *The Time.*

It floureth and bringeth forth his ſeed in Iuly and Auguſt.

¶ *The Names.*

The French men call it *Barbotine* ; the Italians, *Semen Zena* : whereupon alſo the Latine name
Sementina came : the ſeed is called euery where *Semen ſanctum* : Holy ſeed ; and *Semen contra Lum-
bricos :* in Engliſh, Wormſeed ; the herbe it ſelfe is alſo called Wormſeed, or Wormſeed-wort : ſome
name it *Semen Zedoariæ*, Zedoarie ſeed, becauſe it hath a ſmell ſomewhat reſembling that of Zedoarie.

¶ *The Temperature.*

The ſeed is very bitter, and for that cauſe of nature hot and dry.

¶ *The Vertues.*

It is good againſt wormes of the belly and entrailes, taken any way, and better alſo if a little Ru-　A
barbe be mixed withall, for ſo the wormes are not onely killed, but likewiſe they are driuen downe
by the ſiege, which thing muſt alwaies be regarded.

The ſeed mixed with a little *Aloe ſuccotrina*, and brought to the forme of a plaiſter, and applied to　B
the nauell of a childe doth the like.

Cᴴᴬᴘ. 452.　　*Of forreine and Baſtard Wormewoods.*

1 *Abſinthium album.*
White Wormwood.

2 *Abſinthium Ægyptium.*
Wormwood of Ægypt.

¶ *The Deſcription.*

† 1   ABſinthium album hath ſtraight and vpright ſtalkes, a foot high, beſet with broad leaues, but very deeply cut or clouen, in ſhew like vnto thoſe of the great Daiſie, but white of colour: at the top of the ſtalkes, out of ſcaly heads, as in an vmbell grow floures, compact of ſix ſmall white leaues: the root is long, with ſome fibres annexed to it.

2   This kinde of Wormewood *Geſner* and that learned Apothecarie *Valerandus Denraz*, called *Abſinthium Ægyptium* : the leaues of this plant are very like to the leaues of *Trichomanes*, which is our common Maiden haire, of a white colour, euery ſmall leaſe ſtanding one oppoſite againſt another, and of a ſtrong ſauour.

3   This Wormewood, which *Dodonæus* calleth *Abſinthium inodorum*, and *Inſipidum*, is very like vnto the Sea Wormewood, in his ſmall and tender leaues: the ſtalke beareth floures alſo like vnto the foreſaid Sea Wormewood, but it is of a ſad or deepe colour, hauing neither bitter taſte, nor any ſauour at all; whereupon it was called, and that very fitly, *Abſinthium inodorum*, or *Abſinthium inſipidum*. in Engliſh, ooliſh, or vnſauoury Wormwood. ‡ *Dodonæus* ſaith not that his *Abſinthium inſipidum* is like the ſea Wormewood, but that it is very like our common broad leaued Wormwood, and ſo indeed it is, and that ſo like, that it is hard to be diſcerned therefrom, but onely by the want of bitterneſſe and ſmell. ‡

3 *Abſinthium inodorum.*
Vnſauorie Wormewood.

4 *Abſinthium marinum, Abrotani fœminæ facie.*
Small Lauander Cotton.

4   This kinde of Sea-Wormewood is a ſhrubby and wooddy plant, in face and ſhew like to Lauander Cotton, of a ſtrong ſmell; hauing floures like thoſe of the common Wormwood, at the firſt ſhew like thoſe of Lauander Cotton: the root is tough and wooddy.

¶ *The Place.*

Theſe plants are ſtrangers in England, yet we haue a few of them in Herbariſts gardens.

¶ *The Time.*

The time of their flouring and ſeeding is referred to the other Wormwoods.

¶ *The Names.*

The White Wormewood *Conradus Geſnerus* nameth *Seriphium fœmina*, and ſaith that it is commonly called *Herba alba*, or white herbe: another had rather name it *Santonicum* : for as *Dioſcorides* ſaith, *Santonicum* is found in France beyond the Alpes, and beareth his name of the ſame
country

countrey where it groweth ; but that part of Swisserland which belongeth to France is accounted o' the Romans to be beyond the Alps ; and the Prouince of Santon is far from it : for this is a part of Guines, scituate vpon the coaſt of the Ocean, beneath the floud Gerond Northward : therefore Santon Wormewood, if it haue his name from the Santons, groweth farre from the Alps : but if it grow neere adjoyning to the Alps, then hath it not his name from the Santons.

¶ *The Temperature and Vertues.*

White Wormewood is hot and somewhat dry.

Vnsauourie Wormewood, as it is without smell and taſte, so is it scarse of any hot quality, much leſſe hath it any scouring faculty. Theſe are not vſed in Phyſicke, where the others may be had, being as it were wilde or degenerate kindes of Wormwood ; some of them participating both of the forme and smell of other plants.   **A**

† The figure which was here formerly in the firſt place, by the name of *Abſinthium arboreſcens*, is the firſt of the next chapter ſaue one, where you may ſee more thereof. The white Wormewood mentioned here in the Names, but no where elſe in the chapter, is either the ſame with, or one very like our Sea Wormewood. Let ſuch as are curouſlooke into *Edueerarius* his *Hort. Med.* in the title of *Abſinthium Santonicum* : and in *Dodonæus, Pempt. 1. lib. 3. cap. 5.* where the firſt deſcription is of this Wormewood.

C H A P. 453.   *Of Mugwort.*

1 *Artemiſia, mater Herbarum.*
Common Mugwort.

¶ *The Deſcription.*

1   THe firſt kinde of Mugwort hath broad leaues, very much cut or clouen like the leaues of common Wormewood, but larger, of a darke greene colour aboue, and hoarie vnderneath : the ſtalkes are long and ſtraight, and full of branches, whereon do grow ſmall round buttons, which are the floures, smelling like Marjerome when they wax ripe : the root is great, and of a wooddie ſubſtance.

2   The ſecond kinde of Mugwort hath a great thicke and wooddy root, from whence ariſe ſundry branches of a reddiſh colour, beſet full of ſmall and fine jagged leaues, very like vnto ſea Sothernwood : the ſeed groweth alongſt the ſmall twiggie branches, like vnto little berries, which fall not from their branches in a long time after they be ripe. ‡ I know not how this differeth from the former, but only in the colour of the ſtalke and floures, which are red or purpliſh ; whereas the former is more whitiſh. ‡

3   There is alſo another Mugwort, which hath many branches riſing from a wooddie root, ſtanding vpright in diſtances one from another, of an aſhie colour, beſet with leaues not much vnlike ſea Purſlane ; about the lower part of the ſtalkes, and toward the top of the branches they are narrower and leſſer, and cut with great and deep jagges, thicke in ſubſtance, and of a whitiſh colour, as all the reſt of the plant is : it yeeldeth a pleaſant ſmell like *Abrotanum marinum*, and in taſte is ſomewhat ſaltiſh : the floures are many and yellow : which being vaded, there followeth moſſie ſeed like vnto that of the common Wormewood. ‡ The leaues of this plant are of two ſorts ; for ſome of them are long and narrow, like thoſe of Lauander (whence *Cluſius* hath called it *Artemiſia folio Lauendulæ*) otherſome are cut in or diuided almoſt to the middle rib ; as you may ſee it expreſt apart in a figure by it ſelfe, which ſhewes both the whole, as alſo the diuided leaues.

¶ ₄*The*

3 *Artemisia marina.*   ‡ *Artemisia marina ramulus, folia integra &*
Sea Mugwort.   *dissecta exprimens.*
                A branch shewing the cut and vncut leaues.

¶ *The Place.*

The common Mugwort groweth wilde in sundry places about the borders of fields, about high waies, brooke sides, and such like places.

Sea Mugwort groweth about Rie and Winchelsea castle, and at Portsmouth by the Isle of Wight.

¶ *The Time.*

They floure in Iuly and August.

¶ *The Names.*

Mugwort is called in Greeke, Ἀρτεμισία: and also in Latine, *Artemisia*, which name it had of *Artemisia* Queene of Halicarnassus, and wife of noble *Mausolus* King of Caria, who adopted it for her owne herbe: before that it was called παρθένις, *Parthenis* as *Pliny* writeth. *Apuleius* affirmeth that it was likewise called *Parthenion*: who hath very many names for it, and many of them are placed in *Dioscorides* among the bastard names: most of these agree with the right *Artemisia*, and diuers of them with other herbes, which now and then are numbred among the Mugworts: it is also called *Mater Herbarum*: in high Dutch, **Beifuſz**, and **Sant Johanus Gurtell**: in Spanish and Italian, *Artemisia*: in French, *Armoisa*: in low Dutch, **Bijuoet**, **Sint Jans krupt**: in English, Mugwort, and common Mugwort.

¶ *The Temperature.*

Mugwort is hot and dry in the second degree, and somewhat astringent.

¶ *The Vertues.*

A   *Pliny* saith, That Mugwort doth properly cure womens diseases.

B   *Dioscorides* writeth, That it bringeth downe the termes, the birth, and the after-birth.

C   And that in like manner it helpeth the mother, and the paine of the matrix, to bee boyled as bathes for women to sit in; and that being put vp with myrrh, it is of like force that the bath is of. And that the tender tops are boyled and drunke for the same infirmities; and that they are applied in manner of a pultesse to the share, to bring downe the monethly course.

D   *Pliny* saith, That the traueller or wayfaring man that hath the herbe tied about him feeleth no wearisomnesse at all; and that he who hath it about him can be hurt by no poysonsome medicines, nor by any wilde beast, neither yet by the Sun it selfe; and also that it is drunke against *Opium*, or

the

the juyce of blacke Poppy. Many other fantaſticall deuices inuented by Poëts are to be ſeene in the Works of the Antient Writers, tending to witchcraft and ſorcerie, and the great diſhonour of God; wherefore I do of purpoſe omit them, as things vnworthy of my recording, to your reuiewing.

Mugwort pound with oyle of ſweet almonds, and laid to the ſtomacke as a plaiſter, cureth all the   E paines and griefes of the ſame.

It cureth the ſhakings of the joynts, inclining to the palſie, and helpeth the contraction or draw-   F ing together of the nerues and ſinewes.

† There were formerly two deſcriptions of the *Artemiſia marina*; wherefore I omitted the former, being the more vnperfect.

<h2 style="text-align:center">C H A P. 454;   Of Sothernwood.</h2>

¶ The Kindes.

Dioſcorides affirmeth that Sothernwood is of two kindes, the female and the male, which are euery where known by the names of the greater and of the leſſer; beſides theſe there is a third kind, which is of a ſweeter ſmell, and leſſer than the others, and alſo others of a baſtard kinde.

† 1 *Abrotanum fœmina arboreſcens.*   2 *Abrotanum mas,*
    Female Sothernwood.   Male Sothernwood.

 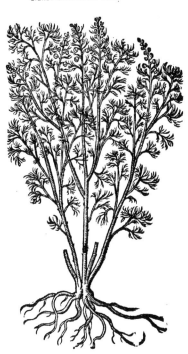

¶ The Deſcription.

1   **T**He greater Sothernwood by carefull manuring doth oftentimes grow vp in manner of a ſhrub, and commeth to be as high as a man, bringing forth ſtalkes an inch thicke, or more, out of which ſpring very many ſprigs or branches, ſet about with leaues diuerſly jagged and finely indented, ſomewhat white, and of a certaine ſtrong ſmell: in ſtead of floures, little ſmall cluſters of buttons doe hang on the ſprigs, from the middle to the very top, of colour yellow, and at the length turne into ſeed. The root hath diuers ſtrings.

2 The

3 *Abrotanum humile.*
Dwarfe Sothernwood.

4 *Abrotanum inodorum.*
Vnſauorie Sothernwood.

5 *Abrotanum campeſtre.*
Wilde Sothernwood.

2   The leſſer Sothernwood groweth low, ful of little ſprigs of a wooddy ſubſtance: the leaues are long, and ſmaller than thoſe of the former, not ſo white : it beareth cluſtering buttons vpon the top of the ſtalkes: the root is made of many ſtrings.

3   The third kind is alſo ſhorter : the leaues hereof are jagged and deeply cut after the maner of the greater Sothernwood, but they are not ſo white, yet more ſweet, wherein they are like vnto Lauander cotton. This kinde is very full of ſeed: the buttons ſtand alongſt on the ſprigs, euen to the very top, and be of a glittering yellow. The root is like to the reſt.

4   The vnſauorie Sothernwood groweth flat vpon the ground, with broad leaues deepely cut or jagged in the edges like thoſe of the common Mugwort : among which riſe vp weake and feeble ſtalkes trailing likewiſe vpon the ground, ſet confuſedly here and there with the like leaues that grow next the ground, of a grayiſh or hoary colour, altogether without ſmell. The floures grow alongſt the ſtalkes, of a yellowiſh colour, ſmall and chaffie: the root is tough and wooddy, with ſome ſtrings annexed thereto.

5   This wilde Sothernwood hath a great long thicke root , tough and wooddy, couered

ouer

ouer with a scaly barke like the scaly backe of an Adder, and of the same colour: from which rise very many leaues like those of Fennell, of an ouerworne greene colour: among which grow small twiggy branches on the tops, and alongst the stalkes doe grow small clustering floures of a yellow colour: the whole plant is of a darke colour, as well leaues as stalkes, and of a strong vnsauourie smell.

¶ The Place.

Theophrastus saith that Sothernwood delighteth to grow in places open to the Sun : *Dioscorides* affirmeth that it groweth in Cappadocia, and Galatia a countrey in Asia, and in Hierapolis a city in Syria : it is planted in gardens almost euery where:that of Sicilia and Galatia is most commended of *Pliny*.

¶ The Time.

The buttons of Sothernwood doe flourish and be in their prime in August, and now and then in September.

¶ The Names.

It is called in Greeke, Κβέιπω : the Latines and Apothecaries keepe the same name *Abrotanum:* the Italians and diuers Spaniards call it *Abrotano* : and other Spaniards, *Yerua lombriguera :* in high Dutch, **Stabwurtz :** in low Dutch, **Auertoone,** and **Auertruijt :** the French, *Aurone,* and *Auroesme:* the English men, Sothernwood: it hath diuers bastard names in *Dioscorides* ; the greater kinde is *Dioscorides* his *Fœmina,* or female Sothernewood ; and *Pliny* his *Montanum,*or mountaine Sothernewood : the mountaine Sothernwood we take for the female,and the champion for the male. There be notwithstanding some that take Lauander Cotton to be the female Sothernwood ; grounding thereupon, becaufe it bringeth forth yellow floures in the top of the sprigs like cluster buttons : but if they had more diligently pondered *Dioscorides* his words, they would not haue beene of this opinion : the lesser Sothernewood is *Mas,*the male, and is also *Plinies* champion Sothernwood; in Latine,*Campestre.* The third,as we haue said, is likewise the female, and is commonly called sweet Sothernwood,becaufe it is of a sweeter sent than the rest. *Dioscorides* seemeth to call this kinde *Siculum,* Sicilian Sotherwood.

¶ The Temperature.

Sothernewood is hot and dry in the end of the third degree : it hath also force to distribute and to rarifie.

¶ The Vertues.

The tops,floures,or feed boyled, and stamped raw with water and drunke, helpe them that can- A not take their breaths without holding their neckes straight vp,and is a remedie for the cramp,and for sinewes shrunke and drawne together ; for the sciatica also, and for them that can hardly make water ; and it is good to bring downe the termes.

It killeth wormes,and driueth them out : if it be drunke with wine it is a remedy against deadly B poysons.

Also it helpeth against the stinging of Scorpions and field spiders,but it hurts the stomacke. C

Stamped and mixed with oyle it taketh away the shiuering cold that commeth by the ague fits, D and it heateth the body if it be annointed therewith before the fits do come.

If it be pouned with barley meale and laid to pushes it taketh them away. E

It is good for inflammations of the eies,with the pulpe of a rosted Quince, or with crummes of F bread,and applied pultis wife.

The ashes of burnt Sothernwood,with some kinde of oyle that is of thin parts,as of *Palma Chri-* G *sti,* Raddish oyle, oyle of sweet Marjerome, or Organie, cure the pilling of the haire of the head, and make the beard to grow quickely : being strewed about the bed,or a fume made of it vpon hot embers,it driueth away serpents : if but a branch be laid vnder the beds head they say it prouoketh venerie.

The seed of Sothernwood made into pouder, or boyled in wine and drunke,is good against the H difficultie and stopping of vrine ; it expelleth,wasteth,consumeth,and digesteth all cold humours, tough slime and flegme,which do vsually stop the spleene,kidnies,and bladder.

Sothernwood drunke in wine is good against all venome and poison. I

The leaues of Sothernwood boyled in water vntill they be soft, and stamped with barley meale K and barrowes greafe vnto the forme of a plaister, dissolue and waste all cold tumors and swellings, being applied or laid thereto.

† The description herein the first place is that of the *Abrotanum fæmina arborescens* of *Dodonæus,*being the very first in his *Pemptades.* The figure which our Author put thereto was that of the Lauander Cotton,which should haue beene in the next chapter faue one : Now the figure that hee should haue put here was put two chapters before,by the name of *Absinthium arborescens,* by which name *Lobel* also calls it : but I haue thought it fitter to put it here, becaufe here was the better description, and the plant is the better referred to this kinde.

CHAP.

Chap. 455.
Of Oke of Ieruſalem, and Oke of Cappadocia.

1 *Botrys.*
Oke of Ieruſalem.

2 *Ambroſia.*
Oke of Cappadocia.

¶ *The Deſcription.*

1   OKe of Ieruſalem, or *Botrys*, hath ſundry ſmall ſtems a foot and a halfe high, diuiding themſelues into many ſmall branches, beſet with ſmall leaues deeply cut or jagged, very much reſembling the leafe of an Oke, which hath cauſed our Engliſh women to call it Oke of Ieruſalem; the vpper ſide of the leafe is of a deepe greene, and ſomewhat rough and haiſy, but vnderneath it is of a darke reddiſh or purple colour: the ſeedie floures grow cluſtering about the branches, like the yong cluſters or blowings of the Vine: the root is ſmall and threddy: the whole herbe is of a pleaſant ſmell and ſauour, and of a feint yellowiſh colour, and the whole plant dieth when the ſeed is ripe.

2   The fragrant ſmell that this kinde of *Ambroſia* or Oke of Cappadocia yeeldeth, hath moued the Poëts to ſuppoſe that this herbe was meate and food for the gods: *Dioſcorides* ſaith it groweth three handfuls high: in my garden it groweth to the height of two cubits, yeelding many weake crooked and ſtreaked branches, diuiding themſelues into ſundry other ſmall branches, hauing from the middeſt to the top thereof many moſſie yellowiſh floures not much vnlike common Worme-wood, ſtanding one before another in good order; and the whole plant is as it were couered ouer with bran or a mealy duſt: the floures doe change into ſmall prickly cornered buttons, much like vnto *Tribulus terreſtris*; wherein is contained blacke round ſeed, not vnpleaſant in taſte and ſmell: the leaues are in ſhape like the leaues of Mugwort, but thinner and more tender: all the whole plant is hoary, and yeeldeth a pleaſant ſauour: the whole plant periſhed with me at the firſt approch of Winter.

¶ *The Place.*
Theſe plants are brought vnto vs from beyond the ſeas, eſpecially from Spaine and Italy.
¶ *The Time.*
They floure in Auguſt, and the ſeed is ripe in September.

¶ *The*

¶ *The Names.*

Oke of Ieruſalem is called in Greeke Βότρυς : in Latine, *Botrys* : in Italian, *Botri* : in Spaniſh, *Bien Granada* : in high-Dutch, **Traubenkraut**, and **Grottenkraut** : in French and low-Dutch, *Pyment* : in Engliſh, Oke of Ieruſalem, and of ſome, Oke of Paradiſe.

Oke of Cappadocia is called in Greeke Κμβροσία : in Latine , *Ambroſia*, neither hath it any other knowne name. *Pliny* ſaith, *Ambroſia* is a wandering name, and is giuen vnto other herbs : for *Botrys* (Oke of Ieruſalem, as we haue written) is of diuers alſo called *Ambroſia*. In Engliſh it is called oke of Cappadocia.

¶ *The Temperature.*

Theſe plants are hot and dry in the ſecond degree, and conſiſt of ſubtill parts.

¶ *The Vertues.*

Theſe plants be good to be boiled in wine, and miniſtred to ſuch as haue their breſts ſtopt, and are ſhort winded and cannot eaſily draw their breath; for they cut and waſt groſſe humors & tough flegm. The leaues are of the ſame force, beeing made vp with ſugar they commonly call it a Conſerue.   A

It giueth a pleaſant taſte to fleſh that is ſodden with it, and eaten with the broth.   B

It is dried and laid among garments, not onely to make them ſmell ſweet, but alſo to preſerue them from moths and other vermin, which thing it doth alſo performe.   C

† There were formerly two more deſcriptions in this Chapter, both which were made by looking vpon the figures in *Lobels Icons* ; the former being of his *Ambroſis ſpontanea ſtigoſi-r*, which is nothing elſe but the *Coronopus Ruelli* or Swines Creſſes. The later was of his *Ambroſia tenuifolia*, which our Authour in the laſt Chapter ſet forth by the name of *Abrotanum campeſtre*.

CHAP. 456.   *Of Lavander Cotton.*

† *Chamæcypariſſus.*
Lavander Cotton.

¶ *The Deſcription.*

Lavander Cotton bringeth forth cluſtred buttons of a golden colour, and of a ſweet ſmel, and is often vſed in garlands and houſes. It hath a wooddy ſtocke , out of which grow forth branches like little boughes, ſlender, very many, a cubit long, ſet about with little leaues, long, narrow, purled or crumpled : on the tops of the branches ſtand vp floures , one alone on euery branch, made vp with ſhort threds thruſt cloſe together, like to the flours of Tanſie, and to the middle buttons of the floures of Cammomill, but yet ſomething broader, of colour yellow , which be changed into ſeed of an obſcure colour : the root is of a wooddy ſubſtance ; the ſhrub it ſelfe is white both in branches and leaues , and hath a ſtrong ſweet ſmell.

‡ There are ſome varieties of this plant, which *Matthiolus*, *Lobel*, and others refer to *Abrotanum fœmina*, and ſo cal it ; and by the ſame name our Author gaue the figure thereof in the laſt Chapter ſaue one, though the deſcription did not belong thereto, as I haue formerly noted. Another ſort thereof our Author, following *Tabernamontanus* and *Lobel*, ſet forth a little before by the name of *Abſinthium marinum Abrotani fœminæ facie*, which *Dodonæus* calls *Santolina prima*; and this here figured, *Santolina altera*. He alſo mentions three other differences thereof, which chiefly conſiſt in the leaues ; for his third hath very ſhort and ſmall leaues like thoſe of Heath ; whence *Bauhine* calls it *Abrotanum fœmina folijs Ericæ*. The fourth hath the leaues leſſe toothed, and more like Cypres, whence it is called in the *Aduerſ. Abrotanum peregrinum cupreſſi folijs*. The fift hath not the ſtalkes

growing vpright,but creeping;the leaues are toothed,more thicke and hoary than the reſt,in other reſpects alike. *Banhine* calls it *Abrotanum fœmina repens caneſcens.*

¶ *The Place.*

Lavander Cotton groweth in gardens almoſt euerie where.

¶ *The Time.*

They floure in Iuly and Auguſt.

¶ *The Names.*

They are called by one name *Santolina,*or Lavander Cotton : of moſt, *Chamæcypariſſus :* but *Pliny* concerning *Chamæcypariſſus* is ſo ſhort and brief,that by him their opinions can neither be rejected nor receiued.

They are doubtleſſe much deceiued that would haue Lavander Cotton to be *Abrotanum fœmina,* or the female Sothernwood;and likewiſe they are in the wrong who take it be *Seriphium,*ſea worm-wood ; and they who firſt ſet it abroch to be a kinde of Sothernwood,we leaue to their errours,be-cauſe it is not abſolutely to be referred to one,but a plant participating both of Wormwood and Sothernwood.

¶ *The Nature.*

The ſeed of Lavander Cotton hath a bitter taſte,being hot and dry in the third degree.

¶ *The Verthes.*

A   *Pliny* ſaith,That the herb *Chamæcypariſſus* being drunke in wine is a good medicine againſt the poiſon of all ſerpents and venomous beaſts.

B   It killeth wormes either giuen greene or dry,and the ſeed hath the ſame vertue againſt wormes, but avoideth them with greater force. It is thought to be equall with the vſuall wormſeed.

† The figure which formerly was in this place was a kinde of Moſſe,which *Tragus* ſet forth by the name of *Sauina ſylueſtris* : *Turner* and *Tabernæmontanus* called it *Chamæcypariſſus.* See more thereof in the Moſſes.

CHAP. 457. *Of Sperage or Aſparagus.*

1 *Aſparagus ſatiuus.*
Garden Sperage.

2 *Aſharagus petraus.*
Stone or Mountaine Sperage.

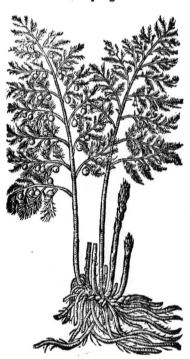 

¶ *The Description.*

1　THe first being the manured or garden Sperage,hath at his first rising out of the ground thicke tender shoots very soft and brittle, of the thicknesse of the greatest swans quill, in taste like the green bean, hauing at the top a certaine scaly soft bud , which in time groweth to a branch of the height of two cubits, diuided into diuers other smaller branches, wheron are set many little leaues like haires, more fine than the leaues of Dill : amongst which come forth small mossie yellowish floures which yeeld forth the fruit, green at the first, afterward as red as Corall, of the bignesse of a small pease ; wherein is contained grosse blackish seed exceeding hard,which is the cause that it lieth so long in the ground after his sowing, before it spring vp: the roots are many thicke soft and spongie strings hanging downe from one head, and spred themselues all about, whereby it greatly increaseth.

2　We haue in our marish and low grounds neere vnto the sea, a Sperage of this kinde which differeth a little from that of the garden,and yet in kinde there is no difference at all. but onely in manuring, by which all things or most things are made more beautifull and larger. This may bee called *Asparagus palustris,* marish Sperage.

4　*Asparagus sylvestris aculeatus.*　　　　5　*Asparagus sylvestris spinosus Clusij.*
Wilde prickly Sperage.　　　　　　　　　　Wilde thorny Sperage.

3　Stone or mountain Sperage is one of the wild ones, set forth vnder the title of *Corruda,* which *Lobel* calls *Asparagus petræus* ; and *Galen, Myacanthinus,* that doth very well resemble those of the garden in stalks, roots, and branches, sauing that those fine hairy leaues which are in the garden Sperage are soft, blunt, and tender ; and in this wild Sperage, sharp hard and pricking thornes, though they be small and slender : the root hereof is round, of the bignesse of a pease, and of a black color : the roots are long, thicke, fat, and very many.

4　This fourth kind differeth from the last described, being a wild Sperage of Spain and Hungarie : the plant is altogether set with sharpe thornes, three or foure comming forth together, as are the branches of Whinnes, Goes, and Furfen : the fruit is black when it is ripe, and full of a greenish pulpe, wherein lie hard and blacke seeds, sometimes one, otherwhiles two in a berry . the roots are like the others, but greater and tougher.

‡ *6 Drypis.*
Sperage Thiſtle.

5   *Carolus Cluſius* deſcribes alſo a certain wild Sperage with ſharp prickles all alongſt the ſtalks, orderly placed at euery joint one, hard, ſtiffe, and whitiſh, the points of the thornes pointing downward: from the which joints alſo doe grow out a few long greene leaues faſtened together, as alſo a little yellow floure and one berry three cornered, of a blacke colour, wherein is contained one black ſeed, ſeldom more: the roots are like the other.

6   *Drypis* beeing likewiſe a kinde hereof, hath long and ſmall roots creeping in the ground like Couch-graſſe; from which ſpring vp branches a cubit high, ful of knotty joints: the leaues are ſmal like Iuniper, not much differing from *Corruda* or *Nepa*: the floures grow at the top of the ſtalke in ſpoky tufts or roundles, of a white colour, cloſely thruſt together: the ſeed before it is taken out of the huske is like vnto Rice; being taken out, like that of Melilot, of a ſaffron colour.

¶ *The Place.*

The firſt being our garden Aſparagus, groweth wilde in Eſſex, in a medow neere to a mill, beyond a village called Thorp; and alſo at Singleton not far from Carby, and in the medowes neere Moulton in Lincolnſhire. Likewiſe it growes in great plenty neere Harwich, at a place called Bandamar lading, and at North Moulton in Holland a part of Lincolnſhire.

The wilde Sperages grow in Portugal and Biſcay amongſt ſtones; one of which, *Petrus Bellonius* maketh mention to grow in Candy, *lib.1. cap.*18. of his Singularities.

¶ *The Time.*

The bare naked tender ſhoots of Sperage ſpring vp in Aprill, at what time they are eaten in ſallads; they floure in Iune and Iuly, the fruit is ripe in September.

¶ *The Names.*

The garden Sperage is called in Greeke Ἀσπάραγος, in Latine likewiſe *Aſparagus*: in ſhops, *Sparagus* and *Speragus*: in high-Dutch, **Spargen**: in low-Dutch, **Aſparges**, and **Coralcruyt**, that is to ſay, *Herba Coralli*, or Coral-wort, of the red berries, which beare the colour of Corall: in Spaniſh, *Aſparragos*: in Italian, *Aſparago*: in Engliſh, Sperage, and likewiſe Aſparagus after the Latine name: in French, *Aſperges*. It is named Aſparagus, of the excellency, becauſe *aſparagi*, or the ſprings hereof are preferred before thoſe of other plants whatſoeuer: for this Latine word *Aſparagus* doth properly ſignifie the firſt ſpring or ſprout of euery plant, eſpecially when it is tender, and before it do grow into an hard ſtalk, as are the buds, tendrels, or yong ſprings of wild Vine or hops, and ſuch like.

Wild Sperage is properly called in Greeke μυάκανθα, which is as much to ſay as Mouſe-prickle; and Ἀσπάραγος πετραῖος, that is to ſay, *Petræus Aſparagus*, or ſtone Sperage: it is alſo named in Latine *Aſparagus ſylveſtris*, or *Corruda*.

¶ *The Temperature.*

The roots of the garden and wilde Sperage do clenſe without manifeſt heate and drineſſe.

¶ *The Vertues.*

A   The firſt ſprouts or naked tender ſhoots hereof be oftentimes ſodden in fleſh broth and eaten; or boiled in faire water, and ſeaſoned with oile, vineger, ſalt, and pepper, then are ſerued vp as a ſallad: they are pleaſant to the taſte, eaſily concocted, and gently looſe the belly.

B   They ſomwhat prouoke vrin, are good for the kidnies and bladder, but they yeeld little nouriſhment to the body, and the ſame moiſt, yet not faulty: they are thought to increaſe ſeed and ſtir vp luſt.

† The *Nepa* formerly mentioned in this Chapter, but now omitted, was againe ſet forth by our Author among the Fulſes, where you may finde it.

CHAP. 458. *Of Horſe-taile or Shaue-graſſe.*

¶ *The Deſcription.*

1 GReat Horſe-taile riſeth vp with a round ſtalke hollow within like a reed, a cubit high, compact as it were of many ſmall pieces one put into the end of another, ſomtimes of a reddiſh colour, very rough, and ſet at euery joint with many ſtiffe Ruſh-like leaues, or rough briſtles, which maketh the whole plant to reſemble the taile of a horſe, whereof it tooke his name : on the top of the ſtalke do ſtand in ſtead of floures cluſtered and thicke catkins, not vnlike to the firſt ſhoots of Sperage, which is called *Myacantha :* the root is jointed , and creepeth in the ground.

2 This ſmall and naked Shaue-graſſe, wherewith Fletchers and Combe-makers doe rub and poliſh their worke, riſeth out of the ground like the firſt ſhoots of Aſparagus, jointed or kneed by certain diſtances like the precedent , but altogether without ſuch briſtly leaues, yet exceeding rough and cutting : the root groweth aſlope in the earth like thoſe of the Couch-graſſe.

1 *Equiſetum majus.*   2 *Equiſetum nudum.*
Great Horſe-taile.   Naked Horſe-taile.

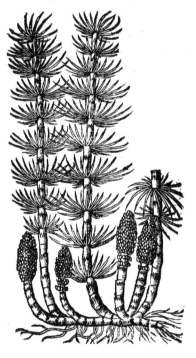 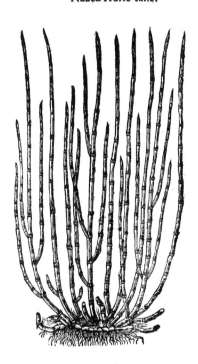

3 Horſe-taile which for the moſt part groweth among corn and where corn hath been, hath a very ſlender root and ſingle ; from which riſe vp diuers jointed ſtalks, whereon doe grow very long rough narrow jointed leaues, like vnto the firſt deſcribed, but thicker and rougher, as is the reſt of the plant.

4 Water Horſe-taile, that growes by the brinks of riuers and running ſtreams, and often in the middeſt of the water, hath a very long root according to the depth of the water, groſſe, thicke, and jointed, with ſome threds anexed thereto : from which riſeth vp a great thicke jointed ſtalk whereon grow long rough ruſhy leaues pyramide or ſteeple-faſhion. The whole plant is alſo tough, hard and fit to ſhaue and rub wooden things as the other.

5 This kinde of Horſe-taile that growes in woods and ſhadowie places, hath a ſmall root and ſingle, from which riſeth vp a rough chamfered ſtalke jointed by certaine ſpaces , hauing at each joint two buſhes of rough briſtly leaues, ſet one againſt another like the other of his kinde.

3 *Equisetum segetale.*
Corne Horse-taile.

4 *Equisetum palustre.*
Water Horse-taile.

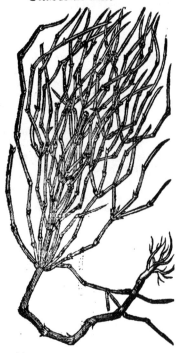

5 *Equisetum sylvaticum.*
Wood Horse-taile.

6 *Cauda equina fæmina.*
Female Horse-taile.

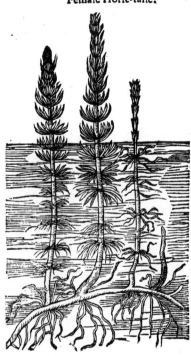

9 *Iuncaria Salmanticenſis*.
Italian ruſhy Horſe-taile.

6　The female Horſe-taile groweth for the moſt part in wateriſh places,and by the brinks of ſmall rills and pirling brooks : it hath a long root like that of Couch-graſſe,from which riſe vp diuers hollow ſtalks,ſet about at certaine diſtances with ſmall leaues in rundles like thoſe of Woodroofe,altogether barren of ſeed and floure, wherof it was called by *Lobel* , *Polygonon fœmina ſemine Vidua*. ‡ This is ſometimes found with ten or more ſeeds at each joint ; whence *Bauhine* hath called it *Equiſetum paluſtre brevioribus folijs polyſpermon*. ‡

‡ 7　In ſome boggy places of this kingdom is found a rare and pretty *Hippuris* or Horſe-taile, which growes vp with many little branches ſome two or three inches high, putting forth at each joint many little leaues, cluſtering cloſe about the ſtalke,and ſet after the maner of other Horſe-tailes:towards the top of the branches the joints are very thicke : the colour of the whole plant is gray,a little inclining to greene,very brittle, and as it were ſtony or grauelly like Coralline , and will craſh vnder your feet as if it were frozen;and if you chew it you ſhall finde it all ſtony and grauelly. My friend M ͬ *Leonard Buckner* was the firſt that found this plant and brought it to me : hee had it three miles beyond Oxford,a little on this ſide Euenſham ferry,in a bog vpon a common by the Beacon hill neere Cumner wood , in the end of Auguſt, 1632. M ͬ *Bowles* hath ſince found it growing vpon a bog not far from Chiſelhurſt in Kent. I queſtion whether this be not the *Hippuris lacuſtris quædam folijs manſu arenoſis* of *Geſner*. But if *Geſners* be that which *Bauhine* in his *Prodromus*, *pag.*24. ſets forth by the name of *Equiſetum nudum minus variegatum*, then I iudge it not to be this of my deſcription : for *Bauhines* differs from this,in that it is without leaues,and oft times bigger: the ſtalks of his are hollow,theſe not ſo. This may be called *Hippuris Coralloides*,Horſe-taile Coralline.

8　Toward the later end of the yere,in diuers ditches,as in S.*Iames* his Parke, in the ditches on the backe of Southwarke toward S.*Georges* fields, &c. you may finde couered ouer with water a kinde of ſtinking Horſe-taile ; it growes ſometimes a yard long, with many joints and branches, and each joint ſet with leaues,as in the other Horſe-tailes,but they are ſomewhat jagged or diuided towards the tops. I take this to be the *Equiſetum fœtidum ſub aqua repens*, deſcribed in the fifth place of *Bauhines Prodromus*. We may call in Engliſh,ſtinking water Horſe-taile. ‡

9　*Cluſius* hath ſet forth a plant that he referreth vnto the ſtocke of Horſe-tailes,which he thus deſcribeth ; It hath many twiggy or ruſhy ſtalks,whereupon it was called *Iuncaria*,and may be Engliſhed,Ruſh weed : the leaues grow vpon the branches like thoſe of flax;on the tops of the ſtalks grow ſmall chaffie floures of a whitiſh colour. The ſeed is ſmall and blacke of colour, the root little and white : the whole plant is ſweetiſh in taſte.

10　*Dodonæus* ſets forth another Horſe-taile,which he called climing Horſe-tail, or Horſe-tail of Olympus. There is(ſaith he)another plant like Horſe-taile,but greater and higher ; it riſeth vp oftentimes with a ſtalke as big as a mans arme, diuided into many branches, out of which there grow long ſlender ſprigs very full of joints like to the firſt Horſe-taile : the floures ſtand about the joints,of a moſſie ſubſtance,ſmall as are thoſe of the Cornel tree ; in places whereof grow vp red fruit full of ſoure juice,not vnlike to little Mulberries,in which is the ſeed. The root is hard and wooddy. This growes now and then to a great height,and ſometimes lower. *Bellonius* writes in his Singularities,That it hath bin ſeen to be equall in height with the Plane tree: it comes vp lower neere to ſhorter and leſſer trees and ſhrubs,yet doth it not faſten it ſelfe to the trees with any tendrels or claſping aglets,much leſſe doth it winde it ſelfe about them, yet doth it delight to ſtand neere and cloſe vnto them.

¶ *The*

¶ *The Place.*

The titles and deſcriptions ſhew the places of their growing. The laſt *Bellonius* reports to grow in diuers vallies of the mountaine Olympus, and not far from Raguſa a city in Sclavonia.

¶ *The Time.*

They floure from Aprill to the end of Summer.

¶ *The Names.*

Horſe-taile is called in Greeke Ἵππουϱις, *Hippuris* : in Latine, *Equiſetum*, and *Equinalis* : of *Pliny, lib.15. cap.28. Equiſetis,* of the likeneſſe of a horſe haire : of ſome, *Salix equina* : in ſhops, *Cauda E-quina* : in high-Dutch, **Schaffthew:** in low-Dutch, **Peerttrert :** in Italian, *Coda di Cavallo* : in Spa-niſh, *Coda de mula* : in French, *Queve de Cheval,* and *Caqueve* : in Engliſh, Horſe-taile, Shave-graſſe, and in ſome places, Ioynts.

Shaue-graſſe is not without cauſe named *Aſprella,* of his ruggedneſſe, which is not vnknowne to women, who ſcoure their pewter and wodden things of the kitchen therewith, which the Germane women call **kannenkraut :** and therefore ſome of our huſwiues do call it Pewter-wort. Of ſome the tenth is called *Ephedra, Anobaſis,* and *Caucon.*

¶ *The Temperature.*

Horſe-taile (as *Galen* ſaith) hath a binding facultie with ſome bitterneſſe, wherefore it mightily drieth, and that without biting.

¶ *The Vertues.*

A　　*Dioſcorides* ſaith, that Horſe-taile being ſtamped and laid to, doth perfectly cure wounds, yea al-though the ſinues be cut in ſunder, as *Galen* addeth. It is of ſo great and ſingular vertue in healing wounds, as that it is thought and reported for truth, to cure the wounds of the bladder and other bowels, and helpeth ruptures and burſtings.

B　　The herb drunke either with water or wine, is an excellent remedy againſt bleeding at the noſe, and other fluxes of bloud: it ſtayeth the ouermuch flowing of womens floures, the bloudy flix, and other fluxes of the belly.

C　　The juice of the herb taken in the ſame manner doth the like, and more effectually.

D　　Horſe-taile with his roots boiled in wine is very profitable for the vlcers of the kidnies & blad-der, the cough, and difficultie of breathing.

Chap. 459.　*Of Sea Cluſter or Sea Raiſon.*

† 1 *Vua marina minor.*
The ſmall ſea Grape.

¶ *The Deſcription.*

1　SMall ſea Grape is not vnlike to Horſe-taile; it bringeth forth ſlender ſtalks almoſt like Ruſhes, ſet with many lit-tle joints, ſuch as thoſe are of the Horſ-taile, and diuided into many wings or branches, the tops whereof are ſharpe pointed, ſomewhat hard and pricking. It is without leaues : the floures grow in cluſters out of the joints, with little ſtemmes, they are ſmall, and of a whitiſh green colour : The fruit conſiſteth of many little pearles, like to the vnripe berries of Raſpis or Hind-berry: when it is ripe it is red, with a ſaffron colour, in taſt ſweet and pleaſant : the ſeed or kernell is hard, three ſquare, ſharp on euery ſide, in taſte binding: the root is jointed, long, and creeps aſlope : the plant it ſelfe alſo doth rather lie on the ground, than ſtand vp. It growes all full of ſmall ſtalks and branches, ca-ſting themſelues all abroad.

2　*Carolus Cluſius* hath ſet forth another ſort of ſea Grape, far different from the precedent. It riſeth vp to the height of a man, hauing many branches of a wooddy ſubſtance, in forme like to Spaniſh Broom, without any leaues at all; where-upon doe grow cluſters of floures vpon ſlender foot-ſtalks of a yellowiſh moſſie or herby colour, like thoſe of the Cornell tree : after which come the fruit like vnto the Mulberry, of a reddiſh co-lour and ſowre taſte, wherein lieth hid one or two

ſeeds

feeds like thoſe of Mullet,blacke without and white within : the root is hard,tough,and wooddy.

2 *Vva marina major.*
Great ſhrubby ſea Grape.

3 *Tragos Matthioli.*
Baſtard ſea Grape.

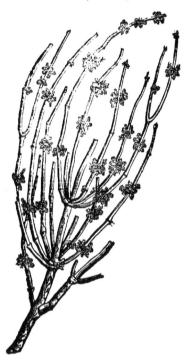

3 *Tragon Matthioli*,or rather *Tragos improbus Matthioli*,which hee vnaduiſedly called *Tragon*, is without controuerſie nothing elſe but a kinde of *Kali* : this plant riſeth vp out of the ground with ſtalks ſeldom a cubit high,diuided into ſundry other groſſe thicke and writhen branches,ſet or armed with many pricking leaues of the colour and ſhape of *Aizoon*,and ſomwhat thick and fleſhy : among which come forth ſuch prickly buds as are to be ſeen in *Tribulus terreſtris*, as that it is hard for a man to touch any part thereof without pricking his fingers : the flours are of an herby colour, bringing forth flat ſeed like vnto *Kali* : the root is ſlender,and ſpredeth vnder the turf of the earth : the whole plant is full of clammy juice,not any thing aſtringent,ſomwhat ſaltiſh,and of no ſingular vertue as yet knowne : wherefore I may conclude,that this cannot be *Tragos Dioſcoridis*,and the rather,for that this *Tragos* of *Matthiolus* is an herb,and not a ſhrub, as I haue before ſpoken in *Vva marina*,neither beareth it any berries or grains like wheat,neither is it pleaſant in taſte or ſmell, or any thing aſtringent,all which are to be found in the right *Tragos* before expreſſed ; which(as *Dioſcorides* ſaith)is without leaues,neither is it thorny,as *Tragos improbus Matthioli* is:this plant I haue found growing in the Iſle of Shepey,in the tract leading to the houſe of Sʳ *Edward Hobby*,called Sherland.

¶ *The Place.*

It loueth to grow vpon dry banks and ſandy places neere the ſea : it is found in Languedoc not far from Montpelier,and in other places by the ſea ſide,but is a ſtranger in England.

¶ *The Time.*

When it growes naturally the fruit is ripe in Autumne : the plant it ſelfe remaines long green, for all the cold in winter.

¶ *The Names.*

It is called of the later herbariſts *Vva marina*: in French, *Raiſin de mer*,of the pearled fruit,and the likeneſſe it hath with the Raſpis berry, which is as it were a Raiſin or Grape, conſiſting of many little ones : it is named in Greeke τράγ, but is not called *Tragus* or *Tragos* of a Goat, (for ſo ſignifieth the Greeke word) or of his ranke and rammiſh ſmell, but becauſe it brings forth fruit fit to be
eaten;

eaten ; of the Verb τρωγω, which ſignifies to eat : it may be called Scorpion, becauſe the ſprigs therof are ſharp pointed like a Scorpions taile.

¶ *The Temperature.*

The berries or Raiſins, and eſpecially the ſeed that is in them, haue a binding qualitie, as wee
haue ſaid, and are dry in the later end of the ſecond degree.

¶ *The Vertues.*

A   *Dioſcorides* writeth, That the Raiſins of ſea grape ſtay the flix, and alſo the Whites in women
when they much abound.

† Our Author, as you ſee, gaue the hiſtorie of the leſſer in the firſt place, but formerly the figure was in the third place, and another figure of the ſame in the
ſecond place, and the figure of the greater was in the firſt place.

<h2 style="text-align:center">CHAP. 460.   Of Madder.</h2>

¶ The Kindes.

THere is but one kinde of Madder only which is manured or ſet for vſe, but if all thoſe that are
like it in leaues and manner of growing were referred thereto, there ſhould be many ſorts, as
Gooſe-graſſe, ſoft Cliuer, our Ladies Bedſtraw, Woodrooſe, and Croſſe-wort; all which are like to
Madder in leaues, and therefore thought to be wilde kindes thereof.

1 *Rubia tinctorum.*
Red Madder.

2 *Rubia ſylveſtris.*
Wilde Madder.

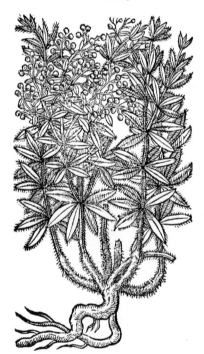

¶ The Deſcription.

1   THe garden or mannured Madder hath long ſtalkes or trailing branches diſperſed farre
abroad vpon the ground, ſquare, rough, and full of joints, at euery joint ſet round with
greene rough leaues in manner of a ſtarre, or like thoſe of Woodrooſe : the floures grow at the top
of

of the branches, of a feint yellow colour: after which come the ſeed, round, green at the firſt, afterward red, and laſtly of a black colour: the root, long, fat, ful of ſubſtance, creepeth far abroad within the vppermoſt cruſt of the earth, and is of a reddiſh colour when it is green and freſh.

2　Wilde Madder is like in forme to that of the garden, but altogether ſmaller, and the leaues are not ſo rough, but ſmooth and ſhining: the floures are white: the root is very ſmall and tender and oftentimes of a reddiſh colour.

‡ Rubia marina　　‡ 4 Rubia ſpicata Cretica.
Sea Madder.　　　　Small Candy Madder.

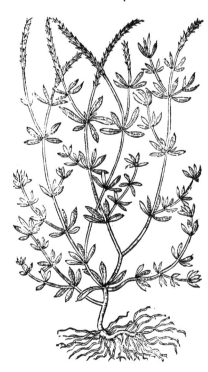

3　Sea Madder hath a root two foot long, with many dry threds hanging thereat, of a reddiſh colour like Alkanet, on the out ſide of the ſame forme and bigneſſe, but within of the colour of the ſcrapings of Iuniper or cedar wood, ſending forth diuers ſlender ſtalks round and ful of joints; from which come forth ſmall thin leaues ſtiffe and ſharp pointed, ſomewhat hairy, in number commonly foure, ſtanding like a Burgonian Croſſe: from the boſome of which come forth certaine tufts of ſmaller leaues thruſt together vpon a heape: the floures grow at the top of the ſtalkes, of a pale yellowiſh colour.

‡ Rubia ſpicata Cretica Cluſij.

4　This hath proceeding from the root many knotty foure ſquare rough little ſtalkes, a foot high, diuided immediatly from the root into many branches, hauing but one ſide branch growing forth of one joint; about which joints grow ſpred abroad, foure or fiue, ſomtimes ſix narrow ſhort ſharpe pointed leaues, ſomewhat rough: the tops of the ſtalks and branches are nothing but long ſmall foure ſquare ſpikes or eares, made of three leafed green huſks: out of the top of each huſke groweth a very ſmall greeniſh yellow floure, hauing foure exceeding ſmal leaues ſcarce to be ſeen: after which followeth in each huſke one ſmall blackiſh ſeed, ſomwhat long, round on the one ſide, with a dent or hollowneſſe on the other. The root is ſmal, hard, wooddy, crooked or ſcragged, with many little branches or threds, red without, and white within, and periſheth when the ſeed is ripe. Iuly 19. 1621.

Synanchica

Synanchica Lug. p.1185.

5   The root is crooked,blackiſh without,yellow vnderneath the skinne,white within that and wooddy, about fiue or ſix inches long, with many hairy ſtrings : from the root ariſe many foure-ſquare branches trailing vpon the ground,ſometimes reddiſh toward the root.The leaues are ſmal and ſharp pointed like thoſe of *Gallium,* and grow along the ſtalke on certain knees or ioints,foure or fiue together, ſometimes fewer : from thoſe ioints the ſtalke diuideth it ſelfe towards the top into many parts,wheron grow many floures, each floure hauing foure leaues,ſomtimes white,ſomtimes of a fleſh colour, and euerie leafe of theſe fleſh coloured leaues is artificially ſtraked in the middle and neere the ſides with three lines of a deeper red,of no pleaſant ſmel : after which comẽs the ſeed,ſomething round,growing two together like ſtones : it floureth all the Summer long, and groweth in dry chalky grounds aboundantly. *Auguſt* 13. 1619. *Iohn Goodyer.* ‡

‡ *6 Rubia minima.*
Dwarfe Madder.

‡   6   *Lobel* thus deſcribes this dwarf Madder : There is another (ſaith he) which I gathered,growing vpon Saint Vincents rocks,not far from Briſtoll : the leaues are of the bigneſſe of thoſe of Rupture-woort , ſharpe pointed , and growing after the manner of thoſe of Madder, vpon little creeping ſtalkes ſome inch and halfe high,wherẽon grow yellowiſh ſmall flours.The root is ſmall,and of the colour of Corall. ‡

¶ *The Place.*

Madder is planted in gardens , and is verie common in moſt places of England. Mʳ *George Bowles* found it growing wilde on Saint *Vincents* rock,and out of the cliffes of the rocks of Aberdovie in Merioneth ſhire.

The ſecond groweth in moiſt medowes in mooriſh grounds,and vnder buſhes almoſt euerie where.

3   This growes by the ſea ſide in moſt places.

‡   The fourth growes only in ſome few gardens with vs,but the ſift may be found wilde in many places. I found it in great plenty on the hil beyond Chattam in the way to Canterbury ‡

¶ *The Time.*

They flouriſh from May vnto the end of Auguſt : the roots are gathered and dried in Autumne, and ſold to the vſe of Diers and medicine.

¶ *The Names.*

Madder is called in Greeke ἐρυθρόδανον, *Erythrodanum :* in Latine,*Rubia,*and *Rubcia :* in ſhops, *Rubia tinctorum : Paulus Ægineta* ſheweth that it is named *Thapſon* which the Diers vſe,and the Romanes call it *Herba rubia :* in Italian, *Rubbia,*and *Robbia :* in Spaniſh, *Ruvia,Roya,*and *Granza :* in French, *Garance :* in high-Dutch, **Rotte :** in low-Dutch, **Mee,** and **Mee Crappen :** in Engliſh, Madder, and red Madder.

¶ *The Temperature.*

Of the temperature of Madder it hath been diſputed among the Learned, and as yet not cenſured,whether it binde or open:ſome ſay both:diuers diuerſly deem.A great Phyſition (I do not ſay the great learned) called me to account as touching the faculties hereof,although he had no commiſſion ſo to doe : notwithſtanding I was content to be examined vpon the point,what the nature of Madder was,becauſe I haue written that it performeth contrarie effects, as ſhall be ſhewed : the roots of Madder,which both the Phyſitians and Diers do vſe,as they haue an obſcure binding

power

power and force; ſo be they likewiſe of nature and temperature cold and dry: they are withall of diuers thin parts, by reaſon whereof their colour doth eaſily pierce: yet haue they at the firſt a certaine little ſweetneſſe, with an harſh binding quality preſently following it; which not onely wee our ſelues haue obſerued, but alſo *Avicen* the prince of phyſitions, who in his 58. chapter hath written, that the root of Madder hath a rough and harſh taſte: now M^r. Doctor, whether it binde or open I haue anſwered, attending your cenſure: but if I haue erred, it is not with the multitude, but with thoſe of the beſt and beſt learned.

¶ *The Vertues.*

The decoction of the root of Madder is euery where commended for thoſe that are burſten, bruſed, wounded, and that are fallen from high places.   A

It ſtencheth bleeding, mitigateth inflammations, and helpeth thoſe parts that bee hurt and bruſed.   B

For theſe cauſes they be mixed with potions, which the later Phyſitians call wound drinkes: in which there is ſuch force and vertue, as *Matthiolus* alſo reporteth, that there is likewiſe great hope of curing of deadly wounds in the cheſt and intrails.   C

Our opinion and iudgement is confirmed by that moſt expert man, ſometimes Phyſition of Louaine, *Iohannes Spiringus*, who in his *Rapſodes* hath noted, that the decoction of Madder giuen with *Triphera*, that great compoſition is ſingular good to ſtay the reds, the hæmorrhoids and bloudy flix, and the ſame approued by diuers experiments: which confirmeth Madder to be of an aſtringent and binding qualitie.   D

Of the ſame opinion as it ſeemeth is alſo *Eros Iulia* her freed man (commonly called *Trotula*) who in a compoſition againſt vntimely birth doth vſe the ſame: for if he had thought that Madder were of ſuch a quality as *Dioſcorides* writeth it to be of, hee would not in any wiſe haue added it to thoſe medicines which are good againſt an vntimely birth.   E

For *Dioſcorides* reporteth, that the root of Madder doth plentifully prouoke vrine, and that groſſe and thicke, and oftentimes bloud alſo, and it is ſo great an opener, that being but onely applied, it bringeth downe the menſes, the birth, and after-birth: but the extreme redneſſe of the vrine deceiued him, that immediately followeth the taking of Madder, which rednes came as he thought from bloud mixed therewith, which notwithſtanding commeth no otherwiſe than from the colour of the Madder.   F

For the root hereof taken any manner of way doth by and by make the vrine extreme red: no otherwiſe than Rubarb doth make the ſame yellow, not changing in the meane time the ſubſtance thereof, nor making it thicker than it was before, which is to bee vnderſtood in thoſe which are in perfect health, which thing doth rather ſhew that it doth not open, but binde, no otherwiſe than Rubarbe doth: for by reaſon of his binding quality the wateriſh humors do for a while keep their colour. For colours mixed with binding things do longer remaine in the things coloured, and do not ſo ſoone vade: this thing they will know that gather colours out of the juyces of floures and herbes, for with them they mixe allume, to the end that the colour may be retained and kept the longer, which otherwiſe would bee quickely loſt. By theſe things it manifeſtly appeareth that Madder doth nothing vehemently either clenſe or open, and that *Dioſcorides* hath raſhly attributed vnto it this kinde of quality, and after him *Galen* and the reſt that followed, ſtanding ſtiffely to his opinion.   G

Pliny ſaith, that the ſtalkes with the leaues of Madder, are vſed againſt ſerpents.   H

The root of Madder boyled in Meade or honied water, and drunken, openeth the ſtopping of the liuer, the milt and kidnies, and is good againſt the jaundiſe.   I

The ſame taken in like manner prouoketh vrine vehemently, inſomuch that the often vſe thereof cauſeth one to piſſe bloud, as ſome haue dreamed.   K

Langius and other excellent Phyſitions haue experimented the ſame to amend the lothſome colour of the Kings-euill, and it helpeth the vlcers of the mouth, if vnto the decoction be added a little allume and hony of Roſes.   L

‡ 5   The fifth being the *Synanchica* of *Daleſchampius*, dries without biting, and it is excellent againſt ſquinancies, either taken inwardly, or applied outwardly, for which cauſe they haue called it *Synanchica, Hiſt. Lugd.* ‡   M

C H A P. 461.   *Of Gooſe-graſſe, or Cliuers.*

¶ *The Deſcription.*

1   A *Parine*, Cliuers or Gooſe-greaſe, hath many ſmall ſquare branches, rough and ſharpe, full of joynts, beſet at euery joint with ſmall leaues ſtar faſhion, and like vnto ſmall Madder:

the floures are very little and white, pearking on the tops of the ſprigs : the ſeeds are ſmall, round, a little hollow in the middeſt in manner of a nauell, ſet for the moſt part by couples : the roots ſlender and full of ſtrings : the whole plant is rough, and his ruggedneſſe taketh hold of mens veſtures and woollen garments as they paſſe by : being drawne along the tongue it fetcheth bloud : *Dioſcorides* reports, that the ſhepheards in ſtead of a Cullender do vſe it to take haires out of milke, if any remaine therein.

2   The great Gooſe-graſſe of *Pliny* is one of the Moone-worts of *Lobel*, it hath a very rough tender ſtalke, whereupon are ſet broad leaues ſomewhat long, like thoſe of Scorpion graſſe, or *Alyſſon Galeni*, *Galens* Moonewort, very rough and hairy, which grow not about the joynts, but three or foure together on one ſide of the ſtalke : the floures grow at the top of the branches, of a blew colour : after which commeth rough cleauing ſeed, that doth ſtick to mens garments which touch it : the root is ſmall and ſingle.

<table>
<tr><td>1 Aparine.
Gooſe-graſſe, or Cleuers.</td><td>2 Aparine major Plinij.
Great Gooſe-graſſe.</td></tr>
</table>

¶ *The Place.*

Gooſe-graſſe groweth neer the borders of fields, and oftentimes in the fields themſelues mixed with the corne, alſo by common waies, ditches, hedges, and among thornes : *Theophraſtus* and *Galen* write, that it groweth among Lentles, and with hard embracing it doth choke it, and by that meanes is burdenſome and troubleſome vnto it.

¶ *The Time.*

It is found plentifully euery where in Summer time.

¶ *The Names.*

It is named in Greeke, ἀφαρίνη : *Aparine* : in Latine, *Lappa minor*, but not properly : *Pliny* affirmeth it to be *Lappaginis ſpeciem* : of ſome, *Philanthropos*, as though he ſhould ſay a mans friend, becauſe it taketh hold of mens garments ; of diuers alſo for the ſame cauſe, *Philadelphos* : in Italian, *Speronella* : in Spaniſh, *Preſera*, or *amor di Hortalano* : in high Dutch, **Kleeb kraut** : in French, *Reble, ou Grateron* : in low Dutch, **Kleefcruyt** : in Engliſh, Gooſe-ſhare, Gooſe-graſſe, Cleuer, or Clauer.

¶ *The Temperature.*

It is, ſaith *Galen*, moderately hot and dry, and ſomewhat of thin parts.

¶ *The*

¶ *The Vertues.*

The juyce which is preſſed out of the ſeeds, ſtalkes, and leaues, as *Dioſcorides* writeth, is a remedie A
for them that are bitten of the poiſonſome ſpiders called in Latine *Phalangia*, and of vipers, if it be
drunke with wine.

And the herbe ſtamped with ſwines greaſe waſteth away the kernels by the throte. B

Pliny teacheth that the leaues being applied doe alſo ſtay the abundance of bloud iſſuing out of C
wounds.

Women do vſually make pottage of Cleuers with a little mutton and Otemeale, to cauſe lank- D
neſſe, and keepe them from fatneſſe.

CHAP. 462. *Of Croſſe-wort.*

¶ *The Deſcription.*

1   CRoſſe-wort is a low and baſe herbe, of a pale greene colour, hauing many ſquare feeble
rough ſtalks full of joynts or knees, couered ouer with a ſoft downe: the leaues are little,
ſhort, & ſmal, alwaies foure growing together, and ſtanding croſſewiſe one right againſt
another, making a right Burgunion croſſe: toward the top of the ſtalke, and from the boſome of
thoſe leaues come forth very many ſmall yellow floures, of a reaſonable good ſauour, each of which
is alſo ſhaped like a Burgunion croſſe: the roots are nothing elſe but a few ſmall threds or fibres.

1 *Cruciata.*
Croſſe-wort.

‡ 2 *Rubia Cruciata læuis.*
Croſſe-wort Madder.

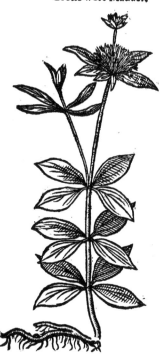

‡ 2   This in mine opinion may be placed here as fitly as any where elſe, for it hath the leaues
ſtanding croſſe-waies foure at a joynt, ſomewhat like thoſe of the largeſt Chickeweed: the ſtalkes
are betweene a foot and a halfe and two cubits high. The white Starre-faſhioned floures ſtand in
roundles about the tops of the ſtalkes. It growes plentifully in Piemont, on the hills not farre from
Turine. *Lobel* ſets it forth by the name of *Rubia Læuis Taurinenſium.* ‡

¶ *The Place.*

Cruciata,or Crossewort,groweth in moist and fertile medowes ; I found the same growing in the Churchyatd of Hampstead neere London, and in a pasture adjoyning thereto, by the mill : also it groweth in the Lane or high way beyond Charlton,a small village by Greenewich, and in sundry other places.

¶ *The Time.*

It floureth for the most part all Summer long.

¶ *The Names.*

It is called *Cruciata*,and *Cruciatis*,of the placing of the leaues in manner of a Crosse : in English, Crosse-wort,or Golden Mugweet.

¶ *The Temperature.*

Crosse-wort seemeth to be of a binding and dry quality.

¶ *The Vertues.*

A   Crosse-wort hath an excellent property to heale, joyne, and close wounds together,yea it is very fit for them,whether they be inward or outward, if the said herbe be boyled in wine and drunke.

B   The decoction thereof is also ministred with good successe to those that are bursten : and so is the herbe,being boyled vntill it be soft,and laied vpon the bursten place in manner of a pultis.

CHAP. 463. *Of Wodrooffe.*

1 *Asperula.*
Woodrooffe.

‡ 2 *Asperula flore caruleo.*
Blew Woodrooffe.

¶ *The Description.*

1   WOodrooffe hath many square stalkes full of joynts,and at euery knot or joynt seuen or eight long narrow leaues,set round about like a star, or the rowell of a spurre:the floures grow at the top of the stems, of a white colour, and of a very sweet smell, as is the rest of the herbe,which being made vp into garlands or bundles, and hanging vp in houses in the heat of Summer,doth very well attemper the aire,coole and make fresh the place,to the delight and comfort of such as are therein.

2 There

‡ 3 *Sagina Spergula.*
Spurrey.

2　There is another fort of Woodrooffe called *Asperula Carulea*,or blew Woodrooffe,it is an herbe of a foot high,soft,hairy,and something branched,with leaues & ftalks like thofe of white Woodrooffe : the floures thereof are blew, ftanding vpon fhort ftems on the tops of the ftalks : the feed is fmall, round,and placed together by couples ; the root is long and of a red colour.

3　There is another herbe called *Sagina spergula,*or Spurry,which is fown in Brabant, Holland, and Flanders, of purpofe to fatten cattel, and to caufe them to giue much milke , and there called Spurrey,and Franke Spurrey:it is a bafe and low herbe, very tender,hauing many jointed ftalks,whereupon do grow leaues fet in round circles like thofe of Woodrooffe, but leffer and fmoother, in forme like the rowell of a fpur:at the top of the ftalks do grow fmall white floures;after which come round feed like thofe of Turneps : the root is fmall and threddie.

‡ 4　There are one or two plants more, which may fitly be here mentioned : the firft of them is the *Spergula marina* of *Dalefchampius,* which from a pretty large wooddy and roughifh root fends vp jointed ftalks fome foot long; at each joint come forth two long thick round leaues, and out of their bofomes other leffer leaues:the top of the ftalks is diuided into fundry branches,bearing floures of a faint reddifh colour,compofed of fiue little leaues,with yellowifh threds in the middle : after which follow cups or feed veffels, which open into foure parts, and containe a little flat reddifh feed : it grows in the falt marfhes about Dartford, and other fuch places ; floures in Iuly and Auguft, and in the meane fpace ripens his feed. We may call this in Englifh,Sea Spurrie.

5　This other hath a large root,confidering the fmalneffe of the plant : from which arife many weak flender branches fome three or foure inches long,fomtimes more,lying commonly flat on the ground,hauing many knots or joints ;at each whereof vfually grow a couple of white fcaly leaues, and out of their bofomes other fmall fharpe pointed little greene leaues : at the tops of the branches grow little red floures,fucceeded by fuch, yet leffer heads than thofe of the former ; it floures in Iuly and Auguft,and growes in fandy grounds,as in Tuthill-fields nigh Weftminfter:the figure fet forth in *Hift.Lug d.p.* 2179,by the title of *Chamapeuce Plinij;Camphorata minor Dalefchampij,*feems to be of this plant,but without the floure : *Bauhine* in his *Prodromus* defcribes it by the name of *Alfine Spergula facie.* This may be called Chickweed Spurrey,or fmall red Spurrey. ‡

¶ *The Place.*
White Woodrooffe groweth vnder hedges,and in woods almoft euery where ; the fecond groweth in many places of Effex,and diuers other parts in fandy grounds.The third in Corne fields.

¶ *The Time.*
They floure in Iune and Iuly.

¶ *The Names.*
Moft haue taken Woodrooffe to be *Pliny* his *Alyffos,*which as he faith,doth differ from *Erythrodanum,*or Garden Madder, in leaues onely, and leffer ftalkes : but fuch a one is not only this,but alfo that with blew floures : for *Galen* doth attribute to *Alyffos,* a blew floure : notwithftanding *Galens* and *Plinies Alyffos* are thought to differ by *Galens* owne words, writing of *Alyffos* in his fecond booke of Counterpoyfons,in *Antonius Cous* his compofition,in this maner : *Alyffos* is an herbe very like vnto Horehound, but rougher and fuller of prickles about the circles : it beareth a floure tending to blew.

Woodrooffe is named of diuers in Latine *Afperula odorata,*and of moft men *Afpergula odorata :* of others,*Cordialis,*and *Stellaria :* in high Dutch, 𝕳𝖊𝖗𝖙𝖟𝖋𝖗𝖊𝖕𝖙 : in low Dutch, 𝕷𝖊𝖚𝖊𝖗𝖐𝖗𝖆𝖚𝖙 : that is to ſay *Iecoraria*, or *Hepatica*, Liuerwort : in French, *Muguet* : in Engliſh, Woodrooffe, Woodrowe, and Woodrowell.

¶ *The Temperature.*

Woodrooffe is of temperature ſomthing like vnto our Ladies Bedſtraw, but not ſo ſtrong, being in a meane betweene heate and drineſſe.

¶ *The Vertues.*

A   It is reported to be put into wine, to make a man merry, and to be good for the heart and liuer : it preuaileth in wounds, as *Cruciata*, and other vulnerary herbes do.

Chap. 464: *Of Ladies Bedſtraw.*

¶ *The Kindes.*

THere be diuers of the herbes called Ladies Bedſtraw, or Cheeſe-renning ; ſome greater, others leſſe ; ſome with white floures, and ſome with yellow.

¶ *The Deſcription.*

1   Ladies Bedſtraw hath ſmall round euen ſtalkes, weake and tender, creeping hither and thither vpon the ground : whereupon do grow very fine leaues, cut into ſmall jags, finer than thoſe of Dill, ſet at certaine ſpaces, as thoſe of Woodrooffe : among which come forth floures of a yellow colour, in cluſters or bunches thicke thruſt together, of ſtrong ſweet ſmell but not vnpleaſant : the root is ſmall and threddy.

1 *Gallium luteum.*   2 *Gallium album.*
Yellow Ladies Bedſtraw.   Ladies Bedſtraw with white floures.

2   Ladies Bedſtraw with white floures is like vnto Cleauers or Gooſe-graſſe, in leaues, ſtalkes, and manner of growing, yet nothing at all rough, but ſmooth and ſoft : the floures be white, the ſeed round : the roots ſlender, creeping within the ground: the whole plant rampeth vpon buſhes, ſhrubs, and all other ſuch things as ſtand neere vnto it : otherwiſe it cannot ſtand, but muſt reele and fall to the ground.

3   This ſmall *Gallium*, or Ladies little red Bedſtraw, hath beene taken for a kind of wilde Madder; neuertheleſſe it is a kinde of Ladies Bedſtraw, or Cheeſe-renning, as appeareth both by his vertues in turning milke to cheeſe, as alſo by his forme, being in each reſpeĉt like vnto yellow *Gallium*, and differs in the colour of the floures, which are of a dark red colour, with a yellow pointal in the middle, conſiſting of foure ſmall leaues : the ſeed hereof was ſent me from a Citiſen of Strauſburg in Germany, and it hath not beene ſeene in theſe parts before this time.

4   There is likewiſe another ſort of *Gallium* for diſtinĉtions ſake called : *Mollugo*, which hath ſtalks that need not to be propped vp, but of it ſelfe ſtandeth vpright, and is like vnto the common white *Gallium*, but that it hath a ſmoother leaſe. The floures thereof be alſo white, and very ſmall. The root is blackiſh.

† 3 *Gallium rubrum*.
Ladies Bedſtrow with red floures.

4 *Gallium, ſive Mollugo montana*.
Great baſtard Madder.

¶ *The Place.*
The firſt groweth vpon ſunnie bankes neere the borders of fields, in fruitfull ſoiles almoſt euery where.
The ſecond groweth in mariſh grounds and other moiſt places.
The third groweth vpon mountaines and hilly places, and is not yet found in England.
The fourth and laſt groweth in hedges among buſhes in moſt places.

¶ *The Time.*
They floure moſt of the Summer moneths.

¶ *The Names.*
The firſt is called in Greeke, τ····: it hath that name of milke, called in Greeke, γ···, into which it is put as cheeſe-renning : in Latine likewiſe *Gallium* : in high Dutch, **Wagerkraut**, **Walſtroo** : in low Dutch, **Walſtroo** : in French, *Petit Muguet* : in Italian, *Galio* : in Spaniſh, *Coaia leche yerua* : in Engliſh, our Ladies Bedſtrow, Cheeſe-renning, Maids haire, and pety Mugwet.

The others are *Species Lappaginis*, or kindes of ſmall Burres, ſo taken of the Antients: The laſt, of the ſoftneſſe and ſmoothneſſe of the leaues, is commonly called *Mollugo* : diuers take it for a kind of wilde Madder, naming it *Rubia ſylueſtris*, or wilde Madder.

¶ *The Temperature.*
Theſe herbes, eſpecially that with yellow floures, is dry and ſomething binding, as *Galen* ſaith.

¶ *The*

¶ *The Vertues.*

A　The floures of yellow Maids haire, as *Dioſcorides* writeth, is vſed in ointments againſt burnings, and it ſtancheth bloud: it is put into the Cerote or Cere-cloath of Roſes: it is ſet a ſunning in a glaſſe, with Oyle Oliue, vntill it be white: it is good to annoint the wearied Traueller: the root thereof drunke in wine ſtirreth vp bodily luſt: and the floures ſmelled vnto worke the ſame effect.

B　The herbe thereof is vſed for Rennet to make cheeſe, as *Matthiolus* reporteth, ſaying, That the people of Tuſcanie or Hetruria doe vſe to turne their milke, that the Cheeſe which they make of Sheeps and Goats milke might be the ſweeter and more pleaſant in taſte, and alſo more wholſome, eſpecially to breake the ſtone, as it is reported.

C　The people in Cheſhire, eſpecially about Namptwich, where the beſt Cheeſe is made, doe vſe it in their Rennet, eſteeming greatly of that Cheeſe aboue other made without it.

D　We finde nothing extant in the antient writers, of the vertues and faculties of the white kinde: but are as herbes neuer had in vſe either for Phyſicke or Surgerie.

† The figure that was formerly in the third place was of the *Gallium album minus* of *Tabern.* which commonly hath but two leaues at a ioynt, yet ſometimes it is found with three.

CHAP. 465. *Of Ferne.*

¶ *The Kindes.*

THere be diuers ſorts of Ferne, differing as well in forme as place of growing; whereof there be two ſorts according to the old writers, the male and the female; and theſe be properly called Ferne: the others haue their proper names, as ſhall be declared.

1 *Filix mas.*　　　　　　　　2 *Filix fœmina.*
Male Ferne.　　　　　　　　Female Ferne or Brakes.

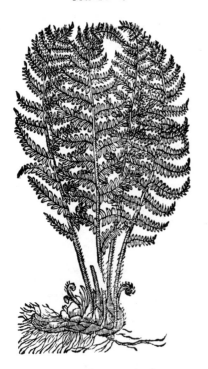

¶ *The*

¶ *The Deſcription.*

1　THe female Ferne bringeth forth preſently from the root broad leaues and rough, ſome-what hard, eaſie to be broken, of a light greene colour, and ſtrong ſmell, more than a cubit long, ſpred abroad like wings, compounded as it were of a great number ſtanding vpon a middle rib, euery one whereof is like a feather, nicked in the edges and on the backe-ſide are ſprinkled as it were with a very fine earthy coloured duſt or ſpots, which many raſhly haue taken for ſeed: the root conſiſteth of a number of tufts or threds, and is thicke and blacke, and is without ſtalke and ſeed, and altogether barren.

‡ *Filicis (vulgo) maris varietas & differentiæ.*
Differences of the male Ferne.

I haue obſerued foure ſorts of Ferne, by moſt writers eſteemed to be the male Ferne of *Dioſcorides*: by *Anguillara, Geſner, Ceſalpinus*, and *Cluſius*, accounted to bee the female, and ſo indeed doe I thinke them to be, though I call them the male with the multitude. If you looke on theſe Fernes according to their ſeuerall growths and ages, you may make many more ſorts of them than I haue done; which I am afraid hath beene the occaſion of deſcribing more ſorts than indeed there are in nature. Theſe deſcriptions I made by them when they were in their perfect growths.

1 *Filix mas ramoſa pinnulis dentatis.*

The roots are nothing but an abundance of ſmall blacke hairy ſtrings, growing from the lower parts of the maine ſtalkes (for ſtalkes I will call them) where thoſe ſtalkes are joyned together. At the beginning of the Spring you may perceiue the leaues to grow forth of their folding cluſters, couered with browniſh ſcales at the ſuperficies of the earth, very cloſely joyned together: a young plant hath but a few leaues; an old one ten, twelue, or more: each ſtalke at his lower end neere the joyning to his fellowes, at his firſt appearing, before he is an inch long hauing ſome of thoſe blacke fibrous roots for his ſuſtenance. The leaues being at their full growth hath each of them a three-fold diuiſion, as hath the Ferne which is commonly called the female: the maine ſtalke, the ſide branches growing from him, and the nerues growing on thoſe ſide branches bearing the leaues: the maine ſtalke of that plant I deſcribe was fully foure foot long (but there are vſually from one foot to foure in length) full of thoſe browniſh ſcales, eſpecially toward the root, firme, one ſide flat, the reſt round, naked fully one and twenty inches, to the firſt paire of ſide branches. The ſide branches, the longeſt being the third paire from the root, were nine inches long, and ſhorter and ſhorter towards the top, in number about twenty paire; for the moſt part towards the root they grow by couples, almoſt oppoſite, the neerer the top the further from oppoſition: the nerues bearing the leaues, the longeſt were two inches and a quarter long, and ſo ſhorter and ſhorter toward the tops of the ſide branches; about twenty in number on each ſide of the longeſt ſide branch. The leaues grow for the moſt part by couples on the nerue, eight or nine paire on a nerue; each leafe being gaſhed by the ſides, the gaſhes ending with ſharpe points, of a deepe greene on the vpper ſide, on the vnder ſide paler, and each leafe hauing two rowes of duſty red ſcales, of a browne or blackeiſh colour: toward the top of the maine ſtalke thoſe ſide branches change into nerues, bearing onely the leaues. When their leaues are at their full growth, you may ſee in the middeſt of them at their roots the ſaid ſcaly folding cluſter; and as the old leaues with their blacke threddy roots wholly periſh, they ſpring vp; moſt yeares you may finde many of the old leaues greene all the Winter, eſpecially in warme places. This groweth plentifully in the boggie ſhadowie moores neere Durford Abbey in Suſſex, and alſo on the moiſt ſhadowie rockes by Mapledurham in Hampſhire, neer Peters-field; and I haue found it often on the dead putrified bodies and ſtems of old rotten okes, in the ſaid moores; neere the old plants I haue obſerued very many ſmall young plants growing, which came by the falling of the ſeed from thoſe duſty ſcales: for I beleeue all herbes haue ſeeds in themſelues to produce their kindes, *Gen.* 1.11.&.12.

The three other haue but a twofold diuiſion, the many ſtalkes and the nerues bearing the leaues. The roots of them all are blacke fibrous threds like the firſt, their maine ſtalkes grow many thicke and cloſe together at the root, as the firſt doth: the difference is in the faſhion of their leaues, and manner of growing, and for diſtinctions ſake I haue thus called them.

2 *Filix mas non ramoſa pinnulis latis denſis minutim dentatis.*
The leaues are of a yellowiſh greene colour on both ſides, ſet very thicke and cloſe together on

the

the nerue, that you cannot see betweene them, with maruellous small nickes by their sides, and on their round tops: each leafe hath also two rowes of dusty seed scales; the figures set forth by *Lobel*, *Tabern.* and *Gerard*, vnder the title of *Filix mas*, do well resemble this Ferne. This growes plentifully in most places in shadowie woods and copses.

3 *Filix mas non ramosa pinnulis angustis, raris, profunde dentatis.*

The leaues are of a deepe greene, not closely set together on the nerue, but you may farre off see betwixt them, deeply indented by the sides, ending with a point not altogether sharpe; each leafe hath also two rowes of dusty seed scales. I haue not seene any figure well resembling this plant. This groweth also in many places in the shade.

4 *Filix mas non ramosa pinnulis latis auriculatis spinosis.*

The leaues are of a deeper greene than either of the two last described, placed on the nerue not very close together, but that you may plainely see betweene them; each leafe (especially those next the stalk) hauing on that side farthest off the stalke a large eare or outgrowing ending, with a sharpe pricke like a haire, as doth also the top of the leafe: some of the sides of the leaues are also nicked, ending with the like pricke or haire. Each leafe hath two rowes of dusty seed scales. This I take to be *Filix mas aculeata major Bauhini.* Neither haue I seene any figure resembling this plant. It groweth abundantly on the shadowie moist rockes by Maple-durham neere Peters-field in Hampshire. *Iohn Goodyer.* Iuly. 4. 1633. ‡

2 The female Ferne hath neither floures nor seed, but one onely stalke, chamfered, something edged, hauing a pith within of diuers colours, the which being cut aslope, there appeareth a certaine forme of a spred-Eagle: about this stand very many leaues which are winged, and like to the leaues of the male Ferne, but lesser: the root is long and blacke, and creepeth in the ground, being now and then an inch thicke, or somewhat thinner. This is also of a strong smell, as is the male.

¶ The Place.

Both the Fernes are delighted to grow in barren dry and desart places: and as *Horace* testifieth;

Neglectis vrenda Filix innascitur agris.

It comes not vp in manured and dunged places, for if it be dunged (as *Theophrastus, lib. 8. cap. 8.* reporteth) it withereth away.

The male ioyeth in open and champion places, on mountaines and stony grounds, as *Dioscorides* saith. ‡ It growes commonly in shadowie places vnder hedges. ‡

The female is often found about the borders of fields vnder thornes, and in shadowie-woods.

¶ The Time.

Both these Fernes wither away in Winter: in the Spring there grow forth new leaues, which continue greene all Summer long.

¶ The Names.

The former is called in Greeke, πτέρις: *Nicander* in his discourse of Treacle nameth it βλάχνον: in Latine, *Filix mas*: in Italian, *Felce*: in Spanish, *Helecho, Falguero*, and *Feyto*: in high Dutch, **walbt farne**: in French, *Fougere,* or *Fenchiere maste*: in low Dutch, **Uaren Manneken**: in English, male Ferne.

The second kinde is called in Greeke, θηλυπτερίς: that is, *Filix fœmina*, or female Ferne: in Latine, as *Dioscorides* noteth among the bastard names *Lingua ceruina*: in high Dutch, **walbt farn weiblin,** and **Grosz farnkraut**: in low Dutch, **Uaren wijfken**: in French, *Fougere femelle*: in English, Brake, common Ferne, and female Ferne.

¶ The Temperature.

Both the Fernes are hot, bitter and dry, and something binding.

¶ The Vertues.

A   The root of the male Ferne being taken to the weight of halfe an ounce, driueth forth long flat wormes out of the belly, as *Dioscorides* writeth, being drunke in Mede or honied water; and more effectually, if it be giuen with two scruples or two third parts of a dram of Scamonie, or of blacke Hellebor: they that will vse it, saith he, must first eate Garlicke. After the same manner, as *Galen* addeth, it killeth the childe in the mothers wombe. The root hereof is reported to be good for them that haue ill spleenes: and being stamped with swines greafe and applied, it is a remedy against the pricking of the reed: for proofe hereof, *Dioscorides* saith the Ferne dieth if the Reed be planted about it; and contrariwise, that the Reed dieth if it bee compassed with Ferne: which is vaine to thinke, that it hapneth by any antipathy or naturall hatred, and not by reason this Ferne prospereth not in moist places, nor the Reed in dry.

B   The female Ferne is of like operation with the former, as *Galen* saith. *Dioscorides* reports, That this bringeth barrennesse, especially to women; and that it causeth women to be deliuered before their time: hee addeth, that the pouder hereof finely beaten is laid vpon old vlcers, and healeth

the

the galled neckes of oxen and other cattell : it is alſo reported, that the root of Ferne caſt into an hogſhead of wine keepeth it from ſouring.

The root of the male Ferne ſodden in Wine is good againſt the hardneſſe and ſtopping of the milt : and being boyled in water, ſtayeth the laske in young children, if they be ſet ouer the decocti-on thereof, to eaſe their bodies by a cloſe ſtoole.

C

Chap. 466. *Of Water-Ferne, or Oſmund the water-man.*

¶ *The Deſcription.*

Water-Ferne hath a great triangle ſtalke two cubits high, beſet vpon each ſide with large leaues ſpred abroad like wings, and dented or cut like Polypody: theſe leaues are like the large leaues of the Aſh-tree ; for doubtleſſe when I firſt ſaw them a far off it cauſed me to wonder thereat, thinking that I had ſeene young Aſhes growing vpon a bog, but beholding it a lit-tle neerer, I might eaſily diſtinguiſh it from the Aſh, by the browne rough and round graines that grew on the top of the branches, which yet are not the ſeed thereof, but are very like vnto the ſeed. The root is great and thicke, folded and couered ouer with many ſcales and interlacing roots, ha-uing in the middle of the great and hard wooddy part thereof ſome ſmall whiteneſſe, which hath beene called the heart of *Oſmund* the water-man.

Filix florida, ſiue Oſmunda Regalis.
Water-Ferne, or Oſmund Royall.

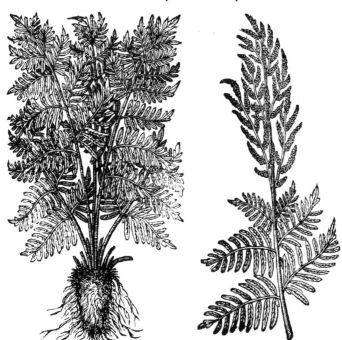

¶ *The Place.*

It groweth in the midſt of a bog at the further end of Hampſted heath from London, at the bot-tome of a hill adjoyning to a ſmall cottage, and in diuers other places, as alſo vpon diuers bogges on a heath or a common neere vnto Bruntwood in Eſſex, eſpecially neere vnto a place there that ſome haue digged, to the end to finde a neſt or mine of gold ; but the birds were ouer fledge, and flowne away before their wings could be clipped. ‡ It did grow plentifully in both theſe places, but of late it is all deſtroyed in the former. ‡

¶ *The*

¶ *The Time.*

It flouriſheth in Summer, as the former Fernes : the leaues decay in Winter : the root continu-eth freſh and long laſting ; which being brought into the Garden proſpereth as in his natiue ſoile, as my ſelfe haue proued.

¶ *The Names.*

It is called in Latine, *Oſmunda* : it is more truly named *Filix paluſtris*, or *aquatilis* : ſome terme it by the name of *Filicaſtrum* : moſt of the Alchimiſts call it *Lunaria major* : *Valerius Cordus* nameth it *Filix latifolia* : it is named in high Dutch, **Groſʒ farn** : in low Dutch, **Groot Waren, wñt Waren** : in Engliſh, Water-Ferne, Oſmund the Water-man : of ſome, Saint Chriſtophers herbe, and Oſmund.

¶ *The Temperature.*

The root of this alſo is hot and dry, but leſſer than they of the former ones.

¶ *The Vertues.*

A   The root and eſpecially the heart or middle part thereof, boyled or elſe ſtamped, and taken with ſome kinde of liquor, is thought to be good for thoſe that are wounded, dry-beaten and bruiſed, that haue fallen from ſome high place : and for the ſame cauſe the Empericks do put it in decoæti-ons, which the later Phyſitians doe call wound-drinkes : ſome take it to be ſo effeætuall, and of ſo great a vertue, as that it can diſſolue cluttered bloud remaining in any inward part of the body, and that it alſo can expell or driue it out by the wound.

B   The tender ſprigs thereof at their firſt comming forth are excellent good vnto the purpoſes a-foreſaid, and are good to be put into balmes, oyles, and conſolidatiues, or healing plaiſters, and into vnguents appropriate vnto wounds, punætures, and ſuch like.

Chap. 467. Of Polypodie or Wall-Ferne.

1 Polypodium.
Wall Ferne, or Polypodie of the wall.

2 Polypodium quercinum.
Polypody of the Oke.

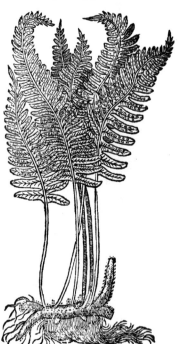

‡ 2 *Polypodium Indicum.*
Indian Polypodie.

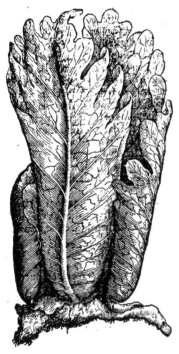

¶ *The Description.*

1 THe leaues of Polypodie might be thought to be like those of male Ferne, but that they are far lesser, and not nicked at all in the edges: these doe presently spring vp from the root, being cut on both the edges with many deepe gashes euen hard to the middle rib: on the vpper side they are smooth, on the nether side they are lightly poudered as it were with dusty marks: the root is long, not a finger thicke, creeping aslope; on which are seen certain little buttons like those pits and dents that appeare in the tailes of cuttle fishes: this hath in it a certaine sweetnesse, with a taste somwhat harsh: this kind of Ferne likewise wanteth not only floures and seed, but stalks also.

2 Polypodie of the oke is much like that of the wall, yet the leaues of it are more finely cut, smooth on the vpper side, of a pale greene colour, together with the stalkes and middle ribs; on the nether side rough like those of Ferne. This Ferne also liueth without a stalke: it groweth without seed: the root hath many strings fastned to it, one folded within another, of a meane bignesse, and sweet in taste: it sendeth forth here & there new dodkins or springs whereby it increaseth.

‡ 3 *Clusius* in his *Exoticks, lib. 4. cap. 7.* giueth vs the historie of an Indian Fern or Polypodie found amongst the papers of Dr *Nicolas Colie* a Dutch physitian, who died in his return from the East Indies: the root of it was six inches long, and almost one thicke, of the same shape & colour as the ordinarie one is: from this came vp three leaues, of which the third was lesser than the other two. the two larger were eleuen inches long, and their bredth from the middle rib (which was very large) was on each side almost fiue inches; the edges were diuided almost like an oken leafe, from the middle rib came other veins that ran to the ends of the diuisions, and betweene these be smaller veines variously diuaricated and netted, which made the leafe shew prettily: the colour of it was like that of a dry oken leafe. Where Dr *Colie* gathered this it was vncertaine, for he had left nothing in writing. ‡

¶ *The Place.*
It growes on the bodies of old rotten trees, and also vpon old walls, and the tops of houses. It is likewise found among rubbish neere the borders of fields, especially vnder trees and thornes, and now and then in woods; in some places it groweth ranke and with a broader leafe, in others not so ranke, and with a narrower leafe.

That which growes on the bodies of old okes is preferred before the rest: in stead of this most do vse that which is found vnder the okes, which for all that is not to be termed *Quircinum*, or Polypodie of the Oke.

¶ *The Time.*
Polypodie is green all the yeare long, and may be gathered at any time: it bringeth forth new leaues in the first Spring.

¶ *The Names.*
The Grecians call it πολυπόδιον, of the holes of the fishes *Polypi* appearing in the roots: it is called in Latine *Polypodium*, after the Greek name, and many times *Filicula*, as though they should say *parua Filix*, or little Ferne: the Italians name it *Polipodio*: the Spaniards, *Filipodio*, and *Polypodio*: in high-Dutch, **Engelfusz, Baumfarn, Dropfoourts:** in low-Dutch, **Boom baren:** in French, *Polypode:* and we of England, Polypodie; that which groweth vpon the wall wee call Polypodie of the wall; and that on the Oke, Polypodie of the Oke.

¶ *The Temperature.*
Polypodie doth dry, but yet without biting, as *Galen* teacheth.

Ccccc   ¶ *The*

¶ *The Vertues.*

A　*Dioſcorides* writeth, That it is of power to purge, and to draw forth choler and flegme. *Actuarius* addeth, That it likewiſe purgeth Melancholy : others ſuppoſe it to be without any purging force at all, or elſe very little ; of which minde is *Iohannes Monardus*, who thinketh it purgeth very gently ; which thing is confirmed by Experience, the Miſtreſſe of things. For in very deed, Polypody it ſelfe doth not purge at all, but only ſerueth a little to make the belly ſoluble, being boiled in the broth of an old cocke, with Beets or Mallowes, or other-like things that moue to the ſtoole by their ſlipperineſſe. *Iohannes Meſue* reckoneth vp Polypodie among thoſe things that do eſpecially dry & make thin : peraduenture he had reſpect to a certain kind of *Arthritis* or ache in the joints, in which not one only part of the body, but many together are moſt commonly touched : for which it is very much commended by the Brabanders and other inhabitants about the riuer Rhene and the Maze. In this kinde of diſeaſe the hands, the feet, and the joints of the knees and elbowes doe ſwell. There is joined withall a feebleneſſe in mouing, through the extremitie of the pain : ſometimes the vpper parts are leſſe grieued, and the lower more ; the humors do alſo eaſily run from one place to another, and then ſettle. Againſt this diſeaſe the Geldres and Cleuelanders doe vſe the decoction of Polypodie, whereby they hope the ſuperfluous humors may be waſted and dried vp, and that not by and by, but in continuance of time ; for they appoint that this decoction ſhould be taken for certain daies together.

B　But this kinde of gout is ſooner taken away either by bloud-letting or purgations, or by both, and afterwards by ſweat : neither is it hard to be cured, if theſe generall remedies be vſed in time ; for the humors do not remaine fixed in thoſe joints, but are rather gathered together, than ſettled about them.

C　Therefore the body muſt out of hand be purged, and then that which remaines is to be waſted and conſumed away by ſuch thiugs as procure ſweat.

D　Furthermore, *Dioſcorides* ſaith that the root of Polypodie is very good for members out of joint, and for chaps between the fingers.

E　The root of Polypody boiled with a little hony, water, and pepper, and the quantity of an ounce giuen, emptieth the belly of cholericke and pituitous humors. Some boile it in water and wine, and giue to the quantitie of three ounces for ſome purpoſes with good ſucceſſe.

C H A P. 468.　*Of Oke-Ferne.*

‡　O Vr Author here (as in many other places) knit knots ſomewhat intricate to looſe ; for firſt he confounds in the Names and Nature the Polypodie of the Oke or leſſer Polypodie, with the *Dryopteris* or Oke-Ferne ; but that I haue now put backe to the former chapter, his fit place. Then in the ſecond place did he giue the Deſcription of the *Dryopteris* of the *Adverſar.* taken from thence, *pag. 363.* Then were the Place, Time, Names, &c. taken out of the Chapter of *Dryopteris candida* of *Dodonæus*, being *Pempt. 3. lib. 5. cap. 4.* but the figure was of the *Filicula fœmina petræa 4.* of *Tabernamontanus.* Now I will in this chapter giue you the *Dryopteris* of the *Adverſaria*, then that of *Dodonæus*, and thirdly that of *Tragus* ; for I take them to be different, and this laſt to be that figured by our Author out of *Tabernamontanus.* ‡

¶ *The Deſcription.*

1　T His kinde of Ferne, called *Dryopteris* or *Filix querna*, hath leaues like vnto the female Ferne before ſpoken of, but much leſſer, ſmaller, and more finely cut or jagged, & is not aboue a foot high, being a very ſlender and delicate tender herb. The leaues are ſo finely jagged, that in ſhew they reſemble feathers, ſet round about a ſmall rib or ſinue ; the backe ſide being ſprinkled not with ruſſet or browne markes or ſpeckes as the other Fernes are, but as it were painted with white ſpots or marks, not ſtanding out of the leaues in ſcales, as the ſpots in the male Ferne, but they are double in each leafe cloſe vnto the middle rib or ſinue. The root is long, brown, and ſomewhat hairy, very like vnto Polypody, but much ſlenderer, of a ſharpe and cauſticke taſte. ‡ *Rondeletius* affirmed that he found the vſe of this deadly, being put into medicines in ſtead of Polypodie, by the ignorance of ſome Apothecaries in Dauphenie in France. Mr *Goodyer* hath ſent me an accurat deſcription, together with a plant of this Ferne, which I haue thought good here alſo to ſet forth.

Dryopteris

‡ 1 *Dryopteris Adverſ.*
True Oke-Ferne.

‡ 2 *Dryopteris alba Dod.*
White Oke-Ferne.

‡ 3 *Dryopteris Tragi.*
Tree-Ferne.

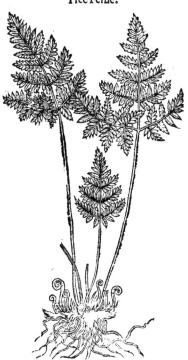

Dryopteris Penæ & Lobelij.

The roots creepe in the ground or myre neere the turfe or vpper part thereof, and fold among ſt themſelues as the roots of *Polypodium* do, almoſt as big as a wheat ſtraw, and about fiue, ſix, or ſeuen inches long, cole blacke without, and white within, of a binding taſte inclining to ſweetneſſe, with an innumerable company of ſmall blacke fibres like haires growing thereunto. The ſtalkes ſpring from the roots in ſundry places, in number variable, according to the length and increaſe of the root: I haue ſeen ſmall plants haue but one or two, and ſome bigger plants fourteene or fifteene: they haue but a twofold diuiſion, the ſtalke growing from the root, and the nerue bearing the leaues: the ſtalk is about fiue, ſix, or ſeuen inches long, no bigger than a bennet or ſmal graſſe ſtalk, one ſide flat, as are the male Fernes, the reſt round, ſmooth, and green. The firſt paire of nerues grow about three inches from the root, and ſo do all the reſt grow by couples, almoſt exactly one againſt another, in number about eight, nine, or ten couples, the longeſt ſeldome exceeding an inch in length. The leaues grow on the nerues alſo by couples, eight or nine couples on a nerue, without any nicks or indentures, of a yellowiſh green colour. The Ferne may be ſaid to be like *Polypodium* in his creeping root, like the male Fern in his ſtalk, and like the female Fern in his nerues & leaues. I could finde no ſeed-ſcales on the back ſides of any of the leaues of this Ferne. Many yeares paſt I found this ſame in a very wet moore or bog, being the land of *Richard Auſten*, called White-row Moore, where Peate is now digged, a mile from Peters-field in Hampſhire: and this ſixth of Iuly, 1633, I digged vp there many plants, and by them made this deſcription. I neuer found it growing in any other place: the leaues periſh at Winter, and grow vp againe very late in the ſpring.   *Iohn Goodyer.*   Iuly 6. 1633.

2   *Dodonæus* thus deſcribes his: *Dryopteris* (ſaith he) doth much reſemble the male Fern, but the leaues are much ſmaller and more finely cut, ſmooth on the fore ſide, and of a yellowiſh greene together with the ſtalks and middle nerues: on the backe ſide it is rough like the other Fernes, and alſo liueth without ſtalke or ſeed. The root conſiſts of fibres intricately folded together, of an indifferent thickneſſe, here and there putting vp new buds. This is the *Adianthum* of the *Adverſ.* who affirme the vſe thereof to be ſafe, and not pernitious and deleterie, as that of *Dryopteris.* It thus differs from the former; the leaues of this are not ſet directly one oppoſit to another, the diuiſions of the leaues are larger and more diuided. The root is more thready, and creeps not ſo far as that of the former.

3   This (which is *Cluſius* his *Filix pumila ſaxatilis prima*, and which I take to be the *Dryopteris*, or *Filix arborea* of *Tragus*) hath blacke ſlender long creeping roots, with few ſmall hard hairy fibres faſtned to them, of a very aſtringent taſte: from theſe riſe vp ſundry ſtalks a foot high, diuided into certaine branches of winged leaues, like to thoſe of the female Ferne, but much leſſe, tenderer, and finer cut, and hauing many blackiſh ſpots on their lower ſides. This differs from the two former, in that the leaues are branched, which is a chiefe difference; and *Bauhinus* did very well obſerue it, if he had as well followed it, when he diuided *Filix* into *ramoſa* and *non ramoſa.* ‡

¶ *The Place.*

It is oftentimes found in ſunny places, in the vallies of mountaines and little hills, and in the tops of the trunks of trees in thicke woods.

¶ *The Time.*

The leaues hereof periſh in Winter, in the Spring new come forth.

¶ *The Names.*

This is called in Greeke δρυοπτερίς: in Latine, *Querna Filix : Oribaſius, lib.* 11. of Phyſicall Collections, calls it *Bryopteris*, of the moſſe with which it is found: for as *Dioſcorides* writeth, it groweth in the moſſe of Okes. The Apothecaries in times paſt miſcalled it by the name of *Adiantum*; but they did worſe in putting it in compound medicines in ſtead of *Adiantum. Valerius Cordus* calleth it *Pteridion :* in low-Dutch, €ijcken baren: the Spaniards, *Helecho de Roble :* it is named in Engliſh, Oke Ferne, Petty Ferne, and it may moſt fitly be called Moſſe-Ferne.

¶ *The Nature and Vertues.*

A   Oke Ferne hath many taſtes, it is ſweet, biting, and bitter, it hath in the root a harſh or choking taſte, and a mortifying qualitie, and therefore it taketh away haires. *Dioſcorides* ſaith further, That Oke Ferne ſtamped roots and all is a remedie to root vp haires, if it be applied to the bodie after ſweating, the ſweat being wiped away.

C H A P. 469.   *Of blacke Oke-Ferne.*

¶ *The Deſcription.*

1   THere is alſo a certaine other kinde of Ferne like to the former Oke-Ferne of *Dodonæus* his deſcription, but the ſtalks and the ribs of the leaues are blackiſh, and the leaues of a deeper green colour: this groweth out alſo immediatly from the root, and is likewiſe diuerſly but not ſo finely indented: the root is made vp of many ſtrings, not vnlike the male Fern, but much leſſer.

2   The female blacke Fern is like to the male, ſauing his leaues are not ſo ſharp at the points, more white and broad than the male, wherein conſiſteth the difference.

¶ *The Place.*

They grow likewiſe vpon trees in ſhadowie woods, and now and then in ſhadowie ſandy bankes and vnder hedges.

¶ *The*

1 *Onopteris mas.*
The male blacke Ferne.

¶ *The Time.*

They remaine greene all the yeare long, other-wiſe than Polypodie and Maidens haire do; yet do they not ceaſe to bring forth new leaues in Sum-mer: they are deſtitute of flours and ſeed, as is the former.

¶ *The Names.*

This is called of diuers of the later herbariſts, *Dryopteris nigra,* or blacke Oke-Ferne, of the like-neſſe it hath with *Dryopteris;* which we haue cal-led in Engliſh, Oke-Ferne or Moſſe-Ferne: of o-thers, *Adiantum nigrum* or black Maiden-hair, that it may differ from the former, which is falſly cal-led *Adiantum.* There are of the later Herbariſts who would haue it to bee *Lonchitis aſpera,* or rough Spleen-wort: but what likeneſſe hath it with the leaues of *Scolopendrium?* none at all: therfore it is nor *Lonchitis aſpera,* much leſſe *Adiantum Plinij,* which differs not from *Adiantum Theophraſti;* for what he hath of *Adiantum* he taketh out of *Theo-phraſtus:* the right *Adiantum* we wil deſcribe here-after. Notwithſtanding blacke Oke-Ferne was vſed of diuers vnlearned Apothecaries of France and Germany for *Adiantum* or Maiden-haire of Lumbardy: but theſe men did erre in doing ſo, yet not ſo much as they who take Polypody of the Oke for the true Maiden-haire.

¶ *The Temperature and Vertues.*

The black Oke-Ferne hath no ſtipticke qualitie at all, but is like in faculty to *Trichomanes* or Eng-liſh Maiden-haire.

CHAP. 470   *Of Harts-Tongue.*

¶ *The Deſcription.*

1 THe common kinde of Harts tongue, called *Phyllitis,* that is to ſay, a plant conſiſting only of leaues, bearing neither ſtalk, floure, nor ſeed, reſembling in ſhew a long tongue, where-of it hath bin and is called in ſhops *Lingua Ceruina,* that is, Harts tongue: theſe leaues are a foot long, ſmooth and plaine vpon one ſide, but vpon that ſide next the ground ſtraked ouerthwart with certain long rough marks like ſmall wormes, hanging on theback ſide thereof. The root is blacke, hairy and twiſted, or ſo growing as though it were wound together,

2 The other kinde of Fern, called *Phyllitis multifida* or *Laciniata,* that is, jagged Harts tongue, is very like the former, ſauing that the leaues therof are cut or iagged like a mans hand, or the palm and browantles of a Deare, bearing neither ſtalke, floure, nor ſeed.

3 There is another kinde of Harts tongue called *Hemionitis,* which hath bred ſome controuer-ſie among writers, for ſome haue taken it for a kind of Harts tongue, as it is indeed: others deſcribe it as a proper plant by it ſelfe, called *Hemionitis,* of *[illegible],* that is *Mulus,* a Mule, becauſe Mules do de-light to feed thereon. It is barren in ſeeds, ſtalks, and floures, and in ſhape it agreeth very well with our Harts-tongue. The roots are compact of many blackiſh haires; the leaues are ſpotted on the back ſide like the common Harts tongue, and differ in that, that this *Hemionitis* in the baſe or low-eſt parts of the leaues is arched after the manner of a new Moon or a forked arrow, the yongeſt and ſmalleſt leaues being like vnto the great Bindweed called *Volubilis.*

4 There is a kinde of Ferne called alſo *Hemionitis ſterilis,* which is a very ſmall and baſe herbe not aboue a finger high, hauing foure or fiue ſmall leaues of the ſame ſubſtance and colour, ſpot-ted on the backe part, and in taſte like vnto Harts-tongue; but the leaues beare the ſhape of them of *Tota-bona,* or good *Henry,* which many of our Apothecaries do abuſiuely take for Mercury. The roots are very many, ſmooth, blacke, and threddy, bearing neither ſtalke, floure, nor ſeed. This plant

1 *Phyllitis.*
Harts-tongue.

2 *Phyllitis multifida.*
Finger Harts-tongue.

‡ 3 *Hemionitis maior.*
Mules Ferne or Moone-Ferne.

‡ 4 *Hemionitis minor.*
Small Moon-Ferne.

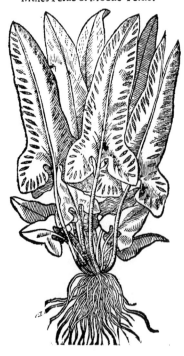

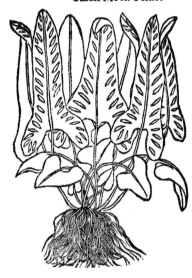

‡ 5 *Hemionitis peregrina.*
Handed Moon-Ferne.

my very good friend Mr *Nicolas Belſon* found in a grauelly lane in the way leading to Oxey park neer Watford, fifteen miles from London. It groweth likewiſe on the ſtone walls of Hampton court, in the garden of Mr *Huggens* keeper of the ſaid houſe.

5   There is a kinde of Fern called alſo *Hemionitis*, but with this addition *peregrina*, that is very ſeldome found, and hath leaues very like vnto Hartstongue, but that it is palmed or branched in the part next the ground in manner of the ſecond *Phyllitis* at the top of the leaues, otherwiſe they reſemble one another in nature and ſome.

¶ *The Place.*

The common Harts-tongue growes by the way ſides in great plenty, as you trauell from London to Exceſter, in ſhadowie places and moiſt ſtony vallies and wells, and is much planted in gardens.

The ſecond I found in the garden of Mr *Cranwich* a ſurgeon dwelling at Much-Dunmow in Eſſex, who gaue me a plant for my garden.

‡   Mr *Goodyer* found it wilde in the bankes of a lane neer Swaneling, not many miles from South-hampton. ‡

It groweth vpon Ingleborow hils, and in diuers other mountaines of the North of England.

¶ *The Time.*

It is green all the yeare long, yet leſſe green in winter: in ſummer it now and then brings forth new leaues.

¶ *The Names.*

It is called in Greeke, ••••••: in Latine alſo *Phyllitis* : in ſhops, *Lingua Cervina*, and falſly *Scolopendria* : for it differs much from the right *Scolopendria* or ſtone Ferne : it is called in high Dutch, **Hirſzong** : in low-Dutch, **Herſtonge** : in Spaniſh, *Lengua cervina* : in French, *Langue de cerf* : in Engliſh, Harts tongue ; of ſome, Stone Harts tongue. *Apuleius, cap.*83. nameth it *Radiolus*.

¶ *The Nature.*

It is of a binding and drying facultie.

¶ *The Vertues.*

Common Harts tongue is commended againſt the laske and bloudy flix : *Dioſcorides* teacheth,   A That being drunke in wine it is a remedie againſt the bitings of Serpents.

It opens the hardneſſe and ſtopping of the ſpleen and liuer, and all other griefes proceeding of   B opilations or ſtoppings whatſoeuer.

CHAP. 471.

Of Spleene-woort or Milt-waſte.

¶ *The Deſcription.*

1   SPleen-wort, being that kinde of Fern called *Aſplenium* or *Ceterach*, and the true *Scolopendria*, hath leaues a ſpan long, jagged or cut vpon both ſides, euen hard to the middle rib, euery cut or inciſure being as it were cut halfe round (whereby it is knowne from the rough Spleenwort) not one cut right againſt another, but one beſides the other, ſet in ſeueral order, being ſlipperie and green on the vpper ſide, ſoft and downy vnderneath ; which when they bee withered are folded vp together like a ſcrole, and hairy without, much like to the rough Bear-worme wherewith men bait their hooks to catch fiſh. The root is ſmall, blacke, and rough much platted or interlaced, hauing neither ſtalke, floure, nor ſeeds.

1 *Aſplenium*

1 *Aſplenium, ſive Ceterach.*
Spleen-wort or Milt-waſte.

2 *Lonchitis aſpera.*
Rough Spleenwort.

† 3 *Lonchitis aſpera major.*
Great rough Spleen-wort.

† 4 *Lonchitis Maranthæ.*
Baſtard Spleen-wort.

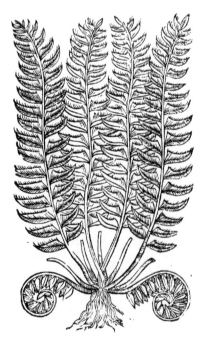

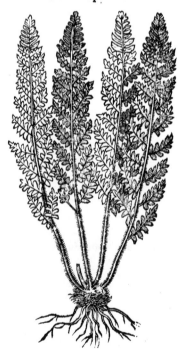

2   Rough Spleenwort is partly like the other Ferns in shew, and beareth neither stalk nor seed, hauing narrow leaues a foot long and somewhat longer, slashed on the edges euen to the middle rib, smooth on the vpper side, and of a swart green colour vnderneath, rough, as is the leaues of Polypodie: the root is blacke, and set with a number of slender strings.

‡ 3   This greater Spleenwort hath leaues like Ceterach, of a span long, somwhat resembling those of Polypodie, but that they are more diuided, snipt about the edges, and sharpe pointed: the root is fibrous and stringy. This growes on the rocks and mountainous places of Italy, and is the *Lonchitis aspera maior* of *Matthiolus* and others. ‡

4   This kinde of Spleenwort is not only barren of stalkes and seeds, but also of those spots and marks wherewith the others are spotted: the leaues are few in number, growing pyramide or steeple-fashion, great and broad below, and sharper toward the top by degrees: the root is thick, black and bushy as it were a Crowes nest.

¶ *The Place.*

Ceterach groweth vpon old stone walls and rocks in darke and shadowie places throughout the West parts of England, especially vpon the stone walls by Bristow, as you go to S. *Vincents* rocke, as likewise about Bathe, Wells, and Salisbury, where I haue seen great plenty thereof.

The rough Spleenwort groweth vpon barren heaths, dry sandy bankes, and shadowie places in most parts of England, but especially on a heath by Lonon called Hampsted heath, where it grows in great aboundance.

¶ *The Names.*

Spleenwort or Milt-waste is called in Greeke Ασπληνου: in Latine likewise *Asplenium*, and also *Scolopendria*: of *Gaza*, *Mula herba*: in shops, *Ceterach*: in high-Dutch, **Steynfarne**: in low-Dutch, **Steynuaren**, and **Miltcruyt**: in English, Spleenwort, Milt-waste, Scale-Ferne, and Stone-Ferne: it is called *Asplenion*, because it is especiall good against the infirmities of the Spleene or Milt; and *Scolopendria*, of the likenesse it hath with the Beare-worme before remembred.

♁ Rough Milt-waste is called of diuers of the later writers, *Asplenium syluestre*, or wild Spleenwort: of some, *Asplenium magnum*, or great Spleenwort. *Valerius Cordus* calls it *Strutiopteris* and *Dioscorides, Lonchitis asper*, or rough Spleenwort: in Latine, according to the same Author, *Longina*, and *Calabrina*: in English, rough Spleenwort or Milt-waste.

¶ *The Nature.*

These plants are of thin parts, as *Galen* witnesseth, yet are they not hot, but in a meane.

¶ *The Vertues.*

Dioscorides teacheth, that the leaues boiled in wine and drunke by the space of forty daies, take   A away infirmities of the spleen, help the strangury and yellow jaundice, cause the stone in the bladder to moulder and passe away; all which are performed by such things as be of thinne or subtile parts. He addeth likewise, that they stay the hicket or yeoxing, and also hinder conception, either inwardly taken, or hanged about the party: and therefore, saith *Pliny*, Spleenwort is not to bee giuen to women, because it bringeth barrennesse.

There be Empericks or blinde practitioners of this age who teach, that with this herb not onely   B the hardnesse and swelling of the spleene, but all infirmities of the liuer also may be effectually, and in very short time remoued, insomuch that the sodden liuer of a beast is restored to his former constitution againe, that is, made like vnto a raw liuer, if it be boiled againe with this herb.

But this is to be reckoned among the old Wiues fables, and that also which *Dioscorides* tells of,   C touching the gathering of Spleene-wort in the night, and other most vain things, which are found here and there scattered in old books: from which most of the later Writers do not abstaine, who many times fillvp their pages with lies and frivolous toyes, and by so doing do not a little deceive yong Students.

† Formerly vnder the title of *Lonchitis Maruscha* was put the figure now in the third place, and the figure which should haue bin there was in the third place of the next Chapter, vnder the title of *Filicula petræa mas.*

CHAP.

Chap. 472: *Of diuers ſmall Fernes.*

¶ *The Deſcription.*

1 THis ſmall or dwarfe Ferne,which is ſeldome found except in the banks of ſtony fountains,wells,and rocks bordering vpon riuers,is very like the common Brakes in leaues, but altogether leſſer. The root is compoſed of a bundle of blacke thready ſtrings.

2 The female,which is found likewiſe by running ſtreams,wells, and fountaines vpon rockes and ſtony places, is like the precedent, but is a great deale ſmaller, blacker of colour, fewer roots, and ſhorter.

1 *Filicula fontana mas.*
The male fountaine Ferne.

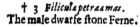
‡ 3 *Filicula petraa mas.*
The male dwarfe ſtone Ferne.

3 The male dwarfe Ferne, that groweth vpon the ſtony mountaines of the North and Weſt parts of England,eſpecially toward the ſea,and alſo in the joints of ſtone walls among the mortar, hath ſmall leaues deeply cut on both ſides,like vnto *Ceterach* or Spleen-wort, barren both of ſeeds and ſtalks,as alſo of thoſe ſpots or marks that are to be ſeen vpon the back part of the other Ferns : the root creepeth along,ſet with ſome few hairy ſtrings reſembling thoſe of the Oke-Ferne called *Dryopteris.*

4 The female ſtone Ferne hath diuers long leaues riſing from a thready root, contrary to that of the male,compoſed of many ſmall leaues finely minced or cut like the teeth of a ſaw, of a whitiſh green colour,without any ſpots or marks at all,ſeeds, or ſtalks, which groweth vnder ſhadowie rockes and craggy mountaines in moſt places. ‡ From a ſmall root compoſed of many blacke hairy and intricately folding ſtrings, come vp many leaues two or three inches high, ſtiſte,thicke, darke greene,and ſhining : in the diuiſion, growth, poſition, ſhape and taſte it reſembles the male Ferne , and hath alſo ruſty ſpots on the backe : the middle ribbe and ſtalke is of a ſhining brownith

4ᵒ *Filicula Petræa fœmina, ſiue Chamæfilix marina*
The female dwarfe ſtone Ferne.   (*Anglica.*

niſh ſilken colour : it growes in the chinkes of
the Rocks by the Sea ſide in Cornewall.

¶ *The Place.*

The place is ſufficiently touched in the deſcription.

¶ *The Time.*

They flouriſh both Winter and Summer,
for when the leaues wither by reaſon of age,
there ariſe young to ſupply the place , ſo that
they are not to be ſeen without greene and withered leaues both at once.

¶ *The Names.*

It ſufficeth what hath bin ſaid of the names
in their ſeuerall titles:notwithſtanding the laſt
deſcribed wee haue called *Chamæfilix marina
Anglica :* which groweth vpon the rockie cliffe
neere Hatwich,as alſo at Douer , among the
Sampire that there groweth.

¶ *The Temperature and Vertues.*

Their temperature and faculties in working A
are referred vnto the kinds of black Oke Ferns,
called *Dryopteris,* and *Onopteris.*

† It is hard to ſay what our Author in this chapter meant,by his figures and deſcriptions,wherefore I haue left his deſcriptions as I found them:the ſecond figure which was very like the firſt I haue omitted : for the third,which was of the *Lonchitis Maranthæ,*mentioned in the foregoing chapter,I haue put *Claſius* his figure of his *Filix ſaxatilis :*which groweth in ſuch pl ces,and reaſonable well fits our Authors deſcription : in the fourth place I haue put Lobels *Chamæfilix marina
Anglica,*and his deſcription,which our Author as I iudge,intended in that place to haue giuen vs.

CHAP. 473.   *Of true Maiden-haire.*

¶ *The Kindes.*

THeophraſtus and Pliny haue ſet downe two Maiden-haires,the blacke and the white, whereunto
may be added another called *Ruta muraria,* or wall Rue,equall to the others in facultie,whereof
we will intreat.

1 *Capillus Veneris verus.*
True Maiden-haire.

¶ *The Deſcription.*

1   WHoſo will follow the variable opinions of Writers concerning the Fern
called *Adianthum verum,* or *Capillus
Veneris verus,*muſt of neceſſitie be brought into a La-
byrinth of doubts , conſidering the diuers opinions
thereof:but this I know that Venus-haire,or Maiden-
haire,is a low herb growing an hand high,ſmooth, of
a darke crimſon colour, and glittering withall : the
leaues be ſmal, cut in ſunder,and nicked in the edges
ſomething like thoſe of Coriander , confuſedly or
without order placed, the middle rib whereof is of a
blacke ſhining colour : the root conſiſteth of manie
ſmall threddy ſtrings.

2   This Aſſyrian Maiden-haire is likewiſe a baſe
or low herbe, hauing leaues, flat, ſmooth,and plaine,
ſet vpon a blackiſh middle rib, like vnto that of the
other Maiden-haire, cut or notched in the edges, na-
ture keeping no certain forme,but making one leafe of this faſhion, and another far different from
it : the root is tough and threddie.

3   This plant which we haue inſerted among the Adianthes as a kinde thereof , may without
error ſo paſſe,which is in great requeſt in Flanders and Germanie, where the practitioners in phyſicke do vſe the ſame in ſtead of *Capillus Veneris,* and with better ſucceſſe than any of the Capillare
herbes,

herbs,although *Matthiolus* and *Diofcorides* himfelfe hath made this wall Rue to be a kind of *Paronychia* or Nailewort : notwithftanding the Germans will not leaue the vfe thereof,but receiue it as the true Adianth,efteeming it equall,if not far better,than either *Ceterach,Capillus Veneris verus,*or *Trichomanes,*called alfo *Polytrichon:* it brings forth very many leaues round and flender, cut into two or three parts,very hard in handling,fmooth and green on the out fide,and of an ilfauored dead color vnderneath,fet with little fine fpots,which euidently fheweth it to be a kind of Fern. The root is blacke and full of ftrings.

    2 *Capillus Veneris Syriaca.*   3 *Ruta muraria,fiue Salvia vitæ.*
     Affyrian Maiden-haire.   Wall Rue or Rue Maiden-haire.

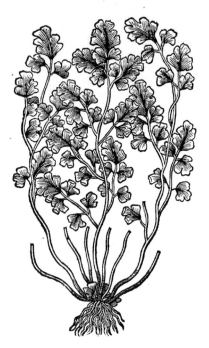 

¶ *The Place.*

    The right Maiden-haire groweth vpon walls in ftony fhadowie and moift places, neere to fountaines and where water droppeth : it is a ftranger in England;yet I haue heard it reported by fome of good credit,that it growes indiuers places of the Weft country of England.

    The Affyrian Maiden-haire taketh name of his natiue country Affyria,being a ftranger in Europe.

    Stone Rue groweth vpon old walls neere vnto waters, Wells, and fountaines: I found it vpon the wall in the Churchyard of Dartford in Kent, hard by the riuer fide where people ride through, and alfo vpon the wals of the Churchyard of Sittingburn in the fame county,in the middle of the towne,hard by a great lake of water,and alfo vpon the Church walls of Railey in Effex,and diuers other places.

¶ *The Time.*

    Thefe plants are green both winter and fummer,and yet haue neither floures nor feed.

¶ *The Names.*

    Maiden-haire is called in Greeke Ἀδίαντον : *Theophraftus* and *Pliny* name it *Adiantum nigrum,* black Maiden-haire ; for they fet downe two Maiden-haires,the blacke and the white, making this the blacke,and the Rue of the wall the white : it is called in Latine,*Polytrichum, Callitrichum, Cincinalis, Terræ Capillus,Supercilium terræ :* of *Apuleius,Capillus Veneris,Capillaris,Crinita :* and of diuers,*Coriandrum putei :* the Italians keep the name *Capillus Veneris :* in Englifh,black Maiden-haire,and *Venus* haire,and it may be called our Ladies haire.

                                                     It

It is called *Adianton* becauſe the leafe, as *Theophraſtus* ſaith, is neuer wet, for it caſteth off water that falleth thereon, or being drowned or couered in water, it remaineth ſtill as if it were dry, as *Pliny* likewiſe writeth ; and is termed *Callitricon* and *Polytricon*, of the effect it hath in dying haire, and maketh it to grow thicke.

Wall Rue is commonly called in Latine, *Ruta muraria*, or *Ruta muralis* : of ſome, *Salvia vitæ*, but wherefore I know not, neither themſelues, if they were liuing : of the Apothecaries of the Low-Countries *Capillus Veneris*, or Maiden haire, and they haue vſed it a long time for the right Maiden-haire ; it is that kind of *Adiantum* which *Theophraſtus* termed *Adiantum Candidum*, or white Maiden-haire, for hee maketh two, one blacke, and the other white, as we haue ſaid. *Pliny* doth likewiſe ſet downe two kindes, one he calleth *Polytricon*; the other, *Tricomanes*, or Engliſh Maiden-haire, whereof we will intreate in the chapter following, which he hath falſely ſet downe for a kinde of *Adiantum*, for *Tricomanes* doth differ from *Adiantum*.

Some there be that thinke Wall-Rue is *Paronichia Dioſcoridis*, or *Dioſcorides* his Whitlow-wort, wherein they haue been greatly deceiued : it is called in high Dutch, **Maurranten:** in low Dutch, **Steencrupt :** in French, *Rue de maraille* : in Engliſh, Wall-Rue, and white Maiden haire.

¶ *The Temperature and Vertues.*

The true Maiden-haire, as *Galen* teſtifieth, doth dry, make thin, waſte away, and is in a meane be-   A
tweene heate and coldneſſe : *Meſue* ſheweth that it conſiſteth of vnlike or diſagreeing parts, and that ſome are watery and earthy, and the ſame binding, and another ſuperficially hot and thinne : And that by this it taketh away obſtructions or ſtoppings, maketh things thinne that are thicke, looſeneth the belly, eſpecially when it is freſh and greene : for as this part is thin, ſo is it quickely reſolued, and that by reaſon of his binding and earthy parts : it ſtoppeth the belly, and ſtaieth the laske and other fluxes.

Being drunke it breaketh the ſtone , and expelleth not onely the ſtones in the kidnies, but alſo   B
thoſe which ſticke in the paſſages of the vrine.

It raiſeth vp groſſe and ſlimie humors out of the cheſt and lungs, and alſo thoſe which ſticke in   C
the conduits of the winde pipe, it breaketh and raiſeth them out by ſpitting, if a loch or licking medicine be made thereof.

Moreouer, it conſumeth and waſteth away the Kings-euill, and other hard ſwellings, as the ſame   D
Author affirmeth, and it maketh the haire of the head or beard to grow, that is fallen and pilled off.

Dioſcorides reckoneth vp many vertues and operations of this Maiden-haire, which do not onely   E
differ, but are alſo contrary one to another. Among others he ſaith, that the ſame ſtancheth bloud : and a little before, that it draweth away the ſecondines, and bringeth downe the deſired ſickeneſſe : which words do confound one another with contrarieties ; for whatſoeuer things do ſtanch bloud, the ſame do alſo ſtay the termes.

He addeth alſo in the end, that it is ſowne about ſheepe-folds for the benefit of the ſheepe, but   F
what that benefit ſhould be he ſheweth not.

Beſides, that it cannot be ſowne, by reaſon it is without ſeed, it is euident, neither can it fitly bee   G
remoued. Therefore in this place it ſeemeth that many things are tranſpoſed from other places, and falſly added to this chapter : and peraduenture ſome things are brought hither out of diſcourſe of *Cytiſus*, or Milke Trefoile, whereof here to write were to ſmall purpoſe.

Wall-Rue is not much vnlike to blacke Maiden-haire in temperature and faculty.   H

Wall-Rue is good for them that haue a cough, that are ſhort winded, and that be troubled with   I
ſtitches and paine in their ſides.

Being boyled, it cauſeth concoction of raw humors which ſticke in the lungs ; it taketh away   K
the paine of the kidnies and bladder, it gently prouoketh vrine, and driueth forth ſtones.

It is commended againſt ruptures in young children, and ſome affirme it to be excellent good,   L
if the pouder thereof be taken continually for forty daies together.

CHAP. 474.   *Of Engliſh, or common Maiden-haire.*

¶ *The Deſcription.*

1   **E**Ngliſh Maiden-haire hath long leaues of a darke green colour, conſiſting of very many ſmall round leaues ſet vpon a middle rib, of a ſhining blacke colour, daſhed on the nether ſide with ſmall rough markes or ſpeckes, of an ouerworne colour : the roots are ſmall and threddy.

1 *Trichomanes mas.*
The Male Englifh Maiden-haire.

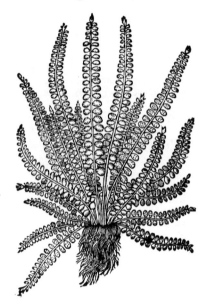

2 The female Englifh Maiden-haire is like vnto the precedent, fauing that it is leffer, and wanteth thofe fpots or markes that are in the other, wherein confifteth the difference. ‡ Our Authors figure was of the *Trichomanes fœmina* of *Tabernamontanus*, which expreffes a variety with branched leaues, and therein only was the difference. ‡

¶ *The Place.*

It growes for the moft part neer vnto fprings and brookes, and other moift places, vpon old ftone walls and rocks: I found it growing in a fhadowie fandie lane in Betfome, in the parifh of Southfleet in Kent, vpon the ground where-as there was no ftones or ftony ground neere vnto it, which before that time I did neuer fee; it groweth likewife vpon ftone walls at her Majefties palace of Richmond, & in moft ftone wals of the Weft and North parts of England. ‡ Mr. *Goodyer* faith, that in Ianuary, 1624. hee faw enough to lade an horfe growing on the bancks in a lane, as he rode betweene Rake and Headly in Hampfhire neere Wollmer For-reft. ‡

¶ *The Time.*

It continueth a long time, the coldneffe of Winter doth it no harme, it is barren as the other Fernes are, whereof it is a kinde.

¶ *The Names.*

It is called in Greeke, τριχομανες: in Latine, *Filicula*, as though we fhould fay, *Parua Filix*, or little Ferne; alfo *Capillaris*: in fhops, *Capillus Veneris*. *Apuleius* in his 51 chapter maketh it all one with *Callitrichon*: of fome it is called *Polytrichon*: in Englifh, common Maiden-haire.

¶ *The Temperature and Vertues.*

A   Thefe as *Diofcorides* and *Galen* do write, haue all the faculties belonging to *Adiantum*, or blacke Maiden-haire.

B   The decoction made in wine and drunke, helpeth them that are fhort winded, it helpeth the cough, ripeneth tough flegme, and auoideth it by fpitting.

C   The lie wherein it hath beene fodden, or laid to infufe, is good to wafh the head, caufing the fcurfe and fcales to fall off, and haire to grow in places that are pild and bare.

CHAP. 475. *Of Thiftles.*

¶ *The Kindes.*

THe matter of the Thiftles is diuers, fome Thiftles ferue for nourifhment, as the Artichoke without prickles, and the Artichoke with prickles; other for medicine, as the root of Carline which is good for many things; the bleffed Thiftle alfo, otherwife called *Carduus benedictus*; Sea Huluer, and diuers others: fome are poifonfome, as *Chamæleon niger*; one fmooth, plaine, and with-out prickles, as the Thiftle called Beares Breech, or *Acanthus fatiuus*, whereof there is another with prickles, which we make the wilde, of the which two we intend to write in this chapter.

¶ *The Defcription.*

1   BEares breech of the garden hath broad leaues, fmooth, fomewhat blacke, gafhed on both the edges, and fet with many cuts and fine nickes: between which rifeth vp in the midft a big ftalke brauely deckt with flours, fet in order from the middle vpward, of color white, of forme long, which are armed as it were with two catkins, one higher, another lower: after them grow forth the husks, in which is found broad feed: the roots be blacke without, and white with-in,

in,and full of clammie juyce, and are diuided into many off. fprings, which as they creepe far,fo do they now and then bud forth and grow afrefh : thefe roots are fo full of life, that how little foeuer of them remaine, it oftentimes alfo bringeth forth the whole plant.

1 *Acanthus fativus.* -
Garden Beares-breech.

‡ 2 *Acanthus fyl.aculeatus.*
Prickly Beares-breech.

2 Wilde Beares-breech,called *Acanthus fylueftris,Pena* fetteth forth for *Chameleonta Monfpelien-fium,* and reporteth that he found it growing amongft the grauelly and moift places neere to the walls of Montpellier, and at the gate of Aegidia, betweene the fountane and the brooke neere to the wall: this thiftle is in ftalke, floures, colour of leaues and feed like the firft kinde, but fhorter and lower,hauing large leaues,dented or jagged with many cuts and incifions, not only in fome few parts of the leaues,as fome other thiftles,but very thickly dented or clouen,and hauing many fharp, large,white and hard prickles about the fides of the diuifions and cuts,not very eafie to be handled or touched without danger to the hand and fingers.

¶ *The Place.*

Diofcorides writeth,that garden Branke Vrfine groweth in moift and ftony places,and alfo in gardens : it were vnaduifedly done to feeke it in either of the Germaines any where, but in gardens onely ; in my garden it doth grow very plentifully.

The wilde was found in certaine places of Italy neer to the fea, by that notable learned man *Alfonfus Pancius,*Phyfition to the Duke of Ferrara,and profeffor of fimples and Phyficke,and is a ftranger in England. ‡ I haue feene it growing in the garden of M^r.*Iohn.Parkinfon.* ‡

¶ *The Time.*

Both the Branke Vrfines do floure in the Summer feafon, the feed is ripe in Autumne : the root remaineth frefh ; yet now and then it perifhes in Winter in both the Germaines,if the weather be too cold : but in England the former feldome or neuer dieth.

¶ *The Names.*

It is called in Greeke,   the Latines keepe the fame name *Acanthus:* yet doth *Acanthus* fignifie generally all kinde of Thiftles,and that is called *Acanthus* by the figure *Antonomafia :* the Englifh name is,Branke Vrfine,and Beares-breech.

The tame or garden Branke Vrfine is named in Latine,*Sativus,*or *Hortenfis Acanthus :* in Greeke,   : and of *Galen, Oribafius,* and *Pliny,*   : *Pliny* alfo calleth this *Acanthus levis,* or fmooth

Branke

Branke Vrfine, and reporteth it to be a city herbe, and to ferue for arbors : fome name it *Branca Vr-fina* (others vfe to call Cow-Parfnep by the name of *Branca Vrfina*, but with the addition *Germanica:*) the Italians call it *Acantho*, and *Branca Orfina* : the Spaniards, *Yerua Giguante* : the Ingrauers of old time were wont to carue the leaues of this Branke Vrfine in pillers, and other workes, and alfo vpon the eares of pots ; as among others *Virgil* teftifieth in the third Eclog of his Bucolicks :

> *Et nobis idem Alcimedon duo pocula fecit,*
> *Et molli circum eft anfas ample xus Acantha.*

‡ I take *Virgils Acanthus* to be that which we now commonly call *Pyracantha*, as I fhall here-after fhew when I come to treat thereof. ‡

The other Branke Vrfine is named in Greeke, ἀγρία ἄκανθα: and in Latine, *Sylueftris Acanthus*, or wild Branke Vrfine, and they may be called properly *Acantha*, or *Spina*, a prickle, by which name it is found called of moft Herbarifts, *Acanthus* : yet there is alfo another *Acanthus* a thorny fhrub : the liquor which iffueth forth of it, as *Herodotus* and *Theophraftus* affirme, is a gumme : for difference whereof peraduenture this kinde of *Acanthus* is named *Herbacantha* : There is likewife found among the baftard names of *Acanthus* the word *Mamolaria*, and alfo *Crepula*, but it is not expreffed to which of them, whether to the wilde or tame it ought to be referred.

¶ *The Temperature.*

The leaues of the garden Branke Vrfine confift in a meane as it were betweene hot and cold, be-ing fomwhat moift, with a mollifying and gentle digefting faculty, as are thofe of the Mallow, and therefore they are profitably boyled in clyfters, as well as Mallow leaues. The root, as *Galen* teach-eth, is of a more drying quality.

¶ *The Vertues.*

A   *Diofcorides* faith, that the roots are a remedy for lims that are burnt with fire, and that haue beene out of joynt, if they be laied thereunto : that being drunke they prouoke vrine, and ftop the belly : that they helpe thofe that be broken, and be troubled with the crampe, and be in a confumption of the lungs.

B   They are good for fuch as haue the ptificke and fpet bloud withall; for thofe that haue faln from fome high place, that are bruifed and dry beaten, and that haue ouerftrained themfelues, and they are as good as the roots of the greater Comfrey, whereunto they are very like in fubftance, tough juyce, and quality.

C   Of the fame root is made an excellent plaifter againft the ache and numneffe of the hands and feet.

D   It is put into clyfters with good fucceffe againft fundry maladies.

Cʜᴀᴘ. 476.   *Of the Cotton Thiftle.*

¶ *The Defcription.*

1   THe common Thiftle, whereof the greateft quantity of down is gathered for diuers pur-pofes, as well by the poore to ftop pillowes, cufhions, and beds for want of feathers, as alfo bought of the rich vpholfters to mix with the feathers and down they do fell, which deceit would be looked vnto : this Thiftle hath great leaues, long and broad, gafhed about the edges, and fet with fharpe and ftiffe prickles all alongft the edges, couered all ouer with a foft cot-ton or downe : out from the middeft whereof rifeth vp a long ftalke about two cubits high, corne-red, and fet with filmes, and alfo full of prickles : the heads are likewife cornered with prickles, and bring forth floures confifting of many whitifh threds : the feed which fucceedeth them is wrapped vp in downe ; it is long, of a light crimfon colour, and leffer than the feed of baftard Saffron : the root groweth deep in the ground, being white, hard, wooddy, and not without ftrings.

2   The Illyrian cotton thiftle hath a long naked root, befet about the top with a fringe of many fmall threds or jags : from which arifeth a very large and tall ftalke, higher than any man, rather like a tree than an annuall herbe or plant : this ftalke is garnifhed with fcroles of thinne leaues, from the bottome to the top, fet full of moft horrible fharpe prickes, and fo is the ftalke and euery part of the plant, fo that it is impoffible for man or beaft to touch the fame without great hurt and dan-ger : his leaues are very great, far broader and longer than any other Thyftle whatfoeuer, couered with an hoarie cotton or downe like the former : the floures doe grow at the top of the ftalkes,

which

which is diuided into sundry branches, and are of a purple colour, set or armed round about with the like, or rather sharper thornes than the aforesaid.

1 *Acanthium album.*
The white Cotton Thistle.

2 *Acanthium Illyricam purpureum.*
The purple Cotton Thistle.

¶ *The Place.*

These Thistles grow by high waies sides, and in ditches almost euery where.

¶ *The Time.*

They floure from Iune vntill August, the second yeare after they be sowne: and in the meane time the seed waxeth ripe, which being thorow ripe the herbe perisheth, as doe likewise most of the other Thistles, which liue no longer than till the seed be fully come to maturity.

¶ *The Names.*

This Thistle is taken for that which is called in Greeke, ἄκανσος, which *Dioscorides* describeth to haue leaues set with prickles round about the edges, and to be couered with a thin downe like a copweb, that may be gathered and spun to make garments of, like those of silke : in high Dutch it is called, 𝔴𝔢𝔦𝔰𝔷𝔴𝔢𝔤𝔢 𝔡𝔦𝔰𝔱𝔦𝔩 : in low Dutch, 𝔴𝔦𝔱𝔱𝔢 𝔴𝔢𝔡𝔥 𝔡𝔦𝔰𝔱𝔢𝔩 : in French, *Chardon argentin* : in English, Cotton-Thistle, white Cotton-Thistle, wilde white Thistle, Argentine or the Siluer Thistle.

¶ *The Temperature and Vertues.*

Dioscorides saith, That the leaues and roots hereof are a remedy for those that haue their bodies drawne backwards; thereby *Galen* supposeth that these are of temperature hot.   A

CHAP. 477.   *Of our Ladies-Thistle.*

¶ *The Description.*

THe leaues of our Ladies-Thistle are as bigge as those of white Cotton-Thistle: for the leaues thereof be great, broad, large, gashed in the edges, armed with a multitude of stiffe and sharpe prickles, as are those of Ote-Thistle, but they are without down, altogether slippery, of a light

greene

Carduus Maria.
Ladies Thiſtle.

A

B

C

D

green and ſpeckled, with white and milky ſpots and lines drawne diuers waies: the ſtalk is high, and as bigge as a mans finger: the floures grow forth of heads full of prickles, being threds of a purple colour: the ſeed is wrapped in downe like that of Cotton Thiſtle: the root is long thicke, and white.

¶ *The Place.*

It groweth vpon waſte and common places by high waies, and by dung-hils almoſt euery where.

¶ *The Time.*

It floureth and ſeedeth when Cotton-Thiſtle doth.

¶ *The Names.*

It is called in Latine, *Carduus Lacteus*, and *Carduus Maria*; in high Dutch, **Onſer Vrouwen Diſtell**: in French, *Chardon de noſtre dame*: in Engliſh, our Ladies Thiſtle: it may properly be called *Leucographus*, of the white ſpots and lines that are on the leaues: *Pliny* in his 27. booke, chap. 11. maketh mention of an herb called *Leucographis*, but what maner of one it is he hath not expreſſed; therefore it would be hard to affirme this to be the ſame that his *Leucographis* is; and this is thought to be *Spina alba*, called in Greeke ἄκανδα λευκὴ, or white Thiſtle, Milke Thiſtle, and *Carduus Ramptarius*: of the Arabians, *Bedoard*, or *Bedeguar*, as *Matthæus Syluaticus* teſtifieth.

¶ *The Temperature and Vertues.*

The tender leaues of *Carduus Leucographus*, the prickles taken off, are ſometimes vſed to bee eaten with other herbes.

Galen writeth, that the roots of *Spina alba* do dry and moderately binde, that therefore it is good for thoſe that be troubled with the laske and the bloudy flix, that it ſtaieth bleeding, waſteth away cold ſwellings; eaſeth the paine of the teeth if they be waſhed with the decoction thereof.

The ſeed thereof is of a thin eſſence and hot facultie, therefore he ſaith that it is good for thoſe that be troubled with cramps.

Dioſcorides affirmeth that the ſeeds being drunke are a remedy for infants that haue their ſinews drawne together, and for thoſe that be bitten of ſerpents: and that it is thought to driue away ſerpents, if it be but hanged about the necke.

Chap. 478. *Of the Globe-Thiſtle.*

¶ *The Deſcription.*

1   GLobe Thiſtle hath a very long ſtalke, and leaues jagged, great, long, and broad, deeply gaſhed, ſtrong of ſmell, ſomewhat greene on the vpper ſide, and on the nether ſide whiter and downy: the floures grow forth of a round head like a globe, which ſtandeth on the tops of the ſtalkes; they are white and ſmall, with blew threds in the midſt: the ſeed is long, with haires of a meane length: the root is thicke and branched.

2   There is another Globe Thiſtle that hath leſſer leaues, but more full of prickles, with round heads alſo: but there groweth out of them beſides the floures, certaine long and ſtiffe prickles.

3   There is likewiſe another kinde reſembling the firſt in forme and figure, but much leſſer, and the floures thereof tend more to a blew.

4   There is alſo another Globe Thiſtle, which is the leaſt, and hath the ſharpeſt prickles of all the reſt: the head is ſmall; the floures whereof are white, like to thoſe of the firſt.

5 There

1 *Carduus globoſus.*
The Globe-Thiſtle.

‡ 3 *Carduus globoſus minor.*
Small Globe-Thiſtle.

‡ 5 *Carduus globoſus capitulo latiore.*
Flat headed Globe-Thiſtle.

5   There is a certaine other kinde hereof, yet the head is not ſo round, that is to ſay, flatter and broader aboue ; out of which ſpring blew floures : the ſtalke hereof is ſlender, and couered with a white thin downe : the leaues are long, gaſhed likewiſe on both ſides, and armed in euery corner with ſharpe prickles.

                                                       6 There

6   There is another called the Down-Thiſtle,which riſeth vp with thicke and long ſtalks. The leaues thereof are jagged, ſet with prickles, white on the nether ſide : the heads be round and many in number, and are couered with a ſoft downe, and ſharpe prickles ſtanding forth on euery ſide,be-ing on the vpper part fraughted with purple floures all of ſtrings : the ſeed is long, and ſhining, as doth the ſeed of many of the Thiſtles.

‡ *6 Cardu**us** eriocephalus.*
Woolly headed Thiſtle.

¶ *The Place.*

Theſe are ſown in gardens,and do not grow in theſe countries that we can finde.
‡   I haue found the ſixth by Pocklington and in other places of the Woldes in Yorke-ſhire. M*r*. *Goodyer* alſo found it in Hamp-ſhire. ‡

¶ *The Time.*
They floure and flouriſh when the other Thiſtles do.

¶ *The Names.*
Fuchſius did at the firſt take it to bee *Cha-mæleon niger* ; but afterwards being better ad-uiſed, he named it *Spina peregrina,* and *Cardu-us globoſus.* *Valerius Cordus* doth fitly call it *Sphærocephalus:*the ſame name doth alſo agree with the reſt, for they haue a round head like a ball or globe. Moſt would haue the firſt to be that which *Matthiolus* ſetteth downe for *Spina alba :* this Thiſtle is called in Engliſh, Globe Thiſtle,and Ball-Thiſtle.
The downe or woolly headed **Thiſtle** is called in Latine, beeing deſtitute of another name,*Eriocephalus,*of the woolly head : in En-gliſh, Downe Thiſtle, or woolly headed Thi-ſtle. It is thought of diuers to be that which *Bartholomæus Vrbeveteranus* and *Angelus Palea,* Franciſcan Friers, report to be called *Corona Fratrum,* or Friers Crowne : but this Thiſtle doth far differ from that,as is euident by thoſe things which they haue written concerning *Corona Fratrum ;* which is thus : In the borders of the kingdome of Aragon towards the kingdome of Caſtile we finde another kinde of Thiſtle, which groweth plentifully there by common wayes, and in wheate fields,&c. *Vide Dod.Pempt.5.lib.5.cap.5.*

¶ *The Temperature and Vertues.*
Concerning the temperature and vertues of theſe Thiſtles we can alledge nothing at all.

Cʜᴀ·ᴘ. 479. *Of the Artichoke.*

¶ *The Kindes.*

THere be three ſorts of Artichokes,two tame or of the garden, and one wilde, which the Italian eſteemeth greatly of,as the beſt to be eaten raw,which he calleth *Cardune.*

¶ *The Deſcription.*

1   **T**He leaues of the great Artichoke,called in Latine *Cinara,* are broad,great, long,ſet with deepe gaſhes in the edges, with a deepe channell or gutter alongſt the middle, hauing no prickles at all, or very few, and they bee of a greene aſh colour : the ſtalke is aboue a cubit high, and bringeth forth on the top a fruit like a globe, reſembling at the firſt a cone or Pine apple, that is to ſay,made vp of many ſcales;which is when the fruit is great or looſed of a greeniſh red colour within, and in the lower part full of ſubſtance and white ; but when it opens it ſelfe there growes
alſo

1 *Cinara maxima Anglica.*
The great red Artichoke.

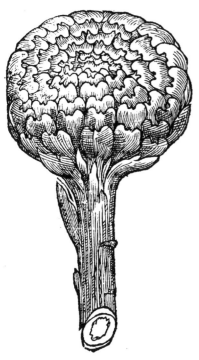

2 *Cinara maxima alba.*
The great white Artichoke.

3 *Cinara sylvestris.*
Wilde Artichocke.

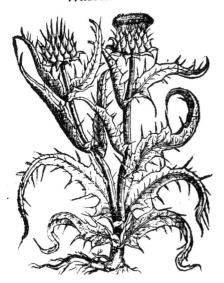

also vpon the cone a floure all of threds, of a gallant purple tending to a blew colour. The feed is long, greater and thicker than that of our Ladies thistle, lying vnder soft and downy haires which are contained within the fruit. The root is thicke, and of a meane length.

2   The second great Artichoke differeth from the former in the colour of the fruit, otherwise there is little difference, except the fruit hereof dilateth it selfe further abroad, and is not so closely compact together, which maketh the difference.

3   The prickly Artichoke, called in Latine *Carduus*, or *Spinosa Cinara*, differeth not from the former, saue that all the corners of the leaues hereof, and the stalkes of the cone or fruit, are armed with stiffe and sharpe prickles, whereupon it beareth well the name of *Carduus*, or Thistle.

¶ *The Place.*

The Artichoke is to be planted in a fat and fruitfull soile : they doe loue water and moist ground. They commit great error who cut away the side or superfluous leaues that grow by the sides, thinking thereby to increase the greatnesse of the fruit, when as in truth they depriue the root from much water by that meanes, which would nourish it to the feeding of the fruit ; for if you marke the trough or hollow channell that is in euery leafe, it shall appeare very euidently, that the

Creator

Creator in his ſecret wiſedome did ordaine thoſe furrows, euen from the extreme point of the leafe to the ground where it is faſtned to the root, for no other purpoſe but to guide and leade that water which falls farre off, vnto the root; knowing that without ſuch ſtore of water the whole plant would wither, and the fruit pine away and come to nothing.

¶ *The Time.*

They are planted for the moſt part about the Kalends of Nouember, or ſomewhat ſooner. The plant muſt bee ſet and dunged with good ſtore of aſhes, for that kinde of dung is thought beſt for planting thereof. Euery yeare the ſlips muſt be torne or ſlipped off from the body of the root, and theſe are to be ſet in April, which will beare fruit about Auguſt following, as *Columella, Paladius,* and common experience teacheth.

¶ *The Names.*

The Artichoke is called in Latine, *Cinara,* of *Cinis,* Aſhes, wherewith it loueth to be dunged. *Galen* calleth it in Greeke, Κάκτος : but with *k* and *v* in the firſt ſyllable : of ſome it is called *Cactos* : it is named in Italian, *Carcioffi, Archiocchi* : in Spaniſh, *Alcarrofa:* in Engliſh, Artichoke: in French, *Artichaux* : in low Dutch, **Artichoken** : whereupon diuers call it in Latine, *Articocalus,* and *Articoca* : in high Dutch, **Strobildorn.**

The other is named in Latine commonly not onely *Spinoſa cinara,* or prickly Artichoke, but alſo of *Palladius, Carduus* : of the Italians, *Cardo,* and *Cardino* : of the Spaniards, *Cardos* : of the Frenchmen, *Chardons* : *Leonhartus Fuchſius* and moſt writers take it to be *Scolymus Dioſcoridis* ; but *Scolymus Dioſcoridis* hath the leafe of Chameleon or *Spina alba,* with a ſtalke full of leaues, and a prickly head: but neither is *Cinara* the Artichoke which is without prickles, nor the Artichoke with prickles any ſuch kinde of herbe ; for though the head haue prickles, yet the ſtalke is not full of leaues, but is. many times without leaues, or elſe hath not paſt a leafe or two. *Cynara* doth better agree with that which *Theophraſtus* and *Pliny* call Κάκτος, *Cactus,* and yet it doth not bring forth ſtalkes from the root creeping alongſt the ground : it hath broad leaues ſet with prickles ; the middle ribs of the leaues, the skin pilled off, are good to be eaten, and likewiſe the fruit, the ſeed and downe taken away ; and that which is vnder it as tender as the braine of the Date tree : which things *Theophraſtus* and *Pliny* report of *Cactus.* That which they write of the ſtalkes, ſent forth immediately from the root vpon the ground, which are good to be eaten, is peraduenture the ribs of the leaues, euery ſide taken away (as they be ſerued vp at the table) may be like a ſtalke, except euen in Sicilia, where they grew only in *Theophraſtus* time. It bringeth forth both certaine ſtalkes that lie on the ground, and another alſo ſtanding ſtraight vp; but afterwards being remoued and brought into Italy or England, it bringeth forth no more but one vpright : for the ſoile and clyme do much preuaile in altering of plants, as not onely *Theophraſtus* teacheth, but alſo euen experience it ſelfe declareth : and of *Cactus, Theophraſtus* writeth thus ; Κάκτος (*Cactus*) groweth onely in Sicilia : it bringeth forth preſently from the root ſtalkes lying along vpon the ground, with a broad and prickly leafe : the ſtalkes being pilled are fit to be eaten, being ſomewhat bitter, which may be preſerued in brine : it bringeth forth alſo another ſtalke, which is likewiſe good to be eaten.

¶ *The Temperature and Vertues.*

A   The nailes, that is, the white and thicke parts which are in the botome of the outward ſcales or flakes of the fruit of the Artichoke, and alſo the middle pulpe whereon the downy ſeed ſtands, are eaten both raw with pepper and ſalt, and commonly boyled with the broth of fat fleſh, with pepper added, and are accounted a dainty diſh, being pleaſant to the taſte, and good to procure bodily luſt : ſo likewiſe the middle ribs of the leaues being made white and tender by good cheriſhing and looking to, are brought to the table as a great ſeruice together with other junkets : they are eaten with pepper and ſalt as be the raw Artichokes : yet both of them are of ill juyce; for the Artichoke containeth plenty of cholericke juyce, and hath an hard ſubſtance, inſomuch as of this is ingendred melancholy juyce, and of that a thin and cholericke bloud, as *Galen* teacheth in his booke of the faculties of nouriſhments. But it is beſt to eate the Artichoke boyled : the ribbes of the leaues are altogether of an hard ſubſtance : they yeeld to the body a raw and melancholy juyce, and containe in them great ſtore of winde.

B   It ſtayeth the inuoluntary courſe of the naturall ſeed either in man or woman.

C   Some write, that if the buds of yong Artichokes be firſt ſteeped in wine, and eaten, they prouoke vrine, and ſtirre vp the luſt of the body.

D   I finde moreouer, That the root is good againſt the ranke ſmell of the arme-holes, if when the pith is taken away the ſame root bee boyled in wine and drunke : for it ſendeth forth plenty of ſtinking vrine, whereby the ranke and rammiſh ſauour of the whole body is much amended.

C H A P.

CHAP. 480. Of Golden Thistle.

¶ The Description.

1 THe stalkes of Golden Thistle rise vp forthwith from the root, being many, round, and branched. The leaues are long, of a beautifull greene, with deepe gashes on the edges, and set with most sharpe prickles: the floures come from the bosome of the leaues, set in a scaly chaffie knap, very like to Succory floures, but of colour as yellow as gold : in their places come vp broad flat and thin seeds, not great, nor wrapped in downe : the root is long, a finger thicke, sweet, soft, and good to be eaten, wherewith swine are much delighted : there issueth forth of the Thistle in what part soeuer it is cut or broken, a juyce as white as milke.

‡ There is some variety of this Thistle ; for it is found much larger about Montpelier than it is in Spaine, with longer branches, but fewer floures : the leaues also are spotted or streaked with white like as the milke Thistle : whence *Clusius*, whom I here follow, hath giuen two figures thereof ; the former by the name of *Scolymus Theophrasti Hispanicus* ; and the other by the title of *Scolymus Theophrasti Narbonensis*. This with white spots I saw growing this yere with Mr. *Tradescant* at South Lambeth. ‡

1 *Carduus Chrysanthemus Hispanicus.*
　The Spanish golden Thistle.

‡ *Carduus Chrysanthemus Narbonensis.*
　The French golden Thistle.

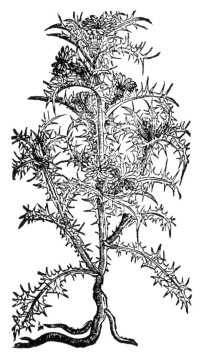 

2 The golden Thistle of Peru, called in the West Indies, *Figue del Inferno*, a friend of mine brought it vnto me from an Island there called Saint Iohns Island, among other seeds. What reason the inhabitants there haue to call it so, it is vnto me vnknowne, vnlesse it bee because of his fruit, which doth much resemble a fig in shape and bignesse, but so full of sharpe and venomous prickles, that whosoeuer had one of them in his throat, doubtlesse it would send him packing either to heauen or to hell. This plant hath a single wooddy root as big as a mans thumbe, but somewhat long : from which ariseth a brittle stalke, full of joynts or knees, diuiding it selfe into sundry other small branches, set full of leaues like vnto the milke Thistle, but much smaller, and straked with many white lines or streakes : and at the top of the stalkes come forth faire and goodly yellow floures, very like vnto the sea Poppy, but more elegant, and of greater beauty, hauing in the midst
　　　　　　　　　　　　　　　　　　　　　　　　　　　　　　thereof

thereof a ſmall knop or boll, ſuch as is in the middle of our wilde Poppy, but full of ſharpe thorns, and at the end thereof a ſtaine or ſpot of a deepe purple : after the yellow floures be fallen, this foreſaid knop groweth by degrees greater and greater, vntill it come to full maturity, which openeth it ſelfe at the vpper end, ſhewing his ſeed, which is very blacke and round like the ſeed of muſtard. The whole plant and each part thereof doth yeeld very great abundance of milky juyce, which is of a golden colour, falling and iſſuing from any part thereof, if it be cut or bruiſed: the whole plant periſheth at the approch of Winter. The vertues hereof are yet vnknowne vnto me, wherefore I purpoſe not to ſet downe any thing thereof by way of conjecture, but ſhall, God willing, be ready to declare that which certaine knowledge and experience either of mine owne or others, ſhall make manifeſt vnto me.

¶ *The Place.*

The golden Thiſtle is ſowne in gardens of the Low-Countries. *Petrus Bellonius* writes, That it groweth plentifully in Candy, and alſo in moſt places of Italy : *Cluſius* reporteth that he found it in the fields of Spaine, and of the kingdome of Caſtile, and about Montpelier, with fewer branches, and of a higher growth.

The Indian Thiſtle groweth in Saint Iohns Iſland in the Weſt Indies, and proſpereth very well in my garden.

¶ *The Time.*

They floure from Iune to the end of Auguſt: the ſeed of the Indian golden Thiſtle muſt bee ſowne when it is ripe, but it doth not grow vp vntill May next after.

¶ *The Names.*

This Thiſtle is called in Latine, *Carduus Chryſanthemus* : in Greeke of *Theophraſtus*, Σκολυμος: for thoſe things which he writeth of *Scolymus* in his ſixth and ſeuenth bookes doe wholly agree with this Thiſtle *Chryſanthemus* : which are theſe ; *Scolymus*, doth floure in the Summer Solſtice, brauely and a long time together ; it hath a root that may be eaten both ſod and raw, and when it is broken it yeeldeth a milky juyce : *Gaza* nameth it *Carduus*. Of this *Pliny* alſo makes mention *lib.21.cap.16*. *Scolymus*, ſaith he, differs from thoſe kindes of Thiſtles, *viz. Acarna*, and *Atractilis*, becauſe the root thereof may bee eaten boyled. Againe, *lib.22.cap.22*. The Eaſt Countries vſe it as a meate : and he calleth it by another name Λιμωνιον. Which thing alſo *Theophraſtus* ſeemeth to affirme, in his ſixt booke ; for when hee reckoneth vp herbes whoſe leaues are ſet with prickles, he addeth *Scolymus*, or *Limonia*.

Notwithſtanding, *Pliny* maketh mention likewiſe of another *Scolymus*, which he affirmeth to bring forth a purple floure, and betweene the middle of the prickes to wax white quickely, and to fall off with the winde ; in his twentieth booke, *cap.23*. Which Thiſtle doubtleſſe doth not agree with *Carduus Chryſanthemus*, that is, with *Theophraſtus* his *Scolymus*, and with that which we mentioned before : ſo that there be in *Pliny* two *Scolymi* ; one with a root that may be eaten, and another with a purple floure, turning into downe, and that ſpeedily waxeth white. *Scolymus* is likewiſe deſcribed by *Dioſcorides* ; but this differs from *Scolymus Theophraſti*, and it is one of thoſe which *Pliny* reckoneth vp, as we will more at large declare hereafter. But let vs come againe to *Chryſanthemus* : This the inhabitants of Candy, keeping the markes of the old name, do call *Aſcolymbros* : the Italians name it *Anconitani Rinci*: the Romans, *Spina borda*: the Spaniards, *Cardon lechar*, and of diuers it is alſo named *Glycyrrhizon*, that is to ſay, *dulcis Radix*, or ſweet root : it is called in Engliſh, golden Thiſtle : ſome would haue it to be that which *Vegetius* in *Arte Veterinaria* calls *Eryngium*: but they are deceiued ; for that *Eryngium* whereof *Vegetius* writeth, is *Eryngium marinum*, or ſea Huluer, of which we will intreat.

The golden Thiſtle of India may be called *Carduus Chryſanthemus*, of his golden colour, adding thereto his natiue country *Indianus*, or *Peruanus*, or the golden Indian Thiſtle, or the golden Thiſtle of Peru : the ſeed came to my hands by the name *Fique del Inferno* : in Latine, *Ficus infernalis*, the infernall fig, or fig of hell.

¶ *The Temperature and Vertues.*

A   The root and tender leaues of this *Scolymus*, which are ſometimes eaten, are good for the ſtomacke, but they containe very little nouriſhment, and the ſame thinne and waterie, as *Galen* teacheth.

B   *Pliny* ſaith that the root hereof was commended by *Eratoſthenes*, in the poore mans ſupper, and that it is reported alſo to prouoke vrine eſpecially ; to heale tetters and dry ſcurfe, being taken with vinegre ; and with wine to ſtir vp fleſhly luſt, as *Heſiod* and *Alcæus* teſtifie ; and to take away the ſtench of the arme-holes, if an ounce of the root, the pith picked out, be boyled in three parts of wine, till one part be waſted, and a good draught taken faſting after a bath, and likewiſe after meat :

which

which later words *Dioscorides* likewise hath concerning his *Scolymus* : out of whom *Pliny* is thought to haue borrowed these things,

✝ The plant our Author here describes in the second place, is that which I described and figured formerly pag.401 .by the name of *Papauer spinosum*. I must confesse, I there should haue omitted it, because it is here set forth sufficiently by our Authour, whereof indeed I had a little remembrance, and therefore at that time sought his Index by all the names I could remember, but not making it a *Carduus*, I at that time missed thereof ; but here finding it, I haue let the history stand as it was, and onely omitted the figure which you may finde before, and something also in the history not here deliuered.

CHAP. 481.   *Of white Carline Thistle of* Dioscorides.

¶ *The Description.*

1   THe leaues of Carline are very full of prickles, cut on both edges with a multitude of deepe gashes, and set along the corners with stiffe and very sharpe prickles ; the middle ribs whereof are sometimes red : the stalke is a span high or higher, bringing forth for the most part onely one head or knap being full of prickles, on the outward circumference or compasse like the Vrchin huske af a chesnut : and when this openeth at the top, there groweth forth a broad floure, made vp in the middle like a flat ball, of a great number of threds, which is compassed about with little long leaues, oftentimes somewhat white, very seldome red : the seed vnderneath is slender and narrow, the root is long, a finger thicke, something blacke, so chinked as though it were split in sunder, sweet of smell, and in taste somewhat bitter.

‡ 1 *Carlina caulescens magno flore.*
Tall Carline Thistle.

2 *Carlina, seu Chamæleon albus* Dioscoridis.
The white Carline Thistle of *Dioscorides*
with the red floure.

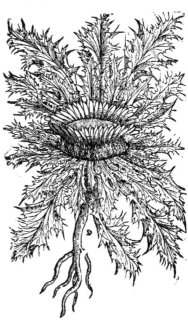

2   There is also another hereof without a stalke, with leaues also very full of prickles, like almost to those of the other, lying flat on the ground on euery side : among which there groweth forth in the middle a round head or knap, set with prickles without after the same manner, but greater: the floure whereof in the middle is of strings, and paled round about with red leaues, and sometimes with white, in faire and calme weather the floures both of this and also of the other lay

themselues

‡ 3 *Carlina acaulos minor flore purp.*
Dwarfe Carline Thiſtle.

themſelues wide open, and when the weather is
foule and miſty, are drawne cloſe together: the
root hereof is long, and ſweet of ſmell, white,
ſound, not nicked or ſplitted as the other.

‡ 3 This ſmall purple Carline Thiſtle hath
a pretty large root diuided oft times at the top
into diuers branches, from which riſe many
greeneleaues lying ſpred vpon the ground, deeply cutand ſet with ſharpe prickles, in the midſt
of theſe leaues come vp ſometimes one, but o-
therwhiles more ſcaly heads, which carry a pretty large floure compoſed of many purple threds
like that of the Knapweed, but larger, and of a
brighter colour, theſe heads grow vſually cloſe
to the leaues, yet ſometimes they ſtand vpon
ſtalkes three or foure inches high: when the
floure is paſt they turne into down, and are carried away with the winde: the ſeed is ſmall and
grayiſh. This growes vpon Blacke-Heath, vpon the chalky hils about Dartford, and in many
ſuch places. It floures in Iuly and Auguſt. *Tra-*
gus calls it *Chamæleon albus, vel exiguus* ; Lobel,
Carduus acaulis, Septentrionalium, and *Chamæleon*
albus, Cordi ; *Cluſius, Carlina minor purpureo flore,*
and hee ſaith, in the opinion of ſome it ſeemes
not vnlike to the *Chamæleon* whereof *Theophra-*
ſtus makes mention, *lib. 6. cap. 3. Hiſt. plant.* ‡

¶ *The Place.*

They both grow vpon high mountaines in
deſart places, and oftentimes by highway ſides:
but that which bringeth forth a ſtalke groweth euery where in Germany, and is a ſtranger in England.

¶ *The Time.*

They floure and ſeed in Iuly and Auguſt, and many times later.

¶ *The Names.*

The former is called in Latine, *Carlina,* and *Cardopatium* ; and of diuers, *Carolina,* of *Charlemaine,*
the firſt Romane Emperor of that name, whoſe army (as it is reported) was in times paſt through
the benefit of this root deliuered and preſerued from the plague : it is called in high Dutch, **Eber-**
wurtz : in low Dutch, French, and other Languages, as likewiſe in Engliſh, *Carline,* and Carline
Thiſtle : it is *Dioſcorides* his *Leucacantha,* the ſtrong and bitter roots ſhew the ſame ; the Faculties
alſo are anſwerable, as forthwith we will declare : *Leucacantha* hath alſo the other names, but they
are counterfeit, as among the Romanes *Gniacardus* ; and among the Tuſcans, *Spina alba,* or White
Thiſtle, yet doth it differ from that Thiſtle which *Dioſcorides* calleth *Spina alba,* of which hee alſo
writeth apart, doth likewiſe attribute to both of them their owne proper faculties and operations,
and the ſame differing.

The later writers do alſo call the other *Carlina altera,* and *Carlina humilis,* or *minor,* low or little
Carline ; but they are much deceiued who goe about to referre them both to the Chamæleons; for
in Italy, Germany, or France, *Chamæleones,* the Chamæleons doe neuer grow, as there is one witneſſe
for many, *Petrus Bellonius,* in his fifth booke of Singularities, who ſufficiently declareth what difference there is betweene the Carlines and the Chamæleons; which thing ſhall be made manifeſt
by the deſcription of the Chamæleons.

¶ *The Temperature and Vertues.*

A   The root of Carline, which is chiefely vſed, is hot in the later end of the ſecond degree, and dry
in the third, with a thinneſſe of parts and ſubſtance; it procureth ſweat, it driueth forth all kinde of
wormes of the belly, it is an enemy to all maner of poyſons, it doth not onely driue away infections
of the plague; but alſo cureth the ſame if it be drunke in time.

B   Being chewed it helpeth the tooth-ache ; it openeth the ſtoppings of the liuer and ſpleene.
C   It prouoketh vrine, bringeth downe the menſes, and cureth the dropſie.
D   And it is giuen to thoſe that haue been dry beaten, and fallen from ſome high place.

The

The like operations *Dioscorides* hath concerning *Leucacantha*, saying, that it hath a root like Cy- E
perus, bitter and ftrong, which being chewed cafeth the pain of the teeth : the decoction of it with
a draught of wine is a remedie againft pains of the fides, and is good againft the Sciatica or ache
in the huskle bones, as alfo the crampe.

The juice alfo being drunke is of like Vertues.   F

Chap. 482.   *Of wilde Carline Thiftle.*

¶ *The Defcription.*

1   THe great wilde Carline Thiftle rifeth vp with a ftalke a cubit high or higher, diuided
into certaine branches : the leaues are long, and very full of prickles in the edges like
thofe of Carline : the floures grow alfo vpon a prickely head, being fet with threds in
the midft, and paled round about with little yellowifh leaues: the root is flender, and hath a twinging tafte.

2   *Clufius* defcribeth a certain other alfo of this kinde, with one only ftalke, flender, fho t and
not aboue an handfull high, with prickly leaues like thofe of the other, but leffer, both of them couered with a certain hoary down : the heads or knaps are for the moft part two , they haue a pale
downe in the midft, and leaues ftanding round about, being fomewhat ftiffe and yellow: the root is
flender, and of a reddifh yellow.

1 *Carlina fylveftris major.*
The great wilde Carline Thiftle.

2 *Carlina fylveftris minor.*
The little wilde Carline Thiftle.

¶ *The Place.*

The great Carline is found in vntoiled and defart places, and oftentimes vpon hils. ‡ It growes
vpon Black-heath, and in many other places of Kent. ‡

The leffer Carline *Clufius* writeth that he found growing in dry ftony and defart places, about
Salmantica a city of Spain,

¶ *The Time.*

They floure and flouriſh in Iune and Iuly.

¶ *The Names.*

It is commonly called in Latine, and that not vnfitly, *Carlina ſylueſtris*, for it is like to Carline in floures, and is not very vnlike in leaues. And that this is *κάκτος*, it is ſo much the harder to affirm, by how much the briefer *Theophraſtus* hath written herof; for he ſaith that this is like baſtard ſafron, of a yellow colour and fat juice: and *Acorna* differs from *Acarna*: for *Acarna*, as *Heſychius* ſaith, is the Bay tree, but *Acorna* is a prickly plant.

¶ *The Nature and Vertues.*

It is hot, eſpecially in the root, the twinging taſte thereof doth declare; but ſeeing it is of no vſe, the other faculties be vnſearched out.

Chap. 483. *Of Chamæleon Thiſtle.*

¶ *The Kindes.*

THere be two Chamæleons, and both black, the vertue of their roots do differ, and the roots alſo differ in kinde, as *Theophraſtus* declareth.

† 1 *Chamæleon niger.*　　　　　　　　　2 *Chamæleon niger Salmanticenſis.*
The blacke Chamæleon Thiſtle.　　　　　The Spaniſh blacke Chamæleon.

 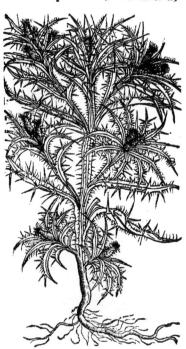

¶ *The Deſcription.*

1　THe leaues of blacke Chamæleon are leſſer and ſlenderer than thoſe of the prickly Artichoke, and ſprinkled with red ſpots: the ſtalke is a cubit high, a finger thicke, & ſomwhat red: it beareth a tufted rundle, in which are ſlender prickly flours of a blew color like the Hyacinth. The root is thicke, black without, of a cloſe ſubſtance, ſometimes eaten away; which being cut is of a yellowiſh colour within, and being chewed it bites the tongue.

2　This

2　This blacke Chamæleon hath many leaues long and narrow,very full of prickles, of a light green in a manner white : the ſtalk is chamfered, a foot high, and diuided into branches, at the tops whereof ſtand purple floures growing forth of prickly heads : the root is blacke, and ſweet in taſt. This is deſcribed by *Cluſius* in his Spaniſh Obſeruations by the name of *Chamæleon Salmanticenſis*, of the place wherein he found it ; for he ſaith that this growes plentifully in the territorie of Salmantica a city of Spain : but it is very manifeſt that this is not blacke Chamæleon, neither doth *Cluſius* affirme it.

¶ *The Place.*

It is very common, ſaith *Bellonius*, in Lemnos, where it beareth a floure of ſo gallant a blew , as it ſeems to contend with the sky in beauty ; and that the floure of Blew-bottle being of this colour, ſeemes in compariſon of it to be but pale. It groweth alſo in the fields neere Abydum, and hard by the riuers of Helleſpont; and in Heraclea in Thracia.

Chamæleon Salmanticenſis growes plentifully in the territorie of Salmantica a city in Spain.

¶ *The Time.*

They floure and flouriſh when the other Thiſtles do.

¶ *The Names.*

The blacke Chamæleon is called in Greeke, χαμαιλέων μέλας : in Latine, *Chamæleon niger* : of the Romans, *Carduus niger*, and *Vernilago* : of ſome, *Crocodilion* : in Engliſh, the Chamæleon Thiſtle, or the Thiſtle that changeth it ſelfe into many ſhapes and colours.

¶ *The Temperature and Vertues.*

The root hereof, as *Galen* ſaith, containeth in it a deadly quality : it is alſo by *Nicander* numbred　A among the poiſonous herbs, in his booke of Treacles ; by *Dioſcorides, lib. 6.* and by *Paulus Ægineta*. Wherefore it is vſed only outwardly, as for ſcabs, morphews, tettars, and to be briefe, for all ſuch things as ſtand in need of clenſing : moreouer, it is mixed with ſuch things as do diſſolue and mollifie, as *Galen* ſaith.

† The figure which was formerly in the firſt place did not agree with the hiſtorie (which was taken out of *Dodonæus*) though Tabern. gàue it for *Chamæleon niger* ; for it is the *Picnomos Creta, &c.* of Lobel. You ſhall finde it hereafter with the *Acarna Valerandi.*

Chap. 484. *Of Sea Holly.*

¶ *The Kindes.*

Dioſcorides maketh mention only of one ſea Holly. *Pliny, lib. 22. cap. 7.* ſeemes to acknowledge two, one growing in rough places, another by the ſea ſide. The Phyſitions after them haue obſerued more.

¶ *The Deſcription.*

1　SEa Holly hath broad leaues almoſt like to Mallow leaues, but cornered in the edges, and ſet round about with hard prickles, fat, of a blewiſh white, and of an aromatical or ſpicy taſte : the ſtalke is thick, about a cubit high, now and then ſomwhat red below : it breaketh forth in the tops into prickly round heads or knops, of the bigneſſe of a Wall-nut, held in for the moſt part with ſix prickly leaues compaſſing the top of the ſtalke round about ; which leaues as well as the heads are of a gliſtering blew : the floures forth of the heads are likewiſe blew, with white threds in the midſt : the root is of the bigneſſe of a mans finger, ſo very long, as that it cannot be all plucked vp but very ſeldome ; ſet here and there with knots, and of taſte ſweet and pleaſant.

2　The leaues of the ſecond ſea Holly are diuerſly cut into ſundry parcels, being all ful of prickles alongſt the edges : the ſtalke is diuided into many branches, and brings forth prickly heads, but leſſer than thoſe of the other : from which there alſo grow forth blew floures, ſeldome yellow : there ſtand likewiſe vnder euerie one of theſe, ſix rough and prickly leaues like thoſe of the other, but thinner and ſmaller : the root hereof is alſo long, blacke without, white within, a finger thick, of taſte and ſmell like that of the other, as be alſo the leaues, which are likewiſe of an aromaticall or ſpicy taſte, and being new ſprung vp and as yet tender, be alſo good to be eaten.

¶ *The*

1 *Eryngium marinum.*
Sea Holly.

2 *Eryngium Mediterraneum.*
Levant ſea Holly.

¶ *The Place.*

Eryngium marinum groweth by the ſea ſide vpon the baich and ſtony ground. I found it growing plentifully at Whitſtable in Kent, at Rie and Winchelſea in Suſſex, and in Eſſex at Landamer lading, at Harwich, and vpon Langtree point on the other ſide of the water, from whence I brought plants for my garden.

Eryngium campeſtre groweth vpon the ſhores of the Mediterranean ſea, and in my garden alſo.

¶ *The Time.*

Both of them do floure after the ſummer Solſtice, and in Iuly.

¶ *The Names.*

This Thiſtle is called in Greeke Ἠρύγγιον: and likewiſe in Latine *Eryngium*: and of *Pliny* alſo *Erynge*: in ſhops, *Eryngus*: in Engliſh, ſea Holly, ſea Holme, or ſea Hulver.

The firſt is called in Latine, *Eryngium marinum*: in low-Dutch euery where, **Cruys diſtl, Eindeloos, Meerwoztele:** in Engliſh, ſea Holly.

The ſecond is named of *Pliny, lib. 22. cap. 8. Centum Capita*, or hundred headed Thiſtle: in high-Dutch, **Manstrew, Brancken diſtil, Radendiſtel:** in Spaniſh, *Cardo corredor*: in Italian, *Eringio* and *Iringo*: it is ſyrnamed *Campeſtre*, or champian ſea Holly, that it may differ from the other.

¶ *The Temperature.*

The roots of them both are hot, and that in a mean, and a little dry alſo, with a thinneſſe of ſubſtance, as *Galen* teſtifieth.

¶ *The Vertues.*

A   The roots of ſea Holly boiled in wine and drunke, are good for them that are troubled with the Collicke, it breaketh the ſtone, expelleth grauel, and helpeth all the infirmities of the kidnies, prouoketh vrine, greatly opening the paſſages, being drunke fifteen daies together.

B   The roots themſelues haue the ſame propertie if they be eaten, and are good for thoſe that bee liuer ſicke, and for ſuch as are bitten with any venomous beaſt: they eaſe cramps, convulſions, and the falling ſickneſſe, and bring down the termes.

The

The roots condited or preſerued with ſugar as hereafter followeth,are exceeding good to be gi- C
uen to old and aged people that are conſumed and withered with age,and which want natural moiſture: they are alſo good for other ſorts of people that haue no delight or appetite to venery,nou-
riſhing and reſtoring the aged,and amending the defects of nature in the yonger.

¶ The manner to condite Eringos.

Refine ſugar fit for the purpoſe,and take a pound of it, the white of an egge,and a pinte of cleer D
water,boile them together and ſcum it,then let it boile vntil it be come to good ſtrong ſyrrup,and
when it is boiled,as it cooleth adde thereto a ſaucer full of roſewater,a ſpoone-full of Cinnamon
water,and a grain of muske,which haue been infuſed together the night before, and now ſtrained :
into which ſyrrup being more than halfe cold, put in your roots to ſoke and infuſe vntill the next
day ; your roots being ordered in manner hereafter following :
Theſe your roots being waſhed and picked, muſt be boiled in faire water by the ſpace of foure
houres,til they be ſoft : then muſt they be pilled clean as ye pil parſneps,& the pith muſt be drawn
out at the end of the root : but if there be any whoſe pith cannot be drawn out at the end,then you
muſt ſlit them and ſo take it out : theſe you muſt alſo keep from much handling,that they may be
clean : let them remain in the ſyrrup till the next day, and then ſet them on the fire in a faire broad
pan vntill they be very hot,but let them not boile at all: let them remain ouer the fire an houre or
more,remoouing them eaſily in the pan from one place to another with a wooden ſlice. This done,
haue in a readineſſe great cap or royall papers, whereupon ſtrow ſome ſugar, vpon which lay your
roots,hauing taken them out of the pan. Theſe papers you muſt put into a ſtouve or hot-houſe to
harden ; but if you haue not ſuch a place,lay them before a good fire : in this maner if you condite
your roots, there is not any that can preſcribe you a better way. And thus you may condite any
other root whatſoeuer,which will not only be exceeding delicat,but very wholſome,and effectual
againſt the diſeaſes aboue named.

Aetius ſaith,a certain man affirmed, that by the continual vſe of ſea Holly he neuer after voided E
any ſtone,when as before he was very often tormented with that diſeaſe.

It is drunke,ſaith <i>Dioſcorides,</i>with Carrot ſeed againſt very many infirmities, in the weight of a F
dram.

The juice of the leaues preſſed forth with wine,is a remedy for thoſe that are troubled with the G
running of the reins.

They report of the herb ſea Holly,if one goat take it into her mouth,it cauſeth her firſt to ſtand H
ſtill,and afterwards the whole flocke, vntill ſuch time as the ſheepheard take it from her mouth.
Plutarch.

CHAP. 485. Of baſtard Sea Hollies.

¶ The Deſcription.

1　THis <i>Eryngium</i>, which <i>Dodonæus</i> in his laſt edition calleth <i>Eryngium planum</i>; and <i>Pena</i>
more fitly and truly,<i>Eryngium Alpinum cæruleum,</i>hath ſtalks a cubit and a half high,ha-
uing ſpaces between euery joint : the lower leaues are greater and broader and notched
about the edges,but thoſe aboue are leſſe, compaſſing or enuironing each joint ſtar-faſhion, beſet
with ſoft tender pricks not much hurtfull to the hands of ſuch as touch them : the knobs or heads
are alſo prickly,in colour blew. The root is bunchy or knotty like that of Elecampane,black without,white within,like the Eringes in ſweetneſſe and taſte.

2　The ſecond baſtard ſea Holly , whoſe picture is ſet forth in <i>Dodonæus</i> his laſt edition verie
gallantly,being alſo a kinde of Thiſtle,hath leaues like the former Erynges, but broader next the
roots,than thoſe that grow next the ſtalks,ſomwhat long,greeniſh,ſoft,and not prickly,but lightly
creviſed or notched about the edges,greater than Quince leaues. The ſtalks grow more than a cubit high,on the tops whereof there hang downwards fiue or ſix knobs or heads, in colour and flours
like the other ; hauing three or foure whitiſh roots of a foot long.

3　This third kinde of baſtard <i>Eryngium</i> hath his firſt leaues (which do grow next the ground)
great,broad,and ſoft,growing as it were in a rundle about the root. The ſtalke is ſmall and ſlender,
diuided into ſome branches,which beare many little leaues turning or ſtanding many waies,which
be

1 *Eryngium cæruleum.*
Blew ſe a Holly.

2 *Eryngium ſpurium primum Dod.*
Baſtard ſea Holly.

3 *Eryngium pumilum Cluſij.*
Dwarfe ſea Holly.

4. *Eryngium Montanum.*
Mountaine ſea Holly.

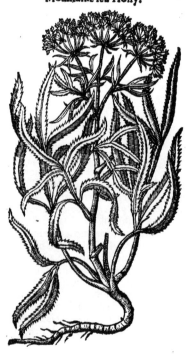

‡ 3 *Eryngium puſillum planum*
Small ſmooth ſea Holly.

be alſo ſlender,prickly,and ſet about the ſtalks ſtar-faſhion. The knobs or heads growing at the tops of the branches are round and prickely, bearing lit-tle blew flours and leaues which compaſſe them a-bout : the root is ſlender,and laſts but one yeare.

4 The fourth kinde of baſtard ſea-Holly,which *Pena* calleth *Eryngium montanum recentiorum*, and is the fourth according to *Dodonæus* his account,is like. the Erynges not in ſhape but in taſte : this beareth a very ſmall and ſlender ſtalke of a meane height, wheron grow three or foure leaues, and ſeldom fiue, made of diuers leaues ſet vpon a middle rib,narrow, long,hard,and of a darke green colour,dented vpon both edges of the leafe like a ſaw : the ſtalke is a cu-bit high,iointed or kneed,and diuiding it ſelfe into many branches, on the tops whereof are round tufts or vmbels wherein are contained the floures,and af-ter they be vaded, the ſeeds, which are ſmall, ſome-what long,well ſmelling,and ſharp in taſte:the root is white and long,not a finger thicke,in taſte ſweer, but afterwards ſomewhat ſharpe,and in ſent and ſa-uor not vnpleaſant : when the root is dried it may be crumbled in pieces,and therefore quickly braied

‡ 5 This is a low plant preſently from the root di-uided into ſundry branches,ſlender,round, & lying on the ground : at each joint grow leaues without any certain order,broad toward their ends, and nar-rower at their ſetting on, ſnipt about their edges. Thoſe next the root were ſome inch broad, and two or more long, of a yellowiſh green colour:the ſtalks are parted into ſundry branches , and at each joynt haue little leaues,and rough green heads with blewiſh flours in them : the roots creep and are ſom-what like thoſe of Aſparagus. This neither *Cluſius* nor *Lobel* found wilde,but it grew in the garden of *Iohn Mouton* of Tournay,a learned Apothecarie very skilfull in the knowledge of plants:wher-upon they both called it *Eryngium puſillum planum Moutoni*. ‡

¶ *The Place.*
Theſe kinds of ſea Holly are ſtrangers in England, we haue the firſt and ſecond in our London gardens.

¶ *The Time.*
They floure and flouriſh when the Thiſtles do.

¶ *The Names.*
Theſe plants be *Eryngia ſpuria,*or baſtard ſea Hollies,being but lately obſerued,& therfore haue no old names.

The firſt may be called in Latine,*Eryngium Boruſſicum*,or *Non ſpinoſum*;ſea Holly without pric-kles.

The ſecond is called by *Matthiolus,Eryngium planum,*or flat ſea Holly : others had rather name it *Alpinum Eryngium,*or ſea Holly of the Alps.

The third is rightly called *Eryngium pumilum,*little ſea Hulver.

Matthiolus makes the fourth to be *Crithmum quartum,*or the fourth kind of Sampier:others, as *Dodonæus* and *Lobel,*haue made it a kinde of ſea Hulver.

¶ *The Nature and Vertues.*
Touching the faculties hereof we haue nothing to ſet downe, ſeeing they haue as yet no vſe in. meat or medicine. But that they be hot the very taſte doth declare.

CHAP.

Cʜᴀᴘ. 486. *Of Starre Thiſtle.*

¶ *The Deſcription.*

1   THe Star Thiſtle, called *Carduus ſtellatus,* hath many ſoft frizled leaues, deepely cut or
gaſht, altogether without prickles : among which riſeth vp a ſtalke, diuiding it ſelf into many other branches, growing two foot high ; on the tops whereof are ſmall knops
or heads like the other Thiſtles, armed round about with many ſharpe prickles, faſhioned like a
blaſing ſtarre, which at the beginning are of a purple colour, but afterwards of a pale bleake or
whitiſh colour : the ſeed is ſmall, flat, and round, the root is long and browne without.

1 *Carduus ſtellatus.*
The Srar-Thiſtle.

† 2 *Carduus Solſtitialis.*
S. Barnabies Thiſtle.

2   S. Barnabies Thiſtle is another kinde of Star-Thiſtle, notwithſtanding it hath prickles no
where ſaue in the head only, and the prickles ſtand forth of it in manner of a ſtar : the ſtalks are two
cubits high, parted into diuers branches ſofter than are thoſe of Star-Thiſtle : which ſtalkes haue
velms or thin skins cleauing vnto them all in length, by which they ſeem to be foure ſquare. The
leaues are ſomewhat long, ſet with deep gaſhes on the edges : the floures are yellow, and conſiſt of
threds, the ſeed is little, the root long and ſlender.

¶ *The Place.*

The two firſt grow vpon barren places neere cities and townes, almoſt euery where.

¶ *The Time.*

They floure and flouriſh eſpecially in Iuly and Auguſt.

¶ *The Names.*

The firſt is called in Latine *Stellaria,* as alſo *Carduus ſtellatus,* and *Carduus Calcitrapa :* but they are
deceiued who take it for *Eryngium,* or ſea Holly, or any kinde thereof. *Matthiolus* ſaith it is called
in Italian, *Calcatrippa :* in high-Dutch, 𝔪𝔞𝔩𝔩𝔢𝔫 𝔡𝔦𝔰𝔱𝔢𝔩 : in low-Dutch, 𝔖𝔱𝔢𝔯𝔯𝔢 𝔡𝔦𝔰𝔱𝔢𝔩 : in French,
Chauſſetrappe : in Engliſh, ſtar Thiſtle.

S. Barnabies Thiſtle is called in Latine, *Solſtitialis ſpina,* becauſe it floureth in the ſummer Solſtice

ſtice, as *Geſner* ſaith, or rather becauſe after the Solſtice the prickles thereof be ſharpeſt: of *Guillan-dinus*, *Eryngium*, but not properly, and *Stellaria Horatij Augerij*, who with good ſucceſſe gaue it againſt the ſtone, dropſies, green ſickneſſe, and quotidian feuers. It is called in Engliſh as aboue ſaid, Saint Barnabies Thiſtle.

¶ *The Temperature.*

The Star-Thiſtle is of a hot temperature.

¶ *The Vertues.*

The ſeed is commended againſt the ſtrangurie : it is reported to driue forth the ſtone, if it bee A drunke with wine.

Baptiſta Sardus affirmeth, that the diſtilled water of this Thiſtle is a remedy for thoſe that are in- B fected with the French Pox, and that the vſe of this is good for the liuer, that it taketh away the ſtoppings thereof.

That it clenſeth the bloud from corrupt and putrified humours.　C

That it is giuen with good ſucceſſe againſt intermitting feuers : whether they be quotidian or D tertian.

As touching the faculties of Saint Barnabies Thiſtle, which are as yet not found out, we haue E nothing to write.

† There were formerly three figures and deſcriptions in this chapter, and all of them out of the 14. and 15. chapter of the fifth booke, and fifth *Pemptas* of *Do-dxaus*, but the firſt and ſecond figures were both of the firſt deſcribed, the third figure was of the *Acanthium peregrinum* of *Tabernæmontanus*, which *Bauhine* knowes not what to make of, but I thinke it was drawne for, and (if the tuberous clogs of the roots were ſomewhat large) might very well ſerue for the *Cirſium maximum Alphadi radice*, whoſe figure as I drew it from the plant I will hereafter giue you: the third deſcription was of the *Iacea maior lutea*, deſcribed in the third place of the 249. Chap. pag. 727.

CHAP. 487.　*Of Teaſels.*

¶ *The Kindes.*

OVr age hath ſet downe two kindes of Teaſels: the tame, and the wilde. Theſe differ not ſaue on-ly in the husbanding ; for all things that are planted and manured doe more flouriſh, and be-come for the moſt part fitter for mans vſe.

1 *Dipſacus ſativus.*	2 *Dipſacus ſylveſtris.*
Garden Teaſell.	Wilde Teaſell.

‡ 3 *Dipſacus minor, ſiue Virga paſtoris.*
Shepheards rod.

¶ *The Deſcription.*

1 GArden Teaſell is alſo of the number of the Thiſtles; it bringeth forth a ſtalke that is ſtraight, very long, iointed, and ful of prickles: the leaues grow forth of the ioynts by couples, not onely oppoſite or ſet one right againſt another, but alſo compaſſing the ſtalke about, and faſtened together; and ſo faſtened, that they hold dew and raine water in manner of a little baſon: theſe be long, of a light greene colour, and like to thoſe of Lettice, but full of prickles in the edges, and haue on the outſide all alongſt the ridge ſtiffer prickles: on the tops of the ſtalkes ſtand heads with ſharpe prickles like thoſe of the Hedge-hog, and crooking backward at the point like hookes: out of which heads grow little floures: The ſeed is like Fennell-ſeed, and in taſte bitter: the heads wax white when they grow old, and there are found in the midſt of them when they are cut, certaine little magots: the root is white, and of a meane length.

2 The ſecond kinde of Teaſell which is alſo a kinde of Thiſtle, is very like vnto the former, but his leaues are ſmaller and narrower: his flours of a purple colour, and the hooks of the Teaſell nothing ſo hard or ſharpe as the other, nor good for any vſe in dreſſing of cloath.

3 There is another kinde of Teaſell, being a wilde kinde thereof, and accounted among theſe Thiſtles, growing higher than the reſt of his kindes; but his knobbed heads are no bigger than a Nutmeg, in all other things elſe they are like to the other wilde kindes. ‡ This hath the lower leaues deeply cut in with one gaſh on each ſide at the bottome of the leafe, which little eares are omitted in the figure: the leaues alſo are leſſe than the former, and narrower at the ſetting on, and hold no water as the two former do: the whole plant is alſo much leſſe. ‡

¶ *The Place.*

The firſt called the tame Teaſell is ſowne in this country in gardens, to ſerue the vſe of Fullers and Clothworkers.

The ſecond kinde groweth in moiſt places by brookes, riuers, and ſuch like places.

The third I found growing in moiſt places in the high way leading from Braintree to Henningham caſtle in Eſſex, and not in any other place except here and there a plant vpon the high way from Much-Dunmow to London. ‡ I found it growing in great plenty at Edgecombe by Croydon, cloſe by the gate of the houſe of my much honoured friend Sir *Iohn Tunſtall.* ‡

¶ *The Time.*

Theſe floure for the moſt part in Iune and Iuly.

¶ *The Names.*

Teaſell is called in Greeke, δίψακος, and likewiſe in Latine, *Dipſacus, Labrum Veneris,* and *Carduus Veneris:* it is termed *Labrum Veneris,* and *Lauer Lauacrum,* of the forme of the leaues made vp in faſhion of a baſon, which is neuer without water: they commonly call it *Virga paſtoris minor,* and *Carduus fullonum:* in high Dutch, **Garden Diſtell:** in low Dutch, **Kaerden:** in Spaniſh, *Cardencha:* and *Cardo Penteador:* in Italian, *Diſſaco,* and *Cardo:* in French, *Chardon de ſoullon, Verge à bergier:* in Engliſh, Teaſell, Carde Teaſell, and Venus baſon.

The third is thought to be *Galedragon Plinij:* of which he hath written in his 27. booke, the tenth Chapter.

¶ *The Temperature.*

The roots of theſe plants are dry in the ſecond degree, and haue a certaine clenſing faculty.

¶ *The*

¶ *The Vertues.*

There is small vse of Teasell in medicines : the heads (as we haue said) are vsed to dresse wool- A
len cloth with.

`Dioscorides` writeth, That the root being boyled in wine, and stamped till it is come to the sub- B
stance of a salue, healeth chaps and fistulaes of the fundament, if it be applied thereunto ; and that
this medicine must bee reserued in a box of copper, and that also it is reported to be good for all
kindes of warts.

It is needlesse here to alledge those things that are added touching the little wormes or magots C
found in the heads of the Teasell, and which are to be hanged about the necke, or to mention the
like thing that *Pliny* reporteth of Galedragon : for they are nothing else but most vaine and trifling
toies, as my selfe haue proued a little before the impression hereof, hauing a most grieuous ague,
aud of long continuance : notwithstanding Physicke charmes, these worms hanged about my neck,
spiders put into a walnut shell, and diuers such foolish toies that I was constrained to take by fan-
tasticke peoples procurement, notwithstanding, I say, my helpe came from God himselfe, for these
medicines and all other such things did me no good at all.

† The figure which formerly was put into the second place, was of the *Dipsacus secundus* of *Tabernamontanus*, which differs from our common one, in that the leaues
are deeply diuided, or cut in on their edges.

CHAP. 488. *Of Bastard Saffron.*

‡ 1 *Carthamus siue Cnicus.*   † 2 *Cnicus alter caruleus.*
    Bastard Saffron.   Blew floured Bastard Saffron.

¶ *The Description.*

1   CNicus, called also bastard Saffron, which may very wel be reckoned among the Thistles,
riseth vp with a stalke of a cubite and a halfe high, straight, smooth, round, hard, and
wooddy, & branched at the top : it is defended with long leaues, somthing broad, sharp
                Fffff   pointed,

pointed; and with prickles in the edges ; from the tops of the stalks stand out little heads or knops of the bignesse of an Oliue or bigger, set with many sharp pointed and prickly scales: out of which come forth floures like threds, closely compact, of a deepe yellow shining colour, drawing neere to the colour of Saffron : vnder them are long seeds, smooth, white, somewhat cornered, bigger than a Barly corne, the huske whereof is something hard, the inner pulpe or substance is fat, white, sweet in taste : the root slender and vnprofitable.

2   There is also another kinde of Bastard Saffron, that may very well be numbred amongst the kindes of Thistles, and is very like vnto the former, sauing that his flockie or threddy floures are of a blew colour : the root is thicker, and the whole plant is altogether more sharpe in prickles : the stalkes are also more crested and hairy.

¶ *The Place.*

It is sowne in diuers places of Italy, Spaine, and France, both in Gardens and in Fields : *Pliny lib.25.cap.15.*saith, that in the raigne of *Vespasian* this was not knowne in Italy : being in Ægypt onely of good account, and that they vsed to make oyle of it, and not meat.

¶ *The Time.*

The floures are perfected in Iuly and August : the root after the seed is ripe, the same yeare it is sowne withereth away.

¶ *The Names.*

It is called in Greeke, κνίκος: in Latine also *Cnicus*, or *Cnecus* : in shops, *Cartamus*, or *Carthamum*: of diuers, *Crocus hortensis*, and *Crocus Saracenicus* : in Italian, *Zaffarano Saracinesco*, and *Zaffarano saluatico*: in Spanish, *Alasor*, and *Semente de papagaios* : in high Dutch, **Wilden Zaffron**: in French, *Safran Sauuage* : in English, Bastard Saffron : of some, Mocke Saffron, and Saffron D'orte, as though you should say, Saffron *de horto*, or of the garden. *Theophrastus* and *Pliny* call it *Cnicus vrbana*, and *satiua*, or tame and garden bastard Saffron, that it may differ from *Atractylis*, which they make to be a kinde of *Cnicus syluestris*, or wilde Bastard Saffron, but rather a *species* of the Holy Thistle.

¶ *The Temperature.*

We vse, saith *Galen*, the seed only for purgations : it is hot, and that in the first degree, as *Mesues* writeth.

¶ *The Vertues.*

A   The juyce of the seed of bastard Saffron bruised and strained into honied water or the broth of a chicken, and drunke, prouoketh to the stoole, and purgeth by siege slimy flegme, and sharp humors: Moreouer it is good against the collicke, and difficulty of taking breath, the cough, and stopping of the brest, and is singular against the dropsie.

B   The seed vsed as aforesaid, and strained into milke, causeth it to curdle and yeeld much cruds, and maketh it of great force to loose and open the belly.

C   The floures drunke with honied water open the liuer, and are good against the jaundise : and the floures are good to colour meat in stead of Saffron.

D   The seed is very hurtfull to the stomacke, causing desire to vomit, and is of hard slow digestion, remaining long in the stomacke and entrails.

E   Put to the same seed things comfortable to the stomacke, as Annise seed, Galingale, or Mastick, Ginger, *Sal gemma*, and it shall not hurt the stomacke at all, and the operation thereof shall be the more quicke and speedy.

F   Of the inward pulpe or substance hereof is made a most famous and excellent composition to purge water with, commonly called *Diacarthamon*, a most singular and effectuall purgation for those that haue the dropsie.

G   The perfect description hereof is extant in *Guido* the Surgion, in his first Doctrine, and the sixt Tractat.

H   We haue not read, or had in vse that Bastard Saffron with the blew floure, and therefore can say nothing of his vertues.

† The figure formerly was of the *Cnicus carnient!*

CHAP. 489.   *Of Wilde Bastard Saffron.*

¶ *The Description.*

1   A *Tractylis*, otherwise called wilde Bastard Saffron, bringeth forth a straight and firme stalke, very fragile or brittle, diuided at the toppe into certaine branches : it hath

long,

long jagged leaues set with prickles: the heads on the tops of the branches are very full of sharpe prickles:out of which grow flours all of threds, like those of bastard Saffron,but they are of a light yellow colour, and sometimes purple: the feed is somwhat great, browne, and bitter, otherwise like that of bastard Saffron: the root is of a meane bignesse.

1 *Atractylis.*
Wilde Bastard Thistle.

2 *Carduus Benedictus.*
The Blessed Thistle.

2   The stalkes of *Carduus Benedictus*, or Blessed Thistle, are round, rough and pliable, and being parted into diuers branches, do lie flat on the ground: the leaues are jagged round about, and full of harmlesse prickles in the edges: the heads on the tops of the stalkes are set with prickles, and inuironed with sharpe prickling leaues, out of which standeth a yellow floure: the feed is long, and set with haires at the top like a beard: the root is white, and parted into strings: the whole herb, leaues and stalkes, and also the heads, are couered with a soft and thin downe.

¶ *The Place.*

Atractylis groweth in Candie, and in diuers prouinces and Islands of Greece, and also in Languedocke: and is an herbe growing in our English gardens.

Carduus Benedictus is found euery where in Lemnos, an Island of the Midland Sea, in Champion grounds, as *Petrus Bellonius* testifieth: it is diligently cherished in Gardens in thefe Northerne parts.

¶ *The Time.*

Atractylis is very late before it floureth and feedeth.

Carduus Benedictus floureth in Iuly and August, at which time it is especially to be gathered for Physicke matters.

¶ *The Names.*

Atractylis is called in Greeke, Ἀτρακτυλὶς ἄγρια: of the Latines likewise, *Atractylis*, and *Cnycus sylvestris*; and becaufe women in the old time were wont to vfe the stiffe stalke therof *pro fuso aut colo*, for a spindle or a distaffe, it is named *Fufus agrestis*, and *Colus Rustica*; which thing *Petrus Bellonius* reporteth the women in Greece do also euen at this day; who call *Atractylis* by a corrupt name *Ardactilis*, diuers of the later herbarists name it *Sylvestris Carthamus*: that is to say in low Dutch, **wilden Car-thamus**: and in English, wilde Bastard Saffron, or Spindle Thistle.

Blessed Thistle is called in Latine euery where *Carduus Benedictus*, and in shops by a compound
Fffff 2   word,

word, *Cardo-benedictus* : it is moſt plaine, that it is *Species Atractylidis*, or a kind of wilde baſtard Saffron: it is called *Atractylis hirſutior*, hairie wilde baſtard Saffron: *Valerius Cordus* nameth it *Cnicus ſapinus*. it is called in high Dutch, **Beſeegnete diſtell, Lardo Benedict** : the later name whereof is known to the Low Country-men: in Spaniſh it is called *Cardo Sancto* : in French, *Chardon benoiſt*, or *benciſt* : in the Iſle Lemnos, *Garderacantha* : in Engliſh, Bleſſed Thiſtle, but more commonly by the Latine name *Carduus Benedictus*.

¶ The Temperature.

Wilde baſtard Saffron doth dry and moderately digeſt, as *Galen* witneſſeth.

As *Cardus benedictus* is better, ſo is it alſo hot and dry in the ſecond degree, and withall clenſing and opening.

¶ The Vertues.

A   The tops, ſeed, and leaues of *Atractylis*, ſaith *Dioſcorides*, being beaten and drunke with pepper and wine, are a remedy for thoſe that are ſtung of the Scorpion.

B   Bleſſed Thiſtle taken in meat or drinke, is good for the ſwimming and giddineſſe of the head, it ſtrengthneth memory, and is a ſingular remedy againſt deafeneſſe.

C   The ſame boiled in wine and drunke hot, healeth the griping paines of the belly, killeth and expelleth wormes, cauſeth ſweat, prouoketh vrine, and driueth out grauell ; clenſeth the ſtomacke, and is very good againſt the Feuer quartaine.

D   The juyce of the ſaid *Carduus* is ſingular good againſt all poyſon, as *Hierome Bocke* witneſſeth, in what ſort ſoeuer the medicine be taken; and helpeth the inflammation of the liuer, as reporteth *Ioachimus Camerarius* of Noremberg.

E   The pouder of the leaues miniſtred in the qnantitie of halfe a dram, is very good againſt the peſtilence, if it be receiued within 24. houres after the taking of the ſickeneſſe, and the party ſweat vpon the ſame : the like vertue hath the wine, wherein the herbe hath beene ſodden.

F   The greene herb pounded and laid to, is good againſt all hot ſwellings, as *Eryſipelas*, plague-ſores, and botches, eſpecially thoſe that proceed of the peſtilence, and is alſo good to be laid vpon the bitings of mad dogs, ſerpents, ſpiders, or any venomous beaſts whatſoeuer ; and ſo is it likewiſe if it be inwardly taken.

G   The diſtilled water thereof is of leſſe vertue.

H   It is reported that it likewiſe cureth ſtubborne and rebellious vlcers, if the decoction be taken for certaine daies together ; and likewiſe *Arnoldus de Villa noua* reporteth, that if it be ſtamped with Barrows greaſe to the form of an vnguent, adding thereto a little wheat floure, it doth the ſame, being applied twice a day.

I   The herbe alſo is good being ſtamped and applied, and ſo is the juyce thereof.

K   The extraction of the leaues drawne according to Art, is excellent good againſt the French diſeaſe, and quartaine ague, as reporteth the foreſaid *Camerarius*.

L   The ſame Author reporteth, that the diſtilled water taken with the water of Louage, and Dodder, helpeth the ſauce-flegme face, if it be drunke for certaine daies together.

CHAP. 490.
Of Thiſtle vpon Thiſtle, and diuers other Wilde Thiſtles.

¶ The Deſcription.

1   AMong all the Thornes and Thiſtles, this is moſt full of prickles ; the ſtalks thereof are very long, and ſeeme to be cornered by reaſon of certaine thin skins growing to them, being ſent downe forth of the leaues : the leaues are ſet round about with many deepe gaſhes, being very full of prickles as well as the ſtalks : the heads are very thicke ſet in euery place with ſtiffe prickles, and conſiſt of a multitude of ſcales ; out of which grow purple floures, as they do out of other Thiſtles, ſeldome white : the root is almoſt ſtraight, but it groweth not deep.

2   To this alſo may be referred that which *Lobel* writeth to be named of the Italians *Leo*, and *Carduus ferox*, for it is ſo called of the wonderfull ſharpe and ſtiffe prickles, wherewith the whole plant aboundeth ; the ſtalke thereof is ſhort, ſcarce a handfull high ; the floure groweth forth of a prickly head, and is of a pale yellow colour, like that of wilde baſtard Saffron, and it is alſo inuironed and ſet round about on euery ſide with long hard thornes and prickles.

3   The third groweth ſeldome aboue a cubit or two foot high : it bringeth forth many round ſtalkes, parted into diuers branches ; the leaues are like thoſe of white Cotton Thiſtle, but leſſer, and blacker, and not couered with downe or Cotton : vpon the tops of the ſtalks grow little heads
like

† 1 *Polyacanthos.*
Thiſtle vpon Thiſtle.

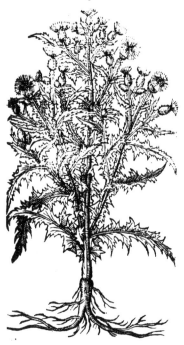

2 *Carduus ferox.*
The cruell Thiſtle.

† 3 *Carduus Aſinius ſive Onopyxos.*
The Aſſes Thiſtle, or Aſſes box.

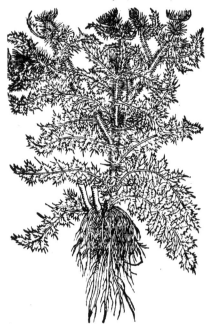

‡ 4 *Carduus vulgatiſſimus viarum.*
The Way Thiſtle.

like Hedge-hogs,out of which fpring gallant purple floures, that at length are turned into downe, leauing feeds behinde them like thofe of the other Thiftles : the root confifteth of many fmall ftrings.

4   The fourth rifeth vp with an higher ftalke,now and then a yard long,round and not fo full of branches nor leaues,which are fharpe and full of prickles, but leffer and narrower : the heads be alfo leffer, longer, and not fo full of ftiffe prickles : the floures are of a white colour, and vanifh into downe : the root is blacke,and of a foot long.

5   This wilde Thiftle which groweth in the fields about **Cambridge**, hath an vpright ftalke, whereon do grow broad prickly leaues : the floures grow at the tops of the branches, confifting of a flockie downe,of a white colour tending to purple, of a moft pleafant fweet fmell, ftriuing with the fauour of muske : the root is fmall, and perifheth at the approch of Winter. ‡ I had no figure directly fitting this ; wherefore I put that of *Dodonæus* his *Onopordon*,which may well ferue for it, if the leaues were narrower,and more diuided. ‡

† 5 *Carduus Mufcatus.*   6 *Carduus lanceatus.*
The musked Thiftle.   The Speare Thiftle.

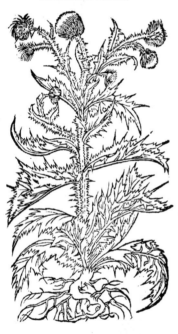 

6   The Speare Thiftle hath an vpright ftalke,garnifhed with a skinnie membrane, full of moft fharpe prickles : whereon do grow very long leaues, diuided into diuers parts with fharp prickles ; the point of the leaues are as the point of a fpeare, whereof it tooke his name : the floures grow on the tops of the branches,fet in a fcaly prickly head, like vnto the heads of Knapweed in forme, confifting of many threds of a purple colour : the root confifteth of many tough ftrings.

7   *Theophraftus* his fifh Thiftle called *Acarna*, which was brought from Illyria to Venice, by the learned *Valerandus Donrez*, defcribed by *Theophraftus*, hath horrible fharpe yellow prickles, fet vpon his greene indented leaues, which are cornered on the backe fide with an hoary downe (as all the reft of the plant) hauing a ftalke of a cubit and a halfe high,and at the top certaine fcaly knops containing yellow thrummie floures,armed or fenced with horrible fharp prickles : the root is long and threddy.

8   The other kinde of fifh Thiftle, being alfo another *Acarna* of *Valerandus* defcription, hath long and large leaues,fet full of fharp prickles,as though it were fet full of pins:all the whole plant is couered with a certain hoarineffe, like the former : there arifeth vp a ftalke nine inches long, yea in fome fertile grounds a cubit high, bearing the floure of *Carduus benedictus*, ftanding thicke together,but leffer.

‡ 9 This

7 *Acarna* Theophraſti.
Theophraſtus his fiſh Thiſtle.

8 *Acarna Valerandi* Donrez.
Donrez his fiſh Thiſtle.

† 9 *Picnomos.*
The thicke or buſh headed Thiſtle.

‡ 9   This Thiſtle in the opinion of *Bau-hine*, whereto I much incline, is the ſame with the former. The root is ſmall, the leaues long, welting the ſtalks at their ſetting on, and armed on the edges with ſharpe prickles : the ſtalkes lie trailing on the ground like thoſe of the ſtar-Thiſtle, ſo ſet with prickles, that one knoweth not where to take hold thereof : it hath many cloſely compaꞇ vmbels, conſiſting of pale yel-lowiſh little floures like thoſe of Groundſwell : the ſeed is like that of *Carthamus*, ſmal and chaf-fie. *Pena* and *Lobel* call this *Picnomos Cretæ Salo-nenſis*, of a place in Prouince where they firſt found it, called the Crau, being not farre from the city Salon. *Tabernamontanus* ſet it forth foɾ *Chamæleon niger*, and our Author formerly gaue the figure hereof by the ſame title, though his hiſtorie belonged to another, as I haue former-ly noted. ‡

¶ *The Place.*

The two firſt grow on diuers banks not farre from mount Apennine, and ſometimes in Italy, but yet ſeldome.

The way Thiſtles grow euery where by high-waies ſides and common paths in great plenty.

The places of the reſt haue beene ſufficiently ſpoken of in their deſcriptions.

¶ *The*

¶ *The Time.*

Thefe kindes of Thiftles do floure from the beginning of Iune vntill the end of September.

¶ *The Names.*

Thefe Thiftles comprehended in this prefent chapter are by one generall name called in Latine *Cardui fylueftres*, or wilde Thiftles ; and that which is the fecond in order is named *Scolymus* : but not that *Scolymus* which *Theophraftus* declareth to yeeld a milky juyce (of which we haue written before) but one of thofe which *Pliny* in his twentieth booke, *cap.23.* defcribeth : of fome they are taken for kindes of Chamæleon : their feuerall titles doe fet forth their feuerall Latine names, and alfo the Englifh.

‡ There was formerly much confufion in this chapter, both in the figures and hiftory, which I will here endeauour to amend, and giue as much light as I can, to the obfcuritie of our Authour and fome others ; to which end I haue made choife of the names as the fitteft place.

1   This defcription was taken out of *Dodonæus*, and the title alfo of *Onopordon* which was formerly put ouer the figure, and they belong to the Thiftle our Author before defcribed by the name of *Acanthium purp. Illyricum, cap.476.* I haue therefore changed the title, yet let the defcription ftand, for it reafonable well agrees with the figure which is of the *Carduus fpinofiſſimus vulgaris* of *Lobel*, and *Polyacantha Theophrafti* of *Tabern.* of this Thiftle I obferue three kindes : the firft is a Thiftle fome two cubits and a halfe high, with many flender ftalkes and branches exceeding prickly, hauing commonly fiue prickly welts running alongft the ftalkes : the leaues on the vpper fides as alfo the ftalkes are of a reafonable frefh greene colour, but the vnderfide of the leafe is fomwhat whitifh : the heads confift of fundry hairy greene threds which looke like prickles ; but they are weake, and not prickly : the floure is of the bigneſſe, and of the like colour and fhape as the common Knapweed, yet fomewhat brighter : it growes on ditch fides, and floures in Iuly. This I take to be the *Aculeofa Gafa* of the *Aduerf.pag.374.* but not that which *Lobel* figures for it in his *Icones.* This is that which *Tabernamontanus* figures for *Polyacantha*, and our Authour gaue his figure in this place. The fecond of thefe I take to be that which *Lobel* hath figured for *Polyacantha*, and *Dodonæus* for *Carduus fyl.3.* (which figure we here giue you) and in the *Hift.Lugd.pag.1473.* it is both figured and defcribed by the name of *Polyacanthos Theophrafti*. In the figure there is little difference : in the things themfelues this; the ftalkes of this are as high as thofe of the laft, but flenderer, with fewer and ftraighter branches, and commonly edged with foure large welts, which haue fewer, yet longer prickles than thofe of the former : the leaues and ftalkes of this are of a grayifh or whitifh colour : the heads are longifh, but much fmaller than thofe of the former, and they feldome open or fpred abroad their floures, but only fhew the tops of diuers reddifh threds of a feint colour. This growes as frequently as the former, and commonly in the fame places. The third, which I thinke may fitly be referred vnto thefe, growes on wet heaths and fuch like places, hauing a ftalke fometimes foure or fiue cubits high, growing ftraight vp, with few branches, and thofe fhort ones : the floures are of an indifferent bigneſſe, and commonly purple, yet fometimes white. I thinke this may be the *Onopyxos alter Lugdunenf.* or the *Carduus paluftris* defcribed in *Bauhinus* his *Prodromus, pag.156.*

2   The fecond which is a ftranger with vs, is the *Phœnix, Leo & Carduus ferox* of *Lobel* and *Dod.* *Bauhine* hath refer'd it to *Acarna*, calling it *Acarna minor caule non foliofo.*

3   The third defcription was alfo out of *Dodonæus*, being of his *Carduus fylueftris primus*, or the *Onopyxos Dodonæi* of the *Hift.Lugd.* The figures formerly both in the third and fourth place of this Chapter were of the *Acanthium Illyricum* of *Lobel* ; or the *Onopordon* of *Dodonæus*, formerly mentioned.

4   This defcription alfo was out of *Dodonæus*, being of his *Carduus fylueftris alter*, agreeing in all things but the colour of the floures, which fhould be purple. *Lobel* in his Obferuations defcribeth the fame Thiftle by the name of *Carduus vulgatiſſimus viarum* : but both he and *Dodonæus*, giue the figure of *Carlina fylueftris* for it : but neither the floures nor the heads of that agree with that defcription. I iudge this to be the Thiftle that *Fabius Columna* hath fet forth for the *Ceanothos* of *Theophraftus* ; and *Tabern.* for *Carduus aruenfis* : and our Author, though vnfitly, gaue it in the next place for *Carduus mufcatus.*

5   The Muske-Thiftle I haue feene growing about Deptford, and (as far as my memory ferues me) it is very like to the third here defcribed : it growes better than a cubit high, with reafonable large leaues, and alfo heads which are a little foft or downy, large, with purple floures : the heads before the floures open fmell ftrong of muske. I haue found no mention of this but only in *Gefner*, *de Collectione in parte*, where hee hath thefe words ; *Carduus aruenfis major purpureo flore (qui flore nondum nato Atofchum olebat) floret Iulio.* Our Author formerly gaue an vnfit figure for this, as I formerly noted.

There is fufficient of the reft in their titles and defcriptions. ‡

¶ *The Temperature and Vertues.*

Thefe wild Thiftles(according to *Galen*)are hot and dry in the fecond degree,and that through the propertie of their effence they driue forth ftinking vrine, if the roots be boyled in Wine and drunke ; and that they take away the ranke fmell of the body and arme-holes. **A**

Diofcorides faith,that the root of the common Thiftle applied plaifterwife correcteth the filthy fmell of the arme-holes and whole body. **B**

And that it workes the fame effect if it be boyled in wine and drunke,and that it expelleth plentie of ftinking vrine. **C**

The fame Author affirmeth alfo, that the herbe being as yet greene and tender is vfed to be eaten among other herbes after the manner of Afparagus. **D**

This being ftamped before the floure appeareth, faith *Pliny*, and the juyce preffed forth,caufeth haire to grow where it is pilled off, if the place be bathed with the juyce. **E**

The root of any of the wilde Thiftles being boyled in water and drunke, is reported to make them dry that drinke it. **F**

It ftrengtheneth the ftomacke; and it is reported (if we beleeue it) that the fame is alfo good for the matrix, that boyes may be engendred : for fo *Chereas* of Athens hath writien, and *Glaucias*, who is thought to write moft diligently of Thiftles. **G**

This Thiftle being chewed is good againft a ftinking breath. Thus farre *Pliny*,in his twentieth booke,*cap*.23. **H**

CHAP. 491.   *Of the* Melon *or Hedge-hog* Thiftle.

Melocarduus Echinatus Pena & Lob.
The Hedge-hog Thiftle.

¶ *The Defcription.*

WHo can but maruel at the rare and fin̄ gular workemanfhip which the Lord God almighty hath fhewed in this Thiftle, called by the name *Echino-Melocactos*,or *Melo-carduus Echinatus?* This knobby or bunchy maffe or lump is ftrangely compact and context together,containing in it fundry fhapes, and formes, participating of a Pepon or Melon, and a Thiftle, both being incorporate in one body ; which is made after the forme of a cock of hay, broad and flat below,but fharp toward the top, as bigge as a mans body from the belly vpward : on the outfide hereof are fourteene hard ribbes, defcending from the crowne to the loweft part, like the bunchy or out fwelling rib of a Melon ftanding out,and chanelled betweene : at the top or crowne of the plant iffueth forth a fine filken cotton, wherewith it is full fraught : within which cotton or flockes lie hid certain fmal fheaths or cods,fharpe at the point,and of a deep fanguine colour, anfwering the cods of *Capficum* or Indian pepper, not in fhew only,but in colour,but the cods are fomewhat fmaller. The furrowed or chanelled ribs on the outfide are garnifhed or rather armed with many prickly ftars, ftanding in a compaffe like fharp crooked hornes or hookes, each ftar confifting of ten or twelue prickes, wherewith the outward barke or pilling is garded , fo that without

hurt to the fingers it cannot be touched : this rinde is hard, thicke and like vnto Aloes, of the colour of the Cucumber : the flefh or inner pulpe is white, fat, waterifh, of tafte foure, vnfauory and cooling, much like vnto the meat of a raw Melon or Pompion. This plant groweth without leafe or ftalke, as our Northerne Thiftle doth,called *Carduus Acaulos*, and is bigger than the largeft
Pompion :

Pompion : the roots are small, spreading farre abroad in the ground, and consisting of blacke and tough twigs, which cannot endure the injury of our cold clymate.

¶ *The Place.*

This admirable Thistle groweth vpon the cliffes and grauelly grounds neere vnto the sea side, in the Islands of the West-Indies, called S. *Margarets* and S. *Iohns* Isle, neere vnto *Puerto rico*, or *Porto rico*, and other places in those countries, by the relation of diuers trauellers that haue iourneyed into those parts, who haue brought me the plant it selfe with his seed ; the which would not grow in my garden by reason of the coldnesse of the clymate.

¶ *The Time.*

It groweth, floureth, and flourisheth all the yeare long, as doe many other plants of those Countries.

¶ *The Names.*

It is called, *Carduus Echinatus*, *Melocarduus Echinatus*, and *Echino Melocactos* : in English, the Hedge-hog Thistle, or prickly Melon Thistle. ‡ Such as are curious may see more hereof in *Clusius* his *Exoticks, lib. 4. cap. 24.* ‡

¶ *The Temperature and Vertues.*

There is not any thing extant set forth of the antient or of the later Writers, neither by any that haue trauelled from the Indies themselues : therefore we leaue it to a further consideration.

C H A P. 492.   *Of the gummie Thistle, called Euphorbium.*

<table>
<tr><td>1 *Euphorbium.*</td><td>2 *Anteuphorbium.*</td></tr>
<tr><td>The poysonous gum Thistle.</td><td>The Antidote against the poysonous Thistle.</td></tr>
</table>

¶ *The Description.*

1   **E**Vphorbium (whereout that liquor or gum called in shops *Euphorbium* is extracted) hath very great thicke grosse and spreading roots, dispersed far abroad in the ground : from which arise long and round leaues, almost like the fruit of a great Cucumber, a foot and a halfe long, ribbed, walled, and furrowed like vnto the Melon : these branched ribs are set or

armed

armed for the moft part with certaine prickles ftanding by couples, the point or fharpe end of one garding one way, and the point of another looking directly a cleane contrary way : thefe prickes are often found in the gumme it felfe, which is brought unto vs from Libya and other parts : the leaues hereof being planted in the ground will take root well, and bring forth great increafe, which thing I haue proued true in my garden: it hath perifhed againe at the firft approch of Winter. The fap or liquor that is extracted out of this plant is of the colour and fubftance of the Creame of Milke; it burneth the mouth extremely, and the duft or powder doth very much annoy the head and the parts thereabout, caufing great and vehement fneefing.

1   This rare plant called *Anteuphorbium* hath a very thicke groffe and farre fpreading root, very like vnto *Euphorbium*; from which rifeth vp many round greene and flefhie ftalkes, whereupon doe grow thicke leaues like Purflane, but longer, thicker, and fatter : the whole plant is full of cold and clammie moifture, which repreffeth the fcortching force of *Euphorbium*; and it wholly feemes at the firft view to be a branch of greene Corall.

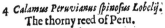

3 *Cereus Peruvianus fpinofus Lobelij.*
The Torch-Thiftle or thorny Euphorbium.

4 *Calamus Peruvianus fpinofus Lobelij.*
The thorny reed of Peru.

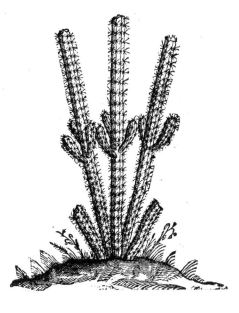

3   There is not among the ftrange and admirable plants of the world any one that giues more caufe of maruell, or more moueth the minde to honor and laud the Creator, than this plant, which is called of the Indians in their mother tongue *Vragua*, which is as much to fay, a torch, taper, or wax candle; whereupon it hath been called in Latine by thofe that vnderftood the Indian tongue, *Cereus*, or a Torch. This admirable plant rifeth vp to the height of a fpeare of twenty foot long, although the figure expreffe not the fame; the reafon is, the plant when the figure was drawne came to our view broken : it hath diuers bunches and vallies, euen as is to be feene in the fides of the Cucumber, that is, furrowed, guttered, or chamfered alongft the fame, and as it were laid by a di-rect line, with a welt from one end vnto the other : vpon which welt or line doe ftand fmall ftar-like Thiftles, (fharpe as needles, and of the colour of thofe of the Melon Thiftle, that is to fay, of a browne colour : the trunke or body is of the bigneffe of a mans arme, or a cable rope; from the middle whereof thruft forth diuers knobby elbowes of the fame fubftance, and armed with the like prickles that the body of the trunke is fet withall : the whole plant is thicke, fat, and full of a fla-fhie fubftance, hauing much juyce like that of Aloes, when it is hardned, and of a bitter tafte : the

floures

floures grow at the top or extreme point of the plant: after which follow fruit in ſhape like a fig, full of a red juyce, which being touched ſtaineth the hands of the colour of red leade: the taſte is not vnpleaſant.

4 There hath been brought from the Indies a prickly reed of the bigneſſe of a good big ſtaffe, of the length of ſix or eight foot, chamfered and furrowed, hauing vpon two ſides growing vnto it an vneuen membrane or skinny ſubſtance, as it were a jag or welt ſet vpon the wing of a garment, and vpon the very point of euery cut or jagge armed with moſt ſharpe prickles: the whole trunke is filled full of a ſpongeous ſubſtance, ſuch as is in the hollownes of the brier or bramble; amongſt the which is to be ſeene as it were the pillings of Onions, wherin are often found liuing things, that at the firſt ſeeme to be dead. The plant is ſtrange, and brought dry from the Indies, therefore wee cannot write ſo abſolutely hereof as we deſire, referring what more might be ſaid to a further conſideration or ſecond edition.

¶ The Place.

Theſe plants grow vpon Mount Atlas, in Lybia, in moſt of the Iſlands of the Mediterranean ſea, in all the coaſt of Barbary, eſpecially in S. Crux neere vnto the ſea ſide, in a barren place there called by the Engliſh-men Halfe Hanneken, which place is appointed for Merchants to confer of their buſineſſe, euen as the Exchange in London is: from which place my friend Mr. *William Martin*, a right expert Surgeon, did procure me the plants of them for my garden, by his ſeruant that he ſent thither as a Surgeon of a ſhip. Since which time I haue receiued plants of diuers others that haue trauelled into other of thoſe parts and coaſts: notwithſtanding they haue not endured the cold of our extreme Winter.

¶ The Time.

They put forth their leaues in the Spring time, and wither away at the approch of Winter.

¶ The Names.

It is called both in Greeke and Latine, Εὐφόρβιον, *Euphorbium* : *Pliny* in one place putteth the herbe in the feminine gender, naming it *Euphorbia* : the juyce is called alſo *Euphorbion*, and ſo it is likewiſe in ſhops: we are faine in Engliſh to vſe the Latine word, and to call both the herbe and juyce by the name of Euphorbium, for other name we haue none: it may be called in Engliſh, the gum Thiſtle.

¶ The Temperature.

Euphorbium (that is to ſay, the congealed juyce which we vſe) is of a very hot, and, as *Galen* teſtifieth, cauſticke or burning faculty, and of thinne parts: it is alſo hot and dry in the fourth degree.

¶ The Vertues.

A An emplaiſter made with the gumme Euphorbium, and twelue times ſo much oyle, and a little wax, is very ſingular againſt all aches of the joynts, lameneſſe, palſies, crampes, and ſhrinking of ſinewes, as *Galen, lib.4. de medicamentis ſecundum genera*, declareth at large, which to recite at this preſent would but trouble you ouermuch.

B Euphorbium mingled with oyle of Bay and Beares greaſe cureth the ſcurfe and ſcalds of the head, and pildneſſe, cauſing the haire to grow againe, and other bare places, being annointed therewith.

C The ſame mingled with oyle, and applied to the temples of ſuch as are very ſleepy, and troubled with the lethargie, doth waken and quicken their ſpirits againe.

D If it be applied to the nuque or nape of the necke, it bringeth their ſpeech againe that haue loſt it by reaſon of the Apoplexie.

E Euphorbium mingled with vinegre and applied taketh away all foule and ill-fauoured ſpots, in what part of the body ſoeuer they be.

F Being mixed with oyle of Wall-floures, as *Meſues* ſaith, and with any other oyle or oyntments, it quickely heateth ſuch parts as are ouer cold.

G It is likewiſe a remedy againſt old paines in the huckle bones, called the Sciatica.

H *Ætius, Paulus, Actuarius*, and *Meſue* doe report, That if it bee inwardly taken it purgeth by ſiege water and flegme; but withall it ſetteth on fire, ſcorcheth and fretteth, not onely the throat and mouth, but alſo the ſtomacke, liuer, and the reſt of the entrals, and inflames the whole body.

I For that cauſe it muſt not be beaten ſmall, and it is to be tempered with ſuch things as allay the heate and ſharpeneſſe thereof, and that make glib and ſlippery; of which things there muſt bee ſuch a quantitie, as that it may bee ſufficient to couer all ouer the ſuperficiall or outward part thereof.

K But it is a hard thing ſo to couer and fold it vp, or to mix it, as that it will not burne or ſcorch. For though it be tempered with neuer ſo much oyle, if it be outwardly applied it raiſeth bliſters, eſpecially in them that haue ſoft and tender fleſh, and therefore it is better not to take it inwardly.

It

It is troublefome to beat it, vnleffe the nofthrils be carefully ftopt and defended; for if it hap- L
pen that the hot fharpneffe thereof do enter into the nofe, it caufeth itching, and moueth neefing,
and after that by reafon of the extremitie of the heate, it drawes out aboundance of flegme and
filth, and laft of all bloud, not without great quantitie of teares.

But againft the hot fharpneffe of Euphorbium, it is reported that the inhabitants are remedied M
by a certaine herb, which of the effect and contrarie faculties is named *Anteuphorbium*. This plant
alfo is full of juice, nothing at all hot and fharpe, but coole and flimy, allaying the heat and fharpneffe of *Euphorbium*. We haue not yet learned That the old Writers haue fet downe any thing as
touching this herb: notwithftanding it feems to be a kinde of Orpine, which is the antidote or
counterpoifon againft the venomous poifon of *Euphorbium*.

---

‡ CHAP. 493. *Of foft Thiftles and Thiftle gentle.*

‡ THete are certain other plants by moft writers referred to the Thiftles: which beeing o-
mitted by our Author, I haue thought fit here to giue you.

‡ 1 *Cirfium maximum Afphodeli radice.*
Great foft bulbed Thiftle.

2 *Cirfium majus alterum.*
Great foft Thiftle.

¶ *The Defcription.*

1   THe firft and largeft of thefe hath leaues confifting of great longifh bulbs like thofe of
the Afphodil; from whence arife many large ftalks three or foure cubits high, crefted
and downy: the leaues are very long and large, juicy, greenifh, cut about the edges, and
fet with foft prickles. At the tops of the ftalks and branches grow heads round and large, out of
which come flours confifting of aboundance of threds of a purple color, which flv away in down.
This growes wilde in the mountainous medowes and in fome wet places of Auftria. I haue feen
it growing in the garden of Mr *Iohn Parkinfon*, and with Mr *Tuggye*. It floures in Iuly. *Clufius* hath
called it *Cirfium maximum mont. incano folio bulbofa radice*. But hee gaue no figure thereof, nor any
elfe, vnleffe the *Acanthium peregrinum* in *Tabernam*. (which our Author formerly, as I before noted,

gaue by the name of *Solſtitialis lutea peregrina*) were intended for this plant, as I verily thinke it was. I haue giuen you a figure which I drew ſome yeares ago by the plant it ſelfe.

2   The root of this is long, yet ſending forth of the ſides creeping fibres, but not bulbous: the leaues are like thoſe of the laſt mentioned, but leſſe, & ſet with ſharp prickles of a greeniſh colour, with the middle rib white: the heads ſometimes ſtand vpright, and otherwhiles hang down; they are very prickely, and ſend forth floures conſiſting of many elegant purple threds. The ſtalkes are thicke, creſted and welted with the ſetting on of the leaues. This groweſ wilde vpon the ſea coaſts of Zeeland, Flanders, and Holland: it floures in Iune and Iuly. It is the *Cirſium tertium* of *Dodonæus*; and *Cirſium majus* of *Lobel*.

3   This, whoſe root is fibrous and liuing, ſendeth forth leſſer, narrower, and ſofter leaues than thoſe of the former, not jagged or cut about their edges, nor hoary, yet ſet about with prickles; the ſtalks are creſted: the heads are ſmaller, and grow three or foure together, carrying ſuch purple floures as the former. This is that which *Matthiolus, Geſner,* and others haue ſet forth for *Cirſium*: *Dodonæus* for *Cirſium 2.* and *Cluſius* hath it for his *Cirſium quartum, or Montanum ſecundum.*

‡ 3 *Cirſium folijs non hirſutis.*
Soft ſmooth leaued Thiſtle.

‡ 4 *Cirſium montanum capitulis parvis.*
Small Burre Thiſtle.

4   The leaues of this are ſomewhat like thoſe of the laſt deſcribed, but larger, and welting the ſtalks further at their ſetting on: they are alſo ſet with prickles about the edges: the ſtalkes are ſome two cubits high, diuided into ſundry long ſlender branches, on whoſe tops grow little rough prickly heads, which after the floures come to perfection, do hang downewards, and at length turne into down; amongſt which lies hid a ſmooth ſhining ſeed. This groweſ wilde in diuers wooddy places of Hungarie and Auſtria. It is the *Cirſium* of *Dodonæus*: the *Cirſium 2.* or *Montanum 1.* of *Cluſius*: and *Cirſium alterum* of *Lobel*. It floures in Iune: the root is about the thickneſſe of ones little finger, fibrous alſo and liuing.

5   This ſends vp long narrow leaues, hairy, and ſet about the edges with ſlender prickles: out of the middeſt of theſe leaues groweth vp a ſtalke, ſomtimes a foot, otherwhiles a cubit high, ſlender, ſtiffe, and downy: vpon which grow leaues ſomwhat broad at their ſetting on, and there alſo a little nicked or cut in. This ſtalke ſometimes hath no branches, otherwhiles to or three long ſlender ones, ot the tops wheriof grow out of ſcaly heads ſuch floures as the common Knap-weed,
                                                which

‡ 5 *Cirsium montanum Anglicum*.
Single headed Thistle.

‡ *Cirsij Anglici alia Icon* Pennei.
Pennies figure of the same.

‡ *Cirsij Anglici Icon* Lobelij.
Lobels figure of the same.

‡ 6 *Carduus mollis folijs dissectis*.
Iagged leaued Thistle gentle.

which at length turne into down : among which lies hid a small shining seed like the other plants of this kinde. The root is made of diuers thicke fibres, which run in the ground, and here and there put vp new heads. This plant wants no setting forth ; for *Clusius* giues vs the figure and historie of it, first by the name of *Cirsium Pannonicum* 1. *pratense* ; then he giues another historie thereof, with a worser figure (which he receiued of Dr *Thomas Penny* of London) by the name of *Cirsium Anglicum* 2. *Lobel* also described it, and set it forth with a figure expressing the floure already faded, by the name of *Cirsium Anglicum*. *Bauhine* in his *Pinax*, deceiued by these seuerall expressions, hath made three seuerall plants of this one ; a fault frequent in many writers of plants. *Clusius* found it growing in the mountainous medowes alongst the side of the Danow in Austria. *Penny*, in the medowes at the foot of Ingleborow hil in Yorkshire : *Lobel*, in the medowes at a place called Acton in Glocester-shire. I found this only once, and that was in a medow on this side High-gate, hauing beene abroad with the Company of Apothecaries, and returning homeward that way in the company of Mr *Iames Walsall*, *William Broad*, and some others. I haue giuen you both the figures of *Clusius*, his owne in the first place, and that of Dr *Penny* in the second, but the former is the better. I haue also giuen you that of *Lobel*.

‡ 7 *Carduus mollis folijs Lapathi.*
Docke leaued Thistle gentle.

6   These also *Clusius* (whom I herein follow) addeth to the kindes of This jagged leafed one, which hee calleth *Carduus mollior primus* , hath many leaues at the root, both spreading vpon the ground, and also standing vpright ; and they are couered with a white soft downinesse, yet greene on the vpper side : they are also much diuided or cut in euen to the middle rib, like to the softer or tenderer leaues of the Star Thistle: they haue no prickles at all vpon them : out of the middest of these leaues grow vp one or two stalkes, round, crested, purplish, hoary, and some cubit or better high. The leaues that grow vpon the lower part of the stalke are diuided, those aboue not so : the tops of the stalks somtimes, yet very seldome, are parted into branches, which carry scaly heads containing elegant floures made of many purple strings. The floure decaying, there succeedeth a cornered seed. The root sometimes equalls the the thicknesse of ones finger, brownish, long, and somewhat fibrous. It floures in May, and growes vpon the hilly places of Hungary.

7   The stalke of this is some foot or better high, thick, crested, and somwhat hairy: the leaues about the root are somewhat large, and in shape like those of *Bonus Henricus* (abusiuely called in English, Mercury) somewhat sinuated about the edges, and set with harmlesse prickles, greene aboue, and very hoary vnderneath, like the leaues of the white Poplar : those that grow vpon the stalke are lesser and narrower, out of whose bosomes toward the tops of the stalke grow out little branches, which carry three, foure, or more little scaly heads like those of the blew Bottle or Knapweed : whereout grow threddy blewish purple floures : the seed is wrapped in downe, and not vnlike that of Blew-Bottle : the root is blacke, hard, and liuing, sending forth shoots on the sides. It growes vpon the highest Austrian Alps, and floures in Iuly. *Clusius* calls this *Carduus mollior Lapathifolio.*

¶ *The Temperature and Vertues.*

These plants seem by their taste to be of a moderately heating and drying facultie, but none of them are vsed in medicine, nor haue their vertues set down by any Author. ‡

                                   C H A P.

Chap. 494. *Of three leafed Graſſe or Medow Trefoile.*

¶ The Kindes.

THere be diuers ſorts of three leafed Graſſes, ſome greater, others leſſer; ſome beare floures of one colour, ſome of another: ſome of the water, and others of the land: ſome of a ſweet ſmel, others ſtinking: and firſt of the common medow Trefoiles, called in Iriſh *Shamrocks*.

1 *Trifolium pratenſe.*
Medow Trefoile.

‡ 3 *Trifolium majus flore albo.*
Great white Trefoile.

¶ The Deſcription.

1　MEdow Trefoile bringeth forth ſtalkes a cubit long, round and ſomething hairy, the greater part of which creepeth vpon the ground: whereon grow leaues conſiſting of three ioined together, one ſtanding a little from another, of which thoſe that are next the ground and roots are rounder, and they that grow on the vpper part longer, hauing for the moſt part in the midſt a white ſpot like a halfe moone. The floures grow at the tops of the ſtalks in a tuft or ſmall Fox-taile eare, of a purple colour, and ſweet of taſte. The ſeed growes in little husks, round and blackiſh: the root is long, wooddy, and groweth deep.

2　There is another of the field Trefoiles, differing from the precedent eſpecially in the color of the floures; for as thoſe are of a bright purple, contrariwiſe theſe are very white, which maketh the difference. The leaues, floures, and all the whole plant is leſſe than the former.

3. 4.　There is alſo a Trefoile of this kinde which is ſowne in fields of the low-Countries, in Italy and diuers other places beyond the ſeas, that comes vp ranker and higher than that which groweth in medowes, and is an excellent food for Cattell, both to fatten them and cauſe them to giue good ſtore of milke.

‡　Of this there is one more with white floures which hath ſtalkes ſome foot high, and narrow hairy leaues, with a root of the thickneſſe of ones little finger. This is *Cluſius* his *Trifolium majus primum,*

‡ 4 *Trifolium majus flo.purpureo.*
Great purple Trefoile.

‡ 5 *Trifolium luteum Lupinum.*
Hop Trefoile.

‡ 6 *Trifolium luteum minimum.*
Little yellow Trefoile.

primum. The other hath stalks some cubit high, with larger joints and leaues: the floure or head of floures is also larger, of an elegant red colour. This *Clusius* calls *Trifolium majus tertium.* ‡

5. 6 Likewise we haue in our fields a smaller Trefoile that brings forth yellow flours, a greater and a lesser, & diuers others also, differing from these in diuers notable points, the which to distinguish apart would greatly inlarge our volume, and yet to small purpose : therefore we leaue them to be distinguished by the Curious, who may at the first view easily perceiue the difference, and also that they be of one stocke or kindred.

‡ The greater of these yellow Trefoiles hath prety large yellow heads, which afterward become of a brownish colour, and somewhat resemble an hop ; whence *Thalius* called it *Lupulus sylvaticus*, or *Trifolium luteum alterum lupulinum : Dodonæus* cals it *Trifolium agrarium.* The leaues are smal, and lightly nickt about the edges: the lesser hath smaller and far lesser yellow heads, which are succeeded by many little crooked clustring seeds. The leaues of this are small, and also snipt about the edges. Both this and the other hath two little leaues close by the fastning of the foot-stalkes of the leaues to the main stalks : wherefore I referre them to the Medickes, and vsually call this later, *Medica sem. racemoso.* It is the *Trifol. luteum minim.* of *Pena* and *Lobel* ; and *Trifol. arvense* of *Taber.* ‡

¶ *The*

¶ *The Place.*

Common medow Treſoile growes in medowes, fertile paſtures, and wateriſh grounds: the others loue the like ſoile.

¶ *The Time.*

They floure from May to the end of Summer.

¶ *The Names.*

Medow Treſoile is called in Latine *Trifolium pratenſe* : in high-Dutch, **weiſenklee**: in low-Dutch, **Claueren** : in French, *Treffle,* and *Trainiere,* and *Viſumarus,* as *Marcellus* an old writer teſtifies: in Engliſh, Common Treſoile, Three leafed graſſe : of ſome, Suckles, Hony-ſuckles, and Cocks-heads : in Iriſh, *Shamrocks.*

¶ *The Temperature.*

The leaues and floures of medow Treſoiles are cold and dry.

¶ *The Vertues.*

The decoction of three leafed graſſe made with hony and vſed as a cliſter, is good againſt frettings and paines of the guts, and driueth forth tough and ſlimy humors that cleaue vnto the guts.   A

The leaues boiled with a little barrows greaſe, and vſed as a pultis, take away hot ſwellings and inflammations.   B

Oxen and other cattell do feed on the herb, as alſo calues and yong lambs. The flours are acceptable to Bees.   C

Pliny writeth and ſetteth it downe for certaine, that the leaues hereof do tremble and ſtand right vp againſt the comming of a ſtorme or tempeſt.   D

The medow Treſoile (eſpecially that with the blacke halfe-moon vpon the leafe) ſtamped with a little hony, takes away the pin and web in the eies, ceaſing the pain and inflammation thereof, if it be ſtrained and dropped therein.   E

CHAP. 495.   *Of ſtinking Treſoile or Treacle Claver.*

Trifolium bituminoſum.
Treacle Claver.

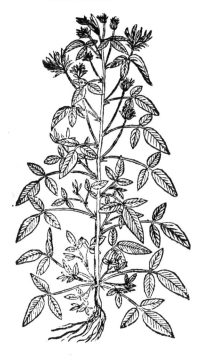

¶ *The Deſcription.*

TReacle Claver growes vpright like a ſhrubbie plant, with ſtalkes of a cubit and a halfe high ; whereupon grow next the ground broad leaues three joined together : thoſe vpon the ſtalks are longer and narrower. The ſtalks are couered ouer with a rough euill coloured hairineſſe : the leaues are of a darke blacke green colour, and of a loathſome ſmell, like the pitch called *Bitumen Iudaicum,* whereof it took the name. The flours grow at the top of the ſtalks, of a dark purpliſh colour tending to blewneſſe, in ſhape like thoſe of Scabious. The ſeed is broad, rough, long, and ſharp pointed : the root is ſmal & tender, and canot endure the coldneſſe of our Winter, but periſheth at the firſt approch thereof.

¶ *The Place.*

It groweth naturally, ſaith *Hippocrates, Hippiatros,* not *Cous,* in rough places, as *Ruellius* tranſlateth it: in Germany, France, and England it neuer commeth vp of it ſelfe, but muſt be ſown in gardens, as my ſelfe haue proued diuers times, and was conſtrained to ſow it yerely, or elſe it would not come vp, either of his owne ſowing or otherwiſe

¶ *The Time.*

It floured not in my garden vntill the end of Auguſt.   ¶ *The*

¶ *The Names.*

Nicander calls this Trefoile ꝺꝛꝟꞁꞃꞁ in Latine, *Trifolium acutum*, or sharpe pointed Trefoile : of *Pliny*, *Trifolium odoratum*, but not properly : of others, *Trifolium Asphaltæum, sive Bituminosum*, or stone Pitch Trefoile.

Avicen calleth it *Tarfilon*, and not *Handacocha* : *Avicen* doth comprehend *Dioscorides* his *Loti*, that is to say, *Lotus urbana sylvestris*, and *Ægyptia*, which *Dioscorides* confoundeth one with another in one Chapter. In English it is called Clauer gentle, Pitch Trefoile, stinking Trefoile, and Treacle Clauer.

¶ *The Temperature.*

This Trefoile, called *Asphaltæum*, as *Galen* saith, is hot and dry in the third degree, as *Bitumen* is.

¶ *The Vertues.*

A   Being drunke, it taketh away the pain of the sides which commeth by obstruction or stoppings, prouoketh vrine, and bringeth downe the desired sickenesse.

B   *Hippocrates* writeth, that it doth not only bring them downe, but likewise the birth, not onely inwardly taken, but also outwardly applied : if a woman, saith he, be not well clensed after her childbearing, giue her this Trefoile to drinke in white wine.

C   *Dioscorides* saith, that the seeds and leaues being drunke in water are a remedie for the pleurisie, difficultie of making water, the falling sickenesse, the dropsie when it first beginneth, and for those that be troubled with the Mother : the quantity to be taken at once is three drams of the seeds, and foure of the leaues.

D   The leaues drunke in Oxymel or a syrrup of vineger made with hony, is good for those that are bitten with serpents.

E   Some affirm, that the decoction of the whole plant, root and leaues, taketh away pain comming of the sting of serpents, if the part be washed therewith : but if any other man hauing an vlcer, bee washed with that water wherewith he was bathed that was bitten of the serpent, they say that hee shall be troubled in the same manner that the stinged party was.

F   Some also giue with wine three leaues, or a small quantitie of the seeds in tertian agues, and in Quartane foure, as a sure remedie against the fits.

G.   The root also is put into antidotes or counterpoisons, saith *Dioscorides* : but other antient Physitions do not only mix the root with them, but also the seed, as we may see in *Galen, lib.* 2. of his antidots, in many compositions ; that is to say, in the Treacles of *Ælius Gallus*, *Zeno Laudocens*, *Claudius Apollonius, Eudemus, Heraclides, Dorotheus*, and *Heras*.

H   The herbe stamped and applied vpon any invenomed wound or made with poisoned weapon, drawes the poison from the depth most apparantly : but if it be applied vpon a wound where there is no venomous matter to work vpon, it doth no lesse infect that part, than if it had bin bitten with some serpent or venomous beast : which wonderful effect it doth not performe in respect of any vitious qualitie that it hath in it selfe, but because it doth not finde that venomous matter to worke vpon, which it naturally draweth (as the Load-stone doth iron) wherupon it is constrained through his attractiue qualitie, to draw and gather together humors from farre vnto the place, whereby the paine is greatly increased.

CHAP. 496.   *Of diuers other Trefoiles.*

¶ *The Description.*

1   TRee leaued Grasse of America hath diuers crooked round stalks leaning this way and that way, and diuided into diuers branches, whereon grow leaues like those of the medow Trefoile, of a blacke greene colour, and of the smell of pitch Trefoile or Treacle Clauer : the floures grow at the top of the branches, made vp in a long spiked chaffie eare, white of colour : after which comes the seed, somwhat flat, almost like to those of Tares : the roots are long strings of a wooddy substance.

2   This three leafed Grasse (which *Dodoneus* in his last edition calleth *Trifolium cochleatum primum* ; and *Lobel, Fænum Burgundiacum*) hath diuers round vpright stalkes of a wooddy rough substance, yet not able of it selfe to stand without a prop or stay : which stalkes are diuided into diuers small branches, whereupon doe grow leaues joined three together like the other Trefoiles, but of a darke swart greene colour : the floures grow at the tops of the stalks, in shape like those of the codded Trefoile, but of a darke purple colour : the seed followeth, contained in small wrinckled

huskes

1 *Trifolium Americum.*
Trefoile of America.

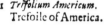

† 2 *Trifolium Burgundiacum.*
Burgondy Trefoile.

3 *Trifolium Salmanticum.*   Portugal Trefoile.

husks turned round,after the maner of a water fnaile. The root is thick,compofed of diuers tough
threddy ftrings,and lafteth long in my garden with great increafe.

3   This three leaued Graffe of Salmanca,a city as I take it in Portugal,differs not much from
our field Trefoile : it hath many branches weak and tender,trailing vpon the ground,of two cubits
and a halfe high : whereupon grow leaues fet together by three vpon a ftemme , from the bofome
whereof thruft forth tender foot-ftalkes,whereon ftand moft fine floures of a bright red tending to
purple : after which come the feeds wrapped in fmall skins,of a red colour

4   The Hart Trefoile hath very many flexible branches, fet vpon a flender ftalke,of the length
of two or three foot , trailing hither and thither : whereupon doe grow leaues joined together by
three

4 *Trifolium Cordatum.*
Heart Treſoile.

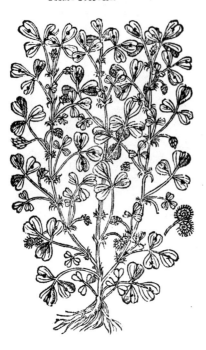

5 *Trifolium ſiliquoſum minus.*
Small codded Treſoile.

‡ 6 *Coronopus ex codice Cæſareo.*
Crow-foot Treſoile.

three on little ſlender foot-ſtalks, euery little
leaſe of the faſhion of a heart, whereof it took
his name : amongſt which come forth ſcaly or
chaffie yellow floures : the root is thicke and
threddy.   ‡ I take the plant which our Authour heere figured and intended to deſcribe
vnto vs, to be of that *Medica* which *Camerarius*
calls *Arabica*, which growes wild in many places with vs, hauing the leaues a little dented in
at the ends, ſo that they reſemble the vulgar figure of an heart ; and each leaf is marked with
a blackiſh or red ſpot : the flours be ſmall and
yellow, the ſeed is contained in rough buttons
wound vp like the other ſnaile Treſoils, wherof it is a kinde. I haue giuen you the figure a
little more exquiſite, by the addition of the
ſpots and cods. ‡

5   This kinde of three leaued Graſſe is a
low herb creeping vpon the ground : the leaues
are like thoſe of the common Treſoile, but leſſer, and of a grayiſh greene colour : the floures
are faire and yellow, faſhioned like thoſe of
Broome, but leſſer : after come three or foure
cods, wherein is contained round ſeed : the root
is long and reddiſh.   ‡ This is the *Trifolium
Corniculatum*, or *Melilotus coronata* of *Lobel* : *Lotus pentaphyllos* of *Geſner*. ‡

This codded Treſoile is like vnto the laſt
deſcribed in euery reſpect, ſauing that this
plant is altogether ‡ larger, hauing ſtalks a cubit

bit and a halfe high : the leaues are alſo foure times as large, two roundiſh leaues growing by the ſtalke,and three longiſh ones growing vpon a ſhort foot-ſtalk comming forth betweene the two roundiſh leaues : both the ſtalke and leaues haue a little ſoft downineſſe or hairineſſe on them:the floures grow cluſtering together on the tops of the ſtalks,in ſhape,bigneſſe,and colour like that of the laſt deſcribed,but commonly more in number : they are alſo ſucceeded by ſuch cods as thoſe of the former.

6   The figure which *Dodonæus* hath ſet forth out of an old Manuſcript in the Emperors Library,being there figured for *Coronopus*,ſeems to be of the laſt deſcribed, or ſome plant very like thereto ; though the fiue leaues at each joint be not put in ſuch order as they ſhould be,yet all the parrs are well expreſt,according to the drawing of thoſe times : for you ſhall finde few antient expreſſions come ſo neere as this doth. ‡

7   There is a kind of Clauer growing about Narbon in France,that hath many twiggy tough branches comming from a wooddy root ; whereon are ſet leaues three together , after the maner of the other Trefoiles,ſomewhat long,hairy, and of an hoary or ouerworn green colour:the floures are yellow,and grow at the tops of the branches like thoſe of Broome.

7 *Lotus incana,ſiue Oxytriphyllon Scribonij largi.*
    Hoary Clauer.

‡ 8 *Trifolium luteum ſiliqua cornuta.*
    Yellow horned Trefoile.

‡   This ſends vp many branches from one root, ſome cubit or more long, commonly lying along vpon the ground,round,flexible,and diuided into ſundry branches: the leaues ſtand together by threes,and are like thoſe of the true *Medica* or Burgondy Trefoile, but much leſſe : the floures grow cluſtering together at the tops of the branches,like in ſhape to thoſe of the former,of a yellow colour,and not without ſmell : they are ſucceeded by ſuch,yet narrower crooked coddes,as the Burgondy Trefoile hath(but the Painter hath not well expreſſed them :) in theſe cods are contained ſeeds like thoſe alſo of that Trefoile,and ſuch alſo is the root, which liues long,and much increaſes.It growes in Hungary,Auſtria,and Moravia.It floures in Iune and Iuly. *Cluſius* calls it *Medica flore flavo. Tabernamontanus,Lens maior repens :* and *Tragus,Meliloti maioris ſpecies tertia. Bauhine* ſaith,That about Nimes in Narbone it is found with floures either yellow,white, green, blew,purple,blacke,or mixt of blew and green ; and he calleth it *Trifolium ſylveſtre luteum ſiliqua cornuta*; or *Medica fruteſcens.* ‡

¶ *The Place.*

The ſeuerall titles of moſt of theſe plants ſet forth their naturall place of growing: the reſt grow in moſt fertile fields of England.

¶ *The Time.*

They floure and flouriſh moſt of the ſummer moneths.

¶ *The Names.*

There is not much to be ſaid as touching their Names, more than hath bin ſet downe.

¶ *The Temperature and Vertues.*

The temperature and faculties of theſe Treſoiles are referred to the common medow Treſoils.

† The figure formerly put in the ſecond place was of the leſſer Treſoile deſcribed in the laſt chapter ſave one.

Chap. 497.

Of the great Trefoiles or winged Clavers.

¶ *The Deſcription.*

† 1   THe great Hares-foot being a kinde of Treſoile, hath a hard and wooddy root, full of black threddy ſtrings: from whence ariſe diuers tough and feeble branches, whereupon grow leaues ſet together by threes, making the whole plant to reſemble thoſe of the Medow Treſoile: the floures grow at the top of the ſtalks, compoſed of a bunch of gray haires: among the which ſoft matter commeth forth ſmall floures of a moſt bright purple colour, ſomewhat reſembling the floures of the common medow Treſoile, but far greater. *Lobel* calls this *Lagopus minimus folio, & facie Trifolij pratenſis* : *Dodonæus, Lagopus maior folio Trifolij.*

‡ 1 *Lagopus maximus.*   ‡ 2 *Lagopus major ſpica longiore.*
The great Hares-foot Treſoile.   Great large headed Hares-foot.

 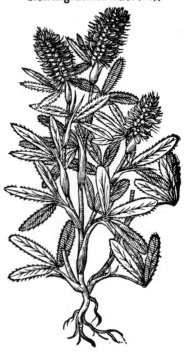

‡ 2 This

‡ 2   This elegant plant(which *Tragus* hath fet forth for *Cytifus*, *Lobel* by the name of *Lagopus altera folio prinnato*, and *Clufius* for his *Trifolij majoris 3.altera fpecies*) hath ftalkes fome foot and better high,whereon grow leaues fet together by threes,long,hoary,and lightly fnipt about the edges, with elegant nerues or veines, running from the middle ribbe to the fides of the leaues,which are moft confpicuous in hot Countries,and chiefly then when the leafe begins to decay. At the tops of the branches,in long and large heads grow the floures,of an elegant fanguine colour.This floures in May and Iune, and growes wilde vpon fome mountaines of Hungary and Auftria, I haue feene them,both this and the former,growing in the gardens of fome of our Florifts.

3   This other great kinde of Hares-foot fends forth one flender, yet ftiffe ftalke,whereon grow leaues whofe foot-ftalkes are large at the ferring on, encompaffing the ftalkes : the leaues themfelues grow by threes,long,narrow,and fharpe pointed, of a grayifh colour like thofe of the common Hares-foot ; the fpike at the tuft is foft and downy, with little reddifh floures amongft the whitifh hairineffe.This growes wild in Spaine:*Clufius* cals it *Lagopus angustifolius Hifpanicus major*.

There is another fort of this defcribed by *Lobel* and *Pena* in the *Aduerf*.whofe leaues are longer and narrower than this, the whole plant alfo is oft times leffer: they call it *Lagopus altera angustifolia*. ‡

‡ 3 *Lagopus angustifolius Hifpanicus.*
Narrow leafed Spanith Hare-foot.

3 *Lagopodium, Pes leporis.*
Little Hares-foot Trefoile.

4   The fmall Hares-foot hath a round rough and hairy ftalke,diuiding it felfe into diuers other branches ; whereupon do grow fmall leaues, three joyned together, like thofe of the fmall yellow Trefoile : the floures grow at the very point of the ftalkes, confifting of a rough knap or bufh of haires or downe,like that of *Alopecuros*, or Fox-taile, of a whitifh colour tending to a light blufh, with little white floures amongft the downineffe : the root is fmall and hard.

¶ *The Place.*

The firft groweth in the fields of France and Spaine,and is a ftranger in England,yet it groweth in my garden.

The fmall Hare-foot groweth among corne, efpecially among Barly, and likewife in barren paftures almoft euery where.

¶ *The Time.*

They floure and flourifh in Iune,Iuly,and Auguft.

¶ *The Names.*

The great Hare-foot Treſoile is called of *Tragus, Cytiſus:* of *Cordus, Trifolium magnum:* of *Lobelius, Lagopum maximum,* and *Lagopodium :* in Greeke, λαγώπους : in Engliſh, the great Hares-foot.

The laſt, being the ſmalleſt of theſe kinds of Treſoiles, is called *Lagopus,* and *Pes Leporis:* in Dutch, **Haſen Pootkens :** in high Dutch, **Haſen fuſz :** in French, *Pied de lieure :* in Engliſh Hare-foot.

¶ *The Temperature and Vertues.*

A   The temperature and faculties are referred vnto the other Treſoiles, whereof theſe are kindes : notwithſtanding *Dioſcorides* ſaith, that the ſmall Hares-foot doth binde and dry. It ſtoppeth, ſaith he, the laske, if it be drunke with red wine. But it muſt be giuen to ſuch as are feueriſh with water.

† Our Author in the firſt place formerly gaue the figure of *Tabern.* his *Lagopodium flore albo,* being only a variety of that plant you ſhall hereafter finde deſcribed by the name of *Anthyllis leguminoſa;* now he made the deſcription ſomewhat in the leaues to agree with the figure, though nothing almoſt with the truth of that hee intended to deſcribe; for(as it is euident by the names) he intended to deſcribe both the firſt and ſecond (which are here now deſcribed) in the firſt place, for he hath confounded them both together in the names.

Chap. 498.   *Of Water Trefoile, or Bucks Beanes.*

Trifolium paludoſum.
Marſh Treſoile.

¶ *The Deſcripton.*

1   THe great Marſh Treſoile hath thicke fat ſtalkes, weake and tender, full of a ſpungious pith, very ſmooth, and of a cubit long: wheron do grow leaues like to thoſe of the garden Beane, ſet vpon the ſtalkes three ioyned together like the other Treſoiles, ſmooth, ſhining, and of a deepe greene colour: among which toward the top of the ſtalkes ſtandeth a buſh of feather-like floures of a white colour, daſht ouer ſlightly with a waſh of light carnation : after which the ſeed followeth, contained in ſmall buttons, or knobby husks, of a browne yellowiſh colour like vnto Millet, and of a bitter taſte : the roots creepe diuers waies in the mirie mariſh ground, being full of ioynts, white within, and full of pores, and ſpungy, bringing forth diuers by-ſhoots, ſtalkes, and leaues, by which meanes it is eaſily increaſed, and largely multiplied.

2   The ſecond differeth not from the precedent, ſauing it is altogether leſſer, wherein conſiſteth the difference, if there be any : for doubtleſſe I think it is the ſelfe ſame in each reſpect, and is made greater and leſſer, according to his place of growing, clymate and countrey.

¶ *The Place.*

Theſe grow in Mariſh and Fenny places, and vpon boggie grounds almoſt euery where.

¶ *The Time.*

They floure and flouriſh from Iune to the end of Auguſt.

¶ *The Names.*

Mariſh Treſoile is called in high Dutch, **Biberklee,** that is to ſay *Caſtoris Trifolium,* or *Trifolium fibrinum :* in low Dutch, of the likeneſſe that the leaues haue with the garden Beanes, **Bocxboonen,** that is to ſay, *Faſelus Hircinus,* or *Boona Hircina :* the later Herbariſts call it *Trifolium paluſtre,* and *Paludoſum:* of ſome, *Iſopyrum :* in Engliſh, marſh-Clauer, marſh-Treſoile, and Buckes-Beanes.

¶ *The Temperature and Vertues.*

A   The ſeed of *Iſopyram,* ſaith *Dioſcorides,* if it be taken with meade or honied water, is good againſt the cough and paine in the cheſt.

It is alſo a remedy for thoſe that haue weake liuers and ſpet bloud, for as *Galen* ſaith it clenſeth and cutteth tough humors, hauing alſo adioyned with it an aſtringent or binding qualitie.

CHAP:

CHAP. 499.   *Of ſweet Trefoile, or garden Clauer.*

Trifolium odoratum.
Sweet Trefoile.

¶ *The Deſcription.*

SWeet Trefoile hath an vpright ſtalk, hollow, and of the height of two cubits, diuiding it ſelfe into diuers branches: whereon do grow leaues by three and three like to the other Trefoiles, ſteightly and ſuperficiouſly nicked in the edges: from the boſom whereof come the floures, euery one ſtanding on his owne ſingle foot-ſtalk; conſiſting of little chaffie husks, of a light or pale blewiſh colour: after which come vp little heads or knops, in which lieth the ſeed, of a whitiſh yellow colour, and leſſer than that of Fenugreeke: the root hath diuers ſtrings: the whole plant is not onely of a whitiſh greene colour, but alſo of a ſweet ſmell, and of a ſtrong aromaticall or ſpicie ſent, and more ſweet when it is dried: which ſmel in the gathered and dried plant doth likewiſe continue long: and in moiſt and rainy weather, it ſmelleth more than in hot and dry weather: and alſo when it is yet freſh and greene it loſeth and recouereth again his ſmell ſeuen times a day; whereupon the old wiues in Germany do call it **Sieuen gezeiten kraut**, that is, the herbe that changeth ſeuen times a day.

¶ *The Place.*
It is ſowne in Gardens not onely beyond the ſeas, but in diuers gardens in England.

¶ *The Time.*
It is ſowne in May, it floureth in Iune and Iuly, and perfecteth his ſeed in the end of Auguſt, the ſame yeare it is ſowne.

¶ *The Names.*

It is called commonly in Latine *Trifolium odoratum*: in high Dutch as we haue ſaid **Sieuen gezetten**: in low Dutch, **Seuenghetiercruit**, that is to ſay, an herbe of ſeuen times: it is called in Spaniſh, *Trebol real*: in French, *Treffle oderiſerant*: in Engliſh, Sweet Treoile, and garden Clauer: it ſeemeth to be *Lotus Vrbana*, or *ſatiua*, of which *Dioſcorides* writeth in his fourth book: neuertheleſſe diuers Authors ſet downe Melilot, for *Lotus vrbana*, and *Trifolium odoratum*, but not properly. ‡ The Gardiners and herbewomen in Cheapſide commonly call it, and know it by the name of Balſam, or garden Balſam. ‡

¶ *The Temperature.*

Galen ſaith, that ſweet Trefoile doth in a meane concoct and dry, and is in a meane temperate facultie betweene hot and cold: the which faculties vndoubtedly are plainely perceiued in this ſweet Trefoile.

¶ *The Vertues.*

The juyce preſſed forth, ſaith *Dioſcorides*, with hony added thereto, clenſeth the vlcers of the eies,   A called in Latine *Argema*, and taketh away ſpots in the ſame, called *Albugines*; and remoueth ſuch things as do hinder the ſight.

The oile wherein the floures are infuſed or ſteeped, doth perfectly cure greene wounds in very   B ſhort ſpace; it appeaſeth the paine of the gout, and all other aches, and is highly commended againſt ruptures, and burſtings in young children.

The juyce giuen in white Wine cureth thoſe that haue fallen from ſome high place, auoideth   C congealed and clotted bloud, and alſo helpeth thoſe that do piſſe bloud, by meanes of ſome great bruiſe, as was prooued lately vpon a boy in Fanchurch ſtreet, whom a cart went ouer, whereupon he did not onely piſſe bloud, but alſo it moſt wonderfully guſhed forth, both at his noſe and mouth.

D   The dried herbe laid among garments keepeth them from Mothes and other vermine.

CHAP. 500. *Of Fenugreeke.*

¶ *The Deſcription.*

1   FEnugreeke hath a long ſlender trailing ſtalke, greene, hollow within, and diuided into diuers ſmall branches : whereon doe grow leaues like thoſe of the medow Trefoile, but rounder and leſſer, greene on the vpper ſide, on the lower ſide tending to an aſh colour : among which come ſmall white floures, after them likewiſe long ſlender narrow cods, in which do lie ſmall vneuen ſeeds, of a yellowiſh colour : which being dryed, haue a ſtrong ſmell, yet not vnpleaſant : the root is ſmall, and periſheth when it hath perfected his ſeed.

1 *Fœnumgrœcum.*
Fenugreeke.

‡ 2 *Fœnumgrœcum ſylueſtre.*
Wilde Fenugreeke.

2   There is a wilde kinde hereof ſeruing for little vſe, that hath ſmall round branches, full of knees or joynts : from each joynt proceedeth a ſmall tender foot-ſtalk, wheron do grow three leaues and no more, ſomewhat ſnipt about the edges, like vnto thoſe of Burgundie Haie: from the boſoms whereof come forth ſmall yellow floures, which turne into little cods : the root is thicke, tough, and pliant.

¶ *The Place.*

Fenugreeke is ſowne in fields beyond the ſeas : in England we ſow a ſmall quantity thereof in our gardens.

¶ *The Time.*

It hath two ſeaſons of ſowing, according to *Columella*, of which one is in September, at what time it is ſowne that it may ſerue for fodder againſt Winter ; the other is in the end of Ianuary, or the beginning of February, notwithſtanding we may not ſow it vntill Aprill in England.

¶ *The*

¶ *The Names.*

It is called in Greeke, ~~~~, or as it is found in *Pliny* his copies *Carpbos* : in Latine, *Fœnum Græ-*
cum : *Columella* ſaith that it is called *Siliqua* : in *Pliny* we reade *Silicia* : in *Varro*, *Siliculа* : in high Dutch,
Bockſhoane : in Italian, *Fiengreco* : in Spaniſh, *Alfornas* : in French, *Fenegrec* : and in Engliſh,
Fenegreeke.

¶ *The Temperature and Vertues.*

It is thought according to *Galen* in his booke of the Faculties of nouriſhments, that it is one of   A
thoſe ſimples which do manifeſtly heat, and that men do vſe it for food, as they do Lupines; for it is
taken with pickle to keep the body ſoluble; and for this purpoſe it is more agreeable than Lupines,
ſeeing it hath nothing in his owne proper ſubſtance, that may hinder the working.

The juyce of boyled Fenegreeke taken with hony is good to purge by the ſtoole all manner of   B
corrupt humors that remaine in the guts, making ſoluble through his ſlimineſſe, and mitigating
paine through his warmeneſſe.

and becauſe it hath in it a clenſing or ſcouring faculty, it raiſeth humors out of the cheſt : but   C
there muſt be added vnto it no great quantity of honie, leaſt the biting quality ſhould abound.

In old diſeaſes of the cheſt without a feuer, fat dates are to be boyled with it, but when you haue   D
mixed the ſame juyce preſſed out with a great quantity of hony, and haue againe boiled it on a ſoft
fire to a meane thickeneſſe, then muſt you vſe it long before meat.

In his booke of the Faculties of ſimple medicines, he ſaith, that Fenegreeke is hot in the ſecond   E
degree and dry in the firſt : therefore it doth kindle and make worſe hot inflammations, but ſuch as
are leſſe hot and more hard are thereby cured by being waſted and conſumed away.

The meale of Fenegreeke, as *Dioſcorides* ſaith, is of force to mollifie and waſte away: being boiled   F
with mead and applied it taketh away inflammations, as well inward as outward.

The ſame being tempered or kneaded with niter and vineger, doth ſoften and waſte away the   G
hardneſſe of the milt.

It is good for women that haue either impoſthume, vlcer, or ſtopping of the matrix, to bathe and   H
ſit in the decoction thereof.

The juyce of the decoction preſſed forth doth clenſe the haire, taketh away dandraffe, ſcoureth   I
running ſores of the head, called of the Græcians ~~~~ : being mingled with gooſe greaſe, and put vp
in manner of a peſſary, or mother ſuppoſitory, it doth open and mollifie all the parts about the
mother.

Greene Fenegreeke bruiſed and pounded with vineger, is a remedy for weake and feeble parts,   K
and that are without skin, vlcerated and raw.

The decoction thereof is good againſt vlcers in the low gut, and foule ſtinking excrements of   L
thoſe that haue the bloudy fliх.

The oile which is preſſed out thereof ſcoureth haires and ſcars in the priuie parts.   M

The decoction of Fenegreeke ſeed, made in wine, and drunke with a litttle vineger, expelleth all   N
euill humors in the ſtomacke and guts.

The ſeed boiled in wine with dates and hony, vnto the form of a ſyrrup, doth mundifie and clenſe   O
the breſt, and eaſeth the paines thereof.

The meale of Fenegreeke boiled in meade or honied water, conſumeth and diſſolueth all cold   P
hard impoſthumes and ſwellings, and being mixed with the roots of Marſh Mallowes and Linſeed
effecteth the ſame.

It is very good for women that haue any griefe or ſwelling in the matrix, or other lower parts, if   Q
they bathe thoſe parts with the decoction thereof made in wine, or ſit ouer it and ſweat.

It is good to waſh the head with the decoction of the ſeed, for it taketh away the ſcurfe, ſcales,   R
nits, and all other ſuch like imperfections.

Chap. 501.   *Of Horned Clauer, and blacke Clauer.*

¶ *The Deſcription.*

1   THe horned Clauer, or codded Trefoile, groweth vp with many weake and ſlender ſtalkes
lying vpon the ground : about which are ſet white leaues, ſomewhat long, leſſer and narrower than any of the other Trefoiles : the floures grow at the tops, of the faſhion of
thoſe of Peaſon, of a ſhining yellow colour : after which come certaine ſtraight cods, bigger than
thoſe of Fenegreeke, but blunter at their ends, in which are contained little round ſeed ; the root is
hard and wooddy and ſendeth forth young ſprings euery yeare.

1 *Lotus trifolia corniculata.*
Horned or codden Caler.

2 *Lotus quadrifolia.*
Foure leaſed graſſe.

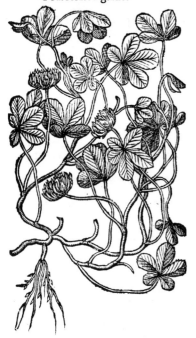

2 This kinde of three leafed graſſe, or rather foure leafed Trefoile, hath leaues like vnto the common Trefoile, ſauing that they bee leſſer, and of a browne purpliſh colour, known by the name of Purple-wort, or Purple-graſſe; whoſe floures are in ſhape like the medow Trefoile, but of a duſty ouerworne colour tending to whitenesse ; the which doth oftentimes degenerate, ſometime into three leaues, ſometimes in fiue, and alſo into ſeuen, and yet the plant of his nature hath but foure leaues & no more. ‡ I do not thinke this to be the purple leaued Trefoile with the white floure,which is commonly called Purple-graſſe,for I could neuer obſerue it to haue more leaues than three vpon a ſtalke. ‡

‡ 3 The root of this is ſmall and white, from which ariſe many weake hairy branches ſome cubit long;wheron grow ſoft hairy leaues three on one foot-ſtalke, with two little leaues at the root therof,& out of the boſoms of theſe vpon like foot-ſtalkes grow three leſſer leaues, as alſo floures of the bignes and ſhape of thoſe of a Vetch, but of a braue deep crimſon veluet colour : after theſe are paſt come cods ſet with foure thinne welts or skins which make them ſeem foure ſquare ; whence *Camerarius* called it *Lotus pulcherrima tetragonolobus*: the ſeed is of an aſh colour, ſomewhat leſſe then a peaſe. It floures moſt of the Summer moneths,and is for the prettineſſe of the floure preſerued in many Gardens by yearely ſowing the ſeed, for it is an annuall plant. *Cluſius* hath it by the name of *Lotus ſiliquoſus rubello flore :* and hee ſaith the ſeeds were diuers times ſent out of Italy by the name of *Sandalida*. It is alſo commonly called in Latine *Piſum quadratum.* ‡

¶ *The Place.*

The firſt groweth wilde in barren ditch bankes,paſtures,and dry Mountaines.

‡ 3 *Lotus ſiliqua quadrata.*   Square crimſon veluet peaſe.

The

The ſecond groweth likewiſe in paſtures and fields, but not ſo common as the other; and it is planted in gardens.

¶ *The Time.*

They floure in Iuly and Auguſt.

¶ *The Names.*

The ſecond is called *Lotus Trifolia*: in Engliſh, horned Clauer, or codded Trefoile.

The other is called *Lotus quadrifolia*, or foure leaſed Graſſe, or Purple-wort: of *Pena* and *Lobel*, *Quadrifolium phæum fuſcum hortorum*.

¶ *The Temperature and Vertues.*

Their faculties in working are referred vnto the medow Trefoiles: notwithſtanding it is repor- A ted, that the leaues of Purple-wort ſtamped, and the juyce giuen to drinke, cureth young children of the diſeaſe called in Engliſh the Purples.

CHAP. 502.   *Of* Medicke Fodder, *or* ſnaile Clauer.

¶ *The Deſcription.*

1   THis kinde of Trefoile, called Medica, hath many ſmall and ſlender ramping branches, crawling and creeping along vpon the ground, ſet full of broad leaues ſlightly indented about the edges: the flours are very ſmall, and of a pale yellow colour, which turne into round wrinkled knobs, like the water Snaile, or the fiſh called Periwinckle: wherein is contained flat ſeed faſhioned like a little kidney, in colour yellow, in taſte like a Vetch or peaſe: the root is ſmall, and dieth when the ſeed is ripe: it growes in my garden, and is good to feed cattell fat.

| 1 *Trefolium Cochleatum.* | ‡ 2 *Medica fructu cochleato ſpinoſo.* |
| Medicke Fodder. | Prickly Snaile Trefoile. |

‡   There are many varieties of theſe plants, and they chiefely conſiſt in the fruit; for ſome are ſmooth and flat, as this firſt deſcribed: other ſome are rough and prickly, ſome with leſſer, and

other-

otherſome with bigger prickles ; as alſo with them ſtanding diuers waies, ſome are onely rough, and of thoſe, ſome are as big as a ſmall nut, otherſome no bigger then a peaſe. I giue you here the deſcriptions of three rough ones (as I receiued them from Mr. *Goodyer*) whereof the laſt is of the ſea, which, as you may ſee, our Author did but ſuperficially deſcribe.

2 *Medica majoris Bæticæ ſpecies prima, ſpinulis intortis.*

This hath foure ſquare reddiſh ſtreaked hairy trailing branches, like the ſmall Engliſh *Medica*, greater and longer, foure or fiue foot long : the leaues are alſo ſmooth, growing three together, neither ſharpe poiuted, nor yet ſo broad at the top as the ſaid Engliſh Medica, but blunt topped, with a ſmall blacke ſpot in the midſt, not crooked : the floures are alſo yellow, three, foure, or fiue on a foot-ſtalke: after commeth a round writhed fruit fully as big as a haſell nut, with ſmall prickles not ſtanding fore-right, but lying flat on the fruit, finely wrapped, plaited, folded, or interlaced together, wherein lieth wrapped the ſeed in faſhion of a kidney, very like a kidney beane, but foure times ſmaller, and flatter, of a ſhining blacke colour without, like poliſhed Ieat ; containing a white kernell within : the root is like the former, and periſheth alſo at Winter.

Medica majoris Bæticæ ſpinoſæ ſpecies altera.

The branches alſo creepe on the ground, and are ſtraked, ſmooth, foure-ſquare, reddiſh here and there, three or foure foot long : the leaues are ſmooth, finely notched about the edges, ſharpe poiuted, without blacke ſpots, very like *Medica pericarpio plana*: the floures are ſmall and yellow like the other : the fruit is round, writhed or twined in alſo, fully as big as a haſell nut, ſomewhat cottonie or woolly, with ſhort ſharpe prickles : wherein lyeth alſo wrapped a ſhining blacke kidney-like ſeed, ſo like the laſt deſcribed, that they are not to be diſcerned apart : the root is alſo alike, and periſheth at Winter.

Medica marina ſpinoſa ſpecies.

The branches of this are the leaſt and ſhorteſt of all the 'reſt , little exceeding a foot or two in length, and are foure ſquare, greene, ſomewhat hairy, and trailing on the ground : the leaues are like to thoſe of *Medica pericarpio plano*, not fully ſo ſharpe pointed, without blacke ſpots, ſoft, hairy, three on a foot-ſtalke: the floures grow alongſt the branches, on very ſmall foot-ſtalkes, forth of the boſomes of the leaues, (not altogether on or neere the tops of the branches) and are very ſmall and yellow, but one on a foot-ſtalke : after commeth ſmall round writhed fruit, no bigger than a peaſe, with very ſhort ſharpe prickles, wherein is contained yellowiſh ſeed of the faſhion of a kidney like the former, and is the hardeſt to be plucked forth of any of the reſt : the root is alſo whitiſh like the roots of the other, and alſo periſheth at Winter.   Aug. 2. 1621. *Iohn Goodyer.* ‡

3 *Trifolium Cochleatum marinum.*
Medick Fodder of the ſea.

3   This kinde alſo of Trefoile, (call *Medica marina* : in Engliſh, ſea Trefoile, growing natu-rally by the ſea ſide about Weſtcheſter, and vpon the Mediterranean ſea coaſt, and about Venice) hath leaues very like vnto the common medow Trefoile, but thicker, and couered ouer with a
flockie

flockie hoarineſſe like *Gnaphalium*, after the manner of moſt of the ſea herbes : the floures are yellow : the ſeeds wrinkled like the former, but in quantitie they be leſſer.

¶ *The Place.*

The firſt is ſowne in the fields of Germany, Italy, and other countries, to feed their cattell, as we in England doe Bucke-wheat : wee haue a ſmall quantity thereof in our Gardens, for pleaſures ſake.

The third groweth neere vnto the ſea ſide in diuers places.

¶ *The Time.*

Medica muſt be ſowne in Aprill; it floureth in Iune and Iuly : the fruit is ripe in the end of Auguſt.

¶ *The Names.*

Medick fodder is called of ſome *Trifolium Cochleatum*, and *Medica* : in French, *L'herbe à Limaſſon* : in Greeke, Μηδική : in Spaniſh, *Mielguas* : of the Valentians and Catalons, *Alfafa*, by a word either barbarous or Arabicke : for the chiefe of the Arabian writers, *Auicen*, doth call *Medica, Cot, Alaſeleti*, and *Alfasfafa*.

The other is called Sea Clauer, and Medick fodder of the ſea.

¶ *The Temperature and Vertues.*

Medick Fodder is of temperature cold, for which cauſe it is applied greene to ſuch inflamma-　A tions and infirmities as haue need of cooling.

CHAP. 503.　*Of Wood Sorrell, or Stubwort.*

1 *Oxys alba.*
White Wood Sorrell.

¶ *The Deſcription.*

1　OXys *Pliniana*, or *Trifolium acetoſum*, being a kinde of three leafed graſſe, is a low and baſe herbe without ſtalke; the leaues immediately riſing from the root vpon ſhort ſtemmes at their firſt comming forth folded together, but afterward they do ſpred abroad, and are of a faire light greene colour, in number three, like the reſt of the Trefoiles; but that each leafe, hath a deep cleft or rift in the middle : among theſe leaues come vp ſmall and weake tender ſtems, ſuch as the leaues do grow vpon, which beare ſmall ſtarre-like floures of a white colour, with ſome brightnes of carnation daſht ouer the ſame : the floure conſiſteth of fiue ſmall leaues ; after which come little round knaps or huſkes full of yellowiſh ſeed : the root is very threddy, and of a reddiſh colour : the whole herbe is in taſte like Sorrell, but much ſharper and quicker, and maketh better greene ſauce than any other herbe or Sorrell whatſoeuer.

‡　My oftmentioned friend Mr. *George Bowles* ſent me ſome plants of this with very faire redde floures, which hee gathered in Aprill laſt, in a wood of Sir *Thomas Walſinghams* at Chiſſelhurſt in Kent, called Stockwell wood, and in a little round wood thereto adjoyning. ‡

The ſecond kinde of *Oxys* or Wood Sorrell is very like the former, ſauing that the floures are

of

2 *Oxys lutea.*
Yellow Wood Sollell.

of a yellow colour, and yeeld for their ſeed veſſels ſmall and long horned cods ; in other reſpeɔts alike.

¶ *The Place.*

Theſe plants grow in woods and vnder buſhes, in ſandie and ſhadowie places in euery countrey. ‡ I haue not as yet found any of the yellow growing with vs. ‡

¶ *The Time.*

They floure from the beginning of Aprill vnto the end of May and midſt of Iune.

¶ *The Names.*

Wood Sorrell or Cuckow Sorrell is called in Latine *Trifolium acetoſum :* the Apothecaries and Herbariſts call it *Alleluya,* and *Panis Cuculi,* or Cuckowes meate, becauſe either the Cuckow feedeth thereon or by reaſon when it ſpringeth forth and floureth the Cuckow ſingeth moſt, at which time alſo *Alleluya* was wont to be ſung in Churches. *Hieronymus Fracaſtorius* nameth it *Lujula. Alexander Benedictus* ſaith that it is called *Alimonia :* in high Dutch, **Saurellklee :** in low Dutch, **Coeckcoeckbrʒoot :** in French, *Pain de Cocu :* in Engliſh, wood Sorrell, wood Sower, Sower Trefoile, Stubwort, Alleluia, and Sorrell du Bois.

It is thought to be that which *Pliny, lib. 27. cap. 12.* calleth *Oxys* ; writing thus : *Oxys* is three leaſed, it is good for a feeble ſtomack, and is alſo eaten of thoſe that are burſten. But *Galen* in his fourth booke of Simples ſaith, that *Oxys* is the ſame which *Oxalis* or Sorrell is : and *Oxys* is found in *Pliny* to be alſo *Iunci ſpecies,* or a kinde of Ruſh.

¶ *The Temperature.*

Theſe herbes are cold and dry like Sorrell.

¶ *The Vertues.*

A Sorrell du Bois or Wood Sorrell ſtamped and vſed for greene ſauce, is good for them that haue
C ſicke and feeble ſtomackes ; for it ſtrengthneth the ſtomacke, procureth appetite, and of all Sorrell ſauces is the beſt, not onely in vertue, but alſo in the pleaſantneſſe of his taſte.

B It is a remedy againſt putrified and ſtinking vlcers of the mouth, it quencheth thirſt, and cooleth mightily any hot peſtilentiall feuer, eſpecially being made with a ſyrrup of ſugar.

CHAP. 504. *Of noble Liuer-wort, or golden Trefoile.*

¶ *The Deſcription.*

1 NOble Liuerwort hath many leaues ſpred vpon the ground, three cornered, reſembling the three leafed graſſe, of a perfect graſſe greene colour on the vpper ſide, but grayiſh vnderneath : among which riſe vp diuers ſmall tender foot-ſtalks of three inches long, on the ends whereof ſtands one ſmall ſingle blew floure, conſiſting of ſix little leaues, hauing in the middle a few white chiues : the ſeed is incloſed in little round knaps, of a whitiſh colour; which being ripe do ſtart forth of themſelues : the root is ſlender, compoſed of an infinite number of blacke ſtrings.

2 The ſecond is like vnto the precedent in leaues, roots, and ſeeds : the floures hereof are of a ſhining red colour, wherein conſiſteth the difference.

This ſtrange three leaued Liuerwort differeth not from the former, ſauing that this brings forth double blew floures tending to purple, and the other not ſo.

There is another in my garden with white floures, which in ſtalkes and euery other reſpect is like the others.

1 *Hepaticum trifolium.*
Noble Liuerwort.

2 *Hepatica trifolia rubra.*
Noble red Liuerwort.

3 *Hepatica multiflora Lobelij.*
Noble Liuerwort with double floures.

¶ *The Place.*

Theſe pretty floures are found in moſt places of Germanie in ſhadowie woods among ſhrubs, and alſo by high-waies ſides: in Italy likewiſe, and that not only with the blew floures, but the ſame with double floures alſo, by the report of *Alfonſus Pancius* Dr. of Phyſick in the Vniuerſity of Ferrara, a man excellently well ſeen in the knowledge of Simples. They do all grow likewiſe in my garden, except that with double floures, which is as yet a ſtranger in England : ‡ It is now plentifull in many gardens. ‡

¶ *The Time.*

They floure in March and April, and perfect their feed in May.

¶ *The Names.*

Noble Liuerwort is called *Hepatica trifolia, Hepatica aurea, Trifolium aureum:* of *Baptiſta Sardus, Herba Trinitatis :* in high Dutch, **Edel Leber kraut** : in low Dutch , **Edel leuer cruijt** : in French, *Hepatique :* in Engliſh, Golden Trefoile, three leaued Liuerwort, noble Liuerwort, and herbe Trinitie.

¶ *The Temperature.*

Theſe herbes are cold and dry, with an aſtringent or binding quality.

¶ *The*

¶ *The Vertues.*

A   It is reported to be good againſt the weakeneſſe of the liuer which proceedeth of an hot cauſe; for it cooleth and ſtrengthneth it not a little.

B   *Baptiſta Sardus* commendeth it, and writeth that the chiefe vertue is in the root; if a ſpoonfull of the pouder thereof be giuen certaine daies together with wine, or with ſome kinde of broth, it profiteth much againſt the diſeaſe called *Enterocele*.

CHAP. 505.   *Of Melilot, or plaiſter Clauer.*

¶ *The Deſcription.*

1   THe firſt kinde of Melilot hath great plenty of ſmall tough and twiggy branches, and ſtalkes full of joynts or knees, in height two cubits, ſet full of leaues three together, like vnto Burgondy hay. The flours grow at the top of the ſtalk, of a pale yellow color, ſtanding thickely ſet and compact together, in order or rowes, very like the floures of *Securidaca altera*: which being vaded there follow certaine crooked cods bending or turning vpward with a ſharpe point, in faſhion not much vnlike a Parrets bill, wherein is contained ſeed like Fenugreeke, but flatter and ſlenderer: the whole plant is of a reaſonable good ſmell, much like vnto hony, and very full of juyce: the root is very tough and pliant.

1 *Melilotus Syriaca odora.*	2 *Melilotus Italica & Patauina.*
Aſſyrian Clauer.	Italian Clauer.

2   The ſecond kinde of Melilot hath ſmall and tender vpright ſtalkes, a cubit high, and ſomewhat more, of a reddiſh colour, ſet full of round leaues three together, not ſnipt about the edges like the other Trefoiles; and they are of a very deepe greene colour, thicke, fat, and full of juyce. The floures grow alongſt the tops of the ſtalkes, of a yellow colour, which turne into rough round ſeeds as big as a Tare, and of a pale colour. The whole plant hath alſo the ſauour of hony, and periſheth when it hath borne his ſeed.

3 The third kind of Melilot hath round stalks and jagged leaues set round about, not much vnlike the leaues of Fenugreeke, alwaies three growing together like the Trefoiles, and oftentimes couered ouer with an hoarinesse, as though meale had been strewed vpon them. The floures be yellow and small, growing thicke together in a tuft, which turne into little cods, wherein the seed is contained : the root is small, tough, and pliant.

4 The fourth kinde of Melilot growes to the height of three cubits, set full of leaues like the common Melilot, and of the same sauour : the floures grow alongst the top of the stalkes, of a white colour, which turne into small soft huskes, wherein is contained little blackish seed : the root is also tough and pliant.

3 *Melilotus Coronata.*
Kings Clauer.

4 *Melilotus Germanica.*
Germane Clauer.

‡ Although our Author intended this last description for our ordinary Melilot, yet hee made it of another which is three times larger, growing in some gardens (where it is onely sowne) aboue two yards high, with white floures and many branches : the whole shape thereof is like the common kinde, as far as I remember. The common Melilot hath weake cornered greene stalkes some two foot and better high ; whereon grow longish leaues snipt and oftentimes eaten about the edges, of a fresh greene colour : out of the bosomes of the leaues come little stalkes some handfull long, set thicke on their tops with little yellow floures hanging downe and turning vp againe. each floure being composed of two little yellow leaues, whereof the vppermost turnes vp againe, and the vndermost seemes to be parted into three. The floures past, there succeed little cods wherein is the seed. ‡

¶ *The Place.*

These plants grow in my garden : the common English Melilot *Pena* setteth forth for *Melilotus Germanica* : but for certainty no part of the World doth enjoy so great plenty thereof as England, and especially Essex; for I haue seene betweene Sudbury in Suffolke, and Clare in Essex, and from Clare to Heningham, and from thence to Ouendon, Pulmare, and Pedmarsh, very many acers of earable pasture ouergrowne with the same ; insomuch that it doth not onely spoyle their land, but the corne also, as Cockle or Darnel, and is a weed that generally spreadeth ouer that corner of the Shire.

¶ *The Time.*

Theſe herbes do floure in Iuly and Auguſt.

¶ *The Names.*

Plaiſter Clauer is called by the generall name, *Melilotus*, of ſome, *Trifolium odoratum*; yet there is another ſweet Trefoile, as hath beene declared. Some call it *Trifolium Equinum*, and *Caballinum*, or Horſe-Trefoile, by reaſon it is good fodder for horſes, who do greedily feed thereon : likewiſe *Trifolium Vrſinum*, or Beares Trefoile : of *Fuchſius, Saxifraga lutea*, and *Sertula Campana* : of *Cato, Serta Campana*, which moſt doe name *Corona Regia* : in high Dutch, **Grote Steenclaueren** : of the Romanes and Hetrurians, *Tribolo*, as *Matthiolus* writeth : in Engliſh, Melilot, and Plaiſter-Clauer : in Yorkeſhire, Harts-Clauer.

¶ *The Temperature.*

Melilote, ſaith *Galen*, hath more plenty of hot ſubſtance then cold (that is to ſay, hot and dry in the firſt degree) it hath alſo a certaine binding quality, beſides a waſting and ripening faculty. *Dioſcorides* ſheweth, that Melilot is of a binding and mollifying quality, but the mollifying quality is not proper vnto it, but in as much as it waſteth away, and digeſteth humors gathered in hot ſwellings, or otherwiſe : for ſo far doth it mollifie or ſupple that thing which is hard, which is not properly called mollifying, but digeſting and waſting away by vapors : which kind of quality the Grecians call διαφορητικί.

¶ *The Vertues.*

A　　Melilote boiled in ſweet wine vntill it be ſoft, if you adde thereto the yolke of a roſted egge, the meale of Fenugreeke and Lineſeed, the roots of Marſh Mallows and hogs greaſe ſtamped together, and vſed as a pultis or cataplaſme, plaiſterwiſe, doth aſſwage and ſoften all manner of ſwellings, eſpecially about the matrix, fundament and genetoires, being applied vnto thoſe places hot.

B　　With the juyce hereof, oile, wax, roſin and turpentine, is made a moſt ſoueraigne healing and drawing emplaiſter, called Melilote plaiſter, retaining both the colour and ſauour of the herbe, being artificially made by a skilfull Surgion.

C　　The herbe boyled in wine and drunke prouoketh vrine, breaketh the ſtone, and aſſwageth the paine of the kidnies, bladder and belly, and ripeneth flegme, and cauſeth it to be eaſily caſt forth.

D　　The juyce thereof dropped into the eies cleereth the ſight, conſumeth, diſſolueth, and cleane taketh away the web, pearle, and ſpot in the eies.

E　　Melilote alone with water healeth *Recentes melicerides*, a kinde of wens or rather apoſtems contayning matter like hony; and alſo the running vlcers of the head, if it be laied to with chalke, wine and galls.

F　　It likewiſe mitigateth the paine of the eares, if the juyce be dropped therein mixed with a little wine, and taketh away the paine of the head, which the Greekes call κεφαλαγίαν, eſpecially if the head be bathed therewith, and a little vineger and oyle of Roſes mixed amongſt it.

---

‡ C H A P. 506.　*Of certaine other Trefoiles.*

‡　THoſe Trefoiles being omitted by our Author, I haue thought good to put into a chapter by themſelues, though they haue little affinity one with another, the two laſt excepted.

¶ *The Deſcription.*

1　THe firſt of thoſe in roots, ſtalkes, and manner of growing is like the Medicke or ſnaile Trefoiles formerly deſcribed : the leaues are hairy : the floures yellow and ſmall : after which follow crooked flat cods, of an indifferent bredth, wherein is contained ſeeds made after the faſhion of little Kidneyes; this the Italians, according to *Lobel*, call *Lunaria radiata*; in the *Hiſt. Lugd.* it is called *Medica ſyl. altera lunata.*

2　The root of this is long and thicke, couered with a yellowiſh rinde, and hauing a white ſweet pith in the inſide, couered with a hairineſſe on the top, and ſending forth ſundry fibres : from this riſe vp many weake long foot-ſtalkes, whereon grow leaues ſet together by threes, long, narrow, ſmooth, lightly nickt on the edges : amongſt theſe riſeth vp commonly one ſtalke (yet ſometimes two) ſmooth and naked, three or foure inches long ; on the top thereof grow ſpike faſhion, eight or ten pretty large light purple floures, each of them being ſet in a cup diuided into fiue parts. This growes vpon diuers parts of the Alps : and *Pona* in his *Mons Baldus* ſet it forth by the name of *Trifolium anguſtifolium Alpinum. Bauhinus* ſaith, the root hereof taſts like liquorice, wherefore it may be called *Glycyrrhiza Aſtragaloides*, or *Aſtragalus dulcis* : and he receiued it out of Spaine by the name of *Glycyrrhiza*. He calls it in his *Prodromus, Trifolium Alpinum flore magno radice dulci.*

The

‡ 1 *Trifolium ſiliqua lunata.*
Moone Trefoile.

‡ 2 *Trifol. anguſtifol. Alpinum.*
Liquorice Trefoile.

‡ 3 *Trifolium ſpinoſum Creticum.*
Prickly Trefoile.

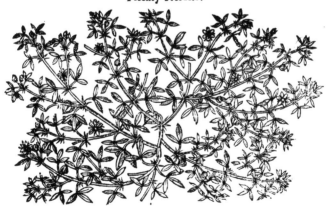

3 This thorny Trefoile hath a long threddy root, from which ariſe many ſhort branched ſtalks ſome two handfuls high, cornered, and ſpred vpon the ground : the joints which are many, are commonly red, & armed with foure ſharp prickles, and out of each of them vpon ſhort foot-ſtalks grow two trifoile leaues, green, longiſh, and ending in a little prickle : out of theſe joints alſo grow little foot-ſtalks which carry ſingle floures made of fiue little leaues of the ſhape and colour of the little Blew-bell floure, with ten chives in the middle tipt with yellow : after theſe follow fiue cornered ſharp pointed heads, containing a ſingle flat red ſeed in each corner. *Cluſius* ſet forth this by the name of *Trifolium ſpinoſum Creticum :* the ſeed was ſent out of Candy by the name of פרא זבד: He queſtions whether it may not be the true *Tribulus terreſtris* of *Dioſcorides.*

4 The roots, ſtalkes, and leaues of this pretty Trefoile doe not much differ from the common

white

‡ 4 *Trifolium fragiferum.*
Straw-berry Trefoile.

white Trefoil, but there is some difference in the floures and seed ; for the floures of this are smal, grow thicke together, and are of a whitish blush colour : after which follow heads made of little bladders or thin skins, after such a maner as they resemble a Straw-berry or Raspas, and they are of a grayish colour here and there marked with red: the stalks seldom grow aboue three inches high. It growes in most salt marishes, as in Dartford salt marish in those below Purfleet, and the like. It floures in Iuly and August. *Clusius* hath set it forth by the name of *Trifolium fragiferum Frisicum :* some had rather cal it *Trifolium vesicarium,* Bladder Trefoile.

5   There are two other Trefoiles with which I thinke good to acquaint you, and those by the similitude of the cups which contain the flours, and become the seed-vessels, may be fitly called *Stellata ;* and thus *Bauhine* calls the first *Trifolium stellatum :* whereto for distinctions sake I add *hirsutum,* calling it *Trisol.stellatum hirsutum,* Rough starry headed Trefoile: it hath a smal long white root, from which arise stalkes some foot high, round, slender, hairy, and reddish, hauing few leaues or branches : the leaues stand three on a stalke, as in other Trefoiles, smooth on the vpper side, and hairy below : the floures are small and red, like in shape to those of the common red Trefoile, but lesser ; and they stand each of them in a cup, reddish, and rough below, and on the vpper part cut into fiue long sharp leaues standing open as they commonly figure a star : the floures fallen, these cups dilate themselues, and haue in the middle a longish transuerse whitish spot. I saw this flouring in May in the garden of Mr *Tradescant,* who first brought plants hereof from Fermentera a small Island in the Mediterranean sea.

6   This other (which for any thing that I know is not figured nor described by any) hath stalks sometimes a foot, otherwhiles little aboue an inch high, hairy, and diuided but into few branches : the leaues, which stand by threes, are fastned to long foot-stalks, and they themselues are somewhat longish, hauing no little sharpe pointed leaues growing at the setting on of the foot-stalks to the stalks : they are green of colour, and not snipt about the edges : the heads that grow on the tops of the stalks are round, short, and green, with small purple or else whitish flours like those of the common Trefoile, but lesser, standing in cups diuided into fiue parts, which when the flours are fallen, become somewhat bigger, harsher, and more prickly, but open not themselues so much as those of the former : the seed is like that of Millet, but somewhat rounder : this flours in Iune, and the seed is ripe in Iuly. I first obserued it in Dartford salt marish, the tenth of Iune, 1633. I haue named this *Trifolium stellatum glabrum,* Smooth starry headed Trefoile.

¶ *The Temperature and Vertues.*

These, especially the three last, seeme to be of the same temper and vertue as the common Medow Trefoiles, but none of them are at this day vsed in physicke, or known, vnlesse to some few. ‡

Chap. 507. *Of Pulse.*

¶ *The Kindes.*

THere be diuers sorts of Pulse, as Beans, Peason, Tares, Chiches, and such like, comprehended vnder this title Pulse : and first of the great Bean or garden Bean.

¶ *The*

¶ The Deſcription.

1 THe great Beane riſeth vp with a foure ſquare ſtalke, ſmooth, hollow, without joynts, long and vpright, which when it is thicke ſowne hath no need of propping, but when it is ſowne alone by it ſelfe it ſoone falleth down to the ground: it bringeth forth long leaues one ſtanding from another, conſiſting of many growing vpon one rib or ſtem, euery one wherof is ſomewhat fat, ſet with veines, ſlippery more long than round. The floures are eared, in forme long, in colour either white with blacke ſpots, or of a blackiſh purple: after them come vp long cods, thicke, full of ſubſtance, ſlenderer below, frized on the inſide with a certaine white wooll as it were, or ſoft flockes; which before they be ripe are green, and afterwards being dry they are black and ſomewhat hard, as be alſo the cods of broome, yet they be longer than thoſe, and greater: in which are contained three, foure, or fiue Beans, ſeldome more, long, broad, flat, like almoſt to a mans naile, great, and oftentimes to the weight of halfe a dram; for the moſt part white, now and then of a red purpliſh colour; which in their vpper part haue a long blacke nauell as it were, which is couered with a naile, the colour whereof is a light greene: the skin of the fruit or beane is cloſely compacted, the inner part being dry is hard and ſound, and eaſily cleft in ſunder; and it hath on the one ſide an euident beginning of ſprouting, as haue alſo the little peaſe, great Peaſe, Ciches, and many other Pulſes. The roots hereof are long, and faſtned with many ſtrings.

1 Faba major hortenſis.
The great garden Beane.

2 Faba ſylueſtris.
The wilde Beane.

2 The ſecond kinde of Beane (which Pena ſetteth forth vnder the title of Sylueſtris Græcorum Faba, and Dodonæus, Bona ſylueſtris; which may be called in Engliſh, Greeke Beanes) hath ſquare hollow ſtalkes like the garden Beanes, but ſmaller. The leaues be alſo like the common Beane, ſauing that the ends of the rib whereon thoſe leaues doe grow haue at the very end ſmall tendrels or claſpers, ſuch as the Peaſe leaues haue. The floures are in faſhion like the former, but of a darke red colour: which being vaded, there ſucceed long cods which are blacke when they be ripe, within which is incloſed blacke ſeed as big as a Peaſe, of an vnpleaſant taſte and ſauour.

‡ 3   The common Beane in ftalkes, leaues, floures, and cods is like the former great garden Beane, but leſſer in them all ; yet the leaues are more, and grow thicker, and out of the boſomes of the leaues vpon little foot-ftalks grow the floures, commonly ſix in number, vpon one ftalke, which are ſucceeded by ſo many cods, leſſer and rounder than thoſe of the former : the Beanes themſelues are alſo leſſe, and not ſo flat, but rounder, and ſomewhat longiſh : their colour are either whitiſh, yellowiſh or elſe blacke. This is ſowne in moſt places of this kingdome, in corne fields, and knowne both to man and beaſt. I much wonder our Author forgot to mention ſo common and vulgarly knowne a Pulſe. It is the *Bana* or *Faſelus minor* of *Dodonæus* ; and the *Faba minor* of *Pena* and *Lobel.* ‡

¶ The Place.

The firſt Beane is ſowne in fields and gardens euery where about London.

This blacke Beane is ſowne in a few mens Gardens who bee delighted in variety and ſtudy of herbes, whereof I haue great plenty in my garden.

¶ The Time.

They floure in Aprill and May, and that by parcels, and they be long in flouring : the fruit is ripe in Iuly and Auguſt.

¶ The Names.

The garden Beane is called in Latine *Faba* : in Engliſh, the garden Beane : the field Beane is of the ſame kinde and name , although the fertilitie of the ſoile hath amended and altered the fruit into a greater forme. ‡ The difference betweene the garden and the field Beane is a ſpecificke diffe-rence, and not an accidental one cauſed by the ſoile, as euery one that knoweth them may well per-ceiue. ‡

The blacke Beane, whoſe figure we haue ſet forth in the ſecond place, is called *Faba ſylueſtris* : of ſome thought to be the true Phyſicke Beane of the Antients ; whereupon they haue named it *Faba Veterum*, and alſo *Faba Græcorum*, or the Greeke Beane. Some would haue the garden Beane to be the true *Phaſeolus*, or Kidney Beane ; of which number *Dodonæus* is chiefe, who hath ſo wrangled and ruffled among his relatiues, that all his antecedents muſt be caſt out of dores : for his long and te-dious tale of a tub wee haue thought meet to commit to obliuion. It is called in Greeke σ´ιαμι, whereupon the Athenians feaſt daies dedicated to *Apollo* were named πυανίψια, in which Beanes and Pulſes were ſodden : in Latine it is called *Faba freſa* or *fracta*, broken or bruiſed Beanes.

‡   *Dodonæus* knew well what he did, as any that are either iudicious or learned may ſee, if they looke into the firſt chapter of the ſecond booke of his fourth *Pemptas*. But our Authors words are too iniurious, eſpecially being without cauſe, and againſt him, from whom he borrowed all that was good in this his booke, except the figures of *Tabernamontanus*. It may be Dr. *Prieſt* did not fit his tranſlation in this place to our Authors capacitie ; for *Dodonæus* did not affirme it to be the *Phaſeo-lus*, but *Phaſelus*, diſtinguiſhing betweene them. ‡

¶ The Temperature and Vertues.

A   The Beane before it be ripe is cold and moiſt : being dry it hath power to bind and reſtraine, ac-cording to ſome Authors : further of the temperature and vertues of *Galen.*

B   The Beane (as *Galen* faith in his booke of the Faculties of Nouriſhments) is windie meate, al-though it be neuer ſo much ſodden and dreſſed any way.

C   Beanes haue not a cloſe and heauy ſubſtance, but a ſpongie and light, and this ſubſtance hath a ſcouring and clenſing faculty ; for it is plainely ſeene, that the meale of Beanes clenſeth away the filth of the skin ; by reaſon of which qualitie it paſſeth not ſlowly through the belly.

D   And ſeeing the meale of Beanes is windie, the Beanes themſelues if they be boyled whole and eaten are yet much more windie.

E   If they be parched they loſe their windineſſe, but they are harder of digeſtion, and doe ſlowly deſcend, and yeeld vnto the body thicke or groſſe nouriſhing juyce; but if they be eaten greene be-fore they be ripe and dried, the ſame thing hapneth to them which is incident to all fruits that are eaten before they be fully ripe ; that is to ſay, they giue vnto the body a moiſt kinde of nouriſh-ment, and therefore a nouriſhment more full of excrements, not onely in the inward parts, but alſo in the outward, and whole body through : therefore thoſe kindes of Beans do leſſe nouriſh, but they do more ſpeedily paſſe thorow the belly, as the ſaid Author in his booke of the Faculties of ſimple medicines faith, that the Beane is moderately cold and dry.

F   The pulpe or meate thereof doth ſomewhat clenſe, the skin doth a little binde.

G   Therefore diuers Phyſitians hath giuen the whole Beane boyled with vineger and ſalt to thoſe that were troubled with the bloudy flix, with laskes and vometings.

H   It raiſeth flegme out of the cheſt and lungs: being outwardly applied it dryet hwithout hurt the watery humors of the gout. We haue oftentimes vſed the ſame being boiled in water, and ſo mixed with ſwines greaſe.

We

We haue laid the meale thereof with Oxymel, or ſyrrup of vineger, both vpon bruiſed and woun-   I
ded ſinewes, and vpon the wounded parts of ſuch as haue been bitten or ſtung, to take away the fiery heat.

It alſo maketh a great plaiſter and pultis for mens ſtones and womens paps: for theſe parts when   K
they are inflamed, haue need of moderate cooling, eſpecially when the paps are inflamed through
the cluttered and congealed milke contained in them.

Alſo milke is dried vp with that pultis.   L

The meale thereof (as *Dioſcorides* further addeth) being tempered with the meale of Fenugreeke   M
and hony, doth take away blacke and blew ſpots, which come by dry beatings, and waſteth away kernels vnder the eares.

With Roſe leaues, Frankincenſe, and the white of an egge, it keepeth backe the watering of the   N
eies; the pin and the web, and hard ſwellings.

Being tempered with wine it healeth ſuffuſions, and ſtripes of the eies.   O

The Beane being chewed without the skin, is applied to the forehead againſt rheumes and fal-   P
ling downe of humours.

Being boyled in wine it taketh away the inflammation of the ſtones.   Q

The skins of Beans applied to the place where the hairs were firſt plucked vp, wil not ſuffer them   R
to grow big, but rather conſumeth their nouriſhment.

Being applied with Barly meale parched, and old oyle, they waſte away the Kings euill.   S

The decoction of them ſerueth to die woollen cloth withall.   T

This Beane being diuided into two parts (the skin taken off) by which it was naturally joyned   V
together, and applied, ſtancheth the bloud which doth too much iſſue forth after the biting of the
horſeleach, if the one halfe be laied vpon the place.

The blacke Beane is not vſed with vs at all, ſeeing, as we haue ſaid, it is rare, and ſowne onely in a   X
few mens gardens, who be delighted in variety and ſtudy of herbes.

Chap. 508.   Of Kidney Beane.

¶ The Kindes.

THe ſtocke or kindred of the Kidney Bean are wonderfully many; the difference eſpecially conſiſteth in the colour of the fruit: there be other differences, wherof to write perticularly would
greatly ſtuffe our Volune with ſuperfluous matter, conſidering that the ſimpleſt is able to diſtinguiſh apart the white Kidney Beane from the blacke, the red from the purple, and likewiſe thoſe of
mixt colours from thoſe that are only of one colour: as alſo great ones from the little ones. Wherefore it may pleaſe you to be content with the deſcription of ſome few, and the figures of the reſt,
with their ſeuerall titles in Latine and Engliſh, referring their deſcriptions vnto a further conſideration, which otherwiſe would be an endleſſe labour, or at the leaſt needleſſe.

¶ The Deſcription.

1   THe firſt kinde of *Phaſeolus* or garden Smilax hath long and ſmall branches growing very high, taking hold with his claſping tendrels vpon poles and ſtickes, and whatſoeuer
ſtandeth neere vnto him, as doth the Hop or Vine, which are ſo weake and tender that
without ſuch props or ſupporters they are not able to ſuſtaine themſelues, but will run ramping on
the ground fruitleſſe: vpon the branches do grow broad leaues almoſt like Iuie, growing together
by three, as in the common Trefoile or three leaued Graſſe: among which come the floures, that do
vary and differ in their colours, according to the ſoile where they grow, ſometimes white, ſomtimes
red, and oftentimes of a pale color: afterwards there come out long cods, whereof ſome are crooked, and ſome are ſtraight, and in thoſe the fruit is contained, ſmaller than the common Beane, ſomwhat flat, and faſhioned like a Kidney, which are of diuers colours, like vnto the floures: whereto for
the moſt part theſe are like.

2   There is alſo another *Dolichus* or Kidney Beane, leſſer, ſhorter, and with ſmall cods, whoſe
floures and fruit are like in forme to the former Kidney Beanes, but much leſſer, and of a blacke colour.

3   There is likewiſe another ſtrange Kidney Beane, which doth alſo winde it ſelfe about poles
and props neere adjoyning, that hath likewiſe three leaues hanging vpon one ſtem, as haue the other
Kidney Beans, but euery one is much narrower and alſo blacker: the cods be ſhorter, plainer, and flatter, and containe fewer ſeeds.

1 *Phaſeolus albus.*
White Kidney Beane.

2 *Phaſeolus niger.*
Blacke Kidney Beane.

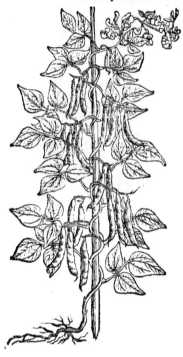

3 *Smilex hortenſis rubra.*
Red Kidney Beane.

4 *Smilex hortenſis flaua.*
Pale yellow kidney Beane.

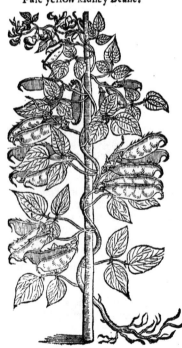

‡ 5 *Phaseolus peregrinus fructu minore albo.*
Indian Kidney bean with a small white fruit.

‡ 6 *Phaseolus peregrinus fructu minore frutescens.*
Indian Kidney bean with a small red fruit.

‡ 7 *Phaseolus peregrinus angustifolius.*
Narrow leafed Kidney bean.

4 This Kidney bean differeth not from the others but only in the colour of the fruit, which are of a pale yellow colour, wherein consisteth the difference.

‡ Besides the varieties of these Kidny beans mentioned by our Author, there are diuers other reckoned vp by *Clusius*, which haue bin brought from the East and West Indies, and from some parts of Africa. I will only giue you the figures of two or three of them out of *Clusius*, with the colour of their floures and fruit.

5 The stalk of this is low and stif: the floures of a whitish yellow on the out side, and of a Violet colour within: the fruit is snow white, with a blacke spot in the eye. This is *Phaseolus peregrinus* 4 of *Clusius*.

6 This hath leaues like the marsh Tresoile, floures growing many together, in shape & magnitude like those of common Pease: the coddes were narrow, and contained three or foure seeds which were smal, no bigger than the seeds of *Laburnum*; the painter expressed two of them in the leafe next vnder the vppermost tuft of floures: this is *Clusius* his *Phaseolus peregrinus* 5.

7 This groweth high, winding about poles or other supporters: the leaues are narrower than the former : the fruit lesser and flatter, of a reddish colour. This is the *Phaseolus peregrinus* 6 of *Clusius*.

8 This windes about poles, and growes to a
great

8 *Phaſcolus Braſilianus.*
Kidney Beane of Braſile.

8 *Phaſeoli Braſiliani ad vivum.*
The Braſile Kidney Bean in his full bignes.

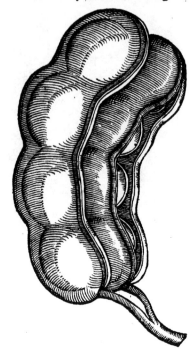

9 *Phaſcolus Ægyptiacus.*   The party coloured Beane of Egypt.

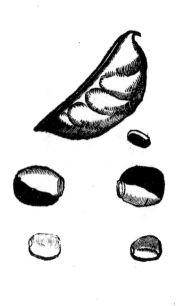

10 *Phaſeoli Americi purgantes.*
Purging Kidney Beane of America.

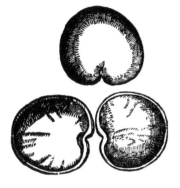

Phaſeoli magni lati albi.

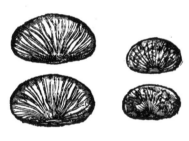

Phaſeoli Braſiliani.

Phaſeoli parui ex America delati.

Phaſeoli parui pallido albi ex America delati.

Phaſeoli rubri.

Phaſeoli rubri Indiani duriſſimi.

great height, with ſoft hairy leaues and large cods, wherein are contained ſeeds of diuers colours; ſometimes they are red, otherwhiles of a whitiſh aſh colour, ſometimes wholly blacke, and otherwhiles ſpotted.

9　The Egyptian Beane is ſomewhat like the other Kidney beanes in his growing : his fruit is of the bigneſſe of a ſmall Haſell nut, blacke on one ſide, and of a golden yellow or Orenge colour on the other.

Beſides theſe you finde here figured, and diuers others deſcribed by *Cluſius*, I think it not amiſſe to mention two more. The firſt of theſe, which was procured by Mr. *Tradeſcant*, and growes in our Gardens, is a large plant, not differing in maner of growth from the former Indian Kidney Beanes, but his floures are large, many, and of an elegant Scarlet colour : whence it is vulgarly termed by our Floriſts, the Scarlet Beane. The other I haue ſeene grow to a little height, but it would not indure; but the cods of it which were brought to vs were ſome three inches long, and couered with a hairy downe of a reddiſh colour, which put vpon the hands or skin in any part of the body would ſting like a Nettle, and this was called the Stinging Beane : I thinke it came from ſome part of the Eaſt-Indies.

¶ The

¶ *The Place.*

Kidney Beanes doe eaſily and ſoone ſpring vp, and grow into a very great length, being ſowne neere to long poles faſtned hard by them, or hard by arbors or banqueting places, otherwiſe they lie flat on the ground, ſlowly come vp, hardly bring forth fruit, and become faulty and ſmitted, as *Theo-phraſtus* writeth.

¶ *The Time.*

It is ſowne in the Spring, eſpecially in the midſt of April, but not before : the fruit is ripe about the end of Summer.

¶ *The Names.*

Hippocrates,Diocles,Theophraſtus, and moſt of the other old Writers do call it διμχοι: diuers of the bigneſſe of the ſeed do name it λίοι and λίσιοι : in Latine, *Siliqua: Dioſcorides* calleth it *Smilax*, becauſe it climeth vp as *Smilax* doth, and taketh hold of props, ſtaies, and ſhrubbes ſtanding neere vnto it : others name it φασίολοι, a Diminitiue deriued from φάσηλοι : for φάσηλοι and φασίολοι are not one and the ſelfe ſame pulſe called by diuers names, as ſome ſuppoſe, but ſundry fruits one differing from the other, as *Galen* in his firſt booke of the Faculties of Nouriſhments doth ſufficiently declare, where he in-treateth of them both. For firſt he diſputeth of *Phaſeli* and *Ochri*, Beanes and Peaſe; then afterward others comming betweene, he writeth of *Dolichus*, which alſo is named *Phaſeolus* : and though hee may be thought to doubt what manner of pulſe that is which *Theophraſtus* calleth *Dolichus* : not-withſtanding he gathereth and concludeth that it is a fruit of a garden plant in Italy, and in Caria, growing in the fields, which is in forme longer than the Cichlings, and was commonly called in his time *Faſeolus*. Of his opinion is *Paulus Ægineta*, writing of *Phaſeolus*, which he nameth *Dolychus*, in the 79.chap. of his firſt booke. Moreouer, *Faſelus* was in times paſt a common pulſe in Italy and Rome, and *Dolichus* a ſtrange pulſe; for *Columella* and *Palladius*, writers of husbandry, haue made men-tion of the ſowing of *Phaſelus* : and *Virgil* calleth it *Vilis* in the firſt of his *Georgicks* : but concer-ning the ſowing of *Dolichus* or Kidney Beane, none of the Latines haue written, by reaſon that the ſame was rare in Italy, and ſowne onely in gardens, as *Galen* hath affirmed, naming it oftentimes a garden plant, and ſhewing that the ſame, as we haue ſaid, is ſowne in Caria; and likewiſe *Dioſcorides* nameth it σμίλαξ κηπαία, that is to ſay, *Smilax hortenſis*, or garden Smilax, becauſe it groweth in gardens : who alſo writing of this in another ſeuerall chapter, ſheweth plainely, that *Smilax hortenſis*, or *Doli-chus* which he nameth *Phaſeolus*, is another plant differing from *Faſelus*.

For which cauſes it is not to be doubted, but that *Phaſelus* with three ſyllables, differeth from *Faſeolus* with foure ſyllables, no otherwiſe than *Cicer, Cicercula*, and *Cicera* differ, which notwithſtan-ding be neere one to another in names : and it is not to be doubted but that they are deceiued, who thinke it to be one and the ſelfe ſame Pulſe called by ſundry names.

This plant is named in Engliſh, Kidney Beane, Sperage Beanes : of ſome, Faſelles, or long Pea-ſon, French Beanes, Garden Smilax, and Romane Beanes : in French, *Feues de Romme* : in Dutch, **Turckſboonen.**

¶ *The Temperature.*

Kidney Beans, as *Dioſcorides* teacheth, do more looſe the belly than Peaſon; they are leſſe windy, and nouriſh well, and no leſſe than Peaſon, as *Diocles* ſaith : they be alſo without ingendring windi-neſſe at all : the Arabian Phyſitions ſay that they are hot and moiſt of nature.

¶ *The Vertues.*

A　The fruit and cods of Kidney Beanes boyled together before they be ripe, and buttered, and ſo eaten with their cods, are exceeding delicate meat, and doe not ingender winde as the other Pulſes doe.

B　They doe alſo gently looſe the belly, prouoke vrine, and ingender good bloud reaſonable well, but if you eat them when they be ripe, they are neither toothſome nor wholeſome. Therefore they are to be taken whileſt they are yet greene and tender, which are firſt boyled vntill they be tender; then is the rib or ſinew that doth run alongſt the cod to be taken away ; then muſt they be put into a ſtone pipkin, or ſome other veſſell with butter, and ſet to the fire againe to ſtew, or boyle gently : which meat is very wholeſome, nouriſhing, and of a pleaſant taſte.

Chap. 509. *Of the flat Beane called Lupine.*

¶ *The Deſcription.*

1　THe tame or Garden Lupine hath round hard ſtems, which of themſelues doe ſtand vp-right without any ſuccour, help or ſtay : the leaues conſiſt of fiue, ſix, or ſeuen joyned to-
　　　　　　　　　　　　　　　　　　　　　　　　　　　gether,

1 *Lupinus ſativus.*
Garden Lupines.

2 *Lupinus flore luteo.*
Yellow Lupines.

3 *Lupinus flore cæruleo.*
Blew Lupine.

‡ 4 *Lupinus major flo.cæruleo.*
The great blew Lupine.

gether, like thoſe of the Chaſt tree, greene on the vpper ſide, and on the nether ſide white and dow-ny, and in the euening about the ſetting of the Sun they hang flagging downwards as though they were withered: among theſe there commeth vp a tuft of flours of a pale or light bluſh colour, which turne into great rough cods, wherein is the fruit, which is flat and round like a cake, of a white co-lour, and bitter in taſte : and where they cleaue vnto the cod, in that part they haue a certaine dent like a little nauell. This Lupine hath but one root, which is ſlender and wooddy, hauing hanging on it a few ſmall threds like haires.

2   The yellow Lupine is like to the garden one in ſtalke and leaues, yet both of theſe leſſer and ſhorter. It hath beautifull floures of an exceeding faire gold yellow colour, ſweet of ſmell, made vp into an eare, of the colour of the yellow Violet, and ſomewhat of the ſmell : the cods are ſmall, hard, ſomewhat hairy : the ſeeds be little, flat, round, in taſte extreme bitter, of ſundry colours, ill-fauoured, far leſſer than the tame one.

3   The blew Lupines are longer than the yellow, and diuided into more wings or branches: the leaues be leſſer and thinner : the floures ſmall: and leſſer than the yellow, of a blew colour: the ſeeds be alſo of diuers colours, bitter, and leſſer than any of them all.

‡   4   There is alſo another blew Lupine, whoſe leaues, ſtalks, floures, and cods are like: but lar-ger then thoſe of the firſt deſcribed : the floures are of colour blew, with ſome whiteneſſe here and there intermixt. ‡

¶ *The Place.*

They require (ſaith *Theophraſtus*) a ſandy and bad ſoile : they hardly come vp in tilled places, be-ing of their owne nature wilde : they grow in my garden, and in other mens gardens about London.

¶ *The Time.*

They are planted in Aprill, and bring forth their fruit at two or three ſundry times, as though it did floure often, and bring forth many crops: the firſt in May, the ſecond in Iuly, the laſt in Septem-ber, but it ſeldome commeth to ripeneſſe.

¶ *The Names.*

This pulſe is named in Greeke, θέρμος ἥμερος : in Latine: *Lupinus*, and *Lupinus ſatiuus*: in high Dutch, Feigbonen : in Italian, *Lupine domeſtico* : in Spaniſh, *Entramocos*: in the Brabanders language, Witch boonen, and Lupinen : in French, *Lupins* : in Engliſh, Garden Lupine, tame Lupine, and of ſome after the German name Fig-beane.

¶ *The Temperature and Vertues.*

A   The ſeed of the garden Lupines is πολύχρησον, that is to ſay, much and often vſed, as *Galen* ſaith in his books of the Faculties of Nouriſhments: for the ſame being boyled and afterwards ſteeped in faire water, vntill ſuch time as it doth altogether loſe his naturall bitterneſſe, and laſtly being ſeaſoned with a reaſonable quantity of ſalt, it is eaten with pickle. The Lupine is of an hard and earthy ſub-ſtance, wherfore it is neceſſarily of hard digeſtion, and containeth in it a thicke juyce; of which be-ing not perfectly concocted in the veines, is engendred a bloud or juyce which is properly called crude, or raw: but when it hath loſt all his bitterneſſe by preparing or dreſſing of it (as aforeſaid) it is like ἄποιον, that is to ſay, to ſuch things as are without reliſh, which is perceiued by the taſte ; and being ſo prepared, it is, as *Galen* writeth in his books of the Faculties of ſimple medicines, one of the emplaiſtickes or clammers.

B   But whileſt the naturall bitterneſſe doth as yet remaine, it hath power to clenſe and to conſume or waſte away ; it killeth wormes in the belly, being both applied in manner of an ointment and gi-uen with hony to licke on, and alſo drunke with water and vineger.

C   Moreouer, the decoction thereof inwardly taken, voideth the wormes; and likewiſe if it be ſun-dry times outwardly vſed as a bath, it is a remedy againſt the morphew, ſore heads, the ſmall Pox, wilde ſcabs, gangrenes, venomous vlcers, partly by clenſing, aud partly by conſuming and drying without biting; being taken with Rue and Pepper, that it may be the pleaſanter, it ſcoureth the li-uer and milt.

D   It bringeth downe the menſes, and expelleth the dead childe if it be laied to with myrrhe and hony.

E   Moreouer, the meale of Lupines doth waſte or conſume away without any biting quality, for it doth not onely take away blacke and blew ſpots that come of dry beatings, but alſo it cureth *Chæ-radas*, and *Phymata* : but then it is to be boyled either in vineger or oxymell, or elſe in water and vi-neger, and that according to the temperature of the grieued parties, and the diuerſities of the diſea-ſes, *Quod ex vſu eſt eligendo*: and it alſo taketh away blew marks, and what thing ſoeuer elſe we haue ſaid the decoction could do, all the ſame doth meale likewiſe performe.

F   Theſe Lupines, as *Dioſcorides* doth furthermore write, being boyled in raine water till they yeeld a certaine creme, are good to clenſe and beautifie the face.

G   They cure the ſcabs in ſheepe with the root of blacke Chameleon Thiſtle, if they be waſhed with the warme decoction.

The

The root boyled with water and drunke, prouoketh vrine.   H

The Lupines being made ſweet and pleaſant, mixed with vineger and drunke, take away the loth-   I
ſomneſſe of the ſtomacke, and cauſe a good appetite to meat.

Lupines boyled in that ſtrong leigh which Barbards do vſe, and ſome Wormwood, Centorie and   K
bay ſalt added thereto, ſtay the running and ſpreading of a *Gangræna*, and thoſe parts that are depriued of their nouriſhment and begin to mortifie, and ſtaieth the ambulatiue nature of running
and ſpreading vlcers, being applied thereto very hot, with ſtuphes of cloth or tow.

Cʜᴀᴘ. 510.   *Of Peaſon.*

¶ *The Kindes.*

THere be diuers ſorts of Peaſon, differing very notably in many reſpects; ſome of the garden, and
others of the field, and yet both counted tame: ſome with tough ſkinnes or membranes in the
cods, and others haue none at all, whoſe cods are to be eaten with the Peaſe when they be young, as
thoſe of the young Kidney Beane: others carrying their fruit in the tops of the branches, are eſteemed and taken for Scottiſh Peaſon, which is not very common, There be diuers ſorts growing wild,
as ſhall be declared.

1 *Piſum majus.*   2 *Piſum minus.*
Rownciuall Peaſe.   Garden and field Peaſe.

¶ *The Deſcription.*

1   THe great Peaſe hath long ſtalks, hollow, brickle, of a whitiſh green colour, branched, and
ſpred vpon the ground, vnleſſe they be held vp with proppes ſet neere vnto them: the
leafe thereof is wide and long, made vp of many little leaues which be ſmooth, white,
growing vpon one little ſtalke or ſtem, and ſet one right againſt another: it hath alſo in the vpper
part long claſping tendrels, wherewith it foldeth it ſelfe vpon props and ſtaies ſtanding next vnto

Kkkkk 2   it:

3 *Pisum vmbellatum.*
Tufted or Scottish Peafe.

4 *Pisum excorticatum.*
Peafe without skins in the cod.

5 *Pisum sylvestre.*
Wilde Peafe.

6 *Pisum perenne sylvestre.*
Euerlasting wilde Peafe.

it : the floure is white and hath about the middle of it a purple ſpot : the cods be long, round *Cilindriforma* : in which are contained ſeeds greater than *Ochri*, or little Peaſon ; which being dry are cornered, and that vnequall, of colour ſometimes white and ſometimes gray : the roots are ſmall.

2   The field Peaſe is ſo very well knowne to all, that it were a needleſſe labour to ſpend time about the deſcription.

3   Tufted Peaſe are like vnto thoſe of the field, or of the garden in each reſpect ; the difference conſiſteth onely in that, that this plant carreth his floures and fruit in the tops of the branches in a round tuft or vmbel, contrary to all other of his kinde, which bring forth their fruit in the midſt, and alongſt the ſtalkes : the root is thicke and fibrous.

4   Peaſe without skins, in the cods differ not from the precedent, ſauing that the cods hereof want that tough skinnie membrane in the ſame, which the hogs cannot eat by reaſon of the toughneſſe ; whereas the other may be eaten cods and all the reſt, euen as Kidney beanes are : which being ſo dreſſed are exceeding delicate meat.

5   The wilde Peaſe differeth not from the common field Peaſe in ſtalke and leaues, ſauing that this wilde kinde is ſomewhat leſſer : the floures are of a yellow colour, and the fruit is much leſſer.

6   The Peaſe whoſe root neuer dies differeth not from the wilde Peaſe, onely his continuing without ſowing, being once ſowne or planted, ſetteth forth the difference.

¶ *The Place.*

Peaſe are ſet and ſowne in gardens, as alſo in the fields in all places of England. The tufted Peaſe are in reaſonable plenty in the Weſt part of Kent, about Sennocke or Seuenock ; in other places not ſo common.

The wilde Peaſe do grow in paſtures and earable fields in diuers places, ſpecially about the field belonging vnto Biſhops Hatfield in Hartfordſhire.

¶ *The Time.*

They be ſowne in the Spring time, like as be alſo other pulſes, which are ripe in Summer : they proſper beſt in warme weather, and eaſily take harme by cold, eſpecially when they floure.

¶ *The Names.*

The great Peaſe is called in Latine, *Piſum Romanum*, or *Piſum majus* : in Engliſh, Romane Peaſe, or the greater Peaſe, alſo garden Peaſe : of ſome, Branch Peaſe, French Peaſe, and Rounſiuals. *Theophraſtus* and other old Writers do call it in Greeke, *πίσον* : in Latine, alſo *Piſum* : in low Dutch, **Roomſche erwiten** : in French, *des Pois*. The little Peaſe is called of the Apothecaries euery where *Piſum*, and *Piſum minus* : it is called in Engliſh, little Peaſe, or the common Peaſe.

¶ *The Temperature and Vertues.*

The Peaſe, as *Hippocrates* ſaith, is leſſe windie than Beans, but it paſſeth ſooner through the belly. A *Galen* writeth, that Peaſon are in their whole ſubſtance like vnto Beanes, and be eaten after the ſame manner that Beanes are, notwithſtanding they differ from them in theſe two things both becauſe they are not ſo windie as be the Beans, and alſo for that they haue not a clenſing faculty and therfore they do more ſlowly deſcend through the belly. They haue no effectuall quality manifeſt, and are in a meane between thoſe things which are of good and bad juyce, that nouriſh much and little, that be windy and without winde, as *Galen* in his booke of the Faculties of Nouriſhments hath written of theſe and of Beanes.

CHAP. 511.   *Of the tame or Garden Ciche.*

¶ *The Deſcription.*

GArden Ciche bringeth forth round ſtalkes, branched and ſomewhat hairy, leaning on the one ſide, the leaues are made of many little ones growing vpon one ſtem or rib, and ſet one right againſt another : of which euery one is ſmall, broad and nicked on the edges, leſſer than the leaues of wilde Germander : the floures be ſmall, of colour either white, or of a reddiſh purple : after which come vp little ſhort cods, puffed vp as it were with winde like little bladders, in which doe lie two or at the moſt three ſeeds cornered, ſmall towards the end, with one ſharpe corner, not much vnlike to a rams head, of colour either white, or of a reddiſh blacke purple ; in which is plainly ſeene the place where they begin firſt to ſprout. The root is ſlender, white and long : for as *Theophraſtus* ſaith, the Ciche taketh deepeſt root of all the Pulſes.

¶ *The Place.*

It is ſowne in Italy, Spaine, and France, euery where in the fields. It is ſowen in our London gardens, but not common.

Cicer ſativum.
Garden Ciche.

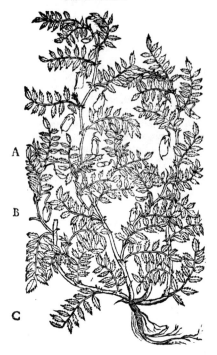

A

B

C

¶ *The Time.*

It is ſowne in April, being firſt ſteeped in water a day before : the fruit is ripe in Auguſt.

¶ *The Names.*

It is called in Greeke, ἐρέβινθος κριὸς : in Latine, *Cicer arietinum*, or Rams Ciches, & of the blackiſh purple colour, *Cicer nigrum*, *vel rubrum* : blacke or red Ciche : and the other is named *Candidum vel album Cicer* : or white Ciche : in Engliſh, Common Cich, or Ciches, red Cich, of ſome, Sheepes Ciche Peaſe, or Sheepes Ciche Peaſon.

¶ *The Temperature and Vertues.*

The Ciche, as *Galen* writeth in his booke of the Faculties of nouriſhments, is no leſſe windie than the true Beane, but it yeeldeth a ſtronger nouriſhment than that doth : it prouoketh luſt, and it is thought to engender ſeed.

Some giue the ſame to ſtalion horſes. Moreouer, Ciches do ſcoure more than doe the true Beanes : inſomuch as certaine of them doe manifeſtly diminiſh or waſte away the ſtones in the Kidneyes : thoſe be the blacke and little Ciches called *Arietina*, or Rams Ciches, but it is better to drinke the broth of them ſodden in water.

Both the Rams Ciches, as *Dioſcorides* ſaith, the white and the blacke prouoke vrine, if the decoction therof be made with Roſemary, and giuen vnto thoſe that haue either the Dropſie or yellow jaundiſe ; but they are hurtfull vnto the Bladder and Kidneies that haue vlcers in them.

CHAP. 512. *Of wilde Ciches.*

¶ *The Kindes.*

THe wilde Ciche is like to the tame (ſaith *Dioſcorides*) but it differeth in ſeed : the later writers haue ſet downe two kindes thereof, as ſhall be declared.

¶ *The Deſcription.*

1   THe firſt wilde Cich bringeth forth a great number of ſtalkes branched, lying flat on the ground : about which be the leaues, conſiſting of many vpon one rib as do thoſe of the garden Cich, but not nicked in the edges, more like to the leaues of Axcich : the flours come forth faſtned on ſmall ſtems, which grow cloſe to the ſtalkes, of a pale yellow colour, and like vnto eares : in their places come vp little cods, in forme and bigneſſe of the fruit of garden Ciches, blacke and ſomething hairy, in which lieth the ſeed, that is ſmall, hard, flat, and glittering, in taſte like that of Kidney Beane : the root groweth deepe, faſtened with many ſtrings.

2   There is another kinde of wild Cich that hath alſo a great number of ſtalkes lying vpon the ground, about which ſtand ſoft leaues, ſomething hairy and white, conſiſting of three broad leaues ſtanding vpon a middle rib, the leaſt of which ſtand neereſt to the ſtem, and the greateſt at the very top : the floures come forth at the bottome of the leaues many together, of colour yellow : after which grow ſmall long huskes, ſoft and hairy, in euery one whereof is a little cod, in which lie two ſeeds like little Cichlings.

¶ *The*

1 *Cicer ſylueſtre.*
The wilde Cich.

2 *Cicer ſylueſtre latifolium.*
Broad leafed wilde Cich.

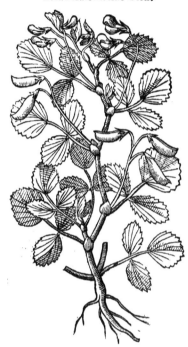

¶ *The Place.*

Theſe plants are ſowne in the parts beyond the ſeas for to feed their cattell with in Winter, as we do Tares, Vetches, and ſuch other baſe pulſe.

¶ *The Time.*

The time anſwereth the Vetch or Tare.

¶ *The Names.*

The wild Cich hath no other name in Latine but *Cicer ſylueſtre* : the later writers haue not found any name at all.

¶ *The Temperature and Vertues.*

Their temperature and vertues are referred to the garden Cich, as *Theophraſtus* affirmes ; and *Galen* ſaith that the wilde Cich is in all things like vnto that of the garden, but in Phyſicks vſe more effectually, by reaſon it is more hotter and drier, and alſo more biting and bitter.

CHAP. 513.   *Of Lentils.*

¶ *The Deſcription.*

1   THe firſt Lentil groweſ vp with ſlender ſtalks, and leaues which be ſomewhat hard, growing aſlope from both ſides of the rib or middle ſtalke, narrow and many in number like thoſe of Tares, but narrower and leſſer : the floures be ſmall, tending ſomewhat towards a purple : the cods are little and broad : the ſeeds in theſe are in number three or foure, little, round, plaine, and flat : the roots are ſmall and thready.

2   The ſecond kinde of Lentill hath ſmall tender and pliant branches a cubit high ; whereon do grow leaues diuided or conſiſting of ſundry other ſmall leaues, like the wilde Vetch, ending at the middle rib with ſome claſping tendrels, wherewith it taketh hold of ſuch things as are neere vnto it : among theſe come forth little browniſh floures mixed with white, which turne into ſmall flat cods, containing little browne flat ſeed, and ſometimes white.

¶ *The*

1 *Lens major.*
Great Lentils.

2 *Lens minor.*
Little Lentils.

¶ *The Place.*

Theſe Pulſes do grow in my garden ; and it is reported vnto me by thoſe of good credit, that a-bout Watford in Middleſex and other places of England the husbandmen doe ſow them for their cattell,euen as others do Tares.

¶ *The Time.*

They both floure and wax ripe in Iuly and Auguſt.

¶ *The Names.*

They are called in Greeke, φακός, or φακῆ : in Latine, *Lens,* and *Lenticula* : in high Dutch, **Linſen** ; in French, *Lentille* : in Italian, *Lentichia* : in Spaniſh, *Lenteia* : in Engliſh, **Lentils.**

¶ *The Temperature and Vertues.*

A　Lentils as *Galen* ſaith, are in a meane betweene hot and cold, yet are they dry in the ſecond de-gree : their skin is aſtringent or binding, and the meate or ſubſtance within is of a thicke and ear-thy juyce,hauing a quality that is a little auſtere or ſomething harſh,much more the skin thereof : but the juyce of them is quite contrary to the binding quality;wherefore if a man ſhall boile them in faire water,and afterwards ſeaſon the water with ſalt and pickle,*aut cum ipſis oleo condiens,*and then take it,the ſame drunke doth looſe the belly.

B　The firſt decoction of Lentils doth looſe the belly ; but if they bee boyled againe, and the firſt decoction caſt away, then doe they binde, and are good againſt the bloudy flix or dangerous laskes.

C　They do their operation more effectually in ſtopping or binding, if all or any of theſe following be boyled therewith, that is to ſay, red Beets, Myrtles, pils of Pomegranats, dried Roſes, Medlars, Seruice berries,vnripe Peares,Quinces,Plantaine leaues,Galls,or the berries of Sumach.

D　The meale of Lentils mixed with hony doth mundifie and clenſe corrupt vlcers and rotten ſores, filling them with fleſh againe ; and is moſt ſingular to be put into the common digeſtiues vſed among our London Surgeons for greene wounds.

E　The Lentil hauing the skin or coat taken off, as it loſeth that ſtrong binding quality,and thoſe accidents that depend on the ſame, ſo doth it more nouriſh than if it had the skin on.

F　It ingendreth thicke and naughty juyce, and ſlowly paſſeth through the belly, yet doth it not ſtay the looſeneſſe as that doth which hath his coat on ; and therefore they that vſe to eat too much
　　　　　　　　　　　　　　　　　　　　　　　　　　　　　　　thereof

thereof doe neceſſarily become Lepers, and are much ſubject to cankers, for thicke and dry nou-riſhments are apt to breed melancholy.

Therefore the Lentill is good food for them that through wateriſh humours be apt to fall into G the dropſie, and it is a moſt dangerous food for dry and withered bodies; for which cauſe it brin-geth dimneſſe of ſight, though the ſight be perfect, through his exceſſiue dryneſſe, whereby the ſpi-rits if the ſight be waſted; but it is good for them that are of a quite contrary conſtitution.

It is not good for thoſe that want their termes; for it breedeth thicke bloud, and ſuch as ſlowly H paſſeth through the veines.

But it is ſingular good to ſtay the menſes, as *Galen* in his booke of the faculties of nouriſhments I affirmeth.

It cauſeth troubleſome dreams (as *Dioſcorides* doth moreouer write) it hurteth the head, ſinewes K and lungs.

It is good to ſwallow downe thirty graines of Lentils ſhelled or taken from their husks, againſt L the ouercaſting of the ſtomacke.

Being boyled with parched barly meale and laid to, it aſſwageth the paine and ach of the gout. M

With hony it filleth vp hollow ſores, it breaketh aſchares, clenſeth vlcers: being boyled in wine N it waſteth away wens and hard ſwellings of the throat.

With a quince, and Melilot, and oyle of Roſes, it helpeth the inflammation of the eies and fun- O dament; but in greater inflammations of the fundament, and great deep vlcers, it is boyled with the ride of a pomegranate, dry Roſe leaues, and hony.

And after the ſame maner againſt eating ſores that are mortified, if ſea water be added; it is alſo P a remedy againſt puſhes, the ſhingles, and the hot inflammation called S. Anthonies fire, and for kibes, in ſuch manner as we haue written; being boyled in ſea water and applied, it helps womens breſts in which the milke is cluttered, and cannot ſuffer too great abundance of milke.

CHAP. 514.   *Of Cich or true Orobus.*

Orobus reptus Herbariorum.
The true Orobus.

¶ *The Deſcription.*

THis Pulſe, which of moſt Herbariſts is taken for the true Orobus, and called of ſome, bit-ter Fitch, is one of the Pulſes whoſe tender bran-ches traile vpon the ground, as *Theophraſtus* ſaith, and whoſe long tender branches ſpred far abroad, whereon doe grow leaues like thoſe of the field Vetch: among which grow white ſloures; after which come long cods, that appeare bunched on the outſide againſt the place where the ſeeds do lie, which are ſmall, round, ruſſet of colour, and of a bitter taſte: the root is ſmall and ſingle.

¶ *The Place.*
It proſpereth beſt in a leane ſoile, according to *Columella*: it groweth in woods and copſes in ſundry places of Spaine and Italy, but here only in gardens.

¶ *The Time.*
This is ſowne early and late, but if it be ſowne in the Spring it eaſily commeth vp, and is plea-ſant; and vnpleaſant if it be ſowne in the fall of the leafe.

¶ *The Names.*
This is called in Greeke, οϊ̈βα: the ſhops of Germany haue kept the name *Orobus*: the Itali-ans cal it *Macho*: the Spaniards, *Yeruo*, and *Yeruos*: in Engliſh it is called bitter Vetch, or bitter Fitch, and Orobus, after the Latine name. Of ſome Ers; after the French name.

¶ *The*

¶ *The Temperature and Vertues.*

A　*Galen* in his firft booke of the Faculties of nourifhments faith, That men do altogether abftaine from the bitter Vetch,for it hath a vnpleafant tafte, and naughty juyce; but Kine in A fia and in moft other countries doe eat thereof, being made fweet by fteeping in water ; notwithftanding men being compelled through neceffity of great famine,as *Hippocrates* alfo hath written,doe oftentimes feed thereof ; and we alfo dreffing them after the manner of Lupines, vfe the bitter Vetches with hony,as a medicine that purgeth thicke and groffe humors out of the cheft and lungs.

B　Moreouer, among the bitter Vetches the white are not fo medicinable, but thofe which are neere to a yellow,or to the colour of Okar ; and thofe that haue been twice boyled, or fundry times foked in warer,lofe their bitter and vnpleafant tafte, and withall their clenfing and cutting quality, fo that there is onely left in them an earthy fubftance, which ferues for nourifhment, that dryeth without any manifeft bitterneffe.

C　And in his booke of the Faculties of fimple medicines he faith, That bitter Vetch is dry in the later end of the fecond degree, and hot in the firft : moreouer, by how much it is bitter,by fo much it clenfeth, cutteth, and remooueth ftoppings : but if it be ouermuch vfed it bringeth forth bloud by vrine.

D　*Diofcorides* writeth, that bitter Vetch caufeth head-ache and heauy dulneffe, that it troubles the belly, and driueth forth bloud by vrine ; notwithftanding being boyled it ferues to fatten Kine.

E　There is made of the feed a meale fit to be vfed in medicine, after this maner : the full and white graines are chofen out and being mixed together they are fteeped in water, and fuffered to lie till they be plumpe, and afterwards are parched till the fkinne be broken ; then are they ground, and fearfed or fhaken through a meale fieue,and the meale referued.

F　This loofeth the belly, prouoketh vrine, maketh one well coloured : being ouermuch eaten or drunke it draweth bloud by the ftoole,with gripings,and alfo by vrine.

G　With hony it clenfeth vlcers, taketh away freckles, fun-burnes, blacke fpots in the fkinne,and maketh the whole body faire and cleane.

H　It ftaieth running vlcers or hard fwellings, and gangrens or mortified fores ; it fofteneth the hardneffe of womens breafts,it taketh away and breaketh eating vlcers,carbuncles,and fores of the head : being tempered with wine and applied it healeth the bitings of dogs, and alfo of venomous beafts.

I　With vineger it is good againft the ftrangury,and mitigateth paine that commeth thereof.

K　It is good for them that are not nourifhed after their meat, being parched and taken with hony in the quantitie of a nut.

L　The decoction of the fame helpeth the itch in the whole body, and taketh away kibes, if they be wafhed or bathed therewith.

M　*Cicer* boyled in fountaine water with fome *Orobus* doth affwage the fwelling of the yard and priuie parts of man or woman, if they be wafhed or bathed in the decoction thereof; and the fubftance hereof may alfo be applied plaifterwife.

N　It is alfo vfed for bathing and wafhing of vlcers and running fores,and is applied vnto the fcurfe of the head with great profit.

Chap. 515.　*Of the Vetch or Fetch.*

¶ *The Defcription.*

1　THe Vetch hath flender and foure fquared ftalkes almoft three foot long : the leaues be long, with clafping tendrels at the end made vp of many little leaues growing vpon one rib or middle ftem ; euery one whereof is greater, broader, and thicker than that of the Lentill : the floures are like to the floures of the garden Beane, but of a blacke purple colour : the cods be broad,fmall,and in euery one are contained fiue or fix graines,not round,but flat like thofe of the Lentill,of colour blacke,and of an vnpleafant tafte.

‡ 2　There is another of this kinde which hath a creeping and liuing root,from which it fendeth forth crefted ftalkes fome cubit and halfe high : the leaues are winged, commonly a dozen growing vpon one rib, which ends in a winding tendrel : each peculiar leaues broader toward the bottome,and fharper towards the top, which ends not flat, but fomewhat round. Out of the bofomes of the leaues towards the tops of the ftalkes, on fhort foot-ftalkes grow two, three, or more pretty large Peafe-fafhioned blewifh purple floures, which are fucceeded by fuch cods as the former,but fomewhat leffer ; which when they grow ripe become blacke, and fly open of themfelues,

and

‡ 1 *Vicia.*
Tare, Vetch, or Fetch.

† 2 *Vicia maxima dumetorum.*
Buſh Vetch.

‡ 3 *Vicia ſyl. flo. albo.*
White floured Vetch.

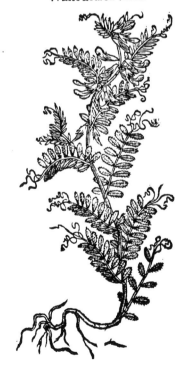

4 *Vicia ſylueſtris, ſiue Cracca major.*
Strangle Tare, Tine, or wild Fetch.

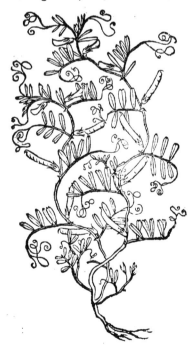

and fo fcatter their feed. This growes in many places wilde among bufhes, both here and in Germany, as appeares by that name, *Bauhine* thence giues it, calling it *Vicia maxima dumetorum*. *Tragus* makes it his *Vicia fyl.altera*; and iudges it to be the *Aphace* of *Diofcorides*; and he faith the Latines call it *Os mundi* : the high Dutch, **S.Criftoffels kraut**, and **Schwartz Linfen.** *Tabern.* calls it *Cracca maior*.

‡ 5 *Vicia fyl.fiue Cracca minima.*
Small wilde Tare.

3   This alfo hath a lafting root, which fendeth vp round crefted branches, a foot and fomtimes a cubit high, whereon grow fuch leaues as thofe of the former, but more white and downie : the floures which grow on fhort footftalkes, out of the bofomes of the leaues, towards the top of the ftalks, are of a whitifh colour, with veines of a duskie colour, diuaricated ouer the vpper leafe : the cods are like thofe of the common Fetch. *Clufius* found this in fome wilde places of Hungarie; it floured in May : he calls it *Vicia fylueftris albo flore.* ‡

4   Strangle Tare, called in fome countries Tine, and of others wilde Vetch, is a ramping herbe like vnto the common Tare, ramping and clymbing among corne where it chanceth, that it pluckerh it downe to the ground, and ouergroweth the fame in fuch fort, that it fpoileth and killerh not only wheat, but all other graine whatfoeuer : the herbe is better known than defired, therefore thefe few lines fhall fuffice for the defcription. ‡ This groweth pretty long, with many flender weake branches : the leaues are much fmaller than the former, and end in clafping tendrels : the floures are of a purple colour, and commonly grow but one at a joynt, and they are fucceeded by flat fharpe pointed cods which containe fome nine or ten feeds apiece.

5   This alfo growes a good height, with flenderer ftalkes than the former, which is diuided into fundry branches : the leaues grow foure or fix vpon foot-ftalkes, ending alfo in clafping tendrels : the floures grow vpon pretty long but very flender foot-ftalkes, fometimes two or three, otherwhiles more, very fmall, and of a whitifh colour inclining to blewneffe : which are fucceeded by little fhort flat cods, containig commonly foure or fiue little feeds of a blackifh colour : this is the *Arachus,fiue Cracca minima* of *Lobel*; but I queftion whether it be that which *Bauhine* in his *Pinax* hath made the fame with it, calling it *Vicia fegetum cum filiquis plurimis hirfutis* : for that which I haue defcribed, and which exactly agrees with this figure of *Lobel* and that defcription in the *Aduerfaria* hath cods very fmooth without any hairineffe at all. This floures moft part of Summer, and growes in moft places both in corne fields and medowes. ‡

¶ *The Place.*
The Tare is fowne in any ground or foile whatfoeuer.

¶ *The Time.*
It floureth in May, and perfecteth his feed toward September.

¶ *The Names.*
It is called in Latine *Vicia à vinciendo*, of binding or wrapping, as *Varro* noteth, becaufe, faith he, it hath likewife clafping tendrels fuch as the Vine hath, by which it crawles vpward vpon the ftalks of the weeds which are next vnto it : of fome, *Cracca*, and *Arachus*, and alfo *Aphaca* : it is called in high Dutch, **Wicken** : in low Dutch, **Witfen** : in French, *Vefce* : in moft fhops it is falfely termed *'Ορόβος* : and *Eruum* ; for *Eruum* doth much differ from *Vicia* : it is called in Englifh, Vetch, or Fetch. The country-men lay vp this Vetch with the feeds and whole plant, that it may be fodder for their cattell.

¶ *The Temperature and Vertues.*
A   Notwithftanding I haue knowne, faith *Galen*, fome, who in time of famine haue fed hereof, efpecially in the Spring, it being but greene ; yet is it hard of digeftion, and bindeth the belly.
Therefore

Therefore ſeeing it is of this kind of nature, it is manifeſt that the nouriſhment that commeth therof hath in it no good juice at all, but ingendreth a thick bloud, and apt to become melancholy.

† The figure of the common Fetch was formerly wanting, and in ſtead thereof was put that of the other deſcribed here in the ſecond place.

CHAP. 516.   *Of Chichlings, Peaſe, and Tare euerlaſting.*

¶ *The Deſcription.*

1 THere is a Pulſe growing in our high and thicke woods, hauing a very thick tough and wooddy root, from which riſe vp diuers long weake and feeble branches, conſiſting of a rough middle rib edged on both ſides with a thin skinny membrane, ſmooth, and of a graſſe green colour; whereon grow at certaine diſtances ſmall flat ſtemmes, vpon which ſtand two broad leaues joining together at the bottome: from betwixt thoſe leaues come forth tough claſping tendrels which take hold of ſuch things as grow next vnto them: from the boſom of the ſtem whereon the leaues do grow, ſhooteth forth a naked ſmooth foot-ſtalke, on which grow moſt beautifull floures like thoſe of the Peaſe, the middle part whereof is of a light red tending to a red purple in grain: the outward leaues are ſomewhat lighter, inclining to a bluſh colour: which beeing paſt, there ſucceed long round cods, wherein is contained ſeed of the bigneſſe of a Tare, but rounder, blackiſh without, yellowiſh within, and of a bitter taſte.

‡ 1 *Lathyrus major latifolius.*
    Peaſe euerlaſting.

‡ 2 *Lathyrus anguſtifolius flore albo.*
    White floured Chichlings.

† 2 Of this kinde there is likewiſe another like vnto the precedent in each reſpect, ſaue that the leaues thereof are narrower and longer, and therfore called of moſt which ſet forth the deſcription, *Lathyrus anguſtifolia*: the floures of this are white, and ſuch alſo is the colour of the fruit: the root is ſmall, and not laſting like that of the former.

‡ 3 The ſtalks, leaues, and floures of this are like thoſe of the precedent, but the floures are of a reddiſh purple colour: the cods are leſſer than thoſe of the former, and in them are contained

lesser,harder,and rounder seeds of a darke or blackish colour. This growes not wilde with vs,but is sometimes sowne in gardens,where it floures in Iune and Iuly.

4   This Egyptian differs not in shape from the rest of his kinde, but the floures are of an elegant blew on the in side , but of an ash colour inclining to purple on the out side : the cods grow vpon long foot-stalks,and are a little welted or winged,and containe but two or three little cornered seeds spotted with blacke spots. This floures in Iune and Iuly,and the seed thereof was sent to *Clusius* from Constantinople,hauing been brought thither out of Egypt.

‡ 3 *Lathyrus angustifol. flo. purp.*
Purple floured Chichlings.

‡ 4 *Lathyrus Ægyptiacus.*
Egyptian Chichlings.

5   The stalks of this are some two or three foot long,winged,weake,and lying on the ground, vnlesse they haue somewhat to support them. Vpon these at certain distances grow winged leaues with two little eares at their setting on to the stalke : these leaues consist of sixe long and narrow greene leaues like those of the other plants of this kinde ; and these six leaues commonly stand vpright by couples one against another; otherwhiles alternatly : the foot-stalke whereon these stand ends in clasping tendrels : the floures are in shape like the former, but the outer leafe is of a faire red or crimson colour,and the inner leafe white: after the floures come the cods,containing some foure or fiue pretty large flat seeds,which swel out of the cods where they lie,and in the spaces betweene each seed are deprest, like that of *Orobus*. This is only a garden plant with vs,and floures in Iune and Iuly,the seed is ripe in August. I haue for this giuen you *Lobels* figure of his *Lathyris angustiore graminco folio*, which may serue, if you but make the leaues and cods to agree with this description. ‡

6   The yellow wilde Tare or Fetch hath diuers very small ramping stalkes, tough, and leaning this way and that way,not able to stand of it selfe without the help of props or things that stand by it : the leaues are very thin and sharp pointed : the floures grow alongst the leaues, in shape of the pease floures,of a bright yellow colour : the roots are very small,long,tough,and in number infinit, insomuch that it is impossible to root it forth,being once gotten into the ground,vnlesse the earth be digged vp with the roots,and both cast into the riuer or burned. Doubtlesse it is the most pernitious and hurtfull weed of all others,vnto all maner of green wholsome herbs or any wood whatsoeuer.

¶ *The*

‡ 5 *Lathyrus annuus ſiliquis Orobi.*
Party coloured Chichling.

‡ 6 *Lathyrus ſylveſtris flo. luteo.*
Tare euerlaſting.

¶ *The Place.*

The firſt growes in ſhadowie woods and amongſt buſhes : there groweth great ſtore thereof in Swainſcombe wood, a mile and a halfe from Greenhithe in Kent, as you go to a village therby called Betſome, and in diuers other places.

The ſixt groweth in moſt graſſie paſtures, borders of fields, and among grain almoſt euery where

¶ *The Time.*

The time anſwereth the other pulſes.

¶ *The Names.*

The firſt is called *Lathyrus*, to make a difference betwixt it and *Lathyris* or Spurge : of *Matthiolus, Clymenum :* of *Cordus, Eruum ſativum :* of *Tragus, Piſum Græcorum:* in Engliſh, Peaſe euerlaſting, great wilde Tare, and Cichling.

‡ The ſecond is the *Eruum album ſativum* of *Fuchſius : Lathyrus* or *Cicercula* of *Dodonæus : Lathyrus anguſtiore gramineo folio* of *Lobel.*

The third is the *Aracus ſiue Cicera* of *Dodonæus :* the *Lathyrus flore purpureo* of *Camerarius.*

The fourth by *Cluſius* is called *Cicercula Ægyptiaca:* by *Camerarius, Aracus Hiſpanicus, ſiue Lathyrus Ægyptiacus.*

The fift is not mentioned by any that I remember, but M* *Parkinſon,* in his Garden of flours, and that by the name I giue it you.

The ſixt is the *Lathyrus ſylveſtris flo. luteus* of *Thalius : Legumen terræ glandibus ſimile* of *Dodonæus : Vicia* of *Tabern.* and it may be the *Aracus flo. luteo* of the *Adverſ.* Howeuer, I haue put *Lobels* figure of the *Aracus* for it, which well enough agrees with it. I vſe for ſome reſemblance it hath to *Aphaca,* to call it *Aphacoides.* ‡

¶ *The Nature and Vertues.*

The temperature and vertues are referred to the manured Tare or Vetch, notwithſtanding they are not vſed for meat or medicine.

CHAP. 517. *Of the oily Pulse called* Sesamum.

Sesamum, sive Sisamum.
The oily Graine.

¶ *The Description.*

† Sesamum hath a thicke and fat vpright stalk a cubit and a half high, garnished with leaues much like the Peach or almond, but rougher, and cut in with somwhat deep gashes on their sides: amongst these leaues come forth large white or else red floures, somwhat shaped like those of Fox-gloues, which turne into round long crested cods containing white flat oilous seed. *Theophrastus* affirmeth that there is a kind therof which is white, bearing only one root. No beast will eat this plant while it is greene, because of his bitternesse: but beeing withered and dried the seed thereof becommeth sweet, and Cattell will feed on the whole plant.

¶ *The Place.*

It groweth both in Egypt and India. *Sesama,* saith *Pliny,* came from the Indies: they make an oile of it. It is a stranger in England.

¶ *The Time.*

It is one of the Summer grains, and is sown before the rising of the seuen stars, as *Pliny* writeth: yet *Columella* saith that *Sesamum* must be sowne after the Autumne Æquinoctial, against the Ides of October: they require for the most part a rotten soile, which the husbandmen of Campania do cal a blacke mold.

¶ *The Names.*

The Grecians call this Grain σήσαμον: the Latines also *Sesamum* and *Sisamum,* and often in the feminine gender *Sesama:* we are constrained for want of an English name to vse the Latine: it is vnknowne to the Apothecaries, especially the plant it selfe; but the seed and oile thereof is to bee found among them in other countries: we may call it Turky Millet.

¶ *The Nature and Vertues.*

A   According to some it is hot and dry in the first degree: the feed thereof, as *Galen* saith, is fat, and therefore being laid vp it commeth to be oily very quickly: wherefore it speedily fills and stuffeth vp those that feed thereof, and ouerthroweth the stomack, being slow of digestion, and yeelding to the body a fat nourishment: therefore it is manifest that it cannot strengthen the stomake or any part thereof, as also no other kind of fat thing; and the juice that comes thereof is thick, and therefore cannot easily passe thorow the veines. Men doe not greedily feed of it alone, but make cakes thereof with hony, which they call σησαμίδες: it is also mixed with bread, and is of an hot temper, for which cause it procureth thirst. And in his booke of the faculties of simple medicines he saith that *Sesamum* is not a little clammy and fat, and therefore it is an emplastick and a softner, and is moderatly hot: the oile which comes thereof is of like temperature, and so is the decoction of the herb also.

B   *Dioscorides* writeth, That *Sesamum* is an enemy to the stomacke, causing a stinking breath if it remain sticking betweene the teeth after it is chewed.

C   It wasteth away grosnesse of the sinues: it is a remedie against bruises of the eares, inflammations, burnings, and scaldings, pains of the joints, and biting of the poisonsom horned serpent called *Cerastes.* Being mixt with oile of roses it takes away the head-ache comming of heate.

D   Of the same force is the herb boiled in wine, but it is especially good for the heat and paine of the eies.

E   Of the herb is made an oile vsed of the Egyptians, which, as *Pliny* saith, is good for the eares.

F   It is a remedie against the sounding and ringing of the eares.

CHAP.

Chap. 518. *Of Hatchet Vetch.*

¶ *The Deſcription.*

1 THe firſt kinde of Hatchet Fetch hath many ſmall branches trailing here and there vp-on the ground : vpon which grow ſmall leaues ſpred abroad like the leaues of the wild Vetch : among which come forth cluſters of ſmal yellow floures,which fade away,and turne into little flat thin and brown cods,wherein is contained ſmall reddiſh ſeed of a bitter taſte.

2 The ſecond kinde of Hatchet Vetch hath many round tough and flexible branches,trailing vpon the ground : whereupon do grow leaues like the former, but more like the leaues of Liquo-rice,and hath the taſte of the Liquorice root ; which cauſed ſome to deeme it a kinde of Liquo-rice : among theſe leaues come forth pale yellow floures : after which there ſucceed ſmal crooked cods turning their points inwardly, one anſwering another like little hornes,containing ſmall flat ſeeds foure cornered,and faſhioned like a little wedge : the root is tough , of a woody ſubſtance, and doth continue fruitfull a very long time.

† 1 *Hedyſarum majus.*
Hatchet Fetch.

‡ 2 *Hedyſarum Glycyrrhizatum.*
Liquorice Hatchet Fetch.

3 There is another kind of *Securidaca* or hatchet Fetch, which hath branches,leaues and roots like the laſt before remembred,and differeth in that,that the floures of this plant are mixed and do vary into ſundry colours, being on the vpper part of a fleſh colour, and on the lower of a white or ſnowy colour,with a purple ſtorks bill in the middle : the leaues are in taſte bitter : the cods are ſmall like thoſe of Birds foot,and not much vnlike the cods of *Orobus.*

4 There is likewiſe another kinde of *Securidaca* or hatchet Fetch, which is dedicated to *Cluſius* by the aforenamed Dr *Penny,*who found it in the North parts of England,hauing leaues,roots,and branches likevnto the former : but the floures of this are white and mixed with ſome purple, and bitter alſo in taſte : the cods are like the claw of a Crab , or (as *Cluſius* ſeth) like the knife which ſhoo-makers do vſe in Flanders ; in which cods are contained ſmal rediſh ſeed : this root is alſo of long continuance. ‡ *Cluſius* doth not ſay that Dr *Penny* found this in the North of England,but in the territorie of Geneva not far from Pontetremile,among the buſhes,and nowhere elſe.

‡ 3 *Hedyfarum majus filiquis articulatis.*
Hatchet Fetch with jointed cods.

‡ 4 *Securidaca minor pallide cærulea.*
Small blew floured Hatchet Fetch.

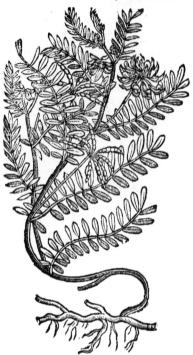

‡ 5 *Securidaca minor lutea.*
Small yellow Hatchet Vetch.

‡ 6 *Securidaca filiquis planis dentatis.*
Indented Hatchet Fetch.

‡ 5   This in the ſtalkes, leaues, colour, and ſhape of the floures is like, yet leſſe than the firſt deſcribed ; the cods are alſo ſmaller, leſſer, and more crooked, and herein onely conſiſts the chiefe difference: it is an annual plant, and growes only in ſome gardens. *Matthiolus, Lobel, Dodonæus,* and others make this their *Hedyſarum,* or *Securidacaminor.*

6   This hath many creſted branches, whereon great winged leaues, that is ſome twenty or more are faſtned to one rib : the floures are like thoſe of the other plants of this kind, but the cods are an inch long, flat, indented or toothed on the ſides : but of what colour the floures and ſeeds are it is not expreſt by *Cluſius,* who only ſet this forth by a picture, and ſome pieces of a dried plant thereof, which he receiued from *Cortuſus,* by the name of *Scolopendria leguminoſa,* or *Hedyſarum peregrinum.* *Cortuſus* had it from *Honorius Bellus,* who obſerued it growing vpon the rockes at Seberico a citie of Illyria. ‡

7   There is alſo another ſort of Hatchet Fetch, which hath very long and tough branches trailing vpon the ground, beſet with leaues like the former, but much greater : the floures grow at the top of the branches, of a pale colour, and turne into rough round and flat cods, faſhioned like little bucklers : the root of this (as of the firſt) dieth at the firſt approch of Winter, as ſoon as the ſeed is ripe. ‡ The ſtalks of this are ſtiffe and creſted, growing to the height of two cubits, with leaues as large as thoſe of Liquorice : the floures are of a faire bright red colour, the cods are made as it were of many rough buckler-like ſeeds, or rather ſeed-veſſels, wherein are contained ſmall browne ſeeds. ‡

‡ 7 *Hedyſarum clypeatum.*
French Honyſuckle.

8 *Ferrum Equinum.*
Horſe-ſhoo.

8   Horſe-ſhoo hath many ſtalks tender and lying vpon the ground : the leaues be thin, and leſſer than thoſe of Axſeed : the floures along the ſtalks are little, after which come vp long cods ſomthing broad, a little bowing, which haue vpon the one ſide deep round and indented cuts, like after a ſort to a horſe-ſhoo : the root is ſomewhat long.

¶ *The Place.*

Theſe plants grow in my garden. The ſecond kinde I found growing in Suffolke in the highway on the right hand as you go from Sudbury to Corner Church, about an hundred paces from the end of the towne, as alſo in ſundry other places of the ſame countrey; and in Eſſex about Dunmow,

mow,and in the townes called Clare and Henningham. ‡ Alſo it growes by Purfleet, about the foot of the hill whereon the winde-mill ſtands,and in diuers parts of Kent. ‡

Horſe-ſhoo commeth vp in certain vntilled and ſunny places of Italy and Languedoc:it groweth alſo in my garden.

¶ *The Time.*

Theſe plants floure in Iune,and their ſeed is ripe in Auguſt.

¶ *The Names.*

The Grecians name this,whether it be a Pulſe,or an infirmitie among corn, *ἀσπρον*: the Latines, of the forme of the ſeed,*Securidaca*,and *Hedyſarum* : in Engliſh,Axſeed,Axwort,Ax-fitch, and hatchet Fitch : it is vnknowne to the Apothecaries.

‡ The ſecond is the *Fænugræcum ſylveſtre* of *Tragus* and *Dodonæus* ; the *Glycyrrhiza ſylveſtris* of *Geſner* ; and the *Glaux vulgaris* of *Lobel* ‡

Horſe-ſhoo is commonly called in Italian,*Sferro de cavallo* : you may name it in Latine, *Ferrum equinum* : in Engliſh,Horſe ſhoo.

¶ *The Nature.*

The ſeeds of theſe plants are hot and dry of complexion.

¶ *The Vertues.*

A   Being drunke it is acceptable to the ſtomacke, and remoueth ſtoppings out of the intrals ; and of like vertue be the new leaues and tender crops of the whole plant.

B   *Dioſcorides* ſheweth that it is alſo good for the ſtomack being taken in drinke,and is mixed with counterpoiſons.

C   And it is thought to hinder conception,if it be applied with hony before the act.

D   The ſeed of Ax-wort openeth the ſtoppings of the liuer,the obſtruction of the ſpleen and of all the inward parts.

E   Horſe-ſhoo is bitter, and like in nature to Axſeed.

† The figure which formerly was in the firſt place agreed with the third deſcription: that which was in the ſecond place was of the *Hedyſarum minus* of *Tabor.* being a kinde of *Ferrum equinum*,which carries the cods many together on the tops of the branches, and growes in Germany: whence *Baukine* calls it *Ferrum Equinum Germanicum ſiliquis in ſummitate.* This growes about Bath,and in diuers places of the Weſt countrey, bearing yellow floures like *Trifolium Corniculatum*,about the middle of Iune. The floures of the indented hatchet Fitch are of a whitiſh blew colour.

Chap. 519. *Of Peaſe Earth-nut.*

¶ *The Deſcription.*

1   THe Peaſe Earth-nut comes vp with ſlender weake ſtalkes : the leaues be thin and little, growing on ſlender ſtems,with claſping tendrels at the ends, with which it imbraceth and taketh hold of ſuch things as ſtand neere it : the flours on the tops of the ſtalks are like thoſe of Peaſe,but leſſe,of a red purple colour,in ſmel not vnpleaſant : in their places come vp long cods in which are three or foure round ſeeds: the roots be thick, long, like after a ſort to Acorns,but much greater,black without,gray within,in taſte like the Cheſnut;out of which beneath doth hang a long ſlender ſtring : there grow out of the ſame alſo other ſtrings hard to the ſetting on of the ſtalke,vnto which creeping aſlope do grow other kernelled roots, whileſt the plant doth thus multiply it ſelfe.

‡ 2   This,which *Thalius* in his *Sylva harcynia* ſets forth by the name of *Aſtragalus ſylvaticus*,was by our Author taken for and confounded with the *Terræ glandes*, and therefore I haue put it with it that the difference may the better appeare,which is not a little to ſuch as heedfully obſerue it. But our Author in this is to be pardoned, ſeeing Dr *Turner*, a man more exquiſite in the knowledge of plants,and who had ſeen the true *Terræ glandes* in Germany,miſtooke this for it,as may appeare by that little Tract of his of the names of plants in Latine and Engliſh, ſet forth *Anno* 1548. for there he ſaith, [I haue ſeen this herb of late in Come parke more aſtringent than it of Germany.] And indeed this growes there,and is much more aſtringent and woody than that of Germany,and in no wiſe fit to be eaten. The root conſiſts of many blacke tuberous particles, here and there ſending forth fibers:from hence ariſe cornered ſtalks ſome foot high, ſmall below, and ſomewhat larger about:the leaues grow forth of the ſtalks,conſiſting ſometimes of two,otherwhiles of 4 longiſh narrow leaues faſtned to one foot-ſtalke,which at the ſetting on hath two little leaues or eares:out of the boſoms of theſe leaues grow ſtalks ſome two inches long,each of which vſually carry a couple of peaſe-faſhioned flours of a purple colour,which fading,vſually become blew : after theſe follow cods ſtraight,round and blacke, and in each of them are commonly contained nine or tenne white

round

1　*Terræ glandes.*
Peaſe Earth-nut.

‡ 2　*Aſtragalus ſylvaticus.*
Wood Peaſe or heath Peaſe.

round ſeeds : it floures in ſummer, and perfects the ſeed in Iuly and Auguſt.

¶ *The Place.*

† 1　This groweth in corne fields, both with the corne it ſelfe, and alſo about the borders of fields among briers and brambles : it is found in diuers places of Germany, but not with vs, that I can yet learne.

2　This is found in the woods and paſtures of England, eſpecially in Hampſtead Wood neere London : it groweth in Richmond heath, and in Come parke likewiſe.

¶ *The Time.*

It floureth in Iune and Iuly ; the nuts after harueſt be digged vp and gathered.

¶ *The Names.*

It is called in high-Dutch, **Erdnuſſen**: in low-Dutch, **Eerdnoten**, **Eerdeeckelen**, and **Muyſen metſteerten**, that is to ſay, tailed Mice, of the ſimilitude or likeneſſe of domeſtick Mice, which the black round and long nuts, with a piece of the ſlender ſtring hanging out behinde do repreſent. The later writers call it in Latine *Terræ glandes*, or *Terreſtres glandes* : in Greeke, χαμαιβαλανοι, *Chamabalani.* in Engliſh, Peaſe Earth-nut.

¶ *The Temperature and Vertues.*

The nuts of theſe Peaſe being boiled and eaten are hardlier digeſted than be either Turneps or Parſneps, yet do they nouriſh no leſſe than the Parſnep : they are not ſo windy as they, but do more ſlowly paſſe through the belly by reaſon of their binding qualitie ; and beeing eaten raw they bee yet harder of digeſtion, and do ſlowlier deſcend.　　A

They be of temperature meanly hot and ſomwhat dry, being withal not a little binding: whereupon alſo they do not only ſtay the fluxes of the belly, but alſo all iſſues of bloud, eſpecially from the mother or bladder.　　B

The root of Peaſe Earth-nut ſtoppeth the belly, and the inordinat courſe of womens ſickneſſe.　　C

Cʜᴀᴘ.

CHAP. 520. *Of Milke-Vetch.*

¶ *The Kindes.*

THere be diuers sorts of herbs contained vnder the title of *Astragalus*; whether I may without breach of promise made in the beginning, insert them amongst the *Legumina*, Pulses, or herby plants, it is doubtfull: but seeing the matter is disputable, I thinke it not amisse to suffer them to passe thus, vntill some other shall finde a place more conuenient and agreeing vnto them in neighborhood.

1 *Astragalus Lusitanicus Clusij.*
Portugal milke-Vetch.

2 *Astragalus Syriacus.*
Assyrian milke Vetch.

¶ *The Description.*

1   THe first kinde of *Astragalus* hath reddish stalks acubit high, a finger thicke, somwhat crested or furrowed, and couered ouer with an hairy mossinesse; which diuide themselues into sundry small branches, beset with leaues consisting of sundry little leaues set vpon a middle rib like the wilde Vetch, placed on the smal pliant branches like feathers, which are likewise couered ouer with a woolly hoarinesse; in taste astringent at the first, but afterwards burning hot: among these leaues come forth many small white floures in fashion like the floures of Lupins, which before their opening seem to be somewhat yellow: the root is maruellous great and large, considering the smalnesse of the plant; for sometimes it groweth to the bignesse of a mans arme, keeping the same bignesse for the space of a span in length, and after diuideth it selfe into two or more forks or branches, blacke without, and wrinkled; white within, hard, and woody, and in taste vnpleasant: which being dried becommeth harder than an horne.

2   The second kinde of *Astragalus* is a rare and gallant plant, and may well be termed *Planta leguminosa*, by reason that it is counted for a kinde of *Astragalus*, resembling the same in the similitude of his stalks and leaues, as also in the thickenesse of his rootes, and the creeping and folding thereof

thereof, and is garnished with a most thicke and pleasant comelinesse of his delectable red floures, growing vp together in great tufts, which are very seemly to behold.

3   There hath been some controuersie about this third kinde, which I am not willing to prosecute or enter into: it may very well be *Astragalus* of *Matthiolus* his description, or else his *Polygala*, which doth exceeding well resemble the true *Astragalus* : his small stalkes grow a foot high, beset with leaues like *Cicer* or *Galega*, but that they are somewhat lesser : among which come forth small pease like flours, of an Orenge colour, very pleasant in sight : the root is tough and flexible, of a finger thicke.

‡ 3 *Astragalus* Matthioli.
Matthiolus his milke Vetch.

‡ 4 *Astragaloides.*
Bastard milke Vetch.

4   The fourth is called of *Mutonus* and other learned Herbarists, *Astragaloides*, for that it resembleth the true *Astragalus*, which groweth a cubit high, and in shew resembleth liquorice: the flours grow at the top of the stalks, in shape like the pease bloome, of a faire purple colour, which turn into small blacke cods when they be ripe : the root is tough and very long, creeping vpon the vpper part of the earth, and of a wooddy substance.

¶ *The Place.*

They grow amongst stones in open places, as *Oribasius* writeth, in places subiect to windes, and couered with snow : *Dioscorides* copies do adde, in shadowie places : it groweth plentifully in Phenea a city in Arcadia, as *Galen* and *Pliny* report : in *Dioscorides* his copies is read, in Memphis a city in Arcadia ; but Memphis is a city in Egypt, and in Arcadia there is none of that name. Some of them grow in my garden, and in sundry places of England wilde : they grow in the medowes neere Cambridge, where the Schollers vse to sport themselues ; and in sundry places of Essex, as about Dunmow and Clare, and many other places of that country.

‡ I should be glad to know which or how many of these our Author here affirms to grow wild in England ; for as yet I haue not heard of nor seen any of them wilde, nor in gardens with vs, except the last described, which growes in some few gardens. ‡

¶ *The Time.*

They floure in Iune and Iuly, and their seed is ripe in September.

¶ *Ti*

¶ *The Names.*

Milke Vetch is called of *Matthiolus, Polygala,* but not properly : of most it is called *Astragalus:* in Spanish, *Garavancillos :* in the Portugals tongue, *Alphabeca :* in Dutch, **Cleyne Ciceren.**

¶ *The Nature and Vertues.*

A   *Astragalus,* as *Galen* saith, hath astringent or binding roots, and therefore it is of the number of those Simples that are not a little drying ; for it glueth and healeth vp old vlcers , and stayeth the flux of the belly, if they be boiled in wine and drunke. The same things also touching the vertues of *Astragalus, Dioscorides* hath mentioned : The root, saith he, drunke in wine, stayeth the laske, and prouoketh vrine, being dried and cast vpon old vlcers it cureth them : it likewise procureth great store of milke in cattell that do eat thereof ; whence it tooke his name.

B   It stoppeth bleeding, but it is with much ado beaten, by reason of his hardnesse.

CHAP. 521. *Of Kidney Vetch.*

1 *Anthyllis leguminosa.*
Kidney Vetch.

2 *Stella leguminosa.*
Starry Kidney Vetch.

¶ *The Description.*

1   KIdney Vetch hath a stalke of the height of a cubit, diuiding it selfe into other branches, whereon grow long leaues made of diuers leaues like those of the Lentil, couered as it were with a soft white downinesse : the floures on the top of the stalks are of a yellow colour, very many joined together as it were in a spoky rundle : after which grow vp little cods in which is contained small seed : the root is slender, and of a wooddy substance. ‡ This is somtimes found with white floures : whereupon *Tabern.* gaue two figures, calling the one *Lagopodium flore luteo,* and the other *Lagopodium flo. albo.* Our Author vnfitly gaue this later mentioned figure in the chapter of *Lagopus,* by the name of *Lagopum maximum.* ‡

2   The starry Kidney Vetch, called *Stella Leguminosa,* or according to *Cortusus, Arcturo,* hath
many

many ſmall flexible tough branches, full of ſmall knots or knees, from each of which ſpringeth forth one long ſmall winged leafe, like birds foot, but bigger: from the boſome of thoſe leaues come forth little tender ſtems, on the ends whereof doe grow ſmall whitiſh yellow floures, which are very ſlender, and ſoone vaded, like vnto them of Birds-foot: theſe floures turne into ſmall ſharp pointed cods, ſtanding one diſtant from another, like the diuiſions of a ſtar, or as though it conſiſted of little hornes; wherein is contained ſmall yellowiſh ſeeds: the root is tough, and deepely growing in the ground.

3   There is another ſort of Kidney Vetch called Birds-foot, or *Ornithopodium*, which hath very many ſmall and tender branches, trailing here and there cloſe vpon the ground, ſet full of ſmall and ſoft leaues, of a whitiſh greene, in ſhape like the leaues of the wilde Vetch, but a great deale leſſer, and finer, almoſt like ſmall feathers: amongſt which the floures do grow, that are very ſmall, yellowiſh, and ſometimes whitiſh; which being vaded there come in place thereof little crooked cods, fiue or ſix growing together, which in ſhew and ſhape are like vnto a ſmall Birds-foot, and each and euery cod reſembling a claw: in which are incloſed ſmall ſeed like that of Turneps.

‡ 3 *Ornithopodium majus.*
The great Birds-foot.

‡ 4 *Ornithopodium minus.*
Small Birds-foot.

‡ 5 *Scorpoides Leguminoſa.*
Small horned pulſe.

4   There is alſo another kinde of *Ornithopodium*, or Birds-foot, called ſmall Birds-foot, which is very like vnto the firſt, but that it is much ſmaller: the branches or ſprigs grow not aboue a hand or halfe an hand in length, ſpreading themſelues vpon the ground with his ſmall leaues and branches, in manner of the leſſer *Arachus*: the floures are like vnto thoſe of the former, but very ſmall, and of a red colour.

‡ 5   This ſmall horned pulſe may fitly here take place: The root thereof conſiſts of many little fibres, from which ariſe two or three little ſlender ſtraight ſtalkes ſome handfull and halfe or foot high: at the tops of theſe grow little ſharpe pointed crooked hornes, rounder and ſlenderer than thoſe of Fenugreeke, turning their ends inwards like the tailes of Scorpions and ſo joynted; the floures are ſmall and yellow; the leaues little, and winged like thoſe of Birds-foot. *Pena* and *Lobel* found this amongſt the corne in the fields in Narbon in France, and they ſet it forth by the name as I haue here giuen you it. ‡

¶ *The Place.*

1, 3. 4. Theſe plants I found growing vpon Hampſtead Heath neere London, right againſt
M m m mm   the

the Beacon, vpon the right hand as you goe from London, neere vnto a grauell pit : they grow alfo vpon blacke Heath, in the high way leading from Greenwich to Charleton, within halfe a mile of the towne.

¶ *The Time.*

They floure from Iune to the middle of September.

¶ *The Names.*

‡ 1 This *Gefner* calls *Vulneraria ruftica: Dodonæus, Lobel,* and *Clufius* call it *Anthyllis,* and *Anthyllis leguminofa.* ‡

3. 4. I cannot finde any other name for thefe plants, but *Ornithopodium :* the firft is called in Englifh, great Birds-foot ; the fecond fmall Birds-foot.

¶ *The Temperature and Vertues.*

Thefe herbes are not vfed either in meate or medicine, that I know of as yet ; but they are very good food for cattell, and procure good ftore of milke, whereupon fome haue taken them for kinds of *Polygala.* ‡

Chap. 522. *Of Blacke milke Tare.*

Glaux Diofcoridis.
Diofcorides his milke Tare.

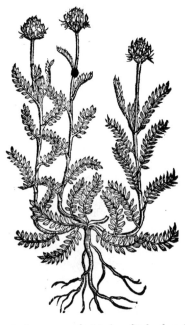

¶ *The Defcription.*

THe true *Glaux* of *Diofcorides* hath very many tough and wooddy branches trailing vpon the ground, fet full of fmall winged leaues, in fhape like the common *Glaux,* but a great deale fmaller, refembling the leaues of Tares, but rather like Birds-foot, of a very gray colour : amongft which come forth knobby & fcaly, or chaffie heads, very like the Medow Trefoile, of a fare purple colour : the root is exceeding long and wooddy, which the figure doth not expreffe and fet forth.

¶ *The Place.*

The true *Glaux* groweth vpon Barton hill, foure miles from Lewton in Bedfordfhire, vpon both the fides of the declination of the hill.

¶ *The Time.*

Thefe plants do floure and flourifh about Midfummer.

¶ *The Names.*

Thefe plant haue in times paft beene called *Glaux, i. folia habens glauca, fiue pallentia,* that is, hauing skie coloured, or pale leaues. Sithens that in times paft, fome haue counted *Glaux* among the kindes of *Polygala,* or Milke-worts, we may therefore call this kinde of *Glaux,* blacke Milke-wort.

¶ *The Temperature.*

Thefe herbes are dry in the fecond degree.

¶ *The Vertues.*

A   The feeds of the common *Glaux* are in vertue like the Lentils, but not fo much aftringent : they ftop the flux of the belly, dry vp the moifture of the ftomacke, and ingender ftore of milke.

† Our Author either not knowing, or forgetting what he had done, againe in this chapter, defcribed the *Glaux Vulgaris,* whofe hiftory he gaue vs but foure chapters before, by the name of *Hedyfarum glycyrhizatum ;* wherefore I haue omitted it here as not neceffary.

CHAP. 523.　　*Of red Fitchling, Medick Fitch, or Cockes-head.*

¶ *The Deſcription.*

1　THe firſt kinde of *Onobrychis* hath many ſmall aud twiggie pliant branches, ramping and creeping through and about buſhes, or whatſoeuer it groweth neere vnto: the leaues and all the reſt of the pulſe or plant is very like to the wilde Vetch or Tare: the floures grow at the top of ſmall naked ſtalkes, in ſhape like the peaſe bloome, but of a purple colour laied ouer with blew, which turne into ſmall round prickly husks, that are nothing elſe but the ſeed.

1　*Onobrychis, ſiue Caput Gallinaceum.*
Medick Fitchling, or Cockes-head.

2　*Onobrychis flore purpureo.*
Purple Cockes-head.

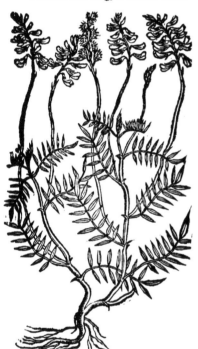

2　The ſecond kind of Fitchling or Coks-head, of *Cluſius* his deſcription, hath very many ſtalks, eſpecially when it is growne to an old plant, round, hard, and leaning to the ground like the other pulſes, and leaues very like *Galega*, or the wilde Vetch, of a biter taſte and lothſome ſauour: among which come forth ſmall ronnd ſtems, at the ends whereof do grow floures, ſpike faſhion, three inches long, in ſhape like thoſe of the great *Lagopus*, or medow Trefoile, but longer, of an excellent ſhining purple colour, but without ſmell: after which there follow ſmall cods, containing little hard and blacke ſeed, in taſte like the Vetch. The root is great and long, hard, and of a wooddy ſubſtance, ſpreading it ſelfe farre abroad, and growing very deepe into the ground.

3　The third kinde of Fitchling or Cocks-head hath from a tough ſmall and wooddy root, many twiggy branches growing a cubit high, full of knots, ramping and creeping on the ground. The leaues are like the former, but ſmaller and ſhorter: among which come forth ſmall tender ſtemmes, whereupon do grow little floures like thoſe of the Tare, but of a blew colour tending to purple: the floures being vaded, there come the ſmall cods, which containe little blacke ſeed like a Kidney, of a blacke colour.

4　The fourth kinde of Fetchling hath firme greene hard ſtalks a cubit and a halfe high, whereupon grow leaues like to the wilde Tare or *Galega*, but ſmaller and ſomewhat hairy, bitter and vnpleaſant in taſte, and in the end ſomewhat ſharpe. At the top of the ſtalkes come forth long ſpiked

3 *Onobrychis 2.Cluſij.*
Blew Medicke Fitch.

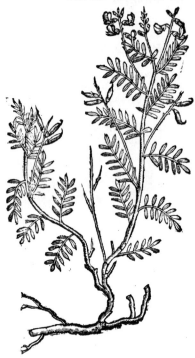

4 *Onobrychis 3.Cluſij flore pallido.*
Pale coloured Medicke Fitch.

5 *Onobrychis montana* 4.*Cluſij.*
Mountaine Medicke Fitch.

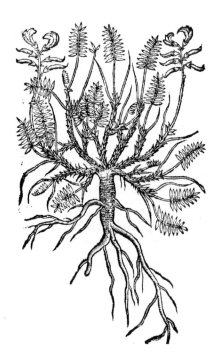

floures of a pale colour, and in ſhape like thoſe
of the ſecond kinde ; which being vaded, there
follow ſmall bottle cods, wherein is contained
little blacke ſeed like the ſeed of Fenugreek, but
ſmaller. The root is thicke and hard, and of a
wooddy ſubſtance, and laſting very long.

5　The fifth kinde of *Onobrychis* hath many
groſſe and wooddy ſtalkes, proceeding immediately from a thick, fat, and fleſhie tough root: the
vpper part of which are ſmall, round, and pliant,
garniſhed with little leaues like thoſe of Lentils, or rather *Tragacantha*, ſomewhat ſoft, and couered ouer with a woolly hairineſſe : amongſt
which come forth little long and naked ſtems,
eight or nine inches long, whereon do grow many ſmall floures of the faſhion of the Vetch or
Lentill, but of a blew colour tending to purple ;
and after them come ſmall cods, wherin the ſeed
is contained.

¶ *The Place.*

The firſt and ſecond grow vpon Barton hill,
foure miles from Lewton in Bedfordſhire, vpon
both the ſides of the hill : and likewiſe vpon the
graſſie balks between the lands of corn two miles
from Cambridge, neere to a water mill towards
London ; and diuers other places by the way
from London to Cambridge : the reſt are ſtrangers in England.

¶ *The Time.*

Theſe plants do floure in Iuly, and their ſeed
is ripe ſhortly after.　　　　　　　　¶ *The*

¶ *The Names.*

It is *ἀνώνυμος*, or without a name among the later writers : the old and antient Phyſitions doe call it *ἄνθρωπος* : for all thoſe things that are found written in *Dioſcorides* or *Pliny* concerning *Onobrychis*, doe eſpecially agree hereunto. *Dioſcorides* writeth thus ; *Onobrychis* hath leaues like a Lentill, but longer; a ſtalk a ſpan high ; a crimſon floure; a little root : it groweth in moiſt and vntilled places : and *Pliny* in like manner ; *Onobrychis* hath the leaues of a Lentill, ſomewhat longer, a red floure, a ſmall and ſlender root : it groweth about ſprings or fountaines of waters.

All which things and euery particular are in this *ἀνώνυμος* or namemeſſe herbe, as it is manifeſt : and therefore it is not to be doubted at all, but that the ſame is the *Onobrychis* of the old Writers : it may be called in Engliſh, red Fetchling, or as ſome ſuppoſe, Medick Fetch, or Cockes-head.

¶ *The Temperature.*

Theſe herbes as *Galen* hath written in his books of the Faculties of ſimple Medicines, doe rarifie or make thinne and waſt away.

¶ *The Vertues.*

Therefore the leaues thereof when it is greene, being yet laied vpon hard ſwellings, or waxen **A** kernels, in manner of a ſalue, do waſt and conſume them away, but being dryed and drunke in wine they cure the ſtrangurie; and laied on with oile it procureth ſweat.

Which things alſo concerning *Onobrychis*, *Dioſcorides* hath in theſe words ſet downe : the herbe **B** ſtamped and applied waſteth away hard ſwellings of the kernels ; but being drunke with wine it helpeth the ſtangury, and rubbed on with oile it cauſeth ſweatings.

Chap. 524. *Of Baſtard Dittanie.*

Fraxinella.
Baſtard Dittanie.

¶ *The Deſcripton.*

BAſtard Dittanie is a very rare and gallant plant, hauing many browne ſtalks, ſomwhat rough, diuided into ſundry ſmall branches, garniſhed with leaues like Liquorice, or rather like the leaues of the Aſh tree, but blacker, thicker, and more ful of juice, of an vnpleaſant ſauor : among which grow floures, conſiſting of fiue whitiſh leaues ſtripped with redde, whereof one which groweth vndermoſt hangeth downe low, but the foure which grow vppermoſt grow more ſtiffe and vpright: out of the midſt of this floure commeth forth a taſſell, which is like a beard, hanging alſo downwards, and ſomewhat turning vp at the lower end : which beeing vaded, there come in place foure huskes joyned together, much like the husks or coddes of Columbines, ſomewhat rough without, ſlimie to handle, and of a lothſome ſauour, almoſt like the ſmell of a goat; whereupon ſome Herbariſts haue called it *Tragium* : in the cods are contained ſmall blacke ſhining ſeeds like Peonie ſeeds in colour : the roots are white, a finger thicke, one twiſting or knotting within another, in taſt ſomwhat bitter.

There is another kinde hereof growing in my garden, not very much differing : the leaues of the one are greater, greener, harder, and ſharper pointed : of the other blacker, not ſo hard, nor ſo ſharpe pointed : the floures alſo hereof be ſomthing more bright coloured, and of the other a little redder.

¶ *The Place.*

Baſtard Dittany groweth wilde in the mountaines of Italy and Germany, and I haue it growing in my garden.

¶ *The Time.*

It floureth in Iune and Iuly : the seed is ripe in the end of August.

¶ *The Names.*

The later Herbarists name it *Fraxinella* : most, χαμαιμηλον, as though they should say *Humilis Fraxinus* or a low ash : in English, bastard, or false Dittanie : the shops call it *Dictamnum*, and *Diptamum*, but not truly, and vse oftentimes the roots hereof in stead of the right Dittany. That it is not the right Dittany it is better knowne than needfull at all to be confuted ; and it is euident that the same is not *Dioscorides* his *Pseudodictamnum*, or bastard Dittanie : but it is plaine to be a kinde of *Tragium* of the old Writers wherewith it seemeth to agree in shew, but not in substance.

‡ The root of this is onely vsed in shops, and there knowne by the name of *Radix Diptamni*, or *Dictamni*. ‡

¶ *The Temperature.*

The root of Bastard Dittany is hot and dry in the second degree, it is of a wasting, attenuating, and opening faculty.

¶ *The Vertues.*

A   It bringeth downe the menses, it also bringeth away the birth and after birth, it helpeth cold diseases of the matrix : and it is reported to be good for those that haue ill stomackes and are short
B   winded.

They also say, that it is profitable against the stingings and bitings of venomous serpents; against deadly poisons, against contagious and pestilent diseases, and that it is with good successe mixed
C   with counterpoisons.

The seed of Bastard Dittanie taken in the quantity of a dram is good against the strangury, pro-
D   uoketh vrine, breaketh the stone in the bladder, and driueth it forth.

The like vertue hath the leaues and juyce taken after the same sort ; and being applied outward-
E   ly, it draweth thornes and splinters out of the flesh.
F   The root taken with a little Rubarb killeth and driueth forth wormes.

Dioscorides reporteth, that the wilde Goats being strikenwith darts or arrowes, will eat *Dictam*, and thereby cause them to fall out of their bodies; which is meant of the right *Dictam*, though *Dodonæus* reporteth that this plant will do the like (which I do not beleeue) ‡ nor *Dodonæus* affirme. ‡

Chap. 525.   *Of Land Caltrops.*

Tribulus Terrestris.
Land Caltrops.

¶ *The Description.*

L And Caltrops hath long branches full of joynts, spred abroad vpon the ground, garnished with many leaues set vpon a middle rib, after the manner of Fetches ; amongst which grow little yellow branches, consisting of fiue small leaues, like vnto the floures of Tormentill : I neuer saw the plant beare yellow, but white floures, agreeing with the description of *Dodonæus* in each respect, saue in the colour of the floures, which doe turne into small square fruit, rough, and full of prickles, wherein is a small kernell or seed : the root is white, and full of strings.

¶ *The Place.*

It groweth plentifully in Spain in the fields : it is hurtfull to corne, but yet as *Pliny* saith, it is rather to bee accounted among the diseases of corne, than among the plagues of the earth : it is also found in most places of Italy and France; I found it growing in a moist medow adjoyning to the wood or Parke of Sir *Francis Carew*, neere Croidon, not farre from London, and not elsewhere ; from whence I brought plants for my garden.

¶ *The*

¶ *The Time.*

It floureth in Iune and Iuly: the fruit is ripe in Auguſt.

¶ *The Names.*

It is called in Greeke, τρίβολος: and in Latine, *Tribulus*: and that it may differ from the other which groweth in the water, it is named τρίβολος χερσαῖος, or *Tribulus terreſtris*: it may be called in Engliſh land Caltrops, of the likeneſſe which the fruit hath with Caltrops, that are inſtruments of Warre caſt in the way to anoy the feet of the enemies horſes, as is before remembred in the Water Saligot.

¶ *The Temperature and Vertues.*

In this land Caltrop there is an earthy and cold quality abounding, which is alſo binding, as   A *Galen* ſaith.

The fruit thereof being drunke waſteth away ſtones in the kidneyes, by reaſon that it is of thinne   B parts.

Land Caltrops, ſaith *Dioſcorides*, being drunke to the quantity of a French crowne weight, and   C ſo applied, cureth the bitings of the Viper.

And if it be drunke in wine it is a remedy againſt poyſons: the decoction thereof ſprinckled   D about killeth fleas.

‡ CHAP. 526.   *Of Spring or mountaine Peaſe or Vetches.*

‡ 1 *Orobus Venetus.*
Venice Peaſe.

‡ 2 *Orobus ſyluaticus vernus.*
Spring Peaſe.

¶ *The Deſcription.*

‡ 1   THis, which *Cluſius* calls *Orobus Venetus*, hath many cornered ſtalkes ſome foot long, whereon grow winged leaues, foure or ſix faſtned to one rib, ſtanding by couples one againſt another, without any odde leafe at the end: theſe leaues are of an indifferent largeneſſe,

and

and of a light greene colour : the flours grow vpon long foot-ſtalks comming forth of the boſoms of the leaues, many together, hanging downe, ſmall, yet ſhaped like thoſe of other Pulſes, and of a purple colour : after theſe follow cods almoſt like thoſe of Fetches, but rounder, red when they be ripe, and containing in them a longiſh white ſeed : the root is hard and wooddy, running diuers waies with many fibres, and liuing ſundry yeares : this varies ſometimes with yellower green leaues and white floures. It floures in May, and growes only in ſome few gardens with vs.

2   The ſtalkes of this alſo are a foot or more high, ſtiffe, cornered, and greene ; on theſe do grow winged leaues ſix or eight on a rib, after the manner of thoſe of the laſt deſcribed : each of theſe leaues hath three veines running alongſt it : the floures in ſhape and manner of growing are like thoſe of the former, but of a moſt elegant purple colour : which fading they become blew. The floures are ſucceeded by ſuch cods as the former, wherein are contained longiſh ſmall variegated ſeed, which ripe, the cods fly open, and twine themſelues round, as in moſt plants of this kinde : the root is blacke, hard, tuberous and wooddy, ſending forth each yeare new ſhoots. This floures in Aprill and May, and ripeneth the ſeed in Iune. This was found by *Cluſius* in diuers mountainous wooddie places of Hungary : he calls it *Orobus Pannonicus* 1.

‡ 3 *Orobus montanus flo.albo.*
White mountaine Peaſe.

‡ 4 *Orobus montanus anguſtifolius.*
Narrow leafed mountaine Peaſe.

3   This hath ſtalkes ſome cubit high, ſtiffe, ſtraight, and creſted ; whereon by turnes are faſt-ned winged leaues, conſiſting of foure ſufficiently large and ſharpe pointed leaues, whereto ſome-times at the very end growes a fifth : the veines in theſe run from the middle rib towards their ed-ges : their taſte is firſt ſomewhat ſouriſh, afterwards bitteriſh. The floures grow vpon ſhort ſtalkes comming forth of the boſomes of the leaues, fiue or ſix together, like thoſe of the Fetch, but of colour white, with ſome little yellowneſſe on the two little leaues that turne vpwards. The cods are like thoſe of the laſt deſcribed, and containe in them a browniſh ſeed, larger than any of the other kindes. This is an annuall plant, and periſheth as ſoone as it hath perfected the ſeed. *Cluſius* giues vs this by the name of *Orobus Pannonicus* 4. *Dodonæus* giues the ſame figure for his *Arachus latifolius :* and *Bauhine* affirmes this to be the *Galega montana*, in the *Hiſt.Lugd.pag.*1139. But theſe ſeeme to be of two ſeuerall plants ; for *Dodonæus* affirmes his to haue a liuing root, and ſuch ſeemes alſo that in the *Hiſt. Lugd.* to be : yet *Cluſius* ſaith expreſſely that his is an annuall, and floureth in
Aprill

Aprill and May, and groweth in ſome wooddy mountainous places of the Kingdome of Hungarie.

4   This fourth hath ſtraight firme cornered ſtales ſome foot or more high, whereupon grow leaues vſually foure on a foot-ſtalke, ſtanding two againſt two, vpright, beeing commonly almoſt three inches long, at firſt of a ſouriſh taſte, but afterwards bitter: it hath no clauicles, becauſe the ſtalkes need no ſupporters: the floures grow vpon long foot-ſtalkes, ſpike-faſhion like thoſe of Peaſe, but leſſe, and white of colour: after theſe follow long blackiſh cods, full of a blacke or elſe ſpotted ſeed: the roots are about the length o ones little finger, faſhioned like thoſe of the Aſphodill or leſſer female Peionie, but leſſer, blacke without, and white within. *Cluſius* found this on the mountainous places nigh the baths of Baden, and in the like places in Hungary: he calls it *Orobus Pannonicus* 3.

¶ *The Temperature and Vertues.*

Theſe are not knowne nor vſed in Phyſicke; yet if the third be the *Galega montana* of the *Hiſtoria Lugd.* then it is there ſaid to be effectuall againſt poyſon, the wormes, the falling ſickeneſſe, and the Plague. ‡

---

‡ CHAP. 527.   *Of ſome other Pulſes.*

‡ 1 *Ochrus, ſiue Erviliᴀ.*
Birds Peaſe.

‡ 2 *Ervum ſylueſtre.*
Crimſon graſſe Fetch.

¶ *The Deſcription.*

‡ 1   THe firſt of theſe hath cornered broad ſtalks like thoſe of euerlaſting Peaſe, and they are weake, and commonly lie vpon the ground, vnleſſe they haue ſomething to ſupport them: the lower leaues are broad, and commonly welt the ſtalke at their ſetting on, and at the end of the firſt leafe do vſually grow out after an vnuſuall manner, two, three, or more other pretty large leaues more long than broad, and the middle rib of the firſt leafe runnes out beyond the ſetting on of the higheſt of the out-growing leaues, and then it ends in two or three claſping tendrels. Thoſe leaues that grow the loweſt vpon the ſtalkes haue commonly the feweſt comming out of
them.

them. The floures are like thoſe of other pulſes,of colour white : the cods are ſome inch and halfe long, containing ſome halfe dozen darke yellow or blackiſh ſmall Peaſe : theſe cods grow one at a joynt,on ſhort foot-ſtalkes comming forth of the boſomes of the leaues, and are welted on their broader ſide,which ſtands towards the maine ſtalke. This growes with vs only in gardens. *Dodonæus, Pena,*and *Lobel* call it *Ochrus ſylueſtris,ſiue Ervilia.*

2   The ſtalkes of this grow vp ſometimes a cubit high, being very ſlender, diuided into branches, and ſet vnorderly with many graſſe-like long narrow leaues : on the tops of the ſtalkes and branches, vpon pretty long foot-ſtalkes grow pretty peaſe-faſhioned floures of a faire and pleaſant crimſon colour : which fallen, there follow cods, long, ſmall, and round, wherein are nine, ten, or more round hard blacke ſhining graines : the root is ſmall, with diuers fibres, but whether it die when the ſeed is perfected, or no as yet I haue not obſerued. This growes wilde in many places with vs,as in the paſture and medow grounds about Pancridge Church. *Lobel* and *Dodon.* call this *Ervum ſylueſtre* ; and they both partly iudge it to be the firſt *Catanance* of *Dioſcorides,* and by that name it is vſually called. It floures in Iune and Iuly,and the ſeed is ripe in Auguſt.

3   This alſo,though it be not frequently found, is no ſtranger with vs ; for I haue found it in the corne fields about Dartford in Kent and ſome other places. It hath long ſlender joynted creeping ſtalkes,diuided into ſundry branches,whereon ſtand pretty greene three cornered leaues two at a joynt, in ſhape and bigneſſe like thoſe of the leſſer Binde-weed. Out of the boſomes of theſe

‡ 3 *Aphaca.*
Small yellow Fetch.

leaues at each joynt comes a claſping tendrel, and commonly together with it a foot-ſtalke ſome inch or more long,bearing a pretty little peaſe-faſhioned yellow flour,which is ſucceded by a ſhort flattiſh cod containing ſix or ſeuen little ſeeds. This floures in Iune,Iuly,and Auguſt, and ſo ripens the ſeed. It is by *Lobel* and others thought to be the *Aphace* of *Dioſcorides, Galen,* and *Pliny* : and the *Pitine* of *Theophraſtus,*by *Anguillara.*

I finde mention in *Stowes* Chronicle,in *Anno* 1555,of a certaine Pulſe or Peaſe,as they term it, wherewith the poore people at that time there being a great dearth,were miraculouſly helped : he thus mentions it ; In the moneth of Auguſt (ſaith he) in Suffolk at a place by the ſea ſide all of hard ſtone and pibble, called in thoſe parts a ſhelfe,lying betweene the townes of Orford and Aldborough,where nether grew graſſe,nor any earth was euer ſeene ; it chanced in this barren place ſuddnely to ſpring vp without any tillage or ſowing, great abundance of Peaſon, whereof the poore gathered (as men iudged) aboue an hundred quarters, yet remained ſome ripe and ſome bloſſoming,as many as euer there were before:to the which place rode the Biſhop of Norwich and the Lord *Willoughby,*with others in great number,who found nothing but hard rockie ſtone the ſpace of three yards vnder the roots of theſe Peaſon: which roots were great and long,and very ſweet.

Geſner alſo,*de Aquatilibus,lib.4.pag.256.*making mention,out of Dr *Cajus* his letters,of the ſpotted Engliſh Whale,taken about that time at Lin in Northfolke,alſo thus mentions thoſe

Peaſe: *Piſa* (ſaith he) *in littore noſtro Britannico quod Orientem ſpectat,certo quodam in loco Suffolciæ, inter Alburnum & Ortfordium oppida,ſaxis inſidentia (mirabile dictu) nulla terra circumfuſa,autumnali tempore Anno* 1555,*ſponte nata ſunt, adeo magna copia, vt ſufficerent vel millibus hominum.* Theſe Peaſe, which by their great encreaſe did ſuch good to the poore that yeare, without doubt grew there for many yeares before, but were not obſerued till   [*Magiſter artis,ingenijque largitor Venter*] ——— hunger made them take notice of them,and quickned their inuention,which commonly in our people is very dull,eſpecially in finding out food of this nature.

My Worfhipfull friend Dr. *Argent* hath told me, that many yeares agoe he was in this place, and cauſed his man to pull away the beach with his hands, and follow the roots ſo long, vntill he got ſome equall in length vnto his height, yet could come to no ends of them : he brought theſe vp with him to London, and gaue them to Dr. *Lobel*, who was then liuing ; and he cauſed them to be drawne, purpoſing to ſet them forth in that Worke which he intended to haue publiſhed, if God had ſpared him longer life. Now whether theſe Peaſe be truly ſo called, and be the ſame with the *Piſum ſyl-ueſtre Perenne*, or different ; or whether they be rather of the ſtocke of the *Lathyrus major*, or of ſome other Pulſe here formerly deſcribed, I can affirme nothing of certainety, becauſe I haue ſeene no part of them, nor could gather by any that had, any certainty of their ſhape or figure : yet would I not paſſe them ouer in ſilence, for that I hope this may come to be read by ſome who liue therea-bour, that may by ſending me the things themſelues, giue me certaine knowledge of them, that ſo I may be made able, as I am alwaies willing, to impart it to others.

¶ The Temperature and Vertues.

I haue not found any thing written of the faculties of the two firſt ; but of *Aphace*, *Galen* ſaith it **A** hath an aſtringent faculty like as the Lentill, and alſo is vſed to be eaten like as it, yet it is harder of concoction, but it dries more powerfully, and heates moderately. The ſeeds (ſaith he) haue an aſtringent faculty ; wherefore parched, broken, and boyled, they ſtay fluxes of the belly. We know (ſaith *Dodonæus*) by certaine experience, that the *Aphace* here deſcribed hath this aſtringent force and faculty. ‡

C H A P . 528.   Of baſtard Rubarb.

1 Thalietrum, ſiue Thalictrum majus.
Great baſtard Rubarb.

2 Thalictrum minus.
Small baſtard Rubarb.

¶ The Deſcription.

1   THe great *Thalietrum* or baſtard Rubarb hath large leaues parted or diuided into diuers other ſmall leaues, ſomewhat ſnipt about the edges, of a blacke or darke greene colour : the

the stalkes are crested or streaked, of a purple colour, growing to the height of two cubits : at the top whereof grow many small and hairy white floures, and after them come small narrow huskes like little cods, foure or fiue growing together : the root is yellow, long, round, and knotty, dispersing it selfe far abroad on the vpper crust of the earth.

2　The small bastard Rubarb is very like vnto the precedent, but that it is altogether lesser : his stalkes are a span or a foot long : his leaues be thin and tender ; the root fine and slender : the little floures grow together in small bundles or tufts, of a light yellow colour, almost white, and are of a gricuous fauour.

‡ 3　There is kept in some gardens a plant of this kinde growing vp with large stalkes to the height of three cubits : the leaues are very like those of Columbines : the floures are made of many white threds : it floures in Iune, and is called *Thalictrum majus Hispanicum*, Great Spanish Bastard Rubarb. ‡

¶ The Place.

These plants doe grow alongst the Ditch sides leading from Kentish street vnto Saint Thomas a-Waterings (the place of Execution) on the right hand. They grow also vpon the bankes of the Thames, leading from Blacke-wall to Woolwich, neere London, and in sundry other places also.

¶ The Time.

They floure for the most part in Iuly and August.

¶ The Names.

Diuers of the later Herbarists do call it *Pigamum*, as though it were πίγανον, that is, Rue ; whereupon most call it *Ruta pratensis*, or Fen Rue : others, *Pseudo-Rhabarbarum*, and *Rhabarbarum Monachorum*, by reason of the yellow colour of the root. But neither of their iudgements is greatly to be esteemed of : they iudge better that would haue it to be *Thalietrum*, which *Dioscorides* describeth to haue leaues something flatter than those of Coriander ; and the stalke like that of Rue : vpon which the leaues do grow. *Pena* calleth it *Thalietrum, Thalictrum*, and *Ruta pratensis* : in English, Bastard Rubarb, or English Rubarb : which names are taken of the colour, and taste of the roots.

¶ The Temperature.

These herbes are hot and dry of complexion.

¶ The Vertues.

A　The leaues of Bastard Rubarb with other pot-herbes do somewhat moue the belly.
B　The decoction of the root doth more effectually.
C　*Dioscorides* saith, That the leaues being stamped do perfectly cure old vlcers. *Galen* addeth, that they dry without biting.

CHAP. 529. Of Goats Rue.

¶ The Description.

G Alega or Goats Rue hath round hard stalkes two cubits or more high, set full of leaues displayed or winged abroad ; euery leafe consisting of sundry small leaues set vpon a slender rib, resembling the leaues of the field Vetch or Tare, but greater and longer. The floures grow at the top of the stalke, clustering together after the manner of the wilde Vetch, of a light skie colour, which turne into long cods small and round, wherein the seed is contained. The root is great, thicke, and of a white colour.

¶ The Place.

It groweth plentifully in Italy euery where in fat grounds and by riuers sides : it groweth likewise in my garden.

¶ The Time.

It floureth in Iuly and August.

¶ The Names.

The Italians call it *Galega*, and *Ruta Capraria* : diuers name it corruptly *Gralega* : *Hieronymus Fracastorius*

Galega.
Goats Rue.

Fracaſtorius calleth it *Herba Galleca*: the Hetruſcians, *Laumeſe*; and it is alſo called by diuers other names in ſundry places of Italy, as are *Geſner* ſaith, as are *Caſtracani*, *Lananna*, *Thorina*, or *Taurina*, *Martanica*, *Sarracena*, *Capraginis*, *Herbaneſa*, *Fœnum grœcum ſylueſtre*, and as *Braſauolus* witneſſeth, *Giarga*. It is named in Engliſh, Italian Fitch, and Goats Rue.

Some iudge that the old Phyſitions were wont to call it *Onobrychis* : others, *Glauce* : diuers would haue it to bee *Polemonium*, but not ſo much *Petr. And. Matthiolus* in his commentaries, as euery one of the deſcriptions mentioned by *Dioſcorides* doe gaineſay them ; as alſo thoſe, who thinke that *Galega*, is *Polygalon*, and that the name of *Galega* came of *Polygalon*, the very deſcription alſo of *Polygalon* is againſt them : for *Galega* is higher and greater than that it may be called a little ſhrub onely of an hand breadth high.

¶ *The Temperature.*

This plant is in a meane temperature betweene hot and cold.

¶ *The Vertues.*

Goats Rue is a ſingular herbe againſt all A venome and poiſon, and againſt wormes, to kill and driue them forth, if the juyce be giuen to little children to drinke.

It is of like vertue if it bee fryed with B Lineſeed oyle, and bound vpon the childes nauell.

It is miniſtred vnto children which are poſſeſſed with a falling euill, a ſpoonfull euery morning C in milke.

Being boiled in vineger, and drunke with a little Treacle, it is very good againſt the infeðtion of D the plague, eſpecially if the medicine be taken within twelue houres.

The herbe it ſelfe is eaten, being boyled with fleſh, as we vſe to eat Cabbage and other worts, E and likewiſe in ſallades, with oyle, vineger and pepper, as we doe eat boyled Spinage, and ſuch like ; Which is moſt excellent being ſo eaten, againſt all poiſon and peſtilence, or any venomous infirmitie whatſoeuer, and procureth ſweat.

It alſo helpeth the bitings and ſtingings of venomous beaſts, if either the juyce or the herbe F ſtamped be laid vpon the wound.

Halfe an ounce of the juyce inwardly taken is reported to helpe thoſe that are troubled with G convulſions, crampes, and all other the diſeaſes aforeſaid.

The ſeeds do feed pullen exceedingly , and cauſe them to yeeld greater ſtore of egges than or- H dinary.

‡ The juyce of the leaues, or the leaues themſelues bruiſed and applied to any part ſwollen I by the ſting of a Bee or waſpe, mitigate the paine, and are a preſent remedy, as M^r. *Cannon* a louer of Plants, and friend of mine, hath aſſured me he hath ſeene by frequent experience. ‡

CHAP. 530. *Of* Pliny *his Leadwort.*

¶ *The Deſcription.*

DEntaria or *Dentillaria* hath offended in the ſuperlatiue degree, in that he hath hid himſelfe like a runnagate ſouldier, when the aſſault ſhould haue beene giuen to the plant *Lepidium*, whereof doubtleſſe it is a kinde. But if the fault be mine, as without queſtion it is, I craue pardon for the ouerſight, and doe intreate the gentle Reader to cenſure me with fauour, whereby I may be more bold to inſert it in this place, rather than to leaue it vntouched. The learned of Narbone (eſpecially *Rondeletius*) haue not without good cauſe accounted this goodly plant for a kinde thereof,

Plumbago Plinij.
Leadwort.

thereof, becaufe the whole plant is of a biting tafte, and a burning faculty, and that in fuch extremity, that it will raife blifters vpon a mans hand: for which caufe fome of the learned fort haue accounted it *Plinies Molybdana*, or *Ægineta* his *Lepidium*: but the new Herbarifts call it *Dentaria*, or *Dentillaria Rondeletij*, who made the like vfe hereof, as he did of *Pyrethrum*; and fuch burning plants, to appeafe the immoderate paine of the tooth-ache and fuch like. This plant hath great thicke tough roots, of a wooddy fubftance, from whence fpring vp long and tough ftalks two cubits high, confufedly garnifhed & befet with long leaues, in color like Woad, of a fharpe and biting tafte. The floures grow at the top of the ftalkes of a purple colour; which being paft there fucceed clofe glittering and hairy huskes, wherein is contained fmall blackifh feed.

¶ *The Place.*

Pena reporteth that *Dentillaria* groweth about Rome, nigh the hedges and corne fields: it likewife groweth in my Garden in great plenty.

¶ *The Time.*

It floureth in Iuly and Auguft.

¶ *The Names.*

Leadwort is called *Molybdana*, *Plumbago Plinij*, & *Dentillaria Rondeletij*: in Italian, *Crepanella*: the Romanes, *Herba S. Antonij*: in Illyria, *Cucurida*: in Englifh Leadwort.

¶ *The Temperature.*

Dentillaria is of a caufticke quality.

¶ *The Vertues.*

A   It helpeth the tooth-ache, and that as fome fay if it be holden in the hand fome fmall while.

C H A P. 531.   *Of Rue, or herbe Grace.*

¶ *The Defcription.*

1   **G**Arden Rue or planted Rue, is a fhrub full of branches, now and then a yard high, or higher: the ftalkes whereof are couered with a whitifh barke, the branches are more greene: the leaues hereof confift of diuers parts, and be diuided into wings, about which are certaine little ones, of an odde number, fomething broad, more long than round, fmooth and fomewhat fat, of a gray colour or greenifh blew: the floures in the top of the branches are of a pale yellow confifting of foure little leaues, fomething hollow: in the middle of which ftandeth vp a little head or button foure fquare, feldome fiue fquare, containing as many little coffers as it hath corners, being compaffed about with diuers little yellow threds: out of which hang pretty fine tips of one colour; the feed groweth in the little coffers: the root is wooddy, and faftned with many ftrings: this Rue hath a very ftrong and ranke fmell, and a biting tafte.

2   the fecond being the wild or mountaine Rue, called *Ruta fylueftris*, is very like to garden Rue, in ftalkes, leaues, floures, feed, colour, tafte, and fauour, fauing that euery little leafe hath fmaller cuts, and is much narrower: the whole plant dieth at the approch of Winter, being an annuall plant, and muft either ftand till it doe fow himfelfe, or elfe muft be fowne of others. ‡ This fecond is a variety of the garden Rue differing from the former onely in fmallneffe. ‡

3   This plant is likewife a wilde kinde of Rue, and of all the reft the fmalleft, and yet more virulent, biting, and ftinking than any of the reft: the whole plant is of a whitifh pale greene, agreeing with the laft before mentioned in each refpect, faue in greatneffe, and in that the venomous fumes or vapors that come from this fmall wild Rue are more noifome & hurtfull than the former. ‡ The leaues lie fpred vpon the ground, and are very finely cut and diuided: the whole plant is of
such

Ruta hortenſis.
Garden Rue.

3 *Ruta ſylveſtris minima.*
The ſmalleſt wilde Rue.

4 *Ruta montana.*
Mountaine Rue.

5 *Harmala.*
Wilde Rue with white flöures.

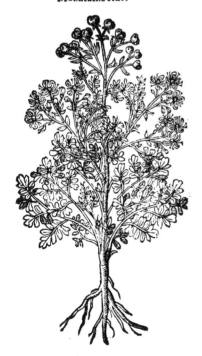

‡ 6 *Ruta Canina.*
Dogs Rue.

such acrimonie, that *Clusius* saith he hath oftener than once obserued it to pierce rhrough three pair of gloues to the hand of the gatherer; and if any one rub his face with his hand that hath newly gathered it, forthwith it will mightily inflame it. He tells a historie of a Dutch Student of Montpelier that went with him a simpling, who putting some of it between his hat and his head to keep him the cooler, had by that means all his face presently inflamed and blistered wheresoeuer the sweat ranne downe. ‡

4   There is another wilde Rue growing on the mountaines of Savoy and other places adioyning, hauing a great thicke root, from which arise great shoots or stalkes, whereon grow leaues very thicke and fat, parted into diuers longish sections, otherwise resembling the leaues of the first described, of a strong and stinking smell: the flours grow on the tops of the stalks, consisting of foure small yellow leaues: the seeds are like the other.

5   *Harmel* is one of the wilde Rues, it bringeth forth immediatly from the root diuers little stalks of a cubit high; whereupon grow green leaues diuersly cut into long pieces, longer and narrower than those of the wilde strong smelling Rue. The flours be white, composed of fiue white leaues: the fruit is three square, bigger than that of the planted Rue, in which the seed lies: the root is thicke, long, and blackish. This Rue in hot countries hath a maruellous strong smell: in cold countries not so.

‡   6   This, which *Matthiolus* gaue for *Sidritis 3.* and *Lobel, Clusius,* and others, for *Ruta Canina,* hath many twiggy branches some cubit and half high, whereon grow leaues resembling those of the *Papauer Rhæas* or *Argemone,* lesser, thicker, and of a blackish green: the floures are of a whitish purple colour, fashioned somewhat like those of *Antirrhinum:* the seed is small, and contained in such vessels as those of Rue, or rather those of *Blattaria:* the whole plant is of a strong and ingratefull smel: it grows in the hot and dry places about Narbon in France, Ravenna, and Rome in Italy. ‡

¶ *The Place.*

Garden Rue ioyeth in sunny and open places: it prospereth in rough and bricky ground and among ashes: it can in nowise away with dung.

The wild are found on mountains in hot countries, as in Cappadocia, Galatia, and in diuers prouinces of Italy and Spain, and on the hills of Lancashire and Yorke.

Pliny saith that there is such friendship between it and the fig-tree, that it prospers no where so well as vnder the fig tree. The best for physicks vse is that which groweth vnder the fig tree, as *Dioscorides* saith: the cause is alledged by *Plutarch, lib.* 1. of his *Symposiacks* or Feasts, for he saith it becomes more sweet and milde in taste, by reason it taketh as it were some part of the sweetnesse of the fig tree, whereby the ouer-ranke qualitie of the Rue is allayed; vnlesse it be that the figge tree whilest it drawes nourishment to it selfe, draweth also the ranknesse away from the Rue.

¶ *The Time.*

They floure in these cold countries in Iuly and August: in other countries sooner.

¶ *The Names.*

The first is *Hortensis Ruta* or garden Rue: in high-Dutch, **Rauten**; in low-Dutch, **Ruijte**: The Italians and Apothecaries keep the Latine name: in Spanish, *Aruda:* in French, *Rue de Iarden*; in English, Rue, and Herb-Grace.

Wilde Rue is called in Greeke, Πήγανον. *Peganon:* in Latine, *Ruta syluestris,* or wilde Rue: in Galatia and Cappadocia, Μῶλυ: of diuers, *Harmala:* of the Arabians, *Harmel:* of the Syrians, *Besara:*

¶ *The Temperature.*

Rue is hot and dry in the later end of the third degree, and wilde Rue in the fourth: it is of thin

and

and ſubtill parts : it waſts and conſumes winde,it cutteth and digeſteth groſſe and tough humors.

¶ The Vertues.

Rue or Herb grace prouokes vrine,brings downe the ſickneſſe,expels the dead childe and after- A birth,being inwardly taken,or the decoction drunk ; and is good for the mother if but ſmelled to. *Pliny,lib.20 cap.13.* ſaith it opens the matrix and brings it into the right place,if the belly all ouer and the ſhare(the breſt ſay the old falſe copies)be anointed therewith : mixt with honey it is a re- medie againſt the inflammation and ſwelling of the ſtones,proceeding of long abſtinence from ve- nerie,called of our Engliſh Mountebanks the Colts euill,if it be boiled with Barrows greaſe, Bay leaues,and the pouders of Fenugreeke and Lineſeed be added thereto,and applied pultis-wiſe.

It takes away crudity and rawneſſe of humors,and alſo windineſſe and old pains of the ſtomack. B

Boiled with vineger it eaſeth paines,is good againſt the ſtitch of the ſide and cheſt,and ſhortnes C of breath vpon a cold cauſe,and alſo againſt the pain in the joints and huckle bones.

The oile of it ſerues for the purpoſes laſt recited : it takes away the colick and pains of the guts, D not only in a cliſter,but alſo anointed vpon the places affected. But if this oile be made of the oile preſſed out of Lineſeed it will bee ſo much the better,and of ſingular force to take away hard ſwel- lings of the ſpleen or milt.

It is vſed with good ſucceſſe againſt the dropſie called in Greeke ὑπωπια, beeing applied to the E belly in manner of a pultis.

The herb a little boiled or ſkalded,and kept in pickle as Sampier,and eaten,quickens the ſight. F

The ſame applied with hony and the juice of Fennell,is a remedie againſt dim eies. G

The juice of Rue made hot in the rinde of a pomegranat and dropped into the eares,takes away H the pain of thereof.

S:Anthonies fire is quenched therewith : it killeth the ſhingles, and running vlcers and ſores in I the heads of yong children,if it be tempered with Ceruſe,white lead,vineger,and oile of roſes,and made into the form of *Nutritum* or *Triapharmacon.*

Dioſcorides ſaith,That Rue put vp in the noſthrils ſtayeth bleeding. K

So ſaith *Pliny* alſo : when notwithſtanding it is of power rather to procure bleeding, through L its ſharpe and biting quality.

The leaues of Rue beaten and drunke with wine are an antidote againſt poiſons,as *Pliny* ſaith. M

Dioſcorides writeth,That a twelue penny weight of the ſeed drunke in wine is a counterpoiſon a- N gaiſt deadly medicines or the poiſon of Wolfes-bane,*Ixia,* Muſhroms or Toad-ſtooles,the biting of Serpents,the ſtinging of Scorpions,Bees,hornets,and waſps ; and is reported,That if a man be anointed with the juice of Rue,theſe will not hurt him ; and that the ſerpent is driuen away at the ſmell thereof when it is burned : inſomuch that when the Weeſell is to fight with the ſerpent,ſhee armeth her ſelfe by eating Rue,againſt the might of the Serpent.

The leaues of Rue eaten with the kernels of Walnuts or figs ſtamped together and made into a O maſſe or paſte,is good againſt all euill aires,the peſtilence or plague,reſiſts poiſon and all venome.

Rue boiled with Dil,Fennel ſeed,and ſome ſugar,in a ſufficient quantitie of wine,ſwageth the P torments and griping pains in the belly,the pain in the ſides and breſt,the difficulty of breathing, the cough,and ſtopping of the lungs,and helpeth ſuch as are declining to a dropſie.

The juice taken with Dill as aforeſaid,helpeth the cold fits of Agues,and alters their courſe : it Q helpeth the inflammation of the fundament,and paines of the gut called *Rectum inteſtinum.*

The juice of Rue drunk with wine purgeth women after their deliuerance,driuing forth the dead R childe,the ſecond ine,and the vnnatural birth.

Rue vſed very often either in meat or drinke,quencheth and drieth vp the naturall ſeed of gene- S ration,and the milke of thoſe that giue ſucke.

The oile wherein Rue hath bin boiled and infuſed many daies together in the Sun,warmeth and T chafeth all cold members if they be anointed therewith : alſo it prouoketh vrine , if the region of the bladder be anointed with it.

If it be miniſtred in cliſters it expels windineſſe, and the torſion or gnawing paines of the guts. V

The leaues of garden Rue boiled in water and drunke cauſeth one to make water,prouoketh the X terms,and ſtoppeth the laske.

Ruta ſylueſtris or wild Rue is more vehement both in ſmel and operation,and therefore the more Y virulent or pernitious ; for ſometimes it fumeth out a vapor or aire ſo hurtfull that it ſcorches the face of him that looketh vpon it,raiſing vp bliſters,wheals,and other accidents : it venometh their hands that touch it,and will infect the face alſo if it be touched before they be clean waſhed:wher- fore it is not to be admitted to meat or medicine.

The end of the ſecond Booke.

THE THIRD BOOKE OF THE
HISTORY OF PLANTS.

Containing the Description, Place, Time, Names, Nature, and Vertues,
of Trees, Shrubs, Bushes, Fruit-bearing Plants, Rosins, Gums, Roses,
Heath, Mosses, some Indian Plants, and other rare Plants not
remembred in the Proeme to the first Booke. Also
Mushroms, Corall, and their seuerall
Kindes, &c.

The Proeme.

Auing finished the treatise of Herbs and Plants in general, vsed for meat, medicine, or sweet smelling vse, only some few omitted for want of perfect instruction; and also being hindered by the slacknesse of the Cutters or grauers of the Figures: these wants we intend to supply in this third and last part. The Tables as well general as particular shall be set forth in the end of this present Volume.

CHAP. I.   *Of Roses.*

¶ *The Kindes.*

He Plant of Roses, though it be a shrub full of prickles, yet it had bin more fit and conuenient to haue placed it with the most glorious floures of the world, than to insert the same here among base and thorny shrubs: for the Rose doth deserue the chief and prime place among all floures whatsoeuer; beeing not onely esteemed for his beauty, vertues, and his fragrant and odoriferous smell; but also because it is the honor and ornament of our English Scepter, as by the conjunction appeareth, in the vniting of those two most Royall Houses of Lancaster and Yorke. Which pleasant floures deserue the chiefest place in crownes and garlands, as *Anacreon Thius* a most antient Greeke Poet (whom *Henricus Stephanus* hath translated in a gallant Latine Verse) affirmes in those Verses of a Rose, beginning thus;

Τὸ ῥόδον τὸ τῶν ἐρώτων, &c.

> *Rosa honos, decusq; florum,*
> *Rosa, cura, amorq; Veris.*
> *Rosa, cælitum voluptas,*
> *Roseis puer Cytheres.*
> *Caput implicat Corollis,*
> *Charitum Choros frequentans.*

The

The Roſe is the honour and beauty of floures,
The Roſe in the care and loue of the Spring:
The Roſe is the pleaſure of th' heauenly Pow'rs.
The Boy of faire *Venus*, *Cythera's* Darling,
Doth wrap his head round with garlands of Roſe,
When to the dances of the Graces he goes.

Augerius Busbequius ſpeaking of the eſtimation and honor of the Roſe, reporteth, That the Turks can by no means endure to ſee the leaues of Roſes fall to the ground, becauſe ſome of them haue dreamed, that the firſt or moſt antient Roſe did ſpring out of the bloud of *Venus*: and others of the Mahumetans ſay that it ſprang of the ſweat of *Mahumet*.

But there are many kindes of Roſes, differing either in the bigneſſe of the floures, or the plant it ſelfe, roughneſſe or ſmoothneſſe, or in the multitude or fewneſſe of the flours, or elſe in colour and ſmell; for diuers of them are high and tall, others ſhort and low, ſome haue fiue leaues, others very many. *Theophraſtus* tells of a certain Roſe growing about Philippi, with an hundred leaues, which the Inhabitants brought forth of Pangæum, and planted in Campania, as *Pliny* ſaith. Which wee hold to be the Holland Roſe, that diuers call the Prouince Roſe, but not properly.

Moreouer, ſome be red, others white, and moſt of them or all ſweetly ſmelling, eſpecially thoſe of the garden.

1 *Roſa alba.*
The white Roſe.

¶ *The Deſcription.*

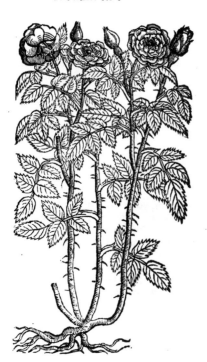

1   F the Curious could ſo be content, one generall deſcription might ſerue to diſtinguiſh the whole ſtock or kindred of the Roſes, being things ſo wel knowne: notwithſtanding I thinke it not amiſſe to ſay ſomthing of them ſeuerally, in hope to ſatisfie all. The white Roſe hath very long ſtalkes of a wooddy ſubſtance, ſet or armed with diuers ſharpe prickles: the branches wherof are likewiſe full of prieckles, whereon grow leaues conſiſting of fiue leaues for the moſt part, ſet vpon a middle rib by couples, the old leaf ſtanding at the point of the ſame, and euery one of thoſe ſmall leaues ſomwhat ſnipt about the edges, ſomewhat rough, and of an ouerworne greene colour: from the boſome whereof ſhoot forth long foot ſtalks, whereon grow very faire double flours of a white colour, and very ſweet ſmell, hauing in the middle a few yellow threds or chiues; which being paſt, there ſucceedeth a long fruit, greene at the firſt, but red when it is ripe, and ſtuffed with a downy choking matter, wherein is contained ſeed as hard as ſtones. The root is long, tough, and of a wooddy ſubſtance.

2   The red Roſe groweth very low in reſpect of the former: the ſtalks are ſhorter, ſmoother, and browner of colour: The leaues are like, yet of a worſe duſty colour: The floures grow on the tops of the branches, conſiſting of many leaues of a perfect red colour: the fruit is likewiſe red when it is ripe: the root is wooddy.

3   The common Damaske Roſe in ſtature, prickely branches, and in other reſpects is like the white

2 *Rosa rubra.*
The red Rose.

3 *Rosa Provincialis, sive Damascena.*
The Province or Damaske Rose.

5 *Rosa sine spinis.*
The Rose without prickles.

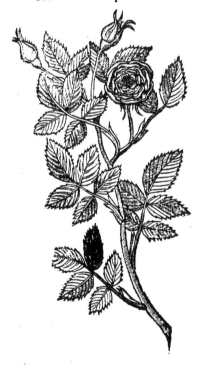

white Rose; the especiall difference consists in the colour and smell of the flours: for these are of a pale red colour, of a more pleasant smel, and fitter for meat and medicine.

4 The *Rosa Provincialis minor* or lesser Province Rose differeth not from the former, but is altogether lesser: the floures and fruit are like: the vse in physicke also agreeth with the precedent.

5 The Rose without prickles hath many young shoots comming from the root, diuiding themselues into diuers branches, tough, and of a wooddy substance as are all the rest of the Roses, of the height of two or three cubits, smooth and plain without any roughnesse or prickles at all: whereon grow leaues like those of the Holland Rose, of a shining deep green colour on the vpper side, vnderneath somewhat hoary and hairy. The flours grow at the tops of the branches, consisting of an infinite number of leaues, greater than those of the Damaske Rose, more double, and of a colour between the red and damask Roses, of a most sweet smell. The fruit is round, red when it is ripe, and stuffed with the like flocks and seeds of those of the damaske Rose. The root is great, wooddy, and far spreading.

6 The Holland or Province Rose hath diuers shoots proceeding from a wooddy root ful of sharpe prickles, diuiding it selfe into diuers branches, wheron grow leaues consisting of fiue

leaues

leaues ſet on a rough middle rib, & thoſe ſnipt about the edges : the flours grow on the tops of the branches, in ſhape and colour like the damaske Roſe, but greater and more double, inſomuch that the yellow chiues in the middle are hard to be ſeene ; of a reaſonable good ſmell, but not fully ſo ſweet as the common damaske Roſe : the fruit is like the other of his kinde.

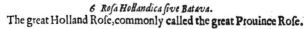

6 Roſa Hollandica ſive Batava.
The great Holland Roſe, commonly called the great Prouince Roſe.

We haue in our London gardens one of the red Roſes, whoſe flours are in quantitie and beautie equall with the former, but of greater eſtimation, of a perfect red colour, wherin it eſpecially diffe-reth from the Province Roſe ; in ſtalks, ſtature, and manner of growing it agrees with our common red Roſe.

¶ *The Place.*
All theſe ſorts of Roſes we haue in our London gardens, except that Roſe without pricks, which as yet is a ſtranger in England. The double white Roſe groweth wilde in many hedges of Lanca-ſhire in great aboundance, euen as Briers do with vs in theſe Southerly parts, eſpecially in a place of the country called Leyland, and at Roughford not far from Latham. Moreouer, in the ſaid Ley-land fields doth grow our garden Roſe wilde, in the plowed fields among the corne, in ſuch aboun-dance, that there may be gathered daily during the time, many buſhels of roſes, equal with the beſt garden Roſe in each reſpect : the thing that giueth great cauſe of wonder, is, That in a field in the place aforeſaid, called Glovers field, euery yeare that the field is plowed for corne, that yeare it wil be ſpred ouer with Roſes, and when it lieth ley, or not plowed, then is there but few Roſes to be ga-thered ; by the relation of a curious gentleman there dwelling, ſo often remembred in our hiſtory.
‡ I haue heard that the Roſes which grow in ſuch plenty in Glovers field euery yere the field is plowed, are no other than Corn Roſe, that is, red Poppies, howeuer our Author was informed. ‡
¶ *The Time.*
Theſe floure from the end of May to the end of Auguſt, and diuers times after, by reaſon the tops and ſuperfluous branches are cut away in the end of their flouring : and then doe they ſomtimes floure euen vntill October and after.

¶ *The*

¶ *The Names.*

The Rose is called in Latine *Rosa* : in Greeke ϱϵιϑι : and the plant it selfe ϱιϑνιϑ : (which in Latine keepeth the same name that the floure hath) and it is called *Rodon* (as *Plutarch* saith) because it sendeth forth plenty of smell.

The middle part of the Roses, that is, the yellow chiues, or seeds and tips, is called *Anthos*, and *Flos Rosæ*, the floure of the Rose : in shops, *Anthera*, or the blowing of the Rose.

The white parts of the leaues of the floure it selfe, by which they are fastened to the cups, be named *Vngues* or nails. That is called *Calix*, or the cup, which containeth and holdeth in together the yellow part and leaues of the floure.

Alabastri are those parts of the cup which are deeply cut, and that compasse the floure close about before it be opened, which be in number fiue, two haue beards and two haue none, and the fifth hath but halfe one : most do call them *Cortises Rosarum*, or the husks of the Roses : the shoots of the plant of Roses, *Strabo Gallus* in his little garden doth call *Viburna*.

The white Rose is called *Rosa alba* : in English, the white Rose : in high Dutch, **Weisz Roosen** : in low Dutch, **witte Roosen** : in French, *Rose Blanche* : of *Pliny, Spineola Rosa*, or *Rosa Campana*.

The red Rose is called in Latine, *Rosa rubro* : the French-men, *Rose Franche, Rose de Pronins*, a towne in Campaigne : of *Pliny, Trachinia*, or *Prænestina*.

The Damaske Rose is called of the Italians *Rosa incarnata* : in high Dutch, **Leibfarbige Roosen** : in low Duth, **Prouincie Roose** : of some, *Rosa Prouincialis*, or Rose of Prouince : in French or some, *Melesia* : the Rose of Melaxo, a city in Asia, from whence some haue thought it was first brought into those parts of Europe.

The great Rose, which is generally called the great Prouince Rose, which the Dutchmen cannot endure ; for say they, it came first out of Holland, and therefore to be called the Holland Rose : but by all likelihood it came from the Damaske Rose, as a kind thereof, made better and fairer by art, which seemeth to agree with truth.

The Rose without prickles is called in Latine, *Rosa sine spinis*, and may be called in English, the Rose without thornes, or the Rose of Austrich, because it was first brought from Vienna, the Metropolitan citie of Austrich, and giuen to that famous Herbarist *Carolus Clusius*.

¶ *The Temperature.*

The leaues of the floures of Roses, because they doe consist of diuers parts haue also diuers and sundry faculties : for there be in them certaine that are earthy and binding, others moist and watery, and sundry that are spirituall and airie parts, which notwithstanding are not all after one sort, for in one kinde these excell, in another those, all of them haue a predominant or ouer-ruling cold temperature, which is neerest to a meane, that is to say, of such as are cold in the first degree, moist, airie, and spirituall parts are predominant in the White roses, Damaske and Muske,

¶ *The Vertues.*

The distilled water of Roses is good for the strengthning of the heart, and refreshing of the spi- A rits, and likewise for all things that require a gentle cooling.

The same being put in junketting dishes, cakes, sauces, and many other pleasant things, giueth a B fine and delectable taste.

It mitigateth the paine of the eies proceeding of a hot cause, bringeth sleep, which also the fresh C roses themselues prouoke through their sweet and pleasant smell.

The juyce of these Roses, especially of Damask, doth mooue to the stoole, and maketh the belly D soluble : but most effectuall that of the Musk Roses : next to them is the juyce of the Damask, which is more commonly vsed.

The infusion of them doth the same, and also the Syrrup made thereof, called in Latine *Drosatum*, E or *Serapium* : the Apothecaries call it Syrrup of roses solutiue, which must be made of the infusion in which a great number of the leaues of these fresh Roses are diuers and sundry times steeped.

It is profitable to make the belly loose and soluble, when as either there is no need of other stron- F ger purgation, or that it is not fit and expedient to vse it : for besides those excrements which stick to the bowels, or that in the first and neerest veines remaine raw, flegmaticke, and now and then cholericke, it purgeth no other excrements, vnlesse it be mixed with certaine other stronger medicines.

This Syrrup doth moisten and coole, and therefore it allayeth the extremity of heat in hot bur- G ning feuers, mitigateth the inflammations of the intrails, and quencheth thirst : it is scarce good for a weake and moist stomacke, for it leaueth it more slacke and weake.

Of like vertue also are the leaues of these preserued in Sugar, especially if they be onely bruised H with the hands, and diligently rempered with Sugar, and so heat at the fire rather than boyled.

¶ *The Temperature of Red Roses.*

There is in the red Roses, which are common euery where, and in the other that be of a deep pur- I ple, called Prouince Roses, a more earthy substance, also a drying and binding quality, yet not without

out certain moiſture joyned, being in them when they are as yet freſh, which they loſe when they be dried: for this cauſe their juyce and infuſion doth alſo make the body ſoluble, yet not ſo much as of the others aforeſaid. Theſe Roſes being dried, and their moiſture gone, do bind and dry; and likewiſe coole, but leſſer than when they are freſh.

¶ *The Vertues.*

I   They ſtrengthen the heart, and helpe the trembling and beating thereof.

K   They giue ſtrength to the liuer, kidnies, and other weake intrails; they dry and comfort a weake ſtomacke that is flaſhie and moiſt; ſtay the whites and reds, ſtanch bleeding in any part of the body, ſtay ſweatings, binde and looſe, and moiſten the body.

L   And they are put into all manner of counterpoyſons and other like medicines, whether they be to be outwardly applied or to be inwardly taken, to which they giue an effectuall binding, and certaine ſtrengthning quality.

M   Hony of Roſes, or *Mel Roſarum*, called in Greeke ϵδωμι, which is made of them, is moſt excellent good for wounds, vlcers, iſſues, and generally for ſuch things as haue need to be clenſed and dried.

N   The oyle doth mitigate all kindes of heat, and will not ſuffer inflammations or hot ſwellings to riſe, and being riſen it doth at the firſt aſſwage them.

¶ *The Temperature and Vertues of the parts.*

O   The flours or bloomings of Roſes, that is to ſay, the yellow haires and tips, do in like manner dry and binde, and that more effectually than of the leaues of the roſes themſelues: the ſame temperature the cups and beards be of; but ſeeing none of theſe haue any ſweet ſmell, they are not ſo profitable, nor ſo familiar or beneficiall to mans nature: notwithſtanding in fluxes at the ſea, it ſhall auaile the Surgion greatly, to carry ſtore thereof with him, which doth there preuaile much more than at the land.

P   The ſame yellow called *Anthera*, ſtayeth not only thoſe lasks and bloudy fluxes which do happen at the Sea, but thoſe at the land alſo, and likewiſe the white flux and red in women, if they bee dried, beaten to pouder, and two ſcruples thereof giuen in red wine, with a little pouder of Ginger added thereto: and being at the Sea, for want of red Wine you may vſe ſuch liquor as you can get in ſuch extremity.

Q   The little heads or buttons of the Roſes, as *Pliny* writeth, doe alſo ſtanch bleeding and ſtop the laske.

R   The nailes or white ends of the leaues of the floures are good for watering eies.

S   The juice, infuſion, or decoction of Roſes, are to be reckoned among thoſe medicines which are ſoft, gentle, looſing, opening and purging gently the belly, which may be taken at all times and in all places, of euery kinde or ſex of people, both old and young, without danger or perill.

T   The Syrrup made of the infuſion of Roſes, is a moſt ſingular and gentle looſing medicine, carrying downewards cholericke humors, opening the ſtoppings of the liuer, helping greatly the yellow jaundiſe, the trembling of the heart, and taking away the extreme heat in agues and burning feuers; which is thus made:

V   Take two pound of Roſes, the white ends cut away, put them to ſteepe or infuſe in ſix pintes of warme water in an open veſſell for the ſpace of twelue houres: then ſtraine them out, and put thereto the like quantitie of Roſes, and warme the water again, ſo let it ſtand the like time: do thus foure or fiue times; in the end adde vnto that liquor or infuſion, foure pound of fine Sugar in pouder; then boyle it vnto the forme of a ſyrrup, vpon a gentle fire, continually ſtirring it vntill it be cold; then ſtraine it, and keepe it for your vſe, whereof may be taken in white wine, or other liquor, from one ounce vnto two.

X   Syrrup of the juyce of Roſes is very profitable for the griefes aforeſaid, made in this manner:

Y   Take Roſes, the white nailes cut away, what quantitie you pleaſe, ſtampe them, and ſtraine out the juyce, the which you ſhall put to the fire, adding thereto Sugar, according to the quantity of the juyce: boyling them on a gentle fire vnto a good conſiſtence.

Z   Vnto theſe ſyrrups you may adde a few drops of oyle of Vitriol, which giueth it a moſt beautifull colour, and alſo helpeth the force in cooling hot and burning feuers and agues: you likewiſe may adde thereto a ſmall quantity of the juyce of Limons, which doth the like.

A   The conſerue of Roſes, as well that which is crude and raw, as that which is made by ebullition or boiling, taken in the morning faſting, and laſt at night, ſtrengthneth the heart, and taketh away the ſhaking and trembling thereof, ſtrengthneth the liuer, kidneies, and other weake intrals, comforteth a weake ſtomacke that is moiſt and raw; ſtaieth the whites and reds in women, and in a word is the moſt familiar thing to be vſed for the purpoſes aforeſaid, and is thus made:

B   Take the leaues of Roſes, the nails cut off, one pound, put them into a cleane pan; then put therto a pinte and a halfe of ſcalding water, ſtirring them together with a woodden ſlice, ſo let them ſtand to macerate, close couered some two or three houres; then set them to the fire slowly to boyle, adding thereto three pounds of sugar in powder, letting them to simper together according to discretion, some houre or more; then keepe it for your vse.

The same made another way, but better by many degrees: take Roses at your pleasure, put them to boyle in faire water, hauing regard to the quantity; for if you haue many Roses you may take more water; if fewer, the lesse water will serue: the which you shall boyle at the least three or foure houres, euen as you would boile a piece of meate, vntill in the eating they be very tender, at which time the Roses will lose their colour, that you would thinke your labour lost, and the thing spoiled. But proceed, for though the Roses haue lost their colour, the water hath gotten the tincture thereof; then shall you adde vnto one pound of Roses, foure pound of fine sugar in pure pouder, and so according to the rest of the Roses. Thus shall you let them boyle gently after the sugar is put therto, continually stirring it with a woodden Spatula vntill it be cold, whereof one pound weight is worth six pound of the crude or raw conserue, as well for the vertues and goodnesse in taste, as also for the beautifull colour.

The making of the crude or raw conserue is very well knowne, as also Sugar roset, and diuers other pretty things made of Roses and Sugar, which are impertent vnto our history, because I intend nether to make thereof an Apothecaries shop, nor a Sugar-Bakers storehouse, leauing the rest for our cunning confectioners.

Chap. 2.   *Of the Muske Roses.*

¶ *The Kindes.*

THere be diuers sorts of Roses planted in gardens, besides those written of in the former chapter, which are of most writers reckoned among the wilde Roses, notwithstanding we thinke it conuenient to put them into a chapter betweene those of the garden and the brier Roses, as indifferent whether to make them of the wilde Roses, or of the tame, seeing we haue made them denizons in our gardens for diuers respects, and that worthily.

1 *Rosa Moschata simplici flore.*
The single Muske Rose.

2 *Rosa Moscata multiplex.*
The double Muske Rose.

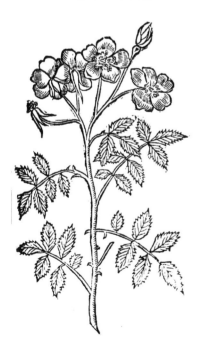

¶ *The Description.*

1   THe single Muske Rose hath diuers long shoots of a greenish colour and wooddy substance, armed with very sharpe prickles, diuiding it selfe into diuers branches: whereon doe grow long leaues, smooth and shining, made of diuers leaues set vpon a middle rib, like the other Roses: the floures grow on the tops of the branches, of a white colour, and pleasant sweet smell, like that of Muske, wherof it tooke his name; hauing certaine yellow seeds in the middle, as the rest of the Roses haue: the fruit is red when it is ripe, and filled with such chaffie flockes and seeds as those of the other Roses: the root is tough and wooddy.

2   The double Muske Rose differeth not from the precedent in leaues, stalkes, and roots, nor in the colour of the floures, or sweetnesse thereof, but onely in the doublenesse of the floures, wherein consisteth the difference.

3   Of these roses we haue another in our London gardens, which of most is called the blush rose, it floureth when the Damaske Rose doth: the floures hereof are very single, greater than the other Muske Roses, and of a white colour, dasht ouer with a light wash of carnation, which maketh that colour which wee call a blush colour: the proportion of the whole plant, as also the smell of the floures, are like the precedent.

3 *Rosæ Moschatæ species major.*
The great Muske Rose.

4 *Rosa Holosericea.*
The Veluet Rose.

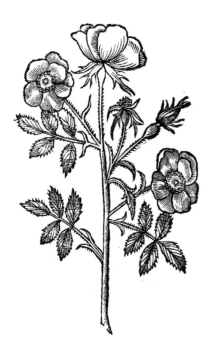

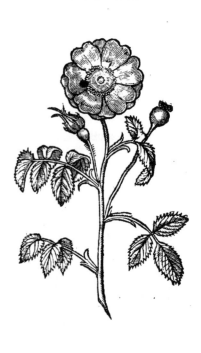

4   The Veluet Rose groweth alwaies very low, like vnto the red Rose, hauing his branches couered with a certaine hairy or prickly matter, as fine as haires, yet not so sharpe or stiffe that it will harme the most tender skin that is: the leaues are like the leaues of the white Rose: the flours grow at the top of the stalkes, doubled with some yellow thrums in the midst, of a deepe and blacke red colour, resembling red crimson Veluet, whereupon some haue called it the Veluet Rose: when the floures be vaded, there follow red berries full of hard seeds, wrapped in a downe or woollinesse like the others.

5   The yellow Rose which (as diuers do report) was by Art so coloured, and altered from his first estate, by grafting a wilde Rose vpon a Broome-stalke; whereby (say they) it doth not onely change his colour, but his smell and force. But for my part I hauing found the contrary by mine owne experience, cannot be induced to beleeue the report: for the roots and off-springs of this Rose haue
brought

brought forth yellow roſes,ſuch as the maine ſtocke or mother bringeth out,which euent is not to be ſeen in all other plants that haue been graffed. Moreouer,the ſeeds of yellow roſes haue brought forth yellow Roſes, ſuch as the floure was from whence they were taken;which they ſhould not do by any conjecturall reaſon, if that of themſelues they were not a naturall kinde of Roſe. Laſtly it were contrary to that true principle,

Nature ſequitur ſemina quodque ſua : that is to ſay.

Euery ſeed and plant bringeth forth fruit like vnto it ſelfe, both in ſhape and nature : but leauing that errour,I will proceed to the deſcription : the yellow roſe hath browne and prickly ſtalks or ſhoots,fiue or ſix cubits high,garniſhed with many leaues, like vnto the Muske roſe,of an excellent ſweet ſmell,and more pleaſant than the leaues of the Eglantine : the floures come forth among the leaues, and at the top of the branches of a faire gold yellow colour : the thrums in the middle, are alſo yellow : which being gone,there follow ſuch knops or heads as the other Roſes do beare.

5 *Roſa lutea.*
The yellow Roſe.

‡ 6 *Roſa lutea multiplex.*
The double yellow Roſe.

‡ 6 Of this kinde there is another more rare and ſet by, which in ſtalks,leaues,and other parts is not much different from the laſt deſcribed,onely the floure is very double,and it ſeldome fairely ſhewes it ſelfe about London,where it is kept in our chiefe gardens as a prime rariety. ‡

7 The Canell or Cinnamon Roſe, or the Roſe ſmelling like Cinnamon, hath ſhots of a brown colour, foure cubits high, beſet with thorny prickles, and leaues like vnto thoſe of Eglantine, but ſmaller and greener,of the ſauour or ſmell of Cinnamon,whereof it tooke his name, and not of the ſmell of his floures (as ſome haue deemed) which haue little or no ſauour at all : the floures be exceeding double, and yellow in the middle, of a pale red colour, and ſometimes of a carnation : the root is of a wooddy ſubſtance.

8 We haue in our London gardens another Cinnamon or Canell Roſe, not differing from the laſt deſcribed in any reſpect, but onely in the floures;for as the other hath very double floures, contrariwiſe theſe of this plant are very ſingle,wherein is the difference.

7 *Rofa Cinnamomea pleno flore.*
The double Cinnamon Rofe.

‡ 8 *Rofa Cinnamomea flore fimplici.*
The fingle Cinnamon Rofe.

¶ *The Place.*

Thefe Rofes are planted in our London Gardens, and elfe-where, but not found wilde in England.

¶ *The Time.*

The Muske Rofe floureth in Autumne, or the fall of the leafe : the reft floure when the Damask and red Rofe do.

¶ *The Names.*

The firft is called *Rofa Mofchata*, of the fmell of Muske, as we haue faid : in Italian, *Rofa Mofchetta*: in French, *Rofes Mufquets*, or *Mufcadelles* : in low Dutch, 𝕸𝖚𝖘𝖐𝖊𝖙 𝖗𝖔𝖔𝖘𝖊𝖓: in Englifh Musk Rofe: the Latine and Englifh titles may ferue for the reft.

¶ *The Temperature.*

The Muske Rofe is cold in the firft degree, wher in airy and fpirituall parts are prodominant:the reft are referred to the Brier Rofe and Eglantine.

¶ *The Vertues.*

A   Conferue or fyrrup made of the Muske Rofe, in manner as before told in the Damaske and Red Rofes, doth purge very mightily waterifh humors, yet fafely, and without all danger, taken in the quantity of an ounce in weight.

B   The leaues of the floures eaten in the morning, in manner of a fallad, with oyle, vineger and pepper, or any other way according to the appetite and pleafure of them that fhall eate it, purge very notably the belly of waterifh and cholericke humors, and that mightily, yet without all perill or paine at all, infomuch as the fimpleft may vfe the quantity, according to their owne fancie : for if they do defire many ftooles, or fieges, they are to eate the greater quantity of the leaues; if fewer, the leffe quantity ; as for example : the leaues of twelue or fourteene floures giue fix or eight ftooles, and fo increafing or diminifhing the quantity, more or fewer, as my felfe haue often proued.

C   The white leaues ftamped in a wooden difh with a piece of Allum and the juyce ftrained forth into fome glafed veffell, dried in the fhadow, and kept, is the moft fine and pleafant yellow colour that may be diuifed, not onely to limne or wafh pictures and Imagerie in books, but alfo to colour meats and fauces, which notwithftanding the Allum is very wholefome.

There

There is not any thing extant of the others, but are thought to be equall with the white Muske Rose, whereof they are taken and holden to be kindes.

CHAP. 3. *Of the wilde Roses.*

¶ *The Description.*

1 THe sweet Brier doth oftentimes grow higher than all the kindes of Roses; the shoots of it are hard, thicke, and wooddy ; the leaues are glittering, and of a beautifull greene colour, of smell most pleasant : the Roses are little, fiue leaued, most commonly whitish, seldom tending to purple, of little or no smell at all : the fruit is long, of colour somewhat red, like a little oliue stone, & like the little heads or berries of the others, but lesser than those of the garden : in which is contained rough cotton, or hairy downe and seed, folded and wrapped vp in the same, which is small and hard : there be likewise found about the slender shoots hereof, round, soft, and hairy spunges, which we call Brier Balls, such as grow about the prickles of the Dog-Rose.

1 *Rosa syluestris odora.*   The Eglantine, or sweet Brier.

2 We haue in our London gardens another sweet Brier, hauing greater leaues, and much sweeter : the floures likewise are greater, and somewhat doubled, exceeding sweet of smell, wherein it differeth from the former.

3 The Brier Bush or Hep tree, is also called *Rosa canina*, which is a plant so common and well knowne, that it were to small purpose to vse many words in the description thereof : for euen children with great delight eat the berries thereof when they be ripe, make chaines and other prettie gewgawes of the fruit : cookes and gentlewomen make Tarts and such like dishes for pleasure thereof, and therefore this shall suffice for the description.

4 The Pimpinell Rose is likewise one of the wilde ones, whose stalks shoot forth of the ground in many places, of the height of one or two cubits, of a browne colour, and armed with sharpe prickles,

‡ 2 *Rosa syl. odora flore duplici.*
The double Eglantine.

3 *Rosa Canina inodora.*
The Brier Rose or Hep-tree.

4 *Rosa Pimpinellæ folio.*
The Pimpinell Rose.

kles, which diuide themselues toward the tops into diuers branches, whereon doe grow leaues consisting of diuers small ones, set vpon a middle rib like those of Burnet, which is called in Latine *Pimpinella*, whereupon it was called *Rosa Pimpinella*, the Burnet Rose. The floures grow at the tops of the branches, of a white colour, very single, and like vnto those of the Brier or Hep tree: after which come the fruit, blacke, contrary to all the rest of the roses, round as an apple; whereupon some haue called it *Rosa Pomifera*, or the Rose bearing apples: wherein is contained seed, wrapped in chaffie or flockie matter, like that of the Brier. the root is tough and wooody.

¶ *The Place.*

These wilde Roses do grow in the borders of fields and woods, in most parts of England. The last growes very plentifully in a field as you go from a village in Essex, called Graies (vpon the brinke of the riuer Thames) vnto Horndon on the hill, insomuch that the field is full fraught therewith all ouer.

It groweth likewise in a pasture as you goe from a village hard by London called Knights brige vnto Fulham, a Village thereby, and in many other places.

We haue them all except the Brier Bush in our London gardens, which we think vnworthy the place.

¶ *The*

¶ *The Time.*

They floure and flouriſh with the other Roſes.

¶ *The Names.*

The Eglantine Roſe, which is *Cynorrhodi*, or *Canina Roſa ſpecies*, a kinde of Dogs Roſe : and *Roſa ſylueſtris*, the wilde Roſe : in low Dutch, **Eglantier :** in French, *Eſglentine* ; and as *Ruellius* teſtifies, *Eglenterium* : who alſo ſuſpects it to be *Cynosbaton*, or *Canirubus* : of which *Dioſcorides* hath written in theſe words ; *Cynosbatus*, or *Canirubus*, which ſome call *Oxycantha*, is a ſhrub growing like a tree, full of prickles, with a white floure, long fruit like an oliue ſtone, red when it is ripe, and downie within : in Engliſh we call it Eglantine or ſweet Brier.

The ſpongie balls which are found vpon the branches are moſt aptly and properly called *Spongiola ſylueſtris Roſe*, the little ſponges of the wilde Roſe. The ſhops miſtake it by the name of *Bedeguar* ; for *Bedeguar* among the Arabians is a kinde of Thiſtle, which is called in Greeke Σκανδα λεωκη that is to ſay, *Spina alba* the white Thiſtle, not the white Thorne, though the word doe import ſo much.

The Brier or Hop tree is called *Sylueſtris Roſa*, the wilde Roſe: in high Dutch, **wilden Roſen :** in French, *Roſes ſauuages* : *Pliny, lib.8.cap.25.* ſaith that it is *Roſa Canina*, Dogs Roſe : of diuers *Canina ſentis*, or Dogs Thorne: in Engliſh, Brier Buſh, and Hep tree : the laſt hath beene touched in the deſcription.

¶ *The Temperature and Vertues.*

The faculties of theſe wilde Roſes are referred to the manured Roſe, but not vſed in Phyſicke　A where the other may be had : notwithſtanding *Pliny* affirmeth, that the root of the Brier Buſh is a ſingular remedy found out by Oracle, againſt the biting of a mad dog, which hee ſets downe in his eight booke, Chap.41.

The ſame Author *lib.25.cap.2.* affirmeth, that the little ſpongie Brier Ball ſtamped with honey　B and aſhes cauſeth haires to grow which are fallen through the diſeaſe called *Alopecia*, or the Foxes euill.

Fuchſius affirmes that the ſpongie excreſcence or ball growing vpon the Brier are good againſt　C the ſtone and ſtrangurie, if they be beaten to pouder and inwardly taken.

They are good not as they be diureticks or prouokers of vrine, or as they are wearers away of the　D ſtone, but as certaine other binding medicines that ſtrengthen the weake and feeble kidnies; which do no more good to thoſe that be ſubject to the ſtone, than many of the diuretickes, eſpecially of the ſtronger ſort ; for by two much vſing of diuretickes or piſſing medicines, it hapneth that the kidnies are ouerweakened, and oftentimes too much heated, by which meanes not onely the ſtones are not diminiſhed, worne away, or driuen forth, but oftentimes are alſo increaſed and made more hard : for they ſeperate and take away that which in the bloud is thin, waterie, and as it were whey-iſh ; and the thicker part, the ſtronger ſorts of diuretickes do draw together and make hard : and in like manner alſo others that are not ſo ſtrong, by the ouermuch vſing of them, as *Galen lib.5.* of the faculties of ſimple medicines reporteth.

The fruit when it is ripe maketh moſt pleaſant meats and banqueting diſhes, as tarts and ſuch　E like ; the making whereof I commit to the cunning cooke, and teeth to eate them in the rich mans mouth.

Chap. 4.　　*Of the Bramble or blacke-berry buſh.*

¶ *The Deſcription.*

1　The common Bramble bringeth forth ſlender branches, long, tough, eaſily bowed, ramping among hedges and whatſoeuer ſtands neere vnto it ; armed with hard and ſharpe prickles, whereon doe grow leaues conſiſting of many ſet vpon a rough middle ribbe, greene on the vpper ſide, and vnderneath ſomewhat white : on the tops of the ſtalkes ſtand certaine floures, in ſhape like thoſe of the Brier Roſe, but leſſer, of colour white, and ſometimes waſht ouer with a little purple : the fruit or berry is like that of the Mulberry, firſt red, blacke when it is ripe, in taſte betweene ſweet and ſoure, very ſoft, and full of grains : the root creepeth, and ſendeth forth here and there young ſprings.

‡ *Rubus repens fructu caſio.*

‡ 2　This hath a round ſtalke ſet full of ſmall crooked and very ſharpe pricking thornes, and creepeth on hedges and low buſhes of a great length, on the vpper ſide of a light red colour, and vnderneath greene, and taketh root with the tops of the trailing branches, whereby it doth mightily

encreaſe :

encreaſe : the leaues grow without order, compoſed of three leaues, and ſometimes of fiue, or elſe the two lower leaues are diuided into two parts, as hop leaues are now and then, of a light greene colour both aboue and vnderneath. The floures grow on the tops of the branches, *racematim*, many together, ſometimes white, ſometimes of a very light purple colour, euery floure containing fiue leaues, which are crompled or wrinkled, and do not grow plaine : the fruit followes, firſt greene, and afterwards blew, euery berry compoſed of one or two graines, ſeldome aboue foure or fiue growing together, about the bigneſſe of corans, wherein is contained a ſtony hard kernell or ſeed, and a iuice of the colour of Claret wine, contrary to the common *Rubus* or Bramble, whoſe leaues are white vnderneath : the berries being ripe are of a ſhining blacke colour, and euery berry containes vſually aboue forty graines cloſely compacted and thruſt together. The root is wooddy and laſting This growes common enough in moſt places, and too common in ploughed fields. Sept. 6. 1619. *Iohn Goodyer.* ‡

3 The Raſpis or Framboiſe buſh hath leaues and branches not much vnlike the common Bramble, but not ſo rough nor prickly, and ſometimes without any prickles at all, hauing onely a rough hairineſſe about the ſtalkes: the fruit in ſhape and proportion is like thoſe of the Bramble, red when they be ripe, and couered ouer with a little downineſſe; in taſte not very pleaſant. The root creepeth far abroad, whereby it greatly encreaſeth. ‡ This growes either with prickles vpon the ſtalks, or elſe without them : the fruit is vſually red, but ſometimes white of colour. ‡

1 *Rubus.*
The Bramble Buſh.

2 *Rubus Idæus.*
The Raſpis Buſh or Hinde-berry.

4   Stone Bramble ſeldome groweth aboue a foot high, hauing many ſmall flexible branches without prickles, trailing vpon the ground, couered with a reddiſh barke, and ſomewhat hairy : the leaues grow three together, ſet vpon tender naked foot-ſtalkes ſomewhat ſnipt about the edges : the floures grow at the end of the branches, conſiſting of foure ſmall white leaues like thoſe of the Cherry-tree : after which come ſmall Grape-like fruit, conſiſting of one, two, or three large tranſparent berries, ſet together as thoſe of the common Bramble, of a red colour when they be ripe, and of a pleaſant taſte, but ſomwhat aſtringent. The roots creepe along in the ground very farre abroad, whereby it greatly increaſeth.

4   *Chamæmorus* (called in the North part of England, where they eſpecially doe grow, Knot-berries, and Knought-berries) is likewiſe one of the Brambles, though without prickles : it brings
forth

forth ſmall weake branches or tender ſtems of a foot high ; whereon do grow at certaine diſtances rough leaues in ſhape like thoſe of the Mallow, not vnlike to the leaues of the Gooſeberry buſh : on the top of each branch ſtandeth one floure and no more, conſiſting of fiue ſmall leaues of a dark purple colour : which being fallen, the fruit ſucceedeth, like vnto that of the Mulberry, whereof it was called *Chamæmorus*, dwarfe Mulberry ; at the firſt white and bitter, after red and ſomewhat pleaſant : the root is long, ſomething knotty ; from which knots or joynts thruſt forth a few threddie ſtrings. ‡ I take that plant to which our Author hereafter hath allotted a whole chapter, and called *Vaccinia nubis*, or Cloud-berries, to be the ſame with this, as I ſhall ſhew you more largely in that place. ‡

4 *Rubus Saxatilis.*
Stone blacke-Berry buſh.

4 *Chamæmorus.*
Knot berry buſh.

¶ *The Place.*

The Bramble groweth for the moſt part in euery hedge and buſh.

The Raſpis is planted in gardens : it groweth not wilde that I know of, except in the field by a village in Lancaſhire called Harwood, not far from Blackeburne.

I found it among the buſhes of a cauſey, neere vnto a village called Wiſterſon, where I went to ſchoole, two miles from the Nantwich in Cheſhire.

The ſtone bramble I haue found in diuers fields in the Iſle of Thanet, hard by a village called Birchinton, neere Queakes houſe, ſomtimes Sir *Henry Criſpes* dwelling place. ‡ I feare our Author miſtooke that which was here added in the ſecond place, for that which he figured and deſcribed in the third (now the fourth) which I know not yet to grow wilde with vs. ‡

Knot-berries do loue open ſnowie hills and mountaines ; they grow plentifully vpon Ingleborow hils among the heath and ling, tweluc miles from Lancaſhire, being thought to be the higheſt hill in England.

They grow vpon Stane-more betweene Yorkeſhire and Weſtmerland, and vpon other wet Fells and mountaines.

¶ *The Time.*

Theſe floure in May and Iune with the Roſes : their fruit is ripe in the end of Auguſt and September.

¶ *The*

¶ *The Names.*

The Bramble is called in Greeke, Βάτος: in French, *Rouges, Loi Duyts Brelmers* : in Latine, *Rubus*, and *Sentis*, and *Vepres*, as *Ouid* writeth in the firſt booke of Metamorphoſis:

> *Aut Lepori, qui vepre latens hoſtilia cernit*
> *Ora canum.* ——

> Or to the Hare, that vnder Bramble cloſely lying, ſpies
> The hoſtile mouth of Dogs. ——

Of diuers it is called *Cynosbatus*, but not properly ; for *Cynosbatus* is the wilde Roſe, as we haue written : in high Dutch, **Bremen**: in low Dutch, **Bremen**: in French, *Rouce* : in Italian, *Garʒa* : in Engliſh, Bramble buſh, and Blacke-berry buſh.

The fruit is named in Latine *Morum rubi*; and as *Fuchſius* thinketh, *Vaccinium*, but not properly : in ſhops, *Mora Bati* : and in ſuch ſhops as are more barbarous, *Mora Baſſi* : in Engliſh, Blacke-berries.

The Raſpis is called in Greeke, Βάτος ιδαία : in Latine, *Rubus Idaa*, of the mountaine Ida on which it groweth: in Engliſh, Raſpis, Framboiſe, and Hinde-berry.

¶ *The Temperature and Vertues.*

A   The yong buds or tender tops of the Bramble buſh, the floures, the leaues, and the vnripe fruit, do very much dry and binde withall : being chewed they take away the heate and inflammation of the mouth, and almonds of the throat : they ſtay the bloudy flix, and other fluxes, and all manner of bleedings : of the ſame force is their decoction, with a little hony added.

B   They heale the eies that hang out, hard knots in the fundament, and ſtay the hemorrhoides, if the leaues be laid thereunto.

C   The juyce which is preſſed out of the ſtalks, leaues, and vnripe berries, and made hard in the Sun, is more effectuall for all thoſe things.

D   The ripe fruit is ſweet, and containeth in it much juyce of a temperate heate, therefore it is not vnpleaſant to be eaten.

E   It hath alſo a certaine kinde of aſtriction or binding quality.

F   It is likewiſe for that cauſe wholeſome for the ſtomacke, and if a man eat too largely therof, ſaith *Galen*, hee ſhall haue the head-ache : but being dried whileſt it is yet vnripe it bindeth and drieth more than the ripe fruit.

G   The root beſides that it is binding containeth in it much thinne ſubſtance, by reaſon whereof it waſteth away the ſtone in the kidnies, ſaith *Galen*.

H   *Pliny* writeth, that the berries and flours do prouoke vrine, and that the decoction of them in wine is a preſent remedy againſt the ſtone.

I   The leaues of the Bramble boyled in water, with honey, allum, and a little white wine added thereto, make a moſt excellent lotion or waſhing water to heale the ſores in the mouth, the priuie parts of man or woman, and the ſame decoction faſtneth the teeth.

K   The Raſpis is thought to be like the Bramble in temperature and vertues, but not ſo much binding or drying. The Raſpis, ſaith *Dioſcorides*, performeth thoſe things which the Bramble doth. The fruit is good to be giuen to thoſe that haue weake and queaſie ſtomackes.

Chap. 5.   *Of Holly Roſes, or* Ciſtus.

¶ *The Deſcription.*

Ciſtus hath beene taken of diuers to be a kinde of Roſe : the old Writers haue made two ſorts thereof, male and female; and likewiſe a third ſort, which is called *Ledum* : the later Herbariſts haue diſcouered diuers more, as ſhall be declared.

¶ *A generall Deſcription, wherein all the ſorts of Ciſtus are compriſed.*

Ciſtus and his kindes are wooddy ſhrubs full of branches, of the height of two or three cubits : ſome haue broad leaues, others rough, vneuen, wrinkled, ſomewhat downy, and moſt like the leaues of Sage ; although ſome haue the leaues of Roſemary, others the forme of thoſe of the Poplar tree : the floures grow on the tops of the branches, like vnto the wild Roſe, yet ſuch as very quickely vade, periſh, and fall away : thoſe of the male are moſt of a reddiſh blew or purple colour; and of the female white : in their places come vp little heads or knops ſomewhat round, in which is contained ſmall ſeed : the roots of them all are wooddy.

There

There groweth vp fometimes vnder the fhrub hard to the roots, a certaine excrefcence or hypo-cift, which is thicke, fat, groffe, full of juyce, without leaues, wholly confifting of many little cafes or boxes, as do thofe of Henbane or of the Pomegranate tree; of a yellowifh red colour in one kind, and in another white, and in certaine other greene or graffie, as Diofcorides faith.

¶ The Defcription.

1   THe firft kinde of *Ciftus* groweth vp like a fmall bufh or fhrub, of a wooddy fubftance, three or foure cubits high, garnifhed with many fmall and brittle branches, fet full of crumpled or rugged leaues very like vnto Sage leaues : at the top of the branches come floures of a purple colour, in fhape like vnto a fingle Brier Rofe, hauing leaues fomewhat wrinkled like a cloth new dried before it be fmoothed, and in the midft a few yellow chiues or thrums : the floures for the moft part do perifh and fall away before noone, and neuer ceafe flouring in fuch ma-ner from the moneth of May vnto the beginning of September, at which time the feed is ripe, be-ing of a reddifh colour, and is contained in an hard hairy huske not much vnlike the huske of Hen-bane.

1 *Ciftus mas anguftifolius.*
The male Holly Rofe.

2 *Ciftus mas cum Hypociftide.*
The male Holly Rofe with his excrefcence.

2   The fecond fort of *Ciftus*, being another kinde of the male *Ciftus*, which *Pena* calls *Ciftus mas cum Hypociftide*, is like vnto the former, but that from the root of this kinde there commeth a cer-taine excrefcence or out-growing, which is fometimes yellow, fometimes greene, and fometimes white ; from which is drawne by an artificiall extraction a certaine juyce called in fhops *Hypociftis*.

3   This kinde of *Ciftus* hath many wooddy ftalkes diuided into diuers brittle branches of a ruf-fet colour : whereon doe grow rough leaues fomewhat cut or toothed on the edges, and of an ouer-worne colour : the floures grow on the tops of the branches, in forme of a Muske Rofe, but of an ex-cellent bright purple colour : after which come round knops, wherein is contained fmall reddifh feed : the root is tough and wooddy.

4   This fourth fort of *Ciftus* hath diuers wooddy branches, whereon are fet, thicke thruft toge-ther, diuers fmall leaues narrow like thofe of Winter Sauorie, but of an ouerworne ruffet colour : the root and floures are like the precedent,

3 *Ciſtus mas dentatus.*
Toothed or ſnipt male Ciſtus.

4 *Ciſtus mas tenuifolius.*
Thin leafed Ciſtus.

5 *Ciſtus fœmina.*
The female Ciſtus.

7 *Ciſtus folio Halimi:*
Ciſtus with leaues like Sea Purſlane.

5   The firſt of the females is like vnto the male Ciſtus in each reſpeſt, ſauing that the floures hereof are of a white colour, with diuers yellow thrums in the middle, and the others purple, wherin conſiſteth the difference.

6   The ſecond female of *Matthiolus* deſcription hath many hard and wooddy ſtalks branched with diuers armes or wings ; whereon are ſet by couples rough hoary and hairy leaues of a dark ruſſet colour, among which come forth ſmall white floures like vnto thoſe of the Iaſmin : the root is rough and wooddy.   ‡ This I iudge all one with the former, and therefore haue omitted the figure as impertinent, although our Aurhour followed it, making the floure ſo little in his Deſcription. ‡

† 7   The ſeuenth ſort of Ciſtus growes vp to the height of a ſmall hedge buſh, hauing diuers brittle branches full of pith : whereon are ſet leaues by couples like thoſe of ſea Purſlane, that is to ſay, ſoft, hoary, and as it were couered oüer with a kinde of mealineſſe : the floures are yellow, and leſſe than thoſe of the former.

8 *Ciſtus folio Lavandula.*
Lavander leaued Ciſtus,

9 *Ciſtus folio Thymi.*
Ciſtus with the leaues of Tyme.

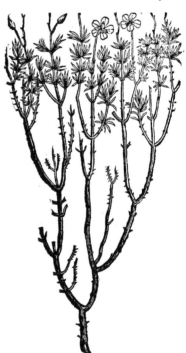

8   The eighth Ciſtus hath likewiſe ſhrubby ſtalks in manner of a hedge tree, whereon grow at certain diſtances diuers leaues cloſe ioined together at the ſtalke like thoſe of the former, but ſomwhat lower and narrower : the floures we haue not expreſt in the figure, for that we haue no certain knowledge of them.

9   This ninth Ciſtus is likewiſe a wooddy ſhrub ſome foot high : the ſtalks are very brittle, as are all the reſt of his kinde, whereon grow very ſmall leaues like thoſe of Tyme : the floures are white, which maketh it one of the females.

10   The low or baſe Ciſtus with broad leaues growes like a ſmal ſhrub, of a woody ſubſtance; the leaues are many, of a darke green colour : the floures are in forme like the other, but of a yellow colour : the roots are likewiſe wooddy.

11   This narrow leafed low Ciſtus hath diuers tough branches leaning to the ground, whereon do grow without order many ſmall narrow leaues ſomwhat long, of a gummy taſte at the firſt, afterwards bitter : the floures grow on the tops of the branches , of a yellow colour, conſiſting of fiue leaues, with certaine chiues in the middle : after which follow three ſquare cods or ſeed-veſſels : the root is tough and wooldy.

10 *Ciſtus humilis latifolius.*
Low Ciſtus with broad leaues.

11 *Ciſtus humilis anguſtifolius.*
Low Ciſtus with narrow leaues.

12 *Ciſtus humilis Auſtriaca Cluſij.*
Low Ciſtus of Auſtria.

13 *Ciſtus humilis Serpillifolio.*
Low Ciſtus with leaues like wilde Tyme.

12 The low or baſe Ciſtus of Auſtria groweth likewiſe leaning to the ground, hauing many wooddy branches very firme and tough, couered with a blackiſh barke ; whereon grow very many rough and hairy leaues in ſhape like thoſe of the ſmall Myrtle, of a ſhining green on the vpper ſide and of an aſtringent taſte : on euery branch ſtandeth one floure, ſeldom two, in form like the other, but conſiſting of one leafe deeply diuided into fiue parts, of a fleſhy white colour.

13 This low ſort of Ciſtus hath many tough branches trailing vpon the ground, of a reddiſh colour, whereon grow ſmall leaues like thoſe of wilde Tyme, of a dark green colour, very thick and fat, and ſomewhat hairy : the floures grow at the top of the branches, of a gold yellow colour, conſiſting of fiue ſmall leaues of a very ſweet ſmell : the root is thicke, hard, and wooddy.

14 This ſtrange and rare plant of *Lobels* obſeruation I haue thought meet to inſert among the kindes of Ciſtus, as a friend of theirs, if not one of the kinde : it hath leaues like the male Ciſtus, (the firſt in this chapter deſcribed) but more hairy, bearing at the top of his branches a ſmall knop in ſhape like a rotten Strawbery, but not of the ſame ſubſtance, for it is compact of a ſcaly or chaffie matter, ſuch as is in the midſt of the Camomill floures, and of a ruſſet colour.

14 *Ciſtus exoticus Lobelij.*
  Lobels ſtrange Ciſtus.

16 *Myrtociſtus Th. Pennei Angli.*
  Dᵣ *Penny* his Ciſtus.

15 The adulterine or counterfeit Ciſtus growes to the height of a hedge buſh : the branches are long and brittle, whereon grow leaues like thoſe of the willow, of an ouerworn ruſſet colour : the floures are ſmall, conſiſting of fiue little yellow leaues : the whole plant being well viewed ſeemes to be a willow, but at the firſt ſight one of the Ciſtus, ſo that it is a plant participating of both : the root is wooddy. ‡ *Bauhine* iudges this (which our Author out of *Tabern.* figured and named *Ciſtus adulterinus*) to be the Ciſtus ſet forth in the eight place of the next chapter ſaue one : but I rather iudge it to be the *Ledum Sileſiacum* ſet forth in the eleuenth place of that chapter, and again in the twelfth, where you may finde more thereof. ‡

16 This kinde of Ciſtus, which Dᵣ *Penny* a famous Phyſition of London deceaſed did gather vpon the Iſland of Majorca, and called by the name μυρτοκιστ, in Latin, *Myrtociſtus Balearica*, is a ſhrub growing to the height of three cubits, hauing a very rough bark, beſet round about with rough and ſcabbed warts, which barke wil of it ſelfe eaſily fall away from the old branches or boughes of the tree.

tree. The leaues of this tree are almoſt like them of *Myrtus*, very rough vnderneath like the branches aforeſaid; but the leaues that grow higher, and towards the top of the branches, are ſmooth, growing about the branches very thicke together, as in the other kinds of Ciſtus. The floures are yellow, growing on the top of the twigs, conſiſting of fiue long leaues ful of many very long chiues within. When the floures be vaded, there followeth a very long and fiue ſquare head or husk ful of ſeed. The whole tree is very ſweet, out of which iſſueth a gum or roſin, or rather a thicke clammie and fat juice, ſuch as commeth forth of the kindes of *Ledum*.

17   This annuall Ciſtus groweth vp from ſeed with one vpright ſtalk to the height of a cubit, oft times diuided into other ſmal branches: whereon grow rough leaues ſomewhat long, of a dark green colour: the floures grow at the top of the ſtalks, conſiſting of fiue ſmall yellow leaues: which being paſt, there followes a three ſquare ſeed veſſel full of ſmall reddiſh ſeed. The root is ſmal and wooddy, and periſheth when the ſeed is perfeðed.

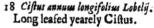

17 *Ciſtus annuus.*
Ciſtus laſting one yeare.

18 *Ciſtus annuus longifolius Lobelij.*
Long leafed yearely Ciſtus.

18   This other Ciſtus which laſteth but a yeare, hath long ſtalks diuided into other branches of the height of two cubits; whereon grow long rough leaues ſet three together at certain diſtances, the middlemoſt whereof is longer than the other two: the flours grow on the ſides of the branches like the female Ciſtus, of a white colour: the root is of a wooddy ſubſtance, as are all the reſt of his kinde.

‡ 19   This growes ſome foot high, with a ſquare rough greeniſh ſtalke, whereon by couples at certain ſpaces ſtand little longiſh rough leaues, yet toward the top of the ſtalke they ſtand ſomtimes three together: vpon the top of the little branches grow floures like thoſe of the other Ciſtus, of colour yellow, with a fine ſanguine ſpot vpon each leafe of the floure. It groweth in ſome parts of France, as a'ſo on the Alps in Italy. *Cluſius* deſcribes it by the name of *Ciſtus annuus 2. Pona* in his *Mons Baldus* :alls it *Ciſtus annuus flore guttato.*

20   This hath many ſlender branches whereon grow ſmall roundiſh leaues, hoary, and ſomewhat like thoſe of Marjerome, ſomewhat leſſe, with the middle rib ſtanding out. The floures grow vpon the tops of the branches, and conſiſt of fiue white leaues, with a darke purple ſpot in the middle of each leafe: the threads in the middle of the floure are of a yellow colour: the ſeed-veſſels

are

are of the bignesse of those of flax, but three square, containing a seed of the bignesse of that of Henbane. *Clusius* found this in diuers parts of Spaine, and sets it forth by the name of *Cistus folio Sampsuchi.* ‡

‡ 19 *Cistus annuus flore maculato.*
Spotted annuall Cistus.

‡ 20 *Cistus folio Sampsuchi.*
Marjerome leafed Cistus.

¶ *The Place.*

Holly Roses grow in Italy, Spain, and Languedoc, and in the countries bordering vpon the riuer Padus, in all Hetruria and Massiles, and in many other of the hotter prouinces of Europe, in dry and stony places, varying infinitely according to the diuersitie of the regions where they grow. Of which I haue two sorts in my garden, the first, and the *Cistus annuus*.

¶ *The Time.*

They floure from May to September.

¶ *The Names.*

The Holly Rose is called in Greeke, κίσος, and κίσθος: in Latine also *Cistus*, and *Rosa syluatica*: of diuers, *Rosa Canina*, as *Scribonius Largus* writes, but not properly: in Spanish, *Estepa*: of the Portugals *Rosella*: in English, Holly Rose, and Cistus after the Greek name. The fungous excrescence growing at the root of Cistus, is called in Greeke ὑποκισθίς, because it growes vnder the shrub Cistus: it is also called *Limodoron*: some call it *κίσος*, among which is *Paulus Ægineta*, who also doth not call that *Hypocistis* which groweth vnder the shrub Cistus, but the juice hereof: whereupon might grow the word *Hypocistis*, by which name the Apothecaries call this juice when it is hardned: of some it is called *Erithanon*, *Citinus*, and *Hypoquistidos*.

¶ *The Nature.*

Cistus, as *Galen* saith, doth greatly dry, almost in the second degree, and it is of that coldnes that it hath withall a temperat heate: the leaues and the first buds being beaten do only dry and bind, in such sort as they may close vp vlcers, and joine together green wounds.

¶ *The Vertues.*

The floures are of most force, which being drunke with wine are good against the bloudy flixe, weaknesse of the stomacke, fluxes, and ouerflowing of moist humors. A

They cure putrified vlcers, being applied in manner of a pultis. *Dioscorides* teacheth, That they B are a remedie for eating vlcers, called in Greeke νομή, being anointed therewith; and that they cure burnings, scaldings, and old vlcers.

C   *Hypociſtis* is much more binding : it is a ſure remedie for all infirmities that come of fluxes , as
voiding of bloud,the whites,the laske,and the bloudy flix : but if it be requiſit to ſtrengthen that
part which is ouerweakned with a ſuperfluous moiſture , it doth notably comfort and ſtrengthen
the ſame.

D   It is excellent to be mixed with fomentations that ſerue for the ſtomack and liuer.

E   It is put into the Treacle of Vipers, to the end it ſhould comfort and ſtrengthen weake bodies,
as *Galen* writeth.

CHAP. 6.

Of other Plants reckoned for dwarfe kindes of Ciſtus.

1. 2. *Helianthemum Anglicum luteum vel album.*
English yellow or white dwarfe Ciſtus.

¶ *The Deſcription.*

1   THe Engliſh dwarfe Ciſtus,called of
Lobel,Panax Chironium (but there is
another *Panax* of *Chirons* deſcription
which I hold to be the true and right *Panax,* notwithſtanding hee hath inſerted it amongſt the
kindes of Ciſtus , as being indifferent to joyne
with vs and others for the inſertion)is a low and
baſe plant creeping vpon the ground , hauing
many ſmall tough branches of a browne colour;
wherupon grow little leaues ſet together by couples,thicke,fat,and ful of ſubſtance,and couered
ouer with a ſoft downe ; from the boſome whereof come forth other leſſer leaues : the floures before they be open are ſmall knops or buttons, of
a browne colour mixed with yellow, and beeing
open and ſpred abroad are like thoſe of the wilde
Tanſie,& of a yellow colour,with ſome yellower
chiues in the middle : the root is thicke,and of a
wooddy ſubſtance.

2   The ſecond is very like to the precedent,
ſauing that the leaues are long,and do not grow
ſo thicke thruſt together, and are more woolly.
the flours are greater,and of a white color,wher-
in the eſpeciall difference conſiſteth.The root is
like the former.

3 *Helianthemum luteum Germanicum.* The yellow dwarfe Ciſtus of Germany.

3   There is found in Germanie a certaine plant like to Ciſtus and *Ledon*,but much leſſer, c. ... ping vpon the ground vnleſſe it be propped vp,hauing a multitude of twiggy branches ſlender and fine,whereupon grow leaues leſſer than thoſe of Ledon or Ciſtus, very like to that of our Engliſh white dwarfe Ciſtus,of a full ſubſtance, ſlightly haired,wherein is contained a tough juice. The floures are ſmall like little Roſes or the wilde Tanſie,of a yellow colour : the roots be ſlender,woody,and ſomething red.

4 *Helianthemum album Germanicum*.   The white dwarfe Ciſtus of Germanie.

5 *Helianthemum Sabaudicum.*
   The dwarfe Ciſtus of Savoy.

6 *Helianthemum anguſtifolium.*
   Narrow leafed dwarfe Ciſtus.

4   This differeth not from the laſt deſcribed,ſauing that the floures hereof are very white,and the others yellow,wherein they eſpecially differ.

5   The dwarf Ciſtus of Savoy hath diuers tough branches of a reddiſh colour, very tough and woody, diuided into diuers other branches ; whereon are ſet ſmal leaues foure together, by certain ſpaces: the floures grow at the top of the branches like thoſe of our yellow dwarfe Ciſtus, of a yellow colour: the root is very wooddy.

6   This dwarfe Ciſtus with narrow leaues hath very many ſmall flexible branches, of a brown colour, very ſmooth, and ramping vpon the ground; whereon grow ſmall long narrow leaues like like thoſe of Tyme of Candy; from the boſome whereof come forth diuers other ſmaller leaues: the floures grow on the tops of the branches, of a bleake yellow colour: the root is likewiſe woody.

‡ 7   To theſe I may fitly adde two more: the firſt hath creeping ſtalks ſome foot or two long, blackiſh, and diuided into ſundry ſmaller branches: the leaues grow thicke and many together, ſet by couples, though the figure do not well expreſſe ſo much: theſe leaues are ſmall, of the bigneſſe of thoſe of Tyme, thicke, green aboue, whitiſh vnderneath, and of a bitter taſte: at the ends of the branches grow two or foure floures neere together, very ſmall, compoſed of fiue little leaues of a kinde of fleſh colour: to theſe ſucceed heads opening themſelues when they come to ripeneſſe into fiue parts, and containing a very ſmal ſeed: the root is hard and wooddy, ſending out certaine fibres; alſo the branches here and there put forth ſome fibres. This plant dried hath a pretty pleaſant ſmell. It growes vpon the higheſt Auſtrian and Styrian Alps, and is ſet forth by *Cluſius* by the name of *Chamæciſtus ſeptimus.*

‡ 7 *Chamæciſtus ſerpillifolius.*   ‡ 8 *Chamæciſtus Friſicus.*
Tyme leafed dwarfe Ciſtus.   Friſian dwarfe Ciſtus.

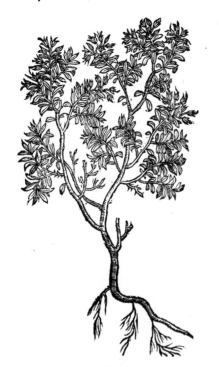 

8   The ſame Author alſo in his *Curæ Poſteriores* giues vs the hiſtorie of this, which he receiued with ſome other rare plants from *Iohn Dortman* a famous and learned Apothecary of Groeningen. This little plant is in leafe and root almoſt like and neere of the ſame bigneſſe with the Celticke Nard, yet the ſtalkes are vnlike, which are ſmall, ſet with a few longiſh leaues, and at the tops they carry fiue or ſix pretty floures like thoſe of Crow-feet, conſiſting of ſix leaues apiece, of a yellow colour, yet with ſome few ſpots of another colour, and theſe ſet in a double ring about the middle: after theſe follow heads or ſeed-veſſels with forked tops, filled with a chaffie ſeed: the whole plant ſmels ſomewhat ſtrong. It growes together with *Gramen Pernaſſi* in rotten mooriſh places about a village in the country of Drent. *Dortman* called this, *Hirculus Friſicus: Cluſius* addes, *qui Chamæciſti genus.* ‡

¶ *The Place.*

Their seuerall titles haue touched their natiue countries; they grow in rough dry and sunny places in plain fields and vpon mountains.

Those of our English growing I haue found in very many places, especially in Kent vpon the chalky banks about Grauesend, Southfleet, and for the most part all the way from thence to Canturbury and Douer.

¶ *The Time.*

They floure from Iuly to the end of August.

¶ *The Names.*

Tragus calls dwarfe Cistus in the high-Dutch tongue 𝕳𝖊𝖞𝖉𝖊𝖓 𝕳𝖞𝖘𝖔𝖕𝖊: in Latine, *Gratia Dei*: but there is another herb called also of the later herbarists *Gratia Dei*, which is *Gratiola: Valerius Cordus* nameth it *Helianthemum*, and *Solis flos* or Sun-floure: of *Clusius*, *Chamæcistus*, or dwarfe Cistus.

Pliny writeth, that *Helianthemum* growes in the champian country Temiscyra in Pontus, and in the mountains of Cilicia neere the sea: saying further, that the wise men of those couutries & the Kings of Persia do anoint their bodies herewith, boiled with Lions fat, a little Saffron, and Wine of Dates, that they may seem faire and beautifull; and therefore haue they called it *Heliocaliden*, or the beauty of the Sun. *Matthiolus* saith, that *Helianthemum* is taken of some to be *Panaces Chironium* or *Chirons* All-heale: but it is nothing likely, as we haue said.

¶ *The Nature and Vertues.*

The faculties and temperature are referred to the kindes of Cistus, for it healeth wounds, stancheth bloud, and stoppeth the spitting of bloud, the bloudy flix, and all other issues of bloud.

The same boiled in wine healeth vlcers in the mouth and privy parts, if they be washed therewith: to be briefe, it ioineth and strengthneth: which things do plainly and euidently shew, That it is not only like to Cistus and Ledon in forme, but in vertues and faculties also, and therefore it is manifest that it is a certain wilde kinde of Cistus and Ledon.

CHAP. 7. *Of Cistus Ledon and Ladanum.*

¶ *The Kindes.*

THere be diuers forts of Cistus, whereof that gummy matter is gathered called in shops *Ladanum*, and *Labdanum*, but vnproperly.

¶ *The Description.*

1　CIstus Ledon is a shrub growing to the height of a man, and sometimes higher, hauing many hard wooddy branches, couered with a blackish barke; whereupon grow leaues set together by couples, one right against another like vnto wings, of an inch broad, of a black swart green on the vpper sides, and whitish vnderneath: whereon is gathered a certain clammic transparent or through-shining liquor, of a very hot sweet smel, which being gathered & hardned is that which in shops is called *Labdanum*: the floures grow at the ends of the branches like little roses, consisting of fiue white leaues, euery one decked or beautified toward the bottome with pretty darke purplish spots tending to blackenesse, hauing in the middle very many yellow chiues such as are in the middle of the Rose; after come the knaps or seed-vessels full of most small reddish seed: the whole plant being dried groweth somewhat whitish, and of a pleasant smell, which it retaineth many yeares.

2　The second groweth likewise to the height of an hedge bush, the branches are long and very fragil or easie to breake, whereon grow leaues greener than any other of his kind, yet vnderneath of a hoary colour, growing toward winter to be somewhat reddish, of a soure and binding taste. The flours are like the precedent, the forme whereof the Grauer hath omitted, in other respects like the former.

3　The third sort of Cistus Ledon groweth vp to the height of a small hedge bush, hauing many twiggy branches, whereon grow leaues like those of the Poplar tree, sharp at the point, couered ouer with that clammy dew that the others are: the floures grow at the tops of the branches, of a white colour like the precedent.

4　The

1 *Cistus Ledon* 1 *Clusij*.
The first Cistus bringing *Ladanum*.

2 *Cistus Ledon* 2 *Clusij*.
The second gum Cistus.

3 *Cistus Ledon populea fronde*.
Cistus Ledon with leaues like Poplar.

4 *Sistus Ledon* 4 *Clusij*.
Cistus Ledon 4 of *Clusius*.

5 *Ciſtus Ledon* 5.*Cluſij*.
The fift Ciſtus Ledon.

6 *Ciſtus Ledon* 6.*Cluſij*.
The ſixth Ciſtus Ledon.

7 *Ciſtus Ledon* 7.*Cluſij*.
The 7. Ciſtus Ledon.

8 *Ciſtus Ledon cum Hypociſtide Lobelij*.
The 8. Ciſtus Ledon, with his excreſcenſe.

9 Ciſtus Ledon 10 Cluſij.
The tenth Ciſtus Ledon.

10 Ciſtus Ledon Myrtifolium.
Ciſtus Ledon with leaues like Myrtle.

11 Ciſtus Ledum Sileſiacum.
The Polonian Ciſtus Ledon.

4 This fourth of *Cluſius* deſcription groweth likewiſe to the height of a ſhrubby buſh,
hauing many branches flexible, hoary, and
hairy : the leaues are like the reſt of his kind,
but ſofter, more hairy, of a ſwart green colour,
daſht ouer with that dewy fatneſſe not onely
in the Spring time, but in the heate of Summer alſo : the floures are white with yellow
thrums in the middle : the reſt anſwereth the
laſt deſcribed.

5 The fifth growes vp like a hedge buſh
with many tough branches, whereon are ſet
long rough leaues hoary vnderneath, ſomewhat daſhed ouer with that fatty dew or humor that the reſt are poſſeſſed of : the floures
are likewiſe of a white colour, with certaine
yellow chiues in the middle. The root is
wooddy.

† 6 The ſixt hath diuers ſmal branches
couered with a blackiſh barke: the floures are
ſet together at the toppes of the branches by
certaine ſpaces : they are yellow, and like the
former in each reſpect.

7 This is a low ſhrub growing to the height
of two cubits, hauing many branches couered with a bark of the colour of aſhes, whereon are confuſedly ſet diuers leaues at certain
diſtances, ſmall, narrow, like thoſe of Winter
Sauorie, of an ouerworne ruſſet colour, verie
thicke, fat, and glutinous: the flours are white,
and differ not, nor the ſeed from the reſt.

                  8 The

8   The eighth groweth vp like a little hedge bush, hauing leaues like the common female Ciftus, fauing that thofe of this plant are fprinkled ouer with that clammy moifture, and the other not fo: the floures and feed are alfo like. From the root of this plant commeth fuch like excrefcence called *Limodoron, Orobanche*, or *Hypociftis*, as there doth from the firft male Ciftus, wherein it differeth from all the reft vnder the name Ledon.

9   The ninth hath diuers brittle ftalkes of an afh colour tending to a ruffet; whereon are fet very may leaues like thofe of Thyme, of an ouerworne colour: the floures are white, with certaine yellow chiues in the middle, which the grauer hath omitted in the figure.

10   The tenth groweth vp like a fmall fhrub, hauing brittle ftalkes, couered with a blackifh barke, and diuided into diuers branches; whereon are fet vpon fhort truncheons or fat foot-ftalkes, foure or fiue like thofe of the Myrtle tree, of a ftrong fmell: the floures are likewife of a white colour.

12 *Ciftus Ledum Rorifmarini folio.*
Ciftus Ledon with leaues like Rofemarie.

13 *Ciftus Ledum Matthioli.*
Ciftus Ledon of *Matthiolus* defcription.

11. 12.   The twelfth kinde of Ciftus Ledon groweth vpright with a ftraight body or ftocke, bringeth at the top many fmall twigs or rods of a cubit long, couered with a barke of the colour of afhes, which diuide themfelues into other branches, of a purplifh colour, befet with long and narrow leaues, not much vnlike to Rofemary, but longer; of a greene colour aboue, but vnderneath hauing as it were a long rib, made or compact of wooll or downe; of a fweet and pleafant fmell, and fomewhat fharpe in tafte: on the tops of the branches grow knops or heads compact as it were of many fcales, of an iron or ruffet colour: out of which commeth and proceedeth a certaine round and long mane, or hairy panickled tuft of floures, with many long, tender, greene, and fomewhat woolly ftalkes or twigs growing vnto them, of a fweet fent and fmell: the floures confift of fiue little white leaues, within which are contained ten white chiues with a long ftile or pointal in the midft of the floure: when the floures be vaded, there fucceed long knops or heads which are fiue cornered, in fhape and bigneffe like vnto the fruite and berries of *Cornus*; which being greene, are befpeckled with many filuer fpots, but being ripe, are of a red colour; containing within them a long yellow feed, which is fo fmall and flender, that it is like to the duft or pouder that falleth out of wormeholes. ‡ This is the *Ledum Silefiacum* of *Clufius*; and the *Ledum Rorifmarini folio* of *Tabernamontanus*: it is alfo the *Rofmarinum fylueftre* of *Matthiolus*; and *Chamæpeuce* of *Cordus*: and I am

Qqqqq   deceiued

deceiued if the figure which *Tabernamontanus* and our Author out of him gaue by the name of *Ci-ſtus adulterinus*, were not of this. ‡

13   Among the ſhrubby buſhes comprehended vnder the title of *Ciſtus Ledum*, *Matthiolus* hath ſet forth one, whereof to write at large were impoſſible, conſidering the Author is ſo briefe, and of our ſelues we haue not any acquaintance with the plant it ſelfe: *Dioſcorides* to helpe what may be, ſaith, that it is a ſhrub growing like vnto the ſtocke or kindred of the *Ciſti*: from whoſe leaues is gathered a clammy dew which maketh that gummie matter that is in ſhops called *Lapdanum*: it groweth, ſaith he, in hot regions (but not with vs :) the Mauritanians call the juyce or clammy mat-ter, *Leden*, and *Laden*: of ſome, *Ladano*, and *Odano*: in Spaniſh, *Xara*: and further ſaith, it groweth in Arabia, where the buſh is called *Chaſus*: thus much for the deſcription. ‡ Our Author here ſeems to make *Dioſcorides* to comment vpon *Matthiolus*, which ſhewes his learning, and how well he was exerciſed in reading or vnderſtanding any thing written of Plants. But of this enough; The Plant here figured which *Matthiolus* iudges to be the true *Ledon*, or *Ciſtus Ladanifera* of *Dioſcorides*, hath large ſtalkes and branches, whereon grow very thicke leaues, broad alſo and long, with the nerues running alongſt the leaues: the floures of this conſiſt of fiue white leaues, and the ſeed is contained in a three cornered ſeed veſſell. ‡

14 *Ciſtus Ledum Alpinum Cluſij.*   ‡ 15 *Ciſtus Ledon folijs Roriſmarini.*
  The Mountaine Ciſtus.   Roſemary leaued Ciſtus Ledon.

4   The foureteenth Ciſtus, being one of thoſe that doe grow vpon the Alpiſh mountaines, which *Lobel* ſetteth downe to be *Balſamum alpinum* of *Geſner*: notwithſtanding I thinke it not amiſſe to inſert it in this place, hauing for my warrant that famous Herbariſt *Carolus Cluſius*: this plant is one of the beautifulleſt, differing in very notable points, yet reſembling them in the wooddy bran-ches and leaues: it riſeth vp hauing many weake branches leaning to the ground, yet of a wooddie ſubſtance, couered ouer with an aſh-coloured barke: the leaues are broad, and very tough, of a ſhi-ning greene colour, and a binding taſte: the floures grow at the tops of the branches like little bels, hanging downe their heads, diuided at the lips or brims into fiue diuiſions, of a deep red colour on the out ſide, and daſht ouer here and there with ſome ſiluer ſpots; on the inſide of a bright ſhining red colour, with certaine chiues in the middle, and of a very ſweet ſmell, as is all the reſt of the plant; after which come ſmall heads or knaps, full of ſeed like duſt, of a very ſtrong ſmell, making the head of them to ake that ſmel therto: the root is long, hard, and very wooddy: oftentimes there is
                           found

found vpon the trunke or naked part of the ſtalkes certain excreſcences, or out-growings in manner of galls, of a fungous ſubſtance, like thoſe of Touchwood, white within, and red without, of an aſtringent or biting taſte.

‡ 15 This growes ſome cubit or better high, and hath long narrow glutinous leaues like in ſhape to thoſe of Roſemarie, ſet by couples but not very thicke: the branches whereon the floures doe grow are ſlender, and the ſeed veſſels are diuided into fiue parts as in other plants of this kinde. This Claſius found in Spaine, and ſets forth for his *Ledum nonum.* ‡

¶ The Place.

Ciſtus Ledon groweth in the Iſland of Candie, as *Bellonius* doth teſtifie, in vntilled places euery where: it is alſo found in Cyprus, as *Pliny* ſheweth, and likewiſe in many places of Spaine that lie open to the Sun: moreouer, both the forme and bigneſſe of the leaues, and alſo of the plants themſelues, as well of thoſe that bring forth *Ladanum,* as the other Ciſtus, do vary in this wonderful maner, according to the diuerſitie of the places and countries where they grow: they are ſtrangers in theſe Northernly parts, being very impatient of our cold clymate.

¶ The Time.

They floure for the moſt part from May to the end of Auguſt: the clammy matter which falleth vpon the leaues, which is a liquid kinde of Roſen of a ſweet ſmell, is gathered in the Spring time, as *Dioſcorides* ſaith: but as *Petrus Bellonius* affirmeth (being an eye witnes of the gathering) in the midſt of ſummer, and in the extreme heat of the Dog-daies, the which in our time not without great care and diligence, and as great labour, is gathered from the whole plant (with certain inſtruments made in manner of tooth-pickes, or eare-pickes, which in their tongue they call *Ergaſtiri*) and not gathered from the beards of Goats, as it is reported in the old fables of the lying Monks themſelues, called *Calohieros,* that is to ſay Greekiſh Monkes, who of very mockery haue foiſted that fable among others extant in their workes.

‡ I thinke it not amiſſe for the better explanation of the matter here treated of, as alſo to ſhew you after what manner our Authour in diuers places gaue the teſtimony of ſundry Writers, and how well hee vnderſtood them, here to ſet downe in Engliſh the words of *Bellonius* concerning the gathering of *Ladanum,* which are theſe. [The Greekes (ſaith he) for the gathering of *Ladanum,* prouide a peculiar inſtrument which in their vulgar tongue they terme *Ergaſtiri:* This is an inſtrument like to a Rake without teeth, to this are faſtened ſundry thongs cut out of a raw and vntanned hide, they gently rub theſe vpon the *Ladanum* bearing ſhrubs, that ſo the liquid moiſture concrete about the leaues may ſticke to them, which afterwards with kniues they ſhaue off theſe thongs in the heat of the day. Wherefore the labour of gathering *Ladanum* is exceeding great, yea intollerable, ſeeing they muſt of neceſſitie ſtay in the mountaines all the day long in the greateſt heat of the Dog-daies: neither vſually ſhall you finde any other who will take the paines to gather it; beſides, the *Calohieroi,* that is, the Greeke Monkes. It is gathered no where in the whole Iſland of Candy in greater plenty, than at the foot of the mountaine Ida at a village called Cogualino, and at Milopotamo. ‡]

¶ The Names.

The ſhrub it ſelfe is called in Greeke, λήδον, or λῆδος: the Latines keepe the name *Ledon* or *Ladon,* and is a kinde of *Ciſtus* or Holly Roſes: the fat or clammie matter which is gathered from the leaues, is named *Ladanon* and *Ledanon,* according to the Greeke: the Apothecaries corruptly call it *Lapdanum: Dioſcorides* counteth that to be the beſt which is ſweet of ſmell, and ſomewhat greene, that eaſily waxeth ſoft, is fat, without ſand, and is not eaſily broken, but very full of Roſin or gumme.

¶ The Temperature.

Ladanum, ſaith *Galen,* is hot in the later end of the firſt degree, hauing alſo a little aſtrictiue or binding quality, it is likewiſe of a thin ſubſtance, and therefore it ſofteneth, and withall doth moderately digeſt, and alſo concoct.

¶ The Vertues.

Ladanum hath a peculiar property againſt the infirmities of the mother, it keepeth haires from falling; for it waſteth away any ſetled or putrified humor that is at their roots.   A

Dioſcorides ſaith, That *Ladanum* doth binde, heate, ſouple, and open, being tempered with wine, Myrrhe, and oyle of Myrtles; it keepeth haires from falling, being annointed therewith; or laied on mixed with wine, it maketh the markes or ſcars of wounds faire and well coloured.   B

It taketh away the paine of the eares if it be poured or dropped therein, mixed with honied water, or with oyle of Roſes.   C

A fume made thereof draweth forth the after-birth, and taketh away the hardneſſe of the matrix.   D

E   It is with good ſucceſſe mixed with mollifying plaiſters that mitigate paine.

F   Being drunke with wine, it ſtoppeth the laske and prouoketh vrine.

G   There is made hereof diuers ſorts of Pomanders, chaines, and bracelets, with other ſweets mixed therewith.

CHAP. 8. *Of Roſemary.*

¶ *The Deſcription.*

1   ROſemarie is a wooddy ſhrub, growing oftentimes to the height of three or foure cubits, eſpecially when it is ſet by a wall : it conſiſteth of ſlender brittle branches, whereon do grow very many long leaues, narrow, ſomewhat hard, of a quicke ſpicy taſte, whitiſh vnderneath, and of a full greene colour aboue, or in the vpper ſide, with a pleaſant ſweet ſtrong ſmell ; among which come forth little floures of a whitiſh blew colour : the ſeed is blackiſh : the roots are tough and wooddy.

1 *Roſmarinum Coronarium.*   2 *Roſmarinum ſylueſtre.*
Garden Roſemarie.   Wilde Roſemarie.

2   The wilde Roſemary *Cluſius* hath referred vnto the kindes of Ciſtus Ledon ; we haue as a poore kinſman thereof inſerted it in the next place, in kindred or neighbourhood at the leaſt. This wilde Roſemary is a ſmall wooddy ſhrub, growing ſeldome aboue a foot high, hauing hard branches of a reddiſh colour, diuiding themſelues into other ſmaller branches of a whitiſh color: whereon are placed without order diuers long leaues greene aboue, and hoarie vnderneath, not vnlike to thoſe of the dwarfe Willow, or the common Roſemary, of a dry and aſtringent taſte, of little ſmell or none at all : the floures ſtand on the tops of the branches, ſet vpon bare and naked foot-ſtalkes, conſiſting of fiue ſmall leaues of a reddiſh colour, ſomewhat ſhining ; after which appeare little knaps full of ſmall ſeed : the root is tough and wooddy.

3   This plant grows vp like an hedge ſhrub of a wooddy ſubſtance, to the height of two or three
cubits ;

3 *Casia Poetica, Lobelij.*
The Poëts Rosemary or Gardrobe.

cubits; hauing many twiggie branches of a green colour:whereupon do grow narrow leaues like vnto *Linaria* or Toad-flax,of a bitter taste; among which come forth small mossie floures, of a greenish yellow colour like those of the Cornell tree, and of the smell of Rosemarie : which hath moued me to place it with the Rosemaries as a kind thereof,not finding any other plant so neere vnto it in kinde and neighbourhood : after the floures be past, there succeed fruit like those of the Myrtle tree,greene at the first,and of a shining red colour when they bee ripe,like Corall,or the berries of *Asparagus*,soft and sweet in taste, leauing a certaine acrimony or sharpe taste in the end : the stone within is hard as is the nut,wherein is contained a small white kernel,sweet in tast:the root is of a wooddie substance : it floureth in the Summer ; the fruit is ripe in the end of October: the people of Granade,Montpelier,and of the kingdom of Valentia,doe vse it in their presses and Wardrobes,whereupon they cal it *Guardalobo*. ‡ This in *Clusius* his time when he liued about Montpelier was called *Osyris* ; but afterwards they called it *Casia*, thinking it that mentioned by the Poët *Virgil*, the which it cannot be, for it hath no sweet smell. *Pena* and *Lobel* iudge it to be the *Casia* of *Theophrastus*,wherewith also it doth not well agree. ‡

¶ *The Place.*

Rosemary groweth in France,Spaine,and in other hot countries ; in woods, and in vntilled places : there is such plenty thereof in Languedocke, that the inhabitants burne scarce any other fuell : they make hedges of it in the gardens of Italy and England,being a great ornament vnto the same : it groweth neither in the fields nor gardens of the Easterne cold countries ; but is carefully and curiously kept in pots, set into the stoues and cellers,against the iniuries of their cold Winters.

Wild Rosemary groweth in Lancashire in diuers places, especially in a field called Little Reed, amongst the Hurtle berries, neere vnto a small village called Maudsley ; there found by a learned Gentleman often remembred in our History (and that worthily) Mr. *Thomas Hesketh.*

¶ *The Time.*

Rosemary floureth twice a yeare,in the Spring, and after in August.
The wilde Rosemary floureth in Iune and Iuly.

¶ *The Names.*

Rosemary is called in Greeke, λιβανωτις στεφανωματικη: in Latine, *Rosemarinus Coronaria* : it is surnamed *Coronaria*,for difference sake betweene it and the other *Libanotides*,which are reckoned for kindes of Rosemary,and also because women haue beene accustomed to make crownes and garlands thereof: in Italian,*Rosemarino coronario* : in Spanish,*Romero* : in French and Dutch,*Rosemarin.*
Wilde Rosemary is called *Rosemarinus syluestris* : of *Cordus*,*Chamæpeuce.*

¶ *The Temperature.*

Rosemary in hot and dry in the second degree,and also of an astringent or binding quality, as being compounded of diuers parts,and taking more of the mixture of the earthy substance.

¶ *The Vertues.*

Rosemary is giuen against all fluxes of bloud ; it is also good,especially the floures thereof, for A all infirmities of the head and braine,proceeding of a cold and moist cause; for they dry the braine, quicken the sences and memory,and strengthen the sinewie parts.

Serapio witnesseth, That Rosemary is a remedy against the stuffing of the head, that commeth B through coldnesse of the braine, if a garland thereof be put about the head, whereof *Abin Mesuai* giueth testimony.

Dioscorides teacheth that it cureth him that hath the yellow iaundice, if it be boyled in water C and drunke before exercise, and that after the taking thereof the patient must bathe himselfe and drinke wine.

The diſtilled water of the floures of Roſemary being drunke at morning and euening firſt and laſt, taketh away the ſtench of the mouth and breath, and maketh it very ſweet, if there be added thereto, to ſteepe or infuſe for certaine daies, a few Cloues, Mace, Cinnamon, and a little Anniſe ſeed.

E   The Arabians and other Phyſitions ſucceeding, do write, that Roſemary comforteth the braine the memorie, the inward ſenſes, and reſtoreth ſpeech vnto them that are poſſeſſed with the dumbe palſie, eſpecially the conſerue made of the floures and ſugar, or any other way confected with ſugar, being taken euery day faſting.

F   The Arabians, as *Serapio* witneſſeth, giue theſe properties to Roſemarie : it heateth, ſay they, is of ſubtill parts, is good for the cold rheume which falleth from the braine, driueth away windines, prouoketh vrine, and openeth the ſtoppings of the liuer and milt.

G   *Tragus* writeth, that Roſemarie is ſpice in the Germane Kitchins, and other cold countries. Further, he ſaith, That the wine boyled with Roſemarie, and taken of women troubled with the mother, or the whites, helpeth them, the rather if they faſt three or foure houres after.

H   The floures made vp into plates with Sugar after the manner of Sugar Roſet and eaten, comfort the heart, and make it merry, quicken the ſpirits, and make them more liuely.

I   The oile of Roſemary chimically drawne, comforteth the cold, weake and feeble braine in moſt wonderfull manner.

K   The people of Thuringia do vſe the wilde Roſemarie to prouoke the deſired ſickneſſe.

L   Thoſe of Marchia vſe to put it into their drinke the ſooner to make their clients drunke, and alſo do put it into cheſts and preſſes among clothes, to preſerue them from moths or other vermine.

† The vertues in the two laſt places properly belong to the *Roſemarinum ſylueſtre* of *Matthiolus*, which is the *Chamæpeuce* of *Cordus*, and is deſcribed in the 11 place of the foregoing chap. er, by the name of *Ciſtus Ledum ſileſiacum.*

Chap. 9.   *Of Vpright Wood-binde.*

1 *Periclymenum rectum Sabaudicum.*
Sauoy Honiſuckles.

2 *Periclymenum rectum Germanicum.*
Germane Honiſuckles.

 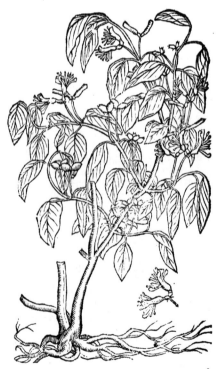

¶ The

¶ *The Deſcription.*

1   THIs ſtrange kinde of Hony-ſuckle,found in the woods of Sauoy,repreſents vnto vs that
ſhrub or hedge-buſh called *Cornus fœmina*, the Dog-berry tree, or Pricke-timber tree,
hauing leaues and branches like the common Wood-binde, ſauing that this doth not
clamber or clymbe as the others do, but contrariwiſe groweth vpright, without leaning to one ſide
or other,like a ſmall tree or hedge-buſh : the flours grow vpon the tender ſprayes or twiggie branches,by couples,not vnlike in ſhape and colour to the common Wood-bind, but altogether leſſer,
and of a white colour, hauing within the ſame many hairy chiues like the other of his kinde : after
which come red berries ioyned together by couples : the root is tough and wooddy.

2   The ſtalkes of the ſecond be oftentimes of a meane thickneſſe,the wooddy ſubſtance ſomewhat whitiſh and ſoft : the branches be round, and couered with a whitiſh barke, notwithſtanding
in the beginning when the ſprayes be young they are ſomewhat reddiſh. The leaues be long, like
thoſe of the common Hony-ſuckle, ſoft, and of a white greene : on the lower ſide they be whiter,
and a little hairy : the floures be leſſer than any of the Wood-bindes, but yet of the ſame faſhion,
and of a whitiſh colour, ioyned together by couples vpon ſeuerall ſlender foot-ſtalkes, like little
wilde Cherries,of a red colour,the one leſſer oftentimes than the other.

3 *Periclymenum rectum fructu cæruleo.*   4 *Periclymenum rectum fructu rubro.*
Vpright Wood-binde with blew berries.   Cherry Wood-binde.

3   This ſtrange kind of Wood-binde, which *Carolus Cluſius* hath ſet forth in his Pannonicke
Obſeruations, riſeth vp oftentimes to the height of a man, euen as the former doth ; which diuides
it ſelfe into many branches,couered with a rough blacke barke,that choppeth and gapeth in ſundry
clefts as the barke of the Oke. The tender branches are of a whitiſh greene colour, couered with a
woolly hairineſſe, or an ouerworne colour whereupon doe grow leaues ſet by couples one againſt
the other,like vnto the common Woodbinde,of a drying bitter taſte : the floures grow by couples
likewiſe, of a whitiſh colour. The fruit ſuccedeth, growing like little Cherries each one in his
owne foot-ſtalke,of a bright and ſhining blew colour ; which being bruiſed, doe die the hands of a
reddiſh colour,and they are of a ſharpe winie taſte,and containe in them many ſmall flat ſeeds. The
root is wooddy,diſperſing it ſelfe far abroad.

4 This

4 This kind of vpright Wood-bind groweth vp likewiſe to the height of a man, and oftentimes more high, like to the laſt deſcribed but altogether greater. The berries hereof are very blacke, wherein eſpecially is the difference. ‡ The leaues of this are as large as Bay leaues, ſharpe pointed, greene aboue, and whitiſh vnderneath, but not hairy, nor ſnipt about the edges : the floures grow by couples, of a whitiſh purple, or wholly purple : to theſe paires of floures there commonly ſucceeds but one berry, larger than any of the former, of the bigneſſe of a little cherry, and of the ſame colour, hauing two markes vpon the top thereof, where the floures ſtood. ‡

Periclymeni 3. & 4. flores.
The floures of the third and fourth.

5 Chamæpericlymenum.
Dwarfe Hony-ſuckle.

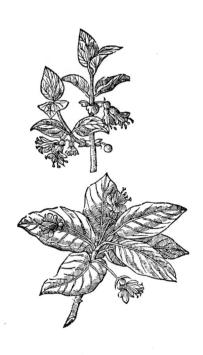

5 To the kinds of Wood-bindes this plant may likewiſe be referred, whoſe picture with this deſcription was ſent vnto Cluſius long ſince by that learned Doctor in Phyſicke Thomas Penny (of our London colledge of famous memory:) it riſeth vp with a ſtalke of a foot high; whereupon are ſet by couples faire broad leaues one right againſt another, ribbed with certain nerues like thoſe of Plantaine, ſharpe pointed, and ſomewhat hollowed in the middle like Spoone-woort : from the boſome of which leaues come forth ſmall floures, not ſeen or deſcribed by the Author: after which commeth forth a cluſter of red berries, thruſt hard together as thoſe of Aaron or Prieſts pint. The root is tough and very ſlender, creeping far abroad vnder the vpper cruſt of the earth, whereby it occupieth much ground.

¶ The Place.

Theſe plants are ſtrangers in England : they grow in the woods and mountaines of Switzerland, Germany, Sauoy, and other thoſe parts tending to the Eaſt, Eaſt North-Eaſt, and Eaſt and by South.

I haue a plant of the firſt kinde in my garden : the reſt as yet I haue not ſeene, and therefore cannot write ſo liberally thereof as I could wiſh.

‡ The dwarfe Hony-ſuckle growes in the maritime parts of Norway and Sweden, and the countries thereabout. ‡

¶ The Time.

They floure for the moſt part when the others doe, that is to ſay, in May and Iune, and their fruit is ripe in September.

¶ The

¶ *The Names.*

Vpright Wood-binde or Hony-ſuckle is called *Periclymenum ſtans*, and *Periclymenum rectum*, or vpright Wood-binde : of *Dodonæus, Xyloſteum* : in high Dutch, **Honds kiſſen**, that is to ſay, *Canum Ceraſa*, or Dog Cherries. The Engliſh names are expreſſed in their ſeuerall titles. It hath beene called *Chamæceraſus*, but not truly.

¶ *The Temperature and Vertues.*

Touching the temperature and vertues of theſe vpright Wood-bindes we haue no experience at all our ſelues, neither haue we learned any thing of others.

Chap. 10.　　*Of Sene.*

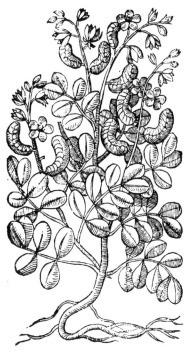

Sena folijs obtuſis.
Italian Sene.

¶ *The Deſcription.*

SEne bringeth forth ſtalkes a cubit high, ſet with diuers branches : the leaues are long, winged, conſiſting of many ſmall leaues like thoſe of Liquorice, or of baſtard Sene: the floures come forth of the bottom of the wings, of colour yellow, ſtanding vpon ſlender foot-ſtalkes; from which after the floures be gone hang forked cods, the ſame bowing inward like a halfe-Moone, plain and flat, in which are contained ſeeds like to the ſeeds or kernels of grapes, of a blackiſh colour. The root is ſlender, long, and vnprofitable, which periſheth when the leaues are gathered for medicine, and the ſeeds be ripe and muſt be ſowne againe the next yeare, euen as we do corne.

There is another kinde of Sene growing in Italy, like the other in each reſpect ſauing that it is greater, and hath not that force in purging that the other hath.

¶ *The Place and Time.*

This is planted in Syria and Ægypt, alſo in Italy, in Prouince, in France, in Languedoc. It hardly groweth in high and low Germany, neither in England : it proſpereth in hot Regions, and cannot away with cold; for that cauſe it is in Italy ſowne in May, and continueth no longer than Autumne : the beſt is brought from Alexandria and Egypt. The Arabians were the firſt that found it out.

¶ *The Names.*

The Perſians call it *Abalzemer*, as *Meſue* his copy teacheth : the Apothecaries *Sena*, by which name it was knowne to *Actuarius* the Grecian, and to the later Latines : it is called in Engliſh, Sene.

¶ *The Temperature.*

Sene is of a meane temperature, neither hot nor cold, yet inclining to heate, and dry almoſt in the third degree : it is of a purging faculty, and that by the ſtoole, in ſuch ſort as it is not much troubleſome to mans nature, hauing withall a certaine binding quality, which it leaueth after the purging.

¶ *The Vertues.*

It voideth forth flegmaticke and cholericke humors, alſo groſſe and melancholike, if it be helped with ſomething tending to that end.　A

It is a ſingular purging medicine in many diſeaſes, fit for all ages and kindes.　B

It purgeth without violence or hurt, eſpecially if it be tempered with Aniſe ſeed or other like　C
ſweet ſmelling things added or with gentle purgers or lenitiue medicines. It may be giuen in pouder, but commonly the infuſion thereof is vſed.

The

D   The quantitie of the pouder is a dram weight, and in the infuſion, foure, fiue, or more. It may be mixed with any liquor.

E   It is in the decoction or in the infuſion tempered with cold things in burning agues and other hot diſeaſes : in cold and long infirmities it is boyled with hot opening ſimples and ſuch like ; or elſe it is ſteeped in wine, in which manner as familiar to mans nature, it draweth forth gently by the ſtoole, almoſt without any kinde of paine, crude and raw humors.

F   Moſt of the Arabians commend the cods, but our Phyſitions the leaues rather ; for vnleſſe the cods be full ripe they ingender winde, and cauſe gripings in the belly. For they are oftentimes gathered before they be ripe, and otherwiſe eaſily fall away being ſhaken downe by the winde, by reaſon of their weake and ſlender ſtalkes.

G   Some alſo thinke that Sene is hurtfull to the ſtomacke, and weakneth the ſame, for which cauſe they ſay that Ginger or ſome ſweet kinde of ſpice is to be added, whereby the ſtomacke may be ſtrengthned. Likewiſe *Meſue* noteth that it is ſlow in operation, and therefore Salgem is to bee mixed with it. Moreouer, Sene purgeth not ſo ſpeedily as ſtronger medicines do.

H   Notwithſtanding it may be helped not only by Salgem, but alſo by other purging things mixed therewith, that is to ſay, with ſimple medicines, as Rubarb, Agaricke, and others ; and with compounds, as that which is called *Catholicon*, or the Electuary *Diaphœnicon*, or that which is made of the juyce of Roſes, or ſome other, according as the condition or quality of the diſeaſe and of the ſicke man requireth.

I   The leaues of Sene are a familiar purger to all people, but they are windie, and do kinde the bodie afterwards, very much diſquieting the ſtomacke with rumbling and belching : for the auoiding of which inconuenience there muſt bee added Cinnamon, Ginger, Anniſe ſeed, and fennell ſeed, Raiſins of the Sun, and ſuch like that do breake winde, which will the better help his purging qualitie.

K   Sene doth better purge when it is infuſed or ſteeped, than when it is boyled : for doubtleſſe the more it is boyled the leſſe it purgeth, and the more windie it becommeth.

L   Take Borage, Bugloſſe, Balme, Fumitory, of each three drams, Sene of Alexandria very well prepared and pouned, two ounces, ſtrow the pouder vpon the herbes and diſtill them : the water that commeth thereof reſerue to your vſe to purge thoſe that liue delicately, being miniſtred in white wine with Sugar, in condited confections, and ſuch dainty waies, wherein delicate and fine people do greatly delight: you may alſo (as was ſaid before) adde hereunto according to the maladie, diuers purgers, as Agaricke, Mirobalans, &c.

M   The pouder of Sene after it is well prepared two ounces, of the pouder of the root of Mechoacan foure drams, pouder of Ginger, Anniſe ſeeds, of each a little, a ſpoonfull of Anniſe ſeeds, but a very little Ginger, and a modicum or ſmall quantity of *Sal gemmæ*: this hath beene proued a very fit and familiar medicine for all ages and ſexes. The patient may take one ſpoonefull or two thereof faſting, either in pottage, ſome ſupping in drinke, or white wine. This is right profitable to draw both flegme and melancholy from the breſt and other parts.

N   The leaues of Sene and Cammomill are put in baths to waſh the head.

O   Sene opens the inward parts of the body which are ſtopped, and is profitable againſt all griefes of the principall members of the body.

P   Take Sene prepared according to art one ounce, Ginger halfe a quarter of an ounce, twelue cloues, Fennell ſeed two drams, or in ſtead thereof Cinnamon and Tartar, of each halfe a dram, pouder all theſe ; which done, take thereof in white wine one dram before ſupper, which doth maruellouſly purge the head.

Q   Handle Sene in manner aboue ſpecified, then take halfe an ounce thereof, which done adde therto ſixty Raiſins of the Sunne with the ſtones pickt out, one ſpoonefull of Anniſe ſeed braied ; boile theſe in a quart of ale till one halfe be waſted, and while it is boiling put in your Sene : let it ſtand ſo till the morning, then ſtraine it and put in a little Ginger : then take the one halfe of this potion and put thereunto two ſpoonfulls of ſyrrup of Roſes : drinke this together, I meane the one halfe of the medicine at one time, and if the patient cannot abide the next day to receiue the other halfe, then let it be deferred vntill the third day after.

R   Sene and Fumitory (as *Raſis* affirmeth, doe purge aduſt humors, and are excellent good againſt ſcabs, itch, and the ill affection of the body.

S   If Sene be infuſed in whey, and then boyled a little, it becommeth good Phyſicke againſt melancholy, clenſeth the braine and purgeth it, as alſo the heart, liuer, milt, and lungs, cauſeth a man to looke young, engendreth mirth, and taketh away ſorrow: it cleereth the ſight, ſtrengthneth hearing, and is very good againſt old feuers and diſeaſes ariſing of melancholy.

†   There were formerly two figures in this chapter, which differed onely in that the firſt, which was the *Sena Orientalis*, had leſſer, narrower, and ſharper pointed leaues than the *Sena Italica*, which was the ſecond.

Chap. II. *Of Bastard Sene.*

¶ *The Description.*

1   **C**Olutea and Sene be so neere the one vnto the other in shape and shew, that the vnskilful Herbarists haue deemed *Colutea* to be the right Sene. This Bastard Sene is a shrubby plant growing to the forme of a hedge bush or shrubby tree: his branches are straight, brittle, and wooddy ; which being carelesly broken off, and as negligently prickt or stucke in the ground, will take root and prosper at what time of the yeare soeuer it be done ; but slipt or cut, or planted in any curious sort whatsoeuer, among an hundred one will scarcely grow : these boughes or branches are beset with leaues like *Sena* or *Securidaca*, not much vnlike liquorice : among which come forth faire broom-like yellow floures, which turn into small cods like the sownd of a fish or a little bladder, which will make a cracke being broken betweene the fingers : wherein are contained many blacke flat seeds of the bignesse of Tares, growing vpon a small rib or sinew within the cod : the root is hard, and of a wooddy substance.

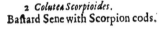

| 1 *Colutea.* | 2 *Colutea Scorpioides.* |
| Bastard Sene. | Bastard Sene with Scorpion cods. |

  2   Bastard Sene with Scorpion cods is a small wooddy shrub or bush, hauing leaues, branches and flours like vnto the former bastard Sene, but lesse in each respect: when his small yellow floures are fallen there succeed little long crooked cods like the long cods or husks of *Matthiolus* his *Scorpioides*, whereof it took his name : the root is like the root of the Box tree, or rather resembling the roots of *Dulcamara* or Bitter-sweet, growing naturally in the shadowie woods of Valena in Narbone ; whereof I haue a small plant in my garden, which may be called Scorpion Sene.

  3   The low or dwarfe *Colutea* of *Clusius* description, hath a thicke wooddy root couered with a yellowish barke, with many fibres annexed thereto, which bringeth forth yearely new shoots; whereby it greatly increaseth, of a cubit and a halfe high, smooth, and of a greene colour ; whereon doe grow leaues composed of six or seuen leaues, and sometimes nine, set vpon a middle rib like those of the common kinde, of a stipticke taste, with some sharpenesse or biting : the floures grow vpon

3 *Colutea ſcorpioides humilis.*
Dwarfe Baſtard Sene.

4 *Colutea ſcorpioides montana Cluſij.*
Mountaine Baſtard Sene.

5 *Colutea minima, ſiue Coronilla.*
The ſmalleſt Baſtard Sene.

ſlender foot-ſtalkes, long and naked like thoſe of the Peaſe, and of a yellow colour, of little or no ſmell at all, and yet that little nothing pleaſant : after which come forth long cods, wherein is contained ſmall ſeed like thoſe of the Strangle Tare.

4 This mountaine baſtard Sene hath ſtalks, leaues, and roots like the laſt deſcribed. The floures grow on the tops of the branches in maner of a crowne ; whereupon ſome haue called it *Coronilla :* in ſhape like thoſe of the Peaſe, and of a yellow colour : the cods as yet we haue not ſeen, and therefore not expreſſed in the figure.

5 This ſmall Baſtard Sene groweth like a ſmall ſhrub creeping vpon the ground, halfe a cubit high, bringing forth many twiggy branches, in maner of thoſe of the Spaniſh broome; wherupon do grow leaues like thoſe of Lentils or the Strangle Tare, with many ſmal leaues ſet vpon a middle rib, ſomwhat fat or full of juice, of the colour of the leaues of Rue or Herbegrace, of an aſtringent and vnpleaſant taſte : the floures grow at the tops of the branches, of a yellow color, in ſhape like thoſe of the ſmalleſt broome : after which come little crooked cods like the clawes or toes of a bird, wherin is contained ſeed ſomewhat long, blacke, and of an vnſauorie taſte : the root is long, hard, tough, and of a wooddy ſubſtance.

6 There

6　There is also found another sort hereof not much differing from the former, sauing that this plant is greater in each respect, wherein especially consisteth the difference.

¶ *The Place.*

Colutea or bastard Sene groweth in diuers gardens, and comneth vp of seed ; it quickly comes to perfection, insomuch that if a sticke thereof be broken off and thrust into the ground, it quickly taketh root, yea although it be done in the middle of summer, or at any other time, euen as the sticks of willow or Elder, as I my selfe haue often proued : the which bring forth flours and fruit the next yeare after.

The second with scorpion cods groweth likewise in my garden : the last growes in diuers barren chalky grounds of Kent toward Sittinburne, Canturbury, and about Southfleet ; I haue not seene them elsewhere : the rest are strangers in England.

¶ *The Time.*

They floure from May til summer be well spent, in the meane season the cods bring forth ripe seed.

¶ *The Names.*

This shrub is called of *Theophrastus* in Greeke κολουτεα, with the dipthong ου in the second syllable : in Latine, as *Gaza* expounds it, *Coloutea*, or *Colutea* : in high-Dutch, **Welsch linsen** : in French, *Baguenaudier* : they are deceiued that thinke it to be *Sena* or any kinde thereof, although wee haue followed others in giuing it to name bastard Sene, which name is very vnproper to it: in low Dutch it is called **Sene boom** : and we may vse the same name Sene tree in English.

This *Colutea* or bastard Sene differs from that plant κολυτεα with *v* in the second syllable, of which *Colutea Theophrastus* writeth, *lib.* 3. ‡ The fift is the *Polygala Valentina* of *Clusius*. ‡

¶ *The Temperature and Vertues.*

Theophrastus nor any other hath made mention of the temperature or faculties of these plants, A more, than that they are good to fatten cattell, especially sheep.

† There were formerly in the fifth and sixth places here two figures no way different, but that which was in the sixth place was a little larger, and Lobels title which he puts in his *Icons* ouer this, was diuided betwcen them : for as you see, *Colutea minima, siue Coronilla*, was ouer in the fift ; and *Colutea, siue Polygala Valentina Clusij* was ouer the sixt.

CHAP. 12. *Of Liquorice.*

¶ *The Description.*

1　THe first kind of Liquorice hath many wooddy branches rising vp to the height of two or three cubits, beset with leaues of an ouerworn green color, consisting of many smal leaues set vpon a middle ribbe like the leaues of *Colutea* or the Mastick tree, somewhat glutinous in handling : among which come small knops growing vpon short stemmes between the leaues and the branches, clustering together, and making a round form or shape: out of which grow small blew flours of the colour of an English Hyacinth : after which succeed round rough prickly heads, consisting of diuers rough and scaly husks closely and thicke compact together, in which is contained a flat seed : the root is straight, yellow within, and browne without, of a sweet and pleasant taste.

2　The common and vsuall Liquorice hath stalkes and leaues very like the former, sauing that his leaues are greener and greater, and the floures of a light shining blew colour : but the floures of this are succeeded by longish cods that grow not so thicke clustering together in round heads as the former, but spike fashion, or rather like the wilde Vetch called *Onobrychis*, or *Galega* : the cods are small and flat like vnto the Tare : the roots are of a brownish colour without, and yellow within like Box, and sweeter in taste than the former.

¶ *The Place.*

These plants grow wilde in sundry places of Germany, France and Spaine, but they are planted in gardens in England, whereof my garden hath plenty : the poore people of the North parts of England do manure it with great diligence, whereby they obtaine great plenty thereof, replanting the same once in three or foure yeares.

¶ *The Time.*

Liquorice floureth in Iuly, and the seed is ripe in September.

¶ *The*

1 *Glycyrrhiza echinata Dioscoridis.*
Hedge hog Licorice.

‡ 2 *Glycyrrhiza vulgaris.*
Common Licorice.

¶ The Names.

The first is called in Greeke, Γλυκύῤῥιζα : in Latine, *Dulcis radix*, or sweet Root: this Licorice is not knowne either to the Apothecaries or vulgar people : we call it in English , *Dioscorides* his Licorice.

It is most euident, that the other is *Glycyrrhiza* or Licorice: the Apothecaries cal it by a corrupt word, *Liquiritia* : the Italians , *Regalitia* : the Spaniards , *Regeliza*, and *Regalitia* : in high-Dutch, **Sußhotz, Sußwurtzel** : in French, *Rigolisse, Raigalisse*, and *Rezlisse*: in low-Dutch, **Callissehout, suethout** : in English, common Licorice : *Pliny* calls it *Scythica herba* : it is named *Scythice*, of the country Scythia where it groweth.

¶ The Temperature.

The nature of *Dioscorides* his Licorice, as *Galen* testifith , is familiar to the temperature of our bodies, and seeing it hath a certain binding qualitie adjoined, the temperature thereof, so much as is hot and binding, is specially of a warm qualitie, comming neerest of all to a meane temper : besides, for that it is also sweet, it is likewise meanly moist.

Forasmuch as the root of the common Licorice is sweet, it is also temperately hot and moist; notwithstanding the barke hereof is somthing bitter and hot, but this must be scraped away : The fresh root when it is full of juice doth moisten more than the dry.

¶ The Vertues.

A　　The root of Licorice is good against the rough harshnesse of the throat and brest ; it opens the pipes of the lungs when they be stuffed or stopt, ripeneth the cough, and bringeth forth flegme.

B　　The juice of Licorice made according to art, and hardned into a lump, which is called *Succus Liquiritiæ*, serueth well for the purposes aforesaid, being holden vnder the tongue, and there suffered to melt.

C　　Moreouer, with the juice of Licorice, Ginger, and other spices, there is made a certaine bread or cakes called Ginger-bread, which is very good against the cough and all infirmities of the lungs and brest ; which is cast into moulds, some of one fashion, and some of another.

D　　The juice of Licorice is profitable against the heate of the stomacke and of the mouth.

The

The same is drunk with wine and Raisins against the infirmities of the liuer and chest, scabs or sores of the bladder, and diseases of the kidnies.

Being melted vnder the tongue it quencheth thirst: it is good for green wounds being laid thereupon, and for the stomacke if it be chewed.

The decoction of the fresh roots serueth for the same purposes.

But the dried root most finely poudered is a singular remedie for a pin and a web in the eie, if it be strewed thereupon.

Dioscorides and *Pliny* also report, that Licorice is good for the stomacke and vlcers of the mouth, being cast vpon them.

It is good against hoarsnesse, difficultie of breathing, inflammation of the lungs, the pleurisie, spitting of bloud or matter, consumption and rottennesse of the lungs, all infirmities and ruggednesse of the chest.

It takes away inflammation, mitigateth and tempereth the sharpnesse and saltnesse of humors, concocteth raw humors, and procureth easie spitting.

The decoction is good for the kidnies and bladder that are exulcerated.

It cureth the stranguric, and generally all infirmities that proceed of sharpe salt and biting humors.

These things concerning Licorice hath also *Theophrastus : viz.* That with this and cheese made of mares milke the Scythians were reported to be able to liue eleuen or twelue dayes.

The Scythian root is good for shortnesse of breath, for a dry cough, and generally for all infirmities of the chest.

Moreouer, with hony it healeth vlcers, it also quencheth thirst if it be held in the mouth : for which cause they say that the Scythians do liue eleuen or twelue daies with it and *Hippace*, which is cheese made of mares milke, as *Hippocrates* witnesseth.

Pliny, lib. 25. cap. 8. hath thought otherwise than truth, That *Hippace* is an herb so called.

† Both the figures formerly were of the first described.

CHAP. 13.　*Of Milke Trefoile or Shrub-Trefoile.*

¶ *The Kindes.*

THere be diuers kindes or sorts of the shrubby Trefoile, the which might very well haue passed among the three leaued Grasses, had it not been for our promise in the proem of our first part, That in the last booke of our historie the shrubby or wooddy plants should be set forth, euery one as neere as might be in kindred and neighborhood.

¶ *The Description.*

1　THe first kinde of *Cytisus* or shrubby Trefoile growes to the forme of a small shrub or wooddy bush two or three cubits high, branching into sundry small boughes or armes, set full of leaues like the small Trefoile, darke greene, and not hairy, three growing alwaies together: among these come forth smal yellow flours like them of French broom, which do turn into long and flat cods, containing small seed of a blackish colour.

2　The second kinde of *Cytisus* is likewise a small shrub, in shape after the manner of the former, but that the whole plant is altogether smaller, and the leaues rounder, set together by couples, and the small cods hairy at the ends, which sets forth the difference. ‡ The leaues of this are almost round, and grow three together close to the stalke : they are smooth, of a fresh greene, and the middlemost leafe of the three is the largest, and ends in a sharp point : the flours are of the bignes and colour of the *Trifolium Corniculatum :* it flours in May. ‡

3　The root of this third kinde is single, from whence spring vp many smooth brittle stalks diuided into many wings and branches, wheron grow green leaues smaller than those of medow Trefoile : the flours are yellow, lesser than Broom floures, otherwise very like, growing about the tops of the twiggy branches, diuided into spoky tufts ; which being vaded, there follow thin long narrow cods lesser than those of the Broome, wherin is contained small blacke seed. The root is long, deeply growing into the ground, and sometimes waxeth crooked in the earth. ‡ This also hath smooth green leaues, and differs little (if any thing at all) from the first described, wherfore I thought it needlesse to giue a figure. Our Author called it *Cytisus siliquosus*, Codded shrub Trefoile, because one of the branches was fairely in the figure exprest with cods ; I know no other reason, for all the *Cytisi* are codded as well as this. ‡

　　　‡ The

1 Cytisus.
The first shrub Trefoile.

2 Cytisus.
The second shrub Trefoile.

4. Cytisus hirsutus.
Hairy shrub Trefoile.

5 Cytisus incanus.
Hoary shrub Trefoile.

4   The fourth kinde of *Cytisus* hath a great number of small branches and stalkes like the former, but it is a low plant and more woolly; whose stalks and branches grow not very high, but yet very plentifully spred about the sides of the plant: the leaues are greater than the former, but lesser than those of medow Trefoile: the floures grow close together, as though they were bound vp or compact into one head or spoky tuft somewhat greater than the former: the cods are also greater, and more hairy: the root groweth very deep into the ground, whereunto are adjoined a few fibres: it falls out to be more hairy or woolly in one place than in another, and the more hairy and woolly that it is, the whiter it waxeth; for the roughnesse bringeth it a certaine whitish colour. ‡ The branches of this oft times lie along vpon the ground: the leaues are smooth, green aboue, and hoary vnderneath: the floures yellow, which fading, somtimes become orange coloured. The cods are round, and seeds brownish. ‡

5   The fifth kinde of *Cytisus* groweth to the height of a cubit or more, hauing many slender twiggy branches like Broom, streaked, and very hard, whereupon grow leaues very like Fenugreek, yet all hoary, three together; from the bosom of which, or between the leaues and the stalks come forth yellow floures very like Broom, *Spartum*, or Pease, but smaller: the cods be like vnto Broome cods, of an ash colour, but slenderer, rougher and flatter; in the seuerall cels or diuisions whereof are contained bright shining seeds like the blacke seeds of Broome: all the whole plant is hoary like *Rhamnus* or *Halymus*.

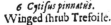

6 *Cytisus pinnatus.*
Winged shrub Trefoile.

7 *Cytisus 7. cornutus.*
The horned shrub Trefoile.

6   The sixt kind of *Cytisus* or bush Trefoile growes to the height of a tall man, with long stalks couered ouer with a blackish barke, and a few boughes or branches beset or garnished with leaues like the common Trefoile, but smaller, growing also three together, wherof the middlemost of the three leaues is twice as long as the two side leaues; the vpper side whereof is greene, and the lower side somwhat reddish and hairy: the floures grow along the stalks almost from the bottom to the top, of a golden yellow colour, fashioned like the Broom floure, but greater than any of the rest of his kinde, and of a reasonable good sauor: the seed hath the pulsie taste of *Cicer*.

7   The seuenth kinde of *Cytisus* hath many tough and hairy branches rising from a wooddie root foure or fiue cubits high; which are diuided into sundry smaller branches, beset with leaues like the medow Trefoiles; amongst which come forth yellow floures like Broome, that turne into

crooked flat cods like a ſickle, wherein is contained the ſeed taſting like *Cicer* or *Legumen*. The whole plant is hoary like *Rhamnus*, and being broken or bruiſed ſmelleth like Rocket.

    8   This eighth kinde of *Cytiſus* which *Pena* ſetteth forth is doubtleſſe another kinde of *Cytiſus*, reſembling the former in leaues, floures, and cods, ſauing that the ſmall leaues (which are alwayes three together) are a little ſnipt about the edges : the whole plant is ſlenderer, ſofter, and greener, rather reſembling an herb than a ſhrub : the root is ſmall and ſingle.

    9   This baſtard or miſ-begotten ſhrub Trefoile, or baſtard *Cytiſus* groweth vp like a ſhrub, but not of a wooddy ſubſtance, hauing tender ſtalks ſmooth and plain, whereon grow hairy leaues like the other, diuers ſet vpon one foot-ſtalke, contrarie to all the reſt : the floures grow along the ſtems like thoſe of the ſtock Gillofloures, of a yellow colour : the root is tough and wooddy.

<table>
<tr><td align="center">8 Cytiſus 8.
The eighth ſhrub Trefoile.</td><td align="center">9 Cytiſus adulterinus, ſiue Alyſſon fruticans.
Baſtard ſhrub Trefoile.</td></tr>
</table>

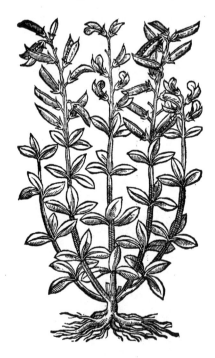

¶ *The Place.*
    Theſe plants were firſt brought into Italy and Greece from one of the Iſles of Cyclades, called Cyntho or Cynthuſa, and ſince found in many places of France, as about Montpelier, Veganium, and other places : they are ſtrangers in England, though they grow very plentifully in Scotland, as it is reported ; whereof I haue two ſorts in my garden, that is to ſay, *Cytiſus marantha*, or the horned *Cytiſus* ; and likewiſe one of the ſmalleſt, that is to ſay, the third in number. ‡ The ſecond groweth in the garden of M ͬ Iohn Tradeſcant. ‡

¶ *The Time.*
    Theſe plants floure for the moſt part in May, Iune, and Iuly, and ſome after : the ſeed is ripe in September.

¶ *The Names.*
    The Grecians and Latines do call this ſhrub κύτισος, of Cynthuſa an Iſland before mentioned, in which place they are in great eſtimation, for that they do ſo wonderfully feed cattell, and encreaſe milke in their dugs, nouriſh ſheep and goats which bring young ones good for ſtore and increaſe. Our Author doth call theſe plants in Greeke κύτισος, that is to ſay in Latine, *Fœcundum fœnum*, fertil or fruitfull Hay, for that the kindes hereof cauſe milke to encreaſe, maketh good bloud and juice, augmenteth ſtrength, and multiplieth the naturall ſeed of generation : they may be called in Engliſh, milke Trefoile, of the ſtore of milke which they increaſe.

                                               ¶ *The*

¶ *The Temperature.*

The leaues of milke Trefoile do coole, as *Dioſcorides* writeth: they aſſwage ſwellings in the beginning, if they be ſtamped and laid vnto them with bread: the decoction thereof drunke prouoketh vrine. *Galen* teacheth, That the leaues of milke Trefoile haue a digeſting or waſting quality, mixed with a waterie and temperat facultie, as haue thoſe of the Mallow.

¶ *The Vertues.*

Women, ſaith *Columella*, if they want milke, muſt ſteepe dry milke Trefoile in faire water, and when it is throughly ſoked, they muſt the next day mix a quart or thereabouts of the ſame preſſed or ſtrained forth with a little Wine, and ſo let it be giuen vnto them to drinke, and by that meanes they themſelues ſhall receiue ſtrength, and their children comfort by aboundance of milke. **A**

Hippocrates reckons vp milke Trefoile among thoſe things that encreaſe milke, in his book of the nature of women and womens diſeaſes. **B**

Alſo *Ariſtomachus* of Athens in *Pliny* commandeth to giue with wine the dry plant, and the ſame likewiſe boiled in water, to nurſes to drinke when their milke is gon. **C**

Democritus and *Ariſtomachus* do promiſe that you ſhall want no Bees, if you haue milke Trefoile for them to feed on: for all writers with one conſent do conclude (as *Galen* ſaith) that Bees gather of the floures of Milke Trefoile very great ſtore of hony. **D**

Columella teacheth, That milk Trefoile is notable good for Hens, Bees, Goats, Kine, and all kind of cattel, which quickly grow fat by eating thereof; and that it yeeldeth very great ſtore of milk. **E**

The people of Betica and Valentia, where there is great ſtore of *Cytiſus*, doe vſe it very much for the Silke-wormes to hang their web vpon after they haue been well fed with the leaues of Mulberries. **F**

Milke Trefoile is likewiſe a maruellous remedie againſt the Sciatica and all other kindes of Gouts. **G**

‡ The deſcription that formerly was in the firſt place belonged to that deſcribed and figured in the ſeuenth.

CHAP. 14.　*Of Baſtard milke Trefoiles.*

¶ *The Deſcription.*

1　THis riſeth vp with little ſtalks from the root, brittle, very many in number, parted into wings & branches, about which grow many leaues leſſer than thoſe of the medow Trefoile, of colour green: the floures about the tops of the twigs be orderly placed in manner like eares, of colour yellow, leſſer than thoſe of Broom, otherwiſe all alike: in their places grow vp ſlender cods, long, narrow, and leſſer than the cods of Broom, rough alſo and hairy; in which do lie little blackiſh ſeeds: the root is long, and groweth deepe, and oftentimes creepeth aſlope.

2　The ſecond kinde of baſtard milke Trefoile is like vnto the former in plentifull ſtalks and twigs, but that it is lower and more downy, neither doe the ſtalks thereof ſtand vpright, but rather incline to the one ſide: the leaues alſo are ſomwhat greater, but yet leſſer than thoſe of the medow Trefoile, wholly white, and they neuer open themſelues out, but keep alwaies folded, with the middle rib ſtanding out: the flours likewiſe be cloſelier ioined together, and compacted as it were into a little head, and be alſo ſomething greater: the cods in like manner are a little bigger and hairy, and of a blackiſh purple or murrey: the root groweth deep in the ground, being diuided into a few ſprigs: it oftentimes hapneth to grow in one place more hairy or downy than in another; the more hairy and downy it is, the more white and hoary it is, for the hairineſſe doth alſo bring with it a certain whitiſh colour.

3　The third kinde of baſtard milke Trefoile brings forth a company of yong ſhoots that are ſomwhat writhed and crooked, long leaues of a faire green colour: the floures are cloſed together, long, white, or elſe galbineous, ſweetly ſmelling, that is to ſay, hauing the ſmell of hony: the ſhrub it ſelfe is alwaies green both Summer and Winter. ‡ This growes ſome foot or better high, with ſlender hoary branches, ſet with leaues three ſtanding together vpon a very ſhort ſtalk, and the middle leafe is as long again as the other two; they are very white and hoary, and the flours grow out of the boſoms of the leaues all alongſt the ſtalks. This is that mentioned in the vertues of the former chapter at *F*, for the ſilke wormes to worke vpon. ‡

4　The fourth ſhrub is likewiſe one of the wilde kinde, though in face and ſtature like the manured

1 *Pſeudocytiſus 1.*
The firſt baſtard ſhrub Trefoile.

2 *Pſeudocytiſus 2.*
The ſecond baſtard ſhrub Trefoile.

3 *Cytiſus ſemper-virens.*
The euer-green ſhrub Trefoile.

4 *Pſeudocytiſus hirſutus.*
The hairy baſtard tree Trefoile.

nured *Cytiſus*. It groweth vp like a ſmall ſhrub or hedge buſh, to the height of two or three yards; on whoſe branches do grow three rough or hairy leaues ſet vpon a ſlender foot-ſtalke, of a griſtle-green colour aboue, with a reddiſh hairineſſe below: the floures grow alongſt the ſtalks, from the middle to the top, of a bright ſhining yellow colour: the root is likewiſe wooddy.

¶ *The Place.*

Theſe kindes of milke Trefoile are found in Moravia, ſo called in our age, which in times paſt was named *Marcomannorum prouincia*, and in the vpper Pannonia, otherwiſe called Auſtria, neere to highwaies and in the borders of fields, for they ſeem after a ſort to ioy in the ſhade. ‡ Theſe grow (according to *Cluſius*) in ſundry parts of Spain. ‡

¶ *The Time,*

They floure eſpecially in Iune and Iuly.

¶ *The Names.*

It is euident enough that they are baſtard kindes of milke Trefoiles, and therefore they may be called and plainly termed *Pſeudocytiſi* or baſtard milke Trefoiles, or *Cytiſi ſylueſtres*, that is to ſay, wilde milke Trefoiles.

¶ *The Nature and Vertues.*

VVhat temperature theſe ſhrubs are of, or what vertues they haue we know not, neither haue wee as yet found out any thing by our owne experience, wherefore they may be referred vnto the other milke-Trefoiles.

Chap. 15.   *Of the venomous tree Trefoile.*

† 1 *Dorycnium Monſpelienſium.*
The venomous Trefoile of Montpelier.

2 *Dorycnium Hiſpanicum.*
The venomous Trefoile of Spain.

¶ *The Deſcription.*

1   THe venomous Tree Trefoile of Montpellier hath many tough and pliant ſtalkes two or three cubits high, diuided into ſundry ſmall twiggy braunches, beſet with leaues three
togethér

together,placed from joint to joint by fpaces,fomewhat hoary,very likevnto the leaues of *Cytifus* or Rue : among which come forth many fmall moffie white floures, tuft fafhion, in fmall bundles like Nofegaies,and very like the floures of the Oliue or Oke tree,which turne into fmall roundifh bladders,as it were made of parchment ; wherein is contained blacke feed like wilde *Lotus*, but in tafte like the wilde Tare : the whole plant is of an vnfauory fmel ; the root is thick,and of a wood-die fubftance.

2   The Spanifh venomous Trefoile hath a wooddy ftalke,rough and hoary, diuided into other fmall branches,whereon grow leaues like the precedent : the floures grow on the tops of the bran-ches,whereon grow leaues like thofe of the Peafe,and of a yellow or rather greenifh color,where-in it differeth from the precedent.

¶ *The Place.*

Thefe venomous Trefoiles grow in Narbon,on the barren artd ftony craggy mountains,at Fron-tignana,and about the fea coafts,and are ftrangers in England.

¶ *The Time.*

They flourifh from May to the end of Iune.

¶ *The Names.*

Dorycnium,or ☌☌☌☌☌☌, is that poifonous or venomous plant wherewith in times paft they vfed to poifon their arrow heads,or weapons, thereby to do the greater hurt vnto thofe whom they did af-faile or purfue, whereupon it tooke his name. Great controuerfie hath been amongft Herbarifts, what manner of plant *Dorycnium* fhould be ; fome faying one thing,and fome another :which con-trouerfies and fundry opinions are very well confuted by the true cenfure of *Rondeletius*, who hath for a definitiue fentence fet down the plant defcribed for the true *Dorycnium*, & none other,which may be called in Englifh,Venomous tree Trefoile. ‡ Thefe plants do not fufficiently anfwer to the defcription of *Diofcorides*,neither can any one fay certainly that they are poifonous. ‡

¶ *The Temperature.*

Dorycnium is very cold without moiftning.

¶ *The Vertues.*

Venomous Trefoile hath not one good qualitie that I can read of,but it is a peftilent venomous plant,as hath been faid in the defcription.

† The figures were formerly tranfpofed.

Chap. 16.   *Of Of the fhrub Trefoile, called alfo* Make-bait.

Polemonium, five Trifolium fruticans.
Shrubby Trefole, or yellow Iafmine.

¶ *The Defcription.*

THis fhruby plant called *Polemonium* hath many wooddy twigges growing into the height of four or fiue cubits,hauing fmal ny twiggy branches of a dark green color,gar-nifhed with fmall leaues of a deep green color, alwaies three ioined together vpon little foot-ftalks,like the *Cytifus* bnfh or field Trefoile,but fmaller : the floures be yellow and round, diui-ded into fiue or fix parts, not much vnlike the yellow Iafmine,which hath caufed many to cal it yellow Iafmine,euen to this day. When the floures be vaded there fucceed fmall round ber-ries as big as a peafe,of a black purplifh colour when they be ripe,which being broken wil die or colour the fingers like Elder berries:within thefe berries are contained a fmall flat feed like vnto Lentils : the root is long and fmall, cree-ping hither and thither vnder the earth,putting forth new fprings or fhoots in fundry places, whereby it greatly increafeth.

¶ *The Place.*

It grows plentifully in the country of Mont-pelier,at New-caftle,vpon the dry hils,and hot banks of the oliue fields,and in the ftony fields and Wood of Gramuntium : it groweth in my garden,and in other herbarifts gardens of Eng-land.

¶ The

¶ *The Time.*

It floureth in Summer : the feed is ripe in Autumne : the fhrub it felfe is alwaies green, & hath a lafting root.

¶ *The Names.*

Moft do call it *Cytifus*, but we had rather name it *Trifolium fruticans :* for it doth not agree with *Cytifus* or Milk Trefoile, as in the chapter before it is plain enough by his defcription, vnleffe it be *Cytifus Marcelli*, or *Marcellus* his Milke Trefoile, with which peraduenture it might be thought to haue fome likeneffe, if the floures which are yellow were white, or *galbineous*, that is to fay, blew.

There be diuers alfo that take this Trefoile to be *Polemonium*, forafmuch as the leaues of it feem to be fomewhat like thofe of common Rue ; but *Polemonium* hath not the leafe of common Rue, otherwife called Herb-grace, but of the other, that is to fay, of S. *Iohns* Rue : it is called in Englifh, fhrubby Trefoile, or Make-bait.

¶ *The Temperature.*

Polemonium is dry in the fecond degree, with fome acrimonie or fharpneffe.

¶ *The Vertues.*

This fhrubby plant hath fo many fingular and excellent vertues contained in it, that fome haue A called it by the name *Chiliodunamis*, that is, hauing a thoufand properties.

It is very effectual againft the ftinging of Scorpions ; and (as fome fay) if a man hold it in his B hand, he cannot be hurt with the biting of any venomous beaft.

Being taken with vineger it is very good for thofe that are fpleneticke, and whofe fpleen or milt C is affected with oppilations or ftoppings.

If the root be taken in wine it helpeth againft the bloudy flix, it prouoketh vrine being drunke D with water, fcoureth away grauell, and eafeth the pain and ache called the Sciatica.

CHAP. 17.   *Of Broome and Broome-Rape.*

1 *Genifta*.
Broome.

2 *Rapum Genifta, fiue Orobanche.*
Broom-Rape, or Orobanch.

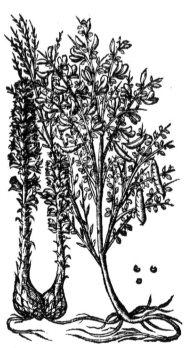

‡ *Orobanche Monſpeliaca flo. oblongis.*
Long floured Broome-Rape.

‡ *Orobanche flore majore.*
Great floured Broome Rape.

‡ *Orobanche Ramoſa.*
Branched Broom-Rape.

¶ *The Deſcription.*

1   BRoom is a buſh or ſhrubby plant, it
hath ſtalks or rather wooddy branches, from which do ſpring ſlender
twigs, cornered, green, rough, and that be eaſily bowed, many times diuided into ſmal branches ; about which do grow little leaues of an
obſcure green colour, & braue yellow floures,
and at the length flat cods, which beeing ripe
are black, as are thoſe of the common Vetch,
in which doe lie flat ſeeds, hard, ſomething
browniſh, and leſſer than Lentils : the root is
hard and wooddy, ſending forth diuers times
another plant of the colour of an Oken leafe,
in ſhape like vnto the baſtard Orchis , called
Birds neſt, hauing a root like a Turnep or
Rape, whereupon it is called *Rapum Geniſtæ*, or
Broom-Rape.

2   This is a certain bulbed plant growing
vnto the roots of Broom, big below, and ſmaller aboue, couered with blackiſh ſcales , and
of a yellowiſh pulp within : from which doth
riſe a ſtalke a ſpan long, hauing whitiſh flours
about the top, like almoſt to thoſe of dead
Nettle. After which grow forth long thicke
and round husks, in which are contained verie
many ſeeds, and good for nothing : the whole
plant is of the colour of the Oken leafe.

‡ Of

3 *Genifta Hifpanica.*
Spanifh Broome.

5 *Chamægenifta Anglica.*
Englifh Dwarfe Broome.

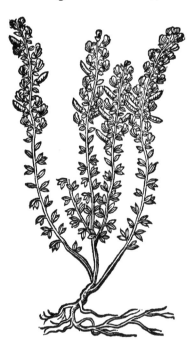

6 *Chamægenifta Pannonica.*
Dwarfe Broome of Hungary.

‡ Of this *Orobanche* or Broome Rape there
are fome varieties obferued and fet forth by *Lobel* and *Clufius*; the firft of thefe varieties hath
longer and fmaller floures than the ordinarie.
The fecond hath larger floures, and thofe of a
blewifh colour, and is fometimes found among
corne. The third is parted towards the top into fundry branches; the floures of this are either blew, purplifh or elfe white, and it willingly growes among Hempe. ‡

3 The Spanifh Broome hath likewife
wooddy ftems, from whence grow vp flender
pliant twigs, which be bare and naked without
leaues, or at the leaft hauing but few fmall
leaues, fet here and there far diftant one from
another, with yellow floures not much vnlike
the floures of common Broome, but greater,
which turne into fmall long cods, wherein is
contained browne and flat feed: the root is
tough and wooddy.

4 Small leafed or thin leafed Broome hath
many tough pliant fhoots rifing out of the
ground, which grow into hard and tough ftalks,
which are diuided into diuers twiggy branches
whereon doe grow very fmall thin leaues, of a
whitifh colour; whereupon fome haue called it
Genifta alba, white Broome: the floures grow at
the top of the ftalkes, in fhape like thofe of the
common Broom, but of a white colour, wherein
it fpecially differeth from the other Broomes.

Sffff   5 Englifh

5   Engliſh Dwarfe Broome hath many twiggy branches, very greene, tough, ſomewhat ſtraked or cornered, leaning toward the ground : whereon doe grow leaues ſet without order, ſometimes two together, and often three or foure growing faſt together, like vnto the common Broome, greene on the vpper ſide, hoary vnderneath, and of a bitter taſte : among which leaues come forth yellow floures like thoſe of common Broome, but leſſer, of little or no ſmell at all : after which appeare ſmall cods ſomewhat hairy, wherein is contained ſmall ſeed : the root is tough and wooddy. ‡ Bauhine iudges theſe two laſt deſcribed to be onely varieties of the common Broome ; to whoſe opinion I do much incline, yet I haue let our Authors deſcription ſtand, together with the figure of this later, which ſeemingly expreſſes the greateſt difference. ‡

6   The Dwarfe Broome of Hungary hath ſtalkes and yellow floures like thoſe of the laſt deſcribed : the leaues hereof are different, they are longer and more in number : the whole plant is altogether greater, wherein eſpecially conſiſteth the difference.

¶ *The Place.*

The common Broome groweth almoſt euery where in dry paſtures and low woods.

The Broome Rape is not to be found but where Broome doth grow ; it groweth in a Broome field at the foot of Shooters hill next to London ; vpon Hampſtead Heath, and in diuers other places.

Spaniſh Broome groweth in diuers kingdomes of Spaine and Italy ; we haue it in our London gardens.

The White Broome groweth likewiſe in Spaine and other hot regions ; it is a ſtranger in England ; of this *Titus Calphurnius* makes mention in his ſecond Eclog of his Bucolicks, writing thus :

Cernis vt, ecce pater, quas tradidit Ornite vacca
Molle ſub hirſuta latus explicuere geniſta.

See Father, how the Kine ſtretch out their tender ſide
Vnder the hairy Broome, that growes in field ſo wide.

¶ *The Time.*

Broome floureth in the end of Aprill or May, and then the young buds of the floures are to bee gathered and laid in pickle or ſalt, which afterwards being waſhed or boyled, are vſed for ſallads, as Capers be, and be eaten with no leſſe delight : the cods and ſeeds be ripe in Auguſt ; the Rape appeareth and is ſeene eſpecially in the moneth of Iune.

The Spaniſh Broome doth floure ſooner, and is longer in flouring.

¶ *The Names.*

This ſhrub is called in Latine, *Geniſta*, or as ſome would haue it *Geneſtra* : in Italian *Geneſtra* : in Spaniſh likewiſe, *Geneſtra*, or *Gieſtra* : in high Dutch, **Pfrimmen** : in low Dutch, **Brem** : in French, *Geneſt* : in Engliſh Broome. ‡ The Spaniſh Broome by moſt Writers is iudged to be the *Spartium* of *Dioſcorides*. ‡

¶ *The Temperature and Vertues.*

A   The twigs, floures and ſeeds of Broome are hot and dry in the ſecond degree : they are alſo of a thin eſſence, and are of force to clenſe and open, and eſpecially the ſeed, which is dryer and not ſo full of ſuperfluous moiſture.

B   The decoction of the twigs and tops of Broome doth clenſe and open the liuer, milt, and kidnies.

C   It driueth away by the ſtoole watery humors, and therefore it is wholeſome for them that haue the dropſie, eſpecially being made with wine ; but better for the other infirmities with water.

D   The ſeed alſo is commended for the ſame purpoſes.

E   There is alſo made of the aſhes of the ſtalkes and branches dryed and burnt, a lie with thin white wine, as Rheniſh wine, which is highly commended of diuers for the greene ſickeneſſe and dropſie, and this doth mightily expell and driue forth thin and watery humors together with the vrine, and that by the bladder ; but withall it doth by reaſon of his ſharpe quality many times hurt and fret the intrailes.

F   *Meſue* ſaith, That there is in the floures and branches a cutting moiſture, but full of excrements, and therefore it cauſeth vomit : and that the plant doth in all his parts trouble, cut, attenuate, and violently purge by vomit and ſtoole, flegme and raw humors out of the joynts.

G   But theſe things are not written of Broome, but of *Spartium*, which purgeth by vomit, after the manner of Hellebor, as both *Dioſcorides* and *Pliny* do teſtifie.

H   *Meſue* alſo addeth, That Broome doth breake the ſtone of the kidnies and bladder, and ſuffereth not the matter whereof the ſtone is made to lie long, or to become a ſtone.

I   The young buds or little floures preſerued in pickle, and eaten as a ſallad, ſtirre vp an appetite to meate and open the ſtoppings of the liuer and milt.

The

The ſame being fully blowne, ſtamped and mixed with ſwines greaſe, doe eaſe the paine of the L
gout.

And *Meſue* writeth, That this tempered with hony of Roſes, or with an egge, doth conſume away M
the Kings euill.

The Rape of the Broome or Broome Rape, being boyled in wine, is commended againſt the pains N
of the kidneyes and bladder, prouoketh vrine, breaketh the ſtone, and expelleth it.

The juyce preſſed forth of Broome rape healeth greene wounds, and clenſeth old and filthy vl- O
cers: the later Phyſitions do affirme that it is alſo good for old venomous, and malicious vlcers.

That worthy Prince of famous memory *Henry* 8. King of England, was wont to drinke the diſtil- P
led water of Broome floures, againſt ſurfets and diſeaſes thereof ariſing.

Sir *Thomas Fitzherbert* Knight, was wont to cure the blacke jaundice with this drinke onely. Q

Take as many handfulls (as you thinke good) of the dried leaues of Broome gathered and brayed R
to pouder in the moneth of May, then take vnto each handfull of the dried leaues, one ſpoonful and
a halfe of the ſeed of Broome brayed into pouder: mingle theſe together, and let the ſicke drinke
thereof each day a quantity, firſt and laſt, vntill he finde ſome eaſe. The medicine muſt be continued and ſo long vſed, vntill it be quite extinguiſhed: for it is a diſeaſe not very ſuddenly cured, but
muſt by little and little be dealt withall.

Orobanch or Broome rape ſliced and put into oyle Oliue, to infuſe or macerate in the ſame, as ye S
do Roſes for oyle of Roſes, ſcoureth and putteth away all ſpots, lentils, freckles, pimples, wheals and
puſhes from the face, or any part of the body, being annointed therewith.

Dioſcorides writeth, That Orobanch may be eaten either raw or boiled, in manner as we vſe to eat T
the ſprigs or young ſhoots of *Aſparagus*.

The floures and ſeeds of Spaniſh Broome are good to be drunke with meade or honied water in V
the quantity of a dram, to cauſe one to vomit with great force and violence, euen as white Hellebor,
or neeſing pouder.

If it be taken alone, it looſeneth the belly, driueth forth great quantitie of waterie and filthy hu- X
mours.

Chap. 18. *Of baſe Broome or greening weed.*

¶ *The Deſcription.*

1   THis baſe kinde of Broom called Greenweed or Diers weed, hath many tough branches
proceeding from a wooddy root: whereon do grow great ſtore of leaues, of a deep green
colour, ſomwhat long like thoſe of Flax: the floures grow at the top of the branches not
much vnlike the leaues of Broome, but ſmaller; of an exceeding faire yellow colour, which turne
into ſmall flat cods, wherein is contained a little flat ſeed.

2   *Carolus Cluſius* ſetteth forth another kinde of Broome, which *Dodonæus* calleth *Geniſta tincto-*
ria, being another ſort of Diers weed: it groweth like the Spaniſh Broome: vpon whoſe branches
do grow long and ſmall leaues like Flax, greene on the vpper ſide, and of an hoary ſhining colour
on the other. The floures grow at the top of the ſtalkes, ſpike faſhion, in forme and colour like the
former: the roots are thicke and wooddy.

3   *Carolus Cluſius* ſetteth forth two kindes of Broome. The firſt is a low and baſe plant, creeping
and lying flat vpon the ground, whoſe long branches are nothing elſe, but as it were ſtalkes conſiſting of leaues thicke in the middeſt, and thinne about the edges, and as it were diuided with ſmall
nicks; at which place it beginneth to continue the ſame leaſe to the end, and ſo from leaſe to leaſe,
vntill it haue increaſed a great ſort, all which doe as it were make one ſtalke; and hath none other
leaues, ſauing that in ſome of the nicks or diuiſions there commeth forth a ſmall leafe like a little
eare. At the end of thoſe flat and leaſed ſtalks come forth the floures, much like the floures of the
common Greening weed, but leſſer, and of a yellow colour, which turne into ſmall cods. The roots
are very long, tough, and wooddy, full of fibres, cloſing at the top of the root, from whence they proceed as from one body.

4   This kinde of Greeneweed called of ſome *Chamæſpartium*, hath a thicke wooddy root: from
which riſe vp diuers long leaues, conſiſting as it were of many pieces ſet together like a paire of
Beads (as may better be perceiued by the figure, than expreſſed by words) greene on the vpper ſide,
and whitiſh vnderneath, very tough, and as it were of a ruſhie ſubſtance: among which riſe vp very
ſmall naked ruſhy ſtalkes; on the top whereof groweth an eare or ſpike of a chaffie matter, hauing
here and there in the ſaid eares diuers yellow floures like Broome, but very ſmall or little.

1 *Geniſtella tinctoria.*
Greeneweed or Diers weed.

2 *Geniſtella infectoria.*
Wooddy Diers-weed.

3 *Geniſtella pinnata.*
Winged Greeneweed.

4 *Geniſtella globulata.*
Globe Greeneweed.

5   The fifth Greenweed hath a wooddy tough root, with certaine ſtrings annexed thereto: from which riſe vp diuers long, flat leaues, tough, and very hard, conſiſting as it were of many little leaues, ſet one at the end of another, making of many one entire leafe, of a greene colour: amongſt which come forth diuers naked hard ſtalkes, very ſmall and ſtiffe, on the tops whereof ſtand ſpikie eares of yellow floures, like thoſe of Broome, in ſhape like that great three leafed graſſe, called *Lagopus*, or like the Fox-taile graſſe: after which come flat cods, wherein is incloſed ſmall ſeed like to Tares both in taſte and forme.

5 *Geniſtella Lagopoides major*.
Hares foot Greeneweed.

5 *Geniſtella Lagopoides minor*.
Small Greeneweed with Hares foot floure.

6   This differeth not from the precedent in ſtalks, roots and leaues: the floures conſiſt of a flockie ſoft matter, not vnlike to the graſſie tuſt of Foxtaile, reſembling the floure of *Lagopus*, or Haresfoot, but hauing ſmall yellow floures leſſer than the former, wherein it chiefely differeth from the other of his kinde.

¶ *The Place*.

The firſt being our common Diersweed, groweth in moſt fertile paſtures and fields almoſt euery where. The reſt are ſtrangers in England.

¶ *The Time*.

They floure from the beginning of Iuly to the end of Auguſt.

¶ *The Names*.

The firſt of theſe Greenweeds is named of moſt Herbariſts *Flos Tinctorius*, but more rightly, *Geniſta Tinctoria*, of this *Pliny* hath made mention [The Greeneweeds, ſaith he, do grow to dye clothes with] in his 18. booke, 6. chapter. It is called in high Dutch, **Ferblumen**, and **Ackerbaum**: in Italian, *Cerretta*, and *Coſaria*, as *Matthiolus* writeth in his chapter of *Lyſimachia*, or Looſe-ſtrife: in Engliſh, Diers Greeneweed, baſe Broome, and Woodwaxen.

The reſt we refer to their ſeuerall titles.

¶ *The Temperature and Vertues*.

Theſe plants are like vnto common Broome in bitterneſſe, and therefore are hot and dry in the ſecond degree: they are likewiſe thought to be in vertues equall; notwithſtanding their vſe is not ſo well knowne, and therefore not vſed at all where the other may be had: we ſhall not need to ſpeake of that vſe that Diers make thereof, being a matter impertinent to our Hiſtory.

A

Chap. 19. *Of Spaniſh baſe Broomes.*

¶ *The Deſcription.*

‡ 1 THis growes to the height of a cubit, and is couered with a creſted and rough barke, and diuided into many longiſh branches creſted and greene, which at their firſt ſpringing vp haue ſome leaues vpon them, which fall away as ſoone as the plant comes to floure: from the ſides of the branches come forth long foot-ſtalkes whereon hang ſome ſmall yellow flours, which are ſucceeded by ſhort round yellowiſh red cods which commonly con-taine but one ſeed, ſeldome two, and theſe hard and blacke, and like a little Kidney, which when it is ripe will rattle in the cod being ſhaken. ‡

1 *Pſeudoſpartum Hiſpanicum Aphyllum.*
Spaniſh Broome without leaues.

2 *Pſeudoſpartum album Aphyllum.*
The white leafe-leſſe Spaniſh Broome.

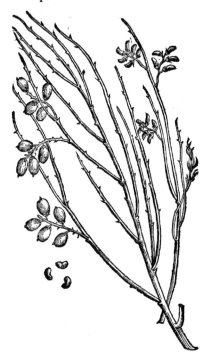

2 This naked broome groweth vp to the height of a man: the ſtalk is rough, and void of leaues, very greene and pliant, which diuideth it ſelfe into diuers twiggie branches, greene and tough, like ruſhes: the floures grow all along the ſtalkes like thoſe of Broome, but of a white colour, wherein it differeth from all the reſt of his kinde.

¶ *The Place.*

Theſe grow in the Prouinces of Spaine, and are in one place higher and more buſhie, and in an other lower.

¶ *The Time.*

‡ The firſt floures in May, and the ſecond in February. ‡

¶ *The Names.*

Theſe baſe Spaniſh broomes may be referred to the true, which is called in Greeke ἀσπνι the La-tines vſe the ſame name, calling it ſometimes *Spartum,* and *Spartium:* in Spaniſh, *Retama:* in Engliſh, Spaniſh Broome, and baſtard Spaniſh Broome.

¶ *The Temperature and Vertues.*

A Both the ſeeds and juyce of the branches of theſe baſe Broomes, wherewith they in Spaine and other hot regions do tie their vines, do mightily draw, as *Galen* writeth.

Dioſcorides

Dioscorides saith, That the seeds and floures being drunke in the quantity of a dram, with Mede or honied water, doth cause one to vomit strongly, as the Hellebor or neesing pouder doth, but yet without jeopardy or danger of life : the seed purgeth by stoole.

The juyce which is drawne from out of the branches steeped in water, being first bruised, is a remedy for those that are tormented with the Sciatica, and for those that be troubled with the Squincie, if a draught thereof be drunke in the morning ; some vse to steepe the branches in Sea water, and to giue the same in a clister, which purgeth forth bloudy and slimy excrements.

† In this chapter formerly in the first place was againe figured and described the true *Spartium* or Spanish Broome; which I haue now omitted, because it was figured and described in the last chapter saue one before. In the second place was described that figured in the third; and in the third place was a description to no purpose, which I therefore omitted, and as you see described anew and put in the first place that which formerly held the second.

CHAP. 20.

Of Furze, Gorsse, Whin, or prickley Broome.

¶ The Kindes.

THere be diuers sorts of prickely Broome, called in our English tongue by sundry names, according to the speech of the countrey people where they doe grow : in some places, Furzes ; in others, Whins, Gorsse, and of some, prickly Broome.

† 1 *Genista spinosa vulgaris*.
Great Furze bush.

2 *Genista spinosa minor*.
The small Furze bush.

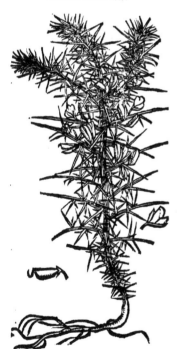

¶ The Description.

1　THe Furze bush is a plant altogether a Thorne, fully armed with most sharpe prickles, without any leaues at all except in the Spring, and those very few and little, and quickly falling away : it is a bushy shrub, often rising vp with many wooddy branches to the height of foure or fiue cubits or higher, according to the nature and soile where they grow : the greatest and highest that I did euer see do grow about Excester in the West parts of England,

where

where the great ſtalks are dearely bought for the better ſort of people, and the ſmall thorny ſpraies for the poorer ſort. From theſe thorny branches grow little floures like thoſe of Broome, and of a yellow colour, which in hot regions vnder the extreme heate of the Sunne are of a very perfect red colour: in the colder countries of the Eaſt, as Danzicke, Brunſwicke, and Poland, there is not any branch hereof growing, except ſome few plants and ſeeds which my ſelfe haue ſent to Elbing otherwiſe called Meluin, where they are moſt curiouſly kept in their faireſt gardens, as alſo our common Broome, the which I haue ſent thither likewiſe, being firſt deſired by diuers earneſt letters: the cods follow the floures, which the grauer hath omitted, as a German who had neuer ſeene the plant it ſelfe, but framed the figure by heare-ſay: the root is ſtrong, tough, and wooddy.

We haue in our barren grounds of the North part of England another ſort of Furze, bringing forth the like prickly thornes that the other haue: the onely difference conſiſteth in the colour of the floures; for the others bring forth yellow floures, and thoſe of this plant are as white as ſnow.

† 2 To this may be ioyned another kinde of Furze which bringeth forth certaine branches that be ſome cubit high, ſtiffe, and ſet round about at the firſt with ſmall winged Lentil-like leaues and little harmeleſſe prickles, which after they haue been a yeare old, and the leaues gone, be armed onely with moſt hard ſharpe prickles, crooking or bending their points downewards. The floures hereof are of a pale yellow colour, leſſer than thoſe of Broome, yet of the ſame forme: the cods are ſmall, in which do lie little round reddiſh ſeeds: the root is tough and wooddy.

† 3 *Geniſta Spinoſa minor ſiliqua rotunda.*   4 *Geniſta aculeata.*
Small round codded Furze.   Needle Furze or petty Whin.

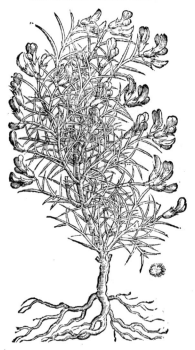

‡ Of this *Cluſius* reckons vp three varieties: the firſt growing ſome cubit high, with deep yellow floures: the ſecond growes higher, and hath paler coloured floures: the third groweth to the height of the firſt, the floures alſo are yellow, the branches more prickly, and the leaues hairy; and the figure I giue you is of this third variety.

3 This ſeldome exceeds a foot in height, and it is on euery ſide armed with ſharpe prickles, which grow not confuſedly, as in the common ſort, but keepe a certaine order, and ſtill grow forth by couples: they are of a lighter greene than thoſe of the common Furze: on the tops of each of the branches grow two or three yellow floures like thoſe of the former, which are ſucceeded by little round rough hairy cods of the bigneſſe of Tares. This floures in March, and groweth in the way betweene Burdeaux and Payone in France, and vpon the Pyrenaean mountaines. *Cluſius* makes it his *Scorpius* 2, or ſecond ſort of Furze: *Lobel* calls it *Geniſta ſpartium ſpinoſum alterum.* ‡

                                                          4 This

4   This ſmall kinde of Furze (growing vpon Hampſtead heath neere London, and in diuers
other barren grounds,where,in manner nothing elſe will grow) hath many weake and flexible branches of a wooddy ſubſtance : whereon doe grow little leaues like thoſe of Tyme : among which are
ſet in number infinite moſt ſharpe prickles, hurting like needles, whereof it tooke his name. The
floures grow on the tops of the branches like thoſe of Broome, and of a pale yellow colour. The
root is tough and wooddy.

‡ 5   This plant (ſaith *Cluſius*) is wholly new and elegant, ſome ſpan high, diuided into many
branches, ſome ſpred vpon the ground, others ſtanding vpright, hauing plentifull ſtore of greene
prickles : the floures in ſhape are like thoſe of Broome, bnt leſſe, and of a blewiſh purple colour,
ſtanding in rough hairy whitiſh cups, two or three floures commonly growing neere together:
ſometimes whileſt it floures it ſendeth forth little leaues, but not very often, and they are few, and
like thoſe of the ſecond deſcribed, and quickely fall away, ſo that the whole plant ſeemes nothing
but prickles or like a hedge-hog when ſhee folds vp her ſelfe : the root is wooddy, and large for the
proportion of the plant. It growes in the kingdome of Valentia in Spaine, where the Spaniards
call it *Erizo*, that is, the Hedge-hog ; and thence *Cluſius* alſo termed it *Erinacea*. It floureth in
Aprill. ‡

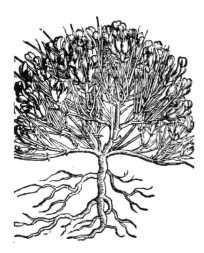
5 *Geniſta ſpinoſa humilis.*
Dwarfe or low Furze.

6 *Geniſta aculeata minor, ſiue Nepa Theophr.*
Scorpion Furze.

6   The ſmalleſt of all the Furze is that of the Antients called *Nepa*, or Scorpion Furze, as the
word *Nepa* ſeemeth to import : it is a ſtranger in England : it hath beene touched of the Antients
in name onely : which fault they haue beene all and euery of them to be complained of, being ſo
briefe that nothing can be gathered from their deſcription : and therefore I refer what might hereof be ſaid to a further conſideration.   ‡   This hath a thicke wooddy blacke root ſome halfe foot
long, from whence ariſe many ſlender branches ſome foot high, which are ſet with many ſtiffe and
ſharpe prickles, growing ſomewhat after the manner of the wilde prickly Sperage : the yong plants
haue little leaues like thoſe of Tragacanth; the old ones none : the flours are ſmall, and come forth
at the bottome of the prickles, and they are ſucceeded by broad cods wherein the ſeed is contained. It growes iu diuers places of France and Spaine, and is thought to be the *Scorpius* of *Theo-*
phraſtus, which *Gaza* tranſlates *Nepa*. ‡

¶ *The Place.*
The common ſort hereof are very well knowne to grow in paſtures and fields in moſt places of
England. The reſt are likewiſe well knowne to thoſe that curiouſly obſerue the difference.

¶ *The Time.*
They floure from the beginning of May to the end of September.

¶ *The Names.*
Furze is commonly called *Geniſta ſpinoſa* : in high Dutch, Gaſpeldozen : in Engliſh, Furze, Fur-
zen buſhes, Whinne, Gorſſe, and Thorne-Broome.
This thorny Broome is taken for *Theophraſtus* his *Scorpius*, which *Gaza* nameth *Nepa* : the name
Scorpius in *Pliny* his πολωνυμον, that is to ſay, ſignifying many things, and common to certaine Plants :

for

for befides this *Scorpius* of which he hath made mention, *lib.25.cap.5.* fetting downe *Theophraftus* his words, where he maketh *Aconitum Thelyphonon* to be *Scorpius, lib.23.cap.10.* and likewife other plants vnder the fame title, but vnproperly.

¶ *The Temperature and Vertues.*

A   There is nothing written in *Theophraftus* concerning the faculties of *Scorpius fpinofus,* or Furze : *Pliny* feemeth to attribute vnto it the fame vertues that *Scorpioides* hath : notwithftanding the later Writers do agree that it is hot and dry of complexion : the feeds are vfed in medicines againft the ftone, and ftaying of the laske.

† This chapter hath vndergone a great alteration : as thus ; the firft, third, and fourth defcriptions belonged to the third figure : the fecond and fifth defcription, to the fifth figure : and the firft, fecond, and fourth figures had no defcriptions belonging to them. The figure that was in the firft place is now in the third : the fecond ftill holds his place : the third is in the firft, belonging thereto of right : and for handfomneffe fake I haue made the fourth and fifth change place. This *Neps* alfo in the fixth place was formerly mentioned by our Author (but now omitted) in the chapter of Afparagus.

<h2 style="text-align:center">CHAP. 21.</h2>

<h3 style="text-align:center">Of Cammocke, Furze, Reſt-Harrow, or Petty Whinne.</h3>

¶ *The Kindes.*

THere be diuers forts of Reſt-Harrow, which fome haue inferted among the fmooth Broomes, others, among thofe with prickles, whereof fome haue purple floures and likewife full of prickles ; others white floures, and fharpe thornes : fome alfo purple floures, others white, and alfo yellow, and euery of them void of prickles.

1 *Anonis, fiue Reſta Bouis.*
Cammocke, or Reſt-Harrow.

3 *Anonis non fpinofa purpurea.*
Purple Reſt-Harrow without prickles.

¶ *The Defcription.*

1   CAmmock or ground Furze rifeth vp with ftalkes a cubit high, and often higher, fet with diuers joynted branches, tough, pliable, and full of hard fharpe thornes : among which do grow leaues in forme like thofe of S. Iohns-wort, or rather of the Lentill, of a

deepe

deepe green colour: from the bofome of which thorns and leaues come forth the floures, like thofe of Peafon, of a purple colour: after which doe come the cods, in which do lie flat feed: the root is long, and runneth far abroad, very tough, and hard to be torne in pieces with the plough, infomuch that the oxen can hardly paffe forward, but are conftrained to ftand ftill; whereupon it was called Reft-Plough, or Reft-harrow.

4 *Anonis, fiue Spina lutea.*
Yellow Reft-Harrow.

2 We haue in our London paftures, and likewife in other places, one of the Reft-Harrowes, not differing from the precedent in ftalkes, leaues, or prickles: the onely differnce is, that this plant bringeth forth white flours, and the others not fo: whence we may call it *Anonis flore albo*, Cammocke with white floures.

3 Reft-harrow without thornes hath a tough hoary rough ftalke, diuided into other rough branches, whereon are fet without order, long leaues fharpe pointed, fleightly cut about the edges, of an hoary colour, and fomewhat hairie: from the bofome whereof commeth forth purple Peafe like floures of a reafonable good fmell: the root is very tough, long, and wooddy.

4 The yellow floured Cammock is a ftranger in thefe parts, it is only found in the cold Eafterne countries, for ought that I can learne; it differs not from the laft defcribed, fauing that the floures hereof are of a darke yellow colour, wherein it differeth from all the other of his kinde.

¶ *The Place.*

Thefe grow in earable grounds in fertile paftures, and in the borders of fields, in a fat, fruitfull, and long lafting foile: it is fooner found than defired of husbandmen, becaufe the tough and wooddie roots are comberfome vnto them, for that they ftay the plough, and make the oxen ftand.

¶ *The Time.*

They fend forth new fhoots in May: they be ful growne in Autumne, and then thofe that of nature are prickly be fulleft of fharpe thornes: they floure in Iuly and Auguft.

¶ *The Names.*

Cammocke is called in Greeke, Ἀνωνίς, or Ὀνωνίς: and likewife in Latine *Anonis*, and *Ononis*: Of Herbarifts commonly *Arefta Bouis*, and *Remora aratri*, becaufe it maketh the Oxen whileft they be in plowing to reft or ftand ftill: it is alfo called *Acutella*, of the ftiffe and fharpe thorns which pricke thofe that paffe by: in French, *Arefte beuf*, and *Boucrande*.

Crateuas nameth it *Ægipyrus*: in high Dutch, **Stalkraut**: in low Dutch, **Prangwoztele**: in Italian, *Bonaga*: in Spanifh, *Gattilhos*: in French, *Arrefte beuf, Beuf & Bouerande*: in Englifh, Cammocke, Reft-harrow, Petty Whinne, and ground Furze.

¶ *The Temperature.*

The root of Cammocke is hot in the third degree, as *Galen* faith: it cutteth alfo and maketh thinne.

¶ *The Vertues.*

The barke of the root drunke with Wine prouoketh vrine, breaketh the ftone, and driueth it forth. A

The root boyled in water and vineger allayeth the paine of the teeth, if the mouth be often wafhed therewith hot. B

Pliny reporteth, That being boyled in Oxymel (or the fyrrup made with hony and vineger) till the one halfe be wafted, it is giuen to thofe that haue the falling fickeneffe. *Matthiolus* reporteth, that he knew a man cured of a ruprure, by taking of the pouder of this root for many monetlis together. C

The tender fprigs or crops of this fhrub before the thornes come forth, are preferued in pickle, and be very pleafant fauce to be eaten with meat as a fallad, as *Diofcorides* teacheth. D

CHAP.

CHAP. 22.
Of Gooſe-berrie, or Fea-berry Buſh.

¶ *The Kindes.*

THere be diuers ſorts of the Gooſe-berries ; ſome greater, others leſſe : ſome round, others long ; and ſome of a red colour : the figure of one ſhall ſerue for the reſt.

‡ I will not much inſiſt vpon diuerſities of fruits, becauſe my kinde friend Mr. *Iohn Parkinſon* hath ſufficiently in his late Worke diſcourſed vpon that ſubject ; onely becauſe I iudge many will be deſirous to know their names, and where to get them, I will briefely name the chiefe varieties our Kingdome affords ; and ſuch as are deſirous of them may finde them with Mr. *Iohn Millen* liuing in Old-ſtreet.

The ſorts of Gooſe-berries are theſe : the long greene, the great yellowiſh, the blew, the great round red, the long red, and the prickly Gooſe-berry. ‡

Vua Criſpa.
Gooſe-berries.

¶ *The Deſcription.*

THe Gooſe-berry buſh is a ſhrub of three or foure cubits high, ſet thicke with moſt ſharpe prickles: it is likewiſe full of branches, ſlender, wooddy, and prickly : whereon do grow round leaues cut with deepe gaſhes into diuers parts like thoſe of the Vine, of a very greene colour : the floures be very ſmall, of a whitiſh greene, with ſome little purple daſhed here and there: the fruit is round, growing ſcatteringly vpon the branches, greene at the firſt, but waxing a little yellow through maturitie, full of a winie iuyce ſomewhat ſweet in taſte when they be ripe ; in which is contained hard ſeed of a whitiſh colour : the root is wooddy, and not without ſtrings annexed thereto.

There is another whoſe fruit is almoſt as big as a ſmall Cherry, and very round in forme : as alſo another of the like bigneſſe, of an inch in length, in taſte and ſubſtance agreeing with the common ſort.

We haue alſo in our London gardens another ſort altogether without prickles: whoſe fruit is very ſmal, leſſer by much than the common kinde, but of a perfect red colour, wherein it differeth from the reſt of his kinde.

¶ *The Place.*

Theſe plants doe grow in our London Gardens and elſe-where in great abundace.

¶ *The Time.*

The leaues come forth in the beginning of Aprill or ſooner : the fruit is ripe in Iune and Iuly.

¶ *The Names.*

This ſhrub hath no name among the old Writers, who as we deeme knew it not, or elſe eſteemed it not : the later Writers call it in Latine, *Croſſularia* : and oftentimes of the berries, *Vua Criſpa, Vua ſpina, Vua ſpinella,* and *Vua Criſpina* : in high Dutch, **Kruſelbeer** : in low Dutch, **Stekelbeſſen** : in Spaniſh, *Vua Criſpa,* or *Eſpina :* in Italian, *Vua ſpina :* in French, *Groiſelles :* in Engliſh, Gooſe-berry, Gooſe-berry buſh, and Fea-berry buſh in Cheſhire, my natiue country.

¶ *The Temperature.*

The berries of this buſh before they be ripe are cold and dry, and that in the later end of the ſecond degree. and alſo binding.

¶ *The Vertues.*

A   The fruit is vſed in diuers ſauces for meat, as thoſe that are skilfull in cookerie can better tell than my ſelfe.

They

They are vſed in broths in ſtead of Verjuice, which maketh the broth not onely pleaſant to the   B
taſte, but is greatly profitable to ſuch as are troubled with an hot burning ague.

They are diuerſly eaten, but they euery way ingender raw and cold bloud : they nouriſh nothing   C
or very little : they alſo ſtay the belly and ſtench bleedings.

They ſtop the menſes or monethly ſicknes, except they happen to be taken into a cold ſtomack,   D
then do they not help, but rather clog or trouble the ſame by ſome manner of flix.

The ripe berries, as they are ſweeter, ſo doe they alſo little or nothing binde, and are ſomething   E
hot, and yeeld a little more nouriſhment than thoſe that be not ripe, and the ſame not crude or raw:
but theſe are ſeldome eaten or vſed as ſauce.

The juice of the green Gooſeberries cureth all inflammations, *Eryſipelas*, and S. *Anthonies* fire.   F

They prouoke appetite, and cure the vehement heate of the ſtomacke and liuer.   G

The young and tender leaues eaten raw in a ſallad prouoke vrine, and driue forth the ſtone and   H
grauell.

Chap. 23.   *Of Barberries.*

¶ *The Kindes.*

There be diuers ſorts of Barberries, ſome greater, others leſſer, and ſome without ſtones.

Spina acida, ſive Oxyacantha.
The Barberry buſh.

¶ *The Deſcription.*

THe Barberry plant is an high ſhrub or buſh,
hauing many yong ſtraight ſhoots & branches very full of white prickly thorns, the rinde
whereof is ſmooth and thin, the wood it ſelf yellow : the leaues are long, very greene, ſleightly
nicked about the edges, and or a ſoure taſte: the
flours be yellow, ſtanding in cluſters vpon long
ſtems : in their places come vp long berries, ſlender, red when they be ripe, with a little hard kernell or ſtone within ; of a ſoure and ſharp taſte :
the root is yellow, diſperſeth it ſelf far abroad,
and is of a wooddy ſubſtance.

Wee haue in our London gardens another
ſort, whoſe fruit is like in forme and ſubſtance,
but one berry is as big as three of the common
kinde, wherein conſiſteth the difference.

We haue likewiſe another without any ſtone,
the fruit is like the reſt of the Barberries both
in ſubſtance and taſte.

¶ *The Place.*

The Barberry buſh growes of it ſelfe in vntoiled places and deſart grounds, in woods and the
borders of fields, eſpecially about a gentlemans
houſe called Mr *Monke*, at a village called Iver
two miles from Colebrooke, where moſt of the
hedges are nothing elſe but Barberry buſhes.

They are planted in moſt of our Engliſh gardens.

¶ *The Time.*

The leaues ſpring forth in Aprill, the floures and fruit in September.

¶ *The Names.*

Galen calleth this Thorne in Greeke, οξυάκανθα, who maketh it to differ from οξυάκανθος, in his book
of the Faculties of ſimple Medicines ; but more plainly in his booke of the Faculties of Nouriſhments ; where he reckoneth vp the tender ſprings of Barberries amongſt the tender ſhoots that are

to be eaten, ſuch as *Oxyacanthus* or the Hawthorne bringeth not forth, wherein he plainely made a difference, *Oxyacantha* the Barberry buſh, and *Oxyacanthus* the Hawthorne tree.

Dioſcorides hath not made mention of this Thorne; for that which he calleth *Oxyacantha*, in the fœminine gender, is *Galens Oxyacanthus* in the maſculine gender.

Auicen ſeemes to containe both theſe herbes vnder the name of *Amyrberis*, but we know they are neither of affinitie nor neighborhood, although they are both prickly.

The ſhrub it ſelfe is called in ſhops Barberries, of the corrupted name *Amyrberis* : of the later writers *Creſpinus* : in Italian, *Creſpino* : in Spaniſh, *Eſpino de maiuelas* : in high-Dutch, **Paiſſelbeer :** in low-Dutch, **Sauſeboom :** in French, *Eſpine vinette* : and thereupon by a Latine name, *Spiniuineta, Spina acida,* and *Oxyacantha Galeni* : ‡ in Engliſh, a Barberry buſh, or Piprige tree, according to Dr *Turner.* ‡

¶ *The Temperature.*

The leaues and berries of this Thorne are cold and dry in the ſecond degree, and as *Galen* alſo affirmeth, they are of thin parts, and haue a certain cutting facultie.

¶ *The Vertues.*

A   The leaues are vſed of diuers to ſeaſon meat with, and in ſtead of a ſallad, as be thoſe of Sorrell.

B   The decoction thereof is good againſt hot burnings and cholericke agues : it allayeth the heat of the bloud, and tempereth the ouermuch heate of the liuer.

C   The fruit or berries are good for the ſame things, and be alſo profitable for hot Laskes, and for the bloudy flix, and they ſtay all manner of ſuperfluous bleedings.

D   The green leaues of the Barberry buſh ſtamped and made into ſauce as that made of Sorel, called Green-ſauce, doth coole hot ſtomacks, and thoſe that are vexed with hot burning agues, & procureth appetite.

E   A conſerue made of the fruit and ſugar performeth all thoſe things before remembred, & with better force and ſucceſſe.

F   The roots of the tree ſteeped for certain daies together in ſtrong lie made with the aſhes of the Aſh tree, and the haire often moiſtned therewith, maketh it yellow.

G   ‡ The bark of the roots is alſo vſed in medicins for the jaundice, and that with good ſucceſſe.‡

CHAP. 24.   *Of the white Thorne or Hawthorne tree.*

¶ *The Kindes.*

THere be two ſorts of the white Thorn trees deſcribed by the later writers, one very common in moſt parts of England, another very rare, and not found in Europ, except in ſome few rare gardens of Germay; which differs not from our common Hawthorne, ſauing that the fruit thereof is as yellow as Saffron. We haue in the Weſt of England one growing at a place called Glaſtenbury which brings forth his floures about Chriſtmaſſe, by the report of diuers of good credit, who haue ſeen the ſame : but my ſelfe haue not ſeen it, and therefore leaue it to be better examined.

¶ *The Deſcription.*

1   THe white Thorn is a great ſhrub growing often to the height of a peare tree, the trunk or body is great, the boughes and branches hard and woody, ſet with long ſharp thorns : the leaues be broad, cut with deep gaſhes into diuers ſections, ſmooth, and of a gliſteting green colour : the floures grow vpon ſpoky rundles, of a pleaſant ſweet ſmell, ſomtimes white, and often daſht ouer with a light waſh of purple, which hath moued ſome to thinke a difference in the plants : after which come the fruit, being round berries, green at the firſt, and red when they be ripe; wherein is found a ſoft ſweet pulpe and certaine whitiſh ſeed : the root growes deepe in the ground, of a hard wooddy ſubſtance.

2   The ſecond and third haue bin touched in the firſt title, notwithſtanding I haue thought it not vnfit to inſert in this place a plant participating with the Hawthorne in floures and fruit, and with the Servis tree in tree in leaues, and not vnlike in fruit alſo.

Theophraſtus hath ſet forth this Tree vnder the name of *Aria*, which groweth vnto the forme of a ſmall tree, delighting to grow in the ſhadowie woods of Cumberland and Weſtmerland, and many other places of the North country, where it is to be found in great quantity : but ſeldome in

Spain,

Spain, Italy, or any hot region. This tree is garniſhed with many large branches beſet with leaues ſike the Peare tree, or rather like the Aller leaf, of a dark green colour aboue, and white vnderneath: among theſe leaues come forth tufts of white floures very like vnto Hawthorn floures, but bigger: after which ſucceed ſmall red berries like the berries of the Hawthorn, in taſte like the Neapolitan Medlar; the temperature and faculties whereof are not yet knowne.

1 *Oxyacanthus.*
The Hawthorne tree.

2 *Aria Theophraſti.*
Cumberland Hawthorne.

¶ *The Place.*

The Hawthorn groweth in woods and in hedges neere vnto highwaies almoſt euery where. The ſecond is a ſtranger in England. The laſt groweth at Glaſtenbury abbey, as it is credibly reported to me. ‡ The *Aria* groweth vpon Hampſted heath, and in many places of the Weſt of England.‡

¶ *The Time.*

The firſt and ſecond floure in May, whereupon many do call the tree it ſelfe the May buſh, as a chiefe token of the comming in of May: the leaues come forth a little ſooner: the fruit is ripe in the beginning of September, and is a food for birds in Winter.

¶ *The Names.*

Dioſcorides deſcribeth this ſhrub, and nameth it οξυάκανθα, in the feminine gender; and *Galen* in his booke of the Faculties of ſimple medicines, οξυάκανθος, in the maſculine gender: *Oxyacanthus* ſaith he, is a tree, and is like to the wilde Peare tree in forme, and the vertues not vnlike, &c. Of *Oxyacantha, Dioſcorides* writeth thus: It is a tree like to the wilde Peare tree, very full of thornes, &c. *Serapio* calleth it *Amyrberis*: and ſome, ſaith *Dioſcorides*, would haue it called Ῥαμνίνη, but the name *Pyrina* ſeemeth to belong to the yellow Hawthorn: it is called in high-Dutch, 𝕳𝕒𝕠𝕘𝕓𝕠𝕫𝕖𝕟: in low-Dutch, 𝕳𝕒𝕘𝕖𝕓𝕠𝕫𝕖𝕟: in Italian, *Bagaia*: in Spaniſh, *Pirlitero*: in French, *Aube-eſpine*: in Engliſh, White-thorn, Hawthorn tree; and of ſome Londoners, May-buſh. ‡ This is not the *Oxyacantha* of the Greeks, but that which is called *Pyracantha*, as ſhall be ſhewed hereafter. ‡

The ſecond is thought to be the *Aria* of *Theophraſtus*, and ſo *Lobel* and *Tabern.* call it. Some, as *Bellonius, Geſner*, and *Cluſius*, referre it to the *Sorbus*, and that not vnfitly: in ſome places of this Kingdome they call it a white Beam tree. ‡

　　　　　　　　　　¶ *The*

¶ *The Temperature.*

The fruit of the Hawthorne tree is very astringent.

¶ *The Vertues.*

The Hawes or berries of the Hawthorne tree, as *Dioscorides* writeth, doe both stay the Laske the menses, and all other fluxes of bloud. Some Authors write, That the stones beaten to pouder and giuen to drinke are good against the stone.

Chap. 25. *Of Goats-Thorne.*

¶ *The Description.*

1   THe first *Tragacantha* or Goats-thorne hath many branchy boughes and twiggs, slender and pliant, so spread abroad vpon euery side, that one plant doth sometimes occupie a great space or room in compasse : the leaues are small, and in shape like Lentill leaues, whitish, and somwhat mossie or hairy, set in rowes one opposit against another: the flour is like the blossom of the Lentill, but much lesser, and of a whitish colour, and somtimes marked with purple lines or streakes : the seed is inclosed in small cods and huskes, almost like vnto the wilde *Lotus* or horned Trefoile : the whole plant on euery side is set full of sharp prickly thornes, hard, white, and strong : the roots run vnder the ground like Licorice roots, yellow within, and black without, tough limmer, and hard to breake ; which being wounded in sundry places with some iron toole, and laid in the Sun at the highest and hottest time of Summer, issueth forth a certaine liquor, which being hardned by the Sun, is that gum called in shops *Tragacantha*, and of some, though barbarously, *Dragagant*.

       1 *Tragacantha, siue Spina Hirci.*   2 *Spina Hirci minor.*
         Goats Thorne.   Small Goats Thorne.

2   The second kinde of *Tragacantha* is a low and thicke Shrub, hauing many shoots growing from one turfe, of a white or grayish colour, about a cubit high, stiffe and wooddy : the leaues are like the former, and garded with most stiffe pricks not very safely to be touched : among the thornie leaues come forth many floures in small tufts like *Genistella*, but that they are white : the cods
                                                           are

are many, ſtraight and thorny like *Geniſtella*, wherin are many ſmall white and three cornered ſeeds as big as muſtard ſeed. ‡ this differs from the former, in that it is ſmaller, and loſeth the leaues euery winter, when as the former keeps on the leaues vntill new ones come on in the Spring. The middle of the winged leaues ends in a pricke, which by the falling of the leaues becomes a long and naked thorne. I haue giuen you a more accurat figure hereof out of *Cluſius*, wherein the leaues, floures, cods, and ſeeds are all expreſſed apart. ‡

3   The Grecians haue called this plant Νινϱῴδια, becauſe it is good for the ſinues: it ſhould ſeem it tooke the name *Poterion* of *Potrix*, becauſe it loueth a waterie or fenny ſoile: it hath ſmall branches, and leaues of *Tragacantha*, growing naturally in the tract of Piedmont in Italy: it ſpreadeth abroad like a ſhrub: the barke or rinde is blackiſh, and dry without great moiſture, very much writhed or wrinkled in and out as that of *Nepa* or *Corruda*: the ſharp prickes ſtand not in order as *Tragacantha*, but confuſedly, and are finer and three times leſſer than thoſe of *Tragacantha*, growing much after the manner of *Aſtragalus*; but the particular leaues are green aboue, and white below, ſhaped ſomwhat like Burnet: the ſeed is ſmall and red, like vnto Sumach, but leſſer.

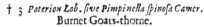

‡ *Tragacanthæ minoris icon accuratior.*
A better figure of the Goats-thorne.

† 3 *Poterion Lob. ſiue Pimpinella ſpinoſa Camer.*
Burnet Goats-thorne.

¶ *The Place.*

Petrus Bellonius, lib.1. of Singularities reports, that there is great plenty hereof growing in Candy vpon the tops of the mountains. *Theophraſtus* ſaith that it was thought to grow nowhere but in Candy; but now it is certain that it is found in Achaia, Peloponeſus, and in Aſia: it growes alſo in Arcadia, which is thought to be inferior to that of Candy. It is thought by *Lobel* to grow in Languedoc in France, whereof *Theophr. lib. 9.* hath written, that the liquor or gum iſſues out of it ſelf, and that it is not needfull to haue the root broken or cut. The beſt is that, ſaith *Dioſcor.* which is through-ſhining, thin, ſmooth, vnmixt, and ſweet of ſmell and taſte.

¶ *The Time.*

They floure and flouriſh in the Summer moneths. I haue ſown the ſeed of *Poterion* in April, hauing receiued it from *Ioach. Camer.* of Noremberg: it grew in my garden two yeares together, and afterwards periſhed by ſome miſchance.

¶ *The Names.*

Goats-thorne is called in Greeke τραγάκανθα : of moft writers likewife *Tragacantha* : we may call it in Latine *Spina hirci* : in French, *Barbe Renard* : and in Englifh for want of a better name, Goats-thorne : the liquor or gum that iffueth forth of the root beareth the name alfo of *Tragacantha* : it is called in fhops *Gumma Tragacantha*, and in a barbarous manner, *Gumma Tragacanthi* : in Englifh, Gum Dragagant.

¶ *The Temperature.*

Each part of this plant is of a drying facultie without biting : it doth confolidate or glew to-gether finues that be cut : but the roots boiled in wine haue that facultie fpecially, being giuen to thofe that haue any griefe or hurt in the finues.

Gum Dragagant hath an emplafticke qualitie, by reafon whereof it dulleth and allayeth the fharpneffe of humors, and doth alfo fomthing dry.

¶ *The Vertues.*

The gum is fingular good to be licked in with hony againft the cough, roughneffe of the throat, hoarfneffe, and all fharp and thin rheumes and diftillations : being laid vnder the tongue it taketh away the roughneffe thereof.

Being drunke with Cute or the decoction of Licorice it taketh away and allayeth the heate of the vrine : it is alfo vfed in medicines for the eies.

The greateft part of thofe artificiall beades, fweet chaines, bracelets, and fuch like pretty fweet things of pleafure are made hard and fit to be worne, by mixing the gum hereof with other fweets, being firft fteeped in Rofe water vntill it be foft.

† The figure which was in the third place was of the plant defcribed in the fecond, which *Matthiolus* and *Tabern.* made their *Poterion*, but it agreed not with the defcription, which was taken out of the *Advorf.*

Chap. 26.   *Of the Ægyptian Thorne.*

‡ 1 *Acacia Diofcoridis.*   † 2 *Acacia altera trifolia.*
The Egyptian Thorne.   Thorny Trefoile.

¶ *The Deſcription.*

1　Dioſcorides makes mention of *Acacia*,whereof the firſt is the true and right *Acacia*, which is a ſhrub or hedge tree,but not growing right or ſtraight vp as other ſmall trees doe: his branches are wooddy, beſet with many hard and long thornes ; about which grow the leaues,compact of many ſmall leaues cluſtering about one ſide,as in the Lentil : the flours are whitiſh, the huskes or cods be plaine and flat, yea very broad like Lupines, eſpecially on that ſide where the ſeed groweth,which is contained ſomtimes in one part,and ſomtimes in two parts of the husk,growing together in a narrow neck : the ſeed is ſmooth and gliſtering.There is a black juyce taken out of theſe huskes, if they be dried in the ſhade when they be ripe ; but if they are not ripe, then it is ſomwhat red : ſome dowring a juice out of the leaues and fruit : there flowes alſo a gum out of this tree,which is the gum of Arabia,called gum Arabick.

2　*Dioſcorides* hauing deſcribed *Spina Acacia*, ſets downe a ſecond kinde thereof, calling it *Acacia altera*,which hath the three leaues of Rue or *Cytiſus*,and coddes like thoſe of *Geniſtella*,but ſomewhat more blunt at the end,and thicke at the backe like a Raſor,and ſtill groweth forward narrower and narrower,vntill it come to haue a ſharpe edge : in theſe cods are contained three or foure flat ſeeds like *Geniſtella*,which before they wax ripe are yellow, but afterwards blacke : the whole plant groweth to the height of *Geniſta ſpinoſa* or Gorſſe, both in ſhape, height, and reſemblance, and not to the height of a tree,as *Matthiolus* would perſuade vs,but full of ſharp thornes like the former.

¶ *The Place.*

The true Acacia groweth in Egypt,Paleſtina,Lombardy,and Syria,as *Dioſcorides* writes:among the ſhrubs and trees that remaine alwaies green, Acacia is noted for one,by *Petrus Bellonius,lib.* 1. *cap.*44. of his Singularities.

The other Acacia groweth in Cappadocia and Pontus, as *Dioſcorides* writeth ; it is alſo found in Corſica, and in diuers mountaines of Italy, and likewiſe vpon all the coaſt of Liguria and Lombardie,and vpon the Narbon coaſt ot the Mediterranean ſea.

¶ *The Time.*

Theſe floure in May,and their fruit is ripe in the end of Auguſt.

¶ *The Names.*

The tree Acacia is named of the Grecians Ἀκακία, yea euen in our time, & likewiſe of the Latines *Acacia* : it is alſo called *Egyptia ſpina* : this ſtrange thorne hath no Engliſh name that I can learne, and therefore it may ſtill keep the Latine name Acacia ; yet I haue named it the Egyptian thorne. The juice alſo is called Acacia. The Apothecaries of Germanie vſe in ſtead hereof the juice that is preſſed out of ſloes or ſnags,which they therefore call *Acacia Germanica*. *Matthiolus* pictureth for *Acacia* the tree which the later Herbariſts call *Arbor Iudæ*, to which he hath vntruly added thornes, that he might belie Acacia ; and yet he hath not made it agree with *Dioſcorides* his deſcription.

They call this Ἐνεγ αγίια : in Latine, *Acacia altera*, or the other Acacia,and *Pontica Acacia*,or Ponticke Acacia.

¶ *The Nature.*

The juice of Acacia,as *Galen* ſaith, conſiſts not of one onely ſubſtance,but is both cold and earthy,to which alſo is coupled a certaine waterie eſſence,and it likewiſe hath thinne and hot parts diſperſed in it ſelfe ; therefore it is dry in the third degree, and cold in the firſt, if it be not waſhed ; and in the ſecond if it be waſhed;for by waſhing it loſeth his ſharp and biting quality,and the hot parts.

¶ *The Vertues.*

The juice of Acacia ſtoppeth the laske,the inordinat courſe of womens termes,and mans invo-　A luntarie iſſue called *Gonorrhæa*, if it be drunke in red wine.

It healeth the blaſtings and inflammations of the eies, and maketh the skin and palmes of the　B hands ſmooth,after the healing of the *Serpigo* : it healeth the bliſters & extreme heat in the mouth, and maketh the haires blacke that are waſhed therewith.

·It is good,ſaith *Dioſcorides*,againſt S.Anthonies fire, the ſhingles,Chimetla,Pterygia,& whit-　C lowes.

The gum doth binde and ſomwhat coole : it hath alſo joined vnto it an emplaſticke qualitie,by　D which it dulleth or alayeth the ſharpneſſe of the medicines wherewith it is mixed. Being applied with the white and yolk of an egge it ſuffers not bliſters to riſe in burned or ſcalded parts. *Dioſc.*

The juice of the other,ſaith *Dioſcorides*,doth alſo binde,but it is not ſo effectuall nor ſo good in　E ₃ye medicines.

† Our Author gaue but formerly one figure,which was that in the ſecond place, and he would haue perſuaded vs that it was of the right *Acacia*,yet in his deſcription he tells vs otherwiſe.

CHAP. 27.

Of Box Thorne, and the juice thereof called Lycium.

¶ *The Deſcription.*

1   BOx Thorne is a rare plant in ſhape not vnlike the Box tree,whereof it hath bin reckoned for a wilde kinde, hauing many great branches ſet full of round and thicke leaues very like that of the common Box tree : amongſt which grow forth moſt ſharpe pricking thornes : the floures grow among the leaues, which yeeld forth ſmall black berries of a bitter taſte, as big as a pepper corne : the juice whereof is ſomwhat oily,and of a reddiſh colour : which bitter juice being ſet on fire, doth burne with a maruellous cracking and ſparkling : the aſhes thereof are of a reddiſh colour : it hath many wooddy roots growing aſlope.

1 *Lycium, ſiue Pyxacantha.*
Box Thorne.

‡ 2 *Lycium Hiſpanicum.*
Spaniſh Box Thorne.

2   The other kinde of *Pyxacantha* or *Lycium* groweth like vnto the common Priuet, hauing ſuch like leaues, but ſomewhat narrower : the tops of the ſlender ſprigs are furniſhed with prickles : the root is tough, and of a wooddy ſubſtance.

¶ *The Place.*

·   They grow in Cappadocia and Lycia, and in many other countries : it proſpers in rough places, it hath likewiſe bin found at Languedoc and Prouince in France. *Bellonius* writeth that he found it in Paleſtina.

Matthiolus pictureth for Box Thorne a plant with Box leaues, with very many boughes and certaine thornes ſtanding among them : but the notable Herbariſt *Anguillara* and others hold opinion that it is not the right ; with whom we alſo agree.

There is drawn out of the leaues and branches of Box Thorn, or as *Pliny* ſaith, out of the boughs and roots being throughly boiled, a juice named *Lycium.*

Dioſcorides ſaith that the leaues and branches muſt be brayed, and the infuſion made many daies

in

in the decoction thereof, after which the feces or wooddy stuffe must be cast away, and that which remaineth boiled againe till it become as thicke as hony. *Pliny* saith that the roots and branches are very bitter, and for three daies together they must be boiled in a copper vessell, and the wood and sticks often taken out till the decoction be boiled to the thicknesse of hony.

¶ *The Time.*

They floure in Februarie and March, and their fruit is ripe in September.

¶ *The Names.*

It is named in Greeke πυξάκανθα, which a man may call in Latine *Buxea spina*; and in English, box Thorne: of some, Asses Box tree, and prickly Box: it is also named *Lycium*, of the juice which is boiled out of it: the juice is properly called λύκιον, and retaineth in Latine the same name *Lycium*: it is termed in English, Thorne Box. But it seemeth to me that the originall name *Lycium* is fitter, being a strange thing, and knowne to very few: the Apothecaries know it not, who in stead thereof do vse amisse the juice of the fruit of Woodbinde, and that not without great error, as we haue already written. ‡ It is vnknown in our shops, neither is there any thing vsed for it, it being wholly out of vse: wherefore our Author might here very well haue spared *Dodonæus* his words. ‡

Dioscorides teacheth to make a ῥοῦ or Sumach which is good for those things that *Lycium* is, and is vsed when *Lycium* is not to be had, and it is fit to be put into all medicines in stead thereof.

¶ *The Temperature.*

Lycium, or the juice of Box Thorne is, as *Galen* teacheth, of a drying qualitie, and compounded of diuers kinds of substances, one of thin parts digesting and hot, another earthy and cold, wherby it enioyeth his binding faculitie: it is hot in a mean, and therefore it is vsed for seuerall purposes.

¶ *The Vertues.*

Lycium cleareth the sight, saith *Dioscorides*, it healeth the scuruy festred sores of the eye lids, the itch, and old fluxes or distillations of humors: it is a remedie for the running of the eares, for Vlcers in the gums, and almonds of the throat, and against the chaps or gallings of the lips and fundament.

† The figure which was in the second place was of the *Lycium Italicum* of *Matthiolus* and others; but the description and title better fitted this *Lycium Hispanicum* of *Lobel*, which therefore I put thereto. The figure also of the *Lycium Italicum* of *Matthiolus* our Author gaue againe in the next chapter: saue two.

Chap. 28.　*Of Ram or Harts Thorne.*

¶ *The Kindes.*

AFter the opinion of *Dioscorides* there bee three forts of *Rhamnus*; one with long flat and soft leaues, the other with white leaues, and the third with round leaues, which are somewhat blackish. *Theophrastus* and *Pliny* affirme that there are but two, the one white, and the other black, both which do beare Thornes. But by the labor and industrie of the new and late writers there are found sundry sorts moe, all which and euery one of them are plants of a woody substance, hauing also many straight twiggy and pliant branches set with most sharp pricking thornes.

¶ *The Description.*

1　THis is a shrub growing in the hedges, and bringing forth straight branches and hard thornes like those of the Hawthorn, with little long leaues something fat and soft: and this hath that notable learned man *Clusius* more diligently described in these words; The Ram is a shrub fit to make hedges of, with straight branches parting it selfe into many twigs, white, and set with stiffe and strong thornes, hauing leaues which for the most part grow by foures or fiues at the root of euery Thorne, long, something fat, like to those of the Oliue tree, somewhat white, but tender and full of juice; which in Autumne do sometimes fall off, leauing new growing in their places: the floures in Autumne are something long, whitish, diuided at the brims into fiue parts: in their places is left a seed in shew as in *Gelsemine*, notwithstanding it was neuer my chance to see the fruit: the root is thicke, and diuersly parted.

‡ I obserued another, saith the same Author, almost like to the former, but lower, and diuided into more branches, with lesser leaues, more thick and salt of taste, and whiter also than the former: the floures are like those of the former in all things but their colour, which in this are purple.

2　This hath more flexible stalks and branches, and these also set with thornes: The leaues are narrow, and not so thicke and fleshy as those of the former, yet remaine alwayes greene like as they do: The floures are small and mossie, of a greenish colour, growing thicke about the branches, and they are succeeded by a round fruit, yellowish when it is ripe, and remaining on the shrubbe all the

Winter,

‡ 1 *Rhamnus* 1. *Cluſij flo. albo.*
White floured Ram-thorne.

‡ *Rhamnus alter Cluſ. flore purpureo.*
Purple floured Ram-thorne.

‡ 2 *Rhamnus* 2. *Cluſij.*
Sallow Thorne.

3 *Rhamnus* 3. *Cluſij.*
Ram or Harts Thorne.

Winter. The whole shrub looks as if it were sprinkled ouer with dust.

3   To these may be added another growing with many branches to the height of a Sloe Tree, or blacke Thorne, and these are couered with a blackish barke, and armed with long prickles. The leaues, as in the first, grow forth of certain knots many together, long, narrow, fleshy, green, and continuing all the yeare: their tast is astringent, somwhat like that of Rhabarb: the flours shew themselues at the beginning of the Spring, of a greenish colour, growing thicke together, and neere the setting on of the leaues: in Summer it carries a blacke fruit almost like a Sloe, round, and harsh of taste.

¶ *The Place.*

The first of these growes in sundry places of Spaine, Portugal, and Province: the other varietie hereof *Clusius* saith he found but only in one place, and that was neere the city Horivela, called by the Antients Orcellis, by the riuer Segura, vpon the borders of the kingdom of Valentia. The second grows in many maritime places of Flanders and Holland, and in some vallies by riuers sides. The third growes in the vntilled places of the kingdom of Granado and Murcia. ‡

¶ *The Time.*

This Ram is euer green together with his leaues: the fruit or berries remain on the shrub euen in Winter.

¶ *The Names.*

The Grecians call this Thorne ῥάμνος: the Latines also *Rhamnus*: diuers name it Παρωθίνη, Ἀκανθαλευθη, that is, *Spina alba*, or White Thorne; *Spina Ceruialis*, or Harts Thorne, as wee finde written amongst the bastard words. *Marcellus* nameth it *Spina salutaris*, and *Herba salutaris*, which hath, saith hee, as it were a grape. It is called in Italian, *Marruco*, and *Rhamno*: in Spanish, *Scambrones*: in English, Ram or Harts Thorne.

¶ *The Temperature.*

The Ram, saith *Galen*, doth dry and digest in the second degree, it cooleth in the later end of the first, and in the beginning of the second degree.

¶ *The Vertues.*

The leaues, saith *Dioscorides*, are laid pultiswise vpon hot cholericke inflammations and S. Anthonies fire, but we must vse them whilest they be yet but tender, as *Galen* addeth.   A

‡   The leaues and buds or young shoots of the first are eaten as sallads, with oile, vineger, and salt, at Salamanca and other places of Castile, for they haue a certain acrimonie and acidity which are gratefull to the taste. A decoction of the fruit of the third is good to foment relaxed and weak or paralyticke members, and to ease the paine of the gout, as the inhabitants of Granado told *Clusius*. ‡   B

† Our Author in this Chapter gaue only the figure of the third, and the description of the first, and the place of the second, with the names and faculties in generall.

CHAP. 29. *Of Christs Thorne.*

¶ *The Description.*

CHrists Thorne or Ram of Lybia is a very tough and hard shrubby bush, growing vp somtimes to the height of a little tree, hauing very long and sharpe prickely branches; but the Thornes that grow about the leaues are lesser, and not so prickly as the former. The leaues are small, broad, and almost round, somewhat sharp pointed, first of a darke green colour, and then somewhat reddish. The floures grow in clusters at the tops of the stalks, of a yellow colour: the husks wherin the seeds be contained are flat and broad, very like vnto small bucklers, as hard as wood, wherin are contained three or foure thin and flat seeds like the seed of Line or Flax.

¶ *The Place.*

This Thorne groweth in Lybia; it is better esteemed of in the countrey of Cyrene than is their Lote tree, as *Pliny* affirmeth. Of this herb *Diphilus Siphnius* in *Athenaus* his 14 booke makes mention, saying, That he did very often eat of the same in Alexandria, that beautifull city.

Petrus Bellonius, who trauelled ouer the Holy land, saith, That this shrubby Thorne *Paliurus* was

the

Palinrus.
Chrifts Thorne.

the Thorne wherewith they crowned our Sauiour Chrift: his reafon for the proofe hereof is this, That in Iudæa there was not any Thorne fo common,fo pliant,or fo fit for to make a crown or garland of,nor any fo full of cruell fharpe prickles.It groweth throughout the whole countrey in fuch abundance,that it is their common fuell to burn; yea, fo common with them there, as our Gorffe, Brakes,and Broome is here with vs. *Iofephus,lib.*1. *cap*.11. of his Antiquities faith, That this thorne hath the moft fharp prickles of any other; wherefore that Chrift might bee the more tormented, the Iewes rather tooke this than any other. Of which I haue a fmall tree growing in my garden, that I haue brought forth by fowing of the feed.

¶ *The Time.*

The leaues fall away and continue not alwayes greene,as doe thofe of the Rams : it buds forth in the Spring,as *Pliny* teftifieth.

¶ *The Names.*

This thorny fhrub is called in Greek, Παλιυρος: the Latines and Italians retaine the fame name *Palinrus .* for want of an Englifh name it may bee termed Ram of Lybia,or Chrifts thorn. *Pliny* faith the feed is called *Zura.*

¶ *The Temperature.*

The leaues and root of Chrifts thorne doe euidently binde and cut.

¶ *The Vertues.*

A   By vertue of this cutting qualitie the feed doth weare away the ftone,and caufe tough and flimy humors to remoue out of the cheft and lungs,as *Galen* faith.

B   The decoction of the leaues and root of Chrifts Thiftle,as *Diofcorides* writes,ftoppeth the belly, prouokes vrine,and is a remedie againft poifons and the bitings of ferpents.

C   The root doth wafte and confume away *Phymata* and *Oedemata*,if it be ftamped and applied.

D   The feed is good for the cough,and weares away the ftone in the bladder.

CHAP. 30.   *Of Buck-Thorne, or laxatiue Ram.*

¶ *The Defcription.*

1   BVck-Thorne growes in manner of a fhrub or hedge tree ; his trunke or body is often as big as a mans thigh ; his wood or timber is yellow within,and the barke is of the colour of a Chefnut,almoft like the barke of a Cherry tree.The branches are befet with leaues that are fomewhat round,and finely fnipt about the edges like the leaues of the Crab or Wilding tree: among which come forth thorns which are hard and prickly : the flonres are white and fmall, which being vaded,there fucceed little round berries,green at the firft,but afterwards black,wherof that excellent green colour is made which the Painters and limners do call Sap-green:but thefe berries before they be ripe do make a faire yellow colour,being fteeped in vineger.

‡   2   Befides the common kinde *Clufius* mentions two other : the firft of which hath branches fome two cubits long,fubdiuided into diuers others, couered with a fmooth barke like that of the former,which,the vpper rinde being taken off,is of a yellowifh greene colour, and bitterifh tafte: the branches haue fome few prickles vpon them, and commonly end in them:the leaues are almoft like vnto thofe of the common kinde, but fmaller,narrower, and fomwhat refembling thofe of the blacke Thorne, hauing fomewhat a drying tafte : the floures confift of foure leaues of a yellowifh

greene

‡ 1 *Rhamnus solutivus.*
Buck-thorne.

‡ 2 *Rhamnus solutivus minor.*
Middle Buck-thorne.

† 3 *Rhamnus solutivus pumilus.*
Dwarfe Buck-thorne.

greene colour : the root is wooddy as in other
shrubs: *Clusius* found this growing in the mountanous places of Austria, and calls it *Spina infe-*
ctoria pumila.

3   This other hath branches some cubit
long, and of the thickenesse of ones little finger,
or lesser, couered with a blacke and shriuelled
barke : and towards the top diuided into little
boughs, which are couered with a thin & smoo-
ther barke, and commonly end in a sharp thorn:
the leaues much resemble those of the Slo-tree
yet are they shorter and lesser, greene also, and
snipt about the edges; first of an astringent, and
afterwards of somewhat a bitterish taste; the
floures which grow amongst the leaues are of
an herby colour, and consist of foure leaues : the
fruit is not much vnlike that of the former; but
distinguished with two, & somtimes with three
crests or dents, first green, and then blacke when
it is ripe : the root is thicke, wooddy and hard.
Clusius found this on the hill aboue the Bathes
of Baden, hee calls it *Spina infectoria pumila* 2.
This *Matthiolus* and others call *Lycium Italicum:*
and our Author formerly gaue the figure of
Matthiolus and *Tabernamontanus* by the name of
Lycium Hispanicum, and here againe another for
his *Rhamnus solutivus*, which made me to keepe
it in this chapter, and omit it in the former, it
being described in neither. ‡

¶ *The Place.*

Buck-thorne groweth neere the borders of fields, in hedges, woods, and in other vntoiled places : it delighteth to grow in riuers and in water ditches : it groweth in Kent in ſundry places, as at Farningham vpon the cony burrowes belonging ſometime to Mr *Sibil*, as alſo vpon cony burrowes in South-fleet, eſpecially in a ſmall and narrow lane leading from the houſe of Mr *Willian Swan* vnto Longfield-downes, alſo in the hedge vpon the right hand at Dartford townes end towards London, and in many places more vpon the chalkie bankes and hedges.

¶ *The Time.*

It floureth in May the berries be ripe in the fall of the leafe.

¶ *The Names.*

The later Herbariſts call it in Latine *Rhamnus ſolutivus*, becauſe it is ſet with thornes, like as the Ram, and beareth purging berries. *Matthiolus* nameth it *Spina infectoria*; *Valerius Cordus*, *Spina Cerui*, and diuers call it *Burgiſpina*. It is termed in high Dutch, **Creukbeer Wegbdorn**; in Italian, *Spino Merlo, Spino Zerlino, Spino Ceruino* : in Engliſh, Laxatiue Ram, Way-thorne, and Buck-thorne : in low Dutch they call the fruit or berries **Rhijnbeſſen**, that is, as though you ſhould ſay in Latine, *Bacca Rhenanæ* : in Engliſh, Rheinberries : in French, *Nerprun*.

¶ *The Temperature.*

The berries of this Thorne, as they be in taſte bitter and binding, ſo be they alſo hot and dry in the ſecond degree.

¶ *The Vertues.*

A   The ſame do purge and void by the ſtoole thicke flegme, and alſo cholericke humors : they are giuen being beaten into pouder from one dram to a dram and a halfe : diuers do number the berries, who giue to ſtrong bodies from fifteene to twenty or more; but it is better to breake them and boile them in fat fleſh broth without ſalt, and to giue the broth to drinke : for ſo they purge with leſſer trouble and fewer gripings.

B   ◀ There is preſſed forth of the ripe berries a juyce, which being boyled with a little Allum is vſed of painters for a deepe greene, which they do call Sap-greene.

C   The berries which be as yet vnripe, being dried or infuſed or ſteped in water, do make a faire yellew colour : but if they be ripe they make a greene.

CHAP. 31.   *Of the Holme, Holly, or Huluer tree.*

Agrifolium.
The Holly tree.

¶ *The Deſcription.*

THe Holly is a ſhrubbie plant, notwithſtanding it oftentimes growes to a tree of a reaſonable bigneſſe : the boughes whereof are tough and flexible, couered with a ſmooth and green barke. The ſubſtance of the wood is hard and ſound, and blackiſh or yellowiſh within, which doth alſo ſinke in the water, as doth the Indian wood which is called *Guaiacum* : the leaues are of a beautifull green colour, ſmooth and glib, like almoſt the bay leaues, but leſſer, and cornered in the edges with ſharp prickles, which notwithſtanding they want or haue few when the tree is old : the floures be white, and ſweet of ſmell : the berries are round, of the bigneſſe of a little Peaſe, or not much greater, of colour red, of taſt vnpleaſant, with a white ſtone in the midſt, which do not eaſily fall away, but hang on the boughes a long time : the root is wooddy.

There is made of the ſmooth barke of this tree or ſhrub, Birdlime, which the birders and country men do vſe to take birds with : they pull off the barke, and make a ditch in the ground, ſpecially in moiſt, boggy, or foggy earth, wherinto they put this bark, couering the ditch with boughes of trees, letting it remaine there til it be rooted and putrified, which will be done in the

the ſpace of twelue daies or thereabout : which done, they take it forth, and beat in morters vntill it be come to the thickeneſſe and clammineſſe of Lime : laſtly, that they may cleare it from pieces of barke and other filthineſſe, they do waſh it very often : after which they adde vnto it a little oyle of nuts, and after that do put it vp in earthen veſſels.

¶ The Place and Time.

The Holly tree groweth plentifully in all countries. It groweth green both winter and ſummer; the berries are ripe in September, and they do hang vpon the tree a long time after.

¶ The Names.

This tree or ſhrub is called in Latine *Agrifolium* : in Italian, *Agrifoglio* : and *Aguifoglio* : in Spaniſh, *Azebo* : in high Dutch, **walddiſtel**, and of diuers **Steepalmen** : in low Dutch, **Huls** : in French, *Hous* and *Houſſon* : in Engliſh, Holly, Huluer, and Holme.

¶ The Temperature.

The berries of Holly are hot and dry, and of thin parts, and waſte away winde.

¶ The Vertues.

They are good againſt the collicke : for ten or twelue being inwardly taken bring away by the A ſtoole thicke flagmaticke humors, as we haue learned of them who oftentimes made triall thereof.

The Birdlime which is made of the barke hereof is no leſſe hurtfull than that of Miſſeltoe, for it B is maruellous clammy, it glueth vp all the intrails, it ſhutteth and draweth together the guts and paſſages of the excrements, and by this meanes it bringeth deſtruction to man, not by any quality, but by his gluing ſubſtance.

Holly beaten to pouder and drunke, is an experimented medicine againſt all the fluxes of the C belly, as the dyſentery and ſuch like.

Chap. 32. Of the Oke.

1 *Quercus vulgaris cum glande & muſco ſuo.*
The Oke tree with his Acornes and Moſſe.

¶ The Deſcription.

1 THE common Oke groweth to a great tree, the trunke or body wherof is couered ouer with a thicke rough barke full of chops or rifts : the armes or boughes are likewiſe great, diſperſing themſelues farre a-broad : the leaues are bluntly indented about the edges, ſmooth, and of a ſhining greene colour, whereon is often found a moſt ſweet dew and ſomewhat clammie, and alſo a fungous excreſcence, which wee call Oke Apples. The fruit is long, couered with a browne, hard, and tough pilling, ſet in a rough ſcaly cup or husk : there is often found vpon the body of the tree, and alſo vpon the branches, a certaine kind of long white moſſe hanging downe from the ſame : and ſometimes another wooddie plant, which we call Miſſeltoe, being either an excreſcence or outgrowing from the tree it ſelfe, or of the doung (as it is reported) of a bird that hath eaten a certaine berry. ‡ Beſides theſe there are about the roots of old Okes within the earth certaine other excreſcences, which *Bauhine* and others haue called *Vuæ quercinæ,* becauſe they commonly grow in cluſters together, after the manner of Grapes and about their bignes, being ſomtimes round, and otherwhiles cornered, of a wooddy ſubſtance, hollow within; and ſometimes of a purple, otherwhiles of a whitiſh colour on the outſide : their taſte is aſtringent, and vſe ſingular in all Dyſenteries

Vuuuu 2   ries

ries and fluxes of bloud, as *Encelius* affirmes, *cap.51.de lapid. & gem.* ‡

3   *Carolus Cluſius* reporteth that hee found this baſe or low Oke not farre from Lisbone, of the height of a cubit, which notwithſtanding did alſo beare an Acorne like that of our Oke-tree, ſauing that the cup is ſmoother, and the Acorne much bitterer, wherein it differeth from the reſt of his kinde.

2 *Quercus vulgaris cum excrementis fungoſis.*   3 *Quercus humilis.*
The common Oke with his apple or greene Gall.   The dwarfe Oke.

 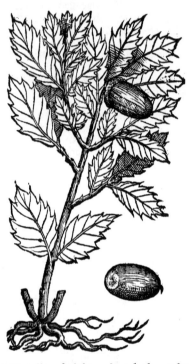

There is a wilde Oke which riſeth vp oftentimes to a maruellous height, and reacheth very far with his armes and boughes, the body wherof is now and then of a mighty thickeneſſe, in compaſſe two or three fathoms: it ſendeth forth great ſpreading armes, diuided into a multitude of boughs. The leaues are ſmooth, ſomething hard, broad, long, gaſhed in the edges, greene on the vpper ſide: the Acornes are long, but ſhorter than thoſe of the tamer Oke; euery one faſtened in his owne cup, which is rough without: they are couered with a thin rinde or ſhell: the ſubſtance or kernell with-in is diuided into two parts, as are Beans, Peaſe, and Almonds: the bark of the yong Okes is ſmooth, glib, and good to thicken skins and hides with, but that of the old Okes is rugged, thicke, hard, and full of chops: the inner ſubſtance or heart of the wood is ſomething yellow, hard and ſound, and the older the harder: the white and outward part next to the barke doth eaſily rot, being ſubiect to the worme, eſpecially if the tree be not felled in due time: ſome of the roots grow deepe into the earth, and otherſome far abroad, by which it ſtiffely ſtandeth.

¶ *The Place.*

The Oke doth ſcarcely refuſe any ground; for it groweth in a dry and barren ſoile, yet doth it proſper better in a fruitfull ground: it groweth vpon hills and mountaines, and likewiſe in vallies: it commeth vp euery where in all parts of England, but it is not ſo common in other of the South and hot regions.

¶ *The Time.*

The Oke doth caſt his leaues for the moſt part about the end of Autumne: ſome keepe their leaues on, but dry all Winter long, vntill they be thruſt off by the new Spring.

¶ *The Names.*

The Oke is called in Greeke δρῦς: in Latine, *Quercus*: of ſome, *Placida*, as *Gaza* tranſlats it. It may be called *Satiua*, *Vrbana*, or *Culta*; ſome alſo *Emeros mudion*, and *Robur*: the Macedonians ἐπυέτρυπι.

as though you ſhould ſay *Viriquercus*, as *Gaza* expoundeth it, or *Vere Quercus*, the true Oke. We may name it in Engliſh, the tamer Oke-tree : in French, *Cheſne* : in Dutch, **Eycken boom.**

The fruit is named in Greeke .... in Latine, *Glans* : in high Dutch, **Eichel :** in low Dutch, **Eckel :** in Spaniſh, *Belloins* : in Italian, *Chiande* : in Engliſh, Acorne, and Maſt.

The cup wherein the Acorne ſtandeth is named in Greeke ...., as *Paulus Ægineta*, in his third booke, 42 chapter teſtifieth, ſaying, *Omphacis* is the hollow thing out of which the Acorne groweth : in Latine, *Calix glandis* : in ſhops, *Capula glandis* : in Engliſh, the Acorne cup.

¶ *The Temperature and Vertues.*

The leaues, barke, Acorne cups, and the Acornes themſelues, do mightily binde and dry in the A the third degree, being ſomewhat cold withall.

The beſt of them, ſaith *Galen*, is the thin skin which is vnder the barke of the tree, and that next, B which lieth neereſt to the pulpe, or inner ſubſtance of the Acorne; all theſe ſtay the whites, the reds, ſpitting of bloud and laskes : the decoction of theſe is giuen, or the pouder of them dried, for the purpoſes aforeſaid.

Acornes if they be eaten are hardly concocted, they yeeld no nouriſhment to mans body, but C that which is groſſe, raw, and cold.

Swine are fatted herewith, and by feeding thereon haue their fleſh hard and ſound. D

The Acorns prouoke vrine, and are good againſt all venome and poyſon, but they are not of ſuch E a ſtopping and binding faculty as the leaues and barke.

The Oke apples are good againſt all fluxes of bloud and lasks, in what manner ſoeuer they be ta- F ken, but the beſt way is to boile them in red wine, and being ſo prepared, they are good alſo againſt the exceſſiue moiſture and ſwelling of the jawes and almonds or kernels of the throat.

The decoction of Oke Apples ſtaieth womens diſeaſes, and cauſeth the mother that is falne G downe to returne againe to the naturall place, if they do ſit ouer the ſaid decoction being very hot.

The ſame ſteeped in ſtrong white wine vineger, with a little pouder of Brimſtone, and the root of H *Ireos* mingled together, and ſet in the Sun by the ſpace of a moneth, maketh the haire blacke, conſumeth proud and ſuperfluous fleſh, taketh away ſun-burning, freckles, ſpots, the morphew, with all deformities of the face, being waſhed therewith.

The Oke Apples being broken in ſunder about the time of their withering, doe foreſhew the ſe- I quell of the yeare, as the expert Kentiſh husbandmen haue obſerued by the liuing things found in them : as if they finde an Ant, they foretell plenty of graine to enſue : if a white worme like a Gentill or Magot, then they prognoſticare murren of beaſts and cattell ; if a ſpider, then (ſay they) we ſhall haue a peſtilence or ſome ſuch like ſickeneſſe to follow amongſt men : theſe things the learned alſo haue obſerued and noted ; for *Matthiolus* writing vpon *Dioſcorides* ſaith, that before they haue an hole through them, they containe in them either a flie, a ſpider, or a worme ; if a flie then warre inſueth, if a creeping worme, then ſcarcitie of victuals; if a running ſpider, then followeth great ſickneſſe or mortalitie.

Chap. 33. *Of the Scarlet Oke.*

¶ *The Kindes.*

ALthough *Theophraſtus* hath made mention but of one of theſe Holme or Holly Okes onely, yet hath the later age ſet downe two kinds thereof; one bearing the ſcarlet grain, and the other only the Acorn: which thing is not contrarie to *Dioſcorides* his opinion, for he intreateth of that which beareth the Acorne, in his firſt booke, among ..... or the Okes : and the other hee deſcribeth in his fourth booke, vnder the title ...., or *Coccus Baphice*.

¶ *The Deſcription.*

THe Oke which beareth the Scarlet graine is a ſmall tree, in manner of a hedge tree, of a meane bigneſſe, hauing many faire branches or boughes ſpread abroad: whereon are ſet leaues, green aboue, white vnderneath, ſnipt about the edges, and at euery corner one ſharpe prickle, in maner of the ſmoother Holly : among which commeth ſometimes, but not often, ſmall Acornes, ſtanling in little cups or husks, armed with prickles as ſharpe as thornes, and of a bitter taſte. Beſides the Acornes, there is found cleauing vnto the wooddy branches, a certaine kind of berries, or rather an excreſcence, of the ſubſtance of the Oke Apple, and of the bigneſſe of a Peaſe, at the firſt white, and of the colour of aſhes when they be ripe, in which are ingendred little Maggots, which ſeeme

Ilex Coccigera.
The Scarlet Oke.

to be without life vntill they feele the heate of the fun, and then they creep, and feeke to flie a-way. But the people of the countrey (which make a gaine of them) doe watch the time of their flying, euen as wee doe Bees, which they then take and put into a linnen bag, wherein they fhake and boult them vp and downe vntil they bee dead, which they make vp into great lumpes oftentimes, and likewife fell them to diers apart, euen as they were taken forth of the bag, whereof is made the moft perfect Scarlet.

¶ *The Place.*

This Oke groweth in Languedocke, and in the countries thereabout, and alfo in Spain: but it beareth not the fcarlet grain in all places, but in thofe efpecially which lie toward the Mid-land fea, and which be fubject to the fcorching heat of the Sun, as *Carolus Clufius* witneffeth; & not there alwaies, for when the tree waxeth old it growes to be barren. Then do the people cut and lop it downe, that after the young fhoots haue attained to two or three yeares growth, it may become fruitfull againe.

Petrus Bellonius in his books of Singularities fheweth, That *Coccus Baphicus* or the Scarlet graine doth grow in the Holy land, and neere to the lake which is called the Sea of *Tiberias*, and that vpon little trees, whereby the inhabi-tants get great ftore of wealth, who feparat the husks from the pulpe or Magots, and fell this being made vp into balls or lumpes, much dea-rer than the empty fhels or husks.

Of this graine alfo *Paufanias* hath made mention in his tenth booke, and fheweth, that the tree which bringeth forth this graine is not great, and alfo groweth in Phocis, which is a countrey in Macedonia neere to the Boetians, not far from the mountaine Parnaffus.

Theophraftus writeth, that ωῖνος, or the Scarlet Oke, is a great tree, and rifeth vp to the height of the common Oke: amongft which writers there are fome contrarietie. *Petrus Bellonius* reporteth it is a little tree, and *Theophraftus* a great one, which may chance according to the foile and clymate; for that vpon the ftonie mountains they cannot grow to that greatneffe as thofe in the fertill grounds.

¶ *The Time.*

The little graines or berries which grow about the boughes begin to appeare efpecially in the Spring, when the Southweft windes do blow: the floures fall and are ripe in Iune, together with the Maggots growing in them, which receiuing life by the heat of the Sun, doe forthwith flie away (in manner of a Moth or Butterflie) vnleffe by the care and diligence of the keepers, they be killed by much and often fhaking them together, as aforefaid.

The tree or fhrub hath his leaues alwaies greene: the Acornes be very late before they be ripe, feldome before new come vp in their place.

¶ *The Names.*

The Scarlet Oke is called in Greeke πρῖνος: in Latine, *Ilex*: the later writers, *Ilex Coccigera*, or *Cocci-fera*; in Spanifh, *Cofcoia*: for want of a fit Englifh name, wee haue thought good to call it by the name of Scarlet Oke, or Scarlet Holme Oke: for *Ilex* is named of fome in Englifh, Holme, which fignifyeth Holly or Huluer. But this *Ilex*, as well as thofe that follow, might be called Holme Oke, Huluer Oke, or Holly Oke, for difference from the fhrub or hedge tree *Agrifolium*, which is fimply called Holme, Holly, and Huluer.

The graine or berry that ferueth to die with is properly called in Greeke κόκκος βαφικὴ: in Latine; *Coccus infectoria*, or *Coccum infectorium*: Pliny alfo nameth it *Cufculium*: or as moft men doe reade it, *Qufquilium*: the fame Author faith, that it is likewife named *Scarlecion*, or Maggot berry.

The Arabians and the Apothecaries doe know it by the name of *Chefmes*, *Chermes*, and *Kermes*: They are deceiued who thinke that *Chefmes* doth differ from *Infectorium Coccum*: it is called in Ita-lian, *Grano de tinctori*: in Spanifh, *Grana de tinctoreros*: in high Dutch, 𝕾𝖈𝖍𝖆𝖗𝖑𝖆𝖈𝖍𝖇𝖊𝖊𝖗: in French, *Vermillon,*

Vermillon, and *Graine d'efcarlate* : in Englifh, after the Dutch, Scarlet Berry, or Scarlet graine, and after the Apothecaries word, *Coccus Baphicus* : the Maggot within is that which is named Cutcho-nele, as moft do deeme.

Theophraftus faith, the Acorne or fruit hereof is called of diuers, κ.λ., *Acylum*.

¶ *The Temperature and Vertues.*

This graine is aftringent and fomwhat bitter, and alfo dry withour fharpeneffe and biting, there-fore, faith *Galen*, it is good for great wounds and finewes that bee hurt, if it be layd thereon : fome temper it with Vineger ; others with Oxymel or fyrrup of Vineger. **A**

It is commended and giuen by the later Phyfitians to ftay the Menfes : it is alfo counted among thofe Simples which be cordials, and good to ftrengthen the heart. Of this graine that noble and famous confection *Alkermes*, made by the Arabians, hath taken his name, which many doe highly commend againft the infirmities of the heart : notwithftanding it was chiefly deuifed in the begin-ning for purging of melancholy ; which thing is plainly perceiued by the great quantitie of *Lapis Lazulus* added thereto : and therefore feeing that this ftone hath in it a venomous quality, and like-wife a property to purge melancholy, it cannot of it felfe be good for the heart, but the other things be good, which be therefore added, that they might defend the heart from the hurts of this ftone, and correct the malice thereof. **B**

This compofition is commended againft the trembling and fhaking of the heart, and for fwou-nings and melancholy paffions, and forrow proceeding of no euident caufe : it is reported to recre-ate the minde, and to make a man merry and joyfull. **C**

It is therefore good againft melancholy difeafes, vaine imaginations, fighings, griefe and forrow without manifeft caufe, for that it purgeth away melancholy humors : after this manner it may be comfortable for the heart, and delightfull to the minde, in taking away the materiall caufe of for-row : neither can it otherwife ftrengthen a weake and feeble heart, vnleffe this ftone called *Lapis Cyaneus* be quite left out. **D**

Therefore he that is purpofed to vfe this compofition againft beatings and throbbings of the heart, and fwounings, and that not as a purging medicine, fhall do wel and wifely by leauing out the ftone *Cyaneus* ; for this being taken in a little weight or fmall quantity, cannot purge at all, but may in the meane feafon trouble and torment the ftomacke, and withall thorow his fharpe and veno-mous quality (if it be oftentimes taken) be very offenfiue to the guts and intrails, and by this means bring more harme than good. **E**

Moreouer, it is not neceffary, no nor expedient, that the briftle died with Cochenele, called *Chefmes*, as the Apothecaries terme it, fhould be added to this compofition ; for this briftle is not died without *Auripigmentum*, called alfo Orpiment, and other pernitious things joyned therewith, whofe poifonfome qualities are added to the juyces together with the colour, if either the briftle or died filke be boyled in them. **F**

The berries of the Cochenele muft be taken by themfelues, which alone are fufficient to die the juyces, and to impart vnto them their vertue : neither is it likewife needfull to boile the raw filke together with the graines, as moft Phyfitians thinke : this may be left out, for it maketh nothing at all for the ftrengthning of the heart. **G**

CHAP. 34. *Of the great Skarlet Oke.*

¶ *The Defcription.*

THe great Skarlet Oke, or the great Holme Oke, groweth many times to the full height of a tree, fometimes as big as the Peare-tree, with boughes far fpreading like the Acorne or com-mon Maft trees : the timber is firme and found : the leaues are fet with prickles round about the edges, like thofe of the former Skarlet Oke : the leaues when the tree waxeth old haue on them no prickles at all, but are fomewhat bluntly cut or indented about the edges, greene on the vpper fide, and gray vnderneath : the Acorne ftandeth in a prickely cup like our common Oke Acorne, which when it is ripe becommeth of a browne colour, with a white kernell within of tafte not vn-pleafant. There is found vpon the branches of this tree a certaine kinde of long hairy moffe of the colour of afhes, not vnlike to that of our Englifh Oke. ‡ This tree is euer greene, and at the tops of the branches about the end of May, here in England, carrieth diuers long catkins of moffie yel-low floures, which fall away and are not fucceeded by the Acornes, for they grow out vpon other ftalks. *Clufius* in the yeare 1581 obferued two trees ; the one in a garden about the Bridge, and the other in the priuat garden at White-Hall, hauing leffer leaues than the former. The later of thefe is yet ftanding, and euery yeare beares fmall Acornes, which I could neuer obferue to come to any maturity. ‡

¶ *The*

Ilex major Glandifera.
The great Skarlet Oke.

‡ *Ilicis ramus floridus.*
The floures of the great Skarlet Oke.

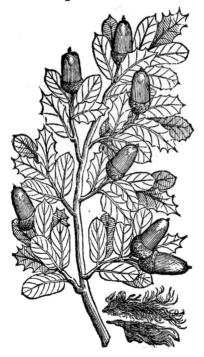

¶ *The Place.*

In diuers places there are great woods of thefe trees, hills alfo and vallies are beautified there-with: they grow plentifully in many countries of Spaine, and in Languedocke and Prouince in great plenty. It is likewife found in Italy. It beareth an Acorne greater, and of a larger fize than doth the tame Oke: in fome countries leffer and fhorter: they are ftrangers in England, notwithftanding there is here and there a tree thereof, that hath been procured from beyond the feas: one groweth in her Majefties priuie Garden at White-Hall, neere to the gate that leadeth into the ftreet, and in fome other places here and there one.

¶ *The Time.*

It is greene at all times of the yeare: it is late before the Acornes be ripe. *Clufius* reporteth, that he faw the floures growing in clufters of a yellow colour in May.

¶ *The Names.*

This Oke is named in Greeke πρῖνος: in Latine, *Ilex*: in Spanifh, *Enzina*: in Italian, *Elize*: in French, *Chefneuerd*: in Englifh, Barren Skarlet Oke, or Holme Oke, and alfo of fome, French or Spanifh Oke.

The Spaniards call the fruit or Acorne *Bellota*, or *Abillota*. *Theophraftus* feemeth to call this tree not *Prinos*, but *Smilax*; for he maketh mention but of one *Ilex* onely, and that is of Scarlet Oke; and he fheweth that the Arcadians do not call the other *Ilex*, but *Smilax*: for the name *Smilax* is of many fignifications: there is *Smilax* among the Pulfes, which is alfo called *Dolichus*, and *Phafcolus*; and *Smilax afpera*, and *Lauis*, amongft the Binde-weeds: likewife *Smilax* is taken of *Diofcorides* to be *Taxus*, the Yew tree. Of *Smilax, Theophraftus* writeth thus in his third booke: the Inhabitants of Arcadia do call a certaine tree *Smilax*, being like vnto the Skarlet Oke: the leaues thereof be not fet with fuch fharpe prickles, but tenderer and fofter.

Of this *Smilax Pliny* alfo writeth, in his fixteenth booke, chap. 6. There be of *Ilex*, faith he, two kindes, *Ex ijs in Italia folio non multum ab oleis diftant*, called of certaine Græcians *Smilaces*, in the prouinces *Aquifolia*: in which words, in ftead of Oliue trees may perchance bee more truely placed *Suberis*, or the Corke-tree; for this kinde of *Ilex* or *Smilax* is not reported of any of the old Writers

to

to haue the leaſe of the Oliue tree : but *Suber* in Greeke, called *Phellos*, or the Corke tree, hath a little leafe.

The leaues of this Oke hath force to coole and repell and keepe backe, as haue the leaues of the A
Acornes or Maſt trees : being ſtamped or beaten, and applied, they are good for ſoft ſwllings, and
ſtrengthen weake members.

The barke of the root boyled in water vntill it be diſſolued, and layd on all night, maketh the B
haire blacke, being firſt ſcoured with *Cimolia*, as *Dioſcorides* ſaith.

Cluſius reporteth, that the Acorne is eſteemed of, eaten, and brought into the market to be ſold, C
in the city of Salamanca in Spaine, and in many other places of that country ; and of this Acorne
Pliny alſo hath peraduenture written, *lib. 16.cap. 5.* in theſe words : Moreouer, at this day in Spaine
the Acorne is ſerued for a ſecond courſe.

CHAP. 35.   *Of the great Holme-Oke.*

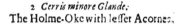

1 *Cerris majore Glande.*	2 *Cerris minore Glande.*
The Holme-Oke with great Acornes.	The Holme-Oke with leſſer Acornes.

¶ *The Deſcription.*

1   AMong the wilder Okes this is not the leaſt, for his comely proportion, although vnprofitable for timber, to make coles, carts, Wainſcot, houſes, or ſhips of: the fruit
is not fit for any man or beaſt to eate, neither any property knowne for the vſe of Phyſicke or Surgerie · it groweth vp to the height of a faire tree, the trunke or body is great, and very
faire to looke vpon : the wood or timber ſoft and ſpongie, ſcarce good to be burned : from which
ſhooteth forth very comely branches diſperſing themſelues farre abroad ; whereon are ſet for
the moſt part by couples very faire leaues, greene aboue, and of an ouerworne ruſſet colour vnderneath, cut or ſnipt about the edges very deepe : the Acorne groweth faſt vnto the boughes, without any foot-ſtalke at all, being very like vnto our common Acorne, ſet in a rough and prickly cup
like an Hedge-hog or the Cheſnut huske, of a harſh taſte, and hollow within : this tree beareth or
bringeth

‡ *Cerri minoris ramulus cum flore.*
A branch of the ſmaller Holme Oke with floures.

bringeth forth oft times a certain ſmooth kind of Gall not altogether vnprofitable. This Oke likewiſe bringeth forth another kind of excreſence,which the Grauer hath omitted in the figure, which is called in Greeke φιμος *Gaza* nameth it *Penis*. This *Penis* or pricke is hollow, moſſie, hanging downe halfe a yard long,like a long rag of linnen cloath.

2   The ſecond is altogether like the firſt, ſauing that this beareth ſmaller A-cornes, and the whole tree is altogether leſſe,wherein conſiſteth the difference.

‡   Both this & the former cary floures cluſtering vpon long ſtalkes,like as in the common Oke;but the fruit doth not ſuc-ceed them, but grow forth in other pla-ces. ‡

¶ *The Place.*

This Oke groweth in vntoiled places, it is ſeldome times found,and that but in Woods onely : it is for the moſt part vn-knowne in Italy, as *Pliny* reporteth.

¶ *The Time.*

They bring forth their fruit or Acornes in the fall of the leafe.

¶ *The Names.*

This Oke is called in Greeke Κηριιςψ: in Latine, *Cerrus* : yet doth *Pliny* make men-tion both of *Ægilops*, and alſo of *Cerrus* : Κηριιςψ is likewiſe one of the diſeaſes of corne,called in Latine *Feſtuca* : in Engliſh,wilde Otes,and far differing from the tree *Ægilops*.

That which hangeth from the boughes, *Pliny,lib.16.cap.8.* calleth *Panus* onely : that acorne tree named *Ægilops* bringeth forth *Panos arentes*, withered prickes, couered with white moſſie jags han-ging downe, not onely in the barke, but alſo from the boughes, halfe a yard in bigneſſe,bearing a ſweet ſmell,as we haue ſaid among ointments.

¶ *The Temperature and Vertues.*

We finde nothing written of the faculties of this tree among the old Writers, neither of our owne experience.

CHAP. 36. *Of the Corke Oke.*

¶ *The Deſcription.*

1   THe Corke tree is of a middle bigneſſe like vnto *Ilex*,or the barren skarlet Oke, but with a thicker body, and fewer boughes : the leaues be for the moſt part greater, broader, rounder,and more nicked in the edges : the barke of the tree is thicke,very rugged,and full of chinkes or crannies that cleaueth and diuideth it ſelfe into pieces, which vnleſſe they be ta-ken away in due time do giue place to another barke growing vnderneath,which when the old is re-moued is maruellous red, as though it were painted with ſome colour : the Acorne ſtandeth in a cup,which is great, briſtled, rough, and full of prickles : this Acorne is alſo aſtringent or binding, more vnpleaſant than the Holme Acorne,greater in one place,and leſſe in another.

2   The Corke tree hath narrow leaues groweth likewiſe to the height and bigneſſe of a great tree ; the trunke or body whereof is couered with a rough and ſcabbed barke of an ouerworne blac-kiſh colour, which likewiſe cleaueth and caſteth his coat when the inner barke groweth ſomewhat thicke: the branches are long, tough, and flexible, eaſie to be blowed any way, like thoſe of the
Oziar,

Oziar; whereupon do grow leaues like thoſe of the precedent, but longer, and little or nothing indented about the edges : the fruit groweth in ſmall cups as the Acornes doe : they are leſſer than
thoſe of the other kinde, as is the reſt of the tree, wherein is the chiefeſt difference. ‡ This varies
in the leafe, (as you may ſee in the figure) which in ſome is ſnipt about the edge, in other ſome no
at all. ‡

1. 2. *Suber latifolium & anguſtifolium.*
The Corke tree with broad and narrow leaues.

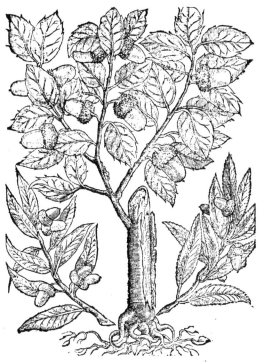

¶ *The Place.*

It groweth in the country of Aquitania, neere to the mountaines called Pyrenæi : it alſo groweth
plentifully in the kingdomes of Spaine, differing ſomewhat from that of Aquitania, as *Cluſius* declareth : it is likewiſe found in Italy, and that in the territory of Piſa, with a longer leafe, and ſharper pointed ; and about Rome with a broader, and cut in the edges like a ſaw, and rougher, as *Matthiolus* teſtifieth.

¶ *The Time.*

The leaues of the firſt are alwaies greene in Spaine and Italy, about the Pyrenæan mountaines
they fall away in Winter.

¶ *The Names.*

This tree is called in Greeke ☙●●●: in Latine, *Suber* : in French, *Liege* : in Italian, *Sugaro* : the ſame
names do alſo belong to the barke : the Spaniards call the tree *Alcornoque* : the Engliſhmen, Corke
tree ; and the barke, *Corcha de Alcornoque* ; whereupon the Low-country-men and Engliſh-men alſo
do call it Corke ; and yet it is called in low Dutch alſo **Wlothout,**

¶ *The Temperature and Vertues.*

This barke doth moſt manifeſtly dry, with a binding faculty.　　　　　　　　　　　　A

Being beaten to pouder and taken in water it ſtancheth bleeding in any part of the body. The　B
Corke which is taken out of wine veſſels, ſaith *Paulus* being burnt, maketh aſhes which doe mightily dry, and are mixed in compoſitions diuiſed againſt the bloudy flix.

Corke is alſo profitable for many things : it is vſed (ſaith *Pliny*) about the anchors of ſhips, Fi-　C
ſhers nets, and to ſtop veſſels with ; and in Winter for womens ſhoos, which vſe remaines with vs
euen to this day : fiſhermen hang this barke vpon the wings of their nets for feare of ſinking : and
ſhoo-makers put it in ſhooes and pantofles for warmneſſe ſake.

CHAP.

Chap. 37.   Of the Gall tree.

¶ *The Kindes.*

OF trees that bring forth Galls there be diuers ſorts, as may appeare by the diuers formes and ſorts of Galls ſet forth in this preſent chapter, which may ſerue for their ſeuerall diſtinctions, whereof ſome bring forth Acornes likewiſe, and ſome nothing but Galls : the figures of ſome few of the trees ſhall giue you ſufficient knowledge of the reſt : for all the Acorne or Maſt trees bring forth Galls, but thoſe trees whoſe figures we haue ſet forth doe beare thoſe galls fit for medicine, and to thicken skins with.

Dioſcorides and *Galen* make but two ſorts of Galls ; the one little, yellow, full of holes, and more ſpongie in the inner part, both of them round, hauing the forme of a little ball, and the other ſmooth and euen on the out ſide : ſince the later writers haue found moe, ſome hauing certaine little knobs ſticking forth, like in forme to the Gall , which doth alſo cleaue and grow without ſtalke to the leafe. There is alſo found a certaine excreſcence of a light greene colour, ſpongie and watery, in the middle whereof now and then is found a little flie or worme : which ſoft ball in hot countries doth oftentimes become hard, like the little ſmooth Gall, as *Theophraſtus* ſaith.

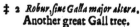

1 *Galla, ſiue Robur majus.*
The great Gall tree.

‡ 2 *Robur, ſiue Galla major altera.*
Another great Gall tree.

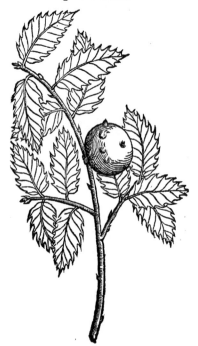

¶ *The Deſcription.*

THe Gall tree growes vp to a ſufficient height, hauing a very faire trunke or body, whereon are placed long twiggie branches bringing forth very faire leaues, broad, and nicked in the edges like the teeth of a ſaw : among which come forth acornes, although the figure expreſſe not the ſame, like thoſe of the Oke, and likewiſe a wooddy excreſcence, which we name the Gall, hauing certaine ſmall eminences or bunches on the out ſide, growing for the moſt part vpon the ſlender branches without ſtalks, and ſometimes they grow at the ends thereof ; which by the heate of the Sun are harder, greater, and more ſollid in one country than another, according to the ſoile and clymat.

‡ 2 This

3 *Galla minor.*
The little Gall tree.

‡ 2 This growes to the height of a tal man, hauing leaues deepely diuided on the edges like the Oke, greene aboue, but hairy and hoary below: it carries a great gall of the bignesse of a little apple, and that in great plenty, and without any order. This growes in diuers parts of old Castile in Spain, and in all the mountainous Woods about Vienna in Austria. ‡

3 The lesser Gal tree differs not from the former, sauing that it is altogether lesser: the fruit or gall is likewise lesser, wherin especially consisteth the difference.

¶ *The Place.*

The Galls are found in Italy, Spain, and Bohemia, and most of the hot regions.

¶ *The Time.*

The Gall, saith *Pliny*, appeareth or commeth forth when the Sunne comes out of the signe *Gemini*, and that generally in one night.

¶ *The Names.*

The Gall tree is called *Quercus, Robur,* and *Galla arbor*: in Greek, ᵂᵃⁿᵃ: the Apothecaries and Italians keepe the name *Galla* for the fruit: in high-Dutch, 𝕲𝖆𝖑𝖔𝖕𝖋𝖊𝖑𝖑: in low-Dutch, 𝕲𝖆𝖑𝖓𝖔𝖙𝖊𝖓: in Spanish, *Agatha, Galha,* and *Bugalha*: in French, *Noix de Galle*: in English, Gaules, and Galls.

¶ *The Temperature and Vertues.*

The Gall called *Omphacitis*, as *Galen* writeth, is dry in the third degree, and cold in the second: it **A** is a very harsh medicine; it fasteneth and draweth together faint and slacke parts, as the ouergrowings in the flesh: it repelleth and keepeth backe rheumes and such like fluxes, and doth effectually dry vp the same, especially when they haue a descent into the gums, almonds of the throat, & other places of the mouth.

The other Gall doth dry and also binde, but so much lesser, by how much the harsh or choking **B** qualitie is diminished: being boiled, beaten, and also applied in manner of a plaister it is laid with good successe vpon the inflammations of the fundament, and falling down thereof. It is boiled in water if there be need of a little astriction; and in wine, especially in austere wine, if more need require.

Galls are very profitable against the Dysenterie and the Cœliack passion, being drunk in wine, **C** or the pouder thereof strewed vpon meats.

Galls are vsed in dying and colouring of sundry things, and in making of inke.

Lastly, burnt Galls do receiue a further faculty, namely to stanch bloud, and are of thinne parts, **D** and of a greater vertue to dry than be those that are not burnt. They must be laid vpon hot burning **E** coles vntill they come to be thorow white, and then they are to be quenched in vineger and wine.

Moreouer, Galls are good for those that be troubled with the bloudy flixe and common laskes, **F** being taken in wine or water, and also applied or vsed in meats: finally, these are to be vsed as oft as need requireth to dry and binde.

Oke apples are much of the nature of Galls, yet are they far inferior to them, and of lesser force. **G**

† Our Author out of *Tabern.* gaue the figures of fourteene varieties of Galls; some being large, others small; some round, others longish, and other forts diuersly tempered.

CHAP. 38.   Of Miſſeltoe or Miſteltoe.

1 *Viſcum.*
Miſſeltoe.

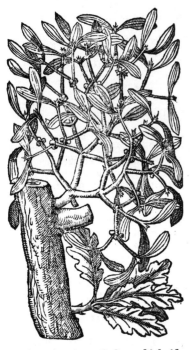

¶ *The Deſcription.*

1 Viſcum or Miſſeltoe hath many ſlender branches ſpread ouerthwart one another, and wrapped or interlaced one within another, the bark wherof is of a light green or Popinjay colour : the leaues of this branching excreſcence be of a brown green colour:the floures be ſmall and yellow : which beeing paſt, there appeare ſmall cluſters of white tranſlucent berries, which are ſo cleare that a man may ſee thorow them, and are of a clammy or viſcous moiſture, whereof the beſt Bird-lime is made, far exceeding that which is made of the Holm or Holly bark:and within this berry is a ſmall blacke kernel or ſeed : this excreſcence hath not any root, neither doth increaſe himſelfe of his ſeed, as ſome haue ſuppoſed ; but it rather commeth of a certaine moiſture and ſubſtance gathered together vpon the boughs and joints of the trees, through the barke whereof this vaporous moiſture proceeding, brings forth the Miſſeltoe. Many haue diuerſly ſpoken hereof: Some of the Learned haue ſet down,that it comes of the dung of the bird called a Thruſh , who hauing fed of the ſeeds thereof, as eating his owne bane,hath voided and left his dung vpon the tree, whereof was ingendred this bery, a moſt fit matter to make lime of to intrap and catch birds withall.

2 Indian Miſſelto groweth likewiſe vpon the branches of trees,running alongſt the ſame in maner of Polypodie : the ſtrings of the roots are like thoſe of Couch-graſſe,from which riſe vp diuers ſtalks ſmooth and euen,ſet with joints and knees at certaine diſtances : toward the top comes forth one leafe ribbed like the Plantain leafe, wheron are marked certain round eies ſuch as are in the haft of a knife ; from the boſome wherof commeth forth a chaffie branch, ſet with ſmall leaues which continue green Winter and Summer.

2 *Viſcum Indicum Lobelij.*
Indian Miſſeltoe.

3 *Viſcum Peruvianum Lobelij.*
Miſſeltoe of Peru.

3 There is found alſo another plant growing vpon the boughes or branches of trees, in maner as our Miſſeltoe doth, and may very well be reckoned as a kind thereof: the plant cleaues vnto the branches, being ſet thereto as it were with the pillings of the ſea Onion, of the bredth of a mans hand toward the bottom, and ſomewhat hollow : the tops whereof are very ſmall and ruſhy, hollow likewiſe, and of a purple colour : among which comes forth a branch like that of *Haſtula Regia*, or Kings ſpeare, reſembling the buſh of Otes, couered with a white ſilk ſuch as is to be found in *Aſcle-pias*, of a ſalt and nitrous taſte, and very vnpleaſant.

¶ *The Place.*

The firſt kinde of Miſſeltoe groweth vpon Okes and diuers other trees almoſt euery where ; as for the other two they are ſtrangers in England.

¶ *The Time.*

Miſſeltoe is alwaies greene as well in winter as ſummer : the berries are ripe in Autumne, they remain all winter thorow, and are a food for diuers birds, as Thruſhes, Black-birds, & Ring-doues.

¶ *The Names.*

Miſſeltoe is called in Greeke 'ιξω, and 'ιξι : in Latine, *Viſcum* : in high-Dutch, **Miſtell** : in Low-Dutch, **Marentacken** : in Italian, *Viſchio* : in Spaniſh, *Liga* : in the Portugal tongue, *Viſgo*: in Engliſh, Miſſel, and Miſſeltoe.

The glue which is made of the berries of Miſſel is likewiſe called *Viſcum*, and *Ixia* : in Engliſh, Bird-lime. *Ixia* is alſo called *Chamæleon albus*, by reaſon of the glue which is oft times found about the root thereof. This word is alſo aſcribed to *Chamæleon niger*, as wee reade amohgſt the baſtard names. *Ixia* is likewiſe reckoned vp by *Dioſcorides, lib.6*. and by *Paulus Ægineta, lib.5*. amongſt the poiſons : but what this poiſonſome and venomous *Ixia* is it is hard and doubtfull to declare:many would haue it to be *Chamæleon niger*:others, the glue or clammy ſubſtance which is made of the berries of Miſſeltoe ; who do truly thinke that *Ixia* differeth from *Chamæleon niger* : for *Paulus Ægine-ta, lib.5.cap.30*. in reckoning vp of ſimple poiſons hath firſt made mention of *Chamæleon niger*; then a little after of *Ixia* : and whileſt he doth particularly diſcourſe of euery one, he treateth of *Chamæ-leon niger, cap.32*. and of *Ixia* which he alſo nameth *Vlophonon, cap.47*. and telleth of the dangerous and far differing accidents of them both. And *Dioſcorides* himſelfe, *lib.6*. where he ſetteth down his iudgment of ſimple poiſons, intreateth firſt of *Chamæleon niger*, and then a little after of *Ixia*. Theſe things declare that *Chamæleon niger* differs from *Ixia*, which is reckoned among the poiſons. More-ouer, it can no where be read, that *Chamæleon niger* beareth Bird-lime, or hath ſo glutinous or clam-my a ſubſtance as that it ought to be called *Ixia*:therfore *Ixia*, as it is one of the poiſons, is the glue that is made of the berries of Miſſeltoe, which becauſe it is ſharp and biting, inflameth and ſetteth the tongue on fire, and with his ſlimy and clammy ſubſtance doth ſo draw together, ſhut, and glue vp the guts, as that there is no paſſage for the excrements, which things are mentioned among the miſchiefes that *Ixia* bringeth.

‡ I can by no means approue of, or yeeld to this opinion here deliuered out of *Dodonæus* by our Author ; which is, That the Bird-lime made of the berries of Miſſeltoe is poiſon ; or that *Ixia* ſet forth by *Dioſcorides* and *Nicander* for a poiſon is meant of this : for this is manifeſtly treated of in *Dioſcorides, lib.3.ca.103*. by the name of ιξι : when as the other is mentioned, *lib.6.cap.21*. by the name of ιξι. Alſo daily experience ſhewes this plant to haue no maligne or poiſonous, but rather a contrarie facultie, being frequently vſed in medicines againſt the Epilepſie. Such as would ſee more concerning *Ixia* or *Ixias*, let them haue recourſe to the firſt chapter of the firſt part of *Fabius Columna, de Stirpib.min.cognitis & rarioribus*, where they ſhall finde it largely treated of. ‡

¶ *The Nature and Vertues.*

The leaues and berries of Miſſelto are hot and dry, and of ſubtill parts : the Bird-lime is hot and biting, and conſiſts of an airy and watery ſubſtance, with ſome earthy quality ; for according to the iudgement of *Galen*, his acrimony ouercommeth his bitterneſſe ; for if it be vſed in outward appli-cations, it draweth humors from the deepeſt or moſt ſecret parts of the body, ſpreading and diſper-ſing them abroad, and digeſting them. A

It ripeneth ſwellings in the groin, hard ſwellings behinde the eares, and other impoſtumes, be-ing tempered with roſin and a little quantitie of wax. B

With Frankincenſe it mollifieth old vlcers and malitious impoſtumes, being boiled with vn-ſlaked lime, or with *Gagate lapide* or *Aſio*, and applied, it waſteth away the hardneſſe of the ſpleen. C

With Orpment or *Sandaracha* it taketh away foule ilfauored nailes, being mixed with vnſlaked lime and wine lees it receiueth greater force. D

It hath been moſt credibly reported vnto me, That a few of the berries of Miſſeltoe bruiſed and ſtrained into oile and drunken, hath preſently and forthwith rid a grieuous and ſore ſtitch. E

Chap. 39. *Of the Cedar Tree.*

¶ *The Kindes.*

THere be two Cedars, one great bearing cones, the other ſmall bearing berries like thoſe of Iuniper.

Cedrus Libani.
The great Cedar tree of Libanus.

¶ *The Deſcription.*

THe great cedar is a very big & high tree, not onely exceeding all other reſinous trees and thoſe which beare fruit like vnto it, but in his talneſſe and largeneſſe farre ſurmounting all other trees: the body or trunk thereof is commonly of a mighty bigneſſe, inſomuch as four men are not able to fathom it, as *Theophraſtus* writeth. The bark of the lower part which proceedeth out of the earth, to the firſt young branches or ſhoots, is rough and harſh ; the reſt which is among the boughes is ſmooth and glib : the boughes grow forth almoſt from the bottom, and not farre from the ground, euen to the very top , waxing by degrees leſſer & ſhorter ſtil as they grow higher, the tree bearing the forme and ſhape of a Pyramide or ſharpe pointed ſteeple : theſe compaſſe the body round about in maner of a circle, and are ſo orderly placed by degrees, that a man may clymbe vp by them to the very top as by a ladder : the leaues be ſmall and round like thoſe of the Pine tree, but ſhorter, and not ſo ſharp pointed : all the cones or clogs are far ſhorter and thicker than thoſe of the Fir tree, compact of ſoft, not hard ſcales, which hang not downewards , but ſtand vpright vpon the boughes, whereunto alſo they are ſo ſtrongly faſtned, as they can hardly be pluckt off without breaking ſome part of the branches, as *Bellonius* writeth. The timber is extreame hard, and rotteth not, nor waxeth old ; thete is no wormes nor rottenneſſe can hurt or take the hard matter or heart of this wood, which is very odoriferous and ſomwhat red. *Solomon* King of the Iewes did therefore build Gods Temple in Ieruſalem of Cedar wood. The Gentiles were wont to make their Diuels or Images of this kinde of wood, that they might laſt the longer.

¶ *The Place.*

The Cedar trees grow vpon the ſnowy mountaines, as in Syria vpon mount Libanus, on which there remaine ſome euen to this day, ſaith *Bellonius*, planted as is thought by *Solomon* himſelf: they are likewiſe found on the mountains Taurus and Amanus, in cold and ſtony places. The merchants of the Factorie of Tripolis told me, That the Cedar tree groweth vpon the declining of the mount Libanus, neere to the hermitage by the city Tripolis in Syria : The inhabitants of Syria vſe it to make boats of, for want of the Pine tree.

¶ *The Time.*

The Cedar tree remaines alwaies green, as other trees which beare ſuch maner of fruit: the timber of the Cedar tree, and the images and other works made thereof, ſeem to ſweat and ſend forth moiſture in moiſt and rainy weather, as do likewiſe all that haue an oily juice, as *Theophraſtus* witneſſeth.

¶ *The Names.*

The huge and mighty tree is called in Greeke κέδρος : in Latine likewiſe *Cedrus* : in Engliſh, Cedar, and Cedar tree. *Pliny, lib. 24. cap. 5.* nameth it *Cedrelate*, as though he ſhould ſay *Cedrus abies*, or *Cedrina abies*, Cedar Firre ; both that it may differ from the little Cedar, and alſo becauſe it is very like the Firre tree.

The

The Roſin hath no proper name, but it may be named *Cedrina*, or Cedar Roſin.

The pitch which is drawne out of this is properly called κίαρα; yet *Pliny* writeth, that the liquor of the Torch tree is alſo named *Cedrium*. The beſt, ſaith *Dioſcorides*, is fat, thorow ſhining, and of a ſtrong ſmell; which being poured out in drops vniteth it ſelfe together, and doth not remain ſeuered.

¶ *The Nature and Vertues.*

Cedar is of temperature hot and dry, with ſuch an excellent tenuitie and ſubtiltie of parts, that A it ſeemeth to be hot and dry in the fourth degree, eſpecially the Roſin thereof.

There iſſueth out of this tree a roſin like to that which iſſueth out of the Fir tree, very ſweet in B ſmell, of a clammy or cleauing ſubſtance, the which if you chew in your teeth it will hardly be gotten forth again, it cleaueth ſo faſt: at the firſt it is liquid and white, but being dried in the Sunne it waxeth hard: if it be boiled in the fire an excellent pitch is made thereof called Cedar pitch.

The Egyptians were wont to coffin and embalme their Dead in Cedar and with Cedar pitch, C although they vſed other means, as *Herodotus* recordeth.

The condited or embalmed body they call in ſhops *Mumia*, but very vnfitly; for *Mumia* among D the Arabians is that which the Grecians call *Piſſaſphalton*, as appeareth by *Avicen, cap.*474 and out of *Serapio, cap.*393.

He that interpreted and tranſlated *Serapio* was the cauſe of this error, who rendred *Mumia* after E his owne fancy, and not according to the ſence and meaning of his author *Serapio*, ſaying, That this *Mumia* is a compoſition of Myrrh and Aloes mingled together with the moiſture of mans body.

The gum of Cedar is good to be put into medicines for the eies; for being anointed therewith F it cleareth and clenſeth the ſight from the Haw and from ſtripes.

Cedar infuſed in vineger and put into the eares, killeth the wormes therein; and being mingled G with the decoction of Hyſſop, appeaſeth the ſounding, ringing, and hiſſing of the eares.

If it be waſhed or infuſed in vineger, and applied to the teeth, it eaſeth the tooth-ache. H

If it be put into the hollowneſſe of the teeth, it breaketh them, and appeaſeth the extreme griefe I thereof.

It preuaileth againſt *Angina's*, and the inflammation of the tonſils, if a gargariſme be made of it. K

It is good to kill nits and lice and ſuch like vermin: it cureth the biting of the ſerpent Ceraſtes L being layd on with ſalt.

It is a remedie againſt the poiſon of the ſea Hare, if it be drunke with ſweet wine. M

It is good alſo for Lepers: being put vp vnderneath it killeth all manner of wormes, and draweth N forth the birth, as *Dioſcorides* writeth.

Chap. 40. *Of the Pitch tree.*

¶ *The Deſcription.*

† 1 **P***icea*, the tree that droppeth pitch, called Pitch tree, groweth vp to be a tall faire and big tree, remaining alwaies green like the Pine tree: the timber of it is redder than that of the Pine or Fir: it is ſet full of boughes not onely about the top, but much lower, & alſo beneath the middle part of the body, which many times hang down, bending toward the ground: the leaues be narrow, not like thoſe of the Pine tree, but ſhorter and narrower, ſharpe pointed like them, yet are they blacker, and withall couer the yong and tender twigs in manner of a circle, like thoſe of the Fir tree; but being many, and thick ſet, grow forth on all ſides, and not onely one right againſt another, as in the Yew tree: the fruit is ſcaly, and like vnto the Pine apple, but ſmaller: the barke of the tree is ſomewhat blacke, tough, and flexible, not brittle as is the barke of the Fir tree; vnder which next to the wood is gathered a roſin, which many times iſſueth forth, and is like to that of the Larch tree.

‡ 2 Of this ſort (ſaith *Cluſius*) there is found another that neuer growes high, but remaineth dwarfiſh, and it caries certain little nugaments or catkins of the bigneſſe of a ſmall nut, compoſed of ſcales lying one vpon another, and ending in a prickly leaf, which in time opening ſhew certain empty cauities or cels: from the tops of theſe ſomtimes grow forth brauches ſet with many ſhort and pricking little leaues: all the ſhrub hath ſhorter and paler coloured leaues than the former. I obſerued neither fruit nor floure on this, neither know I whether it carry any. *Dalechampius* ſeems to haue known this, and to haue called it *Pinus Tubulus* or *Tibulus.* ‡

1 *Picea major.*
The Pitch tree.

‡ 2 *Picea pumila.*
The dwarſe Pitch tree.

¶ *The Place.*

   The Pitch tree groweth in Greece, Italy, France, Germany, and all the cold Regions euen vnto Ruſſia.

¶ *The Time.*

   The fruit of the Pitch tree is ripe in the end of September.

¶ *The Names.*

   The Grecians call this Cone tree πεύκη: the Latines, *Picea,* and not *Pinus* ; for *Pinus,* or the Pine tree, is the Grecians πεύκη, as ſhall be declared : that πεύκη is named in Latine *Picea, Scribonius Largus* teſtifieth in his 201 Compoſition, writing after this manner : *Reſina pituinæ, i. ex Picea arbore,* ſigni-fying in Engliſh, of the roſin of the tree *Pitys,* that is to ſay, of the Pitch tree. With him doth *Pliny* agree, *lib. 16. cap. 10.* where hee tranſlating *Theophraſtus* his words concerning *Peuce* and *Pitys,* doth tranſlate *Pitys, Picea,* although for *Peuce* he hath written *Larix,* as ſhall be declared. *Pliny* writeth thus, *Larix vſtis radicibus non repullulat :* and the Larch tree doth not ſpring vp againe when the roots ate burnt : the Pitch tree ſpringeth vp again, as it hapned in Lesbos, when the wood Pyrthæ-us was ſet on fire. Moreouer, the wormes *Pityocampæ* are ſcarce found in any tree but onely in the Pitch tree, as *Bellonius* teſtifieth. So that they are not raſhly called *Pityocampæ,* or the wormes of the Pitch tree ; although moſt of the Tranſlators name them *Pinorum eruca,* or the Wormes of the Pine trees : and therefore *Pitys* is ſyrnamed by *Theophraſtus,* ἐψῶσαι, becauſe Wormes and magots are bred in it. But foraſmuch as the name *Pitys* is common both to the tame Pine, and alſo to the Pitch tree, diuers of the later writers do for this cauſe ſuppoſe that the Pitch tree is named by *The-ophraſtus,* πεύκη ἀγρία, or the wilde Pine tree. This *Picea* is named in high-Dutch, **Schwartz Tanne-baum,** and **Rot Tannebaum,** and oftentimes alſo **Josenholtz** ; which name notwithſtanding a-greeth alſo with other plants : in Engliſh, Pitch tree : in low-Dutch, **Peck boom.**

¶ *The Temperature and Vertues.*

   The leaues, bark, and fruit of the Pitch tree are all of one nature, vertue, and operation, and of the ſame facultie with the Pines.

C H A P.

CHAP. 41.   Of the Pine Tree.

¶ *The Kindes.*

THe Pine tree is of two ſorts, according to *Theophraſtus* : the one ἥμερος, that is to ſay, tame, or of
the garden ; the other ἀγρία, or wild : hee ſaith that the Macedonians doe adde a third, which is
Ἄκαρπος, barren or without fruit, which is to vs vnknowne : the later writers haue found more, as ſhall
be declared.

Pinus ſativa, ſive domeſtica.
The tame or manured Pine tree.

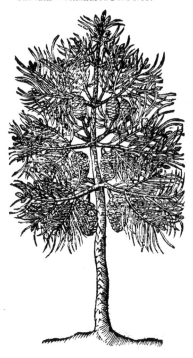

¶ *The Deſcription.*

THe Pine tree growes high and great in the
trunke or body, which below is naked, but
aboue is clad with a multitude of boughes
which diuide themſelues into diuers branches,
whereon are ſet ſmall leaues, very ſtraight, narrow, ſomwhat hard and ſharp pointed : the wood
or timber is hard, heauy, about the heart or middle full of an oilous liquor, and of a reddiſh colour : the fruit or clogs are hard, great, and conſiſt
of many ſound woody ſcales, vnder which are included certaine knobs without ſhape, couered
with a woodden ſhell like ſmall nuts, wherin are
white kernels, long, very ſweet, and couered with
a thin skin or membrane, that eaſily is rubbed off
with the fingers , which kernell is vſed in medicine.

¶ *The Place.*

This tree groweth of it ſelf in many places of
Italy, and eſpecially in the territory of Rauenna,
and in Languedoc, about Marſiles, in Spaine, and
in other hot regions, as in the Eaſt countries. It
is alſo cheriſhed in the gardens of pleaſure, both
in the Low-countries and England.

¶ *The Time.*

The Pine tree groweth greene both Winter
and Summer : the fruit is commonly two yeares
before it be ripe : wherefore it is not to be found
without ripe fruit , and alſo others as yet very
ſmall and not come to ripeneſſe.

¶ *The Names.*

It is called in Latine, *Pinus*, and *Pinus ſativa*, *Vrbana*, or rather *Manſueta* : in Engliſh, tame or garden Pine : of the Macedonians and other Grecians, πεύκη ἥμερος ; but the Arcadians name it πίτυς, for
that which the Macedonians call πεύκη ἥμερος, the Arcadians name πίτυς, as *Theophraſtus* ſaith : & ſo doth
the tame Pine in Arcadia and about Elia change her name : and by this alteration of them it happens that the fruit or nuts of the Pine tree found in the cones or apples, be named in Greek by *Dio-*
ſcorides, Galen, Paulus, and others, πιτυΐδες ; as though they ſhould terme it *Pityos fruĉtus*, or the fruit of
the Pine tree.

There is alſo another πίτυς in Latine, *Picea*, or the Pitch tree, which differeth much from the pine
tree : but *Pytis* of Arcadia differeth nothing from the Pine tree, as we haue ſaid.

The fruit or apples of theſe be called in Greeke, κῶνοι, and in Latine *Coni* : notwithſtanding *Co-*
nos is a common name to all the fruits of theſe kinde of trees : they be alſo named in Latine, *Nuces*
pineæ : by *Mneſitheus* in Greek : ϛρόβιλοι, by *Diocles Cariſtius*, πιτυοϛρόβιλοι, which be notwithſtanding the
fruit or clogs of the tree that *Theophraſtus* nameth πεύκη, or the wilde Pine tree, as *Athenæus* ſaith. It
is thought that the whole fruit is called by *Galen* in his fourth commentary vpon *Hippocrates* books
of diet in ſharp diſeaſes, *Strobilos* : yet in his ſecond book of the Faculties of nouriſhments he doth
not call *Conos* or the apple by the name of *Strobilos*, but the nuts contained in it. And in like manner in his ſeuenth booke of the faculties of ſimple medicines ; The Pine apple fruit, ſaith he, which
they

they call *Coccalus* and *Strobilus*, as we haue ſaid before, that theſe are named in Greeke ϰῶνοι. This apple is called in high-Dutch, **Zyrbel** : in low-Dutch, **Pijn appel** : in Engliſh, Pineapple, Clog, and Cone.

¶ *The Temperature and Vertues.*

A   The kernels of theſe nuts concoct and moderately heate, being in a mean between cold and hot: they make rough parts ſmooth, are a remedy againſt an old cough, and long infirmities of the cheſt, being taken by themſelues or with hony, or elſe with ſome other licking thing.

B   It cureth the Ptyſicke, and thoſe that pine and conſume away through the rottenneſſe of their lungs : it recouereth ſtrength, nouriſheth, and is reſtoratiue to the body.

C   It yeeldeth a thick and good juice, and nouriſheth much, yet is it not altogether of good digeſtion, and therefore it is mixed with preſerues, and boiled with ſugar.

D   The ſame is good againſt the ſtone in the kidnies, and againſt frettings of the bladder, and ſcalding of vrine, for it allayeth the ſharpneſſe, mitigateth pain, and gently prouoketh vrine: moreouer it increaſeth milke and ſeed, and therefore it alſo prouoketh fleſhly luſt.

E   The whole cone or apple being boiled with freſh Horehound, ſaith *Galen*, and afterwards boiled again with a little hony vntill the decoction be come to the thickneſſe of honey, maketh an excellent medicine for to clenſe the cheſt and lungs.

F   The like ſaith *Dioſcor.* The whole cones (ſaith he) newly gathered from the trees, broken and boiled in ſweet wine are good for an old cough, and conſumption of the lungs, if a good draught of the liquor be drunke euery day.

G   The ſcales of the Pine apple, with the barke of the tree, ſtop the laske and bloudy flixe, prouoke vrine, and the decoction alſo hath the like propertie.

Chap. 42.   *Of the wilde Pine tree.*

1 *Pinus ſylueſtris.*
The wilde Pine tree.

2 *Pinus ſylueſtris mugo.*
The low wilde Pine tree.

¶ *The Deſcription.*

1   THe firſt kinde of wilde Pine tree groweth very great, but not ſo high as the former, bee-ing the tame or mannured Pine tree: the barke thereof is glib; the branches are ſpred abroad, beſet with long ſharpe pointed leaues: the fruit is ſomewhat like the tame Pine tree, with ſome roſin therein, and ſweet of ſmell, which doth eaſily open it ſelfe, and quickly falleth from the tree.

2   The ſecond kinde of wilde Pine tree groweth not ſo high as the former, neither is the ſtem growing ſtraight vp, but yet it bringeth forth many branches, long, ſlender, and ſo eaſie to bee bent or bowed, that hereof they make hoops for wine hogſheads and tuns: the fruit of this Pine is grea-ter than the fruit of any of the other wilde Pines.

3   The third kinde of wild Pine tree groweth ſtraight vpright, and waxeth great and high, yet not ſo high as the other wilde kindes: the branches grow like the Pitch tree, the fruit is long and big, almoſt like the fruit of the ſaid Pitch tree; wherein are contained ſmal triangled nuts like the nuts of the Pine apple tree, but ſmaller, and more brittle: in which is contained a kernell of a good taſt, like the kernel of the tame Pine apple: the wood is beautiful, and ſweet of ſmell, good to make tables and other works of.

4   There is another wilde Pine of the mountaine, not differing from the precedent but in ſta-ture, growing for the moſt part like an hedge tree, wherein is the difference.

3 *Pinus ſylueſtris montana.*
The mountain wilde Pine tree.

4 *Pinus montana minor.*
The ſmaller wilde Pine tree.

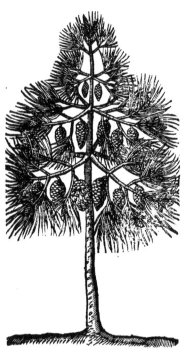 

5   This kinde of Pine, called the ſea Pine tree, groweth not aboue the height of two men, ha-uing leaues like the tame Pine tree, but ſhorter: the fruit is of the ſame form, but longer, ſomwhat faſhioned like a Turnep. This tree yeeldeth very much roſin. ‡ *Bauhine* iudgeth this all one with the third. ‡

6   The ſixt kind of wilde Pine, being one of the ſea Pines, groweth like an hedge tree or ſhrub ſeldome exceeding the height of a man, with little leaues like thoſe of the Larch tree, but alwayes continuing with a very little cone and fine ſmall kernell.

7   The baſtard wilde Pine tree groweth vp to a meane height, the trunke or body, as alſo the
branches

5 *Pinus maritima major.*
The great sea Pine tree.

6 *Pinus sylvestris minor.*
The little sea Pine tree.

7 *Teda, sive Pseudopinus.*
The bastard wild Pine.

‡ 8 *Pinaster Austriacus.*
Dwarfe Pine with vpright cones.

‡ 9 *Pinaster maritimus minor.*
Dwarfe sea Pine.

ches and leaues are like vnto thofe of the manured Pine tree: the onely difference is, That
fome yeares it refembles the Pine it felfe, and
other yeares as a wilde hedge tree, varying often, as Nature lifteth to play and fport her felf
among her delights, with other plants of leffe
moment: the timber is foft, and not fit for building, but is of the fubftance of our Birch tree:
the fruit is like thofe of the other wild Pines,
whereof this is a kinde.

‡ 8   This dwarfe Auftrian Pine exceeds
not the height of a man, but immediatly from
the root is diuided and fpread abroad into
tough bending pretty thick branches, couered
ouer with a rough barke: the leaues, as in the
former, come two out of one hofe, thicker,
fhorter, blunter pointed, and more greene than
the former: the cones or clogs are but fmall,
yet round and compact, and hang not downewards, but ftand vpright: the root is tough and
woody like other plants of this kind. It growes
on the Auftrian and Styrian Alps. *Clufius* fets
it forth by the name of *Pinafter 4. Auftriacus.*

9   The other Dwarfe is of the fame height
with the former: with fuch tough and bending
branches, which are neither fo thicke, nor clad
with fo rough a barke, nor fo much fpred. The
leaues alfo are fmaller, and not vnlike thofe of
the Larix tree, but not fo foft, nor falling euery
yeare as they do. The cones are little and flender, the kernels fmall, blackifh, and winged as
the reft. *Clufius* found this onely in fome few
places of the kingdome of Murcia in Spaine;
wherefore he calls it *Pinafter 3. Hifpanicus.* Dodonæus calls it *Pinus maritima minor.* ‡

¶ *The Place.*

Thefe wilde Pines grow vpon the cold mountaines of Livonia, Polonia, Norvegia, and Ruffia, e-
fpecially vpon the Ifland called Holland within the Sownd, beyond Denmarke, and in the Woods
by Narua vpon the Liefeland fhore, and all the tract of the way, being a thoufand werfts (each werft
containing three quarters of an Englifh mile) from Narua vnto Muicouia, where I haue feen them
grow in infinite numbers.

¶ *The Time.*

The fruit of thefe Pine trees is ripe in the end of September: out of all thefe iffueth forth a
white and fweet fmelling rofin: they are alfo changed into *Teda*, and out of thefe is boiled through
the force of the fire a blacke pitch: the Pitch tree and the Larch tree be alfo fometimes changed
into *Teda*, yet very feldome, for *Teda* is a proper and peculiar infirmitie of the wilde Pine tree. A
tree is faid to be changed into *Teda*, when not only the heart of it, but alfo the reft of the fubftance
is turned into fatneffe.

¶ *The Names.*

All thefe are called in Greeke Πεύκη ἄγρια; and in Latine, *Sylveftres Pini*: of Pliny *Pinaftri*; *Pinafter,*
faith he, *lib.16.cap.10.* is nothing elfe but *Pinus fylveftris*, or the wilde Pine tree, of a leffer height,
and full of boughes from the middle, as the tame Pine tree in the top, (moft of the couples haue
falfly) of a maruellous height: they are farre deceiued who thinke that the Pine tree is called in
Greeke Πεύκη, befides the tame Pine, which notwithftanding is not fo called of all men, but onely of
the Arcadians (as we haue faid before) Πεύκη: all men do name the wilde Πεύκη: and therefore *Teda* or
the Torch Pine hereof is faid to be in Latine, not *Picea*, but *Pinea*, that is, not the Pitch tree, but the
Pine tree, as *Ouid* doth plainly teftifie in his Heroicall Epiftles:

Vt

Vt vidi, vt perij, nec notis ignibus arsi,
Ardet vt ad magnos Pinea Teda deos.

Also in *Fast. 4.*

Illic accendit geminas pro lampade Pinus.
Hinc cereris sacris nunc quoque Teda datur.

The same doth also *Virgil* signifie, *Æneid. 7.*

Ipsa inter medias, flagrantem seruida Pinum.
Sustinet. ───────────

Where in stead of *Flagrantem Pinum, Seruius* admonisheth vs to vnderstand *Teda Pinea. Catullus* also consenteth with them, in the mariage Song of *Iulia* and *Mallius*

─────── *Manu*
Pineam quate tedam.

And *Prudentius, in Hymno Cerei Paschalis,*

Seu Pinus piceam fert alimoniam.

Moreouer, the herbe *Peucedanos*, or Horestrong, so named of the likenesse of ꝏꝏ, is called also in Latine *Pinastellum*, of *Pinus*, the Pine tree : all which things do euidently declare, that ꝏꝏ is called in Latine not *Picea*, but *Pinus*.

The first of these wilde kindes may be *Idæa Theophrasti*, or *Theophrastus* his Pine tree, growing on mount Ida, if the apple which is shorter, were longer : for he nameth two kinds of wilde Pines, one of mount Ida, and the other the sea Pine with the round fruit : but wee hold the contrary ; for the fruit or apple of the wild mountain Pine is shorter, and that of the sea Pine longer. This may more truly be *Macedonum mas*, or the Macedonians male Pine : for they make two sorts of wilde Pines, the male and the female, the male more wrythed and harder to bee wrought vpon, the female more easie ; but the wood of this is more wrythed, & not so much in request for works as the other, wherefore it seems to be the male. This wilde Pine tree is called in high-Dutch, **Hartzbaum**, and **wilder Hartzbaum** : in Gallia Celtica, *Elvo Alevo* : and in Spanish, *Pino Carex.*

The second wilde Pine tree is named commonly of the Italians about Trent and Anagnia, *Cembro*, and *Cirmolo* ; it seems to differ nothing at all from the Macedonians wilde female Pine, for the wood is easie to be wrought on, and serueth for diuers and sundry works.

The third they call *Mugo* : this may be named not without cause, χαμαιπιτυς, that is to say, *Humilis Pinus*, or dwarfe Pine : yet doth it differ from *Chamæpeuce*, the herb called in English, ground Pine.

The fourth wild Pine is named in Greeke, παραλιωτιδα : in Latine, *Maritima*, and *Marina pinus* : in English, sea Pine.

That which the Latines call *Teda*, is named in Greeke δαις, and δαδιον in high-Dutch, **Kynholtz**: it may be termed in English, Torch Pine.

Pliny is deceiued, in that he supposeth Torch Pine to be a tree by it selfe, making it the sixt kind of Cone tree : as likewise he erreth in taking *Larix*, the Larch tree, for πιτυς the Pine tree. And as *Dioscorides* maketh so little difference as scarce any, betweene πιτυς and πευκη, and supposeth them to bee both of one kinde, so likewise he sets downe faculties common to them both.

¶ *The Nature and Vertues.*

A   The barke of them both, saith he, doth binde : being beaten and applied it cureth Merigals, and also shallow vlcers and burnings, if it be laid on with Litharge and fine Frankincense.

B   With the Cerot of Myrtles it healeth vlcers in tender bodies : beeing beaten with Copperas it stayeth tettars and creeping vlcers : it drawes away the birth and after-birth, if it be taken vnder in a fume : being drunke it stoppeth the belly, and prouoketh vrine.

C   *Galen* hath almost the same things ; but he saith that the barke of the Pine tree is more temperat than that of the Pitch tree : the leaues stamped take away hot swellings and the sores that do come thereof.

D   Being stamped and boiled in vineger they asswage the pain of the teeth, if they be washed with this decoction hot : the same be also good for those that haue bad liuers, being drunke with water or mede.

E   Of the same operation is likewise the barke of the Pine nuts ; but *Galen* affirmeth that the Cone or apple, although it seeme to be like these is notwithstanding of lesser force, insomuch as it cannot effectually performe any of the aforesaid vertues, but hath in it a certaine biting quality, which hurteth.

The Torch Pine cut into small pieces and boiled in vineger, is a remedy likewise against the tooth-ache if the teeth be washed with the decoction.   F

Of this there is made a profitable spather or slice to be vsed in making of compound plaisters and pessaries that ease paine.   G

Of the smoke of this is made a blacke which serueth to make inke of, and for eating sores in the corners of eies, and against the falling away of the haire of the eie lids, and for watering and bleere eies, as *Dioscorides* teacheth.   H

Of Rosins.

¶ The Kindes.

1 OVt of the Pine tres, especially of the wilde kinds, there issueth forth a liquid, whitish, and sweet smelling Rosin, and that many times by it selfe; but more plentifully either out of the cut and broken boughes, or forth of the body when the tree commeth to be a Torch Pine.

2 There issueth also forth of the crackes and chinkes of the barke, or out of the cut boughes, a certaine dry Rosin, and that forth of the Pine tree or Firre Tree.

There is likewise found a certaine congealed Rosin vpon the cones or apples.

It is called in Latine *Resina*; in Greeke, ῥητίνη in high Dutch, **Hartz**: in low Dutch, **Herss**: in Italian, *Ragia*: in Spanish, *Resina*: in English, Rosin.

The first is named in Latine, *Liquida Resina*: in Greeke, ῥητίνη ὑγρά, and of diuers, ἀυτόρρυτος, that is to say, issueth out of it selfe: of the Lacedemonians, προπρίππωσις, or *Primiflua*, the first flowing Rosin: and in Cicilia, ἀκταίαν, as *Galen* writeth in his third booke of medicines according to the kindes: in shops, *Resina Pini*, or Rosin of the Pine tree, and common Rosin. It hapneth oftentimes through the negligent and carelesse gathering thereof, that certaine small pieces of wood, and little stones be found mixed with it: this kinde of Rosin *Galen* surnameth συγκεχυμένη, as though he should say, confused, which being melted and clensed from the drosse becommeth hard and brittle.

The like hapneth also to another liquid Rosin, which after it is milted, boyled, and cooled againe, is hard and brittle, and may likewise be beaten, ground, and searced; and this Rosin is named in Greeke φρυκτή: in Latine, *Fricta*, and many times *Colophonia*, in Greeke, Κολοφωνία which name is vsed among the Apothecaries, and may stand for an English name; for *Galen* in his third booke of medicines according to their kindes saith, that it is called *Fricta*, and of some *Colophonia* that, saith he, is the driest Rosin of all, which some call *Fricta*, others, *Colophonia*: because in times past, as *Dioscorides* writeth, it was fetched from *Colophon*, this being yellow or blacke in comparison of the rest, is white when it is beaten: *Pliny* in his 14 booke, 20. chapter.

The second Rosin is named in Greeke ῥητίνη ξηρά, specially that of the Pitch tree without fatnesse, and that soone waxeth dry, which *Galen* in his 6. booke of Medicines according to their kindes, calleth properly ξηρὰ ῥητίνη: that which in Asia is made of the Pitch tree being very white, is called *Spagas*, as *Pliny* testifieth.

The third is called in Greeke ῥητίνη τερβῶθίνη the same is also named ἔυοσμος τερβῶθίνη, that is vnknowne in shops. Yet there is to be sould a certaine dry Rosin, but the same is compounded of the Rosins of the Pine tree, of the cones or clogs, and of the Firre tree mixed altogether, which they call *Garipot*: this is vsed in perfumes in stead of Frankincense, from which notwithstanding it farre differeth.

¶ The Temperature and Vertues.

All the Rosins are hot and dry, but not all after one manner: for there is a difference among them: they which be sharper and more biting, are hotter, as that which commeth of the cones, being of Rosins the hottest, because it is also the sharpest: the Rosin of the Pitch Tree is not so much biting, and therefore not so hot: the Rosin of the Firre tree is in a meane betweene them both; the liquid Rosin of the Pine is moister, comming neere to the quality and facultie of the Larch Rosin.   A

The Rosins which are burnt or dried, as *Dioscorides* testifieth, are profitable in plaisters, and compositions that ease wearisomenesse; for they do not onely supple or mollifie, but also by reason of the thinnesse of their parts and drynesse, they digest: therefore they both mollifie and waste away swellings, and through the same faculty they cure wearisomenesse: being vsed in compound medicines for that purpose.   B

The liquid Rosins are very fitly mixed in ointments, commended for the healing vp of greene wounds, for they both bring to suppuration, and do also glue and vnite them together.   C

D   Moreouer, there is gathered out from the Rosins as from Frankincense, a congealed smoke, called in Latine *Fuligo* ; in Greeke, λιγνυς: and in English, Blacke, which serueth for medicines that beautifie the eie lids, and cure the fretting sores of the corners of the eies, and also watering eies, for it drieth without biting.

E   There is made hereof, saith *Dioscorides*, writing inke, but in our age not that which we write withall, but the same which serueth for Printers to print their bookes with, that is to say, of this blacke, or congealed smoke, and other things added.

Of Pitch and Tar.

The manner of drawing forth of Pitch.

Out of the fattest wood of the Pine tree changed into the Torch Pine, is drawne Pitch by force of fire. A place must be paued with stone, or some other hard matter, a little higher in the middle, about which there must also be made gutters, into which the liquor shall fall ; then out from them other gutters are to be drawne, by which it may be receiued ; being receiued, it is put into barrels. The place being thus prepared, the clouen wood of the Torch Pine must be set vpright ; then must it be couered with a great number of Fir and Pitch boughes, and on euery part all about with much lome and earth : and great heed must be taken, least there be any cleft or chinke remaining, onely a hole left in the top of the furnace, thorow which the fire may be put in, and the flame and smoke may passe ouer : when the fire burneth the Pitch runneth forth, first the thin, and then the thicker.

This liquor is called in Greeke πισσα : in Latine, *Pix* : in English, Pitch, and the moisture, euen the same that first runneth is named of *Pliny* in his 16. booke, 11. chapter, *Cedria* : There is boyled in Europe, saith he, from the Torch Pine a liquid Pitch vsed about ships, and seruing for many other purposes ; the wood being clouen is burned with fire, and set round about the furnaces on euery side, after the manner of making Charcoles : the first liquor runneth through the gutter like water: (this in Syria is called *Cedria*, which is of so great vertue, as in Ægypt the bodies of dead men are preserued being all couered ouer with it) the liquor following being now thicker, is made Pitch. But *Dioscorides* writeth, That *Cedria* is gathered of the great Cedar tree, and nameth the liquor drawne out of the Torch tree by force of fire, πισσα ύγρα: that is, that which the Latines call *Pix liquida*: the Italians, *Pece liquida*: in high Dutch, **Weich bach** : in low Dutch, **Teer** : in French, *Poix foudire*: in Spanish, *Pex liquida*: certaine Apothecaries, *Kitran*: and we in English Tar.

And of this when it is boyled is made a harder Pitch : this is named in Greeke, ξηρα πισσα: in Latine, *Arida*, or *sicca Pix*: of diuers, παλιμπισσα as though they should say, *Iterata Pix*, or Pitch iterated : because it is boyled the second time. A certaine kinde hereof being made clammie or glewing is named βρυτια: in shops, *Pix naualis*, or Ship Pitch : in high Dutch, **Bach**; in low Dutch, **Steenpeck :** in Italian, *Pece secca*: in French, *Poix seche*: in Spanish, *Pez seca*: in English, Stone Pitch.

¶ The Temperature and Vertues.

A   Pitch is hot and dry, Tarre is hotter, and stone Pitch more drying, as *Galen* writeth. Tar is good against inflammations of the Almonds of the throat, and the uvula, and likewise the Squincie, being outwardly applied.

B   It is a remedie for mattering eares with oyle of Roses : it healeth the bitings of Serpents, if it be beaten with salt and applied.

C   With an equall portion of wax it taketh away foule ilfauoured nailes, it wasteth away swellings of the kernels, and hard swellings of the mother and fundament.

D   With barly meale and a boies vrine it consumeth χοιραδας, or the Kings euill : it staieth eating vlcers, if it be laid vnto them with brimstone, and the barke of the Pitch Tree, or with branne.

E   If it be mixed with fine Frankincense, and a cerote made thereof, it healeth chops of the fundament and feet.

F   Stone Pitch doth mollifie and soften hard swellings : it ripens and maketh matter, and wasteth away hard swellings and inflammations of kernels : it filleth vp hollow vlcers, and is fitly mixed with wound medicines.

G   What vertue Tar hath when it is inwardly taken we may reade in *Dioscorides* and *Galen*, but we set downe nothing thereof, for that no man in our age will easily vouchsafe the taking.

H   There is also made of Pitch a congealed smoke or blacke, which serueth for the same purposes as that of the Rosins doth.

CHAP. 43. *Of the Firre or Deale Tree.*

¶ *The Deſcription.*

1　THe Firre tree groweth very high and great, hauing his leaues euer greene; his trunke or body ſmooth, euen, and ſtraight, without joynts or knots, vntill it hath gotten branches; which are many and very faire, beſet with leaues, not much vnlike the leaues of the Ewe tree, but ſmaller: among which come forth floures vpon the taller trees, growing at the bottomes of the leaues like little catkins, as you may ſee them expreſt in a branch apart by themſelues: the fruit is like vnto the Pine Apple, but ſmaller and narrower, hanging downe as the Pine Apple: the timber hereof excelleth all other timber for the maſting of ſhips, poſts, rails, deale boords, and ſundry other purpoſes.

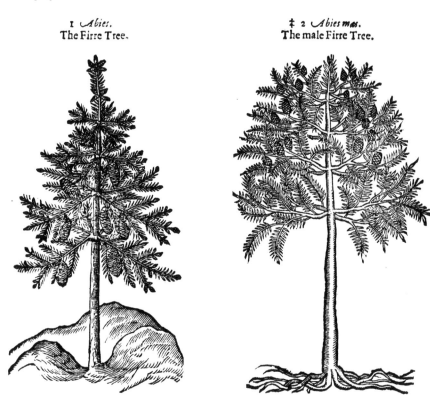

1 *Abies.*
The Firre Tree.

‡ 2 *Abies mas.*
The male Firre Tree.

2　There is another kinde of Firre Tree, which is likewiſe a very high and tall tree, and higher than the Pine: the body of it is ſtraight without knots below, waxing ſmaller and ſmaller euen to the very top: about which it ſendeth forth boughes, foure together out of one and the ſelfe ſame part of the body, placed one againſt another, in manner of a croſſe, growing forth of the foure ſides of the body, and obſeruing the ſame order euen to the very top: out of theſe boughes grow others alſo, but by two and two, one placed right againſt another, out of the ſides which bend downewards when the other beare vpwards: the leaues compaſſe the boughs round about, and the branches thereof: they be long, round, and blunt pointed, narrower, and much whiter than thoſe of the Pitch Tree, that is to ſay, of a light greene, and in a manner of a white colour: the cones or clogs be long, and longer than any others of the cone trees: they conſiſt of a multitude of ſoft ſcales, they hang downe from the end of the twigs, and doe not eaſily fall downe, but remaine on the tree a very long time: the kernels in theſe are ſmall, not greater than the kernels of the Chctric ſtone, with a thinne skin growing on the one ſide, very like almoſt to the wings of Bees, or great Flies: the timber or ſubſtance of the wood is white, and clad with many coats, like the head of an Onion.

‡ *Abietis ramus cum julis.*
A branch with Catkins or floures.

¶ *The Place.*

The Firre trees grow vpon high mountaines, in many woods of Germany and Bohemia, in which it continueth alwaies greene;it is found alfo on hils in Italy,France,& other countries; it commeth downe oftentimes into the vallies: they are found likewife in Pruſe, Pomerania, Liefeland, Ruſſia, and eſpecially in Norway, where I haue ſeene the goodlieſt trees in the world of this kinde, growing vpon the rockie and craggie mountaines,almoſt without any earth about them, or any other thing, ſauing a little moſſe about the roots,which thruſt them ſelues here and there into the chinkes and cranies of the rockes, and therefore are eaſily caſt downe with any extreme gale of winde. I haue ſeen theſe trees growing in Cheſhire, Staffordſhire,and Lancaſhire, where they grew in great plenty,as is reported, before *Noahs* floud : but then being ouerturned and ouerwhelmed haue lien ſince in the moſſes and waterie mooriſh grounds very freſh and ſound vntill this day,& ſo full of a reſinous ſubſtance, that they burne like a Torch or Linke, and the inhabitants of thoſe countries do call it Fir-wood, and Firewood vnto this day:out of this tree iſſueth the roſin called *Thus*,in Engliſh,Frankinſence:but from the young Fir trees proceedeth an excellent cleare and liquid Roſin,in taſte like to the peelings or outward rinde of the Pomecitron.

¶ *The Time.*
The time of the Fir tree agreeth with the Pine trees.

¶ *The Names.*
The tree is called in Latine *Abies* : in Greeke, ἐλάτη: amongſt the Græcians of our time the ſame name remaineth whole and vncorrupt : it is called in high Dutch, **Weiſʒ Thannen** and **weiſʒ Thannen baum** : in low Dutch, **Witte Dennen Boom**, or **Abel-boom**,and **Maſt-boom** : in Italian, *Abete* : in Spaniſh, *Abeto* : in Engliſh, Fir-tree,Maſt-tree,and Deale-tree.The firſt is called in French, *du Sap*,or *Sapin* ; the other is *Suiffe*.

The liquid roſin which is taken forth of the barke of the young Firre-trees, is called in Greeke δάκρυον τῆς ἐλάτης : in Latine, *Lachryma abietis*,and *Lachryma abiegna* : in the ſhops of Germany, as alſo of England, *Terebinthina Veneta*, or Venice Turpentine:in Italian, *Lagrimo* : diuers doe thinke that *Dioſcorides* calleth it ἰλασίας ρητίνη, *Oleaſa Reſina*,or oile Roſin ; but oile Roſin is the ſame that *Pix liquida*,or Tar is.

Arida Abietum Reſina, or dry Roſin of the Firre-trees, is rightly called in Greeke, ξηρὰ ἐλατίνη, and in Latine, *Abiegna Reſina*: it hath a ſweet ſmell,and is oftentimes vſed among other perfumes in ſtead of Frankinſcence.

¶ *The Temperature.*
The barke,fruit,and gums of the Fir-tree,are of the nature of the Pitch tree and his gums.

¶ *The Vertues.*

A   The liquid Roſin of the Fir tree called Turpentine,looſeth the belly,driueth forth hot cholerick humours, clenſeth and mundifieth the kidnies, prouoketh vrine, and driueth forth the ſtone and grauell.

B   The ſame taken with Sugar and the pouder of Nutmegs, cureth the ſtrangurie, ſtaieth the Gonorrhœa or the inuoluntary iſſue of mans nature, called the running of the rains,and the white flux in women.

C   It is very profitable for all green and freſh wounds,eſpecially the wounds of the head:for it healeth and clenſeth mightily,eſpecially if it be waſhed in Plantaine water,and afterward in Roſe water,the yolke of an egge put thereto,with the pouders of *Olibanum* and Maſticke finely ſearced, adding thereto a little Saffron.

C H A P.

CHAP. 44.　Of the Larch Tree.

¶ The Description.

1　THe Larch is a tree of no small height, with a body growing straight vp: the barke wherof in the neither part beneath the boughes is thicke, rugged and full of chinkes, which being cut in sunder is red within, and in the other part aboue smooth, slipperie, something white without : it bringeth forth many boughes diuided into other lesser branches, which be tough and pliable. The leaues are small and cut into many jags, growing in clusters thicke together like tassels, which fall away at the approch of Winter : the floures, or rather the first shewes of the cones or fruit be round, and grow out of the tenderest boughes, being at the length of a braue red purple colour : the cones be small, and like almost in bignesse to those of the Cypresse tree, but longer, and made vp of a multitude of thin scales like leaues : vnder which lie small seeds, hauing a thin velme growing on them very like to the wings of Bees and wasps : the substance of the wood is very hard of colour, especially that in the midst somewhat red, and very profitable for workes of long continuance.

1 *Laricis ramulus.*	2 *Larix cum Agarico suo.*
A brach of the Larnch tree.	The Larch tree with his Agarick.

It is not true that the wood of the Larch tree cannot be set on fire, as *Vitruvius* reporteth of the castle made of Larch wood, which *Cæsar* besieged, for it burneth in chimneies, and is turned into coles, which are very profitable for Smithes, as *Matthiolus* writeth.

There is also gathered of the Larch tree a liquid Rosin, very like in colour and substance to the whiter hony, as that of Athens or of Spaine, which notwithstanding issueth not forth of it selfe, but runneth out of the stocke of the tree, when it hath beene bored euen to the heart with a great and long auger and wimble.

Galen writeth, that there be after a sort two kindes hereof, in his fourth book, of Medicines according to the kindes, one like vnto Turpentine, the other more sharper than this, hotter, more liquid of a ſtronger ſmell, and in taſte bitterer and hotter: but the later is thought not to be the Roſine of the Larch, but of the Fir tree, which *Galen* becauſe it is after a ſort like in ſubſtance, might haue taken for that of the Larch tree.

There groweth alſo vpon the Larch tree a kinde of Muſhrum or excreſcence, not ſuch as is vpon other trees, but whiter, ſofter, more looſe and ſpungie than any other of the Muſhrums, and good for medicine, which beareth the name of *Agaricus*, or Agaricke: I finde that *Pliny* ſuppoſeth all the Maſticke trees, and thoſe that beare Galls, do bring forth this *Agaricum*: wherein he was ſomewhat deceiued, and eſpecially in that he tooke *Glandifera* for *Conifera*, that is, thoſe trees which beare maſt or Acornes, for the Pine apple trees: but among all the trees that beare *Agaricus*, the Larch is the chiefe, and bringeth moſt plenty of Agaricke.

¶ The Place.

The Larch tree groweth not in Greeee, or in Macedon, but chiefely vpon the Alps of Italy, not far from Trent, hard by the riuers *Benacus* and *Padus*; and alſo in other places of the ſame mountains: it is likewiſe found on hils in Morauia, which in times paſt was called the country of the Marcomans: *Fuchſius* writeth, that it groweth alſo in Sileſia: others, in Luſatia, in the borders of Poland: it alſo groweth plentifully in the woods of Gallia Ceſalpina.

Pliny hath ſaid ſomewhat hereof, contradicting the writings of others, in his 16.booke, 8.chapter, where he ſaith, that ſpecially the Acorne trees of France do beare Agaricke, and not only the acorne trees, but the Cone trees alſo, among which, ſaith he, the Larch tree is the chiefe that bringeth forth Agaricke, and that not onely in Gallia, which now is called France, but rather in Lumbardy and Piemont in Italy, where there be whole woods of Larch trees, although they be found in ſome ſmal quantitie in other Countries.

The beſt Agaricke is that which is whiteſt, very looſe and ſpungie, which may eaſily be broken, and is light, and in the firſt taſte ſweet, hard and well compact: that which is heauie, blackiſh, and containing in it little thweds as it were of ſinewes, is counted pernicious and deadly.

¶ The Time.

Of all the Cone trees onely the Larch tree is found to be without leaues in the Winter: in the Spring grow freſh leaues out of the ſame knobs, from which the former did fall. The cones are to be gathered before Winter, ſo ſoon as the leaues are gone: but after the ſcales are looſed and opened, the ſeeds drop away: the Roſine muſt be gathered in the Summer moneths.

¶ The Names.

This tree is called in Greeke, λάριξ: in Latine alſo *Larix*, in Italian and Spaniſh, *Larice*: in high Dutch, **Lerchenbaum**: in low Dutch, **Lozkenboom**: in French, *Meleſe*: in Engliſh, Larch tree, and of ſome Larix tree.

The liquid Roſin is named by *Galen* alſo λάριξ: the Latines call it *Reſina Larigna*, or *Reſina Laricea*, Larch Roſin: the Italians, *Larga*: the Apothecaries, *Terebinthina*, or Turpentine, and it is ſould and alſo mixed in medicines in ſtead thereof: neither is that a thing newly done; for *Galen* likewiſe in his time reporteth, that the Druggers ſold the Larch Roſine in ſtead of Turpentine: and this may bee done without errour; for *Galen* himſelfe in one place vſeth Larch Roſin for Turpentine; and in another, Turpentine for Larch Roſine, in his booke of medicines according to the kindes.

The Agaricke is alſo called in Greeke, ἀγάρικαι and ἀγάρικαι: in Latine, *Agaricum* and *Agaricus*, and ſo likewiſe in ſhops: the Italians, Spaniards, and other nations do imitate the Greeke word; and in Engliſh we call it Agaricke.

¶ The Temperature and Vertues.

A   The leaues, barke, fruit, and kernell, are of temperature like vnto the Pine, but not ſo ſtrong.

B   The Larch Roſin is of a moiſter temperature than all the reſt of the Roſines, and is withall without ſharpeneſſe or biting, much like to the right Turpentine, and is fitly mixed with medicines which perfectly cure vlcers and greene wounds.

C   All Roſins, ſaith *Galen*, that haue this kinde of moiſture and clamminiſſe ioyned with them, do as as it were binde together and vnite dry medicines, and becauſe they haue no euident biting quality, they do moiſten the vlcers nothing at all: therefore diuers haue very well mixed with ſuch compound medicines either Turpentine Roſin, or Larch Roſin: thus far *Galen*. Moreouer, Larch Roſine performeth all ſuch things that the Turpentine Roſin doth, vnto which, as we haue ſaid, it is much like in temperature, which thing likewiſe *Galen* himſelfe affirmeth.

D   Agaricke is hot in the firſt degree and dry in the ſecond, according to the old writers. It cutteth, maketh thin, clenſeth, taketh away obſtructions or ſtoppings of the intrailes, and purgeth alſo by ſtoole.

E   Agaricke cureth the yellow iaundiſe proceeding of obſtructions, and is a ſure remedy for cold ſhakings, which are cauſed of thicke and cold humors.

F   The ſame being inwardly taken and outwardly applied, is good for thoſe that are bit of venomous beaſts which hurt with their cold poiſon.

It

It prouoketh vrine, and bringeth downe the menſes : it maketh the body well coloured, driueth   G forth wormes, cureth agues, eſpecially quotidians and wandring feauers, and others that are of long continuance, if it be mixed with fit things that ſerue for the diſeaſe : and theſe things it performes by drawing forth and purging away groſſe, cold, and flegmaticke humours, which cauſe the diſeaſes.

From a dram weight, or a dram and a halfe, to two, it is giuen at once in ſubſtance or in pouder :   H the weight of it in an infuſion or decoction is from two drams to fiue.

But it purgeth ſlowly, and doth ſomewhat trouble the ſtomacke ; and therefore it is appointed   I that Ginger ſhould be mixed with it, or wilde Carrot ſeed, or Louage ſeed, or Sal gem, in Latine, *Sal foſsilis*.

Galen, as *Meſue* reporteth, gaue it with wine wherein Ginger was infuſed : ſome vſe to giue it with   K Oxymel, otherwiſe called ſyrrup of vineger, which is the faſeſt way of all.

Agaricke is good againſt the paines and ſwimming in the head, or the falling Euill, being taken   L with ſyrrup of vineger.

It is good againſt the ſhortneſſe of breath, called *Aſthma*, the inueterate cough of the lungs, the   M ptyſicke, conſumption, and thoſe that ſpet bloud : it comforteth the weake and feeble ſtomacke, cauſeth good digeſtion, and is good againſt wormes.

Chap. 45.   *Of the Cypreſſe Tree.*

Cupreſſus ſatiua & ſylueſtris.
The Garden and wild Cypreſſe tree.

¶ *The Deſcription.*

THe tame or manured Cypreſſe tree hath a long thicke and ſtraight body ; whereupon many ſlender branches do grow, which do not ſpred abroad like the branches of other trees, but grow vp alongſt the body, yet not touching the top : they grow after the faſhion of a ſteeple, broad below, and narrow toward the top : the ſubſtance of the wood is hard, ſound, well compact, ſweet of ſmell, and ſomewhat yellow, almoſt like the yellow Saunders, but not altogether ſo yellow, neither
ther

ther doth it rot nor wax old, nor cleaueth or choppeth it ſelfe. The leaues are long, round like thoſe of Tamariske, but fuller of ſubſtance. The fruit or nuts do hang vpon the boughes, being in manner like to thoſe of the Larch tree, but yet thicker and more cloſely compact: which being ripe do of themſelues part in ſunder, and then falleth the ſeed, which is ſhaken out with the winde: the ſame is ſmall, flat, very thin, of a ſwart ill fauoured colour, which is pleaſant to Ants or Piſmires, and ſerueth them for food.

Of this diuers make two kindes, the female and the male; the female barren, and the male fruitfull. _Theophraſtus_ reporteth, that diuers affirme the male to come of the female. The Cypreſſe yeelds forth a certaine liquid Roſin, like in ſubſtance to that of the Larch tree, but in taſte maruellous ſharpe and biting.

The wilde Cypreſſe, as _Theophraſtus_ writeth, is an high tree, and alwaies greene, ſo like to the other Cypreſſe, as it ſeemeth to be the ſame both in boughes, body, leaues, and fruit, rather than a certaine wilde Cypreſſe: the matter or ſubſtance of the wood is found, of a ſweet ſmell, like that of the Cedar tree, which rotteth not: there is nothing ſo criſped as the root, and therefore they vſe to make precious and coſtly workes thereof.

‡ I know no difference betweene the wilde and tame Cypreſſe of our Author, but in the handſomeneſſe of their growth, which is helped ſomewhat by art. ‡

¶ _The Place._

The tame and manured Cypreſſe groweth in hot countries, as in Candy, Lycia, Rhodes, and alſo in the territory of Cyrene: it is reported to be likewiſe found on the hils belonging to mount Ida, and on the hils called _Leuci_, that is to ſay, white, the tops whereof be alwaies couered with ſnow. _Bellonius_ denieth it to be found vpon the tops of theſe hills, but in the bottoms on the rough parts and ridges of the hills: it groweth likewiſe in diuers places of England where it hath been planted, as at Sion a place neere London, ſometimes a houſe of Nunnes: it groweth alſo at Greenewich, and at other places, and likewiſe at Hampſted in the garden of Mr _Wade_, one of the Clerkes of her Majeſties priuie Councell.

The wilde kinde of Cypreſſe tree groweth hard by _Ammons_ Temple, and in other parts of the country of Cyrene vpon the tops of mountaines, and in extreme cold countries. _Bellonius_ affirmeth, that there is found a certaine wilde Cypreſſe alſo in Candy, which is not ſo high as other Cypreſſe trees, nor groweth ſharpe toward the top, but is lower, and hath his boughes ſpred flat, round about in compaſſe: he ſaith the body thereof is alſo thicke: but whether this be _Thya_, of which _Theophraſtus_ and _Pliny_ make mention, we leaue it to conſideration.

¶ _The Time._

The tame Cypres tree it alwaies greene; the fruit may be gathered thrice a yeare, in Ianuarie, May, and September, and therefore it is ſyrnamed _Triſera_.

The wilde Cypres tree is late, and very long before it buddeth.

¶ _The Names._

The tame Cypres is called in Greek, κυπάρισσος, or κυπάρισσος: in Latine, _Cupreſſus_: in ſhops, _Cypreſſus_: in Italian, _Cypreſſo_: in French and Spaniſh, _Cipres_: in high Dutch, **Cipreſſenbaum**: in low Dutch, **Cypreſſeboom**: in Engliſh, Cypres and Cypres tree.

The fruit is named in Greeke, σφαίρια τᾶς κυπαρίσσου in Latine, _Pilulæ, Cupreſſi, Nuces Cupreſſi_, and _Galbuli_: in ſhops, _Nuces Cypreſſi_: in Engliſh, Cypres nuts or clogs. This tree in times paſt was dedicated to _Pluto_, and was ſaid to bee deadly; whereupon it is thought that the ſhadow thereof is vnfortunate.

The wilde Cypres tree is called in Greeke, θύα or θεῖον, and θεῖον from this doth differ θυια, being a name not of a plant, but of a mortar in which dry things are beaten: _Thya_ as _Pliny_ writeth, _lib._ 13. _cap._ 16. was well knowne to _Homer_: he ſheweth that this is burned among the ſweet ſmells, which _Circe_ was much delighted withall, whom hee would haue to be taken for a goddeſſe, to their blame that call ſweet and odoriferous ſmells, euen all of them, by that name; becauſe hee doth eſpecially make mention withall in one verſe, of _Cedrus_ and _Thya_: the copies haue falſly _Larix_, or Larch tree, in which it is manifeſt that he ſpake onely of trees: the verſe is extant in the fifth booke of _Odyſſes_, where he mentioneth, that _Mercurie_ by _Iupiters_ commandement went to _Calypſus_ den, and that he did ſmell the burnt trees _Thya_ and _Cedrus_ a great way off.

Theophraſtus attributeth great honor to this tree, ſhewing that the roofs of old Temples became famous by reaſon of that wood, and that the timber thereof, of which the rafters are made is euerlaſting, and it is not hurt there by rotting, cobweb, nor any other infirmitie or corruption.

¶ _The Temperature._

The fruit and leaues of the Cypres are dry in the third degree, and aſtringent.

¶ _The Vertues._

A   The Cypres nuts being ſtamped and drunken in wine, as _Dioſcorides_ writeth, ſtop the laske and bloudy flix, and are good againſt the ſpitting of bloud and all other iſſues of bloud.

They

They glue and heale vp great vlcers in hard bodies : they safely and without harme soke vp and B consume the hid and secret moisture lying deepe and in the bottome of weake and moist infirmities.

The leaues and nuts are good to cure the rupture, to take away the *Polypus*, being an excrescence C growing in the nose.

Some do vse the same against carbuncles and eating sores, mixing them with parched Barley D meale.

The leaues of Cypres boyled in sweet wine or Mede, help the strangury and difficulty of making E water.

It is reported that the smoke of the leaues doth driue away gnats, and that the clogs doe so like- F wise.

The shauings of the wood laid among garments preserue them from the moths : the rosin killeth G Moths, little wormes, and magots.

† Our Author in this chapter hath put together two chapters of *Dodonaus* ; the one of Cypresse, the other of *Thya*, out of *Theophrastus* and others. *Vid. Tempt.6. lib.5.cap.7.& 8.*

CHAP. 46. *Of the tree of Life.*

Arbor Vitæ.
The Tree of Life.

¶ *The Description.*

TH e tree of Life growes to the height of a smal tree, the barke being of a dark reddish colour : the timber very hard, the branches spreading themselues abroad, hanging downe toward the ground by reason of the weakenesse of the twiggie branches surcharged with very oileous and ponderous leaues, casting, and spreading themselues like the feathers of a wing, resembling those of the Sauine tree, but thicker, broader, and more ful of gummie or oileous substance : which being rubbed in the hands do yeeld an aromatick, spicie, or gummy sauor, very pleasant and comfortable : amongst the leaues come forth small yellowish flours, which in my garden fall away without any fruit : but as it hath beene reported by those that haue seene the same, there followeth a fruit in hot Regions, much like vnto the fruit of the Cypres tree, but smaller, compact of little and thinne scales closely pact one vpon another, which my selfe haue not yet seene. The branches of this tree laid downe in the earth will very easily take root, euen like the Woodbind or some such plant; which I haue often proued, and thereby haue greatly multiplied these trees.

¶ *The Place.*
This tree groweth not wile in England, but it groweth in my garden very plentifully.

¶ *The Time.*
It endureth the cold of our Northerne clymat, yet doth it loose his gallant greenes in the winter moneths : it floureth in my garden about May.

¶ *The Names.*
Theophrastus and *Pliny*, as some thinke, haue called this sweet and aromaticall tree *Thuia*, or *Thya*: some call it *Cedrus Lycia* : the new writers doe terme it *Arbor vitæ*: in English, the Tree of Life, I doe not meanethat whereof mention is made, *Gen.3.22.*

¶ *The Temperature.*
Both the leaues and boughes be hot and dry.

¶ *The Vertues.*
Among the plants of the New-found Land, this Tree, which *Theophrastus* calls *Thuia*, or *Thua*,
is

is the moſt principall, and beſt agreeing vnto the nature of man, as an excellent cordial, and of a very pleaſant ſmell.

CHAP. 47. *Of the Yew tree.*

Taxus.
The Yew tree.

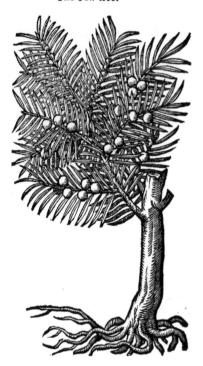

¶ *The Deſcription.*

‡ IN ſtead of the deſcription and place mentioned by our Author (which were not amiſſe) giue me leaue to preſent you with one much more accurate, ſent mee by Mᵗ *Iohn Goodyer.*

Taxus glandifera bacciferáque.
The Yew bearing Acornes and Berries.

THe Yew tree that beareth Acornes and berries is a great high tree remaining alwaies greene, and hath vſually an huge trunke or body as big as the Oke, couered ouer with a ſcabbed or ſcaly barke, often pilling or falling off, and a yong ſmooth barke appearing vnderneath; the timber hereof is ſomewhat red, neere as hard as Box, vniuerſally couered next the barke with a thicke white ſap like that of the Oke, and hath many big limmes diuided into many ſmall ſpreading branches : the leaues be about an inch long, narrow like the leaues of Roſemary, but ſmooth and of a darker greene colour, growing all alongſt the little twigs or branches cloſe together, ſeldome one oppoſite againſt another, often hauing at the ends of the twigs little branches compoſed of many leaues like the former, but ſhorter and broader, cloſely compact or joyned together : amongſt the leaues are to be ſeene at all times of the yeare, ſmall ſlender buds ſomewhat long, but neuer any floures; which at the very beginning of the Spring grow bigger and bigger, till they are of the faſhion of little Acornes, with a white kernell within : after they are of this forme, then groweth vp from the bottoms of the Acornes a reddiſh matter, making beautifull reddiſh berries more long than round, ſmooth on the out ſide, very clammie within, and of a ſweet taſte, couering all the Acorne, only leauing a little hole at the top, where the top of the Acorne is to be ſeene : theſe fallen, or deuoured by birds, leaue behinde them a little whitiſh huske made of a few ſcales : appearing like a little floure, which peraduenture may deceiue ſome, taking it to be ſo indeed : it ſeemes this tree, if it were not hindred by cold weather, would alwaies haue Acornes and berries on him, for hee hath alwaies little buds, which ſo ſoone as the Spring yeelds but a reaſonable heate, they grow into the forme of Acornes : about the beginning of Auguſt, ſeldome before, you ſhall finde them turned into ripe berries, and from that time till Chriſtmaſſe, or a little after, you may ſee on him both Acornes and red berries.

Taxus tantum florens.
The Yew which only floures.

The Yew which onely beareth floures and no berries, is like the other in trunke, timber, barke, and leaues ; but at the beginning of Nouember, or before, this tree doth begin to be very thicke ſet or fraught on the lower ſide or part of the twigs or little branches, with ſmall round buds, very neere as big, and of the colour of Radiſh ſeed, and doe ſo continue all the Winter, till about the beginning or middle of February, when they open at the top, ſending forth one ſmall ſharpe pointall, little longer than the huske, diuided into many parts, or garniſhed towards the top with many
ſmall

ſmall duſty things like floures, of the colour of the huskes; and if you ſhall beate or throw ſtones into this tree about the end of February, or a good ſpace after, there will proceed and fly from theſe floures an abundance of duſty ſmoke. Theſe duſty floures continue on the trees till about harueſt, and then ſome and ſome fall away, and ſhortly after the round buds come vp as aforeſaid.

¶ The Place.

Theſe trees are both very common in England : in Hampſhire there is good plenty of them growing wilde on the chalkie hills, and in Church-yards where they haue been planted.

¶ The Time.

The time is expreſſed in their deſcriptions. Dec. 19. 1621. Iohn Goodyer. ‡

¶ The Names.

This tree is named by Dioſcorides, Σμιλαξ: by Theophraſtus, Μιλος but Nicander in his booke of Coun-terpoyſons, Σμιλος: Galen doth alſo call it Κακτος: it is named in Latine Taxus : in high Dutch. ᴇp= benbaum : in low Dutch, Ibenboom : In Italian, Taſſo : in Spaniſh, Toxo, and Taxo : in French, Yf : in Engliſh, Ewe, or Eue tree : in the vnlearned ſhops of Germany, if any of them remaine, it is cal-led Tamariſcus ; where in times paſt they were wont not without great error, to mix the barke hereof in compound medicines, in ſtead of the Tamariske barke.

¶ The Temperature.

The Yew tree, as Galen reporteth, is of a venomous quality, and againſt mans nature. Dioſcorides writeth, and generally all that heretofore haue dealt in the facultie of Herbes, that the Yew tree is very venomous to be taken inwardly, and that if any doe ſleepe vnder the ſhadow thereof it cau-ſeth ſickneſſe and oftentimes death. Moreouer, they ſay that the fruit thereof being eaten is not onely dangerous and deadly vnto man, but if birds doe eat thereof it cauſeth them to caſt their fea-thers, and many times to die. All which I dare boldly affirme is altogether vntrue : for when I was yong and went to ſchoole, diuers of my ſchoole-fellowes and likewiſe my ſelfe did eat our fils of the berries of this tree, and haue not onely ſlept vnder the ſhadow thereof, but among the branches alſo, without any hurt at all, and that not one time, but many times. Theophraſtus ſaith, That αιφυρα, animalia, Gaza tranſlates them Iumenta, or labouring beaſts die, if they doe eate of the leaues ; but ſuch cattell as chew their cud receiue no hurt at all thereby.

Nicander in his booke of counterpoiſons doth reckon the Yew tree among the venomous plants, ſetting downe alſo a remedy, and that in theſe words, as Gorræus hath tranſlated them.

Parce Venenata Taxo, quæ ſurgit in Oeta
Abietibus ſimilis, lethoque abſumit acerbo
Ni præter morem pleno cratere meraca
Fundere vina pares, cum primum ſentiet æger
Arctari obſtructas fauces animæque canalem.

‡ Shun th'poys'nous Yew, the which on Oeta growes,
Like to the Firre, it cauſes bitter death,
Vnleſſe beſides they vſe pure wine that flowes
From empty'd cups, thou drinke, when as thy breath
Begins to faile, and paſſage of thy life
Growes ſtraight.———

Pena and Lobel alſo obſerued that which our Author here affirmes, and dayly experience ſhewes t to be true, that the Yew tree in England is not poyſonous : yet diuers affirme, that in Prouince in France, and in moſt hot countries, it hath ſuch a maligne quality, that it is not ſafe to ſleepe or long :o reſt vnder the ſhadow thereof. ‡

CHAP. 48.   Of the Iuniper tree.

¶ The Kindes.

AMong the Iuniper trees one is leſſer, another greater, being a ſtrange and forreine tree : one of theſe bringeth forth a floure and no fruit ; the other fruit and no floures.

¶ The Deſcription.

1   THe common Iuniper tree groweth in ſome parts of Kent vnto the ſtature and bignes of a faire great tree, but moſt commonly it growes very low like vnto ground Furres : this

1 *Iuniperus.*
The Iuniper tree.

2 *Iuniperus maxima.*
The great Iuniper tree.

‡ 3 *Iuniperus Alpina minor.*
Small Iuniper of the Alps.

tree hath a thin bark or rinde, which in hot regions will chop and rend it ſelf into many cranies or pieces: out of which rifts iſſueth a certaine gum or liquor much like vnto Frankincenſe: the leaues are very ſmall, narrow, and hard, and ſomwhat prickly, growing euer green along the branches, thicke together: amongſt which come forth round and ſmall berries, greene at the firſt, but afterward blacke, declining to blewneſſe, of a good ſauor, and ſweet in taſte, which do wax ſomwhat bitter after they be dry and withered.

2　The great Iuniper tree comes now and then to the height of the Cypres tree, with a greater and harder leafe, and alſo with a fruit as big as Oliue berries, as *Bellonius* writeth, of an exceeding faire blew colour, and of an excellent ſweet ſauour.

‡ 3　This exceeds not the height of a cubit, but growes low, and as it were creeps vpon the ground, and conſiſts of ſundry thicker and ſhorter branches than the common kind, tough alſo, writhen, and hard to breake; three leaues alwaies growing at equall diſtances, as in the common, but yet broader, ſhorter, and thicker, neither leſſe pricking than they, of a whitiſh greene colour on the inſide, and green without incompaſſe the tender branches. *Cluſius*, who giues vs this figure and hiſtory, obſerued not the floure, but the fruit is like that of the ordinarie,

nary,but yet fomewhat longer : it growes vpon the Auftrian Alps, and ripens the fruit in Auguft and September. ‡

¶ *The Place.*

The common Iuniper is found in very many places,efpecially in the South parts of Eugland.

Bellonius reporteth,That the greater groweth vpon mount Taurus. *Aloyfius Anguillara* writes,that it is found on the fhores of the Ligurian and Adriaticke fea, and in Illyrium, bringing forth great berries : and others fay it groweth in Provence of France : it comes vp for the moft part in rough places and neere to the fea,as *Diofcorides* writeth.

¶ *The Time.*

The Iuniper tree floureth in May ; the floure whereof is nothing elfe but as it were a little yel-lowifh duft or pouder ftrewed vpon the boughes. The fruit is ripe in September, and is feldome found either winter or fummer without ripe and vnripe berries,and all at one time.

¶ *The Names.*

The Iuniper tree is called in Greeke ἄρκευθος : the Apothecaries keepe the Latine name *Iuniperus* : the Arabians call it *Archonas*,and *Archencas* : the Italians, *Ginepro* : in high-Dutch, **Wechholter** : in Spanifh,*Enebro,Ginebro*,and *Zimbro* : the French men and bafe Almanes,*Geneue* : in Englifh,Iu-niper tree.

The leffer is named in Greeke, κάρυος : in Latine *Iuniperus* : the great Iuniper tree is called as fome thinke in Greeke, ἀρκευθίς μεγάλη : in Latine by *Lobel, Iuniperus maximus Illyricus cœrulea bacca*, by reafon of the colour of the berries ; and may be called in Englifh,blew Iuniper.

The berries are called *Grana Iuniperi* : in Greeke, ἄρκευθος, although the tree it felfe is oftentimes called alfo by the fame name ἄρκευθος : it is termed in high-Dutch, **Krametbeer, Wechholterbeer** : in low-Dutch, **Genebzebeffen** : in Spanifh, *Neurinas* : in Englifh, Iuniper berries.

The gum of the Iuniper tree is vfually called of the Apothecaries *Vernix* : in Latine, *Lachrima Iuniperi*. *Serapio* nameth it *Sandarax*, and *Sandaracha* : but there is another *Sandaracha* amongft the Grecians,being a kinde of Orpment, which growes in the fame minerals wherein Orpment doth ; and this doth far differ from *Vernix* or the Iuniper gum. *Pliny,lib.11. cap.7.* makes mention alfo of another *Sandaracha*,which is called *Erithrace*,and *Cerinthus* : this is the meat of Bees whileft they be about their worke.

¶ *The Temperature.*

Iuniper is hot and dry in the third degree,as *Galen* teacheth : the berries are alfo hot,but not al-together fo dry : the gum is hot and dry in the firft degree,as the Arabians write.

¶ *The Vertues.*

The fruit of the Iuniper tree doth clenfe the liuer and kidnies, as *Galen* teftifieth : it alfo makes A thin clammy and groffe humors. It is vfed in countrepoifons and other wholfome medicines. Be-ing ouer largely taken it caufeth gripings and gnawings in the ftomack,and maketh the head hot ; it neither bindeth nor loofeth the belly : it prouoketh vrine.

Diofcorides reporteth, That this being drunke is a remedie againft the infirmities of the cheft, B coughs,windineffe,gripings,and poifons,and that the fame is good for thofe that be troubled with cramps,burftings,and with the difeafe called the Mother.

It is moft certain,That the decoction of thefe berries is fingular good againft an old cough,and C againft that with which children are now and then extremely troubled,called the Chin-cough, in which they vfe to raife vp raw tough and clammy humors that haue many times bloud mixed with them.

Diuers in Bohemia do take in ftead of other drinke the water wherein thofe berries haue beene D fteeped,who liue in wonderfull good health.

This is alfo drunke againft poifons and peftilent feuers,and it is not vnpleafant in the drinking : E when the firft water is almoft fpent,the veffell is again filled vp with frefh.

The fmoke of the leaues and wood driues away ferpents and all infection and corruption of the F aire,which bring the plague or fuch like contagious difeafes : the juice of the leaues is laid on with wine,and alfo drunke againft the bitings of the viper.

The afhes of the burned barke being applied with water take away fcurfe and filth of the skin. G

The pouder of the wood being inwardly taken is pernitious and deadly,as *Diofcorides* his vulgar H copies do affirme ; but the true copies do vtterly deny it,neither do any of the old writers affirm it.

The fume and fmoke of the gum doth ftay flegmaticke humors that diftill out of the head,and I ftoppeth the rheume : the gum doth ftay raw and flegmaticke humors that ftick in the ftomacke and guts,if it be inwardly taken and alfo drunke.

It killeth all manner of wormes in the belly,it ftayeth the menfes and hemorrhoids : it is com- K mended alfo againft fpitting of bloud,it drieth hollow vlcers and filleth them with flefh, if it be caft thereon : being mixed with oile of Rofes it healeth chops in the hands and feet,

L   There is made of this and oile of Lineſeed mixt together,a liquor called Verniſh,which is vſed to beautifie pictures and painted tables with,and to make iron gliſter, and to defend it from ruſt.

Chap. 49.
Of the prickly Cedar, or Cedar Iuniper.

¶ *The Kindes.*

THe prickely Cedar tree is like to Iuniper,and is called the ſmall or little Cedar, for difference from the great and tall Cedar which bringeth Cones: and of this there are two kinds,as *Theophraſtus* and *Pliny* do teſtifie,that is to ſay,one of Lycia,and another crimſon.

¶ *The Deſcription.*

1   THe crimſon or prickly cedar ſeems to be very like the Iuniper tree in body & boughs, which are writhed,knotty,and parted into very many wings:the ſubſtance of the wood is red,and ſweet of ſmel like that of the Cypreſſe: the tree is couered ouer with a rugged barke; the leaues be narrow and ſharp pointed,harder than thoſe of Iuniper, ſharper and more pricking,and ſtanding thinner vpon the branches : the fruit or berry is ſomtimes as big as a haſell nut,or as *Theophraſtus* ſaith,of the bigneſſe of Myrtle beries,and being ripe it is of a reddiſh yellow or crimſon colour,ſweet of ſmell,and ſo pleaſant in taſte, that the countreymen now and then doe eat of the ſame with bread.

1 *Oxycedrus Phænicia.*
Crimſon prickly Cedar.

2 *Oxycedrus Lycia.*
Rough Lycian Cedar.

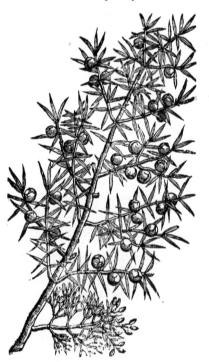 

2   The other low Cedar which growes in Lycia is not ſo high as the former,hauing likewiſe a wrythed body as big as a mans arme,full of boughes : the barke is rough, yellowiſh without, and red within : the leaues ſtand thicker, like at the firſt to thoſe of Iuniper, but yet ſomewhat ſhorter,and in the third or fourth yeare thicker, long and round withall, comming neere to the leaues

of

‡ 3 *Cedrus Lycia altera.*
The other Lycian Cedar.

of the Cypresse tree, or of the second Sauine, that is, blunt and not pricking at all; which being braised betwixt the fingers, doth yeeld a very pleasant smell : so doth one and the selfe same plant bring forth below sharp & prickly leaues, & aboue thicke and blunt ones, as that notable learned Herbarist *Clusius* hath most diligently obserued : the fruit or berry is round like that of Iuniper, of color yellow when it is ripe, inclining to a red, in tast somewhat bitter, but sweet of smell.

‡ 3 This also hath Cypresse-like leaues, not vnlike those of the last described, yet somewhat thicker and broader : the fruit is also much larger, being as big as Hasell nuts, and of a red or scarlet colour ; whence *Lobel* calleth it *Cedrus Phœnicia altera.* ‡

¶ *The Place.*

The prickely Cedar with the Crimson colour comes vp higher and greater in certaine places of Italy, Spain, and Asia, than in other countries. For that which grows on mount Gargarus in Apulia is much higher and broader than those that grow elswhere, and bringeth forth greater beries of the bignes of an hasell nut, and sweeter, as that most diligent writer *Bellonius* reporteth. *Clusius* shewes, that the prickely Cedar and the Iuniper tree be of so great a growth in diuers places of Spaine, that the body of them is as thicke as a man.

The Lycian Cedar is found in Province of France, not far from Massilia, and growes in a great part of Greece, in Illyricum and Epyrum.

¶ *The Time.*

Both of them are alwaies green, and in Winter also full of fruit, by reason that they continually bring forth berries, as when the old do fall new come in their places : in the spring grow vp new buds and beginnings of berries : in Autumne they wax ripe the second yeare, as do the berries of Iuniper.

¶ *The Names.*

They are called in Latine *Minores*, and *Humiles Cedri*, Little and low Cedars, for difference from the tall and great Cedar which beareth Cones.

The former is named in Greeke, οξυκεδρος, and κεδρις μικρα : in Latine, *Oxycedrus*, and *Cedrus Punica* : in English, prickly Cedar, and Crimson Cedar : *Pliny* syrnameth it *Phœnicea*, of the crimson color of the fruit : the Spaniards cal this also *Enebro*, as *Clusius* testifies, euen by the same name they giue the Iuniper ; wherein likewise they are thought to imitate diuers of the old writers, who haue not by names distinguished the Iuniper from the Cedar, but haue, as *Theophrastus* noteth, called them *Cedros*, Cedar trees, yet with an addition οξυκεδρος, or prickly Cedars.

The other with the blunt leafe is named of *Theophrastus*, κεδρις : of *Pliny* also *Lycia Cedrus* : in Province of France, *Morucin* : diuers name this *Sabina*, and vse it in stead of Sauin, which they want : as the apothecaries of Epidaurus, and in diuers cities of Greece, and also in Illyricum & Epyrum, as *Bellonius* testifieth. Some would haue it to be οξα, *Thya* : but *Thya*, according to *Theophrastus*, is like not only in body, leaues, and boughes, but in fruit also, to the Cypresse tree : but the fruit of this is nothing like to the Cypresse cones.

The fruit of this Cedar is named by *Theophrastus*, κεδρις, *Cedrus* notwithstanding *Cedrus*, as hee himselfe doth also testifie (*Gaza* nameth it *Cedrula*) is a certaine little shrub which neuer grows to a tree.

The gum or liquor which issues from the prickly Cedar is also called *Vernix*, and is sold insted thereof.

¶ *The Nature and Vertues.*

The little Cedar, as *Galen* writeth, is hot and dry in a manner in the third degree : the matter or A substance thereof is sweet of smell like that of Iuniper, and is vsed for perfumes and odoriferous smels together with the leaues.

　　　　　　The

The berries or fruit of the low Cedar haue their faculties not ſo ſtrong, as the ſame Author te-ſtifieth, inſomuch as that they may alſo be eaten, yet if they bee taken too plentifully they cauſe head-ache, and breed heate and gnawings in the ſtomacke. Yet there is a difference betweene theſe two Cedar berries; for the crimſon ones are not ſo hot and dry, by reaſon they are ſweeter & plea-ſanter to the taſte, and therefore they are better to be eaten, and do alſo yeeld vnto the body a kind of nouriſhment: but the berries of that of *Lycia* are biting, hotter and drier alſo than thoſe of Iu-niper, from which they differ eſpecially in the biting quality, they bring no nouriſhment at all, and though a man eat neuer ſo few of them, hee ſhall feele gnawings in his ſtomacke, and paine in his head.

     The Peaſants doe feed thereon rather to ſatisfie their hunger, than for any delight they haue in the taſte, or the phyſicall vertues thereof; albeit they be good againſt the ſtrangurie, and prouoke vrine.

Cʜᴀᴘ. 5. *Of Savin.*

¶ *The Kindes.*

THere be two kinds of Savin, one like in leafe to Tamarisk, the other to the Cypres tree; where-of the one beareth berries, the other is barren.

1 *Sabina ſterilis.*   2 *Sabina baccifera.*
Barren Savin.   Savin bearing berries.

¶ *The Deſcription.*

1   THe firſt Savin, which is the common kind, and beſt of all known in this country, grow-eth in manner of a low ſhrub or tree, the ſtem or trunke whereof is ſomtimes as big as a mans arm, diuiding it ſelfe into many branches ſet full of ſmall leaues like Cypreſſe or Tamariske, but thicker, and more ſharp and prickly, remaining green winter and ſummer, in ſmell ranke or very ſtrong, barren both of floures and fruit.

                                                             2 The

‡ 3 *Sabina baccata altera.*
The leſſer berry-bearing Savin.

2   The other Savin is an high tree, as *Bellonius* ſaith, as tall as the Almond tree, and much like to the tame Cypreſſe tree: the body is wrythed, thick, and ſomtimes of ſo great a compaſſe as that it cannot be fathomed ; the ſubſtance of the wood is red within, as is that of Iuniper and of the prickly Cedar : the barke is not very thicke, and it is of a yellowiſh red : the leaues are of a maruellous gallant greene colour, like to thoſe of the Cypres tree, yet thicker or more in number ; in taſte bitter, of a ſpicie ſmell, and like Roſin : the boughes are broader, and thick ſet as it were with wings, like thoſe of the Pitch tree and of the Yew tree : on which grow a great number of beries very round like thoſe of the little Cedars, which at the firſt are green, but when they be ripe they are of a blackiſh blew. Out of the root hereof iſſueth oft times a Roſin, which being hard is like to that of the Iuniper tree, and doth alſo crumble in the chewing.

‡   3   There is another which differs from the laſt deſcribed, onely in that the leaues are ſmaller and leſſe pricking than thoſe of the former, as alſo the branches leſſer. *Lobel* cals this *Savina baccata altera.* ‡

¶ *The Place.*

Both of them grow vpon hils, in woods & otherlike vntoiled places, as in Candy, Myſia, and elſewhere. *P. Bellonius* reporteth, that he found them both vpon the tops of the mountains Taurus, Amanus, and Olympus.

The firſt is planted in our Engliſh gardens almoſt euerie where. The ſecond is planted both by the ſeed and by the ſlip : the ſlips muſt be ſet in a ground that is meanly moiſt and ſhadowie, till they haue taken root : the ſhrubs which grow of theſe decline toward the one ſide, retaining ſtill the nature of the bough : but that Savin which is planted by the ſeed groweth more vpright : this in continuance of time bringeth forth ſeeds, and the other for the moſt part remaines barren. Both theſe grow in my garden.

¶ *The Time.*

They both continue alwaies greene : The one is found to be laden with ripe fruit commonly in Winter ; but it hath fruit at all times, for before the old berries fall, new are come vp.

¶ *The Names.*

Savin is called in Greeke βράθυ, or βράθυ : in Latine, *Sabina.*

The firſt is commonly called in the Apothecaries ſhops by the name *Savina* : of diuers, *Savimera* : the Italians and Spaniards keepe the Latine name : it is called in high-Dutch, **Siben baum :** in low-Dutch, **Sauel boom :** in French, *Savenier* : in Engliſh, common Savin, or garden Savin.

Some name the other *Cupreſſus Cretica,* or Cypres of Candy, as *Pliny* ſaith, *lib.* 12. *cap.* 17. making mention of a tree called *Bruta.* Some there are that take this to be *altera Sabina,* or the ſecond Savin, and to be read *Bruta* for βράθυ, *Brathu,* by altering the vowels : for *Pliny* deſcribes it, *lib.* 12. *cap.* 17. to be like the Cypres tree, in theſe words ; They reeke in the mountain Elimæi the tree *Bruta,* beeing like to the broad Cypres tree, hauing white boughes, yeelding a ſweet ſmell when it is ſet on fire ; whereof mention is made with a miracle, in the ſtories of *Claudius Cæſar.* It is reported, That the Parthians doe vſe the leaues in drinkes ; that the ſmell is very like to that of Cypreſſe, and that the ſmoke thereof is a remedie againſt the ſmell of other woods. It growes beyond Piſitigris, neere to the towne Sittaca, on mount Zagrus. Thus far *Pliny.*

The mountaines Elimæi are deſcribed by *Strabo* in the country of the Aſſyrians, next after the mountain Sagrus aboue the Babylonians ; by *Ptolomæus,* not far from the Perſian gulfe : therefore it is hard to ſay that *Bruta* is *Sabina altera* or the ſecond Savine, ſeeing that ſo great a diſtance of the place may vndoubtedly cauſe a difference, and that it is not largely, but briefely deſcribed. It ſeemeth that *Thya* mentioned by *Theophraſtus* is more like vnto Savin : but yet foraſmuch as *Thya* is like in fruit to the Cypres tree, and not to the fruit or berries of the little Cedars, it is alſo very

manifeſt

manifeſt,that the ſecond Sauin is not *Thya*, neither *Vitæ arbor*,ſo called of the later Herbariſts : it is likewiſe named by *Lobel*, *Sabina genuina baccifera,atro cærulea*, that is, the true Sauine that bears berries of a blackiſh blew colour.

¶ *The Temperature.*

The leaues of Sauine,which are moſt vſed in medicine, are hot and dry in the third degree, and of ſubtill parts, as *Galen* ſaith.

¶ *The Vertues.*

A   The leaues of Sauine boyled in Wine and drunke prouoke vrine,bring downe the menſes with force,draw away the after-birth,expell the dead childe,and kill the quicke: it hath the like vertue receiued vnder in a perfume.

B   The leaues ſtamped with honey and applied, cure vlcers, ſtay ſpreading and creeping vlcers, ſcoure and take away all ſpots and freckles from the face or body of man or woman.

C   The leaues boiled in oile Oliue,and kept therein,kill the worms in children,if you anoint their bellies therewith : and the leaues poudered and giuen in milke or Muſcadell doe the ſame.

D   The leaues dried and beat into fine pouder, and ſtrewed vpon thoſe kindes of excreſcences *ſub præputio*,called Caroles,and ſuch like,gotten by dealing with vnclean women, take them away perfectly,curing and healing them : but if they be inueterate and old, and haue been much tampered withall,it ſhall be neceſſarie to adde vnto the ſame a ſmall quantity of *Auripigmentum* in fine pouder, and vſe it with diſcretion, becauſe the force of the medicine is greatly increaſed thereby and made more corroſiue.

CHAP. 51.   *Of Tamariske.*

1 *Tamariſcus Narbonenſis.*
French Tamariske.

2 *Tamariſcus Germanica.*
Germane Tamariske.

¶ *The*

¶ *The Description.*

1 THe first kinde of Tamariske groweth like a small hedge, couered with a reddish barke, hauing many branches set and bedeckt with leaues much like to Heath: among which come forth small mossie white floures declining to purple, which turn into a pappous or downy seed that flieth away with the winde, as that of Willow doth: the root is woody, as the roots of other shrubs be, and groweth diuers waies.

2 The German Tamariske hath many wooddy branches and shoots rising from the root, with a white barke, hauing his leaues thicker and grosser than the former, and not so finely jagged or cut. The flours are reddish, and larger than the former, growing not vpon footstalks many thick clustering together, as those of the former, but each a pretty distance from another, on the tops of the branches spike-fashion, and begin to flour below: which turn into seed, that is likewise caried away with the winde.

¶ *The Place.*

Tamariske groweth by running streams, and many times by riuers that break forth, and not seldome about fenny grounds, commonly in a grauelly soile, for it best prospereth in moist and stony places: it is found in Germany, Vindelicia, Italy, Spain, and also in Greece.

The Tamarisks do also grow in Egypt and Syria, as *Dioscorides* writeth; and likewise in Tylus an Isle in Arabia, as *Theophr.* noteth, the wood whereof, saith he, is not weake, as with vs in Greece, but strong, like *cyprus* or timber, or any other strong thing: this Tamariske *Dioscorides* cals *ἥμερον*, that is to say, tame or planted; and saith that it bringeth forth fruit very like to Galls, in taste rough and binding.

Petrus Bellonius in his second booke of Singularities reporteth, That he saw in Egypt very high Tamarisks and great like other trees, and that sometimes in moist places by riuers sides, and many times also in dry & grauelly grounds where no other trees did grow, which now and then did beare hanging on the boughes such a multitude of Galls, that the inhabitants call *Chermasel*, as beeing ouerladen they are ready to breake. Both these grow and prosper well in gardens with vs heere in England.

¶ *The Time.*

These trees or shrubs floure in May, and in the later end of August: their seed is caried away with the winde.

¶ *The Names.*

They are called in Greeke μυρίκη, and in Latine also *Myrica* and *Tamarix*: in shops, *Tamariscus*: of *Octauius Horatianus, Murica: Dioscorides* makes that which growes in Italy and Greece to be *ἄγρια*, or wild Tamarisk: it is named in high-Dutch, **Tamarischenholtz, & Porst:** in low-Dutch, **Ibenboom, Tamariscbboom:** in Italian, *Tamarigio*: in Spanish, *Tamarguira*, and *Tamariz*: in French, *Tamaris*: in English, Tamariske.

¶ *The Temperature and Vertues.*

Tamariske hath a clensing and cutting facultie, with a manifest drying; it is also somewhat astringent or binding, and by reason of these qualities it is very good for an hard spleen, being boiled with vineger and wine, either the root, leaues, or tender branches, as *Galen* writeth. A

Moréouer, *Dioscorides* teacheth, That the decoction of the leaues made with wine doth wast the spleen, and that the same is good against the tooth-ache, if the mouth be washed therewith: that it bringeth down the menses, if the patient sit therein; that it killeth lice and nits, if the parts be bathed therewith. B

The ashes of burnt Tamariske haue a drying facultie, and greatly scouring withall, and a little binding. C

The floures and downy seed of the greater Tamariske do greatly binde, insomuch as they come very neere to the Gall named *Galla Omphacitis*, but that the roughnesse of taste is more euident in the gall: the which floures are of an vnequal temper, for there is ioined to the nature therof a great thinnesse of parts, and clensing facultie, which the Gall hath not, as *Galen* writeth. D

These floures we fitly vse (saith *Dioscor.*) in stead of Gall, in medicines for the eies and mouth. E

It is good to stanch bloud, and to stay the laske and womens whites, it helpeth the yellow jaundice, and also cureth those that are bit of the venomous spider called *Phalangium*: the bark serueth for the same purposes. F

The leaues and wood of Tamariske haue great power and vertue against the hardnesse and stopping of the spleen, especially the leaues being boiled in water, and the decoction drunke, or else infused in a small vessell of ale or beere, and continually drunke; and if it be drunk forth of a cup or dish made of the wood or timber of Tamariske, is of greater efficacie. G

CHAP.

C H A P. 52.   *Of Heath, Hather, or Linge.*

¶ *The Kindes.*

THere be diuers sorts of Heath, some greater, some lesser, some with broad leaues, and some narrower ; some bringing forth berries, and others nothing but floures.

¶ *The Description.*

1   THe common Heath is a low plant, but yet wooddy and shrubby, scarce a cubit high: it brings forth many branches, whereupon doe grow sundry little leaues somewhat hard and rough, very like to those of Tamariske or the Cypres tree: the floures are orderly placed alongst the branches, small, soft, and of a light red colour tending to purple : the root is also wooddy, and creepeth vnder the vpper crust of the earth: and this is the Heath which the Antients tooke to be the right and true Heath.

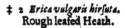

‡ 1 *Erica vulgaris, siue Pumila.*
Common or dwarfe Heath.

‡ 2 *Erica vulgaris hirsuta.*
Rough leafed Heath.

There is another Heath which differeth not from the precedent, sauing that this plant bringeth forth floures as white as snow, wherein consisteth the difference : wherefore we may cal it *Erica pumila alba,* Dwarfe Heath with white floures.

2   The great Heath (which *Clusius* at his being in England found in the barren grounds about Windsor, which in his Spanish trauels he maketh the first kinde) groweth to the height of two cubits, seldome higher, full of branches, couered with a blackish barke : whereon are set in very good order by couples small rough square leaues finer than those of Tamarisk or Cypres. The floures inclose the little twiggy branches round about at certaine distances, from the lower part to the top fashioned like little bottles, consisting of foure parts, of a shining purple colour, very beautifull to behold, and the rather to be esteemed because it floureth twice in the yeare. The root is likewise wooddy.

‡ 3 This

† 3 *Erica maior flore albo Clusij.*
The great Heath with white floures.

4 *Erica major flore purpureo.*
Great Heath with purple floures.

† 5 *Erica cruciata.*
Crossed Heath.

6 *Erica Pyramidalis.*
Steeple Heath.

‡ 3 This,ſaith *Cluſius*,which is the largeſt that I haue ſeen,ſomtimes exceeds the height of a man,very ſhrubby,hauing a hard and blackiſh red wood : the leaues are ſmall and ſhort, growing about the branches by foures,of a very aſtringent taſte: it hath plentifull ſtore of floures growing all alongſt the branches, ſo that ſometimes the larger branches haue floures for a foot long. This floure is hollow and longiſh,well ſmelling,white and beautifull. It growes betwixt Lisbon and the Vniuerſitie of Conimbrica in Portugal,where it floures in Nouember,December,and Ianuary. ‡

† 4 Of this kind there is another ſort with whitiſh purple floures, more frequently found than the other ſort ; which floures are ſomewhat greater than the former,but in forme like,and flouring at the ſame time. ‡ The leaues alſo are hairy,and grow commonly by foures : the hollow ſloures grow cluſtering together at the very tops of the branches, and are to be found in Iuly and Auguſt. It growes on diuers heathy places of this kingdome. ‡

5   Croſſed Heath growes to the height of a cubit and a halfe, full of branches, commonly ly-ing along vpon the ground,of a dark ſwart colour : whereon grow ſmall leaues ſet at certain ſpaces two vpon one ſide,and two on the other, oppoſite,one anſwering another, euen as do the leaues of Croſſewort.The floures in like manner ſtand along the branches croſſe-faſhion,of a dark ouerworn greeniſh colour.The root is likewiſe wooddy,as is all the reſt of the plant.

6   This ſteeple Heath hath likewiſe many wooddy branches garniſhed with ſmall leaues that eaſily fall off from the dried ſtalks ; among which come forth diuers little moſſie greeniſh floures of ſmall moment. The whole buſh for the moſt part groweth round together like a little cock of hay,broad at the lower part and ſharp aboue like a pyramide or ſteeple,whereof it tooke his name.

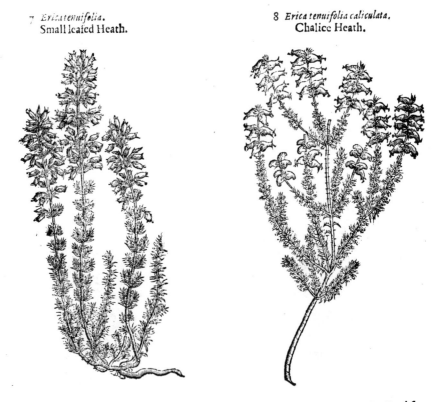

7   *Erica tenuifolia.*
Small leafed Heath.

8   *Erica tenuifolia caliculata.*
Chalice Heath.

7   This ſmall or thinne leaſed Heath is alſo a low and baſe ſhrub,hauing many ſmall and ſlen-der ſhoots comming from the root, of a reddiſh browne colour ; whereupon doe grow very many ſmall leaues not vnlike to them of common Tyme,but much ſmaller and tenderer:the flours grow in tufts at certaine ſpaces,of a purple colour. The root is long,and of a wooddy ſubſtance. ‡ The branches of this are commonly whitiſh,the leaues very green : the flours are ſmalleſt at both ends and biggeſt in the middeſt,hollow, and of a faire purple colour , which doth not eaſily decay. It
floures

floures moſt part of Summer,and growes in many heathy grounds. ‡

8   Chalice heath hath alſo many wooddy branches growing from the roots, ſlender, of a red-diſh brown colour,a foot and half high,garniſhed with very little leaues leſſer than thoſe of Time: the flours grow on the tops and vpper parts of the branches,and be in number fiue, ſix, or more, hanging downward, in faſhion long, hollow within like a little tunnel or open cup or chalice,of a light purpliſh colour : the root creepeth,and putteth forth in diuers places new ſprings or ſhoots.

9   The Heath that bringeth forth berries hath many weake and ſlender branches of a reddiſh colour,which trailing vpon the ground do take hold thereof in ſundry places, whereby it mightily increaſeth : the leaues are ſomewhat broad,of a thicke and fleſhy ſubſtance, in taſte ſomthing drying at the firſt,but afterwards ſomewhat ſharpe and biting the tongue : among which come forth ſmall floures of an herby colour:which being vaded there ſucceed ſmall round berries that at the firſt are green,and afterward blacke,being as big as thoſe of Iuniper, wherein is contained purple juice like that of the Mulbery:within thoſe berries are contained alſo ſmall three cornered grains: the root is hard,and of a woody ſubſtance. ‡ I found this growing in great plenty in Yorkſhire on the tops of the hills of Gisbrough,betwixt it and Roſemary-topin(a round hill ſo called) & ſome of the people thereabouts told me they called the fruit Crake-berries. This is the ſame that Matthiolus calls Erica baccifera ; and it is the Erica Coris folio 11. of Cluſins. ‡

‡ 10   This which our Authour figured as you ſee in the tenth place (putting the deſcription of the former thereto)hath brittle branches growing ſome cubit high,couered with a bark blacker than the reſt : the leaues are like thoſe of the former, but blacker and ſmaller, growing about the ſtalks by threes,of a hottiſh taſte with ſome aſtriction : in September and October it caries a fruit on the tops of the branches different from the reſt,for it is very beautiful,white,tranſparent,reſembling dusky and vneuen pearles in forme and colour,ſucculent alſo and of an acide taſt, commonly containing three little ſeeds in each berrie : in Nouember this fruit becomes dry,and fals away of it ſelfe. Cluſius only obſerued this in Portugal,and at the firſt ſight afar off took the white berries to haue been graines of Manna. He calls it Erica Coris folio 10.

11   I remember(ſaith Dodonæus)that I obſerued another Heath which grew low,yet ſent forth

‡ 11 *Erica pumil. 3. Dod.*
Dodonæus his dwarfe Heath.

‡ 12 *Erica ternis per intervalla ramis.*
Heath with three branches at a joint.

‡ 13 *Erica peregrina Lobelij.*
Lobels ſtrange Heath.

‡ 14 *Erica Coris folio 7. Cluſij.*
Creeping Dutch Heath.

‡ 15 *Erica Coris folio.9.Cluſij.*
Small Auſtrian Heath.

many wooddy and twiggy branches, hauing vpon them little narrow and longiſh leaues ; on theſe ſtalkes ſpike faſhion to the tops of them, yet but on one ſide, grow elegant redde floures, pointed with blacke. This growes in that tract of Germany which leads from Bohemia to Noremberg on dry and vntilled places, and neere woods. It floures in Aprill.

12　This ſhrubby Heath is commonly ſome cubit high, hauing ſlender branches which come out of the maine ſtemmes commonly three together ; and the leaues alſo grow in the ſame order, the tops of the branches are adorned with many floures of a darke purple colour, hollow, round, biggeſt below, and ſtanding vpon long footſtalks. *Cluſius* found this growing in the vntilled places of Portingale aboue Lisbone, where it floured in December ; he calls it *Erica Coris folio*, 5.

13　Beſides all theſe (ſaith *Lobel*, hauing firſt treated of diuers plants of this kinde) there is a certaine rarer ſpecies growing like the reſt after the manner of a ſhrub in pots, in the Garden of Mr *Iohn Brancion*: the leafe is long, and the purple floures, which as far as I remember conſiſted of foure little leaues apiece, grow on the tops of the branches. I know not whence it was brought, and therfore for the rarity I call it *Erica peregrina*, that is, Strange or Forreine Heath.

14　This hath many round blackiſh purple branches ſome foot or cubit high, lying oft times along vpon the ground : theſe are beſet with many narrow little leaues, almoſt like thoſe of the third deſcribed, yet ſomewhat longer, commonly growing foure, yet ſometimes fiue together, of an aſtringent taſte ; the little floures grow on the top of the branches, longiſh, hollow, and of a light purple colour, comming out of foure little leaues almoſt of the ſame colour ; when theſe are ripe and dried they containe a blackiſh and ſmall ſeed; the root is hard, wooddy, and runnes diuers waies; the weake branches alſo that lie vpon the ground now and then take root againe. *Cluſius* found this growing plentifully in diuers mountanous places of Germany, where it floured in Iune and Iuly.

15　The weake ſtalkes of this are ſome foot high, which are ſet with many ſmall greene leaues growing commonly together by threes ; the tops of the branches are deckt with little hollow and longiſh floures diuided at their ends into foure parts, of a fleſh colour, together with the foure little leaues out of which they grow, hauing eight blackiſh little threds in them, with a purpliſh pointall in the middle. The ſeed is blacke and ſmall, the root wooddy as in other plants of this kinde. *Cluſius* found this in ſome mountanous woods of Auſtria, where it floured in Aprill and May. ‡

¶ *The Place.*

Heath groweth vpon dry mountaines which are hungry and barren, as vpon Hampſteed Heath neere London, where all the ſorts do grow, except that with the white floures, and that which beareth berries. ‡ There are not aboue three or foure ſorts that I could euer obſerue to grow there. ‡
Heath with the white floures groweth vpon the downes neere vnto Graueſend.

Heath which beareth berries groweth in the North parts of England, namely, at a place called Crosby Lauenſwaith, and in Cragge cloſe alſo in the ſame country : from whence I haue receiued the red berries by the gift of a learned Gentleman called Mr *Iames Thwaits*.

¶ *The Time.*

Theſe kindes or ſorts of Heath do for the moſt part floure all the Summer, euen vntill the laſt of September.

¶ *The Names.*

Heath is called in Greeke, ἐρείκη in Latine alſo *Erica*: diuers do falſly name it *Myrica* in high and low Dutch, **Heijden :** in Italian, *Erica*: in Spaniſh, *Breſo Querro*: in French, *Bruyre*: in Engliſh, Heath, Hather and Linge.

¶ *The Temperature.*

Heath hath,as *Galen* faith, a digefting faculty, confuming by vapors : the floures and leaues are to be vfed.

¶ *The Vertues.*

A   The tender tops and floures,faith *Diofcorides*, are good to be laid vpon the bitings and ftingings of any venomous beaft : of thefe floures the Bees do gather bad hony.

B   The barke and leaues of Heath may be vfed for,and in the fame caufes that Tamariske is vfed.

† The figure which our Author gaue in the ninth place by the name of *Erica baccifera latifolia* I take to be the *Vitis Idaa,*1, of *Clufius* (which you fhall finde in his due place) and in ftead thereof I haue giuen you our ordinary berry-bearing Heath.

Chap. 53.   Of Heath of Iericho.

1 *Rofa Hiericontea major.*
The Heath Rofe of Ierico.

¶ *The Defcription.*

1   THis kinde of Heath which of the later writers hath beene called by the name *Rofa Hiericontea*; the coiner fpoiled the name in the mint,for of all plants that haue bin written of,there is not any more vnlike vnto the Rofe,or any kinde thereof than this plant:what moued them thereto I know not:but thus much of my owne knowledge, it hath neither fhape, nature, nor facultie agreeing with any Rofe; the which doubtleffe is a kind of Heath, as the barren foile, and that among Heath, doth euidently fhew,as alfo the Heathie matter wherewith the whole plant is poffeffed,agreeing with the kindes of Heath in very notable points. It rifeth vp out of the ground,of the height of four inches, or an hand breadth, compact or made of fundry hard ftickes, (which are the ftalkes) clafping or fhutting it felfe together into a round forme, intricately weauing it felfe one ftick ouerthwart another,like a little net : vpon which wooddy fticks do grow leaus not vnlike to thofe of the Oliue tree,which maketh the whole plant of a round forme,and hollow within ; among the leaues on the infide grow fmall moffie floures,of a whitifh herby colour,which

2 *Rofa Hiericontea ficcata.*   The Heath Rofe of Ierico dried.

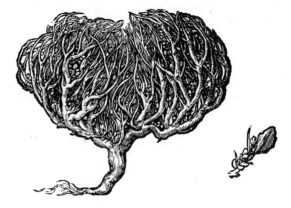

turne into little feed, like the feed of Rocket, but lefser: the whole plant is of the fubftance of heath, and wooddy.

2   The fecond figure fetteth forth the dried plant, as it is brought vnto vs from beyond the feas; which being fet into a difh of warme water, for halfe an houre, openeth it felfe in forme, as when it did grow, and taken forth vntill it be dry, returneth fhut againe as before.

¶ *The Place.*

It groweth in the barren grounds of France, and other hot regions, among the Heath and fuch like plants: it is a ftranger in England, yet dried we haue them in great plenty. ‡ I haue not read nor heard that this growes wilde in France; but *Bellonius* faith it growes in Arabia *deferta : Bauhine* faith it eafily grew and flourifhed many yeares in his garden at Bafill. ‡

¶ *The Time.*

The feed being fowne in our cold climate, is fowne in Aprill; it perifheth when it is fprung vp, and bringeth neither floures nor feed.

¶ *The Names.*

This kind of Heath is called *Rofa Hiericontea*, or *de Hiericho*, the Rofe of Ierico: of fome, the Rofe of Ierufalem, and alfo *Rofa Mariæ* : in Englifh, the Heath Rofe.

¶ *The Temperature and Vertues.*

There is not any of the antient nor later writers that haue fet downe any certainty of this plant as touching the temperature and faculties, but onely a bare picture with a flender defcription.

CHAP. 54.   *Of the Chafte tree.*

1 *Vitex, fiue Agnus caftus.*
The Chafte tree.

‡ 2 *Vitex latiore ferrato folio.*
Chafte tree with cut leaues.

¶ *The Description.*

1　Vitex, or the Chaste tree, groweth after the manner of a bushie shrub or hedge tree, hauing many twiggie branches, very pliant and easie to be bent without breaking, like to the Willow : the leaues are for the most part diuided into fiue or seuen sections or diuisions, much like the leaues of Hemp, whereof each part is long and narrow, very like vnto the willow leafe, but smaller : the floures do grow at the vppermost parts of the branches, like vnto spikie eares, clustering together about the branches, of a light purple or blew colour, and very sweet smell, the fruit is small and round, like vnto the graines or cornes of pepper.

‡ 2　*Lobel* mentions another variety hereof that differs from the former onely in that it hath broader leaues, and these also snipt about the edges. ‡

¶ *The Place.*

Vitex groweth naturally in Italy, and other hot regions, by water courses and running streames : I haue it growing in my garden.

¶ *The Time.*

Vitex beginneth to recouer his lost leaues in May, and the floures come forth in August.

¶ *The Names.*

†　The Grecians call this shrub ἄγνος, and λύγε: *Agnos* (i.) *Castus*, Chaste : because, faith *Pliny* in his 24. booke, 9. chapter, the Athenian Matrons in their feast called *Thesmophoria* dedicated to the honour of *Ceres*, desirous to keepe themselues chaste, doe lay the leaues in their beds vnder them : the Latines name it *Vitex*, and of diuers it is termed, as wee finde among the bastard and counterfeit names, ἄγνος : in Latine, *Salix marina*, or *Salix Amerina*, and *Piper Agreste* : in high Dutch, Schafft-mulle, Keuschbaum : in low Dutch, and also of the Apothecaries, *Agnus Castus* : the Italians *Vitice*, *Agno casto* : in Spanish, *Gattile casto* : in English, Chaste tree, Hempe tree, and of diuers, *Agnus castus*. ‡ The name *Agnus Castus* comes by confounding the Greeke name *Agnos* with *Castus*, the Latine interpretation thereof. ‡

¶ *The Temperature.*

The leaues and fruit of *Agnus castus* are hot and dry in the third degree : they are of very thinne parts, and waste or consume winde.

¶ *The Vertues.*

A　*Agnus Castus* is a singular medicine and remedy for such as would willingly liue chaste, for it withstandeth all vncleannes, or desire to the flesh, consuming and drying vp the seed of generation, in what fort soeuer it bee taken, whether in pouder onely, or the decoction drunke, or whether the leaues be carried about the body ; for which cause it was called *Castus* ; that is to say, chaste, cleane, and pure.

B　The seed of *Agnus Castus* drunken, driueth away, and dissolueth all windinesse of the stomacke, openeth and cureth the stoppings of the liuer and spleen; and in the beginning of dropsies, it is good to be drunke in wine to the quantity of a dram.

C　The leaues stamped with butter, dissolue and asswage the swelling of the genitories and cods, being applied thereto.

D　The decoction of the herbe and seed is good against paine and inflammation about the matrix, if women be caused to sit and bathe their priuie parts therein : the seed being drunke with Pennyroiall bringeth downe the menses, as it doth also both in a fume and in a pessarie : in a pultis it cureth the head-ache, the Phrenticke, and those that haue the Lethargie are wont to be bathed herewith, oile and vineger being added thereto.

E　The leaues vsed in a fume, and also strowed, driue away serpents ; and being laid on do cure their bitings.

F　The seed laied on with water doth heale the clifts or rifts of the fundament; with the leaues, it is a remedy for lims out of joynt, and for wounds.

G　It is reported that if such as journey or trauell do carry with them a branch or rod of *Agnus Castus* in their hand, it will keepe them from Merry-galls, and wearinesse : *Dioscorides.*

C H A P. 55.　*Of the Willow Tree.*

¶ *The Description.*

1　THe common Willow is an high tree, with a body of a meane thicknesse, and riseth vp as high as other trees doe if it be not topped in the beginning, soone after it is planted; the
barke

barke thereof is smooth, tough, and flexible: the wood is white, tough, and hard to be broken: the leaues are long, lesser and narrower than those of the Peach tree, somewhat greene on the vpper side and slipperie, and on the nether side softer and whiter: the boughes be couered either with a purple, or else with a white barke: the catkins which grow on the toppes of the branches come first of all forth, being long and mossie, and quickly turne into white and soft downe, that is carried away with the winde.

1 *Salix.*
The common Willow.

2 *Salix aquatica.*
The Oziar or water Willow.

2   The lesser bringeth forth of the head, which standeth somewhat out, slender wands or twigs, with a reddish or greene barke, good to make baskets and such like workes of: it is planted by the twigs or cods being thrust into the earth, the vpper part whereof when they are growne vp, is cut off, so that which is called the head increaseth vnder them, from whence the slender twigs doe grow, which being oftentimes cut, the head waxeth greater: many times also the long rods or wands of the higher Withy trees be lopped off and thrust into the ground for plants, but deeper, and aboue mans height: of which do grow great rods, profitable for many things, and commonly for bands, wherewith tubs and casks are bound.

3   The Sallow tree or Goats Willow, groweth to a tree of a meane bignesse: the trunke or body is soft and hollow timber, couered with a whitish rough barke: the branches are set with leaues somewhat rough, greene aboue, and hoarie vnderneath: among which come forth round catkins, or aglets that turne into downe, which is carried away with the winde.

4   This other Sallow tree differeth not from the precedent, but in this one point, that is to say, the leaues are greater and longer, and euery part of the tree larger, wherein is the difference. ‡ Both those last described hath little roundish leaues like little eares growing at the bottoms of the footstalkes of the bigger leaues, whereby they may bee distinguished from all other Plnats of this kinde. ‡

5   The Rose Willow groweth vp likewise to the height and bignesse of a shrubby tree; the body whereof is couered with a scabby rough barke: the branches are many, whereupon do grow very many twigs of a reddish colour, garnished with small long leaues, somewhat whitish: amongst which come forth little floures, or rather a multiplication of leaues, joyned together in forme of a

3 Salix Caprea rotundifolia.
The Goat round leafed Willow.

4 Salix Caprea latifolia.
The Goat broad leafed Sallow.

5 Salix Roſea Anglica.
The Engliſh Roſe Willow.

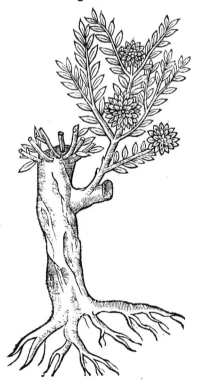

Roſe, of a greeniſh white colour, which doe not only make a gallant ſhew, but alſo yeeld a moſt cooling aire in the heat of Summer, being ſet vp in houſes, for the decking of the ſame.

6   The low or baſe Willow growes but low, and leaneth weakely vpon the ground, hauing many ſmall and narrow leaues, ſet vpon limber and pliant branches, of a darke or blackiſh greene colour: amongſt which come forth long ſlender ſtems full of Moſſie floures, which turne into a light downy ſubſtance that flieth away with the winde.

7   The dwarfe Willow hath very ſmall and ſlender branches, ſeldome times aboue a foot, but neuer a cubit high, couered with a duskie barke, with very little and narrow leaues, of a greene colour aboue, and on the vpper ſide, but vnderneath of a hory or ouerworne greeniſh colour, in bigneſſe and faſhion of the leaues of garden Flax: among which come forth little duskiſh floures, which doe turne into downe that is carried away with the winde: the root is ſmall and threddy, of the bigneſſe of a finger, and of a blackiſh colour.

8   There is another kinde of willow like to the former, and differeth from it in that, the leaues of this kinde are ſmaller and narrower, as big as the leaues of Myrtle, hauing ſmall knobby floures of a duskiſh colour, which

6 Salix humilis.
The low Willow.

7 Chamæitea, ſiue Salix pumila.
The dwarfe Willow.

‡ 8 Salix humilis repens.
Creeping dwarfe Willow.

which turne into downe that flyeth away with the winde: the root is ſmall and limber, not growing deepe, but running along vpon the vpper cruſt of the earth.

¶ The Place.

Theſe Willowes grow in diuers places of England : the Roſe-Willow groweth plentifully in Cambridge ſhire, by the riuers and ditches there in Cambridge towne they grow abundantly about the places called Paradiſe and Hell-mouth, in the way from Cambridge to Grandcheſter : I found the dwarfe Willowes growing neere to a bog or mariſh ground at the further end of Hampſted heath vpon the declining of the hill, in the ditch that incloſeth a ſmall Cottage there, not halfe a furlong from the ſaid houſe or cottage.

¶ The Time.

The Willowes doe floure at the beginning of the Spring.

¶ The Names.

The Willow tree is called in Greeke ιτέα in Latine, Salix : in high Dutch weyden : in low Ditch, wilgen : in Italian, Salice; Salcio : in French, Saux : in Spaniſh, Salgueiro, Salzer, and Sauz : in Engliſh, Sallow, Withie, and Willow.

The

The greater is called in Latine *Salix perticalis*, common Withy, Willow, and Sallow, especially that which being often lopped sendeth out from one head many boughes : the kinde hereof with the red barke is called of *Theophrastus*, blacke Withy; and the other white : *Pliny* calleth the blacke *Græca*, or Greeke Withie (the red, being the Greeke Withy) saith he, is easie to be cleft, and the whiter, *Amerina*.

Theophrastus writeth, That the Arcadians doe call the lesser Ελικι, not Ἰτια, *Pliny* also nameth this *Helice* : both of them doe make this to be *Salicis tertia species*, the third kinde of Sallow : the same is likewise called in Latine, *Salix pumila*, *Salix viminalis*, *Gallica Salix* ; and by *Columella*, *Sabina*, which he saith that many doe terme *Amerina* : in high Dutch, **klein wepden** : in low Dutch, **wijmen** : in English, Osier, small Withy, Twig Withy : *Petrus Crescentius* nameth it *Vincus*.

¶ *The Temperature*.

The leaues, floures, seed, and barke of Willowes are cold and dry in the second degree, and astringent.

¶ *The Vertues*.

A   The leaues and barke of Withy or Willowes do stay the spitting of bloud, and all other fluxes of bloud whatsoeuer in man or woman, if the said leaues and barke be boyled in wine and drunke.

B   The greene boughes with the leaues may very well be brought into chambers and set about the beds of those that be sicke of feuers, for they doe mightily coole the heate of the aire, which thing is a wonderfull refreshing to the sicke Patients.

C   The barke hath like vertues : *Dioscorides* writeth, That this being burnt to ashes, and steeped in vineger, takes away cornes and other like risings in the feet and toes : diuers, saith *Galen*, doe slit the barke whilest the Withy is in flouring, and gather a certain juyce with which they vse to take away things that hinder the sight, and this is when they are constrained to vse a clensing medicine of thin and subtill parts.

CHAP. 56.   *Of the Oliue tree.*

1 *Olea satiua.*
The manured Oliue tree.

2 *Olea syluestris.*
The wilde Oliue tree.

¶ *The*

¶ The Deſcription.

1 THe tame or manured Oliue tree groweth high and great with many branches, full of long narrow leaues not much vnlike the leaues of Willowes, but narrower and ſmaller : the floures be white and very ſmall, growing vpon cluſters or bunches : the fruit is long and round, wherein is an hard ſtone : from which fruit is preſſed that liquor which we call oyle Oliue.

2 The wilde Oliue is like vnto the tame or garden Oliue tree, ſauing that the leaues are ſomething ſmaller : among which ſometimes doe grow many prickly thornes : the fruit hereof is leſſer than of the former, and moe in number, which do ſeldome come to maturity or ripenes in ſo much that the oile which is made of thoſe berries, continueth euer greene, and is called oile Omphacine, or oile of vnripe Oliues.

¶ The Place.

Both the tame and the wilde Oliue trees grow in very many places of Italy, France, and Spaine, and alſo in the Iſlands adjoyning : they are reported to loue the ſea coaſts ; for moſt doe thinke, as *Columella* writeth, that aboue fixty miles from the ſea they either die, or elſe bring forth no fruit : but the beſt, and they that doe yeeld the moſt pleaſant oile are thoſe that grow in the Iſland called Candy.

¶ The Time.

All the Oliue trees floure in the moneth of Iune : the fruit is gathered in Nouember or December : when they be a little dried and begin to wrinkle they are put into the preſſe, and out of them is ſqueezed oyle, with water added in the preſſing : the Oliues which are to bee preſerued in ſalt and pickle muſt be gathered before they be ripe, and whileſt they are greene.

¶ The Names.

The tame or garden Oliue tree is called in Greeke ϵλαία, and Ελαία ημερος in Latine, *Olea ſatiua*, and *Vrbana* : in high Dutch, Oelbaum : in low Dutch, Olijſboome : in Italian, *Oliuo domeſtico* : in French, *Oliuier* : in Spaniſh, *Oliuo*, and *Oliuera* : in Engliſh, Oliue tree.

The berry is called *Oliua* : in Greeke alſo Ελαία : in Spaniſh, *Aʒeytuna* : in French, Dutch, and Engliſh, Oliue.

Oliues preſerued in brine or pickle are called *Colymbades*.

The wilde Oliue tree is named in Greeke, Αγριελαία : in Latine, *Olea ſylueſtris*, *Oleaſter*, *Cotinus*, *Olea Æthiopica* : in Dutch, wald Oelbaum : in Italian, *Oliuo ſaluatico* : in Spaniſh, *Aʒebuche*, *Azambuſ beyro* : in French, *Oliuier ſauuage* : in Engliſh, wilde Oliue tree.

¶ The Temperature and Vertues.

The Oliues which be ſo ripe as that either they fall of themſelues, or be ready to fall, which are **A** named in Greeke, Δρυππαι, be moderately hot and moiſt, yet being eaten they yeeld to the body little nouriſhment.

The vnripe Oliues are dry and binding. **B**

Thoſe that are preſerued in pickle, called *Colymbades*, doe dry vp the ouermuch moiſture of the **C** ſtomacke, they remoue the loathing of meate, ſtirre vp an appetite ; but there is no nouriſhment at all that is to be looked for in them, much leſſe good nouriſhment.

The branches, leaues, and tender buds of the Oliue tree doe coole, dry, and binde, and eſpecial- **D** ly of the wild Oliue ; for they be of greater force than thoſe of the tame : therefore by reaſon they be milder they are better for eye medicines, which haue need of binding things to be mixed with them.

The ſame do ſtay S. Anthonies fire, the ſhingles, epinyctides, night wheales, carbuncles, and ea- **E** ting vlcers : being laid on with honey they take away eſchares, clenſe foule and filthy vlcers, and quench the heate of hot ſwellings, and be good for kernels in the flanke : they heale and skin wounds in the head, and being chewed they are a remedy for vlcers in the mouth.

The juyce and decoction alſo are of the ſame effect : moreouer the juyce doth ſtay all manner of **F** bleedings, and alſo the whites.

The juyce is preſſed forth of the ſtamped leaues, with Wine added thereto (which is better) or **G** with water, and being dried in the Sun it is made vp into little cakes like perfumes.

The ſweat or oyle which iſſueth forth of the wood whileſt it is burning healeth tetters, ſcurfs **H** and ſcabs, if they be annointed therewith.

The ſame which is preſſed forth of the vnripe Oliues is as cold as it is binding. **I**

The old oile which is made of ſweet and ripe Oliues, being kept long, doth withall become hot- **K** ter, and is of greater force to digeſt or waſte away ; and that oile which was made of the vnripe Oliue, being old, doth as yet retaine ſome part of his former aſtriction, and is of a mixt faculty, that is to ſay, partly binding, and partly digeſting ; for it hath got this digeſting or conſuming faculty by age, and the other property of binding of his owne nature.

The

L   The oile of ripe Oliues mollifieth and aſſwageth paine, diſſolueth tumors or ſwellings, is good for the ſtiffeneſſe of the joynts, and againſt cramps, eſpecially being mingled according to art, with good and wholeſome herbes appropriate vnto thoſe diſeaſes and griefes, as *Hypericon*, Cammomill, Dill, Lillies, Roſes, and many others, which do fortifie and increaſe his verrues.

M   The oile of vnripe Oliues, called *Omphacinum Oleum*, doth ſtay, repreſſe, and driue away the beginning of tumors and inflammations, cooling the heate of burning vlcers and exulcerations.

CHAP. 57. *Of Priuet or Prim Print.*

Liguſtrum.
Priuet, or Prim Print.

¶ The Deſcription.

PRiuet is a ſhrub growing like a hedge tree, the branches and twigs hereof be ſtraight, and couered with ſoft gliſtring leaues of a deepe green colour, like thoſe of Peruincle, but yet longer, greater alſo than the leaues of the Oliue tree: the floures be white, ſweet of ſmell, very little, growing in cluſters; which being vaded there ſucceed cluſters of berries, at the firſt greene, and when they be ripe blacke like a little cluſter of grapes, which yeeld a purple juyce: the root groweth euery way aſlope.

¶ The Place.

The common Priuet groweth naturally in euery wood, and in the hedge rowes of our London gardens: it is not found in the countrey of Polonia and other parts adjacent.

¶ The Time.

It floureth in the end of May, or in Iune: the berries are ripe in Autumne or about Winter, which now and then continue all the Winter long; but in the meane time the leaues fall away, and in the Spring new come vp in their places.

¶ The Names.

It is called in Latine, *Liguſtrum*: in Italian at this day, *Guiſtrico*, by a corrupt word drawne from *Liguſtrum*: it is the Grecians ωνιρια, and in no wiſe κιαργι: for Cypres is a ſhrub that groweth naturally in the Eaſt, and Priuet in the Weſt. They be very like one vnto another, as the deſcriptions doe declare; but yet in this they differ, as witneſſeth *Bellonius*, becauſe the leaues of Priuet do fall away in Winter, and the leaues of Cyprus are alwaies greene: Moreouer, the leaues of Cyprus doe make the haire red, as *Dioſcorides* ſaith, and (as *Bellonius* reporteth) doe giue a yellow colour: but the leaues of Priuet haue no vſe at all in dying. And therefore *Pliny, lib. 24. cap. 10.* was deceiued, in that hee judged Priuet to be the ſelfe ſame tree which Cyprus is in the Eaſt: which thing notwithſtanding he did not write as he himſelfe thought, but as other men ſuppoſe; for *lib. 12. cap. 14.* he writeth thus, Some (ſaith he) affirme this, *viz.* Cyprus, to be that tree which is called in Italy, *Liguſtrum*; and that *Liguſtrum* or Priuet is that plant which the Grecians call κιαργι, the deſcription doth declare.

Phylliria, ſaith *Dioſcorides*, is a tree like in bigneſſe to Cyprus, with leaues blacker and broader than thoſe of the Oliue tree: it hath fruit like to that of the Maſtick tree, blacke, ſomething ſweet, ſtanding in cluſters, and ſuch a tree for all the world is Priuet, as we haue before declared.

Serapio the Arabian, *cap. 44.* doth call Priuet *Mahaleb*. There is alſo another *Mahaleb*, which is a graine or ſeed of which *Auicen* maketh mention, *cap. 478.* that it doth by his warme and comfortable heate diſſolue and aſſwage paine. *Serapio* ſeemeth to intreate of them both, and to containe diuers of the *Mahaleb* vnder the title of one chapter: it is named in high Dutch, **Beinholtzleſu, Mundtholtz, Rhein oder Schulweiden**: in low Dutch, **Keelcrupt, Monthout**: in French, *Troeſne*: in Engliſh, Priuet, Primprint, and Print.

Some

Some there be that would haue the berries to be called *Vaccinia*, aud *Vaccinium* to be that of which *Vitruuius* hath made mention in his feuenth booke of Architecture or the art of building, chap.14.of purple colours: after the fame manner, faith hee, they temper *Vaccinium*, and putting milke vnto it doe make a gallant purple : in fuch breuitie of the old Writers what can be certainly determined?

¶ The Temperature.

The leaues and fruit of Priuet are cold,dry,and aftringent.

¶ The Vertues.

The leaues of Priuet doe cure the fwellings, apoftumations, and vlcers of the mouth or throat, A being gargarifed with the iuyce or decoction thereof, and therefore they be excellent good to be put into lotions,to wafh the fecret parts, and the fcaldings with women, cankers and fores in childrens mouthes.

Chap. 58.   Of Mocke-Priuet.

1 *Phillyrea anguftifolia.*
Narrow leaued Mocke-Priuet.

2 *Phillyrea latiore folio.*
The broader leaued Mocke-Priuet.

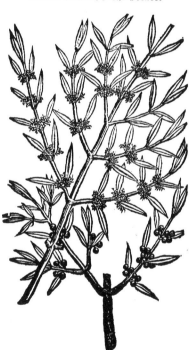
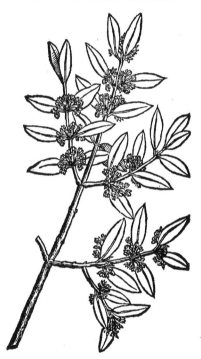

¶ The Defcription.

1   CYprus is a kinde of Priuet, and is called *Phillyrea*, which name all the forts or kindes thereof doe retaine, though for diftinction fake they paffe vnder fundry titles. This plant groweth like an hedge tree, fometimes as big as a Pomegranate tree, befet with flender twiggy boughes which are garnifhed with leaues growing by couples, very like the leaues of the Oliue tree,but broader,fofter,and of a green colour : from the bofomes of thefe leaues come forth great bunches of fmall white floures, of a pleafanr fweet fmell : which being vaded, there fucceed clufters of blacke berries very like the berries of the Alder tree.

2   The fecond Cyprus, called alfo *Phillyrea latifolia*, is very like the former in **body, branches,**
leaues,

3 *Phillyrea ferrata 2.Clufij.*
The fecond toothed Priuet of *Clufius.*

leaues, floures, and fruit ; and the difference is this, that the leaues of this plant are broader, but in faculty they are like.

3   This kinde of Priuet rifeth vp like an hedge bufh, of the height of fiue or fix cubits : the branches are long, fragile or brittle, coue-red with a whitifh barke, whereon are fet leaues fomewhat broad, jagged on the edges like the teeth of a faw, and of a deepe greene colour : among which come forth the flours, which nei-ther my Author nor my felfe haue feene : the berries grow vpon fmall foot-ftalkes, for the moft part three together, being round, and of the bigneffe of pepper graines, or Myrtle ber-ries, of a blacke colour when they be ripe.

¶ *The Place.*

Thefe plants do grow in Syria neere the ci-ty Afcalon, and were found by our induftrious *Pena* in the mountaines neere Narbone and Montpelier in France : the which I planted in the garden at Barne-Elmes neere London, be-longing to the right Honourable the Earle of Effex: I haue them growing in my garden like-wife.

¶ *The Time.*

The leaues fhoot forth in the firft of the Spring : the floures fhew themfelues in May and Iune : the fruit is ripe in September.

¶ *The Names.*

This Priuet is called in Greeke, *κύπρος*, and in Latine alfo *Cyprus* : they may be named in Eng-lifh, Eafterlin Priuet, and Mock-Priuet, for the reafon following: they are deceiued who taking *Pliny* for their Author, do thinke that it is *Liguftrum*, or our Wefterne Priuet, as we haue fhewed in the former chapter, it is the Arabians *Alcanna*, or *Henne* : and it is alfo called of the Turkes *Henne* euen at this prefent time.

¶ *The Temperature.*

The leaues of thefe kindes of Priuet haue a binding quality, as *Diofcorides* writeth.

¶ *The Vertues.*

A   Being chewed in the mouth they heale the vlcers thereof, and are a remedy againft inflammati-ons or hot fwellings.

B   The decoction thereof is good againft burnings and fcaldings.

C   The fame being ftamped and fteeped in the juyce of Mullen and laid on, do make the haire red, as *Diofcorides* noteth. *Bellonius* writeth, that not onely the haire, but alfo the nether parts of mans body and nailes likewife are coloured and died herewith, which is counted an ornament among the Turks.

D   The floures being moiftned in vineger and applied to the temples affwage head-ache.

E   There is alfo made of thefe an oile called *Oleum Cyprinum*, fweet of fmell, and good to heate and supple the finewes.

Chap. 59. *Of baftard Priuet.*

¶ *The Defcription.*

1   THis fhrubby tree, called *Macaleb*, or *Mahaleb*, is alfo one of the Priuets : it rifeth vp like vnto a fmall hedge tree, not vnlike vnto the Damfon or Bulleffe tree, hauing many vp-right ftalkes and fpreading branches : whereon do grow leaues not vnlike to thofe of the *Phillyrea* of *Clufius* defcription : amongft which come forth moffie floures of a white colour, and of a perfect

sweet

ſweet ſmell, growing in cluſters, many hanging vpon one ſtem, which the Grauer hath omitted : after which come the berries, green at the firſt, and blacke when they be ripe, with a little hard ſtone
within, in which lies a kernell.

2   Geſner and Matthiolus haue ſet forth another Macaleb, being alſo another baſtard Priuet. It
groweth to a ſmall hedge tree, hauing many green branches ſet with round leaues like thoſe of the
Elme tree, ſomwhat ſnipt about the edges : the flours are like thoſe of the precedent, the fruit, or
rather the kernell thereof, is as hard as a bead of Corall, ſomwhat round, and of a ſhining blacke colour ; which the cunning French perfumers do bore thorow, making thereof bracelets, chaines, and
ſuch like trifling toyes, which they ſend into England, ſmeared ouer with ſome odde ſweet Compound or other, and they are here ſold vnto our curious ladies and gentlewomen for rare & ſtrange
Pomanders, for great ſummes of mony.

1 Phillyria arbor, verior Macaleb.
Baſtard Priuet.

2 Macaleb Geſneri.
Corall Priuet.

¶ The Place.
Theſe trees grow in diuers places of France, as about Tholouſe and ſundry other places. They
are ſtrangers in England.

¶ The Time.
The floures bud forth in the Spring : the fruit is ripe in Nouember and December.

¶ The Names.
This baſtard Priuet is that tree which diuers ſuſpect to be that Mahaleb or Machaleb of which
Avicen writeth, cap. 478. and which alſo Serapio ſpeaketh of out of Meſue : but it is an hard thing to
affirme any certaintie thereby, ſeeing that Avicen hath deſcribed it without marks. Notwithſtan-
ding this is taken to be the ſame of moſt writers, and thoſe of the beſt. We may call it in Engliſh,
Baſtard Priuet, or Corall, or Pomander Priuet, being without doubt a kinde thereof.

¶ The Nature and Vertues.
Concerning this baſtard Priuet we haue learned as yet no vſe thereof in phyſicke. The kernels A
that are found in the ſtones or fruit, as they be like in taſte to thoſe of Cherries, ſo be they alſo anſwerable to them in temperature ; for they are of a temperat heate, and gently prouoke vrine. and
be therefore good for the ſtone : more wee haue not to write than hath been ſpoken in the deſcription.

CHAP. 60.　*Of the fruitleſſe Privet.*

¶ *The Deſcription.*

1　THis ſhrubby buſh, called of *Pliny* and *Cluſius*, *Alaternus*, growes vp to a ſmal hedge tree, in forme like vnto a baſtard Privet, but the leaues are more like thoſe of *Ilex* or French Oke, yet ſtiffer and rounder than thoſe of *Macaleb* : among which come forth tufts of greeniſh yellow floures like thoſe of the Lentisk tree : vnder and among the leaues come forth the berries, like thoſe of *Laurus Tinus*, in which are contained two kernels like the acines or ſtones of the Grape.

1 *Alaternus Plinÿ.*
Fruitleſſe Privet.

2 *Alaternus humilior.*
The lower fruitleſſe Privet.

2　The ſecond kinde of *Alaternus* is likewiſe a fruitleſſe kinde of Privet, hauing narrow leaues ſomwhat ſnipt about the edges; from the boſoms whereof come forth ſmall herby colored flours, which being vaded, the fruit ſucceedeth, whereof *Auicen* ſpeaketh, calling it *Fagaras*, being a fruit in bigneſſe & form like thoſe in ſhops called *Cocculus Indi*, and may be the ſame, for any thing that hath bin written to the contrary. This fruit hangs as it were in a darke aſh coloured skin or huske, which incloſeth a ſlender ſtiffe ſhell like the ſhell of a nut, couered with a thin or black filme, whether it be the fruit of this plant it is not cenſured; notwithſtanding you ſhall find the figure hereof among the Indian fruits, by the name *Fagaras*.

‡　This hath ſhorter branches and rounder leaues than the former: the floures are larger, and greener, to which ſucceed fruit cluſtering together, firſt green, then red, and afterwards blacke, and conſiſting of three kernels. It floures in February and the beginning of March, and growes in ſundry places of Spain. The fruit of this is not the *Fagaras*, neither doth the *Fagaras* mentioned by our Author any way agree with the *Cocculus Indi* of the ſhops, as ſhal be ſhewed hereafter in their fit places. ‡

¶ *The Place.*
Theſe Plants grow in the ſhadowy woods of France, and are ſtrangers in England.

¶ *The*

¶ *The Time.*

The Time answereth the rest of the Priuets.

¶ *The Names.*

Alaternus of *Pliny* is the same *phillyrea* which *Theophrastus* hath written of by the name *Philyca*: and *Bellonius* also, *lib.* 1. *cap.* 42. of his Singularities, and the people of Candy call it *Eleprinon*: the Portugals, *Casea*: in French, *Dalader*, and *Sangin blanc*: in English, barren or fruitlesse Priuet: notwithstanding some haue thought it to beare fruit, which at this day is called *Fazaras*: with vs, *Cocculus Indi*, as we haue said. ‡ I can by no meanes approue of the English name here giuen by our Author; but iudge the name of Euer-green Priuet (giuen it by Mr *Parkinson*) to be much more fitting to the thing. ‡

¶ *The Temperature and Vertues.*

Whether the plant be vsed in medicine I cannot as yet learn: the fishermen of Portugall vse to seethe the barke thereof in water, with the which decoction they colour their nets of a reddish colour, being very fit for that purpose: the wood also is vsed by Dyers to dye a dark black withall.

CHAP. 61.　*Of the white and blew Pipe Priuet.*

1 *Syringa alba.*
White Pipe.

2 *Syringa cærulea.*
Blew Pipe.

¶ *The Description.*

1　THe white Pipe groweth like an hedge tree or bushy shrub; from the root whereof arise many shoots which in short time grow to be equall with the old stocke, whereby in a little time it increaseth to infinit numbers, like the common English Prim or Priuet, whereof doubtlesse it is a kinde, if wee consider euery circumstance. The branches are couered with a rugged gray barke: the timber is white, with some pith or spongie matter in the middest like Elder, but lesse in quantitie. These little branches are garnished with small crumpled leaues of the shape and bignesse of the Peare tree leaues, and very like in form: among which come forth

the

‡ 3 *Syringa Arabica.*
Arabian Pipe.

4 *Balanus Myrepſica, ſive Glans vnguentaria.*
The oily Acorne.

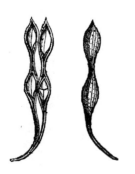

the floure,growing in tufts, compact of four ſmall leaues of a white colour,and of a pleaſant ſweet ſmell ; but in my iudgment they are too ſweet , troubling and moleſting the head in very ſtrange manner. I once gathered the floures and layed them in my chamber window, which ſmelled more ſtrongly after they had lien together a few houres , with ſuch an vnacquainted fauor that they awaked me out of ſleepe, ſo that I could not reſt till I had caſt them out of my chamber. The floures being vaded,the fruit follows,which is ſmall , curled, and as it were compact of many little folds, broad towards the vpper part,and narrow toward the ſtalke,and black when it is ripe , wherein is contained a ſlender long ſeed. The root hereof ſpreadeth it ſelfe abroad in the ground after the manner of the roots of ſuch ſhrubby trees.

2 The blew Pipe groweth likewiſe in manner of a ſmall hedge tree , with many ſhoots riſing from the root like the former, as our common Priuet doth, whereof it is a kind. The branches haue a ſmall quantity of pith in the middle of the wood , and are couered with a darke blacke greeniſh barke or rinde. The leaues are exceeding greene,and crumpled or turned vp like the brimmes of a hat, in ſhape very like vnto the leaues of the Poplar tree : among which come the floures, of an exceeding faire blew colour, compact of many ſmal floures in the form of a bunch of grapes : each floure is in ſhew like thoſe of *Valeriana rubra Dodonæi*,conſiſting of four parts like a little ſtar, of an exceeding ſweet fauour or ſmell, but not ſo ſtrong as the former. When theſe floures be gon there ſucceed flat cods,and ſomwhat long, which being ripe are of a light colour, with a thinne membrane or filme in the middeſt , wherein are ſeeds almoſt foure ſquare, narrow, and ruddy.

‡ 3 This (which *Cluſius* ſetteth forth by the name of *Iaſminum Arabicum,* or *Syringa Arabica*) groweth ſome two or three cubits high, diuided into many ſlender branches, whereon by couples at each joint ſtand leaues like thoſe of the firſt deſcribed, but thinner, and not ſnipt about the edges. On the tops of the branches grow the floures, wholly white, conſiſting of nine, tenne, or twelue leaues ſet in two ranks : theſe floures are ſweet, hauing a ſent as it were compounded of the Spaniſh Iaſmine and Orenge flours. It is a tender plant,and may be graffed on the common Iaſmine,whereon it thriues well,and floures moſt part of the Summer. It groweth plentifully in Egypt ; and *Proſper Alpinus* is thought to mention this by the name of *Sambac Arabum,ſive Gelſeminum Arabicum.* ‡

4 *Glans vnguentaria,* or the oily Acorne, is the fruit of a tree like Tamariske, of the bigneſſe of an Haſell nut ; out of the kernell whereof, no otherwiſe than out of bitter Almonds, is preſſed an oily juyce, which is vſed in pretious Oyntments, as *Dioſcorides* affirmeth. Neither is it in our time wholly rejected ; for the oile of this Fruit mixed with ſweete Odours ſerueth to perfume
gloues

gloues and diuers other things, and is vulgarly known by the name of oile of Ben.

¶ *The Place.*

1 2 These trees grow not wild in England, but I haue them growing in my garden in very great plenty.

¶ *The Time.*

They floure in Aprill and May, but as yet they haue not borne any fruit in my garden, though in Italy and Spain their fruit is ripe in September.

¶ *The Names.*

The later Physitians call the first *Syringa*, or rather χυρνξ, that is to say, a Pipe, because the stalkes and branches therof when the pith is taken out are hollow like a Pipe: it is also many times syrnamed *Candida* or white, or *Syringa Candida flore*, or Pipe with a white floure, because it should differ from *Lillach*, which is somtimes named *Syringa cærulea*, or blew Pipe: in English, white Pipe.

Blew Pipe the later physitians. as we haue said, do name *Lillach* or *Lilach*: of some, *Syringa cærulea*, or blew Pipe: most do expound the word *Lillach*, and call it *Ben: Serapio's* and the Arabians *Ben* is *Glans Vnguentaria*, which the Grecians name Βάλανος Μυρεψικη: from which *Lillach* doth very much differ: amongst other differences it is very apparant that *Lillach* bringeth forth no nut, howsoeuer *Matthiolus* doth falsly picture it with one; for it hath only a little cod, the seed whereof hath in it no oile at all. The figure of the *Balanus Myrepsica* we haue thought good to insert in this Chapter, for want of a more conuenient roome.

¶ *The Temperature and Vertues.*

Concerning the vse and faculties of these herbs neither we our selues haue found out any thing, nor learned ought of others.

‡ *Balanus Myrepsica* taken in the quantity of a dram causeth vomit, drunk with *Hydromel* it pur-   A geth by the stoole, and is hurtfull to the stomacke.

The oile pressed out of this fruit, which is vsually termed oile of Ben, as it hath no good or plea-   B sing smell, so hath it no ill sent, neither doth it become rancide by age, which is the reason that it is much vsed by Perfumers.

The oile smoothes the skin, softens and dissolues hardnesse, and conduces to the cure of all cold   C affects of the sinues: and it is good for the pain and noise in the ears, being mixt with goose grease, and so dropped in warme in a small quantitie. ‡

Chap. 62.   *Of Widow-waile or Spurge Olive.*

¶ *The Description.*

Widow-waile is a small shrub about two cubits high. The stalke is of a woody substance, branched with many small twigs full of little leaues like Privet, but smaller and blacker: on the ends whereof grow small pale yellow floures; which being past, there succedeth a three cornered berry like the *Tithymales*, for which cause it was called *Tricoccos*, that is, three berried *Chamælea*. These berries are green at the first, red afterward, and brown when they be withered, and contain in them an oily fatnesse like that of the Oliue, being of an hot biting tast, and burning the mouth as do both the leaues and rinde. The root is hard and wooddy.

¶ *The Place.*

It is found in most vntilled grounds in Italy and Languedoc in France, in rough and desart places. I haue it growing in my garden.

¶ *The Time.*

It is alwaies green: the seed is ripe in Autumne.

¶ *The Names.*

The Grecians call it χαμελαία, as though they should say low or short Oliue tree: the Latines, *Oleago*, & *Oleastellus*, and likewise *Citocacium*: it is also named of diuers, *Olivella*, as *Matthiolus Sylvaticus* saith: it is called in English, Widow-waile, *quia facit viduas*.

The fruit is named of diuers, κόκκος κνίδιος: in Latine, *Coccus cnidius* · but he is deceiued, saith *Dioscorides*, that nameth the fruit of Spurge Olive, *Coccus cnidius*. *Aviccn* and *Serapio* call *Chamælea* or Spurge Olive, *Mezereon*; vnder which name notwithstanding they haue also contained both the Chamæleons or Carlines, and so haue they confounded *Chamælea* or Spurge Olive with the Carlines, and likewise *Thymælea* or Spurge flax.

Chamælea Arabum Tricoccos.
Widow-waile.

A

B

C

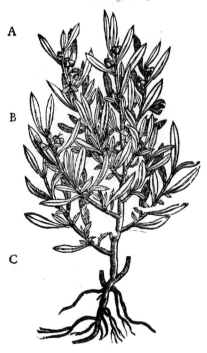

Chamælea Germanica, sive Mezereon.
Spurge Flax, or the dwarfe Bay.

¶ *The Temperature.*

Both the leaues and fruit of Spurge olive, as we haue said, are of a burning and extreme hot temperature.

¶ *The Vertues.*

The leaues, saith *Dioscorides*, purge both flegme and choler, especially taken in pils, so that two parts of Wormwood be mixed with one of Spurge Olive, and made vp into pills with Mede or honied water: they melt not in the belly, but as many as be taken are voided whole.

Mesue also hath a description of pils of the leaues of *Mezereon*, that is, *Chamælea* or spurge Olive (yet *Sylvius* expounds it *Thymælea*, or Spurge flax) but in stead of wormwood he taketh the outward substance of the yellow Mirobalans and *Cepula Mirobalans*, and maketh them vp with *Tereniabin*, that is, with Manna and foure Dates, which they call Tamarinds, dissolued in Endive water; & appointeth the same leaues to be first tempered with very strong vineger, and to be dried.

These pills are commended against the dropsie, for they expell waterie humours, but are violent to nature; therefore wee must vse them as little as may be. Further *Dioscorides* addeth, that the leaues of Spurge olive beaten with hony do clense filthy or crusted vlcers,

CHAP. 63.

Of Germane Olive Spurge.

¶ *The Description.*

THe dwarfe Bay tree, called of Dutchmen *Mezereon*, is a smal shrub two cubits high the branches be tough, limber, and easie to bend, very soft to be cut: whereon grow long leaues like those of Privet, but thicker & fatter: the floures appeare before the leaues, oft times in Ianuarie, clustering together about the stalks at certain distances, of a whitish colour tending to purple, and of a fragrant pleasant sweet smell: after come the small berries, green at the first, but beeing ripe of a shining red, and after of a darke blacke colour, of a very hot and burning tast, inflaming the mouth and throat, with danger of choking. The root is wooddy.

¶ *The Place and Time.*

This plant growes naturally in moist and shadowy woods of most of the East countries especially about Melvin in Poland, from whence I haue had great plenty for my Garden, where they flour in the first of the spring, and ripen their fruit in August.

¶ *The Names.*

It is vsually called in high-Dutch, **Zeilant, Zeidelbast, Lensskraut,** and **Killerhals:** the
Apothe-

Apothecaries of our countrey call it *Mezereon*, but wee had rather name it *Chamælea Germanica*, or Dutch Mezereon, or it may be called Germane Olive Spurge. We haue heard that diuers Italians do name the fruit thereof *Piper montanum*, or Mountain Pepper. Some ſay that *Laureola* or Spurge Laurell is this plant; but there is another *Laureola*, of which we will hereafter treat: but by what name it is called of the antient writers & whether they knew it or no, it is hard to tel. It is thought to be *Cneoron album Theophraſti*, but by reaſon of his breuitie we can affirme no certaintie.

There is, ſaith he, two kindes of *Cneoron*, the white and the blacke, the white hath a long leaf like in forme to Spurge Olive, the blacke is full of ſubſtance like Myrtle; the low one is more white, the ſame is with ſmell, and the blacke without ſmell. The root of both, which groweth deepe, is great: the branches be many, thicke, wooddy, immediatly growing out of the earth, or little aboue the earth, tough; wherefore they vſe theſe to bind with, as with Oziars. They bud and floure when the Autumne Equinoctial is paſt, and a long time after. Thus much *Theophraſtus*.

The Germane Spurge Olive is not much vnlike to the Olive tree in leafe: the floure is ſweet of ſmell, the buds whereof, as we haue written, come forth after Autumne: the branches are wooddy and pliable, the root long, growing deep: all which ſhew that it hath great likeneſſe and affinitie with *Cneoron*, if it be not the very ſame.

¶ *The Temperature.*

This plant is likewiſe in all parts extreme hot: the fruit, leaues, and rinde are very ſharpe and biting; they bite the tongue, and ſet the throat on fire.

¶ *The Vertues.*

The leaues of Mezereon do purge downeward, flegme, choler, and wateriſh humours, with great   A violence.

Alſo if a drunkard doe eat one graine or berry of this plant, hee cannot be allured to drinke any   B drinke at that time, ſuch will be the heate in his mouth and choking in the throat.

This plant is very dangerous to be taken into the body, and in nature like to the ſea Tithymale,   C leauing (if it be chewed) ſuch an heate and burning in the throat, that it is hard to be quenched.

The ſhops of Germany and the Low-countries do when need requires vſe the leaues hereof in   D ſtead of Spurge Olive; which may be done without error, for this Germane Spurge Olive is like in vertue and operation to the other: therefore it may be vſed in ſtead therof, and prepared after the ſelfe ſame manner.

CHAP. 64.   *Of Spurge Flax.*

Thymælea.
Spurge Flax, or mountaine Widow-waile.

¶ *The Deſcription.*

SPurge Flax bringeth forth many ſlender branched ſprigs aboue a cubit high, coue-red-round with long and narrow leaues like thoſe of Flax, narrower and leſſer than the leaues of Spurge Olive. The flours are white, ſmal, ſtanding on the vpper parts of the ſprigs: the fruit is round, greene at the firſt, but red when it is ripe, like almoſt to the round berries of the Hawthorne; in which is a white kernell couered with a blacke skin, very hot, and burning the mouth like Mezereon. The root is hard and wooddy.

¶ *The Place.*

It growes in rough mountaines and vntoiled places in hot regions. It groweth in my garden.

¶ *The Time.*

It is green at any time of the yeare, but the fruit is perfected in Autumne.

¶ *The Names.*

The Grecians cal it κνέωρον the Syrians, as *Dioſcorides* witneſſeth, *Apolinon*: Diuers alſo *Chamælea*, but vnproperly. *Dioſcorides* ſaith the leafe is properly called *Cneoron*, and the fruit *Coccos Cnidios*: notwithſtanding thoſe which *Theophraſtus* calls *Cneora* ſeeme to differ from *Thymælea*, or Spurge flax, vnleſſe *Nigrum Cneo-ron*

ron be *Thymælea* : for *Theophraſtus* ſaith that there be two kindes of *Cneoron*, the one white, the other blacke. This may be called in Engliſh, Spurge flax, or mountain Widow-waile. The ſeed of *Thymalea* is called in ſhops *Granum Cnidium*.

¶ *The Temperature.*

Spurge Flax is naturally both in leaues and fruit extreme hot, biting, and of a burning qualitie.

¶ *The Vertues.*

A   The graines or berries, as *Dioſcorides* ſaith, purge by ſiege choler, flegme, and water, if 20 graines of the inner part be drunke ; but it burneth the mouth and throat : wherefore it is to be giuen with fine floure of Barly meale, or in Raiſins, or couered with clarified hony, that it may be ſwallowed.

B   The ſame being ſtamped with Niter and vineger ſerueth to anoint thoſe with which can hardly ſweat.

C   The leaues muſt be gathered about harueſt, and being dried in the ſhade, they are to be layd vp and reſerued.

D   They that would giue them muſt beat them and take forth the ſtrings: the quantity of two ounces and two drammes put into Wine tempered with water, purgeth and draweth forth watery humors : but they purge more gently if they be boiled with Lentils, and mixed with pot-herbs chopped.

E   The ſame leaues beaten to pouder and made vp into trochiskes or flat cakes with the iuyce of ſoure grapes, are reſerued for vſe.

F   The herb is an enemie to the ſtomacke, which alſo deſtroyeth the birth if it be applied.

†   Our Author formerly following *Tabern.* gaue two figures and deſcriptions in this Chap. but being both of one thing, I omitted the worſer figure and deſcription.

<h2 style="text-align:center">CHAP. 65.   Of Spurge Laurel.</h2>

Laurcola florens.
Laurel or Spurge Laurell flouring.

           Laureola cum fructu.
Laurel with his fruit,

¶ *The Defcription.*

SPurge Laurell is a fhrub of a cubit high,oftentimes alfo of two, and fpreadeth with many little boughes which are tough and lithy,and couered with a thicke rinde : the leaues be long,broad, groffe,fmooth,blackifh green,fhining like the leaues of Laurell,but leffer,thicker,and without fmell,very many at the top cluftering together. The floures bee long,hollow,of a whitifh greene, hanging beneath and among the leaues : the berries when they be ripe are blacke, with a hard kernell within,which is a little longer than the feed of Hemp : the pulpe or inner fubftance is white : the root wooddy,tough,long,and diuerfly parted,growing deep : the leaues,fruit,and bark,as well of the root as of the little boughes,do with their fharpneffe and burning qualitie bite and inflame the tongue and throat.

¶ *The Place.*

It is found on mountaines in vntilled rough fhadowy and wooddy places,as by the lake of Lozanna in Geneva, and in many places neere the riuer of Rhene and of the Maze. ‡ It growes abundantly alfo in the woods in moft parts of England. ‡

¶ *The Time.*

The floures bud very foon,a little after the Autumne Æquinoctial : they are full blown in winter or in the firft Spring : the fruit is ripe in May and Iune ; the plant is alwaies greene, and endureth the cold ftormes of winter.

¶ *The Names.*

It is called in Greeke δαφνοειδὲς, of the likeneffe it hath with the leaues of the Laurel or Bay tree: in Latine likewife *Daphnoides:* the later Latines for the fame caufe name it *Laureola,*as though they fhould fay *Minor Laurus* or little Laurel. It is called χαμεδάφνη, and πήλιον : notwithftanding there is another *Chamædaphne* and another *Peplion.* This fhrub is commonly called in Englifh, Spurge Laurell ; of diuers,Lawrell or Lowry.

Some fay that the Italians name the berries thereof *Piper montanum,*or mountaine Pepper,as alfo the berries of Dutch Mezereon : others affirme them to be called in high-Dutch alfo Zeilant.

It may be *Theophraftus* his *Cneoron* ; for it is very like the Myrtle in leafe , it is alfo a branched plant,tough and pliable,hauing a deep root without fmell,with a blacke fruit.

¶ *The Temperature.*

It is like in temperature and facultie to the German Spurge Oliue, throughout the whole fubftance biting and extreme hot.

¶ *The Vertues.*

The dry or green leaues of Spurge Laurel,faith *Diofcorides,*purgeth by fiege flegmatick humors, it prouoketh vomit and bringeth downe the menfes,and being chewed it draweth water out of the head.   A

It likewife caufeth neefing. Moreouer,fifteen grains of the feed thereof drunk are a purgation. B

C H A P. 66.   *Of Rofe-Bay or Oleander.*

¶ *The Defcription.*

1   ROfe bay is a fmal fhrub of a gallant fhew like the Bay tree,bearing leaues thicker,greater,longer,and rougher than the leaues of the Almond tree : the floures be of a fair red colour,diuided into fiue leaues,not much vnlike a little Rofe : the cod or fruit is long like *Afclepias* or *Vincetoxicum,*and full of fuch white down, among which the feed lieth hidden:the root is long,fmooth,and wooddy.

2   The fecond kinde of Rofe Bay is like the firft,and differs in that, that this plant hath white floures ; but in other refpects it is very like.

¶ *Thi*

1 *Nerium, ſiue Oleander.*
The Roſe Bay.

2 *Nerium flore albo.*
The Roſe Bay with white flours.

¶ *The Place.*

Theſe grow in Italy and other hot regions by riuers and the ſea ſide: I haue them growing in my garden.

¶ *The Time.*

In my garden they floure in Iuly and Auguſt: the cods be ripe afterwards.

¶ *The Names.*

This plant is named in Greeke Νἤιον, by *Nicander*, Νἤιε: in Latine likewiſe *Nerion* and alſo *Rhodo-dendron*,and *Rhododaphne*,that is to ſay, *Roſea arbor*,or *Roſea Laurus* : in ſhops,*Oleander* : in Italian, *O-leandro* : in Spaniſh, *Adelfa, Eloendro*,and *Alendro* : in French,*Roſagine* : in Engliſh, Roſe tree, Roſe-Bay,Roſe Bay tree,and Oleander.

¶ *The Temperature and Vertues.*

A   This tree being outwardly applied,as *Galen* ſaith,hath a digeſting facultie:but if it be inwardly taken it is deadly and poiſonſome,not only to men,but alſo to moſt kindes of beaſts.

B   The flours and leaues kill dogs,aſſes,mules,and very many other foure footed beaſts:but if men drinke them in wine they are a remedy againſt the bitings of Serpents,and the rather if Rue be ad-ded.

C   The weaker ſort of Cattell,as ſheep and goats,if they drinke the water wherein the leaues haue been ſteeped,are ſure to die.

CHAP. 67. *Of dwarfe Roſe Bay.*

¶ *The Deſcription.*

DWarfe *Nerium* or Roſe Bay hath leaues which for the moſt part are alwaies green,rough,and ſmall,of a pale yellow colour like Box,far leſſer than Oleander : the whole plant is of a ſhrub-by ſtature,leaning this way and that way,as not able to ſtand vpright without help; his bran-ches are couered and ſet full of ſmall floures,of a ſhining skarlet or crimſon colour,growing vpon the

Chamærhodendros Alpigena.
Dwarfe Rofe Bay.

Laurus.
The Bay tree.

the hills as you goe from Trent to Verona, which in Iune and Iuly are as it were couered with a fcarlet colored carpet, of an odoriferous fauor and delectable afpect; which being fallen, there commeth feed and faire berries like Afparagus.

¶ *The Place.*
The place and time are expreffed in the defcription.

¶ *The Names.*
This may be called in Englifh, dwarf rofe Bay of the Alps. I find nothing extant of the vertues, but am conftrained to leaue the reft vnto your owne difcretion.

† The other plant of our Author, formerly defcribed in this Chapter in the fecond place, by the name of *Chamærododendros montana*, I haue here omitted, becaufe he fet it forth before by the name of *Ciftus Ledum Silefiacum*, giuing two figures and one defcription, in the 11 and 12 places of the 8 Chap. of this 3 Booke.

CHAP. 68.
Of the Bay or Lawrel Tree.

¶ *The Defcription.*

1   THe Bay or Laurell tree commeth oftentimes to the height of a tree of a mean bigneffe : it is full of boughes, couered with a green barke : the leaues thereof are long, broad, hard, of colour green, fweetly fmelling, and in tafte fomewhat bitter. The flours alongft the boughes and leaues are of a green colour; the beries are more long than round, and be couered with a blacke rinde or pill : the kernell within is clouen into two parts like that of the Peach and almond, and other fuch, of a brown yellowifh color, fweet of fmell, in tafte fomewhat bitter, with a little fharp or biting qualitie.

2   There is alfo a certaine other kinde hereof more like a fhrub, fending forth out of the roots many off-fprings ; which notwithftanding growes not fo high as the former, and the barkes of the boughes be fomewhat red ; the leaues be alfo tenderer, & not fo hard; in other things not vnlike.

Thefe two Bay trees *Diofcorides* was not ignorant of; for he faith that the one is narrow leafed, and the other broader or rather harder leafed, which is more like.

¶ *The Place.*
The Laurell or Bay tree groweth naturally in

ly in Spaine and ſuch hot regions : we plant and ſet it in gardens,defending it from cold at the beginning of March eſpecially.

    I haue not ſeen any one tree thereof growing in Denmarke, Suevia, Poland, Livonia, or Ruſſia, or
in any of thoſe cold countries where I haue travelled.

¶ *The Place.*

    The Bay tree groweth green winter and ſummer : it floureth in the ſpring, and the black fruit is
ripe in October.

¶ *The Names.*

    This tree is called in Greeke *δάφνη* : in Latine, *Laurus* : in Italian, *Lauro* : in high-Dutch , **Looer-**
beerbaum: in low-Dutch, **Laurus boom:** in French, *Laurier:* in Spaniſh, *Laurel, Lorel*, and *Laurie-*
ro : in Engliſh, Laurel or Bay tree.

    The fruit is named in Greeke *δαφνίδες* : in Latine, *Lauri baccæ :* in high-Dutch , **Looerbeeren:** in
low-Dutch, **Bakeleer:** in Spaniſh, *Vayas :* in Engliſh, Bay berries.

    The Poets feigne that it tooke his name of *Daphne, Lado* his daughter, with whom *Apollo* fell in
loue.

¶ *The Temperature and Vertues.*

A   The berries and leaues of the Bay tree, ſaith *Galen*, are hot and very dry, and yet the berries more
than the leaues.

B   The barke is not biting and hot, but more bitter, and it hath alſo a certain aſtrictiue or binding
qualitie.

C   The berries with Hony or Cute are good in a licking medicine, ſaith *Dioſcorides* ; againſt the
Ptyſicke or conſumption of the lungs, difficultie of breathing, and all kindes of fluxes or rheumes
about the cheſt.

D   Bay berries taken in wine are good againſt the bitings and ſtingings of any venomous beaſt, and
againſt all venom and poiſon : they clenſe away the morphew : the juice preſſed out hereof is a remedie for pain in the eares, and deafneſſe, if it be dropped in with old wine and oile of Roſes. This
is alſo mixed with ointments that are good againſt weariſomneſſe, and that heate and diſcuſſe or
waſte away humors.

E   Bay berries are put into Mithridate, Treacle, and ſuch like medicines that are made to refreſh
ſuch people as are growne ſluggiſh and dull by means of taking opiate medicines, or ſuch as haue
any venomous or poiſonous qualitie in them.

F   They are good alſo againſt cramps and drawing together of ſinues.

G   We in our time do not vſe the beries for the infirmities of the lungs or cheſt, but miniſter them
againſt the diſeaſes of the ſtomacke, liuer, ſpleen, and bladder : they warme a cold ſtomacke, cauſe
concoction of raw humours, ſtirre vp a decayed appetite, take away the loathing of meat, open the
ſtopping of the liuer and ſpleene, prouoke vrin, bring downe the menſes, and driue forth the ſecondine.

H   The oile preſſed out of theſe, or drawne forth by decoction, doth in ſhort time take away ſcabs
and ſuch like filth of the skin.

I   It cureth them that are beaten blacke and blew, and that be bruiſed by ſquats and falls : it remoueth blacke and blew ſpots and congealed bloud, and digeſteth and waſteth away the humours gathered about the grieued part.

K   *Dioſcorides* ſaith that the leaues are good for the diſeaſes of the mother and bladder, if a bath be
made thereof to bathe and ſit in : that the greene leaues doe gently binde, that being applied they
are good againſt the ſtingings of waſps and Bees : that with barley meale parched and bread they
aſſwage all kinde of inflammations : and that being taken in drinke they mitigate the paine of the
ſtomacke, but procure vomit.

L   The berries of the Bay tree ſtamped with a little Scammonie and ſaffron, and labored in a mortar with vineger and oile of Roſes to the forme of a liniment, and applied to the temples and fore-
part of the head, do greatly ceaſe the pain of the megrim.

M   It is reported that common drunkards were accuſtomed to eat in the morning faſting two leaues
thereof againſt drunkenneſſe.

N   The later Phyſitions do often vſe to boile the leaues of Laurell with diuers meats, eſpecially fiſhes, and by ſo doing there hapneth no deſire of vomiting, but the meat ſeaſoned herewith becommeth more ſauorie and better for the ſtomacke.

O   The barke of the root of the Bay tree, as *Galen* writeth, drunken in wine prouoketh vrine, breakes
the ſtone and driueth forth grauell : it openeth the ſtoppings of the liuer, the ſpleene, and all other
ſtoppings of the inward parts: which thing alſo *Dioſcorides* affirmeth, adding, That it kils the child
in the mothers wombe.

                                                              It

It helpeth the dropfie and the jaundife, and procureth vnto women their defired fickeneffe.

† Our Author here alfo gaue the two figures of *Tabernæmontanus* ; the firft by the name of *Laurus mas*, or the male Bay tree, and the other by the name of *Laurus fœmina*, the female Bay: the difference in the figures was little or none, wherefore I haue made one ferue.

CHAP. 69.　*Of the Wilde Bay tree.*

¶ *The Defcription.*

‡ 1 **L**Aurus Tinus, or the wilde Bay tree, groweth like a fhrub or hedge bufh, hauing many tough and pliant branches, fet full of leaues very like to the Bay leaues, but fmaller and more crumpled, of a deepe and fhining greene colour : among which come forth tufts of whitifh floures, turning at the edges into a light purple : after which follow fmall berries of a blew colour, containing a few graines or feeds like the ftones or feeds of grapes : the leaues and all the parts of the plant are altogether without fmell or fauour.

1 *Laurus Tinus.*
The wilde Bay tree.　　　　　2 *Laurus Tinus Lufitanica.*
The Portingale wilde Bay tree.

2 *Tinus Lufitanica* groweth very like to *Cornus Fœmina*, or the Dog-berry tree, but the branches be thicker, and more ftiffe, couered with a reddifh barke mixed with greene : the leaues are like the former, but larger, hauing many finewes or veines running through the fame like as in the leaues of Sage : the floures hereof grow in tufts like the precedent, but they are of a colour more declining to purple : the fmall branches are likewife of a purple colour : the leaues haue no fmell at all, either good or bad : the berries are fmaller than the former, of a blew colour declining to blackneffe.

¶ *The Place.*

The wilde Bay groweth plentifull in euery field of Italy, Spaine, and other regions, which differ according to the nature and fcituation of thofe countries : they grow in my garden and profper very well.

Ccccc　　　　　　¶ *The*

¶ *The Time.*

The wilde Laurell is euer greene, and may oftentimes be feene moft part of the Winter, and the beginning of the Spring, with the floures and ripe berries growing both at one feafon.

¶ *The Names.*

It is called in Latine *Tinus*, and *Laurus fylueftris* : in Greeke, δάφνη ἀγρία : Cato nameth it *Laurus fyluatica* : in Italian, *Lauro fyluatico* : in Spanifh, *Vua de perro*, otherwife *Follado* ; and of diuers, *Durillo* : in Englifh, wilde Bay.

¶ *The Temperature and Vertues.*

Pliny nor any other of the Antients haue touched the faculties of this wilde Bay, neither haue we any vnderftanding thereof by the later Writers, or by our owne experience.

Chap. 70. *Of the Box Tree.*

Buxus.
The Box tree.

¶ *The Defcription.*

THe great Box is a faire tree, bearing a great body or trunke : the wood or timber is yellow and very hard , and fit for fundry workes, hauing many boughes and hard branches, befet with fundry fmall hard greene leaues, both Winter and Summer like the Bay tree : the floures are very little , growing among the leaues, of a greene colour : which being vaded there fucceed fmall blacke fhining berries, of bigneffe of the feeds of Corianders, which are inclofed in round greenifh huskes, hauing three feet or legs like a jbraffe or boyling pot : the root is likewife yellow , and harder than the timber, but of geeater beauty, and more fit for dagger hafts, boxes, and fuch like vfes, whereto the trunke or body ferueth, than to make medicines ; though foolifh emperickes and Women leaches, doe minifter it againft the Apoplexie, and fuch difeafes: Turners and Cutlers, if I miftake not the matter, doe call this wood Dudgeon, wherewith they make Dudgeon hafted daggers.

There is alfo a certaine other kinde hereof, growing low, and not aboue halfe a yard high, but it fpreadeth all abroad : the branches hereof are many and very flender : the leaues bee round, and of a light greene.

¶ *The Place.*

Buxus, or the Box tree groweth vpon fundry wafte and barren hils in England, and in diuers gardens.

¶ *The Time.*

The Box tree groweth greene Winter and Summer : it floureth in February and March, and the feed is ripe in September.

¶ *The Names.*

The Grecians call it πύξος : in Latine, *Buxus* : in high Dutch, 𝔅uchßbaum : in low Dutch, 𝔅utboom : in Italian, *Boffo* : in Englifh, Box tree.

The leffer may be called χαμαίπυξος and in Latine, *Humi Buxus*, or *Humilis Buxus* : in Englifh, dwarfe Box, or ground Box, and it is commonly called Dutch Box.

¶ *The Temperature and Vertues.*

A   The leaues of the Box tree aree hot, dry, and aftringent, of an euill and lothfome fmell, not vfed in medicine, but onely as I faid before in the defcription.

CHAP.

CHAP. 71. *Of the Myrtle tree.*

¶ *The Deſcription.*

1   THe firſt and greateſt *Myrtus* is a ſmall tree, growing to the height of a man, hauing many faire and pliant branches, couered with a browne barke, and ſet full of leaues much like vnto the Laurell or Bay leafe, but thinner and ſmaller, ſomewhat reſembling the leaues of Peruinele, which being bruiſed do yeeld forth a moſt fragrant ſmell, not much inferiour vnto the ſmell of Cloues, as all the reſt of the kinds do: among theſe leaues come forth ſmall white floures, in ſhape like the flours of the Cherry tree, but much ſmaller, and of a pleaſant ſauour, which do turne into ſmall berries, greene at the firſt, and afterwards blacke.

1 *Myrtus Laurea maxima.*
The Myrtle tree.

‡ 2 *Myrtus Batica latifolia.*
Great Spaniſh Myrtle.

2   There is alſo another kinde of *Myrtus* called *Myrtus Batica latifolia*, according to *Cluſius, Myrtus Laurea*, that hath leaues alſo like Bay leaues, growing by couples vpon his pleaſant green branches, in a double row on both ſides of the ſtalkes, of a light greene colour, and ſomewhat thicker than the former, in ſent and ſmell ſweet: the floures and fruit are not much differing from the firſt kinde.

3   There is likewiſe another kinde of *Myrtus* called *Exotica*, that is ſtrange and not common: it groweth vpright vnto the height of a man like vnto the laſt before mentioned, but that it is repleniſhed with greater plenty of leaues, which do fold in themſelues hollow and almoſt double, broader pointed, and keeping no order in their growing, but one thruſting within another, and as it were croſſing one another confuſedly; in all other points agreeing with the precedent.

4   There is another ſort like vnto the former in floures and branches, but the leaues are ſmooth, flat and plaine, and not crumpled or folded at all, they are alſo much ſmaller than any of the former. The fruit is in ſhape like the other, but that it is of a white colour, wheras the fruit of the other is blacke.

5   There is alſo another kinde of Myrtle, called *Myrtus minor*, or noble Myrtle, as being the

‡ 3 *Myrtus exotica.*
Strange Myrtle.

‡ 4 *Myrtus fructu albo.*
Myrtle with white berries.

‡ 5 *Myrtus minor.*
The little Myrtle.

‡ 6 *Myrtus Batica ſylueſtris.*
Wilde Spaniſh Myrtle.

chiefe of all the reft) although moft common and beft knowne) and it groweth like a little fhrub or hedge bufh, very like vnto the former, but much fmaller: the leaues are fmall and narrow, very much in fhape refembling the leaues of Mafticke Time called *Marum*, but of a frefher greene colour: the floures be white, nothing differing from the former fauing in greatneffe, and that fometimes they are more double.

‡ 6   This growes not very high, neither is it fo fhrubby as the former: the branches are fmall and brittle: the leaues are of a middle bigneffe, fharpe pointed, ftanding by couples in two rowes, feldome in foure as the former, they are blackifh alfo and well fmelling, the floure is like that of the reft: the fruit is round, growing vpon long ftalkes out of the bofomes of the leaues, firft greene, then whitifh, laftly blacke, of a winy and pleafant tafte with fome aftriction. This growes wilde in diuers places of Portugall, where *Clufius* found it flouting in October: he calls it *Myrtus Batica fylueftris.* ‡

¶ *The Place.*

Thefe kindes of Myrtles grow naturally vpon the wooddy hills and fertill fields of Italy and Spaine. ‡ The two laft are nourifhed in the garden of Miftriffe *Tuggie* in Weftminfter, and in fome other gardens. ‡

¶ *The Time.*

Where they joy to grow of themfelues they floure when the Rofes doe: the fruit is ripe in Autumne: in England they neuer beare any fruit.

¶ *The Names.*

It is called in Greeke μύρτος: in Latine, *Myrtus*: in the Arabicke tongue, *Alas*: in Italian, *Myrto*: in Spanifh, *Arrayhan*: in the Portugale language, *Murta*, and *Murtella*: other Nations doe almoft keepe the Latine name, as in Englifh it is called Myrtle, or Myrtle tree.

Among the Myrtles that which hath the fine little leafe is furnamed of *Pliny*, *Tarentina*; and that which is fo thicke and full of leaues is *Exotica*, ftrange or forreine. *Nigra Myrtus* is that which hath the blacke berries: *Candida*, which hath the white berries, and the leaues of this alfo are of a lighter greene: *Satiua*, or the tame planted one is cherifhed in gardens and orchards: *Sylueftris*, or the wilde Myrtle is that which groweth of it felfe; the berries of this are oftentimes leffer, and of the other, greater. *Pliny* doth alfo fet downe other kindes; as *Patritia*, *Plebeia*, and *Conjugalis*: but what manner of ones they are he doth not declare: he alfo placeth among the Myrtles, *Oxymerfine*, or Kneeholm, which notwithftanding is none of Myrtles, but a thorny fhrub,

Pliny in his 14. booke, 16. chap. faith, that the wine which is made of the wild Myrtle tree is called *Myrtidanum*, if the copy be true. For *Diofcorides* and likewife *Sotion* in his Geoponikes report, that wine is made of Myrtle berries when they be thorow ripe, but this is called *Vinum Murtenm*, or *Myrtites*, Myrtle wine.

Moreouer, there is alfo a wine made of the berries and leaues of Myrtle ftamped and fteeped in Muft, or wine new preffed from the grape, which is called, as *Diofcorides* faith, *Myrfinite vinum*, or wine of Myrtles.

The Myrtle tree was in times paft confecrated to *Venus*. *Pliny* in his 15. booke, 29. chapter, faith thus, There was an old Alter belonging to *Venus*, which they now call *Murtia*.

¶ *The Temperature and Vertues.*

The Myrtle confifteth of contrary fubftances, a cold earthineffe bearing the preheminence; it hath alfo a certaine fubtill heate, therefore, as *Galen* faith, it drieth notably.   A

The leaues, fruit, buds, and juyce do binde, both outwardly applied and inwardly taken: they ftay the fpitting of bloud, and all other iffues thereof: they ftop both the whites and reds in women, if they fit in a bath made therewith: after which manner and by fomenting alfo they ftay the fuperfluous courfe of the hemorrhoides.   B

They are a remedy for laskes, and for the bloudy flix, they quench the fiery heate of the eies, if they be laid on with parched Barly meale.   C

They be alfo with good fucceffe outwardly applied to all inflammations newly beginning, and alfo to new paine vpon fome fall, ftroke or ftraine.   D

They are wholefome for a moift and watery ftomacke: the fruit and leaues dried prouoke vrine: for the greene leaues containe in them a certaine fuperfluous and hurtfull moifture.   E

It is good with the decoction hereof made with wine, to bathe lims that are out of joint, and burftings that are hard to be cured, and vlcers alfo of the outward parts: it helpeth fpreading tetters, fcoureth away the dandriffe and fores of the head, maketh the haires blacke, and keepeth them   F from ſhedding; withſtanding drunkenneſſe, if it be taken faſting, and preuaileth againſt poiſon, and the bitings of any venomous beaſt.

G   There is drawne out of the green berries thereof a juyce, which is dried and reſerued for the foreſaid vſes.

H   There is likewiſe preſſed out of the leaues a juyce, by adding vnto them either old wine or raine water, which muſt be vſed when it is new made, for being once dry it putrifieth, and as *Dioſcorides* ſaith, loſeth his vertues.

Chap. 72.   Of ſweet Willow or Gaule.

Myrtus Brabantica, ſiue Elæagnus Cordi.
Gaule, ſweet willow, or Dutch Myrtle tree.

¶ *The Deſcription.*

GAule is a low and little ſhrub or wooddy plant, hauing many browne & hard branches: whereupon doe grow leaues ſomewhat long, hard, thicke, and oileous, of an hot ſauour or ſmell ſomewhat like *Myrtus*: among the branches come forth other little ones, whereupon do grow many ſpoky eares or tufts, full of ſmall floures, and after them ſucceed great ſtore of ſquare ſeeds cluſtering together, of a ſtrong and bitter taſte. The root is hard, and of a wooddy ſubſtance.

¶ *The Place.*

This Gaule groweth plentifully in ſundry places of England, as in the Ile of Ely, & in the Fenny countries thereabouts, whereof there is ſuch ſtore in that countrey, that they make fagots of it and ſheaues, which they call Gaule ſheaues, to burn and heat their ouens. It groweth alſo by Colebrooke, and in ſundry other places.

¶ *The Time.*

The Gaule floureth in May and Iune, and the ſeed is ripe in Aguſt.

¶ *The Names.*

This tree is called of diuers in Latine, *Myrtus Brabantica*, and *Pſeudomyrſine*; and *Cordus* calleth it *Elæagnus, Chamæleagnus*, and *Myrtus Brabantica*. *Elæagnus* is deſcribed by *Theophraſtus* to be a ſhrubby plant like vnto the Chaſte tree, with a ſoft and downy leafe, and with the floure of the Poplar tree; and that which we haue deſcribed is no ſuch plant. It hath no name among the old writers for ought we know, vnleſſe it be *Rhus ſylueſtris Plinij*, or *Pliny* his wilde Sumach, of which he hath written in his 24. booke, 11. chap. [There is, ſaith he, a wilde herbe with ſhort ſtalkes, which is an enemy to poiſon, and a killer of mothes.] it is called in low Dutch, **Gagel**: in Engliſh, Gaule.

¶ *The Temperature.*

Gaule or the wilde Myrtle, eſpecially the ſeed, is hot and dry in the third degree: the leaues be hot and dry, but not ſo much.

¶ *The Vertues.*

A   The fruit is troubleſome to the brain; being put into beere or ale whileſt it is in boyling (which many vſe to do) it maketh the ſame heady, fit to make a man quickely drunke.

B   The whole ſhrubbe, fruit and all, being laied among clothes, keepeth them from mothes and wormes.

CHAP. 73. *Of Worts or Wortle berries.*

¶ *The Kindes.*

VAccinia, or Worts, of which we treat in this place, differ from Violets, neither are they eſtec-
med for their floures but berries : of theſe Worts there be diuers ſorts found out by the later
Writers.

<table>
<tr>
<td align="center">1 *Vaccinia nigra.*
Blacke Worts or Wortle berries.</td>
<td align="center">2 *Vaccinia rubra.*
Red Worts or Wortle berries.</td>
</tr>
</table>

¶ *The Deſcription.*

1   **V**Accinia nigra, the blacke Wortle or Hurtle, is a baſe and low ſhrub or wooddy plant,
bringing forth many branches of a cubit high, ſet full of ſmall leaues of a darke greene
colour, not much vnlike the leaues of Box or the Myrtle tree : amongſt which come
forth little hollow floures turning into ſmall berries, greene at the firſt, afterward red, and at the
laſt of a blacke colour, and full of a pleaſant and ſweet juyce : in which doe lie diuers little thinne
whitiſh ſeeds : theſe berries do colour the mouth and lips of thoſe that eat them, with a blacke colour : the root is wooddy, ſlender, and now and then creeping.

2   *Vaccinia rubra,* or red Wortle, is like the former in the manner of growing, but that the leaues
are greater and harder, almoſt like the leaues of the Box tree, abiding greene oll the Winter long :
among which come forth ſmall carnation floures, long and round, growing in cluſters at the top of
the branches : after which ſucceed ſmall berries, in ſhew and bigneſſe like the former, but that they
are of an excellent red colour and full of juyce, of ſo orient and beautifull a purple to limme withall, that Indian *Lacca* is not to be compared thereunto, eſpecially when this juyce is prepared and
dreſſed with Allom according to art, as my ſelfe haue proued by experience : the taſte is rough and
aſtringent : the root is of a wooddy ſubſtance.

3   *Vaccinia alba,* or the white Wortle, is like vnto the former, both in ſtalkes and leaues, but the
berries are of a white colour, wherein conſiſteth the difference.

‡   The figure which our Author here giues in the third place hath need of a better deſcription,

for

3 *Vaccinia alba.*
The white Worts or Wortle berries.

4 *Vaccinia Pannonica, siue Vitis Idæa.*
Hungarie Wortle berries.

5 *Vaccinia Vrsi, siue Vua Vrsi apud Clusium.*
Beare Wortle berries.

† 6 *Vitis Idæa folijs subrotundis major.*
Great round leaued Wortle berries.

for the difference is not onely in the colour of the berries. This differs from the former in forme and bigneſſe; for it ſends forth many ſtalks from the root, and theſe three, foure, or fiue cubits high, thicke, and diuided into ſundry branches, couered for the moſt part with a blackiſh barke: at the beginning of the Spring from the buds at the ſides of the branches it ſends forth leaues all horie and hairy vnderneath, and greene aboue: from the midſt of theſe vpon little foot-ſtalkes ſtand cluſtering together many little floures, conſiſting of fiue white leaues apiece without ſmell; and then the leaues by little and little vnfold themſelues and caſt off their downineſſe, and become ſnipt about the edges. The fruit that ſucceeds the floures is round, blacke, ſomewhat like, but bigger than a Haw, full of juyce of a very ſweet taſte; wherein lies ten or more longiſh ſmooth blackiſh ſeeds. It growes vpon the Auſtrian and Stirian Alps, where the fruit is ripe in Auguſt. *Cluſius* calls it *Vitis Idæa* 3. *Pena* and *Lobel, Amelancher : Geſner* by diuers names, as *Myrtomalus, Petromelis, Pyrus ceruinus, &c.* ‡

4   *Carolus Cluſius* in his Pannonicke Obſeruations hath ſet downe another of the Wortle berries, vnder the name of *Vitis Idæa,* which differeth from the other Wortle berries, not onely in ſtature, but in leaues and fruit alſo. ‡ The leaues are long, narrow, ſharpe pointed, full of veines, a little hairy, and lightly ſnipt about the edges, greener aboue than below: the fruit growes from the tops of the branches of the former yeare, hanging vpon long foot-ſtalkes, and being as big as little Cherries, firſt greene, then red, and laſtly blacke, full of juyce, and that of no vnpleaſant taſte, containing no kernels, but ſlat white ſeeds commonly fiue in number : the ſtalks are weake, and commonly lie vpon the ground : *Cluſius* found it vpon the Auſtrian mountain Snealben, with the fruit partly ripe, and partly vnripe in Auguſt. It is his *Vitis Idæa* 1. ‡

5   The ſame Author alſo ſetteth forth another of the Wortle berries, vnder the title of *Vua Vrſi,* which is likewiſe a ſhrubby plant, hauing many feeble branches, whereon grow long leaues blunt at the points, and of an ouerworne greene colour: among which, at the tops of the ſtalks come forth cluſters of bottle-like floures of an herby colour : the fruit followeth, growing likewiſe in cluſters, green at the firſt, and blacke when they be ripe: the root is of a wooddy ſubſtance. ‡ This is alwaies greene, and the floures are of a whitiſh purple colour. ‡

‡ 6   This differs from the ſecond, in that the leaues are thinner, more full of veines, and whiter vnderneath: the floure is like the common kind, whitiſh, purple, hollow, and diuided into fiue parts : the fruit alſo is blacke, and like that of the firſt deſcribed. This growes on diuers mountainous places of Germany, where *Cluſius* obſerued it, who made it his *Vitis Idæa* 2. ‡

¶ *The Place.*

Theſe plants proſper beſt in a leane barren ſoile, and in vntoiled wooddy places : they are now and then found on high hills ſubject to the winde, and vpon mountaines : they grow plentifully in both the Germanies, Bohemia, and in diuers places of France and England ; namely in Middleſex on Hampſted heath, and in the woods thereto adjoyning, and alſo vpon the hills in Cheſhire called Broxen hills, neere Beeſton caſtle, ſeuen miles from the Nantwich ; and in the wood by Highgate called Finchley wood, and in diuers other places.

☛ The red Wortle berry groweth in Weſtmerland at a place called Crosby Rauenſwaith, where alſo doth grow the Wortle with the white berry, and in Lancaſhire alſo vpon Pendle hills.

‡ I haue ſeene none of theſe but onely the firſt deſcribed, growing vpon Hampſted heath. The white formerly mentioned in the third deſcription, and here againe in the place, ſeems only a varietie of the ſecond hauing white berries, as far as I can gather by our Author ; for it is moſt certaine, that it is not that which he figured, and I haue deſcribed in the third place. ‡

¶ *The Time.*

The Wortle berries do floure in May, and their fruit is ripe in Iune.

¶ *The Names.*

Wortle berries is called in high Dutch, **Heydelbeeren:** in low Dutch, **Craakebeſſen,** becauſe they make a certaine cracke whileſt they be broken betweene the teeth : of diuers, **Hauerbeſſen:** the French men, *Airelle,* or *Aurelle,* as *Iohannes de Choul* writeth : and we in England, Wort<, Whortle berries, Blacke berries, Bill berries, and Bull berries, and in ſome places Win berries.

Moſt of the ſhops of Germany do call them *Myrtilli,* but properly *Myrtilli* are the fruit of the Myrtle tree, as the Apothecaries name them at this day. This plant hath no name for ought wee can learne, either among the Greekes or antient Latines ; for whereas moſt doe take it to be *Vitis Idæa,* or the Corinth tree, which *Pliny* ſurnameth *Alexandrina,* it is vntrue ; for *Vitis Idæa* is not onely like to the common Vine, but is alſo a kinde of Vine : and *Theophraſtus* who hath made mention hereof doth call it, without an Epethete, *ραμνος,* ſimply, as a little after we will declare ; which without doubt he would not haue done if hee had found it to differ from the common Vine : For what things ſoeuer receiue a name of ſome plant, the ſame are expreſſed with ſome Epethet added to be known to differ from others; as *Laurus Alexandrina, Vitis alba, Vitis nigra, Vitis ſylueſtris,* and ſuch like.

Moreouer, thoſe things which haue borrowed a name from ſome plant are like thereunto, if not wholly,

wholly, yet either in leafe or fruit, or in fome other thing, *Vitis alba & nigra*, that is, the white and blacke Bryonies, haue leaues and clafping tendrels as hath the common Vine, and clyme alfo after the fame manner : *Vitis fylueftris*, or the wilde Vine, hath fuch like ftalkes as the Vine hath, and bringeth forth fruit like to the little Grapes. *Laurus Alexandrina*, and *Chamædaphne*, and alfo *Daphnoides*, are like in leaues to the Laurell tree : *Sycomorus* is like in fruit to the Fig tree, and in leaues to the Mulberry tree : *Chamædrys* hath the leafe of an Oke; *Peucedanus* of the Pine tree : fo of others which haue taken their names from fome other : but this low fhrub is not like the Vine either in any part, or in any other thing.

This *Vitis Idæa* groweth not on the vppermoft and fnowie parts of Mount Ida (as fome would haue it) but about Ida, euen the hill Ida, not of Candy, but of Troas in the leffer Afia, which *Ptolomie* in his fifth booke of Geographie, chap.3. doth call *Alexandri Troas*, or *Alexander* his Troy : whereupon it is alfo aduifedly named of *Pliny, lib.14.cap.3. Vitis Alexandrina*, no otherwife than *Alexandrina Laurus* is faid of *Theophraftus* to grow there : *Laurus*, fyrnamed *Alexandrina*, and *Ficus quædam*, or a certaine Fig tree, and Άμπελος, that is to fay, the Vine, are reported, faith he, to grow properly about Ida. Like vnto this Vine are thofe which *Philoftratus* in the life of *Apollonius* reporteth to grow in Mæonia, and Lydia, fcituated not far from Troy, comparing them to thofe Vines which grow in India beyond Caucafus : The Vines there, faith he, be very fmall, like as be thofe that doe grow in Mæonia and Lydia, yet is the wine which is preffed out of them of a maruellous pleafant tafte.

This Vine which growes neere to mount Ida is reported to be like a fhrub, with little twigs and branches of the length of a cubit, about which are grapes growing aflope, blacke, of the bigneffe of a beane, fweet, hauing within a certaine winie fubftance, foft : the leafe of this is round, vncut and little.

This is defcribed by *Pliny, lib.14.cap.3.* almoft in the felfe fame words : it is called, faith he, *Alexandrina vitis*, and groweth neere vnto Phalacra : it is fhort with branches a cubit long, with a blacke grape of the bigneffe of the Latines Beane, with a foft pulpe and very little, with very fweet clufters growing aflope, and a little round leafe without cuts.

And with this defcription the little fhrub which the Apothecaries of Germany do call *Myrtillus* doth nothing at all agree, as it is very manifeft; for it is low, fcarce a cubit high, with a few fhort branches not growing to a cubit in length: it doth not bring forth clufters or bunches, nor yet fruit like vnto grapes, but berries like thofe of the Yew tree, not fweet, but fomewhat foure and aftringent ; in which alfo there are many little white flat feeds : the leafe is not round, but more long than round, not like to that of the Vine but of the Box tree. Moreouer, it is thought that this is not found in Italy, Greece, or in the leffer Afia, for that *Matthiolus* affirmeth the fame to grow no where but in Germany and Bohemia ; fo far is it from being called or accounted to be *Vitis Idæa* or *Alexandrina*.

The fruit of this may be thought not without caufe to be named *Vaccinia*, fith they are berries ; for they may be termed of *Baccæ*, berries, *Vaccinia*, as though they fhould be called *Baccinia*. Yet this letteth not that there may be alfo other *Vaccinia's* : for *Vaccinia* is πολύσημος dictio, or a word of diuers fignifications. *Virgil* in the firft booke of his Bucolicks, *Eclog.10.* affirmeth that the written Hyacinth is named of the Latines, *Vaccinium*, tranflating into Latine *Theocritus* his verfe which is taken out of his tenth Eidyl.

Καί τυ ἐστιμίλα ἐπ ἡ ἀσχηπια όλαιδων :

Virgil :

Et nigræ Violæ, funt & Vaccinia nigra.

Vitruvius, lib.7. of his Architecture doth alfo diftinguifh *Vaccinium* from the Violet, and fheweth, that of it is made a gallant purple ; which feeing that the written Hyacinth cannot do, it muft needs be that this *Vaccinium* is another thing than the Hyacinth is, becaufe it ferues to giue a purple die.

Pliny alfo, *lib.16.cap.18.* hath made mention of *Vaccinia*, which are vfed to dye bond-flaues garments with, and to giue them a purple colour.

But whether thefe be our *Vaccinia* or Whortle berries it is hard to affirme, efpecially feeing that *Pliny* reckoneth vp *Vaccinia* amongft thofe plants which grow in watery places ; but ours grow on high places vpon mountaines fubject to windes, neither is it certainely knowne to grow in Italy. Howfoeuer it is, thefe our Whortles may be called *Vaccinia*, and do agree with *Plinies* and *Vitruuius* his *V. cinia*, becaufe garments and linnen cloath may take from thefe a purple die.

The red Whortle berries haue their name from the blacke Whortles, to which they be in form very like, and are called in Latine, *Vaccinia rubra* : in High Dutch, **Rooter Heidelbeere** : in Low Dutch, **Roode Crakebeſſon** : the French-men, *Aurelles Rouges* ; they be named in Englifh, Red

Red Worts, or red Wortle berries. *Conradus Gefnerus* hath called this plant *Vitis Idæa rubris acinis:* but the growing of the berries doth fhow, that this doth farre leffe agree with *Vitis Idæa,* than the blacke; for they do not hang vpon the fides of the branches as do the black (which deceiued them that thought it to be *Vitis Idæa*) but from the tops of the fprings in clufters.

As concerning the names of the other they are touched in their feuerall defcriptions.

¶ *The Temperature.*

Thefe *Vaccinia* or Wortle berries are cold euen in the later end of the fecond degree, and dry alfo, with a manifeft aftriction or binding quality.

Red Wortle berries are cold and dry, and alfo binding.

¶ *The Vertues.*

The juyce of the blacke Wortle berries is boyled till it become thicke, and is prepared or kept **A** by adding hony and fugar vnto it : the Apothecaries call it *Rob*, which is preferred in all things before the raw berries themfelues ; for many times whileft they be eaten or taken raw they are offenfiue to a weake and cold ftomacke, and fo far are they from binding the belly, or ftaying the laske, as that they alfo trouble the fame through their cold and raw quality, which thing the boyled juyce called *Rob* doth not any whit at all.

They be good for an hot ftomacke, they quench thirft, they mitigate and allay the heate of hot **B** burning agues, they ftop the belly, ftay vomiting, cure the bloudy flix proceeding of choler, and helpe the felonie, or the purging of choler vpwards and downewards.

The people of Chefhire do eat the blacke Wortles in creame and milke, as in thefe South parts **C** we eate Strawberries, which ftop and binde the belly, putting away all the defire to vomit.

The Red Wortle is not of fuch a pleafant tafte as the blacke, and therfore not fo much vfed to be **D** eaten ; but (as I faid before) they make the faireft carnation colour in the World.

CHAP. 74.
Of the Marish Worts, or Fenne-Berries.

Vaccinia paluſtria.
Marifh Worts.

¶ *The Defcription.*

THe Marifh Wortle berries grow vpon the bogs in marifh or moorifh grounds, creeping thereupon like vnto wilde Time, hauing many fmall limmer and tender ftalkes laid almoft flat vpon the ground, befet with fmal narrow leaues fafhioned almoft like the leaues of Thyme, but leffer : among which come forth little berries like vnto the common blacke Wortle berrie in fhape, but fomewhat longer, fometimes all red, and fometimes fpotted or fpecked with red fpots of a deeper colour : in tafte, rough and aftringent.

¶ *The Place.*

The Marifh Wortle growes vpon bogs and fuch like waterifh and fenny places, efpecially in Chefhire and Staffordfhire, where I haue found it in great plenty.

¶ *The Time.*

The Berries are ripe about the end of Iuly, and in Auguft.

¶ *The Names.*

They are called in high Dutch, **Mofsbeeren, Veenbeffen :** that is to fay, Fen-Grapes, or Fen-Berries, and Marifh-worts, or Marifh-Berries. *Valerius Cordus* nameth them *Oxycoccon :* we haue called them *Vaccinia paluſtria,* or Marifh Wortle-berries, of the likeneffe they haue to the other berries : fome alfo call them Moffe-Berries, or Moore-berries.

¶ *The Temperature.*

Thefe Wortle berries are cold and dry, hauing withall a certain thinneffe of parts and fubftance, with a certaine binding quality joyned.

¶ *The*

¶ *The Vertues.*

A   They take away the heate of burning agues,and alſo the drought,they quench the furious heate of choler, they ſtay vomiting, reſtore an appetite to meate which was loſt by reaſon of cholericke and corrupt humors,and are good againſt the peſtilent diſeaſes.

B   The juyce of theſe alſo is boyled till it be thicke,with ſuger added that it may be kept,which is good for all things that the berries are,yea,and far better.

    † I haue brought this chapter and the next following from the place they formerly held, and ſeated them here amongſt the reſt of their kindred.

Cʜᴀᴘ. 75. *Of Cloud-berry.*

Vaccinia Nubis.
Cloud-berries.

¶ *The Deſcription.*

THe Cloud-berry hath many ſmall threddy roots, creeping farre abroad vnder the vpper cruſt of the earth, and alſo the moſſe,like vnto Couch-graſſe, of an ouerworn reddiſh colour, ſet here and there with ſmall tufts of hairy ſtrings:from which riſe vp two ſmall ſtalks, hard, tough, and of a wooddy ſubſtance (neuer more nor leſſe) on which doe ſtand the leaues like thoſe of the wilde Mallow,and of the ſame colour, full of ſmall nerues or ſinewes running in each part of the ſame: between the leaues commeth vp a ſtalke likewiſe of a wooddy ſubſtance, whereon doth grow a ſmall floure conſiſting of fiue leaues, of an herby or yellowiſh green colour like thoſe of the wilde Auens. After commeth the fruit, greene at the firſt, after yellow, and the ſides next the Sun red when they be ripe ; in forme almoſt like vnto a little heart, made as it were of two, but is no more but one,open aboue,and cloſed together in the bottom,of a harſh or ſharpe taſte,wherein is contontained three or foure little white ſeeds .

¶ *The Place.*

This plant groweth naturally vpon the tops of two high mountaines (among the moſſie places) one in Yorkeſhire called Ingleborough the other in Lancaſhire called Pendle, two of the higheſt mountaines in all England, where the cloudes are lower than the tops of the ſame all Winter long, whereupon the people of the countrey haue called them Cloud-berries, found there by a curiouſ gentleman in the knowledge of plants,called Mr *Heſketh*,often remembred.

¶ *The Time.*

The leaues ſpring vp in May,at which time it floureth : the fruit is ripe in Iuly.

¶ *The Temperature.*

The fruit is cold and dry and very aſtringent.

¶ *The Vertues.*

A   The fruit quencheth thirſt, cooleth the ſtomacke, and allayeth inflammations, being eaten as Worts are, or the decoction made and drunke.

    † My friend M. *Pimble* of Maribone receiued a plant hereof out of Lancaſhire : and by the ſhape of the leafe I could not iudge it to differ from the *Chamæmorus* (formerly deſcribed, pag. 1173 .neither do the deſcriptions much differ in any materiall point : the figures differ more ;but I iudge this a very imperfect one.

Cʜᴀᴘ. 76. *Of ſhrub Heart-wort of Æthiopia.*

¶ *The Deſcription.*

THis kind of Seſely,being the Æthiopian Seſely,hath blackiſh ſtalkes of a wooddy ſubſtance: this plant diuideth it ſelfe into ſundry other armes or branches, which are beſet with thicke fat and oileous leaues, faſhioned ſomewhat like the Wood-binde leaues, but thicker, and
                                             more

Seseli Æthiopicum frutex.
Shrub Sesely, or Hart-wort of Ethiopia.

more gummy, approching very neere vnto the leaues of Oleander both in shape and substance, being of a deepe or darke greene colour, and of a very good sauor and smell, and continueth green in my garden both winter and summer, like the Bay or Laurell. The floures grow at the tops of the branches, in yellow rundles like the floures of Dill: which being past, there succeedeth a darke or dusky seed resembling the seed of Fennel, and of a bitter taste. The root is thicke, and of a woody substance.

¶ *The Place.*

It is found in stony places and on the sea coast not far from Marsilles, and likewise in other places of Languedoc: it also groweth in Ethyopia in the darke and desart woods. It groweth in my garden.

¶ *The Time.*

It flourisheth, floureth, and seedeth in Iuly and August.

¶ *The Names.*

The Grecians call it Νισαιικαι σιολαι: the Latines likewise *Æthiopicum Seseli*: the Egyptians, κύνος φριαν, that is, Dogs Horror: in English, Sesely of Ethiopia, or Ethiopian Hart-wort.

¶ *The Temperature and Vertues.*

Sesely of Ethiopia is thought to haue the same faculties that the Sesely of Marsilles hath, whereunto I refer it.

Chap. 77. *Of the Elder Tree.*

¶ *The Kindes.*

THere be diuers sorts of Elders, some of the land, and some of the water or marish grounds; some with very jagged leaues, and others with double floures, as shall be declared.

¶ *The Description.*

1   THe common Elder groweth vp now and then to the bignesse of a mean tree, casting his boughes all about, and oftentimes remaineth a shrub: the body is almost all wooddy, hauing very little pith within; but the boughs, and especially the yong ones, which be jointed, are full of pith within, and haue but little wood without: the barke of the body and great armes is rough and full of chinks, and of an ilfauored wan colour like ashes: that of the boughs is not very smooth, but in colour almost like; and that is the outward barke; for there is another vnder it neerer to the wood, of colour green: the substance of the wood is sound, somwhat yellow, and that may be easily cleft: the leaues consist of fiue or six particular ones fastned to one ribbe, like those of the Walnut tree, but euery particular one is lesser, nicked in the edges, and of a ranke and stinking smell. The floures grow on spoky rundles, which be thin and scattered, of a white colour and sweet smell: after them grow vp little berries, green at the first, afterwards blacke, whereout is pressed a purple juice, which being boiled with Allom and such like things, doth serue very well for the Painters vse, as also to colour vineger: the seeds in these are a little flat and somwhat long. There groweth oftentimes vpon the bodies of those old trees or shrubs a certaine excrescence called *Auricula Iudæ* or Iewes eare, which is soft, blackish, couered with a skin, somewhat like now and then to a mans eare, which being plucked off and dried, shrinketh together and becommeth hard. This Elder groweth euery where, and is the common Elder.

Dddddd   2 There

1 *Sambucus.*
The common Elder tree.

‡ 2 *Sambucus fructu albo.*
Elder with white berries.

3 *Sambucus laciniatis folijs.*
The jagged Elder tree.

4 *Sambucus racemosa, vel Cervina.*
Harts Elder, or Cluster Elder.

2   There is another also which is rare and ftrange,for the berries of it are not black,but white: this is like in leaues to the former.

3   The jagged Elder tree growes like the common Elder in body,branches,fhoots,pith,flours, fruit,and ftinking fmell, and differs onely in the fafhion of the leaues, which do fo much difguife the tree,and put it out of knowledge,that no man would take it for a kinde of Elder,vntill hee hath fmelt thereunto,which will quickely fhew from whence he is defcended ; for thefe ftrange Elder leaues are very much jagged,rent or cut euen vnto the middle rib. From the trunke of this tree, as from others of the fame kinde,proceedeth a certain flefhy excrefcence like to the eare of a man,e- fpecially from thofe trees that are very old.

4   This kinde of Elder hath white floures,but red berries, and both are not contained in fpoky fundles,but in clufters,and grow after the manner of a clufter of grapes : in leaues and other things it refembleth the common Elder,faue that now and then it groweth higher.

¶ *The Place.*

The common Elder groweth euery where : it is planted about Cony-boroughs for the fhadow of the Conies ; but that with the white berries is rare : the other kindes grow in like places ; but that with the cluftered fruit groweth vpon mountaines. That with the jagged leaues growes in my garden.

¶ *The Time.*

Thefe kindes of Elders floure in Aprill and May,and their fruit is ripe in September.

¶ *The Names.*

This tree is called in Greeke, *κ..:* in Latine and of the Apothecaries, *Sambucus :* of *Gulielmus Salicetus,Beza :* in high-Dutch, **Holunder, Holder** : in low-Dutch, **Vlier** : in Italian, *Sambuco :* in French,*Hus*,and *Suin :* in Spanifh,*Sauco,Sauch,Sambugueyro :* in Englifh, Elder,and Elder tree:that with the white berries diuers would haue to be called *Sambucus fylveftris*, or wild Elder ; but *Mat-thiolus* calls it *Montana*,or mountaine Elder.

¶ *The Temperature and Vertues.*

Galen attributeth the like facultie to Elder that he doth to Danewoort , and faith that it is of a   A drying qualitie,gluing,and moderatly digefting : and it hath not onely thefe faculties, but others alfo ; for the barke,leaues,firft buds,floures,and fruit of Elder,doe not only dry, but alfo heate, and haue withall a purging qualitie,but not without trouble and hurt to the ftomacke.

The leaues and tender crops of common Elder taken in fome broth or pottage open the belly,   B purging both flimy flegme and cholerick humors : the middle bark is of the fame nature,but ftron-ger,and purgeth the faid humors more violently.

The feeds contained within the berries dried are good for fuch as haue the Dropfie,and fuch as   C are too fat and would faine be leaner,if they be taken in a morning to the quantitie of a dram with wine for a certain fpace.

The leaues of Elder boiled in water vntill they be very foft,and when they are almoft boiled e-   D nough a little oile of fweet Almonds added thereto,or a little Linefeed oile,then taken forth, and laid vpon a red cloath or a piece of skarlet,and applied to the hemmorrhoids or Piles as hot as can be fuffered,and fo left remaining vpon the part affected vntill it be fomewhat cold,hauing the like in a readineffe,applying one after another vpon the difeafed part by the fpace of an houre or more, and in the end fome bound to the place,and the patient put warme a bed ; it hath not as yet failed at the firft dreffing to cure the faid difeafe : but if the Patient be dreffed twice it muft needs help, if the firft faile.

. The green leaues pouned with Deeres fuet or Bulls tallow, are good to be laid to hot fwellings   E and tumors,and doe affwage the pain of the gout.

The inner and green bark doth more forcibly purge ; it draweth forth choler and waterifh hu-   F mors , for which caufe it is good for thofe that haue the dropfie , beeing ftamped,and the liquour preffed out and drunke with wine or whay.

Of like operation are alfo the fiefh floures mixed with fome kind of meat, as fried with egges ;   G they likewife trouble the belly and moue to the ftoole : being dried they lofe as wel their purging qualitie as their moifture,and retain the digefting and attenuating qualitie.

The vineger in which the dried floures are fteeped is wholefome for the ftomacke. Being vfed   H with meat it ftirreth vp an appetite, it cutteth and attenuateth or maketh thin groffe and raw hu-nors.

The facultie of the feed is fomwhat gentler than that of the other parts,it alfo moues the belly,   I and draweth forth watery humors, being beaten to pouder, and giuen to a dram weight: being new gathered, ſteeped in vineger, and afterward dried, it is taken and that effectually in the like weight of the dried lees of wine, and with a few Aniſe ſeeds, for ſo it worketh without any maner of trouble, and helpeth thoſe that haue the dropſie : but it muſt be giuen for certain dayes together in a little wine to thoſe that haue need thereof.

K   The gelly of the Elder, otherwiſe called Iewes eare, hath a binding and drying qualitie : the infuſion thereof, in which it hath bin ſteeped a few houres, taketh away inflammations of the mouth and almonds of the throat in the beginning, if the mouth and throat be waſhed therewith, and doth in like manner help the uvula.

L   *Dioſcorides* ſaith, that the tender green leaues of the Elder tree with parched Barly meale do remoue hot ſwellings, and are good for thoſe that are burnt or ſcalded, and for ſuch as be bitten with a mad dog, and that they glew and heale vp hollow vlcers.

M   The pith of the yong boughes is without qualitie : which being dried, and ſomewhat preſſed or quaſhed together, is good to lay open the narrow orifices or holes of fiſtula's and iſſues, if it be put therein.

Chap. 78.   *Of Mariſh or Water Elder.*

1 *Sambucus aquatilis, ſive paluſtris.*   2 *Sambucus Roſea.*
    Mariſh or water Elder.   The Roſe Elder.

¶ *The Deſcription.*

1   MAriſh Elder is not like to the common Elder in leaues, but in boughes. It groweth after the manner of a little tree ; the boughes are couered with a barke of an ill fauoured Aſh colour, as be thoſe of the common Elder : they are ſet with ioynts by
                                                            certain

certain diſtances, & haue in them great plenty of white pith, therfore they haue leſſe wood, which is white and brittle : the leaues be broad, cornered, like almoſt to Vine leaues, but leſſer and ſofter: among which come forth ſpoked rundles which bring forth little floures, the vttermoſt whereof a-longſt the borders be greater, of a gallant white colour, euery little one conſiſting of fiue leaues: the other in the midſt and within the borders be ſmaller, and it floures by degrees, and the whole tuft is of a moſt ſweet ſmell : after which come the fruit or beries, that are round like thoſe of the common Elder, but greater, and of a ſhining red colour, and blacke when they be withered.

2 *Sambucus Roſea* or the Elder Roſe groweth like an hedge tree, hauing many knotty branches or ſhoots comming from the root, full of pith like the common Elder : the leaues are like the vine leaues, among which come forth goodly floures of a white colour, ſprinkled and daſhed heere and there with a light and thin carnation colour, and do grow thicke and cloſely compact together, in quantitie and bulke of a mans hand, or rather bigger, of great beauty, and ſauoring like the floures of the Hawthorne : but in my garden there groweth not any fruit vpon this tree, nor in any other place, for ought that I can vnderſtand.

3 This kinde is likewiſe an hedge tree, very like vnto the former in ſtalks and branches, which are jointed and knotted by diſtances, and it is full of white pith : the leaues be likewiſe cornered. The flours hereof grow not out of ſpoky rundles, but ſtand in a round thicke and globed tuft, in big-neſſe alſo and faſhion like to the former, ſauing that they tend to a deeper purple colour, wherein only the difference conſiſts.

¶ *The Place.*

Sambucus paluſtris or water Elder growes by running ſtreames and water courſes, and in hedges by moiſt ditch ſides.

The Roſe Elder groweth in gardens, and the floures are there doubled by art, as it is ſuppoſed.

¶ *The Time.*

Theſe kindes of Elders floure in Aprill and May, and the fruit of the water Elder is ripe in Sep-tember.

¶ *The Names.*

The water Elder is called in Latine, *Sambucus aquatica*, and *Sambucus paluſtris* : it is called *Opulus* and *Platanus*, and alſo *Chamæplatanus* or the dwarfe Plane tree, but vnproperly. *Valerius Cordus* ma-keth it to be *Lycoſtaphylos* : the Saxons, ſaith *Geſner*, doe call it *Vua lupina*, from whence *Cordus* in-uented the name *Avæ æφʎ̣ɩ̣ʃ* : it is named in high-Dutch, 𝕸𝖆𝖑𝖙 𝖍𝖔𝖑𝖉𝖊𝖗, and 𝕳𝖎𝖗ſ𝖈𝖍 𝖍𝖔𝖑𝖉𝖊𝖗 : in Low-Dutch, 𝕾𝖙𝖜𝖊𝖑𝖈𝖐𝖊𝖓, and 𝕾𝖙𝖜𝖊𝖑𝖈𝖐𝖊𝖓𝖍𝖔𝖚𝖙 : of certaine French men, *Obiere* : in Engliſh, Mariſh El-der, Whitten tree, Oppel tree, and dwarfe Plane tree.

The Roſe Elder is called in Latine, *Sambucus Roſea*, and *Sambucus aquatica*, being doubtles a kind of the former water Elder, the floures being doubled by art, as we haue ſaid. It is called in Dutch, 𝕲𝖍𝖊𝖑𝖉𝖊𝖗ſ𝖈𝖍𝖊 𝕽𝖔𝖔ſ𝖊 : in Engliſh, Gelders Roſe, and Roſe Elder.

¶ *The Nature and Vertues.*

Concerning the faculties of theſe Elders, and the berries of the Water Elder, there is nothing found in any Writer, neither can we ſet downe any thing hereof of our owne knowledge.

C ʜ ᴀ ᴘ. 79.

Of Dane-wort, Wall-wort, or dwarfe Elder.

¶ *The Deſcription.*

D Anewort, as it is not a ſhrub, neither is it altogether an herby plant, but as it were a plant par-ticipating of both, being doubtleſſe one of the Elders, as may appeare both by the leaues, floures, and fruit, as alſo by the ſmell and taſte.

Wall-wort is very like vnto Elder in leaues, ſpoky tufts, and fruit, but it hath not a woody ſtalk, it bringeth forth only green ſtalks, which wither away in winter : theſe are edged and ful of joints like to the yong branches and ſhoots of Elder ; the leaues grow by couples with diſtances, wide, and conſiſt of many ſmall leaues which ſtand vpon a thicke ribbed ſtalke, of which euery one is long, broad, and cut in the edges like a ſaw, wider and greater than the leaues of the common Elder tree : at the top of the ſtalks there grow tufts of white floures tipt with red, with fiue little chiues in them pointed with blacke, which turne into blacke berries like the Elder, in which be little long ſeed : the root is tough, and of a good and reaſonable length, better for phyſicks vſe than the Elder leaues.

Ebulus, ſive Sambucus humilis.
Dane-wort, or dwarfe Elder.

A

B

C

D

E

F

G

¶ *The Place.*

Dane-wort groweth in vntoiled places neere to common wayes, and in the borders of fields. It groweth plentifully in the lane at Kilbury Abby by London: alſo in a field by S. *Ioans* neer Dartford in Kent; and in the highway at old Branford townes end next London, and in many other places.

¶ *The Time.*

The floures are perfected in Summer, and the berries in Autumne.

¶ *The Names.*

It is called in Greeke χαμαιάκτη, that is, *Humilis Sambucus*, or low Elder: in Latine, *Ebulus*, and *E-bulum*: in high-Dutch, **Attich**: in low-Dutch, **Hadich**: in Italian, *Ebulo*: in French, *Hieble*: in Spaniſh, *Yezgos*: in Engliſh, Wall-wort, Dane-wort, and dwarfe Elder.

¶ *The Temperature.*

Wall-woort is of temperature hot and dry in the third degree, and *Galen* doth attribute a ſingular qualitie vnto it to waſte and conſume. It hath alſo a ſtrange and ſpeciall facultie to purge by the ſtoole: the roots be of greateſt force; the leaues haue chiefeſt ſtrength to digeſt and conſume.

¶ *The Vertues.*

The roots of Wall-woort boiled in wine and drunken are good againſt the dropſie, for they purge downwards watery humors.

The leaues conſume and waſte way hard ſwellings, if they be applied pultiſwiſe, or in a fomentation or bath.

Dioſcorides ſaith, that the roots of Wall-woort doe ſoften and open the matrix, and alſo correct the infirmities thereof, if they be boiled in a bath to ſit in; and diſſolue the ſwellings and pains of the belly.

The juice of the root of Danewort maketh the haire blacke.

The yong and tender leafe quencheth hot inflammations, applied with Barley meale. It is with good ſucceſſe laid vpon burnings, ſcaldings, and the bitings of mad dogs; and with Buls tallow or Goats ſuet it is a remedie for the gout.

The ſeed of Wall-wort drunke in the quantitie of a dram is the moſt excellent purger of waterie humors in the world, and therefore moſt ſingular againſt the dropſie.

If one ſcruple of the ſeed be bruiſed and taken with ſyrrup of Roſes and a little Secke, it cureth the dropſie and eaſeth the gout, mightily purging downwards watery humors, being taken once a weeke.

Chap. 80.　*Of Beane Trefoile.*

¶ *The Deſcription.*

1　THe firſt kinde of *Anagyris* or *Laburnum* groweth like vnto a ſmall tree, garniſhed with many ſmall branches like the ſhoots of Oziars, ſet full of pale greene leaues, alwayes three together, like the *Lotus* or medow Trefoil, or rather like the leaues of *Vitex* or the Cytiſus buſh; among which come forth many tufts of floures of a yellow colour, not much vnlike the floures of Broom: when theſe floures be gon there ſucceed ſmall flat cods, wherein are contained ſeeds like *Galega* or the Cytiſus buſh. The whole plant hath little or no ſauor at all. The root is ſoft and gentle, yet of a wooddy ſubſtance.

2 Stinking

2 Stinking Trefoile is a shrub like to a little tree, rising vp to the height of six or eight cubits or somtimes higher: it sendeth forth of the stalks very many slender branches, the barke whereof is of a deepe greene colour: the leaues stand alwayes three together, like those of *Lotus* or medow Trefoile, yet of a lighter green on the vpper side: the floures be long, as yellow as gold, very like to those of Broom, two or three also joined together: after them come vp the cods, wherein lie hard fruit like Kidny beans, but lesser, at the first white, afterwards tending to a purple, and last of all of a blackish blew. The leaues and floures haue a filthy smell like those of stinking Gladdon, and so ranke withall, as euen the passers by are anoyed therewith.

1 *Anagyris.*
Bean Trefoile.

2 *Anagyris fœtida.*
Stinking Bean Trefoile.

 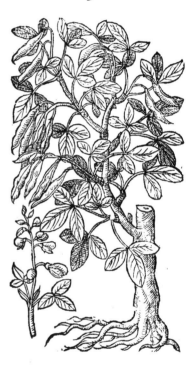

‡ Of *Anagyris* there are foure kinds, two with stinking leaues; the one with longish leaues, the other with rounder.

Two other whose leaues do not stink; the one of these hath somtimes foure or fiue leaues vpon one stalke, and the leaues are long and large. The other hath them lesser and narrower. ‡

¶ *The Place.*

These grow of themselues in most places of Languedoc and Spain, and in other countries also by high-waies, as in the Isle of Candy, as *Bellonius* writeth. The first I haue in my garden; the other is a stranger in England. ‡ Mr *Tradescant* hath two sorts hereof in his garden. ‡

¶ *The Time.*

They floure in Iune, and the seed is ripe in September.

¶ *The Names.*

The Bean Trefoile is called in Greeke Ἀνάγυρις, which name remaineth vncorrupt in Candy euen to this day: in Latine also *Anagyris*, and *Laburnum* of the people of Anagni in Italy, *Eghelo*, which is referred vnto *Laburnum*, whereof *Pliny* writeth, *lib.16.cap.18.* in English, Bean Trefoile, or the pescod tree.

¶ *The Temperature.*

Bean Trefoile, as *Galen* writeth, hath an hot and digesting facultie.

¶ *The*

¶ *The Vertues.*

A   The tender leaues, ſaith *Dioſcorides*, being ſtamped and laid vpon cold ſwellings, doe waſte away the ſame.

B   They are drunke in Cute to the weight of a dram againſt the ſtuffing of the lungs, and do bring downe the menſes, the birth, and the after-birth.

C   They cure the head-ache being drunke with wine; the iuice of the root digeſteth and ripeneth; if the ſeed be taken it procureth vomit, which thing, ſaith *Matthiolus*, the ſeed not only of ſtinking Bean Trefoile doth effect, but that alſo of the other.

Chap. 81. *Of Iudas Tree.*

Arbor Iudæ.
Iudas Tree.

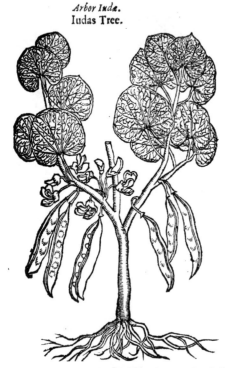

¶ *The Deſcription.*

IVdas Tree is likewiſe one of the hedge-plants, it groweth vp vnto a tree of a reaſonable bigneſſe, couered with a dark coloured barke, whereon grow many tough twiggy branches of a browne colour, garniſhed with round leaues like thoſe of round Birthwort or Sow-bread, but harder, and of a deeper green colour: amongſt which come forth ſmall floures like thoſe of peaſon, of a purple colour mixed with red, which turn into long flat cods preſſed hard together, of a tawny or wan colour, wherein is contained ſmall flat ſeeds like the Lentill, or rather like the ſeed of *Medica*, faſhioned like a little kidney. The root is great and wooddy.

¶ *The Place.*

This ſhrub is found in diuers prouinces of Spain, in hedges, & among briers and brambles: the mountaines of Italy and the fields of Languedoc are not without this ſhrub: it growes in my garden.

¶ *The Time.*

The floures come forth in the Spring, and before the leaues: the fruit or cods be ripe in ſummer.

¶ *The Names.*

It is commonly named in Latine, *Arbor Iudæ*: ſome haue named it *Sycomorus* or Sycomore tree, and that becauſe the flours and cods hang down from the bigger branches; but the right Sycomore tree is like the fig tree in fruit, and in leaues to the Mulberry tree, whereupon it is ſo named. Others take it to be κερκις, of which *Theophraſtus* writeth thus; *Cercis* bringeth forth fruit in a cod: Which words are all ſo few, as that of this no certaintie can be gathered, for there be more ſhrubs that bring forth fruit in cods. The French men call it *Guainier*, as though they ſhould ſay, *Vaginula*, or a little ſheath: moſt of the Spaniards name it *Algorouo loco*, that is, *Siliqua ſylueſtris*, or *fatua*, wild or fooliſh Cod: others, *Arbal d'amor*, for the braueneſſe ſake: it may be called in Engliſh, *Iudas* tree, for that it is thought to be that whereon *Iudas* hanged himſelfe, and not vpon the Elder tree, as it is vulgarly ſaid.

¶ *The Temperature and Vertues.*

The temperature and vertues of this ſhrub are vnknowne; for whereas *Matthiolus* makes this to be *Acacia*, by adding falſly thornes vnto it, it is but a ſurmiſe.

CHAP. 82.   *Of the Carob tree or S. Iohns Bread.*

¶ *The Description.*

THe Carob tree is also one of those that beare cods, it is a tree of a middle bigneſſe, very ful of boughs, the leaues long, and conſiſt of many ſet together vpon one middle rib, like thoſe of the Aſh, but euery particular one of them is broader, harder, and rounder: the fruit or long cods in ſome places are a foot in length, in other places ſhorter by halfe, an inch broad, ſmooth and thicke, in which lie flat and broad ſeeds: the cods themſelues are of a ſweet taſte, and are eaten of diuers, but not before they be gathered and dried; for being green, though ripe, they are vnpleaſant to be eaten, by reaſon of their bad taſte.

Ceratia ſiliqua, ſive Ceratonia.
The Carob tree.

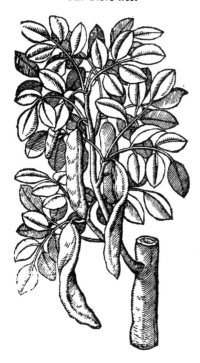

¶ *The Place.*

This groweth in Apulia a Prouince of the kingdome of Naples, and alſo in diuers vn-toiled places in Spain: it is likewiſe found in India & other countries Eaſtward, where the cods are ſo full of ſweet juice as that it is vſed to preſerue Ginger, as *Matthiolus* ſheweth. *Strabo, lib.*15. faith, That *Ariſtobu-lus* reporteth how there is a tree in India of no great bigneſſe, hauing great cods tenne inches long, full of hony, *Quas qui ederint non faſile ſervare*: which thing peraduenture is only to be vnderſtood of the green cods, and thoſe that are not yet dry. It is very well known in the coaſts of Nicea and Liguria in Italy, as alſo in all the tracts and coaſts of the Weſt Indies and Virginea. It groweth alſo in ſundry places of Paleſtine, where there is ſuch plenty of it, that it is left vnto ſwine and other wilde beaſts to feed vpon, as out Acorns and Beech maſt. Moreouer, both yong and old feed thereon for pleaſure, and ſome haue eaten thereof to ſupply and help the neceſſarie nouriſhment of their bodies. This of ſome is called Saint *Iohns* bread, and thought to be that which is tranſlated Lo-cuſts, whereon S. *Iohn* did feed when he was in the wilderneſſe, beſides the wilde Honey whereof hee did alſo eat: but there is ſmall certainty of this: but it is moſt certain, that the people of that country doe feed on theſe cods, called in Greeke ωκέρια; in Latine *Sili-qua*: but Saint *Iohns* food is called in Greek ἀκρίδες, which word is often vſed in the Reue-lation written by Saint *Iohn*, and tranſlated Locuſts. Now we muſt alſo remember that this Greek word hath two ſeuerall interpretations or ſignifications; for taken in the good part it ſignifieth a kinde of creeping creature or fly, which hoppeth or skippeth vp and downe as doth the Graſhop-per; of which kinde of creatures it was lawfull to eat, *Levit.* 22. and *Matth.*3.4. It ſignifieth alſo thoſe Locuſts which came forth of the Bottomleſſe pit, mentioned, *Apoc.9. v.*3,4. &c. which were like vnto horſes prepared for battell. The Hebrew word which the Engliſh Tranſlators haue tur-ned Graſhoppers, *Tremelius* dares not giue the name Locuſt vnto it, but calleth it by the Hebrew name *Arbis*, after the letters and Hebrew name; ſaying thus in the note vpon the 22 verſe of the e-leuenth chapter of *Levit.* Theſe kindes of creeping things neither the Hebrews nor the Hiſtorio-graphers, nor our ſelues do know what they mean; wherefore we ſtill retain the Hebrew words for all the foure kindes thereof. But it is certaine that the Eaſt countrey Graſhoppers and Locuſts were ſometimes vſed in meat; as *Matth.*3.4. and *Marc.*1.6. *Plin. Lib.*11. *Natur.Hiſt.ca.*26.& 29.

Thus

Thus far *Tremelius* and *Iunius*. By that which hath been said it appeareth what S. *Iohn* Baptist fed of, vnder the title Locusts, and that it is nothing like vnto this fruit *Ceratia siliqua*. I rather take the husks or shels of the fruit of this tree to be the cods or husks whereof the Prodigall childe would haue fed, but none gaue them vnto him, though the swine had their fill thereof. These cods being dry are very like Bean cods, as I haue often seen. I haue sowne the seeds in my garden, where they haue prospered exceeding well.

‡ There is no doubt but the ακρίδες or *Siliquæ* mentioned in S. *Lukes* Gospell, *Cap.* 15. *v.* 16. were the cods or fruit of this tree. I cannot beleeue that either the fruit of this or the Locusts were the ακρίδες mentioned by S. *Mathew*, *cap.* 3. *v.* 4. But I am of the opinion of the Greeke Father *Isodor Pelusiota*, who, *lib.* 1. *Epist.* 132. hath these words ; Ἀι ἀκρίδες, ἃς Ἰωάννης ἔφαγεν, οὐ ζῷα ἦσαν, ὡς ποτε ἔνιοι ἀμαθῶς, καὶ δίχα γνωσικῆς, μὴ νοῆσαντες ἀλλ' ἐκριμάσιε Βοτανῶν : ἐπεὶ ἢ πᾶ τις ὅτι πάλιν τὸ μέλι ἄγριον, ἀλλὰ μέλι ἄγριον ὑπὸ μελισσῶν ἀγρίων γινόμενον, &c. That is ; The *Acrides* which *Iohn* fed vpon are not liuing creatures like to Beetles, as some vnlearnedly think; but they are the tender buds of herbs and plants or trees. Neither on the other side is the *Meli agrion* any herb so called, but mountain hony gathered by wilde Bees, &c. ‡

¶ *The Time.*

The Carob tree bringeth forth fruit in the beginning of the Spring, which is not ripe till Autumne.

¶ *The Names.*

The Carob tree is called Κερατωνία : in Latine likewise *Ceratonia* : in Spanish, *Garovo* : in English, Carob tree, and of some, Bean tree, and S. *Iohns* Bread. The fruit or cod is named ακρίδιον : in Latine, *Siliqua*, or *Siliqua dulcis* : in diuers shops, *Xylocaracta* : in other shops of Italy, *Carobe*, or *Carobole* : of the Apothecaries of Apulia, *Salequa* : it is called in Spanish, *Alfarobas*, or *Algarovas* ; and without an article, *Garovas* : in high-Dutch, **S. Johans brot**, that is to say, *Sancti Iohan. panis*, or Saint *Iohns* bread ; neither is it knowne by any other name in the Low-Countries. Some call it in English, Carob.

¶ *The Temperature.*

The Carob tree is dry and astringent, as is also the fruit, and containeth in it a certain sweetnes, as *Galen* saith.

¶ *The Vertues.*

A   The fruit of the Carob tree being eat when it is green doth gently loose the belly ; but beeing dry it is hard of digestion and stoppeth the belly, it prouoketh vrine, is good for the stomacke, and nourisheth well, and much better than when it is green and fresh.

CHAP. 83.   *Of Cassia fistula or Pudding Pipe.*

¶ *The Description.*

Cassia *purgatrix* or *Cassia fistula* groweth vp to be a faire tree, with a tough barke like leather, of the colour of Box, whereupon some haue supposed it to take the Greeke name κίσσος : in Latine, *Coriaceus* : the arms and branches of this are small and limber, beset with many goodly leaues like those of the Walnut tree : among which come forth small flours of a yellow colour, compact or consisting of six little leaues like the floures of *Chelidonium minus* or Pilewort. After these be vaded, there succeed goodly blacke round long cods, whereof some are two foot long, and of a woody substance : in these cods are contained a blacke pulp very sweet and soft, of a pleasant taste, and seruing to many vses in physicke ; in which pulp lieth the seeds couched in little cels or partitions : the seed is flat and brownish, not vnlike the seed of *Ceratia siliqua*, and in other respects very like vnto it also.

¶ *The Place.*

This tree groweth much in Egypt, especially about Memphis and Alexandria, and most parts of Barbary, and is a stranger in these parts of Europe.

¶ *The Time.*

The Cassia tree groweth greene winter and summer, it sheds his old leaues when new are come, by means whereof it is neuer void of leaues : it floureth early in the spring, and the fruit is ripe in Autumne.

¶ *The*

Caſsia fiſtula.
Pudding Pipe tree.

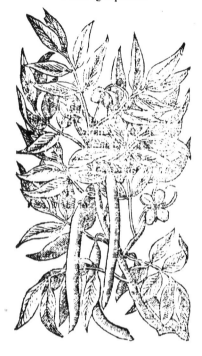

¶ *The Names.*

This tree was vnknown to the old writers, or ſo little accounted of, as that they haue made no mention of it at all: the Arabians were the firſt that eſteemed of it, by reaſon they knew the vſe of the pulp which is found in the pipes; and after them the later Grecians, as *Actuarius* and others of his time, by whom it was named *μαλανεα*, that is to ſay in Latine, *Caſia nigra*: the fruit thereof, ſaith *Actuarius, lib.* 5. is like a long pipe, hauing within it a thicke humour or moiſture, which is not congealed all alike through the Pipe, but is ſeparated and diuided with many partitions, being thin woody ſtrings. The apothecaries call it *Caſia fiſtula*, and with a double ff, *Caſſia fiſtula*. It is called in Engliſh after the Apothecaries word, Caſſia fiſtula, and may alſo be Engliſhed Pudding Pipe, becauſe the cod or pipe is like a pudding. But the old Caſſia fiſtula or *κασια* in Greeke, is that ſweet and odoriferous barke that is rolled together after the manner of a long and round pipe, now named of the Apothecaries *Caſſia lignea*, which is a kinde of Cinamon.

¶ *The Temperature.*

The pulp of this Pipe, which is chiefely in requeſt, is moiſt in the later end of the firſt degree, and little more than temperatly hot.

¶ *The Vertues.*

The pulp of Caſſia fiſtula extracted with violet water is a moſt ſweet and pleaſant medicine, and **A** may be giuen without danger to all weak people, of what age and ſex ſoeuer they be, yea it may be miniſtred to women with childe, for it gently purgeth cholerick humors and ſlimy flegme, if it be taken in the weight of an ounce.

Caſſia is good for ſuch as be vexed with hot agues, pleuriſies, jaundice, or any other inflammati- **B** on of the liuer, being taken as afore is ſhewed.

Caſſia is good for the reins and kidnies, driueth forth grauell and the ſtone, eſpecially if it bee **C** mingled with the decoction of Parſly and Fennell roots, and drunke.

It purgeth and purifieth the bloud, making it more clean than before, breaking therewith the a- **D** crimonie and ſharpneſſe of the mixture of bloud and choler together.

It diſſolueth all phlegmons and inflamations of the breſt, lungs, and the rough artery called *Tra-* **E** *chea arteria*, eaſing thoſe parts exceeding well.

Caſſia abateth the vehemencie of thirſt in agues, or any hot diſeaſe whatſoeuer, eſpecially if it **F** be taken with the iuyce of *Intybum, Cichoreum*, or *Solanum*, depured according to Art. It abateth alſo the intemperat heat of the reins, if it be receiued with diuretick medicines, or with the decoction of Licorice only, and will not ſuffer the ſtone to grow in ſuch perſons as do receiue and vſe this medicine.

The beſt Caſſia for your vſe is to be taken out of the moſt full, heauy, and faireſt cods or canes, **G** and thoſe which ſhine without, and are full of ſoft pulpe within: that pulpe which is newly taken forth is better than that which is kept in boxes, by what art ſoeuer.

Caſſia being outwardly applied taketh away the roughneſſe of the skin, and being laid vpon hot **H** ſwellings it bringeth them to ſuppuration.

Many ſingular compound medicines are made with this Caſſia, which here to recite belongs not **I** to my purpoſe or hiſtorie.

CHAP. 84.

Of the Lentiske or Masticke tree.

Lentiscus.
The Masticke tree.

¶ *The Description.*

THe Masticke tree groweth commonly like a shrub, without any great body, rising vp with many springs and shoots like the Hasell, and oftentimes it is of the height and bignes of a mean tree: the boughes thereof are tough and flexible: the barke is of a yellowish red colour, pliable likewise and hard to be broken. There stand vpon one rib for the most part eight leaues, set vpon a middle ribbe, much like to the leaues of Licorice, but harder, of a deep green colour, and oftentimes somwhat red in the brims, as also hauing diuers veines running along, of a red colour, and somwhat strong of smel: the floures be mossie, and grow in clusters vpon long stems: after them come vp the beries, of the bignesse of Vetches, green at the first, afterward of a purple colour, and last of all blacke, fat, & oily, with an hard blacke stone within, the kernel whereof is white, of which also is made oile, as *Dioscorides* witnesseth: it bringeth forth likewise cods besides the fruit (which may be rather termed an excrescence than a cod) wrythed like an horne; in which lieth at the first a liquor, and afterwards when this waxeth stale, little liuing things like vnto gnats, as in the Turpentine hornes, and in the folded leaues of the Elme tree. There comes forth of the mastick tree a rosin, but dry, called Masticke.

¶ *The Place.*

The Mastick tree groweth in many regions, as in Syria, Candy, Italy, Languedock, and in most Prouinces of Spaine; but the chiefest is in Chios an Iile in Greece, in which it is diligently and specially looked vnto, and that for the Mastick sake, which is there gathered euery yeare most carefully from the husbanded Mastick trees, and sent into all parts of the world.

¶ *The Time.*

The floures be in their pride in the spring time, and the berries in Autumne. The Mastick must be gathered about the time that Grapes be.

¶ *The Names.*

This tree is called in Greeke χ῀ῖος : in Latine, *Lentiscus* : in Italian, *Lentisque* : in Spanish, *Mata*, and *Arcoyra* : in English, Mastick tree, and of some, Lentiske tree.

The Rosin is called in Greeke, ῥητίνη σχίνινη and Μαςίχη : in Latine, *Lentiscina Resina*, and likewise *Mastiche* : in shops, *Mastix* : in Italian, *Mastice* : in high and low Dutch and French also, *Mastic* : in Spanish, *Almastiga, Mastech*, and *Almecega* : in English, Mastick.

Clusius writeth, That the Spaniards call the oile that is pressed out of the beries, *Azeyte de Mata*.

¶ *The Temperature.*

The leaues, barke, and gum of the Mastick tree are of a meane and temperat heate, dry in the second degree, and somwhat stringent.

¶ *The Vertues.*

A   The leaues and barke of the Masticke tree stop the laske, the bloudy flix, spitting of bloud,

bloud, the piſſing of bloud, and all other fluxes of bloud : they are alſo good againſt the falling
ſickeneſſe, the falling downe of the mother, and comming forth of the fundament.

The gum Maſticke hath the ſame vertue, if it be relented in wine and giuen to be drunke.　B

Maſticke chewed in the mouth is good for the ſtomacke, ſtaieth vomiting, increaſeth appetite,　C
comforteth the braines, ſtaieth the falling downe of the rheumes and watery humors, and maketh a
ſweet breath.

The ſame infuſed in Roſewater is excellent to waſh the mouth withall, to faſten looſe teeth,　D
and to comfort the jawes.

The ſame ſpred vpon a piece of leather or veluet, and laid plaiſterwiſe vpon the temples, ſtaieth　E
the rheume from falling into the jawes and teeth, and eaſeth the paines thereof.

It preuaileth much againſt vlcers and wounds, being put into digeſtiues and healing vnguents.　F

It draweth flegme forth of the head gently and without trouble.　G

It is alſo vſed in waters which ſerue to clenſe and make faire the face with.　H

The decoction of this filleth vp hollow vlcers with fleſh if they be bathed therewith.　I

It knitteth broken bones, ſtaieth eating vlcers, and prouoketh vrine.　K

CHAP. 85.　　Of the Turpentine Tree.

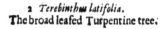

¶ The Deſcription.

1　THe firſt Turpentine Tree groweth to the height of a tall and faire tree, hauing many
long boughes or branches, diſperſed abroad, beſet with long leaues, conſiſting of ſundry
other ſmall leaues, each whereof reſembleth the Bay leafe, growing one againſt another vpon a little ſtem or middle rib, like vnto the leaues of the Aſh tree : the floures be ſmall and reddiſh, growing vpon cluſters or bunches that turne into round berries, which at their beginning are greene,
afterwards reddiſh ; but beeing ripe wax blacke, or of a darke blew colour, clammie, full of fat

　　　and

and oilous in ſubſtance, and of a pleaſant ſauour: this plant beareth an empty cod, or crooked horne ſomewhat reddiſh, wherein are found ſmall flies, wormes or gnats, bred and ingendred of a certaine humorous matter, which cleaueth to the inner ſides of the ſaid cods or hornes, which wormes haue no phyſicall vſe at all. The right Turpentine iſſueth out of the branches of theſe trees, if you doe cut or wound them, the which is faire and cleere, and better than that which is gathered from the barke of the Firre tree.

2   The ſecond kinde of Turpentine tree is very like vnto the former, but that it groweth not ſo great: yet the leaues are greater and broader, and of the ſame faſhion, but very like to the leaues of the Piſtacia tree. The berries are firſt of a ſcarlet colour, and when they be ripe of a skie colour. The great horned cods are ſharpe pointed, and ſomewhat cornered, conſiſting as it were of the ſubſtance of griſtles. And out of thoſe bladders being broken, do creepe and come ſmall flies or gnats, bred of a fuliginous excrement, and ingendred in thoſe bladders. The tree doth alſo yeeld his Turpentine by dropping like the former.

¶ *The Place.*

Theſe trees grow, as *Dioſcorides* ſaith, in Iurie, Syria, Cyprus, Africke, and in the Iſlands called *Cyclades*. *Bellonius* reporteth that there are found great ſtore of them in Syria, and Cilicia, and are brought from thence to Damaſcus to be ſould. *Cluſius* ſaith, That it growes of it ſelfe in Langue-docke, and in very many places of Portingale and Spaine, but for the moſt part like a ſhrub, and without bearing Turpentine.

Theophraſtus writeth, That it groweth about the hill Ida, and in Macedonia, ſhort, in manner of a ſhrub, and writhed; and in Damaſcus and Syria great, in manner of a ſmall tree: he alſo ſetteth downe a certaine male Turpentine, and a female: the male, ſaith he, is barren, and the female fruitfull. And of theſe he maketh the one with a berry red at the firſt, of the bigneſſe of a Lentill, which cannot come to ripeneſſe; and the other with the fruit greene at the firſt, afterwards ſome-what of a yellowiſh red, and in the end blacke, waxing ripe in the Spring, of the bigneſſe of the Græcians Beane, and roſenny.

He alſo writeth of a certaine Indian Turpentine tree, that is to ſay, a tree like in boughes and leaues to the right Turpentine tree, but differing in fruit, which is like vnto Almonds.

¶ *The Time.*

The floures of the Turpentine tree come foorth in the Spring together with the new buds: the berries are ripe in September and October, in the time of Grape gathering. The hornes appeare about the ſame time.

¶ *The Names.*

This tree is called in Greeke τερμινθος, and alſo many times τερμιθος: in Latine, *Terebinthus*: in Itali-an, *Terebintho*: in Spaniſh, *Cornicabra*: in French, *Terebinte*: in Engliſh, Turpentine tree: the Arabi-ans call it *Botin*, and with an article *Albotin*.

The Roſin is ſurnamed τερμιθινη: in Latine, *Terebinthina*: in high Dutch, **Termentijn**: in Eng-liſh, Turpentine, and right Turpentine: in the Arabian language *Albotin*, who name the fruit *Gra-num viride*, or greene berries.

¶ *The Temperature and Vertues.*

A   The barke, leaues, and fruit of the Turpentine tree doe ſomewhat binde, they are hot in the ſe-cond degree, and being greene they dry moderately; but when they are dried they dry in the ſe-cond degree; and the fruit approacheth more neere to thoſe that be dry in the third degree, and alſo hotter. This is fit to be eaten, as *Dioſcorides* ſaith, but it hurteth the ſtomacke.

B   It prouoketh vrine, helpeth thoſe that haue bad ſpleenes, and is drunke in wine againſt the bi-tings of the poiſonſome ſpiders called *Phalangia*.

C   The Roſin of the Turpentine tree excelleth all other Roſins, according to *Dioſcorides* his opini-on: but *Galen* writeth, That the Roſin of the Maſticke tree beareth the preheminence, and then the Turpentine.

D   This Roſin hath alſo an aſtringent or binding faculty, and yet not ſo much as Maſticke; but it hath withall a certaine bitterneſſe joyned, by reaſon whereof it digeſteth more than that of the Maſticke tree: through the ſame quality there is likewiſe in it ſo great a clenſing, as alſo it healeth ſcabs, in his eighth booke of the faculties of ſimple medicines; but in his booke of medicines ac-cording to the kindes, he maketh that of the Turpentine tree to be much like the Roſins of the Larch tree, which he affirmeth to be moiſter than all the reſt, and to be without both ſharpneſſe and biting.

E   The fruit of Turpentine prouoketh vrine and ſtirreth vp fleſhly luſt.

F   The Roſin of this tree, which is the right Turpentine, looſeth the belly, openeth the ſtoppings of the liuer and ſpleene, prouoketh vrine, and driueth forth grauell, being taken the quantity of two or three Beanes.

                                              The

The like quantity washed in water diuers times vntill it be white, then must be put thereto the G like quantity of the yolk of an egge, and laboured together, adding thereto by little and little(continually stirring it) a small draught of possit drinke made of white wine, and giuen to drinke in the morning fasting, it helpeth most speedily the Gonorrhæa, or running of the reines, commonly at the first time, but the medicine neuer faileth at the second time at the taking of it, which giues stooles from foure to eight, according to the age and strength of the patient.

Chap. 86.   Of the Frankincense Tree:

¶ *The Description.*

THe tree from which Frankincense floweth is but low, and hath leaues like the Masticke tree ; yet some are of opinion that the leafe is like the leafe of a Peare tree, and of a grassie colour : the rinde is like that of the Bay tree, whereof there are two kindes : the one growing in mountains and rockie places, the other in the plaine : but those in the plaines are much worse than those of the mountaines : the gum hereof is also blacker, fitter to mingle with Pitch, and such other stuffe to trim ships, than for other vses.

Arbor Thurifera.
The Frankincense tree.

Thuris Limpidi folium Lobelij.
The supposed leafe of the Frankincense tree.

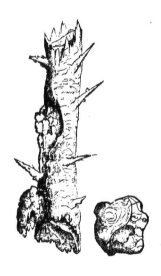

Theuet in his Cosmographie saith, that the Frankincense tree doth resemble a gummy or rosiny Pine tree, which yeeldeth a juyce that in time groweth hard, and is called *Thus,* Frankincense, in which is found sometimes certaine small graines like vnto grauell, which they call the Manna of Frankincense.

Of this there is in Arabia two other sorts, the one, the gum whereof is gathered in the Dog daies when the Sun is in Leo, which is white, pure, cleare, and shining. *Pena* writeth, That he hath seene the cleare Frankincense called *Limpidum,* and yeelding a very sweet smell when it is burnt, but the leafe hath been seldome seene, which the Physition *Launanus* gaue to *Pena* and *Lobel,* together with some pieces of the Rosine, which he had of certaine mariners, but he could affirme nothing of certainty whether it were the leafe of the Frankincense, or of some other Pine tree, yeelding the like juyce or gum. It is, saith he, (which doth seldom happen in other leaues) from the lower part or foot of the leafe to the vpper end, as it were doubled, consisting of two thin rindes or coats, with a sheath a span and a halfe long, at the top gaping open like a hood or fooles coxcombe, and as it were couered with a helmet, which is a thing seldome seene in a leafe, but is proper to the floures of *Napellus,* or *Lonchitis*, as writers affirme; the other is gathered in the Spring, which is reddish, worser than the other in price or value, becaufe it is not fo well concocted in the heat of the Sunne. The Arabians wound this tree with a knife, that the liquor may flow out more abundantly, whereof fome trees yeeld threefcore pound of Frankincenfe.

¶ The Place.

Diofcorides faith it groweth in Arabia, and efpecially in that quarter which is called Thurifera, the beſt in that country is called *Stagonias*, and is round, and if it be broken is fat within, and when it is burned doth quickly yeeld a fmell: next to it in goodneffe is that which groweth in Smilo, leſſer than the other, and more yellow.   **¶ The Time.**

The time is already declared in the defcription.

¶ The Names.

It is called in Greeke λιϐανος: in Latine, *Thus*: in Italian, *Incenfo*: in Dutch, **Weirauch :** in Spanifh, *Encenfo*: in French, *Enceus*: in Englifh, Frankincenfe, and Incenfe: in the Arabian tongue, *Louan*, and of fome few, *Conder*. ‡ The Rofin carries the fame name; but in fhops it is called *Olibanum*, of the Greeke name and article put before it. ‡

¶ The Temperature and Vertues.

A   It hath, as *Diofcorides* faith, a power to heate and binde.

B   It driueth away the dimneffe of the eye-fight, filleth vp hollow vlcers, it clofeth raw wounds, ſtaieth all corruptions of bloud, although it fall from the head.

C   *Galen* writeth thus of it; *Thus* doth heate in the fecond degree, and dry in the firſt, and hath fome fmall aftriction, but in the white there is a manifeſt aftriction; the rinde doth manifeſtly binde and dry exceedingly, and that moſt certainely in the fecond degree, for it is of more groffer parts than Frankincenfe, and not fo fharpe, by reafon whereof it is much vfed in fpitting of bloud, fwellings, in the mouth, the collicke paffion, the flux in the belly rifing from the ſtomacke, and bloudy flixes.

D   The fume or fmoke of it hath a more drier and hotter quality than the Frankincenfe it felfe, being dry in the third degree.

E   It doth alfo clenfe and fill vp the vlcers in the eies, like vnto Myrrhe: thus far *Galen*.

F   *Diofcorides* faith, That if it be drunke by a man in health, it driueth him into a frenfie: but there are few Greekes of his minde.

G   *Auicen* reporteth that it doth helpe and ſtrengthen the wit and vnderſtanding, but the often taking of it will breed the head-ache, and if too much of it be drunke with wine it killeth.

CHAP. 87. *Of Fiſticke Nuts.*

Piſtacia.   The Fiſticke Nut.

¶ The Defcription.

THe tree which beareth Fiſticke nuts is like to the Turpentine tree: the leaues hereof be greater than thofe of the Maſticke tree, but fet after the fame maner, & in like order that they are, being of a faint yellow color out of a green; the fruit or Nuts do hang by their ſtalks in cluſters, being greater than the Nuts of Pine Apples, and much leſſer than Almonds: the huske without is of a grayifh colour, fometimes reddifh, the fhell brickle and white; the fubſtance of the kernell greene; the taſte fweet, pleafant to be eaten, and fomething fweet of fmell.

¶ The Place.

Fiſticke Nuts grow in Perfia, Arabia, Syria, and in India; now they are made free Denizons in Italy, as in Naples and in other Prouinces there.

¶ The Time.

This tree doth floure in May, and the fruit is ripe in September.

¶ The Names.

This Nut is called in Greeke πιϛακιον in *Athenaus*: Nicander Colophonius in his booke of Treacles nameth it φουϛακιον: Poſſidonius nameth it βιϛακιον: others, ψιϛακιον: the Latines obferuing the fame termes, haue named it *Piſtacion, Biſtacion,* or *Phiſtacion:*

ſtacion : the Apothecaries, *Fiſtici :* the Spaniards, *Alhocigos,* and *Fiſticos :* in Italian, *Piſtacchi :* in Engliſh, Fiſticke Nut.

¶ The Temperature and Vertues.

The kernels of the Fiſticke Nuts are oftentimes eaten as be thoſe of the Pine Apples; they be A of temperature hot and moiſt, they are not ſo eaſily concoded, but much eaſier than common nuts: the juyce is good, yet ſomewhat thicke; they yeeld to the body no ſmall nouriſhment, they nouriſh bodies that are conſumed : they recouer ſtrength.

They are good for thoſe that haue the phthiſicke, or rotting away of the lungs. B

They concod, ripen, and clenſe forth raw humors that cleaue to the lights and cheſt. C

They open the ſtoppings of the liuer, and be good for the infirmities of the kidnies, they alſo re- D moue out of the kidnies ſand and grauell ; and aſſwage their paine : they are alſo good for vlcers.

The kernels of Fiſticke nuts condited, or made into comfits, with Sugar, and eaten, doe procure E bodily luſt, vnſtop the lungs and the breſt, are good againſt the ſhortneſſe of breath, and are an excellent preſeruatiue medicine being miniſtred in wine againſt the bitings of all manner of wilde beaſts.

CHAP. 88.　Of the Bladder Nut.

Nux veſicaria.
The Bladder Nut.

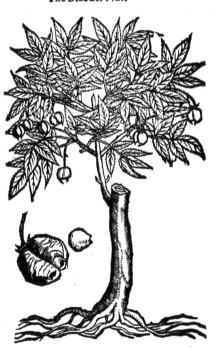

¶ The Deſcription.

THis is a low tree, hauing diuers young ſprings growing forth of the root: the ſubſtance of the wood is white, very hard & ſound; the barke is of a light greene : the leaues conſiſt of fiue little ones, which be nicked in the edges like thoſe of the Elder, but leſſer, not ſo greene nor ranke of ſmell. It hath the pleaſant whitiſh floures of Bryony or *Labruſca,* both in ſmell and ſhape, which turne into ſmall cornered bladders of Winter Cherries, called Alkakengie, but of an ouerworne greeniſh colour : in theſe bladders are contained two little nuts, and ſometimes no more but one, leſſer than the Haſell nut, but greater than the Ram Cich, with a wooddy ſhell and ſomewhat red : the kernell within is ſomething greene; in taſte at the firſt ſweet, but afterwards lothſome and ready to prouoke vomit.

¶ The Place.

It groweth in Italy, Germany and France; it groweth likewiſe at the houſe of Sir *Walter Culpepper* neere Flimmewell in the Weild of Kent, as alſo in the Fryer-yard without Saint Paules gate in Stamford, and about Spalding Abbey, and in the garden of the right Honourable the Lord Treaſurer my very good Lord and Maſter, and by his houſe in the Strand. It groweth alſo in my Garden, and in the Garden hedges of Sir *Francis Carew* neere Croydon, ſeuen miles from London.

¶ The Time.

This tree floureth in May, the Nuts be ripe in Auguſt and September.

¶ The Names.

It is commonly called in high Dutch, 𝔓impernuſʒ: which ſignifieth in low Dutch, 𝔓impernoten : diuers call it in Latine, *Piſtacium Germanicum :* we thinke it beſt to call it *Nux veſicaria.* *Matthiolus* in his Epiſtles doth judge the Turks *Coulcoul* and *Hebulben,* to agree with this: *Gulielmus Quacelbenus* affirmeth, *Coulcoul* to be vſed of diuers in Conſtantinople for a daintie, eſpecially when they be new brought out of Egypt. This plant hath no old name, vnleſſe it be *Staphylodendron Plinij :*

for which it is taken of the later writers: and *Pliny* hath written of it in 16. booke, 16. chapter. There
is alſo (ſaith he) beyound the Alps a tree, the timber whereof is very like to that of white Maple,
and is called *Staphylodendron*, it beareth cods, and in thoſe kernels, hauing the taſt of the Haſel Nut.
It is called in Engliſh, S. Anthonies nuts, wild Piſtacia, or Bladder nuts: the Italians call it *Piſtachio
Saluaticke*: the French-men call it *Baguenaudes a patre noſtres*, for that the Friers do vſe to make beads
of the Nuts.

¶ *The Temperature and Vertues.*

A   Theſe nuts are moiſt and full of ſuperfluous raw humors, and therefore they eaſily procure a readineſſe to vomit, and trouble the ſtomacke, by reaſon that withall they be ſomewhat binding, and
therefore they be not to be eaten.

B   They haue as yet no vſe in Medicine, yet notwithſtanding ſome haue attributed vnto them ſome
vertues in prouoking of Venery.

CHAP. 89. *Of the Haſell tree.*

¶ *The Deſcription.*

1   THe Haſell tree groweth like a ſhrub or ſmall tree, parted into boughes without ioints,
tough and pliable: the leaues are broad, greater and fuller of wrinckles than thoſe of the
Alder tree, cut in the edges like a ſaw, of colour greene, and on the backſide more white,
the barke is thin: the root is thicke, ſtrong, and growing deepe ; in ſtead of floures hang downe catkins, aglets, or blowings, ſlender, and well compact: after which come the Nuts ſtanding in a tough
cup of a greene colour, and jagged at the vpper end, like almoſt to the beards in Roſes. The ſhell is
ſmooth and wooddy: the kernell within conſiſteth of a white, hard, and ſound pulpe, and is couered
with a thin skin, oftentimes red, moſt commonly white ; this kernell is ſweet and pleaſant vnto the
taſte.

1 *Nux Auellana, ſiue Corylus.*   2 *Corylus ſylueſtris.*
The Filberd Nut.   The wilde hedge-Nut.

 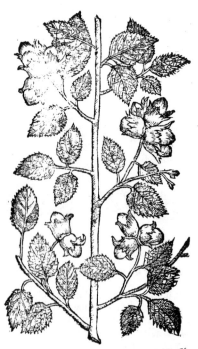

2 *Corylus sylueſtris* is our hedge Nut or Haſell Nut tree, which is very well knowne, and therefore needeth not any deſcription : whereof there are alſo ſundry ſorts, ſome great, ſome little, ſome rathe ripe, ſome later, as alſo one that is manured in our Gardens, which is very great, bigger than any Filberd, and yet a kinde of Hedge Nut : this then that hath beene ſaid ſhall ſuffice for Hedge-Nuts.

‡ 3   The ſmall Turky Nut tree growes but low, and the leaues grow without order, vpon the twigs, they are in ſhape like thoſe of the former, but ſomewhat longer : the chiefe difference conſiſts in the fruit, which is ſmall, and like an Haſell Nut, but ſhorter : the huske, wherein ſometimes one, otherwhiles more Nuts are contained, is very large, tough, and hard, diuided both aboue and below into a great many jags, which on euery ſide couer and hold in the Nuts, and theſe cups are very rough without, but ſmooth on the inſide. *Cluſius* firſt ſet this forth (hauing receiued it from Conſtantinople) by the name of *Auellana pumila Byzantina.* ‡

3 *Auellana pumila Byzantina cum ſuo fructu.*
The Filberd Nut of Conſtantinople.

¶ *The Place.*

The Haſell trees doe commonly grow in Woods and in dankiſh vntoiled places : they are alſo ſet in Orchards, the Nuts whereof are better, and of a ſweeter taſte, and be moſt commonly redde within.

¶ *The Time.*

The catkins or Aglets come forth very timely, before winter be fully paſt, and fall away in March or Aprill, ſo ſoone as the leaues come forth : the Nuts be ripe in Auguſt.

¶ *The Names.*

This ſhrub is called in Latine, *Corylus* : in Greeke, χαρυα Ποντικη, that is, *Nux Pontica*, or Ponticke Nut : in high Dutch, **Haſel ſtrauch** : in low Dutch, **Haſeleer** : in Engliſh, Haſell tree, and Filberd tree ; but the Filberd tree is properly that which groweth in gardens and Orchards, and whoſe fruit is commonly wholly couered ouer with the huske, and the ſhell is thinner.

The Nut is named in Latine, *Nux Pontica, tenuis Nux, parua Nux* : it is alſo called *Nux Praneſtina, Nux Heracleotica,* and commonly *Nux auellana*, by which name it is vſually knowne to the Apothecaries, in high Dutch, **Haſel Nuſz** : in low Dutch, **Haſel Noten** : in Italian, *Nocciuole, Auellane, Necelle* : in French, *Noiſettes, & Noiſelles* : in Spaniſh, *Auellanas* : in Engliſh, Haſell Nut, and Filberd.

Theſe

Theſe Nuts that haue their skins red are the garden and planted Nuts, and the right Ponticke Nuts or Filberds: they are called in high Dutch, **Rhurnuſʒ**, and **Rotnuſʒ**: in low Dutch, **Roode Haſel Noten**: in Engliſh, Filberds, and red Filberds.

The other Nuts which be white are iudged to be wilde.

¶ *The Temperature and Vertues.*

A   Haſell Nuts newly gathered, and not as yet dry, containe in them a certaine ſuperfluous moiſture, by reaſon whereof they are windy: not onely the new gathered Nuts, but the dry alſo, be very hard of digeſtion; for they are of an earthy and cold eſſence, and of an hard and ſound ſubſtance, for which cauſe alſo they very ſlowly paſſe through the belly, therefore they are troubleſome and clogging to the ſtomacke, cauſe head-ache, eſpecially when they be eaten in too great a quantity.

B   The kernells of Nuts made into milke like Almonds do mightily binde the belly, and are good for the laske and the bloudy flix.

C   The ſame doth coole exceedingly in hot feuers and burning agues.

D   The catkins are cold and dry, and likewiſe binding: they alſo ſtay the laske.

E   ‡ The kernels of Nuts rather cauſe than cure the bloudy flix and laske, wherefore they are not to be vſed in ſuch diſeaſes. ‡

CHAP. 90.   *Of the Wall-nut Tree.*

Nux Iuglans.
The Walnut tree.

¶ *The Deſcription.*

THis is a great tree with a thicke and tall body: the barke is ſomewhat greene, and tending to the colour of aſhes, and oftentimes full of clefts: the boughes ſpread themſelues far abroad: the leaues conſiſt of fiue or ſix faſtned to one rib, like thoſe of the aſh tree, and with one ſtanding on the top, which bee broader and longer than the particular leaues of the Aſh, ſmooth alſo, and of a ſtrong ſmell: the catkins or aglets come forth before the Nuts: theſe Nuts doe grow hard to the ſtalke of the leaues, by couples, or by three & three; which at the firſt when they be yet but tender haue a ſweet ſmel, and be couered with a green huske: vnder that is a wooddy ſhell in which the kernell is contained, being couered with a thin skin, parted almoſt into foure parts with a woody skin as it were: the inner pulpe whereof is white, ſweet and pleaſant to the taſt; and that is when it is new gathered, for after it is dry it becommeth oily and rancke.

¶ *The Place.*

The Walnut tree groweth in fields neere common high-wayes, in a fat and fruitfull ground, and in orchards: it proſpereth on high fruitfull bankes, it loueth not to grow in watery places.

¶ *The Time.*

The leaues together with the catkins come forth in the Spring: the Nuts are gathered in Auguſt.

¶ *The Names.*

The tree is called in Greeke, κάρυα: in Latine, *Nux*, which name doth ſignifie both the tree and the fruit: in high Dutch, **Nuſʒbaum**: in low Dutch, **Noote boome**, and **Nootelaer**: in French, *Neiſier*: in Spaniſh, *Nogueyra*: in Engliſh, Walnut tree, and of ſome, Walſh nut tree. The Nut is called in Greeke, κάρυον βασιλικὸν, that is to ſay, *Nux Regia*, or the Kingly Nut: it it likewiſe named *Nux*

Nux Inglans, as though you ſhould ſay *Iouis glans*, Iupiters Acorne ; or *Iuuans glans*, the helping A-corne: and of diuers, *Perſica Nux*, or the Perſian Nut : in high Dutch, **Welſch Nuſz**, and **Baum-nuſz**: in low Dutch, **Ookernoten**, **Walſch Noten**: in Italian, *Noci* : in French, *Noix* : in Spaniſh, *Nuezes*, and *Noues* : in Engliſh, Walnut ; and of ſome, Walſh nut.

¶ The Temperature and Vertues.

The freſh kernels of the nuts newly gathered are pleaſant to the taſte : they are a little cold, and **A** haue no ſmall moiſture, which is not perfectl, concocted : they be hard of digeſtion, and nouriſh little : they ſlowly deſcend.

The dry nuts are hot and dry, and thoſe more which become oily and tanke: theſe be very hurt- **B** full to the ſtomacke, and beſides that they are hardly concocted, they increaſe choler, cauſe head-ache, and be hurtfull for the cheſt, and for thoſe that be troubled with the cough.

Dry Nuts taken faſting with a fig and a little Rue withſtand poyſon, preuent and preſerue the **C** body from the infection of the plague, and being plentifully eaten they driue wormes forth of the belly.

The greene and tender Nuts boyled in Sugar and eaten as Suckad, are a moſt pleaſant and dele- **D** ctable meat, comfort the ſtomacke, and expell poiſon.

The oyle of Walnuts made in ſuch manner as oyle of Almonds, maketh ſmooth the hands and **E** face, and taketh away ſcales or ſcurfe, blacke and blew marks that come of ſtripes or bruiſes.

Milke made of the kernels, as Almond milke is made, cooleth and pleaſeth the appetite of the **F** languiſhing ſicke body.

With onions, ſalt, and honey, they are good againſt the biting of a mad dog or man, if they be **G** laid vpon the wound.

Being both eaten, and alſo applied, they heale in ſhort time, as *Dioſcorides* ſaith, Gangrens, Car- **H** buncles, ægilops, and the pilling away of the haire: this alſo is effectually done by the oyle that is preſſed out of them, which is of thin parts, digeſting and heating.

The outward greene huſke of the Nuts hath a notable binding faculty.   **I**

Galen deuiſed and taught to make of the juyce thereof a medicine for the mouth, ſingular good **K** againſt all inflammations thereof.

The leaues and firſt buds haue a certaine binding quality, as the ſame Author ſheweth, yet there **L** doth abound in them an hot and dry temperature.

Some of the later Phyſitions vſe theſe for baths and lotions for the body, in which they haue a **M** force to digeſt and alſo to procure ſweat.

Chap. 91.   *Of the Cheſtnut tree.*

¶ The Deſcription.

1   THe Cheſtnut tree is a very great and high tree : it caſteth forth very many boughes: the body is thicke, and ſometimes of ſo great a compaſſe as that two men can hardly fa-thom it : the timber or ſubſtance of the wood is ſound and durable : the leaues bee great, rough, wrinkled, nicked in the edges, and greater than the particular leaues of the Walnut tree. The blowings or catkins be ſlender, long, and greene : the fruit is incloſed in a round rough and prickly huſke like to an hedge-hog or Vrchin, which opening it ſelfe doth let fall the ripe fruit or Nut. This nut is not round, but flat on the one ſide, ſmooth, and ſharpe pointed : it is couered with a hard ſhell, which is tough and very ſmooth, of a darke browne colour : the meate or inner ſubſtance of the nut is hard and white, and couered with a thin skin which is vnder the ſhell.

2   The Horſe Cheſtnut groweth likewiſe to be a very great tree, ſpreading his great and large armes or branches far abroad, by which meanes it maketh a very good coole ſhadow. Theſe bran-ches are garniſhed with many beautifull leaues, cut or diuided into fiue, ſix, or ſeuen ſections or di-uiſions like to the Cinkefoile, or rather like the leaues of *Ricinus*, but bigger. The floures grow at the top of the ſtalkes, conſiſting of foure ſmall leaues like the Cherry bloſſome, which turne into round rough prickly heads like the former, but more ſharpe and harder : the Nuts are alſo rounder. ‡ The floures of this, ſaith *Cluſius*, (whoſe figure of them I here giue you) come out of the boſome of the leafe which is the vppermoſt of the branch, and they are many in number growing vpon pret-tie long foot-ſtalkes, conſiſting each of them of foure white leaues of no great bigneſſe ; the two vppermoſt are a little larger than the reſt, hauing round purple ſpots in their middles : out of the middle of the floure come forth many yellowiſh threds with golden pendants. The fruit is con-tained in a prickly huſke that open, in three parts, and it is rounder and not ſo ſharpe pointed as

the

1 *Castanea.*
Chestnut tree.

2 *Castanea Equina cum flore.*
Horse Chestnut tree in floure.

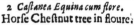

Castanea Equinæ fructus.

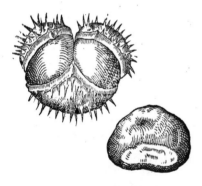

‡ 3 *Castanea Peruana fructus.*

the ordinary Chestnut, neither vnder the vtter coat hath it any peeling within as the other hath, neither is it of so good a taste. ‡

‡ 3   This Americane Chestnut is almost round, but that it is a little flatted on the sides, especially whereas it is fastned to the stalke : the vtter coat is sufficiently thicke, yet brittle, and as it were fungous, of a brownish yellow colour : vnder this are abundance of small, yet stiffe prickes, fast sticking to the shell that contains the kernel : the shell it selfe is brownish, not thick, but tough and hard to breake, smooth and shining on the inside, wherin is contained a kernell of the bignesse and color of an hares kidney, white within, and sweet in taste, like an almond or the common Chestnut. *Clusius* cals this *Castanea Peruana*, or Chestnut of Peru ; and hee saith hee had it from the famous Geographer *Abraham Ortelius*, who had it sent him by *Benedictus Areas Montanus*. The figure is exprest vnder that of the Horse Chestnut. ‡

¶ *The Place.*

The first growes on mountaines and shadowie places, and many times in the vallies : they loue a soft and blacke soile. There be sundry woods of Chestnuts in England, as a mile

mile and a halfe from Feuerſham in Kent, and in ſundry other places: in ſome countries they be greater and pleaſanter : in others ſmaller,and of worſe taſte.

The Horſe Cheſtnut groweth in Italy, and in ſundry places of the Eaſt-countries. ‡ It is now growing with Mr Tradeſcant at South Lambeth. ‡

¶ The Time.

The blowings or aglets come forth with the leaues in Aprill ; but the Nuts later,and be not ripe till Autumne.

¶ The Names.

The Cheſtnut tree beares the name of the Nut both in Greeke and Latine : in high Dutch, **Keaſtenbaum**, and **Kaſtanibaum :** in low Dutch, **Caſtaniboom :** in French, Caſtaignier : in Engliſh, Cheſtnut tree.

The Nut is called in Greeke, κάστανα in Latine, Caſtanea, Iouis glans, Sardinia glans : in high Dutch, **Keſten :** in low Dutch, **Caſtanien :** in Italian, Caſtagne : in French, Chaſtaigne : in Spaniſh, Morones, Caſtanas : in Engliſh, Cheſtnut : the greater Nuts be named of the Italians, Marroni : of the French men and of diuers baſe Almanes, Marons.

The Horſe Cheſtnut is called in Latine, Equina Caſtanea : in Engliſh, Horſe Cheſtnut, for that the people of the Eaſt countries do with the fruit thereof cure their horſes of the cough,ſhortneſſe of breath,and ſuch like diſeaſes.

¶ The Temperature and Vertues.

Our common Cheſtnuts are very dry and binding,and be neither hot nor cold, but in a mean be-   A tweene both : yet haue they in them a certaine windineſſe,and by reaſon of this,vnleſſe the ſhell be firſt cut,they skip ſuddenly with a cracke out of the fire whileſt they be roſting.

Of all the Acornes,ſaith Galen,the Cheſtnuts are the chiefeſt,and do onely of all the wilde fruits   B yeeld to the body commendable nouriſhment;but they ſlowly deſcend,they be hardly concoⓒted, they make a thicke bloud,and ingender winde : they alſo ſtay the belly, eſpecially if they be eaten raw.

Being boiled or roſted they are not of ſo hard digeſtion, they more eaſily deſcend, and are leſſe   C windy,and yet they alſo make the body coſtiue.

Some affirme, That of raw Cheſtnuts dried, and afterwards turned into meale, there is made a   D kinde of bread : yet it muſt needs be, that this ſhould be dry and brittle, hardly concoⓒted and very ſlow in paſſing through the belly ; but this bread may be good againſt the laske and bloudy flix.

An Electuary of the meale of Cheſtnuts and honey is very good againſt the cough and ſpitting   E of bloud.

The barke of the Cheſtnut tree boyled in wine and drunke, ſtops the laske, the bloudy flix, and   F all other iſſues of bloud.

CHAP. 92.   Of the Beech tree.

¶ The Deſcription.

THe Beech is an high tree, with boughes ſpreading oftentimes in manner of a circle, and with a thicke body hauing many armes : the barke is ſmooth : the timber is white, hard, and very profitable : the leaues be ſmooth,thin,broad,and leſſer than thoſe of the blacke Poplar : the catkins or blowings be alſo leſſer and ſhorter than thoſe of the Birch tree and yellow : the fruit or Maſt is contained in a huske or cup that is prickly, and rough briſtled, yet not ſo much as that of the Cheſtnut : which fruit being taken forth of the ſhells or vrchin huskes, be couered with a ſoft and ſmooth skin like in colour and ſmoothneſſe to the Cheſtnuts, but they be much leſſer, and of another forme,that is to ſay,triangled or three cornered : the kernell within is ſweet,with a certaine aſtriction or binding qurlitie : the roots be few,and grow not deepe,and little lower than vnder the turfe.

¶ The Place.

The Beech tree loueth a plaine and open country,and groweth very plentifully in many Forreſts and deſart places of Suſſex, Kent, and ſundry other countries.

¶ The Time.

The Beech floureth in Aprill and May, and the fruit is ripe in September,at what time the Deere do eate the ſame very greedily, as greatly delighting therein ; which hath cauſed forreſters and huntſmen to call it Buck-maſt.

¶ The

Fagus.
The Beech.

¶ *The Names.*

The tree is called in Greeke, φηγὸς in Latine, *Fagus* : in high Dutch, **Buchbaum**, or **Buch**: in low Dutch, **Bukenboom**: in Italian, *Faggi* : in Spanish, *Haia*, *Faia*, and *Fax* : in French, *Fau*, or *Hestre*: in English, Beech tree, Beech-mast, and Buck-mast.

The fruit is called in Latine, *Nuces Fagi*: in Greeke, βάλανοι ἢ φηγοὶ: in low Dutch, **Bueken nootkeng**: in French, *Faine*: in English, Beech-mast. *Dioscorides* reckons the Beech among the Acorne trees, and yet is the mast nothing at all like to an Acorne. Of *Theophrastus* it is called Oxya : of *Gaza*, *Sciscina*.

Pliny also makes mention of this tree, but vnder the name of *Ostrya* (if so be in stead of *Ostrya* we must not reade *Oxya*) *lib*.13.*ca*.21. It brings forth, (saith he, meaning Greece) the tree *Ostrys*, which they likewise call *Ostrya*, growing alone among watery stones, like to the Ash tree in barke and boughes, with leaues like those of the Peare tree, but somwhat longer and thicker, and with wrinkled cuts which run quite through, with a seed like in colour to a Chestnut, and not to barley: the wood is hard and firme, which being brought into the house there followes hard trauell of child, and miserable deaths, as it is reported ; and therefore it is to be forborne, and not vsed as fire wood, if *Plinies* copies be not corrupted.

¶ *The Temperature.*

The leaues of Beech do coole : the kernell of the Nut is somewhat moist.

¶ *The Vertues.*

A   The leaues of Beech are very profitably applied vnto hot swellings, blisters, and excoriations ; and being chewed they are good for chapped lips, and paine of the gums.

B   The kernels or mast within are reported to ease the paine of the kidneys proceeding of the stone, if they be eaten, and to cause the grauell and sand the easier to come forth. With these, mice and Sqirrels are greatly delighted, who do mightily increase by feeding thereon : Swine also be satned herewith, and certaine other beasts : also Deere doe feed thereon very greedily : they be likewise pleasant to Thrushes and Pigeons.

C   *Petrus Crescentius* writeth, That the ashes of the wood is good to make glasse with.

D   The water that is found in the hollownesse of Beeches cureth the naughty scurfe, tetters, and scabs of men, horses, kine, and sheepe, if they be washed therewith.

Chap. 93. *Of the Almond tree.*

¶ *The Description.*

THe Almond tree is like the Peach-tree, yet is it higher, bigger, of longer continuance : the leaues be very long, sharpe pointed, snipt about the edges like those of the Peach tree : the floures be alike : the fruit is also like a Peach, hauing on one side a cleft, with a soft skin without, and couered with a thin cotton ; but vnder this there is none, or very little pulpe, which is hard like a gristle not eaten : the nut or stone within is longer than that of the Peach, not so rugged, but smooth; in which is contained the kernel, in taste sweet, and many times bitter : the root of the tree groweth deepe : the gum which soketh out hereof is like that of the Peachtree.

‡   There are diuers sorts of Almonds, differing in largenes and taste: we commonly haue three or foure sorts brought to vs, a large sweet Almond, vulgarly termed a Iordan Almond ; and a lesser, called a Valence Almond : a bitter Almond of the bignesse of the Valence Almond, and somtimes another bitter one lesse than it. ‡

¶ The

Amygdalus.
The Almond tree.

¶ *The Place.*

The naturall place of the Almond is in the hot regions, yet we haue them in our London gardens and orchards in great plenty.

¶ *The Time.*

The Almond floureth betimes with the Peach : the fruit is ripe in August.

¶ *The Names.*

The tree is called in Greeke, Ἀμυγδάλη in Latine, *Amigdalus :* in French, *Amandier :* in English, Almond tree.

The fruit is called in Greeke, Ἀμύγδαλον in Latine, *Amygdalum :* in shops, *Amygaala :* in high Dutch, **Mandel :** in low Dutch, **Mandelen :** in Italian *Mandole* in Spanish, *Almendras, Amelles,* and *Amendoas :* in French, *Amandes :* in English Almond.

¶ *The Temperature and Vertues.*

Sweet Almonds when they be dry be mo- A derately hot ; but the bitter ones are hot and dry in the second degree. There is in both of them a certain fat and oily substance, which is drawne out by pressing.

Sweet Almonds beeing new gathered are B pleasant to the taste, they yeeld some kinde of nourishment, but the same grosse and earthy, and grosser than those that be dry, and not as yet withered. These doe likewise slowly descend, especially being eaten without their skins ; for euen as the huskes or branny parts of corne doe serue to driue downe the grosse excrements of the belly, so doe likewise the skins or huskes of the almonds : therefore those that be blanched do so slowly descend, as that they do withall binde the belly ; whereupon they are giuen with good successe vnto those that haue the laske or the bloudy flix.

There is drawne out of sweet Almonds, with liquor added, a white juyce like milke, which ouer C and besides that it nourisheth, and is good for those that are troubled with the laske and bloudy flix, it is profitable for those that haue the plurisie and spit vp filthy matter, as *Alexander Trallia-nus* witnesseth : for there is likewise in the Almonds an opening and concocting quality, with a certaine clensing faculty, by which they are medicinable to the chest and lungs, or lights, and serue for raising vp of flegme and rotten humors.

Almonds taken before meate do stop the belly, and nourish but little ; notwithstanding many D excellent meates and medicines are therewith made for sundry griefes, yea very delicat and wholsome meates, as Almond butter, creame of Almonds, marchpane, and such like, which dry and stay the belly more than the extracted juyce or milke ; and they are also as good for the chest and lungs.

They doe serue also to make the Physicall Barley Water, and Barley Creame, which are giuen E in hot Feuers, as also for other sicke and feeble persons, for their further refreshing and nourishments.

The oyle which is newly pressed out of the sweet Almonds is a mitigater of paine and all maner F of aches. It is giuen to those that haue the pleurisie, being first let bloud ; but especially to those that are troubled with the stone in the kidnies : it slackens the passages of the vrine, and maketh them glib or slippery, and more ready to suffer the stone to haue free passage : it maketh the belly soluble, and therefore it is likewise vsed for the collicke.

It is good for women that are newly deliuered ; for it quickly remoueth the throwes which re- G maine after their deliuery.

The oile of Almonds makes smooth the hands and face of delicat persons, and clenseth the skin H from all spots, pimples, and lentils.

Bitter Almonds doe make thin and open, they remoue stoppings out of the liuer and spleene, I therefore they be good against paine in the sides : they make the body soluble, prouoke vrine, bring downe the menfes, helpe the ftrangury, aud clenfe forth of the cheft aud lungs clammie humors: if they be mixed with fome kinde of looch or medicine to licke on: with ftarch they ftay the fpitting of bloud.

L
M   And it is reported that fiue or fix being taken fafting do keepe a man from being drunke.

   Thefe alfo clenfe and take away fpots and blemifhes in the face, and in other parts of the body; they mundifie and make cleane foule eating vlcers.

N   With hony they are laid vpon the biting of mad dogs; being applied to the temples with vineger or oile of Rofes, they take away the head-ache, as *Diofcorides* writeth.

O
P   They are alfo good againft the cough and fhortneffe of winde.

   They are likewife good for thofe that fpit bloud, if they be taken with the fine floure of *Amylum*.

Q   There is alfo preffed out of thefe an oile which pruoketh vrine, but efpecially if a few fcorpions be drowned and fteeped therein.

R   With oile it is fingular good for thofe that haue the ftone, and cannot eafily make water but with extremitie of paine, if the fhare and place betweene the cods and fundament be anoynted therewith.

S   *Diofcorides* faith, That the gum doth heate and binde, which qualities notwithftanding are not perceiued in it.

T   It helpeth them that fpit bloud, not by a binding faculty, but through the clamminesse of his fubftance, and that is by clofing vp of the paffages and pores, and fo may it alfo cure old coughes, and mitigate extreme paines that proceed of the ftone, and efpecially take away the fharpeneffe of vrine, if it be drunke with Baftard, or with any other fweet potion, as with the decoction of Licorice, or of Raifons of the funne. The fame doth likewife kill tetters in the outward parts of the body (as *Diofcorides* addeth) if it be diffolued in vineger.

CHAP. 94. *Of the Peach tree.*

¶ *The Kindes.*

‡ THere are diuers forts of Peaches befides the foure here fet forth by our Author, but the trees do not much differ in fhape, but the difference chiefely confifts in the fruit, whereof I will giue you the names of the choife ones, and fuch as are to be had from my friend Mr *Millen* in Old-ftreet, which are thefe; two forts of Nutmeg Peaches; The Queenes Peach; The Newington Peach; The grand Carnation Peach; The Carnation Peach; The blacke Peach; The Melocotone; The White; The Romane; The Alberza; The Ifland Peach; Peach du Troy. Thefe are all good ones. He hath alfo of that kinde of Peach which fome call *Nuciperfica* or Nectorins, thefe following kindes; the Roman red, the beft of fruits; the baftard Red; the little dainty greene; the Yellow; the White; the Ruffet, which is not fo good as the reft. Thofe that would fee any fuller difcourfe of thefe may haue recourfe to the late worke of Mr *Iohn Parkinfon*, where they may finde more varieties, and more largely handled, and therefore not neceffary for me in this place to infift vpon them. ‡

¶ *The Defcription.*

1 THe Peach tree is a tree of no great bigneffe: it fendeth forth diuers boughes, which be fo brittle, as oftentimes they are broken with the weight of the fruit or with the winde. The leaues be long, nicked in the edges, like almoft to thofe of the Walnut tree, and in tafte bitter: the floures be of a light purple colour. The fruit or Peaches be round, and haue as it were a chinke or cleft on the one fide; they are couered with a foft and thin downe or hairy cotton, being white without, and of a pleafant tafte; in the middle whereof is a rough or rugged ftone, wherein is contained a kernell like vnto the Almond; the meate about the ftone is of a white color. The root is tough and yellowifh.

2 The red Peach tree is likewife a tree of no great bigneffe: it alfo fendeth forth diuers boughes or branches which be very brittle. The leaues be long, and nicked in the edges like to the precedent. The floures be alfo like vnto the former; the fruit or Peaches be round, and of a red colour on the outfide; the meate likewife about the ftone is of a gallant red colour. Thefe kindes of Peaches are very like to wine in tafte, and therefore maruellous pleafant.

3 *Perfica præcocia*, or the d'auant Peach tree is like vnto the former, but his leaues are greater and larger. The fruit or Peaches be of a ruffet colour on the one fide, and on the other fide next vnto the Sun of a red colour, but much greater than the red Peach: the ftones whereof are like vnto the former: the pulpe or meate within is of a golden yellow colour, and of a pleafant tafte.

4 *Perfica*

Persica alba.
The white Peach.

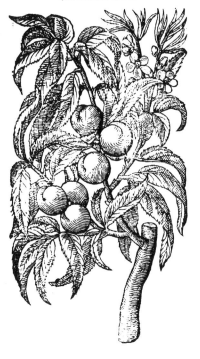

4 *Persica lutea*, or the yellow Peach tree is like vnto the former in leaues and flours, his fruit is of a yellow color on the out side, and likewise on the in side, harder than the rest: in the middle of the Peach is a wooddy hard and rough stone full of crests and gutters, in which doth ly a kernel much like to that of the almond, and with such a like skin: the substance within is white, and of taste somewhat bitter. The fruit hereof is of greatest pleasure, and of best taste of all the other of his kinde; although there be found at this day diuers other sorts that are of very good taste, not remembred of the antient, or set down by the later Writers, whereof to speake particularly would not bee great to our pretended purpose, considering wee hasten to an end.

‡ 5 There is also kept in some of our choice gardens a kind of Peach which hath a very double and beautifull floure, but it is seldom succeeded by any fruit: they call this *Persica flore pleno*, The double blossomed Peach. ‡

¶ *The Place.*

They are set and planted in gardens and Vineyards. I haue them all in my garden, with many other sorts.

¶ *The Time.*

The Peach tree soone comes vp, it beares fruit the third or fourth yeare after it is planted, and it soon decayeth, being not of long continuance. It flours in Aprill, or a little after the leaues appeare, and ripens his fruit in September.

¶ *The Names.*

The Peach tree is called in Greeke, Μηλέα Περσική: in Latine, *Malus Persica*, and *Persica*: in High-Dutch, 𝕻erſichboum: in low-Dutch, 𝕻erſe boom: in French, *Perscher*: in English, Peach tree. ¶ The fruit, as *Galen* testifieth, is named Μῆλα Περσικά, and Περσικόν also, without any addition: in Latine, *Malum Persicum*, and *Persicum*: in high-Dutch, 𝕻ferſing: in low-Dutch, 𝕻erſen: in Italy, *Pesche*: in Spanish, *Pexegos*: in French, *Pisches*: in English, Peach.

¶ *The Nature and Vertues.*

Peaches be cold and moist, and that in the second degree, they haue a juice and also a substance **A** that doth easily putrifie, which yeeldeth no nourishment, but bringeth hurt, especially eaten after other meats, for then they cause the other meats to putrifie. But they are lesse hurtfull if they be taken first; for by reason that they are moist and slippery, they easily and quickly descend, and by making the belly slippery, they cause other meats to slip down the sooner.

The kernels of the Peaches be hot and dry, they open and clense, and are good for the stoppings **B** of the liuer and spleen.

Peaches before they be ripe do stop the laske, but being ripe they loose the belly, and ingender **C** naughty humors, for they are soon corrupted in the stomacke.

The leaues of the Peach tree do open the stoppings of the liuer, and do gently loosen the belly: **D** and being applied plaisterwise vnto the nauel of yong children, they kil the worms, and driue them forth.

The same leaues boiled in milke do kill the wormes in children very speedily.　　　　　　　**E**

The same being dried and cast vpon green wounds cure them.　　　　　　　　　　　　　　**F**

The floures of the Peach tree infused in warme water for the space of ten or twelue houres, and **G** strained, and more floures put to the said liquor to infuse after the same manner, and so iterated six or eight times and strained againe, then as much sugar as it will require added to the same liquor, and boiled vnto the consistence or thicknesse of a syrrup, and two spoonefulls hereof taken, doth so singularly well purge the belly, that there is neither Rubarb, Agarick, nor any other purger comparable vnto it; for this purgeth down waterish humors mightily, and yet without griefe or trouble, either to the stomacke or lower parts of the body.

H   The kernell within the Peach ſtone ſtamped ſmall, and boiled with vineger vntill it be brought to the form of an ointment, is good to reſtore and bring again the haire of ſuch as be troubled with the *Alopecia*.

I   There is drawn forth of the kernels of peaches, with Peniroyall water, a juice like to milk, which is good for thoſe that haue the Apoplexy : if the ſame be oftentimes held in the mouth, it draweth forth water and recouereth ſpeech.

K   The gum is of a meane temperature, but the ſubſtance thereof is tough and clammy, by reaſon whereof it dulleth the ſharpnes of thin humors : it ſerueth in a looch or licking medicine for thoſe that be troubled with the cough, and haue rotten lungs, and ſtoppeth the ſpitting and raiſing vp of bloud, and alſo ſtayeth other fluxes.

Chap. 95.
Of the Aprecocke or Abrecocke tree.

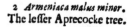

1 *Armeniaca malus major.*   2 *Armeniaca malus minor.*
The greater Aprecocke tree.   The leſſer Aprecocke tree.

¶ *The Deſcription.*

1   THis tree is greater than the Peach tree, and hath a bigger body, it laſteth longer, eſpecially if it be graffed or inoculated: the leaues hereof are broad and ſharp pointed, like thoſe of blacke Poplar, but leſſer, and comming more neere to the leaues of Birch, nicked in the edges : the floures are ſomewhat white, the fruit round like a Peach, yellow within and without, in which doth lie a browne ſtone nothing rough at all as is that of the Peach, ſhorter alſo and leſſer, in which is included a ſweet kernell.

2   We haue another ſort of Aprecock, whoſe trunke or body is equal with the other in greatneſſe, it is like alſo in leaues and brittle branches : his time of flouring, flouriſhing, and manner of growing accordeth : the only point wherein they differ is, that this tree bringeth forth leſſe fruit, and not ſo good in taſte, in euery other reſpect it is like.

‡ Of this alſo Mr *Parkinſon* hath ſet forth diuers varieties, and my fore mentioned friend Mr *Millen* hath theſe fiue ſorts, the common, the long and great, the Muske, the Barbary, and the early Aprecocke. ‡

¶ *The*

¶ *The Place.*

These trees grow in my garden, and now adaies in many other gentlemens gardens throughout England.

¶ *The Time.*

They floure and flourish in Aprill, and their fruit is ripe in Iuly.

¶ *The Names.*

The tree is called in Greeke, μῆλα Ἀρμενιακὰ in Latine, *Malus Armeniaca*: in English, Abrecock tree and Aptecock tree.

The fruit is named μῆλα Ἀρμενιακὰ, and of diuers, πραικόκιον, or Βρικόκκιον, which bee words corrupted from the Latine, for *Praecox* in Latine is diuers times called *Praecoquum*: it is named *Malum Armeniscum*, and commonly *Armeniacum*: in high-Dutch, **Molleten Mollelin, S. Johans Pferling**: in low-Dutch, **Uroege Persen, Auant Persen**: in Italian, *Armeniache, Bacoche, Crisomele, Moniache*: in French, *Abricoz*: in Spanish, *Aluarcoques, Aluarchigas*, and *Albercocs*: in English, Abrecocke, Aprecock, and Aprecox.

Galen seems to make a difference between *Praecocia* and *Armeniaca*, in his book of the Faculties of Nourishments, preferring *Praecocia* before *Armeniaca*: yet he doth confesse that both of them bee called *Armeniaca*. Others pronounce them *Armenia* with foure syllables. And in his book of the Faculties of simple medicines he affirmeth, that both the fruit and the tree are called πραικόκιον: Diuers also of the later physitians do betwixt these make a difference, saying, that the greater ones & those that are graffed be *Armeniaca* (which the French men call *Auant Perses*) and the lesser *Praecosia*: in French, *Abricoz*.

¶ *The Temperature and Vertues.*

Aprecockes are cold and moist in the second degree, but yet not so moist as Peaches, for which cause they do not so soon putrifie; they are also more wholsome to the stomack and pleasant to the taste, yet doe they likewise putrifie and yeeld but little nourishment, and the same cold, moist, and full of excrements. Being taken after meat they corrupt and putrifie in the stomack; being first eaten before meat they easily descend, and cause other meats to passe down the sooner, like as also the Peaches do.

The kernel within it is sweet, and nothing at all in facultie like that of the Peach.

The vertues of the leaues of this tree are not yet found out.

A

B
C

Chap. 96.　*Of the Pomegranat tree.*

¶ *The Kindes.*

AS there be sundry sorts of Apples, Peares, Plums, and such like fruits; so there are two sorts of Pomegranats, the garden and the wilde, and a third sort which is barren and fruitlesse: the fruit of the garden Pomegranat is of three sorts; one hauing a soure juice or liquor, another a very sweet and pleasant liquor, and the third the tast of wine. Of the wild also there be two sorts, and the difference between them is no more than betwixt crabs and weildings, which are both wilde kindes of apples. Therefore the description of the garden Pomegranat shall suffice for the rest.

¶ *The Description.*

1　THe manured Pomegranat tree groweth vp to the height of an hedge tree, being seuen or eight cubits high, hauing many pliant twiggy branches, very limber, tough, and of a browne colour, whereon are set very many leaues in shape like those of the Priuet, but more like those of the Myrtle tree, of a bright shining green colour tending to yellownesse: among which stand certain sharp thorns confusedly set, and likewise hollow floures like to the hedge rose, indented on the edges like a star, of a Carnation colour, and very single: after which commeth the fruit, couered with an hard bark of an ouerworne purplish colour, full of grains and kernells, which after they be ripe are of a crimson colour, and full of juice, which differeth in tast according to the soile, clymat, and country where they grow; some be sweet, others soure, and the third are in a middle between them both, hauing the taste of wine.

† 2　The wild Pomegranat tree is like the other in leaues and twiggy branches, but it is more prickly and horrid. Of this there are two sorts, the one hauing such floures and fruit as the tame Pomegranat, the other bearing floures very double, as may appeare by the figure, which wither and fall away leauing no fruit behind them, as the double floured Cherry doth, and diuers other herbes

　　　　and

1 *Malus granata sive Punica.*
The Pomegranat tree.

2 *Malus Punica sylvestris.*
The wilde Pomegranat.

Balaustia, sive pleni flores Gran syl.
The double floures of wild Pomegranat.

and trees alfo, and it is altogether barren of fruit. Of this *Diofcorides* makes fundry forts, differing in colour; one is white, faith hee, another yellowifh red, and a third fort of the colour of the Rofe. This with red floures is beft known to Apothecaries.

¶ *The Place.*

Pomgranats grow in hot countries toward the South, in Italy, Spain, and chiefely in the kingdome of Granado, which is to be fo named of the great multitude of Pomegranates, which be commonly called *Granata.* They grow in a number of places alfo without mannuring, yet being manured they profper better; for in gardens, vineyards, orchards, & fuch other husbanded grounds they come vp more chearfully. I haue recouered diuers yong trees hereof by fowing the feed or graines, of three or foure cubits high, attending Gods leifure for floures and fruit.

¶ *The Time.*

The Pomegranat floureth in the months of May and Iune. The fruit is ripe in the end of Auguft.

¶ *The Names.*

The Pomegranat tree is called in Latine, *Malus Punica* : in Greeke of the Athenians, Ῥόα
and

and ʒᵘⁱ, as *Galen* ſaith ; in Engliſh, Pomegranat tree.

The Fruit alſo is named ᵒⁱᵃ, and ʼᵖ ⁱᵃ, in Latine, *Malum Punicum* : in ſhops, *Malam*, or *Pomum Granatum* : in high-Dutch, **Granatopffel** : in low-Dutch, **Granatappel** : in Italian, *Melagrano*, and *Pomo Granato*. in Spaniſh, *Granadas*, and *Romanas* : in French, *Pommes Granades* : in Engliſh, Pomegranat.

The floure of the fruitfull Pomgranat tree is called of the Grecians *κύτινος*, which is notwithſtanding properly the cup of the floure : the Latines name it alſo *Cytinus*.

The floure of the wild and barren Pomegranat tree is called *Balauſtion* : the Apothecaries likewiſe term it *Balauſtium*.

The pill or rinde of the Pomegranat, ſo much in vſe, is named in Greeke *σίδιον* : in Latine, *Malicorium*, and *Sidium* : in ſhops, *Cortex Granatorum*, or Pomegranat pill.

¶ *The Nature and Vertues.*

The juicy grains of the Pomegranat are good to be eaten, hauing in them a meetly good juyce ; they are wholſome for the ſtomack, yet contain but thin and ſmall nouriſhment or none at all,   A

The ſweet ones be not ſo cold as the reſt, but they eaſily cauſe hot ſwellings to ariſe, and are not ſo much commended for agues.   B

The ſoure ones, eſpecially if they be withall ſomthing harſh, do euidently coole, dry, and ſomewhat binde.   C

They are good for the heart-burn, they repreſſe and ſtay the ouermuch vomiting of choler, called the Felony ; they help the bloudy flix, aptneſſe to vomit, and vomiting.   D

There is made of the juice of theſe ſoure pomegranates a ſyrrup ſeruing for the ſame purpoſes, and is alſo many times very profitable againſt the longing of women with childe, vnleſſe the coldneſſe of the ſtomack hinder it.   E

The ſeeds of the grains, and eſpecially of the ſoure Pomegranat, being dried, do alſo coole and binde.   F

They ſtop the flix, ſtay vomiting and ſpitting vp of bloud, and ſtrengthen the ſtomacke.   G

Of the ſame effect be the floures both of the tame and wilde Pomegranat tree, being like to the ſeeds in temperature and vertues.   H

They faſten the teeth and ſtrengthen the gums, if they be waſhed therewith.   I

They are good againſt Burſtings that come by falling down of the guts, if they be vſed in plaiſters and applied.   K

The rinde or pill is not only like in facultie to the ſeeds, and both the ſorts of floures, but alſo more auailable ; for it cooleth and bindeth more forcibly ; it bringeth downe the hot ſwellings of the almonds in the throat, being vſed in a gargariſm or lotion for the throat, and is a ſingular remedie for all things that need cooling and binding.   L

Dioſcorides writeth, that there is alſo gathered a juice out of both theſe ſorts of floures, which is very like in facultie to *Hypociſtis*.   M

The bloſſoms of the tame and wild Pomegranats, as alſo the rind or ſhel thereof made into pouder, and drunk in red wine, or boiled in red wine, and the decoction drunk, is good againſt the bloudy flix and all other iſſues of bloud ; yea it is good for women to ſit ouer and bathe themſelues in the decoction hereof. The bloſſoms and ſhels alſo are good to be put into reſtraining pouders, to ſtanch bloud in wounds.   N

The ſeeds or ſtones of Pomegranats dried in the Sun and beaten to pouder, are of like operation with the floures : they ſtop the lask and all iſſues of bloud in man or woman, being taken as aforeſaid.   O

C H A P. 97.   *Of the Quince tree.*

¶ *The Kindes.*

Columella maketh three kindes of Quinces, *Struthia*, *Chryſomeliana* and *Muſtela*, but what maner ones they be he doth not declare. Notwithſtanding we finde diuers ſorts, differing as well in forme, as taſte and ſubſtance of the fruit, whereof ſome haue much core and many kernels, and others fewer.

¶ *The*

Malus Cotonea.
The Quince tree.

¶ *The Description.*

THe Quince tree is not great,but growes low, and many times in maner of a shrub: it is couered with a rugged barke, which hath on it now and then certain scales: it spreadeth his boughes in compasse like other trees, about which stand leaues somewhat round like those of the common Apple tree, greene and smooth aboue, and vnderneath soft and white: the flours be of a white purple colour: the fruit is like an Apple, sauing that many times it hath certain embowed & swelling diuisions:it differeth in fashion and bignesse ; for some Quinces are lesser and round, trust vp together at the top with wrinckles, others longer and greater: the third sort be of a middle manner betwixt both ; they are all of them set with a thinne cotton or freese, and be of the colour of gold, and hurtfull to the head by reason of their strong smell ; they all likewise haue a kinde of choking tast: the pulp within is yellow, and the seed blackish,lying in hard skins as do the kernels of other apples.

¶ *The Place.*

The Quince groweth in gardens and orchards, and is planted oftentimes in hedges and Fences belonging to Gardens and Vineyards:it delighteth to grow on plain and euen grounds, and somwhat moist withall.

¶ *The Time.*

These apples be ripe in the fall of the leafe,and chiefly in October.

¶ *The Names.*

The tree is called in Greek Μηλεα κυδωνια : in Latine,*Malus Cotonea* : in English,Quincetree.
The fruit is named Μηλον κυδωνιον, *Malum Cotoneum,Pomum Cydonium*,and many times *Cydonium* without any addition ; by which name it is knowne to the Apothecaries: it is called in high-Dutch, **Quitten,Quietenopffel**,or **Kuttenopffel :** in low-Dutch,**Queappel :** in Italian, *Mele cotogne* : in Spanish,*Codoyons,Membrilhos*,and *Marmellos* : in French,*Pomme de coing* : in English,**Quince.**

¶ *The Temperature and Vertues.*

A   Quinces be cold & dry in the second degree,and also very much binding,especially when they be raw ; they haue likewise in them a certain superfluous and excremental moisture,which wil not suffer them to lie long without rotting : they are seldom eaten raw,being rosted or baked they are more pleasant.

B   They strengthen the stomack,stay vomiting,stop lasks and also the bloudy flix.

C   They are good for those that spit or vomit bloud, and for women also that haue too great plenty of their monethly courses.

D   *Simeon Sethi* writeth,that the woman with childe that eateth many Quinces during the time of her breeding,shall bring forth wise children and of good vnderstanding.

E   The Marmalad or Cotiniat made of quinces and sugar is good and profitable to strengthen the stomack,that it may retain and keep the meat therein vntill it be perfectly digested ; it also staieth all kindes of fluxes both of the belly and other parts,and also of bloud. Which Cotiniat is made in this manner:

F   Take faire Quinces,paire them,cut them in pieces,and cast away the core, then put vnto euery pound of Quinces a pound of Sugar, and to euery pound of Sugar a pinte of water: these must be boiled together ouer a stil fire till they be very soft,then let it be strained or rather rubbed through a strainer or an hairy Siue, which is better, and then set it ouer the fire to boile againe, vntill it be
stiffe,

ſtiffe, and ſo box it vp, and as it cooleth put thereto a little Roſe water, and a few graines of muske mingled together, which will giue a goodly taſte to the Cotiniat.   This is the way to make Marmalad.

Take whole Quinces and boile them in water vntil they be as ſoft as a ſcalded codling or apple,   G then pill off the skin, and cut off the fleſh, and ſtamp it in a ſtone morter, then ſtraine it as you did the Cotiniat; afterward put it in a pan to dry, but not to ſeeth at all, and vnto euery pound of the fleſh of quinces put three quarters of a pound of ſugar, and in the cooling you may put in roſe water and a little muske, as was ſaid before.

There is boiled with Quinces oile, which therefore is called in Greeke *Melinon* or oile of Quin-   H ees, which we vſe, ſaith *Dioſcorides*, ſo oft as we haue need of a binding thing.

The ſeed of Quinces tempered with water doth make a muſcilage, or a thing like jelly, which   I being held in the mouth, is maruellous good to take away the roughneſſe of the tongue in hot burning feuers.

The ſame is good to be laid vpon burnings and ſcaldings, and to be put into cliſters againſt the   K bloudy flix; for it eaſeth the pain of the guts, and allayeth the ſharpneſſe of biting humors.

Many other excellent dainty and wholſome Confeĉtions are to be made of Quinces, as jelly of   L Quinces, and ſuch like conceits, which for breuities ſake I do now let paſſe.

CHAP. 98.   *Of the Medlar tree.*

¶ *The Kindes.*

THere be diuers ſorts of Medlars, ſome greater, others leſſer ; ſome ſweet, and others of a more harſh taſt : ſome with much core and many great ſtony kernels, others fewer ; and likewiſe one of Naples called Aronia.

1 *Meſpilus ſativa.*
The manured Medlar.

‡ 2 *Meſpilus ſativa altera.*
The other garden Medlar.

¶ *The Description.*

1   THe manured Medlar tree is not great,the body whereof is wrythed,the boughes hard, not easie to be broken : the leaues be longer,yet narrower than those of the Apple tree, darke,green aboue,and somwhat whiter and hairy below : the flours are white & great, hauing fiue leaues apiece : the fruit is small,round,and hath a broad compassed nauel or crowne at the top : the pulp or meat is at the first white,and so harsh or choking,that it canot be eaten vntill it become soft,in which are contained fiue seeds or stones,which be flat and hard.

‡ 2   There is another which differs from the last described, in that the leaues are longer and narrower,the stocke hath no prickles vpon it,the fruit also is larger,and better tasted : in other respects it is like to the last described. This is the *Mespilus fructu praestantiore* of *Tragus* ; and *Mespilus domestica* of *Lobel.* ‡

3   The Neapolitan Medlar tree groweth to the height and greatnesse of an apple tree, hauing many tough and hard boughes or branches,set with sharp thornes like the white Thorne or Hawthorn : the leaues are very much cut or jagged like the Hawthorn leaues,but greater,and more like Smallage or Parsley : which leaues before they fall from the tree do wax red :among st these leaues come forth great tufts of flours of a pale herby colour: which being past,there succeed small long fruit lesser than the smallest Medlar,which at the first are hard,and green of colour,but when they be ripe they are both soft and red, of a sweet and pleasant taste ; wherein is contained three small hard stones as in the former,which be the kernels or seeds thereof.

3 *Mespilus Aronia.*
The Neapolitan Medlar.

‡ 4 *Chamaemespilus.*
Dwarfe Medlar.

4   There is a dwarfe kinde of Medlar growing naturally vpon the Alps and hills of Narbone, and on the rocks of mount Baldus nigh Verona,which hath bin by some of the best Learned esteemed for a kinde of Medlar : others,whose judgements cannot stand with truth or probability,haue supposed it to be *Euonymus* of the Alpes. This dwarfe Medlar groweth like a small hedge tree, of four or fiue cubits high,bearing many smal twiggy wands or crops,beset with many slender leaues greene aboue,and of a skye colour vnderneath, in shew like to a dwarfe Apple tree, but the fruit is
very

very like the Haw, or fruit of the white Thorne, and of a red colour. ‡ The floures come forth in the Spring three or foure together, hollow and of an herby colour, it growes in diuers places of the Alpes : it is the *Chamæmeſpilum* of the *Adu.rſ.* and the *Chamæmeſpilus Geſneri*, of *Cluſius*. ‡

¶ *Th. Place.*

The Medlar trees do grow in Orchards, and oftentimes in hedges among briers and brambles ; being grafted in a White Thorne it proſpereth wonderfull well, and bringeth forth fruit twiſe or thriſe bigger than thoſe that are not grafted at all, almoſt as great as little apples : we haue diuers ſorts of them in our Orchards.

¶ *The Time.*

It is very late before Medlars be ripe, which is in the end of October, but the floures come forth timely enough.

¶ *The Names.*

The firſt is called in Greeke by *Theophraſtus* μεσπίλη· in Latine, *Meſpilus* : in high Dutch, **Neſpel‐baum** : in low Dutch, **Miſpelboome** : in French, *Neſflier* : in Engliſh, Medlar tree.

The Apple or fruit is named in Greeke, μέσπιλον : in Latine likewiſe, *Meſpilum* : in high Dutch, **Neſpel**, in low Dutch, **Miſpele** : in Italian, *Neſpolo* : in French, *Neſfle* : in Spaniſh, *Neſperas* : in Engliſh, Medlar.

Dioſcorides affirmeth, That this Medlar tree is called επιμηλίς, and of diuers, *Sitanion* : *Galen* alſo in his booke of the faculties of ſimple medicines nameth this *Epimelis* : which is called, as he ſaith, by the country-men in Italy, *Vnedo*, and groweth plentifully in Calabria ; for vnder the name of *Meſpi‐lus*, or Medlar tree, he meaneth no other than *Tricoccus*, which is alſo named *Aronia*.

The Neapolitane Medlar tree is called in Greeke μέσπιλος, and μέσπιλη : *Galen* calleth it *Epimelis*. The fruit hereof is called *Tricoccos*, of the three graines or ſtones that it hath: they of Naples call it *Azarolo* : and we may name it in Engliſh, three graine Medlar, or Neapolitane Medlar, or Medlar of Naples.

¶ *The Temperature.*

The Medlars are cold, dry, and aſtringent; the leaues are of the ſame nature : the dwarfe Medlar is dry, ſharpe, and aſtringent.

¶ *The Vertues.*

Medlars do ſtop the belly, eſpecially when they be greene and hard, for after that they haue been kept a while ſo that they become ſoft and tender, they doe not binde or ſtop ſo much, but are then more fit to be eaten.   A

The fruit of the three graine Medlar, is eaten both raw and boyled, and is more wholeſom for the ſtomacke.   B

Theſe Medlars be oftentimes preſerued with ſugar or hony: and being ſo prepared they are plea‐ſant and delightfull to the taſte.   C

Moreouer, they are ſingular good for women with childe : for they ſtrengthen the ſtomacke, and ſtay the lothſomeneſſe thereof.   D

The ſtones or kernels of the Medlars, made into pouder and drunke, doe breake the ſtone, expell grauell, and procure vrine.   E

Chap. 99. *Of the Peare tree.*

¶ *The Kindes.*

TO write of Peares and Apples in particular, would require a particular volume : the ſtocke or kindred of Peares are not to be numbred: euery country hath his peculiar fruit: my ſelfe knowes one curious in graffing & planting of fruits, who hath in one piece of ground, at the point of three-ſcore ſundry ſorts of Peares, and thoſe exceeding good, not doubting but if his minde had beene to ſeeke after multitudes, he might haue gotten together the like number of thoſe of worſe kinds: be-ſides the diuerſities of thoſe that be wilde, experience ſheweth ſundry ſorts : and therefore I thinke it not amiſſe to ſet downe the figures of ſome few, with their ſeuerall titles, as well in Latine as En-gliſh, and one generall deſcription for that, that might be ſaid of many, which to deſcribe apart, were to ſend an owle to Athens, or to number thoſe things which are without number.

‡ Our Author in this chapter gaue eight figures with ſeuerall titles to them, ſo I pluckt a peare from each tree, and put his title to it, but not in the ſame order that he obſerued, for hee made the Katherine peare tree the ſeuenth, which I haue now made the firſt, becauſe the figure expreſſes the whole tree. ‡

¶ *The*

¶ *The generall Description.*

THe Peare tree is for the most part higher than the Apple tree, hauing boughes not spread a-
broad, but growing vp in height; the body is many times great: the timber or wood it selfe is
very tractable or easie to be wrought vpon, exceeding fit to make moulds or prints to be grauen on, of colour tending to yellownesse: the leafe is somewhat broad, finely nicked in the edges,
greene aboue, and somewhat whiter vnderneath: the floures are white: the Peares, that is to say, the
fruit, are for the most part long, and in forme like a Top; but in greatnesse, colour, forme, and taste
very much differing among themselues; they be also couered with skins or coats of sundry colours:
the pulpe or meate differeth, as well in colour as taste: there is contained in them kernels, blacke
when they be ripe: the root groweth straight downe with some branches running aslope.

Pirus superba, siue Katherina.
The Katherine Peare tree.

1 *Pyra Præcocia.*   The Ienneting Peare.
2 *Pyra Iacobæa.*   Saint Iames Peare.
3 *Pyrum regale.*   The Peare royall.

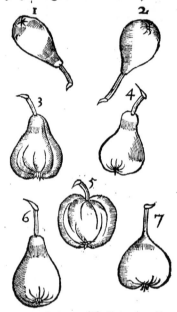

4 *Pyrum Palatinum.*   The Burgomot Peare.
5 *Pyrum Cydonium.*   The Quince peare.
6 *Pyrum Episcopatum.* The Bishops Peare.
7 *Pyrum hyemale.*   The Winter Peare.

¶ *The Place.*

The tame Peare trees are planted in Orchards, as be the apple trees, and by grafting, though vpon wilde stockes, come much variety of good and pleasant fruits. All these before specified, and
many sorts more, and those most rare and good, are growing in the ground of Master *Richard Pointer*, a most cunning and curious graffer and planter of all manner of rare fruits, dwelling in a small
village neere London called Twicknam; and also in the ground of an excellent graffer and painfull
planter, Mr *Henry Banbury*, of Touthill street neere Westminster, and likewise in the ground of a diligent and most affectionate louer of Plants Mr *Warner* neere Horsey downe by London, and in diuers other grounds about London. ‡ Most of the best peares are at this day to be had with Mr *Iohn
Millen* in Old-street, in whose nursery are to be found the choisest fruits this kingdome yeelds. ‡

¶ *The Time.*

The floures doe for the most part come forth in Aprill, the leaues afterwards: all peares are not
ripe at one time: some be ripe in Iuly, others in August, and diuers in September and later.

¶ *The Names.*

The tame or Orchard peare tree is called in Greeke ἄπιος, or with a double ππ ἄππιος: in Latine, *Pyrus Vrbana,*

vrbana,or *Cultiua* : of *Tarentinus* in his Geoponikes ἀγρομιάς:in high Dutch, Birbaum:in low Dutch, Pterboom : in French,*Porrier*.

The Peare or fruit it selfe is called in Greeke ἀπιος:in Latine,*Pyrum* : in high Dutch,Birn : in low Dutch,Peere : in Italian, *Pere* : in French,*Poyre* . in Spanish,*Peras* : in English, Peare.

¶ *The Temperature and Vertues.*

Leauing the diuers and sundry surnames of Peares,let vs come to the faculties which the Physi- A
tions ought to know ; which also vary according to the differences of their tastes : for some Peares
are sweet,diuers fat and vnctious, others soure, and most are harsh, especially the wilde peares, and
some consist of diuers mixtures of tasts, and some hauing no taste at all, but as it were a waterish
taste.

All Peares are cold,and all haue a binding quality and an earthie substance:but the Choke pears B
and those that are harsh be more earthie,and the sweet ones lesse : which substance is so full of superfluous moisture in some, as that they cannot be eaten raw. All manner of Peares do binde and
stop the belly, especially the Choke and harsh ones, which are good to be eaten of those that haue
the laske and the bloudy flix.

The harsh and austere Peares may with good successe be laid vpon hot swellings in the begin- C
ning,as may be the leaues of the tree, which do both binde and coole.

Wine made of the juyce of Peares called in English,Perry,is soluble, purgeth those that are not D
accustomed to drinke thereof, especially when it is new ; notwithstanding it is as wholesome a
drinke being taken in small quantity as wine; it comforteth and warmeth the stomacke,and causeth
good digestion.

CHAP. 100.   *Of the wilde Peare tree.*

¶ *The Kindes.*

AS there be sundry kindes of the manured Peares,so are there sundry wilde,wherof to write apart
were to small purpose : therefore one description with their seuerall titles shall be sufficient for
their distinctions.

1 *Pyrum strangulatorium majus.*
The great Choke Peare.

¶ *The generall Description.*

THe wilde Peare tree growes likewise great,
vpright,full of branches, for the most part
Pyramides like, or of the fashion of a steeple,
not spred abroad as is the Apple or Crab tree :
the timber of the trunke or body of the tree is
very firme and sollid, and likewise smooth, a
wood very fit to make diuers sorts of instruments of, as also the hafts of sundry tooles to
worke withal;and likewise serueth to be cut into many kindes of moulds,not only such prints
as these figures are made of,but also many sorts
of pretty toies,for coifes,brest-plates,and such
like,vsed among our English gentlewomen:the
branches are smooth, couered with a blackish
barke,very fragile or easie to breake,wheron do
grow leaues,in some greater, in other lesser:the
floures are like those of the manured Pear-tree,
yet some whiter than others:the fruit differ not
in shape, yet some greater than others ; but in
taste they differ among themselues in diuers
points, some are sharpe,soure,and of an austere
taste;some more pleasant,others harsh and bitter,and some of such a choking taste, that they
are not to be eaten of hogs & wild beasts,much
lesse of men : they also differ in colour, euery
circumstance whereof to distinguish apart
would greatly enlarge our volume,and bring to
the Reader small profit or commodity.

Gggggg   ¶ The

2 *Pyrum strangulatorium minus.*
The small Choke peare.

3 *Pyrus syluestris.*
The wilde hedge Peare tree.

4 *Pyrus syluestris minima.*
The wilde Crab Peare tree.

5 *Pyrus Pedicularia.*
The Lowsie wilde Peare.

6 *Pyrus Coruina.*
The Crow Peare tree.

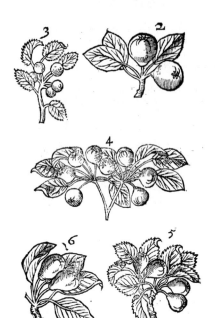

¶ *The Place.*

The wilde Peares grow of themselues without manuring in most places, as woods, or in the borders of fields, and neere to high waies.

¶ *The Time.*

The Time of wilde Peares answereth the tame or manured Peare, notwithstanding for the most part they are not ripe much before Winter.

¶ *The Names.*

The wilde Peare tree is called in Latine, *Pyrus syluestris* and *Pyraster* : in Greeke, ἄχρας by which name both the fruit and tree are known. Peares haue diuers syrnames among the antient Writers, and especially in *Pliny*, in his 15 booke, 15 chapter, none of which are knowne to the later Writers (or not desir-ed :) euery city or euery country haue names of them selues, and Peares haue also diuers names according to the places.

¶ *The Temperature.*

All Peares are of a cold temperature, and the most part of them of a binding quality and an earthy substance.

¶ *The Vertues.*

The vertues of the wilde Peares are referred vnto the garden Peares as touching their binding faculty, but are not to be eaten, because their nourishment is little and bad.

Chap. 101. *Of the Apple Tree.*

¶ *The Kindes.*

THe Latine name *Malus* reacheth far among the old Writers, and is common to many trees, but we will briefely first intreat of *Mali*, properly called Apple trees, whose stocke or kindred is so infinite, that we haue thought it not amisse to vse the same order or method with Apples, that we haue done with Peares ; that is, to giue them seuerall titles in Latine and English, and one generall description for the whole.

¶ *The Description.*

THe Apple tree hath a body or trunke commonly of a meane bignesse, not very high, hauing long armes or branches, and the same disordered : the barke somewhat plaine, and not very rugged : the leaues bee also broad, more long than round, and finely nicked in the edges. The floures are whitish tending vnto a blush colour. The fruit or Apples do differ in greatnesse, forme, colour, and taste ; some couered with a redde skinne, others yellow or greene, varying infinitely

nitely according to the ſoyle and climate, ſome very great, ſome little, and many of a middle ſort ; ſome are ſweet of taſte, or ſomething ſoure; moſt be of a middle taſte betweene ſweet and ſoure, the which to diſtinguiſh I thinke it impoſſible ; notwithſtanding I heare of one that intendeth to write a peculiar volume of Apples, and the vſe of them ; yet when he hath done what he can doe, hee hath done nothing touching their ſeuerall kindes to diſtinguiſh them. This that hath beene ſaid ſhall ſuffice for our Hiſtory.

‡ Our Author gaue foure figures more out of *Tabernamontanus*, with theſe titles. 3. *Malum reginale*, the Qneening or Queene of Apples. 5 *Platomela ſiue Pyra æſtiua :* The Summer Peare-maine. 6 *Platurchapia ſiue Pyra hyemalia :* the Winter Pearemaine. ‡

1 *Malus Carbonaria.*
The Pome-Water tree.

2 *Malus Carbonaria longo fructu.*
The Bakers ditch Apple tree.

¶ *The Place.*

The tame and graffed Apple trees are planted and ſet in gardens and orchards made for that pur-poſe : they delight to grow in good and fertile grounds : Kent doth abound with Apples of moſt ſorts. But I haue ſeene in the paſtures and hedge-rows about the grounds of a worſhipfull Gentle-man dwelling two miles frow Hereford, called Maſter *Roger Bodnome*, ſo many trees of all ſorts, that the ſeruants drinke for the moſt part no other drinke but that which is made of Apples; The quan-tity is ſuch, that by the report of the gentleman himſelfe, the Parſon hath for tithe many hogſheads of Syder. The hogs are fed with the fallings of them, which are ſo many, that they make choiſe of thoſe Apples they do eate, who will not taſte of any but of the beſt. An example doubtleſſe to be followed of Gentlemen that haue land and liuing : but enuie ſaith, the poore will breake downe our hedges, and wee ſhall haue the leaſt part of the fruit ; but forward in the name of God, graffe, ſet, plant and nouriſh vp trees in euery corner of your ground, the labour is ſmall, the coſt is nothing, the commodity is great, your ſelues ſhall haue plenty, the poore ſhall haue ſomwhat in time of want to relieue their neceſſitie, and God ſhall reward your good mindes and diligence.

¶ *The Time.*

They bloome about the end of Aprill, or in the beginning of May. The forward Apples be ripe about the Calends of Iuly, others in September.

¶ *The Names.*

The Apple tree is called in Greeke μηλία: in Latine, *Malus* and *Pomus* : in high Dutch, **Opffel-baum** : in low Dutch, **Appelboom** : in French, *Pommier* : in Engliſh, Apple-tree.

The Grecians name the fruit μῆλον: the Latines, *Malum* or *Pomum* : in high Dutch, **Opfell** : in low Dutch, **Appel** : in French and Spaniſh, *Manſanas* : in Engliſh, Apple.

¶ *The Temperature.*

All Apples be of temperature cold and moiſt, and haue joyned with them a certaine excremen-tall or ſuperfluous moiſture: but as they be not all of like coldneſſe, ſo neither haue they like quan-titie of ſuperfluous moiſture. They are ſooneſt rotten that haue greateſt ſtore of moiſture, and they may be longer kept in which there is leſſe ſtore : for the abundance of excrementall moiſture is the cauſe why they rot.

Sweet Apples are not ſo cold and moiſt, which being roſted or boyled, or otherwiſe kept, retaine or keepe the ſoundneſſe of their pulpe.

They yeeld more nouriſhment, and not ſo moiſt a nouriſhment as do the other Apples, and doe not ſo eaſily paſſe through the belly.

Soure Apples are colder and alſo moiſter : the ſubſtance or pulpe of theſe when they bee boiled doth run abroad, and retaineth not his ſoundneſſe: they yeeld a leſſer nouriſhment, and the ſame raw and cold.

They do eaſily and ſpeedily paſſe through the belly, and therefore they doe mollifie the belly, eſpecially being taken before meat.

Harſh or Auſtere Apples being vnripe, are cold ; they ingender groſſe bloud, and great ſtore of winde, and often bring the Collicke.

Thoſe Apples which be of a middle taſte contain in them oftentimes two or three ſorts of taſts, and yet do they retaine the faculties of the other.

¶ *The Vertues.*

A   Roſted Apples are alwaies better than the raw, the harme whereof is both mended by the fire, and may alſo be corrected by adding vnto them ſeeds or ſpices.

B   Apples be good for an hot ſtomacke : thoſe that are auſtere or ſomewhat harſh doe ſtrengthen a weake and feeble ſtomacke proceeding of heat.

C   Apples are alſo good for all inflammations or hot ſwellings, but eſpecially for ſuch as are in their beginning, if the ſame be outwardly applied.

D   The juyce of Apples which be ſweet and of a middle taſte, is mixed in compoſitions of diuers medicines, and alſo for the tempering of melancholy humours, and likewiſe to mend the qualities of medicines that are dry: as are *Serapium ex pomis Regis Saporis*, *Confectio Alkermes*, and ſuch like com-poſitions.

E   There is likewiſe made an ointment with the pulpe of Apples and Swines greaſe and Roſe wa-ter, which is vſed to beautifie the face, and to take away the roughneſſe of the ſkin, which is called in ſhops *Pomatum* : of the Apples whereof it is made.

F   The pulpe of the roſted Apples, in number foure or fiue, according to the greatneſſe of the Ap-ples, eſpecially the Pome-water, mixed in a wine quart of faire water, laboured together vntill it come to be as Apples and Ale, which wee call Lambes Wooll, and the whole quart drunke laſt at night, within the ſpace of an houre, doth in one night cure thoſe that piſſe by droppes with great anguiſh and dolour ; the ſtrangury, and all other diſeaſes proceeding of the difficulty of making water ; but in twiſe taking, it neuer faileth in any : oftentimes there happeneth with the foreſaid diſeaſes the Gonorrhæa, or running of the Raines, which it likewiſe healeth in thoſe perſons ; but not generally in all ; which my ſelfe haue often proued, and gained thereby both crownes and credit.

G   The leaues of the tree doth coole and binde, and be alſo counted good for inflammations, in the beginning.

H   Apples cut in pieces, and diſtilled with a quantitie of Camphire and butter-milke, take away the markes and ſcarres gotten by the ſmall pockes, being waſhed therewith when they grow vnto their ſtate and ripeneſſe: prouided that you giue vnto the patient a little milke and Saffron, or milke and mithridate to drinke, to expell to the extreme parts that venome which may lie hid, and as yet not ſeene.

C H A P.

CHAP. 102.   *Of the Wilding or Crab tree.*

¶ *The Kindes.*

LIke as there be diuers manured Apples, so are there sundry wilde Apples, or Crabs, whereof to write apart were to small purpose, and therefore one description shall suffice for the rest.

Malus syluestris.
The Wilding or Crab tree.

¶ *The generall Description.*

THere be diuers wilde Apple trees not husbanded, that is to say, not grafted; the fruit whereof is harsh and binding: for by grafting both Apples and Peares become more milde and pleasant. The crab or wilding tree growes oftentimes to a reasonable greatnesse, equall with the Apple tree: the wood is hard, firme, and sollid; the barke rough; the branches or boughes many; the floures and fruit like those of the Apple tree, some red, others white some greater, others lesser: the difference is knowne to all, therefore it shall suffice what hath been said for their seuerall distinctions: we haue in our London gardens a dwarfe kinde of sweet Apple, called *Chamamalus*, the dwarfe Apple tree, or Paradise Apple, which beareth Apples very timely without grafting.

‡ Our Author here also (out of *Tabernæmontanus*) gaue foure figures, whereof I onely retaine the best, with their seuerall titles. 1 *Malus syluestris rubens.* The great Wilding or red Crab tree. 2 *Malus syluestris alba.* The white Wilding or Crab tree. 3 *Malus syluestris minor.* The smaller Crab tree. 4 *Malus duracina syluestris.* The choking leane Crab tree. ‡

¶ *The Place.*
The Crab tree groweth wilde in woods and hedge rowes almost euery where.

¶ *The Time.*
The time answereth those of the garden.

¶ *The Names.*
Their titles doth set forth their names in Latine and English.

¶ *The Temperature.*
Of the temperature of wilde Apples hath been sufficiently spoken in the former chapter.

¶ *The Vertues.*
The juyce of wilde Apples or crabs taketh away the heat of burnings, scaldings, and all inflammations: and being laid on in short time after it is scalded, it keepeth it from blistering.   A

The juyce of crabs or Verjuice is astringent or binding, and hath withall an abstersiue or clensing qualitie, being mixed with hard veest of Ale or Beere, and applied in manner of a cold ointment, that is, spread vpon a cloth first wet in the Verjuyce and wrung out, and then laid to, taketh away the heat of S. Anthonies fire, all inflammations whatsoeuer, healeth scab'd legs, burnings and scaldings wheresoeuer it be.   B

C H A P. 103.

Of the Citron, Limon, Orange, and Affyrian Apple trees.

¶ *The Kindes.*

THe Citron tree is of kindred with the Limon tree, the Orange is of the fame houfe or ftocke, and the Affyrian Apple tree claimeth a place as neereft in kindred and neighbourhood:wherefore I intend to comprehend them all in this one chapter.

1 *Malus medica.*
The Pome Citron tree.

2 *Malus Limonia.*
The Limon tree.

¶ *The Defcription.*

1   THe Citron tree is not very great, hauing many boughes or branches,tough and pliable, couered with a greene barke : whereon do grow greene leaues, long,fomewhat broad,very fmooth,and fweet of fmell like thofe of the Bay tree : among which come forth here and there certaine prickles, fet far in funder : from the bofome whereof come forth fmall floures, confifting of fiue little leaues,of a white colour tending to purple, with certaine threds like haires growing in the middle : the fruit is long,greater many times than the Cucumber, often leffer, and not much greater than the Limon : the barke or rinde is of a light golden colour, fet with diuers knobs or bumps,and of a very pleafant fmell : the pulpe or fubftance next vnto it is thicke, white, hauing a kinde of aromaticall or fpicie fmell,almoft without any tafte at all : the fofter pulpe within that,is not fo firme or follid,but more fpungie,and full of a fower juyce, in which the feed lieth hid, greater and thicker than a graine of Barley.

2   The Limon tree is like vnto the Pome Citron tree in growth,thorny branches,and leaues of a pleafant fweet fmell, like thofe of the Bay-tree : the floures hereof are whiter than thofe of the Citron tree, and of a moft fweet fmell : the fruit is long and thicke, leffer than the Pome Citron :
the

the rinde is yellow, ſomwhat bitter in taſte, and ſweet of ſmell : the pulpe is white, more in quantity than that of the Citron, reſpecting the bigneſſe; in the middle part whereof is contained more ſoft ſpungie pulpe, and fuller of ſoure juyce : the ſeeds are like thoſe of the Pome Citron.

3   The Orenge tree groweth vp to the height of a ſmal Peare tree, hauing many thorny boughes or branches, like thoſe of the Citron tree : the leaues are alſo like thoſe of the Bay-tree, ‡ but that they differ in this, that at the lower end next the ſtalke there is a leſſer leafe made almoſt after the vulgar figure of an heart, whereon the bigger leafe doth ſtand, or is faſtned : ‡ and they are of a ſweet ſmell : the floures are white of a moſt pleaſant ſweet ſmell alſo : the fruit is round like a ball, euery circumſtance belonging to the forme is very well knowne to all : the taſte is ſoure, ſomtimes ſweet, and often of a taſte betweene both : the ſeeds are like thoſe of the Limon.

3 Malus arantia.
The Orange tree.

4 Malus Aßyria.
The Aſſyrian Apple tree.

4   The Aſſyrian Apple tree is like vnto the Orange tree : the branches are like : the leaues are greater : the floures are like thoſe of the Citron tree : the fruit is round, three times as big as the Orange : the barke or peeling is thicke, rough, and of a pale yellow colour, wherein appeare often as it were ſmall clifts or crackes : the pulpe or inner ſubſtance is full of juyce, in taſte ſharpe, as that of the Limon, but not ſo pleaſant : the ſeeds are like thoſe of the Citron.

¶ The Place.

The Citron, Limon, and Orenge trees doe grow eſpecially on the ſea coaſts of Italy, and on the Iſlands of the Adriaticke Tyrrhene, and alſo Ægæan ſeas, and likewiſe on the maine land neer vnto meeres and great lakes : there is alſo great ſtore of them in Spaine, but in places eſpecially joyning to the ſea, or not farre off : they are alſo found in certaine Prouinces of France which lie vpon the midland ſea. They were firſt brought out of Media, as not onely Pliny writeth, but alſo the Poët Virgil affirmeth in the ſecond booke of his Georgickes, writing of the Citron tree after this maner :

Media fert triſtes ſuccos, tardumque ſaporem
Felicis mali, quo non præſentius vllum,
Pocula ſi quando ſæua infecere nouercæ,

Miſcueruntque

Mifcueruntque herbas, & non innoxia verba,
Auxilium venit, ac membris agit atra venena.
Ipfa ingens arbos, faciefque fimillima Lauro ;
Et, fi non alium late jactaret odorem,
Laurus erat ; folia haud vllis labentia ventis ;
Flos apprime tenax. Animas & olentia Medi
Ora fouent illo, & fenibus medicantur anhelu.

The Countrey Media beareth juyces fad,
And dulling taftes of happy Citron fruit,
Than which, no helpe more prefent can he had,
If any time ftepmothers worfe than brute
Haue poyfon'd pots, and mingled herbs of fute
with hurtfull charmes : this Citron fruit doth chafe
Blacke venome from the body in euery place.
The tree it felfe in growth is large and big,
And very like in fhew to th'Laurell tree ;
And would be thought a Laurell, leafe and twig,
But that the fmell it cafts doth difagree :
The floure it holds as faft as floure may be :
Therewith the Medes a remedy do finde
For ftinking breaths and mouthes, a cure moft kinde,
And helpe old men which hardly fetch rheir winde.

¶ *The Time.*

Thefe trees be alwaies greene, and do, as *Pliny* faith, beare fruit at all times of the yeare, fome falling off, others waxing ripe, and others newly comming forth.

¶ *The Names.*

The firft is called in Greeke, Μηλὶ μηδικὰ : in Latine, *Malus Medica*, and *Malus Citria* : in Englifh, Citron tree, and Pomecitron tree.

The fruit is named in Greeke, Μηλον μηδικὸν : in Latine, *Malum Medicum*, and *Malum Citrium* : and *Citromalum. Æmilyanus* in *Athenæus* fheweth, that *Iuba* King of Mauritania hath made mention of the Citron, who faith that this Apple is named among them, *Malum Heffericum : Galen* denieth it to be called any longer *Malum Medicum*, but *Citrium* ; and faith that they who call it *Medicum* doe it to the end that no man fhould vnderftand what they fay : the Apothecaries call thefe Apples *Citrones :* in high Dutch, **Citrin opffell, Citrinaten :** in low Dutch, **Citroenen :** in Italian, *Citroni*, and *Cedri* in Spanifh, *Cidras :* in French, *Citrons :* in Englifh, Citron Apple, and Citron.

The fecond kinde of Citron is called in Latine, *Limonium Malum* ; in fhops, *Limones :* in French, *Limons :* in low Dutch, **Limonen :** in Englifh, Limon, and Lemon.

The third is named in Latine, *Malum anarantium* or *Anerantium :* and of fome, *Aurantium :* of others, *Aurengium*, of the yellow colour of Gold : fome would haue them called *Arantia*, of *Arantium*, a towne in Achaia or Arania, of a country bearing that name in Perfia : it is termed in Italian, *Arancio :* in high Dutch, **Pomeranken :** in low Dutch, **Araengie Appelen :** in French, *Pommes d' Orenges :* in Spanifh, *Naranfas :* in Englifh, Orenges.

The fourth is named of diuers, *Pomum Affyrium*, or the Citron of Affyria, and may be Englifhed Adams Apple, after the Italian name ; and among the vulgar fort of Italians, *Lomie*, of whom it is alfo called *Pomum Adami*, or Adams Apple ; and that came by the opinion of the common rude people, who thinke it to be the fame Apple which *Adam* did eate of in Paradife, when he tranfgreffed Gods commandement ; whereupon alfo the prints of the biting appeare therein, as they fay : but others fay that this is not the Apple, but that which the Arabians do call *Mufa* or *Mofa*, whereof *Auicen, cap. 395.* maketh mention : for diuers of the Iewes take this for that, through which by eating, *Adam* offended, as *Andrew Theuet* fheweth.

¶ *The Temperature and Vertues.*

A   All thefe fruits confift of vnlike parts, and much differing in faculty.

B   The rindes are fweet of fmell, bitter, hot, and dry.

C   The white pulpe is cold, and containeth in it a groffe juyce, efpecially the Citron.

D   The inner fubftance or pap is foure, as of the Citrons and Limons, cold and dry, with thinneffe of parts.

E   The feed becaufe it is bitter is hot and dry.

F   The rinde of the Pomecitron is good againft all poyfons, for which caufe it is put into treacles and fuch like confections.

             It is.

It is good to be eaten against a stinking breath, for it maketh the breath sweet; and being so taken it comforteth the cold stomacke exceedingly.

The white, sound, and hard pulpe is now and then eaten, but very hardly concocted, and ingendreth a grosse, cold, and phlegmaticke juyce, but being condite with sugar, it is both pleasant in taste, and easie to be digested, more nourishing, and lesse apt to obstruction and binding or stopping.

Galen reporteth, That the inner juyce of the Pomecitron was not wont to be eaten, but it is now vsed for sauce; and being often vsed, it represseth choler which is in the stomacke, and procureth appetite: it is excellent good also to be giuen in vehement and burning feuers, and against all pestilent and venomous or infectious diseases: it comforteth the heart, cooleth the inward parts, cutteth, diuideth, and maketh thin, grosse, tough, and slimy humors.

Of this foresaid sharpe juyce there is a syrrup prepared, which is called in shops, *Syrupus de Acetositate Citri*, very good against the foresaid infirmities.

Such a syrrup is also prepared of the sharpe juyce of Limons, of the same quality and operation, so that in stead of the one, the other will serue very well.

A dozen of Orenges cut in slices and put into a gallon of water, adding thereto an ounce of Mercurie sublimate, and boyled to the consumption of the halfe, cureth the itch and manginesse of the body.

Men in old time (as *Theophrastus* writeth in his fourth booke) did not eate Citrons, but were contented with the smell, and to lay them amongst cloathes, to preserue them from Moths.

As often as need required they vsed them against deadly poysons; for which thing they were especially commanded by *Virgils* verses, which we haue before alledged.

Athenæus, lib.3. hath extant a story of some that for certaine notorious offences were condemned to be destroied of Serpents, who were preserued and kept in health and safety by the eating of Citrons.

The distilled water of the whole Limons, rinde and all, drawne out by a glasse Still, takes away tetters and blemishes of the skin, and maketh the face faire and smooth.

The same being drunke prouoketh vrine, dissolueth the stone, breaketh and expelleth it.

The rinde of Orenges is much like in faculty to that of the Citrons and Limons, yet it is so much the more hot as it is more biting and bitter.

The inner substance or soure pap which is full of juyce is of like faculty, or not much inferiour to the faculty of the pap of Citrons or Limons; but the sweet pap doth not much coole or drie, but doth temperately heate and moisten, being pleasant to the taste: it also nourisheth more than doth the soure pap, but the same nourishment is thin and little; and that which is of a middle tast, hauing the smacke of wine, is after a middle sort more cold than sweet, and lesser cold than soure: the sweet and odoriferous floures of Orenges be vsed of the perfumers in their sweet smelling ointments.

Two ounces of the juyce of Limons, mixed with the like quantity of the spirit of wine, or the best *Aqua vitæ* (but the spirit of wine rectified is much better) and drunk at the first approach of the fit of an ague, taketh away the shaking presently: the medicine seldome faileth at the second time of the taking thereof perfectly to cure the same; but neuer at the third time, prouided that the Patient be couered warme in a bed, and caused to sweat.

There is also distilled out of them in a glasse still, a water of a maruellous sweet smell, which being inwardly taken in the weight of an ounce and a halfe, moueth sweat, and healeth the ague.

The seed of all these doth kill wormes in the belly, and driueth them forth: it doth also mightily resist poyson, and is good for the stinging of Scorpions, if it be inwardly taken.

Those which be called Adams Apples are thought to bee like in faculties to the soure juyce, especially of the Limons, but yet they be not so effectuall.

Chap. 104.　*Of the Cornell tree.*

¶ *The Description.*

THe tame Cornell tree groweth somtime of the height and bignesse of a smal tree, with a great number of springs: it is couered with a rugged barke: the wood or timber is very hard and dry, without any great quantity of sap therein: the leaues are like vnto the Dog-berry leaues, crumpled, rugged, and of an ouerworne colour: the floures grow in small bunches before any leaues do appeare, of colour yellow, and of no great value (they are so small) in shew like the floures of the
Oliue

Cornus mas.
The male Cornell tree.

Oliue tree:which being vaded,there come small long berries, which at the first bee greene and red when they be ripe ; of an auftere and harsh tafte, with a certaine fourenesse : within this berry is a small ftone, exceeding hard, white within like that of the Oliue,whereunto it is like both in the fashion and oftentimes in the bignesse of the fruit.

¶ *The Place.*

This groweth in most places of Germanie without manuring:it growes not wild in England. But yet there be fundry trees of them growing in the gardens of fuch as loue rare and dainty plants, wherof I haue a tree or two in my garden.

¶ *The Time.*

The tame Cornell tree floureth sometime in February,& commonly in March, and afterwards the leaues come forth as an vntimely birth : the berries or fruit are ripe in Auguft.

¶ *The Names.*

The Grecians call it κρανία: the Latines, *Cornus* : in high Dutch, **Cornelbaum** : in low Dutch, **Cornoele boom** : the Italians, *Corniolo* : in French, *Cornillier* : in Spanish, *Cornizolos* : in English , the Cornell tree, and the Cornelia tree;of fome,long Cherrie tree.

The fruit is named in Latine, *Cornum* : in high Dutch, **Cornell** : in low Durch, **Cornoele** : in Italian, *Cornole* : in English, Cornell berries and Cornelian Cherries.

This is *Cornus mas Theophrasti*, or *Theophraftus* his small Cornell tree ; for he fetteth downe two forts of the Cornell trees, the male, and the female : he maketh the wood of the male to be found, as in this Cornell tree ; which we both for this canfe and for others alfo haue made to be the male. The female is that which is commonly called *Virga fanguinea*, or Dogs berry tree, and *Cornus fylueftris*,or the wilde Cornell tree,of which we will treat in the next chapter following.

¶ *The Temperature and Vertues.*

A　The fruit of the Cornell tree hath a very harsh or choking tafte : it cooleth,drieth, and bindeth, yet may it alfo be eaten, as it is oftentimes.

B　It is a remedy againft the laske and bloudy flix, it is hurtfull to a cold ftomacke, and increafeth the rawneffe thereof : the leaues and tender crops of the tree are likewife of an harsh and choking tafte,and do mightily dry.

C　They heale greene wounds that are great and deepe, efpecially in hard bodies, but they are not fo good for fmall wounds and tender bodies,as *Galen* writeth.

CHAP. 105.
Of the female Cornell or Dog-berry tree.

¶ *The Defcription.*

THat which the Italians call *Virga fanguinea*,or the bloudy Rod, is like to the Cornell tree, yet it groweth not into a tree,but remaineth a fhrub : the young branches thereof are jointed, and be of an obfcure red purple : they haue within a white fpongie pith like that of Elder, but the old ftalkes are hard and ftiffe, the fubftance of the which is alfo white, and anfwerable to thofe of the Cornell tree : the leaues are alfo like, the middle rib whereof as alfo the brittle foot-ftalkes are fomewhat reddifh : at the top whereof ftand white floures in fpokie rundles, which turne into

berries,

Cornus fœmina.
The Dog-berry tree.

greene at the first, and of a shining blacke colour when they bee ripe, in taste vnpleasant, and not cared for of the birds.

¶ *The Place.*

This shrub groweth in hedges and bushes in euery country of England.

¶ *The Time.*

The floures come forth in the Spring in the moneth of Aprill: the berries are ripe in Autumne.

¶ *The Names.*

The Italians doe commonly call it *Sanguino*, and *Sanguinello. Petrus Crescentius* termes it *Sanguinus*; and *Matthiolus, Virga sanguinea: Pliny, lib.24.cap.10.* hath written a little of *Virga sanguinea*: Neither is *Virga sanguinea*, saith hee, counted more happy; the inner barke doth breake open the scarres which they . . haue healed. It is an hard thing, or perade . ture a rash part, to affirme by these few wor . that *Pliny* his *Virga sanguinea* is the same that the Italian *Sanguino* is. This is called in high Dutch, **Hartriegel:** in low Dutch **Wilde Cornoelle,** that is to say, *Cornus syluestris*, or w . . . Cornell tree: and in French, *Cornillier sauuage:* in English, Hounds tree, Hounds berry, Dogs berry tree, Pricke-Timber: in the North countrey they call it Gaten tree, or Gater tree; the berries whereof seeme to be those which *Chaucer* call th Gater berries: *Valerius Cordus* nameth it ψευδοκρανεια, that is to say, *Falsa* or *Spuria Cornus*, false or bastard Cornell tree: this seemeth also to be *Theophrastus* his θηλυκρανεια, or *Cornus fœmina*, female Cornell tree. This hath little branches hauing pith within, neither be they hard nor sound, like those of the male: the fruit is αβρωτος, that is, not fit to be eaten, and a late fruit which is not ripe till after the Autumne Æquinoctiall; and such is the wilde Cornell tree or Gater tree, the young and tender branches whereof be red, and haue (as we haue written) a pith within: the fruit or berries be vnpleasant, and require a long time before they can be ripe.

¶ *The Temperature.*

The berries hereof are of vnlike parts; for they haue some hot, bitter, and clensing, and very many cold, dry, harsh, and binding, yet they haue no vse in medicine.

¶ *The Vertues.*

Matthiolus writeth, That out of the berries first boyled, and afterwards pressed, there issueth an **A** oyle which the Anagnian country people do vse in lamps: but it is not certaine nor very like, that the barke of this wilde Cornell tree hath that operation which *Pliny* reporteth, of *Virga Sanguinea*; for he saith, as we haue already set downe, that the inner barke thereof doth breake and lay open the scars which they before haue healed.

Chap. 106.   *Of Spindle tree or Pricke-wood.*

¶ *The Description.*

1   PRicke-wood is no high shrub, of the bignesse of the Pomegranat tree: it spreadeth farre with his branches: the old stalks haue their barke somewhat white; the new and those that be lately growne bee greene, and foure square: the substance of the wood is hard, and mixed with a light yellow: the leaues be long, broad, slender, and soft: the floures bee white, many standing vpon one foot-stalke, like almost to a spoked rundle: the fruit is foure square, red, and containing foure white seeds, euery one whereof is couered with a yellow coat, which being taken off giueth a yellow die.

2 This

1 *Euonymus Theophraſti.*
Engliſh Prick-timber tree.

2 *Euonymus latifolius.*
Broad leaſed Spindle tree.

3 *Enonymus Pannonicus.*
Hungary Spindle tree.

2   This other ſort of *Euonymus* groweth
to the forme of an hedge tree, of a meane bigneſſe ; the trunke or body whereof is of the
thickeneſſe of a mans leg, couered with a
rough or ſcabbed barke of an ouerworn ruſſet
colour. The branches thereof are many, ſlender, and very euen, couered with a greene
barke whileſt they be yet young and tender,
they are alſo very brittle, with ſome pith in
the middle like that of the Elder. The leaues
are few in number, full of nerues or ſinewes diſperſed like thoſe of Plantaine, in ſhape like
thoſe of the Pomecitron tree, of a lothſome
ſmell and bitter taſte : amongſt which come
forth ſlender foot-ſtalks very long and naked,
whereon do grow ſmall floures conſiſting of
foure ſmall leaues like thoſe of the Cherry
tree, but leſſer, of a white colour tending to a
bluſh, with ſome yellowneſſe in the middle :
after commeth the fruit, which is larger than
the former, and as it were winged, parted
commonly into foure, yet ſomtimes into fiue
parts, and opening when it is ripe, it ſheweth
the white graines filled with a yellow pulpe.
The root is tough and wooddy, diſperſing it
ſelfe farre abroad vnder the vpper cruſt of the
earth.

3 The

3　The ſame Author ſets forth another ſort, which he found in the mountains of Moravia and Hungary, hauing a trunk or ſtock of the height of three or foure cubits, couered with a bark green at the firſt, afterwards ſprinkled ouer with many blacke ſpots : the boughs are diuided toward the top into diuers ſmall branches, very brittle and eaſie to breake, whereon are placed leaues by couples alſo, one oppoſite to another, ſomewhat ſnipt about the edges, in ſhape like thoſe of the great Myrtle, of an aſtringent taſte at the beginning, after ſomewhat hot and bitter : among which come forth ſmall floures ſtanding vpon long naked footſtalks, conſiſting of foure little leaues of a bright ſhining purple colour, hauing in the middle ſome few ſpots of yellow : after comes the fruit, foure cornered, not vnlike to the common kind, of a ſpongious ſubſtance, and a gold yellow color, wherein is contained not red berries like the other, but black, very like to thoſe of Fraxinella, of a ſhining blacke colour like vnto brandiſhed horne ; which are deuoured of birds when they be ripe, and the rather becauſe they fall of themſelues out of their husks, otherwiſe the bitterneſſe of the huskes would take away the delight.

¶ The Place.

The firſt commeth vp in vntoiled places and amongſt ſhrubs, vpon rough bankes and heapes of earth : it ſerueth alſo oftentimes for hedges in fields, growing amongſt Brambles and ſuch other thornes.

・ ・ The other ſorts Cluſius found in a wood in Hungary beyond the riuer Drauus, and alſo vpon the mountaines of Moravia and other places adiacent.

¶ The Time.

The floures appeare in April, the fruit is ripe in the end of Auguſt, or in the moneth of September.

¶ The Names.

Theophraſtus calleth this ſhrub Εὐώνυμος, and deſcribeth it, lib. 3. of the Hiſtorie of Plants. Diuers alſo falſly reade it Anonymos : Petrus Creſcentius calleth it Fuſanum, becauſe ſpindles be made of the wood hereof ; and for that cauſe it is called in high-Dutch, Spindelbaum, yet moſt of them Hanhoblin : in low-Dutch, Papenhout : in Italian, Fuſano : in French, Fuſin, and Bonnet de preſtre : in Engliſh, Spindle tree, Prickwood, and Prick-timber.

¶ The Temperature and Vertues.

This ſhrub is hurtfull to all things, as Theophraſtus writeth, and namely to Goats : hee ſaith the 　A fruit hereof killeth ; ſo doth the leaues and fruit deſtroy goats eſpecially, vnleſſe they ſcour as wel vpwards as downewards : if three or foure of theſe fruits be giuen to a man they purge by vomit and ſtoole.

C HAP. 107.　Of the blacke Aller Tree.

¶ The Deſcription.

THe blacke Aller tree bringeth forth from the root ſtraight ſtalks diuided into diuers branches, the outward barke whereof is blacke, and that next to the wood yellow, and giueth a colour as yellow as Saffron : the ſubſtance of the wood is white and brittle, with a reddiſh pith in the midſt : the leaues be like thoſe of the Alder tree, or of the Cherry tree, yet blacker, and a litle rounder : the floures be ſomwhat white ; the fruit are round berries, in which appeare a certaine rift or chinke as though two were joined together, at the firſt green, afterward red, and laſtly black ; in this there be two little ſtones : the root runneth along in the earth.

¶ The Place.

The Aller tree groweth in moiſt woods and copſes : I found great plenty of it in a wood a mile from Iſlington, in the way from thence to a ſmall village called Harnſey, lying vpon the right hand of the way ; and in the woods at Hampſted neere London, and in moſt Woods in the parts about London. ・

¶ The Time.

The leaues and floures appeare in the beginning of the ſpring, and the berries about Autumne.

¶ The Names.

This ſhrub is called Alnus nigra, or blacke Alder ; and by others Frangula : Petrus Creſcentius nameth it Avornus : in low-Dutch, Sparkenhout, and oftentimes Pulhout, becauſe boyes make for themſelues arrowes thereof : in high-Dutch, Faulbaum : in Engliſh, blacke Aller tree, and of diuers, Butchers pricke tree.

Alnus nigra ſive frangula.
The blacke Aller tree.

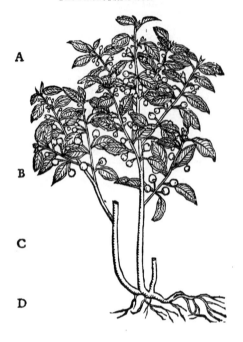

A

B

C

D

¶ *The Tmpereature.*

The inner barke of the blacke Aller tree is of a purging and dry qualitie.

¶ *The Vertues.*

The inner barke hereof is vſed of diuers countrymen, who drinke the infuſion therof when they would be purged: it purges thick flegmatick humors and alſo cholerick, not only by the ſtoole, but oft times alſo by vomit, not without great trouble and paine to the ſtomack: it is therfore a medicine more fit for clownes than for ciuil people, and rather for thoſe that feed groſſely, than for dainty people.

Others affirme the dried barke is more gentle, and cauſeth leſſe pain ; for the green barke (ſay they) which is not yet dried, contains a ſuperfluous moiſture which cauſeth gripings and vomitings, and troubles the ſtomacke.

The ſame bark boiled in wine or vineger makes a lotion for the tooth-ache, and is commended againſt ſcabs and filthineſſe of the skin.

The leaues are reported to be good fodder for cattell, eſpecially for Kine, cauſing them to yeeld good ſtore of milke.

CHAP. 108. *Of the Service Tree.*

¶ *The Deſcription.*

1   THe Seruis tree groweth to the height and bigneſſe of a great tree, charged with many great armes or boughs, which are ſet with ſundry ſmall branches, garniſhed with many great leaues ſomwhat long like thoſe of the Aſh : the flours are white, and ſtand in cluſters, which turne into ſmall brown berries ſomewhat long, which are not good to be eaten til they haue lien a while and are ſoft like the Medlar, whereto it is like in taſte and operation.

2   The common Seruis tree growes likewiſe to the height of a great tree, with a ſtraight body of a browniſh colour, full of branches, ſet with large diſplaied leaues like the Maple or White-thorne, ſauing that they are broader and longer : the floures are white, and grow in tufts ; which being fallen, there come in place thereof ſmal round berries, brown vpon one ſide, and reddiſh toward the Sun, of an vnpleaſant taſte in reſpect of the former, in which are contained little blackiſh kernels.

¶ *The Place.*

Theſe trees are found in woods and groues in moſt places of England: there be many ſmall trees thereof in a little wood a mile beyond Iſlington by London : in Kent it groweth in great aboundance, eſpecially about Southfleet and Graueſend. ‡ The later of theſe I haue ſeen growing wild in diuers places, but not the former at any time as yet. ‡

¶ *The Time.*

They floure in March, and their fruit is ripe in September.

¶ *The Names.*

The firſt is called in Greeke ὄα, and ὄη : in Latine, *Sorbus* : in high-Dutch, **Sperwerbaum :** in low-Dutch, **Sorbedboom :** in French, *Cormier* : in Engliſh, Seruis tree, and of ſome after the Latin Sorb tree.

The

1 *Sorbus.*
The Servis tree.

2 *Sorbus torminalis.*
Common Servis tree.

The common Servis tree is named of *Pliny, Sorbus torminalis* : in high . Dutch, **Arelfel, Efchzo-fel and Wilder Sperwerbaum:** in English, Common Servis tree.

The berries or fruit of the Servis tree is callled 'δη, and δ'ω : in Latine, *Sorbum* : in high-Dutch, **Speierling, Sporopffel** : in low-Dutch, **Sorben** : in Italian, *Sorbe,* and *Sorbole* : in French, *Corme* : in Spanish, *Servas,* and *Sorbas* : in English, Servis : of some, Sorbe Apple.

¶ *The Temperature and Vertues.*

Servis berries are cold and binding, and much more when they be hard, than when they are mild A
and soft : in some places they are quickely soft, either hanged in a place which is not altogether
cold, or laid in hay or chaffe : these Servises are eaten when the belly is too soluble, for they stay
the same ; and if they yeeld any nourishment at all, the same is very little, grosse and cold ; wherefore it is not good to eat of these or other-like fruits, nor to vse them otherwise than in medicines.

They stay all maner of fluxes in the belly, the bloudy flix, and vomiting: they stanch bleeding, if B
they be cut and dried in the sun before they be ripe, and so reserued for vse. These we may vse diuers waies, according to the manner of the griefe and grieued part.

CHAP. 109.   *Of the Ash Tree.*

¶ *The Description.*

1   THe Ash is also an high and tall tree, it riseth vp with a straight body, now and then of
no small bignesse, now and then of a middle sise, and is couered with a smooth barke ;
the wood is white, smooth, hard, and somewhat rough grained : the tender branches
hereof and such as be new grown vp are set with certaine ioynts, and haue within a white and spongie pith ; but the old boughes are wooddy throughout, and be without either ioints or much pith:
the leaues are long and winged, consisting of many standing by couples one right against another
vpon one rib or stalke, the vppermost of all excepted, which standeth alone ; of which euery particular one is long, broad, like to a Bay leafe, but softer, and of a lighter greene, without any sweete

Hhhhhh 2   smell,

ſmell,and nicked round about the edges, out of the yenger ſort of the boughs, hard to the ſetting on of the leaues, grow forth hanging together, many long narrow and flat cods,as it were like almoſt to diuers birds tongues,where the ſeed is perfected, which is of a bitter taſte: the roots bee many,and grow deep in the ground.

Fraxinus.
The Aſh tree.

A

B

¶ *The Place.*

The Aſh doth better proſper in moiſt places,as about the borders of medows and riuers ſides, then in dry grounds.

¶ *The Time.*

The leaues and keyes come forth in Aprill and May,yet is not the ſeed ripe before the fall of the leafe.

¶ *The Names.*

This tree is called in Greeke Μελία. and of diuers, Μελίας: it is named of the Latines, *Fraxinus* : in high-Dutch, **Eſchernbaum, Eſchernholtz, and Stepneſchern**: in low-Dutch, **Eſchen, and Eſchenboom**: in Italian, *Fraſſino* : in French, *Freſne* : in Spaniſh, *Freſno,Fraxino*,and *Freixo*: in Engliſh, Aſh tree.

The fruit like vnto cods is called of the apothecaries,*Lingua auis*,and *Lingua Paſſerina* : it may bee named in Greeke, Ὀρνιθόγλωσσον : yet ſome would haue it called *Orneogloſſum* : others make *Ornus* or wilde Aſh to be called *Orneogloſſum:* it is termed in Engliſh,Aſh-keyes,and of ſome Kite-keyes.

¶ *The Temperature and Vertues.*

The leaues and barke of the Aſh tree are dry and moderately hot, the ſeed is hot & dry in the ſecond degree.

The juice of the leaues or the leaues themſelues being applied, or taken with wine,cure the bitings of vipers,as *Dioſcorides* ſaith.

The leaues of this tree are of ſo great vertue againſt ſerpents,that they dare not ſo much as touch the morning and euening ſhadowes of the tree,but ſhun them afar off,as *Pliny* reports,*lib.16.ca.13.* He alſo affirmeth,that the ſerpent being penned in with boughes laid round about, will ſooner run into the fire,if any be there,than come neere the boughs of the Aſh ; and that the Aſh floureth before the ſerpents appeare,and doth not caſt his leaues before they be gon againe.

C   We write(ſaith he) vpon experience,that if the ſerpent be ſet within a circle of fire & the branches,the ſerpent will ſooner run into the fire than into the boughes. It is a wonderfull courteſie in nature,that the Aſh ſhould floure before the Serpents appeare, and not caſt his leaues before they be gon againe.

D   Both the leaues and barke are reported to ſtop the belly,and being boiled with vineger & water, do ſtay vomiting,if they be laid vpon the ſtomacke.

E   The leaues and barke of the Aſh tree boiled in wine and drunke,do open the ſtoppings of the liuer and ſpleen,and do greatly comfort them.

F   Three or foure leaues of the Aſh tree taken in wine each morning from time to time, doe make thoſe lean that are fat,and keep them from feeding that begin to wax fat.

G   The ſeed or Aſh keyes prouoke vrine, increaſe naturall ſeed,and ſtirre vp bodily luſt,eſpecially being poudered with nutmegs and drunke.

H   The wood is profitable for many things,being highly exalted by *Homer*, and by *Achilles* ſpeare, as *Pliny* writeth.

I   The ſhauings or ſmall pieces being drunke, are ſaid to be pernitious and deadly, as *Dioſcorides* affirmeth.

K   The Lee which is made of the aſhes of the barke cureth the white ſcurfe,and ſuch other roughneſſe of the skin,as *Pliny* teſtifieth.

CHAP.

CHAP. 110.

Of the wilde Ash, otherwise called Quicke-Beam, or Quicken-tree.

Sorbus sylvestris, sive Fraxinus Bubula.
The Quicken-tree,wild Ash,or wild Servis tree.

¶ *The Description.*

THe wilde or Ash Quicken tree *Pena* setteth forth for the wilde Service. This
tree groweth seldom or neuer to the stature
and height of the Ash tree,notwithstanding
it growes to the bignes of a large tree. The
leaues be great and long, and scarce be dif-
cerned from the leaues of the Service Tree.
The floures be white, sweet of smell, and
grow in tuft,which do turn into round berries,green at the first,but when they bee ripe
of a deepe red colour, and of an vnpleasant
taste. The branches are as full of iuice as the
osiar; which is the cause that Boies do make
pipes of the barke thereof, as they doe with
Willowes.

¶ *The Place.*

The wilde Ash or Quicken tree groweth
on high mountains and in thicke Woods in
most places of England, especially about
Namptwich in Cheshire, in the Weilds of
Kent,in Suffex and diuers other places.

¶ *The Time.*

The wild Ash flours in May,and the berries are ripe in September.

¶ *The Names.*

The Latines call this tree *Ornus,* and oftentimes *Sylvestris Fraxinus* or wilde Ash;
and it is also *Fraxini species,*or a kind of Ash:
for the Grecians(as not onely *Pliny* writeth,
but also *Theophrastus*) haue made two kinds of Ash,the one high and tall,the other lower;the high
and tall one is *Fraxinus vulgaris,*or the common Ash; and the lower, *Ornus,* which also is named
Ορεινος, or *Montana Fraxinus,*mountain Ash; as the other, μανια or field Ash, which is also named
Βουμελιος, or as *Gaza* translateth it,*Bubula fraxinus,*but more truly *Magna Fraxinus,* or Great Ash; for
the syllable βου is a signe of bignes. This *Ornus* or great Ash is named in high-Dutch,Walbaum:
in low-Dutch,Hauereschen,or Quereschen,of diuers,Qualster:in French,*Fresne sauvage:* in English,Wilde Ash,Quicken-tree,Quick-beam tree,and Wicken tree. *Matthiolus* makes this to bee
*Sorbus sylvestris,*or wilde Service tree.

¶ *The Temperature and Vertues.*

Touching the faculties of the leaues,barke,or berries,as there is nothing found among the old, **A**
so is there nothiug noted among the later writers: but *Pliny* seems to make this wilde Ash like in
faculties to the common Ash; for *lib.16.cap.13.* where he writes of both the Ashes, hee faith,that
the common Ash is *Crissa,* and the mountain Ash *Spissa.* And forthwith he adds this : The Grecians write that the leaues of them do kill cattel,and yet hurt not those that chew their cud.Which
the old Writers haue noted of the Yew tree, and not of the Ash tree. *Pliny* was deceiued by the
neernesse of the words μιλος and μιλια: Μιλος is the Yew tree,and μιλια the Ash tree: so that hee hath
falsly attributed that deadly facultie to the Ash tree, which belongeth to the Yew tree.

The leaues of the wilde Ash tree boiled in wine are good against the pain in the sides,and stop- **B**
ping of the liuer,and assuage the bellies of those that haue the tympany and dropsie.

Benedictus Curtius Symphorianus is deceiued in the history of *Ornus,*when he thinks out of *Virgils* **C**
*Georgicks,*that *Ornus* hath the floure of the Peare tree ; for out of *Virgils* Verses no such thing at **all**

all can be gathered ; for he intreateth not of the formes of trees, but of the graffing of diuers into others, vnlike and differing in nature; as of the graffing of the Nut tree into the Strawbery tree; the Apple into the Plane tree; the Beech into the Cheſtnut tree; the Peare into the wild Aſh or quick-Beam tree, the Oke into the Elm tree : and in this reſpect he writeth, that the Plane tree bringeth forth an apple, the Beech tree a Cheſtnut ; the wilde Aſh bringeth forth the white floure of the Peare tree ; as is moſt manifeſt out of *Virgils* owne words, after this maner, *lib.* 2. of his *Georgicks*:

Inſeritur vero ex fœtu nucis Arbutus horrida,
Et ſteriles Platani malos geſſere valentes,
Caſtaneæ Fagos : Ornus incanuit albo
Flore Pyri, glandémque ſues fregere ſub Vlmis.

The tree-Strawb'ry on Walnuts ſtock doth grow,
And barren Planes faire Apples oft haue borne ;
Cheſtnuts, Beeſt-maſt ; the Quicken-tree doth ſhew
The Peares white floure ; and Swine oft times th' Aeorn
Haue gather'd vnder Elms. ————

C H A P. III.   *Of Coriars Sumach.*

1 *Rhus Coriaria.*
Coriar Sumach.

2 *Rhus Myrtifolia.*
Wilde or Myrtle Sumach.

¶ *The Deſcription.*

2   COriars Sumach groweth vp vnto the height of an hedge tree, after the manner of the Elder tree, bigger than *Dioſcorides* reporteth it to be, or others, who affirme that *Rhus* groweth two cubits high ; whoſe errors are the greater : but this *Rhus* is ſo like to the
Service

Service tree in shape and manner of growing, that it is hard to know one from the other, but that the leaues are soft and hairy, hauing a red sinue or rib thorow the midst of the leaf: the flours grow with the leaues vpon long stems clustering together like Cats taile or the catkins of the Nut tree, but greater, and of a whitish green colour; after which come clusters of round berries, growing in bunches like Grapes.

2   The Sumach of *Plinies* description groweth like a small hedge tree, hauing many slender twiggy branches garnished with leaues like *Myrtus*, or rather like the leaues of the Iuiube tree. Among which come forth slender mossie flours, of no great account or value, which bring forth smal seeds, inclosed within a cornered case or huske fashioned like a spoon. The trunk or body of both these kindes of Sumach being wounded with some iron instrument, yeeldeth a gum or liquour.

¶ *The Place.*

Sumach groweth, as *Dioscorides* saith, in stony places: it is found in diuers mountains & woods in Spain, and in many places on the mount Apennine in Italy, and also neere vnto Pontus. *Archigenes* in *Galen, lib.8*. of Medicines according to the places affected, sheweth, that it groweth in Syria, making choice of that of Syria.

¶ *The Time.*

The floures of Sumach come forth in Iuly: the seed with the berries are ripe in Autumne.

¶ *The Names.*

This is called in Greeke ῥοῦς: *Rhus*, saith *Pliny*, hath no Latine name, yet *Gaza* after the signification of the Greeke word, feigneth a name, calling it *Fluida*: the Arabians name it *Sumach*: the Italians, *Sumacho*: the Spaniards, *Sumagre*: in low-Dutch, by contracting the word, they cal it 𝕾𝖚𝖒𝖆𝖈 or 𝕾𝖚𝖒𝖆𝖈𝖍 in English, Sumach, Coriars Sumach, and leather Sumach: the leaues of the shrub be called ῥοῦς Κορραιακά: in Latine, *Rhus Coriaria*, or *Rhoe*.

The seed is named *Eruthro*, and ῥοῦς ἐρυθρά: in Latine, *Rhus culinaria*, and *Rhus obsoniorum*: in English, Meat Sumach, and Sauce Sumach.

¶ *The Temperature.*

The fruit, leaues, and seed hereof do very much binde: they also coole and dry: dry they are in the third degree, and cold in the second, as *Galen* saith.

¶ *The Vertues.*

The leaues of Sumach boiled in wine and drunken, do stop the lask, the inordinat course of womens sicknesses, and all other inordinat issues of bloud.   A

The seeds of Sumach eaten in Sauces with meat stop all maner of fluxes of the belly, the bloudie flix, and all other issues, especially the Whites of women.   B

The decoction of the leaues maketh haires blacke, and is put into stooles to fume vpward into the bodies of such as haue the Dysenterie, and is to be giuen them also in drinke.   C

The leaues made into an ointment or plaister with hony and vineger, stayes the spreding nature of gangrens and *Pterygia*.   D

The dry leaues sodden in water vntill the decoction be as thicke as honey, yeeld forth a certaine oilinesse which performeth all the effects of *Licium*.   E

The seed is no lesse effectuall to be strewed in pouder vpon their meats that are *Cœliaci* or *Dysenterici*.   F

The seeds pouned, mixed with hony and the pouder of oke coles, heale the hemorrhoids.   G

There issueth out of the shrub a gum, which being put into the hollownesse of the teeth, taketh away the pain, as *Dioscorides* saith.   H

Chap. 112.   *Of Red Sumach.*

¶ *The Description.*

1   THese two figures are of one and the selfe same plant; the first sheweth the shrub being in floure, the other when it is ful floured with the fruit grown to ripenesse: notwithstanding some haue deemed them to be of two kinds, wherein they were deceiued.

†   This excellent and most beautifull plant *Coggygria* (beeing reputed of the Italians and the Venetians for a kind of *Rhus* or Sumach, because it is vsed for the same purposes whereto *Rhus* serueth,

neth and therein doth farre excell it) is an hedge plant growing not aboue the height of foure or fiue cubits, hauing tough and pliant ſtalkes and twiggy branches like to Oziers, of a brown colour. The leaues be round, thick, and ſtiffe like the leaues of *Capparis*, in colour & ſauor of *Piſtacia* leaues, or *Terebinthus*; among which ariſes a ſmall vpright ſprig, bearing many ſmal cluſtering little greeniſh yellow floures, vpon long and red ſtalks. After which follow ſmall reddiſh Lentill-like ſeeds that carry at the tops a moſt fine woolly or flocky tuft, criſped and curled like a curious wrought ſilken fleece, which curleth and ſoldeth it ſelfe abroad like a large buſh of haires.

1 *Coggygria Theophraſti.*	*vel*	*Cotinus Coriarius Plinij.*
Venice Sumach.	or	Red Sumach.

¶ *The Place.*

Coggygria groweth in Orelans neere Auignion, and in diuers places of Italy, vpon the Alpes of Styria, and many other places. It groweth on moſt of the hils of France, in the high woods of the vpper Pannonia or Auſtria, and alſo of Hungaria and Bohemia.

¶ *The Time.*

They floure and flouriſh for the moſt part in Iuly.

¶ *The Names.*

The firſt is called *Coggygria*, and *Coccygria* : in Engliſh, Venice Sumach, or Silken Sumach ; of *Pliny*, *Cotinus*, in his 16. booke, 18. chapter. There is, ſaith he, on mount Apennine a ſhrub which is called *Cotinus ad lineamenta modo Conchylij colore inſignis*, and yet *Cotinus* is *Oleaſter* or *Olea ſylueſtris*, the wild Oliue tree, from which this ſhrub doth much differ; and therfore it may rightly be called *Cotinus Coriaria*. Diuers would haue named it *Scotinus*, which name is not found in any of the old writers. The Pannonians do cal it *Farblauff* : it is alſo thought that this ſhrub is *Coggygria Plinij*, of which in his 13 book, 22. chapter he writeth in theſe words: *Coggygria* is alſo like to *Vnedo* in leafe, not ſo great, it hath a propertie to looſe the fruit with down, which thing happeneth vnto no other tree.

¶ *The Temperature.*

The leaues and ſlender branches together with the ſeeds are very much binding, cold and dry as the other kindes of Sumach are.

¶ *The Vertues.*

A   The leaues of *Coggygria*, or Silken Sumach, are ſold in the markets of Spaine and Italy for great ſummes

summes of mony, vnto those that dresse Spanish skins, for which purpose they are very excellent.

The root of *Cotinus*, as *Anguillara* noteth, serueth to die with, giuing to wooll and cloth a reddish   B
colour : which *Pliny* knew, shewing that this shrub (that is to say the root) is *ad lineamenta a modo con-chylij colore insignis.*

CHAP. 113.   Of the Alder Tree.

¶ The Description.

1   THe Alder or Aller tree is a great high tree hauing many brittle branches, the bark is of
a brown colour, the wood or timber is not hard, and yet it will last and endure very long
vnder the water, yea longer than any other timber whatsoeuer ; wherefore in fenny and soft marish
grounds they vse to make piles and posts thereof, for the strengthening of the walls and such like.
This timber doth also serue very well to make troughs to conuey water in stead of pipes of Lead.
The leaues of this tree are in shape somewhat like the Hasell, but they are blacker and more wrinkled, very clammy to handle, as though they were sprinkled with hony. The blossoms or floures are
like the aglets of the Birch tree, which beeing vaded, there followeth a scaly fruit closely growing
together, as big as a Pigeons egg ; which toward Autumne doth open, and the seed falleth out and
is lost.

1 *Alnus.*
The Alder tree.

‡ 2 *Alnus hirsuta.*
Rough leafed Alder.

‡ 2   *Clusius* and *Bauhine* haue obserued another kinde of this, which differs from the ordinary,
in that it hath latger and more cut leaues, and these not shining aboue, but hoary vnderneath. The
catkins as also the rough heads are not so large as those of the former: the bark also is whiter. *Clu-*
sius makes it his *Alnus altera :* and *Bauhine*, his *Alnus hirsutus, or folio incano.* ‡

¶ The

¶ *The Place.*

The Aller or Alder tree delighteth to grow in low and moist waterish places.

¶ *The Time.*

The Aller bringeth forth new leaues in Aprill ; the fruit whereof is ripe in September.

¶ *The Names.*

The tree is called in Greeke ꙍꙍꙍ : in Latine *Alnus* : *Petrus Crescentius* nameth it *Amedanus* : it is called in high-Dutch, **Erlenbaum** and **Ellernbaum** : in low-Dutch, **Elsen**, and **Elsen boom** : in Italian, *Alno* : in French, *Aulne* : in English, Alder, and Aller.

¶ *The Temperature.*

The leaues and barke of the Alder are cold, dry, and astringent.

¶ *The Vertues.*

A   The leaues of Alder are much vsed against hot swellings, vlcers, and all inward inflammations, especially of the almonds and kernels of the throat.

B   The bark is much vsed of poore country Diers for the dying of course cloath, caps, hose, & such like into a blacke colour, whereto it serueth very well.

CHAP. 114. *Of the Birch Tree.*

Betula.
The Birch tree.

¶ *The Description.*

THe common Birch tree waxeth likewise a great tree, hauing many boughs beset with many small rods or twigs, very limber and pliant, the barke of the yong twigs and branches is plain, smooth, and full of sap, in colour like the chestnut, but the rind of the body or trunk is hard without, white, rough, and vneuen, full of chinks or creuises : vnder which is found another fine barke, plaine, smooth, and as thin as paper, which heretofore was vsed in stead of paper to write on, before the making of paper was knowne : in Russia and these cold countries it serueth in stead of tiles and slate to couer their houses withall. This tree beareth for his flours certaine aglets like the Hasel tree, but smaller, wherein the seed is contained.

¶ *The Place.*

This common Birch tree grows in woods, fenny grounds, and mountains, in most places of England.

¶ *The Time.*

The catkins or aglets do first appear, and then the leaues, in Aprill or a little later.

¶ *The Names.*

Theophrastus calleth this tree in Greeke, ꙍꙍꙍ : diuers, ꙍꙍꙍ : others ꙍꙍꙍ it is named in Latine, *Betula* : diuers also write it with a double *ll Betulla*, as some of *Plinies* Copies haue it : it is called in high Dutch, **Birkenbaum** : in low-Dutch, **Berckenboom** : in Italian, *Betula* : by them of Trent, *Bedallo* : in French, *Boulean* : in English, Birch tree.

¶ *The Nature and Vertues.*

Concerning the medicinable vse of the Birch tree, or his parts, there is nothing extant either in the old or new writers.

This tree, saith *Pliny, lib. 16. cap. 18. Mirabili candore & tenuitate terribilis magistratuum virgis* : for in times past the magistrats rods were made thereof ; and in our time also Schoolmasters and Parents do terrifie their children with rods made of Birch.

It serueth well to the decking vp of houses and banqueting rooms, for places of pleasure, and for beautifying of streets in the Crosse and Gang weeke, and such like.

CHAP.

Chap. 115.

Of the Horne-beame or hard Beame Tree.

Betulus, ſive Carpinus.
The Horn-beam tree.

¶ The Deſcription.

BEtulus or the Horn-beame tree growes great, and very like vnto the Elme or Wich-Haſell tree, hauing a great body, the wood or timber whereof is better for arrowes and ſhafts, pulleyes for mils, and ſuch like deuices, than Elm or Wich Haſell; for in time it waxeth ſo hard, that the toughneſſe and hardnes of it may be rather compared to horn than vnto wood, and therfore it was called Horne-beam or Hard-beam: the leaues of it are like the Elme, ſauing that they be tenderer; among thoſe hang certaine triangled things, vpon which be found knaps or little heads of the bigneſſe of Ciches, in which is contained the fruit or ſeed: the root is ſtrong and thicke.

¶ The Place.

Betulus or the Horn-beam tree grows plentifully in Northampton ſhire, and in Kent by Graueſend, where it is commonly taken for a kinde of Elme.

¶ The Time.

This tree ſprings in Aprill, and the ſeed is ripe in September.

¶ The Names.

The Horn-beame tree is called in Greeke, ζυγία: which is as much to ſay as Conjugalis, or belonging to the yoke, becauſe it ſerues well to make ζυγία of, in Latine Inga, yokes wherwith oxen are yoked together; which are alſo euen at this time made thereof, as witneſſeth Benedictus Curtius Symphorianus, and our ſelues haue ſufficient knowledge thereof in our owne country, and therefore it may be Engliſhed, Yoke Elm. It is called of ſome Carpinus, and Zugia: it is alſo called Betulus, as if it were a kinde of Birch; but my ſelfe better like it ſhould be one of the Elmes: in high-Dutch, ℈ Hozn: in French Carne: in Italian, Carpino: in Engliſh, Horn-beam, Hard-beam, Yoke-Elme, and in ſome places Witch Haſell.

¶ The Temperature and Vertues.

This tree is not vſed in medicine, the vertues are not expreſſed of the Antients, neither haue we any certain experiments of our owne knowledge, more than hath been ſaid for the vſe of Husbandry.

Chap. 116.   Of the Elme tree.

‡ OVr Author onely deſcribed two Elmes, and thoſe not ſo accuratly but I think I ſhall giue the Reader content, in exchanging them for better, receiued from Mr Goodyer, the which are theſe.

Vlmus vulgatiſſima folio lato ſcabro. The common Elme.

1 THis Elme is a very great high tree, the bárk of the yong trees, and boughes, of the elder, which are vſually lopped or ſhred, is ſmooth aud very tough, and will ſtrip or pill from the wood a great length without breaking: the bark of the body of the old tree as the trees grow in bigneſſe teares or rents, which makes it very rough. The innermoſt wood of the tree is of a reddiſh
yellow

yellow or browniſh colour, and curle d, and after it is dry, very tough, hard to cleaue or rent, wherof aves of carts are commonly made: th e wood next the bark, which is called the ſap, is white. Before the leaues come forth, the floures appeare about the end of Marʧ, which grow on the twigges or branches, cloſely compacted or thruſt together, and are like to the chiues growing in the middle of moſt floures, of a red colour : after which come flat ſeed, more long than broad, not much vnlike the garden Arach ſeed in form and bigneſſe, and do for the moſt part fal away before or ſhortly after the leaues ſpring forth, and ſome hang on a great part of ſummer: the leaues hang on the twigs, of a darke green colour, the middle ſiſe whereof are two inches broad and three inches long; ſome are longer and broader, ſome narrower and ſhorter, rough or harſh in handling on both ſides, nickt or indented about the edges, and many times crumpled, hauing a nerue in the middle, and many ſmaller nerues growing from him : the leafe on one ſide of the nerue is alwaies longer than on the other. On theſe leaues oftentimes grow bliſters or ſmall bladders, in which at the ſpring are little wormes about the bigneſſe of bed-fleas. This Elme is common in all parts of England where I haue trauelled.

1 *Vlmus vulgatiſſ. folio lato ſcabro.*
The common Elme tree.

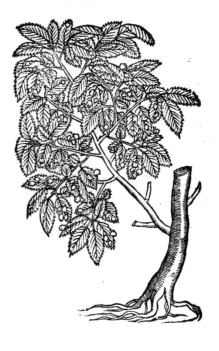

‡ 2 *Vlmus minor folio anguſto ſcabro.*
The narrow leaued Elme.

Vlmus minor folio anguſto ſcabro.
The narrow leafed Elme.

    2 This tree is like the other, but much leſſer and lower ; the leaues are vſually about two inches and a halfe long, and an inch or an inch and a quarter broad, nickt or indented about the edges, and hath one ſide longer than the other, as the firſt hath, and are alſo harſh or rough on both ſides, the barke or Rinde will alſo ſtrip as the firſt doth.   Hitherto I haue not obſerued either the floures or ſeed, or bliſters on the leaues, nor haue I had any ſight of the Timber, or heard of any vſe thereof.   This Kinde I haue ſeene growing but once, and that in the hedges by the high-way, as I rode betweene Chriſt Church and Limmington in the New Forreſt in Hampeſhire,
abou t

about the middle of September 1624. from whence I brought some small plants of it, not a foot
in length, which now, 1633. are risen vp ten or twelue foot high, & grow with me by the first kind,
but are easily to be discerned apart, by any that will looke on both.

‡ 3 *Vlmus folio latißimo scabro.*
Witch Hasell, or the broadest leaued Elme.

4 *Vlmus folio glabro.*
Witch Elme, or smooth leaued Elme.

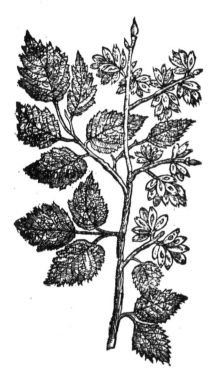

Vlmus folio latißimo scabro.
Witch Hasell, or the broadest leaued Elme.

3   This groweth to be a very great tree, and also very high, especially when he groweth in woods
amongst other trees: the barke on the outside is blacker than that of the first, and is also very tough,
so that when there is plenty of sap it will strip or peele from the wood of the boughes from the one
end to the other, a dozen foot in length or more, without breaking, whereof are often made cords or
ropes : the timber hereof is in colour neere like the first ; it is nothing so firme or strong for naues
of Carts as the fruit is, but will more easily cleaue ; this timber is also couered with a white sappe
next the barke : the branches or young boughes are grosser and bigger, and doe spread themselues
broader, and hang more downewards than those of the first ; the floures are nothing but chiues, very
like those of the first kinde : the seed is also like, but something bigger : the leaues are much broader and longer than any of the kindes of Elme, vsually three or foure inches broad, and fiue or six
inches long, also rough or harsh in handling on both sides, snipt or indented about the edges, neere
resembling the leaues of the Hasell : the one side of the leaues are also most commonly longer
than the other, also on the leaues of this Elme are sometimes blisters or bladders like those of the
first kinde. This prospereth and naturally groweth in any soile moist or dry, on high hills, and in
low vallies in good plenty in most places in Hampshire, where it is commonly called Witch Hasell. Old men affirme, that when long boughes were in great vse, there were very many made of the
wood of this tree, for which purpose it is mentioned in the statutes of England by the name of
Witch Hasell, as 8. *El.* 10. This hath little affinitie with *Carpinus*, which in Essex is called Witch
Hasell,

Vlmus folio glabro.
Witch Elme, or smooth leauen Elme.

4   This kinde is in bignesse and height like the first, the boughes grow as those of the Witch Hasell do, that is hanged more downewards than those of the common Elme, the barke is blacker, than that of the first kinde, it will also peele from the boughes : the floures are like the first, and so are the seeds : the leaues in forme are like those of the first kind, but are smooth in handling on both sides. My worthy friend and excellent Herbarist of happy memory Mr *William Coys* of Stubbers in the Parish of Northokington in Essex told me, that the wood of this kinde was more desired for naues of Carts than the wood of the first. I obserued it growing very plentifully as I rode betweene Rumford and the said Stubbers, in the yeare 1620. intermixed with the first kinde, but easily to be discerned apart, and is in those parts vsually called Witch Elme. ‡

¶ The Place.
The first kinde of Elme groweth plentifully in all places of England. The rest are set forth in their descriptions.

¶ The Time.
The seeds of the Elme sheweth it selfe first, and before the leaues ; it falleth in the end of Aprill, at what time the leaues begin to spring.

¶ The Names.
The first is called in Greeke, πτελέα: in Latine, *Vlmus* in high Dutch, **Ruſt holtz, Ruſtbaum, Ulm-baum**: in low Dutch **Olmen**: in French, *Orme*, and *Omeau*: in Italian, *Olmo*: in Spanish, *Vlmo* in English, Elme tree.

The seed is named by *Pliny* and *Columella*, *Samera*. The little wormes which are found with the liquor within the small bladders be named in Greeke, κώνωπι: it Latine *Culices*, and *Muliones*.

The other Elme is called by *Theophrastus*, ὀρειπτελέα; which *Gaza* translateth *Montiulmus* or mountaine Elme. *Columella* nameth it *Vernacula*, or *Nostras Vlmus*, that is to say, *Italica*, or Italian Elme: it is called in low Dutch, **Herseleer**, and in some places, **Heerenteer.**

¶ The Temperature and Vertues.
A   The leaues and barke of the Elme bee moderately hot, with an euident clensing faculty ; they haue in the chewing a certaine clammy and glewing quality.

B   The leaues of Elme glew and heale vp greene wounds, so doth the barke wrapped and swadled about the wound like a band.

C   The leaues being stamped with vineger do take away scurffe.

D   *Dioscorides* writeth, That one ounce weight of the thicker barke drunke with wine or water purgeth flegme.

E   The decoction of Elme leaues, as also of the barke or root, healeth broken bones very speedily, if they be fomented or bathed therewith.

F   The liquor that is found in the blisters doth beautifie the face, and scoureth away all spots, freckles, pimples, spreading tetters, and such like, being applied thereto.

G   It healeth greene wounds, and cureth ruptures newly made, being laid on with Spleenwoort and the trusse closely set vnto it.

CHAP. 117. *Of the Line or Linden Tree.*

¶ The Description.

1   THe female Line or Linden tree waxeth very great and thicke, spreading forth his branches wide and farre abroad, being a tree which yeeldeth a most pleasant shadow, vnder and within whose boughes may be made braue summer houses and banqueting arbors, because the more that it is surcharged with weight of timber and such like, the better it doth flourish. The barke is brownish, very smooth, and plaine on the outside, but that which is next to the timber is white, moist and tough, seruing very well for ropes, trases, and halters. The timber is whitish, plaine and without knots, yea very soft and gentle in the cutting or handling. Better gunpouder is made of the coales of this wood than of Willow coales. The leaues are greene, smooth,

smooth, shining, and large, somewhat snipt or toothed about the edges : the floures are little, whitish, of a good sauour, and very many in number, growing clustering together from out of the middle of the leafe : out of which proceedeth a small whitish long narrow leafe : after the floures succeed cornered sharpe pointed Nuts, of the bignesse of Hasell Nuts. This tree seemeth to be a kinde of Elme, and the people of Essex about Heningham (wheras great plenty groweth by the way sides) do call it broad leafed Elme.

1 *Tilia fæmina.*
The female Line tree.

2 *Tilia mas.*
The male Line tree.

2   The male *Tilia* or Line tree groweth also very great and thicke, spreading it selfe far abroad like the other Linden tree his barke is very tough and pliant, and serueth to make cords and halters of. The timber of this tree is much harder, more knotty, and more yellow than the timber of the other, not much differing from the timber of the Elme tree : the leaues hereof are not much vnlike Iuy leaues, not very greene, somewhat snipt about the edges : from the middle whereof come forth clusters of litle white floures like the former : which being vaded, there succeed small round pellets, growing clustering together like Iuy berries, within which is contained a little round blackish seed, which falleth out when the berry is ripe.

¶ *The Place.*

The female Linden tree groweth in some woods in Northampton shire ; also neere Colchester, and in many places alongst the high way leading from London to Henningham, in the county of Essex.

The male Linden tree groweth in my Lord Treasurers garden in the Strand, and in sundry other places, as at Barn-elmes, and in a garden at Saint Katherines neere London. ‡ The female growes in the places here named, but I haue not yet obserued the male. ‡

¶ *The Time.*

The trees floure in May, and their fruit is ripe in August.

¶ *The Names.*

The Linden tree is called in Greeke ···· : in Latine, *Tilia* : in high Dutch, **Linden,** and **Lindenbaum :** in low Dutch, **Linde,** and **Lindenboom :** the Italians, *Tilia* : the Spaniard, *Tera* : in French, *Tilet* and *Tilleul* : in English, Linden tree, and Line tree.

¶ *The Temperature.*

The barke and leaues of the Linden or Line tree, are of a temperate heate, somewhat drying and astringent.

¶ *The Vertues.*

A   The leaues of *Tilia* boyled in Smithes water with a piece of Allum and a little honey, cure the sores in childrens mouthes.

B   The leaues boiled vntill they be tender ; and pouned very small with hogs grease, and the pouder of Fenugreeke and Lineseed, take away hot swellings and bring impostumes to maturation, being applied thereto very hot.

C   The floures are commended by diuers against paine of the head proceeding of a cold cause, against disinesse, the Apoplexie, and also the falling sicknesse, and not onely the floures, but the distilled water thereof.

D   The leaues of the Linden (saith *Theophrastus*) are very sweet, and be a fodder for most kind of cattle: the fruit can be eaten of none.

Chap. 118.   *Of the Maple tree.*

‡ 1 *Acer majus.*   † 2 *Acer minus.*
The great Maple.   The lesser Maple.

¶ *The Description.*

THe great Maple is a beautifull and high tree, with a barke of a meane smoothnesse : the substance of the wood is tender and easie to worke on ; it sendeth forth on euery side very many goodly boughes and branches, which make an excellent shadow against the heat of the Sun ; vpon
which

which are great, broad, and cornered leaues, much like to thoſe of the Vine, hanging by long red-diſh ſtalkes ; the floures hang by cluſters, of a whitiſh greene colour; after them commeth vp long fruit faſtened together by couples, one right againſt another, with kernels bumping out neere to the place in which they are combined : in all the other parts flat and thin like vnto parchment, or reſembling the innermoſt wings of graſhoppers : the kernels be white and little.

2   There is a ſmall Maple which doth oftentimes come to the bigneſſe of a tree, but moſt com-monly it groweth low after the manner of a ſhrub : the barke of the young ſhoots hereof is likewiſe ſmooth; the ſubſtance of the wood is white, and eaſie to be wrought on: the leaues are cornered like thoſe of the former, ſlippery, and faſtened with a reddiſh ſtalke, but much leſſer, very like in big-neſſe and ſmoothneſſe to the leafe of Sanicle, but that the cuts are deeper : the floures be as thoſe of the former, greene, yet not growing in cluſters, but vpon ſpoked rundles : the fruit ſtandeth by two and two vpon a ſtem or foot-ſtalke.

¶ *The Place.*

The ſmall or hedge Maple groweth almoſt euery where in hedges and low woods.

The great Maple is a ſtranger in England, onely it groweth in the walkes and places of pleaſure of noble men, where it eſpecially is planted for the ſhadow ſake, and vnder the name of Sycomore tree.

¶ *The Time.*

Theſe trees floure about the end of March, and their fruit is ripe in September.

¶ *The Names.*

This tree is called in Greeke σφένδαμνος : in Latine, *Acer* : in Engliſh, Maple, or Maple tree.

The great Maple is called in high Dutch, **Ahozne**, and **walberſcherne :** the French-men, *Grand Erable,* and *Plaſne* aouſiuely, and this is thought to be properly called σφένδαμνος : but they are far decei-ued that take this for *Platanus,* or the Plane tree, being drawne into this errour by the neereneſſe of the French word; for the Plane tree doth much differ from this. ‡ This is now commonly (yet not rightly) called the Sycomore tree. And ſeeing vſe will haue it ſo, I thinke it were not vnfit to call it the baſtard Sycomore. ‡

The other is called in Latine, *Acer minor:* in high Dutch, **waſſholder:** in low Dutch, **Booghout:** in French, *Erable :* in Engliſh, ſmall Maple, and common Maple.

¶ *The Temperature and Vertues.*

What vſe the Maple hath in medicine we finde nothing written of the Grecians, but *Pliny* in his A 14 booke, 8 chapter affirmeth, That the root pouned and applied, is a ſingular remedy for the paine of the liuer. *Serenus Sammonicus* writeth, that it is drunke with wine againſt the paines of the ſide.

> *Si latus immeritum morbo tentatur acuto,*
> *Accenſum tinges lapidem ſtridentibus vndis.*
> *Hinc bibis : aut Aceris radicem tundis, & vna*
> *Cum vino capis : hoc praſens medicamen habetur.*

> Thy harmeleſſe ſide if ſharpe diſeaſe inuade,
> In hiſſing water quench a heated ſtone :
> This drinke. Or Maple root in pouder made,
> Take off in wine, a preſent med'cine knowne.

CHAP. 119.   *Of the Poplar Tree.*

¶ *The Kindes.*

THere be diuers trees vnder the title of Poplar, yet differing very notably, as ſhall be beclared in the deſcriptions, whereof one is the white, another the blacke, and a third ſort ſet downe by *Pli-ny,* which is the Aſpe, named by him *Lybaca;* and by *Theophraſtus, Kerkis :* likewiſe there is another of America, or of the Indies, which is not to be found in theſe regions of Europe.

¶ *The*

¶ *The Deſcription.*

1   THe white Poplar tree commeth ſoone to perfection,and growes high in ſhort time,full of boughes at the top : the barke of the body is ſmooth,and that of the boughes is like-wiſe white withall : the wood is white,eaſie to be cleft : the leaues are broad, deeply ga-ſhed,and cornered like almoſt to thoſe of the Vine,but much leſſer, ſmooth on the vpper ſide, glib, and ſomewhat greene ; and on the nether ſide white and woolly : the catkins are long,downy,at the firſt of a purpliſh colour : the roots ſpread many waies,lying vnder the turſe, and not growing deep, and therefore it happeneth that theſe trees be oftentimes blowne downe with the winde.

1 *Populus alba.*   2 *Populus nigra.*
The white Poplar tree.   The blacke Poplar tree.

2   The blacke Poplar tree is as high as the white,and now and then higher,oftentimes fuller of boughes, and with a thicker body : the barke thereof is likewiſe ſmooth, but the ſubſtance of the wood is harder,yellower,and not ſo white,fuller of veines,and not ſo eaſily cleft:the leaues be ſom-what long,and broad below toward the ſtem, ſharpe at the point, and a little ſnipt about the edges, neither white nor woolly, like the leaues of the former, but of a pleaſant greene colour : amongſt which come forth long aglets or catkins, which do turne into cluſters : the buds which ſhew them-ſelues before the leaues ſpring out,are of a reaſonable good ſauour,of the which is made that profi-table ointment called *Vnguentum Populeon.*

3   The third kinde of Poplar is alſo a great tree : the barke and ſubſtance of the wood is ſome-what like that of the former : this tree is garniſhed with many brittle and tender branches, ſet full of leaues, in a manner round, much blacker and harder than the blacke Poplar, hanging vpon long and ſlender ſtems,which are for the moſt part ſtill wauering,and make a great noiſe by being beaten one to another,yea though the weather be calme,and ſcarce any winde blowing; and it is known by the name of the Aſpen tree : the roots hereof are ſtronger, and grow deeper into the ground than thoſe of the white Poplar.

4   This ſtrange Poplar,which ſome do cal *Populus rotundifolia,*in Engliſh,the round leafed Pop-lar of India,waxeth a great tree,bedeckt with many goodly twiggie branches,tough and limmer like
the

3 *Populus Libyca.*
The Aſpen tree.

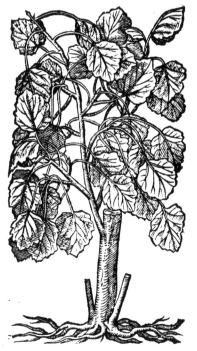

4 *Populus Americana.*
The Indian Poplar tree.

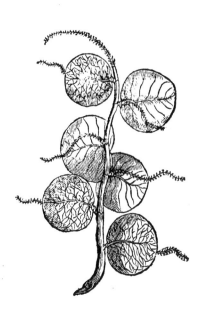

‡ 5 *Populus alba folijs minoribus.*
The leſſer leaued white Poplar.

the Willow, full of joynts where the leaues
doe grow, of a perfect roundneſſe, ſaue where
it cleaueth or groweth to the ſtalke: from the
boſoms or corners of theſe leaues come forth
ſmall aglets, like vnto our Poplar, but ſmaller: the leafe is thicke, and very like the leaues
of *Arbor Indæ*, but broader, of an aſtringent
taſte, ſomewhat heating the mouth, and ſaltiſh.

5 There is alſo another ſort of Poplar
which groweth likewiſe vnto a great tree, the
branches whereof are knotty and bunched
forth as though it were full of ſcabs or ſores:
the leaues come forth in tufts moſt commonly at the end of the boughes, not cut or jagged, but reſembling the leaues of that *Atriplex* called *Pes Anſerinus*; in colour like the
former, but the aglets are not ſo cloſely pac-
ked together, otherwiſe it is like.

¶ *The Place.*

Theſe trees doe grow in low moiſt places,
as in medowes neere vnto ditches, ſtanding
waters and riuers.

The firſt kinde of white Poplar groweth
not very common in England, but in ſome
places here and there a tree: I found many
both ſmall & great growing in a low medow
turning

turning vp a lane at the farther end of a village called Blacke-wall, from London ; and in Eſſex at a place called Ouenden, and in diuers other places.

The Indian Poplar groweth in moſt parts of the Iſlands of the Weſt-Indies.

¶ *The Time.*

Theſe trees do bud forth in the end of March and beginning of Aprill, at which time the buds muſt be gathered to ſerue for *Vnguentum Populeon.*

¶ *The Names.*

The white Poplar is called in Greeke, Λευχή: in Latine, *Populus alba* : of diuers, *Farſarus,* as of *Plautus* in his Comedie *Penulus,* as you may ſee by his words ſet downe in the chapter of Colts-foot, *pag.* 813.

It is called in high Dutch, **Poppelbaum, Weiſz Albarbaum** : in low Dutch, **Abeel,** of his horie or aged colour, and alſo **Abeelboome** ; which the Grammarians doe falſly interpret *Abies,* the Firre tree : in Italian, *Popolo nero* : in French, *Peuplier blanc, Aubel, Obel,* or *Aubeau* : in Engliſh, white Poplar tree, and Abeell, after the Dutch name.

The ſecond is called in Greeke, Κεγχρος: in Latine, *Populus nigra* : by *Petrus Creſcentius, Albarus :* in high Dutch, **Aſpen** : in low Dutch, **Populier** : in Italian, *Popolo nero* : in French, *Peuplier noir :* in Spaniſh, *Alamo nigailho :* in Engliſh, Poplar tree, blacke Poplar, and Peplar. The firſt or new ſprung buds whereof are called of the Apothecaries, *Oculi Populi,* Poplar buds: others chuſe rather to call it *Gemma Populi :* ſome of the Grӕcians name it Σκήμα: whereupon they grounded their error, who raſhly ſuppoſed that thoſe roſenny or clammy buds are not to be put or vſed in the compoſition of the ointment bearing the name of the Poplar, and commonly called in Engliſh, Popilion and Pompillion ; but the berries that grow in cluſters, in which there is no clammineſſe at all.

They are alſo as farre deceiued, who giuing credit to Poëts fables, doe beleeue that Amber commeth of the clammy roſin falling into the riuer Poo.

The third is called of diues, *Populus tremula,* which word is borrowed of the French-men, who name it *Tremble :* it alſo receiued a name amongſt the low-Country men, from the noiſe and ratling of the leaues, *viz.* **Rateeler** : this is that which is named of *Pliny, Libica :* and by *Theophr.* Κερκις. which *Gaʒa* calleth *Populus montana :* in Engliſh, Aſpe, and Aſpen tree, and may alſo be called Tremble, after the French name, conſidering it is the matter whereof womens tongues were made, (as the Poets and ſome others report) which ſeldome ceaſe wagging.

¶ *The Temperature and Vertues.*

A   The white Poplar hath a clenſing faculty, ſaith *Galen,* and a mixt temperature, conſiſting of a watery warme eſſence, and alſo a thin earthy ſubſtance.

B   The barke as *Dioſcorides* writeth, to the weight of an ounce (or as others ſay, and that more truly, of little more than a dram) is a good remedy for the Sciatica or ache in the huckle bones, and for the ſtrangury.

C   That this barke is good for the Sciatica, *Serenus Sammonicus* doth alſo write :

> *Sæpius occultus victa coxendice morbus*
> *Perfurit, & greſſus diro languore moratur :*
> *Populus alba dabit medicos de cortice potus.*

> An hidden diſeaſe doth oft rage and raine,
> The hip ouercome and vex with the paine,
> It makes with vile aking one tread ſlow and ſhrinke;
> The barke of white Poplar is helpe had in drinke.

D   The ſame barke is alſo reported to make a woman barren, if it be drunke with the kidney of a Mule, which thing the leaues alſo are thought to performe, being taken after the ſloures or reds be ended.

E   The warme juice of the leaues being dropped into the eares doth take away the paine thereof.

F   The Roſin or clammy ſubſtance of the blacke Poplar buds is hot and dry, and of thin parts, attenuating and mollifying : it is alſo fitly mixed *acopis & malegmatis :* the leaues haue in a manner the like operation for all theſe things, yet weaker and not ſo effectuall, as *Galen* teacheth.

G   The leaues and young buds of blacke Poplar aſſwage the paine of the gout in the hands or feet, being made into an ointment with May butter.

H   The ointment made of the buds is good againſt all inflammations, bruſes, ſquats, falls, and ſuch like : this ointment is very well knowne to the Apothecaries.

I   *Paulus Ægineta* teacheth to make an oile alſo hereof, called *Ægyrinum,* or oile of blacke Poplar.

CHAP.

CHAP. 120. Of the Plane tree.

Platanus.
The Plane tree.

¶ *The Deſcription.*

THe Plane is a great tree, hauing very long and farre ſpreading boughes caſting a wonderfull broad ſhadow, by reaſon wherof it was highly commended and eſteemed of among the old Romans: the leaues are cornered like thoſe of *Palma Chriſti,* greater than Vine leaues, and hanging vpon little red foot-ſtalkes: the floures are ſmall and moſſie, and of a pale yellowiſh colour: the fruit is round like a ball, rugged, and ſomewhat hairy; but in Aſia more hairy and greater, almoſt as big as a Walnut: the root is great, diſperſing it ſelfe far abroad.

¶ *The Place.*

The Plane tree delighteth to grow by ſprings or riuers: *Pliny* reports that they were wont to bee cheriſhed with wine: they grew afterward (ſaith he) to bee of ſo great honour (meaning the Plane trees) as that they were cheriſhed and watered with wine: and it is found by experience that the ſame is very comfortable to the roots, and wee haue already taught, that trees deſire to drinke Wine. This tree is ſtrange in Italy, it is no where ſeene in Germany, nor in the low-Countries: in Aſia it groweth plentifully: it is found alſo in Candy, growing in vallies, and neere vnto the hill Athos, as *Petrus Bellonius* in his Singularities doth declare: it groweth in many places of Greece, and is found planted in ſome places of Italy, for pleaſure rather than for profit. My ſeruant *William Marſhall* (whom I ſent into the Mediterranean ſea as Surgeon vnto the Hercules of London) found diuers trees hereof growing in Lepanto, hard by the ſea ſide, at the entrance into the towne, a port of Morea, beeing a part of Greece, and from thence brought one of theſe rough buttons, being the fruit thereof. ‡ There are one or two yong ones at this time growing with M.r *Tradeſcant.* ‡

The Plane trees caſt their leaues in Winter, as *Bellonius* teſtifieth, and therefore it is no maruell that they keepe away the Sun in Summer, and not at all in Winter: there is, ſaith *Pliny,* no greater commendation of the tree, than that it keepeth away the Sunne in Summer, and entertaineth it in Winter.

¶ *The Names.*

This tree is called in Greeke, πλάτανος: and likewiſe in Latine *Platanus:* it beareth his name of the bredth: the French-mens *Plaſne* doth far differ from this, which is a kinde of Maple: this tree is named in Engliſh, Plane tree.

¶ *The Temperature and Vertues.*

The Plane tree is of a cold and moiſt eſſence, as *Galen* ſaith: the greene leaues are good to be laid **A** vpon hot ſwellings and inflammations in the beginning.

Being boiled in wine they are a remedy for the running and the watering of the eies, if they bee **B** applied.

The barke and balls do dry: the barke boyled in vineger helpeth the tooth-ache. **C**

The fruit of the Plane tree drunke with Wine helpeth the biting of mad dogs and ſerpents, and **D** mixed with hogs greaſe it maketh a good ointment againſt burning and ſcalding.

The burned barke doth mightily dry, and ſcoureth withall; it remoueth the white ſcurfe, and cu- **E** reth moiſt vlcers.

The

F   The duſt or downe, ſaith *Galen*, that lieth on the leaues of the tree is to be taken heed of, for if it be drawne in with the breath, it is offenſiue to the winde-pipe by his extreme dryneſſe, and making the ſame rough, and hurting the voice, as it doth alſo the ſight and hearing, if it fall into the eyes or eares. *Dioſcorides* doth not attribute this to the duſt or downe of the leaues onely, but alſo to that of the balls.

C H A P. 121.   *Of the Wayfaring tree.*

Lantana, ſiue Viburnum.
The Wayfaring tree.

¶ *The Deſcription.*

THe Wayfaring mans tree growes vp to the height of an hedge tree, of a meane bignes: the trunke or body thereof is couered with a ruſſet barke : the branches are long, tough, and eaſie to be bowed, and hard to be broken, as are thoſe of the Willow, couered with a ſoft whitiſh barke, whereon are broad leaues thicke and rough, ſleightly indented about the edges, of a white colour, and ſomewhat hairy whileſt they be freſh and greene; but when they begin to wither and fall away, they are reddiſh, and ſet together by couples one oppoſit to another. The floures are white, and grow in cluſters : after which come cluſters of fruit of the bigneſſe of a peaſe, ſomewhat flat on both ſides, at the firſt greene, after red, and blacke when they be ripe : the root diſperſeth it ſelfe far abroad vnder the vpper cruſt of the earth.

¶ *The Place.*

This tree groweth in moſt hedges in rough and ſtony places, vpon hils and low woods, ſpecially in the chalky grounds of Kent about Cobham, Southfleet, and Graueſend, and in all the tract to Canterbury.

¶ *The Time.*

The floures appeare in Summer : the berries are ripe in the end of Autumne, and new leaues come forth in the Spring.

¶ *The Names.*

This hedge tree is called *Viurna* of *Ruellius* : in French, *Viorne*, and *Viorna* : in Italian, *Lantana* : it is reputed for the tree *Viburnum*, of which *Virgill* maketh mention in the firſt Eclog, where hee commendeth the City Rome for the loftineſſe and ſtatelineſſe thereof, aboue other Cities, ſaying, that as the tall Cypres trees do ſhew themſelues aboue the low and ſhrubby Viorne, ſo doth Rome aboue other Cities lift vp her head very high ; in theſe verſes :

> *Verum hæc tantum alias inter caput extulit vrbes,*
> *Quantum lenta ſolent inter viburna cupreſſi.*

> But this all other cities ſo excels,
> As Cypreſſe, which 'mongſt bending Viornes dwels.

‡ I judge *Viburnum* not to be a name to any particular plant, but a generall name to all low and bending ſhrubs ; amongſt which this here deſcribed may take place as one. I enquired of a country man in Eſſex, if he knew any name of this : he anſwered, it was called the Cotton tree, by reaſon of the ſoftneſſe of the leaues. ‡

¶ *The Temperature.*

The leaues and berries of Lantana are cold and dry, and of a binding quality.

¶ *The*

¶ *The Vertues.*

The decoction of the leaues of Lantaua is very good to be gargled in the mouth against all swellings and inflammations thereof, against the scuruie and other diseases of the gums, and fastneth loose teeth.   A

The same boyled in lee doth make the haires blacke if they be bathed or washed therewith, and suffered to dry of it selfe.   B

The berries are of the like faculty, the pouder whereof when they be dried stay the laske, all issues of bloud, and also the whites.   C

It is reported that the barke of the root of the tree buried a certaine time in the earth, and afterwards boyled and stamped according to Art, maketh good Bird-lime for Fowlers to catch Birds with.   D

CHAP. 122.   *Of the Beade tree.*

1 *Zizypha candida.*
The Beade tree.

‡ 2 *Zizypha Cappadocica.*
The Beade tree of Cappadocia.

¶ *The Description.*

1   THis tree was called *Zizypha candida* by the Herbarists of Montpellier ; and by the Venetians and Italians, *Sycomorus*, but vntruly : the Portugals haue termed it *Arbor Paradiso* : all which and each whereof haue erred together, both in respect of the fruit and of the whole tree : some haue called it *Zizypha*, though in faculty it is nothing like ; for the taste of this fruit is very vnpleasant, virulent and bitter. But deciding all controuersies, this is the tree which *Auicen* calleth *Azederach*, which is very great. charged with many large armes, that are garnished with twiggie branches, set full of great leaues consisting of sundry small leaues, one growing right opposite to another like the leaues of the ash-tree or Wicken tree, but more deepely cut about the edges like the teeth of a saw : among which come the floures, consisting of fiue small blew leaues layd abroad in manner of a starre : from the middest whereof groweth forth a small hollow cup
resembling

resembling a Chalice: after which succeedeth the fruit, couered with a brownish yellow shell, very like vnto the fruit of Iuiubes (whereof *Dodonæus* in his last edition maketh it a kinde) of a rancke, bitter, and vnpleasant taste, with a six cornered stone within, which being drawne on a string, serueth to make beades of, for want of other things.

2　*Ziziphus Cappadocica* groweth not so great as the former, but is of a meane stature, and full of boughes: the barke is smooth and euen, and that which groweth vpon the trunke and great boughes is of a shining scarlet colour: out of these great armes or boughes grow slender twigges, white and soft, which are set full of whitish leaues, but more white on the contrary or backe part, and are like to the leaues of Willow, but narrower, and whiter: amongst these leaues come forth small hollow yellowish floures, growing at the joynts of the branches, most commonly three together, and of a pleasant sauour, with some few threds or chiues in the middle thereof. After which succeedeth the fruit, of the bignesse and fashion of the smallest Oliue, white both within and without, wherein is contained a small stone which yeeldeth a kernell of a pleasant taste and very sweet.

¶ The Place.

Matthiolus writeth, that *Ziziphus candida* is found in the cloisters of many monasteries in Italy; *Lobel* saith that it groweth in many places in Venice and Narbon; and it is wont now of late to be planted and cherished in the goodliest orchards of all the low-Countries.

Ziziphus Cappadocica groweth likewise in many places of Italy, and specially in Spaine: it is also cherished in gardens both in Germany and in the low-Countries. ‡ It groweth also here in the garden of Mr *Iohn Parkinsou*. ‡

¶ The Time.

These trees floure in Iune, in Italy and Spaine; their fruit is ripe in September; but in Germany and the low-Countries there doth no fruit follow the floures.

¶ The Names.

Ziziphus candida Auicen calleth *Azedcrach*, or as diuers read it, *Azederaeth*: and they name it, saith he, in Rechi, *Arbor Mirobalanorum*, or the Mirobalane tree, but not properly, and in Tabrasten, and Kien, and Thihich. The later writers are farre deceiued in taking it to be the Sycomore tree; and they as much, that would haue it to be the Lote or Nettle tree: it may be named in English, Beade tree, for the cause before alledged.

The other is *altera species Ziziphi*, or the second kinde of Iuiube tree, which *Columella* in his ninth booke and fourth chapter doth call *Ziziphus alba*, or white Iuiube tree, for difference from the other that is syrnamed *Rutila*, or glittering redde. *Pliny* calleth this *Ziziphus Cappadocica*, in his 21 booke, ninth chapter, where he entreateth of the honour of Garlands, of which hee saith there be two sorts, whereof some be made of floures, and others of leaues: I would call the floures (saith he) broomes, for of those is gathered a yellow floure, and *Rhododendron*, also *Zizipha*, which is called *Cappadocica*. The floures of these are sweet of smell, and like to Oliue floures. Neither doth *Columella* or *Pliny* vnaduisedly take this for *Ziziphus*, for both the leaues and floures grow out of the tender and yong sprung twigs, as they likewise do out of the former: the floures are very sweet of smell, and cast their sauor far abroad: the fruit also is like that of the former.

¶ The Temperature.

Auicen writing and intreating of *Azadaraeth*, saith, that the floures thereof be hot in the third degree, and dry in the end of the first.

Ziziphus Cappadocica is cold and dry of complexion.

¶ The Vertues.

A　The floures of *Zyzyphus*, or *Azadaraeth* open the obstructions of the braine.

B　The distilled water thereof killeth nits and lice, preserueth the haire of the head from falling, especially being mixed with whitewine, and the head bathed with it.

C　The fruit is very hurtfull to the chest, and a troublesome enemy to the stomacke; it is dangerous, and peraduenture deadly.

D　Moreouer, it is reported, That the decoction of the barke and of Fumitory, with Mirobalans added, is good for agues proceeding of flegme.

E　The juyce of the vppermost leaues with hony is a remedy against poyson.

F　The like also hath *Rhasis*: the Beade tree, saith hee, is hot and dry: it is good for stoppings of the head: it maketh the haire long; yet is the fruit thereof very offensiue to the stomacke, and oftentimes found to be pernitious and deadly.

G　*Matthiolus* writeth, that the leaues and wood bringeth death euen vnto beasts, and that the poyson thereof is resisted by the same remedies that *Oleander* is.

H　*Ziziphus Cappadocica* preuaileth against the diseases aforesaid, but the decoction therof is very good for those whose water scaldeth them with the continuall issuing thereof, as also for such as are the running of the reines and the exulcerations of the bladder and priuy parts.

A looch

A looch or licking medicine made of it or the ſyrrup is excellent good againſt the ſpitting of G bloud proceeding of the diſtillations of ſharp or ſalt humors.

† The figure that formerly was in the ſecond place was of the narrow leafed kind of *Guaiacum Patauinum*, which you ſhall finde in the ſecond place of the next chapter ſaue one.

Chap. 123.   *Of the Lote or Nettle tree.*

Lotus arbor.
The Nettle tree.

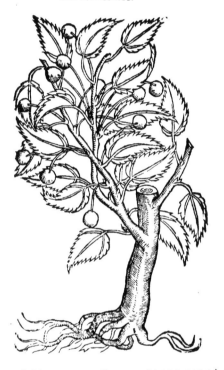

¶ The Deſcription.

THis Lote whereof we write is a tree as big as a Pear tree, or bigger & higher: the body and arms are very thick, the bark whereof is ſmooth, of a gallant greene colour tending to blewneſſe, the boughs are long, and ſpred themſelues all abroad: the leaues be like thoſe of the Nettle, ſharpe pointed and nickt in the edges like a ſaw, and daſht here and there with ſtripes of a yellowiſh white color: the beries be round and hang vpon long ſtalks like cherries, of a yellowiſh white colour at the firſt, and afterward red, but being ripe they are ſomwhat black.

¶ The Place.

This is a rare and ſtrange tree in both the Germanies; it was brought out of Italy, where there is found ſtore thereof, as *Matthiolus* teſtifieth. I haue a ſmall tree thereof in my garden. There is alſo a tree thereof in the garden vnder London wall, ſomtime belonging to Mr *Gray* an apothecarie of London; and another great tree in a garden neere Colemanſtreet in London, belonging to the Queens Apothecarie at the impreſſion hereof, called Mr *Hugh Morgan*, a curious Conſeruer of rare Simples. The Lote tree doth alſo grow in Africke, but it ſomewhat differs from the Italians Lote in fruit, as *Pliny* in plain words ſheweth, *lib.* 13. *cap.* 17. That part of Africk, ſaith he, that lieth toward vs, brings forth the famous Lote tree which they call *Celtis*, and the ſame well known in Italy, but altered by the ſoile: it is as big as the Peare tree, although *Nepos Cornelius* reporteth it to be ſhorter; the leaues are full of fine cuts, otherwiſe they be thought to be like thoſe of the Holme tree. There be many differences, but the ſame are made ſpecially by the fruit, which is as big as a beane, of the colour of Saffron, but before it is thorow ripe it changeth his colour as doth the grape. It growes thick among the boughs after the maner of Myrtle, not as in Italy, after the maner of the Cherry: the fruit of it is there ſo ſweet, as it hath alſo giuen a name to that Country and land, Too hoſpitable to ſtrangers, and forgetfull of their own country.

It is reported that they are troubled with no diſeaſes of the belly that eat it. The better is that which hath no kernell, which in the other kind is ſtony; there is alſo preſſed out of it a Wine like to a ſweet wine, which the ſame *Nepos* denieth to endure aboue ten daies, and the berries ſtamped with *Alica* are reſerued in veſſels for food. Moreouer, we haue heard ſay that armies haue beene fed therewith as they haue paſſed to and fro thorow Africk. The colour of the wood is black, they vſe to make flutes and pipes of it; the root ſerueth for knife hafts and other ſhort works; this is there the nature of the tree. Thus far *Pliny*. In the ſame place he ſaith that this renowned tree groweth about Syrtes and Naſamonæ. And *lib.5.ca.7.* he ſheweth, that there is not far from the leſſer Syrtes the Iſland Menynx, ſyrnamed Lotophagitis, of the plenty of Lote trees.

Strabo,li. 17. affirmeth,that not only Menynx but alſo the leſſer Syrtis is thought to be *Lotopha-*
gitis : Firſt,ſaith he,lieth Syrtis,a certain long Iſland by the name Cercinna,and another leſſer,cal-
led Cercinnitis ; next to this is the leſſer Syrtis,which they cal Lotophagitis Syrtis:the compaſſe
of this gulfe is almoſt 1600 furlongs, the bredth of the mouth 600. By both the capes there be
Iſlands ioined to the main land,that is, Circinna and Menynx,of like bigneſſe:they think that Menynx is the country of the Lotophagi,or thoſe that feed of the Lote trees. Of which country *Homer*
maketh mention,and there are certain monuments to be ſeen,and *Vlyſſes* altar,and the fruit it ſelfe,
for there be in it great plenty of Lote trees,whoſe fruit is wonderfull ſweet. So ſaith *Strabo.*

This Lote is alſo deſcribed by *Theophraſtus,li.*4.where he ſaith there are very many kinds which
be ſeuered by the fruit : the fruit is of the bignes of a bean, which when it waxeth ripe altereth his
colour as Grapes do.The fruit of which the Lotophagi eat is ſweet, pleaſant,harmleſſe,and wholſome for the belly,but that is pleaſanter which is without kernels,whereof they make their wine.

This Lote tree,as the ſame author affirmeth,is by nature euerlaſting,witneſſe thoſe whereof *Pliny*
writ,*lib.*16.*cap.* 44. At Rome(ſaith he)the Lote tree in *Lucina's* court, how much elder it was,than
the church of the city (built in the yeare which was without magiſtrats, 469) is vncertaine. But
doubtleſſe it was elder , becauſe *Lucina* bare the name of that *Lucus* or groue. This is now about
450 yeares old. That is elder which is ſyrnamed *Capillata* or hairy,becauſe the haire of the Veſtal
Virgins was brought vnto it:but the other Lote tree in *Vulcans* church,which *Romulus* built by the
victorie of tenths,is taken to be as old as the city,as *Maſſurius* witneſſeth.

They loſe their leaues at the firſt approch of winter, and recouer them again in Aprill:the fruit
is ripe in September.

The tree is called in Greek, Λωτὸς : in Latine by *Pliny,Celtis* : in Italian, *Perlaro:* by thoſe of Trent,
Bagolaro : in Engliſh,Lote tree,and Nettle tree.

A   The Lote tree is not greatly binding,as *Galen* ſaith,but of thin parts,and of a drying nature.
B   The decoction of the wood beaten ſmall , being either drunke or vſed cliſter-wiſe,helpeth the
bloudy flix,the whites and reds.
C   It ſtops the lask,and maketh the haire yellow,and as *Galen* ſaith keepeth haires from falling.
D   The ſhiuers or ſmall pieces thereof,as the ſame Author ſaith,are boiled ſomtimes in water,ſom-
times in wine,as need ſhall require.

C H A P. 124.

Of Italian Wood of Life,or Pock-wood , vulgarly called Lignum vitæ.

1   ITalian *Lignum vitæ* or Wood of life grows to a faire and beautifull tree,hauing a ſtraight
      and vpright body couered ouer with a ſmooth and dark green barke, yeelding forth many
      twiggy branches ſet forth of goodly leaues like thoſe of the Pear tree,but of greater beau-
ty,and ſomwhat broader ; amongſt which comes forth the fruit, growing cloſe to the branches,al-
moſt without ſtalks ; this fruit is round,at the firſt green,but black when it is ripe, as big as Cherries,of an excellent ſweet taſt when it is dried. But this is not the Indian *Lignum ſanctum* or *Guaia-*
cum whereof our bowles and phyſical drinks be made,but it is a baſtard kinde thereof, firſt planted
in the common garden at Padua by the learned *Fallopius,*who ſuppoſed it to be the right *Guaiacum.*
‡ 2   The leaues of this are longer and narower than the former,but firm alſo and neruous like
as they are ; the fruit is in ſhape like Sebeſtens,but much leſſe, of a blewiſh colour when it is ripe,
with many little ſtones within ; the taſte hereof is not vnpleaſant. *Matthiolus* cals this *Pſeudolotus;*
and *Tabern. Lotus Africana ;* whoſe figure our Author in the laſt chapter ſaue one gaue vnfitly for
the *Zizyphus Cappadocica.* ‡

Guaiacum Patavinum groweth plentifully about Lugdunum or Lyons in France. I planted it in
                                               the

the garden of Barn-Elmes neere London two trees:beſides,there groweth another in the garden of Mr Gray an Apothecary of London,and in my garden likewiſe.

1 *Guaiacum Patavinum latifolium.*
Broad leafed Italian Wood of life.

2 *Guaiacum Patavinum anguſtifól.*
Narrow leafed Italian Guaiacum.

¶ *The Time.*
It floureth in May,and the fruit is ripe in September.
¶ *The Names.*
Guaiacum Patavinum hath bin reputed for the *Lotus* of *Theophraſtus* : in Engliſh it is called the baſtard Meuynwood.
‡   This hath no affinity with the true Indian Guajacum which is freqnently vſed in medicine.
¶ *The Nature and Vertues.*
The fruit of this is thought to be of the ſame temper and quality with that of the Nettle tree.‡

Chap. 125.   *Of the Strawberry tree.*

¶ *The Deſcription.*

THe Strawberry tree groweth for the moſt part low,very like in bigneſſe vnto the Quince tree (whereunto *Dioſcorides* compareth it.) The body is couered with a reddiſh barke both rough and ſcaly : the boughes ſtand thicke on the top,ſomwhat reddiſh. The leaues be broad,long and ſmooth like thoſe of Bayes,ſomwhat nicked in the edges,and of a pale green colour:the flours grow in cluſters,being hollow and white,and now and then on the one ſide ſomwhat of a purple co-lour : in their places come forth certain berries hanging down vpon little long ſtems like to Straw-berries,but greater,without, a ſtone within, but onely with little ſeeds,at the firſt greene,and being ripe they are of a gallant red colour,in taſte ſomwhat harſh,and in a manner without any relliſh;of which Thruſhes and Black-birds do feed in winter.

Arbutus.
The Strawberry tree.

¶ *The Place.*

The Strawberry tree groweth in most countries of Greece, in Candy, Italy, and Spaine, also in the vallies of the mountaine Athos, where being in other places but little, they become huge trees, as *P. Bellonius* writeth. *Iuba* also reporteth, that there be in Arabia of them 50 cubits high. They grow only in some few gardens with vs.

¶ *The Time.*

The Strawberry tree floureth in Iuly & August, and the fruit is ripe in September, after it hath remained vpon the tree by the space of an whole yeare.

¶ *The Names.*

This tree is called in Greek, Κόμαρος : in Latine, *Arbutus* : in English, Strawberry tree, & Arbute tree.

The Fruit is named in Greeke, Μιμαίκυλον, or as others reade it, Μεμαίκυλον : in Latine, *Memæcylum*, and *Arbutus*. *Pliny* calls it *Vnedo*; ground Strawberries (saith he) haue one body, and *Vnedo*, much like vnto them, another body, which only in apple is like to the fruit of the earth. The Italians call this Strawbery *Albatro* : the Spaniards, *Madrono*, *Medronheyro*, and *Medronho* : in French, *Arboutes*, *Arbous*. In English, Tree Strawberry.

¶ *The Temperature and Vertues.*

The fruit of the Strawberry tree is of a cold temper, hurting the stomack and causing headache, wherefore no wholsome food, though it be eaten in some places by the poorer sort of people.

Chap. 126. *Of the Plum Tree.*

¶ *The Kindes.*

TO write of Plums particularly would require a peculiar Volume, and yet the end not be attained vnto, nor the stock or kindred perfectly known, neither to be distinguished apart: the numbers of the sorts or kinds are not known to any one Country, euery clymat hath his own fruit, farre differing from that of other places : my selfe haue sixty sorts in my garden, and all strange and rare: there be in other places many more common, and yet yearly commeth to our hands others not before known; therefore a few figures shall serue for the rest. ‡ Let such as require a larger history of these varieties haue recourse to the oft mentioned Work of Mr *Parkinson* : and such as desire the things themselues may find most of the best with Mr *Iohn Millen* in Old street. ‡

¶ *The Description.*

1 THe Plum or Damson tree is of a mean bignesse, it is couered with a smooth barke : the branches are long, whereon do grow broad leaues more long than round, nicked in the edges : the floures are white; the plums do differ in colour, fashion, and bignesse, they all consist of pulp and skin, and also of kernell, which is shut vp in a shell or stone. Some plums are of a blackish blew, of which some be longer, others rounder, others of the colour of yellow wax, diuers of a crimson red, greater for the most part than the rest. There be also green plums, and withall very long, of a sweet and pleasant taste : moreouer, the pulp or meat of some is drier, and easilier separated from the stone ; of other-some it is moister, and cleaueth faster. Our common Damson is known to all, and therefore not to be stood vpon.

2 The

1 *Prunus domeſtica.*
The Damſon tree.

2 *Prunus Mirobalana.*
The Mirobalan Plum tree.

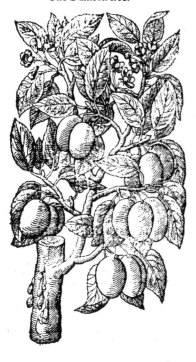

3 *Prunus Amygdalina.*
The Almond Plum tree.

5 *Prunus ſylveſtris.*
The Sloe tree.

2   The Mirobalan Plum tree groweth to the height of a great tree, charged with many great armes or boughes, which diuide themſelues into ſmall twiggy branches, by means whereof it yeeldeth a goodly and pleaſant ſhadow : the trunke or body is couered with a finer and thinner barke than any of the other Plum trees : the leaues do ſomewhat reſemble thoſe of the Cherrie tree, they are very tender, indented about the edges : the ſlours be white: the fruit is round, hanging vpon long foot-ſtalks pleaſant to behold, greene in the beginning, red when it is almoſt ripe, and beeing full ripe it gliſtereth like purple mixed with blacke : the fleſh or meat is full of juice, pleaſant in taſt : the ſtone is ſmall, or of a meane bigneſſe : the tree bringeth forth plenty of fruit euery other yeare.

3   The Almond tree groweth vp to the height of a tree of a meane bigneſſe : the branches are long, ſmooth, and euen : the leaues are broad, ſomthing long, and ribbed in diuers places, with ſmal nerues running through the ſame: the ſlours are white, ſprinkled with a little daſh of purple ſcarcely to be perceiued: the fruit is long, hauing a cleft downe the middle, of a brown red colour, and of a pleaſant taſte.

4 ! The Damaſcen Plum tree groweth likewiſe to a meane height, the branches very brittle: the leaues of a deep green colour: the fruit is round, of a blewiſh blacke colour: the ſtone is like vnto that of the Cherry, wherein it differeth from all other Plums.

5   The Bulleſſe and the Sloe tree are wilde kindes of Plums, which do vary in their kind, euen as the greater and manured Plums do. Of the Bulleſſe, ſome are greater and of better taſte than others. Sloes are ſome of one taſte, and ſome of others, more ſharp ; ſome greater, and others leſſer; the which to diſtinguiſh with long deſcriptions were to ſmall purpoſe, conſidering they be all and euery of them knowne euen vnto the ſimpleſt: therefore this ſhall ſuffice for their ſeuerall deſcriptions.

¶ *The Place.*

The Plum trees grow in all knowne countries of the world : they require a looſe ground, they alſo receiue a difference from the regions where they grow, not only of the forme or faſhion, but eſpecially of the faculties, as we will forthwith declare.

The Plum trees are alſo many times graffed into trees of other kindes, and being ſo ingraffed, they *faciem parentis, ſuccum adoptionis, vt Plinius dicit, exhibent.*

The greateſt variety of theſe rare Plums are to be found in the grounds of Mr *Vincent Pointer* of Twicknam before remembred in the chapter of Apples : although my ſelfe am not without ſome, and thoſe rare and delicate.

The wilde Plums grow in moſt hedges through England.

¶ *The Time.*

The common and garden Plum trees do bloome in April : the leaues come forth preſently with them : the fruit is ripe in Summer, ſome ſooner, ſome later.

¶ *The Names.*

The Plum tree is called in Greek, Κοκκυμηλέα : in Latine, *Prunus* : in high-Dutch, **Pflaumenbaum**: in low-Dutch, **Pruymen** : in Spaniſh, *Ciruelo* : in French, *Prunier* in Engliſh, Plum tree.

The fruit is called in Greeke, Κοκκυμᾶλον : in Latine, *Prunum* in high-Dutch, **Pflaumen** : in low-Dutch, **Pruymen** : in Italian and French, *Prune* : in Spaniſh, *Prunas* : in Engliſh, Prune, and Plum. Theſe haue alſo names from the regions and countries where they grow.

The old Writers haue called thoſe that grow in Syria neere vnto Damaſcus, *Damaſcena Pruna* : in Engliſh, Damſons, or Damask Prunes : and thoſe that grow in Spain, *Hiſpanica*, Spaniſh Prunes or Plums. So in our age we vſe to call thoſe that grow in Hungarie, *Hungarica*, or *Pannonica*, Plums of Hungarie: ſome, *Gallica Pruna*, or French Prunes, of the country of France. *Clearcus Peripateticus* ſaith, that they of Rhodes and Sicilia do call the Damaske Prunes *Brakula*.

¶ *The Temperature and Vertues.*

A   Plummes that be ripe and new gathered from the tree, what ſort ſoeuer they are of, do moiſten and coole, and yeeld vnto the body very little nouriſhment, and the ſame nothing good at all : for as Plummes do very quickly rot, ſo is alſo the juice of them apt to putrifie in the body, and likewiſe to cauſe the meat to putrifie which is taken with them : onely they are good for thoſe that would keep their bodies ſoluble and coole ; for by their moiſture and ſlipperineſſe they do mollifie the belly.

B   Dried Plums, commonly called Prunes, are wholſomer, and more pleaſant to the ſtomack, they yeeld more nouriſhment and better, and ſuch as cannot eaſily putrifie. It is reported, ſaith *Galen* in his booke of the faculties of Nouriſhments, that the beſt doe grow in Damaſcus a city of Syria ; and next to thoſe, they that grow in Spaine : but theſe do nothing at all binde, yet diuers of the Damaſke Damſon Prunes very much ; for Damaſke Damſon Prunes are more aſtringent, but they of Spaine be ſweeter. *Dioſcorides* ſaith, that Damaſke Prunes dried do ſtay the belly ; but *Galen* affirmeth in his books of the faculties of ſimple medicines, that they do manifeſtly looſe the belly,

yet

yet leſſer than they that be brought out of Spain,beeing boiled with Mede or honied water which hath a good quantity of hony in it, they looſe the belly very much(as the ſame Author ſaith) although taken by themſelues, much more if the Mede be ſupped after them.We commend thoſe of Hungary,which be long and ſweet,yet more thoſe of Moravia,the chiefe and principall city in times paſt of the prouince of the Marcomans; for theſe after they be dried,that the watery humor may be conſumed away,be moſt pleaſant to the taſte, and do eaſily without any trouble ſo mollifie the belly,as that in that reſpect they go beyond Caſſia and Manna,as *Tho. Iordanus* affirmeth.

The leaues of the Plum tree are good againſt the ſwelling of the Vvula, the throat, gums, and C kernels vnder the throat nnd jawes; they ſtop the rheum and falling down of humours,if the decoction thereof be made in wine,and gargled in the mouth and throat.

The gum which comes out of the Plum tree doth glue and faſten together,as *Dioſcorides* ſaith. D

Being drunk in wine it waſteth away the ſtone and heales Lichens in infants and yong children : E if laid on with vineger, it works the ſame effects that the gum of the Peach & Cherry tree doth.

The wilde Plums do ſtay and binde the belly, and ſo doe the vnripe plums of what ſort ſoeuer, F while they are ſharp and ſoure,for then they are aſtringent.

The juice of Sloes doth ſtop the belly,the lask and bloudy flix,the inordinat courſe of womens G termes,and all other iſſues of bloud in man or woman,and may very wel be vſed in ſtead of Acatia, which is a thorny tree growing in Egypt,very hard to be gotten,and of a deare price,and therefore the better for wantons,albeit the Plums of this country are equall vnto it in vertues.

CHAP. 127. Of Sebeſten or the Aſſyrian Plum.

Sebeſtena, Myxa, ſive Myxara.
Aſſyrian Plums.

¶ *The Deſcription.*

SEbeſtens are alſo a kinde of Plums,the tree whereof is not vnlike to the Plum tree, ſaving it groweth lower than moſt of the manured Plumme trees : the leaues be harder and rounder: the floures grow at the tops of the branches,conſiſting of fiue ſmall white leaues, with pale yellow threds in the middle like to thoſe of the Plumme tree. After followeth the fruit like little Plums,faſtned in little skinny cups, which when they be ripe are of a greeniſh black colour, wherein is contained a ſmal hard ſtone. The fruit is ſweet in taſte,the pulp or meat is very tough and clammy.

¶ *The Place.*

The Sebeſten trees grow plentifully in Syria and Egypt;they were in times paſt forrein and ſtrange in Italy, now they grow almoſt in euery garden, beeing firſt brought thither in *Plinies* time : Now do the Sebeſten trees, ſaith hee, *lib.*15.*cap.*18. begin to grow in Rome among the Service trees.

¶ *The Time.*

The time doth anſwer the other common Plums.

¶ *The Names.*

Pliny calleth the tree *Myxa* : it may be ſuſpected that this is the tree which *Matron Paraſitus* in his Attick banquet in *Athenaus*,calls κοκκυμηλεα : but we canot certainly affirm it,and eſpecially becauſe diuers haue diuerſly deemed thereof. The berry or fruit is named *Myxon,* and *Myxarion,* neither haue the Latines any other name. The Arabians and the Apothecaries doe call it Sebeſten,which

is

is alſo made an Engliſh name. We may call it the Aſſyrian Plum.

¶ *The Temperature and Vertues.*

A   Sebeſtens be very temperatly cold and moiſt, and haue a thick and clammy ſubſtance, therefore they nouriſh more than moſt fruits do ; but withall they eaſily ſtop the intrals, and ſtuf vp the narrow paſſages, and breed inflammations.

B   They take away the ruggedneſſe of the throat and lungs, and alſo quench thirſt, being taken in a looch or licking medicine, or taken any other way, or by themſelues.

C   Ten drams or an ounce and halfe of the pap or pulp hereof being inwardly taken doth looſe the belly.

D   There is alſo made of this fruit a purging Electuarie, but ſuch an one as quickly mouldeth, and therefore not to be vſed but when it is new made.

Chap. 128.   *Of the Indian Plums or Mirobalans,*

¶ *The Kindes.*

THere be diuers kinds of Mirobalans, as *Chebula*, *Bellirica*, *Emblica*, &c. They likewiſe grow vpon diuers trees, and in countries far diſtant one from another; and *Garcias* the Portugal phyſitian is of opinion that the fiue kindes grow vpon fiue diuers trees.

¶ *The Deſcription.*

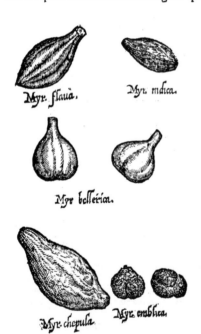

Myr. flaua.   *Myr. indica.*

Myr belleriea.

Myr. chepula.   *Myr. emblica.*

1   THe firſt of the Mirobalan trees, called *Chebula*, is a ſhrubby tree altogether wild (which the Indians call *Aretca*) in ſtature not vnlike to the Plum tree: the branches are many, and grow thick together, wheron are ſet leaues like thoſe of the Peach tree. The fruit is greater than any of the reſt, ſomwhat long, faſhioned like a Peare.

2   This ſecond kind of Mirobalan, called *Flaua*, or *Citrina*, which ſome do call *Aritiqui*, but the common people of India, *Arare*, groweth vpon a tree of meane ſtature, hauing many boughes ſtanding finely in order, and ſet full of leaues like to the Seruice tree.

3   The third kinde of Mirobalans, called *Emblica*, the Indians call *Amiale*; which grow vpon a tree of mean ſtature like the former, but the leaues are very much jagged, in ſhape like the leaues of Fern, but that they be ſomwhat thicker; the Indians put not the fruit hereof to phyſicall vſes, but occupie it for the thickning and tanning of their leather in ſtead of *Rhus* or Coriars Sumach, as alſo to make inke and bletch for other purpoſes.

4   *Mirobalani Bellirica*, called of the Sauages, *Gotni* and *Guti*, groweth vp to a mean ſtature, garniſhed with leaues like Laurel or the bay tree, but ſomwhat leſſer, thinner, and of a pale green color.

5   The fift kinde of Mirobalans is called *Indica*, which the Indians call *Rez annale* : it groweth vpon a tree of mean ſtature, or rather vpon a ſhrub or hedge plant, bearing leaues like the Willow, and a fruit eight ſquare. There is a fift kinde, the tree whereof is not mentioned by Authors.

¶ *The Place and Time.*

The laſt foure kinds of Mirobalans grow in the kingdome of Cambaia : they grow likewiſe in Goa, Batecala, Malanor, and Dabul : the *Kebula* in Biſnager, Decan, Guzarat, Bengala, and in many other places of the Eaſt Indies. The time agrees with other fruits in thoſe countries.

¶ *The Names.*

Thoſe which we haue ſaid to be yellow, the Inhabitants of thoſe countries where they grow, do

call

call them *Arare*: those that be black they call *Rezennale*: the *Bellerica*, *Gotim*; the *Chebula*, *Aretca*; the *Emblica* are called *Aretiqui*.

¶ *The Temperature and Vertues.*

All the kinds of Mirobalans are in taste astringent and sharpe, like to the vnripe Sorbus or Seruice berries, and therefore they are of complexion cold and dry.

The Indians vse them rather to bind than purge; but if they vse them for a purge, it is onely the **A** decoction much conserued in sugar, and especially the *Chebula*: the yellow and black be good that way likewise.

The yellow and *Bellerica* taken before meat stop the laske, and help the weak stomack, as *Garcias* **B** writeth.

The yellow and blacke, or *Indica* and the *Chebula*, purge lightly, if two or three drams be taken, **C** and draw superfluous humors from the head.

The yellow, as some write, purge choler, *Chebula* flegme, *Indica* melancholy, and strengthen the **D** inward parts; but rosted in the embers, or otherwise wasted, they dry more than they purge.

There are two sorts especially brought into these parts of the world conserued, the *Chebula*, and **E** of them the best are somwhat long like a small Limon, with a hard rind and black pith of the taste of a conserued Walnut; and the *Bellerica*, which are round and lesser, and tenderer in eating.

Lobel writeth, that of them the *Emblica* do meanely coole, some do dry in the first degree, they **F** purge the stomack of rotten flegme, they comfort the brain, the sinues, the heart and liuer, procure appetite, stay vomit, and coole the heate of choler, helpe the vnderstanding, quench thirst and heate of the intrals. The greatest and heauiest are the best.

They purge best and with lesse pain, if they be laid in water in the sun vntill they swell, and sod **G** on a soft fire, & after they haue sod and be cold, preserued in foure times so much white honey, put to them.

Garcias found the distilled water very helpful against the French disease & such like infections. **H**

The *Bellerica* are also of a mild operation, and do comfort: they are cold in the first degree, and **I** dry in the second: the others come neere to the *Emblica* in operation.

† I haue in this Chapter contented my selfe with the expressing of the fruits out of *Clusus* and *Lobel*, and omitted the figures of the three Mirobalan trees, which our Author gue vs out of *Tabera*, because I iudge them rather drawne by fancy than by the things themselues.

CHAP. 129. *Of the Iujube tree.*

Iujube Arabum, siue Ziziphus Dodonæi.
The Iujube tree.

¶ *The Description.*

THe Iujube tree is not much lesse than *Ziziphus Candida*, hauing a wreathed trunk or body, and a rough barke full of rifts or crannies, and stiffe branches beset with strong and hard prickles, from whence grow out many long twigs or little stalks halfe a foot or more in length, in shew like Rushes, limmer, and easily bowing themselues, and very slender like the twigs of *Spartum*: about which come forth leaues one aboue another, which are somewhat long, not very great, but hard and tough like to the leaues of *Peruinca* or Peruincle, & among these leaues come forth pale & mossie little flours: after which succeed long red wel tasted sweet berries as big as Olives (of a mean quantitie) or little Prunes, or small plums, wherein there are hard stones, or in which a small kernell is contained.

¶ *The Place.*

There be now at this day Iujube trees growing in very many places of Italy, which in times past were newly brought thither out of Syria, and that about *Plinies* time, as hee himselfe hath written, *lib.17.cap.10.*

¶ *The Time.*

It floureth in Aprill at which time the seeds or stones are to be set or sown for increase.

¶ *The*

¶ *The Names.*

This tree is called in Greeke ζίζυφος and ζίζυφος, with *Iota* in the ſecond ſyllable : in Latine likewiſe *Zizyphus*; and of *Petrus Creſcentius, ZeZulus* : in Engliſh, Iuiube tree.

The fruit or plums are named in Greeke, ζίζυφα,ζίζυφα : *Galen* calleth them σηρικά, as *Auicen* plainely ſheweth, *cap. 369.* intreating of the Iujube, in which be ſet down thoſe things that are mentioned concerning *Serica* in *Galens* books of the faculties of nouriſhments: in Latine likewiſe *Zizypha* and *Serica* : in ſhops, *Iujuba* : in Engliſh, Iujubes.

¶ *The Temperature.*

Iujubes are temperat in heate and moiſture.

¶ *The Vertues.*

A   The fruit of the Iujube tree eaten is of hard digeſtion and nouriſheth very little; but being taken in ſyrrups, electuaries, and ſuch like confections, it appeaſeth and mollifieth the roughneſſe of the throat, the breſt and lungs, and is good againſt the cough, but very good for the reins of the back, the kidnies and bladder.

CHAP. 130.   Of the Cherry tree.

¶ *The Kindes.*

THe antient Herbariſts haue ſet down four kinds of Cherry trees; the firſt is great and wild, the ſecond tame or of the garden, the third hath ſour fruit, the fourth is that which is called in Latine *Chamæceraſus*, or the dwarfe Cherry tree. The later writers haue found diuers ſorts more, ſome bringing forth great fruit, others leſſer ; ſome with white fruit, ſome with blacke, others of the colour of black bloud, varying infinitely according to the clymat and country where they grow.

     1 *Ceraſus vulgaris.*   3 *Ceraſus Hiſpanica.*
   The common Engliſh Cherry tree.   The Spaniſh Cherry tree.

¶ *The Description.*

1   THe English Cherry tree groweth to an high and great tree, the body whereof is of a mean bignesse, which is parted aboue into very many boughes, with a barke somewhat smooth, of a brown crimson colour, tough and pliable; the substance or timber is also brown in the middle, and the outer part is somwhat white: the leaues be great, broad, long, set with veins or nerues, and sleightly nicked about the edges: the floures are white, of a mean bignes, consisting of fiue leaues, and hauing certain threds in the middle of the like colour. The Cherries be round, hanging vpon long stems or footstalks, with a stone in the middest which is couered with a pulp or soft meat; the kernell thereof is not vnpleasant to the taste, though somwhat bitter.

2   The Flanders Cherry tree differeth not from our English Cherry tree in stature or form of leaues or floures, the only difference is, that this tree brings forth his fruit sooner and greater than the other, wherefore it may be called in Latine, *Cerasus præcox, siue Belgica.*

5 *Cerasus Serotina.*
Late ripe Cherry tree.

6 *Cerasus vno pediculo plura.*
The Cluster Cherry tree.

3   The Spanish Cherry tree groweth vp to the height of our common Cherry tree, the wood or timber is soft and loose, couered with a whitish scaly barke, the branches are knotty, greater and fuller of substance than any other Cherry tree; the leaues are likewise greater and longer than any of the rest, in shape like those of the Chestnut tree: the floures are like the others in form, but whiter of colour; the fruit is greater and longer than any, white for the most part all ouer, except those that stand in the hottest place where the sun hath some reflexion against a wall: they are also white within, and of a pleasant taste.

4   The Gascoin Cherry tree groweth very like to the Spanish cherry tree in stature, flours and leaues: it differeth in that it bringeth forth very great Cherries, long, sharp pointed, with a certain hollownesse vpon one side, and spotted here and there with certain prickles of purple color as smal as sand. The taste is most pleasant, and excelleth in beauty.

5   The late ripe Cherry tree groweth vp like vnto our wild English Cherry tree, with the like leaues:

7 *Ceraſus multiflora fructus edens.*
The double floured Cherry tree bearing fruit.

8 *Ceraſus multiflora pauciores fructus edens.*
The double floured barren Cherry tree.

9 *Ceraſus avium nigra & racemoſa.*
Birds Cherry, and black grape Cherry tree.

2 *Ceraſus racemoſa rubra.*
Red Grape Cherry tree.

leaues, branches, and floures, fauing that they are somtimes once doubled : the fruit is small, round, and of a darke bloudy colour when they be ripe, which the French-men gather with their stalkes, and hang them vp in their houses in bunches or handfulls against Winter, which the Physitions do giue vnto their patients in hot and burning feuers, being first steeped in a little warme water, that causeth them to swell and plumpe as full and fresh as when they did grow vpon the tree.

6　The Cluster Cherry tree differeth not from the last described either in leaues, branches, or stature : the floures are also like, but neuer commeth any one of them to be double. The fruit is round, red when they be ripe, and many growing vpon one stem or foot-stalke in clusters, like as the Grapes do. The taste is not vnpleasant although somewhat soure.

7　This Cherrie-tree with double floures growes vp vnto a small tree, not vnlike to the common Cherrie-tree in each respect, fauing that the floures are somewhat double, that is to say, three or foure times double ; after which commeth fruit (though in small quantitie) like the other common Cherry.

8　The double floured Cherry-tree growes vp like vnto an hedge bush, but not so great nor high as any of the others ; the leaues and branches differ not from the rest of the Cherry-trees. The floures hereof are exceeding double, as are the flours of Marigolds, but of a white colour, and smelling somewhat like the Hawthorne floures ; after which come seldome or neuer any fruit, although some Authors haue said that it beareth sometimes fruit, which my selfe haue not at any time seen, notwithstanding the tree hath growne in my Garden many yeeres, and that in an excellent good place by a bricke wall, where it hath the reflection of the South Sunne, fit for a tree that is not willing to beare fruit in our cold climat.

　　11 Cerasus nigra.　　　　　　　　　　12 Chamæcerasus.
The common blacke Cherry-tree.　　　　The dwarfe Cherry-tree.

9　The Birds Cherry-tree, or the blacke Cherry-tree, that bringeth forth very much fruit vpon one branch (which better may be vnderstood by sight of the figure, than by words) springeth vp like an Hedge-tree of small stature, it groweth in the wilde woods of Kent, and are there vsed for stockes to graft other Cherries vpon, of better tast, and more profit, as especially those called the Flanders Cherries : this wilde tree growes very plentifully in the North of England, especially at a place called Heggdale, neere vnto Rosgill in Westmerland, and in diuers other places about Crosbie Rauenswaith, and there called Hegberrie-tree ; it groweth likewise in Martome Parke, foure miles from Blackeburne, and in Harward neere thereunto; in Lancaſhire almoſt in euery hedge: the leaues and branches differ not from thoſe of the wilde Cherry-tree: the floures grow alongſt the ſmall branches, conſiſting of fiue ſmall white leaues, with ſome greeniſh and yellow thrums in the middle: after which come the fruit, greene at the firſt, blacke when they be ripe, and of the bigneſſe of Sloes; of an harſh an vnpleaſant taſte.

10   The other birds Cherry-tree differeth not from the former in any reſpect, but in the colour of the berries; for as they are blacke; ſo on the contrary, theſe are red when they be ripe, wherein they differ.

11   The common blacke Cherry-tree growes vp in ſome places to a great ſtature: there is no difference betweene it and our common Cherry-tree, ſauing that the fruit hereof is very little in reſpect of other Cherries, and of a blacke colour.

12   The dwarfe Cherry-tree groweth very ſeldome to the height of three cubits: the trunke or body ſmall, couered with a darke coloured blacke: wherupon do grow very limber and pliant twiggie branches: the leaues are very ſmall, not much vnlike to thoſe of the Priuite buſh: the floures are ſmall and white: after which come Cherries of a deepe red colour when they be ripe, of taſte ſomewhat ſharpe, but not greatly vnpleaſant: the branches laid downe in the earth, quickely take root, whereby it is greatly increaſed.

My ſelfe with diuers others haue ſundry other ſorts in our gardens, one called the Hart Cherry, the greater and the leſſer; one of the great bigneſſe, and moſt pleaſant in taſte, which we call *Luke Wardes* Cherry, becauſe he was the firſt that brought the ſame out of Italy; another we haue called the Naples Cherry, becauſe it was firſt brought into theſe parts from Naples: the fruit is very great, ſharpe pointed, ſomewhat like a mans heart in ſhape, of a pleaſant taſte, and of a deepe blackiſh colour when it is ripe, as it were of the colour of dried bloud.

We haue another that bringeth forth Cherries alſo very great, bigger than any Flanders Cherrie, of the colour of Iet, or burniſhed horne, and of a moſt pleaſant taſte, as witneſſeth Mr. *Bull*, the Queenes Maieſties Clockemaker, who did taſte of the fruit (the tree bearing onely one Cherry, which he did eate; but my ſelfe neuer taſted of it) at the impreſſion hereof. We haue alſo another, called the Agriot Cherry, of a reaſonable good taſte. Another we haue with fruit of a dun colour, tending to a watchet. We haue one of the dwarfe Cherries, that bringeth forth fruit as great as moſt of our Flanders Cherries, whereas the common ſort hath very ſmall Cherries, and thoſe of an harſh taſte. Theſe and many ſorts more we haue in our London gardens, whereof to write particularly would greatly enlarge our volume, and to ſmall purpoſe: therefore what hath beene ſaid ſhall ſuffice. ‡ I muſt here (as I haue formerly done, in Peares, Apples, and other ſuch fruites) refer you to my two friends Mr. *Iohn Parkinſon*, and Mr. *Iohn Millen*, the one to furniſh you with the hiſtory, and the other with the things themſelues, if you deſire them. ‡

¶ *The Time.*

The Cherrie-treee bloome in Aprill; ſome bring forth their fruit ſooner; ſome later: the red Cherries be alwaies better than the blacke of their owne kinde.

¶ *The Names.*

The Cherry-tree is called in Greeke, κέραϲος: and alſo in Latine, *Ceraſus*: in high Dutch, **Kirſchenbaum**: in low Dutch, **Kerſenboome** and **Crieckenboom**: in French, *Ceriſier*: in Engliſh, Cherry-tree.

The fruit or Cherries be called in Greeke, κεράϲια: and κέραϲα: and in Latine likewiſe, *Ceraſa*: in Engliſh, Cherries: the Latine and Engliſh names in their ſeuerall titles ſhall ſuffice for the reſt that might be ſaid.

¶ *The Temperature and Vertues.*

A   The beſt and principall Cherries be thoſe that are ſomewhat ſower: thoſe little ſweet ones which be wild and ſooneſt wripe be the worſt: they contain bad iuyce, they very ſoone putrifie, and doe ingender ill bloud, by reaſon whereof they do not onely breed wormes in the belly, but trouble ſome agues, and often peſtilent feuers: and therefore in well gouerned common wealths it is carefully prouided that they ſhould not be ſold in the markets in the plague time.

B   Spaniſh Cherries are like to theſe in faculties, but they doe not ſo ſoone putriſie: they be likewiſe cold, and the iuyce they make is not good.

C   The Flanders or Kentiſh Cherries that are through ripe, haue a better iuice, but watery, cold and moiſt: they quench thirſt, they are good for an hot ſtomacke, and profitable for thoſe that haue the ague: they eaſily deſcend, and make the body ſoluble: they nouriſh nothing at all.

D   The late ripe Cherries which the French-men keepe dried againſt Winter, and are by them called *Morelle*, and wee after the ſame name call them Morell Cherries, are dry and do ſomwhat binde, theſe being dried are pleaſant to the taſte, and wholeſome for the ſtomacke, like as Prunnes be, and do ſtop the belly.

Generally

Generally all the kindes of Cherries are cold and moiſt of temperature, although ſome more E cold and moiſt than others: which beeing eaten before meate doe ſoften the belly very gently, they are vnwholeſome either vnto moiſt and rheumaticke bodies, or for vnhealthie and cold ſtomackes.

The common blacke Cherries do ſtrengthen the ſtomacke, and are wholeſomer than the redde F Cherries, the which being dried do ſtop the laske.

The diſtilled water of Cherries is good for thoſe that are troubled with heate and inflammati- G ons in their ſtomackes, and preuaileth againſt the falling ſickeneſſe giuen mixed with wine.

Many excellent Tarts and other pleaſant meats are made with Cherries, ſugar, and other delicat H ſpices, whereof to write were to ſmall purpoſe.

The gum of the Cherry tree taken with wine and water, is reported to helpe the ſtone; it may do I good by making the paſſages ſlippery, and by tempering and alaying the ſharpneſſe of the humors; and in this manner it is a remedy alſo for an old cough. *Dioſcorides* addeth, that it maketh one well coloured, cleareth the ſight, and cauſeth a good appetite to meat.

CHAP. 131.   *Of the Mulberrie Tree.*

1 *Morus.*
The Mulberrie tree.

2 *Morus alba.*
The white Mulberrie tree.

¶ *The Deſcription.*

1   THe common Mulberrie tree is high, and ful of boughes: the body wherof is many times great, the barke rugged; and that of the root yellow: the leaues are broad and ſharpe pointed, ſomething hard, and nicked on the edges; in ſtead of floures, are blowings or catkins, which are downy: the fruit is long, made vp of a number of little graines, like vnto a black-Berrie, but thicker, longer, and much greater, at the firſt greene, and when it is ripe blacke, yet is the juyce (whereof it is full) red: the root is parted many waies.

2　Thewhite Mulberrie tree groweth vntill it be come vnto a great and goodly ſtature, almoſt as big as the former: the leaues are rounder, not ſo ſharpe pointed, nor ſo deeply ſnipt about the edges, yet ſometimes ſinuated or deeply cut in on the ſides, the fruit is like the former, but that it is white and ſomewhat more taſting like wine.

¶ The Place.

The Mulberrie trees grow plentifully in Italy and other hot regions, where they doe maintaine great woods and groues of them, that their Silke wormes may feed thereon. The Mulberry tree is fitly ſet by the ſlip; it may alſo be grafted or inoculated into many trees, being grafted in a white Poplar, it bringeth forth white Mulberries, as *Beritius* in his Geoponickes reporteth. Theſe grow in ſundry gardens in England.

¶ The Time.

Of all the trees in the Orchard the Mulberry doth laſt bloome, and not before the cold weather is gone in May (therefore the old VVriters were wont to call it the wiſeſt tree) at which time the Silke wormes do ſeeme to reuiue, as hauing then wherewith to feed and nouriſh themſelues, which all the winter before do lie like ſmall graines or ſeeds, or rather like the dunging of a fleſh flie vpon a glaſſe, or ſome ſuch thing, as knowing their proper times both to performe their duties for which they were created, and alſo when they may haue wherewith to maintaine and preſerue their owne bodies, vnto their buſineſſe aforeſaid.

The berries are ripe in Auguſt and September. *Hegeſander* in *Athenæus* affirmeth, that the Mulbery trees in his time did not bring forth fruit in twenty yeares together, and that ſo great a plague of the gout then raigned and raged ſo generally, as not onely men, but boies, wenches, eunuches, and women were troubled with that diſeaſe.

¶ The Names.

This tree is named in Greeke μορία, and συκάμινα: in Latine, *Morus*: in ſhops, *Morus Celſi*: in high Dutch, Maulberbaum: in low Dutch, Moerbeſie boom: in French, *Meurier*: in Engliſh, Mulberry tree.

The fruit is called μόρον, and συκάμινον: in Latine, *Morum*: in ſhops, *Morum Celſi*: in high Dutch, Moerbeſie: in Italian, *Moro*: in French, *Meure*: in Spaniſh, *Moras* and *Mores*: in Engliſh Mulberry.

¶ The Temperature and Vertues.

A　Mulberries being gathered before they be ripe, are cold and dry almoſt in the third degree, and do mightily bind; being dried they are good for the lask and bloudy flix, the pouder is vſed in meat, and is drunke with wine and water.

B　They ſtay bleeding, and alſo the reds; they are good againſt inflammations or hot ſwellings of the mouth and jawes, and for other inflammations newly beginning.

C　The ripe and new gathered Mulberries are likewiſe cold and be full of juyce, which hath the taſt of wine, and is ſomething drying, and not without a binding quality: and therefore it is alſo mixed with medicines for the mouth, and ſuch as helpe the hot ſwellings of the mouth, and Almonds of the throat; for which infirmities it is ſingular good.

D　Of the juyce of the ripe berries is made a confection with ſugar, called *Diamorum*: that is, after the manner of a ſyrrup, which is exceeding good for the vlcers and hot ſwellings of the tongue, throat, and almonds, or Vvula of the throat, or any other malady ariſing in thoſe parts.

E　Theſe Mulberries taken in meat, and alſo before meat, do very ſpeedily paſſe through the belly, by reaſon of their moiſture and ſlipperineſſe of their ſubſtance, and make a paſſage for other meats, as *Galen* ſaith.

F　They are good to quench thirſt, they ſtir vp an appetite to meate, they are not hurtfull to the ſtomacke, but they nouriſh the body very little, being taken in the ſecond place, or after meate, for although they be leſſe hurtfull than other like fruits, yet are they corrupted and putrified, vnleſſe they ſpeedily deſcend.

G　The barke of the root is bitter, hot, and dry, and hath a ſcouring faculty: the decoction hereof doth open the ſtoppings of the liuer and ſpleene, it purgeth the belly and driueth forth wormes.

H　The ſame barke being ſteeped in vineger helpeth the tooth-ache: of the ſame effect is alſo the decoction of the leaues and barke, ſaith *Dioſcorides*, who ſheweth that about harueſt time there iſſueth out of the root a juyce, which the next day after is found to be hard, and that the ſame is very good againſt the tooth-ache; that it waſteth away *Phyma*, and purgeth the belly.

I　*Galen* ſaith, that there is in the leaues and firſt buds of this tree a certaine middle faculty, both to binde and ſcoure.

C H A P.

CHAP. 132.   Of the Sycomore Tree.

Sycomorus.
The Sycomore tree.

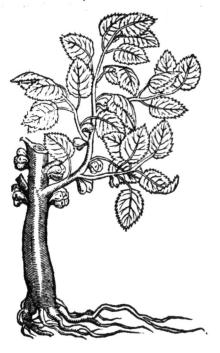

¶ *The Description.*

THE Sycomore tree is of no small height, being very like to the mulberie tree in bignesse & shew, as also in leafe: the fruit is as great as a Fig, and of the same fashion, very like in juyce and taste to the wilde Fig, but sweeter, and without any graines or seeds within, which groweth not forth of the tender boughes, but out of the body and great old armes very fruitfully: this tree hath in it plenty of milkie juyce, which so soone as any part is broken or cut, doth issue forth.

¶ *The Place.*

It groweth, as *Dioscorides* writeth, very plentifully in Caria and Rhodes, and in sundry places of Ægypt, as at the great Cayre or Alkaire, and in places that doe not bring forth much wheat, in which it is an helpe, and sufficeth in stead of bread & corne where there is scarsitie of victuals. *Galen* writeth, that hee saw a plant of the Sycomore tree like to the wilde Fig tree, fruit and all.

¶ *The Place.*

It bringeth forth fruit three or foure times in one yeare, and oftner if it be scraped with an iron knife, or other like instrument.

¶ *The Names.*

This tree is called in Greeke, σνκάμνος, of the Fig tree and the Mulbery tree: in Latine, *Sycomorus*: *Cornelius Celsus* nameth it backeward *Morosycos*: the Egyptians of our time do call it *Ficus Pharaonus*, or *Pharao* his Fig tree, as witnesseth *Bellonius*: and it is likewise termed *Ficus Ægyptia*, Ægyptian Fig tree, and also *Morus Ægyptia*, or Ægyptian Mulbery tree. We call it in English, Sycomore tree after the Greeke and Latine, and also Mulberry Fig tree, which is the right Sycomore tree, and not the great Maple, as we haue said in the chapter of Maple.

The fruit is named in Greeke *Sycomoron*, and in Italian, *Sycomoro* and *Fico d'Egitto*.

¶ *The Temperature and Vertues.*

The fruit of the Sycomore tree hath no sharpnesse in it at all, as *Galen* saith. It is somwhat sweet A in taste, and is of temperature moist after a sort, and cold as be Mulberries.

It is good, saith *Dioscorides*, for the belly; but it is ἄτροφος, that is, without any nourishment, and trou- B blesome to the stomacke.

There issueth forth of the barke of this tree in the beginning of the Spring, before the fruit ap- C peareth a liquour, which beeing taken vp with a spunge, or a little wooll, is dried, made vp into fine cakes, and kept in gallie pots: this mollifieth, closeth wounds together, and dissolueth grosse humours.

It is both inwardly taken and outwardly applied against the biting of serpents, hardnesse of the D milt or spleene: and paine of the stomacke proceeding of a cold cause: this liquor doth very quickly putrifie.

Chap. 133. *Of the Fig Tree.*

¶ *The Description.*

1   THe garden Figtree becommeth a tree of a meane stature, hauing many branches full of white pith within, like Elderne pith, and large leaues of a darke greene colour, diuided into sundry sections or diuisions. The fruit commeth out of the branches without any floure at all that euer I could perceiue, which fruit is in shape like vnto Peares, of colour either whitish, or somewhat red, or of a deepe blew, full of small graines within, of a sweet and pleasant taste; which being broken before it be ripe, doth yeeld most white milke, like vnto the kindes of Spurge, and the leaues also being broken doe yeeld the like liquor; but when the Figges be ripe, the juyce thereof is like hony.

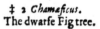

1 *Ficus.*
The Fig tree.

‡ 2 *Chamæficus.*
The dwarfe Fig tree.

2   The dwarfe Fig tree is like vnto the former in leaues and fruit, but it neuer groweth aboue the height of a man, and hath many small shoots comming from the roots, whereby it greatly increaseth.

There is also another wilde kinde, whose fruit is neuer ripe; *Theophrastus* nameth it *Erineos*; *Pliny*, *Caprificus.*

¶ *The Place.*

The Fig trees do grow plentifully in Spaine and Italy, and many other countries, as in England; where they beare fruit, but it neuer commeth to kindely maturity, except the tree be planted vnder an hot wall, whereto neither North, nor North-east windes can come.

¶ *The Time.*

The dwarfe Fig tree groweth in my Garden, and bringeth forth ripe and very great fruit in the moneth of August, of which Figs sundry persons haue eaten at pleasure.

In England the Fig trees put not forth their leaues vntill the end of May, where oftentimes the fruit commeth before the leaues appeare.

¶ *The*

¶ *The Names.*

The fig tree is called in Greeke, ☙, and of diuers, for difference ſake betweene it and the wilde
Fig tree, ☙: in Latine, *Ficus*, and *Ficus ſatiua*, and *Vrbana*: in high Dutch, **Feygenbaum**: in
low Dutch, **Wijgheboom**: in French, *Figuier*: in Italian, *Fico*: in Spaniſh, *Higuera*: in Engliſh,
Fig tree.

The fruit is named in Greeke, ☙: in Latine, *Ficus*: and the vnripe fruit, ☙: in Latine, *Groſ-
ſus*: that which is dried is called in Greeke, ☙: in Latine, *Carica*: in high Dutch, **Feygen**: in low
Dutch, **Wijghen**: in French, *Figues*: in Italian, *Fichi*: in Spaniſh, *Higos*: in Engliſh Fig: the little
ſeeds which are found in them are named by *Galen*, ☙, *Cechramides*.

¶ *The Temperature.*

The greene Figs new gathered are ſomewhat warme and moiſt: the dry and ripe Figs are hot almoſt in the third degree, and withall ſharpe and biting.

The leaues alſo haue ſome ſharpeneſſe, with an opening power, but not ſo ſtrong as the juice.

¶ *The Vertues.*

The dry Figs doe nouriſh better than the greene or new Figs; notwithſtanding they ingender   A
not very good bloud, for ſuch people as do feed much thereon become low ſie.

Figs be good for the throat and lungs, they mitigate the cough, and are good for them that bee   B
ſhort winded: they ripen flegme, cauſing the ſame to be eaſily ſpit out, eſpecially when they bee
ſodden with Hyſſop, and the decoction drunke.

Figges ſtamped with Salt, Rew, and the kernels of Nuts withſtand all poyſon and corruption   C
of the aire. The King of Pontus, called *Mithridates*, vſed this preſeruatiue againſt all venom and
poyſon.

Figs ſtamped and made into the forme of a plaiſter with wheate meale, the pouder of Fenugreek,   D
and Lineſeed, and the roots of Mariſh Mallowes, applied warme, doe ſoften and ripen impoſtumes,
phlegmons, all hot and angry ſwellings and tumors behind the eares: and if you adde thereto the
roots of Lillies, it ripeneth and breaketh Venerious impoſtumes that come in the flanke, which
impoſtume is called *Bubo*, by reaſon of his lurking in ſuch ſecret places: in plaine Engliſh termes
they are called botches.

Figs boiled in Wormewood wine with ſome Barly meale are very good to be applied as an im-   E
plaiſter vpon the bellies of ſuch as haue the dropſie.

Dry Figs haue power to ſoften, conſume and make thinne, and may be vſed both outwardly and   F
inwardly, whether it be to ripen or ſoften impoſtumes, or to ſcatter, diſſolue and conſume them.

The leaues of the Fig tree doe waſte and conſume the Kings Euill, or ſwelling kernels in the   G
throat, and doe mollifie, waſte, and conſume all other tumors, being finely pouned and laid thereon:
but after my practiſe, being boyled with the roots of Mariſh Mallowes vntill they bee ſoft, and ſo
incorporated together, and applied in forme of a plaiſter.

The milkie juyce either of the Figs or leaues is good againſt all roughneſſe of the skin, lepries,   H
ſpreading ſores, tetters, ſmall pockes, meaſels, puſhes, wheales, freckles, lentils, and all other ſpots,
ſcuruineſſe, and deformity of the body and face, beeing mixed with Barley meale and applied: it
doth alſo take away warts and ſuch like excreſcences, if it bee mingled with ſome fatty or greaſie
thing.

The milke doth alſo cure the tooth-ache, if a little lint or cotten be wet therein, and put into the   I
hollowneſſe of the tooth.

It openeth the veines of the hemorrhoids, and looſeneth the belly, being applied to the funda-   K
ment.

Figs ſtamped with the pouder of Fenugreeke, and vineger, and applied plaiſterwiſe, doe eaſe the   L
intollerable paine of the hot gout, eſpecially the gout of the feet.

The milke thereof put into the wound proceeding of the biting of a mad dog, or any other veno-   M
mous beaſt, preſerueth the parts adjoyning, taketh away the paine preſently, and cureth the hurt.

The greene and ripe Figs are good for thoſe that be troubled with the ſtone of the kidneies, for   N
they make the conduits ſlippery, and open them, and doe alſo ſomewhat clenſe: whereupon after
the eating of the ſame, it happeneth that much grauell and ſand is conueighed forth.

Dry or barrell Figs, called in Latine *Carica*, are a remedy for the belly, the cough, and for old in-   O
firmities of the cheſt and lungs: they ſcoure the kidneies, and clenſe forth the ſand, they mitigate
the paine of the bladder, and cauſe women with childe to haue the eaſier deliuerance, if they feed
thereof for certaine daies together before their time.

Dioſcorides ſaith, that the white liquor of the Fig tree, and juyce of the leaues, do curdle milke as   P
rennet doth, and diſſolue the milke that is cluttered in the ſtomacke, as doth vineger.

It bringeth downe the menſes, if it be applied with the yolke of an egge, or with yellow wax.   Q

Cʜᴀᴘ.

CHAP. 134. *Of the prickly Indian Fig Tree.*

Ficus Indica.
The Indian Fig tree.

Fructus.
The fruit.

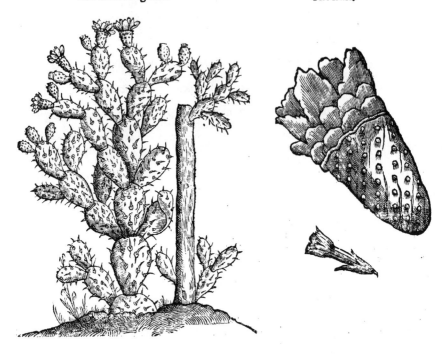

¶ *The Deſcription.*

THis ſtrange and admirable plant, called *Ficus Indica*, ſeemes to be no other thing than a multiplication of leaues, that is, a tree made of leaues, without body or boughes ; for the leafe ſet in
the ground doth in ſhort ſpace take root, and bringeth out of it ſelfe other leaues, from which
do grow others one after another, till ſuch time as they come to the height of a tree, hauing alſo in
the meane ſeaſon boughes as it were comming from thoſe leaues, ſomtimes more, otherwhiles fewer, as Nature liſt to beſtow, adding leafe vnto leafe, whereby it occupieth a great piece of ground :
theſe leaues are long and broad, as thicke as a mans thumbe, of a deepe green colour, ſet full of long,
ſlender, ſharpe and whitiſh prickles : on the tops of which leaues come forth long flours not vnlike
to thoſe of the manured Pomegranate tree, of a yellow colour : after which commeth the fruit like
vnto the common Fig, narrow below, and bigger aboue, of a greene colour, and ſtuffed full of a red
pulpe and juyce, ſtaining the hands of them that touch it, as doe the Mulberries, with a bloudy or
ſanguine colour : the top of which Figs are inuitoned with certain ſcaly leaues like a crowne, wherein are alſo contained ſmall graines that are the ſeeds : the which being ſowne, do bring forth plants
round bodied, like vnto the trunke of other trees, with leaues placed thereon like the other ; which
being ſet in the ground bring forth trees of leaues, as we haue ſhewed.

‡ Vpon this plaut in ſome parts of the Weſt-Indies grow certaine excreſcences, which in continuance of time turne into inſects ; and theſe out-growings are that high prized Cochenele wherewith they die colours in graine. ‡

¶ *The Place.*

This plant groweth in all the tract of the Eaſt and Weſt-Indies, and alſo in the countrey Norembega, now called Virginia, from whence it hath beene brought into Italy, Spaine, England, and
other countries ; in Italy it ſometimes beareth fruit, but more often in Spaine, and neuer as yet in
England, although I haue beſtowed great paines and coſt in keeping it from the iuiury of our cold
clymate.

It

It groweth alſo at S. Crux and other places of Barbary, and alſo in an Iſland of the Mediterra-nean ſea, called Zante, about a day and nights ſailing with a meane winde from Petraſſe a port in Morea, where my ſeruant *William Marſhall* (before remembred) did ſee not onely great ſtore of thoſe trees made of leaues, but alſo diuers other round bodied plants of a woody ſubſtance: from whence he brought me diuers plants thereof in tubs of earth, very freſh and greene, which flouriſhed in my garden at the impreſſion hereof.

¶ The Time.

Theſe plants doe grow greene and freſh both Winter and Summer, by the relation of my fore-ſaid ſeruant: notwithſtanding they muſt bee very carefully kept in theſe countries from the extre-mitie of Winter.

¶ The Names.

This is thought to be the plant called of *Pliny, Opuntium* ; whereof he hath written, *lib.* 2 1.*ca.*17. in this manner: About Opuns is the herbe *Opuntia*, to mans taſte ſweet, and it is to be maruelled, that the root ſhould be made of the leaues, and that it ſhould ſo grow. Opuns is a City neere vnto Phocis in Greece, as *Pauſanias, Strabo,* and *Pliny* teſtifie: but it is commonly called in Latine, *Ficus Indica:* of the Indians, *Tune,* and *Tunas,* and alſo *Anapallas,* as teſtifieth *Bellonius* : in Engliſh, Indian Fig tree.

There is a certaine other deſcribed for the Indian Fig tree, by *Theophraſtus, lib.* 4. which *Pliny, lib.* 12. *cap.* 5. doth eloquently expreſſe almoſt in the ſame words, but turned into Latine, whereof we intend to ſpeake in the next chapter.

¶ The Temperature and Vertues.

We haue no certaine inſtruction from the Antients, of the temperature or faculty of this plant, **A** or of the fruit thereof: neither haue we any thing whereof to write of our owne knowledge, more than that we haue heard reported of ſuch as haue eaten liberally of the fruit thereof, that it changed their vrine to the colour of bloud ; who at the firſt ſight thereof ſtood in great doubt of their life, thinking it had beene bloud, whereas it proued afterwards by experience to be nothing but the tin-cture or colour the vrine had taken from the juyce of the fruit, and that without all hurt or griefe at all.

It is reported of ſome, that the juyce of the fruit is excellent good againſt vlcers of long conti- **B** nuance.

‡ Cocheneſe is giuen alone, and mixed with other things, in maligne diſeaſes, as peſtilent fe- **C** uers and the like, but with what ſucceſſe I know not. ‡

CHAP. 135.   Of the arched Indian Fig tree.

¶ The Deſcription.

THis rare and admirable tree is very great, ſtraight, and couered with a yellow barke tending to tawny : the boughes and branches are many, very long, tough, and flexible, growing very long in ſhort ſpace, as doe the twigs of Oziars, and thoſe ſo long and weake, that the ends thereof hang downe and touch the ground, where they take root and grow in ſuch ſort, that thoſe twigs become great trees : and theſe being growne vp vnto the like greatneſſe, doe caſt their bran-ches or twiggy tendrels vnto the earth, where they likewiſe take hold and root; by meanes whereof it commeth to paſſe, that of one tree is made a great wood or deſart of trees, which the Indians doe vſe for couerture againſt the extreme heat of the Sun, wherewith they are grieuouſly vexed : ſome likewiſe vſe them for pleaſure, cutting downe by a direct line a long walke, or as it were a vault, through the thickeſt part, from which alſo they cut certaine loope-holes or windowes in ſome pla-ces, to the end to receiue thereby the freſh coole aire that entreth thereat, as alſo for light, that they may ſee their cattell that feed thereby, to auoid any danger that might happen vnto them ei-ther by the enemy or wilde beaſts : from which vault or cloſe walke doth rebound ſuch an admi-rable eccho or anſwering voice, if one of them ſpeake vnto another aloud, that it doth reſound or anſwer againe foure or fiue times, according to the height of the voice, to which it doth anſwer, and that ſo plainely, that it cannot be known from the voice it ſelfe : the firſt or mother of this wood or deſart of trees is hard to bee knowne from the children, but by the greatneſſe of the body, which three men can ſcarſely fathom about: vpon the branches whereof grow leaues hard and wrinkled, in ſhape like thoſe of the Quince tree, greene aboue, and of a whitiſh hoary colour vnderneath, whereupon the Elephants delight to feed: among which leaues come forth the fruit, of the bigneſſe of a mans thumbe, in ſhape like a ſmall Fig, but of a ſanguine or bloudy colour, and of a ſweet taſt,

but

but not ſo pleaſant as the Figs of Spaine ; notwithſtanding they are good to be eaten, and withall very wholeſome.

Arbor ex Goa, ſiue Indica.
The arched Indian Fig tree.

¶ *The Place.*

This wondrous tree groweth in diuers places of the Eaſt-Indies, eſpecially neere vnto
Goa, and alſo in Malaca: it is a ſtranger in
moſt parts of the World.

¶ *The Time.*
This tree keepeth his leaues greene Winter and Summer.

¶ *The Names.*
This tree is called of thoſe that haue trauelled, *Ficus Indica* ; the Indian Fig ; and *Arbor Goa,* of the place where it groweth in greateſt plenty : we may call it in Engliſh, the arched Fig tree.
‡ Such as deſire to ſee more of this Fig-
tree, may haue recourſe to *Cluſius* his *Exoticks,
lib. 1. cap. 1.* where he ſhewes it was mentioned
by diuers antient Writers, as *Q.Curtius, lib. 9.
Pliny, lib. 12. cap. 5. Strabo, lib. 5.* and *Theophraſt.
Hiſt. Plant. lib. 4. cap. 5.* by the name of *Ficus
Indica.* ‡

¶ *The Temperature and Vertues.*
We haue nothing to write of the temperature or vertues of this tree, of our owne knowledge : neither haue we receiued from others
more, than that the fruit hereof is generally
eaten, and that without any hurt at all, but rather good, and alſo nouriſhing.

Chap. 136.
Of Adams Apple tree, or the Weſt-Indian Plantane.

¶ *The Deſcription.*

WHether this plant may be reckoned for a tree properly, or for an herby Plant, it is diſpu-
table, conſidering the ſoft and herby ſubſtance whereof it is made ; that is to ſay, when it
hath attained to the height of ſix or ſeuen cubits, and of the bigneſſe of a mans thigh,
notwithſtanding it may be cut downe with one ſtroke of a ſword, or two or three cuts with a knife,
euen with as much eaſe as the root of a reddiſh or Carrot of the like bigneſſe : from a thicke fat
threddy root riſe immediately diuers great leaues, of the length of three cubits and a halfe, ſometimes more, according to the ſoile where it groweth, and of a cubit and more broad, of bigneſſe ſufficient to wrap a childe in of two yeares old, in ſhape like thoſe of Mandrake, of an ouerworne green
colour, hauing a broad ribbe running through the middle thereof : which leaues, whether by reaſon
of the extreme hot ſcorching Sun, or of their owne nature, in September are ſo drie and withered,
that there is nothing thereof left or to bee ſeene but onely the middle rib. From the middeſt of
theſe leaues riſeth vp a thicke trunke, whereon doth grow the like leaues, which the people do cut
off, as alſo thoſe next the ground, by meanes whereof it riſeth vp to the height of a tree, which o-
therwiſe would remaine a low and baſe plant. This manner of cutting they vſe from time to time,
vntill it came to a certaine height, aboue the reach of the Elephant, which greedily ſeeketh after
the fruit. In the middeſt of the top among the leaues commeth forth a ſoft and fungous ſtumpe,
whereon do grow diuers apples in forme like a ſmall Cucumber, and of the ſame bigneſſe, couered
with

with a thin rinde like that of the Fig, of a yellow colour when they be ripe : the pulpe or substance of the meate is like that of the Pompion, without either seeds, stones or kernels, in taste not greatly perceiued at the first, but presently after it pleaseth, and entiseth a man to eate liberally thereof, by a certaine entising sweetnesse it yeelds : in which fruit, if it be cut according to the length (saith mine Author) oblique, transuerse, or any other way whatsoeuer, may be seene the shape and forme of a crosse, with a man fastned thereto. My selfe haue seene the fruit, and cut it in pieces, which was brought me from Aleppo in pickle ; the crosse I might perceiue, as the forme of a spred-Egle, in the root of Ferne ; but the man I leaue to be sought for by those which haue better eies and judgement than my selfe.

Musa Serapionis.
Adams Apple-tree.

Musa Fructus.
Adams Apple.

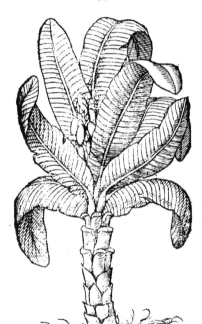

‡ Aprill 10. 1633. my much honoured friend Dr. Argent (now President of the Colledge of Physitions of London) gaue me a plant hee receiued from the Bermuda's : the length of the stalke was some two foot ; the thickenesse thereof some seuen inches about, being crested, and full of a soft pith, so that one might easily with a knife cut it asunder. It was crooked a little or indented, so that at each two or three inches space it put forth a knot of some halfe inch thicknesse, and some inch in length, which encompassed it more than halfe about, and vpon each of these joints or knots, in two rankes one aboue another, grew the fruit, some twentie, nineteene, eighteene, &c. more or lesse, at each knot : for the branch I had, conteined nine knots or diuisions, and vpon the lowest knot grew twenty, and vpon the vppermost fifteene. The fruit which I receiued was not ripe, but greene, each of them was about the bignesse of a large Beane ; the length of them some fiue inches, and the bredth some inch and halfe : they all hang their heads downewards, haue rough or vneuen ends, and are fiue cornered ; and if you turne the vpper side downeward, they somewhat resemble a boat, as you may see by one of them exprest by it selfe : the huske is as thicke as a Beane, and will easily shell off it : the pulpe is white and soft : the stalke whereby it is fastned to the knot is very short and almost as thicke as ones little finger. Tste stalke with the fruit thereon I hanged vp in my shop, where it became ripe about the beginning of May, and lasted vntill Iune : the pulp or meat was very soft and tender, and it did eate somewhat like a Muske-Melon. I haue giuen you the figure of the whole branch, with the fruit thereon, which I drew as soone as I receiued it, and it is marked with this figure 1. The figure 2 sheweth the shape of one particular fruit, with the lower
side

ſide vpwards. 3 The ſame cut through the middle long waies. 4 The ſame cut ſide waies, I haue been told (but how certainely it is I know not) that the floures which precede the fruit are bell-faſhioned, and of a blew colour. I could obſerue no ſeed in the fruit ; it may be it was becauſe it had beene cut from the ſtocke ſo long before it came to maturity. This plant is found in many places of Aſia, Africke, and America, eſpecially in the hot regions : you may finde frequent mention of it amongſt the ſea voiages to the Eaſt and Weſt Indies, by the name of Plantaines, or *Platanus*, *Bannanas*, *Bouanas*, *Dauanas*, *Poco*, &c. ſome (as our Author hath ſaid) haue judged it the forbidden fruit ; otherſome, the Grapes brought to *Moſes* out of the Holy-land. ‡

Muſa fructus exactior Icon.
An exacter figure of the Plantaine fruit.

¶ *The Place.*

This admirable tree groweth in Ægypt, Cyprus, and Syria, neere vnto a chiefe city there called Alep, which we call Aleppo ; and alſo by Tripolis, not far from thence : it groweth alſo in Canara, Decan, Guzarate, and Bengala, places of the Eaſt-Indies.

¶ *The Time.*

From the root of this tree ſhooteth forth young ſprings or ſhoots, which the people take vp and plant for the increaſe of the Spring of the yeare. The leaues wither away in September, as is aboue ſaid.

¶ *The Names.*

It is called *Muſa* by ſuch as trauell to Aleppo : by the Arabians, *Muſa Maum* : in Syria, *Moſe* : The Grecians and Chriſtians which inhabit Syria, and the Iewes alſo, ſuppoſe it to be that tree of whoſe fruit *Adam* did taſte ; which others thinke it to be a rediculous fable : of *Pliny*, *Opuntia*. It is called in the Eaſt-Indies (as at Malauar where it alſo groweth) *Palan* : in Malayo, *Pican* : and in that part of Africa which we call Ginny, *Bananas* : in Engliſh, Adams Apple tree.

¶ *The Temperature.*

Serapio judges, that it heateth in the end of the firſt degree, and moiſtneth in the end of the ſame.

¶ *The Vertues.*

A   The fruit hereof yeeldeth but little nouriſhment : it is good for the heate of the breaſt, lungs, and bladder : it ſtoppeth the liuer, and hurteth the ſtomacke if too much of it be eaten, and procureth

cureth loosnesse in the belly : whereupon it is requisit for such as are of a cold constitution, in the eating thereof to put vnto it a little Ginger or other spice.

It is also good for the reins or kidnies, and to prouoke vrine : it nourisheth the child in the mothers wombe, and stirreth to generation. B

CHAP. 137.   *Of the Date tree.*

Palma.
The Date tree.

Palma cum fructus & flores cum Elat.
The fruit and floures of the Date tree.

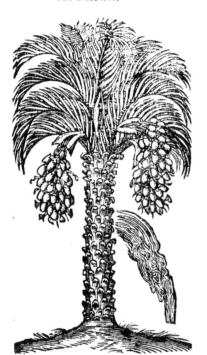
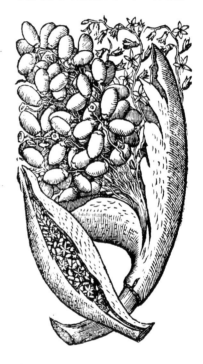

¶ *The Description.*

THe Date tree growes very great and high ; the body or trunke thereof is thicke, and couered with a scaly rugged bark, caused by the falling away of the leaues : the boughs grow only on the top, consisting of leaues set vpon a wooddly middle rib like those of reeds or flags : the inner part of which rib or stalk is soft, light, hollow, and spongie. Among the leaues come forth the floures, included in a long skinny membrane as it were a sheath or hose, like that which couereth the floure de-luce before it be blown ; which being opened of it selfe, white flours start forth, standing vpon short and slender footstalks, which are fastned with certain small filaments or threddie strings like vnto little branches : after which spring out from the same branches the fruit or dates, which be in fashion long and round, in taste sweet, and many times somewhat harsh, of a yellowish red colour ; wherein is contained a long hard stone, which is in stead of kernel and seed ; the which I haue planted many times in my garden, and haue grown to the height of three foot : but the first frost hath nipped them in such sort, that soon after they perished, notwithstanding my industry by couering them, or what else I could do for their succour.

¶ *The Place.*
The Date trees grow plentifully in Affrica and Ægypt ; but those which are in Palestina and Syria,

Syria be the beſt : they grow likewiſe in moſt places of the Eaſt and Weſt Indies, where there be diuers ſorts, as well wild, as tame or manured.

¶ *The Time.*

The Date tree is alwaies greene, and floureth in the ſpring time : the fruit is ripe in September, and being then gathered they are dried in the Sunne, that they may be the better both tranſported into other countries far diſtant, as alſo preſerued from rotting at home.

¶ *The Names.*

The tree is called in Greeke, ✦✦✦ : in Latine *Palma* : in Engliſh, Date tree.

The fruit is named in Greeke, ✦✦✦✦ ✦✦✦✦✦ : that is to ſay, *Glans Palmarum*, or the fruit of the Date trees : and by one word, ✦✦✦✦✦✦✦✦✦ in Latine *Palmula* : in ſhops, *Dactylus* : in high-Dutch, **Datteᵃ len** : in low-Dutch, **Dadelen** : in Italian, *Dattoli* : in French, *Dattis* : in Spaniſh, *Tamaras*, and *Dattiles* : in Engliſh, Date.

The cod or ſheath wherein the floures and Dates are wrapped, is called ✦✦✦, and of ſome, ✦✦✦✦✦.

¶ *The Temperature and Vertues.*

A   All manner of Dates whatſoeuer are hard of digeſtion, and cauſe head-ache : the worſer ſort be thoſe that be dry and binding, as the Egyptian Dates ; but the ſoft moiſt and ſweet ones are leſſe hurtfull.

B   The bloud which is ingendred of Dates in mans body is altogether groſſe, and ſomewhat clammy : by theſe the liuer is very quickly ſtopped, eſpecially being inflamed and troubled with ſome hard ſwelling : ſo is the ſpleen likewiſe.

C   The Dates which grow in colder regions, when they cannot come to perfect ripeneſſe, if eaten too plentifully, they fill the body full of raw humors, ingender wind, and oftentimes cauſe the Leproſie.

D   The drier ſorts of Dates, as *Dioſcorides* ſaith, be good for thoſe that ſpit bloud, for ſuch as haue bad ſtomacks, and for thoſe alſo that be troubled with the bloudy flix.

E   The beſt Dates, called in Latine *Caryota*, are good for the roughneſſe of the throat and lungs.

F   There are made hereof both by the cunning Confectioners & Cooks, diuers excellent, cordial, comfortable and nouriſhing medicines, that procure luſt of the body very mightily.

G   They do alſo refreſh and reſtore ſuch vnto ſtrength as are entring into a conſumption, for they ſtrengthen the feebleneſſe of the liuer and ſpleen, being made into conuenient broths and phyſical medicines directed by a learned Phyſition

H   Dry Dates do ſtop the belly and ſtay vomiting, and the wambling of womens ſtomacks that are with childe, if they be either eaten in meats or otherwiſe, or ſtamped and applied vnto the ſtomack as a pectoral plaiſter.

I   The aſhes of the Date ſtones haue a binding qualitie and emplaſtick facultie, they heal puſhes in the eies, *Staphylomata*, and falling away of the haire of the eie-lids, beeing applied together with Spikenard : with wine it keeps proud fleſh from growing in wounds.

K   The boughs and leaues do euidently bind, but eſpecially the hoſe or caſe of the floures : wherefore it is good to vſe theſe ſo oft as there is need of binding.

L   The leaues and branches of the Date tree doe heale green wounds and vlcers, refreſh and coole hot inflammations.

M   *Galen* in his booke of medicines according to the kinds, mentioneth a compoſition called *Diapalma*, which is to be ſtirred with the bough of a date tree in ſtead of a ſpature or thing to ſtir with, for no other cauſe than that it may receiue thereby ſome kind of aſtriction or binding force.

C H A P. 138.   *Of the wilde Date trees.*

¶ *The Deſcription.*

1   THeophraſtus maketh this plant to be a kinde of Date tree, but low and of ſmall growth, ſeldome attaining aboue the height of a cubit ; on the top whereof ſhoot forth for the moſt part long leaues like thoſe of the Date tree, but leſſer and ſhorter : from the ſides whereof breake forth a buſh of threddy ſtrings, amongſt which riſeth vp ſmall branches garniſhed with cluſters of white floures, in which before they be opened are to bee ſeene vnperfect ſhapes of leaues, cloſely compaſſed about with an innumerable ſort of thinne skinny huls ; which rude ſhapes with the flours are ſerued vp and eaten at the ſecond courſe among other junkets, with a little ſalt and pepper, being pleaſant to the taſte. ‡ The ſtalke is about the thickneſſe of ones

1 *Palmites, sive Chamarriphes.*
The little wilde Date tree.

2 *Palmapinus, sive Palma conifera.*
The wilde Date tree bearing cones.

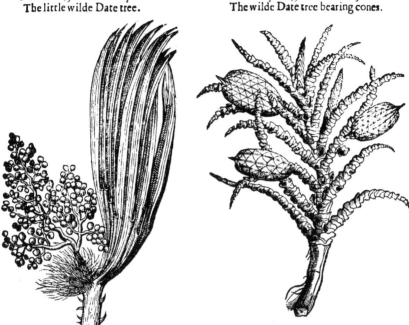

‡ *Fructus Palmapini.*
The fruit of the Cone Date.

little finger, here & there set with a few crooked prickes: the leaues within some handfull or two of the stalke are cut vp and made into little besomes, which are sold in many glasse shops here in London.

2   The wilde Date tree that brings forth cones or key clogs, is of most trauellers into the Indies thought to be barren of Dates, except sometimes it yeeldeth forth some small berries like vnto Dates, but dry, and nothing worth. This groweth vp to the height and bignesse of a low tree, the trunke or body whereof is soft, of a fungous or pithy substance, vnfit for building, as is the mannured Date tree. The branch it selfe was brought vnto vs from the Indies, dry & void of leaues, wherefore we must describe the leaues by report of the bringer. The branches, saith my Authour, are couered ouer with long flaggie leaues, hanging downe of a great length like those of the Date tree : the branches are also couered with a scaly or scabbed barke, very rough, one scale or plate lying ouer another, as tiles vpon a house : the fruit growes at the end of the branches, not vnlike a great Pine Apple cone, couered ouer with a skinne like the Indian Nut, wherein is contained a shell, within which shell lieth hid an acorn or long

Mmmmmm 2   kernell

kernell of an inch long, and ſometimes longer, very hard to be broken, in taſte like the Cheſtnut; which the ſauage people do grate and ſtamp to pouder to make them bread.

¶ *The Place.*

Theophraſtus ſaith, the firſt growes in Candy, but much more plentifully in Cilicia, and are now found in certaine places of Italy by the ſea ſide, and alſo in diuers parts of Spaine.

The other hath been found by trauellers into the Weſt-Indies, from whence haue been brought the naked branches with the fruit.

¶ *The Time.*

The time anſwereth that of the manured Date tree.

¶ *The Names.*

The little Date tree or wilde Date tree is named of *Theophraſtus*, χαμαιριφος : in Naples, *Cephaglione* : in Latine commonly *Palmites*. That which is found in the midſt of the yong ſprings, and is vſed to be eaten in banquets, is called in Greeke, ἐγκεφαλος φοινικος : in Latine, *Palma cerebrum*, the braine of the Date tree.

¶ *The Temperature and Vertues.*

A   *Galen* ſuppoſeth that the braine of the Date tree conſiſteth of ſundry parts, that is to ſay, of a cer- taine watery and warme ſubſtance, and of an earthy and cold ; therefore it is moiſt and cold, with a certaine aſtriction or binding quality.

B   Being taken as a meate it ingendreth raw humors and winde, and therefore it is good to be eaten with pepper and ſalt.

Chap. 139.   *Of the drunken Date tree.*

Areca, ſiue Faufel.
The drunken Date tree.

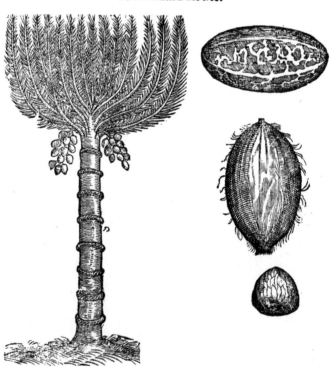

¶ *The Deſcription.*

THe drunken Date tree, which *Carolus Cluſius* calleth *Faufel*, is an Indian tree of a great bigneſſe, the timber whereof is very ſoft and ſpongious, exceeding ſmooth and plaine vnto a great
height,

height, not poſſible to be climbed vp; and therefore the Indians for their eaſier aſcending vp, at ſome diſtances doe tie round about the tree certaine wyths or ropes made of the barkes of trees, as may be perceiued by the figure,wherby very eaſily they go vp and down to gather the fruit at their pleaſure. The top of the tree is diuided into ſundry branches, in ſubſtance like to the great cane; whereupon do grow faire flaggy leaues like thoſe of the Palme or Date tree,whereof doubtleſſe it is a wild kind : from the bottom of which branches comes forth fruit in long branches like traces of Onions,couered with a ſoft pulp like vnto the walnut,rough,and very ful of haire of a yellowiſh colour, and like the dried date when it is ripe : within which huske is contained fruit like vnto the Nutmeg,but greater,very hard,and ſtriped ouer with red and white veins or ſinues.

¶ *The Place, Time, and Names.*

This Date tree,which the Arabians call *Fauſel*,that is by interpretation,*Auellana Indica*, the Indian nut or filberd,*Auicen* and *Serapio* call *Filfel* and *Fufel*. It groweth in the Eaſt Indies in diuers and ſundry places,as in Malauar,where vulgarly it is called *Pac* : and of the Nobles & gentlemen, *Areca*; which name is vſed among the Portugals that dwell in thoſe Indies : in Guzarat & Decan it is called *Cupare* : in Zeilan,*Poaz* : in Malaca,*Pilan* : in Cochin,*Chacani* : in Engliſh,the drunken Date tree,which name we haue coined from his qualitie,becauſe the fruit makes one drunke that eats thereof.

¶ *The Temperature.*

It is cold and dry in the ſecond degree.

¶ *The Vertues.*

The fruit of *Areca* before it be ripe is reckoned among the ſtupefactiue and aſtoniſhing medi- **A** cines,for whoſoeuer eateth thereof waxeth drunk,becauſe it doth exceedingly amaſe and aſtoniſh the ſences.

When the Indians are vexed with ſome intolerable ache or paine , or muſt of neceſſitie endure **B** ſome great torment or torture,they eat of this fruit,wherby the rigor of that pain which otherwiſe they ſhould feele,is very much mitigated.

The iuice of the fruit of *Areca* ſtrengthens the gums, faſtens the teeth, comforts the ſtomacke, **C** ſtayes vomiting and looſneſſe of the belly : it doth alſo purge the body from congealed or clotted bloud gathered within the ſame.

CHAP. 140.   *Of the Indian Nut tree.*

¶ *The Deſcription.*

1   THE Grecians haue not known, but the Arabians haue mentioned this Indian Nut tree, the body wherof is very great,ſmooth,and plaine,void of boughs or branches of a great height ; wherefore the Indians do wrap ropes about the body thereof,as they do about the tree laſt deſcribed,for their more eaſe in gathering the fruit : the timber wherof is very ſpongy within,but hard without,a matter fit to make their Canoos & boats of : on the top of the tree grow the leaues like thoſe of the Date tree,but broad,and ſharpe at the point as thornes,whereof they vſe to make needles,bodkins,and ſuch like inſtruments,wherewith they ſow the ſailes of their ſhips,& do ſuch buſineſſe : among theſe leaues come forth cluſters of floures like thoſe of the Cheſtnut tree,which turn into great fruit of a round form,and ſomwhat ſharp at one end ; in that end next vnto the tree is one hole,ſomtimes two bored through : this Nut or fruit is wrapped in a couerture conſiſting of a ſubſtance not vnlike to hemp before it be beaten ſoft ; there is alſo a finer and gentler ſtuffe next vnto the ſhell, like vnto Flax before it bee made ſoft : in the middle whereof is contained a great nut couered with a very hard ſhel,of a brown colour before it be poliſhed,afterward of a black ſhining colour like burniſhed horne. Next vnto the ſhell vpon the inſide there cleaueth a white cornelly ſubſtance firm and ſollid,of the colour and taſt of a blanched almond : within the cauitie or hollowneſſe thereof is contained a moſt delectable liquour like vnto milke,and of a moſt pleaſant taſte.

2   We haue no certain knowledge from thoſe that haue trauelled into the Indies, of the tree which beareth this little Indian nut;neither haue we any thing of our own knowledge, more than that we ſee by experience the fruit hereof is leſſer,wherein conſiſteth the difference.

‡   The other, expreſſed in the ſame table with the former, by the name of *Meheuhethene*, *Cluſius* receiued it by the ſame name from *Cortuſus* of Padua; yet it doth not (as hee ſaith) well agree with the deſcription; and he rather approues of their opinion who refer it to the *Nux vnguentaria*,

1 *Nux Indica arbor.*
The Indian Nut tree.

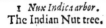

Nux Indica.
The Indian Nut.

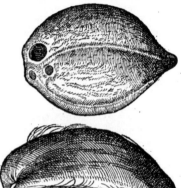

2 *Nucula Indica.*
The little Indian Nut.

or *Ben.* It is fome inch long, of a triangular figure, with a hard and woody fhell : which broken fhewes 3 cels or partitions, in each of which is contained a long kernell white and fweet. ‡

¶ *The Place.*

This Indian Nut groweth in fome places of Africa, and in the Eaft Indies, & in all the Iflands of Weft India, efpecially in Hifpaniola, Cuba, and S. Iohns Ifland, and alfo vpon the continent by Carthagena, Nombre de dios, and Panama, and in Virginea, otherwife called Norembega, part of the fame Continent, for the moft part neere vnto the fea fide and in moift places, but it is feldom found in the vplandifh countries.

¶ *The Time.*

It groweth green Winter and Summer.

¶ *The Names.*

    The fruit is called in Latine *Nux Indica*: of the Indians, *Cocus*: of the Portugals that dwel in the Eaft Indies, *Cocco*, taken from the end, wherein are three holes reprefenting the head of a monkey: *Serapio* and *Rhafis* cal this tree, *Iarralnare*, i. *Arborem nuciferam*, the tree bearing nuts: of *Auicen*, *Glaucial hend* : of the vulgar people *Maro*, and the fruit *Narel* ; which name *Narel* is common among the Perfians and Arabians. It is called in Malauar, *Tengamaran* : the ripe fruit *Tenga*, and the greene fruit *Eleri* : in Goa it is called *Lanhan* : in Malaio, *Triccan* : and the nut *Nihor*.
    The diftilled liquor is called *Sula* ; and the oile that is made thereof, *Copra*.

¶ *The*

¶ *The Temperature.*

It is of a mean temper betwixt hot and cold.

¶ *The Vertues and Vse.*

The Indians vse to cut the twigs and tender branches toward the euening, at the ends whereof A
they haue bottle gourds, hollow canes, and such like things, fit to receiue the water that droppeth
from the branches thereof; which pleasant liquor they drinke in stead of wine: from the which is
drawn a strong & comfortable Aqua vitæ, which they vse in time of need against all maner of sicknesses. Of the branches and boughs they make their houses, of the trunke or body of the tree ships
and boats; of the hemp on the outward part of the fruit they make ropes and cables, and of the finer stuffe failes for their ships.

Likewise they make of the shell of the nut cups to drink in, which wee likewise vse in England, B
garnished with siluer for the same purposes. The kernel serues them for bread and meat: the milky
juice doth serue to coole and refresh their wearied spirits: out of the kernell when it is stamped is
pressed a most pretious oile, not only good for meat, but also for medicine, wherewith they annoint
their feeble lims after their tedious trauel, by means wherof the ach and pain is mitigated, & other
infirmities quite taken away proceeding of other causes.

Chap. 141.   *Of the Dragon Tree.*

1 *Draco arbor.*
The Dragon tree.

Draconis fructus.
The Dragon tree fruit.

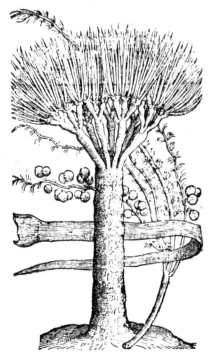

¶ *The Description.*

THis strange and admirable tree groweth very great, resembling the Pine tree, by reason it doth
alwaies flourish, and hath his boughs or branches of equall length and bignesse, which are bare
and naked, of eight or nine cubits long, and of the bignesse of a mans arme: from the ends of which
do shoot out leaues of a cubit and a halfe long, and full two inches broad, somewhat thick, & raised
vp in the middle, then thinner and thinner like a two edged sword: among which come forth little
mossie floures, of small moment, and turn into berries of the bignesse of Cherries, of a yellowish
colour,

·colour,round,light,and bitter,couered with a threefold skin or film,wherin is to be ſeen,as *Monar-* *dus* and diuers others report,the form of a dragon, hauing a long neck and gaping mouth,the ridge or back armed with ſharp prickles like the Porcupine;with a long taile and foure feet very eaſie to be diſcerned : the figure of it we haue ſet forth vnto you according to the greatnes thereof,becauſe our words and meaning may be the better vnderſtood ; and alſo the leafe of the tree in his ful big-neſſe,becauſe it is impoſſible to be expreſſed in the figure:the trunk or body of the tree is couered with a tough bark,very thin and eaſie to be opened or wounded with any ſmall toole or inſtrument; which being ſo wounded in the dog days,bruiſed or bored,yeelds forth drops of a thick red liquor, of the name of the tree called Dragons tears,or *Sanguis draconis*,Dragons bloud : diuers haue doub-ted whether the liquor or bloud were all one with *Cinnabaris* of *Dioſcorides* (not meaning that *Cina-ber* made of Quickſiluer)but the receiued opinion is,they differ not,by reaſon their quality & tem-perature worke the like effect.

¶ *The Place.*

This tree groweth in an Iſland which the Portugals call Madera,and in one of the Canary Iſles called *Inſula portus ſancti* ; and as it ſeemes it was firſt brought out of Africke,though ſome are of a contrarie opinion,and ſay that it was firſt brought from Carthagena in America,by the Biſhop of the ſame prouince.

¶ *The Time.*

The time of his growing we haue touched in the deſcription,where wee ſaid that it flouriſheth and groweth green all the yeare.

¶ *The Names.*

The names haue been ſufficiently ſpoken of in the deſcription,and in their ſeuerall titles.

¶ *The Nature and Vertues.*

A   ‡   The *Sanguis Draconis* which is thought to proceed from this tree,hath an aſtringent faculty, and is with good ſucceſſe vſed in the ouermuch flowing of the courſes,in fluxes, dyſenteries, ſpit-ting of bloud,faſtning looſe teeth,and ſuch other affects which require aſtriction.

B   Smiths alſo vſe it to verniſh ouer their works, to giue them a ſanguine colour, and keepe them from ruſt. ‡

C H A P. 142.   *Of the Saſſafras or Ague tree.*

¶ *The Deſcription.*

THe Saſſafras tree growes very great much like to the Pine tree : the trunk or body is ſtraight, ſmooth,and void of boughs,of a great height :it is couered with a twofold groſſe rind,the vp-permoſt of the colour of aſhes,that next the wood of a tawny colour: on the top come forth many goodly branches like thoſe of the Palm tree,whereon grow green leaues ſomwhat like thoſe of the fig tree,of a ſweet ſmell when they be greene,but much ſweeter when they be dry,declining to the ſmell of fennel,with much ſweetneſſe in taſt :they are green Winter and Summer, neither bearing fruit nor floures,but is altogether barren as it is ſaid : the roots are groſſe,conformable to the greatneſſe of the tree,of a tawny colour,diſperſing themſelues far abroad vnder the vpper cruſt of the earth,by means whereof they are often caſt down with mean blaſts of wind. ‡ The wood of the tree is very ſtrong,hard,and brittle,it hath not ſo ſtrong & pleaſant a ſmell as that of the root, neither is it in ſuch vſe. The leaues are of two ſorts,ſome long and ſmooth,and not ſnipt about the edges ; otherſome,and thoſe chiefely on the end of the branches, are deepely gaſhed in,as it were diuided into three ſeuerall parts. I haue giuen the figure of a branch taken from a little tree,which grew in the garden of M*r Wilmot* at Bow;who died ſome few yeares ago.

¶ *The Place.*

This tree groweth in moſt parts of the Weſt Indies,ſpecially about the cape of Florida, Win-gandico,and Virginia,otherwiſe named Norembega.

¶ *The Time.*

It flouriſheth and keepeth green Winter and Summer.

¶ *The*

Saſſafras.
The Saſſafras tree.

¶ *The Names.*

The Spaniards and French men haue named this tree, *Saſſafras*, the Indians in their tongue, *Pauame*: for want of an Englith name we are contented to call it the Ague tree, of his vertue in healing the Ague.

¶ *The Temperature.*

The boughes and branches hereof are hot and dry in the ſecond degree; the rinde is hotter, for that it entreth into the third degree of heate and dryneſſe, as is manifeſtly perceiued in the decoction.

¶ *The Vertues.*

The beſt of all the tree is the root, and that A worketh the beſt effect, the which hath the rinde cleauing very faſt to the inner part, and is of colour tawny, and much more ſweet of ſmell than all the tree and his branches.

The rinde taſteth of a more ſweet ſmell B than the tree: and the water beeing ſod with the root is of greater and better effects than any other part of the tree, and is of a more ſweet ſmell, and therefore the Spaniards vſe it, for that it worketh better and greater effects.

It is a tree that groweth neere vnto the ſea, C and in temperate places that haue not much drouth, nor moiſture. There be mountaines growing full of them, and they caſt forth a moſt ſweet ſmell, ſo that at the beginning when they ſaw them firſt, they thought they had been trees of Cinnamon, and in part they were not deceiued: for that the rinde of this tree hath as ſweet a ſmell as Cinnamon hath, and doth imitate it in colour and ſharpneſſe of taſte, and pleaſantneſſe of ſmell: and ſo the water that is made of it is of a moſt ſweet ſmell and taſte, as the Cinamon is, and procureth the ſame workes and effects as Cinnamon doth.

The wood hereof cut in ſmall pieces and boyled in water, to the colour of Claret wine, and drunk D for certain daies together, helpeth the dropſie, remoueth oppilation or ſtopping of the liuer, cureth quotidian and tertian agues, and long feuers

The root of Saſſafras hath power to comfort the liuer, and to free from oppilations, to comfort E the weake and feeble ſtomacke, to cauſe good appetite, to conſume windineſſe, the chiefeſt cauſe of cruditie and digeſtion, ſtay vomiting, and make ſweet a ſtinking breath.

It prouoketh vrine, remoueth the impediments that doe cauſe barrenneſſe, and maketh women F apt to conceiue.

Chap. 143. *Of the Storax tree.*

¶ *The Deſcription.*

THe Storax tree groweth to the height and bigneſſe of the Quince tree: the trunke or body is couered with a barke or rinde like vnto the Birch tree: the branches are ſmall and limmer wheron do grow leaues like thoſe of the Quince tree, greeniſh aboue, and whitiſh vnderneath: among which come forth white floures, like thoſe of the Orange tree, of an vnpleaſant ſmell: after commeth the fruit or berries, ſtanding vpon long and ſlender foot-ſtalkes, couered ouer with a little woollineſſe, of the bigneſſe of a bladder nut, of of the ſame colour; wherein is contained ſmall

ſeed,

Styrax arbor.
The Storax tree.

feed, whereto alfo cleaue certain gummy tears bearing the name of the tree, and which iffue from the trunk or body when it is wounded.

¶ *The Place.*

This tree groweth in diuers places of France, Italy, and Spaine, where it bringeth forth little or no gum at all. It groweth in Iudæa, Pamphilia, Siria, Pifidia, Sidon, and many other places of Iury or Paleftine, as alfo in diuers Iflands in the Mediterranean fea, namely Cyprus, Candy, Zant, and other places, where it bringeth forth his gummy liquour in full perfection of fweetneffe, and alfo in great plenty, where it is gathered and put into great Canes or Reeds, whereof as fome deem it took the name *Calamita* ; others thinke of the leaues of reeds wherein they wrap it. Hereof I haue two fmal trees in my garden, the which I raifed of feed.

¶ *The Time.*

It flours in May, and the fruit is ripe in September.

¶ *The Names.*

This tree, as may be gathered by fome, was called *Styrax*, by reafon of the gum or liquor which droppeth out of the fame, being like to the hollow pipes of ice that hang at the eaues of houfes in winter, called *Styria*, or of the Canes or leaues of Reeds fpoken of before: in Latine, *Storax Calamita* : in Englifh, Storax, which is kept in canes or the leaues of Reeds. There floweth from fome of thefe trees a certaine gummie liquour, which neuer groweth naturally hard, but remaineth alwayes thinne, the which is called liquid Styrax or Storax.

¶ *The Temperature.*

The gum of this tree is of an heating, mollifying, and concocting qualitie.

¶ *The Vertues.*

A   It helpeth the cough, the falling down of rheums and humors into the cheft, and hoarfeneffe of the voice, It alfo helpeth the noife and founding of the ears, preuaileth againft *Strumas* or the kings Euill, nodes on the nerues, and hard fwellings proceeding of a cold caufe, as alfo againft all cold poifons, as Hemlock and the like.

B   Of this gum there are made fundry excellent perfumes, pomanders, fweet waters, fweet bags, and fweet wafhing balls, and diuers other fweet chaines and bracelets, whereof to write were impertinent to this hiftorie.

CHAP. 144.

Of the Sorrowfull Tree, or Indian Mourner.

¶ *The Defcription.*

ARbor Triftis, the Sad or forrowfull tree, waxeth as big as an Olive tree, garnifhed with many goodly branches fet full of leaues like thofe of the Plum tree ; among which come forth moft odoriferous and fweet fmelling floures, whofe ftalkes are of the colour of Saffron, which flourifh and fhew themfelues only in the night time, and in the day time looke withered and with a mourning cheere ; the leaues alfo at that time fhrink in themfelues together, much like a tender Plant that is froft-bitten, very fadly lumping, lowring, and hanging downe the head, as though it loathed the light, and could not abide the heate of the Sunne. I fhould but in vain lofe labor in repeating a foolifh fanfie of the poëtical Indians, who would make fooles beleeue that this tree was once a fair daughter of a great Lord or King, and that the Sun was in loue with her ; with other toies which I omit.

Arbor triſtis.
The ſorrowfull tree.

omit. ‡ The floures are white, ſomewhat like thoſe of Iaſmine, but more double, and they are of a very ſweet ſmell. There ſucceed them many little cods, containing ſome ſixe ſeeds apiece ſomewhat like thoſe of *Stramonium.* ‡

¶ *The Place, Time, and Names.*

This tree growes in the Eaſt Indies, eſpecially in Goa and Malayo : in Goa it is called *Pariȝataco* : in Malayo, *Singadi* : in Decan, *Pul* : of the Arabians, *Guart*: of the Perſians and Turks, *Gnl* : in Engliſh, the ſad or ſorowfull tree, or the Indian mourner. The time is ſpecified in the deſcription.

¶ *The Nature and Vertues.*

Wee haue no certaine knowledge of the A temperature hereof ; neuertheleſſe we reade, that the Indians doe colour their broths and meats with the ſtalks of the floures hereof in ſtead of Saffron ; or whatſoeuer they deſire to haue of a yellow colour.

It is reported, That if a linnen cloath bee B ſteeped in the diſtilled water of the flours, & the eies bathed and waſhed therwith, it helpeth the itching and pain thereof, and ſtayes the humors that fall down to the ſame. C

There is made of the ſplinters of the wood certaine tooth-picks, and many pretty toies for pleaſure.

CHAP. 145. *Of the Balſam tree.*

¶ *The Kindes.*

THere be diuers ſorts of trees from which do flow Balſams, very different one from another, not only in form, but alſo in fruit, liquor, and place of growing ; the which to diſtinguiſh would require more time and trauell than either our ſmall time will afford, or riches for our maintenance to diſcouer the ſame in their natural countries : which otherwiſe by report to ſet downe certain matter by incertainties, would diſcredit the Author, and no profit thereby ſhall ariſe to the Reader : notwithſtanding we wil ſet down ſo much as we haue found in the works of ſome trauellers, which beſt agree with the truth of the hiſtorie.

¶ *The Deſcription.*

1 THere be diuers trees growing in the Indies, whoſe fruits are called by the name of the fruit of the Balſam tree : among the reſt this whoſe figure wee here ſet forth vnto your view, we our ſelues haue ſeen & handled, and therfore are the better able to deſcribe it. It is a fruit very crooked, & hollowed like the palm of an hand, two inches long, half an inch thick, couered with a thick ſmooth rind, of the colour of a dry oken leafe ; wherein is contained a kernell (of the ſame length and thickneſſe, apt to fil the ſaid ſhel or rinde) of the ſubſtance of an almond, of the colour of aſhes, fat and oily, of a good ſmell, but very vnpleaſant in taſt.

2 The wood we haue dry brought vnto vs from the Indies for our vſe in phyſicke (a ſmall deſcription may ſerue for a dry ſticke) neuertheleſſe wee haue other fruits brought from the Indies, whoſe figures are not ſet forth, by reaſon they are not ſo well knowne as deſired ; whereof one is of the bigneſſe of a walnut, ſomewhat broad on the vpper ſide, with a rough or rugged ſhell, vneuen, blacke of colour, and full of a white kernell, with much iuice in it, of a pleaſant taſt and ſmell, like the oile of Mace : the whole fruit is exceeding light, in reſpect of the quantity or bigneſſe, euen as

ᴛᴇ

it were a piece of corke,which notwithſtanding ſinketh to the bottom when it fals into the water, like as doth a ſtone.

3   This tree,ſaith *Garcias*,that beareth the fruit *Carpobalſamum*,is alſo one of the Balſam trees: it groweth to the height and bigneſſe of the Pomegranat tree,garniſhed with very many branches, whereon do grow leaues like thoſe of Rue,but of colour whiter, alwaies growing greene : amongſt which come forth floures,whereof wee haue no certaintie : after which commeth forth fruit like that of the Turpentine tree, which in ſhops is called *Carpobalſamum*,of a pleaſant ſweet ſmell ; but the liquor which floweth from the wounded tree is much ſweeter,which liquor is called of ſome, *Opobalſamum*.

1 <i>Balſami fructus.</i>
The fruit of the Balſam tree.   ‡ 3 <i>Balſamum Alpini cum Carpobalſamo.</i>
The Balſam tree with the fruit.

‡ *Proſper Alpinus* hath writ a large Dialogue of the Balſam of the Antients,and alſo figured and deliuered the hiſtorie thereof in his book *De Plant. Ægypti,cap.*14.whither I refer the curious. I haue preſented you with a ſlip from his tree,& the *Carpobalſamum* ſet forth by our Author,which ſeems to be of the ſame plant. The leaues of this are like to thoſe of *Lentiſcus*,alwaies greene, and winged,growing three,fiue,or ſeuen faſtned to one footſtalke : the root is gummy, reddiſh,and wel ſmelling : the floures are ſmal and white like thoſe of *Acacia*,growing vſually three nigh together. The fruit is of the ſhape and bigneſſe of that of the Turpentine tree, containing yellow and well-ſmelling ſeeds filled with a yellowiſh moiſture like hony. Their taſt is bitteriſh,and ſomwhat biting the tongue. ‡

Of theſe Balſam trees there is another ſort , the fruit whereof is as it were a kernell without a ſhell,couered with a thin skin ſtraked with many veins,of a brown colour ; the meat is firm and ſolid like the kernell of the Indian nut,of a white colour,and without ſmell,but of a gratefull taſte ; and it is thought to be hot in the firſt degree,or in the beginning of the ſecond.

There be diuers ſorts more,which might be omitted becauſe of tediouſneſſe,neuertheles I will trouble you with two ſpecial trees worth the noting;There is,ſaith my author, in America a great tree of monſtrous hugeneſſe,beſet with leaues and boughes euen to the ground, the trunk whereof is couered with a twofold bark,the one thicke like vnto Cork,aed another thin next the tree : from betweene which barks doth flow (the vpper barke being wounded)a white Balſam like vnto teares

or

or drops,of a moſt ſweet ſauor and ſingular effects,for one drop of this which thus diſtilleth out of the tree;is worth a pound of that which is made by decoction. The fruit herof is ſmal in reſpect of the others ; it ſeldome exceedeth the bigneſſe of a peaſe, of a bitter taſt, incloſed in a narrow huſk of the length of a finger,ſomthing thin,and of a white colour, which the Indians vſe againſt head-ache: which fruit of moſt is that we haue before deſcribed,called *Carpobalſamum*.

It is alſo written,that in the Iſland called Hiſpaniola,there growes a ſmall tree of the height of two men,without the induſtry of man,hauing ſtalks and ſtems of the colour of aſhes ; whereon do grow green leaues,ſharp at both ends,but more green on the vpper ſide than on the lower,hauing a middle rib ſomewhat thicke and ſtanding out : the footſtalke whereon they grow is ſomwhat red-diſh : among which leaues comes fruit growing by cluſters,as long as a mans hand. The ſtones or graines in the fruit be few, and greene, but growing to redneſſe more and more as the fruit waxeth ripe : from the which is gathered a juice after this manner ; they take the yong ſhoots and buds of the tree,and alſo the cluſters of the fruit, which they bruiſe and boile in water to the thickneſſe of hony,which being ſtrained,they keep for their vſes.

They vſe it againſt wounds and vlcers,it ſtops and ſtancheth the bloud,maketh them clean,bring-eth vp the fleſh,and healeth them mightily and with better ſucceſſe than true Balſam. The bran-ches of the tree being cut,do caſt forth by drops a certain cleare water, more worth than *Aqua vita*, moſt wholſome againſt wounds and all other diſeaſes proceeding from cold cauſes, being drunke ſome few daies together.

¶ *The Place.*

Theſe trees grow in diuers parts of the world,ſome in Egypt and moſt of thoſe countries adja-cent.There groweth of them in the Eaſt and Weſt Indies,as trauellers in thoſe parts report.

¶ *The Time.*

Theſe trees for the moſt part keepe green Winter and Summer.

¶ *The Names.*

Balſam is called in Greek, Βαλσαμον : in Latine alſo *Balſamum* : of the Arabians,*Balſeni,Baleſina,*and *Belſan* : in Italian,*Balſamo* : in French, *Baume*.

The liquor that flowes out of the tree when it is wounded, is called *Opobalſamum* : the wood, *Xy-lobalſamum* : the fruit,*Carpobalſamum*:and the liquor which naturally flowes from the tree in Egypt, *Balſamum*.

¶ *The Temperature.*

Balſam is hot and dry in the ſecond degree,with aſtriction.

¶ *The Vertues.*

Natural Balſam taken in a morning faſting,with a little Roſe water or Wine,to the quantity of five or ſix drops, helpeth thoſe that be aſthmatick or ſhort winded : it preuaileth againſt the pains of the bladder and ſtomack,comforting the ſame mightily : it alſo amendeth a ſtinking breath,and takes away the ſhaking fits of the quotidian ague,if it be taken two or three times. **A**

It helpeth conſumptions,clenſeth the barren wombe, eſpecially being annointed vpon a peſſary or mother ſuppoſitorie,and vſed. **B**

The ſtomack being anointed therewith,digeſtion is helped thereby ; it alſo preſerueth the ſto-mack from obſtruction and windineſſe;it helps the hardneſſe of the ſpleen,eaſeth the griefs of the reins and belly,proceeding of cold cauſes. **C**

It alſo takes away all manner of aches occaſioned by cold, if they be annointed therewith ; but more ſpeedily,if a linnen cloth be wet therein and laid thereon : vſed in the ſame manner it diſſol-ueth hard tumors called *œdemata*;and ſtrengthneth the weak members. **D**

The ſame refreſheth the brain,and comforteth the parts adioyning ; it helpeth the palſie, Con-vulſions,and all griefes of the ſinues,being anointed therewith. **E**

The maruellous effects it worketh in new and green wounds,were here too long to ſet down,and alſo ſuperfluous,conſidering the skilfull Surgeon,whom it moſt concernes,knoweth the vſe there-of : and as for the beggerly Quackſaluers,Runnagates,and knauiſh Mountibankes,we are not wil-ling to inſtruct them in things ſo far aboue their reach, capacitie,and worthineſſe. **F**

CHAP. 146.   *Of a kinde of Balme or Balſam tree.*

¶ *The Deſcription.*

THis tree which the people of the Indies do call *Molli*, groweth to the bigneſſe of a great tree, hauing a trunke or bodie of a darke greene colour, ſprinkeled ouer with many aſh-coloured ſpots : the branches are many, and of very great beauty , whereupon grow leaues not vnlike thoſe of the aſh tree,conſiſting of many ſmal leaues ſet vpon a middle ribbe, growing narrower euer toward the point, euery particular one jagged on the ſides like the teeth of a Saw ; which beeing plucked from the ſtem,yeelds forth a milky juice tough and clammy, ſauouring like the bruiſed leaues of fenell,and in taſt ſeems ſomewhat aſtringent : the floures grow in cluſters on the twiggy branches , like thoſe of the Vine a little before the grapes are formed : after followes the fruit or berries,ſomwhat greater than pepper corns,of an oily ſubſtance, green at the firſt,and of a dark reddiſh colour when they be ripe. ‡ The firſt of the figures was taken from a tree only of three years growth,but the later from a tree come to his full growth,as it is affirmed by *Cluſius* in his *Cur.Poſt*. It differs only,in that the leaues of the old tree are not at all ſnipt or diuided about the edges. ‡

1 *Molli,ſiue Molly Cluſij & Lobelij*. 　　　　　‡ 2 *Molle arboris adulti ramus.*
The Balſam tree of *Cluſius* and *Lobels* deſcription. 　　　A branch of the old tree of Molle.

¶ *The Place.*

This tree,ſaith a learned Phyſition *Ioh.Fragoſus*,growes in the king of Spains garden at Madril, which was the firſt that euer he had ſeen : ſince which time,*Iohn Ferdinando,* Secretarie to the ſaid King,did ſhew vnto the ſaid *Fragoſus* in his own garden a tree ſo large,and of ſuch beauty, that he was neuer ſatisfied with looking on it,and meditating vpon the vertues thereof. Which words I receiued from the hands of a famous learned man called Mr *Lancelot Browne,* Dr in phyſick,and phiſition to the Queenes Maieſtie at the impreſſion hereof ; faithfully tranſlated out of the Spaniſh tongue,without adding or taking any thing away.

They grow plentifully in the vales and low grounds of Peru,as all affirm that haue trauelled to the Weſt Indies ; as alſo thoſe that haue deſcribed the ſingularities thereof. My ſelfe with diuers others,as namely Mr *Nicholas Lete* a Worſhipfull Merchant of the Citie of London ; and alſo a moſt ſkilful Apothecarie Mr *Iames Garret,* who haue receiued ſeeds hereof from the right honorable the Lord *Hunſdon* Lord high Chamberlaine of England ; who is worthy of triple honor,for his care in getting,as alſo for his curious keeping ſuch rare and ſtrange things brought from the fartheſt parts of the world. Which ſeeds we haue ſown in our gardens, where they haue brought forth plants of a foot high, and alſo their beautifull leaues ; notwithſtanding our care, diligence,
and

and induſtry,they haue periſhed at the firſt approch of winter,as not being able by reaſon of their tenderneſſe to endure the cold of our winter blaſts.

¶ The Time.

As touching the time of his flouriſhing and bringing his fruit to maturitie,wee haue as yet no certain knowledge,but is thought to be green both winter and ſummer.

¶ The Names.

This moſt notable tree is called by the Indian name *Molle:*of ſome,*Molly* and *Muelle,*taken from his tender ſoftneſſe,as ſome haue deemed : it may be called the Fenel tree, or one of the Balme or Balſam trees.

¶ The Temperature.

This tree is thought to be of an aſtringent or binding qualitie,whereby it appeares,beſides the hot temperature it hath,to be compounded of diuers other faculties.

¶ The Vertues.

The Indians ſe to ſeeth the fruit or berries hereof in water,and by a ſpeciall ſkill they haue in　A the boiling,do make a moſt wholſome wine or drinke,as alſo a kinde of vineger, and ſomtimes hony;which are very ſtrange effects,theſe three things being ſo contrary in taſte.

The leaues boiled and the decoction drunke,helpeth them of any diſeaſe proceeding of a cold　B cauſe.

The gum which iſſueth from the tree,being white like vnto Manna, diſſolued in milke, taketh　C away the web of the eies,and cleareth the ſight,being wiped ouer with it.

The barke of this tree boiled,and the legs that be ſwoln and full of paine bathed with the deco-　D ction diuers times,taketh away both the infirmities in ſhort ſpace.

This tree is of ſuch eſtimation among the Indians, that they worſhip it as a god, according to　E their ſavage rites and ceremonies. Like as *Pliny* reporteth of *Homers Moly,* the moſt renowned of all plants,which they of old time had in ſuch eſtimation and reuerence, that, as it is recorded, the gods gaue it the name of *Moly,*and ſo writeth *Ovid,*

> *Pacifer huic dederat florem* Cyllenius album,
> *Moly vocant Superi,nigra radice tenetur.*

If any be deſirous to ſee more hereof, they may reade a learned diſcourſe of it ſet forth in La-　F tine by the learned *Lobel,* who hath at large written the hiſtorie thereof,dedicated to the right honorable the Lord Chamberlain at the impreſſion hereof,faithfully examined by the aforeſaid learned phyſitian Dr *Browne,*and his cenſure vpon the ſame : ‡ together with *Lobels* Reply, who iudged this plant (and not without good reaſon) to be a kind of the true Balſam of the Antients, and not much differing from that ſet forth by *Proſper Alpinus,*whereof I haue made mention in the foregoing chapter. ‡

CHAP. 147.　*Of the Canell or Cinnamon tree.*

¶ The Deſcription.

1　THe tree which hath the Cinamon for his bark is of the ſtature of an Olive tree,hauing a body as thick as a mans thigh,from whence the Cinamon is taken;but that taken from the ſmaller branches is much better : which branches or boughes are many and very ſtraight,wheron grow beautifull leaues in ſhape like thoſe of the Orenge tree, and of the colour of the Bay leaf, not as it hath been reported,like vnto the leaues of Hags or Flour-de-Lys. Among theſe pleaſant leaues and branches come forth many faire white floures,which turn into round black fruit or berries of the bigneſſe of an Haſell nut or the Olive berry,and of a black colour ; out of which is preſſed an oile that hath no ſmell at all vntill it be rubbed and chafed between the hands : the trunk or body with the greater arms or boughs of the tree are couered wtth a double or twofold barke like that of the Corke tree, the innermoſt whereof is the true and pleaſant Cinnamon,which is taken from the tree and caſt vpon the ground in the heate of the Sun,through whoſe heate it turneth and

Canella folium Bacillus, & Cortex.
The leafe, barke, and trunke of the Cinamon tree.

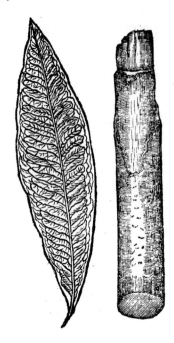

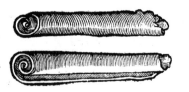

foldeth it felfe round together, as we daily fee by viewing the thing it felfe. This tree being thus peeled recouereth a new bark in the fpace of three yeares, and is then readie to be disbarked as afore. That Cinnamon which is of a pale colour hath not bin well dried in the fun: that of a fair brown color is beft, and that which is blackeft hath bin too much dried, and alfo hath taken fome wet in the time of drying.

‡ 2 Befides the Cinnamon vulgarly known & vfed, there is another fort, which alfo is commonly receiued for the *Caſſia* of *Diofcorides* and the Antients. Now this differs from the former, in that it is of a redder colour, of a more hard follid and compact fubftance, commonly alfo thicker, and if you chew it, more clammy and viſcous. The tafte and fmell are much like Cinnamon, yet not altogether fo ftrong as that of the beft Cinnamon. There is much controuerfie in late Writers, concerning both the true Cinnamon and Caſſia of the Antients: the which I haue not time nor fpace here to mention, much leffe to infift vpon. I haue obferued, that both the Cinnamon & Caſſia that we haue are couered ouer with a rough grayifh barke like that of an Oke or other fuch tree, which is clean fcraped off and taken away before it be brought vnto vs. ‡

¶ *The Place.*

The chiefeft places where the trees doe grow that beare Cinnamon, are Zeilan and Malavar ; but thofe of Zeilan are the beft. They grow in other of the Molucca Ifles, as Iaoa and Iava, the greater and the leffe, and alfo in Mindanoa, for the moft part vpon mountaines.

¶ *The Time.*

The Cinnamon tree groweth green Winter and Summer, as do all the others of the Molucca's and Eaft Indies for the moft part ; the boughs whereof are cut off at feafonable times, by expreffe commandement of the King of the country, and not before he appoints the time.

There hath bin fome controuerfie among writers concerning the tree whofe bark is Caſſia and that tree that beareth Cinnamon, making them both one tree ; but that opinion is not to be receiued, for there is a great difference betwixt them, as there is betwixt an Oke and a Cheftnut tree; for the tree whofe bark is Caſſia is doubtleffe a baftard kinde of Canell or Cinnamon, in fhew it is very like, but in fweetneffe of fmell and other circumftances belonging to Cinnamon far inferior.

¶ *The Names.*

Cinnamon is called in Italian, *Canella* : in Spanifh, *Canola* : in French, *Canelle* : in high-Dutch, **Zimmet coezlin** : in Greek, Κιναμωμον : in Latine likewife *Cinnamomum* : the Arabians, *Darfeni*, and as fome fay, *Querfaa*, others, *Querfe* : in Zeilan, *Cuurde* : in the Ifland Iava they name it *Cameaa* : in Ormus, *Darchini* (i.) *Lignum Chinenfe*, Wood of China : in Malavar, *Cais mains*, which fignifies *Dulce lignum*, or Sweet wood : in Englifh, Cinnamome, Cinnamon, and Canel. The other is called *Caſſia*, and *Caſſia lignea.*

¶ *The Temperature and Vertues.*

Diofcorides writeth, that Cinnamon hath power to warme, and is of thinne parts. It is alfo dry and

and aſtringenr,it prouoketh vrine,cleereth the eies and maketh ſweet breath.

The decoction bringeth downe the menſes, preuaileth againſt the bitings of venomous beaſts, B the inflammotions of the inteſtines and reines.

The diſtilled water hereof is profitable to many, and for diuers infirmites , it comforteth the C weake,cold.and feeble ſtomacke, eaſeth the paines and frettings of the guts and entrailes procee-ding of cold cauſes,it amendeth the euill colour of the face,maketh ſweet breath,and giueth a moſt pleaſant taſte vnto diuers ſorts of meates, and maketh the ſame not onely more pleaſant, but alſo more wholeſome for any bodies or what conſtitution ſoeuer they be, notwithſtanding the binding quality.

⚘ The oile drawne chimically preuaileth againſt the paines of the breſt,comforteth the ſtomacke, D breaketh windineſſe, cauſeth good digeſtion, and being mixed with ſome hony, taketh away ſpots from the face being annointed therewith.

The diſtilled water of the floures of the tree, as *Garcias* the Luſitanian Phyſition writeth,excel- E leth far in ſweetneſſe all other waters whatſoeuer, which is profitable for ſuch things as the barke it ſelfe is.

Out of the berries of this tree is drawn by expreſſion,as out of the berries of the Oliue tree,a cer- F taine oile,or rather a kinde of fat like butter,without any ſmell at all,except it be made warme,and then it ſmelleth as the Cinnamon doth, and is much vſed againſt the coldneſſe of the ſinewes, all paines of the joynts,and alſo the paines and diſtemperature of the ſtomacke and breaſt.

To write as the worthineſſe of the ſubject requireth,would aske more time than we haue to be- G ſtow vpon any one plant ; therefore theſe few ſhall ſuffice, knowing that the thing is of great vſe a-mong many,and knowne to moſt.

‡ *Caſſia* vſed in a larger quantity ſerueth well for the ſame purpoſes which Cinnamon H doth. ‡

Chap. 148.　*Of Gum Lacke and his rotten tree.*

Lacca cum ſuis bacillis.
Gum Lacke with his ſtaffe or ſticke.

¶ *The Deſcription.*

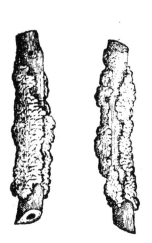

THe tree that bringeth forth that excremen-tal ſubſtance called *Lacca*,both in the ſhops of Europ and elſewhere,is called of the Arabi-ans, Perſians, and Turkes, *Lac Sumutri*, as who ſhould ſay,*Lacca* of Sumutra:ſome which haue ſo termed it haue thought that the firſt plenty thereof came from Sumutta,but herein they haue erred,for the abundant ſtore thereof came from Pegu,where the inhabitants therof do cal it *Lac*, & others of the ſame prouince,*Tree*:the hiſtory of which tree,according to that famous Herbariſt *Cluſius* is as followeth. [There is in the country of Pegu and Malabar, a great tree, whoſe leaues are like them of the Plum tree,ha-uing many ſmall twiggy branches ; when the trunk or body of the tree waxeth old, it rotteth in ſundry places, wherin do breed certain great Ants or Piſmires,which continually work and labour in the time of Harueſt and Summer,a-gainſt the penury of Winter : ſuch is the dili-gence of thoſe Ants,or ſuch is the nature of the tree wherein they harbour, or both, that they prouide for their winter food,a lumpe or maſſe of ſubſtance,which is of a crimſon colour, ſo beautifull and ſo faire, as in the whole World the like is not ſeene,which ſerueth not onely to phyſicall vſes,but is a perfect and coſtly colour for Painters,called by vs,Indian Lack. The Piſmires (as I ſaid) worke out this colour, by ſucking the ſubſtance or matter of Lacca from the tree, as bees doe make hony and wax, by ſucking the matter thereof from all herbes, trees, and floures, and the inhabitants of that country, do as diligently ſeeke for this Lacca, as we in England and in other countries ſeeke in the woods for hony; which Lacca after they haue found, they take from the tree, and dry it into a lumpe; among which ſometimes there come ouer ſome ſtickes and pieces of the tree with the wings of the Ants, which haue fallen among it, as we daily ſee.

‡ The Indian Lacke or Lake which is the rich colour vſed by Painters, is none of that which is vſed in ſhops, nor here figured or deſcribed by *Cluſius*, wherefore our Author was much miſtaken in that hee here confounds together things ſo different; for this is of a reſinous ſubſtance, and a faint red colour, and wholly vnfit for painters, but vſed alone and in compoſition to make the beſt hard ſealing wax. The other ſeemes to bee an artificiall thing, and is of an exquiſite crimſon colour, but of what it is, or how made, I haue not as yet found any thing that carries any probabilitie of truth. ‡

¶ *The Place.*

The tree which beareth Lacca groweth in Zeilan and Malauar, and in other parts of the Eaſt-Indies.

¶ *The Time.*

Of the time we haue no certaine knowledge.

¶ *The Names.*

Indian Lacke is called in ſhops *Lacca*: in Italian, *Lachetta*: *Auicen* calleth it *Luch*: *Paulus* and *Dioſcorides*, as ſome haue thought, *Cancamum*: the other names are expreſſed in the deſcription.

¶ *The Temperature and Vertues.*

A   Lacke or Lacca is hot in the ſecond degree, it comforteth the heart and liuer, openeth obſtructions, expelleth vrine, and preuaileth againſt the dropſie.

B   There is an artificiall Lacke made of the ſcrapings of Braſill and Saffron, which is vſed of Painters, and not to be vſed in Phyſicke as the other naturall Lacca.

Chap. 149.   *Of the Indian leafe.*

Tamalapatra.
The Indian leafe.

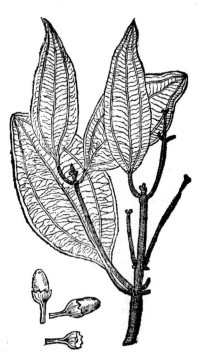

¶ *The Deſcription.*

TAmalapatra, or the Indian leafe grows vpon a great tree like the Orenge tree, with like leaues alſo, but broader, a little ſharp pointed, of a greene gliſtering colour, and three ſmall ribs running through each leafe, after the manner of Ribwort, wherby it is eaſie to be known: it ſmelleth ſomewhat like vnto Cloues, but not ſo ſtrong as Spikenard or Mace (as ſome haue deemed) nor yet of ſo ſubtill and quick a ſent as Cinnamon. There was ſent or added vnto this figure by *Cortuſus* a certaine fruit like vnto a ſmall Acorn, with this inſcription, *Fructus Canella*, the fruit of the Canell tree, which may be doubted of, conſidering the deſcription of the forenamed tree holden generally of moſt to be perfect.

¶ *The Place.*

The Indian leafe groweth not fleeting vpon the water like vnto *Lens Paluſtris*, as *Dioſcorides* and *Pliny* doe ſet downe, (though learned and painfull writers) but is the leafe of a great tree, a branch whereof we haue ſet forth vnto your view, which groweth in Arabia and Cambaya, far from the water ſide.

¶ *The Time.*

Of the time we haue no certain knowledge, but it is ſuppoſed to bee greene Winter and Summer.

¶ *The*

¶ *The Names.*

Tamalapatra is called of the Indians in their mother tongue, eſpecially of the Arabians, *Cadegi Indi*, or *Ladegi Indi*, that is, *Folium Indicum*, or *Indum*, the Indian leafe : but the Mauritanians doe call it *Tembul*. The Latines and Græcians following ſome of the Arabians, haue called it *Malabathrum*.

¶ *The Temperature and Vertues.*

The Indian leafe is hot and dry in the ſecond degree, agreeing with Nardus in temperature, or as **A** others report with Mace : it prouoketh vrine mightily, warmeth and comforteth the ſtomacke, and helpeth digeſtion.

It preuaileth againſt the pin and web in the eies, the inflamed and waterie eies, and all other in- **B** firmities of the ſame.

It is laid among cloathes, as well to keepe them from mothes and other vermine, as alſo to giue **C** vnto them a ſweet ſmell.

CHAP. 150.　　*Of the Cloue tree.*

Caryophylli veri Cluſij.
The true forme of the Cloue tree.

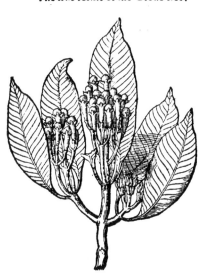

¶ *The Deſcription.*

THe Cloue tree groweth great in forme like vnto the Bay tree, the trunke or body whereof is couered with a ruſſet barke : the branches are many, long, and very brittle, whereupon doe grow leaues like thoſe of the Bay tree, but ſomewhat narrower : amongſt which come the floures, white at the firſt, after of a greeniſh colour, waxing of a darke red colour in the end ; which floures are the very cloues when they grow hard : after when they bee dried in the Sunne they become of that dusky black colour which we daily ſee, wherein they continue. For thoſe that wee haue in eſtimation are beaten downe to the ground before they be ripe, and are ſuffered there to lie vpon the ground vntill they bee dried throughly, where there is neither graſſe, weeds, nor any other herbes growing to hinder the ſame, by reaſon the tree draweth vnto it ſelfe for his nouriſhment all the moiſture of the earth a great circuit round about, ſo that nothing can there grow for want of moiſture, and therfore the more conuenient for the drying of the Cloues. Contrariwiſe, that groſſe kinde of Cloues which hath beene ſuppoſed to be the male, are nothing elſe than fruit of the ſame tree tarrying there vntill it fall downe of it ſelfe vnto the ground, where by reaſon of his long lying, and meeting with ſome raine in the meane ſeaſon, it loſeth the quicke taſte that the others haue. Some haue called theſe *Fuſti*, whereof we may Engliſh them Fuſſes. Some affirme that the floures hereof ſurpaſſe all other floures in ſweetneſſe when they are greene ; and hold the opinion, that the hardned floures are not the Cloues themſelues, (as we haue written) but thinke them rather to be the ſeat or huske wherein the floures doe grow : the greater number hold the former opinion. And further, that the trees are increaſed without labour, graffing, planting, or other induſtrie, but by the falling of the fruit, which beare fruit within eight yeares after they be riſen vp, and ſo continue bearing for an hundred yeares together, as the inhabitants of that country do affirme.

¶ *The Place.*

The Cloue tree groweth in ſome few places of the Molucca Iſlands, as in Zeilan, Iaua the greater and the leſſer, and in diuers other places,

¶ The

¶ *The Time.*

The Cloues are gathered from the fifteenth of September vnto the end of February, not with hands, as we gather Apples, Cherries, and ſuch like fruit, but by beating the tree, as Wall-nuts are gotten, as we haue written in the deſcription.

¶ *The Names.*

The fruit hereof was vnknowne to the antient Grecians : of the later writers called Καρύφυλλοι : in Latine alſo *Caryophyllus*, and *Clauus* : in French, *Clou de Gyroſle* : the Mauritanians, *Charumfel* : in Italian, *Carofano* : in high Dutch, 𝕹𝖆𝖊𝖌𝖊𝖑 : in Spaniſh, *Clauo de eſpecia* : of the Indians, *Calafur* : in the Molucca's, *Changue* : of the Pandets, *Arumfel*, and *Charumfel* : in Engliſh, Cloue tree, and Cloues.

¶ *The Temperature.*

Cloues are hot and dry in the third degree.

¶ *The Vertues.*

A   Cloues ſtrengthen the ſtomacke, liuer, and heart, helpe digeſtion, and prouoketh vrine.

B   The Portugall women that dwell in the Eaſt Indies draw from the Cloues when they bee yet greene, a certaine liquor by diſtillation, of a moſt fragrant ſmell, which comforteth the heart, and is of all Cordials the moſt effectuall.

C   Cloues ſtop the belly : the oile or water thereof dropped into the eies, ſharpens the ſight, and clenſeth away the cloud or web in the ſame.

D   The weight of foure drams of the pouder of Cloues taken in milke procureth the act of generation.

E   There is extracted from the Cloues a certaine oile or rather thicke butter of a yellow colour; which being chafed in the hands ſmelleth like the Cloues themſelues, wherewith the Indians doe cure their wounds and other hurts, as we do with Balſam.

F   The vſe of Cloues, not onely in meate and medicine, but alſo in ſweet pouder and ſuch like, is ſufficiently knowne : therefore this ſhall ſuffice.

‡ There were formerly three figures in this chapter : wherefore I omitted two as impertinent.

Chap. 151.   *Of the Nutmeg tree.*

1 *Nux Muſcata rotunda, ſiue fœmina.*
The round or female Nutmeg.

2 *Nux Myriſtica oblonga, ſiue mas.*
The longiſh or male Nutmeg.

Nux Moſchata cum ſua Maci.
The Nutmeg with his Mace about him.

1　THe tree that beareth the Nutmeg and the Mace is in form like to the Peare tree, but the leaues of it are like thoſe of the Orenge tree, alwaies green on the vpper ſide, and more whitiſh vnderneath ; amongſt which come forth the Nut and Mace as it were the floures. The Nut appeares firſt, compaſſed about with the Mace, as it were in the middle of a ſingle roſe, which in proces of time doth wrap and incloſe the Nut round on euery ſide : after commeth the huske like that of the Wallnut, but of an harder ſubſtance, which incloſeth the Nut with his Mace a the Walll-nut husk doth couer the Nut, which in time of ripeneſſe cleaueth of it ſelf as the walnut huske doth, and ſheweth his Mace, which then is of a perfect crimſon colour, and makes a moſt goodly ſhew, eſpecially when the tree is well laden with fruit : after the Nut becomes dry, the Mace likewiſe gapeth and forſaketh the nut, euen as the firſt husk or couerture, and leaueth it bare as we all doe know. At which time it getteth to it ſelfe a kinde of darke yellow colour, and loſeth that braue crimſon dye which it had at the firſt.

‡　2　The tree which carrieth the male Nutmeg (according to *Cluſius*) thus differs from the laſt deſcribed : the leaues are like thoſe of the former in ſhape, much bigger, being ſometimes a foot long, and three or foure inches broad; their common length is ſeuen or eight inches, and bredth two and a halfe : they are of a whitiſh colour vnderneath, and green and ſhining aboue. The Nuts alſo grow at the very ends of the branches, ſomtimes two or three together, and not only one, as in the common kinde. The nut it ſelfe is alſo larger and longer : the Mace that incompaſſeth it is of a more elegant colour, but not ſo ſtrong as that of the former.

‡　I can ſcarſe beleeue our Authors aſſertion in the foregoing deſcription, that the Nut appeareth firſt, compaſſed about with the Mace as it were in the midſt of a ſingle Roſe &c. But I rather thinke they all come forth together, the Nutmeg, Mace, the green outward husk and all, iuſt as we ſee Walnuts do, and only open themſelues when they come to full maturitie. In the third figure you may ſee expreſt the whole manner of the growing of the Nutmeg, together with both ſorts of Nutmegs taken forth of their ſhels. ‡

¶ *The Place.*
The Nutmeg tree groweth in the Indies, in an Iſland eſpecially called Banda, in the Iſlands of Molucca, and in Zeilan, though not ſo good as the firſt.

¶ *The Time.*
The fruit is gathered in September in great aboundance, all things beeing common in thoſe countries.

¶ *The Names.*
The Nutmeg tree is called of the Grecians Κάρυον μυρισικόν : of the Latines, *Nux Moſchata,* and *Nux Myriſtica* : in Italian, *Noce Moſcada* : in Spaniſh, *Nuez de eſcette* : in French, *Noix Muſcade* : in high-Dutch, Moſchat Nuſz : of the Arabians, *Leuzbane,* or *Granziban* : of the country people where they grow, *Palla* ; the Maces, *Bunapalla.* In Decan the Nut is called *Iapatri* ; and the Maces, *Iaiſol, of Auicen, Iauſiband* (i.) *Nux Bandenſis.* The Mace he calleth *Beſbaſe.* In Engliſh, Nutmeg.

¶ *The Temperature.*
The Nutmeg, as the Mauritanians write, is hot and dry in the ſecond degree compleat, and ſomewhat aſtringent.

¶ *The Vertues.*
Nutmegs cauſe a ſweet breath, and mend thoſe that ſtinke, if they be much chewed and holden A in the mouth.

The

B   The Nutmeg is good agaiſt freckles in the face, quickneth the ſight, ſtrengthens the belly and feeble liuer, taketh away the ſwelling in the ſpleen, ſtayeth the laske, breaketh wind, and is good a-gainſt all cold diſeaſes in the body.

C   Nutmegs bruiſed and boiled in Aqua vitæ vntill they haue waſted and conſumed the moiſture, adding thereunto hony of Roſes, gently boiling them, being ſtrained to the form of a ſyrrup, cure all pains proceeding of windy and cold cauſes, if three ſpoonfulls be giuen faſting for certain daies together.

D   The ſame bruiſed and boiled in ſtrong white Wine vntill three parts be ſodden away, with the roots of Motherwort added thereto in the boiling, and ſtrained, being drunke with ſome ſugar, cu-reth all gripings in the belly proceeding of windineſſe.

E   As touching the choice, there is not any ſo ſimple but knoweth that the heauieſt, fatteſt and ful-leſt of juice are beſt, which may eaſily be found out by pricking the ſame.

C H A P. 152.   *Of the Pepper Plant.*

¶ *The Kindes.*

THere be diuers ſorts of Pepper, that is to ſay, white, blacke, and long Pepper, one greater and longer than the other ; and alſo a kinde of Ethiopian Pepper.

1 *Piper nigrum.*   2 *Piper album.*
Blacke Pepper.   White Pepper.

¶ *The Deſcription.*

1   THe Plant that beareth the blacke Pepper groweth vp like a Vine, amongſt buſhes and brambles, where it naturally groweth ; but where it is manured it is ſowne at the bot-tom of the tree *Fauſel* and the Date trees, whereon it taketh hold and clymbeth vp euen vnto the top, as doth the Vine, ramping and taking hold with his claſping tendrels of any other
thing

thing it meets withall. The leaues are few in number, ‡ growing at each joint one, firſt on one ſide of the ſtalk, then on the other, like in ſhape to the long vndiuided leaues of Iuy, but thinner, ſharp pointed, and ſometimes ſo broad that they are foure inches ouer, but moſt commonly two inches broad and foure long, hauing alwaies fiue pretty large nerues running along them. The fruit grow cluſtering together vpon long ſtalks, which come forth at the joints againſt the leaues, as you may ſee in the figure: the root (as one may conie cture) is creeping, for the branches that ly on the ground do at their joints put forth new fibres or roots. Wee are beholden to *Cluſius* for this exa ct figure and deſcription, which hee made by certaine branches brought home by the Hollanders from the Eaſt Indies. The Curious may ſee more hereof in his Exoticks and notes vpon *Garcias*.

† 3 *Piper longum.*
Long Pepper.

4 *Piper Æthiopicum, ſiue Vita longa.*
Pepper of Æthiopia.

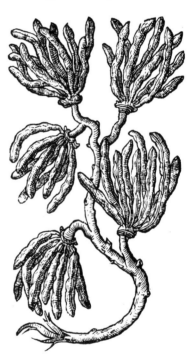

2 The plant that brings white Pepper is not to be diſtinguiſhed from the other plant, but only by the colour of the fruit, no more than a Vine that beareth blacke grapes, from that which brings white: and of ſome it is thought, that the ſelfe ſame plant doth ſomtimes change it ſelf from black to white, as diuers other plants doe. ‡ Neither *Cluſius*, nor any other elſe that I haue yet met with, haue deliuered vs any thing of certaine, of the plant whereon white Pepper growes; *Cluſius* onely hath giuen vs the maner how it growes vpon the ſtalks, as you may ſee it here expreſt. ‡

There is alſo another kind of Pepper, ſeldome brought into theſe parts of Europe, called *Piper Canarium*: it is hollow within, light, and empty, but good to draw flegme from the head, to helpe the tooth-ache and cholericke affe cts.

3 The tree that beareth long pepper hath no ſimilitude at all with the plant that brings black and white Pepper: ſome haue deemed them to grow all on one tree; which is not conſonant to truth: for they grow in countries far diſtant one from another, and alſo that countrey where there is blacke Pepper, hath not any of the long Pepper. And therefore *Galen* following *Dioſcorides*, were together both ouerſeene in this point. This tree, ſaith *Monardus*, is not great, yet of a wooddy ſubſtance, diſperſing here and there his claſping tendrels, wherewith it taketh hold of other trees and ſuch other things as do grow neere vnto it. The branches are many and twiggy; whereon grow eth the fruit, conſiſting of many graines growing vpon a ſlender foot-ſtalke, thruſt or compa ct cloſe

toge ther,

‡ 5 *Piper Caudatum.*
Tailed Pepper.

together, green at the firſt, and afterward blackiſh; in taſt ſharper and hotter than common blacke pepper yet ſweeter and of better taſt. ‡ For this figure alſo I acknowledge my ſelfe beholden to the learned and diligent *Cluſius*, who cauſed it to be drawne from a branch of ſome foot in length, that he receiued from Dr *Lambert Hortenſius*, who brought it from the Indies. The order of growing of the leaues and fruit is like that of the black, but the joints ſtand ſomewhat thicker together; the leaf alſo doth little differ from that of the black, only it is thinner, of a lighter green and (as *Cluſius* thought) hath a ſhorter footſtalke, the veins or nerues alſo were leſſe eminent, more in number, and run from the middle rib to the ſides, rather than alongſt the leafe. ‡

4   This other kinde of Pepper, brought vnto vs from Æthiopia, called of the country where it groweth, *Piper Æthiopicum*: in ſhops, *Amomum*, and *Longa Vita*. It growes vpon a ſmal tree in maner of an hedge buſh, wherupon grow cods in bunches, a finger long, of a brown colour, vneuen, and bunched or puſt vp in diuers places, diuided into fiue or ſix lockers or cels, each wherof containeth a round ſeed ſomwhat long, leſſe than the ſeeds of Peony, in taſte like common Pepper or *Cardamomum*, whoſe facultie and temper it is thought to haue, whereof we hold it a kind.

5   Another kind of Pepper is ſomtimes brought, which the Spaniards cal *Pimenta de rabo*, that is, Pepper with a taile: it is like vnto Cubebes, round, full, ſomewhat rough, black of colour, and of a ſharp quick taſt like the common pepper, of a good ſmell: it growes by cluſters vpon ſmal ſtems or ſtalks, which ſome haue vnaduiſedly taken for *Amomum*. The king of Portugal forbad this kind of Pepper to be brought ouer, for feare leſt the right Pepper ſhould be the leſſe eſteemed, and ſo himſelfe hindred in the ſale thereof.

¶ *The Place.*

Black and white Pepper grow in the kingdome of Malauar, and that very good; in Malaca alſo, but not ſo good; as alſo in the Iſlands Sunde and Cude: there is great ſtore growing in the Kingdome of China, and ſome in Cananor, but not much.

Pepper of Æthiopia growes alſo in America, in all the tract of the country where Nat & Carthago are ſituated. The reſt haue bin ſpoken of in their ſeueral deſcriptions. The white Pepper is not ſo common as the blacke, and is vſed there in ſtead of ſalt.

¶ *The Time.*

The plant riſeth vp in the firſt of the ſpring: the fruit is gathered in Auguſt.

¶ *The Names.*

The Grecians who had beſt knowledge of Pepper, do cal it ~~~~ the Latines, *Piper*: the Arabians, *Fulfel*, and *Fulful*: in Italian, *Pepe*: in Spaniſh, *Pimenta*: in French, *Poiure*: in high-Dutch, Pfeffer: in Engliſh, Pepper.

That of Æthiopia is called *Piper Æthiopicum*, *Amomum*, *Vita longa*, and of ſome *Cardamomum*. I receiued a branch hereof at the hands of a learned Phyſition of London, called Mr *Stephen Bredwel*, with his fruit alſo.

¶ *The Temperature.*

The Arabian and Perſian Phyſitions iudge, that Pepper is hot in the third degree. But the Indian Phyſitions (which for the moſt part are Emperickes) hold; that Pepper is cold, as almoſt all other ſpice, which are hot indeed: the long Pepper is hot alſo in the third degree, and as wee haue ſaid, is thought to be the beſt of all the kinds.

¶ *The Vertues.*

A   *Dioſcorides* and others agreeing with him affirme, that Pepper reſiſteth poiſon, and is good to be put into medicaments for the eies.

All Pepper heateth, prouoketh vrin, digesteth, draweth, disperseth, and clenseth the dimnesse of E the sight, as *Dioscorides* noteth.

‡ I haue omitted in this chapter *Matthiolus* his counterfeit figure which was formerly here.

CHAP. 153. *Of bastard Pepper, called Betle or Betre.*

Betle siue Betre.
Bastard Pepper.

¶ *The Description.*

THis plant climeth and rampeth vpon trees, bushes, or whatsoeuer else it meets withall, like to the Vine, or the black Pepper, whereof some hold it for a kinde. The leaues are like those of the greater Binde weed, but somewhat longer, of a dusty colour, with diuers veins or ribs running through the same. The fruit groweth among the leaues, very crookedly writhed, in shape like the taile of a Lizard, of the taste of Pepper, yet very pleasant to the palate.

¶ *The Place.*

It groweth among the Date trees, and *Areca*, in most of the Molucca Islands, especially in the marish grounds.

¶ *The Time.*

The time answereth that of Pepper.

¶ *The Names.*

It hath been taken for the Indian leafe, but not properly : of most it is called *Tembul*, and *Tambul*: in Malavar, *Betre* : in Decan, Guzarat, and Canam it is called *Pam* : in Molaio, *Siri*.

¶ *The Vse and Temperature.*

The leaues chewed in the mouth are of a bitter taste ; whereupon (saith *Garcias*) they put thereto some Areca, with the lime made of oister shells ; whereunto they also adde some Amber greece, *Lignum Aloes*, and such like, which they stamp together, making it into a paste which they rolle vp into round balls, keepe dry for their vse, and carry the same in their mouthes vntill by little and little it is consumed ; as when we carry sugar-Candy in our mouthes, or the juice of Licorice: which is not only meat to the silly Indians, but also drink in their tedious trauels, refreshing their wearied spirits, and helping memory ; which is esteemed amongst the Empericke physitians to be hot and dry in the second degree. ‡ *Garcias* doth not affirm that the Indians eat it for meat, or in want of drinke, but that they eat it after meat, and that to giue the breath a pleasant sent, which they count a great grace, so that if an inferior person that hath not chewed Betre or some such thing, chance to speak with any great man, he holds his hand before his mouth, lest his breath should offend him. ‡

CHAP. 154. *Of Graines, or Graines of Paradice,*

¶ *The Kindes.*

THere be diuers sorts of Graines, some long, others Peare-fashioned ; some greater, and others lesser.

¶ *The Description.*

†   THe figure hereof setteth forth to your view the cod where in the hot spice lies, which we cal Graines; in shops, *Grana Paradisi* : it growes, by the report of the Learned, vpon a low herby plant ; the leaues are some foure inches long, and three broad, with somewhat a thicke middle rib, from which run transuerse fibres, they much in shape resemble those of Cloues. The fruit is like a great cod or huske, in shape like a fig when it groweth vpon the tree, but of colour russet, thrust full of small seeds or grains of a darke reddish colour (as the figure sheweth which is diuided) of an exceeding hot taste.

Cardamomi genera.
The kindes or sorts of Graines.

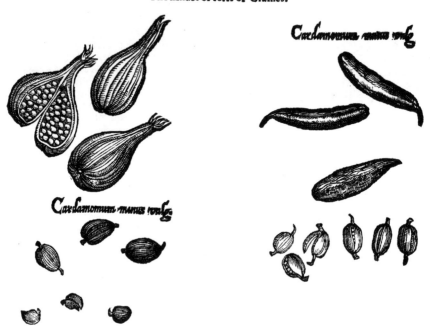

The other sorts may be distinguished by the sight of the picture, considering the only difference consists in forme.

¶ *The Place.*

Graines grow in Ginny, and the Cardamones in all the East Indies, from the port of Calecute vnto Cananor : it groweth in Malauar, in Ioa, and in diuers other places.

¶ *The Time.*

They spring vp in May being sowne of seed, and ripen their fruit in September.

¶ *The Names.*

Graines are called in Greeke Καρδάμωμον : in Latine, *Cardamomum* : of the Arabians, *Corrumeni* : of diuers Gentiles, *Dore* : of *Auicen, Saccolaa quebir* (i.) *magnum* : the other, *Saccolaa cequer* (i.) *minus*. It is called in Malauar *Etremelli* : in Zeilan, *Ençal* : in Bengala, Guzarat, and Decan, *Hil*, and *Eluchi* : The first and largest sort are called of some *Milegnetta*, and *Milegetta* ; in English, Grains, & Grains of Paradice.

¶ *The Temperature.*

Auicen writeth, that *Saccolaa, Cardamomum*, or *Grana Paradisi*, are hot and dry in the third degree, with astriction.

¶ *The Vertues.*

A   The Graines chewed in the mouth draw forth from the head and stomacke waterish and pituitous humors.

B   They also comfort and warm the weake cold and feeble stomack, help the ague, and rid the shaking fits, being drunke with Sacke.

CHAP.

Chap. 155.   Of Yucca or Iucca.

Yucca, ſive Iucca Peruana.
The root whereof the bread Caſſua or
Cazava is made.

¶ *The Deſcription.*

THe Plant of whoſe root the Indian bread called Cazava is made, is a low herbe conſiſting onely of leaues and roots : it hath neither ſtalks, flours, nor fruit, that I can vnderſtand of others, or by experience of the plant it ſelfe, which hath grown in my garden four yeres together, and yet doth grow and proſper exceedingly ; neuertheleſſe without ſtalke, fruit, or floures, as aforeſaid. It hath a very great root, thick and tuberous, and very knobby, ful of iuice ſomwhat ſweet in taſte, but of a pernitious qualitie, as ſaith my Author. From which root riſeth vp immediatly forth of the ground very many leaues ioyned vnto the head of the root in a round circle; the which are long, of the length of a cubit, hollowed like a gutter or trough, very ſmooth, and of a green colour like that of woad: the edges of the leaues are ſharpe like the edge of a knife, and of a brown colour ; the point of the leafe is a prick as ſharpe as a Needle, which hurteth thoſe that vnaduiſedly paſſe by it. The leafe with aduiſed eye viewed is like to a little Wherry or ſuch like boat ; they are alſo very tough, hard to break, and not eaſie to be cut, except the knife be very ſharp.

‡ *Lobel* in the ſecond part of his *Aduerſaria* largely deſcribes and figures this plant; & there he affirmes hee wrot a Deſcription (which hee there ſets down) for our Author, but hee did not follow it, and therefore committed theſe errors: firſt, in that he ſaith it is the root wherof Cazava bread was made; when as *Lobel* in his deſcription ſaid he thought it to be *Alia ſpecies à Yucca Indica ex qua panis communis fit.* Secondly, in that he ſet down the place out of the *Hiſt. Lugd.* (who tooke it out of *Theuet*) endeauoring by that meanes to confound it with that there mentioned, when as he had his from Mr *Edwards* his man: and thirdly (for which indeed he was moſt blame-worthy, and wherein he moſt ſhewed his weakneſſe) for that he doth confound it with the *Manihot* or true *Yucca*, which all affirme to haue a leaf like that of Hemp, parted into ſeuen or more diuiſions; and alſo for that he puts it to the *Arachidna* of *Theophraſtus*, when as he denies it both floure and fruit ; yet within ſome few yeares after our Authour had ſet forth this Worke it floured in his garden.

This ſome yeares puts forth a pretty ſtiffe round ſtalke ſome three cubits high, diuided into diuers vnequall branches carrying many pretty large floures, ſhaped ſomwhat like thoſe of *Fritillaria*, but that they are narrower at their bottoms : the leaues of the floure are ſix, the colour on the inſide white, but on the out ſide of an ouerworn reddiſh colour from the ſtalke to the middeſt of the leafe ; ſo that it is a floure of no great beauty, yet to be eſteemed for the raritie. I ſaw it once floure in the garden of Mr *Wilmot* at Bow, but neuer ſince, though it hath bin kept for ſundry yeres in many other gardens, as with Mr *Parkinſon* and Mr *Tuggy*. This was firſt written of by our Author and ſince by *Lobel* and Mr *Parkinſon*, who keepe the ſame name, as alſo *Bauhine*, who to diſtinguiſh it from the other calls it *Yucca folijs Aloes*. ‡

¶ *The Place.*

This Plant groweth in all the tract of the Indies, from the Magellane ſtraights vnto the cape of Florida, and in moſt of the Iſlands of the Canibals, and others adioining, from whence I had that plant brought me that groweth in my garden, by a ſeruant of a learned and skilfull Apothecary of Exceſter, named Mr *Tho. Edwards*.

¶ *The Time.*

It keepeth green both Winter and summer in my garden, without any couerture at all, notwith-standing the injurie of our cold clymat.

¶ *The Names.*

It is reported vnto me by Trauellers, that the Indians do call it in some parts, *Manihot*, but gene-rally *Yucca* and *Iucca*. It is thought to be the plant called of *Theophrastus*, *Arachidna*: of *Pliny*, *Ara-cidna*.

¶ *The Temperature.*

This plant is hot and dry in the first degree, which is meant by the feces or drosse, when the poi-sonous juice is pressed or strained forth; and is also dry in the middle of the second degree.

Chap. 156.

Of the Fruit *Anacardium*, and *Caious* or *Caiocus*.

¶ *The Description.*

THe antient writers haue bin very briefe in the historie of *Anacardium*: the Grecians haue tou-ched it by the name of *Anacardion*, taking the name from the likenesse it hath of an heart both in shape and colour, called of the Portugals that inhabit the East Indies, *Fava de Malaqua*, the bean of Malaca; for being green, and as it hangeth on the tree, it resembleth a Bean, sauing that it is much bigger: but when they be dry they are of a shining blackish colour, containing betweene the out-ward rinde and the kernell, which is like an almond, a certain oile of a sharp caustick or burning fa-cultie, called *Mel Anacardinum*, although the kernell is vsed in meats & sauces, as we do Oliues and such like, to procure appetite.

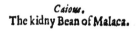

| *Anacardium.* | *Caious.* |
| The Bean of Malaca. | The kidny Bean of Malaca. |

The other fruit groweth vpon a tree of the bignesse of a Peare tree: the leaues are much like to those of the Olive tree, but thicker and fatter, of a feint greene colour: the floures are white, consi-sting of many small leaues much like the floures of the Cherry tree, but much doubled, without smell: after commeth the fruit, according to *Clusius*, of the forme and magnitude of a goose egge, full of juice, in the end whereof is a nut, in shape like an Hares kidny, hauing two rindes, betweene which is contained a most hot and sharp oile like that of *Anacardium*, whereof it is a kinde.

The bean or kernell it selfe is no lesse pleasant and wholesome in eating, than the *Pistacia* or Fi-stick nut, whereof the Indians do eat with great delight, affirming that it prouoketh venery, where-in is their chiefest felicitie. The fruit is contained in long cods like those of beanes, but greater; neere vnto which cods commeth forth an excrescence like vnto an apple, very yellow, of a good smell, spungeous within, and full of juice, without any seeds, stones, or grains at all, somwhat sweet in taste, at the one end narrower than the other, Peare-fashion, or like a little bottle, which hath bin reputed of some for the fruit, but not rightly; for it is rather an excrescence, as is the Oke apple.

¶ *The Place.*

The first growes in most parts of the East Indies, especially in Cananor, Galecute, Cambaya, and Decan. The later in Brasile.

¶ *The*

¶ *The Time.*

Theſe trees floure and flouriſh Winter and Summer.

¶ *The Names.*

Their names haue bin touched in their deſcriptions. The firſt is called *Anacardium*, of the like-neſſe it hath with an heart : of the Arabians *Balador* : of the Indians *Bibo*.

The ſecond is called *Caious* ; and is written *Caious* ; and *Caius* : of ſome, *Cauocus*.

¶ *The Temperature and Vertues.*

The oile of the fruit is hot and dry in the fourth degree; it hath alſo a cauſtick or corroſiue qua- **A** litie : it taketh away warts, breaketh impoſtumes, preuaileth againſt lepry, *alopecia*, and eaſeth the pain of the teeth, being put into the hollowneſſe thereof.

The people of Malavar do vſe the ſaid oile migled with chalke, to marke their cloathes or any **B** other thing they deſire to be coloured or marked, as we vſe chalke, okar, and red marking ſtones, but their colour will not be taken forth again by any art whatſoeuer.

They alſo giue the kernell ſteeped in whay to them that be aſthmaticke or ſhort winded ; and **C** when the ſame is green they drinke the ſame ſo ſteeped againſt the wormes.

The Indians for their pleaſure will giue the fruit vpon a thorne or ſome other ſharpe thing, and **D** hold it in the flame of a candle or any other flame, which there will burn with ſuch crackings, light-nings, and withall yeeld ſo many ſtrange colors, that it is great pleaſure to the beholders who haue not ſeen the like before.

Cʜᴀᴘ. 157.

Of Indian Morrice bells, and diuers other Indian Fruits.

† 1 *Ahouay Theueti.*
Indian Morrice bells.

† 2 *Fructus Higuero.*
Indian Moroſco bells.

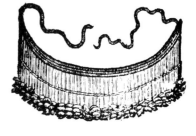
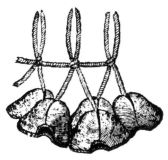
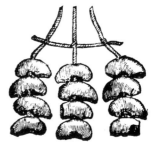

¶ *The Deſcription.*

THis fruit groweth vpon a great tree of the bigneſſe of a peare tree, full of branches, garniſhed with many leaues which are alwayes greene, three or foure fingers long, and in bredth two : when the branches are cut off there iſſueth a milky iuyce not inferior to the fruit in his venomous

qualitie.

qualitie. The trunk or body is couered with a grayiſh barke : the timber is white and ſoft, not fit
to make fire of, much leſſe for any other vſe ; for being cut and put to the fire to burne, it yeeldeth
forth ſuch a loathſome and horrible ſtinke, that neither man nor beaſt are able to endure it, wherefore the Indians haue no vſe thereof, but only of the fruit, which in ſhape is like the Greeke letter Δ,
of the bigneſſe of a Cheſtnut, and couered with a moſt hard ſhell, wherein is contained a kernell of
a moſt venomous and poiſonſome qualitie, wherewith the men beeing angry with their wiues doe
poiſon them, and likewiſe the women their husbands : they likewiſe vſe to dip or anoint and inue-
nom their arrowes therewith, the more ſpeedily to diſpatch their enemies. Which kernel they take
forth with ſome conuenient inſtrument, leauing the ſhel as whole as may be, not touching the kernell with their hands becauſe of its venomous qualitie, which would ſpoile their hands, and ſometimes take away their life alſo. In which ſhells they put ſome little ſtones and tie them on ſtrings,
(as you may perceiue by the figure) which they dry in the Sun, and after tie them about their legs
as we do bels, to ſet forth their dances and Moroſco Matachina's, wherein they take great pleaſure,
by reaſon they think themſelues to excell in thoſe kindes of dances. Which ratling ſound doth
much delight them, becauſe it ſetteth forth the diſtinction of ſounds, for they tune and mix them
with great ones and little ones, in ſuch ſort as we do chimes or bells.

  2 There is alſo another ſort herof, differing only in forme, being of the like venomous quality,
and vſed for the ſame perpoſe. ‡ The fruit of *Higuerro* is like that of a Gourd in pulpe, and may
be eaten : the ſhape of the fruit is round, whereas the other is three cornered. ‡

¶ The Place.

Theſe grow in moſt parts of the Weſt Indies, eſpecially in ſome of the Iſlands of the Canibals,
who vſe them in their dances more than any of the other Indians. ‡ You may ſee theſe on ſtrings
as they are here figured, among many other rarities, with Mr *Iohn Tradeſcant* at South Sambeth. ‡

¶ The Time.

We haue no certain knowledge of the time of flouring or bringing the fruit to maturitie.

¶ The Names and Vſe.

We haue ſufficiently ſpoken of the names and vſe hereof, therefore what hath beene ſayd may
ſuffice.

  † The figures were tranſpoſed.

CHAP. 158.   Of the vomiting and purging Nuts.

<table>
<tr><td>1 Nuces vomica.
Vomiting Nuts.</td><td>2 Nuces purgantes.
Purging Nuts.</td></tr>
</table>

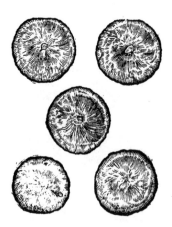 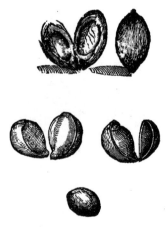

¶ The

¶ The Deſcription.

1　A Vicen and Serapio make Nux vomica and Nux Methel to be one, whereabout there hath been much cauilling : yet the caſe is plain, if the text be true, that the Thorn Apple is Nux Methel. Of the tree that beareth the fruit called in ſhops Nux vomica and Nux methel, we haue no certain knowledge : ſome are of opinion that the fruit is the root of an herb, & not the nut of a tree : wherefore ſince the caſe among the Learned reſteth doubtfuil, we leaue what els might be ſaid to a further conſideration. The fruit is round, flat like a little cake, of a ruſſet ouer-worne colour, fat and firme, in taſte ſweet, and of ſuch an oily ſubſtance, that it is not poſſible to ſtamp it in a mortar to pouder, but when it is to be vſed it muſt be grated or ſcraped with ſome inſtrument for that purpoſe.

2　There be certain nuts brought from the Indies, called Purging nuts, of their quality in purging groſſe and filthy humors : for want of good inſtruction from thoſe that haue trauelled to the Indies, we can write nothing of the tree it ſelfe : the Nut is ſomwhat long, oual, or in ſhape like an Egge, of a browne colour : within the ſhell is contained a kernell, in taſte ſweet, and of a purging facultie.

¶ The Place and Time.

Theſe Nuts do grow in the deſarts of Arabia, and in ſome places of the Eaſt Indies : we haue no certain knowledge of their ſpringing or time of maturitie.

¶ The Names.

Avicen affirmeth the Vomiting nut to be of a poiſonous qualitie, cold in the fourth degree, hauing a ſtupifying nature, and bringing deadly ſleepe.

¶ The Vertues.

Of the phyſicall vertues of the Vomiting Nuts we thinke it not neceſſarie to write, becauſe the　A danger is great, and not to be giuen inwardly, but mixed with other Compoſitions, and that very curiouſly by the hands of a faithfull Apothecarie.

The pouder of the nut mixed with ſome fleſh and caſt vnto Crowes and other rauenous Fowles,　B doth kill, at the leaſt ſo dull their ſences, that you may take them with your hands.

They make alſo an excellent ſallet, mixed with ſome meat or butter, and laid in a garden where　C Cats vſe to ſcrape to bury their excrements, ſpoiling both herbs and alſo ſeeds new ſowne.

CHAP. 159.　Of diuers ſorts of Indian Fruits.

¶ The Kindes.

THeſe fruits are of diuers ſorts and kinds, wherof we haue little knowledge, more than the fruits themſelues, with the names of ſome of them : wherefore it ſhall ſuffice to ſet forth vnto your view the form only, leauing vnto Time, and thoſe that ſhall ſucceed, to write of them at large, who in time may know that which at this time is vnknowne.

‡　OVr Author formerly in this chapter ſet forth diuers ſorts of Indian fruits, and among the reſt, Beritinas, Cacao, Coces Orientales, Buna, Fagaras, Cububæ, &c. but he gaue but only three deſcriptions; and theſe either falſe, or to no purpoſe ; wherefore I haue omitted them, and in this chapter giuen you moſt of theſe fruits which were formerly figured therein, together with an addition of ſundry other out of Cluſius his Exoticks, whoſe figures I haue made vſe of, and here giuen you al thoſe which came to my hands, though nothing ſo many as are ſet forth in his Exoticks ; neither if I ſhould haue had the figures, would the ſhortneſſe of my time, nor bigneſſe of the book (being already grown to ſo large a volume) ſuffer me to haue inſerted them : therefore take in good part thoſe I here giue, together with the briefe hiſtories of them.

¶ The Deſcription.

1　THe fitſt and one of the beſt knowne of theſe fruits, are the Cubibæ, called of the Arabian Phyſitions, Cubibe, and Quabeb ; but of the Vulgar, Quabeb chini; in Iaoa where they plentifully grow, Cumue : the other Indians (the Malayans excepted) call them Cubas ſini, not becauſe they grow in China, but becauſe the Chinois vſe to buy them in Iaoa and Sunda, and ſo carry them to the other ports of India. The Plant which carrieth this fruit hath leaues

like

Cubibæ. Cubi bs.
2 *Cocci Orientales. Cocculus Indi.*

‡ 6 *Amomum verum.*

cubibe

Cocci.

‡ 7 *Amomum spurium.*

‡ 8 *Amomis.*

3 *Fagara.*

9 *Beritinus.*

4 *Mungo.*

5 *Buna.*

‡ 10 *Nuces insanæ.*
Mad Nuts.

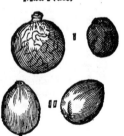

like those of Pepper,but narrower,and it also windes about trees like as Ivy or Pepper doth. The fruit hangs in clusters like as those we call red Currans, and not close thrust together in bunches, as Grapes:the fruit or berries are of the bignesse of Pepper cornes,wrinkled,and of a brownish colour;they are of a hot and biting aromaticke taste,and oft times hollow within,but if they be not hollow,then haue they a pretty reddish smooth round seed vnder their rough vtter huske, each of these berries commonly hath a piece of his footstalke adhering to it. It is reported,That the Natiues where it growes first gently boile or scald these berries before they set them, that so none els may haue them,by sowing the seeds. Some haue thought these to haue been the *Carpesium* of the Antients ; and other-some haue iudged them the seeds of *Agnus Castus*,but both these opinions are erronious.

Thefe are hot and dry in the beginning of the third degree, wherfore they are good against the   A cold and moist affects of the stomacke,and flatulencies ; they help to clense the breft of tough and thicke humors,they are good for the spleen,for hoarfeneffe and cold affects of the womb ; chewed with Maftick they draw much flegmatick matter from the head,they heate and comfort the brain. The Indians vse them macerated in wine to excite venery.

2   The plant which caries this fruit is vnknowne,but the berry is well knowne in shops by the name of *Cocculus Indicus* , some call them *Cocci Orientales* ; others,*Coccula Orientales* ; others,as *Cordus* for one,thinke them the fruit of *Solanum furiosum* ; others iudge them the fruit of a Tithymale, or of a *Clematis*. Thefe beries are of the bignesse of Bay beries,commonly round,and growing but one vpon a stalke; yet somtimes they are a little cornered, and grow two or three clustering together : their outer coat or shel is hard,rough,and of a brownish dusky colour ; their inner substance is very oily,of a bitter taste.

They are vsed with good succeffe to kill lice in childrens heads,being made into pouder and so   B ftrowed amongft the haire. They haue also another facultie which our Author formerly set down in the chapter of *Alaternus*(where he confounded these with *Fagaras*)in these words,which I haue there omitted,to infert here :

In England we vse the fruit called *Cocculus Indi* in pouder,mixed with floure,hony,and crums of   C bread,to catch fish with,it being a numming, foporiferous,or sleeping medicine, causeth the fish to turn vp their bellies,as being fenfeleffe for a time.

3   *Fagara* is a fruit of the bignesse of a Cich pease, couered with a thin coat of a blackish Ash colour,vnder which outer coat is a slender shell containing a sollid kernel involved in a thin black filme.The whole fruit both in magnitude,form,and colour,is so like the *Cocculus Indus* last described,that at the first sight one would take it to be the same. *Avicen* mentions this,*cap.*266.after this manner : What is *Fagara?* It is a fruit like a Chich,hauing the seed of *Mahaleb*,and in the hollowneffe is a black kernell as in *Schehedenegi*,and it is brought out of Sofale.

He places it amongst those that heate and dry in the third degree, and commends it against the coldneffe of the stomacke and liuer,it helps concoction,and bindes the belly.

4   This,which *Clufius* thinks to be *Mungo*(which is vsed in the East Indies about Guzarat and Decan for prouender for horfes)is a small fruit of the bignesse of Pepper,crested,very like Coriander seed,but that it is bigger and blacke,it is of an hot taste.

5   *Buna* is a fruit of the bignesse of *Fagara*, or somewhat bigger or longer, of a blackish Ash colour,couered with a thinne skin furrowed on both sides longwife,whereby it is easily diuided into two parts,which contain each a kernell longish and flat vpon one side,of a yellowish colour, and acide taste.They say that in Alexandria they make a certain very cooling drinke hereof. *Rauwolfius* in his Iournall seems to describe this fruit by the name of *Bunu* ; and by the appellation,form,and faculties,he thinks it may be the *Buncho* of *Avicen*,and *Buncha* of *Rhafis*,to *Almanfor*. *Cluf.*

6   This is a kinde of Cardamome, and by diuers it is thought to be the true *Amomum* of the Antients :and to this purpose *Nicolas Morogna* a phyfitian of Verona hath written a treatife which is set forth at the later end of *Pona's* description of Mount Baldus, to which I refer the Curious. Thefe cods or berries(whether you pleafe to call them) grow thicke clustering together, they are round,and commonly of the bignesse of a cherry;the outer skin is tough,fmoother,whiter, & leffe crested than that of the Cardamome:within this filme lie the feeds clustering together,yet with a thin filme parted into three : the particular feeds are cornered, fomwhat fmoother and larger than those of Cardamomes,but of the same aromatick tafte,and of a brown colour. Their temperature and faculties may be referred to thofe of Cardamomes.

7.   8.   This with the next infuing are by *Clufius* set forth by the names I here giue you them, though,as he faith,neither of them agree with the *Amomum* of *Diofcorides*, they were only branches set thick with leaues,hauing neither any obferuable fmell or tast:they were fent to the learned and diligent Apothecarie *Walarandus Dourez* of Lyons, from Ormuz the famous mart and port towne in the Perfian Bay.

9   Thofe that accompanied the renowned Sir *Francis Drake* in his Voyage about the World, light

11 *Cacao*.   Small Coco's.   ‡ 14 *Guanobanus*.   Tree Melon.

12 *Cucciophora*.   Quince Dates.

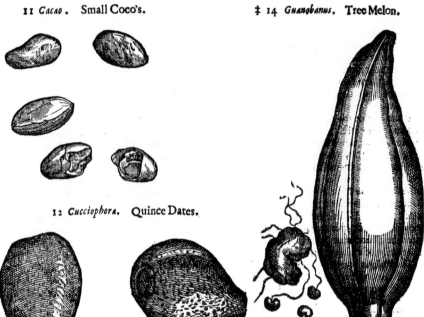

‡ 13 *Baruce. Arara. Orukoria. Cropiet*.   ‡ 15 *Ananas*. The Pinia or Pine Thistle.

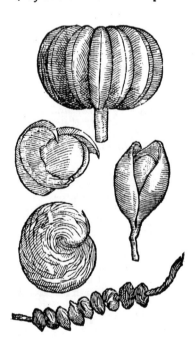

light vpon a certaine defart Iſland, wherein grew many very tall trees, and looking for ſomething a-
mongſt theſe to refreſh themſelues, amongſt others they obſerued ſome bigger than Okes, hauing
leaues like thoſe of the Bay tree, thicke and ſhining, not ſnipt about the edges, their fruit was longiſh like to the ſmall acornes of the Ilex or Holme-Oke, but without any cup, yet couered with a
thin ſhel of an aſh colour, and ſomtimes blacke, hauing within it a longiſh white kernell wrapped
in a thin peeling, being without any manifeſt taſt. They when they found it, though much oppreſt
with hunger, yet durſt not taſte thereof, leſt it ſhould haue bin poiſonous : but afterward comming
to the Iſland Beretina, not far from this, they found it to abound with theſe trees, and learned that
the fruit was not poiſonous, but might be eaten : whereupon afterwards they in want of other victuals, boiled ſome as they doe peaſe, and ground others into floure, wherewith they made puddings.
They found this tree alſo in the Molucco's.

10   The firſt expreſſed in this table is the Mad plum, or as *Cluſius* had rather term it, the Mad
Nut, for he cals them *καπυε Μανικα*, or *Inſana Nuces*. The Hollanders finding them in their return from
the Eaſt Indies, and eating the kernels, were for a time diſtracted, and that variouſly, according to
the particular temperature of each that ate of them ; as you may ſee in *Cluſius Exot. lib. 2. cap. 26.*
This was round, little more than two inches about, with a ſhell not thicke, but ſufficiently ſtrong,
browniſh on the out ſide, and not ſmooth, but on the in ſide of a yellowiſh colour, and ſmooth, containing a membranous ſtone or kernell couered with a black pulp, in form and bigneſſe not much
vnlike a Bullas or Sloe, hauing a large white ſpot on the lower part whereas it was faſtned to the
ſtalke : vnder the pulp lay the kernell, ſomwhat hard, and of an Aſh colour : the footſtalk was ſhort
and commonly caried but one fruit, yet ſometimes they obſerued two growing together : the tree
whereon this fruit grew was of the bigneſſe of a Cherry tree, hauing long narrow leaues like thoſe
of the Peach tree : the other fruit figured in the ſecond place was of a browniſh yellow color, ſomwhat bigger, but not vnlike a ſmall nut, an inch long, and ſomewhat more about, ſmaller below, and
bigger aboue, and as it were parted into foure, beeing very hard and follid. Of this ſee more in the
fourteenth place.

11   The *Cacao* is a fruit well knowne in diuers parts of America, for they in ſome places vſe it
in ſtead of mony, and to make a drink, of which, though bitter, they highly eſteem. The trees which
beare them are but ſmall, hauing long narrow leaues, and will onely grow well in places ſhadowed
from the Sun. The fruit is like an Almond taken out of the huske, and it is couered with a thinne
black skin, wherein is contained a kernell obliquely diuided into two or three parts, browniſh, and
diſtinguiſhed with aſh coloured veines, of an aſtringent and vngratefull taſte.

12   This, which *Cluſius* had from *Cortuſus* for the fruit of *Bdellium*, is thought to be the *Cuci* of
Pliny, and is the *Cuciophera* of *Matthiolus*, and by that name our Author had it in this Chapter. The
whole fruit is of the bigneſſe of a Quince, and of the ſame colour, with a ſweet and fibrous fleſh ,
vnder which is a nut of the bignes of a large walnut or ſomwhat more, almoſt of a triangular form,
bigger below, and ſmaller aboue, well ſmelling, of a darke aſh colour, with a very hard ſhell, which
broken, there is therein contained an hard kernell of the colour and hardneſſe of marble, hauing an
hollowneſſe in the middle, as much as may contain an Haſell nut.

13   In this table are foure ſeuerall fruits deſcribed by *Cluſius, Exot. lib. 2. ca. 21.* The firſt is called *Baruce*, and is ſaid to grow vpon a high tree in Guyana called Hura. It conſiſted of many Nuts
ſome inch long, ſtrongly faſtned or knit together, each hauing a hard wooddy ſhel, falling into two
parts, containing a round and ſmooth kernell couered with an aſh coloured filme.

They ſay the Natiues there vſe this fruit to purge and vomit.   A

The ſecond called Arara growes in Kaiana, but how it is not knowne : it was ſome inch long, couered with a skin ſufficiently hard and blacke, faſtened to a long and rugged ſtalke that ſeemed to
haue carried more than one fruit : the kernell is black, and of the bigneſſe of a wilde Olive.

The Natiues vſe the decoction hereof to waſh maligne vlcers, and they ſay the kernel will looſe B
the belly.

The third named *Orukoria* is the fruit of a tree in Wiapock called *Iurnwa*. They vſe this to cure
their wounds, dropping the juice of the fruit into them. This fruit is flat, almoſt an inch broad and
two long, writhen like the cod of the true Cytiſus, but much bigger, very wrinkled, of an aſh color,
containing a ſmooth ſeed.

The fourth called *Cropiot* is a ſmall and ſhriveled fruit, not much vnlike the particular joints of
the Ethiopian Pepper.

The Savages vſe to take it mixt amongſt their Tabaco, to aſſwage head-ache: there were diuers C
of them put vpon a ſtring (as you may ſee in the figure) the better to dry them.

14   This, which by *Cluſius* and *Lobel* is thought to be the *Guanabanus* mentioned by *Scaliger, Exerc. 281. part. 6.* is a thick fruit ſome foot and halfe long, couered with a thick and hard rind, freezed
ouer with a ſoft downineſſe like as a Quince is, but of a greeniſh colour, with ſome veines, or rather
furrowes running alongſt it as in Melons: the lower end is ſomewhat ſharpe, at the vpper end it is
faſtned

faſtned to the boughes with a firme hard and fibrous ſtalke: this fruit containeth a whitiſh pulpe, which the Ethiopians vſe in burning feuers to quench the thirſt, for it hath a pleaſant tartneſſe: this dried becomes friable, ſo that it may be brought into pouder with ones fingers, yet retains it's aciditie: in this pulp lie ſeeds like little kidnies or the ſeeds of the true *Anagyris*, of a blacke ſhining colour, with ſome fibres comming out of their middles: theſe ſowne brought fotth a plant hauing leaues like the Bay tree, but it died at the approch of Winter. *Cluſ.*

15   *Ananas Pinias*, or Pine Thiſtle, is a plant hauing leaues like the *Aizoon aquaticum*, or Water

‡ 16 *Fabæ Ægyptiæ affinis.*

‡ 19 *Fructus tetragonus.* The ſquare Coco.

‡ 17 *Coxeo Cypote. Amygdalæ Peruanæ.*
Almonds of Peru.

‡ 20 *Arboris laniferæ ſiliqua.*
A cod of the Wooll-bearing tree.

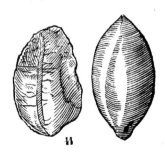

‡ 18 *Buenas Noches.*

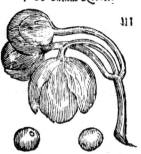

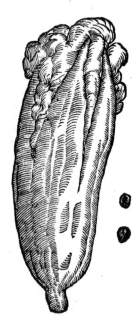

Sengreene,

Sengreene, ſomewhat ſharpe and prickly about the edges : the ſtalke is round, carrying at the top therof one fruit of a yellowiſh colour when it is ripe, of the bignes of a Melon, couered with a ſcale-like rinde : the ſmell is gratefull, ſomewhat like that of the Malecotone : at the top of the fruit, and ſometimes below it come forth ſuch buds as you ſee here preſented in the figure, which they ſet in the ground and preſerue the kind by in ſtead of ſeed:the meat of this fruit is ſweet & very pleaſant of taſte, & yeelds good nouriſhment; there are certaine ſmall fibres in the meat thereof, which though they do not offend the mouth, yet hurt they the gums of ſuch as too frequently feed thereon.

16   The forme of this is ſomwhat ſtrange, for it is like a large Poppy head cut off nigh the top: the ſubſtance thereof was membranous and wrinckled, of a browniſh colour, very ſmooth : the cir-cumference at the top is about nine inches, and ſo it growes ſmaller and ſmaller euen to the ſtalke, which ſeems to haue carried a floure whereto this fruit ſucceeded:the top of the fruit was euen, and in it were orderly placed 24. concauities, in each whereof was contained a little Nut like an Acorn, almoſt an inch long, and as much thick; the vpper part was of a browniſh colour, and the kernel with-in was all rank and mouldy. *Cluſius* could learne neither whence this came, nor how it grew, but with a great deale of probability thinks it may be that which the Antients deſcribed by the name of *Fa-ba Ægyptia.*

17   The former of theſe two *Cluſius* receiued by the name of *Coxco Cypote*, that is, the Nut Cypote: It is of a dusky browne colour, ſmooth, and ſhining, but on the lower part of an aſh colour, rough, which the Painter did not well expreſſe in drawing the figure. The 2. he receiued by the name of *Almendras del Peru*, (i.) Almonds of Peru : the ſhell was like in colour and ſubſtance to that of an almond, and the kernell not vnlike neither in ſubſtance nor taſte : yet the forme of the ſhell was dif-ferent, for it was triangular, with a backe ſtanding vp, and two ſharpe ſides, and theſe very rough.

18   This was the fruit of a large kinde of *Convolvulus* which the Spaniards called *Buenas noches*, or Good-night, becauſe the floures vſe to fade as ſoone as night came. The ſeeds were of a ſooty co-lour as big as large Peaſe, being three of them contained in a skinny three cornered head. You may ſee more hereof in *Cluſius Exot.lib.2.cap.*18.

19   This is the figure of a ſquare fruit which *Cluſius* conjectures to haue been ſome kind of In-dian Nut or Coco : it was couered with a ſmooth rinde, was ſeuen inches long, and a foot and halfe about, being foure inches and a halfe from ſquare to ſquare.

20   About Bantam in the Eaſt-Indies growes a tall tree ſending forth many branches, which are ſet thicke with leaues long and narrow, bigger than thoſe of Roſemary:it carries cods ſix inches long, and fiue about, couered with a thin skin, wrinkled and ſharp pointed, which open themſelues from below into fiue parts, and are full of a ſoft woolly or Cottony matter, wherewith they ſtuffe cuſhions, pillowes and the like, and alſo ſpin ſome for certaine vſes : amongſt the downe lie blacke ſeeds like thoſe of Cotton, but leſſe and not faſtned to the downe.

21   This which *Cluſ.*calls *Palma ſaccifera*, or the Bag Date, becauſe it carries the figure of an Hip-pocras bag, was found in a deſart Iſland in the Antlanticke ocean, by certaine Dutch mariners who obſerued whole woods thereof : theſe bags were ſome of them 22. or more inches long, and ſome ſeuen inches broad in the broadeſt place, ſtrongly woue with threds croſſing one another, of a brow-niſh yellow colour. Theſe ſachels (as they report who cut them from the tree) were filled with fruit of the bigneſſe of a Walnut, huske and all:within theſe were others, as round as if they had bin tur-ned, and ſo hard that you could ſcarce breake them with a hammer:in the midſt of theſe were white kernels, taſting at the firſt ſomewhat like pulſe, but afterwards bitter like a Lupine.

22   The tree which carries this rough cod is very large, as I haue been told by diuers:ſome who ſaw it in Perſia, and others that obſerued it in Mauritius Iſland. *Cluſius* alſo notes that they haue bin brought from diuers places : the cod is ſome three inches long, and ſome two inches broad, of a du-skie red colour, and all rough and prickly : in theſe cods are contained one, two, or more round nuts or ſeeds of a grayiſh aſh-colour, hauing a little ſpot on one ſide, where they are faſtened to the cod they are exceeding hard, and difficult to break, but broken they ſhew a white kernel very bitter and vnpleaſant of taſte. I haue ſeen very many and haue ſome of theſe, and ſome haue offered to ſel them for Eaſt-Indian *Beazor*, whereto they haue ſome ſmall reſemblance, though nothing in faculty like them (if I may credit report, which I had rather doe than make triall) for I haue been told by ſome that they are poyſonous : and by others, that they ſtrongly procure vomit.

23   The long cod expreſſed in this figure is called in the Eaſt-Indies (as *Cluſius* was told) *Kayo baka*, it was round, the thickneſſe of ones little finger, and ſix inches long:the rinde was thick, black, hard and wrinkled, and it contained a hard pulpe of a ſouriſh taſte, which they affirme was eatable.

The other was a cod of ſome inch and halfe long, and ſome inch broad, membranous, rough, and of a browniſh colour, ſharp pointed, and opening into two parts, and diſtinguiſhed with a thin filme into foure cels, wherein were contained ſcarlet Peare faſhioned little berries, hauing golden ſpots eſpecially in the middles. This growes in Braſile, and as *Cluſius* was informed was called *Daburi.*

24   In the ſecond place of the tenth figure and deſcription in this chapter you may finde the

‡ 21 *Palmaſaccifera.*
The Sachell Date.

‡ 22 *Lobus Echinatus.*
Beazor Nuts.

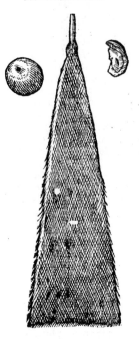

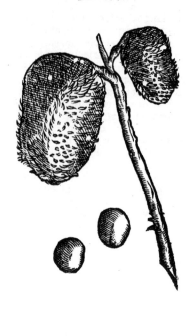

‡ 23 *Kaic baka.*
Daburi.

‡ 24 *Nucula Indica racemoſa.*
The Indian, or rather Ginny Nut.

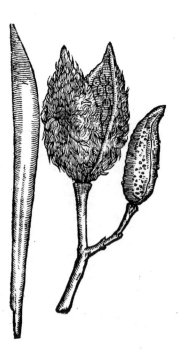

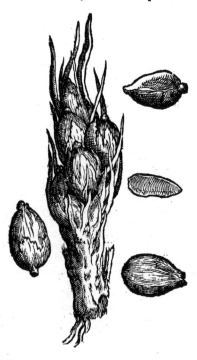

‡ 25 *Fructus squamosi.*
Scalie fruits.

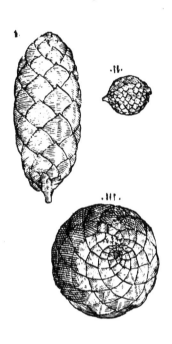

‡ 26 *Fructus alij Exotici.*
Other strange fruits.

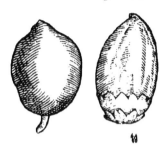

‡ 26 *Fructus alij Exotici.*
Other strange fruits.

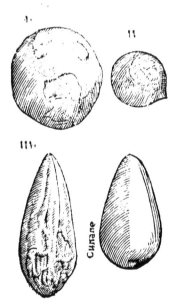

single Nut here figured, described, & set forth; but this figure better expresses the manner of growing therof : for first it presents to the view the nuts in their outer husks growing close together, then the single nuts in and out of their husks, and lastly their kernell : the shell of this nut containes in it a certaine oilie substance, somwhat resembling the oile of sweet almonds: the tree wherof this nut is the fruit growes in Ginny, & is much vsed by the people there, for they presse a liquor forth of the leaues, or else boile them in water, and this serues them in stead of wine and beere, or at least for a common drinke, of the fruit they make bread of a sweet and pleasant taste.

25 These scaily fruits are set forth by *Clusius, Exot.l.2.c.3*. The first was three inches long and two inches about, and had it in a longish hard sollid kernell, with many veines dispersed ouer it, & such kernels are somtimes polished, whereby they become white, and then their blacke veines make a fine shew, which hath giuen occasion to some imposters to put them to saile for rare and pretious stones.

The second was small, round and scaily, and the scailes turned their points downewards towards the stalke.

The third was also scaily, of the bignesse of a Walnut in his huske, with the scales very orderly placed, and of brownish colour : it had a kernel which ratled in it when it was shaken.

26 The first of the two in the former table was brought from Ginny, it was of the bignesse and shape of a plum, two inches long, and one and halfe broad, of a thicke fungous substance, somewhat wrinckled, and blackish on the outside, and within containing a certaine whitish insiped friable pulpe, wherein lay a few small seeds.

The second was so ne inch and halfe long, an inch thicke, couered with an ash coloured skin, composed within of many fibres almost like the huske of the Nut Faufell, at the lower end it stood in a double cup, and it was sharp pointed at the vpper end : in this skin was contained a kernell, or rather nut, blacke, hard, and very wrinckled not much vnlike to that of Faufell, whereto I refer it as a kinde thereof. Theſe two are treated of by *Cluſius, Exot. lib. 2. cap. 23.*

The firſt of the ſecond table (wherein are contained foure figures) was of a round forme, yet a little flat on one ſide, diſtinguiſhed vnder the black and ſhining coat wherewith it was couered, with furrowes running euery way, not vnlike to the Nut Faufell taken forth of his couer : the inner pulpe was hard and whitiſh, firſt of a ſalt, and then of an aſtringent taſte.

The ſecond of theſe was an inch long, but rather the kernell of a fruit, than a fruit it ſelfe; it was round except at the one end, and all ouer knobby (though the picture expreſſe not ſo much) there was alſo ſome ſhew of a triangular forme at the each end.

The third was two inches and a halfe long, and in the broadeſt part ſome inch and more broad: it was ſomewhat crooked, the backe high and riſing, the top narrow, and the loweſt part ſharp pointed, of an aſh-colour, with thicke and eminent nerues running alongſt the back from the top to the lower part, exptreſt with ſuch art, as if they had been done by ſome curious hand : it ſeemeth to haue bin couered ouer with another rinde, but it was worne off by the beating of the waues of the ſea vpon the ſhore.

The fruit *Cunane* figured in the fourth place of this table, was two inches long, and an inch broad at the head, and ſo ſmaller by little and little, with a backe ſtanding out, ſmooth, black, and ſhining, hauing three holes at the top, one aboue, and two below : they ſaid it grew vpon a ſmall tree called Morremor, and was yet vnripe, but when it was ripe it would be as big againe, and that the natiues where it grew (which was as I take it about Wiapock) roſt it vpon the coles, and eate it againſt the head-ache. *Cluſius* ſets forth theſe foure in his *Exot. l. 2. c. 22.* he deſcribes *Cunane cap. 21.* ‡

Chap. 160. *Of Sun-Dew, Youth wort, Ros Solis.*

1 *Ros Solis folio rotundo.*
Sun-Dew with round leaues.

2 *Ros Solis folio oblongo.*
Sun Dew with longiſh leaues.

¶ *The Description.*

1   SVn-Dew is a little herbe, and groweth very low, it hath a few leaues standing vpon slender
stems very small, something round, a little hollow, and like an eare picker, hairy and reddish as be also the stems, hauing dew and moisture vpon them at the driest time of the
yeare, and when the Sun shineth hottest euen at high noone ; and a moneth after there spring vp sub-
tle stalks, a hand bredth high, on which stand small whitish floures : the roots are very slender, and
like vnto haires.

2   The second kinde is like vnto the former, in stalkes and floures, but larger, and the leaues are
longer, and not so round, wherein consisteth the difference,

¶ *The Place.*

They grow in desart, sandie, and sunnie places, but yet watery, and seldome other where than a-
mong the white marish mosse which groweth on the ground and also vpon bogs.

¶ *The Time.*

Sun-Dew flourisheth in Summer, it floureth in May and Iune : it is to be gathered when the weather is most dry, and calme. The distilled water hereof that is drawne forth with a glasse still, is of
a glittering yellow colour like gold, and coloureth siluer put therein like gold.

¶ *The Names.*

It is called in Latine, *Ros Solis*: of diuers, *Rorella*: it is named of others, *Salsi Rosa*, of the dew which
hangeth vpon it, when the Sun is at the hottest : it is called in high Dutch, **Sondaw,** and **Sundew:**
in low Dutch, **Looptchecruit,** which in English signifieth Lustwoort, because sheepe and other cattell, if they do but only tast of it, are prouoked to lust. It is called in English, Sun Dew, Ros Solis,
Youth-woort : in the Northerne parts, Red Rot, because it rotteth sheepe ; and in Yorkshire, Moore
grasse.

¶ *The Temperature.*

It is a searing or causticke herbe, and very much biting, being hot and dry in the fourth degree.

¶ *The Vertues.*

The leaues being stamped with salt do exulcerate and raise blisters, to what part of the body so-   A
euer they be applied.

The later Physitions haue thought this herbe to be a rare and singular remedy for all those that   B
be in a consumption of the lungs, and especially the distilled water thereof : for as the herbe doth
keep and hold fast the moisture and dew, and so fast, that the extreme drying heate of the Sun cannot consume and waste away the same : so likewise men thought that herewith the naturall and ra-
dical humidity in mens bodies is preserued and cherished. But the vse therof doth otherwise teach,
and reason sheweth the contrary : for seeing it is an extreme biting herbe, and that the distilled water is not altogether without this biting quality, it cannot be taken with safety, for it hath also been
obserued, that they haue sooner perished that vsed the distilled water hereof, than those that abstai-
ned from it, and haue followed the right and ordinary course of diet.

Cattell of the female kind are stirred vp to lust by eating euen of a small quantity : which thing   C
hath greatly increased their vain opinion, without sence or reason ; for it doth not moue nor prouoke
cattell to lust, for that it increaseth the substance of the seed, but because through his sharp and biting quality it stirreth vp a desire to lust, which before was dulled and as it were asleepe.

It strengthneth and nourisheth the body, especially if it be distilled with wine, and that liquor   D
made thereof which the common people do call Rosa Solis.

If any be desirous to haue the said drinke effectuall for the purposes aforesaid, let them lay the   E
leaues of Rosa Solis in the spirit of wine, adding thereto Cinnamon, Cloues, Maces, Ginger, Nutmegs, Sugar, and a few graines of Muske, suffering it so to stand in a glasse close stopt from the aire,
and set in the Sun by the space of ten daies, then straine the same and keep it for your vse.

CHAP. 161.   *Of the Mosse of Trees.*

¶ *The Description.*

TRee Mosse hath certaine things like haires, made vp as it were of a multitude of slender leaues,
now and then all to be jagged, hackt, and finely carued, twisted and interlaced one in another,
which cleaue fast to the barkes of trees, hanging downe from the bodies : one of this kinde is more

slender

Muſcus quernus.
The Moſſe of the Oke and of other trees.

ſlender and thinne, another more thicke, another ſhorter, another longer ; all of them for the moſt being of a whitiſh colour, yet oftentimes there is a certaine one alſo which is blacke, but leſſer and thinner : the moſt commendable of them all, as *Pliny* ſaith, be thoſe that are whitiſh, then the reddiſh, and laſtly ſuch as be blacke.

¶ *The Place.*

This Moſſe is found on the Oke tree, the white and blacke Poplar tree, the Oliue tree, the Birch tree, the Apple tree, the Peare tree, the Pine tree, the wilde Pine tree, the Pitch tree, the Firre tree, the Cedar tree, the Larch tree, and on a great ſort of other trees. The beſt, as *Dioſcorides* ſaith, is that of the Cedar tree, the next of the Poplar, in which kinde the White and the ſweet ſmelling Moſſe is the chiefeſt ; the blackiſh ſort is of no account. *Matthiolus* writeth, that in Italy that Moſſe is ſweet which groweth on the Pine tree, the Pitch tree, the Fir tree, and the Larch tree, and the ſweeteſt, that of the Larch tree.

¶ *The Time.*

Moſſe vpon the trees continueth all the yeare long.

¶ *The Names.*

It is called of the Grecians βρύον: of the Latines, *Muſcus*: the Arabians and ſome Apothecaries in other countries call it *Vſnea*: in high Dutch, Moſz : in low Dutch, Moſch: the French-men, *Lu Mouſch* : the Italians, *Muſgo* : in Spaniſh, *Muſco de los arbores*: in Engliſh, Moſſe, tree Moſſe, or Moſſe of trees.

¶ *The Temperature.*

Moſſe is ſomewhat cold and binding, which notwithſtanding is more or leſſe according vnto the nature and faculty of that tree on which it groweth, and eſpecially of his barke : for it taketh vnto it ſelfe and alſo retaineth a certaine property of that barke, as of his breeder of which hee is ingendred : therefore the Moſſe which commeth of the Oke doth coole and very much binde, beſides his owne and proper faculty, it receiueth alſo the extreme binding quality of the Oke barke it ſelfe.

The Moſſe which commeth of the Cedar tree, the Pine tree, the Pitch tree, the Fir tree, the Larch tree, and generally all the Roſine trees are binding, and do moreouer digeſt and ſoften.

¶ *The Vertues.*

A   *Serapio* ſaith, that the wine in which Moſſe hath been ſteeped certain daies, bringeth ſound ſleep, ſtrengthneth the ſtomacke, ſtaieth vomiting, and ſtoppeth the belly.

B   *Dioſcorides* writeth, That the decoction of Moſſe is good for women to ſit in, that are troubled with the whites ; it is mixed with the oile of Ben, and with oiles to thicken them withall.

C   It is fit to be vſed in compoſitions which ſerue for ſweet perfumes, and that take away weariſomneſſe ; for which things that is beſt of all which is moſt ſweet of ſmell.

CHAP. 162.   *Of ground Moſſe.*

¶ *The Kindes.*

THere groweth alſo on the ſuperficiall or vppermoſt part of the earth diuers Moſſes, as alſo vpon rocks and ſtony places, and mariſh grounds, differing in forme not a little.

¶ *The Deſcription.*

1   THe common Moſſe groweth vpon the earth, and the bottome of old and antient trees, but ſpecially vpon ſuch as grow in ſhadowie woods, and alſo at the bottom of ſhadowie hedges,

1 *Muscus terrestris vulgaris.*
Common ground Mosse.

2 *Muscus terrestris scoparius.*
Beesome ground Mosse.

3. 4. *Muscus capillaris, siue Adianthum
aureum majus & minus.*
Goldilockes or golden Maiden-haire
the bigger and the lesser.

hedges and ditches, and such like places : it is very well knowne by the softnesse and length thereof, being a mosse most common, and therefore needeth not any further description.

2   Beesome Mosse, which seldome or neuer is found but in bogs and marish places, yet sometimes haue I found it in shadowie dry ditches, where the Sunne neuer sheweth his face : it groweth
vp halfe a cubit high, euery particular leafe consisting of an innumerable sort of hairy threds set vpon a middle rib, of a shining blacke colour like that of Maiden-haire, or the Capillare Mosse *Adianthum aureum,* whereof it is a kinde.

3   This kinde of Mosse, called *Muscus capillaris,* is seldome found but vpon bogs and moorish
places, and also in some shadowie dry ditches where the Sun doth not come. I found it in great
aboundance in a shadowie ditch vpon the left hand neere vnto a gate that leadeth from Hampsted
heath,

5 *Muſcus ramoſus floridus.*
Flouring branched Moſſe.

heath toward Highgate; which place I haue ſhewed vnto diuers expert Surgeons of London, in our wandering abroad for our farther knowledge in Simples. This kind of Moſſe, the ſtalkes thereof are not aboue one handfull high, couered with ſhort haires ſtanding very thicke together, of an obſcure yellow green colour; out of which ſtalkes ſpring vp ſometimes very fine naked ſtems, ſomewhat blacke, vpon the tops of which hang as it were little graines like wheat cornes. The roots are very ſlender and maruellous fine.

‡ Of this *Adianthum aureum* there are three kindes, different onely in magnitude, and that the two bigger haue many hairie threds vpon their branches, when as the leaſt hath onely three or foure cloſe to the root; and this is the leaſt of plants that I euer yet ſaw grow. ‡

4 Of this there is alſo another kinde altogether leſſer and lower. This kind of moſſe groweth in moiſt places alſo, commonly in old moſſie and rotten trees, likewiſe vpon rocks, and oftentimes in the chinks and crannies of ſtone walls.

† 5 There is oftentimes found vpon old Okes and Beeches, and ſuch like ouer-growne trees, a kind of Moſſe hauing many ſlender branches, which diuide themſelues into other leſſer branches; whereon are placed confuſedly very many ſmall threds like haires, of a greeniſh aſh-colour: vpon the ends of the tender branches ſometimes there commeth forth a floure in ſhape like vnto a little buckler or hollow Muſhrom, of a whitiſh colour tending to yellownes, and garniſhed with the like leaues of thoſe vpon the lower branches.

6 *Muſcus Pyxidatus.*
Cup or Chalice Moſſe.

6 Of this Moſſe there is another kinde, which *Lobel* in his Dutch Herbal hath ſet forth vnder the title of *Muſcus Pyxidatus*, which I haue Engliſhed, Cup Moſſe or Chalice Moſſe: it groweth in the moſt barren dry and grauelly ditch bankes, creeping flat vpon the ground like vnto Liuerwort, but of a yellowiſh white colour: among which leaues ſtart vp here and there certaine little things faſhioned like a little cup called a Beaker or Chalice, and of the ſame colour and ſubſtance of the lower leaues, which vndoubtedly may be taken for the floures: the pouder of which Moſſe giuen to children in any liquor for certaine daies together, is a moſt certaine remedy againſt that perillous malady called the Chin cough.

7 There is likewiſe found in the ſhadowie places of high mountaines, and at the foot of old

and

and rotten trees, a certaine kinde of moſſe in face and ſhew not vnlike to that kinde of Oke Ferne called *Dryopteris*. It creepeth vpon the ground, hauing diuers long branches, conſiſting of many ſmall leaues, euery particular leafe made vp of ſundry little leaues, ſet vpon a middle rib one oppoſit to another.

7 *Muſcus Filicinus.*
Moſſe Ferne.

8 *Muſcus corniculatus.*
Horned or knagged Moſſe.

9 *Muſcus denticulatus.*
Toothed Moſſe.

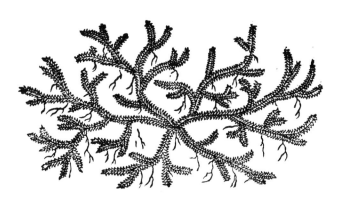

8　There is found vpon the tops of our moſt barren mountaines, but eſpecially where ſea Coles are accuſtomed to be digged, ſtone to make iron of, and alſo where ore is gotten for tinne and lead, a certaine ſmall plant: it riſeth forth of the ground with many bare and naked branches, diuiding themſelues at the top into ſundry knags like the forked hornes of a Deere, euery part whereof is of an ouerworne whitiſh colour.

‡　Our Author formerly gaue another figure and deſcription of this plant, by the name of *Holoſteum petræum*, which I haue omitted, thinking this the better. *Tragus, Lonicerus*, and *Bauhine* refer this to the Fernes; and the laſt of them calleth it *Filex ſaxatilis corniculata : Pena* and *Lobel* made it their *Holoſtium alterum : Thalius* calls it *Adianthum acroſchiſton, ſeu furcatum.* ‡

9　There is found creeping vpon the ground a certaine kinde of moſſe at the bottome of Heath and Ling, and ſuch like buſhes growing vpon barren mountaines, conſiſting as it were of ſcales made vp into a long rope or cord, diſperſing it ſelfe far abroad into ſundry branches, thruſting out

here

here and there certaine roots like threds, which take hold vpon the vpper cruſt of the earth, whe reby it is ſent and diſperſed far abroad : the whole plant is of a yellowiſh greene colour.

10　This other kinde of moſſe is found in the like places : it alſo diſperceth it ſelfe far abroad, and is altogether leſſer than the precedent, wherein conſiſts the difference.

10 *Muſcus minor denticulatus.*
Little toothed Moſſe.

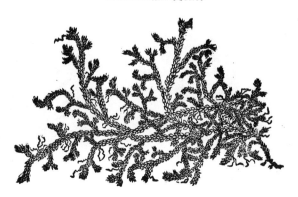

11 *Muſcus clauatus, ſiue Lycopodium.*
Club Moſſe, or Wolfe claw Moſſe.

† 12 *Muſcus clauatus folijs Cupreſſi.*
Heath Cypres.

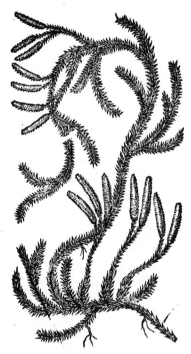

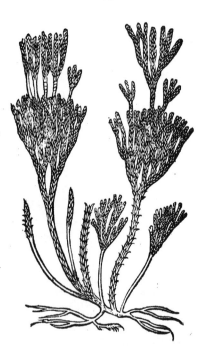

11　There is likewiſe another kinde of Moſſe, which I haue not elſewhere found than vpon Hampſted heath, neere vnto a little cottage, growing cloſe vpon the ground amongſt buſhes and brakes, which I haue ſhewed vnto diuers Surgeons of London, that haue walked thither with mee for their further knowledge in Simples, who haue gathered this kinde of moſſe, whereof ſome haue
made

13 *Muscus ex cranio humano.*
Mosse growing vpon the **skull of a man**.

Muscus ex Cra- neo Hu- maua

made them hat-bands, girdles, and also bands to tye such things as they had before gathered, for the which purpose it most fitly serued; some pieces whereof are six or eight foot long, consisting as it were of many hairie leaues set vpon a tough string, very close couched and compact together, from which is also sent forth certain other branches like the first: in sundry places there be sent down fine little strings, which serue in stead of roots, wherewith it is fastened to the vpper part of the earth, and taketh hold likewise vpon such things as grow next vnto it. There spring also from the branches bare & naked stalkes, on which grow certaine eares as it were like the catkins or blowings of the Hasell tree, in shape like a little club or the reed Mace, sauing that it is much lesser, and of a yellowish white colour, very wel resembling the claw of a Wolfe, whereof it tooke his name; which knobby catkins are altogether barren, and bring forth neither seed nor floure.

‡ 14 *Muscus paruus stellaris.*
Small Heath Mosse.

‡ 12 This, whose figure in the former edition was by our Author vnfitly put for Lauander Cotton (hauing more regard to the title of the figure in *Tabernamontanus*, than to see whether it were that which he there described) it is no other than a kind of *Muscus Clauatus*, or Club-Mosse. I, is thought to be the *Selago* mentioned by *Pliny, lib. 25. cap. 11. Tragus* and some others call it *Sauina syluestris: Turner* and *Tabernamontanus, Chamæcyparissus*: but *Bauhine* the most fitly nameth it *Muscus clauatus folijs Cypressi*: and *Turner* not vnfitly in English, Heath Cypresse. This is a low plant, and keepes greene Winter and Summer: the leaues are like those of Cypresse, bitter in taste, but without smell: it carries such eares or catkins as the former, and those of a yellowish colour: it is found growing in diuers wooddy mountainous places of Germany, where they call it **walb Sewenbaum**, or wilde Sauine. ‡

13 This kinde of Mosse is found vpon the skulls or bare scalpes of men and women, lying long in charnell houses or other places, where the bones of men and women are kept together: it growes very thicke, greene, and like vnto the short mosse vpon the trunkes of old Okes: it is thought to be a singular remedie against the falling Euill and the Chin-cough in children, if it be poudered, and then giuen in sweet wine for certaine daies together.

‡ 14 Vpon diuers heathy places in the moneth of May is to be found growing a little short Mosse not much in shape different from the first described, but much lesse, and parted at the top into star-fashioned heads. *Lobel* calls this, *Muscus in Ericetis proueniens*. ‡

¶ The

¶ *The Place.*

Their feuerall defcriptions fet forth their naturall places of growing.

¶ *The Time.*

They flourifh efpecially in the Summer moneths.

¶ *The Names.*

Goldilocks is called in high Dutch, **Widertodt, golden Wedertodt, Jung Urauwen har:** in low Dutch, **Gulden Wederdoot:** *Fuchſius* nameth it *Polytrichon Apuleij,* or *Apuleius* his Maiden-haire; neuerthelefſe *Apuleius* Maiden-haire is nothing elfe but *Diofcorides* his *Trichomanes,* called Englifh Maiden-haire; and for that caufe we had rather it fhould be termed *Muſcus Capillaris,* or hairy Moffe. This is called in Englifh, Goldilockes: it might alfo be termed Golden Moffe, or Hairy Moffe.

Wolfes claw is called of diuers Herbarifts in our age, *Muſcus terreſtris:* in high Dutch, **Beer-clap, Surtelkraut, Seilkraut:** in low Dutch, **wolffs Clauwen;** whereupon we firft named it *Lycopodium,* and *Pes Lupi* : in Englifh, Wolfes foot, or Wolfes claw, and likewife Club moffe. Moft fhops of Germany in former times did falfly terme it *Spica celtica:* but they did worfe, and were very much too blame, that vfed it in compound medicines in ftead of *Spica Celtica,* or French Spikenard : as touching the reft, they are fufficiently fpoken of in their defcriptions.

¶ *The Temperature.*

The Moffes of the earth are dry and aftringent, of a binding quality, without any heate or cold.

Goldilockes and the Wolfes clawes are temperate in heate and cold.

¶ *The Vertues.*

A   The Arabian Phyfitians doe put moffe among their cordiall medicines, as fortifying the ſtomacke, to ſtay vomit, and to ſtop the laske.

B   Moffe boiled in Wine and drunke ſtoppeth the ſpitting of bloud, piffing of bloud, the termes, and bloudy flix.

C   Moffe made into pouder is good to ſtanch the bleeding of greene and frefh wounds, and is a great helpe vnto the cure of the fame.

D   Wolfes claw prouoketh vrine, and as *Hieronymus Tragus* reporteth, waſteth the ſtone, and driueth it forth.

E   Being ſtamped and boiled in wine and applied, it mitigateth the paine of the gout.

F   Floting wine, which is now become ſlimie, is reſtored to his former goodneffe, if it be hanged in the veffell, as the fame Author teſtifieth.

† The figure formerly in the firſt place was of the *Muſcus Momanus* of *Tabern* being a ſmall kinde of *Muſcus denſiouiatus.* The fifth and fixth were both of one; and fo of the two defcriptions I haue made one more accurate, and referued the better figure.

Cʜᴀᴘ. 163. *Of Liuerwort.*

¶ *The Defcription.*

1   Liuerwort is alfo a kinde of Moffe which ſpreadeth it felfe abroad vpon the ground, hauing many vneuen or crumpled leaues lying one ouer another, as the fcales of Fifhes do, greene aboue, and browne vnderneath : amongſt thefe grow vp ſmall fhort ſtalkes, ſpred at the top like a blaſing ſtarre, and certaine fine little threds are fent downe, by which it cleaueth and ſticketh faſt vpon ſtones, and vpon the ground, by which it liueth and flourifheth.

2   The fecond kinde of Liuerwort differeth not but in ſtature, being altogether leffe, and more ſmooth or euen : the floures on the tops of the ſlender ſtems are not fo much laid open like a ſtar; but the efpeciall difference confiſteth in one chiefe point, that is to fay, this kinde being planted in a pot, and fet in a garden aboue the ground, notwithſtanding it ſpitteth or caſteth round about the place great ſtore of the fame fruit, where neuer any did grow before.

‡   Of this fort which is ſmall, and oftentimes found growing in moiſt gardens among Beares-eares, and fuch plants, when they are kept in pots, there are two varieties, one hauing little ſtalkes fome inch long, with a ſtarre-fafhioned head at the top : the other hath the like tender ſtalke, and a round head at the top thereof. ‡

3   This is found vpon rockes and ſtony places, as well neere vnto the fea, as further into the land : it groweth flat vpon the ſtones, and creepeth not far abroad as the ground Liuerwort doth, it only reſteth it felfe in ſpots and tufts fet here and there, of a duſty ruffet colour aboue, and blackifh vnderneath : among the crumpled leaues rife vp diuers ſmall ſtems, whereupon do grow little ſtar-like floures of the colour of the leaues : it is often found at the bottom of high trees growing vpon

high

1 Hepatica terreſtris.
Ground Liuer-wort.

2 Hepatica ſtellata & vmbellata.
Small Liuerwort with ſtarry and round heads.

3 Hepatica petræa.
Stone Liver-wort.

high mountains, eſpecially in ſhadowy places.

¶ *The Place.*

This is often found in ſhadowy and moiſt places, on rockes and great ſtones laid by the highway, and in other common paths where the ſun beams do ſeldome come, and where no tra-ueller frequenteth.

¶ *The Time.*

It brings forth his blaſing ſtarres and leaues oftentimes in Iune and Iuly.

¶ *The Names.*

It is called of the Grecians, Λειχήν: of the La-tines, *Lichen* : of ſome, βρύον, that is to ſay, *Muſ-cus* or Moſſe, as *Dioſcorides* witneſſeth. It is na-med in ſhops *Hepatica* , yet there are alſo many other herbs named *Hepatica*, or Liverworts ; for difference whereof this may fitly be called *He-patica petræa* or ſtone Liverwort , hauing taken that name from the Germans, who call this Li-ver-woort, 𝕾𝖙𝖊𝖞𝖓 𝕷𝖊𝖇𝖊𝖗𝖐𝖗𝖆𝖚𝖙 : and in Low-Dutch, 𝕾𝖙𝖊𝖊𝖓 𝕷𝖊𝖚𝖊𝖗𝖈𝖗𝖚𝖕𝖙, In Engliſh, Liver-wort.

¶ *The Temperature.*

Stone Liverwort is cold and dry, and ſomwhat binding.

¶ *The Vertues.*

It is ſingular good againſt the inflammations of the liuer, hot and ſharp agues, and tertians that **A**
proceed of choler.

B   *Diofcorides* teacheth, That Liverwort being applied to the place ftancheth bleeding, takes away all inflammations, and is good for a tettar or ringworme called in Greek Ἀιχὴς: and that it is a reme-die for them that haue the yellow jaundice, euen that which commeth by inflammation of the li-uer; and that it alfo quencheth the inflammations of the tongue.

Chap. 164.

Of Lung-wort or wood Liver-wort, and Oifter-greene.

1 *Lichen arborum.*
Tree Lung-wort.

2 *Lichen marinus.*
Sea Lung-wort, or Oifter-greene.

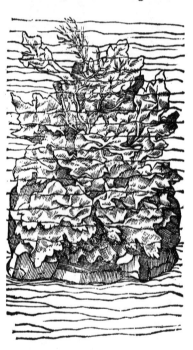

¶ *The Defcription.*

1   TO Liver-wort there is ioyned Lung-wort, which is alfo another kinde of Moffe, drier, broader, of a larger fize, and fet with skales: the leaues hereof are greater, and diuerfly folded one in another, not fo fmooth, but more wrinckled, rough and thicke almoft like a Fel or hide, and tough withall; on the vpper fide whitifh, and on the nether fide blackifh or dufty: it feems after a fort to refemble lungs or lights.

2   . This kind of sea Moffe is an herby matter much like vnto Liver-wort, altogether without ftalke or ftem, bearing many green leaues very vneuen or crumpled, and full of wrinkles, fomewhat broad, not much differing from leaues of crifpe or curled Lettuce. This growes vpon rocks within the bowels of the fea, but efpecially amongft oifters, and in great plenty among thofe oifters called Wal-fleet oifters. It is very well known euen to the poore oifter women which carry oifters to fell vp and down, who are greatly defirous of the faid moffe for the decking and beautifying of their oifters, to make them fell the better. This moffe they call Oifter-green.

‡ 3   The branches of this elegant plant are fome handfull or better high, fpred abroad on e-uery fide, and only confifting of fundry fingle roundifh leaues, whereto are faftned fometimes one,

                                           fometimes

ſometimes two or more ſuch leaues, ſo that the whole plant conſiſts of branches made vp of ſuch round leaues, faſtned together by diuers little and very ſmall threds : the lower leaues which ſticke faſt to the rockes are of a browniſh colour, the other of a whitiſh or light greene colour, ſmooth and ſhining. This growes vpon rocks in diuers parts of the Mediterranean. *Cluſius* ſets it forth by the name of *Lichen marinus*; and he receiued it from *Imperato* by the name of *Sertuloria*: and *Cortuſus* had it from Corſica by the title of *Corallina latifolia*; and he called it *Opuntia Marina*, hauing reference to that mentioned by *Theophraſt.lib.1.cap.12.Hiſt.plant.* ‡

‡ 3 *Lichen marinus rotundifolius*.   Round leaued Oiſter-weed.

4 *Quercus marina*.
Sea Oke or Wrack.

‡ 4 *Quercus marina varietas*.
A varietie of the ſea Oke or Wrack.

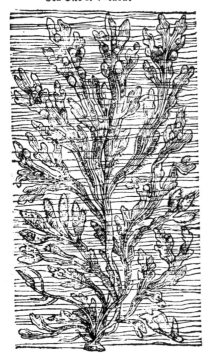
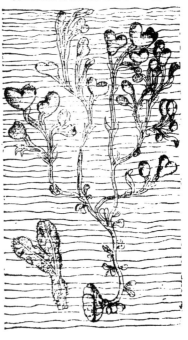

4   There is alſo another ſort of Sea Weed found vpon the drowned rocks which are naked and bare of water at euery tyde. This ſea Weed groweth vnto the rocke, faſtned vnto the ſame at one   end,

end,being a ſoft herby plant very ſlippery, inſomuch as it is an hard matter to ſtand vpon it with̄-
out falling : it rampeth far abroad,and here and there is ſet with certaine puſt vp tubercles or bladders full of winde,which giue a cracke being broken : the leafe it ſelfe doth ſomwhat reſemble the
Oken leafe,whereof it took his name *Quercus marina* or ſea Oke : of ſome,Wrack,and Crow gall.
His vſe in phyſick hath not bin ſet forth,and therefore this bare deſcription may ſuffice.

‡ 5 *Quercus marina ſecunda.*   ‡ 6 *Quercus marina tertia.*
     Sea Thongs.   The third ſea Wrack.

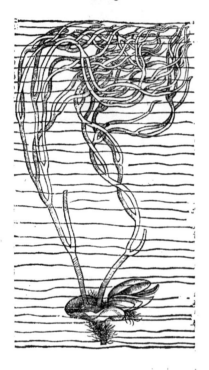 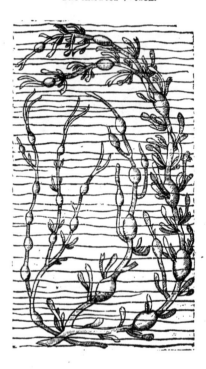

‡   Of this *Quercus marina* or *Fucus* there are diuers ſorts,whereof I will giue you the figures and
a briefe hiſtory. The firſt of theſe is onely a varietie of the laſt deſcribed, differing therefrom in
the narrowneſſe of the leaues,and largeneſſe of the ſwolne bladders.

5   This growes to the length of fiue or ſix foot,is ſmooth and membranous, being ſome halfe
fingers bredth,and variouſly diuided like wet parchment or leather cut into thongs. This hath no
ſwoln knots or bladders like as the former ; and is the *Fucus marinus ſecundus* of *Dodonæus.*

6   This Wrack or Sea Weed hath long and flat ſtalks like the former,but the ſtalks are thicke
ſet with ſwolne knots or bladders,out of which ſomtimes grow little leaues ; in other reſpects it
is not vnlike the former kindes. *Dodonæus* makes this his *Fucus marinus* 3.

7   The leaues of this other Wracke,which *Dodonæus* makes his *Fucus marinus* 4. are narrower,
ſmaller,and much diuided ; and this hath either none or very few of thoſe ſwolne bladders which
ſome of the former kindes haue.

8   This,which *Lobel* cals *Alga marina*,hath jointed black branched creeping roots of the thickneſſe of ones finger,which end as it were in diuers eares or hairy awnes, compoſed of whitiſh hairy
threds ſomwhat reſembling Spikenard ; from the tops wherof come forth leaues long,narrow,ſoft,
and graſſe-like,firſt greene,but white when they are dry. It growes in the ſea as the former. They
vſe it in Italy and other hot countries to pack vp glaſſes with, to keepe them from breaking.

9   Of this Tribe are diuers other plants ; but I will onely giue you the hiſtory of two more,
which I firſt obſerued the laſt yeare,going in company with diuers London Apothecaries to finde
out Simples,as far as Margate in the Iſle of Tenet ; and whoſe figures (not before extant that I
know of) I firſt gaue in my Iournall or enumeration of ſuch Plants as wee there and in other places found. The firſt of theſe by reaſon of his various growth is by *Bauhine* in his *Prodromus* diſtinguiſhed

guiſhed into two, and deſcribed in the ſecond and third places. The third he calls *Fucus longiſsimo latiſsimo, craſsoque folio*, and this is marked with the figure 1. The ſecond he calls *Fucus arboreus Polyſchides*, and this you may ſee marked with the figure 2. This Sea weed (as I haue ſaid) hath a various face; for ſomtimes from a fibrous root, which commonly groweth to a pibble ſtone, or faſtned to a rock, it ſends forth a round ſtalk ſeldom ſo thick as ones little finger, and about ſome half foot in length, at the top whereof grows out a ſingle leafe, ſomtimes an ell long, and then it is about the bredth of ones hand, ending in a ſharpe point, ſo that it very well reſembles a two edged ſword. Somtimes from the ſame root come forth two ſuch faſhioned leaues, but then commonly they are leſſer. Otherwhiles at the top of the ſtalke it diuides it ſelfe into eight, nine, ten, twelue, more or fewer parts, and that iuſt at the top of the ſtalk, and theſe neuer come to that length that the ſingle leaues do. Now this I iudge to be the *Fucus polyſchides* of *Bauhine*. That theſe two are not ſeueral kinds I am certain; for I haue marked both theſe varieties from one and the ſame root, as you may ſee them here expreſt in the figure. At Margate where they grow they cal them ſea Girdles: which name well befits the ſingle one; and the diuided one they may call ſea Hangers, for if you hang the tops downward they do reaſonable wel reſemble the old faſhioned ſword hangers. Thus much for their ſhape: now for their colour, which is not the ſame in all; for ſome are more greene, and theſe can ſcarce be dried; otherſome are whitiſh, and theſe do quickly dry, and then both in colour and ſubſtance are ſo like Parchment, that ſuch as know them not would at the firſt view take them to be nothing els. This is of a glutinous ſubſtance and a little ſaltiſh taſt, and diuers haue told me they are good meat, being boiled tender, and ſo eaten with butter, vineger and pepper.

‡ 7 *Quercus marina quarta.*
Iagged ſea Wrack.

† 8 *Alga.*
Graſſe Wrack.

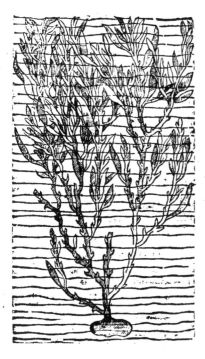 

10　This which I giue you in the tenth place is not figured or deſcribed by any that as yet I haue met with; wherefore I gaue the Figure and Deſcription in the fore-mentioned journal, which I will here repeat. This is a very ſucculent and fungous plant, of the thickneſſe of ones thumb; it is of a dark yellowiſh colour, and buncheth forth on euery ſide with many vnequall tuberoſities or knots: whereupon Mr *Tho. Hicks* being in our company did fitly name it ſea Ragged ſtaffe. We did not obſerue it growing, but found one or two plants thereof ſome foot long apiece.

‡ 9 *Fucns phaſganoides & polyſchides.*
Sea Girdle and Hangers.

‡ 10 *Fucus ſpongioſus nodoſus.*
Sea ragged Staffe.

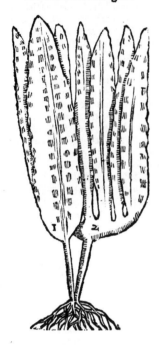

‡ 11 *Conferva.*   Hairy Riuer-weed.

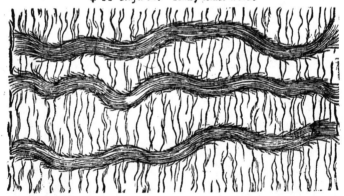

11   In ſome ſlow running waters is to be found this long green hairy weed, thought to be the *Conferva* of *Pliny* : it is made vp only of long hairy green threds, thick thrummed together without any particular ſhape or faſhion, but only following the current of the ſtream. ‡

¶ *The Place, Time, and Names.*

It growes vpon the bodies of old Okes, Beech, and other wild trees, in dark and thick woods: it is oftentimes found growing vpon rocks, and alſo in other ſhadowy places.

It flouriſheth eſpecially in the ſummer moneths.

It taketh his name *Pulmonaria* of the likeneſſe it hath with lungs or lights, called in Latine *Pulmones*, of ſome *Lichen* : it is called in high-Dutch, **Lungenkraut** : in low-Dutch, **Longhencruyt**: in French, *Herbe à Poulmon* : in Engliſh, Lungwort, and wood Liverwort.

¶ *The Temperature.*

This ſeemeth to be cold and dry.

¶ *The*

¶ *The Vertues.*

It is reported, That ſhepheards and certain horſeleeches do with good ſucceſſe giue the pouder   A
hereof with ſalt vnto their ſheep and other cattell that be troubled with the cough and be broken
winded.

Lungwort is much commended of the learned Phyſitions of our time, againſt the diſeaſes of the   B
lungs, eſpecially for the inflammations and vlcers of the ſame, being brought into pouder & drunk
with water.

It is likewiſe commended for bloudy and green wounds, and for vlcers in the ſecret parts, and   C
alſo to ſtay the reds.

Moreouer, it ſtops the bloudy flix and other flixes and ſcourings, either vpwards or downwards   D
eſpecially if they proceed of choler : it ſtayeth vomiting, as ſome ſay, and alſo ſtops the belly.

Oiſter-green fried with egges and made into a tanſie and eaten, is a ſingular remedie to ſtreng-   E
then the weakneſſe of the backe.

Chap. 165.   *Of Sea Moſſe or Coralline.*

¶ *The Kindes.*

THere be diuers ſorts of Moſſe, growing as well within the bowels of the ſea, as vpon the rocks,
diſtinguiſhed vnder ſundry titles.

1 *Muſcus marinus, ſiue Corallina alba.*
White Coralline or Sea moſſe.

† 2 *Muſcus marinus albidus.*
White Sea Moſſe.

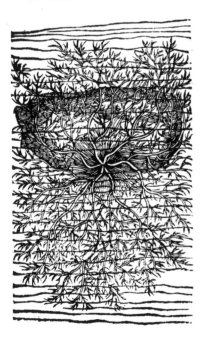

3 *Corallina Anglica.*
Engliſh Coralline.

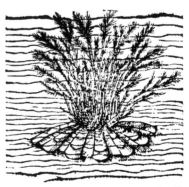

¶ *The*

¶ *The Deſcription.*

1   THis kinde of Sea moſſe hath many ſmall ſtalks finely couered or ſet ouer with ſmall leaues, very much cut or jagged euen like the leaues of Dill, but hard, and of a ſtony ſubſtance.

2   The ſecond is much like the former, yet not ſtony, but more finely cut, and growing more vpright, branching it ſelfe into many diuiſions at the top, growing very thicke together, and in great quantity, out of a piece of ſtone, which is faſhioned like an hat or ſmall ſtony head, whereby it is faſtned vnto the rocks.

3   This third kind of Sea moſſe is very well known in ſhops by the name *Corallina* : it yeeldeth forth a great number of ſhoots in ſhape much like to Corall, being full of ſmall branches diſperſed here and there, diuerſly varying his colour according to the place where it is found, beeing in ſome places red, in others yellow and of an herby colour; in ſome gray or of an aſh colour, in otherſome very white.

4   This Sea moſſe is ſomwhat like the former, but ſmaller, and not ſo plentiful where it groweth, proſpering alwaies vpon ſhels, as of Oiſters, Muſcles, and Scallops, as alſo vpon rolling ſtones in the bottom of the water, which haue tumbled down from the high clifs and rocks, notwithſtanding the old Prouerb, That rolling ſtones neuer gather moſſe.

4 *Corallina minima.*
The ſmalleſt Coralline.

5 *Muſcus Corallinus, ſiue Corallina montana.*
Corall moſſe, or mountain Coralline.

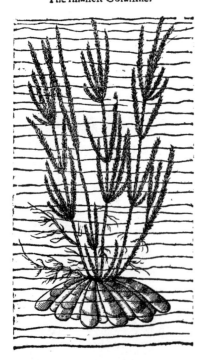

5   There is found vpon the rocks and mountains of France bordering vpon the Mediterranian ſea, a certain kind of Coralline which in theſe parts hath not bin found : it growes in maner like a branch of Corall, but altogether leſſer, of a ſhining red colour, and of a ſtony ſubſtance.

‡   I know not what our Author meant by this deſcription, but the plant which here is figured out of *Tabern.* (and by the ſame title he hath it) is of a moſſe growing vpon Hampſtead heath, and moſt ſuch places of England: it growes vp ſome two or three inches high, and is diuided into very many little branches ending in little threaddy chiues ; all the branches are hollow, and of a very light white dry ſubſtance, which makes it ſomwhat reſemble Coralline, yet is it not ſtony at all. ‡

6   There is found vpon the Rocks neere vnto Narbone in France, and not far from the ſea, a

kinde

6 Fucus marinus tenuifolius.
Fenell Coralline or Fenell Moſſe.

‡ *7 Fucus ferulaceus.*
Sea Fenell.

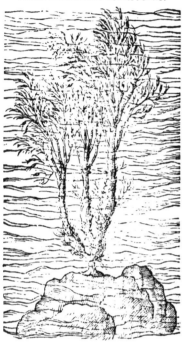

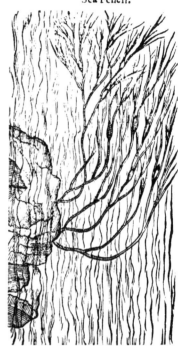

‡ 8 *Fucus tenuifolius alter.*
Bulbous ſea Fenell.

‡ *9 Muſcus marinus Cluſij.*
Branched ſea Moſſe.

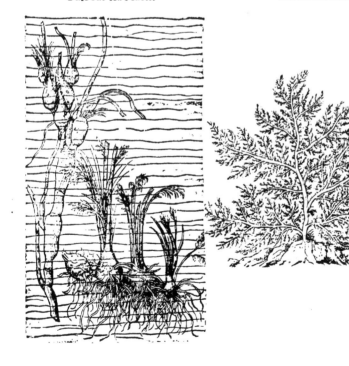

kinde of Coralline : it groweth vp to the form of a ſmall ſhrub,branched diuerſly,wheron do grow ſmall graſſe-like leaues very finely cut or jagged like to Fenell,yet are they of a ſtony ſubſtance as are the reſt of the Corallines,of a darke ruſſet colour.

‡ 7   This growes alſo in the like places,hauing many ſmall long Fennel-like diuided leaues vpon ſtalks ſome foot long,with ſome ſwelling eminences here and there ſet in the diuiſions of the leaues. This by *Lobel* is called by the name I here giue you it.

8   This alſo hath fine cut leaues like thoſe of Fenel,but much leſſe and ſhorter,of a faire green colour : theſe grow vp from round tuberous roots,which, together with the fibres they ſend forth, are of a blackiſh colour : the ſtalks alſo are tuberous and ſwoln as in other plants of this kinde.It growes in the ſea with the former. *Dodonæus* calls this *Fucus marinus virens tenuifolius.*

9   This kind of Sea moſſe grows ſome foure or more inches long,diuided into many branches which are ſubdiuided into ſmaller,ſet with leaues finely jagged like thoſe of Camomil;at firſt ſoft flexible and tranſparent,green below,and purpliſh aboue : being dried it becommeth rough & fragile like as Coralline. It growes in the Mediterranian ſea.

10   This Sea moſſe is a low little excreſcence,hauing ſomwhat broad cut leaues growing many from one root : in the whole face it reſembles the moſſe that growes vpon the branches of okes and other trees,and is alſo white and very like it,but much more brittle. This by *Dodonæus* is called *Muſcus marinus tertius.*

‡ 10 *Muſcus marinus 3.Dod.*   Broad leafed Sea Moſſe.

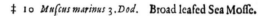
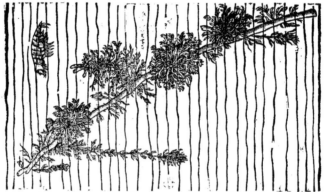

‡ 11 *Abies marina Belgica Cluſ. Cluſius* his ſea Firre

11   Vpon the rocks and ſhels of ſea fiſhes are to be found diuers ſmall plants reſembling others that grow on the land ; and *Cluſius* ſaith,vpon the coaſt of the Low-countries he obſerued one that very much reſembled the Fir tree,hauing branches growing orderly on both ſides, but thoſe very brittle and ſmall,ſeldome exceeding an handfull in height,and couered as it were with many ſmall ſcales. He obſerued others that reſembled Cypreſſe trees,and other branches that reſembled Tamariske or Heath. ‡

¶ *The*

¶ *The Place.*

These Mosses grow in the sea vpon the rocks, and are often found vpon oister shels, muskle shels, and stones. I found great plenty thereof vnder Reculuers and Margate, in the Isle of Tenet, and in other places alongst the sands from thence to Dover.

¶ *The Time.*

The time answereth the other Mosses, and are found at all times of the yeare.

¶ *The Names.*

Sea mosse is called in Greeke, ϲϖοϲϲϋϋϖϋ; in Latine, *Muscus marinus*: of the Apothecaries, Italians, and French men, *Corallina*: in Spanish, *Atalharquiana yerva*: in high-Dutch, 𝔐eermoſz: in low-Dutch, 𝖅ee 𝔐oſch: in English, Sea mosse, and of many Corallina, after the Apothecaries word, and it may be called Corall mosse. The titles distinguish the other kindes.

¶ *The Temperature.*

Corallina consisteth, as *Galen* saith, of an earthy and waterish essence, both of them cold: for by his taste he bindeth, and being applied to any hot infirmitie it also euidently cooleth. The earthy essence of this mosse hath in it also a certain saltnesse, by reason wherof it likewise dries mightily.

¶ *The Vertues.*

Dioscorides commendeth it to be good for the gout which hath need of cooling.   A

The later physitians haue found by experience that it killeth wormes in the belly. It is giuen to B this purpose to children in the weight of a dram or thereabouts.

That which cleaueth to Corall, and is of a reddish colour, is of some preferred and taken for the C best: they count that which is whitish to be the worser, notwithstanding in the French ocean, the Britain, the Low-country, and the German ocean sea there is scarce found any but the whitish Coalline, which the nations neere adioyning do effectually vse.

Chap. 166.   *Of Corall.*

1 *Corallium rubrum.*
Red Corall.

2 *Corallium nigrum, siue Antipathes.*
Blacke Corall.

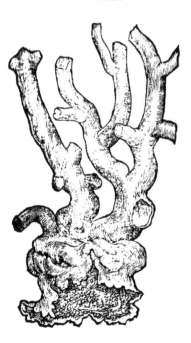

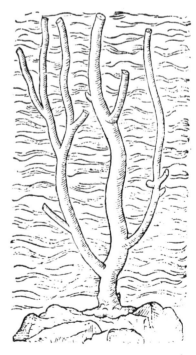

3 *Corallium album.*   White Corall.

4 *Corallium album alterum.*   ‡ 5 *Coralloides albicans.*
The other white or yellow Corall.   Baſtard whitiſh Corall.

¶ *The Deſcription.*

1   ALthough Corall be a matter or ſubſtance euen as hard as ſtones; yet I think it not amiſſe
to inſert it here next the moſſes, and the rather for that the kinds thereof do ſhew themſelues as well in the maner of their growing, as in their place & form, like to the Moſſes. This later
age wherein we liue, hath found more kinds therof than euer were known or mentioned among the
old Writers. Some of theſe Corals grow in the likeneſſe of a ſhrub or ſtony matter, others in a
ſtraight form with crags and joints, ſuch as we ſee by experience ; which being ſo well known, and
in ſuch requeſt for phyſick, I will not ſtand to deſcribe ; only remember this, That ſome Corall is
of a pale colour, others red, and ſome white.

2   The black Corall growes vpon rocks neere the ſea about Maſſilia, in maner of the former, ſauing that this is of a ſhining black colour, very ſmooth, growing vp rather like a tree than a ſhrub.

3   The white Corall is like the former, growing vpon the rocks neere the ſea, and in the Weſt
parts of England about S. *Michaels* mount ; but the branches hereof are ſmaller and more brittle,
finelier diſperſed into a number of branches, of a white colour.   4 The.

‡ 6 *Coralloides rubens.*
Reddish bastard Corall.

‡ 7 *Spongia marina alba.*
White Spunge.

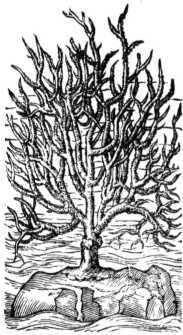

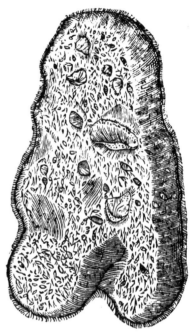

‡ 8 *Spongia infundibuli forma.*
Funnell fashioned Spunge.

‡ 9 *Spongia ramosa.*
Branched Spunge.

4   The fourth and last groweth also vpon the Westerne rocks of the sea, and in the place afore-named, and varieth his colour, sometimes waxing white, sometimes yellow, and sometimes red.

‡   5   This growes vp with many branches some two or three handfulls high : the inner part is a hard wooddy substance, which is couered ouer with a white and hard stony matter, so that it much resembleth White Corall, but that it is neither so thicke, hard, nor smooth, but is rough

and bends eaſily without breaking,which Corall will not do. *Lobel* calls this *Corallina alba*, it growes in the Mediteranian ſea,and vpon the Coaſts of Spaine.

6   This in all reſpects is like the laſt deſcribed, the colour excepted, which is a darke red, and therefore better reſembles the red Corall. *Cluſius* refers both theſe to the *Quercus marina* mentioned by *Theophraſtus,Hiſt.plant.7.cap.4.* ‡

7   There is found growing vpon the rockes neere vnto the ſea,a certaine matter wrought together,of the forme or froth of the ſea,which we call ſpunges, after the Latine name,which may very fitly be inſerted among the ſea Moſſes,wherof to write at large would greatly increaſe our volume, and little profit the Reader,conſidering we haſten to an end, and alſo that the vſe is ſo well knowne vnto all : therefore theſe few lines may ſerue vntill a further conſideration,or a ſecond Edition. ‡ Spunges are not like the *Alcyonium*, that is, an accidentall matter wrought together of the froth of the ſea,as our Author affirmes,but rather of a nobler nature than plants, for they are ſaid to haue ſence,and to contract themſelues at the approach of ones hand that comes to cut them vp, or for feare of any other harme-threatning object, and therefore by moſt Writers they are referred to the ζωσφυτον:which ſome render *Plantanimalia*, that is, ſuch as are neither abſolute plants, nor liuing creatures,but participate of both : they grow of diuers ſhapes and colours vpon the Rockes in the Mediterranean,as alſo in the Archipelago,or Ægean ſea.

8   *Cluſius* obſerued one yet adhering to the ſtone whereon it grew, which in ſhape reſembled a funnell,but in ſubſtance was like another Spunge.

9   There is alſo to be found vpon our Engliſh coaſt a ſmall kinde of ſpunge caſt vp by the ſea, and this is alſo of different ſhapes and colour, for the ſhape it is alwaies diuided into ſundry branches,but that after a different manner ; and the colour is oft times browniſh, and otherwhiles gray or white. *Lobel* makes it *Conſerua marina genus*. ‡

¶ *The Place.*

The place of their growing is ſufficiently ſpoken of in their ſeuerall deſcriptions.

¶ *The Time.*

The time anſwereth the other kindes of ſea Moſſes.

¶ *The Names.*

Corallium rubrum is called in Engliſh, red Corrall. *Corallium nigrum*, blacke Corrall. *Corallium album*,white Corrall.

¶ *The Temperature.*

Corrall bindeth, and meanely cooleth : it clenſeth the ſcars and ſpots of the eies, and is very effuall againſt the iſſues of bloud,and eaſeth the difficulty of making water.

¶ *The Vertues.*

A   Corrall drunke in wine or water,preſerueth from the ſpleene; and ſome hang it about the neckes of ſuch as haue the falling ſickeneſſe,and it is giuen in drinke for the ſame purpoſe.

B   It is a ſoueraigne remedy to drie, to ſtop, and ſtay all iſſues of bloud whatſoeuer in man or woman,and the Dyſentery.

C   Burned Corrall drieth more than when it is vnburned, and being giuen to drinke in water, it helpeth the gripings of the belly,and the griefes of the ſtone in the bladder.

D   Corrall drunke in wine prouoketh ſleepe : but if the patient haue an ague, then it is with better ſucceſſe miniſtred in water, for the Corrall cooleth, and the water moiſtneth the body, by reaſon whereof it reſtraineth the burning heate in agues,and repreſſeth the vapours that hinder ſleepe.

Chap. 167.   *Of Muſhrumes, or Toadſtooles.*

¶ *The Kindes.*

SOme Muſhrumes grow forth of the earth ; other vpon the bodies of old trees, which differ altogether in kindes. Many wantons that dwell neere the ſea,and haue fiſh at will, are very deſirous for change of diet to feed vpon the birds of the mountaines; and ſuch as dwell vpon the hills or champion grounds,do long after ſea fiſh;many that haue plenty of both,do hunger after the earthy excreſcences, called Muſhromes : whereof ſome are very venomous and full of poyſon, others not ſo noiſome ; and neither of them very wholeſome meate ; wherefore for the auoiding of the venomous quality of the one, and that the other which is leſſe venomous may be diſcerned from it, I haue thought good to ſet forth their figures with their names and places of growth. ‡ Becauſe the booke is already growne too voluminous,I will only giue you the figures of ſuch as my Author hath here mentioned, with ſome few others,but not trouble you with any more hiſtory, yet diſtinguiſh betweene ſuch as be eatable,and thoſe that be poiſonous, or at leaſt not to be eaten ; for the firſt figured among the poiſonous ones, is that we call Iewes-eare, which hath no poyſonous facultie in it. *Cluſius* (all whoſe figures I could haue here giuen you) hath written a peculiar tract of theſe baſtard plants, or excreſcences, where ſuch as deſire it may finde them ſufficiently diſcourſed of. ‡

¶ *The*

1　*Fungi vulgatiſſimi eſculenti.*　Common Muſhrums to be eaten.

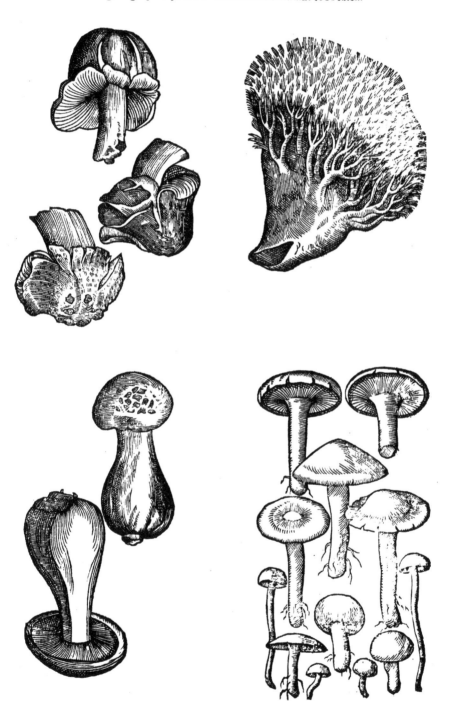

¶ *The Deſcription.*

1   GRound Muſhrums grow vp in one night, ſtanding vpon a thicke and round ſtalke, like vnto a broad hat or buckler, of a very white colour vntil it begin to wither, at what time it loſeth his faire white, declining to yellowneſſe : the lower ſide is ſomewhat hollow, ſet or decked with fine gutters, drawne along from the middle centre to the circumference or round edge of the brim.

2   All Muſhroms are without pith, rib, or veine : they differ not a little in bigneſſe and colour, ſome are great, and like a broad brimmed hat ; others ſmaller, about the bigneſſe of a ſiluer coine called a doller : moſt of them are red vnderneath; ſome more, ſome leſſe; others little or nothing red at a'l : the vpper ſide which beareth out, is either pale or whitiſh, or elſe of an ill-fauoured colour like aſhes (they commonly call it Aſh-colour) or elſe it ſeemeth to be ſomewhat yellow.

There is another kinde of Muſhroms called *Fungi parui lethales galericulati :* in Engliſh, deadly Muſhrums, which are faſhioned like vnto an hood, and are moſt venomous and full of poyſon.

There is a kinde of Muſhrom called *Fungus Clypeiformis lethalus,* that is alſo a deadly Muſhrum, faſhioned like a little buckler.

There is another kinde of Muſhrum, which is alſo moſt venomous and full of poyſon, bearing alſo the ſhape of a buckler, being called *Fungus venenatus Clypeiformis :* in Engliſh, the ſtinking venomous Muſhrum.

2 Fungi lethales, aut ſaltem non eſculenti.
Poyſon Muſhrums, or at the leaſt ſuch as are not vulgarly eaten.

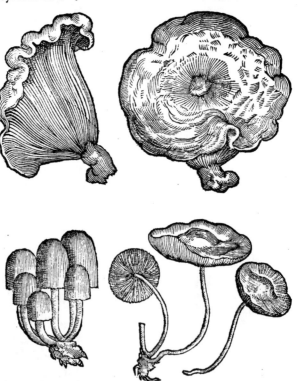

The Muſhrums or Toodſtooles which grow vpon the trunkes or bodies of old trees, very much reſembling *Auricula Iudæ,* that is, Iewes eare, doe in continuance of time grow vnto the ſubſtance of wood, which the Fowlers doe call Touchwood, and are for the moſt halfe circuled or halfe round, whoſe vpper part is ſomewhat plaine, and ſometimes a little hollow, but the lower part is plaited or purſed together. This kinde of Muſhrum the Grecians do call *ἀγαρικ* and is full of venome or poyſon as the former, eſpecially thoſe which grow vpon the Ilex, Oliue, and Oke trees.

There is likewiſe a kinde of Muſhrum called *Fungus Fauaginoſus,* growing vp in moiſt and ſhadowie woods, which is alſo venomous, hauing a thicke and tuberous ſtalke, an handfull high, of a duskiſh colour; the top whereof is compact of many ſmall diuiſions, like vnto the hony combe.

There

Fungus ſambucinus, ſiue Auricula Iudæ.   Iewes eares.

Fungi lethales, ſiue non eſculeuti.   Poyſonous Muſhrums.

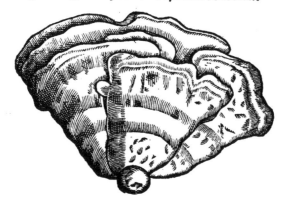

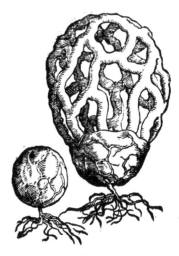

There is alſo found another, ſet forth vnder the title *Fungus virilis penis erecti forma*, which wee Engliſh, Prick Muſhrum, taken from his forme.

3   *Fungus orbicularis*, or *Lupi crepitus*, ſome do call it *Lucernarum fungus*: in Engliſh, Fuſſe balls, Pucke Fuſſe, and Bulfiſts, with which in ſome places of England they vſe to kill or ſmolder their Bees, when they would driue the Hiues, and bereaue the poore Bees of their meat, houſes and liues: theſe are alſo vſed in ſome places where neighbours dwell far aſunder, to carry and reſerue fire from place to place, wherof it tooke the name, *Lucernarum Fungus*: in forme they are very round, ſticking and cleauing vnto the ground, without any ſtalks or ſtems; at the firſt white, but afterwards of a duskiſh colour, hauing no hole or breach in them, whereby a man may ſee into them, which being troden vpon doe breath forth a moſt thin and fine pouder, like vnto ſmoke, very noiſome and hurtfull vnto the eies, cauſing a kinde of blindneſſe, which is called Poor-blinde, or Sandblinde.

Fungi lethales, ſiue non eſculenti.   Poiſonous Muſhrums.

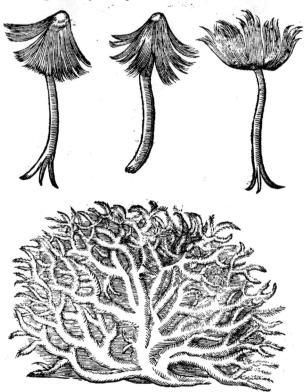

There is another kinde of *Fungus*, or Muſhrum, which groweth in moiſt medowes, and by ditch ſides, fiue or ſix inches high, couered ouer with a skin like a piece of ſheepes leather, of a ruſſet colour; which being taken away there appeareth a long and white ſtumpe, in forme not much vnlike to an handle, mentioned in the title, or like vnto the white or tender ſtalke of Aron, but greater: this kinde is alſo full of venome and poiſon.

There is likewiſe a kinde of Muſhrum, with a certaine round excreſcence, growing within the earth, vnder the vpper cruſt or face of the ſame, in dry and grauelly grounds in Pannonia and the Prouinces adjoining which do cauſe the ground to ſwel, and be full of hils like Mole-hils. The people where they grow, are conſtrained to dig them vp and caſt them abroad like as we do Mole-hils, ſpoiling their grounds, as Mole-hils are hurtfull vnto our ſoile: theſe haue neither ſtalkes, leaues, fibres nor ſtrings annexed or faſtened vnto them, and for the moſt part are of a reddiſh colour, but within of a whitiſh yellow: the Grecians haue called this tuberous excreſcence *Idna*, and the Latines *Tubera*: the Spaniards doe call them *Turmas tierra*: in Engliſh wee may call them Spaniſh Fuſſe-bals.

¶ *The Place.*

Muſhrums come vp about the roots of trees, in graſſie places of medowes, and Ley Land newly turned

Fungus fauiginosus.
Hony-comb'd Mushrome.

Fungus Virilis Penis effigie.
Pricke Mushrome.

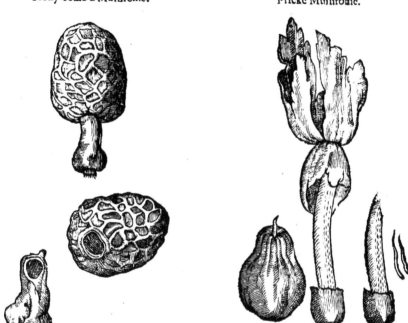

Tubera terræ.
Fusse-balls,or Puck-fists.

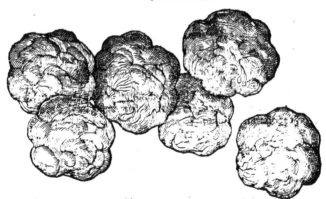

turned; in woods also where the ground is sandy, but yet dankish : they grow likewise out of wood,
forth of the rotten bodies of trees, but they are vnprofitable and nothing worth. Poisonsome mushroms, as *Dioscorides* saith, groweth where old rusty iron lieth, or rotten clouts, or neere to serpents
dens, or roots of trees that bring forth venomous fruit. Diuers esteeme those for the best which
grow in medowes, and vpon mountaines and hilly places, as *Horace* saith, *lib.fer.2.satyr.4.*

———— *pratensibus optima fungis.*
Natura est, alijs male creditur.

The medow Mushroms are in kinde the best,
It is ill trusting any of the rest.

N The

¶ *The Time.*

Diuers come vp in Aprill, and laft not till May, for they flourifh but whileft Aprill continues: others grow later, about Auguft; yet all of them after raine, and therefore they are found one yeare fooner, and another later. Mufhroms, faith *Pliny*, grow in fhoures of raine: they come of the flime of trees, as the fame Author affirmeth.

¶ *The Names.*

They are called in Latine, *Fungi*: in Greeke, μύκητη: in Italian, *Fonghi*: in Spanifh, *Hungos, Cguumenos*: in French, *Campinion*, which word the Low-Country men alfo vfe, and call them **Camper-noellen**: in high Dutch, **Schwemme, Pfifferling**: in Englifh, Mufhroms, Toadftooles, and Paddock-ftooles.

The mufhroms that come vp in Aprill are called in Latine of fome, *Spongiola*: of the Italians, *Prignoli*: and in high Dutch, **Morchel**.

They that are of a light red are called of fome *Boleti*, among the later ones which rife and fall awae in feuen daies. The white, or thofe which bee fomewhat yellow, are called in Latine, *Suilli*: which the later Phyfitions name *Porcini*, or Swine mufhrums. *Suilli*, faith *Pliny*, are dried, being hanged vpon rufhes, which are thruft through them. The dry ones are in our age alfo eaten in Bohemia and Auftria: they that grow by the roots of Poplar trees are called of the Latines, *Populnei*, Poplar mufhrums.

Puffes-fifts are commonly called in Latine, *Lupi crepitus*, or Wolfes fifts: in Italian, *Vefcie de Lupo*: in Englifh, Puffes-fifts, and Fuffe-balls in the North. *Pliny* nameth them *Pezica*, as though he fhould fay, flat.

Tree mufhroms be called in Greeke, μύκητη: in Latine, *Fungi arborum*, and *Fungi arborei*: in Englifh, tree Mufhroms, or Touch-wood: in high Dutch alfo **Schwemme.** They are all thought to be poifonfome, being inwardly taken. *Nicander* writeth, that the mufhroms of the Oliue tree, the Ilex tree, and of the Oke tree bring death.

¶ *The Temperature and Vertues.*

A   *Galen* affirmes, that they are all very cold and moift, and therefore to approach vnto a venomous and mutthering faeultie, and ingender a clammy, pituitous, and cold nutriment if they be eaten. To conclude, few of them are good to be eaten, and moft of them do fuffocate and ftrangle the eater. Therefore I giue my aduice vnto thofe that loue fuch ftrange and new fangled meates, to beware of licking honey among thornes, leaft the fweetneffe of the one do not counteruaile the fharpneffe and pricking of the other.

B   Fuffe-balls are no way eaten: the pouder of them doth dry without biting: it is fitly applied to merigals, kibed heeles, and fuch like.

C   In diuers parts of England where people dwell farre from neighbours, they carry them kindled with fire, which lafteth long: whereupon they were called *Lucernarum Fungi*.

D   The duft or pouder thereof is very dangerous for the eies, for it hath been often feene, that diuers haue beene pore-blinde euer after, when fome fmall quantitie thereof hath beene blowne into their eies.

E   The country people do vfe to kill or fmother Bees with thefe Fuffe-balls, being fet on fire, for the which purpofe it fitly ferueth.

F   ‡ The fungous excrefcence of the Elder, commonly called a Iewes eare, is much vfed againft the inflammations and all other foreneffes of the throat, being boyled in milke, fteeped in beere, vineger, or any other conuenient liquor. ‡

CHAP. 168.
Of great Tooth-wort, or Clownes Lung-wort.

¶ *The Defcription.*

1   THere is often found among the Mufhroms a certaine kinde of excrefcence confifting of a jelly or foft fubftance, like that of the Mufhroms, and therefore it may the more fitly be here inferted: it rifeth forth of the ground in forme like vnto *Orobanche*, or the Broome-Rape, and alfo in fubftance, hauing a tender, thicke, tuberous, or mif-fhapen body, confifting as it were of fcales like teeth (whereof it tooke his name) of a dufty fhining colour tending to purple. The ftalke rifeth vp in the middle, garnifhed with little gaping hollow floures like thofe of Satyrion; on the outfide of an ouerworne whitifh colour: the whole plant refembleth a rude forme of that

jelly,

gellie, or slimie matter, found in the fields, which we call the falling of stars : the root is small and tender.

  2   There is also another sort hereof found, not differing from the precedent : the chiefe difference consisteth in that, that this plant is altogether lesser ; ‡ and hath a root diuersly diuaricated like Corall, white of colour, full of juyce, and without any fibers annexed therto ‡ ; in other respects like.

  1 *Dentaria major Matthioli.*
Great Toothwort, or Lungwort.

                         2 *Dentaria minor.*
                         Little Lungwort.

¶ *The Place.*

  These plants do grow at the bottom of Elme trees, and such like, in shadowie places : I found it growing in a lane called East-lane, on the right hand as ye go from Maidstone in Kent vnto Cockes Heath, halfe a mile from the towne; and in other places thereabout : it doth also grow in the fields about Croidon, especially about a place called Groutes, being the land of a Worshipfull Gentleman called Mr *Garth*, and also in a Wood in Kent neere Crayfoot, called Rowe, or Rough-hill : it groweth likewise neere Harwood in Lancashire, a mile from Whanley, in a wood called Talbot banke.

¶ *The Time.*

They flourish in May and Iune.

¶ *The Names.*

There is not any other name extant, more than is set forth in the description.

¶ *The Temperature and Vertues.*

  There is nothing extant of the faculties hereof, either of the antient or later writers: neither haue we any thing of our owne experience ; onely our country women do call it Lungwort, and do vse it against the cough, and all imperfections of the lungs : but what benefit they reape thereby I know not ; neither can any of judgement giue me further instruction thereof.

CHAP. 169. *Of Saunders.*

¶ *The Kindes.*

THe antient Greekes haue not knowne the sorts of Saunders : *Garcias* and others describe three ; *Album*, *Rubrum*, and *Pallidum :* which in shops is called *Citrinum*.

                                        ¶ *The*

¶ *The Deſcription.*

1 THe Saunders tree groweth to the bigneſſe of the Walnut-tree, garniſhed with many goodly branches; whereon are ſet leaues like thoſe of the Lentiske tree, alwaies greene; among which come forth very faire floures, of a blew colour tending to blackeneſſe; after commeth the fruit of the bigneſſe of a Cherry, greene at the firſt, and blacke when it is ripe; without taſte, and ſteady to fall downe with euery little blaſt of winde: the timber or wood is of a white colour, and a very pleaſant ſmell.

2 There is likewiſe another which groweth very great, the floures and fruit agree with the other of his kinde: the wood is of a yellowiſh colour, wherein conſiſteth the difference.

‡ 3 The third ſort which we call Red-Saunders is a very hard and ſollid wood, hauing little or no ſmell, the colour thereof is very red, it groweth not in thoſe places where the other grow, neither is the forme of the tree deſcribed by any that I know of, it is frequently vſed to colour ſauces, and for ſuch like vſes. ‡

¶ *The Place.*

The white and yellow Saunders grow naturally, and that in great abundance, in an Iſland called Timor, and alſo in the Eaſt-Indies beyond the riuer Sanges or rather Ganges, which the Indians call *Hanga*, and alſo about Iaua, where it is of better odour than any that growes elſe-where.

The red Saunders growes within the riuer Ganges, eſpecially about Tanaſarim, and in the marriſh grounds about Charamandell: *Auicen Scrapio*, and moſt of the Mauritanians call it by a corrupt name, *Sandal*: in Timor, Malaca, and in places neere adjoyning, *Chandama*: in Decan and Guzarate, *Sercanda*: in Latine, *Sandalum* and *Santalum*, adding thereto for the colour *album*, *flauum*, or *Citrinum*, and *Rubrum*, that is, white, yellow, and red Saunders.

¶ *The Time.*

Theſe trees which are the white and yellow Saunders grow greene Winter and Summer, and are not one knowne from another, but by the Indians themſelues, who haue taken very certaine notes and markes of them, becauſe they may the more ſpeedily diſtinguiſh them when the Mart commeth.

¶ *The Names.*

Their names haue been ſufficiently ſpoken of in their deſcriptions.

¶ *The Temperature.*

† Yellow and white Saunders are hot in the third degree, and dry in the ſecond. The red Saunders are not ſo hot.

¶ *The Vertues.*

A   The Indians do vſe the decoction made in water, againſt hot burning agues, and the ouermuch flowing of the menſes, *Eryſipelas*, the gout, and all inflammations, eſpecially if it be mixed with the juyce of Night-ſhade, Houſleeke, or Purſlane.

B   The white Saunders mixed with Roſe-water, and the temples bathed therewith, ceaſeth the pain of the megrim, and keepeth backe the flowing of humors to the eies.

C   *Auicen* affirmeth it to be good for all paſſions of the heart, and maketh it glad and merry, and therefore good to be put into colliſes, jellies, and all delicate meates which are made to ſtrengthen and reuiue the ſpirits.

D   ‡ Red Saunders haue an aſtrictiue and ſtrengthtning faculty, but are not cordiall as the other two, they are vſed in diuers medicines and meates both for their faculty and pleaſing red colour which they giue to them. ‡

CHAP. 170.   *Of Stony wood, or wood made Stone.*

¶ *The Deſcription.*

AMong the wonders of England this is one of great admiration, and contrary vnto mans reaſon and capacitie, that there ſhould bee a kinde of Wood alterable into the hardneſſe of a ſtone called Stonie Wood, or rather a kinde of water, which hardneth Wood and other things, into the nature and matter of ſtones. But wee know that the Workes of God are

wonderfull,

Lignum Lapideum, ſiue in Lapides conuerſum.
Stonie wood, or wood made ſtones.

wonderfull, if we doe but narrowly ſearch the leaſt of them, which wee daily behold ; much more if wee turne our eies vpon thoſe that are ſeldome ſeene, and knowne but of a few, and that of ſuch as haue painfully trauelled in the ſecrets of Nature. This ſtrange alteration of Nature is to bee ſeene in ſundry parts of England and Wales, through the qualities of ſome waters and earth, which change ſuch things into ſtones as do fall therein, or which are of purpoſe for triall put into them. In the North part of England there is a Well neere vnto Knaesborough, which will change any thing into ſtone, whether it be wood, timber, leaues of trees, moſſe, leather gloues, or ſuch like. There be diuers places in Bedfordſhire, Warwickſhire, and Walls, where there is ground of that qualitie, that if a ſtake be driuen into it, that part of the ſtake which is within the ground will be a firme and hard ſtone, and all that which is aboue the ground retaineth his former ſubſtance and nature. Alſo my ſelfe being at Rougby (about ſuch time as our fantaſticke people did with great concourſe and multitudes repaire and run headlong vnto the ſacred Wells of *Newnam Regis*, in the edge of Warwickſhire, as vnto the water of life, which could cure all diſeaſes) I went from thence vnto theſe Wells, where I found growing ouer the ſame a faire Aſh-tree, whoſe boughes did hang ouer the ſpring of water, whereof ſome that were ſeare and rotten, and ſome that of purpoſe were broken off, fell into the water and were all turned into ſtones. Of theſe boughes or parts of the tree I brought into London, which when I had broken in pieces, therein might be ſeene, that the pith and all the reſt was turned into ſtones ; yea many buds and flourings of the tree falling into the ſaid water, were alſo turned into hard ſtones, ſtill retaining the ſame ſhape and faſhion that they were of before they were in the water. I doubt not but if this water were proued about the hardening of ſome Confections Phyſicall, for the preſeruation of them, or other ſpeciall ends, it would offer greater occaſion of admiration for the health and benefit of mankinde, than it doth about ſuch things as already haue beene experimented, tending to very little purpoſe.

<hr />

CHAP. 171.

Of the Gooſe tree, Barnacle tree, or the tree bearing Geeſe.

Britanica Conchæ anatifera.
The breed of Barnacles.

¶ *The Deſcription.*

HAuing trauelled from the Graſſes growing in the bottome of the fenny waters, the Woods, and mountaines, euen vnto Libanus it ſelfe ; and alſo the ſea, and bowels of the ſame, wee are arriued at the end of our Hiſtory ; thinking it not impertinent to the concluſion of the ſame, to end with one of the maruels of this land (we may ſay of the World.) The hiſtory whereof to ſet forth according to the worthineſſe and raritie thereof, would not only require a large and peculiar volume, but alſo a deeper ſearch into the bowels of Nature, than my intended purpoſe will ſuffer me to wade into, my ſufficiencie alſo conſidered ; leauing the Hiſtory thereof rough hewen, vnto ſome excellent man, learned in the ſecrets of nature, to be both fined and refined : in the meane ſpace take it as it falleth out, the naked and bare truth, though vnpoliſhed. There are found in the North parts of Scotland and the Iſlands adiacent, called Orchades, certaine trees whereon do grow certaine ſhells of a white colour tending to ruſſet, wherein are contained little liuing creatures : which ſhells in time of maturity doe open, and out of them grow thoſe little liuing things, which falling into the water do become fowles, which we call Barnacles ; in the North of England, brant Geeſe ; and in Lancaſhire, tree Geeſe : but the other that do fall vpon the land periſh and come to nothing. Thus much by the writings of others, and alſo from the mouthes of people of thoſe parts, which may very well accord with truth.

But what our eies haue ſeene, and hands haue touched we ſhall declare. There is a ſmall Iſland in Lancaſhire called the Pile of Foulders, wherein are found the broken pieces of old and bruiſed ſhips, ſome whereof haue beene caſt thither by ſhipwracke, and alſo the trunks and bodies with the branches of old and rotten trees, caſt vp there likewiſe ; whereon is found a certaine ſpume or froth that in time breedeth vnto certaine ſhells, in ſhape like thoſe of the Muskle, but ſharper pointed, and of a whitiſh colour ; wherein is contained a thing in forme like a lace of ſilke finely wouen as it were together, of a whitiſh colour, one end whereof is faſtned vnto the inſide of the ſhell, euen as the fiſh, of Oiſters and Muskles are : the other end is made faſt vnto the belly of a rude maſſe or lumpe, which in time commeth to the ſhape and forme of a Bird : when it is perfectly formed the ſhell gapeth open, and the firſt thing that appeareth is the foreſaid lace or ſtring ; next come the legs of the bird hanging out, and as it groweth greater it openeth the ſhell by degrees, til at length it is all come forth, and hangeth onely by the bill : in ſhort ſpace after it commeth to full maturitie, and falleth into the ſea, where it gathereth feathers, and groweth to a fowle bigger than a Mallard, and leſſer than a Gooſe, hauing blacke legs and bill or beake, and feathers blacke and white, ſpotted in ſuch manner as is our Magpie, called in ſome places a Pie-Annet, which the people of Lancaſhire call by no other name than a tree Gooſe : which place aforeſaid, and all thoſe parts adioyning do ſo much abound therewith, that one of the beſt is bought for three pence. For the truth hereof, if any doubt, may it pleaſe them to repaire vnto me, and I ſhall ſatisfie them by the teſtimonie of good witneſſes.

Moreouer, it ſhould ſeeme that there is another ſort hereof ; the Hiſtory of which is true, and of mine owne knowledge : for trauelling vpon the ſhore of our Engliſh coaſt betweene Douer and Rumney, I found the trunke of an old rotten tree, which (with ſome helpe that I procured by Fiſhermens wiues that were there attending their husbands returne from the ſea) we drew out of the water vpon dry land : vpon this rotten tree I found growing many thouſands of long crimſon bladders, in ſhape like vnto puddings newly filled, before they be ſodden, which were very cleere and ſhining ; at the nether end whereof did grow a ſhell fiſh, faſhioned ſomewhat like a ſmall Muskle, but much whiter, reſembling a ſhell fiſh that groweth vpon the rockes about Garnſey and Garſey, called a Lympit : many of theſe ſhells I brought with me to London, which after I had opened I found in them liuing things without forme or ſhape, in others which were neerer come to ripeneſſe I found liuing things that were very naked, in ſhape like a Bird : in others, the Birds couered with ſoft downe, the ſhell halfe open, and the Bird ready to fall out, which no doubt were the Fowles called Barnacles. I dare not abſolutely auouch euery circumſtance of the firſt part of this hiſtory, concerning the tree that beareth thoſe buds aforeſaid, but will leaue it to a further conſideration ; howbeit, that which I haue ſeene with mine eies, and handled with mine hands, I dare confidently auouch, and boldly put downe for verity. Now if any will object that this tree which I ſaw might be one of thoſe before mentioned, which either by the waues of the ſea or ſome violent wind had beene ouerturned as many other trees are ; or that any trees falling into thoſe ſeas about the Orchades, will of themſelues beare the like Fowles, by reaſon of thoſe ſeas and waters, theſe being ſo probable conjectures, and likely to be true, I may not without prejudice gaineſay, or indeauour to confute.

‡ The Barnakle whoſe fabulous breed my Author here ſets downe, and diuers others haue alſo

also deliuered,were found by some Hollanders to haue another originall,and that by egs as other birds haue : for they in their third voiage to find out the North-East passage to China & the Molucco's,about the 80 degree and eleuen minutes of Northerly latitude, found two little Islands, in one of which they found aboundance of these geese sitting vpon their eggs,of which they got one Goose,and tooke away sixty eggs,&c. *Vide Pontani,rerum & vrb. Amsteloram.hist.lib.2.cap.22*.Now the shels out of which these birds were thought to fly,are a kind of *Balanus marinus* ; and thus *Fabius Columna*,in the end of his *Phytobasanos*,writing *piscium aliquot historia*, iudiciously proueth. To whose opinion I wholly subscribe, and to it I refer the Curious. His asseueration is this ; *Conchas vulgò Anatiferas,non esse fructus terrestres,neque ex iis Anates oriri ; sed Balani marinæ speciem*. I could haue said somthing more hereof,but thus much I think may serue,together with that which *Fabius Columna* hath written vpon this point. ‡

¶ *The Place.*

The bordes and rotten plankes whereon are found these shels breeding the Barnakle, are taken vp in a small Island adioyning to Lancashire,halfe a mile from the main land, called the Pile of Foulders.

¶ *The Time.*

They spawn as it were in March and Aprill ; the Geese are formed in May and Iune, and come to fulnesse of feathers in the moneth after.

And thus hauing through Gods assistance discoursed somewhat at large of Grasses, Herbes, Shrubs, Trees, and Mosses, and certaine Excrescences of the earth, with other things moe, incident to the historie thereof,we conclude and end our present Volume, with this wonder of England. For the which Gods Name be euer honored and praised.

FINIS.

AN APPENDIX OR ADDITION OF
certaine Plants omitted in the former Hiſtory.

The Preface.

Auing run through the hiſtory of plants gathered by Maſter *Gerard,* and much inlarged the ſame, both by the addition of many figures and hiſtories of plants not formerly contained in it, and by the amending and increaſing the hiſtory of ſundry of thoſe which before were therein treated of; I finde that I haue forgotten diuers which I intended to haue added in their fitting places : the occaſion hereof hath been, my many buſineſſes, the troubleſomneſſe, and aboue all, the great expectation and haſte of the Worke, whereby I was forced to perform this task within the compaſſe of a yeare. Now being conſtant to my firſt reſolution, I here haue, as time would giue me leaue, and my memorie ſerue, made a briefe collection and addition (though without method) of ſuch as offered themſelues to me; and without doubt there are ſundry others which are as fitting to be added as thoſe; and I ſhould not haue bin wanting if time had permitted me to haue entred into further conſideration of them. In the meane time take in good part thoſe that I haue here preſented to your view.

Chap. I.   Of the *Maracoc or Paſſion Floure.*

¶ The Deſcription.

His Plant, which the Spaniards in the Weſt Indies call *Granadilla,* becauſe the fruit ſomwhat reſembles a Pomgranat, which in their tongue they term *Granadas,* is the ſame which the Virginians cal *Maracoc.* The Spaniſh Friers for ſome imaginarie reſemblances in the floure, firſt called it *Flos Paſſionis,* The Paſſion floure, and in a counterfeit figure, by adding what was wanting, they made it as it were an Epitome of our Sauiors paſſion. Thus ſuperſtitious perſons *ſemper ſibi ſomnia fingunt. Bauhine* deſirous to refer it to ſome ſtock or kindred of formerly known plants, giues it the name of *Clematis trifolia :* yet the floures and fruit pronounce it not properly belonging to their tribe; but *Clematis* being a certain genericke name to all wooddy winding plants, this as a ſpecies may come vnder the denomination, though little in other reſpects participating with them. The roots of this are long, ſomwhat like, yet thicker than thoſe of *Sarſa parilla,* running vp and down, and putting vp their heads in ſeueral places : from theſe roots riſe vp many long winding round ſtalks, which grow two, three, foure, or more yards high, according to the heate and ſeaſonableneſſe of the yeare and ſoile whereas they are planted. Vpon theſe ſtalks grow many leaues diuided into three parts, ſharp pointed, and ſnipt about the edges : commonly out of the boſoms of each of the vppermoſt leaues there growes a claſping tendrel and a floure, the floure groweth vpon a little foot-ſtalk ſome two inches long, and is of a longiſh cornered forme, with fiue little crooked hornes at the top, before ſuch time as it opens it ſelfe, but opened, this longiſh head diuides it ſelfe into ten parts, and ſuſtains the leaues of the floure, which are very many, long, ſharp pointed, narrow, and orderly ſpred open one by another, ſome lying ſtraight, others crooked. Theſe leaues are of colour whitiſh, but thick ſpotted with a peach colour, and toward the bottom it hath a ring of a perfect peach colour, and aboue and beneath it a white circle, which giue a great grace to the floure, in the midſt whereof riſeth an vmbone, which parts it ſelfe into foure or fiue crooked ſpotted hornes, with broadiſh heads; from the middeſt of theſe riſes another roundiſh head which carries three nailes or hornes, biggeſt aboue, and ſmalleſt at their lower end. This floure with vs is neuer ſucceeded by any fruit, but in the Weſt Indies, whereas it naturally growes, it beares a fruit,

S ſſſſſ 2   when

when it is ripe of the bignesse and colour of Pomegranats, but it wants such a ring or crown about the top as they haue ; the rinde also is much thinner and tenderer, the pulp is whitish, and without taste, but the liquor is somwhat tart : they open them as they do egges, and the liquor is supped off with great delight both by the Indians and Spaniards (as *Monardus* witnesseth) neither if they sup off many of them shall they finde their stomack opprest, but rather their bellies are gently loose-ned. In this fruit are contained many seeds somewhat like Peare kernels , but more cornered and rough.

Clematis trifolia, siue Flos Passionis.
The Maracoc or Passion-floure.

. This growes wilde in most of the hot countries of America, from whence it hath been brought into our English gardens, where it growes very well, but flours onely in some few places, and in hot and seasonable yeares. It is in good plenty growing with Mistresse *Tuggy* at Westminster, where I haue some yeares seen it beare a great many floures.

Chap. 2. *Of Ribes or red Currans.*

¶ *The Description.*

1   THe plant which carries the fruit which we commonly terme red Currans, is a shrubbie bush of the bignesse of a Gooseberry bush, but without prickles : the wood is soft and white, with a pretty large pith in the middle ; it is couered with a double barke, the vndermost be-ing the thicker, is greene, and the vppermost, which sometimes chaps and pills off, is of a brownish
colour,

colour, and ſmooth : the barke of the yongeſt ſhoots is whitiſh and rough : the leaues which grow vpon foot-ſtalkes ſome two inches long, are ſomewhat like Vine leaues, but ſmaller by much, and leſſe cornered, being cut into three, and ſometimes, but ſeldome, into fiue parts, ſomewhat thicke, with many veines running ouer them, greener aboue than they are below : out of the branches in Spring time grow ſtalkes hanging downe ſome ſix inches in length, carrying many little greeniſh floures, which are ſucceded by little red berries, cleare and ſmooth, of the bigneſſe of the Whortle berries, of a pleaſant tart taſte. Of this kinde there is another, onely differing from this in the fruit, which is twice ſo big as that of the common kinde.

2   The buſh which beares the white Currans is commonly ſtraighter and bigger than the former : the leaues are leſſer, the floures whiter, and ſo alſo is the fruit, being cleare and tranſparent, with a little blackiſh rough end.

<table>
<tr><td>1 Ribes vulgaris fructu rubro.
Red Currans.</td><td>2 Ribes fructu albo.
White Currans.</td></tr>
</table>

3   Beſides theſe there is another, which differs little from the former in ſhape, yet grows ſomewhat higher, and hath leſſer leaues : the floures are of a purpliſh greene colour, and are ſucceeded by fruit as big againe as the ordinary red, but of a ſtinking and ſomewhat loathing ſauour : the leaues alſo are not without this ſtinking ſmell.

¶ *The Place, Time, and Names.*

None of theſe grow wild with vs, but they are to be found plentifully growing in many gardens, eſpecially the two former, the red and the white.

The leaues and floures come forth in the Spring, and the fruit is ripe about Midſommer.

This plant is thought to haue been vnknowne to the Antient Greekes : ſome thinke it the *Ribes* of the Arabian *Serapio. Fuchſius, Matthiolus,* and ſome other deny it ; notwithſtanding *Dodonæus* affirmes it : neither is the controuerſie eaſie to be decided, becauſe the Author is briefe in the deſcription thereof, neither haue we his words but by the hand of a barbarous Tranſlator. Howeuer the ſhops of late time take it (the faculties conſenting thereto) for the true Ribes, and of the fruit hereof prepare their *Rob de Ribes. Dodonæus* calls it *Ribeſium, groſſularia rubra, & Groſſularia tranſmarina* ; and they are diſtinguiſhed into three ſorts, *Rubra, Alba, Nigra Ribeſia,* red, white, and blacke Currans : the Germans call them **S. Johans traubell,** or **traublin,** and **S. Johans Beerlen** : the Dutch, **Beſtkins ouer Zee** : the Italians, *Vuetta roſſa* : the French, *groiſſes, Groiſelles d outre mer* : the Bohemians, **Jahodi S. Jana** : the Engliſh, Red Currans : yet muſt they not be confounded with thofe Currans which are brougbt from Zant,and the continent adjoyning thereto, and which are vulgarly fold by our Grocers ;,for they are the fruit of a fmall Vine,and differ much from thefe.

¶ *The Temperature and Vertues.*

A   The berries of red Currans,as alfo of the white,are cold and dry at the end of the fecond degree, and haue fome aftricîion,together with tenuity of parts.

B   They extinguifh and mitigate feuerifh heates,represfe choler, temper the ouer-hot bloud, refift putrefaction,quench thirft,helpe the dejeîion of the appetite,ftay cholericke vomitings and fcou-rings,and helpe the Dyfentery proceeding of an hot caufe.

C   The juyce of thefe boyled to the height of hony, eitherwith, or without fugar (which is called *Rob de Ribes*) hath the fame qualities, and conduces to the fame purpofes.

Chap. 3.   *Of Parfley Breake-ftone, and baftard Rupturewort.*

1 *Percipier Anglorum Lob.*
Parfley Breake-ftone.

2 *Polygonum Herniariæ facie.*
Baftard rupture-wort.

¶ *The Defcription.*

1   I Thought that it was not altogether inconuenient to couple thefe two Plants together in one Chapter; firft,becaufe they are of one ftature; and fecondly,taken out of one, and the fame Hiftory of Plants,to wit,the *Aduerfaria* of *Pena* and *Lobel.*

The firft of thefe,which the Authors of the *Aduerfaria* fet forrh by the name of *Percepier*, (and ra-ther affert, than affirme to be the *Scandix* of the Antients) is by *Tabernamontanus* called *Scandix mi-nor* : and by *Fabius Columna*, *Alchimilla montana minima* : it hath a fmall wooddy yellowifh fibrous root, from which rife vp one,two,or more little ftalks, feldome exceeding the height of an handfull, and thefe are round and hairy, and vpon them grow little roundifh leaues, like the tender leaues of Cheruil,but hairy,& of a whitifh green color,faftned to the ftalks with fhort foot-ftalks,& hauing little eares at their fetting on : the floures are fmall,greene,and fiue cornered,many cluftering toge-ther at the fetting on of the leaues : the feed is fmall,fmooth,and yellowifh : the ftalks of this grow fometimes vpright, and otherwhiles they leane on the ground : it is to be found vpon dry and bar-ren grounds,as in Hide Parke, Tuthill fields,&c. It floures in May, and ripens the feed in Iune and Iuly. It feemes by the Authors of the *Aduerfaria*,that in the Weft countrey about Briftow they call this Herbe Percepier:but our herbe women in Cheapfide know it by the name of Parfley Breakeftone.

D   This is hot and dry,and of fubtill parts : it vehemently and fpeedily moues vrine,and by fome is kept in pickle; and eaten as a fallad.

E   The diftilled water is alfo commended to be effeîuall to moue vrine, and clenfe the kidnies of grauell.

2   The hiftory of this, by the forementioned Authors, *Aduerf.pag.404.* is thus fet forth vnder this title,*Polygonium Herniaria folijs & facie, perampla radice Aftragalitidi* : Neither (fay they) ought this to be defpifed by fuch as are ftudious of tbe knowledge of Plants; for it is very little knowne, being a very fmall herbe lying along vpon the ground, and almoft ouerwhelmed or couered with the graffe,hauing little branches very full of joints : the little leaues and feeds are whitifh,and very like thofe of *Herniaria* or Rupture-wort: the whole plant is white, hauing a very fmall and moffie floure : the root is larger than the fmalneffe of the plant feemeth to require,hard,branched, diuerfly turning and winding, and therefore hard to be plucked vp : the tafte is dry and hottifh. It growes vpon a large Plaine in Prouince, betweene the cities Arles and Selon. Thus much *Pena* and *Lobel*. I am deceiued, if fome few yeares agone I was not fhewed this plant, gathered in fome part of this kingdome,but where,I am not able to affirme.

Chap. 4. *Of Heath Spurge and Rocke Roſe.*

¶ *The Deſcription.*

1 THeſe Plants by right ſhould haue followed the hiſtory of *Thymelæa*, for in ſhape and faculties they are not much vnlike it. The firſt is a low ſhrub, ſending from one root many branches of ſome cubit long, and theſe bending, flexible, and couered with an outer blackiſh barke, which comprehends another within, tough, and which may bee diuided into fine threds: the leaues are like thoſe of *Chamælea*, yet leſſer, ſhorter, and thicker, a little rough alſo, and growing about the branches in a certaine order: if you chew them they are gummy, bitter at the firſt, and afterward hot and biting: the floures grow among the leaues, longiſh, yellowiſh, and diuided at the end into foure little leaues: the fruit is ſaid to be like that of *Thymælea*, but of a blackiſh colour, the root is thicke and wooddy. It growes frequently in the kingdome of Granado and Valentia in Spaine, it floures in March and Aprill. The Herbariſts there terme it *Sanamunda*, and the common people, *Mierda-cruz*, by reaſon of the purging faculty.

1 *Sanamunda* 1. *Cluſ.*
Heath Spurge.

2 *Sanamunda* 2. *Cluſ.*
The ſecond Heath Spurge.

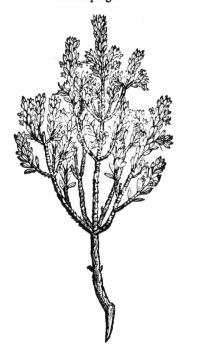

2 The other is a ſhrub ſome cubit high, hauing tough flexible branches couered with a denſe and thick barke, which the outward rinde being taken away, ouer all the plant, but chiefely next the root, may be drawne into threds like Flax or Hemp: the vpper branches are ſet with thick, ſhort, fat, rough ſharp pointed leaues, of ſomewhat a ſaltiſh taſte at the firſt, afterwards of a hot and biting taſte: the floures are many, little and yellow: the root is thicke and wooddy like as that of the former: this growes vpon the ſea coaſt of Spaine, and on the mountaines nigh Granado, where they call it *Sanamunda*, and the common people about Gibralter call it Burhalaga, and they only vſe it to heate their ouens with. It floures in February. *Anguillara* called this, *Empetron*: *Caſilpinus*, *Cneoron*, and in the *Hiſtoria Lugd.* it is the *Cneoron nigrum Myconi*: *Seſamoides minus*: *Daleſchampij*, and *Phicoides, Oribaſij quibuſdam.*

3 This

3   This is bigger than either of the two former, hauing whiter and more flexible branches, whose barke is vnmeasurably tough and hard to breake : the vpper branches are many, and those very downy, and hanging downe their heads, set thicke with little leaues like Stone-crop, and of the like hot or burning faculty : the floures are like those of the former; sometimes greenish, otherwhiles yellow : *Clusius* did not obserue the fruit, but saith, it floured at the same time with the former, and grew in all the sea coast from the Straits of Gibralter, to the Pyrenæan mountaines. *Alfonsus Pantius* called this *Cneoron* : *Lobel* and *Tabernamontanus* call it *Brita Alexandrina.*

3 *Sanamunda 3.Cluf.*
The third Heath Spurge.

4 *Cneoron.Matthioli.*
Rocke Rose.

4   This also may not vnfitly bee joyned to the former, for it hath many tender flexible tough branches commonly leaning or lying along vpon the ground, vpon which without order grow leaues greene, skinny, and like those of the true *Thymelæa*; at first of an vngratefull, and afterwards of a bitter taste, yet hauing none or very little acrimony (as far as may be perceiued by their taste:) the floures grow vpon the tops of the branches six, seuen, or more together, consisting of foure little leaues of a reddish purple colour, very beautifull and well smelling, yet offending the head if they be long smelt vnto : these are succeded by small berries, of colour white, containing a round seed, couered with an ash coloured skin.   The root is long, of the thicknesse of ones little finger, sometimes blackish, yet most commonly yellowish, tough, and smallest at the top where the branches come forth.   It floures in Aprill or May, and ripens the fruit in Iune : it floures sometimes thrice in the yeare, and ripens the fruit twise ; for *Clusius* affirmes that twice in one yeare he gathered ripe berries from one and the same plant.   It growes plentifully vpon the mountainous places of Austria about Vienna ; whither the countrey women bring the floures to the market in great plenty to sell them to deck vp houses : it growes also in the dry medowes by Frankford on the Mœne, where there is obserued a variety with white floures. *Matthiolus* would haue this to be the *Cneoron album* of *Theophrastus* : *Cordus* calls it *Thymelæa minor* : it is the *Cneoron alterum Matthioli*, and *Oleander syl. Auicennæ Myconi*, in the *Hist.Lugd.* The Germanes call it **Stein Roselin:** and we may call it Rocke Rose, or dwarfe Oleander.

5   This plant by *Bauhine* is called *Cneorum album folio oleæ argenteo molli* : and by *Daleschampius*, *Cneorum album*, which hath been the reason I haue put it here, although *Casalpinus*, *Imperatus*, and *Plateau,*

5 *Cneorum album folijs argenteu.*
White rocke Rose.

Chamæbuxus flore Colutea.
Bastard dwarfe Box.

teau who sent it to *Clusius*, would haue it to
be and call it *Dorycnium*. It is a shrubbie
herbe, sending from one root many single
stalks som halfe cubit or better high. The
leaues,which grow vpon the stalks without
order,are like those of the Oliue, but somewhat narower, and couered ouer with a soft
siluer-like downinesse : At the top of the
stalkes grow many floures clustering together,in shape like those of the lesser Binde-
weed,but white of colour. This grows wild
in some parts of Sicily, whence *Casalpinus*
calls it *Dorycnium ex Sicilia*.

¶ *The Temperature and Vertues.*

The three first are very hot, and two first A
haue a strong purging facultie, for taken in
the weight of a dram with the decoction of
cicers they mightily purge by stoole, both
flegme, choler, and also waterish humours,
and they are often vsed for this purpose by
the country people in some parts of Spain.
The faculties of the rest are not known,
nor written of by any as yet.

CHAP. 5.
Of bastard dwarfe Box.

¶ *The Description.*

THis which *Clusius* for want of a name
calls *Anonymos flore Coluteæ, Gesner* called *Chamæbuxus* ; to which *Bauhine* addeth *flore Coluteæ*; and *Besler* in his *hortus Ey-
stettensis*, agreeable to the name I haue giuen it in English,cals it *Pseudochamæbuxus*.
It is a small plant , hauing many creeping
woody tough roots, here and there sending
forth small fibres : From these arise many
tough bending branches some span long,
hauing thicke sharpe pointed green leaues
almost like those of Box , and these grow
vpon the stalks without any order, & when
you first chew them they are of an vngratefull taste,afterwards bitter and hot. At the
tops of the branches do come forth among
the leaues three or foure longish floures for
the most part without smell , yet in some
places they smell sweet like as some of the
Narcisses : they consist of three leaues a-
piece,two whereof are white, and spred a-
broad as wings,a whitish little hood couering their lower ends ; the third is wrapt vp
in form of a pipe, with the end hollow and
crooked,

crooked,and this is of a yellow colour,which by age often times becomes wholly red : after thofe floures fucceed cods broad and flat,little leffe than thofe of the broad leafed *Thlaſpi*,and greene of colour,rough,and in each of thefe cods are commonly contained a couple of feeds,of the bigneffe of little Chichlings,of a blackifh afh colour,rough,and refembling a little dug.

This is fometimes found to vary , hauing the two winged leaues yellow or red, and the middle one yellow.

It floures in Aprill and May,and ripens the feed in Iune. It growes vpon moft of the Auftrian and Styrian Alps,and in diuers places of Hungary. It is neither vfed in phyficke,nor the faculties thereof in medicine known.

CHAP. 6.   Of winged Binde-weed or Quamoclit.

Quamoclit, ſive Convolvulus Pennatus.
Winged Binde-weed.

¶ *The Defcription.*

THe firft that writ of & defcribed this plant was *Cæſalpinus*,and that by the name or *Gelfeminum rubrum alterum*. After him *Camerarius* gaue a figure and defcription thereof in his *hortus Medicus*,by the name of *Quamoclit*. And after him *Fabius Columna* both figured and defcribed it more accuratly,whofe defcription is put to the figure of it we here giue, in *Cluſ*. his *Curæ Poſteriores*. It is fo tender a plant that it wil not come to any perfe&ction with vs,vnleffe in extraordinary hot yeres,& by other artificial helps ; wherefore I wil borrow the defcription thereof out of *Fabius Columna*. This exotick plant,faith he, cannot more fitly be referred to any Kinde, than to the family of the *Convolvuli* or Bindeweeds, for in the nature and whole habit it is almoft like them, except in the fhape of the winged leaues : it is ftored with leffe milke;the floures are long,hollow,but parted into fiue at the top, of a pleafing red colour, with ftreaked lines or folds, ftanding vpon long ftalks one or two together comming out of the bofomes of the leaues at each joint of the branches,& they haue in them fiue yellowifh pointals:then fucceeds a longifh fruit ftanding in a fcaly cup, ending in a fharpe pointall, and couered with a tough skin,as that of the common *Convolvulus*, but leffer, hauing within it foure longifh black hard feeds,of a biting tafte. The leaues grow alternatly out of the joints of the purple winding branches,being winged and finely diuided,twice as fmall as the common *Rheſeda*,of a darke greene colour,but the yong ones are yellowifh,firft hauing a few diuifions, but afterwards more, till they come to haue thirteen on a fide,and one at the top ; but the lower ones are often times forked : by reafon of the great plenty of leaues and flouring ftalks or branches winding themfelues about artificial hoops,croffings,or other fafhioned works of Reeds,or the like,fet for winding herbs to clime vpon,it much delights the eye of the beholder, and is therfore kept in pots in gardens of pleafure. The feed fown in the beginning of the Spring growes vp in Iune,and the firft leaues refemble the winged fruit of the Maple : it floures in the end of Auguft, and ripens the feed in the end of September.

Chap. 7.   Of the Senſitiue herbe.

Herba mimoſa.
The Senſitiue herbe.

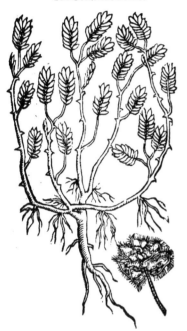

Eius exactior Icon.
A perfect figure thereof.

¶ The Deſcription.

THis which I here call the Senſitiue herbe, is that which Chriſtopher Acoſta ſets forth by the name of Herba mimoſa, or the Mocking herbe, becauſe when one puts his hand thereto, it forthwith ſeemes to wither and hang downe the leaues ; but when you take it away againe, it recouers the priſtine greeneſſe and vigor. I will here giue you that which Acoſta writes thereof, and the figure & hiſtorie which Cluſius giues in his notes vpon him ; and alſo another figure better expreſſing the leaues and manner of growing. There is found, ſaith Acoſta, in ſome gardens another plant ſome fiue handfulls long, reſting vpon the neighboring ſhrubs or wals, hauing a ſlender ſtalke of freſh greene colour, not very round, ſet at certaine ſpaces with ſmal & pricking thornes : the leaues are not vnlike the former, [That is, the Herba viua, which in condition is little different from this] being ſomwhat leſſer than thoſe of the female Fern. It loues to grow in moiſt and ſtony places, and is called Herba mimoſa for the reaſon formerly giuen. The nature hereof is much different from that of Arbor triſtis ; for euery night at Sunne ſet it as it were withers and dries, ſo that one would thinke it were dead, but at Sunne-riſe it recouers the former vigor, and by how much the Sun grows hotter, by ſo much it becomes the greener, and all the day it turns the leaues to the Sun.

This plant hath the ſmel and taſte of liquorice, A, and the leaues are commonly eaten by the Indians againſt the cough, to clenſe the cheſt, and clear the voice. It is alſo thought good againſt the paines of the kidnies, and to heale greene wounds. Thus much Acoſta.

Now, ſaith Cluſius, the leaues of many plants, eſpecially Pulſes, vſe to contract or ſhrink vp their leaues in the night time. Now I receiued a drie plant, which was ſent to mee by the name of Herba mimoſa, by Iames Garret in the end of October, 1599. which he writ he had from the right honorable the Earle of Cumberland, who returning from Saint Iohn de puerto rico in the Weſt Indies, brought it put in a pot with ſome earth, but could not preſerue it aliue. But I cauſed the figure of that dried plant to be expreſſed as well as it might, ſo to fit it to the deſcription following, made alſo by the dried plant. This plant which was wholly dry and without leaues, had a ſingle root, and that not thick, but hard and wooddy, with few fibres, from whence aroſe three or foure ſhort ſtalks, which ſtraight diuided themſelues into ſlender branches which ſpred themſelues round about vpon the ground, at each joint putting forth many long and ſlender fibres, like as in the branches of the common Wood-bind, which lie vpon the ground : theſe branches were a cubit long, and ſomtimes more, round, tough, with ſome prickles, broader at their ſetting on, as you may ſee in the common bramble, yet leſſer, fewer, & leſſe firm ; theſe again were diuided into other more ſlender branches ſet with many little prickles, out of whoſe joints betwixt 2 little leaues grew forth foot-ſtalks bedeckt with their little leaues, which were many, ſet in order, with other to anſwer to them on the other ſide, but hauing no ſingle leafe at the end : they were tender & green, not vnlike the little leaues of Acacia, & theſe at their firſt comming out couered with a thin whitiſh hairines, as I gathered by a little branch retaining the foot-ſtalk and leaues thereon (which he ſent with the former) and it had alſo ſome fibres comming forth therof. He alſo added to the former two little heads, which growing on the ſame plant, he writ he receiued of the

foremeentioned

forementioned right honourable Earle,with fome branches yet retaining the leaues. Thefe little heads confifted of many flender narrow and as it were prickly little leaues;amongft which lay hid round feeds,fmooth,black,and fomwhat fwoln in the middle : the floures I faw not, neither know I whether they were brought with the reft:but whether the leaues of this plant being green,& yet growing on the ground,do wither at the approch of ones hand, as *Chriftopher Acofta* writes, and for that caufe impofes the name thereon, they beft know who haue feene the greene and yet growing plant:for the faculties you may haue recourfe to that which *A Cofta* hath fet down.Thus much out of *Clufius*.

Novemb.7.1632.I being with M*r Iob Beft* at the Trinity houfe in Ratcliffe,among other rarities hee fhewed me a dry plant hereof, which I heedfully obferued, and carefully opening out fome of the broadeft leaues,which(as alfo the whole plant befides)were carelefly dried,I found the leaues grew vfually fome dozen or more on a foot-ftalke,iuft as many on one fide as on the other;& they were couered ouer with a little downineffe,which ftanding out on their edges made them looke as if they had bin fnipt about the edges,which they were not : alfo I found at euery ioint two little hooked prickles,& not two little leaues or appendices at the fetting on of the foot-ftalks,but 3 or foure little leaues,as the rudiment of a young branch,comming forth at the bofome of each foot-ftalke:the longeft branch(as far as I remember) was not aboue a fpan long;I then drew as perfect a figure as I could of the perfecteft branch thereof,drawing as neer as I could the leaues to their ful bigneffe,the which I here prefent you withall. There are two figures formerly extant,the one this of *Clufius*,which I here giue you;and the other in the 18 book & 144 chap.of the *hift.Lugd.* which is out of *Acofta*,and this feemes to be fo far different from that of *Clufius*, that *Bauhine* in his *Pinax* faith,*Clufius notis fuis in Acoftam diuerfam planè figuram propofuit, herbam mimofam nominans:*but he did not well confider it,for if he had,he might haue found thefe fo much different,thus farre to agree; they both make the branches prickly and weak;the leaues many on one rib,one oppofit to another without any odd one at the end:but *Clufius* figures the leaues fo clofe together that they feeme but one leafe,and *Acofta* makes them too far afunder, and both of them make them too fharpe pointed. *Clufius* made his to be taken from a dried plant, and *Acofta* I iudge made his by the Idæa thereof which he had in his memory;and after this maner,if my iudgement faile me not,are moft of the figures in him expreft:but of this enough,if not too much.

CHAP.8.   *Of the Staffe-tree,and euer-greene Priuet.*

1 *Celaftrus Theophrafti.*
The Staffe-tree.

2 *Phillyrea* 1.*Cluf.*
Clufius his firft Mock-Priuet.

¶ *The Deſcription.*

1 THe hiſtorie and figure of this tree are ſet forth in *Cluſius* his *Curæ Poſt.* and there it is
aſſerted to be ſuppe or ſuppe of *Theophraſtus* ; for by diuers places in *Theophraſtus* there
collected, it is euident, that his *Celaſtus* was euer greene, grew vpon very high and cold
mountains, yet might be tranſplanted into plain and milder places ; that it floured exceeding late
and could not perfect the fruit by reaſon of the nigh approch of Winter, and that it was fit for no
other vſe but to make ſtaues for old men.

Now this tree growes but to a ſmall height, hauing a firme and hard body, diuiding it ſelf at the
top into ſundry branches, which being yong are couered with a green bark, but waxing old, with a
brownith one ; it hath many leaues growing alwaies one againſt another, and thicke together, of a
deep ſhining green aboue, and lighter vnderneath, keeping their verdure both winter and ſummer :
they are of the bignes of thoſe of *Alaternus*, not ſnipt about the edges, but only a little nickt, when
they are yet yong : at the top of the tendereſt branches among the leaues, vpon foot ſtalkes of ſome
inch long , grow fiue or ſix little floures, conſiſting commonly of fiue little leaues of a yellowiſh
green colour, and theſe ſhew themſelues in the end of Autumne or the beginning of winter, and alſo in the beginning of the Spring ; but if the Summer be cold and moiſt, it ſhewes the buds of the
floures in October. The fruit growes on a ſhort ſtalke, and is a berry of the bigneſſe of a myrtle, firſt
green, then red of the colour of that of Aſparagus, and laſtly blacke when it is withered : the ſtone
within the berry is little and as it were three cornered, containing a kernell couered with a yellow
filme. Where this growes wild I know not, but it was firſt taken notice of in the publique garden
at the Vniuerſitie of Leyden, from whence it was brought into ſome few gardens of this Kingdome.

2 The firſt *Phyllyria* of *Cluſius* may fitly be referred to the reſt of the ſame Tribe and name, deſcribed formerly, *Lib.3. Cap.59*. It growes ſomewhat taller than the Skarlet Oke, and hath branches of the thickneſſe of ones thumbe and ſomewhat more, and thoſe couered with a greene barke
marked with whitiſh ſpots : the leaues ſomewhat reſemble thoſe of the Skarlet Oke, but greater,
greener, thicker, ſomewhat prickly about the edges, of an aſtringent taſt, but not vngratefull. The
floure thereof *Cluſius* did not ſee : the fruit is a little blacke berry, hanging downe out from the boſome of the leaues, and containing a kernel or ſtone therein. It growes wild in many places of Portugal, where they call it Azebo.

The temper and vertues are referred to thoſe ſet down in the formerly mentioned chapter.

CHAP. 9. *Of Mock-Willow.*

Speirai Theophraſti, Cluſ.
Mock-Willow.

¶ *The Deſcription.*

THis Wilow-leaued ſhrub, which *Cluſius* conjectures may be referred to the *Speiræa* mentioned by *Theophraſtus lib.1.cap.23.hiſt.plant*. I haue
named in Engliſh, Mock-willow, how fitly I know
not, but if any will impoſe a fitter name, I ſhall be
well pleaſed therewith. But to the thing it ſelf. It
is a ſhrub (ſaith *Cluſius*) ſome two cubits high, hauing ſlender branches or twigs couered ouer with a
reddiſh barke, whereon grow many leaues without
order, long, narrow, like thoſe of the Willow, ſnipt
about the edges, of a light greene aboue, and of a
blewiſh green vnderneath, of a drying taſt conjoi-
ned with ſome bitterneſſe. The tops of the branches for ſome fingers length carry thicke ſpikes of
ſmall floures cluſtering together, and conſiſting of
fiue leaues apiece ; out of whoſe middle come
forth many little leaues of a whitiſh red or fleſh
colour, together with the floure, hauing no peculiar ſmell , but ſuch as is in the floure of the Oliue
tree : theſe floures fading, there ſucceed ſmall fiue

Tttttt cornered

cornered heads,which comming to full maturitie contain a fmall & yellowifh dufty feed:it flours in Iuly, and ripens the feed in the end of Auguft. *Clufius* had this plant from *Fredericke Sebizius* Phyſition to the duke of Briga,and that from Briga in Silefia,and hee (as I faid) referreth it to the Σπιγμια of *Theophraftus*,which he reckons among the fhrubs that carry fpike-fafhioned floures.

This is not vfed in medicine,nor the temperature and faculties thereof as yet known.

CHAP. 10.　*Of the Strawberry Bay.*

Adrachne Theophrafti.
The Strawberry Bay.

¶ *The Defcription.*

THe figure and hiftorie of this was fent by *Honor. Bellus* out of Candy to *Clufius*, from whom I haue it. It is that which *Theophraftus* calleth *Adrachne* or (as moft of the printed books haue it) *Andrachne :* but the former feems the righter,and is the better liked by *Pliny, Lib.* 13. *cap.*22. At this day in Candy where it plentifully growes, it is called *Adracla.* It is rather a fhrub than a tree,delighting in rocky and mountainous places,and keeping greene Winter and Summer, hauing leaues fo like thofe of Bayes, that they are diftinguifhable only by the fmell, which thefe are deftitute of. The barke of the bole and all the branches is fo fmooth, red, and fhining,that they fhew like branches of Coral ; this barke crackes or breaks off in Summer, and pills off in thin fleakes ; at which time it is neither red nor fhining,but in a mean betweene yellow and afh-colour. It hath floures twice in the yeare like as the *Arbutus* or Strawberry tree, and that fo like it, that you can fcarfe know the one from the other , yet this differs from it in that it groweth onely in the mountaines, hath not the leaues jagged, neither a rough barke : the wood hereof is very hard,and fo brittle that it will not bend,and they vfe it to burn and to make whorls for their womens fpindles. *Theophraftus* reckons vp this tree amongft thofe which dye not when their barks are taken off, and are alwaies greene, and retaine their leaues at their tops all Winter long :which to be fo *Honorius Bellus* obferued. *Bellonius* alfo obferued this tree in many places of Syria.

The fruit in temperature,as in fhape,is like that of the Strawberry tree.

CHAP. 11.　*Of the Cherry Bay.*

¶ *The Defcription.*

THe Chery-bay is one of the euer-greene trees:it rifes vp to an indifferent height,& is diuided into fundry branches couered ouer with a fwart green bark:that of the yonger fhoots is wholly greene.

green,the leaues alternatly ingirt the branches,and they are long,ſmooth,thick, greene & ſhining, ſnipt alſo lightly about the edges:when the tree is grown to ſome height, at the tops of the bran-ches amongſt the leaues of the former years growth,vpon a ſprig of ſome fingers length it putteth forth a great many little white floures conſiſting of fiue leaues a piece, with many little chiues in them. Theſe floures quickly fall away,and the fruit that ſucceeds them is a berry of an oual figure, of the bigneſſe of a large Cherry or Damſon, and of the ſame colour, and of a ſweet and pleaſant

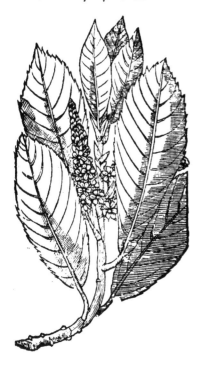

Laurocraſi flos.
The Cherry-bay in floure.

Laurocraſi fructus.
The Cherry-bay with the fruit.

taſte,with a ſtone in it like to a Cherry ſtone. This floures in May,and ripens the fruit in Auguſt or September. It was firſt ſent to *Cluſius* from Conſtantinople, and that by the name of *Trabiſon curmaſi* 1. *Trapazuntina dactylus*,the Date of Trapezon ; but it hath no affinitie with the Date. *Da-lechampius* refers it to the ſecond *Lotus* mentioned by *Theophraſtus,Hiſt.plant.lib.4.ca.4.* but there-with it doth not agree. *Cluſius* and moſt ſince call it fitly *Lauroceraſus*,or *Ceraſus folio Laurino.* It is now got into many of our choice Engliſh gardens,where it is well reſpected for the beauty of the leaues,and their laſting or continuall greenneſſe.

The fruit hereof is good to be eaten,but what phyſicall vertues the tree or leaues thereof haue it is not yet knowne.

Chap. 12.   *Of the euer-greene Thorne.*

THis plant,which *Lobel* and ſome other late Writers haue called by the name of *Pyracantha*,is the *Oxyacantha* mentioned by *Theophraſtus*, *lib.1.cap.15. lib.3.cap.4.hiſt.plant.* amongſt the euer-green trees ; and I thinke rather this than our white Thorn to be the *Oxyacantha* of Dio-ſcorides,lib.1.ca.123. And certainly it was no other than this Thorn which *Virgil* mentioneth by name of *Acanthus,lib.2.Georg.*in theſe words,*Et bacchas ſemper frondentis Acanthi* : That is, And the berries of the E're-green Thorn.

Oxyacantha Theophrasti.
The euer-green Thorne.

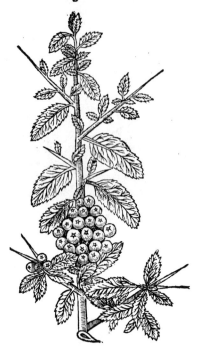

¶ *The Description.*

THis grows vp like a bush, vnles you keep it with pruning, and then it will in time grow to the height of a small tree, as the Hawthorne, whereto it is of affinitie, for the wood is white and hard, like it, and couered ouer with the like barke; but the leaues are somewhat like those of the Damson tree, longish, sharpe pointed, and snipt about the edges: they grow alongst the branches, without any order, yet somtimes they keep this maner of growing, at each knot, where commonly there is a sharp prickle, growes out one of the larger leaues, which may be some inch & half long, & some three quarters of an inch broad: then vpon the prickle & at the comming out thereof are three or foure, more or lesse, much smaller leaues: now these leaues are of a faire & shining green aboue, but paler vnderneath, and they keepe on all the yeare. At the ends and oft times in the middle of the branches come forth clusters of vmbels of little whitish Blush-coloured floures consisting of fiue leaues apiece, with some little chiues in their middls: then follow clusters of beries, in shape tast and bignes like those of Hawthorn, & of the same, but much more orient and pleasing colour, and containing in them the like seed. Now these berries hang long vpon the tree, and make a gallant shew amongst the greene leaues, but chiefely then when as the autumne blasts haue depriued other trees of their wonted verdure. This flours in May and Iune, and ripens the fruit in September and October. It grows wilde in sundry places of Italy and Province in France, but is kept in gardens with vs, where it is held in good esteem for his euer-greenesse and pliablenesse to any work or forme you desire to impose vpon him.

The fruit haue the same faculties that are formerly attributed to Haws, in the third book, *p.*1328. and therefore I will not here repeat them.

Chap. 13.   *Of the Egyptian Nap, or great Iujube tree.*

¶ *The Description.*

THis tree, which for his leaues and manner of growing I thinke may fitly be referred to the Iujubes tree, is of two sorts, that is, the one prickly, and the other not prickly, in other respects they are both alike, so that one figure and history may serue for them both; which I will giue you out of *Clusius*, who recciued this figure together with a description thereof from *Honorius Bellus*, and also added thereto that which *Prosper Alpinus* hath written of it, *cap.*5. *de plant. Ægypt.* It grows to the height of an indifferent Pear tree, and the body and branches thereof are couered with a whitish ash coloured barke: the leaues are like those of the Iujubes tree, two inches long, & one broad, with three nerues running along them, of a deepe shining greene aboue, and more whitish vnderneath, and they grow alternately vpon the branches, and at their comming forth grow tufts of little white floures hanging vpon single long foot-stalks: after these followes the fruit, like vnto a small apple, of the bignesse for the most part of a large Cherry, and sometimes as big as a Wallnut, of a sweet tast, containing therein a kernell or stone like that of an Olive. It beares fruit twice a yeare, for it hath ripe fruit both in the Spring and Fall; yet the vernal fruit seldom comes to good,

by

Oenoplia non spinosa.
The great Iujubes tree.

by reason of the too much moisture of the season, which causes it to become worme-eaten. The Thorny kind is described by *Alpinus*, who rightly iudges it the *Connarus* of *Athenæus*, but the figure he giues is not very accurate. That which wants prickles growes (as well as the prickly one) in Ægypt and Syria, as also in the city Rhetimo in Candy, whither it was brought out of Syria.

The history of both these trees is in *Serapio*, by the name of *Sadar* : but he, according to his custome confounds it with the *Lotus* of *Dioscorides*, from which it very much differs. *Bellonius* in his second booke, and 79. chap. of his Obseruations, reckons vp *Napeca* amongst the trees that are alwaies greene : which is true, in those that grow in Egypt and Syria, but false in such as grow in Candy. That tree in Ægypt and Syria is called *Nep*, or *Nab*. *Alpinus* calls it *Paliurus Athenæi*, or *Nabca Ægyptiorum*, thinking it (as I formerly said) the *Connarus* mentioned in the 14. booke of *Athenæus*, his Deipnosophists.

¶ *The Vertues out of Alpinus.*

The fruit is of a cold and dry facultie, and A
the vnripe ones are frequently vsed to strengthen the stomacke, and stop lasks : the iuyce of them being for this purpose either taken by the mouth, or iniected by clysters : of the same fruit dried and macerated in water, is made an infusion profitable against the relaxation and vlceration of the guts.

The decoction or infusion of the ripe dried B
fruit, is of a very frequent vse against all pestilent feuers : for they affirme that this fruit hath a wonderfull efficacie against venenate qualities, and putrifaction, and that it powerfully strengthens the heart.

Also the iuyce of the perfectly ripe fruit is very good to purge choler forth of the stomacke and C
first veines : and they willingly vse an infusion made of them in all putride feuers to mitigate their heate or burning.

CHAP. 14: *Of the Persian Plum.*

¶ *The Description.*

1 THis tree is thought by *Clusius* (to whom I am beholden for the history and figure) to be the *Persea arbor* mentioned by *Pliny* and *Plutarch*, but he somwhat doubts whether it be that which is mentioned by *Theophrastus*. *Dioscorides* also, *Galen* and *Strabo* make mention of the *Persea arbor*, and they all make it a tree alwaies greene, hauing a longish fruit shut vp in the shell and coat of an Almond : with which how this agrees you may see by this description of *Clusius*.

This tree (saith he) is like to a Peare tree, spreading it selfe far abroad, and being alwaies greene, hauing branches of a yellowish green colour. The leaues are like those of the broadest leaued Baytree, greene aboue, and of a grayish colour vnderneath, firm, hauing some nerues running obliquely, of a good taste and smell, yet biting the tongue with a little astriction. The floures are like those of the Bay, growing many thicke together, and consist of six small whitish yellow leaues. The fruit at the first is like a Plum, and afterwards it becomes Peare-fashioned, of a blacke colour, and pleasant taste : it hath in it a heart-fashioned kernell, in taste not vnlike a Chesnut, or sweet Almond. I found it flouring in the Spring, and I vnderstood the fruit was ripe in Antumne, by the relation of Sig*.

Iohn

Perſea arbor.
The Perſian Plum.

Iohn Placa, Phyſition and Profeſſor of Valentia, who ſhewed me the tree growing in the
Garden of a Monaſterie a mile from Valentia, brought thither, as they ſay, out of America, and he ſaid they called it *Memay*: but
the Spaniards who haue deſcribed America
giue this name to another tree. But diuers
yeares after, I vnderſtood by the moſt learned *Simon de Touar*, a Phyſition of Ciuil, who
hath the ſame tree in his Garden, with other
exoticke plants, that it is not called *Mamay*,
but *Aguacate*. Thus much out of *Cluſius*;
where ſuch as are deſirous, may finde more
largely handled the queſtion, whether this
be the *Perſea* of the Antients or no? *Rariorum
plan. Hiſt. l. 1. c. 2.*

Chap. 15.

Of Geſners wilde Quince.

¶ *The Deſcription.*

Cotonaſter Geſneri.
Geſners wilde Quince.

THe ſhrub which I here figure out of
Cluſius, is thought both by him and o-
thers, to be the *Cotanaſtrum* or *Cidonago*,
mentioned by *Geſner* in his Epiſtles, *lib. 3.
pag.*88. It hath branches ſome cubit long,
tough, and bare of leaues in their lower parts,
couered with a blacke barke: and towards
the tops of the branches grow leaues ſomewhat like thoſe of Quinces: of a darke green
aboue, and whitiſh vnderneath, ſnipt about
the edges: at the tops of the branches grow
vſually many floures, conſiſting of fiue purpliſh coloured leaues apiece, with ſome
threds in their middles: theſe decaying, vnder them grow vp red dry berries without
any pulpe or juyce, each of them containing
foure triangular ſeeds. *Cluſius* found this
flouring in Iune vpon the tops of the Auſtrian Alpes, and he queſtions whether it were
not this which *Bellonius* found in the mountains of Candy, and called *Agriomalea, lib. 1.
cap.* 17. This is not vſed in Phyſicke, nor the
faculties thereof knowne.

CHAP.

CHAP. 16. Of Tamarindes.

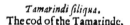

Tamarindus.
The Tamarinde.

Tamarindi siliqua.
The cod of the Tamarinde.

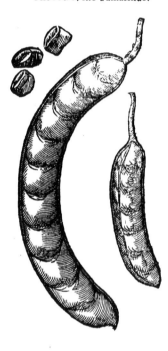

¶ The Description.

TAmarinds, which at this day are a medicine frequently vsed, and vulgarly knowne in shops, were not knowne to the antient Greekes, but to some of the later, as *Actuarius*, and that by the name of *Oxyphœnicæ*, that is, foure Dates, drawne as it may seeme from the Arabicke appellation, *Tamarindi*, that is, Indian Date: but this name is vnproper, neither tree nor fruit being of any affinitie with the Date, vnlesse the Arabicke *Tamar* be a word vsed in composition for fruits of many kindes, as the Greeke μῆλον, the Latine *Malum*, and Apple with vs in English; for we call the Cone of the Pine, and excrescence of the Oke leafe, by the name of Pine Apple, and Oke Apple. But howsoeuer it be, it is no matter for the name, whether it be proper or no, if so be that it serue to distinguish the thing from others, and we know what is denoted by it. In Malauer they call it *Puti*: in Guzarat, *Ambili*, by which name it is knowne in most parts of the East-Indies. This tree is thus described by *Prosper Alpinus, de Plant. Ægypti, cap.* 10. The Tamarind (saith he) is a tree of the bignesse of a Plum tree, with many boughes and leaues like those of the Myrtle, many standing vpon one rib [one against another, with a single one at the end:] it carrieth white floures very like those of the Orange tree: out of whose middle comes forth foure white and very slender threds: after these come thicke and large cods, at first greene, but when they are ripe of an ash-colour; and within these are contained thicke, hard, brownish, cornered seeds, and a blacke acide pulpe. These trees grow in some few gardens of Ægypt, whither they haue been brought out of Arabia and Ethiopia. This plant hath this strange quality that the leaues alwaies follow the Sun, and when it sets th y all contract themselues, and open out themselues againe at the rising thereof; and there is obserued to be such force in this motion, that they closely shut vp and hold their cods (if any be on the tree) and then at the rising of the Sun they forgoe them againe. But I haue obserued this folding vp of the leaues to be common to diuers other Ægyptian plants, as *Acacia, Abrus, Absus*, and *Sesban*. Thus much out of *Alpinus*.

The

The figure I here giue in the firſt place, out of *Lobel*, is of a plant ſome ſix moneths old, ariſen of a ſeed : and ſuch by ſowing of ſeeds I haue ſeene growing in the garden of my diſceaſed friend Mr. *Tuggy*, but they ſtill died at the firſt approach of Winter. The other figure expreſſeth the cods, and ſome of the ſeeds apart, taken forth of the cods : now the cods are neuer brought whole to vs, but the vtter rindes are taken off ; and the ſtrings or nerues that run alongſt the cods, the pulpe and ſeeds in it are cloſe thruſt together, and ſo are brought to vs in pots and ſuch like veſſels.

¶ *The Temperature and Vertues.*

A　　The fruit or pulpe of Tamarindes is cold and dry in the third degree : it is of good vſe in chole-ricke diſeaſes, and burning Feuers, Tertians, and the like : it is a lenitiue and very gently purging medicine and therefore vſed to be put into medicines ſeruing to that purpoſe.

B　　They vſe (ſaith *Alpinus*) the leaues of Tamarindes to kill wormes in young children ; and alſo their infuſion or decoction to looſen the belly ; the leaues are acide, and not vnpleaſant vnto the taſte.

C　　The Arabians preſerue the ſmall and yet greene cods of this tree, as alſo the ripe ones, either with ſugar, or the hony boyled out of the fruit of the Carob tree : they alſo mix the pulpe with ſu-gar, which trauellers carry with them in their journies through the deſart places of Africke, where-with they being dry or ouerheated, may quench their thirſt, coole and refreſh themſelues, and alſo euacuate many hot humors by ſtoole.

D　　In peſtilent and all other burning putrid feuers they drinke the water with ſugar, wherein a good quantitie of Tamarinds haue beene infuſed ; for it is a drinke very pleaſant to ſuch as are thirſty by reaſon of too much heate, for it powerfully cooles and quencheth thirſt.

E　　They are alſo vſed in all putrid feuers, cauſed by cholericke and aduſt humors, and alſo againſt the hot diſtempers and inflammations of the liuer and reines, and withall againſt the Gonorrhæa.

F　　Some alſo commend them againſt obſtructions, the dropſie, jaundiſe, and the hot diſtempers of the ſpleene : they conduce alſo to the cure of the itch, ſcab, leproſie, tetters, and all ſuch vlcerations of the skin which proceed of aduſt humors.

G　　They are not good for ſuch as haue cold ſtomackes, vnleſſe their coldneſſe be corrected by put-ting to them Mace, Aniſe ſeeds, Squinanth, or ſuch like.

Chap. 17.

Of the Mamoera, the Male and Female.

¶ *The Deſcription.*

THe hiſtorie of theſe two trees, together with the figures I here giue you, are it the *Curæ Poſteri-ores* of *Cluſius*, from whence I will take as much as concernes their hiſtory, and briefely here giue it you.

That of the Poët (ſaith he) is moſt true, *Non omnis fert omnia tellus* : for I thinke there is no Pro-uince to be found, which produces not ſome peculiar plant not growing in other regions, as they can teſtifie who haue trauelled ouer forreine countries, eſpecially if they haue applied themſeles to the obſeruations of plants. Amongſt ſuch I thinke I may reckon that honeſt and courteous man *Iohn Van vſele*, who returning out of that part of America called Braſile, ſhewed mee in the yeare 1607. a booke, wherein he in liuely colours had expreſt ſome plants and liuing creatures : for as he told me, when he purpoſed to trauell he learned to paint, that ſo he might expreſſe in colours, for his memorie and delight after he was returned home, ſuch ſingularities as he ſhould obſerue abroad. Now amongſt thoſe which he in that booke had expreſſed, I obſerued two very ſingular, and of a ſtrange nature, whoſe figures without any difficulty he beſtowed vpon me, as alſo the following hi-ſtory.

Theſe two trees, whoſe figures you ſee here expreſt, are of the ſame kinde, and differ only in ſex ; for the one of them, to wit, the male, is barren, and only carries floures, without any fruit ; but the female onely fruit, and that without floure : yet they ſay they are ſo louing, and of ſuch a nature, that if they be ſet far aſunder, and the female haue not a male neere her, ſhee becomes barren and beares no fruit : of which nature they alſo ſay the Palme is.

Now the bole or trunke of that tree which beares the fruit is about two foot thicke, and it grow-eth ſome nine foot high before it begin to beare fruit, but when it hath acquired a juſt magnitude, then ſhall you ſee the vpper part of the tree laden with fruit, and that it will bee as it were thicke

girt

girt about therewith for some nine foot high more : the fruit is round and globe-fashioned, of the shape and magnitude of a small gourd, hauing when it is ripe a yellowish pulpe, which the Inhabitants vse to eat to loosen their bellies. This fruit contains many kernels of the bignesse of a small pease, blacke and shining, of no vse that he could learne, but which were cast away as vnneceffary. The leaues come forth amongst the fruit, growing vpon long footstalks, and in shape much resemble the Plane tree or great Maple.

Mamoera mas.
The male Dug tree.

Mamoera fæmina.
The female Dug tree.

What name the Brasilians giue it he could not tell, but of the Portugals that dwelt there it was called *Mamoera*, and the fruit *Mamaon*, of the similitude I thinke they haue with dugs, which by the Spaniards are called *Mamas* and *Tetas*.

There is no difference in the forme of the trunke or leaues of the male and female, but the male only carries floures hanging down, clustering together vpon long stalkes like the floures of Elder, but of a whitish yellow colour, and these vnprofitable, as they affirme.

Both these trees grow in that part of America wherein is scituat the famous Bay called by the Portugals, *Baya de todos los sanctos*, lying about thirteen degrees distant from the Equator towards the Antartick Pole.

CHAP. 18. *Of the Clove-berry tree.*

¶ *The Description.*

I Must also abstract the historie of this out of the works of the learned and diligent *Clusius*, who set it forth in his *Exot.lib.1.cap.17.*in the next chapter after Cloues.

I put (saith he) the description of this Fruit next after the historie of Cloues , both for the affinity

Amomum quorundam, fortè Garyophyllon Plinij.
The Clove-berry tree.

affinitie of fmell it hath with Cloves, as alfo for another caufe, which I will fhew
hereafter. *Iames Garret* in the yeare 1601,
fent mee from London this round Fruit,
commonly bigger than Pepper cornes, yet
fome leffe, wrinkled, of a brownifh colour,
fufficiently fragil; which opened, I found
contained a feed, round, blacke, which
might be diuided into two parts, of no leffe
aromatick tafte and fmell than the fruit it
felfe, and in fome fort refembling that of
Cloves. It grows in bunches or clufters, as
I conjectured by many berries which yet
kept their ftalkes, and two or three which
ftuck to one little ftalke. To thefe were added leaues of one form, but of much different bignes, for fome of them were 7 inches
long and 3 broad; fome only 5 inches long,
and two and a halfe broad ; others did not
exceed 3 inches in length, and thefe were
not two inches broad; and fome alfo were
much leffe and narrower than thefe, efpecially thofe that were found mixed with the
berries, differing according to the place, in
the boughs or branches which they poffeft.
I obferued none amongft them which had
fnipt leaues, but fmooth, with many fmall
veins running obliquely from the middle
rib to the fides, with their points now narrower, otherwhiles broader and roundifh;
they were of a brownifh afh colour, of a
fufficient acride tafte : the branches which were added to the reft were flender, quadrangular, couered with a bark of an afh colour, and thofe were they of a yeares growth ; for thofe that were of an
after-growth were brownifh, and they had yet remaining the prints where the leaues had growne,
which for the moft part were one againft another, and thefe alfo were of an acride tafte, as well as
the leaues, and of no vngratefull fmell.

I receiued the fame fruit fome yeres before, but without the ftalks, with this queftion propoun-
ded by him which fent it, *An Amomum ?* And certainly the faculties of this fruit are not very much
vnlike thofe which *Diofcorides* attributes to his *Amomum*; for it hath an heating aftrictiue and drying facultie, and I think it may perform thofe things whereto *Diofcorides, lib. 1. cap. 14.* faith his is
good : yet this wanteth fome notes which he giues vnto his, as the leaues of Bryonie, &c.

But I more diligently confidering this Exotick fruit, finde fome prime notes which doe much
moue me (for I will ingenioufly profeffe what I thinke) to judge it the *Garyophyllon* of *Pliny*; for he,
Hift. Nat. lib. 12. cap. 7. after he hath treated of Pepper, addes thefe words; [There is befides in the
Indies a thing like to the Pepper corn, which is called *Garyophyllon*, but more great and fragil; they
affirm it growes in an Indian groue, it is brought ouer for the fmells fake.] Though this defcription be briefe and fuccinct, neither containes any faculties of the fruit it felfe, yet it hath manifeft
notes, which, compared with thofe which the fruit I here giue you poffeffe, you fhal find them very
like; as comparing them to Pepper corns, yet bigger and more fragil, as for the moft part thefe berries are : their fmell is alfo very pleafing, and commeth very neere to that of Cloves ; and for the
fmels fake only they were brought ouer in *Plinies* time. I found, this fruit being chewed made the
breath to fmell wel ; and it is credible, that it would be good for many other purpofes, if trial were
made.

CHAP.

Chap. 19. Of *Guaiacum* or *Indian Pock-wood*.

Guaiaci arboris ramul:
A branch of the Guaiacum tree.

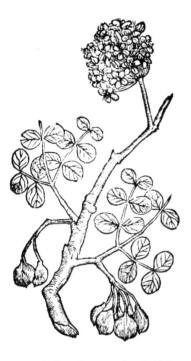

¶ *The Description.*

GVaiacum,which some call *Lignum San-ctum*: others, *Lignum vitæ*, is a well knowne wood,though of a tree vnknown, or at least not certainly knowne; for this fi-gure which I here giue you out of *Clusius*,was gotten, and the history framed as you shall hear by his own words, taken out of his *Scho-lia* vpon the 21 Chapter of *Monardus*. About the beginning(saith he)of the yeare 1601, I receiued from *Peter Garret* a branch of a foot long, which he writ was giuen him by a cer-tain Surgeon lately returned from America, for a branch of the tree *Guaiacum:* which if it be a branch of the true Guaiacum,then hath *Nicolas Monardus* sleightly enough set downe the history of this tree. I thus described the branch that was sent me.

This branch was a foot long, very writhen and distinguished with many knots,scarse at the lower end equalling the thickenesse of a writing pen or goose quil,hauing an hard and yellowish wood,and a wrinkeled barke of an ash colour:at the vpper end it was diuided into slender branches,wherof some yet retai-ned their leaues, and other some the floures and the rudiment of the fruit : the leaues, or more truly the wings or foot-stalkes of the leaues grew on slender branches one against another, each winged leafe hauing foure or six little leaues, alwaies growing by couples one against another,as in the Masticke tree; and these are thickish, round, and distinguished with many veins,which by reason of their drinesse(as I obserued)would easily fall off, leauing the foot-stalks naked,and only retaining the markes whereas the leaues had bin. In the knots or vpper bran-ches there grew as it were swellings,out of which together grew six,eight,ten,or more slender foot-stalks,some inch long,each carrying a floure not great,consisting of six little leaues (but whether white,yellow,or blew,I could not by reason of the drinesse iudge:)out of the middest of the floure grew many little threds, and in some the rudiment of the fruit began to appeare, hauing two cells, almost shaped like the seed-vessell of the common Shephears purse.

Thus much *Clusius*; who afterwards receiued the fruit from two or three, but the most perfect from the learned Apothecarie *Iohn Pona* of Verona: they are commonly parted into two parts or cels,yet hee obserued one with three : he found longish stones in them almost like those of *Euony-mus*,and they consisted of a very hard and hairy substance like to that of the Date stones, contai-ning a smooth kernell of a yellowish colour.

Now will I giue you the descriptions of *Monardus* : then what I haue obserued my selfe of this wood,which I must confesse is very little,yet which may giue some light to the ignorant. Of this wood (saith *Monardus*)many haue written many waies,saying that it is either Ebonie,or a kinde of Box,or calling it by some other names. But as it is a new kind of tree not found in these Regions, or any other of the whole world described by the Antients, but onely those of late discouered ; so this shal be a new tree to vs : howeuer it be,it is a large tree of the bignesse of Ilex, ful of branches, hauing a great matrix or blackish pith, the substance of the wood being harder than Ebonie : The bark is thick,gummy or fat,and when the wood is dry falls easil off: the leaues are smal and hard : the floure yellow;the which is followed by a round sollid fruit,containing in it seeds like those of the Medlar.

It growes plentifully in the Isles of *Sancto Domingo.*

Another

Another kind of this was afterwards found in the Iſland of S. Iohn de Puerto rico, neere to the former: it is alſo like the laſt deſcribed, but altogether leſſe, and almoſt without matrix or pith, ſmelling ſtronger, and being bitterer than the former: which being left, this is now in vſe, and of the wondrous effects it is called *Lignum ſanctum*; neither without deſert, being (experience giuing teſtimonie) it excels the other; yet both their faculties are admirable in curing the French diſeaſe, and therefore the water or decoction of both of them are drunk, either mixed together, or ſeuerally, both for the cure of the forementioned diſeaſe, as alſo againſt diuers other affects. Thus much for *Monardus* his deſcription.

The wood which is now in vſe with vs is of a large tree, whoſe wood is very heauy, ſollid, and fit to turne into bowles or the like, and all that I haue yet ſeen hath beene wholly without matrix or pith, and commonly it is of a dark browniſh colour ſomwhat inclining to yellow, hauing a ring of white ingirting it next to the barke: I haue obſerued a tree whoſe diametre hath bin two foot and a quarter, to haue had as little or leſſe of this white wood, than one whoſe diametre was 13 inches; and this which had thirteen inches had only a white circle about it of one inch in bredth. I thinke the yonger the tree is, the bigger the white circle is: the beſt wood is denſe, heauy, browniſh, leauing a quick and biting taſt in the decoction, as alſo his ſmell and colour. The bark of this wood is alſo denſe and heauy, of a hard ſubſtance and yellowiſh colour within, but rough and greeniſh, or elſe grayiſh without, and of ſomewhat a bitteriſh taſt. Thus much for the deſcription of the wood and his bark. Now let me ſay ſomewhat briefly of the temperature and qualities.

¶ *The Temperature and Vertues.*

A   It is iudged to be hot and dry in the ſecond degree: it hath a drying, attenuating, diſſoluing, and clenſiug facultie, as alſo to moue ſweat, and reſiſt contagion and putrefaction.

B   The decoction of the bark or wood of Guaiacum, made either alone or with other ingredients, as ſhall be thought moſt fit for the temper and age of the patient, is of ſingular vſe in the cure of the French Poxes, and it is the moſt antient and powerfull antidote that is yet knowne againſt that diſeaſe. I forbeare to ſpecifie any particular medicine made thereof, becauſe they are well enough knowne to all to whom this knowledge belongs, and they are aboundantly ſet downe by all thoſe that haue treated of that diſeaſe.

C   It alſo conduceth to the cure of the Dropſie, Aſthma, Epilepſie, the diſeaſe of the bladder and reines, paines of the joints, flatulencies, crudities, and laſtly all Chronicall diſeaſes proceeding from cold and moiſt cauſes: for it oftentimes works ſingular effects whereas other medicines little preuaile.

D   It doth alſo open the obſtructions of the liuer and ſpleen, warms and comforts the ſtomack and all the intrals, and helps to free them from any groſſe viſcous matter that may be apt to breed diſcaſes in them.

CHAP. 20.

Of Guayaua or Orange-Bay.

¶ *The Deſcription.*

SImon de Touar ſent *Cluſius* a branch of the tree which the Spaniards cal *Guayauas*, from which he drew this figure, thus deſcribing it. This branch, (ſaith *Cluſius*) whoſe vpper part together with the fruit I cauſed to be drawne, was ſome foot long, foure ſquare, alternately ſet with leaues growing by couples, being foure inches long, and one and a halfe or two broad, of the forme of Bay leaues, very firm, hauing a ſwelling rib running alongſt the lower ſide, with veins running obliquely from thence to the ſides, of an aſh or grayiſh colour beneath, but ſmooth aboue, with the veins leſſe appearing: which broken, though old, yet retained the ſmel of Bay leaues, and alſo after ſome ſort the taſte: the fruit was ſmooth, yet ſhriueled, becauſe peraduenture it was vnripe, of the bigneſſe of a ſmall apple, longiſh, blackiſh on the out-ſide like a ripe plumme, but within full of a reddiſh pulpe, of an acide taſt; and in the middle were many whitiſh ſeeds of the bignes of Millet, or thoſe that are in Figs.

Nicolas Monardus (as hee is turned into Latine by *Cluſius*) thus giues vs the hiſtory of *Guayauas*, in his ſixty fourth Chapter. It is a Tree (ſaith he) of an indifferent bigneſſe, and hath ſpreading branches, the leafe of the Bay, and a white floure like that of the Orange, yet ſomewhat bigger,

                                                                    and

Guayaua arboris ramus.
The Orange Bay.

and well ſmelling. It eaſily grows whereſoeuer it be ſowne, and ſo ſpreds and creepes that it is accounted as a weed, for it ſpoiles the ground of many paſtures with the too much ſpreding as brambles do. The fruit is like to our apples, of the bigneſſe of thoſe the Spaniards call *Camueſas*, green at the firſt, and of a golden colour when they be ripe, with their inner pulp white, and ſomtimes red; diuided: it hath foure cels, wherein lie the ſeeds, like thoſe of the Medlars very hard, of a browniſh colour, wholly ſtony, without kernell and taſte.

The fruit is vſually eaten, the rind being firſt **A** taken off; it is pleaſing to the palat, wholſome & eaſie of concoction: being green it is good in fluxes of the belly, for it powerfully bindes; and ouer or throughly ripe it looſeth the belly: but between both, that is, neither too green nor ouer-ripe, if roſted it is good both for ſound and ſick; for ſo handled it is wholeſomer, and of a more pleaſing taſte: that alſo is the better which is gathered from domeſticke and huſbanded trees: the Indians profitably bath their ſwolne legs in the decoction of the leaues, and by the ſame they free the ſpleene from obſtruction. The fruit ſeemes to be cold, wherefore they giue it roſted to ſuch as are in Feuers. It growes commonly in all the Weſt Indies. So much *Monardus*.

Chap. 21. *Of the Corall tree.*

¶ *The Deſcription.*

THe ſame laſt mentioned, *Simon de Touar* a learned and prime phyſition of Seuill ſent *Cluſius* three or foure branches of this tree, from whence he framed this hiſtory and figure. He writ (ſaith *Cluſ.*) that this tree grew in his garden, ſprung vp of ſeeds ſent from America, which had the name of Corall impoſed on them, by reaſon the floures were like Corall, but he did not ſet down their ſhape, writing only this in his letter: That hee had two little ſhrubs which had borne floures, and that the greater of them bore alſo cods full of large beans, but in the extreme Winter, which they had the yeare before, he loſt not only that tree and others ſprung vp of Indian ſeed, but alſo many other plants. Now ſeeing that this tree carries cods, I conjecture the floures were in form not vnlike to thoſe of Peaſe, or of the tree called *Arbor Iuda*, but of another colour, to wit, red like Corall; eſpecially ſeeing that in the Catalogue of his garden which he ſent me the yeare before, he had writ thus: [*Arbor Indica dicta Coral, ob ejus florem ſimilem Corallo, &c.* That is, An Indian tree called Corall, by reaſon of the floure like to Corall, whoſe leaues are very like thoſe of *Arbor Iuda*, but this hath thorns, which that wants.] And verily the branches which he ſent (for he writ he ſent the branches with the leaues, but the tree brought out ſome twice or thrice as big) had leaues not much vnlike thoſe of *Arbor Iuda*, but faſtned to a ſhorter foot-ſtalke, and growing one againſt another, with a ſingle one at the end of the branch, which was here and there ſet with ſharp & crooked prickles; but whether theſe branches are onely the ſtalkes of the leaues, or perfect branches, I doubt, becauſe all that he ſent had three leaues apiece: I could eaſily perſuade my ſelfe they were only leaues, ſeeing the vpper part ended in one leafe, and the lower end of one amongſt the reſt, yet ſhewed the place where it ſeemed it grew to the bough. But I affirme nothing, ſeeing there was none whereof I could enquire, by reaſon of his death who ſent them me, which hapned ſhortly after; yet I haue made the form of the leaues, with the maner as I conjectured they grow, to be delineated

neated

Coral arboris ramus.
A branch of the Coral tree.

neated in the figure which I heere giue you. Whether *Matthiolus* in the laſt edition of his Commentaries vpon *Dioſcorides* would haue expreſt this by the *Icon* of his firſt *Acacia*, which is prickly, and hath leaues reſembling thoſe of *Arbor Iudæ*, I know not : but if hee would haue expreſſed this tree, the painter did not well play his part.

After that *Cluſius* had ſet forth thus much of this tree, in his *Hiſt. rarior. plant.* the learned Dr. *Caſtaneda* a Phyſitian alſo of Sevill certified me (ſaith he) that the flours of this tree grow thick together at the tops of the branches, ten, twelue, or more hanging vpon ſhort footſtalks, growing out of the ſame place : whoſe figure he alſo ſent, but ſo rudely drawne, that I could not thereby haue come to any knowledge of the floures, but that he therewith ſent me two dried floures, by which I partly gathered their form. Now theſe floures were very narrow, 2 inches long or more, confiſting of three leaues, the vppermoſt wherof much exceeded the two narrow ones on the ſides both in length and bredth, and it was doubled ; but before the flour was opened it better reſembled a horne or cod, than a floure, and the lower end of it ſtood in a ſhort green cup in the midſt of the floure vnder the vpper leafe that was folded, but open at the top ; there came forth a ſmooth pointall, diuided at the top into nine parts or threds, whoſe ends of what colour they were, as alſo the threds, I know not, becauſe I could not gather by the dried flour, whoſe colour was quite decayed, and the picture it ſelfe expreſſed no ſeparation of the leaues in the floure, no forme of threds, but only the floures ſhut, and reſembling rather cods than floures, and thoſe of a deep red colour. But if I could haue ſeen them freſher, I ſhould haue been able to haue giuen a more exact deſcription : wherefore let the Reader take in good part that which I haue here performed. Thus much *Cluſius*.

Chap. 22. *Of the Sea Lentill.*

¶ *The Deſcription.*

SOme call this *Vna marina* ; and others haue thought it the *Lenticula marina* of *Serapio*, but they are deceiued ; for his *Lenticula marina* deſcribed in his 245 chap. is nothing els but the *Muſcus marinus* or *Bryon thalaſsion* deſcribed by *Dioſcorides, lib. 4. cap. 99.* as any that compare theſe two places together may plainly ſee.

1 The former of theſe hath many winding ſtalks, whereon grow ſhort branches ſet thick with narrow leaues like thoſe of Beluidere or Beſom flax, and amongſt theſe grow many skinny hollow empty round berries of the bigneſſe and ſhape of Lentils, whence it takes the name. This growes in diuers places of the Mediterranian and Adriatick ſeas.

2 This differs little from the former, but that the leaues are broader, ſhorter, and ſnipt about the edges. But this being in probabilitie the *Sargazo* of *Acoſta*, you ſhall heare what he ſays thereof. In that famous and no leſſe to be feared nauigation del Sergazo (for ſo they which ſaile into the Indies call all that ſpace of the Ocean from the 18 to the 34 degree of Northerly latitude) is ſeen a deepe and ſpatious ſea couered with an herb called Serguazo, being a ſpanne long, wrapped with the tender branches as it were into balls, hauing narrow and tender leaues ſome halfe inch long,

much

1 *Lenticula marina angustifolia.*
Narrow leafed sea Lentill.

2 *Lenticula marina serratis folijs.*
Cut leafed sea Lentill.

much snipt about the edges, of colour reddish, of taste insipid, or without any sensible biting, but what is rather drawn from the salt water, than naturally inherent to the plant. At the setting on of each leafe growes a seed round like a pepper corne, of a whitish colour, and somtimes of white and red mixt, very tender when it is first drawn forth of the water, but hard when it is dried, but by reason of the thinnesse very fragil, and full of salt water: there is no root to be obserued in this Plant, but only the marks of the breaking off appears, and it is likely it growes in the deep and sandy bottom of the sea, and hath small roots: yet some are of opinion that this herb is plucked vp and caried away by the rapid course of waters that fall out of many Islands into the Ocean. Now the Master of the ship wherein I was did stiffely maintain this opinion, and in sailing here we were becalmed; but as far as euer we could see we saw the sea wholly couered with this plant; and sending down some yong sailers which should driue the weeds from the ship and clense the water, we plainly saw round heaps thereof rise vp from the bottom of the sea, where by sounding we could find no bottom.

This plant pickled with salt and vineger hath the same tast as Sampier, and may be vsed in stead **A** thereof, and also eaten by such as saile, in stead of Capers. I willed it should be giuen newly taken forth of the sea, to Goats which we caried in the ship, and they fed vpon it greedily.

I found no faculties thereof; but one of the Sailers troubled with a difficulty of making water, **B** casting out sand & grosse humors, ate thereof by chance both raw and boiled, only for that the tast thereof pleased him; after a few daies he told me that he found great good by the eating thereof, and he tooke some of it with him, that so he might vse it when he came ashore. Hitherto *Acosta.*

Chap. 23. *Of the Sea Feather.*

Myriophyllum marinum.
The Sea Feather.

¶ *The Description.*

THis elegant plant, which *Clusius* receiued from *Cortusus* by the name of *Myriophyllum Pelagicum*, is thus described by him. As much (saith he) as I could conjecture by the picture, this was some cubit high, hauing a straight stalk, sufficiently slender, diuided into many branches or rather branched leaues, almost like those of Fern, but far finer, bending their tops like the branches of the Palme, of a yellowish colour: the top of the stalke adorned with lesser leaues ended in certain scales or cloues framed into an head; which are found to contain no other seed than tender plants already formed, shaped like the old one: which falling sink to the bottom of the sea, and there take root and grow, and so become of the same magnitude as the old one from whence they came. The stalk is fastned with most slender and more than capillary fibres, in sted of a root, not vpon rocks and oister shells, as most other sea plants are, but on sand or mud in the bottom of the sea: this stalk when it is dry is no lesse brittle than Coralline or glasse; but greene & yet growieg it is as tough and flexible as *Spartum*, or Matweed.

¶ *The Place.*

It groweth in the deepest streames of the Illyrian sea, whence the fishermen draw it forth with hooks or other instruments, which they call Spern. The whole plant though dried retains the faculties.

¶ *The Names.*

The Italian fishermen call it *Penachio delle Ninfe*, and *Palma de Nettuno*: some also, *Scettro di Nettuno.*

¶ *The Vertues.*

A   They say it is good against the virulent bites of sea Serpents, and the venomous stings or pricks of fishes.

B   Applied to small green wounds it cures them in the space of 24 houres.

C   *Cortusus* writ, that he had made triall thereof for the killing and voiding of worms, and found it to be of no lesse efficacie than any Coralline, and that giuen in lesse quantitie.

Chap. 24. *Of the Sea Fan.*

¶ *The Description.*

THis elegant shrub groweth vpon the rocks of the sea (where it is sometimes couered with the water) in diuers places; for it hath been brought both from the East and West Indies, and as I haue bin informed it is to be found in great plenty vpon the rocks at the Burmuda Isles. *Clusius* calls

Frutex marinus reticulatus.   Sea Fan.

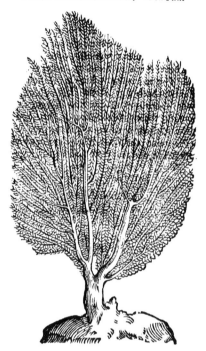

calls it *Frutex marinus elegantißimus*, & thinks it may be referred to the *Palma marina* of *Theophraſtus*. *Bauhine* hath referred it to the *Corallines*, calling it *Corallina cortice reticulato maculoſopurpuraſcente*. It growes ſometimes to the height of three foot, hauing a ſtalk ſome handfull or two high before it part into branches; then is it diuided into three, foure, or more branches, which are ſubdiuided into infinite other leſſer ſtrings, which are finely interwouen & ioyned together as if they were netted, yet leauing ſomtimes bigger, otherwhiles leſſer holes; and theſe twiggy branches become ſmaller and ſmaller, the farther they are from the root, and end as it were in ſmall threds. Theſe branches grow not vp on euery ſide, as in other plants, but flat one beſides another, ſo that the whole plant reſembles a fan, or a cabbage leafe eaten full of holes; yet ſometimes vpon the ſides come forth other ſuch fan like branches, ſome bigger, ſome leſſe, ſometimes one or two, otherwhiles more. The inner ſubſtance of this Sea Fan is a blackiſh tough and hard wood, and it is all couered ouer with a rough Coral-like ſtony matter, of a reddiſh or purpliſh colour, and this you may with your naile or a knife ſcrape off from the ſmooth and black wood.

I know no vſe of this, but it is kept for the beauty and raritie thereof, by many louers of ſuch Curioſities, amongſt which for the rareneſſe of the ſtructure this may hold a prime place.

Chap. 25. *Of China, and baſtard China.*

¶ The Deſcription.

THis root which is brought from the remoteſt parts of the world, and is in frequent vſe with ns, hath not bin known in Europe little aboue ninety yeares: for *Garcias ab Orta* the Portugall Phyſitian writes, That he came to the firſt knowledge thereof in the Eaſt Indies in the yeare 1535, and that by this means, as he relates it: It hapned (ſaith he) that about this time a merchant in the Iſle Diu told the noble gentleman Sᵣ *Mart. Alfonſo de Souſa* my Patron, by what meanes he was cured of the French Poxes, which was by a certain root brought from China; whoſe faculties he much extolled, becauſe ſuch as vſed it needed not obſerue ſo ſtrict a dyet as was requiſite in the vſe of Guajacum; but ſhould only abſtain from beefe, porke, fiſh, and crude fruits; but in China they do not abſtain from fiſh, for they are there great gluttons. When the report of this root was divulged abroad, euery man wonderfully deſired to ſee and vſe it, becauſe they did not wel like the ſtrict dyet they were forced to obſerue in the vſe of Guajacum. Beſides, the inhabitants of theſe countries by reaſon of their idle life are much giuen to gluttony. About this time the China ſhips arriue at Malaca, bringing a ſmal quantity of this root for their own vſe; but this little was ſought for with ſuch earneſtneſſe that they gaue an exceſſiue rate for it: but afterward the Chinois bringing a greater quantity, the price fell, and it was ſold very cheape. From this time Guajacum began to be out of vſe, and baniſhed the Indies, as a Spaniard that would famiſh the Natiues. Thus much *Garcias*, concerning the firſt vſe thereof in the Eaſt Indies.

1   The China now in vſe is a root of the largeneſſe of that of the ordinary Flag, or *Iris paluſtris*, and not much in ſhape vnlike thereto, but that it wants the rings or circles that are imprinted in the other : the outer coat or skinne of this root is thin, ſometimes ſmooth, otherwhile rugged, of a browniſh red colour, and not to be ſeparated from the ſubſtance of the root, which is of an indiffe-rent firmeneſſe, being not ſo hard as wood, but more ſollid than moſt roots which are not of ſhrubs or trees : the colour is ſometimes white, with ſome very ſmall mixture of redneſſe ; otherwhiles it hath a greater mixture of red, and ſome are more red than white : it is almoſt without taſte, yet that it hath is dry, without any bitterneſſe or acrimony at all. The beſt is that which is indifferently ponderous, new, firme, not worme-eaten, nor rotten, and which hath a good and freſh colour, and that either white, or much inclining thereto. The plant whoſe root is this (if we may beleeue *Chri-ſtopher A Coſta*) hath many ſmall prickly and flexible branches, not vnlike the *Smilax aſpera*, or the prickly Binde-weed : the biggeſt of theſe exceedeth not the thickeneſſe of ones little finger. The leaues are of the bigneſſe of thoſe of the broad leaued Plantaine : the roots as large as ones hand, ſometimes leſſe, ſollid, heauie, white, and alſo ſometimes redde, and many oft times growing to-gether.

1   *China vulgaris Officinarum.*
     True China.

2   *Pſeudo-China.*
     Baſtard China.

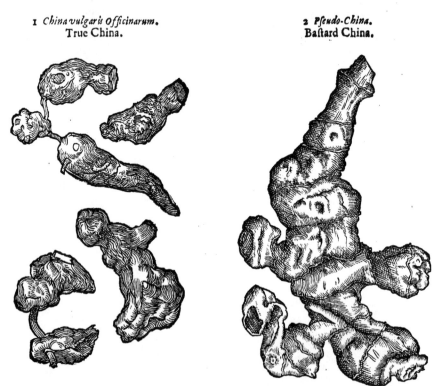

It groweth abundantly in the territory of China, and is alſo found in Malaber, Cochin, Cranga-nor, Coulan, Tanor and other places.

The Chinois call it *Lampatan* : in Decan they call it *Lampatos* : in Canarin, *Bouti* : the Arabians, Perſians and Turks terme it *Choph-China*.

2   This other root, whoſe figure you ſee here expreſt, was ſent from London to *Cluſius* in the yeare 1591, by *Iames Garret*, being brought out of Wingandecaow, or Virginia, with this inſcripti-on, *China ſpecies*, A kind of China. *Cluſius* cauſed this figure thereof to be drawne, and thus deſcri-beth it. This root (ſaith he) was very knotty, and formed with out-growings, or bunches ſtanding out, of a reddiſh colour, and it yet retained at the top ſome part of the ſtalke, being ſomewhat like vnto that of *Smilax aſpera*, or common rough Binde-weed, hard, wooddy, and full of veines, as the ſtalkes of *Smilax aſpera* : the ſubſtance of the root was alſo reddiſh, as the root of the common Flag, at the firſt of a ſaltiſh taſte, it being old, (for ſo it was when I receiued it) and then drying. Now
                                                        I iudge

I iudge this is the ſame that the writer of the Virginian Hiſtory mentions in his chapter of roots, and ſaith, it was brought into England for China, though the Natiues knew no vſe thereof: but they vſe another root very like China, which they call *Tſinaw*, of which being cut, beaten, and preſſed out with water, they draw a iuice wherewith they make their bread. Thus much *Cluſius*; to whoſe words I thinke it not amiſſe to adde that which Mʳ *Thomas Hariot* (who was the Writer of the Virginian hiſtory, here mentioned by *Cluſius*) hath ſet downe concerning this thing.

Tſinaw (ſaith he) is a kinde of root much like vnto that which in England is called the China root, brought from the Eaſt-Indies. And we know not any thing to the contrary but that it may be of the ſame kinde. Theſe roots grow many together in great cluſters, and doe bring forth a Brier ſtalke, but the leaſe in ſhape is far vnlike: which being ſupported by the trees it groweth neereſt vnto, wil reach or clime to the top of the higheſt. From theſe roots while they be new or freſh, being chopt into ſmall pieces and ſtampt, is ſtrained with water a iuyce that maketh bread, and alſo being boiled, a very good ſpoonemeat in manner of a gelly, and is much better in taſte, if it be tempered with oyle. This *Tſinaw* is not of that ſort which by ſome was cauſed to be brought into England for the China root; for it was diſcouered ſince, and is in vſe as is aforeſaid; but that which was brought hither is not yet knowne, neither by vs, nor by the inhabitants, to ſerue for any vſe or purpoſe, although the roots in ſhape are very like. Thus much *Hariot*.

❡ The Temperature and Vertues.

China is thought to be moderately hot and dry: the decoction therof made alone or with other 　**A** things, as the diſeaſe and Symptoms ſhall require, is much commended by *Garcias*, for to cure the French Pox, but chiefly that diſeaſe which is of ſome ſtanding: yet by moſt it is iudged leſſe powerfull than *Guajacum*, or *Sarſapatilla*.

It attenuates, moues ſweat, and dries, and therefore reſiſts putrifaction: it ſtrengthens the liuer, 　**B** helpes the dropſie, cures maligne vlceres ſcabbes, and lepry. It is alſo commended in conſumptions.

The decoction of this root, ſaith *Garcias*, beſides the diſeaſes which haue community with the 　**C** Pox, conduces to the cure of the Palſie, Gout, Sciatica, ſchirrous and œdematous tumours. It alſo helps the Kings-Euill. It cureth the weakeneſſe of the ſtomacke, the inueterate head-ache, the ſtone and vlceration of the bladder; for many by the vſe of the decoction hereof haue beene cured, which formerly receiued helpe by no medicine.

C H A P. 26. 　Of Coſtus.

❡ The Deſcription.

THis ſimple medicine was brieſely deſcribed by *Dioſcorides*, who mentions three kindes hereof, but what part of a plant, whether root, wood or fruit, he hath not expreſt: but one may probably conjecture it is a root, for that hee writes toward the end of the Chapter where he treats thereof, *lib. 1. cap. 15.* that it is adulterated by mixing therewith the roots of *Helenium commagenum*; now a root cannot well be adulterated but with another. Alſo *Pliny, lib. 12. cap. 12.* calls it a root; but neither any of the antient or moderne Writers haue delineated the plant, whoſe root ſhould be this *Coſtus. Dioſcorides* makes three ſorts, as I haue ſaid: the Arabian being the beſt, which was white, light, ſtrong, and well ſmelling: the Indian, which was large, light, and blacke: the Syrian, which was heauie, of the colour of Box, and ſtrong ſmelling. Now *Pliny* makes two kindes, the blacke, and the white, which he ſaith is the better; ſo I iudge his blacke to be the Indian of *Dioſcorides*, and his white, the Arabian. Much agreeable to theſe (but whether the ſame or no, I do not determine) are the two roots whoſe figures I here preſent to your view, and they are called by the names of *Coſtus dulcis* (I thinke they ſhould haue ſaid *odoratus*) and *Coſtus amarus*.

1 The firſt of theſe, which rather from the ſmell than taſte, is called ſweet, is a pretty large root, light, white, and well ſmelling, hauing the ſmell of Orris, or a violet, but ſomewhat more quick and piercing, eſpecially if the root be freſh, and not too old: it is oft times diuided at the top into two, three, or more parts, from whence ſeuerall ſtalkes haue growne, and you ſhall ſomtimes obſerue vpon ſome of them pieces of theſe ſtalkes ſome two or three inches long, of the thickeneſſe of ones
little

little finger,crefted and filled with a foft pith like the ftalks of Elder, or more like thofe of the bur
Docke : the taft of the root is bitter,with fome acrimony,which alfo *Diofcorides* requires in his, for
he faith the taft fhould be biting and hot.Thus much for the firft,being *Coftus dulcis* of the fhops.

1 *Coftus Indicus five odoratus.*
Indian or fweet fmelling Coftus.

2 *Coftus officinarum Lobelij.*
Bitter Coftus.

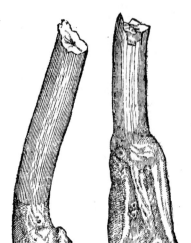

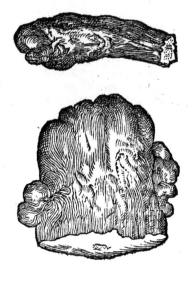

2   The fecond,which is the *Coftus amarus*, and it may be the Indian of *Diofcorides*,and *Niger* of
Pliny,is a root black both within and without,light,yet very denfe.   It feemeth to be of fome large
root,for that it is brought ouer cut into large piéces,of the bigneffe of ones finger,fometimes bigger,fomtimes leffer,which it feems is for the more conuenient drying thereof,for a large root,vnles
it be cut into pieces can fcarfe be well dried:the tafte of this is bitter,fomwhat clammy & ingrate:
the fmell is little or none.

There are fome other roots which haue bin fet forth by late writers for *Coftus*, but becaufe they
are neither in vfe,known here with vs,nor more agreeable to the defcription of the antients,I haft-
ning to an end,am willing to paffe them ouer in filence.

¶ *The Temperature and Vertues.*

A   It hath a heating and attenuating faculty, and therefore was vfed in oile to annoint the bodie a-
gainft the cold fits of Agues,the Sciatica,and when it was needful to draw any thing to the fuper-
ficies of the body
B   It is alfo conuenient to moue vrine,to procure the termes,to help ftrains,convulfions,or cramps
and pains in the fides,and by reafon of the bitterneffe it kils wormes.
C   It is good to be drunk againft the bite of the viper,againft pains of the cheft, and windineffe of
the ftomack,taken in wine and wormwood;and it is vfed to be put into fundry Antidotes.

C H A P.

CHAP. 27.   *Of Drakes Root or Contra-yerva.*

¶ *The Deſcription.*

THat root which of late is known in ſome ſhops by the Spaniſh name *Contra-yerva*, is the ſame which *Cluſius* hath ſet forth by the title of *Drakena radix :* wherfore I will giue you the hiſto-rie of *Cluſius*, and thereto adde that which *Monardus* writes of the *Contra-yerva*. For although *Bauhine* and the Author of the *Hiſtoria Lugdunenſis* ſeeme to make theſe different; yet I finde that both *Cluſius* his figure and hiſtorie exactly agree with the roots ſent vs from Spaine by that title. Wherefore I ſhall make them one, till ſome ſhal ſhew me how they differ : and *Cluſius* ſeems to be of this minde alſo, who deſired but the degree of heate which *Monardus* giues theſe, and that is but the ſecond degree : now theſe haue no taſte at the firſt, vntill you haue chewed them a pretty while, and then you ſhall finde a manifeſt heate and acrimonie in them, which *Cluſius* did alſo obſerue in his.

In the yeare (ſaith *Cluſius*) 1581, the generous Knight Sr *Francis Drake* gaue me at London cer-tain roots, with three or foure Peruvian Beazor ſtones, which in the Autumne before (hauing fini-ſhed his voiage, wherein paſſing the ſtraights of Magellan he had incompaſſed the world) hee had brought with him, affirming them to be of high eſteem amongſt the Peruvians. Now for his ſake that beſtowed theſe roots vpon me, I haue giuen them the title of *Drakena radix*, or *Drakes* root, and haue made them to be expreſſed in a table, as you may here ſee them.

1 *Drakena radix.* Contra-yerva.	2 *Radix Drakenæ affinis.* Another ſort of Contra-yerva.

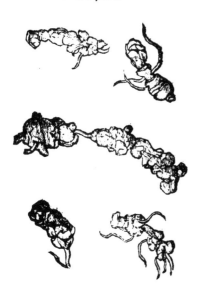 

Theſe roots were for the moſt part ſome halfe inch thicke, longiſh, now and then bunching out into knots and vnequall heads, and their tops looked as if they were compoſed of thicke ſcales, al-moſt like thoſe of the *Dentaria enneaphyllos*; blackiſh without, wrinckled and hard, becauſe dried: their inner part was white, they had ſlender fibres here and there growing out of them, and ſome more thick and large, hard alſo and tough, at which hung other knots. I obſerued no manifeſt ſmel they had, but found them to haue a taſte ſomwhat aſtringent, and drying the tongue at the firſt; but being long chewed they left a quick and pleaſing acrimony in the mouth.

It ſeemed to haue great affinitie with the *Radix S . Helenæ*, whereof *Nic. Monardus* ſpeakes in his bookes of the Simple Medicines brought from the Weſt Indies: but ſeeing *N . Eliot* (who accompanied

nied S.r *Francis Drake* in that voyage,said that the Spaniards in Peru had them in great request,and they could not easily be got of them , and that he had learned by them,that the leaues were present poison,out the root an antidote,and that not only against the same poison, but also against other ; and that it strengthned the heart and vitall faculties,if it were beaten to pouder, and taken in the morning in a little wine ; and giuen in water it mitigated the heate of Feuers. By reason of these faculties it should much agree with the *Radix Contra yerva*,whereof *Monardus* writes in the same booke: yet in these I required the aromatick tast and degree of heate which he attributes to these roots. Thus much *Cluf.*

A   From Charchis a prouince of Peru(saith *Monard.*) are brought certain roots very like the roots of *Iris*,but lesse,and hauing the smell of Fig leaues. The Spaniards that liue in the Indies cal them *Contra-yerva*,as if you should say an antidote against poison ; because the pouder of them taken in white wine is a most present remedie against all poison of what kinde soeuer it be(only sublimate excepted,whose malignity is only extinguished by the drinking of milk)it causes them to be cast vp by vomit,or euacuated by sweat. They also say that Philtres or amorous potions are cast forth by drinking this pouder.It also killeth wormes in the belly. The root chewed hath a certain aromatick taste ioyned with acrimony:wherfore it seems hot in the second degree:Thus far *Monardus.*

2   *Clusius,Exot.l.4.c.11.*being the next after *Drakena radix*,describes this root, whose figure, I giue you in the second place,and that by the same title as it is here set forth. These roots,saith hee, seemed somewhat like the *Drakena radix* which were found in the great ship which brought backe the Viceroy from the East Indies,&was taken by the English:for they were tuberous, & as much as one may gather by their form,crept vpon the surface of the earth,hauing vpon them many hairs and fibres,and being of a sooty colour,yet somwhat inclining to yellow,dying the spittle in chewing them,and being bitter. They as yet retained foot-stalks of the leaues,but of what fashion they were no man can easily guesse. But it is very likely they were of great vse among the Indians,seeing the Viceroy brought them together with other pretious medicines growing in the East Indies. *Iames Garret* sent this to *Clusius*,with the little plant dried,whose figure you see exprest by it.

Chap. 28.   *Of* Lignum Aloes.

Lignum Aloes vulgare.

¶ *The Description.*

IT is a Question,whether the *Agallochum* described *ca.21.li.1.*of *Dioscorides*,be the same which the later Greekes and shops at this time call *Xylo-aloe* or *Lignum aloes.* Many make them the same;others, to whose opinion I adhere, make them different : yet haue not the later shewed vs what *Agallochum* should be;which I notwithstanding wil do. And though I do not now giue you my arguments,yet I will point at the things, and shew positiuely my opinions of them.

The first & best of these is that which some cal *Calumbart*:others,*Calumba*,or *Calambec* : This is of high esteem in the Indies, and seldome found but among the Princes and persons of great qualitie ; for it is sould oft times for the weight in gold. I haue not seen any thereof but in beads.It seemeth to be a whiter wood than the ordinary , of a finer grain,not so subiect to rot,of a more fragrant smel and but light.

The second sort, which is vsually brought ouer, and called in Shops by the name of *Lignum Aloes*, is also a precious and odoriferous wood,especially burnt:the sticks of this are commonly knotty and vnsightly, some parts of them being white, soft, and doted ; other-some dense, blackish,

blackiſh,or rather intermixt with black and white veins,but far more black than white ; which put to the fire will ſweat out an oily moiſture,and burnt yeeld a moſt fragrant odor. This I take for the true *Xyloaloe* of the late Greeks; and the *Agalugen* of *Auicen* ; and that they in the Indies call *Agula.*

The third is a wood of much leſſe price than the former,and I conjecture it may well be ſubſtituted for *Thus* ; which I take to be the *Agallochum* of *Dioſcorides* ; the *Lignum Aloes ſylueſtre* of *Garcias,* and the *Agula braua* of *Linſcoten.* It is a firm ſollid wood ſomwhat like that of Cedar,not ſubject to rot or decay : the colour thereof is blackiſh, eſpecially on the out ſide, but on the inſide it is oft times browniſh and ſpeckled,containing alſo in it an oily ſubſtance,and yeelding a ſweet pleaſing ſmel when it is burnt,but not like that of the two former: the taſte alſo of this is bitterer than that of the former;and the wood(though denſe and ſollid) may be eaſily cleft longways.It is alſo a far handſomer and more ſightly wood than the former,hauing not many knots in it.

Garcias ab Orta thus deſcribes the tree that is the *Lignum Aloes* (I judge it is that I haue ſet forth in the ſecond place) It is,ſaith he,like an olive tree, ſomtimes larger:the fruit or floure I could not yet ſee,by reaſon of the difficulties and dangers which are to be vndergone in the accurat obſeruation of this tree(Tigers frequently there ſeeking their prey.) I had the branches with the leaues brought me from Malaca. They ſay the wood new cut down hath no fragrant odor,nor til it be dried;neither the ſmell to be diffuſed ouer the whole matter of the wood, but in the heart of the tree; for the bark is thick,and the matter of the wood without ſmell. Yet may I not deny,but the barke and wood putrifying that oily fat moiſture,may betake it ſelfe to the heart of the tree and make it the more odoriferous. But there is no need of putrifaction to get a ſmell to the *Lignum Aloes* ; for there are ſundry ſo expert and skilfull in the knowledge thereof,that they will iudge whether that new cut down will be odoriferous or no.For in all ſorts of wood ſome are better than other ſome. Thus far *Garcias* ; where ſuch as are deſirous may ſee more vpon this ſubiect.

¶ *The Nature and Vertues.*

It is moderatly hot and dry,and alſo of ſomwhat ſubtil parts. Chewed it makes the breath ſmel **A** ſweet,and burnt it is a rich perfume.

Taken inwardly it is good to help the too cold and moiſt ſtomack,as alſo the weak liuer. **B**

It is commended likewiſe in dyſenteries and pleuriſies ; and put alſo into diuers cordiall medi- **C** cines and antidotes as a prime ingredient.

Chap. 29. *Of Gedwar.*

1 *Gedwar, aut Geidnar.*   2 *Zedoaria exactior icon.* A better figure of Zedoary.

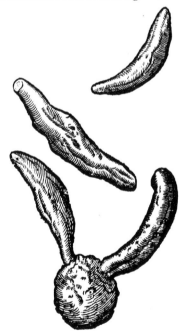

¶ *The Description.*

IN the chapter of Zedoary (which I made the 28 of the first booke) I might fitly haue giuen you this history of Gedwar, which is thought to be that described by *Auicen,li.2.c.734.* and a kind of Zedoary. *Garcias* saith, Gedwar is at a high rate, and not easie to be found, vnlesse with the Indian Mountibanks and juglers, called *Iogues*, which go vp and down the country like rogues, and of these the Kings and noblemen buy *Geiduar.* It is good for many things, but chiefly against poisons and the bites and stings of venomous creatures. Now *Clusius* in his *Auctarium*, at the end thereof giues this figure, with the following history.

1 Because *Garcias* (saith he, *cap.42. lib.1. Aromat.hist.*) treating of Zedoary, writes that *Auicen* called it Gedwar ; and saith that it is of the magnitude of an acorn, and almost of the same shape ; I in my notes at the end of that Chapter affirmed that it was not knowne in Europe, and hard to be known. But in the yere 1605, *Iohn Pona* sent me from Verona together with other things, two roots written on by the name of *Gedwar verum*: they were not much vnlike a longish acorn, or (that I may more truly compare them) the smaller bulbs of an Asphodil or *Anthora*: the one of them was whole and not perished, the other rotten and broken, yet both of them very hard and sollid, of an Ash colour without, but yellowish within; which tasted, seemed to possesse a heating faculty & acrimony.

But although I can affirm nothing of certaintie of this root, yet I made the figure of the wholler of them to be exprest in a table, that so the form might be conceiued in ones mind more easily, than by a naked description. Let the Studious thank *Pona* for the knowledge hereof. Thus far *Cluf.*

2 In the 28 chapter of the first book I gaue the figure of Zedoary out of *Clusius*, hauing not at that time this figure of *Lobel*, which presents to your view both the long and the round, with the manner how they grow together, being not seuerall roots, but parts of one and the same.

CHAP. 30.   *Of Rose-wood.*

Aspalathus albicans torulo citreo.
White Rose-wood.

Aspalathus rubens.
Reddish Rose wood.

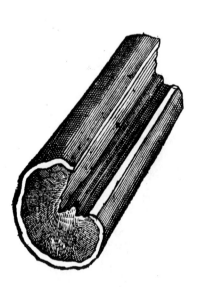

¶ *The Description.*

BOth these as also some other woods are referred to the *Aspalathus* described by *Dioscorides, lib.* 1. *c.* 19. But the later of these I take to be the better of the two sorts there mentioned. The first of them is whitish without, hauing a yellowish or citrine coloured round in the middle: the taste is hottish, and smell somewhat like that of a white-Rose.

The other hath also a small ring of white, next the thicke and rugged barke, and the inner wood is of a reddish colour, very dense, follid and firme, as also indifferent heauie: the smell of this is also like that of a Rose, whence they vulgarly call it *Lignum Rhodium*, Rose-wood, rather than from Rhodes the place where the later of them is said to grow.

¶ *The Faculties out of* Dioscorides.

It hath a heating faculty with astriction, whence the decoction thereof made in wine is conue- A nient to wash the vlcers of the mouth, and the eating vlcers of the priuities and such vncleane sores as the *Ozæna* (a stinking vlcer in the nose so called.)

Put vp in a pessary it drawes forth the childe, the decoction thereof staies the loosenesse of the B belly, and drunke it helpes the casting vp of bloud, the difficulty of making water, and windinesse.

AT the end of this Appendix I haue thought good to giue you diuers descriptions of Plants, which I receiued from my often mentioned friend Mr *Goodyer*, which also were omitted in their fitting places, partly through haste, and partly for that I receiued some of them after the printing of those chapters wherein of right they should haue beene inserted. They are most of them of rare and not written of Plants, wherefore more gratefull to the curious.

Hieracium stellatum Boelij.

THis plant is in round, hairy, straked, branched stalks, and long, rough, blunt indented leaues, like to *Hieracium falcatum*, but scarce a foot high: the floures are also yellow, three times smaller: which past, there succeed long crooked slender sharpe pointed cods or huskes, neere an inch long, spreading abroad, star-fashion, wherein a long seed is contained: this hath no heads or woolly down like any of the rest, but onely the said crooked coddes which doe at the first spread abroad. The root is small, threddy, full of milkie juyce, as is also the whole plant, and it perisheth when the seed is ripe.

Hieracium medio nigrum flore majore Boelij.

This hath at the first spreading vpon the ground many long, narrow, green, smooth leaues, bluntly indented about the edges, like those of *Hieracium falcatum*, but smaller: amongst which rise vp three, foure, or more, small, smooth, straked round stalkes, diuided into other branches, which grow longer than the stalkes themselues, leaning or trailing neere the ground: the flours grow on the tops of the stalkes, but one together, composed of many pale yellow leaues, the middle of each floure being of a blackish purple colour.

Hieracium medio nigrum flore minore Boelij.

This is altogether like the last before described in stalkes and leaues: the floures are also of a blackish purple in the middle, but they are three times smaller.

Hieracium lanosum.

There groweth from one root three, foure or more round vpright soft cottony stalkes, of a reasonable bignesse, two foot high, diuided into many branches, especially neere the top, whereon groweth at each diuision one broad sharpe pointed leafe, diuided into corners, and very much crumpled, and also very soft cottony and woolly, as is the whole plant: the floures are small, double, of a pale yellow colour, very like those of *Pilosella repens*, growing clustering very many together at the tops of the stalkes and branches, forth of small round soft cottony heads: these foure plants grew from

seed

ſeed which I receiued from Mr *Coys*, 1620. and I made theſe deſcriptions by the Plants the 22. of Auguſt, 1621.

Blitum ſpinoſum : eſt Beta Cretica ſemine Bauhini, Matth.
pag. 371.

This ſendeth forth from one root many round greene ſtrailing, joynted, ſmall branches, about a foot long : the leaues are of a light greene colour, and grow at euery joynt one, ſomewhat like the leaues of great Sorrell, but they are round topped without barbes or eares below, or any manifeſt taſte or ſmell, very like the leaues of Beets, but much ſmaller: the floures grow cluſtering together about the joynts, and at the tops of the branches ſmall and greeniſh, each floure conteining fiue or ſix very ſmall blunt topped leaues, and a few duſty chiues in the middle : which paſt, there commeth great prickly ſhriuelled ſeed, growing euen cloſe to the root, and vpwards on the joynts, each ſeed hauing three ſharpe prickes at the top growing ſide-waies, which indeed may be more properly called the huſke; which huſke in the in-ſide is of a darke reddiſh colour, and containeth one ſeed in forme like the ſeed of *Flos Adonis*, round at the lower end, and cornered towards the top, and ſharp pointed, couered ouer with a darke yellowiſh skin ; which skin pulled away, the kernell appeareth yellow on the outſide, and exceeding white within, and will with a light touch fall into very ſmall pouder like meale.

Geranij Batici ſpecies, Boelij.

This hath at the beginning many broad leaues, indented about the edges, ſomwhat diuided, like thoſe of *Geranium Creticum*, but of a lighter greene colour, and ſmaller : amongſt which grow vp many round hairy kneed trailing branches, diuided into many other branches, bearing leaues like the former but ſmaller, and no more diuided. The floures are ſmall like thoſe of *Geranium Moſchatum*, but of a deeper reddiſh colour, each floure hauing fiue ſmall round topped leaues : after followeth ſmall long hairy ſeed, growing at the lower end of a ſharpe pointed beake like that of *Geranium Moſchatum* : the whole plant periſheth when the ſeed is ripe.

Boelius a Low-countrey man gathered the ſeeds hereof in Bætica a part of Spaine, and imparted them to Mr *William Coys*, a man very skilfull in the knowledge of Simples, who hath gotten plants thereof, and of infinite other ſtrange herbes, and friendly gaue me ſeeds thereof, and of many other, *Anno*, 1920.

Antirrhinum minus flore Linariæ, luteum inſcriptum.

This hath at the firſt many very ſmall, round, ſmooth branches from one root, trayling on the ground, about foure or fiue inches long, ſet with many ſmall greene ſhort ſharp pointed leaues, like thoſe of *Serpillum*, but that theſe are longer, ſmooth, and three or foure growing oppoſite one againſt another : amongſt which riſe vp fiue or ſix, ſometimes ten or twelue vpright round ſmooth little ſtalkes a cubit high, diuided into branches bearing ſmall long ſmooth greene leaues, growing without order, as narrow as the vpper leaues of *Oenanthe Anguſtifolia* : at the toppes of the ſtalkes and branches grow cluſtering together fiue ſix or more ſmall yellow floures, flouring vpwards, leauing a long ſpike of very ſmall huſkes, each huſke hauing a ſmall line or chinke as though two huſkes were joyned together, the one ſide of the huſke being a little longer than the other, wherein is contained exceeding ſmall blackiſh ſeed. The root is very ſhort, ſmall, and white, with a few threds, and periſheth at Winter.

This plant is not written of that I can finde. I receiued ſeed thereof from Mr *William Coys* often remembred.

Linaria minor æſtiua.

The ſtalkes are round, ſmooth, of a whitiſh greene colour, a foot high, weake, not able to ſtand vpright : whereon grow long narrow ſharpe pointed leaues, moſt commonly bending or turning downewards. The floures grow in ſpikes at the tops of the branches, yet not very neere together, and are very ſmall and yellow, with a ſmall taile : the ſeed of this plant is ſmall, flat, and of a blackiſh gray colour, incloſed in ſmall round huſkes, and you ſhall commonly haue at one time floures and ripe ſeed all on a ſtalke. The whole plant is like to the common *Linaria*, but that it is a great deale leſſer, and the floures are ſix times as ſmall, and periſh at Winter. I alſo receiued ſeeds thereof from Mr *William Coys*.

Scor-

Scorpioides multiflorus Boëlij.

This Plant is in creeping branches and leaues like the common *Scorpioides Bupleuri folio* : the floures are alſo alike, but a little bigger, and grow foure or fiue together on one foot-ſtalke : the cods are rougher, and very much turned round, or folded one within another : in all things elſe alike.

Scorpioides ſiliqua craſſa Boëlij.

This is alſo like the other in creeping branches and leaues : the flouree are ſomething bigger than any of the reſt, and grow not aboue one or two together on a foot-ſtalke : the cods are crooked, without any rough haire, yet finely checkquered, and ſeuen times bigger than any of the reſt, fully as big as a great Palmer-worme, wherein is the difference : the ſeed is almoſt round, yet extending ſomewhat in length, almoſt as big as ſmall field Peaſon, of a browne or yellowiſh colour. This alſo periſheth when the ſeed is ripe.   *Sept.* 1. 1621.

Silibum minus flore nutante Boëlij.

This Thiſtle is in ſtalkes and leaues much ſmaller than our Ladies Thiſtle, that is to ſay, The ſtalkes are round, ſtraked, ſomewhat woolly, with narrow skinny prickly edges three or foure foot high, diuided into many branches, whereon grow long leaues, deeply diuided, full of white milke-like ſtreakes and ſharpe prickles by the edges : the floures grow on the top of the ſtalkes and branches full of ſmall heads, commonly turning downewards, of the bigneſſe of an Oliue, ſet with very ſmall ſlender ſharpe prickes, containing nothing but ſmall purple chiues, ſpreading abroad like thoſe of *Iacea*, with ſome blewiſh chiues in the middle : the ſeed followeth, incloſed in downe, and is ſmall and grayiſh like the ſeed of other Thiſtles, but it is as clammy as Bird-lime. The whole plant periſheth at Winter, and reneweth it ſelfe by the falling of the ſeed. I finde not this written of. It was firſt gathered by *Boelius* in Spaine, and imparted vnto Mr *William Coys*, who friendly gaue me ſeeds thereof.

Aracus major Baticus Boëlij.

It hath ſmall weake foure ſquare ſtraked trailing branches, two foot high, leſſer but like thoſe of Fetches ; whereon grow many leaues without order, and euery ſeuerall leafe is compoſed of ſix ſeuen, or more ſmall ſharp pointed leaues, like thoſe of Lentils, ſet on each ſide of a middle rib, which middle rib endeth with claſping tendrels : the floures grow forth of the boſomes of the leaues, but one in a place, almoſt without any foot-ſtalkes at all, like thoſe of Vetches, but of a whitiſh colour, with purple ſtreakes, and of a deeper colour tending to purple towards the nailes of the vpper couering leaues : after which follow the cods, which are little aboue an inch long, not fully ſo big as thoſe of the wilde beane, almoſt round, and very hairy : wherein is contained about foure peaſon, ſeldome round, moſt commonly ſomewhat flat, and ſometimes cornered, of a blackiſh colour, neere as big as field peaſon, and of the taſte of Fetches : the whole herbe periſheth when the ſeed is ripe. This plant *Boelius* ſent to Mr *William Coys*, who hath carefully preſerued the ſame kind euer ſince, and friendly imparted ſeeds to me in *Anno* 1620.

Legumen pallidum Vliſſiponenſe, Nonij Brandonij.

This plant is very like, both in ſtalkes, leaues, and cods, to *Aracus major Baticus*, but the floures of this are of a pale yellow or Primroſe colour, and the whole herbe ſmaller, and nothing ſo hairy. It periſheth alſo when the ſeed is ripe. I receiued the ſeeds likewiſe from Mr *Coys*.

Vicia Indica fructu albo.   Piſum Indicum Gerardo.

This Vetch differeth not in any thing at all, either in ſtalkes, leaues, cods, faſhion of the floures, or colour thereof, from our common manured Vetch, but that it groweth higher, and the fruit is bigger and rounder, and of a very cleare white colour, more like to Peaſen than Vetches. Mr *Gerrard* was wont to call this Vetch by the name of *Piſum Indicum*, or Indian Peaſe, gotten by him after the publiſhing of his Herball, as Mr *Coys* reported to me. But the ſaid Mr *Coys* hath in my judgement more properly named it *Vicia fructu albo* : which name I thought moſt fit to call it by, onely adding *Indica* to it, from whence it is reported to haue been gotten.   *Iuly*, 30. 1621.

Aſtragalus marinus Luſitanicus Boelij.

This hath fiue, ſix, or more round ſtraked reddiſh hairy ſtalkes or branches, of a reaſonable bigneſſe, proceeding from one root, ſometimes creeping or leaning neere the ground, and ſometimes ſtanding vpright, a cubit high, with many greene leaues, ſet by certaine diſtances, out of order like thoſe of *Glaux vulgaris*, but leſſer, euery leafe being compoſed of fourteene or more round op-

ped

ped leaues, a little hairy by the edges, set on each side of a long middle rib, which is about nine or ten inches in length, without tendrels : the floures grow forth of the bosomes of the leaues, neere the tops of the stalkes, on long round streaked hairy foot-stalkes, of a very pale yellow colour, like those of *Securidaca minor*, but bigger, growing close together in short spikes, which turne into spikes of the length of two or three inches, containing many smal three cornered cods about an inch long, growing close together like those of *Glaux vulgaris*, each cod containing two rowes of small flat foure cornered seeds, three or foure in each row, of a darke yellowish or leadish colour, like to those of *Securidaca minor*, but three or foure times as big, of little taste : the root is small, slender, white, with a few threds, and groweth downe right, and perisheth when the seed is ripe. I first gathered seeds of this plant in the garden of my good friend Mr *Iohn Parkinson* an Apothecary of London, *Anno*, 1616.

Faba veterum serratis folijs Boelij.

This is like the other wilde Beane in stalks, floures, cods, fruit, and clasping tendrels, but it diffe-reth from it in that the leaues hereof (especially those that grow neere the tops of the stalkes) are notched and indented about the edges like the teeth of a saw. The root also perisheth when the seed is ripe. The seeds of this wilde Beane were gathered by *Boelius* a Low-country man, in Bæti-ca a part of Spaine, and by him sent to Mr *William Coys*, who carefully preserued them, and also im-parted seeds thereof to me, in *Anno* 1620. *Iuly* 31. 1621.

Pisum maculatum Boelij.

They are like to the small common field Peason in stalkes, leaues, and cods; the difference is, the floures are commonly smaller, and of a whitish green colour : the peason are of a darke gray colour, spotted with blacke spots in shew like to blacke Veluet; in taste they are also like, but somewhat harsher. These peason I gathered in the garden of Mr *Iohn Parkinson*, a skilfull Apothecary of Lon-don; and they were first brought out of Spaine by *Boelius* a Low-country man.

Lathyrus æstiuus flore luteo. Iuly, 28. 1621

This is like *Lathyris latiore folio Lobelij*, in stalkes, leaues, and branches, but smaller : the stalks are two or three foot long, made flat with two skins, with two exceeding smal leaues growing on the stalkes, one opposite against another : betweene which spring vp flat foot-stalkes, an inch long, bea-ring two exceeding narrow sharpe pointed leaues, three inches long : betweene which grow the tendrels, diuided into many parts at the top, and taking hold therewith : the floures are small, and grow forth of the bosomes of the leaues, on each foot-stalke one floure, wholly yellow, with purple strakes. After each floure followeth a smooth cod, almost round, two inches long, wherein is con-tained seuen round Peason, somewhat rough, but after a curious manner, of the bignesse and taste of field Peason, and of a darke sand colour.

Lathyrus æstiuus Baticus flore cæruleo Boelij.

This is also like *Lathyris latiore folio Lobelij*, but smaller, yet greater than that with yellow floures, hauing also adjoyning to the flat stalkes, two eared sharpe pointed leaues, and also two other slen-der sharpe pointed leaues, about foure inches long, growing on a flat foot-stalke betweene them, an inch and an halfe long, and one tendrel betweene them diuided into two or three parts: the flours are large, and grow on long slender foure-square foot-stalkes, from the bosomes of the leaues, on each foot-stalke one : the vpper great couering leafe being of a light blew, and the lower smaller leaues of a deeper blew : which past, there come vp short flat cods, with two filmes, edges, or skins on the vpper side, like those of *Eruilia Lobelij*, containing within, four or fiue great flat cornered Pea-son, bigger than field Peason, of a darke sand colour.

Lathyrus æstiuus edulis Baticus flore albo Boelij.

This is in flat skinny stalkes, leaues, foot-stalkes, and cods, with two skins on the vpper side, and all things else like the said *Lathyrus* with blew floures, only the floures of this are milke white : the fruit is also like.

Lrthyrus æstiuus flore miniato.

This is also in skinnie flat stalkes and leaues like the said *Lathyris latiore folio*, but far smaller, not three foot high : it hath also small sharp pointed leaues growing by couples on the stalke, betweene which grow two leaues, about three inches long, on a flat foot-stalk halfe an inch long: also between those leaues grow the tendrels: the floures are coloured like red lead, but not so bright, growing on

smooth

ſmooth ſhort foot-ſtalks one on a foot-ſtalke : after which follow cods very like thoſe of the common field peaſon,but leſſer,an inch and a halfe long, containing foure,fiue,or ſix cornered Peaſon, of a ſand colour, or darke obſcure yellow,as big as common field peaſon,jand of the ſame taſte.

Lathyrus paluſtris Luſitanicus Boelij

Hath alſo flat skinny ſtalks like the ſaid *Lathyrus latiore folio*, but the paire of leaues which grow on the ſtalke are exceeding ſmall as are thoſe of *Lathyrus flore luteo*,and are indeed ſcarce worthy to be called leaues:the other paire of leaues are about two inches long,aboue halfe an inch broad,and grow from betweene thoſe ſmall leaues,on flat foot-ſtalkes,an inch long:betweene which leaues alſo grow the tendrels : the floures grow on foot-ſtalks which are fiue inches long,commonly two on a foot-ſtalke,the great vpper couering leaues being of a bright red colour, and the vnder leaues are ſomewhat paler : after commeth flat cods,containing ſeuen or eight ſmall round peaſon, no bigger than a Pepper corne,gray and blacke,ſpotted before they are ripe, and when they are fully ripe of a blacke colour,in taſte like common Peaſon : the ſtalkes,leaues,foot-ſtalkes and cods are ſomewhat hairy and rough.

Lathyrus aſtivus dumetorum Bæticus Boelij

Hath alſo flat skinny ſtalkes like the ſaid *Lathyrus latiore folio*, but ſmaller, and in the manner of the growing of the leaues altogether contrary. This hath alſo two ſmall ſharp pointed leaues, adjoyning to the ſtalke : betweene which groweth forth a flat middle rib with tendrels at the top hauing on each ſide(not one againſt another)commonly three blunt topped leaues, ſometimes three on the one ſide,and two on the other,and ſometimes but foure in all,about an inch and a halfe longg the floures grow on foot-ſtalks,about two or three inches long,each foot-ſtalke vſually bearing two floures,the great couering leafe being of a bright red colour;and the two vnder leaues of a blewiſh purple colour: after which follow ſmooth cods,aboue two inches long, containing fiue ſix or ſeuen ſmooth Peaſon, of a browne Cheſtnut colour, not round but ſomewhat flat, more long than broad,eſpecially thoſe next both the ends of the cod, of the bigneſſe and taſt of common field Peaſon.

Iuniperus ſterilis.

This ſhrub is in the manner of growing altogether like the Iuniper tree that beareth berries,only the vpper part of the leaues of the youngeſt and tendereſt bowes and branches are of a more reddiſh greene colour : the floures grow forth of the boſoms of the leaues,of a yellowiſh colour,which neuer exceed three in one row, the number alſo of each row of leaues : each floure is like to a ſmall bud,more long than round, neuer growing to the length of a quarter of an inch, being nothing elſe but very ſmall ſhort crudely chiues, very thicke and cloſe thruſt together,faſtened to a very ſmall middle ſtem,in the end turning into ſmall duſt, which flieth away with the winde, not much vnlike that of *Taxus ſterilis* : on this ſhrub is neuer found any fruit. 15.*Maij*.1621.

VVHen the laſt ſheets of this worke were on the Preſſe,I receiued a letter from M.r *Roger Bradſhaghe*, wherein he ſent me incloſed a note concerning ſome plants mentioned by our Author which I haue thought fitting here to impart to the Reader : he writ not then who it was that writ it, but ſince hath certified me that it was one M.r *Iohn Redman* a skilfull Herbariſt, to whom, though vnknown,I giue thankes,for his deſire to manifeſt the truth and ſatisfie our doubts in theſe particulars,

BEcauſe you write that *Gerards* Herball is vpon a review,I haue thought good to put you in mind what I haue obſerued touching ſome plans which by him are affirmed to grow in our Northern parts : firſt the plant called *Pyrola*,which he ſaith groweth in Lanſdale,I haue made ſearch for it the ſpace of twenty yeares,but no ſuch is to be heard of.

Sea Campion with a red floure was told him groweth in Lancaſhire : no ſuch hath euer beene ſeene by ſuch as dwell neere where they ſhould grow.

White Fox-gloues grow naturally in Lanſdale,ſaith he, it is very rare to ſee one in Lanſdale.

Garden Roſe he writes groweth about Leiland in Glouers field wilde : I haue learned the truth from thoſe to whom this Glouers field did belong, and I finde no ſuch thing, onely aboundance of red wilde poppie,which the people call Corne-Roſe,is there ſeene.

White Whortles,as he ſaith,grow at Crosbie in Weſtmerland, and vpon Wendle hill in Lancaſhire : I haue ſought Crosbie very diligently for this Plant and others, which are ſaid to grow there, but none could I finde, nor can I here of any of the countrey people in theſe parts,who dayly are labouring vpon the mountaines where the Whortle berries abound,that any wihte ones hue

ſeene

beene feene, fauing that thofe which *Gerard* calls red Whortles, and they are of a very pale white greene vntill they be full ripe, fo as when the ripe ones looke red, the vnripe ones looke white.

Cloud-berry affuredly is no other than Knout-berry.

Heskets Primrofe groweth in Clap-dale. If Mʳ *Hesket* found it there it was fome extraord˙ry luxurious floure, for now I am well affured no fuch is there to be feene, but i˙ ˙y che˙...ed in our gardens.

Gerard faith many of thefe Northerne plants do grow in Crag-clofe. In the North euery towne and village neere vnto any craggie ground both with vs and in Weftmerland haue clofes fo callea, whereby *Gerards* Crag-clofe is kept clofe from our knowledge.

Chamamorus, fex Vaccinia nubis.　　Knot, or Knout-berry, or Cloud-berry.

THis Knot, Knout, or Cloud-berry (for by all thefe names it is knowne by vs in the North, and taketh thefe names from the high mountaines whereon it groweth, and is perhaps, as *Gerard* faith, one of the brambles, though without any prickles) hath roots as fmall as packe-thred, which creepe far abroad vnder the ground, of an ouerworne red colour, here and there thrufting more faftly into the moffie hillockes tufts of fmall threddy ftrings, and at certaine joynts putting vp fmall ftalks rather tough than wooddy, halfe a foot high, fomthing reddifh below, on which do grow two or three leaues of a reafonable fad greene colour, with foot-ftalks an inch long, one aboue another without order: the higheft is but little, and feldome will fpread open, they are fomething rugged, crifpie, full of nerues in euery part, notched about the edges, and with fome foure gafhes a little deeper than the reft, wherby the whole leafe is lightly diuided into fiue portions. On the top of the ftalke commeth one floure confifting of foure, fometimes of fiue leaues apiece, very white and tender, and rather crumpled than plaine, with fome few fhort yellow threds in the midft: it ftandeth in a little greene huske of fiue leaues, out of which when the floure fades, commeth the fruit, compofed of diuers graines like that of the bramble, as of eight, ten, or twelue, fometimes of fewer, and perhaps through fome mifchance but of three or two, fo joyned, as they make fome refemblance of a heart, from whence (it may be) hath growne that errour in *Gerard* of diuiding this plant into two kindes: the fruit is firft whitifh greene, after becommeth yellow, and reddifh on that fide next the Sun.

It groweth naturally in a blacke moift earth or moffe, whereof the countrey maketh a fewell wee call Turfe, and that vpon the tops of wet fells and mountaines among the Heath, moffe, and brake: as about Ingleborough in the Weft part of Yorke-fhire, on Graygreth a high fell on the edge of Lancafhire, on Stainmor fuch a like place in Weftmerland, and other fuch like high places.

The leaues come forth in May, and in the beginning of Iune the floures: the fruit is not ripe till late in Iuly.

The berries haue a harfh and fomething vnpleafant tafte.

THis Worke was begun to be printed before fuch time as we receiued all the figures from beyond the Seas, which was the occafion I omitted thefe following in their fitting places: but thinking it not fit to omit them wholly, hauing them by me, I will giue you them with their titles, and the reference to the places whereto they belong.

* In Auguft laft whiles this worke was in the Preffe, and drawing to an end, I and Mʳ *William Broad* were at Chiffel-hurft with my oft mentioned friend Mʳ *George Bowles*, and going ouer the heath there I obferued this fmall *Spartum* whofe figure I here giue, and whereof you fhall find mention, in the place noted vnder the title of the figure; but it is not there defcribed, for that I had not feen it, nor could finde the defcription therof in any Author, but in Dutch, which I neither had, nor vnderftood. Now this little Matweed ...h fome fmall creeping ftringy roots: on which grow fomwhat thicke heads, confifting of three or foure leaues, a ...ere wrapt together in one skin, biggeft below, and fo growing fmaller vpwards, as in *Schananth*, vntill they grow vp to the height of halfe an inch, then thefe rufhie greene leaues (whereof the longeft fcarce exceeds two inch . breake out of thefe whitifh skins wherein the.. are wrapp... ... li. ...g vpon the ground, and amongft thefe growes vp a fmall graffie ftalke, fome handfull or better high, bending backe the top, which carries two rowes of fmall chaffie feeds. It is in the perfeᶜtion about the beginning of Auguft.

FINIS.

Cyperus Indicus, siue Curcuma.
Turmericke.
Pag. 33. Lib. 1. Cap. 17.

Iuncus minor capitulis Equiseti.
Club-Rush.
Pag. 35. Lib. 1. Cap. 19. the first.

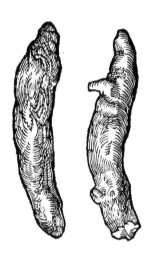

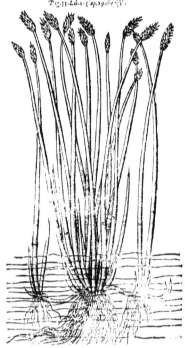

* *Spartum nostras parvum Lobelij.*
Heath Mat weed.
Pag. 41. lib. 1. Cap. 34. the fifth.

Schœnanthi flores.
The Floures of Camels Hay.
Pag. 43. Lib. 1. Cap. 35. the first.

INDEX LATINVS STIRPIVM IN HOC

opere descriptarum nec non nomina quædam Græca,

Arabica, Barbara, &c.

Yyyyy

Index Latinus.

Index Latinus.

Index Latinus.

Index Latinus.

Index Latinus.

NOMINVM ET OPINIONVM HAR-
monia & consensus.

A

Abel, id est, Populus alba
Abbel,i.Sauina
Abrono Serapionis, i. Pisum cor-
datum
Abrotanum fœmina, id est, Chamæcy-
parissus
Abrugi,i.Pisum cordatum
Acanophora,i.Iacea
Acanthus Germanica, i. Sphondylium
Acanthium,i.Onopordon
Acarna Theoph,i.Scolymos
Acetabulum,i.Cotyledon
Accipitrina,i.Hieracium, vel Sophia
Acinaria palustris Gesneri,i. Vaccinia
palustris
Acinus,i.Clinopodium
Achillea,i. Millefolium
Aconitum bacciferum, id est, Christo-
phoriana
Aconitum Pardalianches Dodonæi,i.
Doronichium
Acrocorion,i. Crocus veruus
Acutella,i.Resta bonis
Acus pastoris,i.Geranium
Adianthum album,i. Capillus Veneris
Adianthum album Plinij, id est, Ruta
muraria
Ador est quoddam frumenti genus
Ageratum,i.Balsamina minor
Agnus castus,i. Uitex
Agilensz,i.Auellana
Aglaophotis,i. Pæonia
Ægilops Plinij, id est, Cerris maior
Lobel.
Ægilops,i. Avena sterilis
Ægopogon Tragi,i. Ulmariæ species
Ægelethron,i. Tota bona
Æschinomenen Plin.i. Noli me tan-
gere, vel potius herba mimosa
Aizoon,i. Sempervivum
Aiuga,i.Chamæpitys
Vaseleti Auicennæ,i. Medica
u Daleschampy,i. Phillyrea

Alectorolophos Plinij,i.Fistularia
Alleluia,i.Trifolium acetosum
Alchilel Serapionis,i. Polemonium
Alcibiadion,i.Echium
Alchimelech,i. Melilotus
Alfasfasa Auicennæ,i.Medica
Algosarel Auicennæ,i. Daucus
Alkakengi,i. Halicacabum
Alimonia,i.Trifolium acetosum
Alnam,i.Pulegium
Alnus nigra,i.Frangula
Althæa,i.Ibiscus
Albucus,i.Asphodelus albus
Altercum Plin.i.Hyoscyamus
Alphesera Arabibus,i.Bryonia nigra
Alsaharateia,i. Parthenium
Alscebran,i.Esula
Alsenefu,i.Absinthium
Aluesen,i.Peucedanum
Amaracus,i. Maiorana
Amellus Virgilij, i. Aster Atticus
Amellus,i.Caltha palustris
Ambubeia,i.Chicorium sylvestre
Ampelos Plinij,i. Bryonia nigra
Amyrberis,i.Berberis
Anapallus Bellonij,i.Ficus Indica
Anas,i. Prunus
Anablatum Cordi, i. Dentaria maior
Matthioli
Anblatum Dodonæi,i.Nidus auis
Anchusa,i. Buglossum sylvestre
Andrian Rhasis,i. Fabago
Andration Auerrois,i. Peucedanum
Androsæmum Dodonæi, i. Clymenum
Italorum
Anctum tortaosam i. Meum
Anguria,i.Pepo oblongus
Anonis,i.Resta bovis
Apiastellum Apuleij,i. Bryonia nigra
Apiastellum Dodonæi,i.Melissa
Apocynum Dios.i.Periploca
Apolinaris,i.Hyoscyamus
Apronia,i.Bryonia nigra
Aprus,i.Pisum rubrum
Argentino,i.Potentilla
Aria Theophrasti,i. Sorbus Alpinus
Gesneri

Arcium,i.Bardana
Argentina,i.Ulmaria
Arthritica,i. Primula veris
Artemisia marina,i. Cineraria
Armoracia,i. Raphanus rusticus
Arornas,i.Iuniperus
Arzi,i.Oryza
Asfruntia,i.Imperatoria
Asterion,i. Aster Atticus
Astergis Rhasis, id est, Azaradach
Auicennæ
Asphalathus,i. Acatia Math.
Asplenium sylvestre,i. Lonchitis
Asplenium,i.Ceterach
Asturca,i.Stæchas
Athanasia,i. Tanacetum
Aulitica,i.Chamæmelum
Aureumolus,i.Atriplex
Auornus Petri Placentij,i.Frangula
Azabaser,i. Meum
Azez,i.Lichen.

B

Baaras, id est, Pæonia
Baccharis,i.Conyza maior Matth.
Baccharis officinarum,i. Asarum
Eamia,i. Althæa palustris
Baptisecula,i.Cyanus
Barbahirci,i.Tragopogon
Barba Iovis,i. Sempervivum
Barba Aron,i. Sempervivum
Batis Plinij,i.Crithmum
Baton,i.Terebinthus
Bazari Chichen,i.Linum
Bazara Contana,i.Psyllium
Beccabunga,i. Anagallis aquatica
Bechion,i.Tussilago
Bedoara arabibus,i.Spina alba
Bedeguar Math.Sylvat.i. Spina alba
Bederangi,i.Melissa
Belvidere Italica,i.Scoparia
Beiahalalen,i.Aizoon
Betonica Pauli,i.Veronica
Bihar,i.Buphthalmum
Bifaria,i.Dracunculus
Bifachium,i.Gingidium

Basiatrahagi,

Bagatrahagi, i. Polygonum
Bismalua, i. Ibiscus
Bislingua, i. Hippoglossum
Bombax, i. Gossypium
Botonaria, i. Aphyllanthes
Boutomen, i. Plat.naria
Branca ursina, i. Acanthus
Britannica, i. Bistorta quorund.
Brassica canina, i. Cynocrambe
Broeggia Plinij, i. Helxine Cissampelos
Bruscus, i. Ruscus
Bulbus agrestis, i. Colchicum
Bubonium, i. Aster Atticus
Buccinum, i. Consolida regalis
Bugia, i. Cortex Berberis
Butomon Theoph. i. Iris palustris
Bulef, i. Salix

C

Cachry: marinum, i. Crithmum
Cachrifera, i. Libanotis
Cahade, i. Polium
Cafal, i. Agrimonia
Calabrina, i. Lonchitis
Calchochrum, i. Fumaria
Calcifraga, i. Crithmum
Caltha, i. Calendula
Callionymum Gesn. i. Lilium convallium
Callitrichum, i. Capillus Ven.
Callion Plin. i. Alkakengi
Calicularia, i. Hyoscyamus
Caluegia, i. Galanga
Camphoratum, i. Abrotanum magn.
Candelaria, i. Tapsus barbatus
Cania Plin. i. Vrtica
Cancum Avicen. i. Cheledonium minus
Cantabrica Plin. i. Rapunculus
Cantabrica Turneri, i. Cariophyllus
Capillaris, i. Capillus Ven.
Capnos, i. Fumaria
Capnitis, i. Fumaria
Caprificus Plin. i. Ezula Tragi
Caransul, i. Cariophyllus
Cardamum, i. Pseudobunium
Cardopatium, i. Carlina
Carduus fullonum. i. Dipsacus
Curobia Aliuar. i. Ceratia siliqua
Carica, i. Apios
Carnabadion Semionis Sethi, i. Carui
Cartamus syluest. i. Atractylis
Carpentaria, i. Pseudobunium
Carpesium, i. Cubeba
Carpesium Gal. i. Ruscus
Carnabadion Sim. Sethi, i. Carum
Cradel, i. Sinapis
Casebar, i. Coriandrum
Cassilago Matth. syl. i. Hyoscyamus
Cassutha, i. Cuscuta
Castoris trifol. i. Trifolium paludosum
Casignete, i. Pimpinella

Castrangula, i. Scrophularia
Caleitrapa Matth. i. Carduus stellatus
Catef, i. Atriplex
Cathsium, i. Abrotanum
Cataputia maior, i. Ricinus
Cacon Plin. i. Equisetum
Cauda Vulpium, i. Alopecuros
Celiros, i. Gramen Manna
Caciliana Plin. i. Androsemum Dodo
Centummorbia, i. Nummularia
Centrum Galli, i. Horminum
Centunculus, i. Gnaphalium
Cepaa, i. Anagallis aquat.
Ceratonia, i. Ceratia siliqua
Cercis Theop. i. Arbor Iuda
Cesis, i. Daucus
Cerua maior, i. Ricinus
Ceraunia, i. Crassula
Ceruicalia, i. Trachelium
Circa, i. Gracis, i. Mandragora
Cicrunalis, i. Capillus Ven.
Cicerbita, i. Sonchus
Cnicus, i. Carthamus
Cnicum supin. Cordi, i. Carduus Benedictus
Cnicus syluest. i. Atractylis
Citrago, i. Melissa
Cidromela, i. Malus Medica
Cineraria, i. Iacobea marina
Charantia, i. Balsamina mas
Chamacissus, i. Hedera terrest.
Chamaleuce, i. Tussilago
Chamepeuce (id est) Rosmarinum syluest.
Chamexylon Plin. i. Gnaphalium
Chamalea Germanica, (i.) Mezerean
Chamadaphne, i. Laureola
Chamadaphne Plin. i. Vinca Pervinca
Chamaplium, i. Erysimum
Chamarrhipes, i. Palmites
Chamaleagnus, id est, Myrtus Brabantica
Chamaops Plin. i. Palma humilis
Chaynuba, i. Ceratia siliqua
Charfi, i. Apium
Chastara, i. Betonica
Choth, i. Cucumer
Chironia, i. Brionia nigra
Chitini, i. Althea
Choeradoleihron, Xanthium
Chabece, i. Malua
Chiliodynamus, i. Polemonium aut Scrophoron
Chrysolachanon Plin. i. Atriplex sat.
Chrysanthemum Peruu. i. Flos solis
Chrysocome Gesn. i. Linaria aurea
Chrysogonon, i. Blattaria
Chrysomela Athenai, i. Malus Med.
Citrago, i. Melissa
Clauum Ven. i. Nymphea
Clavicula, i. Hedera Helix
Clematis peregr. i. Flammula Iovis

Clematis Daphnoides, i. Vinca, Pervinca
Clematis alt. Matth. i. Viorna
Cleome Horatii, i. Erysimum
Colubrina, i. Arum Ægypt.
Cocior, i. Fœniculus
Colocasi. i. Faba Ægypt.
Columnaru, i. Campana Lad.
Columbina Actuar. i. Verbena
Combul, i. Nardus
Condisi, i. Saponaria
Condurdum Plin. i. Vaccaria
Consolida media Fuch. i. Bellis maior
Consolida media, i. Bugula
Consolida minor Matth. i. Prunella
Consolida minor Ruell. i. Bellis minor
Couzambuch Turcor. i. Hemerocallis Valent.
Connarus Athenai, i. Iuiuba species maior
Concordia, i. Agrimonia
Conila, i. Myrrhis
Consecratrix. i. Iris nostras
Coralloides Cordi, i. Dentaria Matth.
Corcorus Melochia, i. Olus Iudaic.
Corcorus Marcelli vet. i. Anagal mas
Corydalis, i. Radix cava
Cor Indum, i. Pisum cordatum
Coriziola Rhasis, i. Scamonea
Corona fratrum, i. Carduus Eriocephalus
Corion, i. Coriandrum
Coronopus, i. Cornu Cervi
Corigiola, i. Centumnodia
Corona Monachis, i. Dens Leonis
Corona regia, i. Melilotus
Coroneola, i. Lysimachia
Corydalium, i. Fumaria
Costus spurius Matth. i. Panax Chironium
Costus hortorum, i. Balsamita maior
Costa canina. i. Quinquenervia
Corydalpio cium, i. Consol. regalis
Cotinus Dodo, i. Coccygria Plin.
Cos Avicenna, i. Medica
Crepanella, i. Dentillaria Rondel.
Creta marina i. Crithmum
Crinita, i. Capillus Ven.
Cuculata, i. Pinguicula
Cucurida, i. Dentillaria Rondelet.
Cuminum Æthiop. i. Ammi
Cunilago, i. Conyza
Cunegundis herba, i. Eupatorium Avicenna
Cunila Columella, i. Satureia
Cunophoron, i. Nux Indica
Cander Avicenna, i. Saponaria
Curcuma, i. Cyperus Indus
Curcas Clus. i. Ricinus
Cusbar, i. Coriandrum
Cyanus Hierosolymitana, i. Ptarmica Austriaca
Cyso, i. Hyssopus
Cyminalis, i. Gentiana
Cynanthemis, i. Cotula fœtida

Cynoſpaſtos,s,Pæonia
Cynoſorchis,i,Satyrium
Cynocephalus Apul.i,Antirrhinum
Cynocephalus,i,Anetum
Cynosbatos Dioſc.i.Roſa ſylueſt.
Cynosbatos Tragi,i,Oxyacanthus
Cynosbatos Ruellij,i,Rubus
Cynosbatos Daleſch.i,Capparis
Cyprus,i,Liguſtrum Lob.
Cyprus Dod.i,Phillyrea

D

Damedrios,i,Chamædrys
Danebalchil,i,Equiſetum
Darach,i,Palma
Daracht mous,i,Muſa
Datura,i,Stramonia peregr.
Debonigi,i,Chamomilla
Deſta,i,Beta
Delphinium,i,Conſolida reg.
Diapenſia,i,Sanicula
Didar Arab.i,Vlmus
Digitus Ven.i,Nymphæa
Dili,i,Iſatis
Diocallia Apul.i,Chamomelum
Dioſpiron,i,Milium ſolis
Dioſpiros Plin.i,Lachrima Iob.
Diſanthos Theop. i,Superba Fuchſij
Dochon,i,Panicum
Dolichus Theop.i,Phaſeolus Lob.
Draco herba,i,Tarcon
Draco ſylueſt.i,Ptarmica
Droſutum Haliabbi,i Aniſum
Droſion Cordi,i Alchimilla
Drypis Guillandini,i Tragon Matth.
Drypis Theop. i,Eryngium mar.
Dryopteris Tragi,i,Pieridion Cordi
Dryopteris,i. Adianthum nigrum
Dryophonon Plin.,i,Arabis Dod.
Dulb Arab. i,Platanus
Dulciſida,i,Pæonia
Dulcichinum Guillandini, i,Cyperus
  dulcis Tragi
Dumbebe,i,Endiuia

E

Echium paluſtre Cordi, i,Scorpioi-
  des Dod.
Eghelo Dod.i,Laburnum Lob.
Eleagnus Mat.i,Zizaphus alb.Lob.
Eleoſelinum,i,Paludapium
Elkiageber,i,Roſmarinus
Empetron Dod.i.Kali
Empetron Rondeletij,i.Crithmum
Empetron Tragi,i.herba Turca
Empetron,i,Herniaria
Eneaphyllum Lob.i,Dentaria
Eneaphyllum,i,Lingua ſerpentina
Enneadinamen Geſn. i,Gramen Par-
  naſſi
Ephemum, i,Ranunculus

Euphroſynum Plin.i,Borago
Epilobion Geſn. i,Chamænerium Lo-
  bel.
Epimetron,i Epimedium
Epipaſtis Recentiorum, i, Helleborine
Epipaſtis Rondeletij,i,Herniaria
Eranthemum,i,Flos Adonis
Ericoides,i,Euphraſia lut. Dod.
Eringium Vegetij,i,Acarna Theop.
Eringium Guiland. i,Carduus Stell.
Erinus,i,Corcorus Plin.
Eruum ſylueſt.i,Catanance
Eryphion Apul.i,Ruta
Eryſimum Theop.i,Camelina
Erythrodannum,i,Rubia tinctorum
Exupera,i,Verbena

F

Farfara,i,Tuſſilago
Farfugium, i,Tuſſilago
Farfium Auicenne, i, Thora Valden-
  ſis
Farfrugum, i.Caltha paluſt.
Faranum, i,Tuſſilago
Faudegeni,i,Origanum
Fel terræ,i,Centaureum minus
Feniculus porcinus,i,Peucedanum
Ferulacois,i,Thapſia
Ferraria,i,Agrimonia
Ferraria Lob.i,Scrophularia
Feſtuca Plin.i, Auena ſterilis
Flammula Iouis, i, Clematis Perigri-
  na
Filicaſtrum,i,Oſmunda reg.
Filix Florida,i,Oſmunda regalis
Filix latifolia Cordi,i.Oſmunda rega-
  lis
Ficaria,i,Chelidonium minus
Flos regius,i,Conſolida reg.
Flos Abarnalis,i,Polygala
Fontalis,s,Potamogeiton
Folium Indum, i,Malabathrum Gar-
  ciæ
Fuga Demonis,i,Hypericum
Fuſus,i,Atractylis
Fumaria Corydalis, i,Radix Caua

G

Galedragon Plin. i,Dipſacus
Garoſmus, i,Atriplex Olida
Gallitrichum,i,Horminum
Gelſeminum Indicum,i.Mirab. Peru-
  viana
Genitura,i.Anetum
Genicularis,i,Phu magnum
Genecanthe,i,Bryonia nigra
Geum,i,Cariophyllata
Giezar,i,Daucus
Giezar Aridras,i,Belbunac
Gith,i,Nigella
Githago Plin.i,Lolium

Gladiolus aquat.i,Iuncus Floridus
Globularia,i,Aphyllanthos
Gloſſographe Auicen,i,Fumaria
G.ſſypium,i.Xylon
Granum reg.Meſuæ,i,Ricinus
Gromphena Plin i.Calendula
Groſſularia,i,Ribes
Groſſularia,i,Vua criſpa

H

Habonog Auer.i,Chamæmela
Habal nil Serap.i,Conuolulus
  cæruleus
Halcaſmeg Auer.i,Fœniculus
Halgazar Auer.i,Paſtinaca
Halicacabum veſicarium,i,Alkaken-
  gi
Halicacabum Peregrinum, i, Piſum
  Cordatum
Hameſiteos,i,Chamapitys
Hanab.Althaleb,i,Solanum
Haraha,i,Cucurbita
Harbatum,i,Peucedanum
Hauenaria,i,Cornu ceruu
Haronigi,i, Doronicum
Haſmiſen Syriac. i,Conuoluulus cæ-
  ruleus
Haſtula reg.i,Aſphodelus albus
Haſce,i,Thymus
Handachocha Plin. i, Trifol. bitumi-
  noſum
Haur.Arab.i,Populus alba
Hebene,i,Xylaloes Offic.
Herculania,i.Verbena
Hederalis Ruell.i,Aſclepias
Hedera ſpinoſa,i,Smilax aſpera
Helxine,i,Parietaria
Helice, i, Hedera helix
Hemen,i,Serpillum
Hepatica alba Cordi, i,Gra.Parnaſſi
Herba Scythica, i,Glycyrrhiza vulg.
Herba caſta,i,Pæonia
Herba Leonis,i,Palma
Herba Lucis,i,Chelidonium maius
Herba Impetiginaria,i,Chelidon.ma-
  ius
Herba Vulcani,i,Ranunculus
Herba ſcelerata,i,Ranunculus
Herba ulceraria,i,Ranunculus
Herba cauſtica,i,Ranunculus
Herba Dineotilie, i.Conſolida Rega-
  lis
Herba ſtella,i,Cornu ceruu
Herba Cancri,i,Heliotropium
Herba radioli Apul.i,Polypodium
Herba Leon.Dod.i,Aquilegia
Herba S. Anth.i,Dentillaria Rondele-
  tij
Herba coxendicum.i,Cotyledon
Herba Benedicta,i,Caryophyllata
Herba Fortis id eſt,Solidago Sarace-
  nica
Herba paralyſis, i.Primula veris
Herba Clauellata,i.Viola tricolor
                Herba

Herba Laſſulata, id eſt, Balſamita ma-
ior
Herba Pinnula, i. Hyoſcyamus
Herba Turca, i. Herniaria
Herba Hungarica Dodon, i. Alcea
Herba Simeonis Dodon. i. Alcea
Herba Vrbana, i. Acanthus
Herba Tunica Gordony, id eſt, Ocyma-
ſtrum
Herba Tunica Dodon. id eſt, Caryo-
phyllata
Herba Gallica Fracaſtor, i. Galega
Herba Rutinalis, i. Sphondylium
Herba Sardoa, id eſt, Ranunculus aqua-
ticus
Herba Sacra, i. Tabaco
Herba Sacra Agrippa, i. Meliſſa
Hermodaɗylus Dodon, id eſt, Colchi-
cum
Hermodaɗl. Italorum, i. Iris Tuberoſa
Lobel.
Heſperis Cluſij, i. Leucoium marinum
Lobel.
Hippia, i. Alſine
Hirundinaria, i. Aſclepias
Hortus Veneris, i. Cotyledon
Hormanum Tridentinum, id eſt, Colus
Iouis
Humadh, i. Lapathium
Hunen, i. Iuiube
Humure, i. Vrtica
Hydroſelinum, i. Paludapium
Hydroſelinum Camerary, id eſt, Lauer
maius
Hyoſcyamus Peruuianus, i. Tabaco
Hippogloſſum, Boniſacia, id eſt, Laurus
Alex
Hyoſyris, Plin, i. Iacea nigra
Hyophthalmon, i. Aſter Atticus
Hypecoon Dodon, id eſt, Cuminum ſyl-
ueſtre
Hypecoon Cluſij, i. Alcea Veneta
Hippoſelinon, i. Olus atrum

I

Iarus, id eſt, Arum
Iackuiak, i. Anemone
Iaſione, i. Campanula
Için, i. Enula
Iemuri, i. Nux Moſcata
Ikiza, i. Chamapitys
Iezar Serapionis, i. Paſtinaca
Imperatrix, i. Meum
Inula Ruſtica Scribon. Largi, i. Con-
ſolida maior
Inguinalis, i. Aſter atticus
Intybus, i. Cichorium
Iouis Faba, i. Hyoſcyamus
Iouis Glans, i. Caſtanea
Iouis Flos, i. Lychnis
Iouis Arbor, i. Quercus
Iorgir, i. Eruca
Irio, i. Eryſimum
Iuncus quadratus Celſi, i. Cyperus
Iua Muſcata, i. Chamapitys
Inſaɗti, i. Sambucus
Inſqniamus, i. Hyoſcyamus
Ixopus Cordi, i. Chondrilla

K

Kantmron, i. Centaurenm
Kanz, i. Amygdalus
Kaper, i. Capparis
Katuroch, i. Chelidonium maius
Kebikengi, i. Ranunculus
Keiri, i. Leucoium
Kemetri, i. Pyrus
Kemun, i. Cuminum
Kenne, i. Liguſtrum
Kernagh, i. Ricinus
Kerngha, i. Ricinus
Kermes, i. Coccus infeɗtoria
Kulb, i. Milium ſolis
Knſbera Auerroy, i. Coriandrum
Kuſbor, Coriandrum

L

Labruſca, i. Bryonia nigra
Labrum Veneris, i. Dipſacus
Laburnum, i. Anagyris
Laɗtaria, i. Tithymalus
Laɗtuca leporina, i. Sonchus
Laɗtucella, i. Sonchus
Lanata Cordi, i. Aria Theoph.
Lancea Chriſti, id eſt, Lingua Serpen-
tina
Lantana, i. Viburnum
Lanaria, i. Radicula
Lanceola, i. Quinquenervia
Laudata Nobilium, i. Veronica
Lathyrus, i. Cataputia
Lathyrus, i. Piſum ſylueſtre
Lauer Lauacrum, i. Dipſacus
Laurus Alexandrina, id eſt, Hippo-
gloſſum
Laurus roſea, i. Oleander
Laurus Sylueſtris, id eſt, Laurus Ti-
nus
Laurentia Matthioli, i. Eupula
Leo Columella, i. Aquilegia
Leontoſtomum Geſneri, id eſt, Aquile-
gia
Leo Herba Dodon, i. Aquilegia
Lepidium Plin, i. Piperitis
Leſen Arthaur, i. Bugloſſum
Leucacantha, i. Carlina
Leucacanthemum, i. Chamameum
Lipadion Plin, i. Centaureum
Libanum Apulei, i. Borago
Limodoron Dodon, i. Orobanche
Lingua auis, i. Fraxini ſemen
Lingua Pagana, i. Hippogloſſum
Liliago Cordi, i. Phalangium Lobel.
Liſen, i. Plantago
Lotus Vrbana, i. Trifolium odoratum
Lobel.
Longius, i. Lonchitis
Lichen, i. Hepatica officinarum
Lunaria Arthritica Geſneri, i. Auricu-
la Vrſi
Luciola, i. Lingua ſerpentina
Lunaria Graca, i. Bolbonac
Lunaria maior Dioſcor. id eſt, Alyſ-
ſon
Lupi Cordi, i. dracunculus
Luiula, i. Trifolium Acetoſum

Lycoſtaphylus Cordi, id eſt, Sumi ucus
aquatica
Lutopſis, i. Bugloſſum ſylueſtre
Lycoperſicum, i. Poma amoris

M

Machla, id eſt, Palma
Madon Plinij, id eſt, Bryonia
alba
Mahaleb Auicenna, id eſt, Pſeudoligu-
ſtrum
Mahaleb, i. ſpecies Phillyreæ
Magydaris Theoph. i. Laſerpitium
Malacocciſſos, id eſt, Hedera Terre-
ſtris
Malinathalla Theop, i. Mala inſana vel
potius Cyperus Eſculentus
Malacciſſus Caſſani Baſſi, i. Caltha pa-
luſtris
Maluauiſcus, i. Ibiſcus
Manus martis, i. Quinquefolium
Marana, i. Stramonia
Marathrum, i. Fœniculum
Maru herba Dodon. id eſt, Cerinthe
Plin
Marinella, i. Phu magnum
Marmarites, i. Fumaria
Marmorilla, i. Agrimonia
Maſtaſtes, i. Laſerpitium
Maſton Plinij, i. Scabioſa
Mater Herbarum, i. Artemiſia
Materfilon, i. Iacea nigra
Matriſaluia, i. Horminum
Matryſylua, i. Periclimenum
Maurohebra Caput, id eſt, Antirrh-
num
Medium Dioſcor. id eſt, Viola mar-
ina
Medium Lobel. i. Iris maritima Nar-
bonenſi
Melochia, i. Corcorus
Melampodium, i. Helleborus niger
Melfrugum Dioclis, i. Panicum
Melampyrum, id eſt, Triticum Vacci-
num
Melaſpermum, i. Nigella
Melich Arab, id eſt, Trifolium fruti-
cans
Meleagris Flos, i. Fritillaria
Melanibium, i. Nigella
Meloſpinum, i. Pomum Spinoſum
Memiran Andr. Bellunenſis, i. Chelid.
maius
Memireſin Auicen. idem
Meud Hendi Arabibus, id eſt, Sca-
monea
Memitha Arabibus, id eſt, Papauer
Cornutum
Memacylum, i. Arbutus
Merogonion, i. Pæonia
Mentha Saracenica, id eſt, Balſamita
maior
Men, id eſt, Meum
Memiran Serapionis, i. Chelidonium
minus
Methel, i. Strammonea
Merxenins, i. Maiorana
Meſcatremſir, id eſt, Dictam-
num

Mille grana,i.Herniaria

Menianthe Theop. id est, Trifolium palustre.

Militaris,i.Millefolium.

Miha,i.Styrax

Millemorbia,i.Scrophularia

Mixa,i.Sebesten

Molochia Serapionis, id est, Corcoros Matthioli.

Molybdena,id est, Dentillaria Rondeletij

Momordica,i.Balsamita mas

Morghani Syriaca,id est Fabago Belgarum.

Mochus Dodon. id est, Orobus Lobel.

Morella,i.Solanum hortense

Mula Herba Gaza,i.Ceterach.

Multibona,i.Petroselinum

Mumeiz,i.Sycomorus

Muralia Plin.i. Helxine

Myophonon,i. Doronicum

Myrtus sylvestris,i.Ruscus

Myrica,i.Tamariscus

Myriophyllum,i.Viola aquatilis.

N

Nabatnaho, id est, Mentha

Nanachach,i. Ammi

Nard & Naron Arab,i.Rosa

Nardus Cretica,i.Phu magnum

Nardus Rustica Plinij,i. Conyza vel potius Asarum

Narf.i.Nasturtium

Nargol,i.Palma

Nasturtium hibernum,i. Barbarea

Nenupher,i.Nymphaea

Neottia,i.Nidus auis

Nepa Gaza,i. Genista spinosa

Nerium,i.Oliander

Nicophoron Plinij,i.Smilax Aspera

Nicosiana,i. Tabaco

Nigellastrum,i.Pseudomelanthium

Nilofer,i.Nymphaea

Nil Auicennae, id est, Convolvulus Caeruleus

Nola Culinaria,i. Anemone

Noli me tangere,i.Impatiens herba

Noli me tangere,i.Cucumis sylvestris

Nux Metel,i.Stramonia Fuchsii

Nux Vesicaria, id est, Staphylodendron

Nymphaea minima,i.Morsus Rana

O

Oculus Christi, id est, Horminum syl.

Odontis,i.Dentillaria Rondeletij

Olualidia,i.Chamaemelum

Olea Bohemica,i.Ziziphus alba

Oteagnos,i.Chamelaea

Oleastellum,i.Chamelaea

Olus Indicum,i.Corcoros

Olus album Dodon.i. Valeriana Campestris,vel Lactuca agrina

Onagra Veterum,i.Chamaenerium

Onitis Plinij,i.Origanum

Ononis,i.Resta Bonis

Onobrychis, id est, Caput Gallinaceum

Onobrychis Belgarum, i. Campanula Arvensis

Onosma,i.Buglossum sylvestre

Onopordon, id est, Acanthium Illyricum

Ordelion Nicandri,i.Tordylion

Ophris,i. Bifolium

Ophioglossum , id est , Lingua serpentis

Opuntia Plinij,i. Ficus Indica

Opsago,i.Solanum somniferum

Orbicularis,i.Cyclamen

Orvala,i. Horminum

Oreoselinum,i.Petroselinum

Ornus,i. Fraxinus Bubula

Orontium,i. Antirrhinum

Ostria Cordi,i.Ornus Tragi

Osteocollon,i.Consolida maior

Ostrutium,i.Imperatoria

Osyris,i.Linaria

Othonna,i.Flos Africanus

Oxyacantha,i.Berberis

Oxyacanthus,i.Spina appendix, vel pyracantha

Oxys,i.Trifolium Acetosum,

Oxymyrsine,i.Ruscus

Oxycoccos Cordi, id est, Vaccinia palustis

P

Palma Christi, id est,Ricinus

Palalia,i.Cyclamen

Paderota,i.Acanthus

Panis Cucuis,id est, Trifolium Acetosum

Pancratium,i.Squilla

Panis porcinus,i.Cyclamen

Papaver Spumeum,i.Ben album

Paronichia Dioscor.id est,Ruta Muraria

Passerina,Ruellij,id est, Morsus Gallinae

Pedicularis herba,i.Staphisagria

Peduncularia Marcelli, id est, Staphisagria

Peganon,i.Ruta sylvestris

Pentadactylon,i.Ricinus

Peponella Gesneri, id est, Pimpinella

Perlaro,i.Lotus arbor

Perforata,i.Hypericon

Perdicion,i.Helxine

Peristerion,i.Scabiosa minima

Personata,i.Bardana

Pezice Plinij sunt fungi species

Pes auis,i.Ornithopodium

Pes Leonis,i. Alchimella

Pes vituli,i.Arum

Pes Leporinus,i.Lagopus

Petum Americe,i.Tabaco

Petilius Flos,i. Flos Africanus

Pharnaceum,i.Costus Spurius

Phasganon Theop.i.Gladiolus

Phalangitis,i.Phalangium

Phellos,i.Suber

Phellandrium,i.Cicutaria palustris

Phellandrium Guillandini,i. Angelica

Phoenix,i. Lolium

Philomedium,i.Chelidonium maius

Phleterium,i.Ben Album

Phleos,i.Sagittaria

Phthirion,i.Pedicularis

Phylateria,i. Polemonium

Phyllirea Dodon,i. Ligustrum

Phyllon Theophrasti,i. Mercurialis

Philanthropos,i. Aparine

Pycnacomon Anguill.i.Rheseda

Pimpinella spinosa Camerary,i.Poterion Lobel.

Pinastella,i. Peucedanum

Piper aquaticum,i.Hydropiper

Piper Calecuthium, Indum, Brasilianum,i.Capsicum

Piper agreste,i.Vitex

Pistacia sylvestris, id est, Nux Vesicaria

Pistana,i,Sagittaria

Planta leonis,i,Alchimilla

Pneumonanthe Lobelij,i.Viola Calathiana Dodonei

Podagraria Germanica, id est, Herba Gerardi

Polytricum,i. Capillus Veneris

Polytricum Fuchsij, id est, Muscus capillaris

Polygonatum,id est, Sigillum Salomonis

Polygonoides Dioscoridis, id est,Vinca pervinca

Polyanthemum, i. Ranunculus aquaticus

Pologonum,i. Centumnodia

Populago,i.Tussilago, vel Caltha palustris

Potentilla maior,i. Vlmaria

Pothos Coster,i. Aquilegia

Pothos Theophrasti,i. Aquilegia

Proserpina herba,i.Chamomelum

Protomedia,i.Pimpinella

Pseudorchis,i. Bifolium

Pseudobunium,i. Barbarea

Pseudocapsicum,i.Strichnodendron

Pyrethrum sylvestre,i. Ptarmica

Pteridion Cordi,i. Dryopteris Tragi

Pustech,i.Pistatia

Pulicaria,i. Conyza

Q

Quexai,id est,Nigella

R

Radix Naronica,id est,Iris

Ramel,i.Cistus

Rapum terrae,i.Cyclamen

Raginigi,i.Foeniculum

Raledialemen Haliabbi,id est,Fumaria

Reginaprati,i.Vlmaria

Rosa fatuina,i.Paonia

Rosa Iunonis,i.Lilium

Rorastrum,i.Bryonia

Rorella,i. Ros solis

Rotula solis,i. Chamaemelum

Rhododophne,i.Oleander

Rhodadendros,

Rhododendron,i.Oleander
Rhuselinum Apuleia,i.Ramunculus
Rima Maria,i.Alliaria
Rincus Marinus,i.Crithmum
Rubus Cervinus,i Smilax aspera
Rumex,i.Lapathum
Ruta Capraria,i.Galega
Ruta palustris,i.Thalietrum

S

Sabeteregi,id est,Fumaria
Sabaler,i.Satureia
Sadeb,id est,Ruta
Sacra herba Agrippa,i.Salvia
Saffargel,i.Malus Cydonia
Safarhtramon,i.Sparganium
Salicaria,i.Lysimachia
Salinunca Gesneri,i.Nardus Celtica
Salsirora,i Ros solis
Salicastrum Plin.i.Amara dulcis
Salix Amerina,i.Salix humilis
Salivaria,i.Pyrethrum
Salvia vita,i.Ruta muraria
Salvia agrestis,id est,Scordium alterum
Salvia Romana,i.Balsamita maior
Salusandria,i.Nigella
Samalum Plin.i.Pulsatilla
Samolum Plin.i.Anagallis Aquatica
Sampsucum,i.Amaracus
Sanguis Herculis,id est, Helleborus albus
Sanguinaria,i.Cornu cervi
Sanamunda,id est, Caryophylata quibusdam
Sarix,i.Filix
Sardinia glans,i.Castanea
Sauch,i.Malus Persica
Saxifragia Lutea Fuchsij,id est, Millefolium
Saxifragia rubra,i.Philipendula
Sagitta,i.Sagittaria
Scandix,i.Pecten Veneris
Scarlea,i.Horminum
Scaunix Auerr.i.Nigella
Scissima Gaza,i.Fagus
Schehedenegi,i.Cannabis
Scheiteregi,i.Fumaria
Scoparia,i.Osyris
Scolopendria,i.Lingua cervina
Scorodonia, i.Scordium alterum, vel salvia agrestis
Scorpio Theophrasti, i.Genista spinosa
Scolymos Dioscor.i.Cinara
Scuck Syriaca,i.Papaver Rhœas
Secacul Monardi,i.Sigillum Salomonis
Selago Plin,i.Sauina syluestris Tragi
Selk m,i.Rapum
Seligonion,i.Pæonia
Selanion,i.Crocus vernus
Selliga,i.Nardus Celtica
Seminalis,i.Equisetum
Seneffigi,i.Viola martia
Serpentaria,i.Dracunculus
Sertula Campana,i.Melilotus
Serapias mas,i.Orchis fœmina Tragi

Seygar,i.Nux moscata
Sida Theoph.i.Althæa palustris
Sideritis tertia Matth.i.Ruta canina Monspeliensium
Sideritis,i.Marrubium aquaticum
Siciliana Camerarij,i.Androsemum Dodonei
Siger Indi,i.Palma
Siringa cærulea Dodon.id est, Lilac Matthioli
Siliqua dulcis,i.Ceratie siliqua
Silicula Varronis,i.Fœnugrecum
Siliquastrum Plinij,i.Capsicum
Sigillum Mariæ,i.Bryonia nigra
Sin,id est,Ficus
Sinnasbarium,i.Mentha aquatica
Sinapi Persicum,i.Thlaspi
Siser,i.Sisarum
Silaus Plin.i.Thysselinum
Sison Syriacum,i.Ammi
Sissitiepteris Plin,i.Pimpinella
Siler Plin,i.Alnus nigra
Sithim,i.Larix
Smilax Lænis, i. Convolvulus maior flo.albo
Smyrhiza Plin.i.Myrrhis
Sorbus aucuparia,id est, Fraxinus bubula
Sorbus Alpina Gesn.id est,Aria Theophrasti
Sorbus syluestris, id est, Fraxinus bubula
Solanum rubrum,i.Capsicum
Solanum lignosum Plinij,id est, Amaradulcis
Solanum Tetraphyllum, id est, Herba Paris
Solanum vesicarium,i.Alkakengi
Solanum,i.Solanum hortense
Scilla Stella,i.Pimpinella
Solleo Theoph.i.Anemone
Sparganion Matthioli,id est,Platanaria
Spina acuta,i.Oxyacanthus
Spina acida,i.Oxyacantha
Spina herti,i.Tragacantha
Spina infectoria, id est, Rhamnus solutreus
Spina Indaica,i.Palurus
Spiræa Theoph.i.Viburnum
Sponsa solis,i.Ros solis
Sphacelus Dodonei,Scordium alterum Lobelij
Splyte,i radix cava
Spicata,i.Potamogeiton
Staphylodendron Plin.i. Nux vesicaria
Statice Dalescamp.i.Caryophillus marinus Lobelii
Stataria,i.Peucedanum
Stellaria Horat. Augerii, i. Carduus stellatus
Struthiopteris Cordi,i.Lonchitis
Struthium,i.Saponaria
Strumaria Galeni,i.Lappa minor
Strangulatoria Auicennæ,id est, Doronicum
Sucaram,i.Cicuta
Succisa,i.Mo sus Diaboli
Surum Auicennæ,i. Nigella

Symphitum,i.Consolida maior
Symphoniaca,i. Hyosciamus
Supercilium Veneris, id est, Viola aquatilis
Supercilium terra, id est, Capillus Veneris
Sus,i.Liquiritia

T

Tagotes Indica, id est, Flos Africanus
Tahaleb,i.Lens palustris
Tamecnemum Cordi,i. Vaccaria
Tarifilon Auicennæ,i. Trifolium bituminosum
Tatula Clusii,i.Stramonia
Tatoula Turcis,i.Pomum spinosum
Tamus.Dodon,i.Bryonia Nigra
Taraxacon,i.Dens Leonis
Tarfa,i.Tamariscus
Teda arbor,i.Pinus syluestris
Terzola, Baptista Sardi,i.Eupatorium Cannabinum
Tetrahit,i.Herba Iudaica
Terdina Paracelsi,i.Phu magnum
Terpentaria,i.Betonica Aquatica
Thina,i.Larix
Thut,i.Morus
Thuia Theophrasti,i.Arbor vita
Thysselinum,i.Apium syluestre
Thymbra,i.Satureia
Tornsol bobo,i Heliotropium
Topiaria,i.Acanthus
Trapezuntica Dactylus, id est, Laurocerasus
Tragium,i.Fraxinella
Tragium Germanicum,i. Atriplex olida
Tremula,i.Populus Lybica
Trifolium fibrinum,id est Trifolium palustre
Trifolium cochleatum,i.Medica
Trifolium fruticans,i.Polemonium
Trifolium Asphaltites, i.Trifolium bituminosum
Tuber terra,i.Cyclamen
Turbith,i.Thapsia
Turbith Auicennæ,i.Tripolium
Typhium Theoph.i.Tussilago

V

Vesicaria peregrina, i. Pisum cordatum
Veelguitta,Dod i.Petroselinum
Veratrum,i.Helleborus
Veratrum,nig.Dios.i. Astrantia nigra
Verbascula,i.Primula veris
Verdelbel Haliah,i.Ranunculus
Victoriola,i.Hippoglossum
Vitis alba,i.Bryonia
Vitis Idæa,i.Vaccinia
Virga sanguinea Matthioli,i. Cornus fœmina
Virga pastoris,i.Dipsacus
Vitalis,i.Crassula
Vitalba,i.Viorna
Viticella,i.Momordica

Vincetoxicum

A Table of such English names as are attributed to the Herbes, Shrubs, and Trees mentioned in this History.

A Table

A Supplement or Appendix vnto the generall Table, and to the
Table of English Names, gathered out of antient written
and printed Copies, and from the mouthes of plaine
and simple country people.

A

APer is Dill
Ameos, Ameos
Argentil, Perexpler
Ache, Smallage.
Alliaria, in written copies Cardiaca.

B

BAldmoin, Gentian
Baldmony, Meum
Baldwein, Gentian
Belwed, Iacea nigra.
Bishops wozts, Betony
Birds nest, wilde Parsley
Birds tongue, Stitchwozt
Bigold, Crysanthemum segetum
Blew ball, Blew bottle
Bolts, Ranunculus globosus
Bow wood, knapweed
Brown begle, Bugle
Brookwozt, Consolida minor
Brotherwozt, Puliol mountain
Biderwozt, Vlmaria
Bight, Chelcdonia
Brokeleake, water Dragons
Bulruwozt, Hopewozt
Bucks beans, Trifolium paludosum
Buckram, Iron

C

CArbiack, Alliaria
Carles, Cresses
Catmint, Nepe a
Eenclesse, Daffodill
Chaffeweed, Cottonweed
Cheruelor Cheuerel was called (though
vntruly) Apium risus.
Churles treacle, Allium
Churchwozt, Pennyroyall
Ciderage, Arsmart
Clithe, the Burre docks
Clitheren, Gosgrasse oz Cliuers
Clitte, Lappa
Ciouetongue, Ellebor niger
Cocks foot, Columbine
Cock foot, Cheledonia major
Cowfat, Cow Basll
Criftalyc the lesser Centory
Coznberries, Vaccinea palustris
Crowbell, yellow Daffodill
Crowberries, Erica baccifera
Crowfoot is Ozchis, in Lincolnshire and
Yozkshire
Crow sope, Hopewozt
Crow leek, Hyacinthus Anglicus.
Cropweed, Iacea nigra
Cuiuerwozt, Columbine
Culrage, Arsmart
Cutberdole ? Cutbertill is Blank vrine

D

DIlnote, Cyclamen
Donninethel, wilde Hempe
Dragons female, water Dragons

Drapwozt, Filipendula
Dancedown, Catstaile
Divale is Nightshade

E

EDderwozt, Dracontium
Elleber, Alliaria
Elfdock, Enula Campana
Earthgall, great oz rather small Centozy
Euerfern is wall Fern
Exan, Crosswozt, yet not our Cruciata

F

FIne, white Flour de luce
Fauerel, Cepea
Field Cypres is Chamæpitys
Fieldwozt, Felwozt oz Gentian
Filewozt, Filago minor
Fleabock, Petasites
Fleawozt, Psyllium
Forget me not, Chamæpitys
Forbitten moze, diuels bit
Fauerole, water Dragons
Franke, Spurry
Freiser is the herb that beareth Strawber-
ries, Strawberries

G

GAlingal meke is Aristol. rotunda
Gater tree oz Gater tree is Dogges-
berry tree
Gandergosses is Jecks
Geckbor, Aparine
God kinng Harry, English Mercury
Gosechite, Agrimony
Gosgrasse was somtime called Argentum
Goselbill, Aparine
Garten ginger, Pirentic
Glond, Cow Basll
Grace of God, S. Johns wozt
Grien mustard, Dittander
Groundwill, Groundswell
Ground nædle, Geranium mustatum
Ground Enel, Venus combe

H

HIrene, Cliuers.
Hammerwozt, Pellitozy of the wall
Hardhow Marygolds
Hares eye, Lychnis sylvestris
Harebell, Crow læke
Herb Jvy, Chamæpitys
Henbel Henbane
Herholw, Hedera terrest.
Herb Benner, Hemlock
Herb Peter, Cowslip
Herba martis, Martagon
Hertelowze, Chamædrys
Hertwozt, Fraxinus
Hilfwozt Puliol mountaine
Hirpia major, Common Pimpernel
Holy rope wilde Hemp
Houndberry Solanum
Horfewozt, Filago.
Horfechire, Germander.

I

IAcea alb, wild oz white Tansie
Jmbieb, Housleke
Joan silver pin, double Poppy.

K

KIndlegosts, Gosegrasse
kings crowne, Melilotus
King cob oz king cup is Crowfoot
Kille meete Arise Pansies
Kidnywozt, Nauelwozt

L

LAgwozt, Helleborus albus
Little Wale Gromel
Lichwort is Pellitozy of the wall
Longwozt, Pellitozy of Spain
Lilip læke, Moly
Liliryall, Pennyroyal
Lobewozt water Crowfoot
Lousewozt, Staphisacre
Lustwozt, Sundew
Lyngwozt, Helleborus albus

M

MAns Motherwozt, Palma Christi
May blossoms, Conbal Lillies
Mawroll white Horehound
Mauthen oz Mathes, Cotula foetida
March, Smallage
March beetle, Catstaile
Mæbles, Arage
Morecrop, Pimpernell
Mozel, Nightshade
Mousfpeale, Orobus
Mugwet, woodrosse

N

NEle, Lollium
Nespite Calamint
Nay, Cats mint
Nofeblead, Yarrow

O

ORbal, Orpin
Oxan Cruciata
Oxtongue, Lingua bovis

P

PIgle, Stitchwozt.
Palma de Dieu, *Palma Christ.*
Papwozt, Mercury
Pastel, Woad
Pedelion, Helleborus niger
Peters Baf, Tapsus barbatus
Pewterwozt, Horstaile
Pimentary, Baulme
Pownnædle, Stozkbill
Primrose, Ligustrum
Pygis, Gramen Leucanthemum

Iaaaaad 3

R

Rams foot is water Crowfoot
Red knees is Hydropiper
Robin in the hose is Lychnis sylvest.
Rods gold is Marygold

S

Scabwort is Enula Campana
Sea Dock is Brank vrsine
Seggrom is Ragwort
Self heal was somtimes called Pimpernel
Sheep killing is Cotyledon aquatica
Sleepwort is Lettuce
Staggerwort & Stauerwort is Iacobea
Stanmarch is Alisander
Standelwelks is Satyrion
S. Maries seed is Sow-thistle seed
Small honesty is Pinks
Sowerwort is Aristolochia
Stikepile is Storks bill
Stedfast is Palma Christi.

Stobwort is Oxys
Sparrow tongue is Knot grasse
Stonnord & Stonchore is Stonecrop
Stubwort is wood Sorrell
Swines grasse is Knot grasse
Swine Carse is knot grasse
Swichen is Groundswell
Sowthwort is Columbine

T

Tletwort is wilde Borage
Tank is wilde Parsnep
Tentwort is Ruta muraria
Tetterwort is great Celandine
Toothwort is Shepheards purse
Cutsane is Clymenum Italorum

W

Wallwort is Ebulus, which was somtime called Filipendula
Warence is Madder
Warmot is Wormwood

Wapwort is Pimpernel
Wapbread is Plantago
Wapwort is Hippia maior
Waterwort is Maidenhaire
Weythernop is Feuerfew
White Bothen is great Daisy
Wilde Sauager is Cockle
Wilde Nardus is Asarum
White Golds is great Daisy
Woodmarch is Saniclet
Woodsower is Oxys
Woodbhoney is Fraxinus
Woodnep is Ameos
Wolfs thistle is Chamæleon
Wineberry is Vaccinea
Wormot is Ibiscus
Wit is Hyoscyamus luteus

Y

Yron head is Knapweed

Z

Zeskes was counted Satyrion minor, and is that which Lobel calls Serapias foemina pratensis

A Cata-

A Table of British Names.

A Catalogue of the Brittish Names of Plants, sent me by Master Robert Dauyes of Guissaney in Flint-Shire.

A

A Mes, Dill.
Awrd danadl, Red Archangell Nettles.
Aſuranadl, vide Hwb yr yeben.

B

B Anadyl, Broome.
Banatlos, Furze.
Berw yr'Frengic, Creſſes.
Berw yr awr, water Creſſes.
Bedwen, a Birch tree.
Bactus, Beets.
Blaen yr Twrch, Mercury.
Blaen y gwayw, Spearewort.
Bleidda dug, Wolfes bane.
Brialbu Mair, Cowſlips.
Biwynen, a Ruſh.
Bylwg, Cockle or field Nigella.
Buſt yl y Ddayar, Centorie.

C

C Arn yr ebol, Foleſoot.
Cas gan gythrel, Veruaine
cacamwaei, Butre.
Caliwlyn y inel, Agrimony.
Cancwlwm, Knot graſſe.
Camamill, Camomill.
Ceirch, Oats.
Cenaln, Leekes.
Cennin Pedr, Daffodill.
Cedor y wrach, Horſetaile.
Cegid, Hemlocke.
Celynen, Holly.
Chwerwlys yr kithin, Wood Sage.
Cluſt yr ewic, Lautell.
Cloſilops, Gilloflowres.
Cluſtieu yr Derw, vide Galladr.
Cluſt llygoden, Mouſe eare.
Ciatarlys y dwr, Brookelime.
Coed Ceri, Seruice tree.
Cowarch, Hempe.
Cawer y llaeth, Caliwlyn y mei.
Coed kirin, Plum trees.
Corſen, a Poole reed.
Cribe y Bleiddieu, vide Cacamwaei.
Craith vnnos, Prunel or Selfe-heale.
Crafanke y vran, Crowfoot.
Cribe fan Fraid, Betony.
Cynglennydd, white Mullen.
Cyafon y Celior, Selfwell.

D

D Aily gwaed, Penny royall.
Danadl, Nettles.
Danadten wina, White Atchangell Nettle.
Danit y llew, Dandeleon.
Danadlen ddaս, dead Nettle.

E

E Polgarn yr ardd, Aſtrabacca.
Efrtu, Darnell.
Eiddew, Iuy.
Eiddew y ddayar,
Eidral, ground Iuy.
Eathin yr ieinvide Hwb yr yeben.
Erienlys, S. Iohns wort.
Erbin, Calamint.
Ealun perſis, baſtard Parſley.

F

F A, Beanes.
Fenich y cwn, wilde Cammomill.
Fenich, Fenell.
Fetter, Fitches.

G

G Alladr, Lungwort like Liuerwort.
Garllec, Garlicke.
Gleſyn y Coed, Bugle.
Gladyn, Gladiol or Corne Flag.
Geltndrem, vide Llyſie Ewſtas.
Gold Mair, Marigold.
Gruc, vide Banatlos.
Grayanlys y dwr, Brookelime.
Gwlydd, ſmall Chickeweed.
Gwlydd Mair, Pimpernell.
Gwenynaddail, Gwenynoc, Balme.
Gwyddfyd, Woodbinde or Honiſeckle.
Gwrden y coed, Smooth Bindweed.
Gwallt gwener, Venus haire.
Gwallt y ſorwin, Maiden haire.
Gwayw yr Brenhin, Daffodill.
Gwenith, Wheat.
Gwinwydden, Vine.

H

H Ad y gramandi, Gromel.
Haidd, Barly.
Heſe metſedec, Water torch, or Typha palaſt:
Hoccyt, Mallowes.
Hoccyt gors, Mariſh Mallowes.
Hwb yr yeben, Camock, or reſt harrow.

LL

L Laeth bron Mair, Sage of Ieruſalem
Llaulys, Staueſacre.
Llawenlys, Borage.
Llewic ychwannen, vide y Benſelan.
Llewic yr iâr, Henbane.
Llewpard dug, Aconitum.
Llyſie Juan, Mugwort.
Llyſie llwydion, v. de Llſie Juan.
Llyſie llewelys, Pauls Betony.
Llyſi y wennol, Celandine.
Llym y llygaid, vide Llyſie wennal.
Llyſie Effras, Eyebright.
Llyſie yr Crymman, vide Gwylydd Mair
Llyſie lliw, vide Diers weed.
Llyſie pen câ, Houſleek.
Llyſie yr gwaedlin, Yarrow or Milfoile.
Llyſie Mair, vide Gold mair.
Llyſie Amor, Floure gentle.
Llygaid y Dydd, Dailies.
Llyſie y pwdin, vide Daily gwaed.
Llyſie yr gâth, vide Erbin.
Llyſie Blaidd, vide Bleid dûg.
Llyſie y moch, Nightſhade.
Llyſie y Cribeu, Teaſell.
Llyſie Simion, vide Cas gan gythrel
Llyſie y Cyrph, Periwinkle.
Llyſie Eua.
Lyriaidy mor, Sea banke horne.
Llyſie yr meddaglyn, wilde Carrot.
Llwyfen, Elme tree.
Llwynlys, Scuruy graſſe.

M

M Aſod, Raſplis.
Marchalan, Elecampane.
March rbedya y derw, Polypody, Oke Ferne
Maip, Turneps.

Mare

Marcb vſgal y gerddi. Artichoke.
Mefys. Strawberries.
Menig elluſsion. Fox gloues.
Meirw. Iuniper tree.
Meillionen y meirch. Right trefoile.
Mintas. Mintes.
Moron. Parſneps.
Moron y maes. Wilde Parſneps.
Mwg y ddayar. Fumetory
Mwſsogl. Moſſe.
Mynawyd y bigail. Storks bill.

N

Nyddoes. Spinage.

O

Onnen. An Aſh tree.

P

Pawen yr Arth. Beares breech.
Padere Mair. Croſſewort.
Ierſti y dwr. water Parſley.
Perſli Ffrengic. Smallage.
Phionffrwyth,v. Menig y elluſsion
Pidny y goc. Aron, or Cuckow pint.
Poerlys,v.y ludlys.
Poplys. a Poplar.
Pwrs y Bigail. Shepheards purſe.
Pys y Ceirw. Tares.

R

Rhedyn. Ferne.
Rhedegat y derw,v.Galladr.
Rhâg. Rie.
Rhoſyn. a Roſe.

S

Saeds gwyllt,v.Chwerwlys.
Siwdrmwt. Sothernwood.
Siarcked y melnydd,v.Cynffon Ewynog.
Sirian. Cherries.
Snedan Fair. Engliſh Galengale.
Sewdl y Crydd,v. Blaen yr yswrch.
Suran y gôr. Wood Sorrell.
Suran. Sorrell.
Syſi,v. Mefys.

T

Tafod y ki. Dogs tongue.
Tafod y neidr. Adders tongue.
Tafod yr Hydd. Harts tongue.
Tafol, a Docke.
Tafol Mair. Biſtort.
Tagaradr,v. Hwb yr ychen.
Tafod yr edn. Birds tongue.
Tafod yr ych. Bugloſſe.
Telephin. Orpin.
Tormaen. Filipendula.
Tryw,v.Caliwlyn y mêl.
Troed y glomen. Columbine.
Triacl y tylodion. Tormentilla.
Troed y dryw. Parſley Breakſtone, or ſmall Saxifrage.
Triacl y Cymro. Germander.
Troed yr bedydd. Larke heele.

W

wilffraew,v. Llyſie yr gwaedlin.
wiuniwn. Onions.

Y

Y Bewſelen. Fleabane.
  Y Benlas wenn,v. Claſlys.
  Y bengaled. Red Scabious.
Y Benlas. Blewbottle, or Corneſloure.
Y bengoch. Horehound.
Y Claſlys. Scabious.
Y Derfagl. Medow three leaſed graſſe.
Y Droedydd. Herbe Robert.
Y Drwynſawr. Caliwlyn y mêl.
Y Ddwy geonioc. herbe Twopence or Moneywort.
Y Dorfwyd. wild Tanſy or ſiluerweed.
Y dew bannoc,v. Cynffon Llwynoc.
Y Dinboeth. Arſmart.
Y Ddayarlys. Peony.
Y Doddedige wenn. Pilewort.
Y fendigedi. Tutſan or Parkeleaues.
Y Fabgoll. Poppy.
Y fiolud. Violet.
Y fyſen Y frouwys. ſmall Celandine.
Y ſtidioc lâs,v. Llyſie Jvan.
Y fyddarlys. Prickmadam.
Y fyddygyn,v. Craith un nos.
Y fyw fyth. Llyſſew pentû.
Y gaurs goch,v. Buſtl y Ddayar.
Y gynga,v. Llyſie yr bidl.
Y gloria. wilde Roſe or Spargwort.
Y gâs wenwyn. Diuelsbit.
Y gyſog. a kinde of Spurge.
Y glaiarlys. Y greulys. Groundſwell.
Y gyſgadur. Nightſhade or Morell.
Y gyugroen. Todeflax.
Y llew gwyn dôf. Garden Orach.
Y llew gwyun gwyllt. wilde Orach.
Y lliwlys,v.Llyſiew lliw.
Y llwynbidydd. Ribwort.
Y lindro. Doder.
Y llyſiewyn benegidedic. Valerian.
Y lleuadlyt. Lunaria.
Y Môr gelyn. Sea Holly.
Y Mûrlys. Pellitory of the wall.
Y Papi coch,v. red Poppy, or corne Roſe.
Tr Eſcarlys { Hir geon { Atiſtolochia, long.
           { bychan { or Birthwort, round.
           {   { or Hartwort, ſmall.
Yr Alaw. Water Lilly.
Yr hên lydan,i.fforedd. Waybread.
Yr rhût. Rue or herbe Grace.
Yr vchelfe. Miſſeltoe.
Yr yſcallen Fraith. our Ladies thiſtle.
Yr yſcallen Fendigedic. Card.Benedict.
Yr hokiach. Clownes wort.
Yſcall drain gwynn. Carline thiſtle.
Yſcall. Wilde thiſtles.
Yſcal y moch. Sow thiſtle.
Yſcolfair. Peters wort or ſquare S.Iohns wort.
Yſcaw. Elder trees.
Yſcaw Mair. Walwort.
Yſpaddaden. Whitethorne.
Yſuiab. Muſtard.
Y wermod. Wormwood
Y wermod wenn. Feuerfew.
Y winwidden wenn. white Brionie.
Y winwydden ddû. blacke Brionie.
Y willffrae. Llyſie yr gwaedlin.
Y wenwlydd. Great Thickw, eed.

A TABLE, WHEREIN IS CONTAINED
THE NATVRE AND VERTVES OF ALL THE
Herbes, Trees, and Plants, described in
this present Herbal.

Good

To

Y

An Aduertifement to the Readers.

Courteous Readers, I haue thought fit to aduertife you, that in this second Edition of reuifed Gerrard, you must not expect any Additions or Alterations, otherwife than an Amendment of thofe few Errata I noted in the former Edition : which were chiefly of Figures tranfpofed and Verball efcapes ; of which later fort, you may perhaps finde here and there one , yet fuch as the meaneft Reader may without any difficultie amend. I know it will be expected, that I fhould haue giuen the Figures of fuch things, as I formerly deliuered the Hiftory of without them ; as alfo an Addition of more Plants : both which (I muft confeffe) I could haue done, and the later in great number, yet vpon thefe following confiderations (whereof I would not haue you ignorant) I forbore the performance. Firft, for that I haue determined with my felfe (by Gods fauourable Affiftance) by the joint help of fome of my friends (of whom mention is made in my Epiftle to the Reader) to trauell ouer the moft parts of this Kingdome, for the finding out of fuch Plants, as grow naturally in England; which how farre we haue already performed may be found by diuers places in Gerrard, but chiefly by my Mercurius Botanicus, fet forth Anno. 1634. betweene the time of the former Impreffion and this. For I iudge it requifite that we fhould labour to know thofe Plants which are, and euer are like to be Inhabitants of this Ifle; for I verily beleeue that the diuine Prouidence had a care in beftowing Plants in each part of the Earth, fitting and conuenient to the foreknowne neceffities of the future Inhabitants; and if wee throughly knew the Vertues of thefe, we needed no Indian nor American Drugges.

Secondly, I haue fome friends, addicted to this ftudy, gone into forreigne parts ; from whom I expect to receiue fome things to the further Augmentation of my intended Worke.

Thirdly, there will bee required more time for the performance hereof, than I had to fet forth both thefe Editions, for one may more eafily repaire an old Building, than reare a new one.

Laftly, fuch as haue bought the former Edition fhall receiue no injury ; which they would haue done, if I had added fome few things to this (for the Booke is already fo bigge, it would not haue admitted much.) But I haue thought it more conuenient to fet forth apart fuch Figures and other Additions, as I fhall iudge fit for compleating of this Hiftory of Plants ; and then any that haue this former, may if they pleafe haue the later alfo, and none be injured.

Thefe with fome other confiderations needleffe here to declare, moued mee to forbeare the Enlargement of this fecond Edition.

T. I.